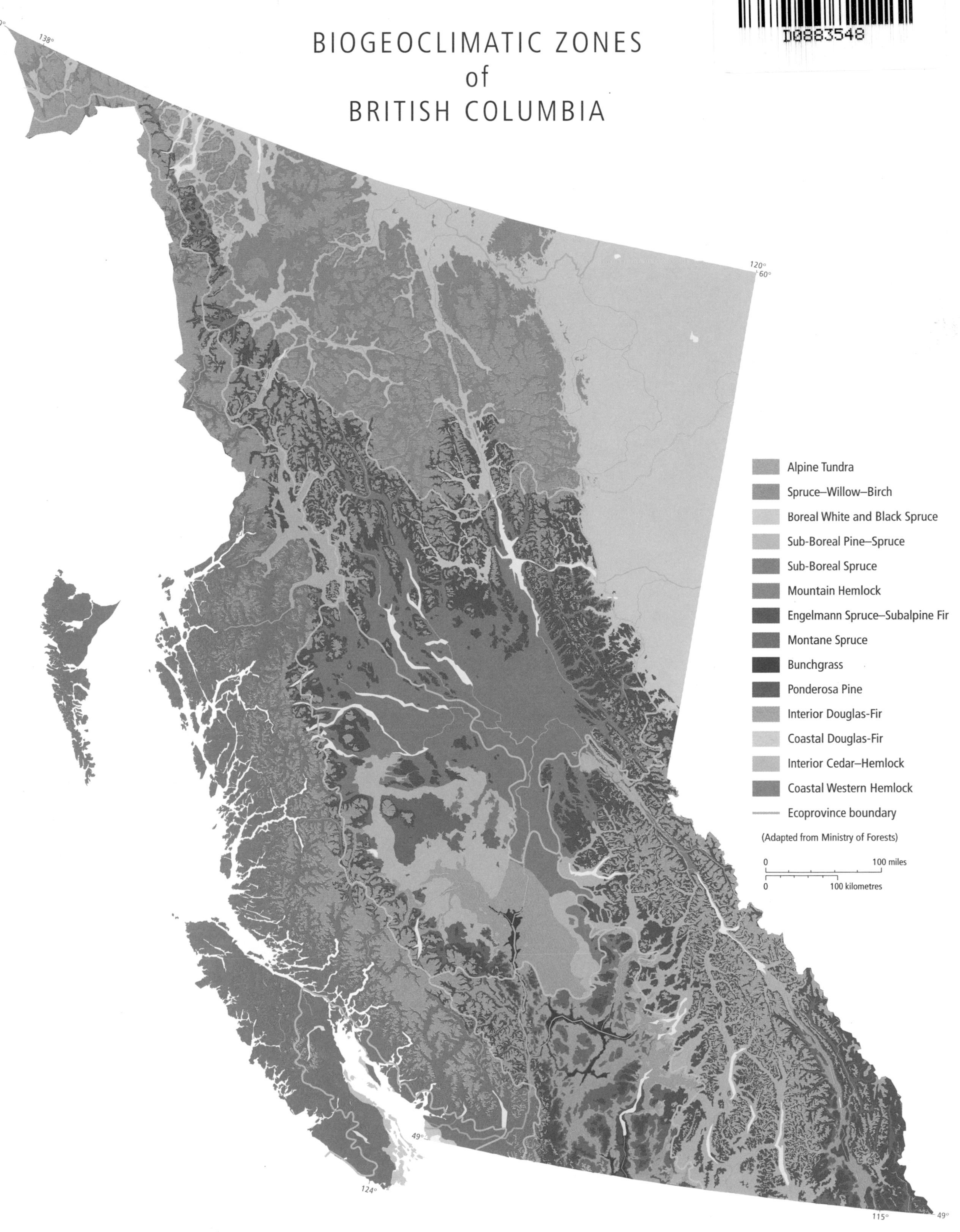

BIOGEOCLIMATIC ZONES
of
BRITISH COLUMBIA
Alpine Tundra
Spruce–Willow–Birch
Boreal White and Black Spruce
Sub-Boreal Pine–Spruce
Sub-Boreal Spruce
Mountain Hemlock
Engelmann Spruce–Subalpine Fir
Montane Spruce
Bunchgrass
Ponderosa Pine
Interior Douglas-Fir
Coastal Douglas-Fir
Interior Cedar–Hemlock
Coastal Western Hemlock
Ecoprovince boundary
(Adapted from Ministry of Forests)
0
100 miles
0
100 kilometres
60°
138°
120°
60°
49°
124°
115°
49°

THE Birds OF BRITISH COLUMBIA

VOLUME 3 PASSERINES

FLYCATCHERS THROUGH VIREOS

THE BIRDS OF BRITISH COLUMBIA

Volume 1 Nonpasserines: Introduction, Loons through Waterfowl

Volume 2 Nonpasserines: Diurnal Birds of Prey through Woodpeckers

Volume 3 Passerines: Flycatchers through Vireos

Volume 4 Passerines: Wood Warblers through Old World Finches

Environment
Canada

Canadian Wildlife
Service

Ministry of Environment,
Lands and Parks
Wildlife Branch

THE Birds OF BRITISH COLUMBIA

VOLUME 3 **PASSERINES**

FLYCATCHERS THROUGH VIREOS

by
R. Wayne Campbell, Neil K. Dawe,
Ian McTaggart-Cowan, John M. Cooper,
Gary W. Kaiser, Michael C.E. McNall,
and G.E. John Smith

UBCPress / Vancouver

Published in cooperation
with Environment Canada
(Canadian Wildlife Service) and the
British Columbia Ministry of
Environment, Lands and Parks
(Wildlife Branch)

Printed in Canada on acid-free paper ∞

ISBN 0-7748-0572-2

Canadian Cataloguing in Publication Data

Main entry under title:
The Birds of British Columbia

Includes bibiliographical references and index.
"Published in cooperation with the Canadian Wildlife Service and the British Columbia Wildlife Branch."
Previous ed. of vols. 1-2 published by: Royal British Columbia Museum.
Partial contents: v. 1. Nonpasserines: introduction and loons through waterfowl – v. 2. Nonpasserines: diurnal birds of prey through woodpeckers – v. 3. Passerines: flycatchers through vireos.
ISBN 0-7748-0622-2 (set) – ISBN 0-7748-0618-4 (v. 1) – ISBN 0-7748-0619-2 (v. 2) – ISBN 0-7748-0572-2 (v. 3)

1. Birds – British Columbia. I. Campbell, R. Wayne (Robert Wayne), 1942- II. Canadian Wildlife Service. III. British Columbia. Wildlife Branch.

QL685.5.B7B574 1997 598.29711 C96-910903-2

UBC Press gratefully acknowledges the ongoing support to its publishing program from the Canada Council, the British Columbia Arts Council, and the Department of Canadian Heritage of the Government of Canada. The Press also acknowledges the assistance and cooperation of Gerald L. Truscott of the Royal British Columbia Museum.

Printed and bound in Canada by Friesens
Set in Palatino and Frutiger by Irma Rodriguez
Book design: DesignGeist; Irma Rodriguez
Cover design: Chris Tyrrell
Copy-editor: Francis J. Chow
Cartographer: Eric Leinberger
Bird illustrations: Michael Hames
Front cover photograph: Antoinette Alexander
Back cover photograph: Mike H. Symons

UBC Press
University of British Columbia
6344 Memorial Road
Vancouver, BC V6T 1Z2
(604) 822-5959
Fax: 1-800-668-0821
E-mail: orders@ubcpress.ubc.ca
http://www.ubcpress.ubc.ca

CONTENTS

British Columbia Field Ornithologists

MacMillan
Bloedel Limited

FBCN
Federation of British Columbia Naturalists

BC hydro

SHEARWATER
SCALING
AND
GRADING

FOREST SERVICE
BRITISH COLUMBIA

BRITISH COLUMBIA
WATERFOWL SOCIETY

WBT
WILD BIRD TRUST
of British Columbia

HABITAT
CONSERVATION
TRUST FUND

Ducks Unlimited Canada

WWF
CANADA

ACKNOWLEDGEMENTS

To complete a project of this complexity and magnitude requires a high level of support from friends and colleagues and a working environment free from the normal demands associated with most government offices. Fortunately, through the years that we have worked on the passerine volumes of *Birds of British Columbia,* many people with the British Columbia Ministry of Environment, Lands and Parks (Wildlife Branch), Environment Canada (Canadian Wildlife Service), and The Nature Trust of British Columbia have cleared our path of many distractions and have allowed us unusual freedom to focus our creative energies on the task at hand. We hope that these volumes reflect the confidence and the total support we received from the following: Michael Dunn, Donald S. Eastman, D. Ray Halladay, Richard W. McKelvey, Arthur M. Martell, William T. Munro, Chris Pharo, Rod S. Silver, Jim H.C. Walker, Stephen P. Wetmore, and Brian Wilson.

We are also grateful to Dennis A. Demarchi for cooperation and assistance in updating the "Ecoregion Classification for British Columbia" used in this volume.

The many contributors who provided continued support to these volumes are not too numerous to mention, and are listed under the following categories.

CONTRIBUTORS TO DATA BASE

A total of 6,498 people provided information on some aspect of the life history of passerine birds in the province by contributing to the British Columbia Nest Records Scheme, British Columbia Wildlife Records Scheme, and British Columbia Photo-Records File, as well as participating in provincial Breeding Bird Surveys, Christmas Bird Counts, and other regional bird surveys. They are all listed in Appendix 4.

Among the general contributors is a group of individuals who have made major contributions in the form of serving as regional or sub-regional editors for quarterly reports to *American Birds/Field Notes,* operating Rare Bird Alerts, coordinating Birder's Nights, coordinating recordkeeping regionally, providing access to detailed field diaries, and consistently submitting meticulous and complete observations over the 20 years this project has been in progress. These major contributors are: David Allison, Jerry and Gladys Anderson, Gerry Ansell, E. Derek Beacham, Alice Beals, Barbara Begg, Winnifred Bennie, Mike Bentley, Jack Bowling, Ken C. Boyce, Jan Bradshaw, Len Brown, Elmer Callin, Richard J. Cannings, Stephen R. and Jean Cannings, Donald G. Cecile, Chris Charlesworth, Stewart Clow, John Comer, Gary S. Davidson, Lyndis Davis, Brent Diakow, Adrian Dorst, Linda Durrell, Michael C.R. Edgell, Kyle Elliott, Maurice Ellison, Anthony J. Erskine, Michael P. Force, David F. Fraser, Jeff Gaskin, Brian R. Gates, Martin Gebauer, Les and Violet Gibbard, J.E. Victor Goodwill, Hilary and Orville Gordon, Al Grass, Tony Greenfield, Larry R. Halverson, Peter Hamel, Willie Haras, Margaret Harris, Robert B. Hay, W. Grant Hazelwood, Margo Hearne, Charles W. Helm, Werner and Hilde Hesse, Dennis Horwood, Richard R. Howie, Douglas and Marian Innes, John Ireland, Pat Janzen, Dale Jensen, Sandra Kinsey, Helen Knight, W. Douglas Kragh, Nancy Krueger, Violet F. Lambie, Laird Law, Douglas Leighton, Jo Ann and Hue MacKenzie, Alan MacLeod, Derrick Marven, Hylda Mayfield, Barb and Mike McGrenere, Karen McLaren, Martin K. McNicholl, William J. Merilees, Alexander Muir, Mark Nyhof, Elsie Nykyfork, Mary Pastrick, Jim Perry, Connie Phillip, Mark Phinney, Douglas Powell, G. Allen Poynter, George and Bea Prehara, Al Preston, D. Michael Price, Roy Prior, Nina Raginsky, Diane Richardson, Anna Roberts, I. Laurie Rockwell, Michael S. Rodway, Manfred Roschitz, Glen R. Ryder, Ron Satterfield, Barbara Sedgewick, Brian Self, Chris R. Siddle, Pam Sinclair, Marion Smith, Andrew C. Stewart, David Stirling, Harvey Thommasen, John Toochin, Rick Toochin, Ruth Travers, Danny Tyson, Linda M. Van Damme, Ronald P. Walker, Sydney and Emily Watts, Wayne C. Weber, Bruce Whittington, and Karen Willies.

DATA ENTRY

The entry of the occurrence and breeding data into electronic data base files was a massive effort and would not have been accomplished without the assistance of many people and organizations.

The Arrowsmith Naturalists (Parksville) obtained a series of grants from Employment and Immigration Canada under the Unemployment Insurance Job Creation Program (Section 25) that allowed most of the passerine data from the Wildlife Records Scheme to be entered into data base files. Pauline Tranfield and Betty Barnes of the Arrowsmith Naturalists helped administer the program.

The following people participated in this program over the three and a half years it took to complete the task; those with an asterisk after their names supervised the data entry: Lone Anderson, Anne Ashmore, Danielle Bras, Susan L. Bridge, Shirley Brittain, Kathryn J. Cole*, Heather Ann Colton, Deborah Cote, Avon Fitch*, Roseanne Marie Harrison, Michael E.J. Hayward, Karen Koehn, Renate L. Liddell*, Terri Martin, Jeanette W. Pryor, Dianna Reed, Marilyn Reed, Rosemarie Sather, Marlene Slawson, Carol J. Strynadka, Jodi Waites*.

The following people entered other portions of the data, particularly the British Columbia Nest Records Scheme, over the years: D. Sean Campbell, Eileen C. Campbell, Tessa N. Campbell, Darren R. Copley, Jordan T. Dawe, Karen E. Dawe, Scott Dickin, Carolyn Hamilton, Renate L. Liddell, Andrew MacDonald, Edward L. Nygren, Charlene Pearce, Cynthia and Michael G. Shepard, and Bernice Smith.

Len Thomas, a graduate student at the University of British Columbia, organized and summarized computer files for trends in coastal and interior Breeding Bird Survey routes.

VOLUNTEERS

The following people assisted in producing initial distribution maps from raw data, searching journals and government reports for bird records, copying articles, and filing references: Shari Baker, Jeniffer Barlow, Colin Barnfield, Marienne Becevel, Elizabeth Brooke, Eileen C. Campbell, Brian Chapel, Vivian Clark, Holly Clermont, Lyndis Davis, Jordan T. Dawe, Dianne Demarchi, Scott Dickin, Peggy Dyson, John Elliott, Maureen L. Funk, Tracey D. Hooper, Steve Madson, Alistar Marr, Anita McKenzie, John D. McIntosh, Heather Melchior, Joan Mogenson, Cindy Moore, Phyllis Mundy, Meg Philpot, Bernice Smith, Calvin Tolkamp, and Lillian Weston.

In 1986, Margaret Harris took over the position of coordinator of the Brtish Columbia Nest Records Scheme from Violet Gibbard, and over the next decade maintained its existence and profile in the province. The scheme has been the major source of breeding information for birds in the province. Margaret's patient and constant encouragement and her reminders to participants that the quality of information on nest cards is more important than the number of cards submitted are reflected in the data analysis for each species account.

FIELD SUPPORT

Many active field naturalists unselfishly shared with the senior authors their favourite, remote, and little-known habitats in the province, which helped fill gaps in the occurrence and breeding information for certain species. These people included: Cathy A. Antoniazzi, Barbara Begg, Jan Bradshaw, Arnie Chaddock, Al Charbonneau, Gary S. Davidson, Adrian Dorst, Violet and Les Gibbard, Larry R. Halverson, Margaret Harris, Katie Hayhurst, Margo Hearne, Richard R. Howie, Sandra J. Kinsey, Dennis Kuch, Laird E. Law, Ron Mayo, Douglas Powell, Al Preston, Anna and Gina Roberts, Glen R. Ryder, Chris Siddle, Andrew C. Stewart, Ruth and Charles Travers, Linda Van Damme, Al and Irene Whitney, and Jim Young.

FINANCIAL AND ADMINISTRATIVE SUPPORT

The following organizations and individuals provided important support to the project, either directly or indirectly: British Columbia Field Ornithologists; British Columbia Hydro and Power Authority; British Columbia Ministry of Environment, Lands and Parks, Wildlife Branch (Ron S. Silver); British Columbia Ministry of Forests, Research Branch (Alison Nicholson); British Columbia Waterfowl Society; Canada–British Columbia Partnership Agreement of Forest Resources: FRDA II; Centennial Wildlife Society of British Columbia; Ducks Unlimited Canada; Employment and Immigration Canada; Environment Canada, Canadian Wildlife Service, Pacific and Yukon Region; Environment Canada, Parks; Federation of British Columbia Naturalists; Frank M. Chapman Memorial Fund (American Museum of Natural History); Habitat Conservation Trust; MacMillan Bloedel Limited; The Nature Trust of British Columbia (L. Ronald Erickson and Helen N. Torrance); Northwest Wildlife Preservation Society; Saltspring Island Garden Club; James R. Slater; Bernice Smith; Win Speechley; William Taylor; and The Wild Bird Trust of British Columbia; World Wildlife Fund, Canada.

PHOTOGRAPHS

We thank the 29 photographers, in addition to the authors, who contributed images to this volume, especially Mark Nyhof, Anna Roberts, Linda M. Van Damme, and Tim Zurowski. Antoinette Alexander provided the cover photograph of the Steller's Jay and Mike H. Symons the photograph of the authors.

REVIEWERS

All species accounts were reviewed from a regional and provincial perspective by the following 24 individuals who reside in British Columbia: **Cathy Antoniazzi** (Prince George area), **Jack Bowling** (Prince George area and B.C. general), **Jan Bradshaw** (Harrison and Shuswap areas), **Richard J. Cannings** (Okanagan valley and B.C. general), **Michael J. Chutter** (B.C. general), **Gary S. Davidson** (west Kootenay), **Dennis A. Demarchi** (B.C. general), **Adrian Dorst** (Tofino-Ucluelet area), **Bryan R. Gates** (southern Vancouver Island), **Tony Greenfield** (Sunshine Coast and B.C. general), **Margo Hearne** (Queen Charlotte Islands), **Richard R. Howie** (Thompson-Nicola valleys and east Kootenay), **Sandra Kinsey** (Prince George area), **Nancy Krueger** (Prince George area), **Laird Law** (Prince George area), **Mark Phinney** (Prince George area), **Douglas Powell** (Revelstoke area), **Anna Roberts** (Chilcotin-Cariboo), **Chris R. Siddle** (Peace River region and B.C. general), **Andrew C. Stewart** (Peace River region and B.C. general), **David Stirling** (B.C. general), **Ruth E. Travers** (central Chilcotin-Cariboo), **Linda M. Van Damme** (west Kootenay), and **Ellen Zimmerman** (Golden area).

The following 86 individuals reviewed at least one species account in their field of expertise: **John W. Aldrich** (Tucson, Arizona – American Robin), **Robert C. Beason** (State University of New York, Geneseo – Horned Lark), **Donald L. Beaver** (Michigan State University, East Lansing – Pacific-slope Flycatcher), **Barbara Begg** (Victoria, British Columbia – Sky Lark), **Louis B. Best** (Iowa State University, Ames – Gray Catbird), **Carl E. Bock** (University of Colorado, Boulder – Black-billed Magpie, Red-breasted Nuthatch, and Golden-crowned Kinglet), **Jeffrey D. Brawn** (Illinois Natural History Survey, Champaign – Western Bluebird), **Leonard A. Brennan** (Mississippi State University, Mississippi – Chestnut-backed and Mountain chickadees), **Richard Brewer** (Western Michigan University, Kalamazoo – Black-capped Chickadee), **James V. Briskie** (Queen's University, Kingston, Ontario – Least Flycatcher), **Jerram L. Brown** (State University of New York, Albany – Steller's Jay), **Fred L. Bunnell** (University of British Columbia, Vancouver – Winter Wren), **Robert W. Butler** (Canadian Wildlife Service, Delta, British Columbia – Tree Swallow, Northwestern Crow, and Bushtit), **Tom J. Cade** (Boise, Idaho – Northern Shrike), **Jeff N. Davis** (University of California, Santa Cruz – Hutton's Vireo), **David F. DeSante** (Point Reyes Station, California – Western Wood-Pewee), **Michael J. DeJong** (University of St. Thomas, St. Paul, Minnesota – Northern Rough-winged Swallow), **Richard A. Dolbeer** (Denver Wildlife Research Center, Sandusky, Ohio – European Starling), **Donald S. Eastman** (British Columbia

Ministry of Environment, Lands and Parks, Victoria – Olive-sided Flycatcher, Western Wood-Pewee, and Alder and Willow flycatchers), **John T. Emlen** (University of Wisconsin, Madison – Cliff Swallow), **Anthony J. Erskine** (Sackville, New Brunswick – Olive-sided Flycatcher, Say's Phoebe, and Barn Swallow), **Millicent S. Ficken** (University of Wisconsin, Milwaukee – Black-capped Chickadee), **Kathleen E. Franzreb** (Clemson University, Clemson, South Carolina – Brown Creeper), **W. Earl Godfrey** (Ottawa, Ontario – House Wren), **Joseph A. Grzybowski** (Norman, Oklahoma – Barn Swallow), **D. Paul Hendricks** (George M. Sutton Avian Research Center, Bartlesville, Oklahoma – American Pipit), **Rachel F. Holt** (University of British Columbia, Vancouver – Mountain Bluebird), **Tracey D. Hooper** (University of British Columbia, Vancouver – Sprague's Pipit), **Daryl Howes-Jones** (Kitchener, Ontario – Warbling Vireo), **Jocelyn Hudon** (Provincial Museum of Alberta, Edmonton – Alder and Willow flycatchers), **Frances C. James** (Florida State University, Tallahassee – American Robin), **Ross D. James** (Royal Ontario Museum, Toronto – Eastern Phoebe, Solitary Vireo, and Warbling Vireo), **Ned K. Johnson** (University of California, Berkeley – Pacific-slope Flycatcher), **Stephen R. Johnson** (LGL Limited, Sidney, British Columbia – European Starling and Crested Myna), **E. Dale Kennedy** (Manhattan, Kansas – House Wren), **Brina Kessel** (University of Alaska, Fairbanks – Gray-cheeked Thrush), **Lawrence Kilham** (Lyme, New Hampshire – White-breasted Nuthatch), **Frank Kime** (Tappen, British Columbia – Siberian Accentor), **Hugh E. Kingery** (Denver, Colorado – American Dipper), **Richard L. Knight** (Colorado State University, Boulder – Black-billed Magpie), **Don Kroodsma** (University of Massachusetts, Amherst – Rock and Bewick's wrens), **Mark G. Lewis** (The Whale Museum, Friday Harbor, Washington – Sky Lark), **David J. Low** (British Columbia Ministry of Environment, Lands and Parks, Kamloops – Pygmy Nuthatch), **William A. Lunk** (Ann Arbor, Michigan – Northern Rough-winged Swallow), **Jo Ann and Hue MacKenzie** (South Surrey, British Columbia – Black-capped Chickadee and Northern Mockingbird), **David A. Manuwal** (University of Washington, Seattle – Hammond's Flycatcher), **Jina Mariani** (United States Forest Service, Butte, Montana – Brown Creeper), **Katherine Martin** (Canadian Wildlife Service, Delta, British Columbia – American Robin), **Stephen G. Martin** (S.G. Martin and Associates Inc., Wellington, Colorado – Varied Thrush), **William J. Merilees** (British Columbia Parks Branch, Parksville – Crested Myna), **Eugene S. Morton** (Smithsonian Institution, Washington, D.C. – Purple Martin), **Michael T. Murphy** (Hartwick College, Oneonta, New York – Eastern Kingbird), **Barry R. Noon** (Redwood Sciences Laboratory, Arcata, California – Swainson's Thrush), **Henri Ouellet** (Hull, Quebec – Say's Phoebe and Gray Jay), **Cynthia A. Paszkowski** (University of Alberta, Edmonton – Hermit Thrush), **Harold S. Pollock** (Victoria, British Columbia – Mountain and Western bluebirds), **David R.C. Prescott** (University of Calgary, Alberta – Alder and Willow flycatchers), **Martin G. Raphael** (Forest Sciences Laboratory, Olympia, Washington – Hammond's Flycatcher), **Stephan Reebs** (University of Moncton, New Brunswick – Black-billed Magpie), **Robert E. Ricklefs** (University of Pennsylvania, Philadelphia – Northern Rough-winged Swallow), **Don Roberson** (Pacific Grove, California – Gray Flycatcher and Brown Thrasher), **Scott K. Robinson** (Illinois Natural History Survey, Urbana – Red-eyed Vireo), **Stephen I. Rothstein** (University of California, Santa Barbara – Cedar Waxwing), **Howard F. Sakai** (Arcata, California – Pacific-slope and Hammond's flycatchers), **John M. Schukman** (Leavenworth, Kansas – Say's Phoebe), **Spencer G. Sealy** (University of Manitoba, Winnipeg – Least Flycatcher, Eastern and Western kingbirds), **Gary F. Searing** (LGL Limited, Sidney, British Columbia – European Starling), **David M. Scott** (University of Western Ontario, London – Gray Catbird), **Brian E. Sharp** (Portland, Oregon – Purple Martin), **Joanne Siderius** (Nelson, British Columbia – Eastern Kingbird), **Jamie N.M. Smith** (University of British Columbia, Vancouver – American Robin), **Susan M. Smith** (Mount Holyoke College, South Hadley, Massachusetts – Chestnut-backed and Black-capped chickadees), **R. Dan Strickland** (Ontario Ministry of Natural Resources, Whitney – Gray Jay), **Bridget J. Stutchbury** (York University, North York, Ontario – Purple Martin and Tree Swallow), **William J. Sydeman** (Point Reyes Bird Observatory, Stinson Beach, California – Pygmy Nuthatch), **Diana F. Tomback** (University of Colorado, Denver – Clark's Nutcracker), **John Toochin** (Vancouver, British Columbia – Crested Myna), **Stephen B. Vander Wall** (University of Nevada–Reno, Reno – Clark's Nutcracker), **Nicolaas A.M. Verbeek** (Simon Fraser University, Burnaby, British Columbia – Horned Lark, American Pipit, and Northwestern Crow), **Jared Verner** (Forestry Sciences Laboratory, Fresno, California – Marsh Wren), **Terence R. Wahl** (Bellingham, Washington – Eurasian Skylark), **Wayne C. Weber** (British Columbia Forest Service, Merritt – Olive-sided, Least, and Yellow-bellied flycatchers, Western Wood-Pewee, Acadian Flycatcher, and Black Phoebe), **John A. Wiens** (Colorado State University, Fort Collins – Horned Lark), **John G. Woods** (Parks Canada, Revelstoke, British Columbia – Steller's Jay and Common Raven), and **John L. Zimmerman** (Kansas State University, Manhattan – Gray Catbird).

APPENDICES

The regional migration dates listed in Appendix 1 were completed by the following people: **J.E. Victor Goodwill** and **Bryan R. Gates** (Victoria), **Adrian Dorst** (Tofino), **Wayne C. Weber** (Vancouver), **Jan Bradshaw** (Harrison), **Tony Greenfield** (Sechelt), **Margo Hearne** and **Peter J. Hamel** (Masset), **Richard J. Cannings** (Okanagan), **Gary S. Davidson** (Nakusp and Fort Nelson), **Larry R. Halverson** (Radium), **Richard R. Howie** (Kamloops), **Anna Roberts** (Williams Lake), **Manfred Roschitz** (Quesnel), **Jack Bowling** (Prince George and Fort Nelson), **Chris R. Siddle** (Fort St. John).

Nearly 40 years of Christmas Bird Counts published in *Audubon Field Notes, American Birds,* and *Field Notes* and summarized in Appendix 2 were compiled onto species spreadsheets by Eileen C. Campbell, Tessa N. Campbell,

Maureen L. Funk, Laura Gretzinger, Edward L. Nygren, and Pam Stacey.

PRODUCTION

We very much enjoyed our affiliation with UBC Press and their associates. Professional help and advice were provided to the authors by Peter Milroy (Director), Holly Keller-Brohman (Managing Editor), and George Maddison (Production Manager). Francis Chow of F & M Chow Consulting copy-edited the book with efficacy; Eric Leinberger and Vincent Kujala designed and prepared the maps and technical figures; and Irma Rodriguez of Artegraphica Design Co. Ltd. typeset and completed the layout for much of the book. Chris Tyrrell of the Royal British Columbia Museum designed the jacket.

INTRODUCTION

GENERAL

Nearly six years have passed since the publication of Volumes 1 and 2 of *The Birds of British Columbia* (Campbell et al. 1990a, 1990b). While preparing those volumes, the authors became convinced that the extensive data on the passerine birds could yield much more detailed information if the move was made from the cumbersome paper card files to a computerized data base. The new technology would not only facilitate the ordering and analysis of the data but also provide a data base that could easily be kept current and used to supplement the information in the published volumes. During the past 6 years, the task of entering into our computers the approximately 500,000 occurrence records and 50,000 breeding records covering all 156 regularly occurring species of the passerine birds of the province was undertaken, and it is now complete.

Thanks to the generosity of the many thousands of cooperators who have provided us with data (Fig. 1), we have been able to interpret the patterns of distribution, migration, habitat association, and breeding chronology for the passerine birds of British Columbia. Meanwhile, the extensive body of data from which the generalizations in the species accounts and elsewhere in this volume have been derived remains in the data base for examination in many contexts that we cannot now imagine. They will thus serve the changing needs of those entrusted with conserving the richest assemblage of bird species of any province in Canada.

For the authors, creating and managing this vast body of data, and moulding it into the texture of this volume, has been a new experience. We have learned much; the task was not easy, but we are confident that the refinements in detail that were possible have improved the accuracy, usefulness, and interest of the volume.

This is the third of what will be 4 volumes. Volume 3 includes the first half of the Order Passeriformes (Perching Birds), from the flycatchers through the vireos. It includes the Families Tyrannidae (Tyrant Flycatchers, 22 species); Alaudidae (Larks, 2 species); Hirundinidae (Swallows, 7 species); Corvidae (Jays, Magpies, and Crows, 9 species); Paridae (Titmice, 4 species); Aegithalidae (Bushtits, 1 species); Sittidae (Nuthatches, 3 species); Certhiidae (Creepers, 1 species); Troglodytidae (Wrens, 6 species); Cinclidae (Dippers, 1 species); Muscicapidae (Kinglets, Bluebirds, and Thrushes, 14 species); Mimidae (Catbirds, Mockingbirds, and Thrashers, 4 species); Prunellidae (Accentors, 1 species); Motacillidae (Wagtails and Pipits, 5 species); Bombycillidae (Waxwings, 2 species); Laniidae (Shrikes, 2 species); Sturnidae (Starlings and Allies, 2 species); and Vireonidae (Vireos, 5 species). Volume 4 will contain accounts of the Emberizidae (Wood Warblers, Sparrows, Buntings, Blackbirds, and their allies), Fringillidae (finches), and Passeridae (Old World Sparrows).

The team of 6 authors that prepared Volumes 1 and 2 has grown by one. While a statistician with the Canadian Wildlife Service, G.E. John Smith assisted in producing the maps and computer programs for the first 2 volumes, and did likewise for this volume.

The format of each species account follows that of the first 2 volumes so that the information you seek will be where you expect it to be. The same is true of the common appendices. The use of computer technology, however, has made it possible to generate information with greater depth and detail than was possible for the nonpasserine volumes. This is particularly noticeable in the treatment of the elements in the breeding cycle and the analysis of changes in populations with time. However, the new detail is not without constraints, and we urge users to study the section "Methods, Terms, and Abbreviations."

Ornithologists and naturalists alike have been fascinated by the rich variety of birds that inhabit the widely diverse ecosystems of British Columbia. During the period covered by this volume, there was intensive public scrutiny

Figure 1. Ornithologists and naturalists have contributed more than 550,000 occurrence and breeding records of passerine birds from British Columbia (near Port Clements, Queen Charlotte Islands, 8 June 1990; R. Wayne Campbell).

Figure 2. The passerine birds include many of the most habitat-sensitive species in British Columbia, and can be an important indicator of our performance as conservators (Creston Wildlife Management Area, 31 May 1995; R. Wayne Campbell).

of the manner in which we, in British Columbia, have been exploiting the natural resources of the province, and the impacts on the land and its biota. Needed changes have been identified, sustainability has emerged as an urgent objective of resource use, and the maintenance of biodiversity is seen as a major criterion of progress towards this objective. Effective application of the strategy that is developing requires detailed information on the fauna and flora of the province, including the presence, numbers, movements, biological associations, and habitat associations of the species within the various ecoprovinces. The body of data available on the birds of British Columbia exceeds that of any other animal group and provides a valuable entry point for studies of biodiversity.

The passerine birds include many of the most habitat-sensitive species in the avifauna of the province, and can be useful indicators of our performance as conservators (Fig. 2). We have assembled all existing information on the geographic, seasonal, and biological distribution; the timing and biological features of the reproductive period; and changes in abundance with time of the passerine species as they occur in British Columbia. While the development of sound conservation plans for many species will require more detail than is now available, the information presented here will provide for the immediate and continuing needs of naturalists, professional biologists, and resource administrators, and at the same time serve as an effective starting point for further exploration and research.

Figure 3. Northwestern Crow, Victoria, 26 December 1995 (R. Wayne Campbell). More than one-half of the world's population breeds in British Columbia.

The Bird Resource: Its National and International Significance

In the introduction to Volume 1 (p. 4), we noted that, among the bird species breeding in Canada, 65 species nested in one province only, and more than one-half of those occur in British Columbia. Of these, 25 are nonpasserine species, to which we can add 12 passerine species:

Gray Flycatcher
Pacific-slope Flycatcher

Figure 4. Say's Phoebe, Atlin, 7 July 1996 (R. Wayne Campbell). British Columbia supports most of the Canadian breeding population.

Figure 5. Townsend's Solitaire, Hat Creek, 13 July 1996 (R. Wayne Campbell). Up to 20% of the world's population breeds in British Columbia.

Northwestern Crow (Fig. 3)
Chestnut-backed Chickadee
Bushtit
Pygmy Nuthatch
Canyon Wren
Bewick's Wren
Western Bluebird
Crested Myna
Hutton's Vireo
Black-throated Gray Warbler

British Columbia also supports most of the Canadian population of another 27 species of passerine birds. The survival of these species as part of the Canadian fauna depends largely upon the maintenance in the province of the special habitat conditions they require, in large enough areas to support viable populations. These species are:

Western Wood-Pewee
Willow Flycatcher
Hammond's Flycatcher
Dusky Flycatcher
Say's Phoebe (Fig. 4)
Violet-green Swallow
Northern Rough-winged Swallow
Steller's Jay
Clark's Nutcracker
Mountain Chickadee
Rock Wren
American Dipper
Townsend's Solitaire (Fig. 5)
Varied Thrush
Sage Thrasher
Warbling Vireo
Townsend's Warbler
MacGillivray's Warbler
Western Tanager
Black-headed Grosbeak
Lazuli Bunting
Brewer's Sparrow
Golden-crowned Sparrow
Rosy Finch
Cassin's Finch
House Finch
Red Crossbill

In Volume 1 we listed 10 species of nonpasserine birds whose breeding population in British Columbia has international significance because it represents a substantial proportion of the world population. There are 11 species of passerine birds in the same category. These species, with estimates of the proportion of their world population found in British Columbia, are as follows:

Hammond's Flycatcher (30-40%)
Northwestern Crow (50-70%; Fig. 3)
Chestnut-backed Chickadee (25-40%)
Townsend's Solitaire (10-20%; Fig. 5)
Varied Thrush (30-40%)
American Dipper (15-30+%)
Townsend's Warbler (40-60%)
MacGillivray's Warbler (30-40%)
Western Tanager (10-20%)
Golden-crowned Sparrow (40-60%)
Rosy Finch (20-40%)

The nesting habitat that these 11 species occupy in British Columbia is vitally important, but the role of the province as an essential part of the migration routes of many passerine species is of at least equal consequence. Because many species of passerine birds are less obvious than the nonpasserines, and because many of them migrate at night, the details of their passage are less well known. It is reasonably certain that the entire Alaska- and Yukon-nesting populations of such species as Varied Thrush, Hermit Thrush, and Townsend's Warbler pass through the province, along with those of many distinctive western subspecies of such widespread species as Swainson's Thrush, Orange-crowned and Wilson's warblers, Fox Sparrow, Lincoln's Sparrow, Song Sparrow, and others.

Figure 6. Mountain Bluebird, 24 March 1996 (R. Wayne Campbell). Records of Mountain Bluebirds wintering in British Columbia are the highest in Canada.

Along the coastal strip of southern British Columbia, the winter climate is moderated by the ocean. Here extended periods of subfreezing temperatures are infrequent, and of modest intensity when they occur. Christmas Bird Counts from the coast reveal that many passerine species are winter residents. For 8 passerine species, British Columbia has reported the highest numbers on Christmas Bird Counts in North America. These are all from the Georgia Depression Ecoprovince.

Sky Lark
Northwestern Crow (see Fig. 3)
Chestnut-backed Chickadee
Golden-crowned Kinglet
Dusky Thrush
Red-throated Pipit
Crested Myna
Brambling

For another 51 species, southern British Columbia has produced the highest Christmas Bird Counts in Canada:

Say's Phoebe (see Fig. 4)
Western Kingbird
Violet-green Swallow
Cliff Swallow
Barn Swallow
Steller's Jay
Clark's Nutcracker
Mountain Chickadee
Bushtit
Pygmy Nuthatch
Rock Wren
Canyon Wren
Bewick's Wren
House Wren
Winter Wren
Marsh Wren
American Dipper
Ruby-crowned Kinglet
Western Bluebird
Mountain Bluebird (Fig. 6)
Townsend's Solitaire (see Fig. 5)
Swainson's Thrush
Hermit Thrush
American Robin
Varied Thrush
American Pipit
European Starling (Fig. 7)
Hutton's Vireo
Orange-crowned Warbler
Yellow Warbler
Black-throated Gray Warbler
Townsend's Warbler
Black-and-White Warbler
MacGillivray's Warbler
Wilson's Warbler
Western Tanager
Green-tailed Towhee
Spotted Towhee
Fox Sparrow

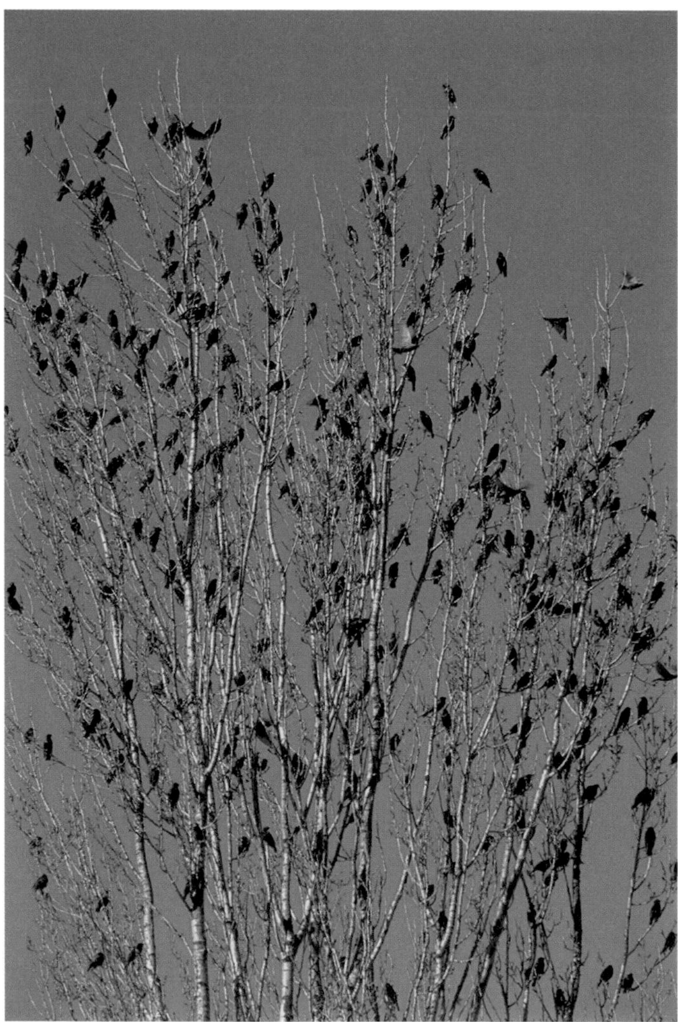

Figure 7. European Starlings, Vancouver, 26 February 1996 (R. Wayne Campbell). The highest numbers reported on Christmas Bird Counts in Canada are found in southern British Columbia.

Figure 8. Red Crossbill, Burnaby, 24 March 1970 (R.Wayne Campbell). Although this species is considered "irruptive" in its migratory habits in British Columbia, the province has produced the highest numbers on Christmas Bird Counts in Canada.

Song Sparrow
Lincoln Sparrow
Golden-crowned Sparrow
White-crowned Sparrow
Harris' Sparrow
Western Meadowlark
Yellow-headed Blackbird
Brewer's Blackbird
Rosy Finch
Purple Finch
Cassin's Finch
Red Crossbill (Fig. 8)

While these highest species totals from the Christmas Bird Counts are an interesting element in the hobby of bird-watching, many have little biological significance. Included in the list are 17 species, largely swallows and warblers, represented by one or a few laggard or dysfunctional individuals from the host of migrants that passed through southern British Columbia. Another 4 species are from the nesting avifauna of northeastern Asia – individuals that migrated down the wrong side of the Pacific Ocean. A few others are single records of wanderers from ranges in other parts of the continent. Birds in these three categories are evidence that the remarkable behavioural endowment of the migratory birds has occasional imperfections. The majority of the list, however, is drawn from the 59 passerine species for which southern British Columbia is part of the normal winter range.

Of the 59 species for which the Christmas Bird Counts in British Columbia have produced the highest totals in Canada, 76% of the highest counts were recorded in the Georgia Depression, 19% in the Southern Interior, 3% in the Queen Charlotte Islands, and 2% in the Southern Interior Mountains.

THE ENVIRONMENT AND CHANGES TO ECOPROVINCES OF BRITISH COLUMBIA

The system of ecological classification presented in detail in Volume 1 (Demarchi et al. 1990, pp. 57-142) has become the standard in British Columbia and has been found to be adaptable for describing regional ecosystems in adjacent provinces and states (Demarchi 1994). In this volume, we have followed that system with a few minor but important refinements that have been adopted by biogeographers of the province (Demarchi 1993, 1995).

While the central concept of the Ecoregion Classification remains the same, some of the boundaries have been redefined (see inside covers). For example, 10 ecoprovinces are still recognized, but the Boreal Plains Ecoprovince no longer extends to the Muskwa Plateau Ecosection; rather, this ecosection is now placed in the Taiga Plains Ecoprovince (Demarchi 1993).

For almost all regions of the province, the distribution of the birds is well reflected in the ecoprovinces. Along the coast, however, the birds of the extensive region designated as the Coast and Mountains Ecoprovince reveal patterns of distribution that are best reflected by further subdividing that ecoprovince. These units include Western Vancouver Island, the Queen Charlotte Islands, the Northern Mainland Coast, and the Southern Mainland Coast (see inside covers). In the text, and on our graphs of seasonal fluctuations and occurrence and breeding chronology, these appear as though they were ecoprovinces.

Further changes to the Ecoregion Classification for British Columbia may be necessary in the future as mapping of the province's ecosystems becomes both more detailed and more extensively applied; however, the changes will probably be confined to boundary definitions and will probably include few new units (Demarchi 1995).

TAXONOMY AND NOMENCLATURE

The general stipulations on this topic given on p. 145 of Volume 1 are equally applicable to the present volume. We have continued to follow the *Check-list of North American Birds*, 6th edition, and its supplements (American Ornithologists' Union 1982, 1983, 1985, 1987, 1989, 1993, 1995).

Recent studies of the DNA of birds by Ahlquist et al. (1987) and Sibley and Ahlquist (1981, 1982a, 1982b, 1982c, 1983, 1984, 1985, 1986, 1987, 1990) have shed new light on the relationships within the passerine genera and species. These have been summarized in the monumental volume *Distribution and Taxonomy of Birds of the World* by Sibley and Monroe (1990). The changes they have found necessary in order to reflect the new knowledge of the relationships within the passerine species of North America have not yet stood the test of scrutiny by the Committee on Classification and Nomenclature of the American Ornithologists' Union, so we have not adopted them here. There is little doubt, however, that this new approach to the taxonomy of birds will in the near future substantially influence our understanding of this complex aspect of ornithology.

METHODS, TERMS, AND ABBREVIATIONS

This section discusses changes to the methodology used in preparing the 2 volumes on the passerine birds; the reader is also referred to pp. 146-51 of Volume 1.

Data Sources and Limitations

The species accounts in Volumes 3 and 4 are based on data from early historical times until 31 December 1995.[1] Except for the following changes, the sources for the data have been previously discussed (Volume 1, p. 146).

The Christmas Bird Count summaries in the passerine species accounts and in Appendix 2 include only official counts (see Volume 1, p. 146) for the 37 years from 1957 to 1993. The locations of the 60 Christmas Bird Counts in British Columbia, along with a complete summary of each locality, are given in Appendix 2.

The Breeding Bird Survey summaries in the passerine species accounts and in Appendix 3 were compiled from data obtained from the Bird Banding Office, Canadian Wildlife Service, Ottawa. The locations of the 100 Breeding Bird Survey routes in British Columbia, along with a complete summary of each route, are given in Appendix 3. Within the Noteworthy Records for each species, the 3 highest counts recorded on coast and interior surveys are included.

For each account with sufficient data, we have also included in the BREEDING section the trend that the species is undergoing on both coastal and interior Breeding Bird Surveys. These data were generated by Len Thomas (pers. comm.) using a compromise between the trend analysis methods of Collins (1990) and Sauer and Geissler (1990).

For the purposes of trend analysis, British Columbia was divided into 2 strata, the Coast and the Interior (see Volume 1, Fig. 176). Smaller strata would contain insufficient data for analysis for most species.

Because of the many factors affecting migratory bird populations in British Columbia, we concluded that the Breeding Bird Survey data were important information that could flag a warning to wildlife managers about species showing significant downward trends in numbers. Thus, we chose to accept as significant those species with a test statistic that had an associated probability of 10% or less. We are not saying that these populations in British Columbia are in trouble; we are saying, however, that the data indicate that numbers for the species show a downward trend within the region(s) noted, and wildlife managers may want to look into the reasons for these apparent declines.

For species and regions where the trend was significant, we have displayed a smooth line to show the variation over the period between 1968 and 1994. This was computed using the LOWESS (locally weighted regression) method described in Cleveland (1979) and Kennedy (1981).

For a complete description and the purpose of the Breeding Bird Surveys, see Robbins et al. (1986).

Reports of birds banded in British Columbia were obtained from the Bird Banding Office, Canadian Wildlife Service, Ottawa. These data, covering the period 1923 through 1991, were incorporated into our data sets. Where there was enough information for a particular species, the data formed the basis for a number of figures showing banding and recovery locations.

Data Management and Summarization

Computer Analysis System

The assembly and analysis of the massive amount of information presented in a work of this size require a great deal of careful planning and effort. Much greater use was made of electronic data processing in the final 2 volumes than in Volumes 1 and 2. Putting the raw data in electronic form accomplished several objectives. It provided a permanent and readily accessible record of over *500,000* observations of all types. It also enabled rigorous editing and checking of the data so that errors and omissions were minimized in the final product.

The approach to writing the accounts differed greatly from that followed in the first 2 volumes. With the raw data computerized, numerous data summaries were routinely produced, and a number of interactive programs were developed so that the authors could search for specific information in the data base. In addition, some of the standard information in the accounts – such as clutch sizes, composition of nesting materials, frequencies of nests by habitat, and heights above ground – were automatically written in sentence form by the computer programs.

These processes greatly reduced the time spent by the authors in compiling standard data summaries, and enabled them to devote more time to considering important information available through the data base as well as from other sources.

Overview of the Data System

The data system was programmed almost entirely in Borland International's data base software, dBASE. We initially used dBASE III Plus and finished with dBASE IV. The distribution maps were produced using the same system as for Volumes 1 and 2, with the mapping programs written using Microsoft BASIC to produce plotting files in Houston Instruments' Hiplot plotting language for output on a Houston Instruments DMP-42 plotter.

The programs summarized data from 3 types of data files: occurrence files, single breeding files, and colony files. The output consisted of (1) working printouts and electronic files for use by the authors to write the species accounts, (2) electronic text files with standard information in sentence form, which could be placed directly in the species account with at most minor grammatical changes, and (3) a map file from which a map consisting of 3 overlays was produced by WestCad Graphics Services. Further, a series of interactive programs were written so that the authors could search for specific information as they wrote the species accounts.

The data base files and data entry screens were designed by Neil K. Dawe. The dBASE IV programs used to summarize the data were written by Neil K. Dawe, G.E. John Smith, and Jordan T. Dawe. The dBASE IV interactive programs were written by Neil K. Dawe, and the Microsoft BASIC programs were written by G.E. John Smith.

[1] Significant records up to the date of publication have been included as postscripts or addenda.

We used personal computers running under Windows 3.1 to facilitate efficient writing. This system was configured to allow the 3 main programs used by the authors – dBASE IV, WordPerfect 5.1, and the Provincial Gazetteer – to run simultaneously. The authors were able to switch from one program to another instantly with a simple series of keystrokes; they could, for example, summarize specific information with a dBASE interactive program and then move immediately to WordPerfect to complete a particular section of text.

The Raw Data

Occurrence Data Base

This data base contains data transferred from occurrence cards in the British Columbia Wildlife Records Scheme filled out by both volunteers and professionals. The majority of these records were verified against the original cards. The data base covers all observations of a species except reports or evidence of nesting. The structure for this data base is shown in Appendix 5.

Breeding Data Base

This data base, a summarization of the British Columbia Nest Records Scheme, consists of nesting information for various species. For a particular year, there is 1 record for each nest regardless of how many times the nest was observed. The complete data base for each species consists of a primary and secondary file and, in the case of a colonial nesting species, a tertiary file. The structure of this data base is shown in Appendix 5.

The primary data base file contains all the constant information pertaining to a nesting record, such as the site location, nest habitat, and observer.

The secondary data base file consists of information about individual observations of a nest, such as date, number of eggs, or number of young. Each record corresponds to an observation. Records in this data base are linked to those in the primary breeding file by the field PRIMARY_NO, thus making it unnecessary to enter the constant nest information in the record for each observation at the nest.

The tertiary data base file contains information about individual visits to a nesting colony and exists only for colonial species. Records in this file are also linked to those in the primary breeding file by the field PRIM_RECNO.

Programs to Summarize the Data

The program system was written in dBASE and consisted of several components. The first produced a standard set of tables containing basic information about each species that was useful for writing the species accounts. The system also prepared part of the information as an electronic text file that could be entered directly into the species account, making transcription unnecessary. In the case of a breeding or colonial species, additional information was provided.

The second purpose of the program system was to provide a list of all the observers whose records were used to compile this volume (Appendix 4, Contributors). A set of computer files used to produce the distribution maps was also prepared.

Finally, a set of interactive programs was written to allow the authors to gather or check information quickly; produce a variety of standard figures, such as the seasonal fluctuations or occurrence and breeding chronology figures; and calculate summaries such as coastal and interior nest success or the variety of nest predators.

Auxiliary Files for the Analysis

Because the data were in electronic form, it was possible to summarize them by ecoprovince in each species account. Because the ecoprovince data were constant, the ecoprovince was not entered in the species data bases. Rather, a separate dBASE file was created that allowed the species file to be linked to the ecoprovince file (containing constant information about the ecoprovinces) by the GRID field. Data from this separate file were used to electronically attach the ecoprovince to every record for each species in the occurrence or breeding files. This facilitated the production of summaries by ecoprovince.

Two other data base files were used by the programs to provide the summarized information: PASSSORT.DBF and WEEKEND.DBF. The structure of these files is shown in Appendix 5.

A list of all the programs used to summarize the data, along with a brief description of their action, is found in Appendix 6.

FORM AND CONTENT

The presentation of the species accounts in the passerine volumes adheres to the pattern established in Volumes 1 and 2 (see Volume 1, pp. 148-51). Because the passerine data were computerized, however, the interpretation of some data was more refined than was possible in the previous volumes. We were able to provide more detail regarding movements of birds and their numbers in various ecoprovinces and ecodivisions, and have included additional figures showing this detail where warranted.

Annual occurrence and breeding chronology. In the first 2 volumes, each map page carried in the upper right corner a bar graph representing the period during which the species is present in British Columbia (in black) and the period during which there is evidence of breeding (in red). With the passerine data on computer, it has been possible to present this information for each ecoprovince, or, in the coastal sector of the province, by ecodivision. Thus a more sensitive and regionally useful presentation of the data is available to those who need it. This graph is presented separately from the map. It should be noted that all records are shown for the week in which they occurred.

Regions of highest numbers in summer and winter. For each species there is a figure showing the ecoprovince, or, on the coast, the ecodivision, with the highest numbers in winter and summer. To avoid excessive complexity, the detail has been kept to the ecoprovince (ecodivision on the coast) level, but this does not imply that the species is most likely to occur throughout that ecoprovince. Several ecoprovinces are geographically extensive and ecologically diverse, and a given species will occur only in those parts where its special requirements are met.

This map was prepared by considering such factors as the number of records, the mean number of birds per record, Christmas Bird Count, and Breeding Bird Survey data.

Seasonal fluctuations. From the data we could derive the accumulated total of all birds reported as seen in each ecoprovince month by month, as well as the number of records that have provided these observations. These data are useful as an indication of a species' arrival or its seasonal changes in numbers within an ecoprovince, although some interpretation is required. It should be obvious that all data below 1 are zero.

Seasonal fluctuations in total number of birds recorded can be dramatic. For example, over the study period total numbers of Violet-green Swallows recorded in February in the Georgia Depression Ecoprovince is 54, in March nearly 20,000, and in April nearly 54,000 birds. In the same months the number of records in February is 54, in March 708, and in April 1,250. In order to display these numbers in a way that permits comparison between ecoprovinces, the seasonal fluctuation data are presented on a logarithmic scale.

Changes in abundance with time. In the case of some species, the Breeding Bird Surveys have indicated a change in numbers present on the survey routes through the 10 or more years that counts have been made. All such apparent changes were subjected to a statistical test for probability of significance. The level of 90% or more was selected for any figure given.

We also supplemented the passerine accounts with two new sections: Brown-headed Cowbird Parasitism and Nest Success.

Brown-headed Cowbird Parasitism. The ratio of parasitized nests was determined for each species both at the coast and in the interior by relating the number of nests in which eggs or young of the cowbird were recorded to the number of nests examined with sufficient care that the presence of cowbird eggs or young would have been detected. Only records that include a statement of the numbers of eggs or young present were used in the calculation. Records of young cowbirds being fed by foster parents were not included in the data used to calculate the percentage of parasitism of a host species. They were, however, summed and included as supplementary information on cowbird parasitism. They also helped to establish the range of hosts used by the Brown-headed Cowbird.

Nest Success. This was calculated from the total number of nests first discovered with eggs that were followed to a known fate. A nest was deemed successful if at least 1 young fledged. In some cases this was assumed if on one visit the young were fully feathered and ready to leave the nest and on a subsequent visit there was no sign of predation.

Nests were not included in the data for calculating success where a nest or eggs were collected for museum purposes, or where desertion was probably the result of the observer's visit. In some instances where nests were parasitized by the Brown-headed Cowbird, the observers noted that they had removed the eggs or young or punctured the cowbird eggs. Such records were not included in the calculations of nest success.

CHECKLIST OF BRITISH COLUMBIA BIRDS

Passerines: Flycatchers through Vireos

This phylogenetic list includes 91 species of birds, flycatchers through vireos, that have been documented in British Columbia through 31 December 1995.

Order PASSERIFORMES: Perching Birds

Family TYRANNIDAE: Tyrant Flycatchers
- Olive-sided Flycatcher
- Western Wood-Pewee
- Yellow-bellied Flycatcher
- Acadian Flycatcher
- Alder Flycatcher
- Willow Flycatcher
- Least Flycatcher
- Hammond's Flycatcher
- Dusky Flycatcher
- Gray Flycatcher
- Pacific-slope Flycatcher ("Western Flycatcher" Complex)
- Black Phoebe
- Eastern Phoebe
- Say's Phoebe
- Ash-throated Flycatcher
- Great-crested Flycatcher
- Tropical Kingbird
- Thick-billed Kingbird
- Western Kingbird
- Eastern Kingbird
- Gray Kingbird
- Scissor-tailed Flycatcher

Family ALAUDIDAE: Larks
- Sky Lark
- Horned Lark

Family HIRUNDINIDAE: Swallows
- Purple Martin
- Tree Swallow
- Violet-green Swallow
- Northern Rough-winged Swallow
- Bank Swallow
- Cliff Swallow
- Barn Swallow

Family CORVIDAE: Jays, Magpies, and Crows
- Gray Jay
- Steller's Jay
- Blue Jay
- Western Scrub-Jay
- Clark's Nutcracker
- Black-billed Magpie
- American Crow
- Northwestern Crow
- Common Raven

Family PARIDAE: Titmice
- Black-capped Chickadee
- Mountain Chickadee
- Boreal Chickadee
- Chestnut-backed Chickadee

Family AEGITHALIDAE: Bushtits
- Bushtit

Family SITTIDAE: Nuthatches
- Red-breasted Nuthatch
- White-breasted Nuthatch
- Pygmy Nuthatch

Family CERTHIIDAE: Creepers
- Brown Creeper

Family TROGLODYTIDAE: Wrens
- Rock Wren
- Canyon Wren
- Bewick's Wren
- House Wren
- Winter Wren
- Marsh Wren

Family CINCLIDAE: Dippers
- American Dipper

Family MUSCICAPIDAE: Kinglets, Bluebirds, and Thrushes
- Golden-crowned Kinglet
- Ruby-crowned Kinglet
- Blue-gray Gnatcatcher
- Northern Wheatear
- Western Bluebird
- Mountain Bluebird
- Townsend's Solitaire
- Veery
- Gray-cheeked Thrush
- Swainson's Thrush
- Hermit Thrush
- Dusky Thrush
- American Robin
- Varied Thrush

Family MIMIDAE: Catbirds, Mockingbirds, and Thrashers
Gray Catbird
Northern Mockingbird
Sage Thrasher
Brown Thrasher

Family PRUNELLIDAE: Accentors
Siberian Accentor

Family MOTACILLIDAE: Wagtails and Pipits
Yellow Wagtail
Black-backed Wagtail
Red-throated Pipit
American Pipit
Sprague's Pipit

Family BOMBYCILLIDAE: Waxwings
Bohemian Waxwing
Cedar Waxwing

Family LANIIDAE: Shrikes
Northern Shrike
Loggerhead Shrike

Family STURNIDAE: Starlings and Allies
European Starling
Crested Myna

Family Vireonidae: Vireos
Solitary Vireo
Hutton's Vireo
Warbling Vireo
Philadelphia Vireo
Red-eyed Vireo

Regular Species

Olive-sided Flycatcher

OSFL

Contopus borealis (Swainson)

RANGE: Breeds from western and central Alaska east across north-central Canada to central Newfoundland, and south to northern Baja California in the west; in the east from central Saskatchewan, Manitoba, and the northern Great Lakes states south through New York in mountainous areas to eastern Tennessee and western North Carolina. Winters in South America from Colombia and Venezuela to Peru, and casually in southern California.

STATUS: *Uncommon* migrant and summer visitant throughout most of the province; absent from the Queen Charlotte Islands.

Breeds.

NONBREEDING: The Olive-sided Flycatcher is widely distributed throughout British Columbia. It occurs from Vancouver Island and the mainland coast east across the province to the Rocky Mountains, and in suitable habitat throughout the interior to the extreme northern portions of the province. It is absent from the Queen Charlotte Islands.

The Olive-sided Flycatcher has been reported from near sea level to the subalpine and alpine regions near 2,400 m elevation. It prefers the edges of mature coniferous forests, both pure and mixed, especially adjacent to water (Fig. 13). These edges may be natural (ponds, sloughs, lakes, wetlands, estuaries, blowdowns, and rivers) or human-made (burns, clearcuts, transmission-line corridors, highways, and recreational clearings such as ski areas and campsites). It also frequents open woodlands on steep mountain slopes (Munro 1935a; Van Tighem and Gyug 1983) and deciduous woodlands, especially in swamp and floodplain forests. Within these habitats, tall trees with dead tops are necessary perch sites for maintenance, singing, and foraging activities. Erskine (1977) mentions that as long as snags and stubs remain, the Olive-sided Flycatcher will be a characteristic bird of burned and clearcut regions in boreal Canada.

This species shows flexibility in its choice of habitats. It has been found in old-growth and climax forests near Sproat Lake (Bryant et al. 1993) and Terrace (Webster 1969a), and in immature and young second-growth forests near Port McNeill, Port Alice (Buckner et al. 1975), and Garibaldi Park (Dow 1963). On the Sunshine Coast, the Olive-sided Flycatcher occurs on the Tetrahedron Plateau in climax mountain hemlock and yellow cedar forests, but only where there are openings created by upland bogs with standing snags (T. Greenfield pers. comm.).

On the south coast, spring migration begins as early as the third week of April, while on the north coast the earliest spring migrants arrive in early May (Figs. 9 and 10). The main coastal movement occurs from mid to late May. In the southern portions of the interior, spring arrivals also begin to appear about the third week of April but the main movement is not evident until mid to late May. By early May, spring migrants have reached the Central Interior and Boreal Plains ecoprovinces, and by mid-May have reached the far northern portions of the province (Fig. 10).

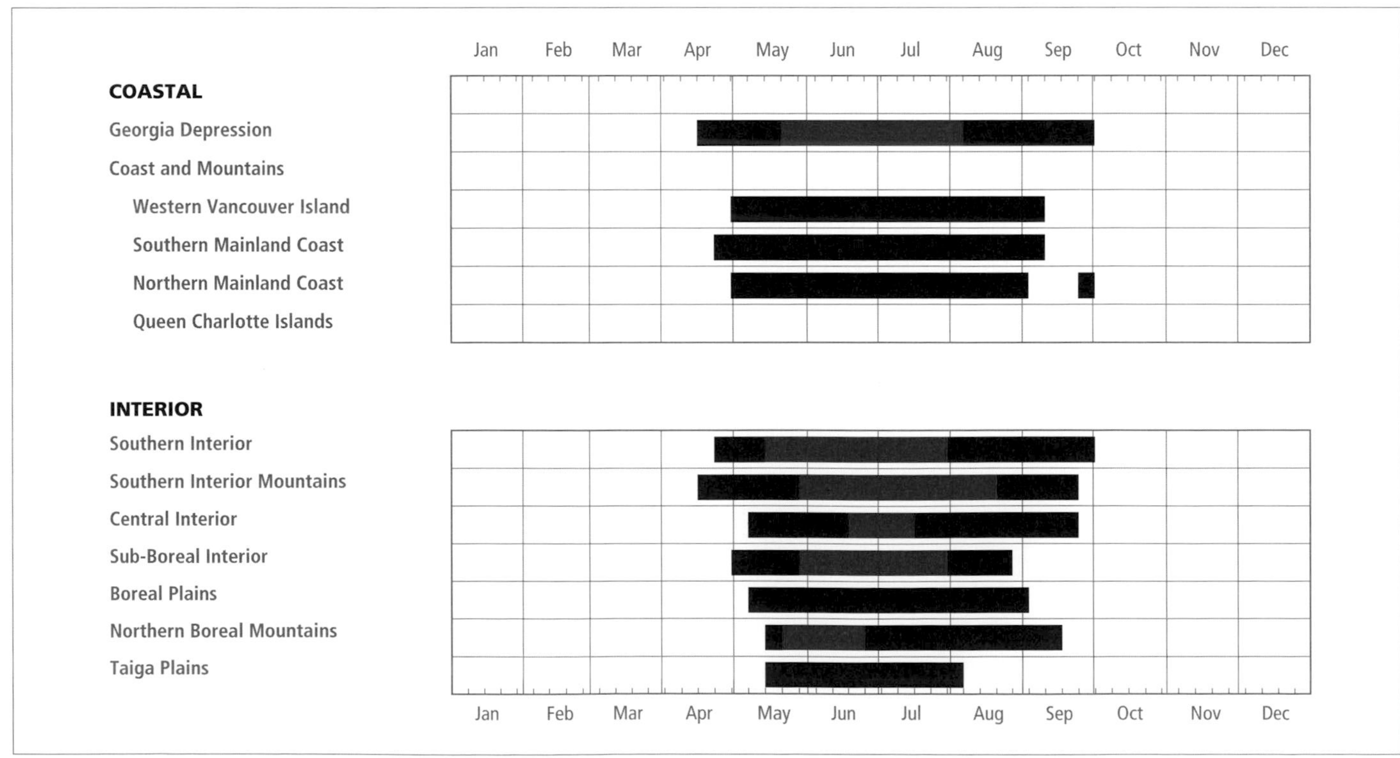

Figure 9. Annual occurrence (black) and breeding chronology (red) for the Olive-sided Flycatcher in ecoprovinces of British Columbia.

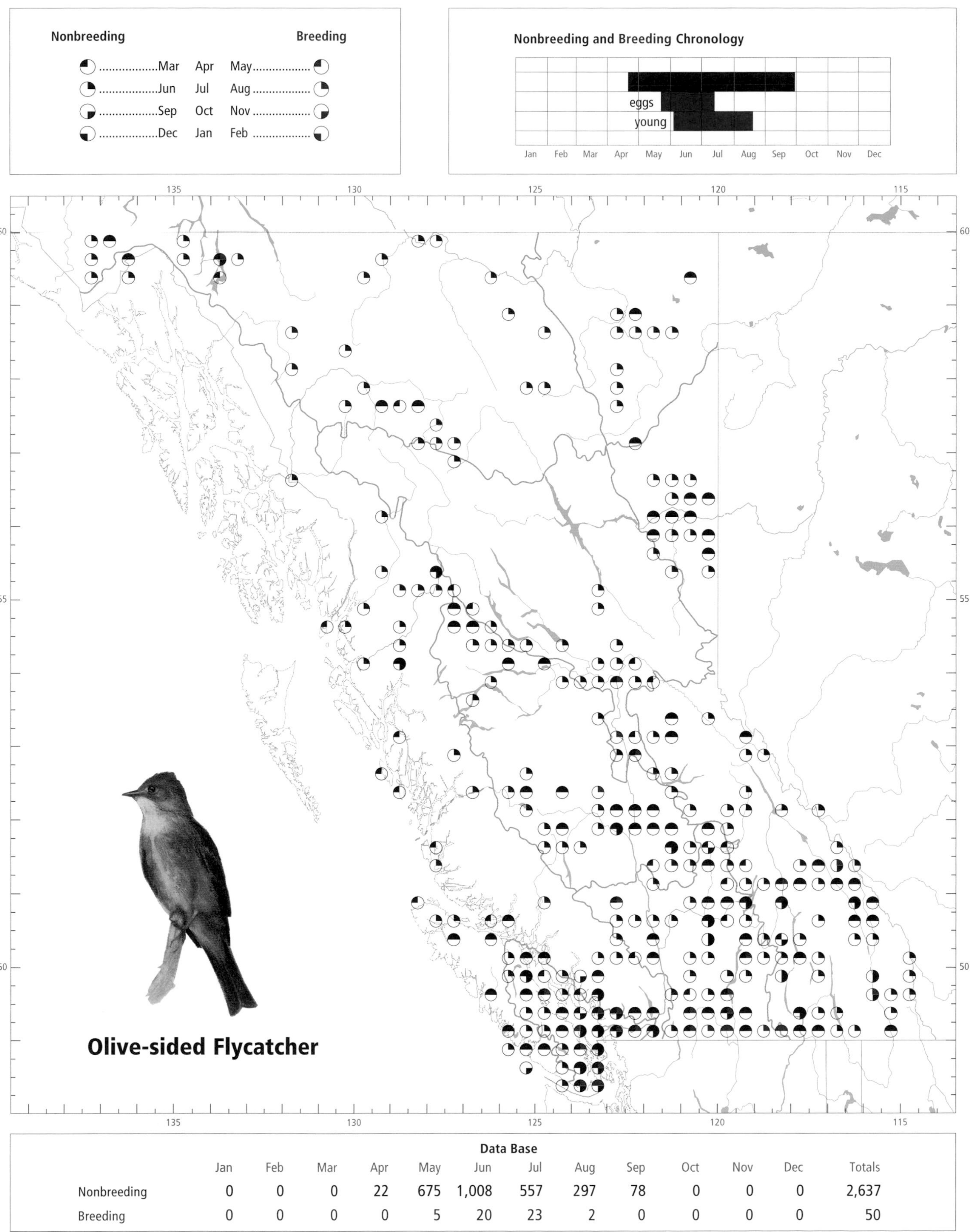

Data Base

	Jan	Feb	Mar	Apr	May	Jun	Jul	Aug	Sep	Oct	Nov	Dec	Totals
Nonbreeding	0	0	0	22	675	1,008	557	297	78	0	0	0	2,637
Breeding	0	0	0	0	5	20	23	2	0	0	0	0	50

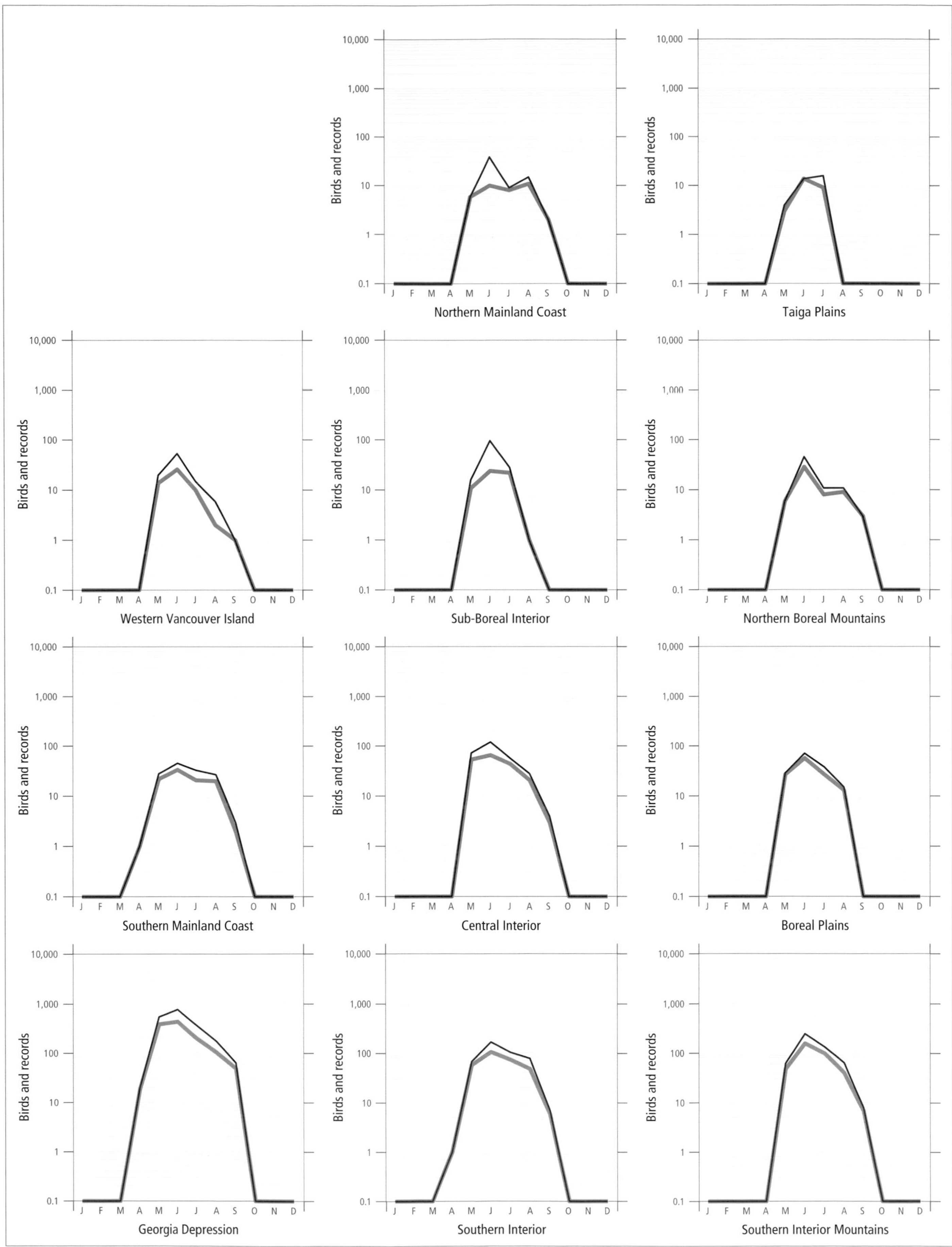

Figure 10. Fluctuations in total number of birds (purple line) and total number of records (green line) for the Olive-sided Flycatcher in ecoprovinces of British Columbia. Nest record data have been excluded.

In autumn, the migration is less pronounced. Most birds have left the far northern interior regions of the province by the end of August. In the southern portions of the interior and along the coast, a few birds can be found through September (Figs. 9 and 10). By October, all birds have left the province.

On the coast, the Olive-sided Flycatcher has been recorded from 20 April to 29 September; in the interior, it has been recorded from 20 April to 27 September (Fig. 9).

BREEDING: The Olive-sided Flycatcher has a widespread breeding distribution across the southern portions of the province, including Vancouver Island, although it likely breeds throughout most of forested British Columbia except the Queen Charlotte Islands.

Its highest numbers occur mainly on southeastern Vancouver Island, in the Georgia Depression Ecoprovince, and in the southwestern portion of the Sub-Boreal Interior Ecoprovince (Fig. 11). It appears to be sparsely but evenly distributed elsewhere in suitable habitats. An analysis of Breeding Bird Surveys for the period 1968 through 1993 shows that the mean number of birds on coastal routes decreased at an average annual rate of 5% (Fig. 12a); the mean number of birds on interior routes also decreased at an average annual rate of 5% over the same period (Fig. 12b).

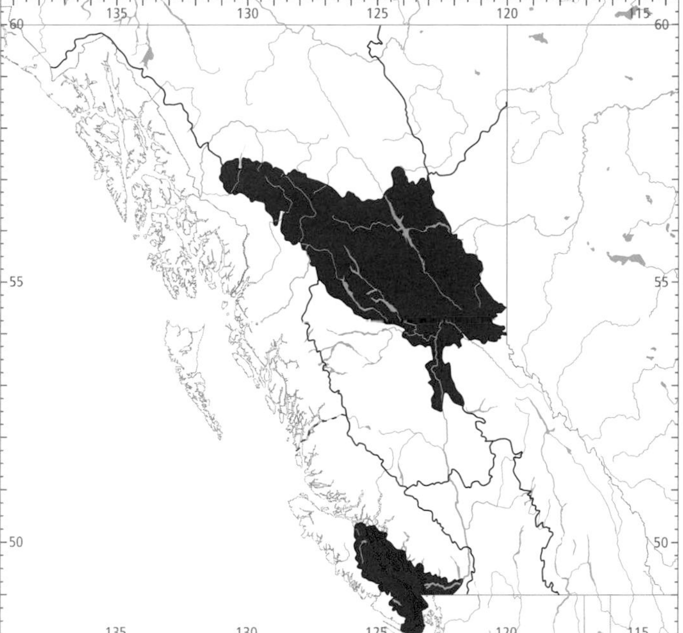

Figure 11. In British Columbia, the highest numbers for the Olive-sided Flycatcher in summer occur in the Georgia Depression and Sub-Boreal Interior ecoprovinces.

Summaries of North American Breeding Bird Surveys indicate that between 1965 and 1985 western continental populations of the Olive-sided Flycatcher declined sharply, with British Columbia populations showing the strongest decrease (Robbins et al. 1986; Droege and Sauer 1987).

Possible reasons for the decline include destruction of tropical wintering habitats, loss of suitable breeding and foraging habitats (e.g., suitable perch sites in tall mature trees), logging activities that alter age classes of forests, and use of forest pesticides and herbicides (Marshall 1988).

The Olive-sided Flycatcher breeds from near sea level to about 2,200 m elevation. On the south coast, it was 10 times more abundant at 1,100 m than between 100 and 400 m elevation (Weber 1975). In the interior, it is a scarce breeder in valley bottoms but becomes more abundant at higher elevations. In the Okanagan valley, most breeding records were from above 900 m elevation, within the Douglas-fir zone; however, there are records from lower elevations at Richter Lake (600 m) and Okanagan Landing (Cannings et al. 1987).

Breeding habitat includes the edges of semi-open mature coniferous forests and mixed woodlands, usually near water (Figs. 13 and 14). One nest was found in a shrub thicket and 3 nests were in human-influenced habitats. Considering the widespread disturbance of habitats in this province (from logging, agriculture, urbanization, and transportation corridors), it is significant that 79% of the nests ($n = 19$) were found in pristine forests. Half of the Olive-sided Flycatcher nests were found in semi-open coniferous forests. Most of the rest were in mixed forests.

On the coast, the Olive-sided Flycatcher has been recorded breeding from 30 May to 5 August; in the interior, it

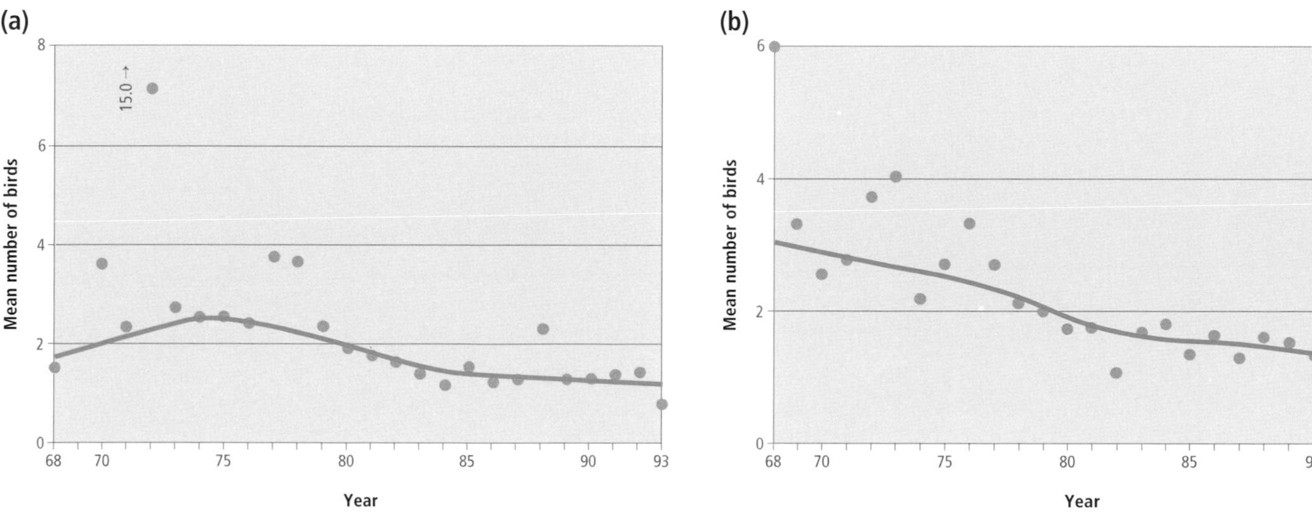

Figure 12. An analysis of Breeding Bird Surveys for the Olive-sided Flycatcher in British Columbia shows that the mean number of birds on coastal routes (a) decreased at an average annual rate of 5.0% over the period 1968 through 1993 ($P < 0.01$); the mean number of birds on interior routes (b) also decreased at an average annual rate of 5.0% over the same period ($P < 0.05$).

Figure 13. In the central interior of British Columbia, the Olive-sided Flycatcher frequents mixed woodland habitats of white spruce, lodgepole pine, Sitka alder, and water birch, usually adjacent to water (Shane Lake, 14 June 1996; Neil K. Dawe).

has been recorded from 20 May (calculated) to 19 August (Wade 1977) (Fig. 9).

Nests: Most nests (76%; *n* = 25) were situated in coniferous trees, including Douglas-fir (45%), ponderosa pine, white spruce, and Engelmann spruce. Several nests were found in trembling aspen and 1 was found in a willow shrub. All of the described nests (*n* = 17) were attached to the upper surface of a horizontal branch, generally well out on the branch but occasionally near the trunk. Nests were bulky structures of interwoven twigs, sticks, and rootlets. The nest cup was lined with beard-lichens, grass, and other plant fibres. Nest heights (*n* = 27) ranged from 3 to 15 m; 14 were between 6 and 12 m.

Eggs: Dates for 14 clutches ranged from 22 May to 15 July. Calculated dates indicate that nests may have eggs as early as 20 May. Sizes of 8 clutches ranged from 1 to 4 eggs (1E-1, 2E-1, 3E-3, 4E-3). Bent (1942) states that the Olive-sided Flycatcher almost invariably lays 3 eggs (Fig. 15), occasionally 4. Bendire (1895) states that about 1 nest in 20 contains 4 eggs. The incubation period is between 10 and 17 days (Walkinshaw 1957); Ehrlich et al. (1988) give it as 14 days.

Young: Dates for 20 broods ranged from 3 June to 19 August. Sizes of 14 broods ranged from 1 to 4 young (1Y-1, 2Y-2, 3Y-9, 4Y-2). The nestling period is 21 to 23 days (Ehrlich et al. 1988).

Nest Success: Of 5 nests found with eggs and followed to a known fate, 2 produced at least 1 fledgling.

Brown-headed Cowbird Parasitism: In British Columbia, only 1 of 26 nests recorded with eggs or young was parasitized by the cowbird. That nest was discovered on southern Vancouver Island. It was only the fourth reported instance in North America of the Olive-sided Flycatcher acting as host for the cowbird (Friedmann 1963; Friedmann et al. 1977; Friedmann and Kiff 1985).

REMARKS: There are no recognized subspecies of the Olive-sided Flycatcher. Bent (1942) gives general life-history information for this species.

Figure 14. In the Cariboo-Chilcotin region of British Columbia, the Olive-sided Flycatcher breeds in mixed, mature coniferous forests of white spruce and lodgepole pine (Stum Lake, 25 May 1993; R. Wayne Campbell).

Figure 15. Nest with 3 eggs of the Olive-sided Flycatcher (Mount Douglas Park, 12 June 1995; R. Wayne Campbell).

NOTEWORTHY RECORDS

Spring: Coastal – Metchosin 20 Apr 1987-1; near Victoria 30 May 1970-4 eggs; near Fulmore Lake 29 Apr 1976-1; Tugwell Lake 19 May 1985-8; Long Beach (Tofino) 2 May 1985-1; Vancouver 20 Apr 1989-1; Gold River 2 May 1974-6; Miracle Beach Park 7 Apr 1961-1; Cape Scott 11 May 1987-2; McInnes Island 1 May 1964-1; Gnarled Islands 17 May 1987-1 (Rodway and Lemon 1991); Hazelton 1 May 1921-1 (MVZ 42181). **Interior** – Richter Lake 22 May 1971-2 eggs just laid; Christina Lake 23 May 1988-adult on eggs; McLean Creek (Okanagan Falls) 8 May 1913-1 (Anderson 1914); 8 km e Princeton 28 Apr 1990-1; Nakusp 10 May 1974-1; Invermere 20 Apr 1981-1; MacArthur Slough (Kamloops) 15 May 1982-1; Revelstoke 3 May 1976-1, 31 May 1976-6; Lac la Hache 27 May 1993-5; 50 km w Williams Lake 10 May 1978-1 at Junction Sheep Range; Quesnel 28 May 1982-1; 24 km s Prince George 4 May 1987-1; Dawson Creek 11 May 1992-1; Taylor 12 May 1987-1; Fort St. John 31 May 1986-1; near Clarke Lake 21 May 1979-1; Parker Lake 10 Jun 1979-2; Atlin 17 May 1934-1 (Swarth 1936).

Summer: Coastal – Sooke Hills 1 Aug 1983-12; near Victoria 10 Jun 1967-4 eggs, 5 Aug 1967-3 fledglings; Tugwell Lake 6 Jul 1983-15; Duncan 15 Jun 1969-3 eggs and 1 Brown-headed Cowbird egg; Kennedy Lake 18 Jul 1972-5; Lighthouse Park 2 Jun 1979-10; Royston 5 Jul 1991-2 nestlings; near Alta Lake 1 Jul 1975-3; Nass River 21 Jun 1975-3. **Interior** – Manning Park 22 Jun 1986-4; Barrett Creek (Nelson) 2 Jun 1983-7; Brookmere 19 Jun 1974-9; Grindrod 28 Aug 1953-1 (UBC 7912); Amiskwi River 1 Jul 1976-7 (Wade 1977); Emerald Lake 19 Aug 1976-adult feeding a recently fledged young (Wade 1977); Vermilion River 28 Jun 1972-10; Pearson Creek 18 Aug 1985-1 fledgling being fed; Kleena Kleene 19 Aug 1962-2 (Paul 1964); Stum Lake 15 Jul 1973-3 fledglings being fed; Bowron Slough 10 Aug 1975-1, last seen (Bell 1975); Prince George 16 Jun 1979-5, 19 Aug 1982-1; St. John Creek 31 Aug 1985-1; Iskut 25 Jun 1990-6; n Stikine River 18 Jun 1975-1; Clarke Lake 30 Jul 1978-1; 22 km ne Fort Nelson 2 July 1982-2 (Campbell and McNall 1982); Liard River 10 Aug 1943-1 (Rand 1944); Wilson Creek 21 Jun 1914-4 eggs (Anderson 1915).

Breeding Bird Surveys: Coastal – Recorded from 25 of 27 routes and on 64% of all surveys. Maxima: Elsie Lake 17 Jun 1970-22; Port Hardy 19 Jun 1982-17; Alberni 10 Jun 1969-15. **Interior** – Recorded from 60 of 73 routes and on 56% of all surveys. Maxima: Ferndale 9 Jun 1968-30; Telkwa High Road 23 Jun 1974-20; Lac la Hache 24 Jun 1973-17.

Autumn: Interior – Atlin 13 Sep 1931-1 (RBCM 5690); Wineglass Ranch (Riske Creek) 10 Sep 1985-1; Golden 2 Sep 1977-2; Kootenay National Park 6 Sep 1965-1 (Seel 1965); Mount Assiniboine 1 Sep 1976-1; Kinnaird Park 22 Sep 1968-1; Scotch Creek 27 Sep 1962-1; Sugar Lake 1 Sep 1927-1; Castlegar 23 Sep 1968-1; Vaseux Lake 16 Sep 1959-1 (Cannings et al. 1987). **Coastal** – Kispiox Valley 2 Sep 1921-1, last seen (Swarth 1924); Kitimat River 29 Sep 1974-1; Egmont 4 Sep 1977-1; Capilano River 9 Sep 1970-1 (Campbell et al. 1972a); Vancouver 9 Sep 1972-4; Pitt Meadows 27 Sep 1964-1 (Hesse and Hesse 1965a); Pachena Point 6 Sep 1974-1 (Hatler et al. 1978); Saanich 28 Sep 1979-1.

Winter: No records.

Western Wood-Pewee

WWPE

Contopus sordidulus Sclater

RANGE: Breeds from central Alaska and southern Yukon east to south-central Manitoba, North Dakota, and Kansas, south to Baja California, Guatemala, and Honduras. Winters from southern Central America to Peru and Bolivia.

STATUS: On the coast, an *uncommon* to *fairly common* migrant and summer visitant to the Georgia Depression Ecoprovince, becoming *rare* to *uncommon* elsewhere on the coast, including Western Vancouver Island. Absent from the Queen Charlotte Islands.

In the interior, an *uncommon* to *fairly common* (locally *common*) migrant and summer visitant throughout the southern and central regions of the province north to the Boreal Plains Ecoprovince; further north and west it becomes *uncommon*.

Breeds.

NONBREEDING: The Western Wood-Pewee (Fig. 16) is widely distributed in forested areas throughout much of the southern two-thirds of the province. It is sparsely distributed in the Northern Boreal Mountains and Taiga Plains ecoprovinces. There are also large areas on the coast where the species has not been reported, including much of the exposed coastal areas of Western Vancouver Island, most of the southern and northern regions of the Coast and Mountains Ecoprovince between Rivers Inlet and Prince Rupert, and all of the Queen Charlotte Islands. In the interior, there is a large area on both sides of the Rocky Mountain Trench from Mackenzie north to Lower Post where records of the species are lacking.

Figure 16. Adult Western Wood-Pewee (Okanagan Falls, August 1989; Phil Gehlen).

The Western Wood-Pewee occurs from near sea level to about 1,900 m elevation. It frequents a wide variety of open coniferous, deciduous, and mixed forests, usually near water or around the edges of clearings (Fig. 21). Habitats include cottonwood riparian lowlands, beaver meadows, open coniferous forests, aspen and birch parklands, Garry oak woodlands, burns, brushy meadows, swamps, orchards, farmsteads, and gardens.

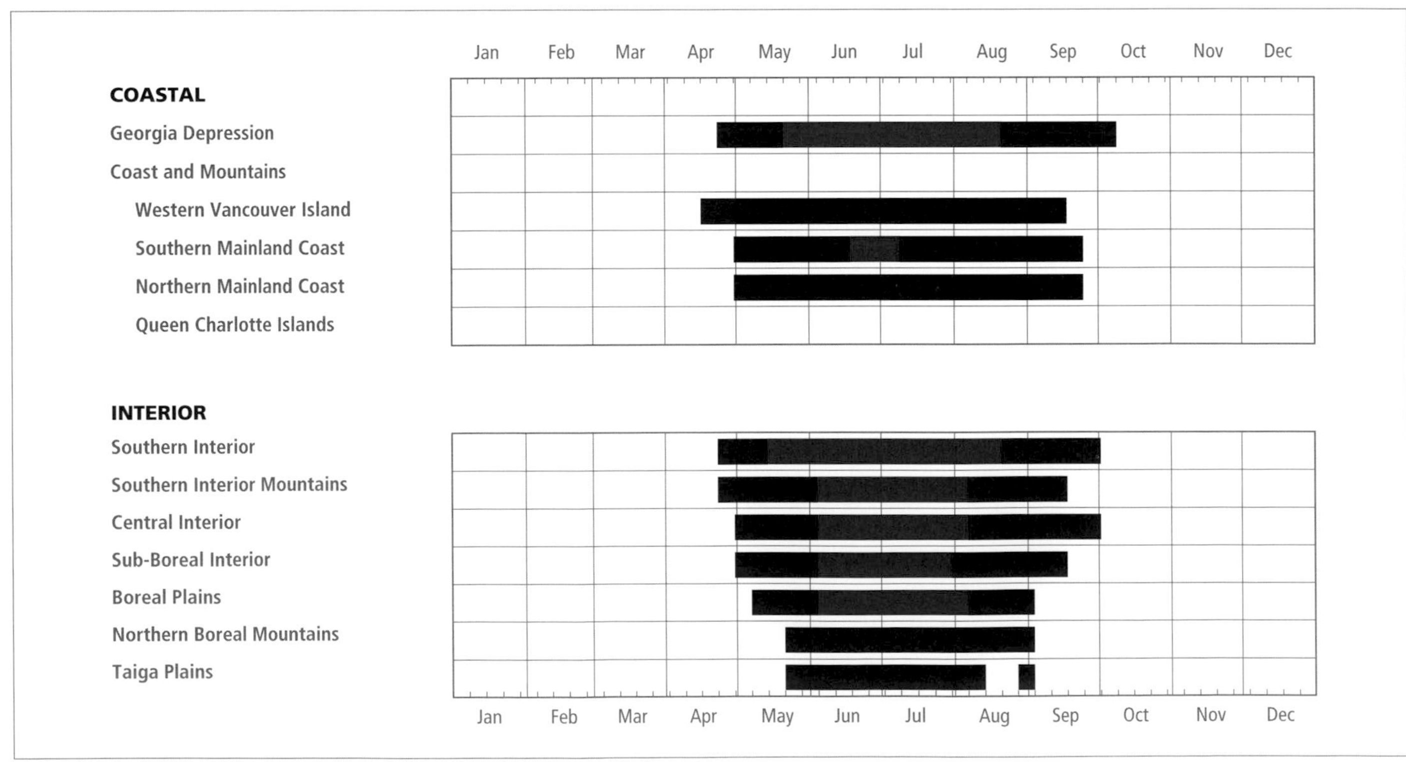

Figure 17. Annual occurrence (black) and breeding chronology (red) for the Western Wood-Pewee in ecoprovinces of British Columbia.

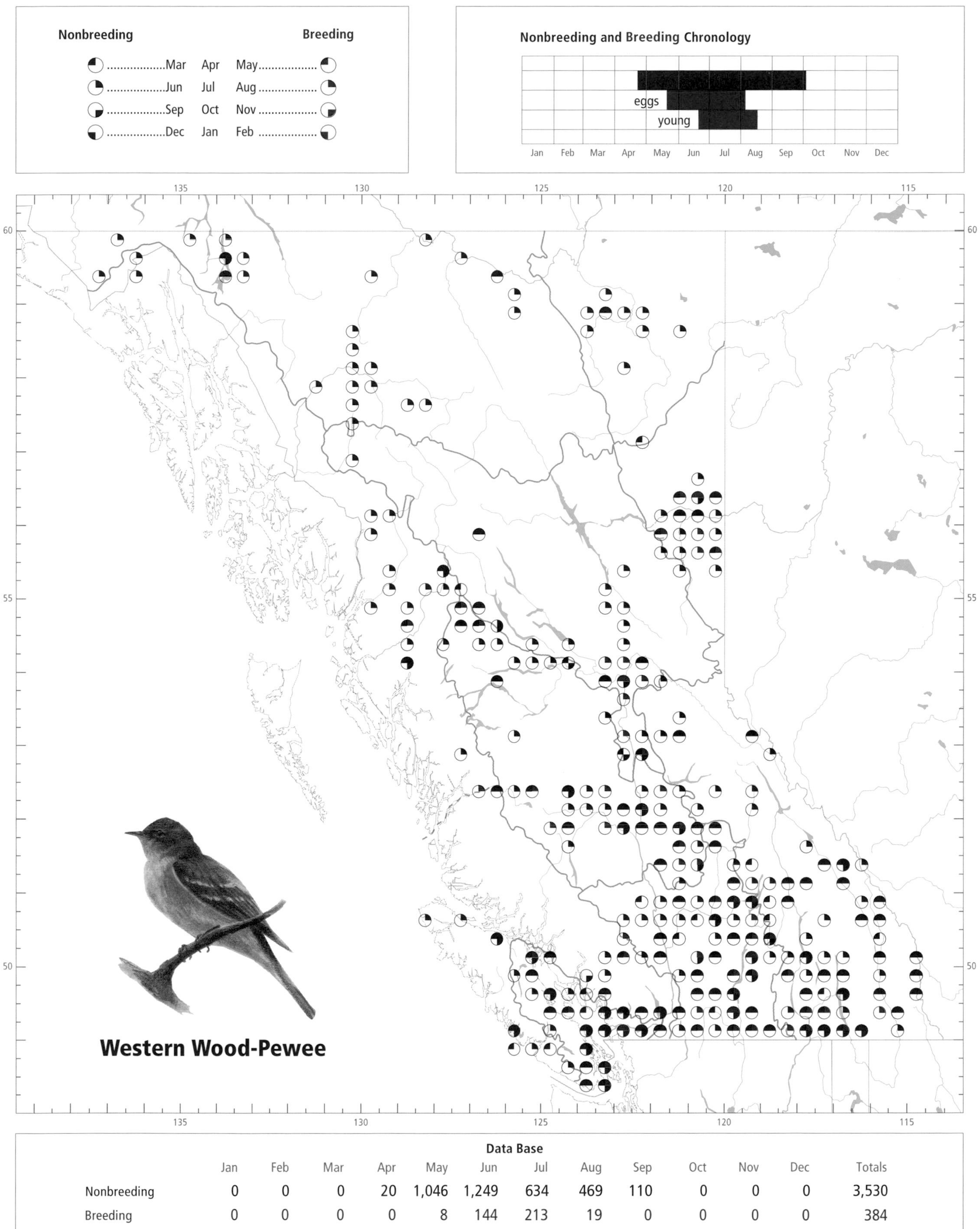

Data Base

	Jan	Feb	Mar	Apr	May	Jun	Jul	Aug	Sep	Oct	Nov	Dec	Totals
Nonbreeding	0	0	0	20	1,046	1,249	634	469	110	0	0	0	3,530
Breeding	0	0	0	0	8	144	213	19	0	0	0	0	384

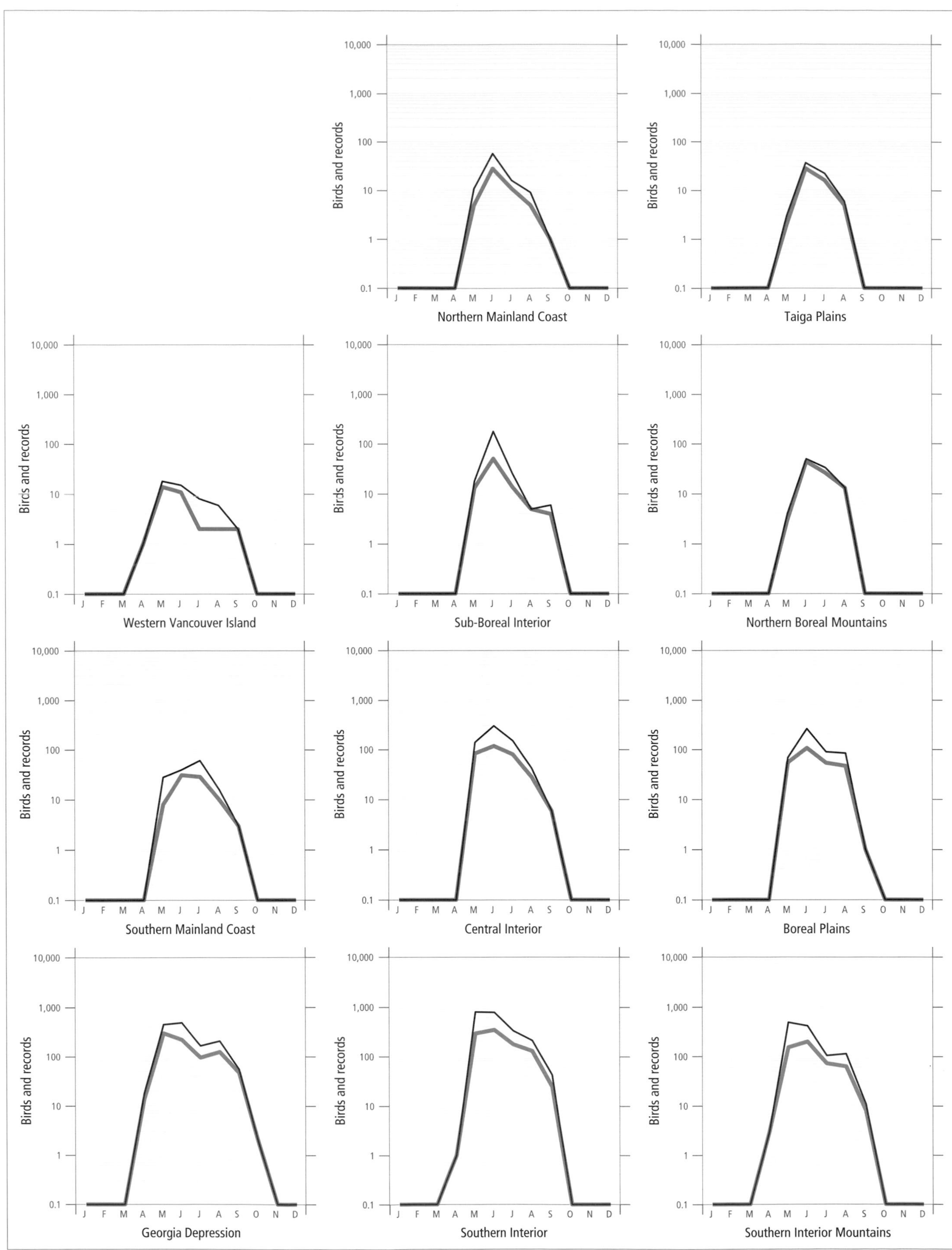

Figure 18. Fluctuations in total number of birds (purple line) and total number of records (green line) for the Western Wood-Pewee in ecoprovinces of British Columbia. Nest record data have been excluded.

Rand (1944) found the Western Wood-Pewee common at Muncho Lake in the scattered spruces of old alluvial fans. In the Peace River area, Cowan (1939) found it abundant in mature aspen forests. In the Okanagan valley, it is most common in ponderosa pine forests, orchards, gardens, and deciduous stands around lakes below 1,000 m (Cannings et al. 1987). In Mount Revelstoke and Glacier national parks, Van Tighem and Gyug (1983) report the highest densities in mixed forests and wetlands in the Interior Cedar-Hemlock forest. In the Smithers area, Pojar (1993) found the highest densities for this species in mature aspen stands, although it also occurred in old and mixed seral stages. Favourite perches for hunting and calling are the dead lower branches of aspens or the tops of smaller trees (Verbeek 1975).

In the Coast and Mountains (Western Vancouver Island), Georgia Depression, Southern Interior, and Southern Interior Mountains ecoprovinces, early spring migrants may arrive in the third or fourth week of April (Figs. 17 and 18), with a peak movement occurring about the second or third week of May. By early May, migrants have reached the Central Interior and Sub-Boreal Interior ecoprovinces, reaching the Boreal Plains by the second week of May; the main movement peaks in those areas by about the third week of that month. Further north, birds arrive about the last 2 weeks of May.

In the north, the autumn migration begins shortly after the young are on the wing, and most birds have left by early September (Fig. 18). In the south, both on the coast and in the interior, the main movement peaks in late August and early September. By mid-September most birds have left the province, and all birds have gone by the second week of October (Figs. 17 and 18).

Migrants are occasionally grounded by poor weather. In late May 1991, 50 Western Wood-Pewees were reported from the Nelson and Castlegar area, including a single flock of 25 birds.

On the coast, the Western Wood-Pewee has been reliably recorded in the province from 22 April to 3 October; in the interior, it has been recorded from 23 April to 30 September (Fig. 17; see REMARKS).

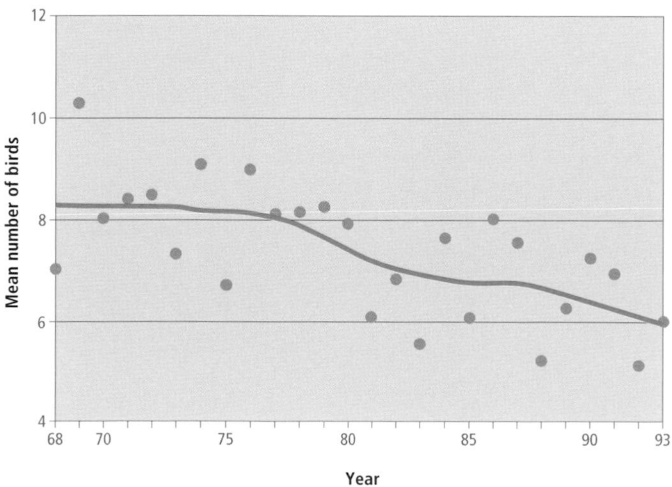

Figure 20. An analysis of Breeding Bird Surveys for the Western Wood-Pewee in British Columbia shows that the mean number of birds on interior routes decreased at an average annual rate of 1% over the period 1968 through 1993 ($P < 0.05$).

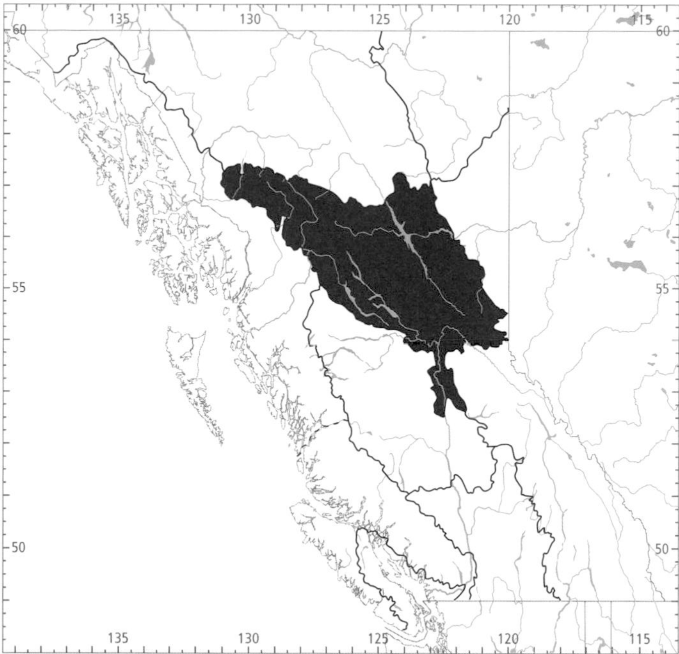

Figure 19. In British Columbia, the highest numbers for the Western Wood-Pewee in summer occur in the Sub-Boreal Interior Ecoprovince.

BREEDING: The Western Wood-Pewee has a widespread breeding distribution from southeastern Vancouver Island east across the province to the west Kootenay and north through the Central Interior to the Skeena River valley in the west and the Peace River area of the Boreal Plains in the east. Breeding records are lacking from Western Vancouver Island, most of the east Kootenay region, all but the extreme southern portions of the Sub-Boreal Interior, and the Northern Boreal Mountains and Taiga Plains. However, the species probably breeds throughout its summer range in the province.

The highest numbers for the Western Wood-Pewee in summer occur in the southern portions of the Sub-Boreal Interior (Rising and Schueler 1980) (Fig. 19). An analysis of Breeding Bird Surveys for the period 1968 through 1993 shows that the number of birds on interior routes has decreased at an average annual rate of 1% (Fig. 20); analysis of coastal routes for the same period could not detect a net change in numbers. As early as the mid-1940s, however, Pearse (1946) noted that the Western Wood-Pewee was "seldom heard now" in the vicinity of Comox on Vancouver Island.

The Western Wood-Pewee has been recorded breeding from near sea level to 1,700 m elevation. It breeds in forested habitats (45%; n = 170; Fig. 21) and human-altered habitats (39%) such as gardens, farms, orchards (Fig. 22), parks, playgrounds, campgrounds, and transmission-line corridors. The majority of nests were reported from the Southern Interior, where there was frequent use of gardens close to water. Similar sites were also used in the Georgia Depression and Southern Interior Mountains, where about one-third of all nests were found in altered habitats. These data may be biased towards nests that are more easily found in sparse growth and semi-open, agricultural areas rather than nests that are in forests or woodlands. In the north, this species frequently nests near beaver ponds. Occasionally mixed woods are used for nesting.

Most nests (43%; n = 142) were reported from cultivated environments. The remaining nests were associated with 24 other classes of open habitat, including swamps, marshes, beaver ponds with drowned trees, small lakes (Fig. 21), rangeland, sagebrush, and talus.

Cannings et al. (1987) state that breeding wood-pewees in the Okanagan were most abundant in aspen groves around ponds and lakes, often preferring areas with water-killed trees. In Kootenay National Park, breeding habitat is restricted to open Douglas-fir bottomlands along the Kootenay River and mixed successional forests of spruce, aspen, and alder (Poll and Porter 1984).

On the coast, the Western Wood-Pewee has been recorded breeding from 25 May to 14 August; in the interior, it has been recorded from 20 May to 17 August (Fig. 17).

Nests: All nests (n = 198) were found in trees, usually living but occasionally dead. Most nests (75%) were found in deciduous trees; trembling aspen (17%), poplar (15%), and black cottonwood (7%) were the most frequently reported native species. A wide variety of garden and orchard species were also used. In the Boreal Plains and Central Interior, up to 50% of the nests were found in trembling aspen, while in the Southern Interior ponderosa pine and poplar were reported with equal frequency (19%; n = 120), followed by trembling aspen (13%).

Almost all nests (97%; n = 172) were situated on horizontal tree branches, usually well out from the trunk; 25% were in natural forks or saddled the limb.

The heights of 187 nests ranged from 0.4 to 18 m above the ground (or water), with 65% between 2.4 and 8.0 m.

Nests were elegantly constructed cups composed mainly of grasses (75%). Other nest materials included lichens (26%), animal hair (15%), feathers (9%), bark strips (6%), and leaves (2%). In addition, mosses and plant down were used, and cobwebs were often incorporated into the outer surface of the basic structure. Lining material consisted of grasses, plant fibres and down, mosses, bark strips, and lichens.

Eggs: Dates for 136 clutches ranged from 20 May to 5 August, with 50% recorded between 20 June and 8 July. Calculated dates indicate that eggs could be found as early as 17 May. Sizes of 90 clutches ranged from 1 to 5 eggs (1E-1, 2E-19, 3E-45, 4E-24, 5E-1), with 77% having 3 or 4 eggs (Fig. 22). The incubation period is 12 or 13 days (Ehrlich et al. 1988).

Young: Dates for 94 broods ranged from 20 June to 17 August, with 52% recorded between 8 and 22 July. Calculated dates indicate that young could be found as early as 29 May. Sizes of 59 broods ranged from 1 to 5 young (1Y-8, 2Y-10, 3Y-27, 4Y-13, 5Y-1), with 68% having 3 or 4 young. The nestling period is 14 to 18 days (Ehrlich et al. 1988).

Figure 21. In central British Columbia, the Western Wood-Pewee breeds in mixed woods of willow, Sitka alder, and black spruce (Tabor Lake, 14 June 1996; Neil K. Dawe).

Brown-headed Cowbird Parasitism: In British Columbia, 6% of 139 nests found with eggs or young were parasitized by the cowbird. Interior nest parasitism was 5% (n = 133); only 1 of 6 coastal nests was parasitized. Friedmann and Kiff (1985) list only 8 instances of the Western Wood-Pewee serving as host for the cowbird in North America; 1 was from British Columbia (Friedmann 1934).

Nest Success: Of 21 nests found with eggs and followed to a known fate, 10 produced at least 1 fledgling.

REMARKS: Two subspecies occur in Canada: *C. s. veliei* Coues and *C. s. saturatus* Bishop, the latter being the race found in British Columbia (American Ornithologists' Union 1957).

All records of this species in British Columbia earlier than 22 April (see Cannings et al. 1987) and later than 3 October have been excluded from the analysis because they lack convincing details.

A Western Wood-Pewee banded as a fledgling in Vernon in May 1934 returned in May 1935 and "nested not only in the same apple tree but on top of its old nest" (Fowle 1940). Bent (1942) provides additional life-history information.

Figure 22. Nest and eggs of the Western Wood-Pewee in a cherry orchard (Keremeos, June 1964; John K. Cooper).

NOTEWORTHY RECORDS

Spring: Coastal – Victoria 6 May 1972-1 (Tatum 1973); Colwood 1 May 1986-1; Koksilah River 22 May 1971-10 (Tatum 1972); Chesterman Beach 22 Apr 1984-1; Surrey 25 May 1963-1 egg; Stanley Park (Vancouver) 1 May 1959-6; Burnt Hill (Pitt Meadows) 24 Apr 1982-1; Sheridan Hill 25 Apr 1980-1; Harrison 1 May 1981-1; Green Lake (Whistler) 11 May 1975-11; Port Neville 22 May 1975-2; Kitimat 15 May 1975-1, 29 May 1975-3. **Interior** – n end Osoyoos Lake 2 May 1974-2 (Cannings 1974); Kinnaird 23 Apr 1972-1; Castlegar to Nelson area 25 May 1991-50 grounded migrants, including 1 flock of 25; Horn Lake (Okanagan) 20 May 1905-4 fresh eggs collected; Penticton 1 May 1950-observed (Bowman 1950); Vernon 25 Apr 1962-1 (Cannings et al. 1987), 30 May 1980-constructing nest; Radium 12 May 1985-1; Nakusp 7 May 1979-1; Kamloops 24 Apr 1971-1; Riske Creek 2 May 1987-1; Williams Lake 3 May 1968-1, 3 May 1981-1; Quesnel 7 May 1982-1; Puntchesakut Lake 11 May 1944-1, first seen (Munro 1947a); 24 km s Prince George 4 May 1975-1; Vanderhoof 19 May 1945-1 (Munro 1949); Tupper Creek (Peace River) 18 May 1938-1 (Cowan 1939); Charlie Lake 30 May 1982-nest half complete; Cecil Lake 10 May 1987-1 (Campbell 1987c); Mile 325 Alaska Highway 23 May 1987-1 (McEwan and Johnston 1987); Liard Hot Springs 27 May 1981-1; Atlin 22 May 1932-1 (RBCM 5677).

Summer: Coastal – Victoria 14 Aug 1921-young in nest; Cultus Lake 21 Jun 1959-30; Ladner 30 Jun 1968-2; Pitt Meadows 20 Jun 1965-10; Mitlenatch Island 31 Aug 1974-1; Campbell River 25 Jun 1976-4 eggs; Bella Coola 3 Jul 1940-1 male juvenile (FMNH 173618); Kimsquit River 3 Aug 1985-5; Terrace 10 Jun 1977-building nest; New Hazelton 24 Aug 1917-1 (NMC 10937); Nass River 21 Jun 1975-13. **Interior** – Blue Lake (Richter Pass) 17 Jun 1960-3 eggs (Campbell and Meugens 1971); Castlegar 14 Jun 1980-2 eggs; Jaffray 22 Jun 1968-16; Barrett Creek (Nelson) 2 Jun 1983-14; Cranbrook 24 Jun 1945-3 eggs (Johnstone 1949); Naramata 5 Aug 1969-eggs, 17 Aug 1962-young (Cannings et al. 1987); Skookumchuck 8 Aug 1976-5 young being fed in nest; Lytton 12 Jul 1963-13; Nicola Lake 1 Aug 1976-12; Little Big Bar Lake 14 Jun 1987-25; Radium 31 Aug 1979-1; Hemp Creek 4 Jun 1962-31 (Dow 1962); Watson Lake 8 Jun 1958-adult incubating; Williams Lake 3 Aug 1971-young on nest flapping wings; Frost Lake (Prince George) 16 Jun 1972-4 eggs; Prince George 3 Aug 1992-fledglings from renest, 31 Aug 1990-1; Seymour Lake 13 Jul 1944-nest with young (Munro 1947a); Tupper 27 Jun 1982-18; Charlie Lake 6 Jun 1986-adult incubating, 14 Aug 1986-7; Beatton Park 4 Aug 1988-young in nest being fed; St. John Creek 28 Aug 1986-1, latest fall record; near Clarke Lake (Fort Nelson) 28 Aug 1978-1; Atlin 27 Aug 1924-1 (MVZ 44759).

Breeding Bird Surveys: Coastal – Recorded from 20 of 27 routes and on 40% of all surveys. Maxima: Kispiox 20 Jun 1993-19; Seabird 2 Jul 1989-17; Kitsumkalum 11 Jun 1978-14. **Interior** – Recorded from 68 of 73 routes and on 90% of all surveys. Maxima: Telkwa High Road 29 Jun 1989-39; Ferndale 7 Jun 1969-32; Fraser Lake 22 Jun 1969-29.

Autumn: Interior – Atlin 1 Sep 1931-1, last seen (Swarth 1936); Beatton Park 2 Sep 1982-1; Kispiox 4 Sep 1921-1, last seen (Swarth 1924); Vanderhoof 1 Sep 1945-1 (Munro 1949); Quesnel 1 Sep 1989-1; Williams Lake 30 Sep 1970-1; Lac la Hache 14 Sep 1942-1 (USNM 413237); Tranquille 9 Sep 1985-1; Monck Park 9 Sep 1971-1; Nakusp 7 Sep 1985-4; Crawford Bay 10 Sep 1989-1; Okanagan Landing 26 Sep 1933-1 (ROM 82796); Trout Creek (Okanagan) 25 Sep 1971-2. **Coastal** – Kitimat River 21 Sep 1991-1, on estuary; Port Neville 7 Sep 1975-1; Discovery Passage 5 Sep 1975-3; Egmont 5 Sep 1977-1; Goose Lake (Maple Ridge) 25 Sep 1964-2; Harrison 20 Sep 1969-1; Long Beach (Tofino) 7 and 11 Sep 1983-1; Richmond 25 Sep 1985-1 (Hunn and Mattocks 1986); Victoria 3 Oct 1954-1, last seen of year (Flahaut and Schultz 1955a).

Winter: No records.

Yellow-bellied Flycatcher

YBFL

Empidonax flaviventris (Baird and Baird)

RANGE: Breeds from west-central and southern Mackenzie and northeastern and east-central British Columbia across the northern and central Prairies, and east as far as Newfoundland and south through the northern Great Lakes states to North Dakota and New York. Winters from Mexico to South America.

STATUS: In the interior, *rare* migrant and *rare* to *uncommon* local summer visitant in east-central and northeastern British Columbia; *casual* elsewhere, except the southern Coast and Mountains Ecoprovince, where it is *accidental*. Breeds.

NONBREEDING: The Yellow-bellied Flycatcher (Fig. 23) is sparsely distributed in the northern two-thirds of the province, where it occupies much of the Taiga and Boreal Plains ecoprovinces; it also occurs locally in the southern portions of the Sub-Boreal Interior Ecoprovince east of Prince George. It has been recorded irregularly elsewhere in the province. Eleven specimens were collected at Indianpoint Lake, near Bowron Lakes, between 1929 and 1934 (Dickinson 1953), which suggests that the Yellow-bellied Flycatcher may be more common there than in other regions south and west of the Fort Nelson Lowland.

The Yellow-bellied Flycatcher occurs at between 13 and 975 m elevation. It prefers forest edges, second-growth woodlands, and swampy areas, including edges of open trembling aspen stands, second-growth deciduous woodlands, white birch near beaver ponds, and flooded willow swamps. In the Taiga Plains, Erskine and Davidson (1976) heard a Yellow-bellied Flycatcher singing from the edge of a black spruce muskeg; others have reported singing males in mature white spruce, balsam poplar, white birch, and young to middle-aged trembling aspen (C. Siddle and M. Phinney pers. comm.). The Yellow-bellied Flycatcher stays fairly close to the ground, perching in the lower branches; it rarely uses exposed or high perch sites.

Figure 23. Adult male Yellow-bellied Flycatcher on territory (52 km east of Prince George, 20 July 1993; Jack Bowling).

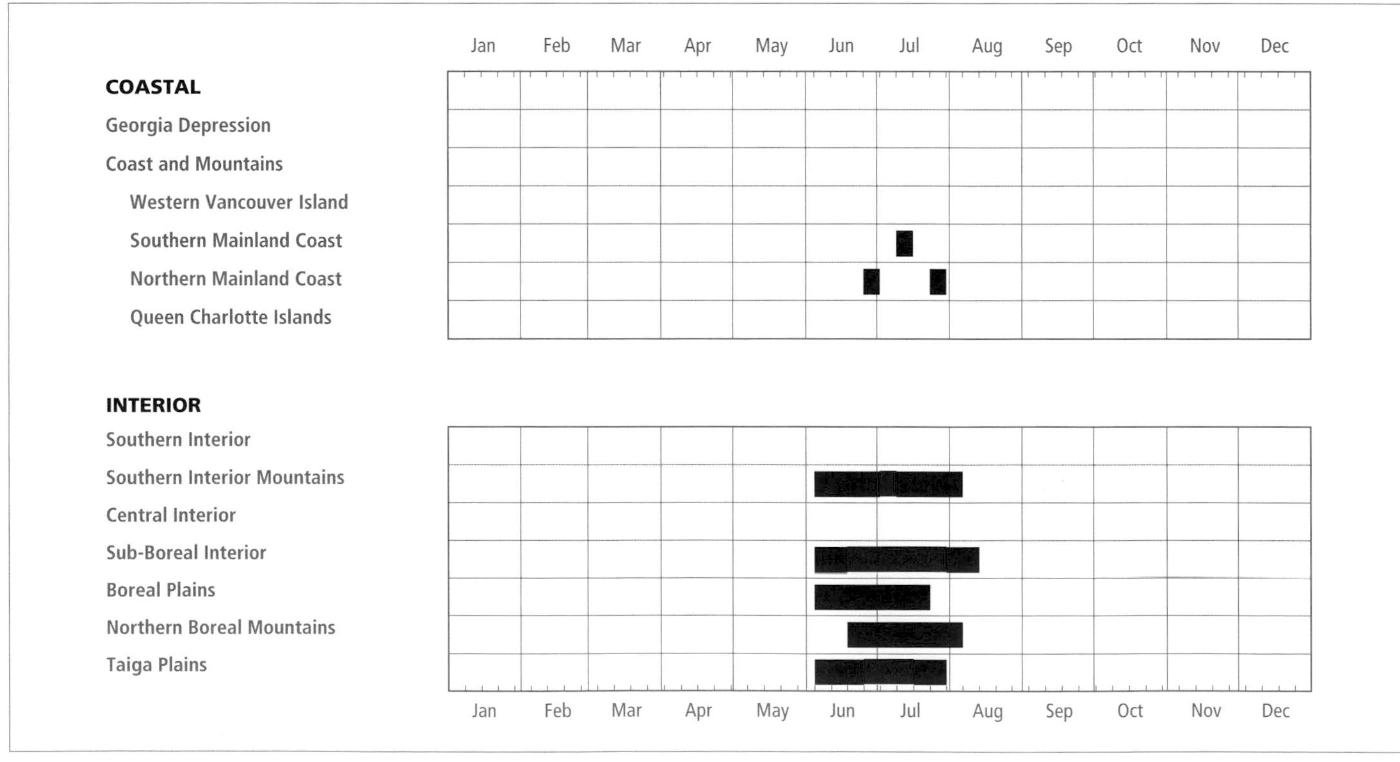

Figure 24. Annual occurrence (black) and breeding chronology (red) for the Yellow-bellied Flycatcher in ecoprovinces of British Columbia.

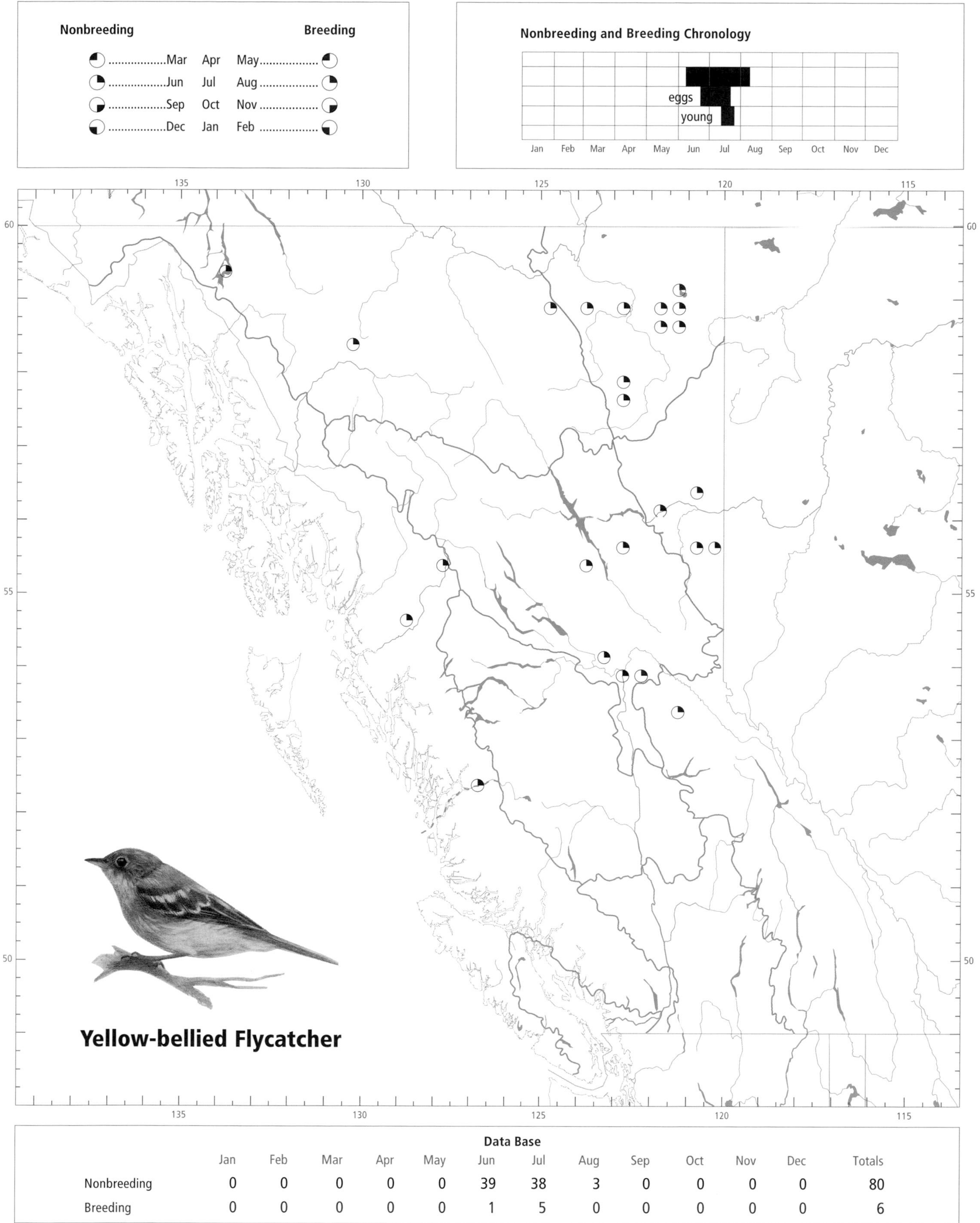

Data Base

	Jan	Feb	Mar	Apr	May	Jun	Jul	Aug	Sep	Oct	Nov	Dec	Totals
Nonbreeding	0	0	0	0	0	39	38	3	0	0	0	0	80
Breeding	0	0	0	0	0	1	5	0	0	0	0	0	6

S.S. Dickey (*in* Bent 1942) eloquently describes the flycatcher's habitat in Canada, in terms that are applicable to some situations in British Columbia. He says, "they are met with in the shadowy underwoods of evergreens, paper birches, and mountain ashes, where cranberries, trailing white snowberry, rare orchids, and an array of slightly emerald mosses carpet the forest floor and cover crumbling logs."

Migratory movements of the species in British Columbia are poorly known. Because it migrates quietly and singly (Hussell 1982a) and appears to be locally distributed, the species may often be overlooked. The spring movement probably occurs in early June (Figs. 24 and 25) and peaks about the second week of June. In southern Ontario the spring movement peaks in early June, with males preceding females by about 4 days (Hussell 1982a). After the nesting season, this species is silent and easily overlooked. The autumn movement in British Columbia probably occurs from late July to mid-August. Although there are no records after 9 August, the species is probably present until late August or early September (Figs. 24 and 25). In Ontario, adults begin migrating in late July, about 3 weeks earlier than young of the year (Hussell 1982b). They probably leave shortly after their young fledge. The same author states that adults usually spend no more than 66 days on the breeding grounds.

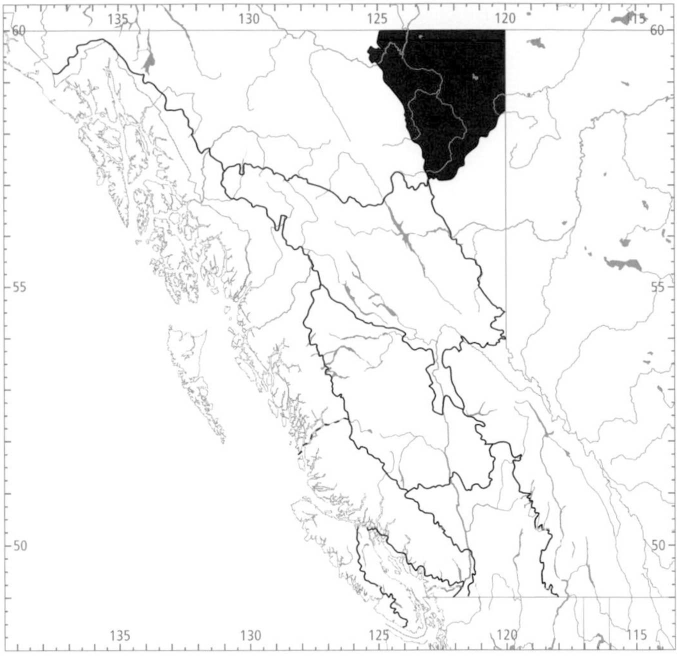

Figure 26. In British Columbia, the highest numbers for the Yellow-bellied Flycatcher in summer occur in the Taiga Plains Ecoprovince.

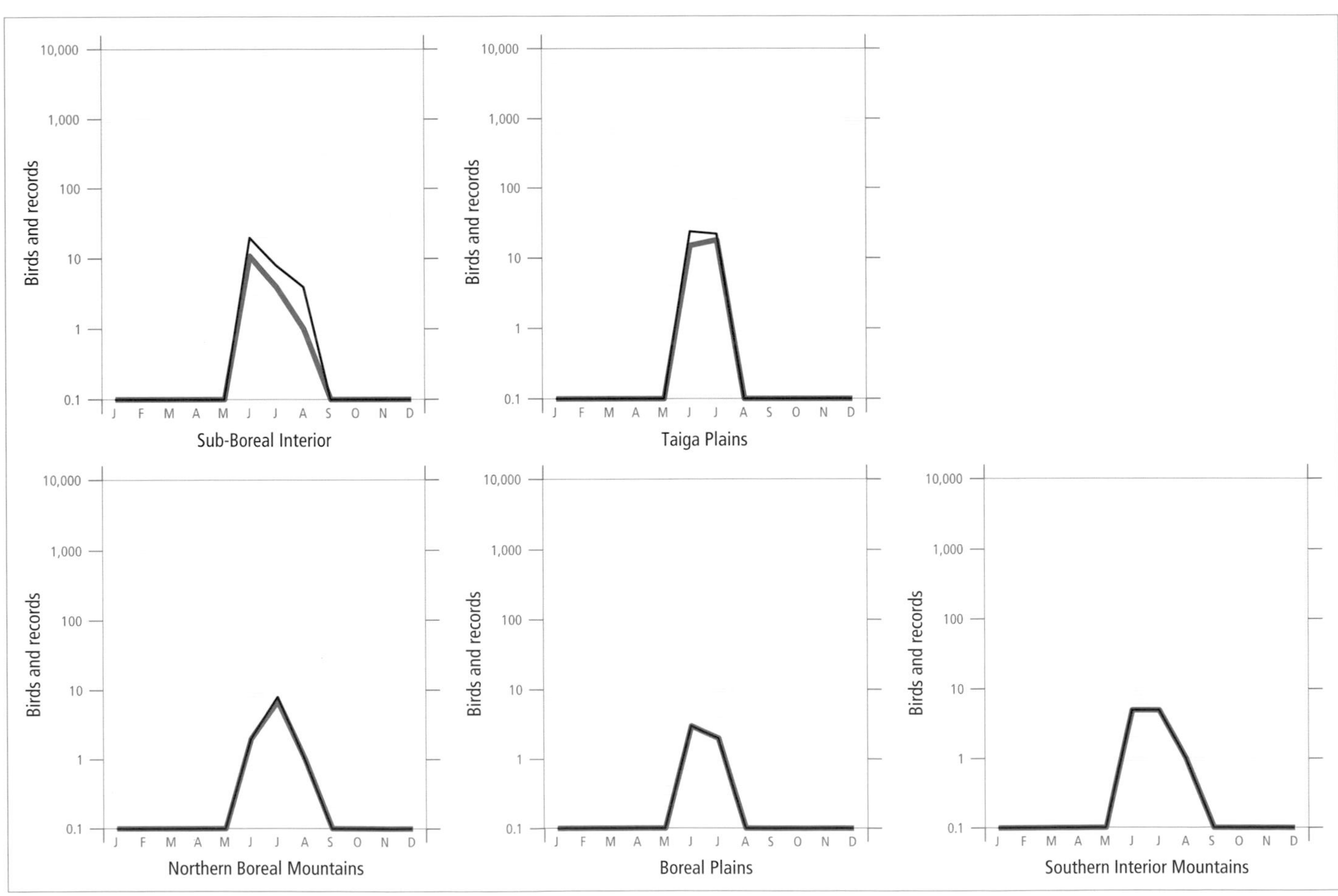

Figure 25. Fluctuations in total number of birds (purple line) and total number of records (green line) for the Yellow-bellied Flycatcher in ecoprovinces of British Columbia. Nest record data have been excluded.

The Yellow-bellied Flycatcher has been recorded in the province from 7 June to 9 August (Fig. 24).

BREEDING: The Yellow-bellied Flycatcher probably breeds locally throughout its range in the northeastern and east-central regions of the province. However, nesting has been confirmed only in the vicinity of Indianpoint Lake and near Prince George. Adults with recently fledged young have been reported from Kotcho Lake. The species is known to breed at between 400 and 600 m elevation.

The highest numbers for the Yellow-bellied Flycatcher in summer occur in the Taiga Plains (Fig. 26). Breeding Bird Surveys on interior routes for the period 1968 through 1993 contained insufficient data for analysis.

In the vicinity of Kotcho Lake, breeding habitat included dry second-growth trembling aspen within mature stands. South of Dawson Creek, the species preferred higher ground in young to middle-aged (20 to 40 years old) pine-birch-aspen forests (M. Phinney pers. comm.), but has also been found in black spruce muskegs and willow swamps (T. Greenfield pers. comm.). Near Prince George, 1 site included a dry mixed woodland of willow, trembling aspen, paper birch, and black cottonwood, with occasional lodgepole pine, Douglas-fir, and white spruce. The understorey was dominated by fireweed. A second site consisted of a 23-year-old plantation of white spruce and lodgepole pine with patches of same-age trembling aspen and willow. Throughout most of its range in North America, this species nests in muskegs and similar wet areas (Bull 1974; Peck and James 1987; Salt and Salt 1966).

The Yellow-bellied Flycatcher has been recorded breeding in the province from 18 June (calculated) to 25 July (Fig. 24).

Nests: The only nest for which details were given was situated on the ground, sunk into a mossy hummock. It was lined with dead grasses.

Eggs: Only 2 nests, each containing 4 eggs, have been located: Indianpoint Lake 5 July 1930 (Dickinson 1953) and near Vama Vama Creek (east of Prince George) 13 to 21 July 1993. Calculated dates indicate that nests may hold eggs as early as 18 June. The incubation period is 12 to 15 days (Walkinshaw 1961, 1967).

Young: There is but 1 record of nestlings for British Columbia: 4 young found near Vama Vama Creek on 25 July 1993. In addition, 2 recently fledged broods, each with 2 young being fed by adults, were reported: 12 and 13 July 1982 (near Kotcho Lake) and 17 July 1992 (32 km e Prince George) (Campbell and Dawe 1992). The nestling period is 13 days (Walkinshaw and Henry 1957).

Brown-headed Cowbird Parasitism: Cowbird parasitism was not found in British Columbia in 2 nests recorded with eggs or young. Walkinshaw (1961), Friedmann (1963), and Friedmann and Kiff (1985) mention that the Yellow-bellied Flycatcher is rarely victimized by the cowbird; they list only 3 occurrences for North America.

REMARKS: There are 2 published records, both lacking details, that we suspect may have been misidentified birds and probably refer to the "Western Flycatcher." In the first of these, Taverner (1919) suggested that the Yellow-bellied Flycatcher remained near the Hazelton area until 30 August. The second was by Cooke (1937), who reported that a Yellow-bellied Flycatcher was banded in Summerland on 23 June 1931 and shot at the place of banding on 30 June a year later.

Very little is known about the distribution and abundance of the Yellow-bellied Flycatcher in British Columbia. It appears to be concentrated in the Fort Nelson Lowland (Fig. 26), and as the forests it inhabits are now being harvested, there may be reason for concern. There are records of the species in second-growth forests elsewhere, but there too it may be vulnerable to logging of its preferred nesting habitat.

DeSante et al. (1985), Pyle et al. (1987), and Kaufman (1990) provide guidance in identification of adult and immature Yellow-bellied Flycatchers in the field and in museum collections.

NOTEWORTHY RECORDS

Spring: No records.

Summer: Coastal – Bella Coola (Nusatsum) 11 Jul 1942-1 (RBCM 10083); Terrace 22 Jun 1948-1 (ROM 81991); Hazelton 24 Jul 1913-1 (Brooks and Swarth 1925). **Interior** – Indianpoint Lake 8 Jun 1934-1 (MVZ 65633), 5 July 1930-1 adult female, nest, and 4 eggs collected (MCZ 282922), 2 Aug 1930-1 (MCZ 282925); Prince George 7 Jun 1993-1 singing male, first arrival; 31.8 km e Prince George 17 Jul 1992-1 adult feeding 2 young (Campbell and Dawe 1992); 30 km e Prince George 22 Jun 1992-5 near Vama forest road; Vama Vama Creek (50 km e Prince George) 19 Jun 1992-2 singing males, 1 photographed (Siddle 1992c), 13 Jul 1993-nest with 4 eggs; Mount Tabor 9 Aug 1992-1 adult and 3 to 4 immatures (Siddle 1993a); 50 km nw Prince George 17 Jun 1993-1 singing male; Hazelton 24 Jul 1913-1 adult male (Brooks and Swarth 1925); Donna Creek (80 km wnw Mackenzie) 13, 16, 26, and 29 Jun 1993-several singing males; Chetwynd 16 Jun 1969-1 (Webster 1969b); Charlie Lake 10 Jun 1986-1; near Farrell Creek 21 Jul 1992-1 (Siddle 1992c); Trutch 14 Jul 1943-1 (Rand 1944); Dease Lake 22 Jun 1983-1 (Grunberg 1983d); near Kotcho Lake 15 Jun 1982-1 (Campbell and McNall 1982), 12 Jul 1982-1 adult feeding 2 recently fledged young; e Fort Nelson 23 Jun 1982-6 to 8 calling (Campbell and McNall 1982); Parker Lake 24 Jul 1988-1 (Siddle 1988d); Kledo Creek 10 Jul 1992-1; McDonald Creek (Mile 114 Alaska Highway) 21 Jul 1943-2 (Rand 1944); Pike River (Atlin) 3 Aug 1914-1 with Hammond's Flycatcher (RBCM 3045; Anderson 1915).

Breeding Bird Surveys: Coastal – Not recorded. **Interior** – Recorded from 1 of 73 routes and on less than 1% of all surveys. Maxima: Ferndale 19 Jun 1992-2.

Autumn: No records.

Winter: No records.

Alder Flycatcher ALFL

Empidonax alnorum Brewster

RANGE: Breeds from central Alaska, central Yukon, northwestern and southern Mackenzie and across the Canadian Prairie provinces to central and eastern Quebec and southern Newfoundland; south to southern Alaska, south-central British Columbia, southern Alberta, Saskatchewan, and south-central Minnesota, east into Appalachia and south in the east to Tennessee and North Carolina; winters in South America.

STATUS: On the coast, *very rare* in the northern Coast and Mountains and *casual* in the southern Coast and Mountains and Georgia Depression ecoprovinces; absent from Vancouver Island and the Queen Charlotte Islands.

In the interior, an *uncommon* to *fairly common* migrant and summer visitant from about the latitude of Lac la Hache in the Central Interior Ecoprovince northeastward into the Taiga Plains Ecoprovince; *rare* westward across the Northern Boreal Mountains Ecoprovince. It is *very rare* in the southern portions of the Southern Interior Mountains Ecoprovince, becoming *fairly common* in the Mount Robson area; *casual* in the Southern Interior Ecoprovince.

Breeds.

CHANGE IN STATUS: Since the early 1970s, the Alder Flycatcher appears to have extended its range southward into the Southern Interior Mountains of southeastern British Columbia. There were reports of the species in the early 1970s from the Mount Robson and Golden areas in the northern parts of that ecoprovince, but it was not reported with regularity in the southern portions (Arrow Lakes area) until the late 1970s and early 1980s. While the Alder Flycatcher was split only recently from the Traill's Flycatcher (American Ornithologists' Union 1983), and would not likely have been distinguished by most observers before 1973, a number of collectors in the east Kootenay probably would have identified the Alder Flycatcher (then a part of the Traill's Flycatcher, *E. traillii traillii*) had it been there. It is not mentioned as occurring in the Kootenay region by Munro and Cowan (1947), Johnstone (1949), Munro (1950, 1958), or Godfrey (1955).

NONBREEDING: The Alder Flycatcher has a widespread distribution through the interior of the province from the Cariboo area northward. It occurs sporadically in the northern Coast and Mountains. It is not as common in the northwestern quarter of British Columbia as it is in the northeast. It is also now found with some regularity in the southeastern portion of the province. It has not been recorded from Vancouver Island. On the Queen Charlotte Islands, Darcus (1930) notes that he observed the Alder Flycatcher in suitable places on Graham Island. However, he did not offer sufficient evidence to confirm his observations, and there has been no independent confirmation subsequently.

The Alder Flycatcher has been reported from near sea level to about 1,300 m elevation. Nonbreeding habitat is similar to breeding habitat (see Figs. 31 and 32).

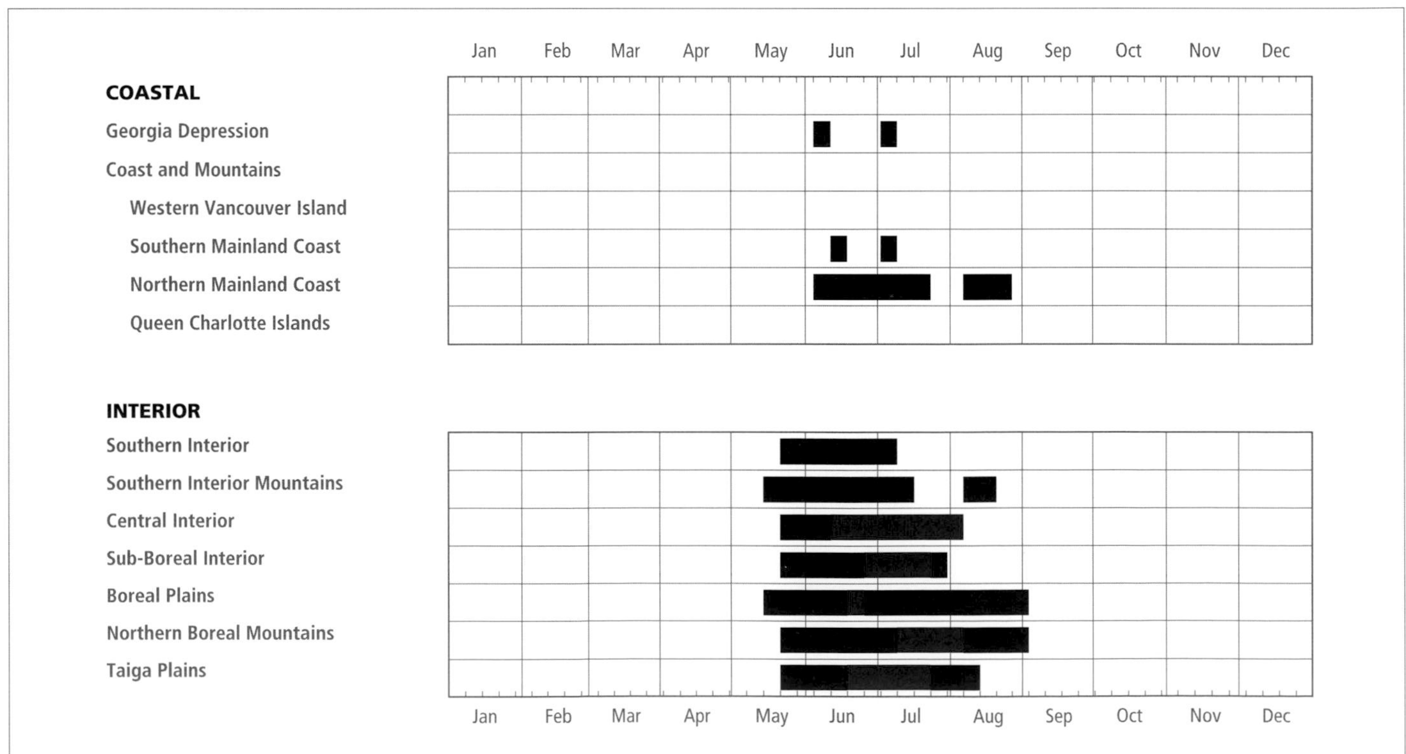

Figure 27. Annual occurrence (black) and breeding chronology (red) for the Alder Flycatcher in ecoprovinces of British Columbia.

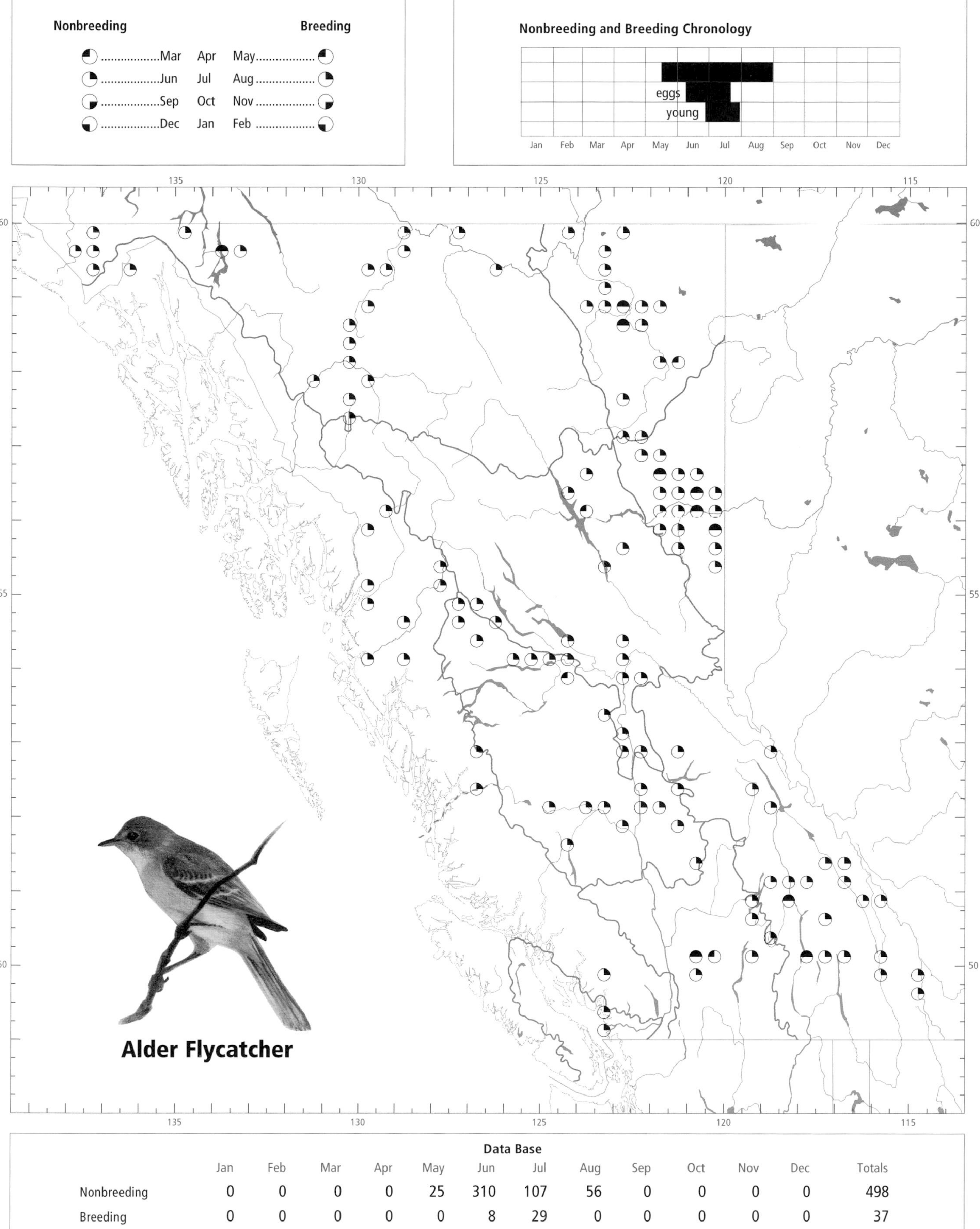

Data Base

	Jan	Feb	Mar	Apr	May	Jun	Jul	Aug	Sep	Oct	Nov	Dec	Totals
Nonbreeding	0	0	0	0	25	310	107	56	0	0	0	0	498
Breeding	0	0	0	0	0	8	29	0	0	0	0	0	37

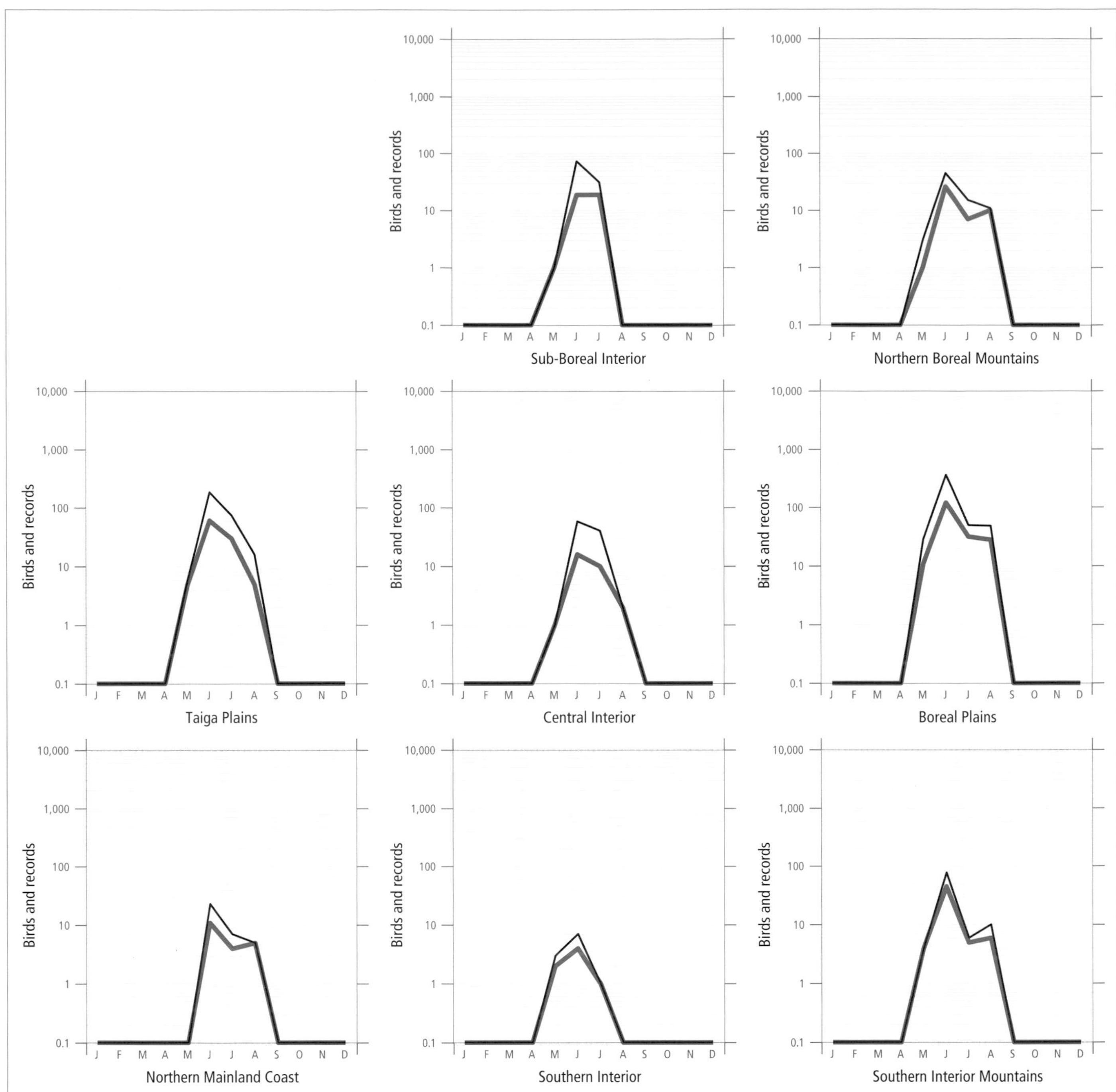

Figure 28. Fluctuations in total number of birds (purple line) and total number of records (green line) for the Alder Flycatcher in ecoprovinces of British Columbia. Nest record data have been excluded.

The Alder and Willow flycatchers are the last of the *Empidonax* flycatchers to arrive in British Columbia. The Alder Flycatcher may arrive in the interior as early as the third week of May; however, the peak interior movement occurs about the first 2 weeks of June (Figs. 27 and 28).

Autumn migration is well under way by the second or third week of August, and birds appear to have left the province by the end of that month (Figs. 27 and 28).

The Alder Flycatcher has been recorded from 16 May to 30 August in the interior, and from 5 June to 24 August on the coast (Fig. 27).

BREEDING: The Alder Flycatcher is known to breed in the Central Interior (Williams Lake area), Sub-Boreal Interior (Fraser Lake area, Quesnel, and near Mackenzie), Boreal Plains (Tupper and Fort St. John), and Taiga Plains (Fort Nelson area) ecoprovinces. There is 1 breeding record from Nuttlude Lake in the Northern Boreal Mountains. However, the Alder Flycatcher likely breeds throughout its range in the province.

The Alder Flycatcher reaches its highest numbers in the Fort Nelson Lowland of the Taiga Plains in northeastern British Columbia (Fig. 29). An analysis of Breeding Bird Surveys

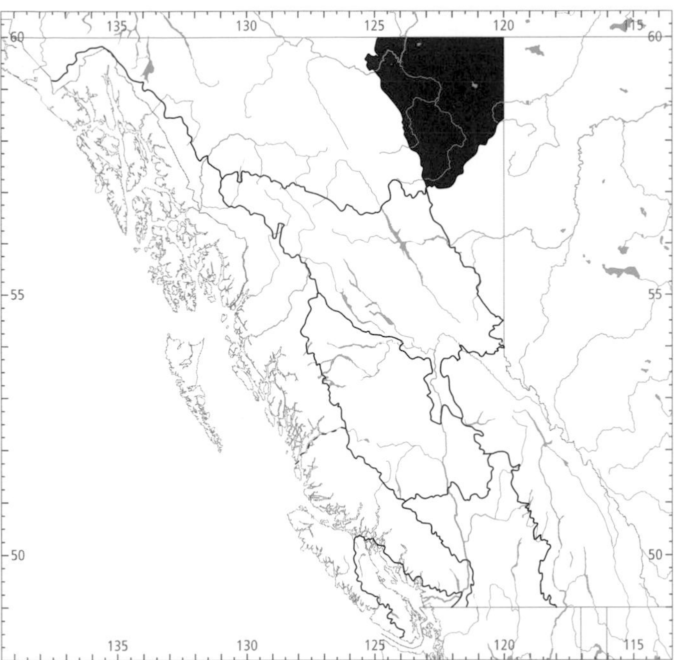

Figure 29. In British Columbia, the highest numbers for the Alder Flycatcher in summer occur in the Taiga Plains Ecoprovince.

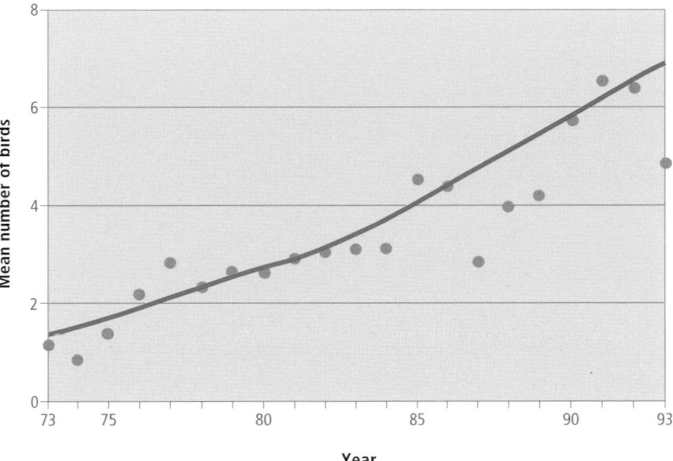

Figure 30. An analysis of Breeding Bird Surveys for the Alder Flycatcher in British Columbia shows that the mean number of birds on interior routes increased at an average annual rate of 10% over the period 1973 through 1993 ($P < 0.10$).

for the period 1973 through 1993 shows that the mean number of birds on interior routes increased at an average annual rate of 10% (Fig. 30); coastal routes for the same period had insufficient data for analysis.

In British Columbia, the Alder Flycatcher has been found breeding at elevations ranging from 340 to 1,300 m. Breeding habitat usually includes shrub thickets and stands of young deciduous trees, usually in close proximity to water such as lakes, ponds, rivers, floodplains, creeks, swamps (Fig. 31), marshes, and sewage lagoons. Willow, alder, black cottonwood, poplar, and trembling aspen stands, particularly 4- to 8-year-old regenerating cuts, have been most often reported as preferred nesting sites (Fig. 32). In addition to riparian habitat, edges created by roadsides, airstrips, avalanche chutes, hedgerows, and even clearcuts are sometimes used.

In the vicinity of Smithers, Pojar (1993) found the Alder Flycatcher in "thickets of sapling aspen or aspen mixed with young cottonwood or willow." The species was absent from stands where trees were older and less dense. This observation indicates a possible preference for the high density of deciduous stems found in younger seral stages. Westworth et al. (1984) and Morgan and Freedman (1986) report similar habitat preferences in Alberta and Nova Scotia, respectively. Near Dawson Creek, in the Boreal Plains, the Alder Flycatcher was most abundant in shrub-dominated habitats of willow and small trembling aspen (Lance and Phinney 1993).

The Alder Flycatcher has been recorded breeding in the province from 12 June (calculated) to 30 July (Fig. 27).

Nests: Most nests (13 of 14) were found in deciduous shrubs, including willow, rose, dogwood, saskatoon, and honeysuckle. A single nest was found in a cow-parsnip. Nine nests were attached to a vertical fork or stem of the plant. Nest heights ranged from ground level to 1.2 m, with 8 nests recorded between 0.6 and 0.8 m.

The nests were loosely woven and bulky, and were formed of plant fibres, grasses, stems, fine rootlets, mosses, and pieces of slender bark strips (Fig. 33). The grass cup was lined with hair, plant down, or fine grass. One nest was unusual in that feathers were included as part of the lining (Harrison 1979). Long grass culms, or streamers, were often found dangling below the nest, and sometimes reached the ground (Stein 1963).

Eggs: Dates for 15 clutches ranged from 21 June to 21 July, with 9 recorded between 29 June and 9 July. Calculated dates indicate that eggs can occur as early as 12 June. Clutch size ranged from 2 to 4 eggs (2E-3, 3E-8, 4E-4), with 8 having 3 eggs. The incubation period ranges between 11 days (Peck and James 1987) and 13 days (Ehrlich et al. 1988).

Young: Dates for 8 broods ranged from 27 June to 30 July, with 6 recorded between 16 July and 26 July. Brood size ranged from 1 to 4 young (1Y-1, 3Y-5, 4Y-2), with 5 having 3 young. The nestling period is 13 to 14 days (Ehrlich et al. 1988).

Brown-headed Cowbird Parasitism: In British Columbia, 10% of 19 nests found with eggs or young were parasitized by the cowbird. In Ontario, Peck and James (1987) found cowbird parasitism in 15% of 47 nests.

Nest Success: Of 2 nests found with eggs and followed to a known fate, 1 produced at least 1 fledgling. Holcomb (1972) found that of 91 nests he studied in Ohio and Nebraska, about 40% produced at least 1 fledgling.

REMARKS: In the aspen parklands of the Central Interior, the Alder and Willow flycatchers are known to breed sympatrically only in the small area from Watson Lake, near 100 Mile House, north to Deep Creek Indian Reserve. Stein (1963) studied the Traill's Flycatcher (*E. traillii*) there in 1958 and 1959 to test the hypothesis that its 2 song morphs were actually distinct species. At that time, the Traill's Flycatcher was

Figure 31. Typical Alder Flycatcher breeding habitat in mixed woodlands of willow and trembling aspen, with some black spruce (Tabor swamp, east of Prince George, 10 May 1991; R. Wayne Campbell).

Figure 32. Alder Flycatcher breeding habitat in mixed deciduous woodlands of willow, Sitka alder, and balsam poplar, with red-osier dogwood understorey (Peace Island Park, near Taylor, 23 June 1996; R. Wayne Campbell).

the exception to the "one song for each species" condition. The results from Stein's work in British Columbia led to the splitting of the Traill's Flycatcher in 1973 into the Alder and Willow flycatchers; they are now recognized as a superspecies (= "*traillii*" complex; American Ornithologists' Union 1973, 1983).

As Kaufman (1990) notes, the song of the Alder Flycatcher is "often described as a 3-syllabled *fee-bee-oh*, but actually sounds like a 2-syllabled *rrree-beep;* a faint third syllable, as in *rrree-beea*, may or may not be audible." That is certainly true in British Columbia, where the 2-syllabled song can often sound like *rrree-peet*. Stein (1963) also found that the Alder Flycatcher may give a *pit* and a *wee-oo* in this order, which sounds superficially like a *fitz-bew*. He notes that the difference between the two is audible, "though not easily distinguishable." Experience and caution are required when reporting observations from areas where both species occur (see other cautionary comments under REMARKS in **Willow Flycatcher**).

In an area of overlap between the Alder and Willow flycatchers in southern Ontario, Barlow and McGillivray (1983) found that the 2 flycatchers showed similar habitat preferences. Despite that, they could find no evidence of competition between them.

Prescott (1987) suggests that where the Alder and Willow flycatchers coexist, they learn to recognize each other as potential competitors. He also finds that the responses of the Willow Flycatcher both to the song of the Alder Flycatcher and to songs of conspecifics are more aggressive than those of the Alder Flycatcher under similar circumstances. He suggests that the recent range expansions of the Willow Flycatcher into areas historically occupied by the Alder Flycatcher may be due in part to interspecific dominance (see also REMARKS in **Willow Flycatcher**). This is an area for further study, particularly in British Columbia, where the 2 species breed sympatrically and the Alder Flycatcher appears to be expanding its range into southeastern British Columbia, an area occupied by the Willow Flycatcher.

Figure 33. Nest and eggs of the Alder Flycatcher (Prince George, 12 June 1985; R. Wayne Campbell).

NOTEWORTHY RECORDS

Spring: Coastal – No records. **Interior** – Rush Lake 21 May 1993-1 calling (Campbell and Stewart 1993); 19 km w Merritt 31 May 1980-2 (Rogers 1982c); Revelstoke 19 May 1986-1; Nulki Lake 24 May 1945-1, spring arrival (Munro and Cowan 1947); Donna Creek (Williston Lake) 25 May 1993-1, spring arrival (Price 1993); Dawson Creek 16 May 1992-1, earliest spring arrival; Beatton Park 30 May 1990-15; Cecil Lake 26 May 1986-1, spring arrival; Jackfish Creek 25 May 1975-2; Kenai Creek 26 May 1995-1; Atlin 21 May 1981-3.

Summer: Coastal – Iona Island (Richmond) 9 Jun 1992-1 (Cooper 1993); Maplewood flats 3 Jul to 7 Aug 1981-1 (Daly 1982); near Squamish 17 Jun 1984-1 recorded on tape; Hagensborg 6 Jul 1938-1 (Rand 1943); Terrace 22 Jun 1947-1 (RBCM 10082); Hazelton 5 Jun 1921-1, first arrival (Swarth 1912); Kispiox Valley 24 Aug 1921-1 (MVZ 42192). **Interior** – Brookmere 27 Jun 1982-1 (Rogers 1982d); 19 km w Merritt 22 Jun 1980-2 (Rogers 1982d); Sparwood 19 Aug 1983-1; Lavington 8 to 28 Jun 1984-1 (Cannings et al. 1987); 23.4 km n Revelstoke 3 Jul 1991-2 calling from mixed woodlands (Campbell and Dawe 1991); Leanchoil 30 Jun 1975-1 (Wade 1977); n Revelstoke 21 Jun 1984-12; Lac la Hache 30 Jul 1959-3 young in nest; Clearwater Lake 18 Jun 1987-7 territorial birds along Highway 20 (Tweit and Mattocks 1987); Quesnel 14 Jul 1900-1 (MVZ 102488), 19 Jul 1984-3 young in nest; 50 km e Likely 12 Jun 1993-4, calling simultaneously at 1,300 m; Prince George 17 Jun 1969-8, 11 Jul 1991-2 calling in Wilkins Park (Campbell and Dawe 1991); Tabor Lake 17 July 1992-2 in wetland (Campbell and Dawe 1992); Francois Lake 4 Aug 1944-1 (Munro 1947a); Vanderhoof 25 Jul 1945-3 young in nest, 5 Aug 1919-1 (NMC 13731); Fraser Lake 21 Jun 1959-3 eggs, 27 Jun 1959-1 egg, 1 young; Bouchie Lake 7 Jun 1944-4 (Munro 1947a); Smithers 1 Jul 1974-8; Fort St. James 1 Jul 1889-2 eggs; Donna Creek (Williston Lake) 3 Jun 1993-noticeable influx (Price 1993); near Mackenzie 3 Jul 1969-3 eggs; Tupper Creek 22 Jun 1941-4 eggs; Tupper Creek to Pouce Coupe 15 Jun 1974-27; Fort St. John 2 Aug 1987-4 recently fledged young; Cecil Lake to Fort St. John 14 Jun 1974-20; Boundary Lake 30 Aug 1982-2; Aline Lake 26 Jun 1980-4 eggs; Nuttlude Lake 30 Jul 1957-4 large young in nest; Fort Nelson 15 June 1975-65 on Breeding Bird Survey, 21 Jul 1986-1 recently fledged young; Parker Lake 12 Aug 1979-3 along road; sw Kotcho Lake 13 Jun 1982-6 in damp willow-spruce area (Campbell and McNall 1982), 29 Jun 1982-4 eggs; Atlin 29 Aug 1924-1 (MVZ 44762).

Breeding Bird Surveys: Coastal – Recorded from 9 of 27 routes and on 4% of all surveys. Maxima: Kispiox 20 Jun 1993-19; Nass River 21 Jun 1975-8; Meziadin 17 Jun 1975-5. **Interior** – Recorded from 41 of 73 routes and on 27% of all surveys. Maxima: Fort Nelson 15 Jun 1975-65; Steamboat 14 Jun 1976-54; McLeod Lake 16 Jun 1992-38.

Autumn: No records.

Winter: No records.

Willow Flycatcher

Empidonax traillii (Audubon)

WIFL

RANGE: Breeds from central and south coastal British Columbia, southern Alberta, Saskatchewan, and southwestern Manitoba east to central Maine and Nova Scotia; south in the west to northern Baja California; east across the southern states to North Carolina and Virginia. Winters in Middle America from Veracruz and Oaxaca south to Panama.

STATUS: On the coast, an *uncommon* to *fairly common* migrant and summer visitant to the Georgia Depression Ecoprovince and southern portions of the Coast and Mountains Ecoprovince; *very rare* on Western Vancouver Island; *casual* in the northern Coast and Mountains; absent from the Queen Charlotte Islands.

In the interior, an *uncommon* to *fairly common* migrant and summer visitant to the Southern Interior, Southern Interior Mountains, and the southern portions of the Central Interior ecoprovinces.

Breeds.

CHANGE IN STATUS: The Willow Flycatcher (Fig. 34) has increased its range eastward in British Columbia. Munro and Cowan (1947) do not mention the species from anywhere in the Southern Interior Mountains, nor do Johnstone (1949) from his work in the east Kootenay or Munro (1947, 1958) in the Creston region. Kelso (1931) has a single record of a Traill's

Figure 34. Adult Willow Flycatcher (Christina Lake, 28 June 1983; Mark Nyhof). Note the bird's distinctive white throat.

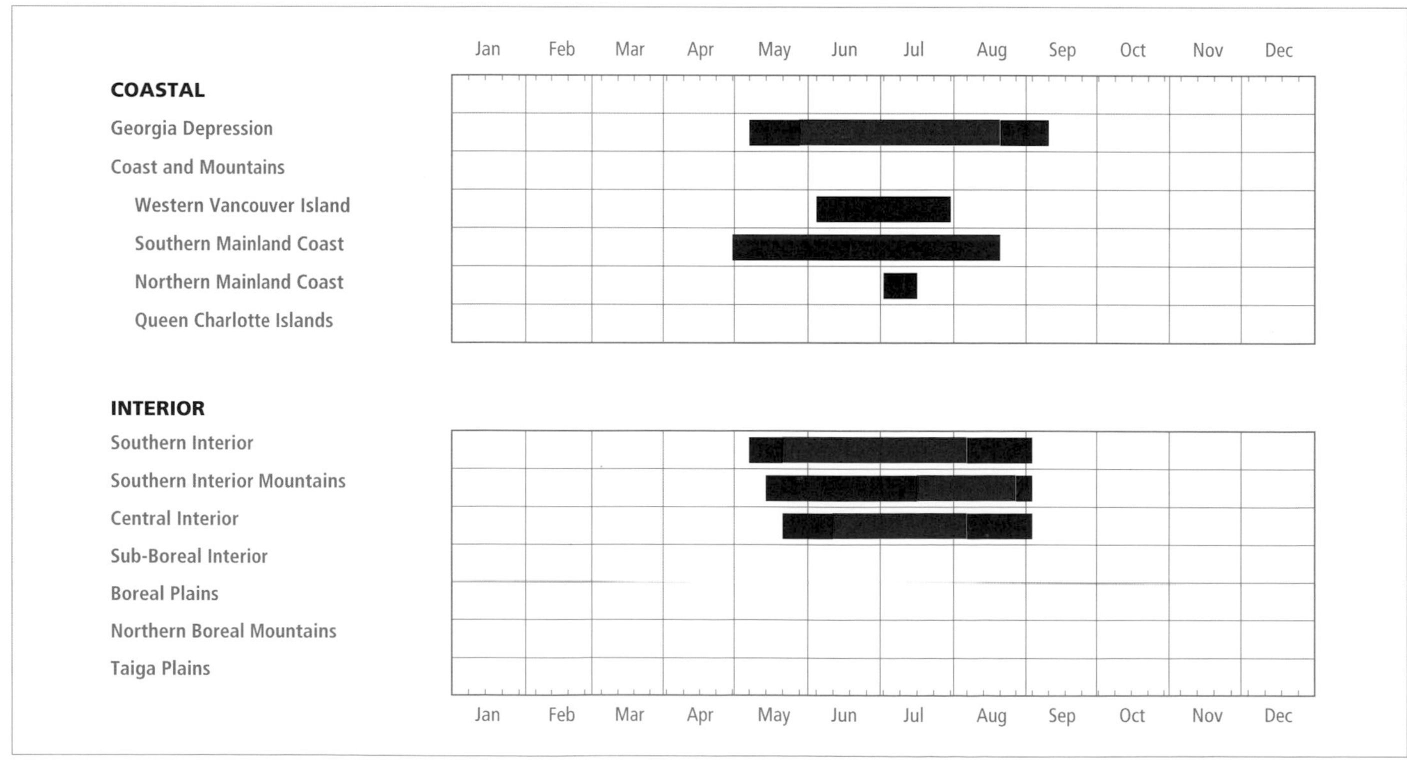

Figure 35. Annual occurrence (black) and breeding chronology (red) for the Willow Flycatcher in ecoprovinces of British Columbia.

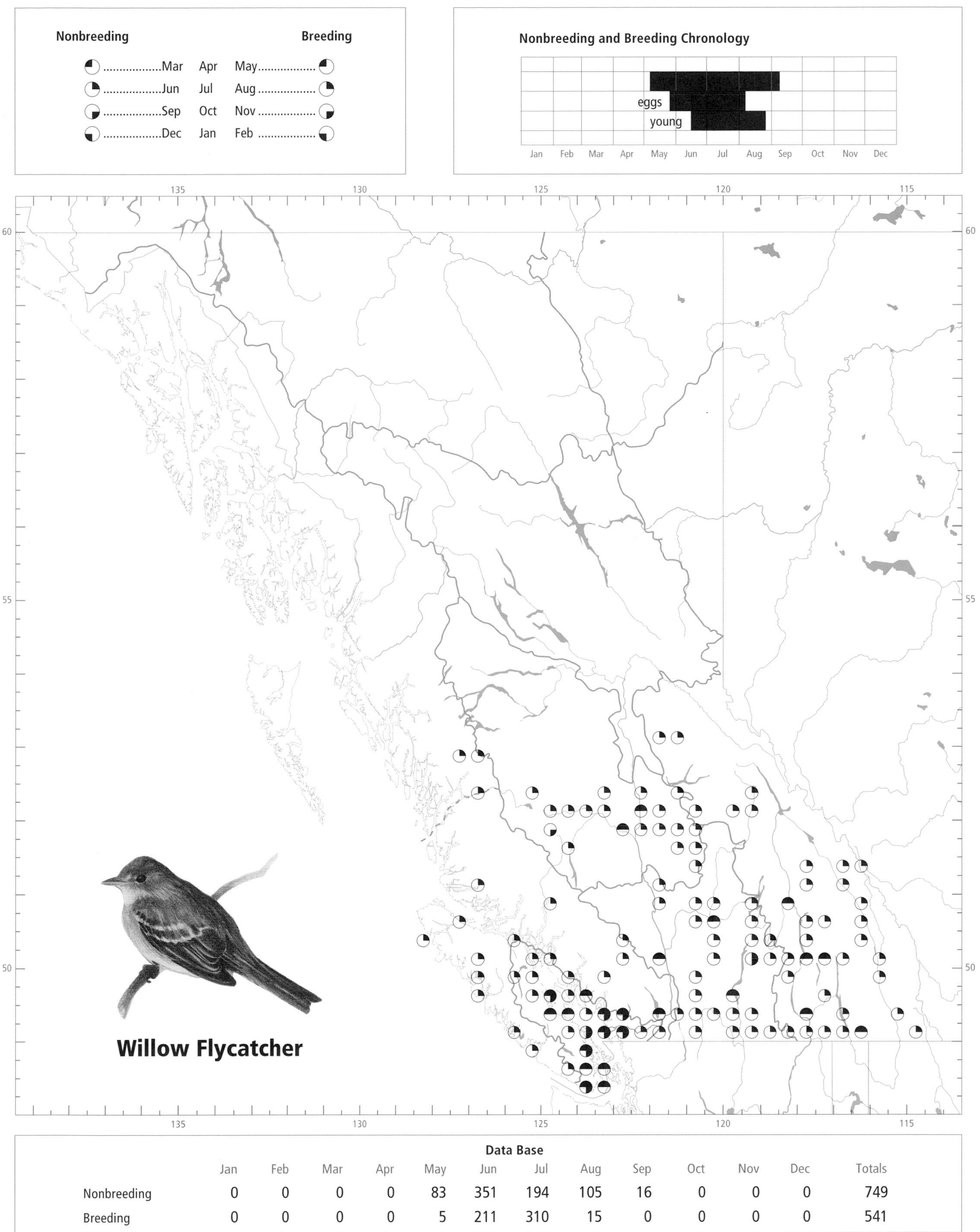

Data Base

	Jan	Feb	Mar	Apr	May	Jun	Jul	Aug	Sep	Oct	Nov	Dec	Totals
Nonbreeding	0	0	0	0	83	351	194	105	16	0	0	0	749
Breeding	0	0	0	0	5	211	310	15	0	0	0	0	541

Flycatcher from Upper Whatshan Lake, which he ascribes to the subspecies *E. t. traillii*. This record is of a nesting pair with 4 large young, and was likely the Willow Flycatcher. The next report of this species in the region is of a specimen taken in the Flathead Valley on 10 July 1952 and identified as *E. t. adastus* (Godfrey 1955). Godfrey (1955) notes that the species is "apparently scarce in most of the East Kootenay." Following a report of the species 2 decades later in Wasa Park (Dawe 1971), sightings became a regular occurrence in the region. There are now more than 230 records of the Willow Flycatcher in the Southern Interior Mountains from Creston in the south to Wells in the north.

NONBREEDING: The Willow Flycatcher (Fig. 34) is widely distributed in suitable habitat across the southern third of the province, including Vancouver Island, north along the coast to Kimsquit, and in the interior to Wells. It has been reported rarely from the west coast of Vancouver Island and the central mainland coast. In the Southern Interior Mountains, where both the Willow and Alder flycatchers occur, the Wil-

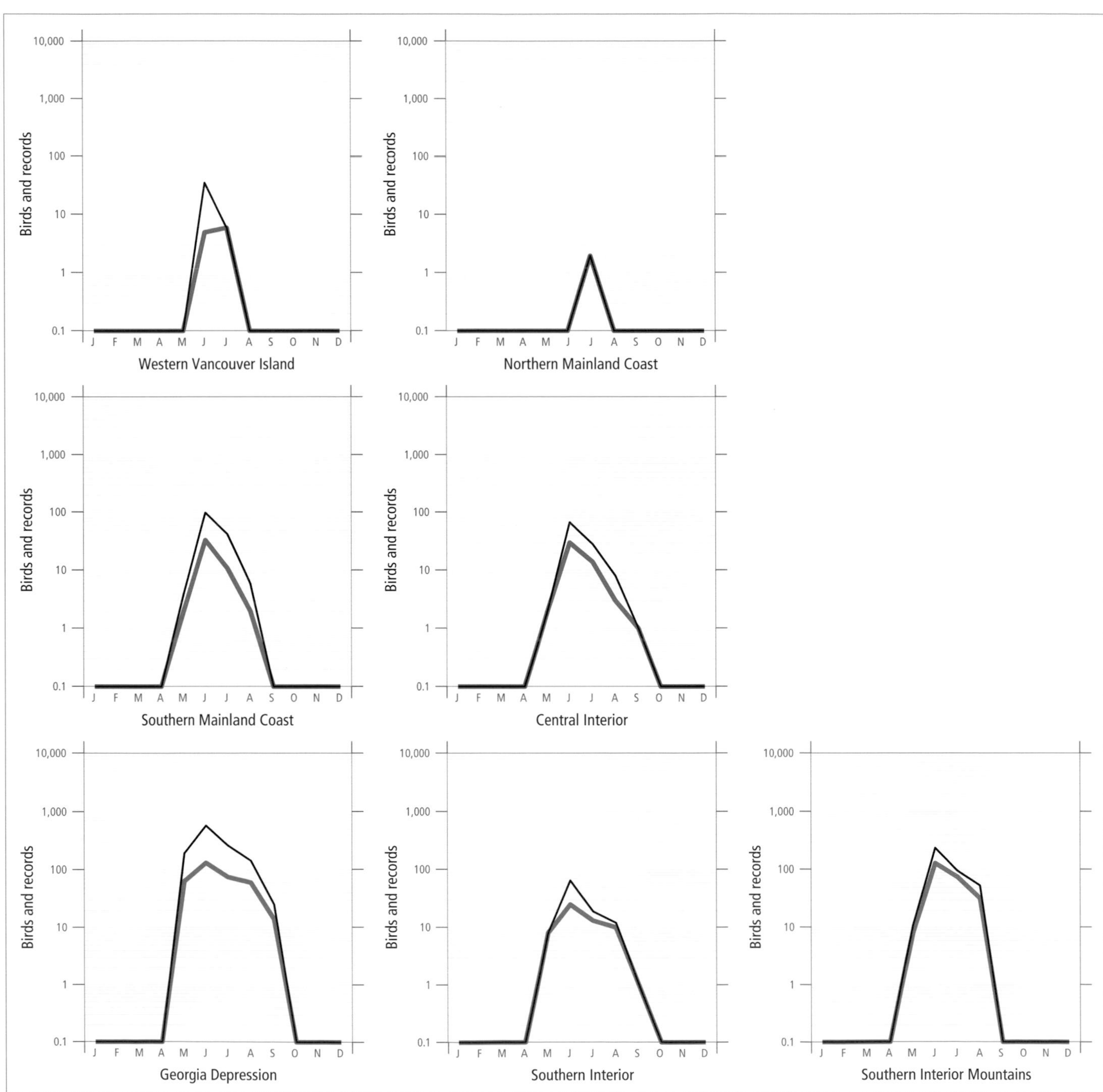

Figure 36. Fluctuations in total number of birds (purple line) and total number of records (green line) for the Willow Flycatcher in ecoprovinces of British Columbia. Nest record data have been excluded.

low Flycatcher is the more common of the two south of the Mount Robson area. Further north the Alder Flycatcher becomes the predominant species. In the Cariboo and Chilcotin areas, where both species breed sympatrically, the Willow Flycatcher is more numerous in the Williams Lake area, and is gradually replaced by the Alder Flycatcher just north of Soda Creek.

On the coast, the Willow Flycatcher occurs from near sea level to 1,070 m elevation. Weber (1975), in his study of altitudinal zonation of birds on Mount Seymour, noted the Willow Flycatcher as common at the lower altitudes (below 600 m), reaching peak numbers below 370 m elevation. In the interior, it has been found from 170 to 1,700 m elevation. The nonbreeding habitat of the Willow Flycatcher in British Columbia is similar to its breeding habitat (see Figs. 39 and 40).

As with all the *Empidonax* flycatchers of British Columbia, identification of silent migrants is difficult (see REMARKS). Reliable records indicate that the Willow Flycatcher may arrive in the first week of May on the coast and in the interior, although the peak movement for both areas appears to be about the last week of May or the first week of June (Figs. 35 and 36).

The autumn movement is also difficult to ascertain, because by then the birds have stopped vocalizing. It probably begins in early August and is at its height from mid to late August or early September. Most birds have left the province by the end of August, although a few may remain into the second week of September (Figs. 35 and 36).

On the coast, the Willow Flycatcher has been recorded from 6 May to 8 September; in the interior, it has been recorded from 7 May to 2 September (Fig. 35; see REMARKS).

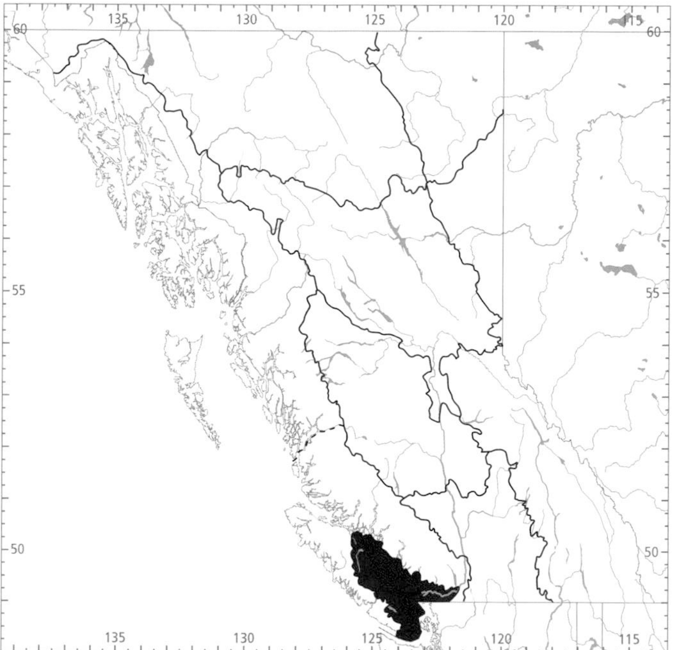

Figure 37. In British Columbia, the highest numbers for the Willow Flycatcher in summer occur in the Georgia Depression Ecoprovince.

BREEDING: The Willow Flycatcher probably breeds throughout most of its range in British Columbia. On the coast, breeding has been documented only from the east coast of Vancouver Island and the Fraser Lowland of the Georgia Depression. In the interior, breeding has been established in the Okanagan valley north to the Shuswap area, in 2 locations in the Southern Interior Mountains, and in the Cariboo and Chilcotin areas of the Central Interior. The northernmost breeding record is from Stum Lake.

In British Columbia, the Willow Flycatcher reaches its highest numbers in the Georgia Depression (Fig. 37). Robbins et al. (1986) note that, based on North American Breeding Bird Surveys, this species reaches its greatest abundance in the coastal rain forest of Washington and British Columbia. An analysis of Breeding Bird Surveys shows that over the period 1973 through 1993, the mean number of birds on coastal routes decreased at an average annual rate of 4% (Fig. 38); analysis of interior routes for the same period could not detect a net change in numbers.

On the coast, nests have been found from near sea level to 150 m elevation. Over this altitudinal range, the Willow Flycatcher frequents riparian habitats adjacent to sewage lagoons, swamps, lakes, rivers, creeks, sloughs, ponds, and other bodies of water. Deciduous shrubs or young trees, such as willow, red-osier dogwood, wild rose, and red alder (Fig. 39) are used. Near Sproat Lake, on central Vancouver Island, the Willow Flycatcher is consistently found in 30- to 35-year-old forests, especially in red alder thickets near small streams (Bryant et al. 1993). It also occurs on drier sites adjacent to estuaries where former marine and fluvial gravel deposits support Nootka rose and Pacific crab apple. Woodlands adjacent to agricultural land and wet areas are also used. Daly (1982) found the Willow Flycatcher to be one of the commonest passerine species around grassy field habitats where red

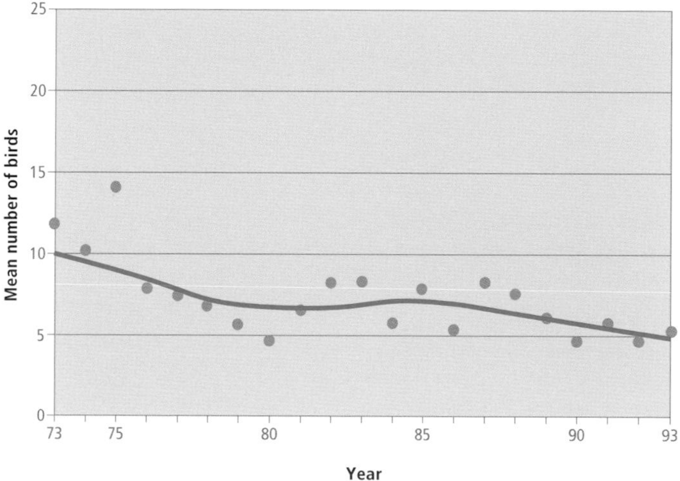

Figure 38. An analysis of Breeding Bird Surveys for the Willow Flycatcher in British Columbia shows that the mean number of birds on coastal routes decreased at an average annual rate of 4% over the period 1973 through 1993 ($P < 0.01$).

Figure 39. On the south coast of British Columbia, the Willow Flycatcher is found locally in young stands of red alder (Vancouver, 6 May 1990; R. Wayne Campbell).

alder and broom thickets lie adjacent to tidal mudflats. Clearcuts in the earlier stages of regeneration can also provide suitable habitat for this species. Laing (1942) noted that "the characteristic call of this bird is now a common sound in the slashings on the eastern slope of Vancouver Island." In a recent study of early-growth clearcuts in Oregon, the Willow Flycatcher was a species with a moderate density estimate (0.7 birds/ha) (Morrison and Meslow 1983).

In the interior, the Willow Flycatcher has been recorded breeding from 275 to 1,225 m elevation. Habitat use is similar to that on the coast and includes riparian willow scrub and thickets (Fig. 40), along with alder, birch, maple, and red-osier dogwood adjacent to lakes, ponds, creeks, rivers, and wet meadows. In Mount Revelstoke and Glacier national parks, habitat is described as shrubby wetlands (Van Tighem and Gyug 1983). In the aspen parklands of the Cariboo and Chilcotin areas, Stein (1963) found both Alder and Willow flycatchers nesting alongside streams and lakes where suitable brush occurred. He also found the Willow Flycatcher nesting on relatively dry hillsides. In the lower grasslands of the same area, A. Roberts (pers. comm.) found the species nesting wherever there was riparian vegetation; the Alder Flycatcher was absent from those sites. The Willow Flycatcher also occurs in scattered rose thickets and woodlands adjacent to (often dry) pasture or agricultural lands and country gardens.

On the coast, the Willow Flycatcher has been recorded breeding from 29 May (calculated) to 13 August; in the interior, it has been recorded from 25 May to 25 August (Fig. 35).

Nests: In the interior, most nests were found in wild rose bushes (56%; n = 147; Fig. 41) followed by willow (13%), red-osier dogwood (11%), poplar, birch, saskatoon, aspen, choke cherry, alder, Oregon-grape, honeysuckle, raspberry, and a small pine tree. On the coast, most nests (43%; n = 44) were found in bracken fern followed by wild rose (18%), vine maple (11%), salmonberry, blackberry, hardhack, red-berry elder, plum, and alder. Our data indicate results similar to those of McCabe (1991): "If there is only willow, or it overwhelmingly dominates a marshland, then the [Willow] flycatchers are likely to select it. In situations where a choice of nesting bushes is available, willow is not preferred."

Most nests (90%; n = 31) were placed on an upright stem or fork, with the remainder attached to a branch (Fig. 41). The heights of 187 nests ranged from 0.6 to 20.0 m, with 73% between 0.9 and 1.5 m. Our data support Stein's (1963) findings that the Willow Flycatcher builds nests farther from the ground and in larger bushes than does the Alder Flycatcher.

The nest of the Willow Flycatcher is usually more compact than that of the Alder Flycatcher (Stein 1963). It is made of grass, plant fibre, moss, bark strips, twigs, and rootlets; the cup is lined with hair, plant down, feathers, and fine grass (Fig. 41). Often the outside of the nest contains cottony material such as plant down or feathers. Streamers are rarely present (see **Alder Flycatcher**).

Eggs: Dates for 208 clutches ranged from 25 May to 6 August, with 53% recorded between 22 June and 9 July. Sizes of 191 clutches ranged from 1 to 5 eggs (1E-17, 2E-40, 3E-73, 4E-57, 5E-4), with 68% having 3 or 4 eggs (Fig. 41). The incubation period in British Columbia from 8 nests (not parasitized by the Brown-headed Cowbird) ranged from 12 to 17 days, with a median of 14 days. The incubation period may be prolonged in the presence of cowbird eggs: incubation periods in British Columbia for 7 nests parasitized by the Brown-headed Cowbird ranged from 14 to 20 days, with a median of 16 days. McCabe (1991) reports a mean incubation period for the Willow Flycatcher of 14.8 days.

Young: Dates for 63 broods ranged from 13 June to 25 August, with 56% recorded between 8 July and 22 July. Sizes of 60 broods ranged from 1 to 4 young (1Y-10, 2Y-13, 3Y-24, 4Y-13), with 83% having 2 to 4 young. The nestling period is about 14 days (McCabe 1991).

Brown-headed Cowbird Parasitism: In British Columbia, 36% of 210 nests found with eggs or young were parasitized by the cowbird. Parasitism on the coast was 7% (n = 45); parasitism in the interior was 44% (n = 165). There were 3 additional reports from the interior of adult Willow Flycatchers feeding cowbird young. Over half of the interior nests that suffered parasitism were from a single observer's property in the Okanagan valley. This probably resulted from the intensive nesting studies in progress on the property, which revealed the nests to the cowbirds. Excluding those nests, the British Columbia rate drops to 13% (n = 114) and the interior rate drops to 18% (n = 68). In McCabe's (1991) Wisconsin study of 537 nests with at least 1 egg, 9% contained 1 or more cowbird eggs; Harris (1991), on the other hand, found an exceptionally high rate (68%) of brood parasitism by Brown-headed Cowbirds in California. Sedgwick and Knopf (1988), in Colorado, also report a very high rate of 40.7% of Willow Flycatcher nests parasitized, but 75% of first nests. McCabe (1991) notes that many of the losses his nests suffered came from abandonment after he removed the cowbird eggs. He surmises that such removal is regarded by the flycatcher as predation, and the flycatcher responds by abandoning the nest. Friedmann et al. (1977) report that 8.4% of 237 sets of eggs in museum collections were parasitized. They note that western populations of *E. traillii* experience only about half as much parasitism as do eastern ones, although Sedgwick and Knopf (1988) raise some doubts about that assertion.

Nest Success: Of 96 nests found with eggs and followed to a known fate, 27 produced at least 1 fledgling, for a nest success rate of 28%. Coastal nest success was 33% (n = 6); interior nest success was 28% (n = 90). These rates are low compared with other studies (McCabe 1991), and may be cause for concern. Besides cowbird parasitism, a number of predators were observed or suspected to have destroyed nests (n = 60), including red squirrel (45%), striped skunk (15%), House Wren, chipmunk, northern flying squirrel, Common Raven, snake, Clark's Nutcracker, American Crow, American black bear, Steller's Jay, and domestic dog.

Figure 40. In the Okanagan valley, the Willow Flycatcher nests in riparian thickets of birch, willow, and red-osier dogwood (north end of Vaseux Lake, 9 August 1996; R. Wayne Campbell).

REMARKS: In their account of the Traill's Flycatcher, Munro and Cowan (1947) ascribe all interior birds to the subspecies *E. t. traillii;* coastal birds are ascribed to the subspecies *E. t. brewsteri.* This has caused some difficulties in trying to determine the historical status of the Alder and Willow flycatchers in the province. Part of the former subspecies *E. t. traillii* is now the Alder Flycatcher, while the remainder of *E. t. traillii* and *E. t. brewsteri* are now the Willow Flycatcher.

Reports of the Willow Flycatcher in British Columbia before 6 May and after 8 September were not accompanied by convincing details and have not been included in this account. Under good conditions, the Willow Flycatcher can be distinguished visually from other small flycatchers in British Columbia (except the Alder Flycatcher) by its white throat and inconspicuous eye ring; few observers recorded those distinctive field marks. In areas of sympatry with the Alder Flycatcher, there appears to be no way at present to distinguish the 2 species by visual clues alone. The song or call must be noted. Thus it is up to the meticulous observer to provide adequate documentation that will allow the determination of arrival and departure dates for this species. Kaufman (1990) and Pyle et al. (1987) provide identification aids.

Care should be used by observers when reporting Alder or Willow flycatchers from areas of sympatry. Stein (1963) notes 2 cases where a bird apparently sang both *fee-bee-o* and *fitz-bew* songs, although he concluded that the *fitz-bews* were "combinations of *pits* and *wee-oos*" of the Alder Flycatcher. Stewart (1975) found that some songs in North Dakota appeared to be intermediate in form between the two. In British Columbia, 2 of the present authors heard a song under good conditions in the Revelstoke area, where both species occur, that appeared to be intermediate between the two, leaning more towards the Willow than the Alder song. D. Prescott

Figure 41. Willow Flycatcher nest and eggs in a rose bush (Williams Lake, June 1969; Anna Roberts).

(pers. comm.) notes that when he was working extensively on Willow and Alder flycatchers, he had several quite experienced birders point out birds singing hybrid songs. In all cases the bird was an Alder Flycatcher singing a 2-note song. This tended to happen later in the year, when the birds started to lose their singing intensity and the last syllable was not produced or was inaudible.

Prescott (1987) showed that the Alder Flycatcher can be displaced from its breeding territory by the more aggressive Willow Flycatcher, and suggested that this could be a factor in the recent range contraction of Alders in Ontario and the accompanying northward expansion of Willows. That does not appear to be occurring in Wisconsin; indeed, the Alder Flycatcher may have extended its range into Willow Flycatcher areas (Robbins 1974), as appears to be happening in British Columbia. In British Columbia, both the Alder and Willow flycatchers appear to have recently moved into the southeastern portion of the province. Although the latter is currently the more common there, only the Alder Flycatcher seems to be increasing its numbers in the province, as suggested by interior Breeding Bird Surveys. Further studies in British Columbia may shed some light on the effects of the range extensions of these 2 species on each other.

The Willow Flycatcher appears on the *American Birds* "Blue List" from 1980 to 1982 (Tate 1981; Tate and Tate 1982). It was first listed because of declines in a number of regions covered by *American Birds,* including the Northern Pacific coast. By 1982, however, it appeared to be increasing or at least stable in some of those areas. In 1986 it was delisted, but retained as a species of "special concern" due to declines in the Northern Great Plains, Middle Pacific Coast, and Southern Pacific Coast regions, the latter area reporting serious declines. In California, the Willow Flycatcher has been extirpated from most of its range (Harris et al. 1988; Unitt 1987). Two reasons suggested for the decline are the alteration and loss of riparian habitat and Brown-headed Cowbird parasitism. These can be interactive, as a host species relegated to marginal habitats may become more vulnerable to its parasites and predators.

One of the principal reasons for habitat loss is cattle grazing in the riparian habitats. Cattle affect the hydrology of the meadows, and their activities directly disturb the flycatcher nests (Taylor 1986; Harris et al. 1988). In Oregon, dramatic population increases in the Willow Flycatcher have been documented following a reduction in cattle grazing and the elimination of willow cutting and spraying (Taylor and Littlefield 1986).

Although the Willow Flycatcher is normally monogamous, polygyny has been reported in this species (Prescott 1986; Sedgwick and Knopf 1989).

Additional information on the biology and systematics of the Willow Flycatcher can be found in Barlow and McGillivray (1983), Frakes and Johnson (1982), King (1955), McCabe (1991), and Sanders and Flett (1989).

See also **Alder Flycatcher**.

NOTEWORTHY RECORDS

Spring: Coastal – Saanich 10 May 1985-1 (Mattocks 1985), 30 May 1982-14 along Munn Road; Cowichan Bay 15 May 1970-10; South Langley 10 May 1967-3; Sea Island 12 May 1974-1; Harrison 6 May 1986-1 at hot springs; Vancouver 12 May 1959-1, 13 May 1984-1; Pitt Meadows 31 May 1981-12; Dudley Marsh (Coombs) 9 May 1985-1; Egmont 12 May 1979-5 singing in area. **Interior** – Castlegar 18 May 1980-1; Naramata 25 May 1969-2 eggs; Lytton 7 May 1966-1; Fountain Valley near Lytton 23 May 1970-1; Revelstoke 19 May 1986-1; near Riske Creek 26 May 1989-1; Williams Lake 30 May 1981-1.

Summer: Coastal – Victoria 31 Aug 1973-1; Saanich 5 Jun 1982-22 between Munn Road and Purple Martin pond; Saanich 15 Jun 1984-13 along powerline near Francis Park; Central Saanich 2 Jul 1983-20 along powerline; Kent 4 Jun 1984-4; Huntingdon 21 Jul 1946-11; Langley 3 Aug 1982-20 at Campbell Valley Park; Mission 31 Aug 1965-10; Sahtlam 6 Jun 1915-4 eggs, incubation started (RBCM 860); Sandhill Creek 19 Jul 1972-1 (Hatler et al. 1978); Little Qualicum River estuary 13 Aug 1982-1 young in nest; Woss 12 Jun 1991-1; Matthews Island 22 Jun 1979-1; Port Hardy 12 Jul 1985-1; Purcell Point 16 Aug 1936-4; Kimsquit River valley 2 July 1986-1, 13 Jul 1939-1 (NMC 29071). **Interior** – White Lake (Okanagan) 3 Jun 1922-1 (NMC 17823); Vaseux Lake 1 Aug 1969-3 eggs at Venner Meadows (Cannings et al. 1987); Naramata 9 Aug 1976-3 young sitting on edge of nest; Upper Whatshan Lake 1 Aug 1931-4 large young in nest; Wasa Park 1 Jun 1971-2 (Dawe 1971); Brouse 4 Aug 1980-2; Wilmer 18 Aug 1976-2; Yoho National Park 25 Aug 1975-4 adults feeding 2 young in the nest at Wapta Marsh; Revelstoke 31 Aug 1985-1; Canim Lake 1 Aug 1978-6; 100 Mile House 12 Aug 1979-1; Lac la Hache 31 Jul 1959-3 young in nest at Disaster Lake; Williams Lake 13 Jun 1978-3 young in nest; Babcock Creek 6 Jul 1990-1; Wells 29 Jul 1990-1 (Siddle 1990c); Lightning Lake (Wells) 26 Jul 1988-4.

Breeding Bird Surveys: Coastal – Recorded from 21 of 27 routes and on 77% of all surveys. Maxima: Gibsons Landing 23 Jun 1974-45; Port Renfrew 27 Jun 1973-31; Squamish 16 Jun 1985-31; Saltery Bay 22 Jun 1975-28. **Interior** – Recorded from 54 of 73 routes and on 62% of all surveys. Maxima: Golden 19 Jun 1994-32; Succour Creek 17 Jun 1994-29; Ferndale 7 Jun 1969-25; Summerland 20 Jun 1987-23.

Autumn: Interior – Kleena Kleene 2 Sep 1962-1 (Paul 1964); Yoho National Park 3 Sep 1967-1 (Wade 1977); Okanagan Landing 2 Sep 1943-1 (MVZ 102496); Madeline Lake (Penticton) 2 Sep 1979-1 (Cannings et al. 1987); Okanagan Falls 7 Sep 1977-1 (Cannings et al. 1987). **Coastal** – Jericho Park (Vancouver) 8 Sep 1990-2 (Weber 1991); Sea Island 4 Sep 1989-2; Somenos Lake 6 Sep 1980-1.

Winter: No records.

Least Flycatcher

LEFL

Empidonax minimus (Baird and Baird)

RANGE: Breeds from southern Yukon and southwestern Mackenzie east across Canada to Nova Scotia; south through central and southeastern British Columbia to Wyoming, Ohio, and Georgia; winters on coastal slopes from northern Mexico south to Honduras and Nicaragua.

STATUS: On the coast, a *very rare* migrant and summer visitant to the Georgia Depression Ecoprovince; *casual* in both the southern and northern Coast and Mountains. Absent from Western Vancouver Island and the Queen Charlotte Islands.

In the interior, an *uncommon* to *fairly common* migrant and summer visitant to the Boreal Plains and the Southern Interior Mountains ecoprovinces; *uncommon* in the Taiga Plains and in the extreme southern portions of the Sub-Boreal Interior and the Central Interior ecoprovinces; *rare* migrant and summer visitant in the Northern Boreal Mountains and the Southern Interior ecoprovinces.

Breeds.

CHANGE IN STATUS: The first documented occurrences of the Least Flycatcher in British Columbia were specimens collected in 1891 and 1892 from Chilliwack and Ashcroft, respectively. These specimens were overlooked by Brooks and Swarth (1925) in their summary of the province's avifauna, as was a specimen taken at Telkwa in 1919 (Rand 1948). Williams (1933a, 1933b) visited the Peace River area in the summers of 1921 and 1922. At that time he found this flycatcher to be common at Nig Creek and Moberly Lake, and along the Sikanni Chief River (Fig. 42). He also encountered it occasionally, further north, along the Liard River.

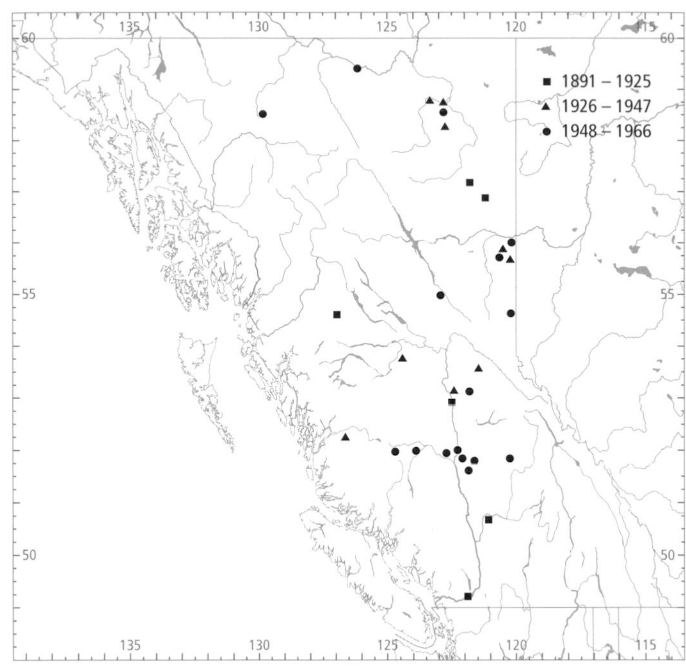

Figure 42. Distribution of the Least Flycatcher in British Columbia, 1891-1966 (Brooks and Swarth 1925; Munro and Cowan 1947; Godfrey 1966).

By 1930 the Least Flycatcher was reported nesting in the Peace River area (Racey 1930), and Cowan (1939) considered

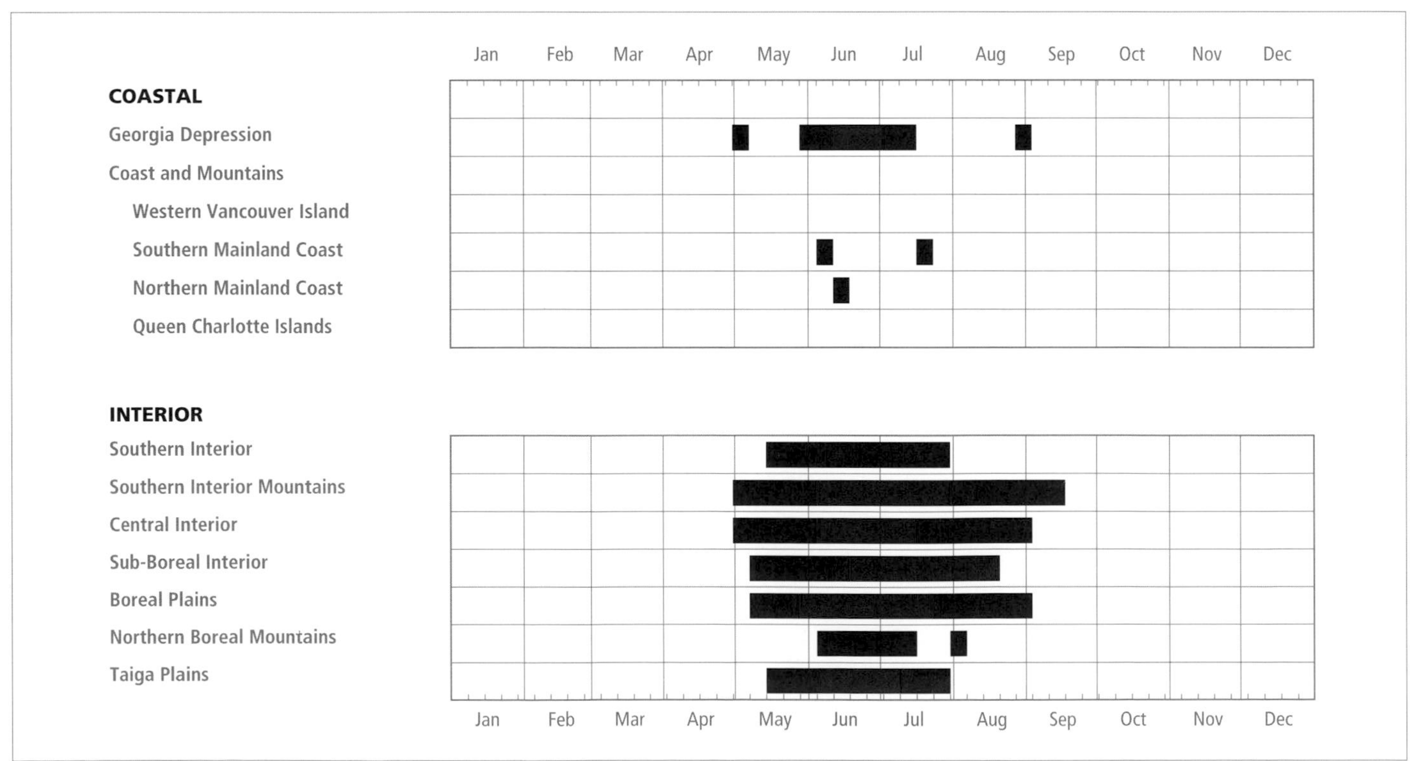

Figure 43. Annual occurrence (black) and breeding chronology (red) for the Least Flycatcher in ecoprovinces of British Columbia.

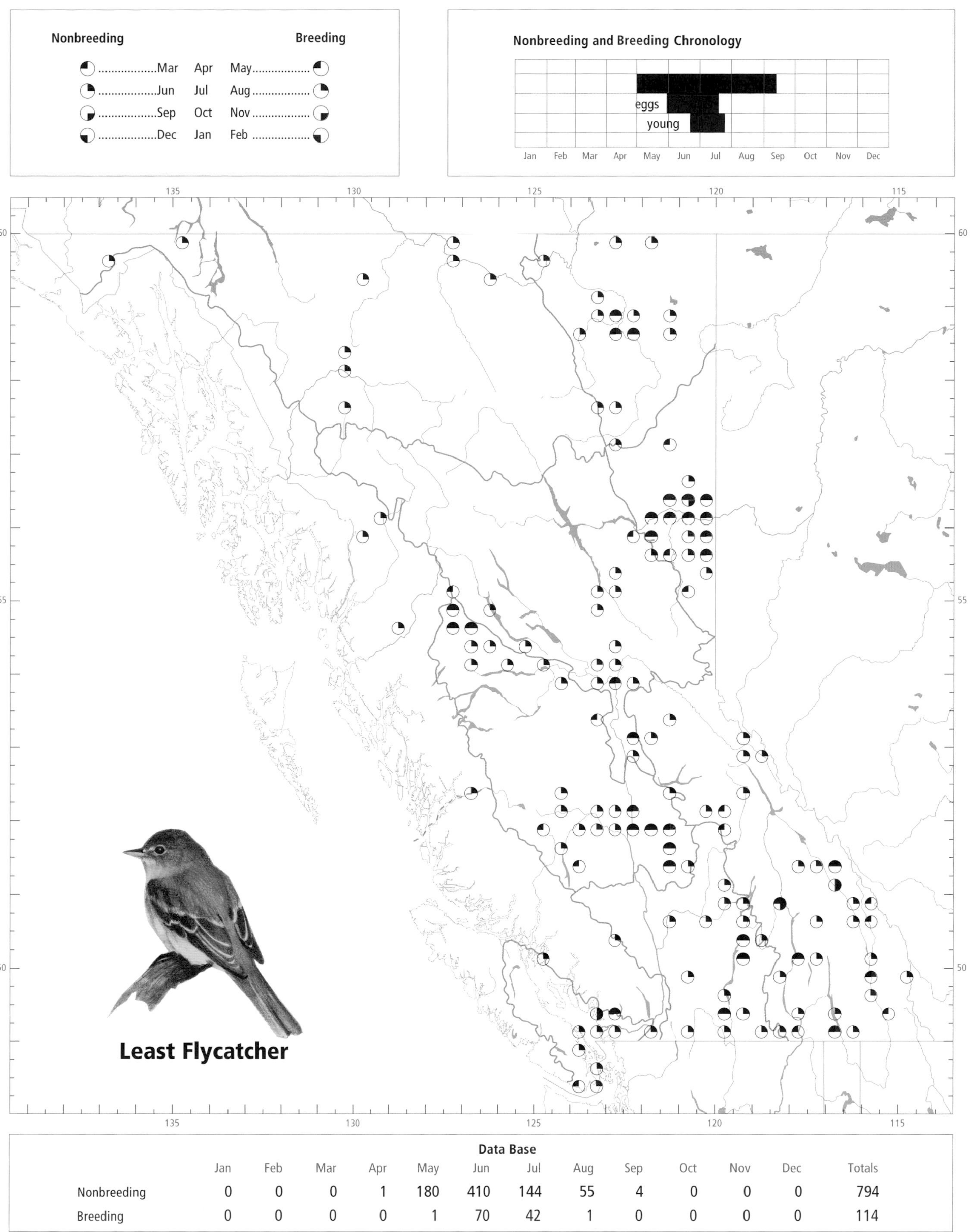

Data Base

	Jan	Feb	Mar	Apr	May	Jun	Jul	Aug	Sep	Oct	Nov	Dec	Totals
Nonbreeding	0	0	0	1	180	410	144	55	4	0	0	0	794
Breeding	0	0	0	0	1	70	42	1	0	0	0	0	114

it to be the most abundant *Empidonax* flycatcher there. The southward and westward expansion of its range had probably begun by 1942, when it was found breeding at Lac la Hache (Fig. 42), although in 1947 it was still considered only a summer bird of the parkland areas of the Peace River region and areas north (Munro and Cowan 1947).

By the late 1950s, the Least Flycatcher had become established in the Cariboo and Chilcotin areas of the Central Interior. It continued its southward expansion (Stein 1961; Erskine and Stein 1964; Paul 1964; Edwards and Ritcey 1967) and also moved into the northwestern part of the province (Fig. 42). It had arrived in the southern portions of the Southern Interior Mountains (e.g., Wasa) by the early 1970s, and had reached the northern portions of the Southern Interior (e.g., Adams and Shuswap lakes) by about the same time. It was first reported in the Okanagan valley in 1975 (Cannings et al. 1987). By 1992 the Least Flycatcher had been reported from much of the southern portions of the interior of the province.

On the south coast, the first record was from Vancouver in 1969 (Weber 1974), and it was found breeding near Victoria the following year (Taylor 1970).

NONBREEDING: The Least Flycatcher is widely distributed throughout the northeastern, central, and southeastern interior of the province; it is sparsely and locally distributed in the rest of the interior. In some years, it occurs in small numbers in the Fraser River valley and on extreme southeastern Vancouver Island; however, coastal observations account for only 6% of the records (Fig. 45). This flycatcher has not been reported from the Queen Charlotte Islands.

On the coast, the Least Flycatcher occurs near sea level; in the interior it is found from valley bottoms to about 1,000 m elevation.

In the interior, the Least Flycatcher frequents pure and mixed deciduous forests (Fig. 44), including aspen groves, alders and willows over swampy ground (Miller 1955), willow thickets, birch tangles, mixed spruce-aspen-birch and aspen–lodgepole pine forests (Stein 1961), and shrubby edges along lakes, ponds, creeks, cutlines, and avalanche slopes. It is particularly common near water, such as small lakes and beaver ponds, where there is a heavy growth of aspen, willow, or water birch. On the coast, it occurs in red alder stands and shrubby thickets. The Least Flycatcher frequents forest understorey shrubs and smaller trees or the open mid-canopy of mature forest stands. It forages at random in the available tree species (Rogers 1985).

Spring migration in the interior may begin as early as late April (Williams Lake; Figs. 43 and 45), but most birds arrive during the second and third weeks of May. Spring migrants may arrive suddenly. For example, in the Peace Lowland, Cowan (1939) reported the first arrivals on 18 May and 2 days later found Least Flycatchers "everywhere." In southern Ontario, peak spring movements occur in mid-May, with males arriving about a week earlier than females (Hussell 1981).

In the Peace River region, the autumn movement reaches a peak during the third week of August, with most birds gone by the end of the month (Figs. 43 and 45). In the Central Interior and Southern Interior Mountains, a similar trend occurs;

Figure 44. Least Flycatcher habitat in pure deciduous woodlands of trembling aspen (Fraser Lake, 4 June 1995; R. Wayne Campbell).

in the Southern Interior Mountains, birds may be found as late as mid-September. In southern Ontario, adults migrated south even earlier, apparently immediately after completion of breeding activities, between mid-July and early August, with juveniles following 1 month later (Hussell 1981). Asynchronous departure of age groups also occurs in Manitoba (Sealy and Biermann 1983). Whether or not a similar trend occurs in British Columbia remains to be determined.

On the coast, the Least Flycatcher has been recorded from 4 May to 1 September; in the interior, it has been recorded from 30 April to 12 September (Fig. 43).

BREEDING: The Least Flycatcher likely breeds throughout most of its summer range in the province; however, it is presently known to breed only in the Peace River region of the Boreal Plains, the Taiga Plains, the eastern Central Interior (west to Hanceville), and the east and west Kootenay districts. There is 1 nesting record from the coast.

The Least Flycatcher reaches its highest numbers in the Peace Lowland of the Boreal Plains in northeastern British Columbia (Fig. 46). An analysis of Breeding Bird Surveys for the period 1968 through 1973 could not detect a net change in numbers on interior routes; coastal routes for the same period contained insufficient data for analysis. The Least

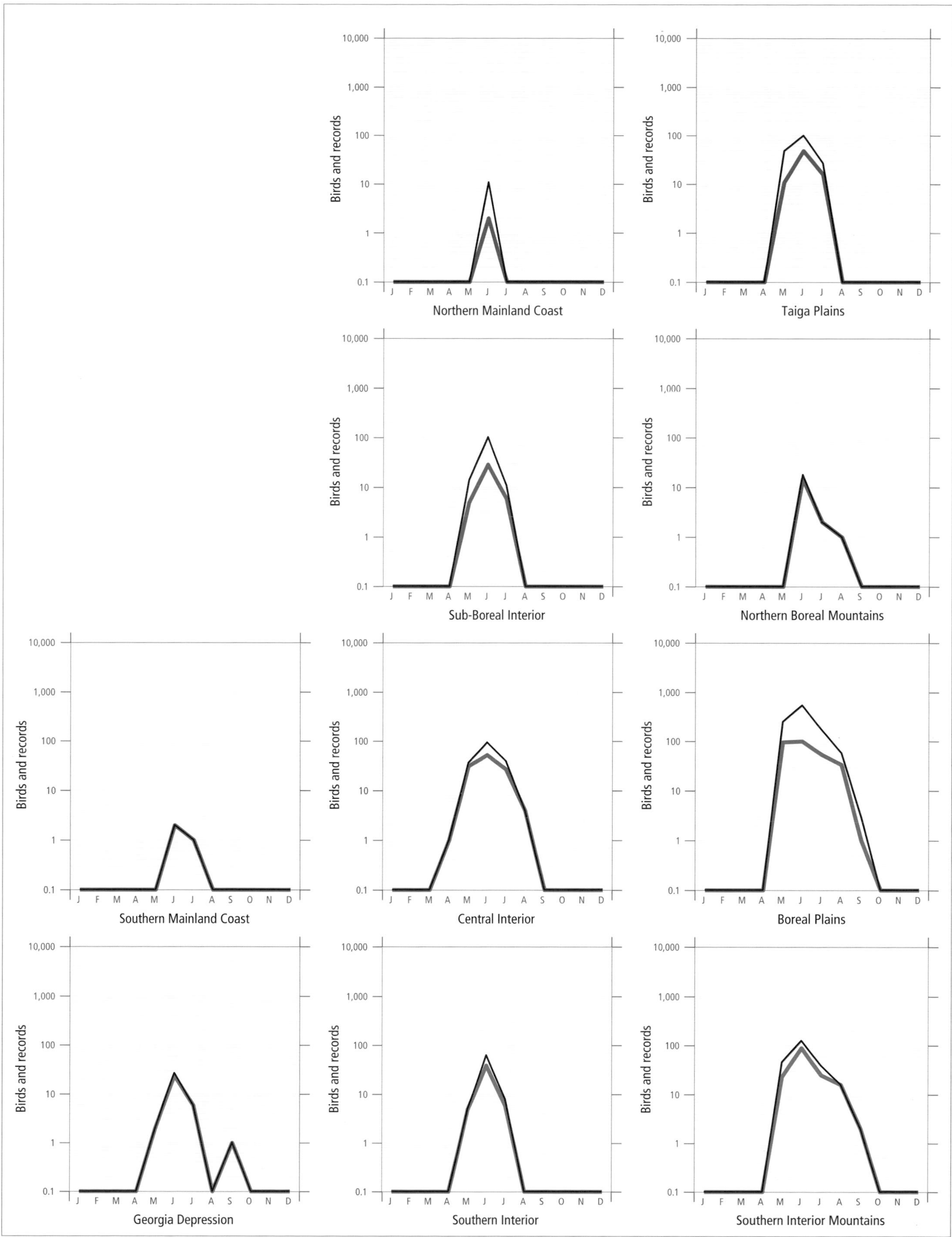

Figure 45. Fluctuations in total number of birds (purple line) and total number of records (green line) for the Least Flycatcher in ecoprovinces of British Columbia. Nest record data have been excluded.

Flycatcher seems to be maintaining stable populations throughout its range in North America (Robbins et al. 1986; Sauer and Droege 1992). In the Western Region of the Breeding Bird Surveys, populations of this flycatcher appear to be increasing (Dobkin 1992; Sauer and Droege 1992), and similar increases are reported for the central and southern Prairies of Canada, although there are declines in parts of its range (e.g., central Quebec and central Ontario) (Erskine et al. 1992).

The Least Flycatcher has been reported breeding from near sea level to 900 m elevation. Almost all nesting takes place in semi-open, mature deciduous and mixed forest habitats dominated by trembling aspen (middle-aged and 50+ years old), willow, black cottonwood, or balsam poplar. Douglas-fir, lodgepole pine, and ponderosa pine are often present in lesser numbers. Many nest sites have open spaces nearby, including lakeshores, city parks, roads, and clearings around houses. In the Peace Lowland, mature trembling aspen or aspen-birch stands (Fig. 47), sometimes with a few spruce, are characteristic of breeding territories. Near Smithers, Pojar (1993) found the species more abundant in stands with less foliage in the mid-canopy but with high foliage cover in the upper canopy. In New Hampshire, Sherry (1979) showed that the Least Flycatcher has higher nesting success in hardwood forests where the mid-canopy is open and affords good visibility.

In the interior, the Least Flycatcher has been recorded breeding from 30 May to 24 July; on the coast, it has been recorded breeding from 15 June (calculated) to at least 27 June (Fig. 43).

Nests: Most nests (94%; n = 48) were in live deciduous trees (Fig. 48); other sites included saskatoon and choke cherry shrubs. Trembling aspen (50%; n = 48), willow (29%), and alder (10%) were the most frequently used nest trees; other trees included paper birch, balsam poplar, and Garry oak. Most nests (71%) were in the fork or crotch of a branch, or on a branch adjacent to the trunk (29%). Distances of 6 nests from the trunk ranged from 0.3 to 2.2 m. The heights of 39 nests ranged from 1.0 to 13.5 m from the ground, with 56% recorded between 2.1 and 4.8 m.

Nests were compact cups, composed mainly of plant down, grass, plant fibre, and bark strips. Other materials included hair, lichens, moss, feathers, leaves, and human-made items such as string and paper.

In British Columbia, 1 Least Flycatcher nest was reused for 3 successive years (A. Stewart pers. comm.). Briskie and Sealy (1988) indicate that Least Flycatchers occasionally reuse old nests.

Eggs: Dates for 32 clutches ranged from 30 May to 19 July, with 57% recorded between 11 and 27 June. Sizes of 22 clutches ranged from 2 to 5 eggs (2E-3, 3E-3, 4E-14, 5E-2), with 14 having 4 eggs. In Manitoba (Briskie and Sealy 1989) and in Ontario (Peck and James 1987), 4-egg clutches were also the most frequent. One clutch in British Columbia had an incubation period of 14 days, which agrees with the 12 to 15 days (mean = 13.6 days) noted by Walkinshaw (1966) and Briskie (1994).

Young: Dates for 20 broods ranged from 21 June to 24 July, with 52% recorded between 30 June and 8 July. Hoffmann (1901) and Walkinshaw (1966) report double-brooding for the Least Flycatcher; but, as Briskie and Sealy (1987a) rightly state, the length of the available nesting season at northern

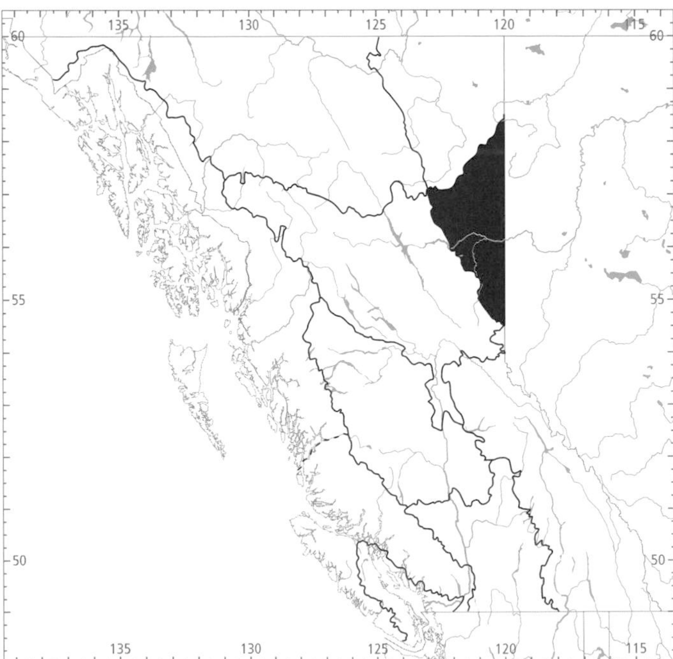

Figure 46. In British Columbia, the highest numbers for the Least Flycatcher in summer occur in the Boreal Plains Ecoprovince.

Figure 47. Least Flycatcher breeding habitat in mixed woodlands of mainly trembling aspen and wild rose understorey with some willow, balsam poplar, young black cottonwood, and paper birch (Pouce Coupe, 12 June 1996; R. Wayne Campbell).

latitudes normally precludes more than 1 successful nesting per season. Sizes of 19 broods ranged from 1 to 6 young (1Y-1, 3Y-9, 4Y-8, 6Y-1), with 17 having 3 or 4 young. The brood of 6 young was accompanied by an adult female and 2 other adults. One brood in British Columbia had a nestling period of 17 days, which agrees with the nestling period of 12 to 17 days noted by Briskie (1994).

Brown-headed Cowbird Parasitism: In British Columbia, 19% of 37 nests found with eggs or young were parasitized by the cowbird. Nest parasitism in the interior was 17% (n = 35); 1 of 2 coastal nests was parasitized. In the early 1960s, Friedmann (1963) concluded that the Least Flycatcher "appears to be molested rather seldom," but a decade later there was an increase of over 100% in numbers of nests parasitized. Parasitism rates in British Columbia are higher than those reported for Manitoba (2.7%; Briskie 1985), Ontario (5%; Peck and James 1987) and Michigan (11.9%; Southern and Southern 1980).

The Least Flycatcher is a typical "acceptor" of cowbird eggs (Rothstein 1975; Briskie and Sealy 1987b). Since the incubation period for cowbirds (10 to 11 days) is about 3 days shorter than that of the Least Flycatcher, most flycatcher eggs fail to hatch. Those that do are quickly outcompeted by the young cowbird. Studies in Manitoba (Briskie and Sealy 1987b) showed that in nests in which the eggs of both Least Flycatcher and Brown-headed Cowbird hatched, no flycatcher chick survived longer than 2 days. As Walkinshaw (1961) noted, successful cowbird parasitism probably results in total reproductive failure for the host.

Figure 48. Adult Least Flycatcher on nest in willow (Williams Lake, 26 June 1968; Anna Roberts).

Nest Success: Of 13 nests found with eggs and followed to a known fate, 6 produced at least 1 fledgling.

REMARKS: The 1966 records of nesting near Penticton (Rogers 1966d) are without documentation and have been excluded from the account.

Freedman et al. (1981) showed that the density of breeding Least Flycatchers decreased with intensity of logging operations. DellaSala and Rabe (1987) also demonstrated that this species retreated further into the forest interior as the adjacent disturbed area increased in size.

For details on the life history of the Least Flycatcher in North America, see Briskie (1994).

NOTEWORTHY RECORDS

Spring: Coastal – Triangle Mountain (Sooke) 4 May 1980-1. **Interior** – Creston 25 May 1986-1; Lac du Bois 20 May 1985-1 singing; 103 Mile Lake (Cariboo) 17 May 1959-1; Watson Lake 17 May 1969-1, first spring arrival (Erskine and Stein 1964); Kleena Kleene 20 May 1961-1 (Paul 1964); Williams Lake 30 Apr 1968-1 banded; Hemp Creek 10 May 1959-1 (Edwards and Ritcey 1967); Taylor 25 May 1985-9; Beatton Park 11 May 1980-1; Charlie Lake 27 May 1978-1 adult building nest, 30 May 1982-1 adult at nest; Nig Creek 27 May 1922-1 (Williams 1933a).

Summer: Coastal – Victoria 27 Jun 1970-4 eggs (RBCM 1691); Mount Tolmie (Victoria) 17 Jun 1970-3 eggs, 25 Jun 1970-eggs broken and replaced by 2 Brown-headed Cowbird eggs (Taylor 1970); Cassidy 23 Jun 1985-1 singing; Chilliwack 4 Jun 1891-1 (MCZ 35989); Vancouver 7 Jun 1969-1 (Weber 1974), 5 and 6 Jun 1984-1 singing (Campbell 1984c); Pitt Meadows 19 Jun 1988-1 male singing; Cortes Island 19 Jun 1976-1; Pemberton 9 Jun 1985-1 (Harrington-Tweit and Mattocks 1985); near Hagensborg 18 Jul 1938-1; Stewart 17 Jun 1975-1. **Interior** – Allison Pass 22 Jun 1986-1 (Mattocks 1986b); Kettle River 4 Jul 1973-1; Kimberley 24 Jul 1975-2 young left nest; Wasa 9 Jun 1971-4 eggs (Dawe 1971); Brouse 12 Jun 1980-1; Coldstream (Vernon) 12 Jun 1975-1; Enderby 18 Aug 1941-1 male (UBC 7882); Ashcroft 8 Jun 1892-1 (ANS 30878), 11 Jun 1892-1 (ANS 30879); Adams Lake 1 Jul 1963-1; Lac la Hache 12 Jul 1942-4 eggs (RBCM 1376); Watson Lake (Cariboo) 9 Jul 1958-1 (Stein 1961); Timothy Lake 8 Aug 1958-1 (Stein 1961); Williams Lake 3 Jun 1965-building nest, 11 Jun-4 eggs; 9 km w Quesnel 16 Jul 1947-2 males (UBC 1833-1818); Mount Robson Park 8 Jun 1970-1 (Stirling 1970); Indianpoint Lake 7 Jul 1930-1 (MCZ 283001), 2 Jun 1934-1 (MCZ 65636; Miller 1955); Meldrum Lake 5 Jul 1948-1; Nulki Lake 29 Aug 1945-1 (ROM 86614); Prince George 22 Jun 1949-1 male (UBC 2113); 14 Jun 1969-12, 12 Aug 1975-4 fledged young; near Burns Lake 11 Jun 1975-1; Smithers 1 Jul 1974-1; Parsnip River 26 Jun 1963-1 (UBC 11756); Pouce Coupe 24 Jun 1930-1 (UBC 5346), nest and 3 eggs and 1 cowbird egg (Racey 1930); Charlie Lake 4 Jun 1986-4 eggs; Trutch 13 Jul 1943-1 (Rand 1944); Dease Lake 15 Jun 1962-1 (NMC 39848); Sikanni Chief River 6 Jun 1922-1 (Williams 1933b); Fort Nelson 29 Jun 1982-1 male (RBCM 17612); 18 Jul 1943-1 (Rand 1944); s Kotcho Lake 16 Jun 1982-1 singing (Campbell and McNall 1982); Cassiar 17 Jun 1975-1; Liard Hot Springs 14 Jul 1956-1; Fireside 20 Jun 1974-1 male (WSM 28205); Tutshi Lake 24 Jun 1979-1 singing.

Breeding Bird Surveys: Coastal – Recorded from 7 of 27 routes and on 3% of all surveys. Maxima: Kitsumkalum 12 Jun 1977-10; Kispiox 20 Jun 1993-7; Pemberton 22 Jun 1986-3. **Interior** – Recorded from 49 of 73 routes and on 39% of all surveys. Maxima: Tupper 10 Jun 1976-67; Hudson's Hope 7 Jun 1976-42; Telkwa High Road 27 Jun 1990-40.

Autumn: Interior – Charlie Lake 2 Sep 1984-3; Nicholson 12 Sep 1976-1; Revelstoke 2 Sep 1986-1. **Coastal** – Vancouver 1 Sep 1986-1.

Winter: No records.

Hammond's Flycatcher

HAFL

Empidonax hammondii (Xantus de Vesey)

RANGE: Breeds from east-central and southeastern Alaska, southern Yukon, northern and central British Columbia, and southwestern Alberta south to California and New Mexico. Winters in highland areas from southern Arizona (rarely) and northern Mexico south to Honduras and Nicaragua.

STATUS: On the coast, an *uncommon* to locally *common* migrant and summer visitant to the Georgia Depression Ecoprovince; elsewhere along the coast it is *uncommon* to *fairly common* except on the Queen Charlotte Islands, where it is *very rare*.

In the interior, an *uncommon* to locally *common* migrant and summer visitant to the Southern Interior and Southern Interior Mountains ecoprovinces, becoming *uncommon* to *fairly common* in the Central Interior and Sub-Boreal Interior ecoprovinces. Locally *rare* in the Boreal Plains, Northern Boreal Mountains, and Taiga Plains ecoprovinces.

Breeds.

NONBREEDING: The Hammond's Flycatcher is widely distributed in forested areas throughout the province south of the latitude of the Skeena River valley and Prince George (54°30'N). It is found only occasionally on the Queen Charlotte Islands. In the northern interior and in the northeastern region east of the Rocky Mountains, it has a rather sparse and local distribution. The Hammond's Flycatcher is most abundant in mountainous areas across southern British Columbia, including Vancouver Island.

On the coast, it occurs from sea level up to the timberline but is more abundant at middle elevations. On southern Vancouver Island, surveys indicate that it is the most abundant *Empidonax* flycatcher in mid-spring up to at least 800 m elevation. In the Okanagan valley it is the common *Empidonax* at moderate to higher elevations (Cannings et al. 1987), although spring migrants may use the valley bottom early, before moving to breeding areas at higher elevations (Taverner 1922). It is also the most common flycatcher on the western slope of the Rocky Mountains south of Kickinghorse Pass (Poll et al. 1984). It is fairly numerous in lowland forests and upper river valleys of the Stikine River (Swarth 1922) and Skeena River (Swarth 1924).

The Hammond's Flycatcher frequents several types of woodland, from open to dense coniferous, deciduous, or mixed (Fig. 50) forests. This species inhabits almost all coniferous forest types (Fig. 51), ranging from lowland Douglas-fir, western hemlock, and western redcedar to subalpine lodgepole pine and Engelmann spruce and white spruce in boreal forests. It is a common spring and autumn migrant in ponderosa pine stands, usually in cooler, moister areas where Douglas-fir predominates. The Hammond's Flycatcher also frequents mixed or deciduous forests and woodlands, where it forages along edges of ponds, lakes, streams, bogs, and swamps in clumps of willow, alder, cottonwood, or aspen. In all habitats, it typically forages in the forest canopy or just beneath the canopy, rather than in shrubs and smaller trees

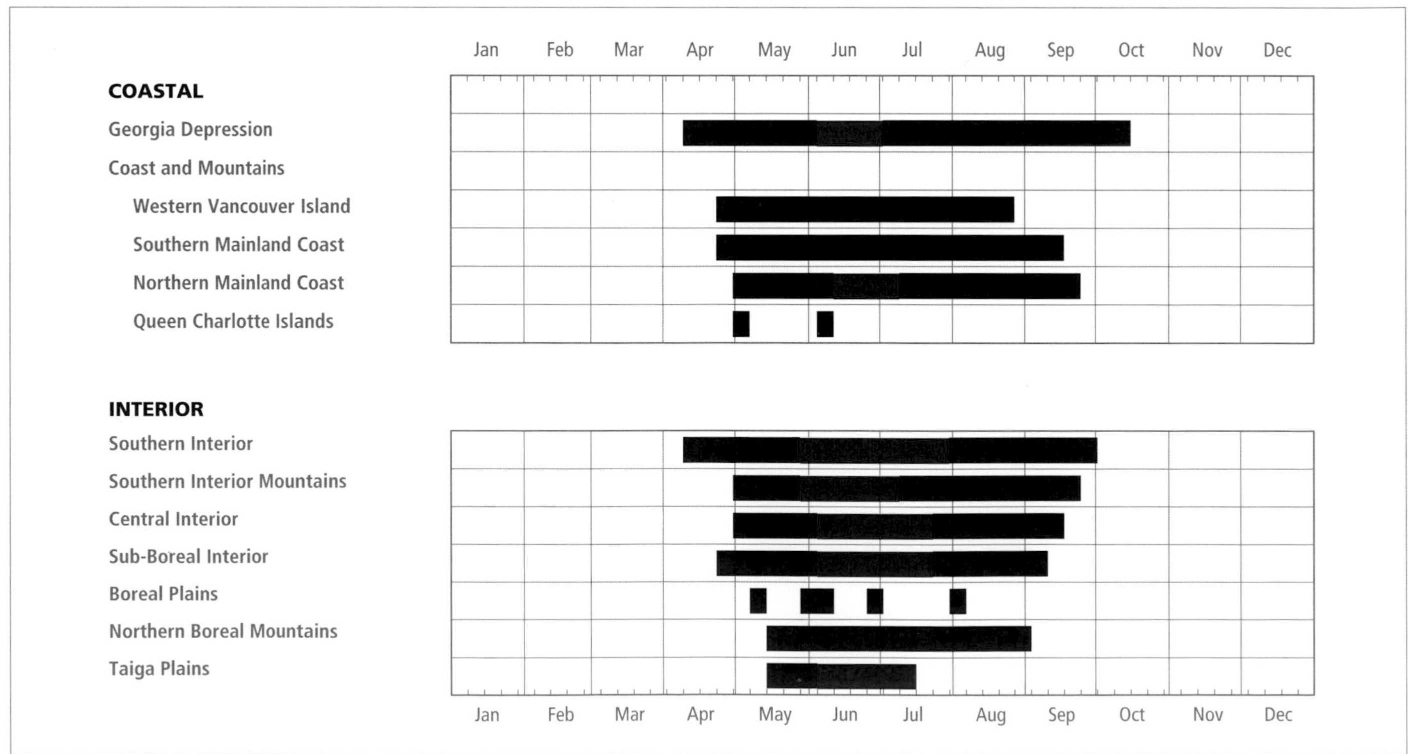

Figure 49. Annual occurrence (black) and breeding chronology (red) for the Hammond's Flycatcher in ecoprovinces of British Columbia.

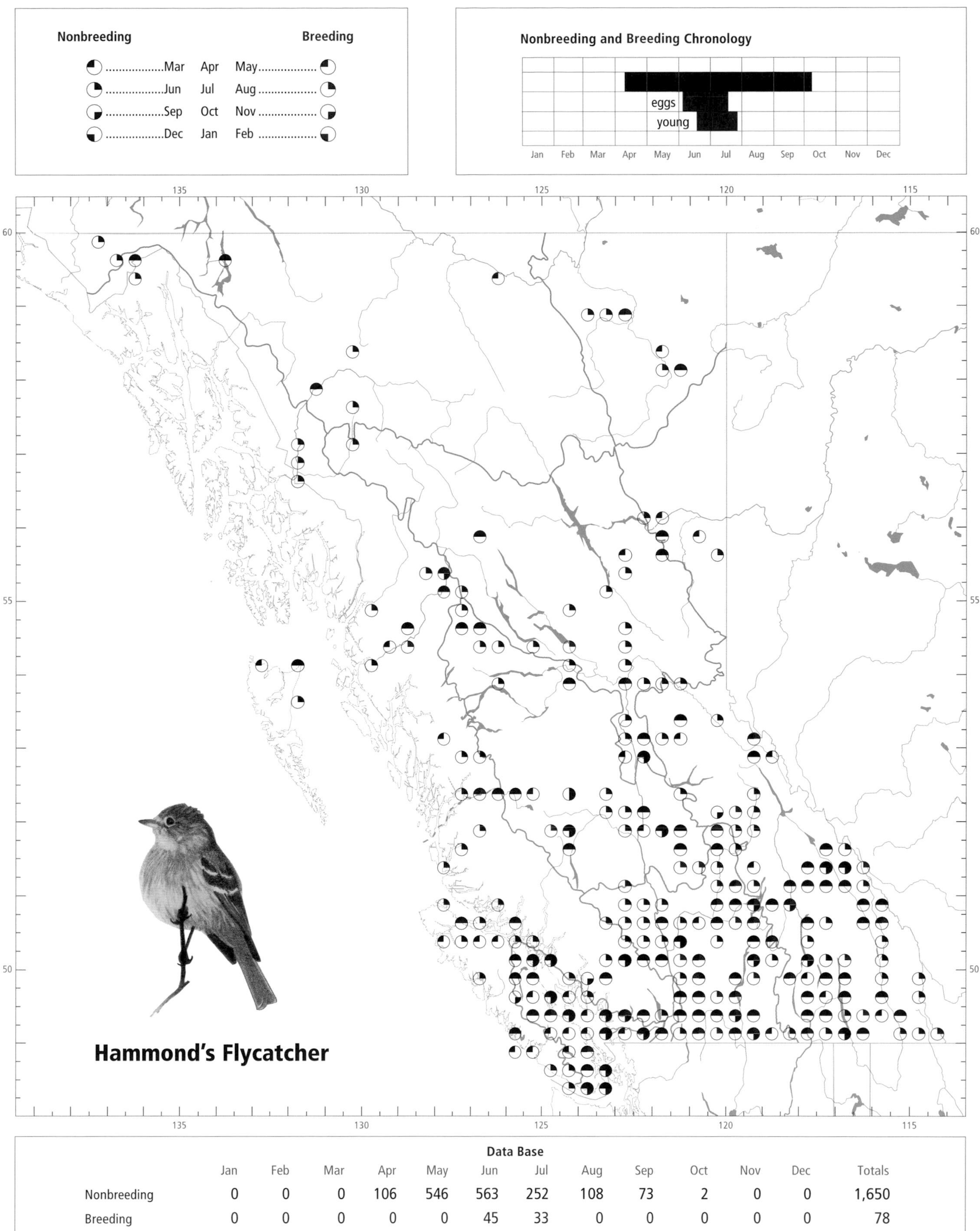

Data Base

	Jan	Feb	Mar	Apr	May	Jun	Jul	Aug	Sep	Oct	Nov	Dec	Totals
Nonbreeding	0	0	0	106	546	563	252	108	73	2	0	0	1,650
Breeding	0	0	0	0	0	45	33	0	0	0	0	0	78

Figure 50. On the coast, Hammond's Flycatcher nonbreeding and breeding habitats include mixed woodlands of black cottonwood, mountain-ash, red alder, western redcedar, and Douglas-fir (sparse), with wild rose and thimbleberry as understorey vegetation (Hagensborg, 11 July 1992; R. Wayne Campbell).

Figure 51. Hammond's Flycatcher breeding habitat in a mixed forest of Douglas-fir, lodgepole pine, and some trembling aspen, with saskatoon as the understorey shrub (Slocan valley, 16 June 1993; R. Wayne Campbell).

of the understorey. It is seldom reported from agricultural and urban habitats, and then only during migration.

On Western Vancouver Island, Bryant et al. (1993) found the Hammond's Flycatcher in all ages of forests, but it was most consistent and abundant in old-growth and 50- to 60-year-old forests. The same attraction of old-growth for this flycatcher was described by Manuwal (1991) in southern Washington. There it was more abundant in old-growth and 95- to 190-year-old forests than in those of 55 to 80 years. In northwestern California, Sakai (pers. comm.) also found the Hammond's Flycatcher in old-growth and large mature stands (about 91 to 199 years old).

In Mount Revelstoke and Glacier national parks, spring migrants frequented forest edges of mature coniferous and mixed forests (especially cedar-hemlock). The Hammond's Flycatcher was less common in deciduous forests and tall shrubbery (Van Tighem and Gyug 1983).

Spring migrants arrive early compared with other *Empidonax* flycatchers. In the southern portions of the province, the spring movement may begin in the second week of April (Figs. 49 and 52) and is most obvious from the end of April to mid-May. The earliest specimen records are 13 April on the coast and 15 April in the interior. In the interior, peak movements at higher elevations in Kootenay National Park occur in late May (Poll et al. 1984). Specimen data indicate that the migration of male Hammond's Flycatchers in Washington state and southern British Columbia is at its height between 27 April and 5 May (Johnson 1965). In the Central Interior, spring migrants are first heard in late April, with the peak movement occurring about 3 weeks later. In the far north, the first spring arrivals are encountered in mid-May.

The autumn movement in the north occurs in August; in the south, it occurs in late August and early September. Most birds have left the province by the third week of September (Figs. 49 and 52).

Migrants move mainly as individuals or in mixed groups of other songbirds. Johnson (1970) notes that in coastal regions the spring migration is early and rapid, in contrast to a later, more protracted movement in the interior. An opposite pattern is shown by the autumn movement, in which there is a leisurely migration through coastal areas compared with an early and comparatively rapid passage southward from the interior.

The Hammond's Flycatcher has been recorded on the coast from 9 April to 8 October; in the interior, it has been recorded from 13 April to 28 September (Fig. 49).

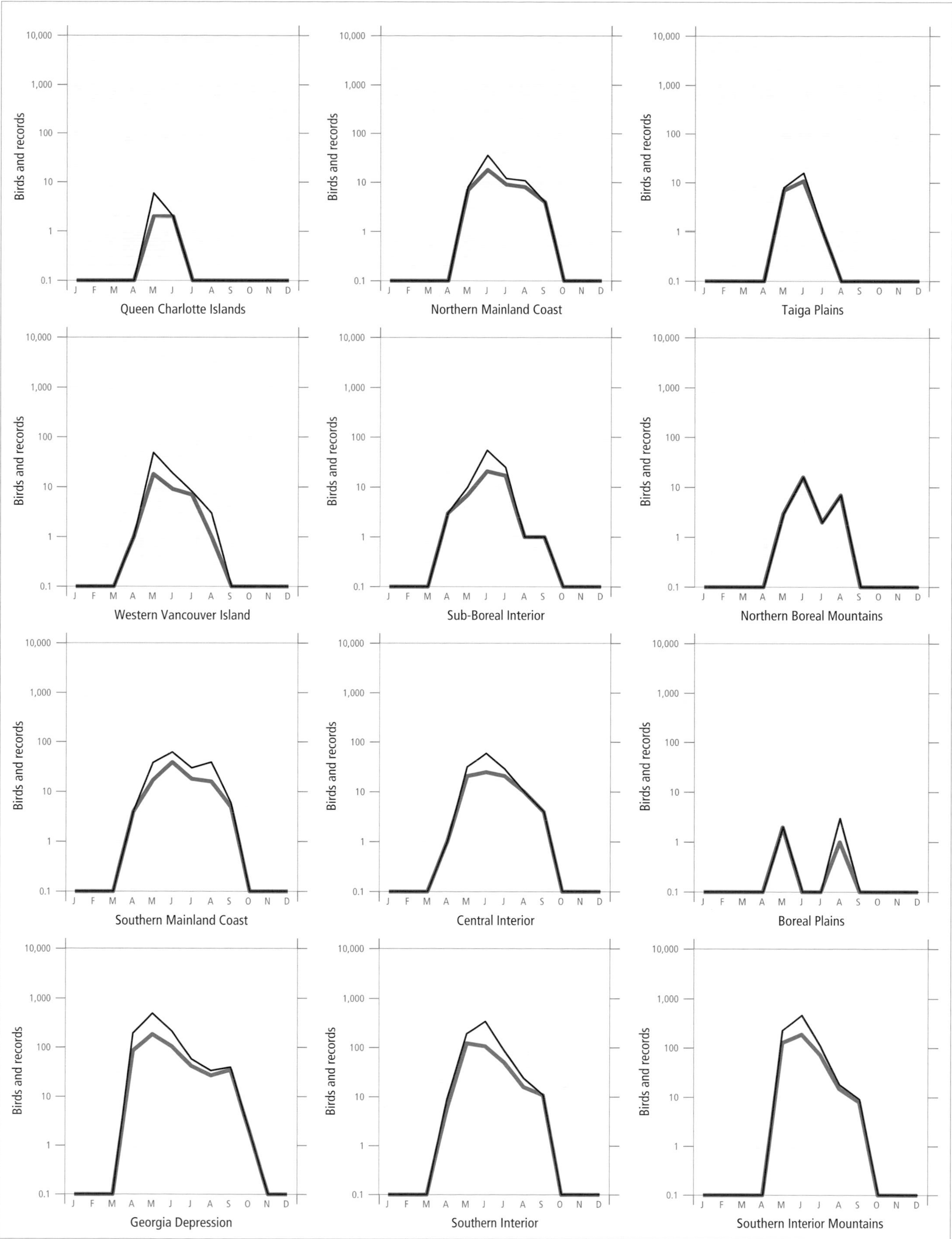

Figure 52. Fluctuations in total number of birds (purple line) and total number of records (green line) for the Hammond's Flycatcher in ecoprovinces of British Columbia. Nest record data have been excluded.

BREEDING: The Hammond's Flycatcher breeds throughout the southern half of the province, including Vancouver Island. There are only 4 known breeding locations north of Prince George. The northernmost location, Fort Nelson, is the only British Columbia nesting record east of the Rocky Mountains. In Alberta, the few known occurrences of this species are on the eastern slope of the Rocky Mountains from the vicinity of Banff southward (Semenchuk 1992).

There are just 2 breeding occurrences on the coast north of Campbell River, both from near Hazelton; in addition, fledged young have been reported from Bella Coola and Lakelse Lake Park. The Hammond's Flycatcher is not known to breed on the Queen Charlotte Islands.

The Hammond's Flycatcher reaches its highest numbers in summer in the Southern Interior Mountains Ecoprovince (Fig. 53). An analysis of Breeding Bird Surveys for the period 1968 through 1993 shows that the number of birds on coastal routes increased at an average annual rate of 8% (Fig. 54); analysis of interior routes for the same period could not detect a net change in numbers.

Breeding habitat is similar to that used by nonbreeding birds (Figs. 50 and 51) except in southern regions, where the Hammond's Flycatcher tends to breed in cooler, shadier, higher-elevation forests than those used during its spring migration. Although the Hammond's Flycatcher breeds over a wide range of elevations from near sea level on the coast to 1,500 m in the interior, in general it nests at higher elevations than the other *Empidonax* flycatchers in the province (Wade 1977). In Kootenay National Park, the Hammond's Flycatcher is very common in the montane spruce and lower Engelmann spruce–subalpine fir forests (Poll and Porter 1984). This habitat is dominated by Engelmann spruce and lodgepole pine, with Douglas-fir at lower elevations, interspersed with willow, birch, and sedge swamps. The highest densities of breeding birds were found in the wetter, spruce-dominated forests along river floodplains. In Yoho National Park, the Hammond's Flycatcher is the most common of the *Empidonax* flycatchers and occupies similar habitats, except that it extends its nesting habitat to include mixed conifer-aspen forests (Wade 1977).

In Montana (Davis 1954; Manuwal 1970), Oregon (Mannan and Meselow 1984), and northwestern California (Sakai 1988), the Hammond's Flycatcher uses much the same habitat as it uses in British Columbia.

On the coast, the Hammond's Flycatcher has been recorded breeding from 11 June to 7 July; in the interior, it has been recorded from 2 June (calculated) to 27 July (Fig. 49).

Nests: About half of 40 nests were situated in live coniferous trees, including Douglas-fir, western redcedar, ponderosa pine, and spruce. The remainder were mostly in live, but some dead, deciduous trees – including trembling aspen, willow, vine maple, poplar, and birch – or in such shrubs as wild rose or saskatoon. In 25 nests where the placement was noted, all were attached to the upper surface of a horizontal branch either in a fork, adjacent to the trunk, or saddled directly on the branch. Nest heights (n = 40) ranged from 0.5 to 14 m, with 60% recorded between 2 and 6 m. On the Olympic Peninsula of Washington, nests have been found as high as 18 m (Bent 1942).

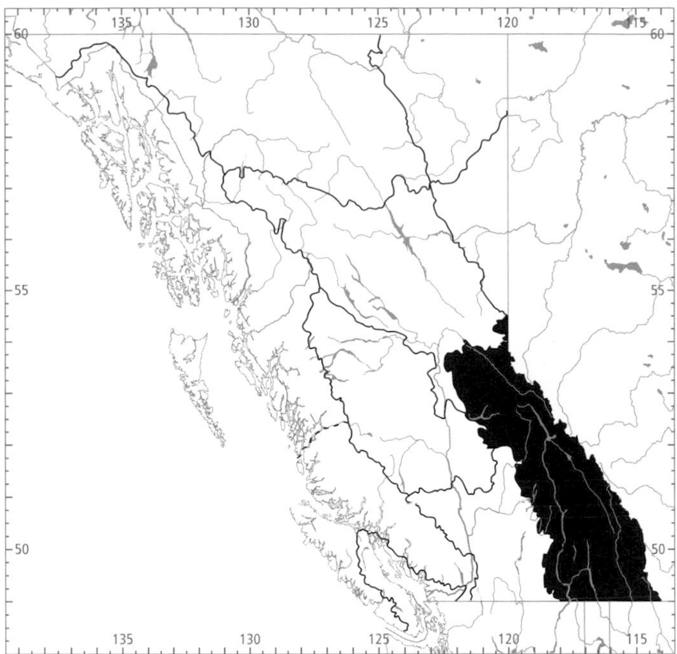

Figure 53. In British Columbia, the highest numbers for the Hammond's Flycatcher in summer occur in the Southern Interior Mountains Ecoprovince.

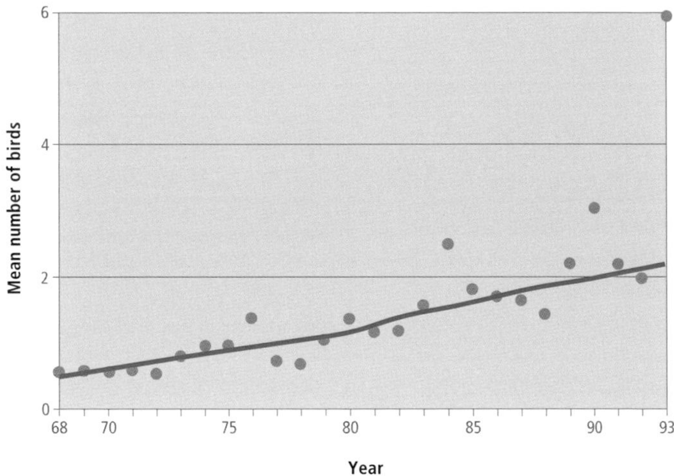

Figure 54. An analysis of Breeding Bird Surveys for the Hammond's Flycatcher in British Columbia shows that the number of birds on coastal routes increased at an average annual rate of 8% over the period 1968 through 1993 ($P < 0.01$).

The nest was a compact cup, composed of grass stems, bark strips, fine rootlets, small twigs, needles, moss, lichens, plant fibres, leaves, and plant down, and was usually lined with hair, fine grass, and feathers.

Eggs: Dates for 33 clutches ranged from 4 June to 18 July, with 57% recorded between 10 June and 1 July. Calculated dates indicate that nests could contain eggs by 2 June. Sizes of 25 clutches ranged from 2 to 4 eggs (2E-7, 3E-7, 4E-11), with 72% having 3 or 4 eggs. Incubation period is about 15 to 16 days (Davis 1954; Sakai 1988).

Young: Dates for 20 broods ranged from 17 June to 27 July, with 63% between 28 June and 18 July. Sizes of 13 broods ranged from 1 to 4 young (1Y-1, 3Y-7, 4Y-5), with 7 having 3 young. The nestling period is 16 to 18 days (Davis 1954; Sakai 1988).

Brown-headed Cowbird Parasitism: In British Columbia, 10% of 41 nests found with eggs or young were parasitized by the cowbird. Parasitism was not reported from coastal nests (n = 4); interior nest parasitism was 11% (n = 37). Friedmann et al. (1977) and Friedmann and Kiff (1985) list only a single instance of cowbird parasitism of the Hammond's Flycatcher.

Nest Success: Of 7 nests found with eggs and followed to a known fate, 2 produced at least 1 fledgling.

REMARKS: The Hammond's Flycatcher is easy to confuse with other *Empidonax* flycatchers, particularly the Dusky Flycatcher and, to a lesser degree, the Pacific-slope Flycatcher. During migration and after the breeding season, when males are usually silent, these species are especially difficult to separate. During the breeding season, knowledge of the characteristic songs and calls of each species is necessary for certain identification. Kaufman (1990) is a good identification aid for the *Empidonax* flycatchers.

In British Columbia, the Pacific-slope and Hammond's flycatchers are sympatric in old-growth stands. While this habitat overlap appears to be widespread here, it is infrequent elsewhere in the ranges of the 2 species. In southern Colorado, the 2 species were studied by Beaver and Baldwin (1975) in areas of sympatry. There they had completely overlapping territories, and used much the same species of insects, without aggressive interaction. A partial separation of the foraging heights led to some differences in the prey secured. Under these conditions the Hammond's Flycatcher was about half as abundant as the Pacific-slope Flycatcher, and was less successful in fledging its young.

The Hammond's Flycatcher is more dependent on old-growth coniferous forests than most other *Empidonax* flycatchers (Mannan 1984). Several studies in northern California have shown that it occurs in higher abundance in old-growth forests than in younger-aged stands (Raphael 1984; Sakai 1987). This preference for old-growth conifers renders the species vulnerable to the logging of old-growth forests (Sakai and Noon 1991).

For further information on the systematics of this species, see Johnson (1963) and Johnson and Marten (1991). Additional life-history information can be found in Davis (1954), Sedgwick (1975, 1994), and Sakai (1988).

NOTEWORTHY RECORDS

Spring: Coastal – Thetis Lake 10 Apr 1988-1 (Campbell 1988b); Spectacle Lake (Cowichan Bay) 9 Apr 1987-1; Kennedy River 28 May 1989-1 male; Tofino 1 May 1983-1 collected; Clayoquot valley 12 May 1989-2 males; Vancouver 13 Apr 1990-1, 16 Apr 1931-1 (RBCM 7794); Stanley Park (Vancouver) 13 Apr 1979-1; Comox 13 Apr 1942-1 (RBCM 13183); Courtenay 12 Apr 1990-1; Cortes Island 21 Apr 1977-1; Green Lake (Whistler) 29 Apr 1973-1 singing; Bella Coola 24 Apr 1933-1 (MVZ 282966); Fife Point 1 May 1979-1; Yakan Point 1 May 1979-5 calling; Hazelton 23 May 1912-1 (UBC 151). **Interior** – Okanagan 15 Apr 1912-2 (FMNH 140895 and 140896); White Lake (Okanagan) 20 Apr 1985-1; Nakusp 1 May 1981-1; Lillooet 13 Apr 1968-1; Tranquille 26 Apr 1986-1; Stein Lake 28 May 1986-2; Yoho National Park 10 May 1976-1 (Wade 1977); 100 Mile House 20 Apr 1985-1; Lac la Hache 30 Apr 1943-1 (ROM 83286); Williams Lake 3 May 1958-1 (Erskine and Stein 1964); Clearwater 2 May 1935-2 (MCZ 282980-282981); Quesnel 7 May 1983-1; 25 km e Prince George 29 Apr 1969-1; Tetana Lake 27 Apr 1938-1 (RBCM 8310; Stanwell-Fletcher and Stanwell-Fletcher 1943); Bear Mountain (near Dawson Creek) 12 May 1993-1; 5 km s Hudson's Hope 30 May 1976-2 (ROM 126463-126464); Kenai Creek 23 May 1995-1; Fort Nelson 15 May 1987-1; Liard Hot Springs 17 May 1981-1; Atlin 17 May 1934-1, first spring arrival (Swarth 1936).

Summer: Coastal – Duncan 1 Jul 1970-3 young in nest; Chemainus 25 Jun 1979-9; Coquitlam 11 Jun 1940-3 fresh eggs (UBC 1080); West Vancouver 31 Aug 1986-2; Egmont 15 Jun 1982-9; Little Qualicum Falls Park 10 Jul 1975-3 young being fed by adults; Quadra Island 24 Jun 1975-16; Quatse Lake 9 Jun 1978-6; Stuie 12 Aug 1982-7; Lakelse Lake 23 Aug 1977-4; Terrace 23 Aug 1977-2 or 3 young near nest; New Hazelton 30 Aug 1917-1 (NMC 10978); Cranberry Junction 21 Jun 1975-11; Stikine River 10 Aug 1919-1 (MVZ 39831). **Interior** – Trail 14 Jun 1902-eggs (Macoun 1900 to 1904); Oliver 2 Jun 1976-8; Horn Lake 5 Jun 1915-4 fresh eggs; Naramata 27 July 1971-1 young ready to leave nest; Waldo 9 Jun 1953-6; Edgewood 17 Jun 1926-2 young in nest; Sorrento 14 Jul 1971-3 young in nest; Spillimacheen 26 Jun 1976-10; Revelstoke 3 Jul 1991-8; Glacier National Park 8 Aug 1942-female with brood patch (Munro and Cowan 1947); Murtle Lake 6 Jun 1977-12; Kleena Kleene 24 Jul 1965-4 young left nest; Prince [Fort] George 5 Jun 1889-2 eggs (USNM 238410); Tomslake 4 Aug 1975-3; Fort St. James 4 Jun 1889-1 egg (USNM 238440); Dease Lake 27 Aug 1962-1 (NMC 49853); Fort Nelson 9 Jul 1986-1 young recently out of nest; Kledo Creek 29 Jun 1987-3; Atlin 3 Jun 1924-1 (MVZ 44763), 31 Aug 1924-1 (MCZ 44764).

Breeding Bird Surveys: Coastal – Recorded from 17 of 27 routes and on 32% of all surveys. Maxima: Kispiox 20 Jun 1993-25; Pemberton 17 Jun 1984-14; Courtenay 17 Jun 1993-12. **Interior** – Recorded from 56 of 73 routes and on 62% of all surveys. Maxima: Beaverdell 12 Jun 1971-79; Ferndale 7 Jun 1969-45; Wingdam 17 Jun 1969-41; Creighton Valley 14 Jun 1985-41.

Autumn: Interior – Chezacut Lake 1 Sep 1933-1 (MVZ 282972); Quesnel 3 Sep 1900-1 (MVZ 102505); Williams Lake 5 Sep 1973-1; Lac la Hache 14 Sep 1942-1 (ROM 86616); Clearwater 13 Sep 1959-1 (UKMU 38360); Yoho National Park 13 Sep 1967-1 (Wade 1977); Revelstoke 18 Sep 1986-2; Nakusp 20 Sep 1983-1; Spences Bridge 15 Sep 1896-1 (ROM 49463); Okanagan Landing 18 Sep 1926-1 (MVZ 102524), 28 Sep 1920-1; Creston 13 Sep 1947-1 (ROM 86615). **Coastal** – 37 km n Hazelton 15 Sep 1921-1 (MVZ 42200), 21 Sep 1921-1 (Swarth 1924); Strathcona Park 7 Sep 1986-1; Mansons Landing 21 Sep 1975-1; Egmont 5 Sep 1977-2; Alta Lake 16 Sep 1937-2 (UBC 5376-5377); Errington 3 Sep 1910-1 (Swarth 1912); Stanley Park (Vancouver) 1 Sep 1975-3; Vancouver 30 Sep 1986-1; Sea Island 8 Oct 1990-1 (Weber 1991), 29 Sep 1995-1 adult banded (R.J. Cannings pers. comm.); Huntingdon 5 Oct 1901-1 (NMC 2690); Victoria 28 Sep 1981-1.

Winter: No records.

Dusky Flycatcher

DUFL

Empidonax oberholseri Phillips

RANGE: Breeds from southwestern Yukon south through northeastern and central British Columbia, southwestern Alberta, southwestern Saskatchewan, and western South Dakota, south to southern California and northern New Mexico. Winters in southern California (casually), southern Arizona, and Mexico south to Guatemala.

STATUS: On the coast, *casual* in the Georgia Depression Ecoprovince, becoming *very rare* further north along the coast; absent from the Queen Charlotte Islands.

In the interior, a *fairly common* to *common* migrant and summer visitant to the southern and central portions of the province north to about Williams Lake in the Central Interior Ecoprovince; *uncommon* to *fairly common* in the upper Skeena and Nechako valleys between Smithers and Prince George. *Very rare* in the Northern Boreal Mountains and Taiga Plains ecoprovinces; *casual* in the Boreal Plains Ecoprovince.

Breeds.

CHANGE IN STATUS: The Dusky Flycatcher (Fig. 55) was unknown in the northeastern corner of the province at the time Munro and Cowan (1947) completed their review. Since 1985, the species has been recorded from 6 localities in the Boreal Plains and Taiga Plains ecoprovinces, north at least to Fort Nelson. Fledged young have been recorded near Fort Nelson.

Figure 55. Adult Dusky Flycatcher at nest with single young (near Chilcotin River, 45 km southwest of Williams Lake, 26 July 1993; R. Wayne Campbell).

NONBREEDING: The Dusky Flycatcher (Fig. 55) is widely distributed throughout the central and southern interior of the province. It occurs less commonly in northern areas, with few reports from north of latitude 56°N. It has a very sparse distribution east of the Rocky Mountains in the Boreal Plains and Taiga Plains. On the coast, the Dusky Flycatcher occurs infrequently at the heads of inlets; it has been reported only rarely from the extreme southwestern mainland coast, and only once on Vancouver Island.

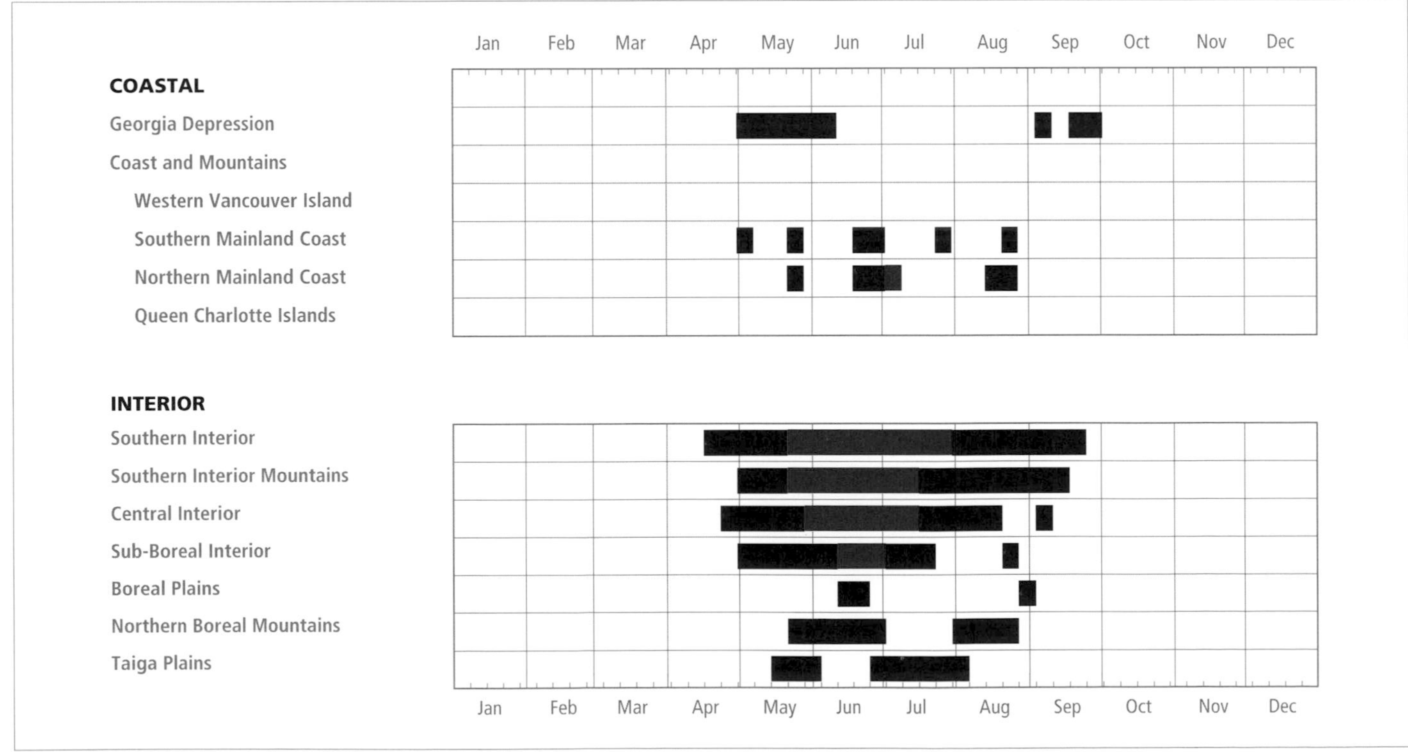

Figure 56. Annual occurrence (black) and breeding chronology (red) for the Dusky Flycatcher in ecoprovinces of British Columbia.

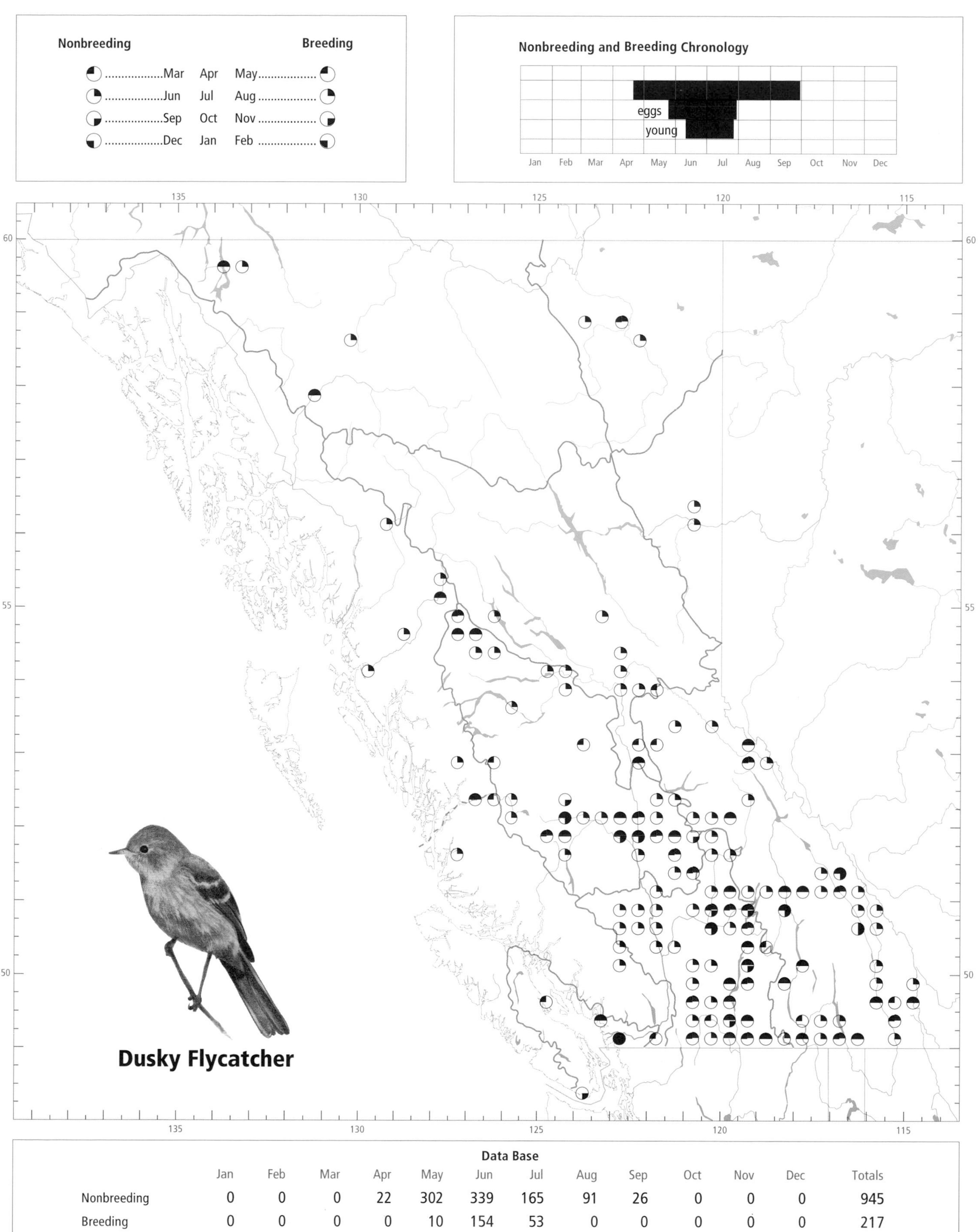

Data Base

	Jan	Feb	Mar	Apr	May	Jun	Jul	Aug	Sep	Oct	Nov	Dec	Totals
Nonbreeding	0	0	0	22	302	339	165	91	26	0	0	0	945
Breeding	0	0	0	0	10	154	53	0	0	0	0	0	217

The Dusky Flycatcher occurs from near sea level to 1,500 m elevation. In the Okanagan valley, it was usually found at the lower to middle elevations compared with the similar Hammond's Flycatcher (Cannings et al. 1987). Recently, however, clearcut logging has opened the high elevational forests and totally changed the elevational separation of the 2 species in the Southern Interior Ecoprovince. There the Dusky Flycatcher now occurs at least to 1,500 m elevation quite commonly (R.J. Cannings pers. comm.). Wade (1977) mentions that in Yoho National Park the Dusky Flycatcher occurs at lower elevations and is rarely found above 1,400 m. Conversely, near Atlin this species was found at higher elevations than the Hammond's Flycatcher (Swarth 1926).

The Dusky Flycatcher frequents a wide range of open woodland and shrub habitats. In the interior, it is typically found in the dry, open coniferous forests on benchlands and in valley bottoms, where it occupies riparian growth along lakes, ponds, streams, brush-filled draws and meadows, or mixed open coniferous-deciduous forests. Willow thickets at the edge of water or forest openings are particularly favoured habitats. It also frequents burns, logged areas (Fig. 57), and avalanche slopes where brushy vegetation occurs. The morphologically similar Dusky and Hammond's flycatchers can usually be distinguished by differences in habitat preferences (Johnson 1963); the former occurs in drier and more open forest habitats, whereas the latter occurs in moister and denser forest types.

In the southern portions of the interior, spring migration may start during the third week of April (Figs. 56 and 58), but usually begins later in the month and reaches its peak in mid-May. In the Prince George area of the southern Sub-Boreal Interior, spring migrants appear during the second week of May, with the peak movement occurring a week later.

In the autumn, the southward movement is less noticeable but begins in August, with most birds gone from southern areas by early September (Figs. 56 and 58).

The Dusky Flycatcher has been recorded in the province from 20 April to 29 September (Fig. 56).

BREEDING: The Dusky Flycatcher probably nests throughout its summer range in British Columbia. Known nesting localities are almost entirely confined to the southern half of the province, north to about Prince George. Although the American Ornithologists' Union (1983) and Godfrey (1986) state that breeding occurs throughout northwestern British Columbia, nesting has not been documented north of the Kispiox valley. There is 1 record from northeastern British Columbia, in the vicinity of Fort Nelson, of adults with fledged young. In addition, Swarth (1926) collected a male in juvenile plumage near Atlin, which suggests that breeding had occurred there. The westernmost breeding record is from the Kispiox valley (Swarth 1924).

The Dusky Flycatcher reaches its highest numbers in summer in the Southern Interior Ecoprovince (Fig. 59). An analysis of Breeding Bird Surveys for the period 1968 through 1993 could not detect a net change in numbers on interior routes; surveys of coastal routes for the same period contained insufficient data for analysis.

Figure 57. In south-central British Columbia, the Dusky Flycatcher occurs in selectively logged areas of mixed Douglas-fir and ponderosa pine, with saskatoon and wild rose understorey (Pruden Pass, near Kamloops, 30 June 1991; R. Wayne Campbell).

The Dusky Flycatcher breeds at low to moderate elevations, with nesting records between 650 and 2,300 m. It selects its nesting territories in relatively open forested environments (see Fig. 57), where it frequently nests in shrub habitats (Fig. 60); less often, it can be found in human-influenced habitats such as gardens, orchards, and farmsteads. In the Okanagan, it is the common breeding *Empidonax* species in the ponderosa pine forests of the valley. It used to have a sparser distribution in the Douglas-fir/trembling aspen woodlands up to about 1,200 m (Cannings et al. 1987); now it is a common breeder in clearcuts with aspen and willow as high as 1,500 m (R.J. Cannings pers. comm.). Near Smithers, Pojar (1993) found the Dusky Flycatcher to be one of the most widespread bird species in all seral stages of trembling aspen forests. In Mount Revelstoke and Glacier national parks, this species occurs in open deciduous shrubbery and thickets, and generally occupies drier habitats than the similar Hammond's Flycatcher. South-facing slopes with aspen thickets and rocky avalanche slopes at low elevations are favoured (Van Tighem and Gyug 1983).

The Dusky Flycatcher has been recorded breeding in British Columbia from 24 May to 29 July (Fig. 56).

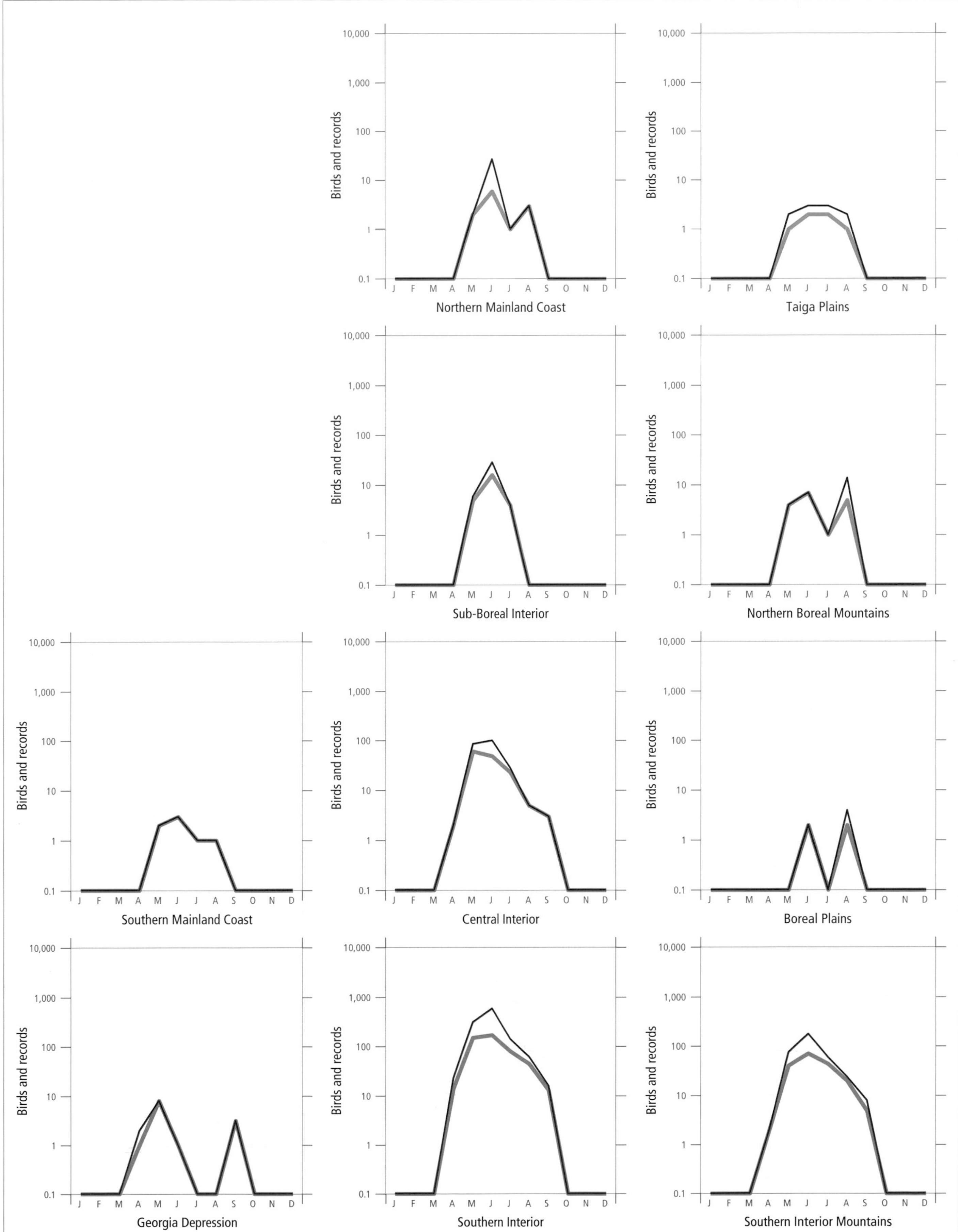

Figure 58. Fluctuations in total number of birds (purple line) and total number of records (green line) for the Dusky Flycatcher in ecoprovinces of British Columbia. Nest record data have been excluded.

Nests: Most nests (95%; n = 73; Fig. 55) were found in living deciduous vegetation. Even in coniferous forests, the actual nest site was usually in deciduous trees and shrubs. Of 90 nests where the nest tree or shrub was noted, most were found in willow (20%), followed by wild rose (16%), trembling aspen (11%), saskatoon (8%), alder (6%), and Douglas maple (3%). At least 15 additional shrub or small-tree species were used by this flycatcher. In the Okanagan valley, wild rose bushes and trembling aspen were used most frequently (Cannings et al. 1987).

Most nests (80%; n = 60) were situated in forks or crotches of branches, more often on vertical than horizontal branches, at distances up to 0.3 m from the centre of the nest tree. The heights of 90 nests ranged from 0.5 to 9.0 m from the ground, with 62% between 1.2 and 3.0 m.

Nests were usually compact cups woven of grasses (68%), plant fibres, plant down, bark strips, spider webs, rootlets, mosses, leaves, lichens, twigs, and needles. Nests were lined with fine grasses, hair, plant down, mosses, feathers, needles, and spider webs.

Eggs: Dates for 78 clutches ranged from 24 May to 29 July, with 53% recorded between 9 June and 25 June. Sizes of 70 clutches ranged from 1 to 5 eggs (1E-4, 2E-10, 3E-17, 4E-38, 5E-1), with 54% having 4 eggs (Fig. 61). The incubation period is 15 or 16 days (Morton and Pereyra 1985; Sedgwick 1993a).

Young: Dates for 42 broods ranged from 10 June to 26 July, with 55% recorded between 21 June and 6 July. Sizes of 37 broods ranged from 1 to 4 young (1Y-4, 2Y-2, 3Y-14, 4Y-17), with 82% having 3 or 4 young. The nestling period is 15 to 20 days (Grinnell et al. 1930; Sedgwick 1993a).

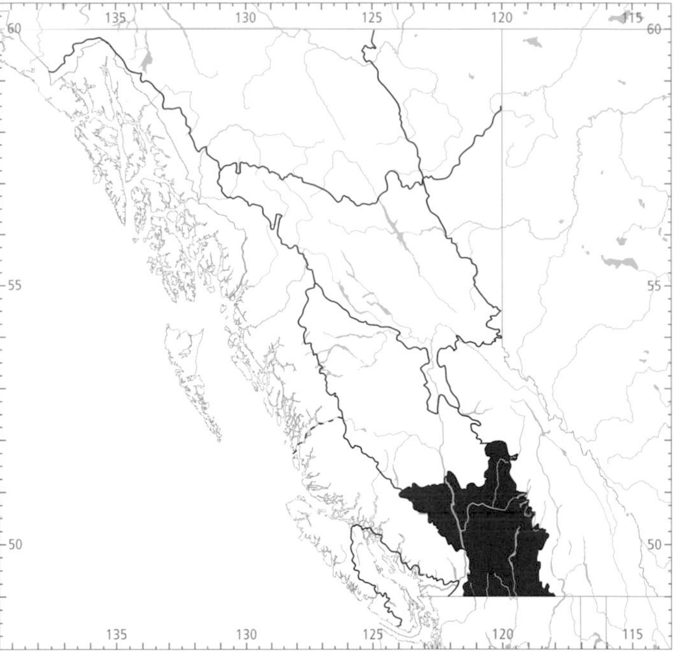

Figure 59. In British Columbia, the highest numbers for the Dusky Flycatcher in summer occur in the Southern Interior Ecoprovince.

Figure 60. Tall willows with black twinberry and fireweed, and an occasional black cottonwood, spruce, and lodgepole pine, are nesting habitat for the Dusky Flycatcher in central British Columbia ("Forests for the World," near Prince George, 14 June 1996; Neil K. Dawe).

Brown-headed Cowbird Parasitism: In British Columbia, 20% of 93 nests found with eggs or young were parasitized by the cowbird. All were reported from the interior of the province. Friedmann and Kiff (1985) consider the Dusky Flycatcher a rare cowbird host.

Nest Success: Of 27 nests found with eggs and followed to a known fate, 10 produced at least 1 fledgling, for a nest success rate of 37%.

REMARKS: The Dusky Flycatcher was formerly known as Wright's Flycatcher (*Empidonax wrightii*). The latter scientific name is now used for the Gray Flycatcher.

Although differences in the habitat occupied may aid in separating the Dusky Flycatcher from other *Empidonax* species, particularly Hammond's and Gray flycatchers, positive identification remains inconclusive by habitat criteria alone. In addition, the songs of the Dusky and Hammond's flycatchers are similar and require "careful concentration to hear and learn the differences" (Kaufman 1990). The Dusky Flycatcher can be positively identified by one of its calls, *ter ... terheet*, which is never given by the Hammond's Flycatcher (Johnson 1963). Kaufman (1990) also notes that the call note is diagnostic, "as the sharp *peep* of the Hammond's is unlike the *whit* of the Dusky."

The Gray Flycatcher has recently invaded Dusky Flycatcher habitat in the southern Okanagan valley, which further complicates the identification problem there (Cannings et al. 1987). Fortunately the tail-dipping behaviour of the Gray Flycatcher is perhaps "the most distinctive behavioural trait of any Empid" (Kaufman 1990), and the songs of the 2 species are readily distinguishable. This author discusses identifying characteristics of the Dusky Flycatcher in comparison with other species of *Empidonax* flycatchers.

Figure 61. Nest with typical clutch of 4 all-white eggs of the Dusky Flycatcher (Sorrento, 24 June 1993; R. Wayne Campbell).

There are a number of reports of this flycatcher from the Vancouver area, some of them published (e.g., Shepard 1976a); only 1 was adequately documented. Others have been excluded from the account.

Johnson (1966b) reviews the impact of competition for resources between these 2 species. See Sedgwick (1993b) for a comprehensive summary of life-history information.

NOTEWORTHY RECORDS

Spring: Coastal – Agassiz 30 Apr 1890-2 (NMC 864-865); Vancouver 18 May 1986-1; Comox 12 May 1934-1 (CMNH 115913); Bella Coola 6 May 1933-1 (NMC 283009); Kitimat Mission 11 May 1975-1 (Hay 1976). **Interior** – Kilpoola Lake 14 May 1972-10; Grand Forks 28 Apr 1984-1; Kearns Creek (Okanagan) 20 Apr 1985-1; Jaffray 25 May 1975-11; Summerland 26 May 1929-4 eggs collected; Edgewood 28 May 1931-3 eggs; Okanagan Landing 24 May 1926-eggs; Tranquille 1 May 1988-2; Shuswap Lake 1 May 1977-2 singing; Leanchoil 17 May 1975-5 (Wade 1977); 100 Mile House 28 Apr 1934-1 (MCZ 283011); Soda Lake 9 May 1983-1, first of spring (Graham 1983); Williams Lake 3 May 1968-1 specimen, 3 May 1976-1 calling; Wells Gray Park 31 May 1962-1 (NMC 50618); Fort Nelson 15 May 1974-1, 23 May 1987-2; Atlin 22 May 1934-1 (CAS 42006).

Summer: Coastal – West Vancouver 4 Jun 1986-1; Indian River 27 Aug 1937-3 (RBCM 4782); Alta Lake 26 Aug 1944-1; Hagensborg 28 Jun 1938-1 (NMC 28733); Lucy Island 15 Jul 1930-1 (RBCM 7818); Hazelton 25 Aug 1921-1 (MVZ 42205); New Hazel-ton 25 Aug 1917-1 (NMC 10942; Taverner 1919); Kispiox valley 4 Jul 1921-2 eggs hatching (Swarth 1924). **Interior** – Manning Park 4 Aug 1962-2; Creston 11 Jul 1984-4 young about 10 days old in nest; Jaynes Lake (Richter Pass) 2 Jul 1960-3 eggs (Campbell and Meugens 1971); near Vaseux Lake 10 Jun 1987-3 eggs and 1 young in nest, 2 Aug 1986-3 young being fed near nest; Brouse 28 Aug 1991-1; Eagle Bay (Shuswap Lake) 23 Jun 1991-3 young near fledging, 3 Jul 1991-nest with 3 eggs, male singing; 100 Mile House 18 Aug 1985-1 young being fed by adult; Prince George 14 Jun 1974-4 eggs, 28 Aug 1981-1; Vanderhoof 14 Jun 1946-4 newly hatched young (Munro 1949); n Ootsa Lake 6 Jun 1986-1 egg; Mile 335 Alaska Highway 2 Aug 1985-1; Peace and Pine rivers 12 Jun 1988-1 at junction; Taylor 28 Aug 1985-3; Atlin 17 Aug 1934-1 (CAS 42009); 24 Aug 1924-1 (MVZ 44765).

Breeding Bird Surveys: Coastal – Recorded from 5 of 27 routes and on 12% of all surveys. Maxima: Pemberton 2 Jul 1977-14; Kwinitsa 29 Jun 1973-13; Kispiox 19 Jun 1994-6. **Interior** – Recorded from 58 of 73 routes and on 69% of all surveys. Maxima: Telkwa High Road 29 Jun 1991-54; Adams Lake 24 Jun 1990-47; Chu Chua 10 Jun 1994-46.

Autumn: Interior – Chezacut 5 Sep 1933-1 (MCZ 283010); Riske Creek 5 and 8 Sep 1978-1; Invermere 3 Sep 1977-2; Tranquille 18 Sep 1983-1; Coldstream 21 Sep 1921-1 (RBCM 9841); Okanagan Landing 7 Sep 1931-2 (UMMZ 68174-68175), 18 Sep 1935-1 (MVZ 102525); Needles 19 Sep 1986-1. **Coastal** – Vancouver 4 Sep 1932-1 (RBCM 7800), 17 Sep 1985-1; Rocky Point (Victoria) 29 Sep 1994-1 banded (Shepard 1995a).

Winter: No records.

Gray Flycatcher

GRFL

Empidonax wrightii Baird

RANGE: Breeds from extreme south-central British Columbia and south-central Idaho south through the Great Basin region of the western United States to south-central California, southern Nevada, central Arizona, south-central New Mexico, and locally in western Texas. Winters from central Arizona and northern Mexico south to central Mexico and Baja California.

STATUS: In the interior, an *uncommon* to locally *fairly common* spring migrant and summer visitant to the southern Okanagan valley in the Southern Interior Ecoprovince. Breeds.

CHANGE IN STATUS: During the past 3 decades, the Gray Flycatcher (Fig. 62) has slowly expanded its range northward from central Oregon into central Washington and south-central British Columbia (American Ornithologists' Union 1957, 1983; Larrison 1971; Yaich and Larrison 1973; Lavers 1975; Cannings 1987). As the last author mentions, "expansion seems to have taken place in an almost linear fashion through ponderosa pine forests along the east side of the Cascade Mountains."

In British Columbia, the Gray Flycatcher was first reported in 1984 along the Camp McKinney Road, 10 km east of Oliver. Two years later, 13 males were on territory and a nest was found in the same locality (Cannings 1987). Since 1986, small numbers have been found in pockets of suitable habitat in the southern Okanagan valley north to Summerland. It appears to be steadily expanding its range in the province.

Figure 62. Adult Gray Flycatcher on nest in crotch of ponderosa pine (near Oliver, 28 May 1987; Tim Zurowski).

NONBREEDING: The Gray Flycatcher (Fig. 62) has a local distribution in the southern Okanagan valley, between Oliver and Summerland, at elevations between 300 and 600 m. Its preferred habitat includes areas where ponderosa pine forests meet grasslands in a narrow elevational band along benchlands above the valley (see BREEDING).

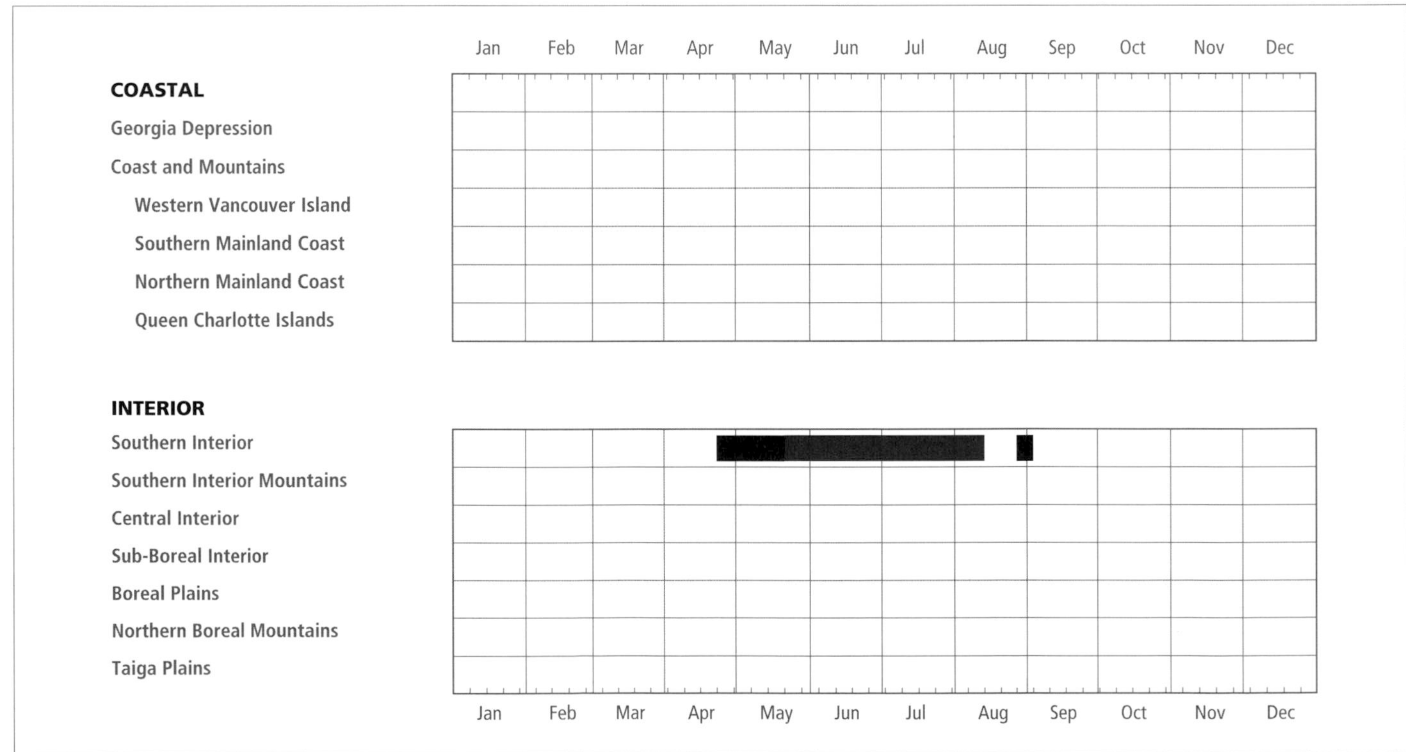

Figure 63. Annual occurrence (black) and breeding chronology (red) for the Gray Flycatcher in ecoprovinces of British Columbia.

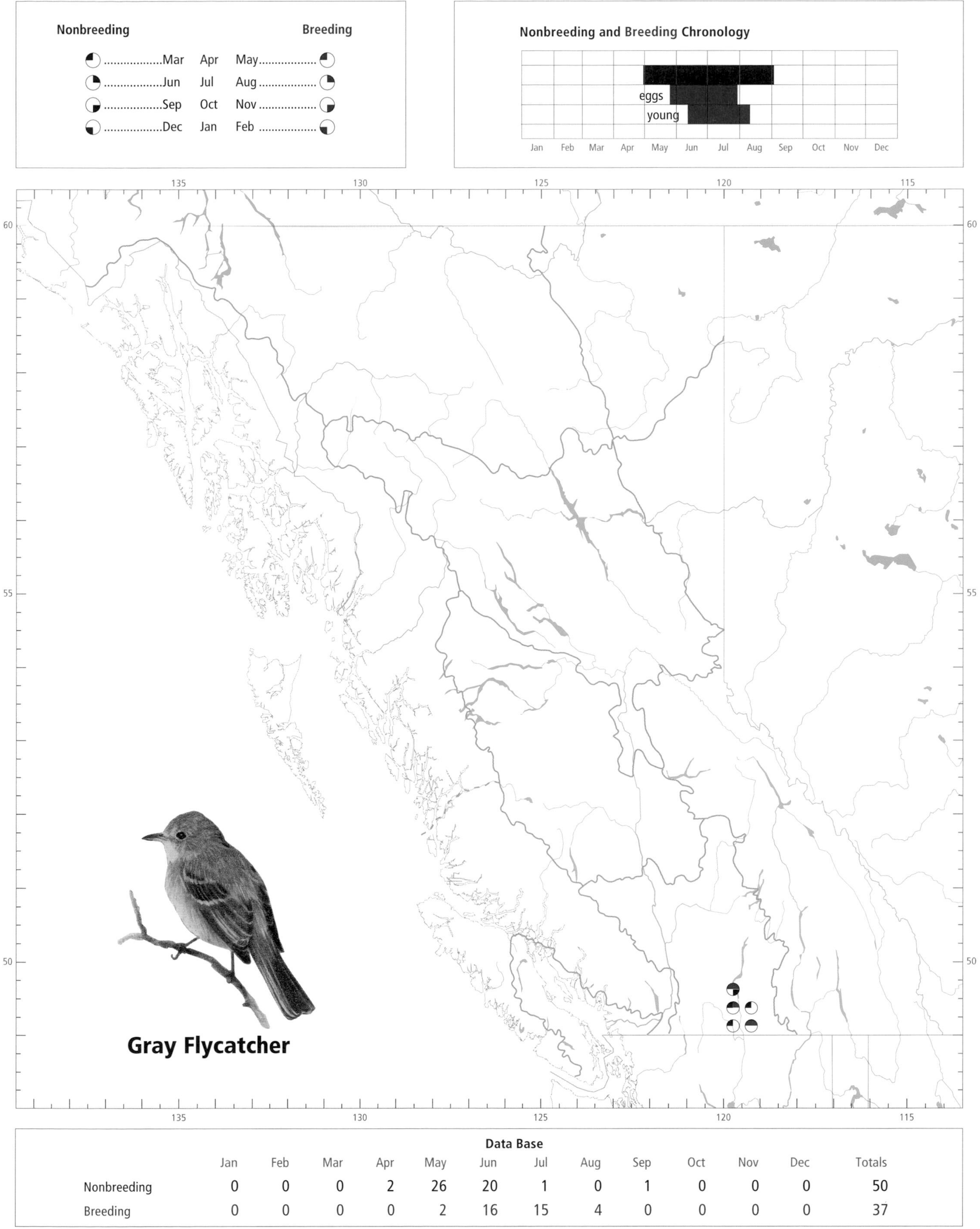

Data Base

	Jan	Feb	Mar	Apr	May	Jun	Jul	Aug	Sep	Oct	Nov	Dec	Totals
Nonbreeding	0	0	0	2	26	20	1	0	1	0	0	0	50
Breeding	0	0	0	0	2	16	15	4	0	0	0	0	37

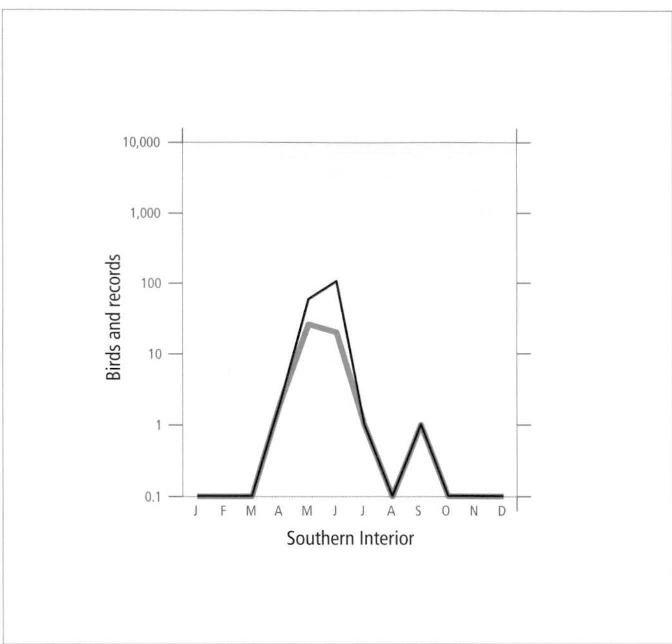

Figure 64. Fluctuation in total number of birds (purple line) and total number of records (green line) for the Gray Flycatcher in the Southern Interior Ecoprovince of British Columbia. Nest record data have been excluded.

Spring arrival can begin as early as the last week of April (Figs. 63 and 64), but most birds return in early May, with the peak movement from mid to late May. The southward movement occurs in August and is complete by early September (Fig. 64).

The Gray Flycatcher has been recorded in the province from 29 April to 2 September (Fig. 63).

BREEDING: The Gray Flycatcher breeds only in the southern Okanagan valley. During a recent survey, Preston (1990) found 51 birds at 8 sites in 3 general localities, as follows: 1, Summerland area (44 birds – Meadow Valley Road, Shingle Creek Ranch Road, sites north of Mt. Nkwala, and Ecological Reserve No. 7 near Upper Trout Creek); 2, east of Vaseux Lake (2 birds); 3, 10 km east of Oliver (5 birds – Camp McKinney Road area). Those birds represented at least 47 territorial males. Singing males were not found in 1990 between Summerland and Peachland, but in 1993 singing males were heard on the west side of Vaseux Lake and at Mahoney Lake. Cannings (1991) suggests that "suitable habitat probably exists west to Princeton and Merritt, north to the Kamloops area, and to the east in the Rocky Mountain Trench near Cranbrook." Also, the Okanagan valley north of Peachland has not been adequately investigated for this species.

In British Columbia, the Gray Flycatcher occurs only in the Southern Interior Ecoprovince (Fig. 65). Breeding Bird Surveys for interior routes for the period 1968 through 1993 contained insufficient data for analysis. Local populations may fluctuate between years, as 13 singing males were found in a 70 ha area at the Oliver site in 1986 (Cannings 1987), where only 5 were found by Preston (1990). The breeding population probably consists of more than the 50 pairs estimated by Preston (1990).

Breeding habitat in British Columbia is restricted to open ponderosa pine forests with an understorey of grasses and scattered shrubs or young pines that the flycatcher uses for perching. The pines are of small or moderate size (10 to 15 m in height; 25 cm diameter at breast height [dbh]), with 10 to 15 m wide clearings scattered throughout the stand (Cannings 1991).

Habitat characteristics are described by the same author at the 3 known breeding localities:

> At Oliver, the understorey is quite shrubby, mainly consisting of antelope-bush (*Purshia tridentata*) and threetip sagebrush (*Artemisia tripartita*). The pine woodland along the Shingle Creek road (Summerland) has an understorey of scattered buckbrush (*Ceanothus velutinus*). Other Summerland sites have little or no shrubs, the ground cover being dominated by bluebunch wheatgrass (*Agropyron spicatum*).

Many of the Summerland breeding territories were in young ponderosa pine stands that had been thinned for silvicultural purposes. This may help create forest openings that are required by this flycatcher for foraging. Also, piles of cut trees are used as preferred low perch sites.

Eleven nests have been found in British Columbia. All but 1 were in second-growth ponderosa pine forests, as described above. Cannings (1987) describes the first nest found:

> A pair was found building a nest on 6 July 1986, against the trunk of a ponderosa pine, on a horizontal branch 1.9 m above the ground. The nest was constructed of sagebrush twigs, grass, feathers, thistledown, and pine needles. The nest contained two eggs

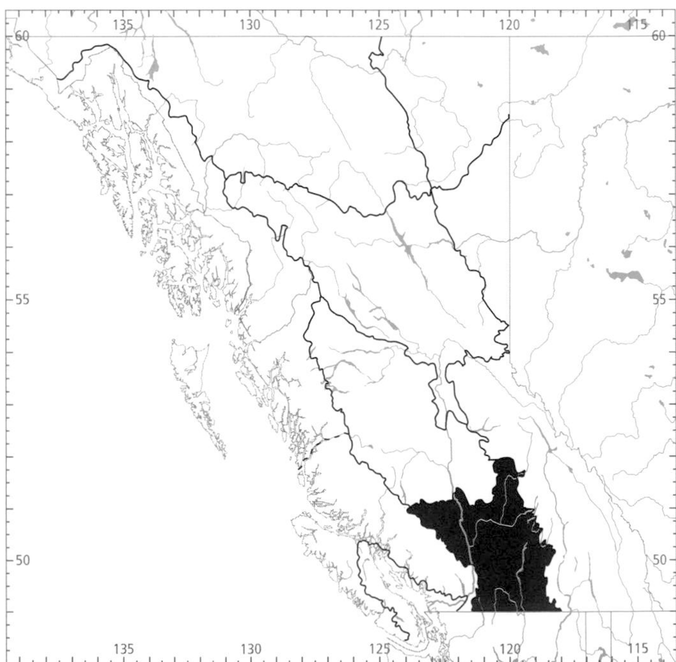

Figure 65. In British Columbia, the highest numbers for the Gray Flycatcher in summer occur in the Southern Interior Ecoprovince.

and a cowbird egg on 10 July and four eggs and a cowbird egg on 18 July. The cowbird egg was destroyed with a needle puncture but was left in the nest. Subsequently, one flycatcher egg was found on the ground and two others failed to hatch. One egg hatched on 27 July and the young fledged on 10 August.

The Gray Flycatcher is known to be double-brooded over much of its range, and this nest, hatching in late July, may have been a successful second nest or a renesting after the failure of an earlier attempt.

In British Columbia, the Gray Flycatcher has been recorded breeding from 28 May (calculated) to 10 August (Fig. 63).

Nests: All nests were in living conifers, either ponderosa pine (10 nests; Fig. 62) or Douglas-fir (1 nest) and were placed in crotches or on horizontal branches next to, or up to 1.5 m from, the trunk. Nest heights ranged from 1.5 to 9.0 m above the ground.

Nests were cups composed mainly of grasses, pine needles, plant stems and fibres, small twigs, and leaves, and lined with plant down and feathers (Fig. 62). Nest building takes 3 days to complete (Russell and Woodbury 1941).

Eggs: Dates for 4 clutches ranged from 5 June to 29 July. Calculated dates indicate that eggs may occur as early as 28 May. Clutch size was 3 or 4 eggs (3E-1, 4E-3). The incubation period from 1 nest in British Columbia was 16 days; however, that nest was parasitized by a cowbird and suffered other disturbance that could have prolonged the incubation period. It is normally reported as 14 days (Russell and Woodbury 1941; Ehrlich et al. 1988).

Young: Dates for 7 broods ranged from 11 June to 10 August. Brood size ranged from 1 to 3 young (1Y-2, 2Y-2, 3Y-3). The nestling period from 1 nest in British Columbia was between 15 and 17 days; Russell and Woodbury (1941) report it as 16 days.

Brown-headed Cowbird Parasitism: In British Columbia, 2 of 9 nests recorded with eggs or young were parasitized by the cowbird. Friedmann et al. (1977) report that the Gray Flycatcher was first found as a host species in 1972, in Washington state, and has become regularly parasitized by the Brown-headed Cowbird. They also note that in 1970 and 1971 the Gray Flycatcher was the most abundant breeding bird in central Oregon, averaging about 25 pairs per 100 ha (cowbirds averaged about 3 "pairs" per 100 ha). Of 28 nests found, 7 were parasitized (25%); there was a 30% parasitism rate in 1970 and 20% in 1971. Of the 7 parasitized nests, only 3 fledged any young, in each case a single cowbird apiece. In none of the 7 did any of the host young survive. For details, see Yaich and Larrison (1973).

It is probable that this flycatcher will continue to be heavily parasitized, especially since it breeds in semi-open and open habitats preferred by cowbirds.

Nest Success: Of 2 nests found with eggs and followed to a known fate, 1 produced at least 1 fledgling.

REMARKS: Because of its small Canadian population, the Gray Flycatcher has been declared a "Vulnerable Species" in Canada by the Committee on the Status of Endangered Wildlife in Canada (Cannings 1991, 1995). However, British Columbia is at the northern edge of a fairly widespread and expanding population in western North America, and there is little cause for concern regarding its overall status.

In British Columbia, the Gray and Dusky flycatchers both breed in dry, open ponderosa pine forests on the benchlands above the valley floor. These similar species prefer different microhabitats and exhibit interspecific territorial behaviour where their preferred habitats join to form vegetational mosaics (Johnson 1966b).

Cannings (1987) notes that the Dusky Flycatcher occurs in moister habitats in small draws in the Okanagan valley. Habitat relationships between the 2 species in British Columbia remain to be investigated.

Cannings (1995) discusses management recommendations and options for this species in British Columbia. They include efforts to determine the distribution of the species, particularly in the Kelowna, Kamloops, and east Kootenay regions, as well as determination of the habitat parameters of the species in the province.

NOTEWORTHY RECORDS

Spring: Coastal – No records. **Interior** – Camp McKinney Road (e Oliver) 11 May 1990-4, 17 May 1987-1 (Campbell 1987c), 18 May 1986-1 singing (Cannings 1987), 28 May 1987-adult on nest (BC Photo 1185; Fig. 62); Inkaneep Indian Reserve 18 May 1986-1; Mahoney Lake 25 May 1993-1 male singing; Irrigation Creek 19 May 1991-1; Trout Creek (Shingle Creek) 20 May 1989-1, 22 May 1988-1; Summerland 29 Apr 1990-1 adult, 30 Apr 1990-1 (Siddle 1990b), 26 May 1991-adult on nest, 3 May 1991-2 (Campbell 1991b); Meadow Valley 19 May 1991-1.

Summer: Coastal – No records. **Interior** – Camp McKinney Road (e Oliver) 1 Jun 1986-13 singing in 70 ha area, 5 Jun 1987-3 eggs; 11 Jun 1989-adult feeding 2 or 3 young in nest, 19 Jun 1984-1 (Cannings 1987), 10 July 1986-2 eggs, 21 July 1993-1 adult on nest, 29 Jul 1986-adult on nest, 10 Aug 1986-1 addled egg and 1 young in nest that fledged this day (Cannings et al. 1987; BC Photo 1103), 6 Aug 1987-3 young, 2 out of nest; Vaseux Lake 13 June 1993-4 eggs; 2.5 km se Trout Creek 24 Jul 1988-1; Summerland 14 Jun 1991-building nest, 22 Jun 1991-nest complete.

Breeding Bird Surveys: Coastal – Not recorded. **Interior** – Recorded from 2 of 73 routes and on 2% of all surveys. Maxima: Summerland 19 Jun 1991-10; Oliver 20 Jun 1986-2.

Autumn: Interior – Summerland 2 Sep 1991-1.

Winter: No records.

"Western Flycatcher" Complex WEFL

Pacific-slope Flycatcher PSFL
Empidonax difficilis Baird

Cordilleran Flycatcher COFL
Empidonax occidentalis Nelson

RANGE: Breeds from southeastern Alaska, coastal and central British Columbia, southwestern Alberta, and South Dakota south to northern Baja California west of the Sierra Nevada, and south to central Mexico through mountainous parts of the western United States. Winters from southern Baja California and northern Mexico south to Yucatan.

STATUS: On the coast, an *uncommon* to locally *common* migrant and summer visitant to the Georgia Depression Ecoprovince, Western Vancouver Island, and the Queen Charlotte Islands; *uncommon* to *fairly common* in the southern and northern Coast and Mountains Ecoprovince. *Casual* on the coast north of Stewart.

In the interior, a locally *uncommon* to *fairly common* migrant and summer visitant to the Southern Interior and Southern Interior Mountains ecoprovinces, becoming *very rare* to *rare* in the Central Interior and Sub-Boreal Interior ecoprovinces. *Very rare* in the Boreal Plains Ecoprovince; *accidental* in the Northern Boreal Mountains Ecoprovince.

Breeds.

CHANGE IN STATUS: The "Western Flycatcher" has expanded its range eastward and northward in the province since the mid-1940s. Munro and Cowan (1947) record only 3 specimens from the interior, all from the Okanagan valley, where the species was listed as occasional. It is now a fairly common summer resident there (Cannings et al. 1987).

Other than an isolated record at Revelstoke in 1890, the first record for the "Western Flycatcher" in the Kootenays was from 1960, and its occurrence was sporadic there until the 1970s. Johnstone (1949) in the east Kootenay, and Munro (1950, 1958) in the Creston region, did not mention this flycatcher. It is now a regular breeder in both areas.

In the Central Interior, other than an isolated record from 1931, reports of the "Western Flycatcher" do not appear until the 1950s. Munro (1947a) does not mention it; however, it is now considered a rare breeding bird in the Cariboo (Roberts and Gebauer 1992).

The first record for the Sub-Boreal Interior was from Prince George in 1969. Not until 1974 was it reported from the Peace River region of the Boreal Plains, where there are

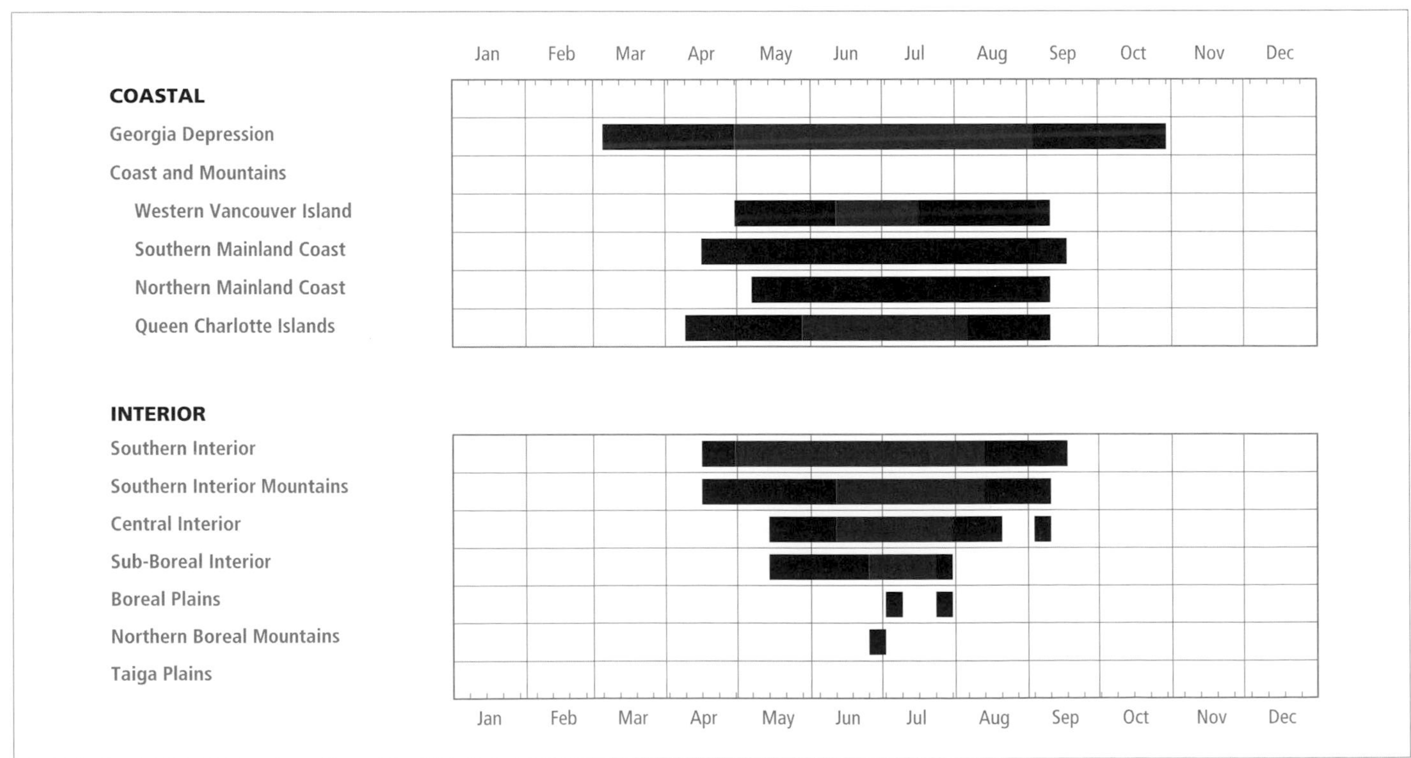

Figure 66. Annual occurrence (black) and breeding chronology (red) for the "Western Flycatcher" in ecoprovinces of British Columbia.

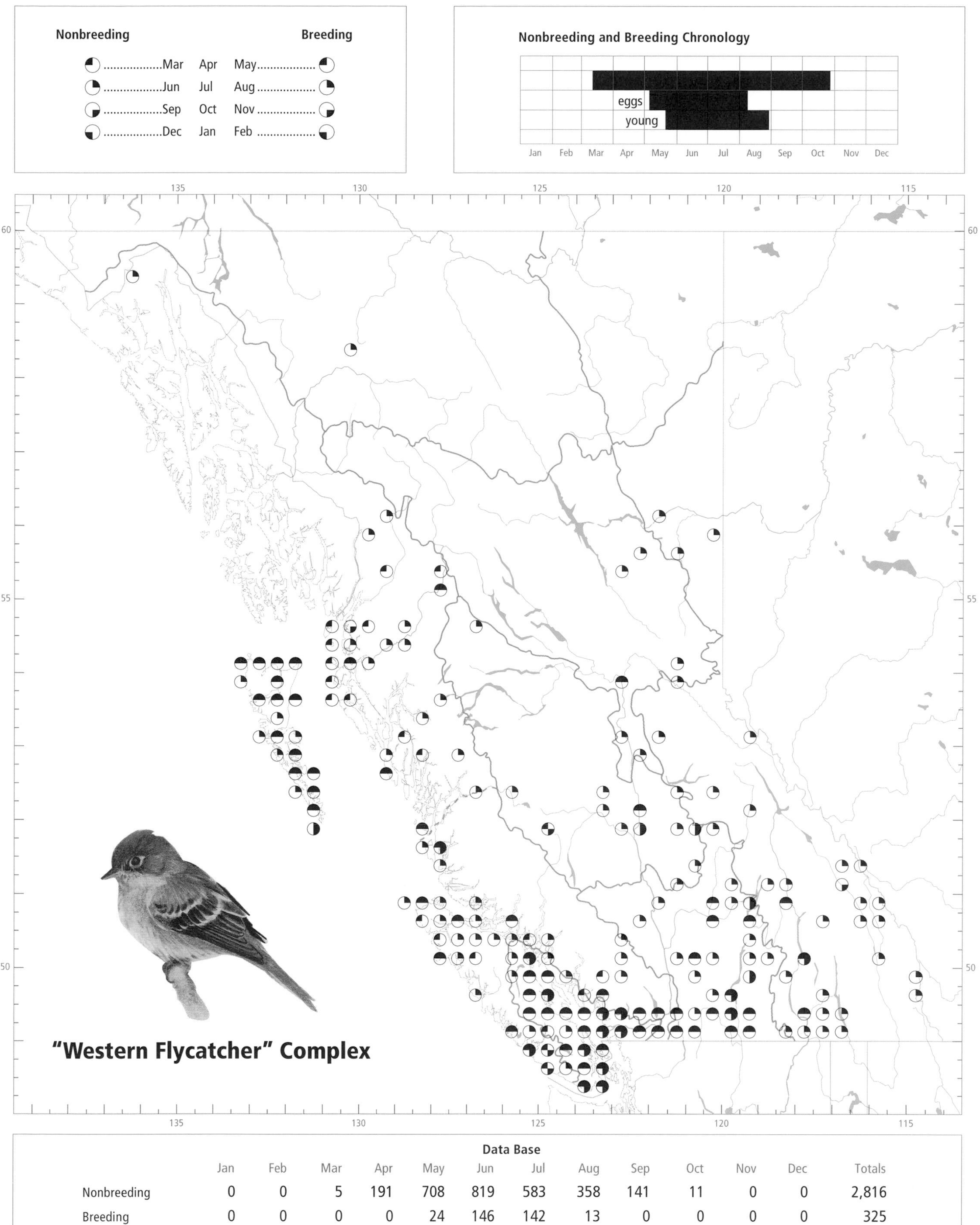

Data Base

	Jan	Feb	Mar	Apr	May	Jun	Jul	Aug	Sep	Oct	Nov	Dec	Totals
Nonbreeding	0	0	5	191	708	819	583	358	141	11	0	0	2,816
Breeding	0	0	0	0	24	146	142	13	0	0	0	0	325

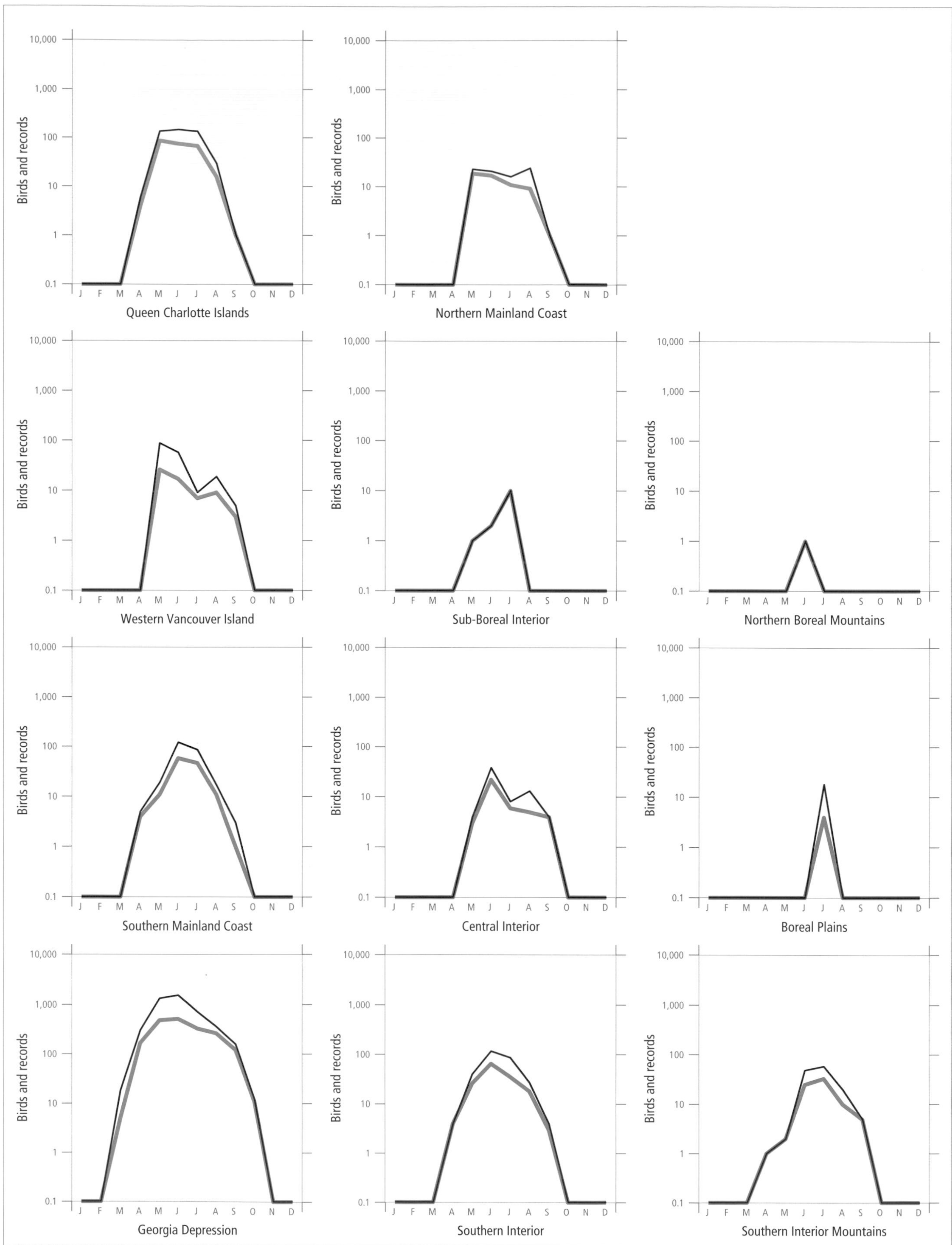

Figure 67. Fluctuations in total number of birds (purple line) and total number of records (green line) for the "Western Flycatcher" in ecoprovinces of British Columbia. Nest record data have been excluded.

now several records, including 9 in July 1992 (C. Siddle pers. comm.).

This range extension is further supported by evidence from Alberta, where the "Western Flycatcher" was not found until 1954. It now occurs there regularly (Salt and Salt 1976; Semenchuk 1992).

NONBREEDING: The "Western Flycatcher" is widely distributed from Vancouver Island and the lower Fraser River valley north along the coast to the Queen Charlotte Islands and the northern Coast and Mountains. In the interior, it has a widespread but localized distribution through the Southern Interior and Southern Interior Mountains, becoming more sparsely distributed through the Central Interior, Sub-Boreal Interior, and Boreal Plains ecoprovinces. Further north this flycatcher is an unusual find.

On the coast, the "Western Flycatcher" occurs from near sea level to about 1,250 m elevation. It is generally restricted to the lower coastal forest zones, giving way to the Hammond's Flycatcher at higher elevations. On Mount Seymour, near Vancouver, it is most abundant at low elevations (100 to 400 m), less abundant at middle elevations (400 to 800 m), and absent above 800 m elevation (Weber 1975). On the Sunshine Coast, however, it is fairly common from sea level through the middle elevations to old-growth mountain hemlock–yellow cedar–fir forests at 1,050 m elevation in the Caren Range and 1,100 m in the Tetrahedron Plateau (T. Greenfield pers. comm.).

The "Western Flycatcher" frequents old-growth and mature second-growth Douglas-fir–western hemlock and mixed coniferous-deciduous forests, where it forages in the lower and middle canopy, often in riparian growth along the edge of openings near creeks and ponds. It is one of the few birds in British Columbia that inhabits extremely dense, shady forests (see Fig. 70). On Western Vancouver Island it occurs in all forest age classes, but is consistently found in old-growth and 50- to 60-year-old forests (Bryant et al. 1993). Peterson and Peterson (1983) note an absence of this species in the years immediately following logging, and its abundance in stands over 20 years of age. It occasionally occurs in suburban gardens with ornamental trees, usually near forest edges.

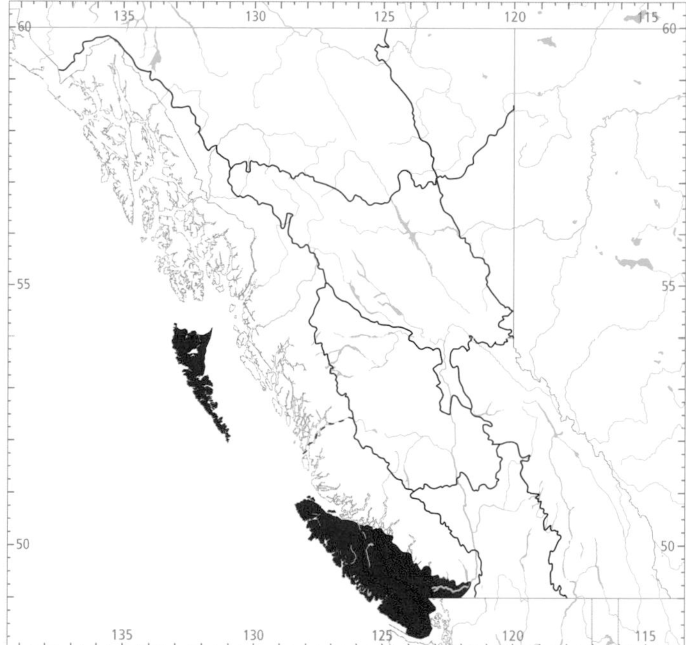

Figure 68. In British Columbia, the highest numbers for the "Western Flycatcher" in summer occur in the Georgia Depression Ecoprovince and on Western Vancouver Island and the Queen Charlotte Islands in the Coast and Mountains Ecoprovince.

In the interior, the "Western Flycatcher" is locally distributed at elevations between 280 and 1,310 m. It is usually restricted to moister, shadier habitats where western redcedar is found. This habitat occurs in heavily wooded creek valleys and ravines, and around lakes and ponds. At middle elevations, where it is sympatric with the Hammond's Flycatcher, interior "Western Flycatchers" are usually found in wooded canyon bottoms.

On the south coast, spring migrants begin arriving in the second week of March (Figs. 66 and 67), but the main movement begins about mid-April and is greatest about the second week of May. On the Queen Charlotte Islands, the first

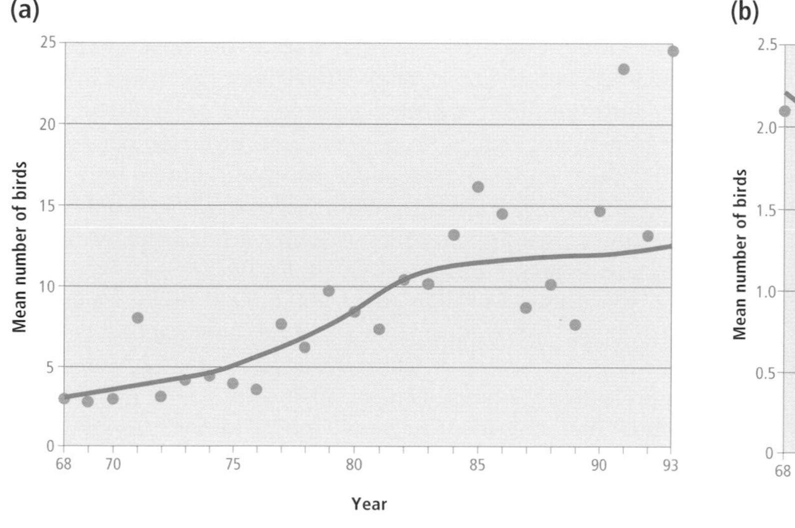

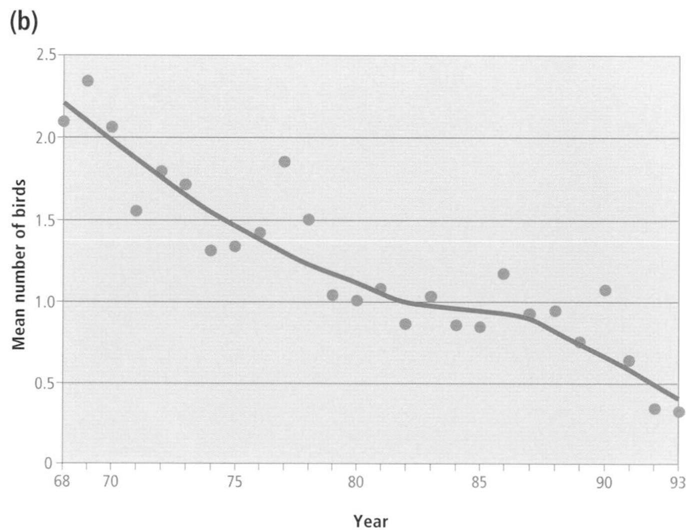

Figure 69. An analysis of Breeding Bird Surveys for the "Western Flycatcher" in British Columbia shows that the mean number of birds on coastal routes (a) increased at an average annual rate of 8% over the period 1968 through 1993 ($P < 0.01$); the mean number of birds on interior routes (b) decreased at an average annual rate of 7% over the same period ($P < 0.10$).

Figure 70. On the Queen Charlotte Islands, the "Western Flycatcher" (= Pacific-slope Flycatcher) breeds in dense Sitka spruce–western hemlock forests (Bag Harbour, 3 June 1990; R. Wayne Campbell).

migrants arrive in the second week of April (Fig. 67), and numbers build in early to mid-May. Ainsley (1991) found that on the Queen Charlotte Islands, males arrive on territory first, with females appearing 1 to 2 weeks later.

In the interior, birds arrive much later than those at the coast (Figs. 66 and 67). The spring movement in the Southern Interior and Southern Interior Mountains begins in the third week of April. In the Okanagan valley, most birds appear in mid to late May (Cannings et al. 1987), while in the Southern Interior Mountains, most arrive in early June, nearly a full month later than the main movement in coastal areas.

Most "Western Flycatchers" have left the province by the end of August, although a few birds may remain until mid-September (Figs. 64 and 65). An exception occurs in the Georgia Depression, where the peak autumn movement is in the first 2 weeks of September. Most birds have left that area by the end of September, although stragglers may be found until the end of October in some years.

On the coast, the "Western Flycatcher" has been recorded from 11 March to 27 October; in the interior, it has been recorded from 16 April to 13 September (Fig. 66).

BREEDING: The "Western Flycatcher" probably breeds throughout its range in British Columbia. On the coast, nesting has been confirmed only in the Georgia Depression, on the west coast of Vancouver Island, and on the Queen Charlotte Islands. In the interior, it breeds locally throughout much of its range north to the Prince George area. There are a number of reports of nesting in the southern west Kootenay and the Okanagan valley; elsewhere in the interior, nesting records are scarce. The northernmost breeding record is from McGregor, northeast of Prince George.

The "Western Flycatcher" reaches its highest numbers in the rainforests of Western Vancouver Island and the Queen Charlotte Islands in the Coast and Mountains, and in the Georgia Depression (Fig. 68). Johnson (1980) also notes that this flycatcher reaches its most dense and continuous breeding populations in the lush forests of the Pacific coast and on adjacent islands. He indicates that along the west coast of North America, populations increase in density from south to north, and reach their maximum numbers in coastal British Columbia and southeastern Alaska. An analysis of Breeding Bird Surveys for the period 1968 through 1993 shows that the mean number of birds on coastal routes increased at an average annual rate of 8% (Fig. 69a); the mean number of birds on interior routes decreased at an average annual rate of 7% over the same period (Fig. 69b). This decrease in numbers on interior routes appears inconsistent with the northeasterly expansion noted under CHANGE IN STATUS.

Elevations and breeding habitats used by the "Western Flycatcher" are similar to those of the nonbreeding period, but may be more restricted. Throughout the province this complex is usually associated with forests, although over 50% of our breeding records come from rural and urban areas, pastures, orchards, and other cultivated farmlands.

The forest types occupied in coastal areas vary from the mature western hemlock–western redcedar and Sitka spruce–

western hemlock stands (Fig. 70) through mixed forests of arbutus, Garry oak, and Douglas-fir, to the early successional stages of red alder, vine maple, and birch.

In the interior, breeding habitat is quite variable. It includes farms, pastures, and suburban parks and gardens, as well as trembling aspen, black cottonwood, riparian creek bottoms (Fig. 71), dry ponderosa pine, interior Douglas-fir, Engelmann spruce–subalpine fir, and sub-boreal spruce forests. In the interior forested areas, breeding "Western Flycatchers" are very localized. They are usually associated with water courses, and thus openings, in the forest. The moist ravine bottoms offer airways under the canopy that are necessary for foraging as well as for providing nest sites along the mossy streambanks (Johnson 1980). This microhabitat is usually shadier and moister than the surrounding forest.

There is a distinct difference between habitat types used by the coastal nesting populations of the "Western Flycatcher" and those used by interior populations. On the coast, about 70% of nests were found in forested habitats while 30% of the nests were associated with human habitation; virtually the reverse was true in the interior.

On the coast, the "Western Flycatcher" has been recorded breeding from 6 May (calculated) to 29 August; in the interior, it has been recorded breeding from 7 May to 8 August (Fig. 66).

Nests: Over half the nests (56%; n = 154) were found in either deciduous or coniferous trees and shrubs, including red alder, wild rose, Douglas-fir, western redcedar, willow, vine maple, Garry oak, poplar, and spruce. About 30% of the nests were on buildings, bridges, and other artifacts. Nests were also found on cliff banks; 1 was in a cave. Compared with those in the interior, nest sites selected at the coast reflect the difference in habitat types noted above. In coastal areas, although many nests were built on ledges of occupied or unoccupied buildings, most were placed in vegetation (in clumps of sucker growth, attached to a fork or crotch of a branch or saddled on the branch, in bark crevices and roots of overturned trees, in hollows or cavities in trees), whereas in the interior most nests were placed on ledges in abandoned buildings and on bridge timbers.

The heights of 129 nests ranged from 0.3 to 13.0 m, with 52% between 1.3 and 2.8 m.

Nests were large and often bulky, composed of moss, lichens, conifer needles, grass, plant fibres, and rootlets within which the nest cup was constructed. They were usually lined with moss, plant down, hair, fine grass, and feathers. Nest construction is by the female alone and takes 5 to 6 days (Ainsley 1991).

The "Western Flycatcher" in coastal British Columbia (= Pacific-slope Flycatcher) appears to be a persistent renester after destruction of an early nest. It also frequently produces 2 broods a year. Pairs under observation have been seen to fledge young from a first nest in May and a second nest in July. These observations have been made in the Douglas-fir–arbutus–Garry oak arid forests of southeastern Vancouver Island and the Gulf Islands; Ainsley (1991) noted similar findings on the Queen Charlotte Islands.

Figure 71. In the south-central interior of British Columbia, the "Western Flycatcher" has been found breeding in a shrubby 10 m gully with choke cherry, saskatoon, wild rose, and big sagebrush interspersed with Douglas-fir and ponderosa pine (Kamloops, 15 June 1996; Neil K. Dawe).

Eggs: Dates for 146 clutches ranged from 14 May to 8 August, with 53% recorded between 10 June and 3 July. Calculated dates indicate that eggs may be present as early as 6 May. Sizes of 123 clutches ranged from 1 to 5 eggs (1E-2, 2E-8, 3E-33, 4E-79, 5E-1), with 64% having 4 eggs. The incubation period in British Columbia is 13 to 16 days (Ainsley 1991; Davis et al. 1963). This period may be prolonged by 2 to 4 days when the nest is parasitized by the Brown-headed Cowbird.

A male in 1 of 7 territories studied on the Queen Charlotte Islands courted a second female while his first mate was incubating. This resulted in the male being associated with 2 nesting females simultaneously (Ainsley 1991).

Young: Dates for 64 broods ranged from 20 May to 29 August, with 54% recorded between 1 and 23 July. Sizes of 48 broods ranged from 1 to 4 young (1Y-3, 2Y-6, 3Y-18, 4Y-21), with 81% having 3 or 4 young. The nestling period in British Columbia is 15 to 16 days (Ainsley 1991).

Brown-headed Cowbird Parasitism: In British Columbia, 7% of 167 nests found with eggs or young were parasitized by the cowbird. Coastal nest parasitism was 3% (n = 99); interior nest parasitism was 12% (n = 68). For North America, Friedmann et al. (1977) reported that 3.9% of 157 nests contained cowbird eggs.

Nest Success: Of 41 nests found with eggs and followed to a known fate, 26 fledged at least 1 young, for a nest success rate of 63%. Coastal success was 50% (n = 14); interior success was 70% (n = 27).

REMARKS: The former Western Flycatcher (*E. difficilis*; American Ornithologists' Union 1983) was recently recognized as including 2 species (American Ornithologists' Union 1989), the coastal Pacific-slope Flycatcher (*E. difficilis*) and the interior Cordilleran Flycatcher (*E. occidentalis*). The decision was based largely on the studies of the birds in an area in northern California in which the 2 forms appeared to be living sympatrically with little evidence of interbreeding (Johnson and Marten 1988; Johnson 1994).

There is no doubt that the Pacific-slope Flycatcher occurs on the coastal mainland and islands. However, the status of the Pacific-slope and Cordilleran flycatchers east of the Coast Mountains needs to be clarified.

According to Johnson (1980), the southeastern interior of British Columbia is occupied by the interior form (*E. occidentalis*) of the former Western Flycatcher, while the coastal form (*E. difficilis*) breeds along the coast and in the Okanagan valley. He suggests a distributional gap between the 2 species in southeastern and central British Columbia. There are, however, good populations of the "Western Flycatcher" in these alleged gaps, particularly in the west Kootenay along the east side of Kootenay Lake, from the south end of the lake north to Crawford Bay.

The "Western Flycatcher" in some regions occurs sympatrically with other, similar-looking *Empidonax* flycatchers. In southeastern Washington, "Western" and Willow flycatchers occur together during the breeding season. There they occupy quite different habitat niches within the same general locality: the "Western" favours shady, forested habitat whereas the Willow inhabits brushy, open habitat. In floodplain forest habitat, where tall trees and open brush occur in patches, both species are present and occupy the same foraging niche (Frakes and Johnson 1982).

"Western" and Hammond's flycatchers also occur sympatrically in western North America (Beaver and Baldwin 1975), although when 1 species is abundant the other is usually scarce. The 2 are thought to be ecological equivalents, with the "Western Flycatcher" occurring at lower elevations and the Hammond's Flycatcher at higher elevations, presumably because of interspecific competition for food between the species. These authors found interior "Western Flycatchers" and Hammond's Flycatchers breeding sympatrically in Colorado.

Coexistence between the "Western" and Hammond's flycatcher populations in Colorado was accompanied by differences in foraging niche, with the Hammond's Flycatcher feeding higher than the "Western Flycatcher" in the aspen-conifer habitat. Sympatry seemed to occur where the habitat was intermediate in composition. In preferred habitat, one species presumably is able to exclude the other. This may explain the low breeding success of the Hammond's Flycatcher in the presence of a high "Western" population, even though nest site preference is so different (Beaver and Baldwin 1975). However, Johnson (*in* Beaver and Baldwin 1975) suggests that behavioural relationships between other *Empidonax* flycatchers and the "Western Flycatcher" may be different on the coast.

Ainsley (1991) detected differences between the calls and songs of the "Western Flycatcher" on the Queen Charlotte Islands and other coastal populations. This is an area in need of further study in the province.

Identification of Pacific-slope or Cordilleran flycatchers in the field is extremely difficult; it should not be based on nonvocal characteristics such as colouring of throat and breast plumage. According to Kaufman (1990), the only way to separate the Pacific-slope Flycatcher from the Cordilleran Flycatcher in the field is by the call notes of the males. However, Johnson (1994) points out that the male position note may be used to distinguish the Cordilleran Flycatcher if a 2-part note is heard – *pit-weet!* Unfortunately, both species in the northwest give the "boat-shaped" or slurred sinusoidal note – *peewhitt!* Thus, contrary to Kaufman (1990), the slurred call is not necessarily evidence of a Pacific-slope male. However, Johnson (1994) notes that the advertising song of *E. difficilis* and *E. occidentalis* differ profoundly when relatively remote populations are compared. Even where the nesting distributions of the 2 taxa approach more closely, the songs approach in structure but do not overlap. Johnson (1994) describes the final syllable as "low-high" in *E. difficilis* and "high-low" in *E. occidentalis*. He states that the difference is audible in the field at sympatric localities in California.

For more information on the life history of the "Western Flycatcher," see Davis et al. (1963), Sakai (1988), Sakai and Noon (1991), and Ainsley (1991).

NOTEWORTHY RECORDS

Spring: Coastal – Beacon Hill 23 Mar 1969-8 (Tatum 1970); Goldstream River 20 May 1985-14; Sidney Island 18 May 1984-25 on survey; Mayne Island 22 April 1991-2, 20 May 1983-4 nestlings; Huntingdon 26 Apr 1931-1 (UBC 5398); Surrey 11 Mar 1966-1, 15 May 1963-12; Tofino 1 May 1931-1; Burnaby 11 May 1958-4 eggs; Denman Island 11 Mar 1958-1; Comox 13 Apr 1942-1 (RBCM 13191); Klaskish Inlet 24 May 1978-7; Swanson Bay 12 May 1936-1 (MCZ 283033); Delkatla Inlet 27 Apr 1979-2; Sapsucker Creek 23 May 1985-8; McClinton Bay 15 to 21 May 1985-13 (RBCM 18406-18). **Interior** – Oliver 21 May 1984-3 eggs; White Lake (Okanagan) 20 Apr 1985-1; Penticton 27 Apr 1903-1; Kinnaird 16 Apr 1969-1; Revelstoke 23 May 1890-1 (NMC 913); Doc English Bluff (Williams Lake) 20 May 1979-2; Kleena Kleene 20 May 1961-1; 24 km s Prince George 15 May 1985-1.

Summer: Coastal – Beaver Lake (Victoria) 10 Jun 1973-70 on survey; Chemainus 25 Jun 1979-29 on survey; Quatse Lake 9 Jun 1978-5; Triangular Hill 4 Jun 1987-21; Mayne Island 19 Jul 1983-4 nestlings; Vancouver 20 Aug 1964-1 (UBC 11896); Alouette Lake 1 Jun 1963-12, 20 Jul 1975-14; Mitlenatch Island 24 Aug 1971-3 (Sirk and Sirk 1971); Gaultheria Lake (Brooks Peninsula) 1 Aug 1981-8; Alert Bay 29 Aug 1930-1 (RBCM 9471); Port Hardy 28 Aug 1935-1 (NMC 26139); Goose Island 29 June 1988-12; Reef Island 11 Jul 1977-8; Fairfax Inlet 1 Jun 1977-12; Ship Island 2 Jul 1977-7; Skidegate Inlet 2 Aug 1960-3 eggs; Masset 22 Jun 1946-21; Pleasant Camp 2 Jun 1976-2 singing, 2 Jun 1981-4; Lakelse Lake 30 Jul 1946-1, 30 Aug 1934-1. **Interior** – Summerland 14 Jun 1986-7 on survey; Christina Lake 8 Jul 1991-25; South Slocan 1 Aug 1971-4; Kuskonook 22 Jun 1978-8 on survey; Nelson 1 Aug 1971-4 nestlings; Grindrod 24 Aug 1953-1 (UBC 7907); Merritt 28 Jul 1983-2 eggs, 29 Jul 1980-3 nestlings; Bridge Lake 20 Jun 1977-6; Williams Lake 9 Aug 1978-6; Prince George 20 Jun 1969-1 (NMC 56933); Pine Pass 2 Jul 1972-1; Bijou Provincial Park 1 Jul 1992-1 singing at picnic site; Moberly Lake Provincial Park 2 Jul 1992-1 singing; s bank Peace River 75 km e Farrell Creek 21 Jul 1992-a pair singing and foraging; 1 to 3 km e Farrell Creek 21 Jul 1992-4 along s bank Peace River; Lynx Creek (Peace River) 28 Jul 1986-7 singing; e Hudson's Hope 26 Jul 1985-8; Stoddart Creek (Fort St. John) 20 May 1993-1 singing male tape-recorded; McGregor 5 Jul 1969-4 eggs; Dease Lake 28 Jun 1962-1 (NMC 49854).

Breeding Bird Surveys: Coastal – Recorded from 24 of 27 routes and on 64% of all surveys. Maxima: Masset 6 Jun 1993-69; Queen Charlotte City 25 Jun 1994-65; Port Hardy 1 Jul 1983-48. **Interior** – Recorded from 42 of 73 routes and on 30% of all surveys. Maxima: Kuskonook 17 Jun 1985-18; Mabel Lake 27 Jun 1990-12; Summerland 17 Jun 1985-10.

Autumn: Interior – Alkali Lake 3 Sep 1953-1 (RBCM 15704); Kleena Kleene 9 Sep 1961-1 (Paul 1964); Nicholson 5 Sep 1975-1; Anglemont 1 Sep 1977-2; Brouse 2 Sep 1991-1; Kelowna 13 Sep 1971-1 and Summerland 13 Sep 1972-1 (Cannings et al. 1987). **Coastal** – Burnt Cliff Island 4 Sep 1969-1; Cape St. James 7 Sep 1978-1, latest seen; Calvert Island 15 Sep 1934-3 (MCZ 283030-32); Vancouver 13 Sep 1932-1 (RBCM 7815), 19 Sep 1971-1 (Campbell et al. 1972b); Reifel Island 30 Sep 1984-1; Wascana 2 Oct 1990-1; Nitinat Lake 6 Sep 1985-2; Saanich 2 Oct 1981-1, 27 Oct 1981-1; Oak Bay 15 Sep 1985-8; Victoria 13 Sep 1914-1 (RBCM 3048), 24 Sep 1973-1 (RBCM 12139).

Winter: No records.

Eastern Phoebe EAPH

Sayornis phoebe (Latham)

RANGE: Widespread breeder east of the Rocky Mountains from northeastern British Columbia, southern Mackenzie, and northern Saskatchewan east to New Brunswick and south to New Mexico, central Texas, and Georgia. Winters from the Gulf states south through southern and eastern Mexico.

STATUS: On the coast, *accidental* in the Georgia Depression and northern Coast and Mountains ecoprovinces.

In the interior, an *uncommon* migrant and summer visitant northeast of the Rocky Mountains in the Boreal Plains and Taiga Plains ecoprovinces; *very rare* in the Southern Interior Mountains; *accidental* in the Sub-Boreal Interior.

Breeds.

CHANGE OF STATUS: At the time Munro and Cowan (1947) completed their review of the bird fauna of the province, the Eastern Phoebe (Fig. 72) was known only from the Peace River Parkland of the Boreal Plains north to the Sikanni Chief River in the southern Taiga Plains. Since that time it has expanded its range through the Taiga Plains north to Fort Nelson and the Petitot River area near the Yukon boundary. In addition, between 1962 and 1989 it was reported from 5 areas in the Southern Interior Mountains (1962, Hemp Creek in Wells Gray Park; 1976, Spillimacheen and Brisco; 1985, Revelstoke; 1989, Creston). In all but the Creston instance, 2 birds were observed through the summer; the pair at Spillimacheen built a nest and laid eggs (R.R. Howie pers. comm.).

Figure 72. Adult Eastern Phoebe (south of Dawson Creek, 16 June 1996; Linda M. Van Damme).

NONBREEDING: The Eastern Phoebe (Fig. 72) occurs regularly only in the northeastern portion of the province east of the Rocky Mountains. It is a sporadic migrant to the Southern Interior Mountains from Hemp Creek in the north to Creston in the south. Elsewhere in the province, its occurrence is unusual. It frequents semi-open, low-elevation habitats, usually near water. Unlike the Say's Phoebe, it rarely occurs at high

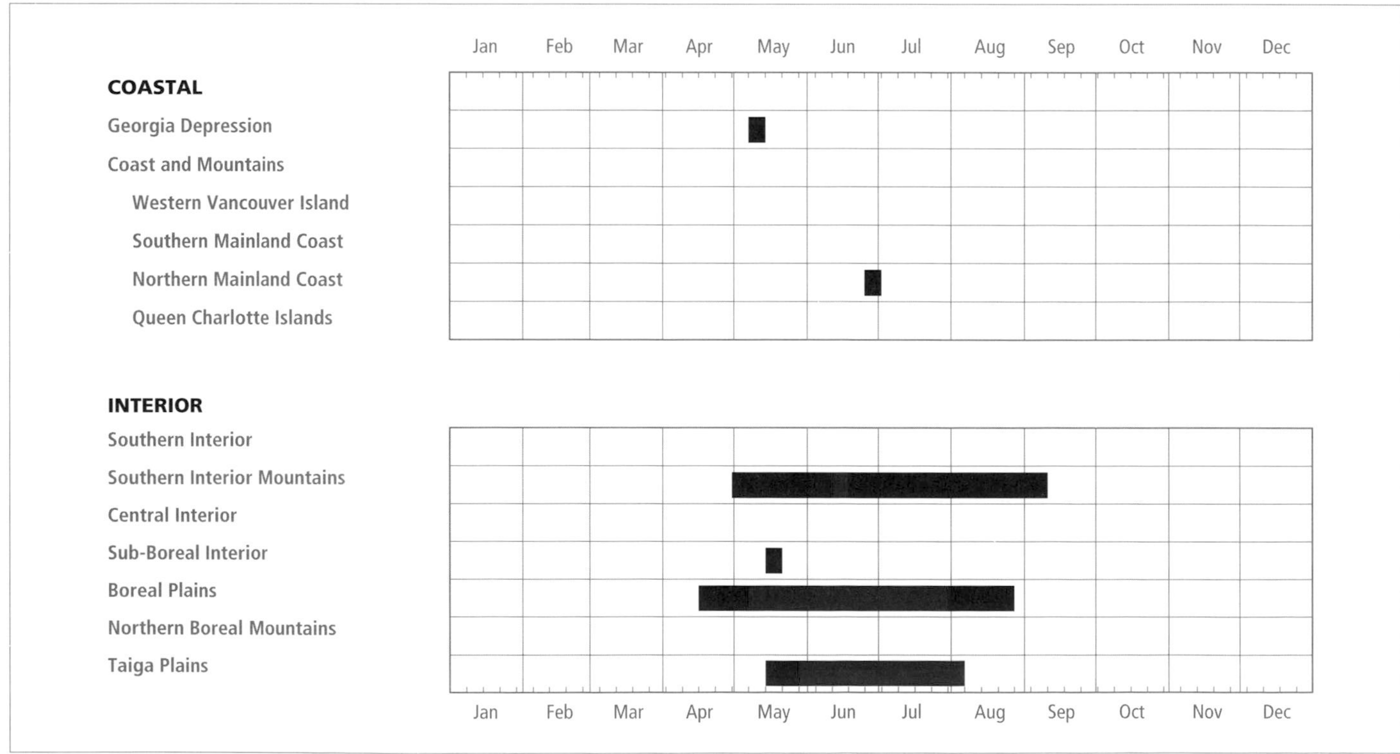

Figure 73. Annual occurrence (black) and breeding chronology (red) for the Eastern Phoebe in ecoprovinces of British Columbia.

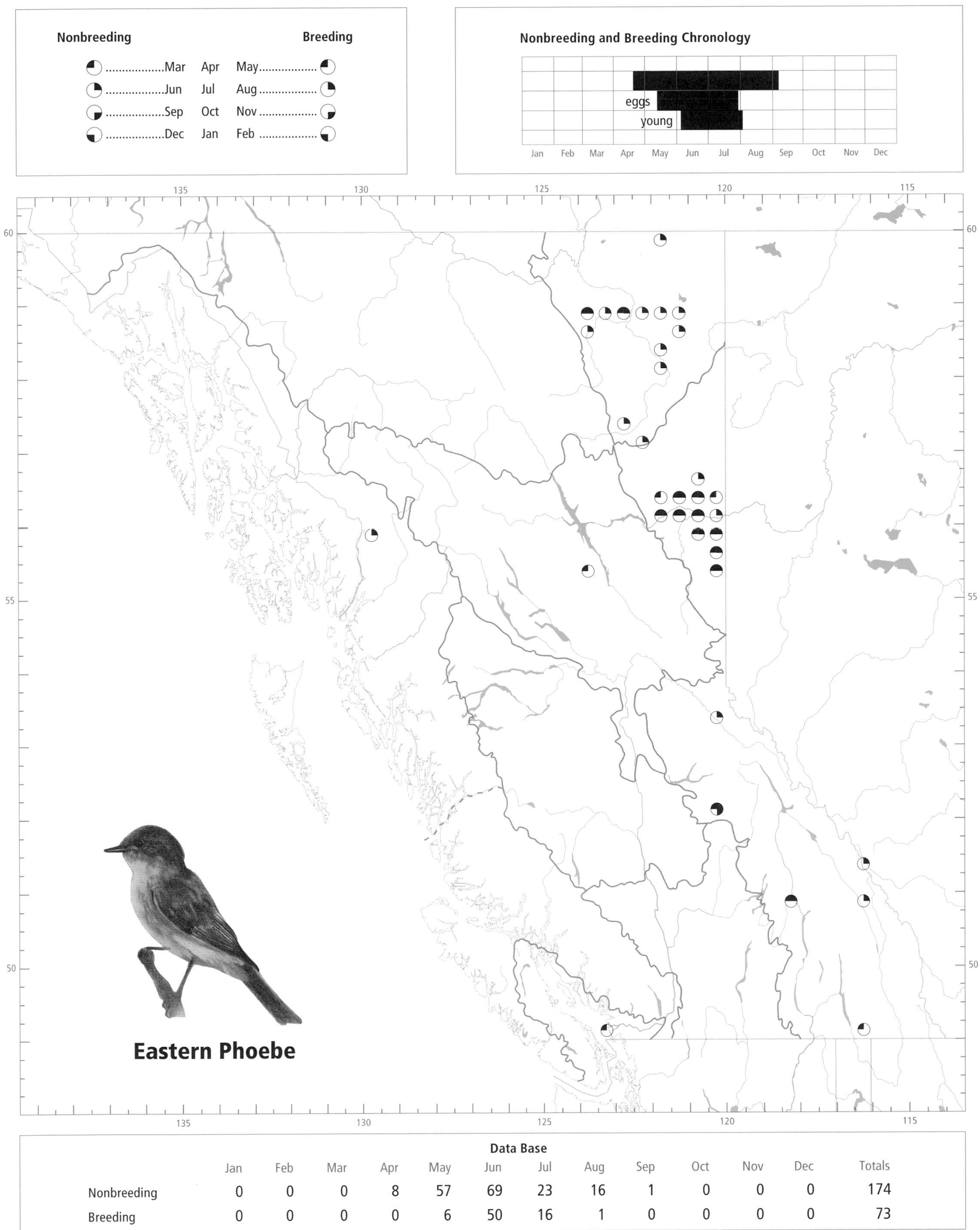

Data Base

	Jan	Feb	Mar	Apr	May	Jun	Jul	Aug	Sep	Oct	Nov	Dec	Totals
Nonbreeding	0	0	0	8	57	69	23	16	1	0	0	0	174
Breeding	0	0	0	0	6	50	16	1	0	0	0	0	73

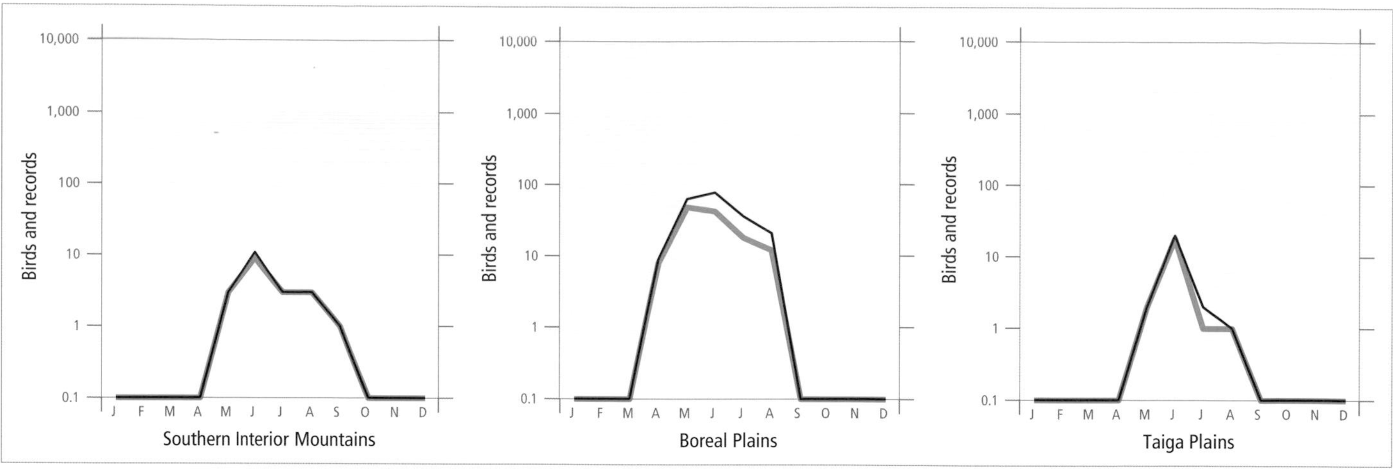

Figure 74. Fluctuations in total number of birds (purple line) and total number of records (green line) for the Eastern Phoebe in ecoprovinces of British Columbia. Nest record data have been excluded.

elevations, even in late summer. It has been recorded from near sea level to 1,400 m elevation.

Spring migrants frequent riparian growth along river banks, lakes, marshes, and beaver ponds, and occur also on the edges of clearings in deciduous or mixed forests and agricultural lands. Dead branches of trembling aspen and willow, or snags, are used as perches from which they hawk insects. The species is often most abundant in human-influenced habitats such as rural clearings with abandoned or seldom-used buildings. It also occurs around farm buildings and ranch houses.

Spring migration consistently begins in the last 2 weeks of April, when the first migrants arrive in the Peace Lowland (Figs. 73 and 74). The peak movement occurs in the second and third week of May; by the second week of May the species is fairly common there (Cowan 1939). Further north, at Fort Nelson, spring migrants arrive during the second week of May, with the movement peaking towards the end of the month.

The southward migration probably begins in late July and early August, peaking in mid-August. All birds have left the Peace River region by the end of August (Figs. 73 and 74). A sighting at Wells Gray Park is the only autumn record for the province, and there are no winter records.

The Eastern Phoebe has been recorded in the province from 19 April to 6 September (Fig. 73).

BREEDING: The Eastern Phoebe breeds in the northeastern corner of the province from Tupper Creek north to Mile 335 on the Alaska Highway. There is 1 nesting record for the Southern Interior Mountains, at Spillimacheen.

The Eastern Phoebe reaches its highest numbers in the Boreal Plains (Fig. 75). Its centre of breeding abundance is in the Peace Lowland between Tupper and Charlie Lake. Breeding Bird Surveys of interior routes for the period 1968 through 1993 contained insufficient data for analysis. Results from North American Breeding Bird Surveys since 1965 (Robbins et al. 1986; Erskine et al. 1992) indicate that there has been a steady but slight decline in the numbers of the Eastern Phoebe over most of its range in North America.

The Eastern Phoebe has been reported nesting between 400 and 800 m elevation. Most nests (67%; $n = 43$) were associated with forested habitat; 28% were associated with human-influenced areas. Only 2 nests were found in grassland areas. Of the 31 nests reported from forest stands, 58% occurred among trembling aspen or other deciduous species, and 13% occurred in mixed spruce-aspen forest. Human-influenced habitats included residential areas, parks, and farmyards. Nests were often described as being associated with edge habitats. In all these habitats, however, nests were most often associated with human-made structures (Fig. 76).

The Eastern Phoebe often nests near water, especially if clay or rocky banks are present. Some areas have become traditional nesting sites: the banks of the Peace River at Farrell Creek and at Hudson's Hope, the ledges of clay banks of a

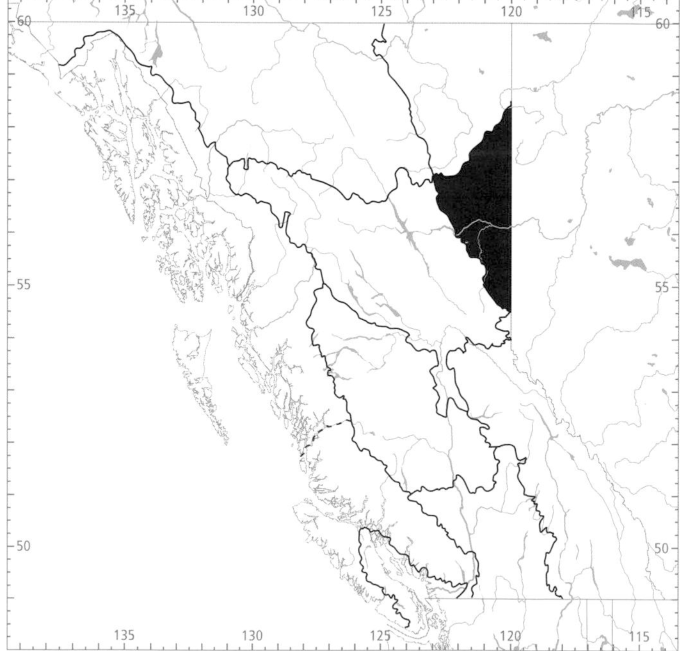

Figure 75. In British Columbia, the highest numbers for the Eastern Phoebe in summer occur in the Boreal Plains Ecoprovince.

seasonal stream just north of the Beatton Recreational Area on the Beatton River, the shores of Swan and Charlie lakes, St. John Creek north of Fort St. John, the south shore of the Peace River across from Taylor, and Stoddart Creek. In such habitats, the Eastern Phoebe is a bird of the woodland edges, catching insects over water or over small open spaces and nesting beneath the overhanging roots at the top of the banks.

The Eastern Phoebe has been recorded breeding in the province from 12 May to 2 August (Fig. 73).

Nests: Most nests (90%; n = 49) were built on human-made structures such as houses (in active use or abandoned), bridges, sheds, barns, picnic shelters, culverts, and similar structures. There they were situated under the eaves (48%; n = 41), or on rafters, girders, or beams (38%) of the structure. A few were found in natural sites such as among tree roots or in niches on clay or dirt banks. The heights of 43 nests ranged from 1.2 to 6.9 m, with 63% recorded between 1.9 and 3.0 m.

Nests were usually bulky cups, often shallow and incompletely circular with 1 flat side. They were composed mainly of mud, mosses, and grasses, and, less often, fine rootlets, lichens, leaves, plant stems, and feathers. Linings consisted of tightly woven fine grasses and hair.

Eggs: Dates for 27 clutches ranged from 12 May to 29 July, with 57% recorded between 2 June and 5 July. Sizes of 21 clutches ranged from 1 to 6 eggs (1E-1, 2E-1, 3E-1, 4E-11, 5E-6, 6E-1), with 11 having 4 eggs. The average clutch size in British Columbia is similar to that in Ontario (Peck and James 1987). Published incubation periods range from 13 to 18 days (Fannes 1980; Peck and James 1987).

Young: Dates for 27 broods ranged from 5 June to 2 August, with 58% recorded between 14 and 28 June. Sizes of 20 broods ranged from 1 to 5 young (1Y-1, 2Y-1, 3Y-5, 4Y-11, 5Y-2), with 11 having 4 young. Calculated dates indicate that young could be found as early as late May. The nestling period is 13 to 16 days (Weeks 1979; Fannes 1980). Two nests suggest that the Eastern Phoebe is double-brooded in British Columbia. Conrad and Robertson (1993) discuss parental provisioning by this species in first and second broods.

Figure 76. Nest site of the Eastern Phoebe under the overhang of an abandoned building (Peace Island Park, south of Taylor, 16 June 1986; R. Wayne Campbell).

Brown-headed Cowbird Parasitism: Cowbird parasitism was not found in British Columbia in 37 nests recorded with eggs or young. The Eastern Phoebe is a frequent cowbird host elsewhere in North America (Friedmann 1963; Peck and James 1987). Southern and Southern (1980) also report a freedom from parasitism in northern Michigan, although cowbirds were common in the area.

Nest Success: Of 5 nests found with eggs and followed to a known fate, 3 produced at least 1 fledgling.

REMARKS: Reports of the Eastern Phoebe outside its normal range in the northeastern portion of the province are infrequent. Several summer reports in the Okanagan valley were not included in the account for lack of adequate documentation (Cannings et al. 1987). In addition, single reports from Vancouver, Jaffray, Lac la Hache, Williams Lake, and Quesnel, and on Breeding Bird Surveys from Prince George, Fraser Lake, and Mount Morice, lack supporting details and have also been excluded from the account.

See Hill and Gates (1988) and Weeks (1994) for additional information on the nesting ecology and life history of the Eastern Phoebe.

NOTEWORTHY RECORDS

Spring: Coastal – Reifel Island 13 May 1989-1 (BC Photo 1251; Weber 1992; Campbell 1989c). **Interior** – Creston 6 and 7 May 1989-1 male singing under bridge; Revelstoke 24 May 1989-1 (Campbell 1989c); Hemp Creek (Wells Gray Park) 12 May 1962-1 collected (Edwards and Ritcey 1967); 45 km e Mackenzie 19 May 1993-1 (Siddle 1993d); Tupper 5 May 1938-common in area (Cowan 1939), 6 May 1938-1 male (RBCM 8163), 12 May 1938-eggs (Cowan 1939); sw Dawson Creek 19 Apr 1991-1; Rolla 12 May 1922-1 (Williams 1933a); Fort St. John 25 Apr 1987-1, 30 May 1962-5 eggs; Beatton Park 11 May 1983-3; Fort Nelson 12 May 1979-1; Mile 335 Alaska Highway 23 May 1987-1 male (McEwan and Johnston 1987b).

Summer: Coastal – Stewart 27 Jun 1993-1 (Siddle 1993d). **Interior** – Brisco 12 Jul 1976-1; Spillimacheen Jun 1976-1, incubating eggs; Revelstoke 21 Aug 1986-1; Wapta Marsh (Yoho National Park) 31 Jul and 4 Aug 1976-1 (Wade 1977); Hemp Creek (Wells Gray Park) 25 Aug 1962-2 (Edwards and Ritcey 1967); McBride 28 Jun 1970-1; Tupper 27 Jun 1938-fully feathered young (Cowan 1939); Dawson Creek 25 Jun 1930-1 (UBC 5426); Swan Lake (Peace River) 21 Jun 1988-3; Fort St. John 5 Jun 1962-4 nestlings; Charlie Lake 20 Jul 1974-1 egg and 3 nestlings; Beatton Park 12 Jul 1983-3; Stoddart Creek (Fort St. John) 30 Aug 1985-1; Sikanni Chief River 7 Jun 1922-1 (Williams 1933a); Kledo Creek 2 Aug 1985-3 fully feathered nestlings; Fort Nelson 2 Aug 1985-1 (Grunberg 1986); s Kotcho Lake 16 Jun 1982-1 male (RBCM 17614; Campbell and McNall 1982).

Breeding Bird Surveys: Coastal: Not recorded. **Interior** – Recorded from 7 of 73 routes and on 3% of all surveys. Maxima: Fort St. John 27 Jun 1986-3; Tupper 10 Jun 1976-2; Steamboat 28 Jun 1980-2; Fort Nelson 19 Jun 1974-1.

Autumn: Interior – Hemp Creek (Wells Gray Park) 6 Sep 1962-1 (Edwards and Ritcey 1967). **Coastal** – No records.

Winter: No records.

Say's Phoebe

Sayornis saya (Bonaparte)

SAPH

RANGE: Widespread in western North America. Breeds from western Alaska and northern Yukon east to northern Mackenzie District and south through the interior of British Columbia, central Alberta, central Saskatchewan, and southwestern Manitoba, and throughout the central Plains states east of the coastal mountains, to Baja California and central Mexico. Winters from the southwestern United States to southern Mexico.

STATUS: On the coast, a *rare* transient to the Georgia Depression Ecoprovince on the south coast, *accidental* in winter; *very rare* transient in the Coast and Mountains Ecoprovince, including the Queen Charlotte Islands.

In the interior, an *uncommon* migrant and summer visitant to the Southern Interior (locally *fairly common* in spring in the Okanagan valley), Southern Interior Mountains, and Central Interior ecoprovinces; *rare* to locally *fairly common* in the Sub-Boreal Interior and Northern Boreal Mountains ecoprovinces; *rare* transient to the Boreal Plains Ecoprovince; *very rare* transient in the Taiga Plains Ecoprovince; in winter, *casual* in the Southern Interior and *accidental* in the Southern Interior Mountains ecoprovinces.

Breeds.

NONBREEDING: The Say's Phoebe (Fig. 77) is widely distributed throughout the interior of British Columbia from the extreme southern portions of the province to the Yukon boundary in the north. It is concentrated in the southern third of the province north to about Williams Lake. In northern areas, its known distribution is closely associated with highway routes and human habitation.

Figure 77. Adult Say's Phoebe with food for nestlings (Surprise Lake Road, east of Atlin, 7 July 1996; R. Wayne Campbell).

On the coast, it occurs irregularly; most records from the lower Fraser River valley and southeastern Vancouver Island are in March, April, May, August, and September (Fig. 78). It has a spotty distribution elsewhere along the coast.

The Say's Phoebe occurs from near sea level to 3,200 m elevation. It is a bird of open country, often near human habitation, where it typically perches on fencelines, utility wires

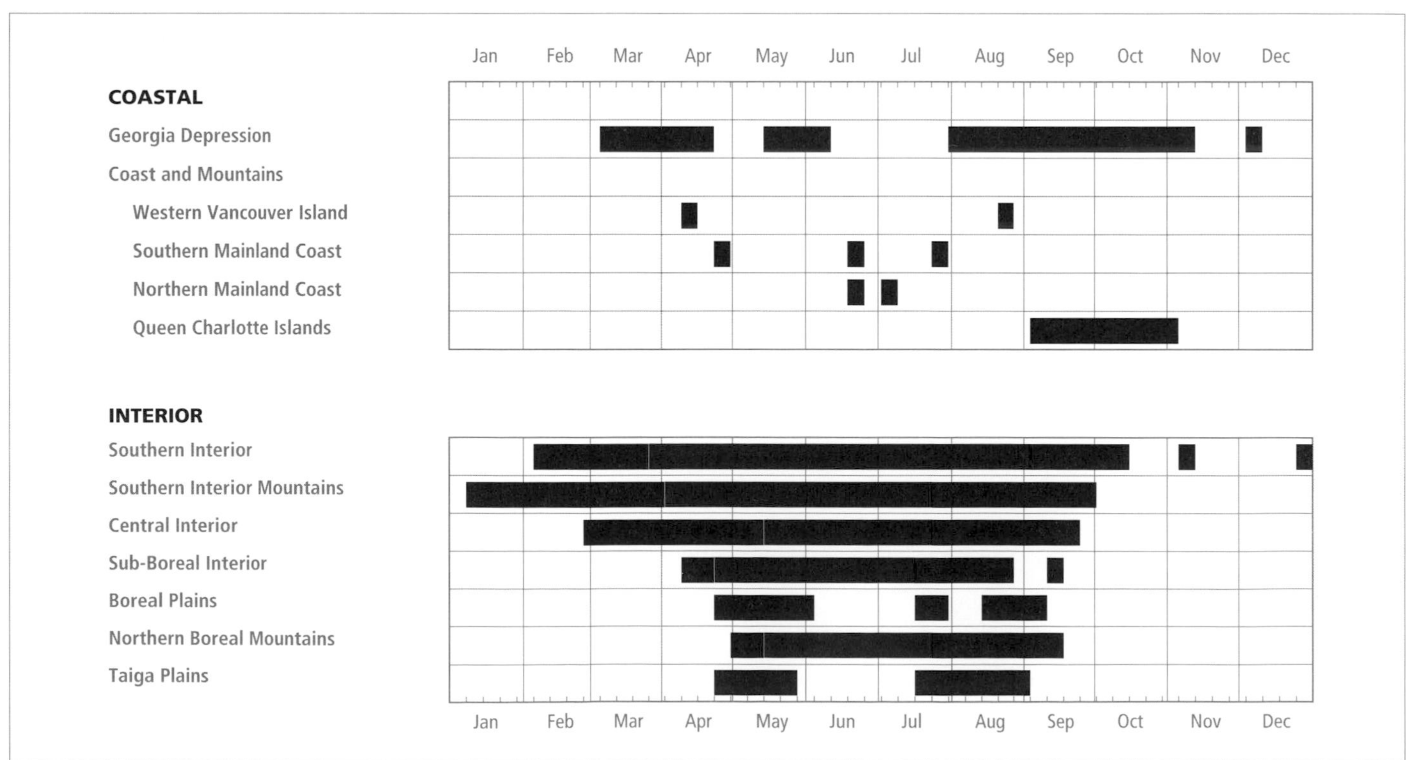

Figure 78. Annual occurrence (black) and breeding chronology (red) for the Say's Phoebe in ecoprovinces of British Columbia.

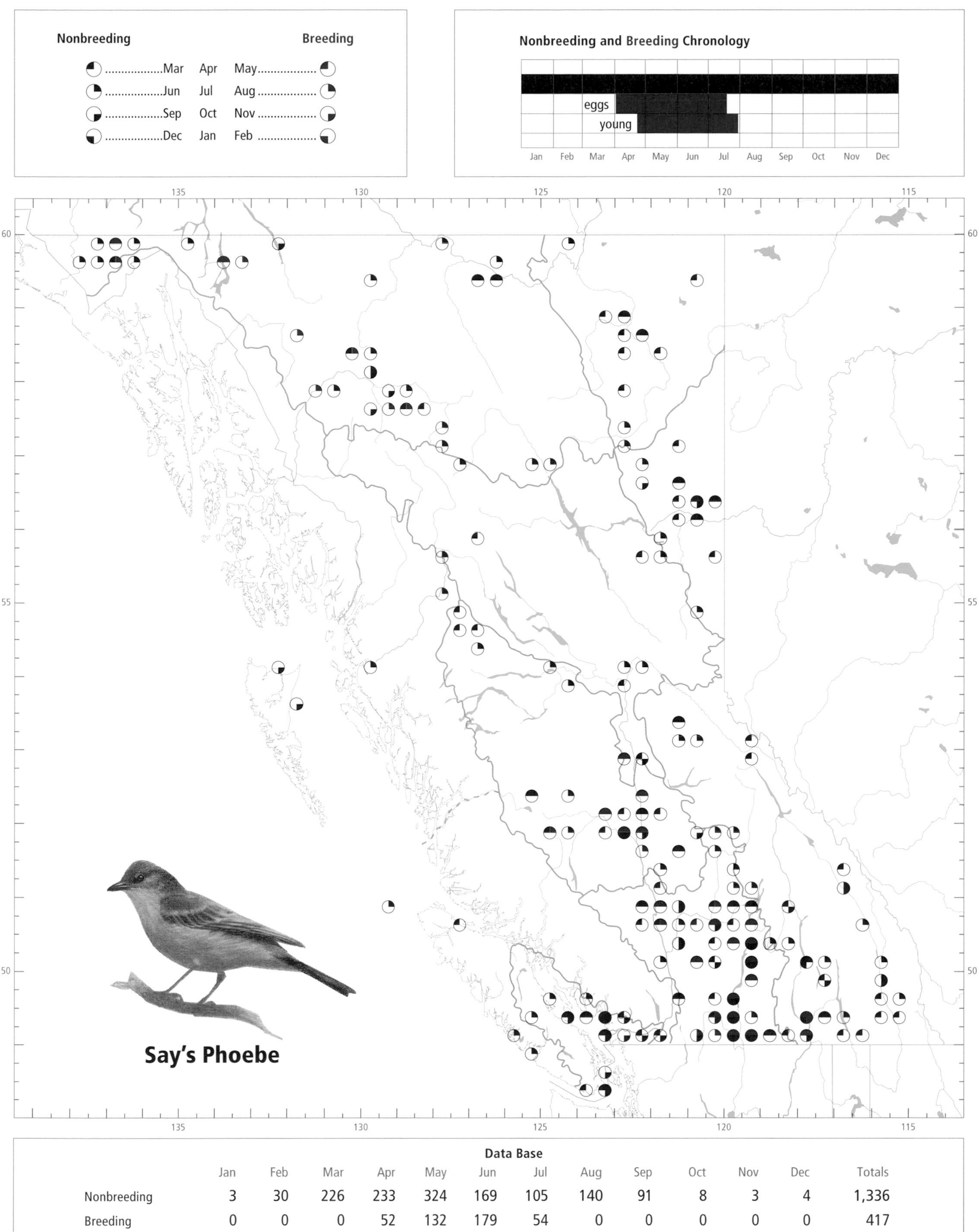

Data Base

	Jan	Feb	Mar	Apr	May	Jun	Jul	Aug	Sep	Oct	Nov	Dec	Totals
Nonbreeding	3	30	226	233	324	169	105	140	91	8	3	4	1,336
Breeding	0	0	0	52	132	179	54	0	0	0	0	0	417

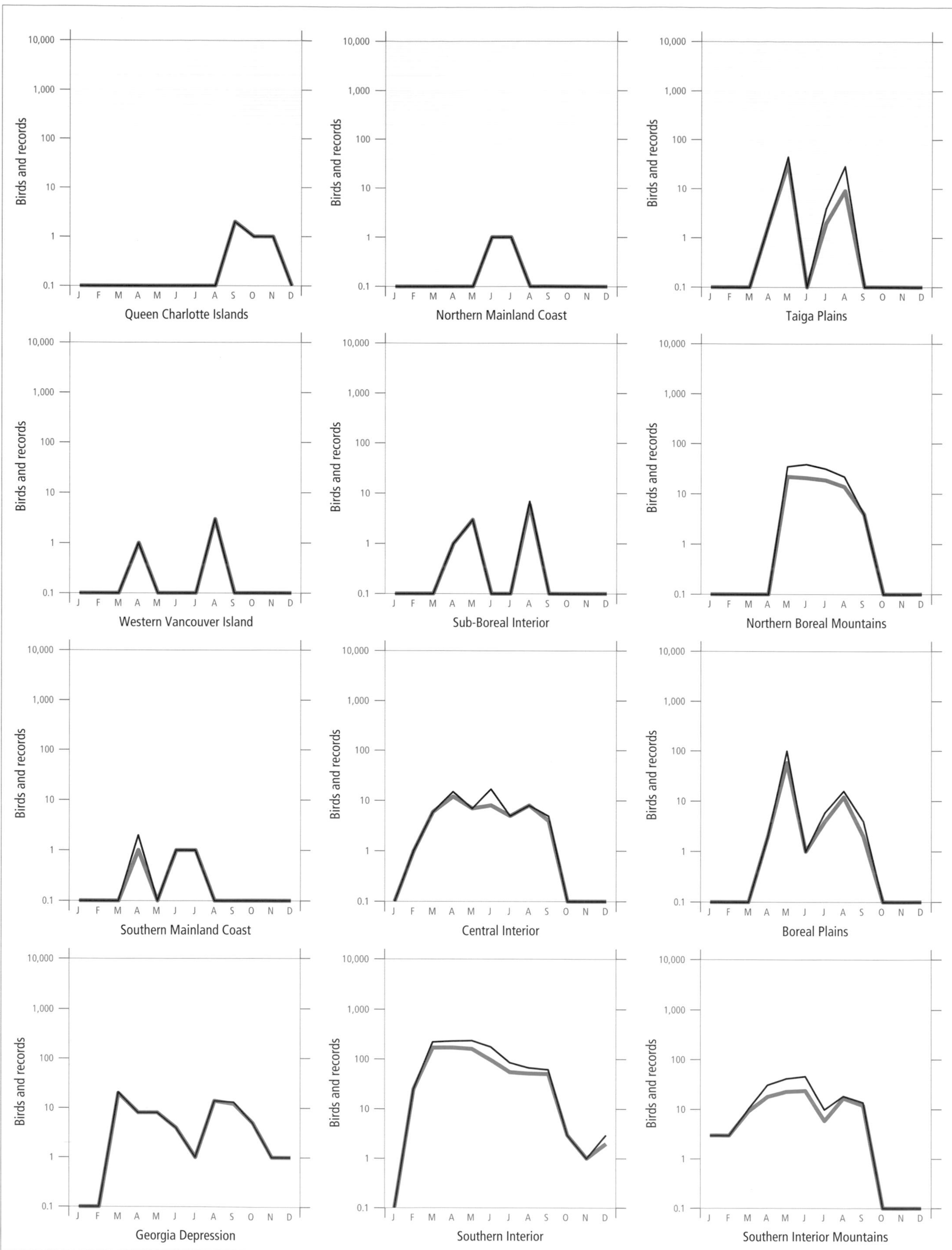

Figure 79. Fluctuations in total number of birds (purple line) and total number of records (green line) for the Say's Phoebe in ecoprovinces of British Columbia. Christmas Bird Count and nest record data have been excluded.

or poles, and shrubs. It frequents open rangelands, agricultural country, semi-arid sites, and benches along rivers in the southern and central portions of the interior. In forested areas, it occurs in recently logged sites, natural clearings, burns, and swamps, as well as on the edges of lakes, ponds, or streams, roadsides, and cliff faces. In the Peace Lowland, it frequents open, rolling farmlands and edges of marshes, meadows, fields, and landings in clearcuts. Further north, it is usually found near buildings and bridges, or along fencelines. In the mountainous northwest, it occurs in river bottom meadows, along rocky bluffs or canyons, and near mountain peaks above the timberline (Fig. 80).

The Say's Phoebe is among the first migrant passerines to appear in spring, and is by far the earliest-arriving flycatcher species. In the Okanagan valley, the first migrants may arrive by the second week of February (Figs. 78 and 79); however, the main movement does not normally begin until the end of February or beginning of March and peaks between the end of March and mid-April. Before spring migration ends in the Okanagan valley, the early arrivals are already nesting (Cannings et al. 1987). Birds may arrive in the Southern Interior Mountains and the Central Interior by the end of February or early March; however, the migration in these areas is not as noticeable as in the Southern Interior. The few birds found in the Sub-Boreal Interior usually arrive in April.

Migrants, believed to be from another source, arrive in the Boreal Plains and Taiga Plains ecoprovinces at the end of April. These birds, likely from Alberta, pass rapidly through the northeastern corner of the province, and have left the area by the end of May (Figs. 78 and 79). On the extreme south coast, the Say's Phoebe appears irregularly in spring from the second week of March through early June. In mid to late summer, some Say's Phoebes disperse to subalpine and alpine areas (Fig. 80).

The Say's Phoebe begins its southward movement from the northern interior by the third week of July, and most birds have left the area by the end of August, although stragglers may be found as late as the third week of September (Figs. 78 and 79). In the southern portions of the interior, the main autumn movement occurs in late August and early September, with most phoebes gone from the province by the end of

Figure 80. Some Say's Phoebes disperse to subalpine and alpine areas in midsummer to forage prior to autumn migration (Three Guardsmen Mountain, 7 August 1980; R. Wayne Campbell).

September. There are 4 winter occurrences of this flycatcher from the extreme southern interior region and 1 from Sechelt on the south coast.

In the interior, the Say's Phoebe has been recorded regularly from 8 February to 30 September. On the coast, it has been recorded mainly during migration periods: 9 March to 5 June, and 4 August to 6 November (Fig. 78).

BREEDING: There are 2 discrete breeding populations of the Say's Phoebe in British Columbia (American Ornithologists' Union 1957). A southern population (*S. s. saya*) nests throughout much of the central-southern interior south of the Williams Lake area (latitude 53°N). It occupies the region from Manning Park east to the extreme southern portion of the west Kootenay, and extends northward through the Okanagan valley and Thompson Basin into the Cariboo and Chilcotin areas. A northern population (*S. s. yukonensis*) nests north of latitude 57°30′N and west of longitude 128°W. This area encompasses the northwestern portion of the province from the Spatsizi Mountains northwest to Atlin Lake, Chilkat Pass, and the Tatshenshini valley. The 2 populations are separated by at least 600 km, over which the species is not known to nest.

Although migrants regularly pass through, there is no confirmed breeding in the east Kootenay or in the vast northeastern portion of the province north of Quesnel and east of Spatsizi Provincial Park.

The highest numbers for the Say's Phoebe in British Columbia in summer occur in the Southern Interior and Northern Boreal Mountains (Fig. 81). Breeding Bird Surveys of interior routes for the period 1968 to 1993 contain insufficient data for analysis. Those data may be biased by the phoebe's very early nesting period, as most birds have stopped singing in June and early July, when the Breeding Bird Surveys are conducted. In western North America, populations appear to have increased between 1965 and 1979 (Robbins et al. 1986).

The Say's Phoebe breeds between 270 and 1,860 m elevation. It has become closely associated with human-influenced habitats. Breeding usually takes place in open country, in and around buildings or other structures. Most nests (77%; n = 186; Fig. 83) were found in human-influenced areas, and of these about two-thirds were in rural, agricultural, or cultivated areas. There, human activity has greatly increased available foraging habitats and nesting sites. Natural habitats included those associated with grasslands, semi-arid areas, open forests, alpine cliffs (Weeden 1960), shrublands, and rocky bluffs. Nests were built adjacent to open areas, often along banks of rivers and lakes, canyons, rocky outcroppings, meadows, and mountainsides.

The Say's Phoebe has been recorded breeding in the province from 1 April to 22 August (Fig. 78).

Nests: Most nests (89%; n = 199) were in human-made structures, including sheds, houses, abandoned buildings, garages, barns, bridges, mine shafts, and culverts. Natural sites included cliffs, caves (Fig. 82), sinkholes, clay or dirt banks, and rock crevices. On structures, nests were positioned on rafters, girders, beams, shelves, electrical fixtures (Fig. 83), ledges, drainpipes, and eaves. The heights of 167 nests ranged from 2.4 m below ground to 24 m above ground, with 55% recorded between 2 and 2.8 m above ground.

Nests were described as neat to bulky cups that varied greatly in shape because they were frequently placed against a vertical surface. The most frequently used materials were dry grasses, animal hair, and feathers. Some nests included a variety of other local material such as mosses, lichens, plant fibres, cloth, string, fibrous furniture padding, and – rarely – mud. The nest lining was of finer material, including fine grasses, hair, and plant or bird down.

The Say's Phoebe often builds its nest in the same site year after year; in 1 case, a site was reportedly used for up to 7 consecutive years. If the old nest is present, the bird will frequently repair it; if not, it builds a new nest or occasionally uses the old nest of another species (Fig. 83).

Eggs: Dates for 123 clutches ranged from 1 April to 9 July, with 53% recorded between 7 May and 12 June. Sizes of 106 clutches ranged from 1 to 6 eggs (1E-5, 2E-9, 3E-11, 4E-48, 5E-29, 6E-4), with 73% having 4 or 5 eggs. The Say's Phoebe is double-brooded in southern British Columbia; complete clutches found after the end of May are likely second clutches (Fig. 84). There is 1 record of a pair successfully raising 3 broods in a single season at Williams Lake. The incubation period in British Columbia is 14 to 17 days ($n = 5$).

Young: Dates for 128 broods ranged from 22 April to 28 August, with 53% recorded between 22 May and 29 June. Sizes of 114 broods ranged from 1 to 6 young (1Y-4, 2Y-11, 3Y-21, 4Y-49, 5Y-24, 6Y-5), with 64% having 4 or 5 young. The nestling period in British Columbia is 17 to 21 days ($n = 5$).

Brown-headed Cowbird Parasitism: In British Columbia, 1 of 169 nests found with eggs or young was parasitized by the cowbird. Friedmann et al. (1977) and Friedmann and Kiff (1985) indicate that the Say's Phoebe is seldom parasitized.

Nest Success: Of 55 nests found with eggs and followed to a known fate, 39 produced at least 1 fledgling, for a nest success rate of 71%. Known and suspected causes of nest failure included predation by a red squirrel (1 nest with young), chipmunk (1 nest with eggs), Merlin (1 nest with young), Clark's Nutcracker (1 nest with young), and domestic cat (1 nest, destroyed); parasitism by a Brown-headed Cowbird (1 nest); adult mortality (1 nest abandoned); and human disturbance (2 nests).

REMARKS: The American Ornithologists' Union (1957) recognizes 3 races of Say's Phoebe: *Sayornis saya quiescens* of Baja California, *S. s. yukonensis* of Alaska and northwestern British Columbia, and *S. s. saya* of the remaining part of the species' range. The race *saya* is a pale, "scorched"-appearing bird with a shorter tail and longer bill than *yukonensis*, which is a darker, clearer gray bird (Bishop 1931). Browning (1976), however, considers the name *yukonensis* to be a synonym of nominate *saya* on the basis of the "great amount of variation in colour samples" among specimens of *yukonensis* and "lack of mensural differences" between it and samples of *saya*.

The breeding distribution listed for British Columbia by the American Ornithologists' Union (1983) is incorrect and should not include the northeastern portion of the province.

The Say's Phoebe has the widest north-to-south breeding distribution of any North American tyrannid. It breeds from the deserts of Mexico to the Arctic tundra, where it has recently (1960) expanded its range in Alaska to North Slope rivers with rocky cliff habitat (Cade and White 1973).

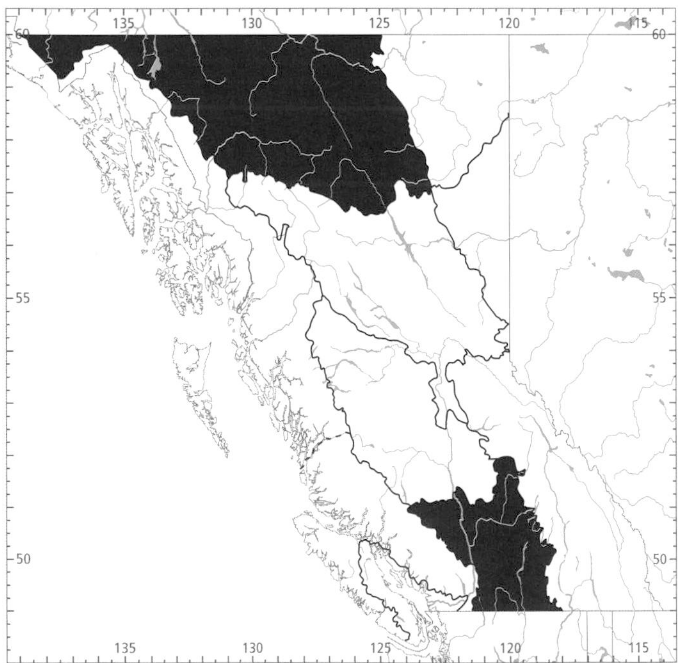

Figure 81. In British Columbia, the highest numbers for the Say's Phoebe in summer occur in the Southern Interior and Northern Boreal Mountains ecoprovinces.

Figure 82. A Say's Phoebe nest site in a natural cave (Ord Road, Kamloops, 30 June 1991; R. Wayne Campbell).

Figure 83. A brood of 3 young Say's Phoebes in a Barn Swallow nest on a light fixture in a cabin (Farwell Canyon, south of Riske Creek, 23 June 1987; Anna Roberts).

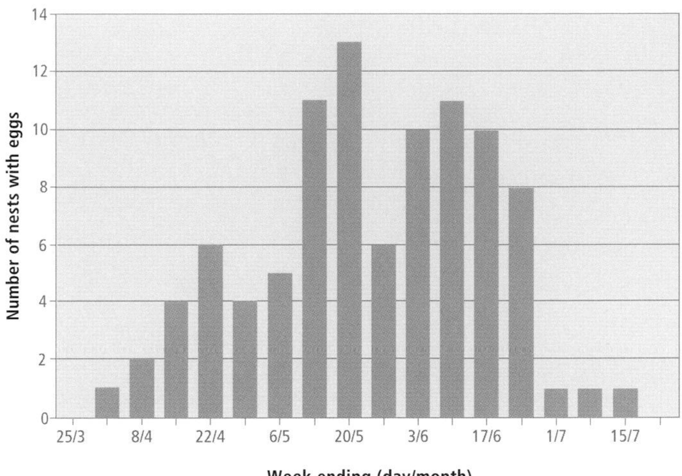

Figure 84. Weekly number of clutches for the Say's Phoebe in the Southern Interior Ecoprovince (n = 94). The figure is based on the week the maximum number of eggs was recorded. The figure suggests that complete clutches after the end of May likely represent second clutches.

NOTEWORTHY RECORDS

Spring: Coastal – Whiffin Spit (Sooke) 19 and 20 Mar 1976-1 (Shepard 1976c); Kirby Creek (Sooke) 14 May 1960-1 (Stirling 1960); Tsawwassen 9 Mar 1980-1; Chilliwack 26 Apr 1896-1 (Brooks 1917); Sea Island (Richmond) 16 Mar 1974-1 (Shepard 1975a); Pitt Meadows 17 to 20 Mar 1972-1 (Crowell and Nehls 1972c), 31 Mar 1971-2 (Campbell et al. 1972b); Rathtrevor Beach Park 3 Apr 1993-1; Sechelt 15 Mar 1972-1; Pulteney Point 11 Apr 1976-1 (BC Photo 506). **Interior** – Osoyoos 4 May 1985-3 half-grown nestlings; Kilpoola Lake 21 Apr 1962-4 eggs (Campbell and Meugens 1971); near Oliver 28 Mar 1976-7; Castlegar 10 Apr 1975-6 eggs; Moyie 7 Mar 1942-1 (RBCM 10994); Naramata 22 Apr 1966-1 egg and 5 nestlings, 5 Apr 1967-1 egg; Vernon 1 Mar 1964-1; Tranquille 1 Mar 1986-1; Shuswap Park 27 May 1967-1 (Anonymous 1967); Riske Creek 5 Mar 1984-1; Fort St. John 17 May 1987-10; Charlie Lake 28 Apr 1981-2; Dease Lake 5 May 1980-1; Mile 306 Alaska Highway 28 Apr 1975-1; Atlin 10 May 1930-1 (Swarth 1936), 16 May 1981-1 (Campbell 1981), 31 May 1981-1 egg; Chilkat Pass 15 May 1977-3.

Summer: Coastal – Clover Point (Victoria) 13 Aug 1985-1; Victoria Jul 1890-1 (Munro and Cowan 1947); Cedar Hill (Victoria) 20 Aug 1990-1 at golf course (Siddle 1991a); Saanich 13 Aug 1983-1; Stubbs Island 25 Aug 1893-1 (Macoun and Macoun 1909); North Vancouver 3 and 4 Jun 1990-1 (Siddle 1991a); Mount Steele 23 Aug 1992-1 in alpine habitat at 1,700 m; Qualicum Beach 5 Jun 1984-1 (RBCM 18053); Yale 12 Jul 1924-1; Triangle Island (Scott Islands) 23 Aug 1976-1 (BC Photo 474); Kispiox River valley 23 Aug 1921-1 (Swarth 1924). **Interior** – Kilpoola Lake 14 Jun 1968-5 nestlings (Campbell and Meugens 1971); Trail 20 Jul 1983-4 nestlings; Castlegar 7 May 1973-4 eggs; Kaleden 22 Aug 1985-5 ready to fledge; Trout Creek 14 Jul 1967-2 eggs plus 1 cowbird egg; Lillooet 28 Aug 1968-female and 2 new fledglings; Lytton 12 Jul 1963-3; Crater Lake (Shuswap) 21 Aug 1970-4 (Sirk 1970); Farwell Canyon 27 Jul 1986-4 fledglings, 23 Jun 1987-3 nestlings (Fig. 83); Alexis Creek 28 May 1946-eggs (Munro and Cowan 1947); Cecil Lake 21 Jul 1987-2, first autumn migrant; Tatlatui Lake 21 Jun 1986-1; Marion Creek (Spatsizi Provincial Park) 3 Aug 1976-numerous along warm cliffs facing southeast (Hodson 1976); Parker Lake (Fort Nelson) 2 Aug 1985-1; Fort Nelson 19 and 22 Aug 1986-8; Clarke Lake 12 Aug 1979-1; Level Mountain 2 Jun 1978-3 nestlings; Mount McDame 21 Jun 1956-6 pairs; Mount Vaughan 18 Jul 1980-3 cold eggs; Chilkat Pass (Mile 83) 28 Jun 1958-6 nestlings with eyes not open (Weeden 1960); Liard River valley 14 Aug 1943-7 (Rand 1944); Atlin 1 Jul 1980-2 new fledglings; Three Guardsmen Mountain 12 Jun 1979-1 (Fig. 80).

Breeding Bird Surveys: Coastal – Recorded from 1 of 27 routes and on less than 1% of all surveys. Maxima: Kwinitsa 19 Jun 1969-1. **Interior** – Recorded from 19 of 73 routes and on 9% of all surveys. Maxima: Mount Morice 26 Jun 1968-5; Kamloops 27 Jun 1982-5; Columbia Lake 26 Jun 1973-4; Lavington 10 Jun 1978-4; Fraser Lake 22 Jun 1969-3; Kuskonook 30 Jun 1977-3; Pennington 30 Jun 1991-3.

Autumn: Interior – Teslin Lake 10 Sep 1924-1, last autumn departure (Swarth 1926); Atlin 2 Sep 1929-1 (Swarth 1936); McBride River 7 Sep 1977-1; Tatogga Lake 3 Sep 1977-1; Halfway River (Fort St. John) 9 Sep 1984-3; Chilcotin River bridge at Highway 20 (w Williams Lake) 22 Sep 1978-1; Nicholson 28 Sep 1976-1; Rose Hill (Kamloops) 7 Oct 1981-1; Summerland 30 Sep 1971-1; Okanagan Landing 15 Sep 1914-1 (Munro and Cowan 1947), 7 Nov 1912-1; Princeton 1 Oct 1927 (RBCM 13197). **Coastal** – Masset 18 Sep 1943-1 (RBCM 10387); Tlell 8 Sep 1989-1, Oct 1989-1, 1 Nov 1989-1 (Campbell 1990a); Qualicum Beach 6 Sep 1976-1 (Dawe 1976); Jericho Park (Vancouver) 25 Oct 1982-1; Chilliwack 1 Oct 1887-1 (Brooks 1917); Vedder Canal 23 Sep 1980-1; Victoria 15 Oct 1965-1, 6 Nov 1965-1 (Davidson 1966).

Winter: Interior – Wineglass Ranch (Riske Creek) 29 Feb 1988-1, first spring arrival; Vernon 19 Dec 1982-1 (Cannings et al. 1987); Nakusp 10 Jan to mid-Feb 1972-1; Lavington 12 Feb 1980-1; Okanagan Landing 12 Feb 1924-1, 29 Feb 1924-1; Penticton 27 and 28 Dec 1976-2 (Cannings et al. 1987); Vaseux Lake 8 Feb 1991-1 (Siddle 1991b); Skaha Lake 31 Dec 1980-1 (Cannings et al. 1987); Keremeos 27 Feb 1965-1 (UBC 12584). **Coastal** – Sechelt 5 Dec 1984-1 (T. Greenfield pers. comm.).

Christmas Bird Counts: Interior – Recorded from 2 of 27 localities and on less than 1% of all counts. Maxima: Penticton 27 Dec 1976-**2**, all-time Canadian high count (Anderson 1977); Vernon 19 Dec 1982-1. **Coastal** – Not recorded.

Ash-throated Flycatcher

ATFL

Myiarchus cinerascens (Lawrence)

RANGE: Breeds from southeastern Washington and southwestern Oregon, Idaho, and Wyoming, south and east through California and the southwestern states to Baja California and central Mexico. Winters from extreme southwestern United States south throughout Mexico to El Salvador.

STATUS: On the coast, *very rare* early summer to late autumn vagrant in the Georgia Depression Ecoprovince, including southeastern Vancouver Island; *casual* on Western Vancouver Island; *accidental* in the southern Coast and Mountains Ecoprovince.

In the interior, *accidental* in the Southern Interior Mountains Ecoprovince.

CHANGE IN STATUS: The Ash-throated Flycatcher (Fig. 85) was first reported in British Columbia in 1944, at Ainsworth in the west Kootenay, on the basis of a well-described sighting by K. Racey. That report was not considered by Munro and Cowan (1947). The occurrence of this flycatcher was subsequently confirmed in the province in 1953 when 1 bird was collected, and a second bird banded, along the edge of the Fraser River at Marpole (Hughes 1954). Erskine (1960) found another individual at Jericho Beach in 1958. By 1971, the Ash-throated Flycatcher was considered a "casual transient" in the Vancouver area (Campbell et al. 1972b). Since the 1980s, 1 or 2 birds have appeared in 12 of the last 14 years ending in 1993.

Figure 85. Ash-throated Flycatcher (Adrian Dorst).

OCCURRENCE: The Ash-throated Flycatcher (Fig. 85) is a rarity north of Washington state and is considered an irregular straggler to southern British Columbia. On the south coast, reliable documentation is limited to reports of 36 individual birds: 5 from southeastern Vancouver Island, 4 from Western Vancouver Island, 25 from the Greater Vancouver area and lower Fraser River valley east to Langley, and 2 from the mainland coast south of Alice Lake. A single report

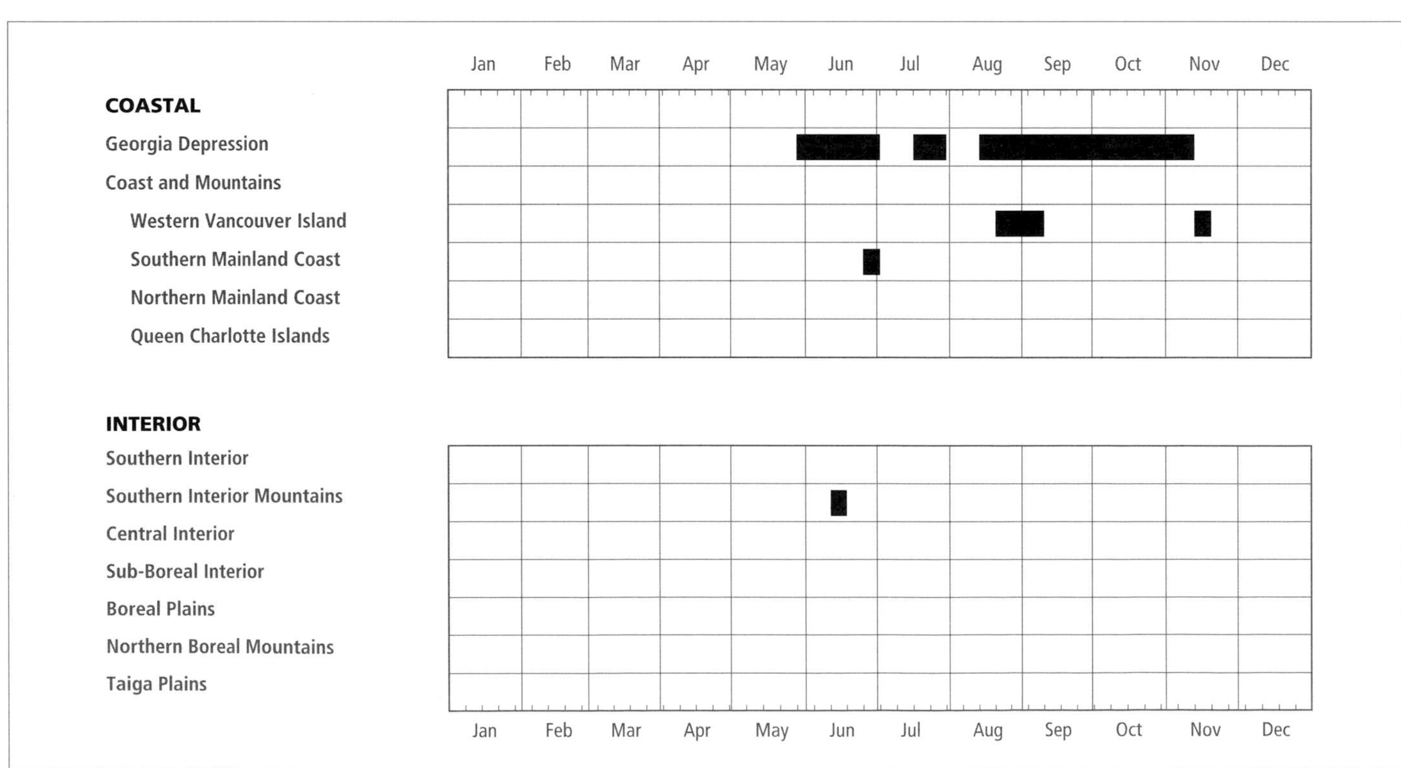

Figure 86. Annual occurrence for the Ash-throated Flycatcher in ecoprovinces of British Columbia.

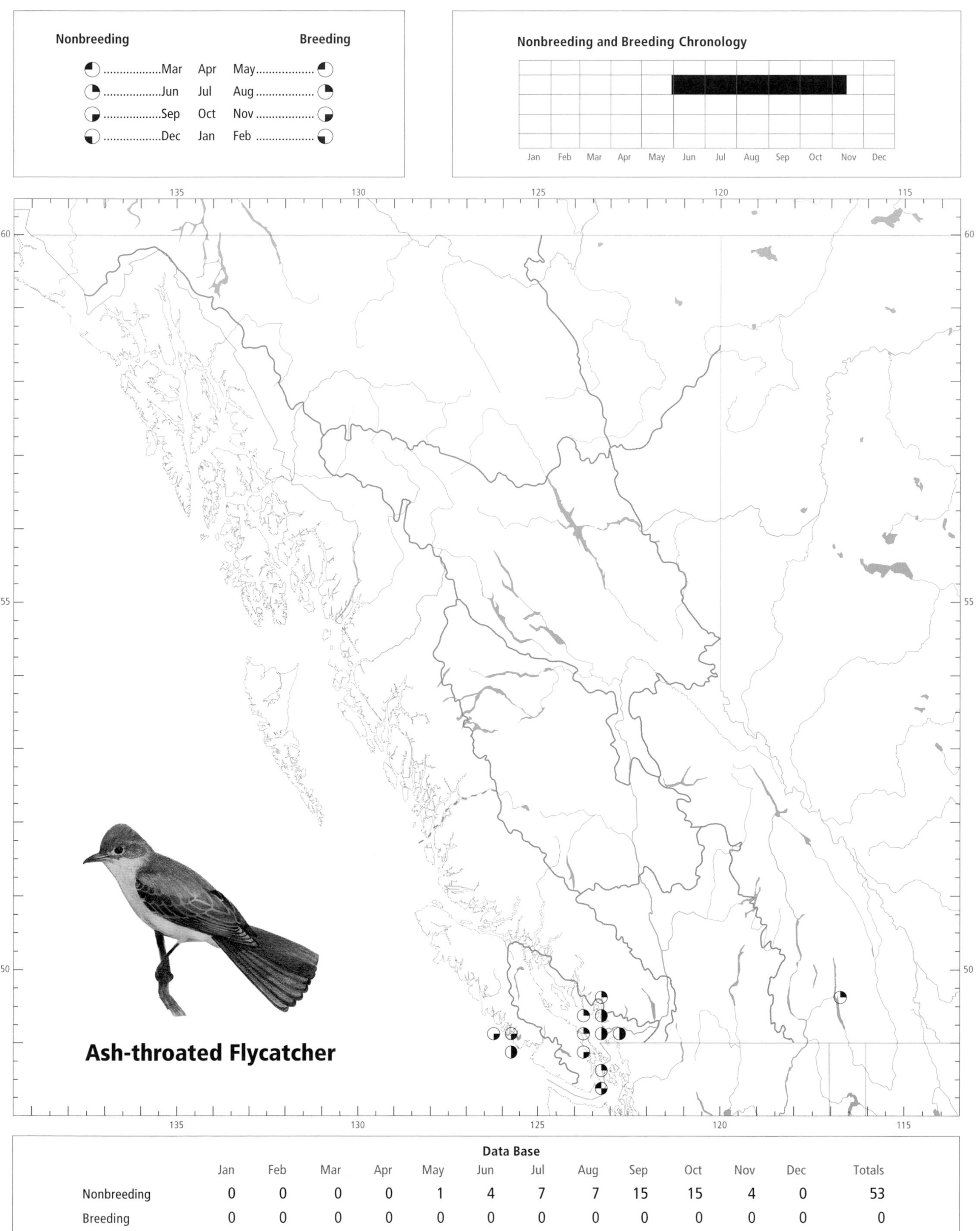

Data Base

	Jan	Feb	Mar	Apr	May	Jun	Jul	Aug	Sep	Oct	Nov	Dec	Totals
Nonbreeding	0	0	0	0	1	4	7	7	15	15	4	0	53
Breeding	0	0	0	0	0	0	0	0	0	0	0	0	0

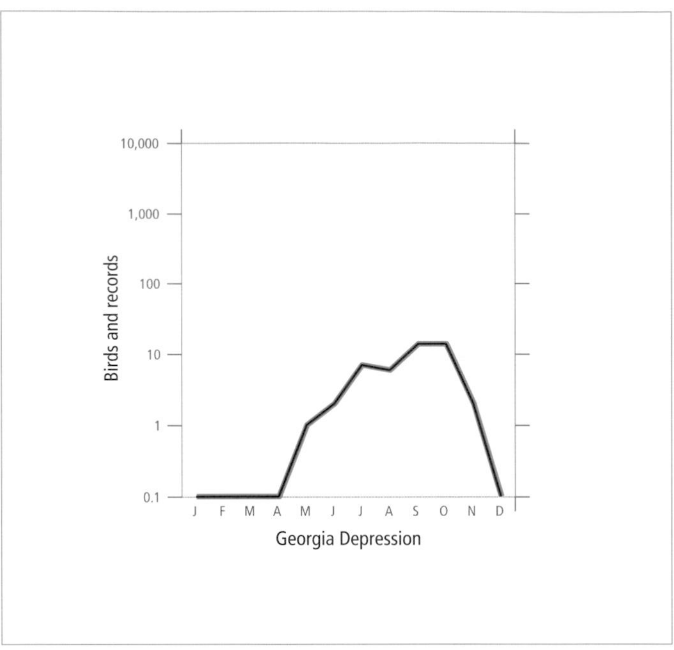

Figure 87. Fluctuation in total number of birds and records (purple and green lines) for the Ash-throated Flycatcher in the Georgia Depression Ecoprovince of British Columbia.

from the west Kootenay is the only acceptable record for the interior.

The Ash-throated Flycatcher has been recorded in British Columbia from near sea level to 450 m elevation (interior). It has occurred in suburban and rural parklands with open areas (Fig. 88), and in brushy patches (Fig. 89), riparian habitats, mouths of creeks with shrubs, and open deciduous woodlands. It avoids thick forests.

There is no regular movement of the Ash-throated Flycatcher into British Columbia (Fig. 86). As Roberson (1980) suggests, our records are "spring overshoots and early fall vagrants." The majority of records are from September and October (Fig. 87). A small number of individuals evidently disperse northward along the coast from breeding populations in southern Oregon, with a few reaching southwestern British Columbia.

All records have been of single birds seen for only 1 or 2 days. The longest documented stay was of an individual sighted at Blackie Spit (Fig. 89) over a period of 52 days, between 18 August and 8 October 1991.

On the coast the Ash-throated Flycatcher has been recorded, in most weeks, from 29 May to 14 November (Fig. 86).

REMARKS: Since the *Myiarchus* flycatchers are difficult to identify (Phillips and Lanyon 1970; Phillips et al. 1966), sightings must be well documented with specimens, photographs, or comprehensive field notes. We have excluded 5 interior reports that lacked sufficient documentation: Vaseux Lake and Peachland (Cannings et al. 1987), Paul Lake near Kamloops (Leckie 1970), Revelstoke, and Kleena Kleene.

Figure 88. On the southwestern mainland coast of British Columbia, the Ash-throated Flycatcher has been found in suburban parklands with open spaces (Jericho Park, Vancouver, 7 August 1996; R. Wayne Campbell).

Figure 89. In British Columbia, the Ash-throated Flycatcher frequents brushy patches rather than dense forests (Blackie Spit, Delta, 7 August 1996; R. Wayne Campbell).

NOTEWORTHY RECORDS

Spring: Coastal – Mount Tolmie (Victoria) 29 May 1990-1 (Taylor 1990). **Interior** – No records.

Summer: Coastal – Ucluelet 23 Aug 1987-1; Blackie Spit 18 to 31 Aug 1991-1 stayed into autumn (Siddle 1992a); Pitt Meadows 1 Jul 1983-1; Maple Ridge 28 Aug 1976-1 (Shepard 1976e); Jericho Park (Vancouver) 15 Jun 1980-1, 24 Aug 1958-1 (Erskine 1960), 29 to 30 Aug 1983-1; Kitsilano Point (Vancouver) 26 Aug 1987-1; Maplewood (North Vancouver) 1 Jul 1991-1, 27 Jul 1991-1 photographed; Nanaimo 23 Jun 1991-1 (Gillespie 1992); Wilson Creek (Sechelt) 27 Jul 1991-1 (Greenfield 1991); Alice Lake Park 27 Jun 1971-1. **Interior** – Ainsworth 15 Jun 1944-1.

Breeding Bird Surveys: Coastal – Not recorded. **Interior** – Not recorded.

Autumn: Interior – No records. **Coastal** – Langley 3 Oct 1981-1 photographed, 12 Oct 1981-1 photographed, 4 Oct 1982-1; Ambleside Park (West Vancouver) 9 to 11 Sep 1971-1 (Campbell et al. 1972b; BC Photo 176), 12 Sep 1982-1 (Campbell 1982d); Stanley Park (Vancouver) 13 Sep 1980-1 (Hunn and Mattocks 1981); University of British Columbia (Vancouver) 2 Sep 1975-1 (Shepard 1976a); Vancouver 13 Sep 1971-1 (Campbell et al. 1972b); Marpole (Vancouver) 7 Oct 1953-1 adult male in delayed moult (UBC 10547), 11 Oct 1953-1 banded (Hughes 1954); Sea Island 3 to 18 Oct 1980-1 (Hunn and Mattocks 1981); Alaksen National Wildlife Area (Reifel Island) 9 and 10 Nov 1992-1 (Siddle 1993a); Delta 2 Oct 1984-1 (Hunn and Mattocks 1985); Blackie Spit 1 Sep to 8 Oct 1991-1, first seen 18 Aug (Siddle 1992a), 11 Sep 1981-1 photographed, 12 Sep 1982-1 photographed (Campbell 1982d); Beach Grove 2 Nov 1985-1 (Campbell 1986a); Copper Canyon 16 Oct 1983-1; Vargas Island 14 Nov 1971-1 (Hatler et al. 1978); Sandhill Creek (Long Beach) 7 Sep 1983-1; Ucluelet 14 Nov 1972-1 (Hatler et al. 1978); Saanich 23 Sep 1984-1 (Hunn and Mattocks 1985).

Winter: No records.

Tropical Kingbird

TRKI

Tyrannus melancholicus Vieillot

RANGE: Breeds locally in southern Arizona and from northern Mexico south to central Argentina. Winters from Sonora and northeastern Mexico south through its South American breeding range.

STATUS: *Very rare* autumn vagrant on the extreme south coast, including southern Vancouver Island; *casual* in winter.

CHANGE IN STATUS: Munro and Cowan (1947) report the Tropical Kingbird in British Columbia from a single specimen collected near Sooke, west of Victoria, in February 1923 (Kermode 1928). In 1972, 49 years later, a second bird was recorded, again on southern Vancouver Island. Since then the Tropical Kingbird has been recorded, only in autumn, in 10 of the 20 years through 1992. This apparent increase in occurrence may reflect an expanding population or simply increased observational effort and an increased ability of observers to identify this species. Although it has been reported with some regularity over the past decade, it remains an infrequent visitor in the province.

OCCURRENCE: The Tropical Kingbird has been found only in coastal areas of southeastern Vancouver Island, on southwestern Vancouver Island between Ucluelet and Tofino, and on the Fraser River delta.

All records but one (Kermode 1928) are in autumn between late September and late November, with most (68%) occurring in October. A similar pattern has been documented in Oregon (Schmidt 1989) and Washington (Roberson 1980). British Columbia occurrences involve at least 17 individual birds. Birds arriving in the province are probably overshoots from a much larger autumn movement in California that occurs, with some regularity, every year (Roberson 1980). Most of those birds may be immatures that wander northward using the immediate coast as a corridor.

The Tropical Kingbird has been recorded in British Columbia from 26 September to 23 November except for its first appearance on 23 February 1923.

All records, listed in chronological order, are as follows:

(1) French Beach (Sooke) 23 February 1923-1 (RBCM 4776; Kermode 1928; Slipp 1942).
(2) Cadboro Bay 17 to 22 October 1972-1 seen and photographed at Gyro Park (Crowell and Nehls 1974a; Tatum 1973; BC Photo 240).
(3) Tofino 11 October 1976-1 perched on television antenna (Mattocks and Hunn 1978a).
(4) Sea Island 20 to 23 October 1976-1 in bushes along Ferguson Road (Shepard 1977; Mattocks and Hunn 1978a).
(5) Colwood 26 to 30 October 1977-1 on telephone wires (Mattocks and Hunn 1978a).
(6) Wier Beach (Metchosin) 22 and 23 October 1978-1 hawking flies from various perches, including telephone wires, in area along Sandgate Road (Hunn and Mattocks 1979b).
(7) Colwood 14 to 23 November 1982-1 caught numerous insects on the wing and sang several times. It frequented Lagoon Road, Fairview Road, Millburn Drive, Metchosin Road, and Wickheim Road, and was photographed (Hunn and Mattocks 1983).
(8) Ucluelet 22, 26, and 30 October 1985-1 frequenting parking lot (Hunn and Mattocks 1986; BC Photo 1043); Tofino 22, 28, and 30 October 1985-1 on telephone wires (Campbell 1986a; BC Photo 1043); Long Beach (Vancouver Island) 2 and 4 November 1985-1 hunting from piles of kelp on beach (Campbell 1986a). These all appear to be the same bird and are listed under 1 record even though they involve 3 locations.
(9) Nanaimo 24 October 1986-1.
(10) Tofino 5 November 1987-1 on telephone wires near Maquinna Hotel (Campbell 1988a).
(11) Tofino 31 October and 1 November 1989-1 (Weber and Cannings 1990).
(12) Jordan River 2 November 1990-1.
(13) Tofino 24 to 28 October and 1, 6, and 8 November 1990-1 hawking from wires and trees in 4 square block area.
(14) Roberts Bank (Delta) 28 October and 1 November 1990-1.
(15) Iona Island (Richmond) 26 September 1992 (Siddle 1993a).
(16) Tofino 10, 16, 20, and 21 October 1992-1.
(17) Central Saanich 27 October and 4 to 8 November 1992-1 (Siddle 1993a).

REMARKS: The American Ornithologists' Union (1957) lists 3 subspecies of the Tropical Kingbird: *T. m. couchii, T. m. occidentalis,* and *T. m. chloronotus*. Recently, Traylor (1979) split the 2 sibling species into the Couch's Kingbird (*Tyrannus couchii*) and the Tropical Kingbird (*T. melancholicus*). Kermode (1928) listed the first British Columbia record as a "Lichtenstein Kingbird" (*T. m. satrapa*), a name that is no longer valid (van Rossem 1929; Slipp 1942). That specimen should be referred to the race *occidentalis*.

One of the authors (JMC) saw a "Western" type kingbird, which lacked white outer tail feathers, along the Upper Muskwa River on 23 August 1979 (Cooper and Adams 1979). Close observation could not be made, however, and this report must remain hypothetical.

Rea (1969), Collins (1974), Dunn (1979), and Kaufman (1993) offer advice for separating the Tropical Kingbird from similar species, such as the Western Kingbird (*T. verticalis*) and Cassin's Kingbird (*T. vociferans*). Kaufman (1993) is particularly well illustrated.

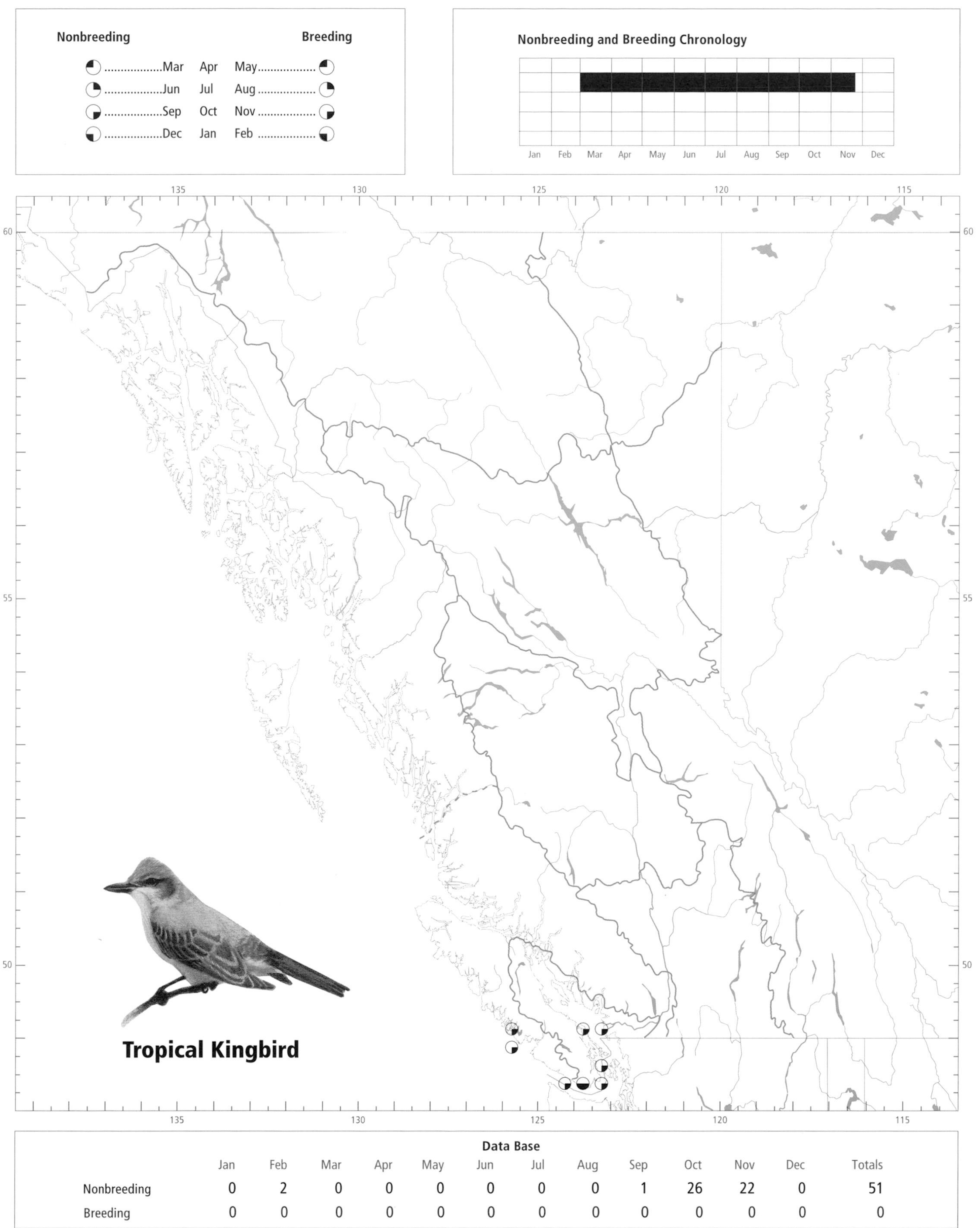

Data Base

	Jan	Feb	Mar	Apr	May	Jun	Jul	Aug	Sep	Oct	Nov	Dec	Totals
Nonbreeding	0	2	0	0	0	0	0	0	1	26	22	0	51
Breeding	0	0	0	0	0	0	0	0	0	0	0	0	0

Western Kingbird
Tyrannus verticalis Say

WEKI

RANGE: Breeds from the southern interior of British Columbia to southern Manitoba, south throughout the western half of the United States to northern Baja California, southern Texas, and northwestern Mexico; occasionally east to southern Ontario, Arkansas, and Louisiana. Winters mainly from central coastal Mexico south to Costa Rica, with small numbers in the states bordering the Gulf of Mexico.

STATUS: On the coast, a *rare* migrant and summer visitant to the upper Fraser River delta of the Georgia Depression Ecoprovince; *uncommon* spring transient and *very rare* autumn transient elsewhere in the ecoprovince, including southeastern Vancouver Island; *accidental* in winter. *Very rare* migrant and summer visitant to the southern Coast and Mountains Ecoprovince; *very rare* transient to Western Vancouver Island; *accidental* in the northern Coast and Mountains.

In the interior, a *fairly common* migrant and summer visitant to the Southern Interior Ecoprovince; *uncommon* to locally *fairly common* in the Southern Interior Mountains Ecoprovince; *uncommon* in the Central Interior Ecoprovince; *rare* in the southern Sub-Boreal Interior Ecoprovince; *accidental* in the Boreal Plains and Northern Boreal Mountains ecoprovinces.

Breeds.

CHANGE IN STATUS: In North America, the breeding range of the Western Kingbird (Fig. 90), formerly known as the Arkansas Kingbird, has expanded since 1900 (Taverner 1927; Bent 1942). Planted trees, buildings, and networks of utility poles probably allowed its spread across the plains (Nice 1924). The increase in abundance and distribution in British Columbia since the mid-1940s (Munro and Cowan 1947) is best documented in the Okanagan basin. There the Western Kingbird found suitable habitat in the residential and orchard areas that replaced much of the original sagebrush and antelope-brush. It also followed highway corridors and farm clearings into higher elevations that were once dense forests (Cannings et al. 1987).

Figure 90. Adult Western Kingbird near Oliver (June 1981; Mark Nyhof).

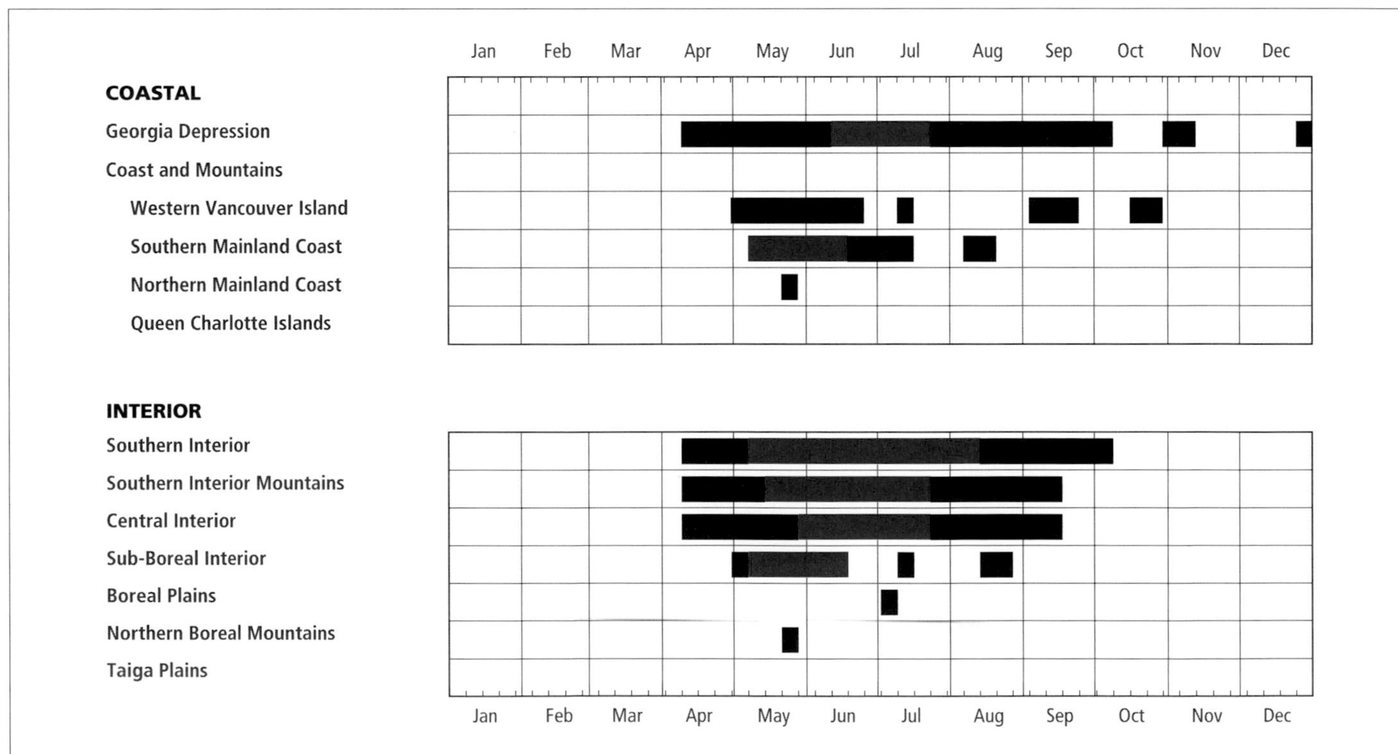

Figure 91. Annual occurrence (black) and breeding chronology (red) for the Western Kingbird in ecoprovinces of British Columbia.

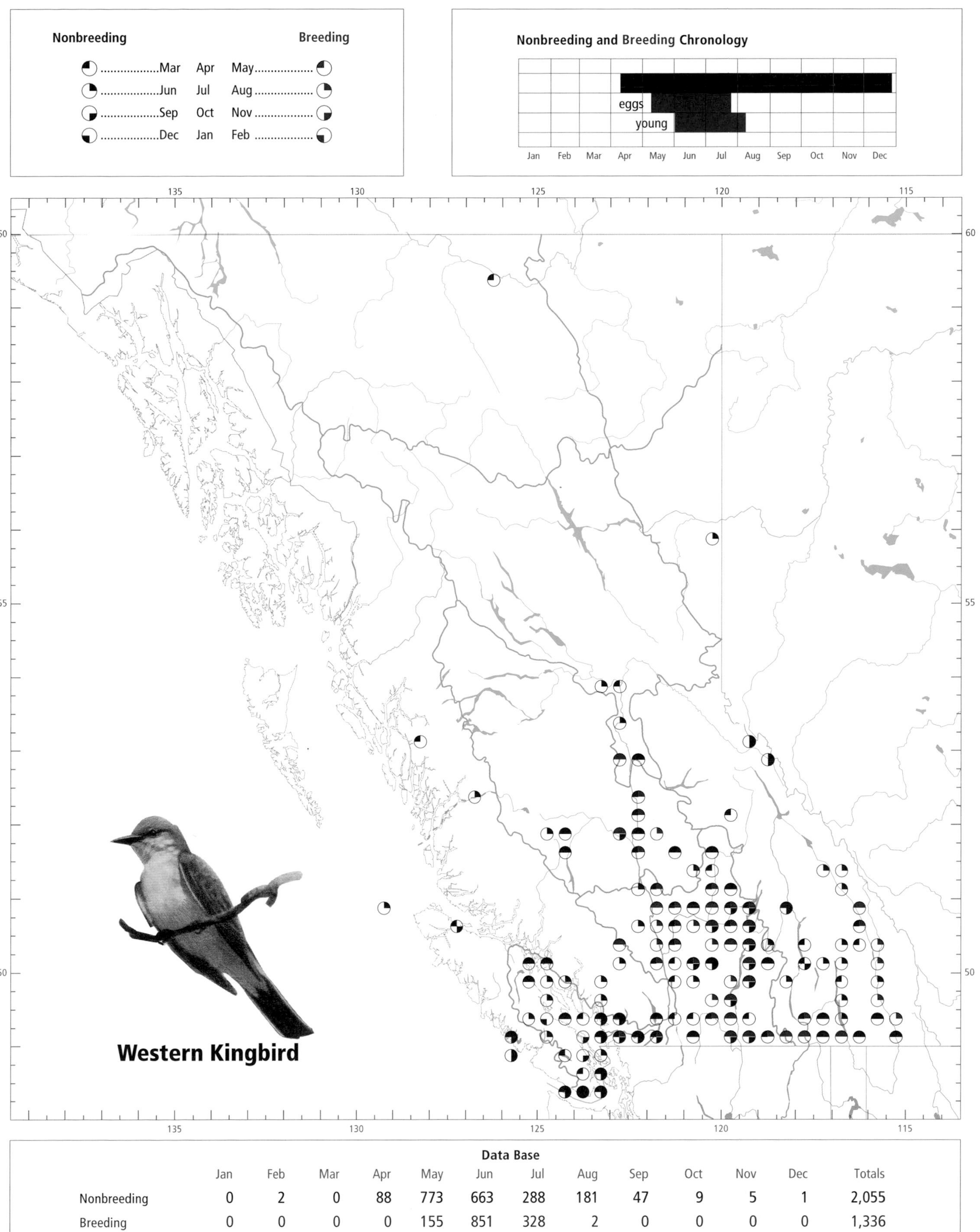

Data Base

	Jan	Feb	Mar	Apr	May	Jun	Jul	Aug	Sep	Oct	Nov	Dec	Totals
Nonbreeding	0	2	0	88	773	663	288	181	47	9	5	1	2,055
Breeding	0	0	0	0	155	851	328	2	0	0	0	0	1,336

By the 1950s, the Western Kingbird was slowly expanding its range eastward across southern British Columbia (Munro 1950; Butler et al. 1986) into the east Kootenay, where it was considered a "scarce summer visitant" (Johnstone 1949). Over the next several decades it expanded its range northward along clearings and corridors into the Columbia River valley as far north as Golden. It now occurs there regularly but locally, in agricultural areas and around settlements.

Although some northward range expansion has taken place into the Central Interior and southern Sub-Boreal Interior, the species still remains an uncommon to rare summer visitant there (Munro 1945a, 1946, 1947a; Erskine and Stein 1964; Roberts and Gebauer 1992).

On the south coast, the Western Kingbird has increased its status over the past 25 years from being of irregular occurrence to one that occurs annually but in small numbers (Campbell 1969b, 1972; Weber et al. 1990).

NONBREEDING: The Western Kingbird (Fig. 90) is widely distributed across southern British Columbia, including Vancouver Island. It is fairly widespread in open valley bottoms in the Okanagan, Thompson, and Kootenay regions, but is scarcer in northern portions of the west Kootenay. It is more sparsely distributed in the Cariboo and Chilcotin areas, reaching its usual northern limit south of Prince George. On the coast, transients occur infrequently along the west and east coasts of Vancouver Island, and on the southwestern mainland from the lower Fraser River valley along the Sunshine Coast north to Powell River. The species now occurs regularly in the eastern lower Fraser River valley, near Chilliwack and Agassiz.

The Western Kingbird frequents relatively low elevations, from near sea level to 200 m on the coast and up to 1,300 m in the interior. In the interior it occurs in open, hot, and dry habitats, including grasslands, sagebrush flats (see Fig. 95), open rangelands, and farmlands within the Ponderosa Pine and Interior Douglas-fir zones. It is most abundant along forest edges in the Ponderosa Pine and Bunchgrass biogeoclimatic zones. In grassland areas, it requires a few large trees or human-made structures for perch sites. It is particularly numerous along highway and road corridors, where it typically perches on fences, utility poles, or telephone lines, often in the vicinity of farm buildings (Fig. 91). It also inhabits treed suburbs and villages. On the coast, it frequents the edges of pastures, airports, farmlands, sewage lagoons, beaches, parks, and other open areas.

Figure 92. Throughout British Columbia, the Western Kingbird forages and breeds in farmland habitats (Castlegar, 13 June 1993; R. Wayne Campbell).

Spring arrival of the Western Kingbird is 2 to 3 weeks earlier than that of the Eastern Kingbird. Migration patterns are similar for both the interior and the coast. The first migrants may arrive as early as the second week of April (Figs. 91 and 93), but birds do not normally arrive in numbers until the end of April or the first week of May. Over 55 years, the mean and median date of first arrival in the Okanagan valley has been 29 April (Cannings et al. 1987). Peak numbers are reached by the second or third week of May.

The Western Kingbird is an early autumn migrant, with the southward movement beginning in early August and reaching a peak from mid to late August. Most birds have left the province by late August, with only a few stragglers remaining into September or later (Figs. 91 and 93). Most migrants travel as individuals or in family groups. There is 1 winter record (see Christmas Bird Counts and REMARKS).

In the interior, the Western Kingbird has been recorded from 9 April to 4 October; on the coast, it has been recorded regularly from 15 April to 27 December (Fig. 91).

BREEDING: The Western Kingbird breeds mainly in the southern interior of the province from Keremeos to Wardner, north to Spillimacheen in the east Kootenay and Castlegar in the west Kootenay, and through the Southern Interior, Central Interior, and southern Sub-Boreal Interior to Quesnel in the north and the vicinity of Riske Creek in the west. Local breeding populations occur on the coast in the Douglas-fir forests of the Pemberton valley, and in farmlands and sloughs of the upper Fraser River valley.

The Western Kingbird reaches its highest numbers in summer in the ponderosa pine forests of the Southern Interior, from Osoyoos north to Kamloops (Fig. 94). An analysis of Breeding Bird Surveys for the period 1968 to 1993 could not detect a net change in numbers on interior routes; surveys on coastal routes for the same period contained insufficient data for analysis. Robbins et al. (1986) note that British Columbia populations showed the only persistent decrease between 1965 and 1979, although that appears to have since changed. From the western region of the Breeding Bird Surveys, population trends of the Western Kingbird show an increase in numbers between 1966 and 1988 at an average annual rate of 1.6% (Sauer and Droege 1992).

Few Western Kingbirds breed above 1,200 m elevation. In the Southern Interior, the open coniferous forests of the valley bottoms and adjacent benchlands are dominated by ponderosa pine. The dominant ground cover is bluebunch wheatgrass. In that environment, 50% of the Western Kingbird nests were in woodlands that had been altered by

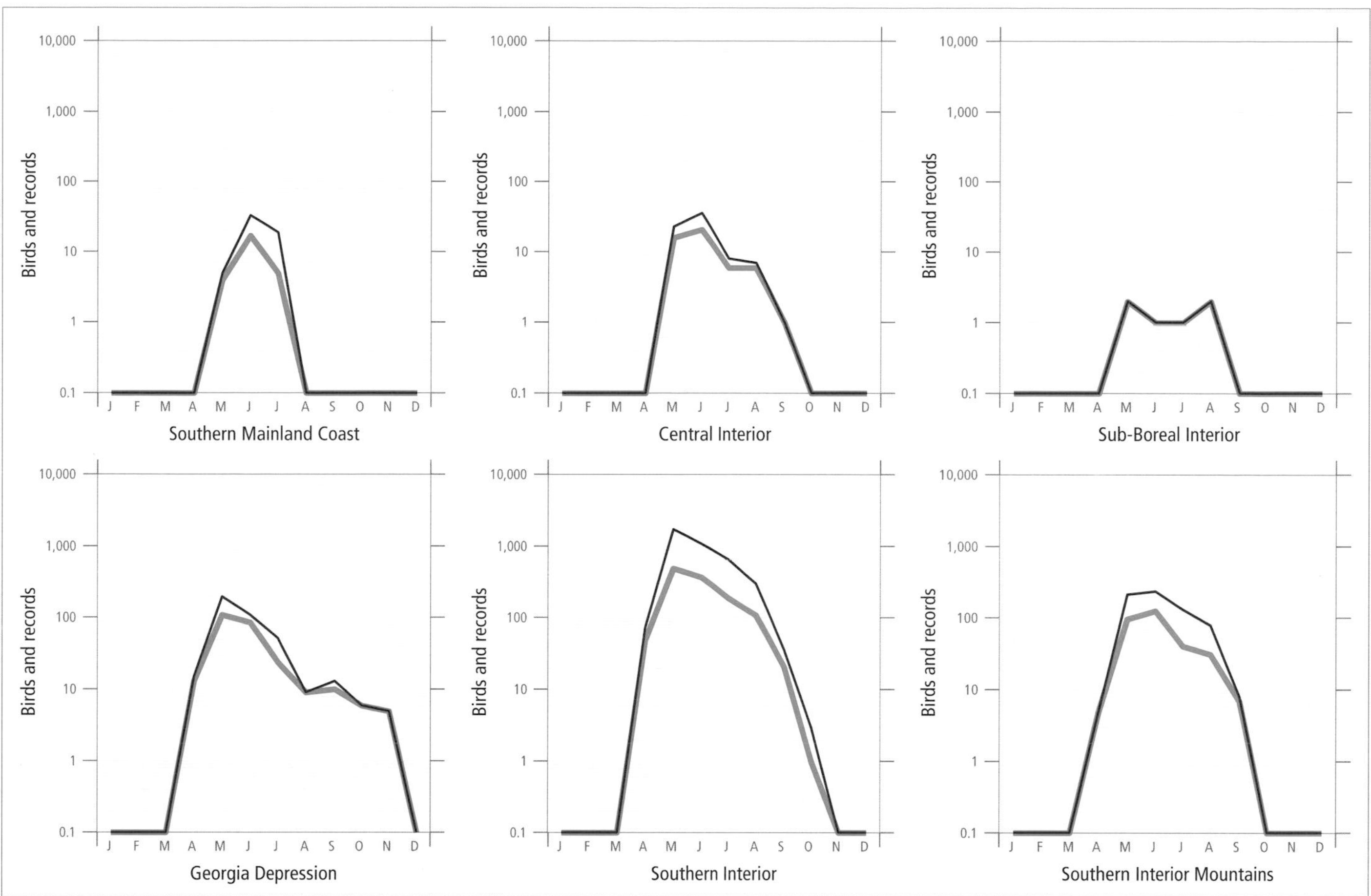

Figure 93. Fluctuations in total number of birds (purple line) and total number of records (green line) for the Western Kingbird in ecoprovinces of British Columbia. Christmas Bird Count and nest record data have been excluded.

humans, while another 30% were in open forest situations (n = 922).

The adaptability of the Western Kingbird is revealed by its use of 40 habitat classes within the basic habitat types. The most frequently used classes were cultivated farmland (21%; n = 778; Fig. 92), orchards or vineyards (14%), big sage shrubland (9%; Fig. 95), arid rangeland (9%), and open ponderosa pine stands (8%), followed by rural and suburban sites (8%; Fig. 96). In British Columbia, most nest sites were found in association with roadsides (31%; n = 407), gardens (13%), and open hillsides (10%).

In the interior, the Western Kingbird has been recorded breeding from 9 May (calculated) to 11 August; on the coast, it has been recorded from 11 May (calculated) to 11 July (Fig. 91).

Nests: Most nests (84%; n = 1,138) were situated on power poles (Fig. 96). Nests were built behind transformers; on crossbars between brackets (Fig. 97), wires, and insulators; or on other fixtures on the pole itself. Other sites included a wide variety of living and dead coniferous and deciduous trees and bushes; ledges in buildings, including houses, barns, sheds, garages, and cabins, as well as bridges; and cliff faces. Munro (1919, 1927) and Green (1928), referring to the Okanagan valley in British Columbia, state that the Western Kingbird used abandoned Northern Flicker nest holes (also Pinkowski 1982) and American Robin nests, the decayed top

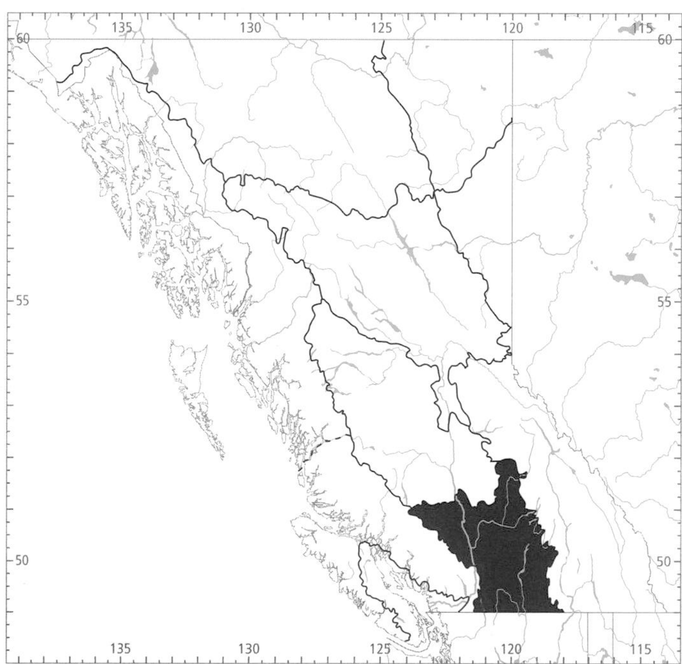

Figure 94. In British Columbia, the highest numbers for the Western Kingbird in summer occur in the Southern Interior Ecoprovince.

of fence posts and trembling aspens, and the eaves troughs of houses. The heights of 1,075 nests ranged from ground level to 45 m, with 73% recorded between 6 and 9 m. Nests were mainly bulky, untidy cups of grasses (93% of nests; Fig. 97), string (15%), twigs (13%), plant fibres, and remains of rootlets and forbs, along with at least 24 other natural and human-made materials.

Eggs: Dates for 659 clutches ranged from 9 May to 25 July, with 52% recorded between 4 and 26 June. Sizes of 107 clutches ranged from 1 to 5 eggs (1E-8, 2E-8, 3E-28, 4E-57, 5E-6), with 53% having 4 eggs (Fig. 97). Bent (1942) reports occasional clutches of 6 or 7 eggs in the United States. The incubation period in British Columbia is 18 to 20 days.

Young: Dates for 443 broods ranged from 1 June to 11 August, with 55% recorded between 25 June and 6 July. Sizes of 157 broods ranged from 1 to 5 young (1Y-8, 2Y-31, 3Y-63, 4Y-53, 5Y-2), with 74% having 3 or 4 young. The nestling period in British Columbia is 16 to 19 days. There is evidence of double-clutching. In the Okanagan valley, 1 brood fledged on 19 June and a second brood, in the same tree but in a different nest, fledged on 11 August (Cannings et al. 1987).

Brown-headed Cowbird Parasitism: Cowbird parasitism was not found in British Columbia in 259 nests recorded with eggs or young. Friedmann and Kiff (1985) and Smith (1972) note that the Western Kingbird is rarely recorded as a host. British Columbia data further support the tentative designation of the Western Kingbird as a rejector species (Rothstein 1975). As Friedmann et al. (1977) mention, the "paucity of observed cases of natural parasitism agrees well with this rejector behaviour of the Western Kingbird."

Nest Success: Of 29 nests found with eggs and followed to a known fate, 21 produced at least 1 fledgling, for a nest success rate of 71%.

REMARKS: Anderson (1914) reported an early brood of 3 young Western Kingbirds in the Okanagan valley on 3 May 1913. It is likely this nest was a misidentified Say's Phoebe (Cannings et al. 1987).

The Western Kingbird and Eastern Kingbird occur sympatrically over much of the southern interior of British Columbia. However, the Eastern Kingbird tends to be found in thicker, wetter riparian forest and at higher elevations, while the Western Kingbird favours lower, drier, more open areas with larger and taller trees. In Manitoba, the Western Kingbird usually nests in sites having fewer but larger trees than the sites chosen by the Eastern Kingbird, and its nests

Figure 95. In parts of British Columbia, the Western Kingbird breeds in big sage shrubland habitats (12 km south of Ashcroft, 10 June 1990; R. Wayne Campbell).

Figure 96. Power poles in suburban and rural areas provide a suitable nesting site for the Western Kingbird in British Columbia (Radium Hot Springs, 7 July 1991; R. Wayne Campbell).

are placed higher than those of the latter species (MacKenzie and Sealy 1981). The 2 species will nest close to each other in gardens, along road corridors (e.g., Richter Pass; Campbell and Meugens 1971), and around farmhouses where a mixture of favourable habitats occur.

Since the Western Kingbird also breeds in tall bushes and trees adjacent to transportation corridors, it is frequently a victim of highway mortality (Barkley 1966; Campbell 1984f).

In late autumn, especially October and November, Western Kingbirds should be carefully scrutinized because of the possibility of confusion with the Tropical Kingbird (*q.v.*), which occurs as a vagrant from the south at that time. We have not included the record of a single bird seen during the following Christmas Bird Count because convincing details were lacking: Victoria 21 December 1963-1 (Anderson 1976b).

For additional life-history and breeding habitat information, see Bent (1942), Blancher and Robertson (1984), and MacKenzie and Sealy (1981).

Figure 97. The Western Kingbird frequently builds its untidy nest between wooden crossbars of power poles in British Columbia (near Merritt, 25 June 1993; R. Wayne Campbell).

NOTEWORTHY RECORDS

Spring: Coastal – Jordan River 16 May 1959-1 perched on driftwood (Stirling 1960a); Saanich 15 Apr 1984-1, 13 May 1973-1 (Crowell and Nehls 1973c); Sumas 25 Apr 1905-1 (FMNH 140266); Blackie Spit 4 to 7 May 1990-1; Westham Island 24 May 1973-1 (BC Photo 469); Chilliwack 22 May 1986-30 counted in area (Mattocks 1986); Tofino 2 May 1985-1; Agassiz 2 May 1976-2; University of British Columbia (Vancouver) 21 May 1970-1 (Campbell 1970c); Jericho Park (Vancouver) 27 Apr 1985-1; Hope 19 May 1986-1; Wilson Creek (Sechelt) 19 May 1987-1; Qualicum Beach 25 Apr 1977-1 (Dawe 1980); Miracle Beach Park 26 May 1982-1 (BC Photo 797); Oyster Bay (Campbell River) 6 May 1973-1; Mansons Landing (Cortes Island) 15 May 1975-1; Port Hardy 22 May 1936-1 collected; Khutze Inlet 24 May 1936-1 (MCZ 282913). **Interior** – Osoyoos 1 May 1966-1; Creston 10 May 1928-1 (Mailliard 1932), 17 May 1983-adult incubating; Trail 18 May 1975-adult on nest; Oliver 17 May 1964-40 counted in area; Kinnaird 14 Apr 1971-1; Princeton 14 Apr 1975-4; Penticton 10 May 1969-eggs; Kalamalka Lake 9 Apr 1962-2; Nakusp 4 May 1977-1; Vernon 19 May 1980-adult incubating; Kamloops 9 Apr 1988-1, 27 Apr 1926-1 (UBC 5540); Radium 15 May 1980-4 on powerline (Halverson 1981); Lytton 8 May 1966-2; Murtle River 26 May 1951-1 collected (Edwards and Ritcey 1967); Clearwater 3 May 1935-2 (NMC 282910-11); Alkali Lake 25 Apr to 18 May 1946-1 (Munro 1955b); Williams Lake 13 Apr 1980-1; Quesnel 3 May 1985-1; Mile 496 Alaska Highway 21 to 23 May 1981-1.

Summer: Coastal – Albert Head (Victoria) 1 Jun 1936-1 (RBCM 5317); Central Saanich 10 Jul 1974-1 (Crowell and Nehls 1974d); Colquitz 5 to 12 Jun 1957-1; South Pender Island 4 Jun 1957-1; Saturna Island 4 Jun 1991-1 (Campbell 1992a); Chilliwack 28 Jun 1929-1 (RBCM 6435); Iona Island 15 Aug 1970-1 (Crowell and Nehls 1970d); Ucluelet 19 Jun 1909-1 (NMC 4167; Rand 1943); Tofino 3 Jun 1990-1; McKenzie Beach (Long Beach) 14 Jul 1983-1 photographed; Cheam Slough 26 Aug 1976-1 at nest; Agassiz 11 Jul 1985-2 nestlings; Port Alberni 20 Jun 1974-1 (Crowell and Nehls 1974d); Qualicum Beach 14 Jun 1978-1 (Dawe 1980); Cranberry Lake (Powell River) 19 Aug 1979-1; Squamish 6 Jun 1975-1 feeding near dump; Brackendale 24 Jun 1916-1 (NMC 9440; Taverner 1917); Pemberton 14 Jun 1971-2 fledglings with parents; Mount Currie 1 Jun 1986-1; Anderson Lake 23 Jun 1968-1; Triangle Island 22 Jun 1975-1 (Vermeer et al. 1976); Bella Coola 15 Jun 1989-1 photographed (Campbell 1992a). **Interior** – Keremeos 9 Jul 1984-1 dead on highway (Campbell 1984f); Cranbrook 14 Jul 1937-young present (Johnstone 1949); Vernon 11 Aug 1980-3 fledged (Cannings et al. 1987); Skihist (Lytton) 1 Aug 1964-2; Windermere 22 Jul 1982-2 nestlings; Lillooet 26 Jul 1916-1 adult (NMC 9630); Spillimacheen 22 Jul 1960-2 nestlings, 6 Jul 1972-4 fledged; Chase to Kamloops 11 Jul 1963-25 counted on wires; Shuswap Park 23 Aug 1966-1 killed on road (Barkley 1966); Strathnaver 6 Jul 1966-1 (Grant 1966); Dog Creek 15 Aug 1953-1 male (NMC 47991); Kleena Kleene 12 Jun 1965-1; Williams Lake 1 Jun 1972-1 adult on eggs, 16 Jul 1977-3 young fledged, 27 Aug 1984-1; Soda Creek 10 Jul 1978-adult feeding nestling; Dale Lake (Quesnel) 10 Jul 1979-1; Beverly Lake 22 Aug 1959-1; Farmington 4 Jul 1975-1.

Breeding Bird Surveys: Coastal – Recorded from 4 of 27 routes and on 7% of all surveys. Maxima: Pemberton 2 Jul 1977-7; Seabird 2 Jul 1989-2; Chilliwack 10 Jun 1973-1; Saltery Bay 23 Jun 1974-1. **Interior** – Recorded from 34 of 73 routes and on 33% of all surveys. Maxima: Lavington 9 Jun 1979-20; Mabel Lake 6 Jul 1974-18; Oliver 28 Jun 1981-14.

Autumn: Interior – Riske Creek 15 Sep 1986-1; Brouse 10 Sep 1978-1; Celista 4 Oct 1964-1; e of Kamloops 1 Sep 1960-1; Enderby 11 Sep 1889-1 (NMC 557); Swan Lake (Vernon) to O'Keefe 9 Sep 1961-some seen (Cannings et al. 1987); Summerland 9 Sep 1978-1; Okanagan Centre 1 Sep 1931-1 (MVZ 102392); Keremeos 2 Sep 1974-1. **Coastal** – Pulteney Point 7 Sep 1976-1 flycatching around lightstation (BC Photo 495); Tofino 23 Oct 1978-1; Ucluelet 20 Oct 1973-1; Iona Island (Richmond) 3 Oct 1981-1; Crescent Beach 2 Sep 1971-2 (Campbell et al. 1972b); Blackie Spit 5 and 6 Sep 1971-1 (Campbell et al. 1972b); Duncan 2 Oct 1971-1 (Tatum 1972; Crowell and Nehls 1972a); Saanich 5 Oct 1984-1, 6 Nov 1983-1, 6 Nov 1984-1 (Hunn and Mattocks 1985); Victoria 1 and 11 Nov 1963-1.

Winter: Interior – No records. **Coastal** – see Christmas Bird Counts and REMARKS.

Christmas Bird Counts: Coastal – Recorded from 1 of 33 localities and on less than 1% of all counts. Maximum: Deep Bay 27 Dec 1992-**1**, all-time Canadian high count (Monroe 1993). See REMARKS.

Eastern Kingbird

EAKI

Tyrannus tyrannus (Linnaeus)

RANGE: Breeds from southern Mackenzie and northeastern and southern British Columbia east across southern and central Canada to Nova Scotia, south to northeastern California, central Texas, the Gulf coast, and southern Florida. Winters in South America, from Colombia to northern Chile and Argentina.

STATUS: On the coast, a *rare* to locally *uncommon* migrant and summer visitant to the Georgia Depression Ecoprovince; *very rare* in the southern Coast and Mountains Ecoprovince; a *very rare* vagrant on the west coast of Vancouver Island and the northern Coast and Mountains; *casual* on the Queen Charlotte Islands.

In the interior, a *fairly common* to *common* migrant and summer visitant in the Southern Interior and Southern Interior Mountains ecoprovinces; *fairly common* in the Cariboo and Chilcotin areas of the Central Interior Ecoprovince; *rare* in the Sub-Boreal Interior Ecoprovince. An *uncommon* to *fairly common* migrant and summer visitant to the Peace Lowland of the Boreal Plains Ecoprovince; *uncommon* in the Fort Nelson Lowland of the Taiga Plains Ecoprovince; *casual* in the Northern Boreal Mountains Ecoprovince.

Breeds.

NONBREEDING: The Eastern Kingbird (Fig. 98) is widely distributed throughout lower elevations in the southern interior of the province, from the eastern slope of the Cascade Mountains at Princeton and the coastal mountains at Lillooet east to the Rocky Mountain Trench and north to the Cariboo and Chilcotin areas. It is less numerous further north as far as Prince George and Smithers. A discrete population occupies the northeastern portion of the province, and this population may follow a separate migration route. On the coast, the distribution of the Eastern Kingbird is sparse; it is localized in the lower Fraser River valley and of intermittent and scattered occurrence on southern Vancouver Island. A number

Figure 98. Adult Eastern Kingbird (Castlegar, 17 June 1993; R. Wayne Campbell). This conspicuous bird is easily identified by its black upperparts, white underparts, and all-black tail with a white tip.

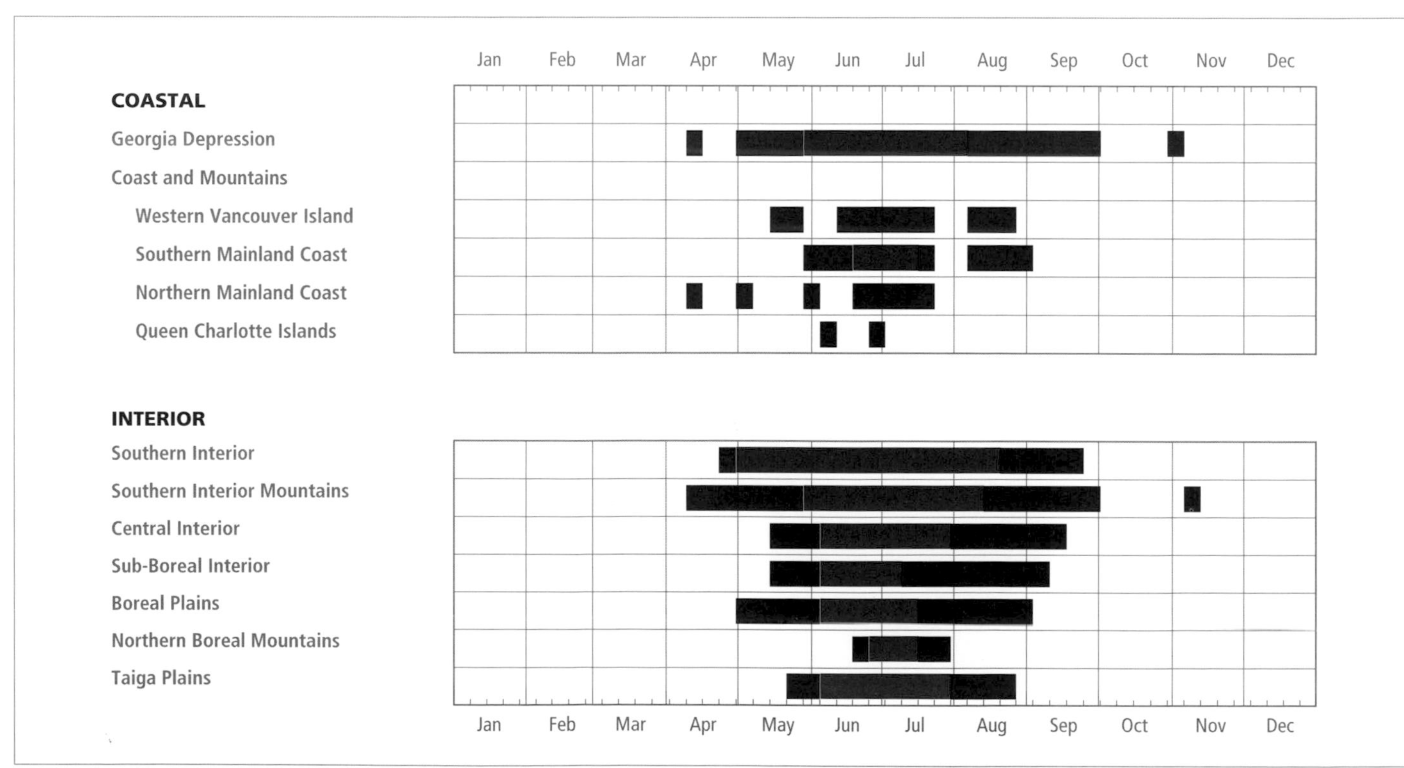

Figure 99. Annual occurrence (black) and breeding chronology (red) for the Eastern Kingbird in ecoprovinces of British Columbia.

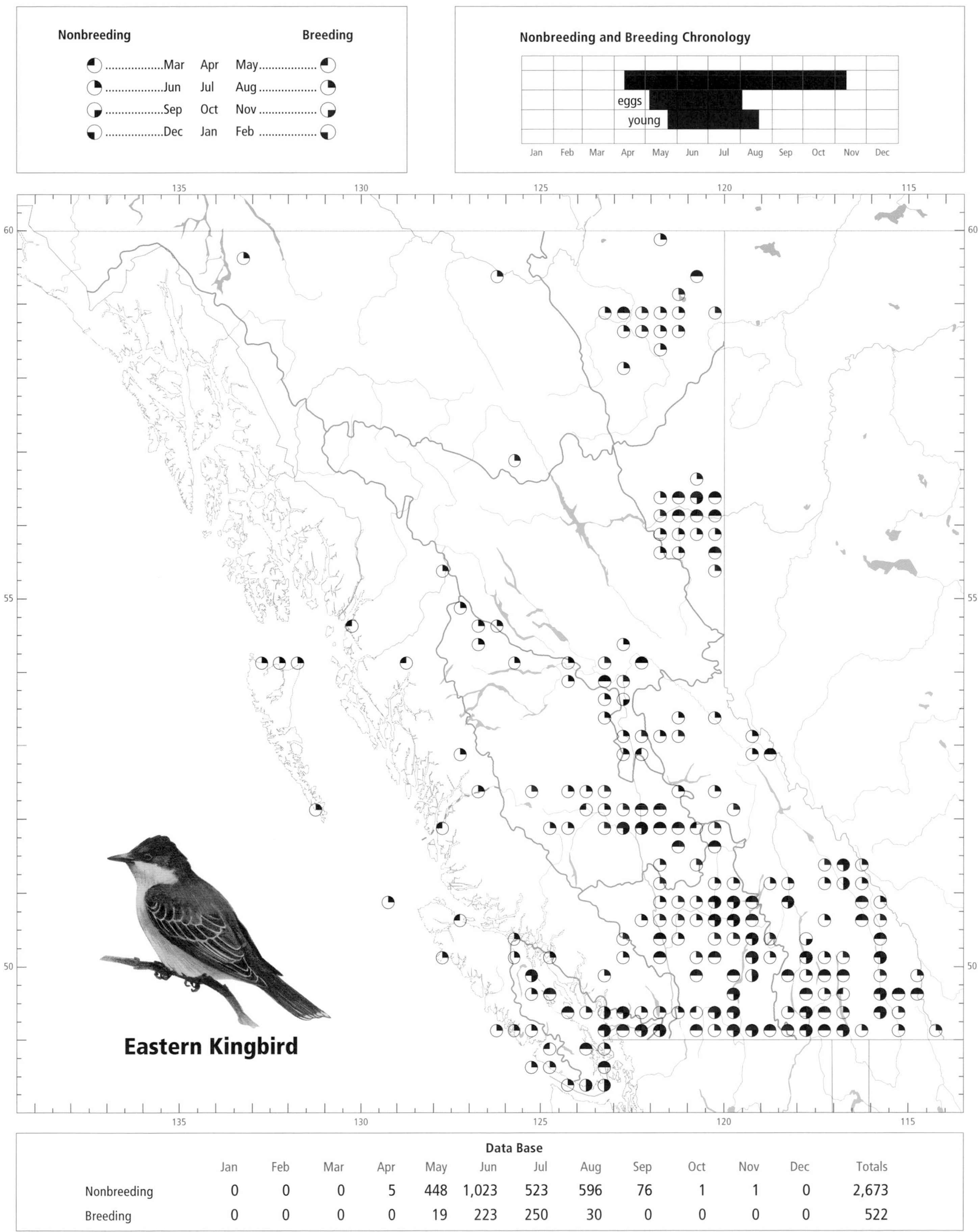

Data Base

	Jan	Feb	Mar	Apr	May	Jun	Jul	Aug	Sep	Oct	Nov	Dec	Totals
Nonbreeding	0	0	0	5	448	1,023	523	596	76	1	1	0	2,673
Breeding	0	0	0	0	19	223	250	30	0	0	0	0	522

of the few records along the coast, north of Vancouver Island, are at the heads of deep fiords, where river valleys originate in the interior.

The Eastern Kingbird occurs at elevations from near sea level to 1,300 m. Like the Western Kingbird, it is essentially a bird of valley bottoms and lower slopes and rarely occurs in mountainous terrain. It prefers dense riparian woodlands, particularly those with many dead or bare deciduous trees and shrubs (Figs. 100 and 103). It also uses similar edges of sloughs, marshes, ponds, wet meadows, lakes (Fig. 100), and sewage lagoons. Beaver ponds with flood-killed trees and shrubs are favoured in forested areas. It also frequents hedgerows along ditches and pastures; open, dry fields with tall grass and scattered trees; tall plants or posts that provide perches for foraging; orchards (especially in the Okanagan valley and west Kootenay); or open industrial areas such as airports. On the coast, it is associated with river sloughs, estuaries, logged areas, shrubs bordering fields (Butler and Campbell 1987), open headlands, islands, beaches, city parks, and golf courses. In the lower Fraser River valley, it frequents the backwater sloughs of rivers.

Spring migrants can arrive in the southern portions of the province as early as the second week of April, although birds do not usually arrive in numbers until the second or third week of May (Figs. 99 and 101). Numbers continue to build through May into the first week of June. In the northeast, early migrants can appear by the first week of May, but they do not usually arrive in numbers until the end of May and beginning of June. Stragglers arrive until mid-June in the south and late June in the northeast (Fig. 101).

Soon after breeding, the Eastern Kingbird loses its solitary, aggressive behaviour, and family groups form loose flocks as they prepare to move southward. This flocking behaviour is unique among the flycatchers in British Columbia. The Eastern Kingbird is most conspicuous in mid-August, when southbound migrants may merge into flocks of up to 300 birds.

A southward movement is noticeable by the second week of August in the northeast, and most birds have left the area by the end of August or the first week of September (Figs. 99 and 101). In the south, numbers begin to build by the last week of July (Southern Interior) or the first week of August (Southern Interior Mountains and Georgia Depression), and the movement peaks by the third week of that month. The main movement is over by the first week of September, with a few stragglers remaining through early September (Figs. 99 and 101); after mid-September, sightings are unusual. There are no winter records.

The Eastern Kingbird has been recorded regularly in the province from 1 May to 10 September on the coast and from 2 May to 11 September in the interior (Fig. 99).

BREEDING: The Eastern Kingbird breeds in suitable habitats throughout the interior, from the international boundary north through the Cariboo and Chilcotin areas to the Prince George region. There is an isolated breeding record from the Ingenika

Figure 100. Typical foraging and nesting habitat for the Eastern Kingbird in parts of the Peace River area includes stands of bare deciduous trees in lakes (Swan Lake, south of Dawson Creek, 14 June 1996; R. Wayne Campbell). Note territorial adult in upper left corner.

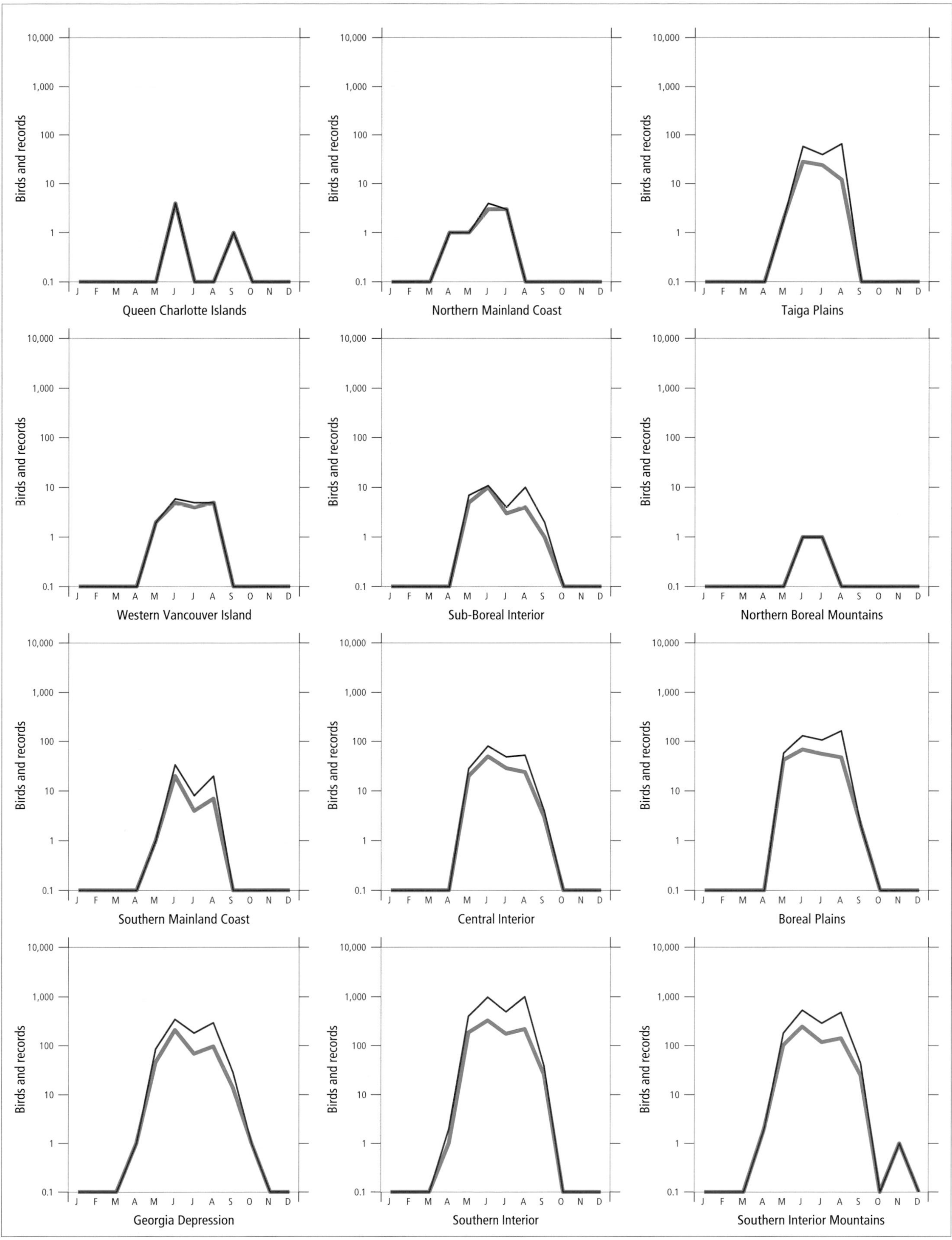

Figure 101. Fluctuations in total number of birds (purple line) and total number of records (green line) for the Eastern Kingbird in ecoprovinces of British Columbia. Nest record data have been excluded.

River valley in the Northern Boreal Mountains. The northeastern population nests in the Peace Lowland and locally in the Fort Nelson Lowland. On the coast, it breeds only in the lower Fraser River valley, northern Gulf Islands, and southeastern Vancouver Island, and at Pemberton in the southern mainland of the Coast and Mountains.

Its highest numbers in summer occur in the Southern Interior and Southern Interior Mountains (Fig. 102). An analysis of Breeding Bird Surveys for the period 1968 to 1993 could not detect a net change in numbers on interior routes; coastal surveys for the same period contained insufficient data for analysis.

The Eastern Kingbird breeds from near sea level on the coast to about 1,300 m in the interior. Breeding habitat is similar to nonbreeding habitat (Figs. 100 and 103) except at outer coastal sites. Almost 40% of the nests (n = 360) were recorded in human-influenced habitats that have large open areas. The typical natural breeding habitat is the forested edge of open areas, often riparian habitat along the shores of ponds and lakes (Fig. 100). This kingbird prefers sites with quantities of dead trees and shrubs at the water's edge (Fig. 103) or actually in the water (Mailliard 1931) (Fig. 104c).

The Eastern Kingbird is a versatile species; nest records (n = 305) include territories in at least 42 habitat classes. There are regional habitat preferences, however. In the semi-arid Southern Interior, 57% of 172 nests were in human-associated habitats, 37% were in various woodland and grassland classes, and only 6% were associated with wetlands. Throughout the rest of the province, wetlands were a preferred habitat and accounted for 48% of the nest sites (n = 133). Preference seemed to be given to sites where dead shrubs and trees stood along the wetland edges or extended over water (Fig. 103).

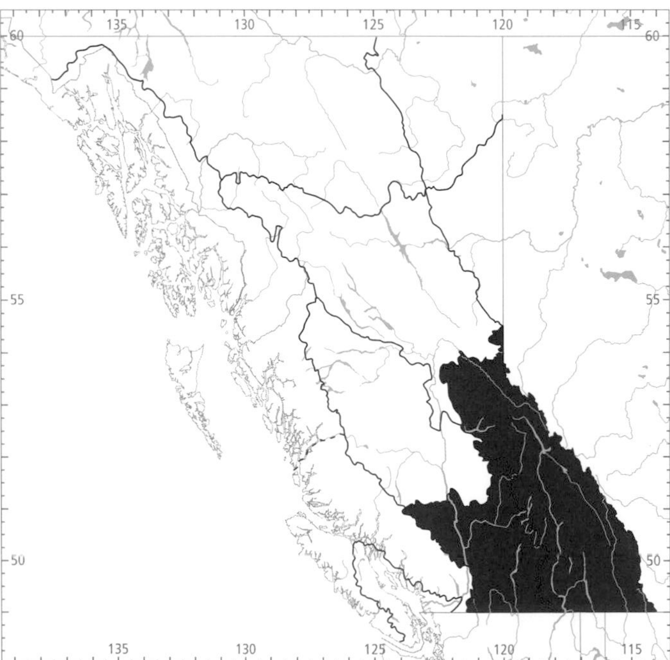

Figure 102. In British Columbia, the highest numbers for the Eastern Kingbird in summer occur in the Southern Interior and Southern Interior Mountains ecoprovinces.

Proximity to water was a specific habitat component for 41% of 222 nests. At Creston, Siderius (1994) found that the more water near the nest (Fig. 104c), the greater the chance of success; incubating birds spent more time on the nest and visited it more often when the nest was near water.

The Eastern Kingbird has been recorded breeding on the coast from 1 June to 1 August and in the interior from 4 May (calculated) to 18 August (Fig. 99).

Nests: In more natural sites, most nests (74%; n = 335) were located in dead or living, erect or fallen trees or shrubs (Fig. 104a, c, d). Twenty-three different species were reported. Willows were the single most frequent choice (11%), followed by birch (5%) and aspen (3%). About 7% of all nests were found in conifers. The remaining nests were found on utility poles (23%) as well as posts, pilings, abandoned cars, bridges, buildings, and water towers.

Tree and shrub nests were usually found on horizontal or occasionally up-slanting branches, in forks or crotches but occasionally against the trunk (73%; n = 222). Other nests were found on a shelf or ledge (23%), on girders or beams of artificial structures, in nest boxes and culverts, on cliffs and banks (Fig. 104b), or in old nests. The Eastern Kingbird vigorously defends its nest site from intruders but may nest within 90 m of a pair of Western Kingbirds (Fig. 107). However, there are a number of incidences of nest trees being shared with the

Figure 103. In forested areas of British Columbia, beaver ponds with flood-killed trees and shrubs are favoured sites for Eastern Kingbirds (2 km north of 150 Mile House, 25 April 1990; R. Wayne Campbell).

following species: American Robin (4), Cedar Waxwing (2), Tree Swallow (1), and Bullock's Oriole (1). Siderius (1994) found that the kingbird destroyed eggs in artificial nests that were placed near its own. The heights of 335 nests ranged from ground level to 27 m, with 52% recorded between 1.5 and 7 m.

Nests were bulky, untidy, woven cups (Fig. 105). In 2 instances, nests were built on top of old nests of the American Robin. Grasses were the major constituent of most nests (83%; n = 266), followed by twigs (27%), plant stems and fibres (22%), plant down (9%), feathers (8%), string (7%), hair (7%), and rootlets (7%). Other materials included strips of bark, moss, leaves, paper, lichen, wood chips, sedges, and down. Linings consisted of fine grasses, plant down, mammal hair, rootlets, plant fibres, and other soft and easily shaped plant material.

Figure 104. The Eastern Kingbird is a versatile and adaptable species that selects a wide variety of situations in which to build its nest. These may include: (a) saskatoon bushes in fields and along highways, Richter Pass, 18 June 1993; (b) natural cavities in river banks, Creston, 2 August 1993; (c) dead trees bordering sloughs and slow-moving rivers, Creston, 4 June 1994; and (d) the tops of broken trees and posts, Creston, 30 May 1995 (all R. Wayne Campbell).

Eggs: Dates for 204 clutches ranged from 22 May to 2 August, with 51% recorded between 18 June and 3 July. Calculated dates indicate that nests may contain eggs as early as 4 May. Sizes of 138 clutches ranged from 1 to 5 eggs (1E-3, 2E-11, 3E-30, 4E-81, 5E-13), with 59% having 4 eggs (Fig. 106). Siderius (1994) found similar results in her study at Creston. Bent (1942), Davis (1941), and Murphy (1983) report that the Eastern Kingbird's most frequent clutch size is 3, but Blancher and Robertson (1985) reported a modal clutch size of 4 in Ontario. The incubation period for 1 nest in British Columbia was 14 days (Cannings et al. 1987). Siderius (1994), at Creston, found that incubation lasted about 14 days after the first egg was laid.

Young: Dates for 159 broods ranged from 22 May to 18 August, with 53% between 3 and 23 July. Sizes of 99 broods ranged from 1 to 5 young (1Y-8, 2Y-22, 3Y-31, 4Y-32, 5Y-6), with 64% having 3 or 4 young. The nestling period for 1 nest in British Columbia was 15 days (Cannings et al. 1987), which is within the range of 15 to 19 days given by Murphy (1983). Siderius (1994), at Creston, reports a nestling period of from 14 to 21 days after the first young hatches.

Brown-headed Cowbird Parasitism: Of 213 nests reported with eggs or young in British Columbia, none was parasitized by the cowbird. There was, however, 1 report of adult kingbirds feeding a fledged cowbird. The Eastern Kingbird usually ejects cowbird eggs from the nest (Rothstein 1975; Hamas 1980), so few have been reported (Friedmann and Kiff 1985). The kingbird may increase its reproductive costs by accidentally ejecting its own eggs with the parasitic egg (Murphy 1986).

Nest Success: Of 36 nests found with eggs and followed to a known fate, 22 produced at least 1 fledgling, for a nest success rate of 61%. Murphy (1983, 1986) and Blancher and Robertson (1985) report nest success varying between 30% and 50%. Siderius (1994) found that vigorous defence near the nest was effective, but birds who chased the predator too far increased the risk of predation.

REMARKS: The late date, listed by Cannings et al. (1987), of a nest with eggs at Oliver on 7 August 1968 should read 8 July 1968.

There is no significant morphometric variation within the species' geographic range. Male and female Eastern Kingbirds have similar body size; however, there is significant sexual dimorphism. Males have longer tails, wings, and bills than females (van Wynsberghe et al. 1992).

Figure 105. The Eastern Kingbird's nest is a bulky cup consisting mainly of grasses, twigs, plant fibres, plant stems, and mosses (Creston, 30 May 1995; R. Wayne Campbell).

Figure 106. The eggs of the Eastern Kingbird, which are indistinguishable from those of the Western Kingbird, are creamy-white and heavily but irregularly spotted (Castlegar, 17 June 1995; R. Wayne Campbell). Four eggs are usually laid in British Columbia.

Figure 107. In the Richter Pass region of the Southern Interior Ecoprovince in British Columbia, both Eastern and Western kingbirds nest in roadside bushes, sometimes within 90 m of each other (Richter Pass, 18 June 1993; R. Wayne Campbell).

Siderius (1994) banded 62 nestlings from 20 nests at Creston. Only 2 were recaptured there, and only 1 was found nesting there in following years.

The Eastern Kingbird is unusual among passerines in British Columbia in that it has a period of post-fledging parental care that extends for 5 weeks or more (Morehouse and Brewer 1968). Siderius (1994) found that parents fed nestlings for up to 24 days after fledging.

Several observers have recorded the Eastern Kingbird eating fruit during its southward migration, notably berries of the red-osier dogwood (Beal 1912; Cannings et al. 1987; Siderius 1994). Siderius (1994) notes that dogwood berries are easier than insects for fledglings to collect. Like many North American insectivores, kingbirds take advantage of abundant supplies of small figs on the tropical wintering grounds (Fitzpatrick 1982).

For additional information on breeding biology, foraging behaviour, and habitat selection by the Eastern Kingbird, see Blancher and Robertson (1985), Murphy (1987, 1988), and MacKenzie and Sealy (1981).

NOTEWORTHY RECORDS

Spring: Coastal – Island View Beach 19 May 1964-1; Bamfield 17 May 1981-1; Terra Nova (Richmond) 5 May 1923-1; Langley 3 May 1968-1 incubating; Lulu Island 5 May 1924-1 (RBCM 7140); Pitt Meadows 14 Apr 1973-1; Qualicum Beach 28 May 1986-1; Courtenay 27 May 1934-1; between Courtenay and Comox 28 May 1931-1 (Pearse 1931); Wolf Lake (Campbell River) 23 May 1960-1; Namu 30 May 1983-1 (Campbell 1983c); s Kitimat River 1 May 1960-1 (Hay 1976); Port Simpson 10 Apr 1887-1 (RBCM 613). **Interior** – Richter Pass 2 May 1968-1 (Campbell and Meugens 1971); Creston 12 May 1928-1 (Mailliard 1932), 23 May 1928-1 incubating; 4 km e Grand Forks 23 May 1972-1 incubating; Salmo 11 Apr 1981-1; Naramata 15 May 1969-1 incubating (Cannings et al. 1987); Okanagan Landing 6 May 1940-1; Crescent Bay 16 May 1981-1; Lytton 20 May 1986-1; Chase to Pritchard 26 May 1980-10; Sorrento to Kamloops 26 May 1970-10; Kamloops 18 May 1982-1; Celista 24 Apr 1948-1; Radium 15 May 1980-4 (Halverson 1981); Yoho National Park 25 Apr 1975-1 (Wade 1977); Williams Lake 16 May 1970-1; Riske Creek 17 May 1989-1; Willow River 24 May 1979-1; Tupper Creek 5 May 1938-1 (Cowan 1939); Fort St. John 9 May 1982-1; Charlie Lake 15 May 1982-1; Parker Lake 30 May 1975-1.

Summer: Coastal – Mount Tolmie 18 Jun 1970-1 (Tatum 1971); Saanich 1 Jun 1986-1; Carmanah Point 18 Jun 1978-1, 23 Aug 1979-1; Pachena Point 20 Jul 1975-1 (Shepard 1976a; BC Photo 407); Sumas Prairie 28 Jun 1936-4 eggs collected, incubation advanced, 4 Aug 1923-1; Cultus Lake 17 Jul 1927-2 (RBCM 13161-2); Ladner 10 Jun 1949-1 female (UBC 2120); Iona Island 1 Jun 1983-1; Chesterman Beach 6 Aug 1983-1; Tofino 29 Jun 1983-2; University of British Columbia (Vancouver) 1 July 1970-2 (Campbell 1970a); Jericho Park (Vancouver) 1 Jun 1983-1; Haney 1 Jun 1979-2; Port Coquitlam 1 Jun 1962-4 fresh eggs (RBCM 307); Pitt Meadows 7 Jun 1964-10 in area, 13 Jun 1964-nesting (BC Photo 40), 30 Jul 1961-first nestling (Boggs and Boggs 1961a); Alouette Lake 19 Jun 1963-1 at picnic site (Dow 1963); Cleland Island 14 Jun 1976-1; Qualicum Beach 3 to 28 Jun 1980-1; near Lasqueti Island 3 Jun 1978-1; Courtenay 9 Jun 1929-2 (Pearse 1930); between Courtenay and Comox 7 Aug 1929-3 recent fledglings (Bishop 1930); Miracle Beach Park 6 Jun 1971-1; Mitlenatch Island 6 Jun 1977-1; Cortes Island 6 Jun 1978-1 (Crowell and Nehls 1971d); Northey Lake 16 Aug 1951-1 adult with 2 immatures; Brooks Peninsula 6 Aug 1981-1 (Campbell and Summers, in press); Fry Lake 8 Jun 1959-2; Acous Peninsula 13 Jun 1990-1 (Siddle 1990c); Pemberton 4 Jun 1967-5; Triangle Island 5 Jul 1974-1 (Vermeer et al. 1976); Kimsquit River estuary 7 Jul 1986-1; 10 km e Bella Coola 8 Aug 1977-1 hawking insects in garden; Anthony Island 4 Jun 1982-1 in small spruce; McIntyre Bay 27 Jun 1989-1; Langara Island 10 Jun to 13 Jul 1927-1 (Darcus 1930); Delkatla Slough 27 Jun 1991-1; Hazelton 1 Jun 1960-2, 22 Jun 1921-1 male (MVZ 42175). **Interior** – 2 km s Creston valley 31 Jul 1981-3 near fledging; Trail 31 Jul 1979-2 fledglings, 1 nestling; Manning Park 16 Jun 1961-eggs, 11 Jul 1962-1 incubating; Oliver 7 Aug 1963-2 fledglings; Vaseux Lake 11 Aug 1964-50; Waneta 4 Aug 1982-1 nestling; Penticton 18 Aug 1966-2 nestlings; Kelowna 18 Aug 1983-300; Lytton 4 Jun 1964-1 (NMC 52412); Revelstoke 23 Aug 1989-40; Watson Lake (100 Mile House) 10 Aug 1958-1; 10 km w Williams Lake 28 Aug 1977-3 fledglings being fed by adults (Murphy 1983); n Riske Creek 11 Jun 1958-5 eggs; Hanceville 10 Aug 1958-15 in area; Stum Lake 21 Jun 1973-1, 6 Jul 1973-1 (Ryder 1973); Quesnel 10 Jun 1978-eggs; Bowron Lake Park 2 Jun 1971-1 (Runyan 1971); Indianpoint Lake 1 Jun 1929-1 male (MCZ 282894); 40 km w Prince George 8 Jun 1969-1; Tiltzarone Lake 31 Aug 1944-1 (Munro 1947a); Aleza Lake 19 Jun 1963-1 male (UBC 11702); Bouchie Creek 31 Aug 1944-1 (Munro and Cowan 1947); McLure Lake (Smithers) 1 June 1980-1; One Island Lake 28 Jun 1978-4 nestlings (Campbell 1978c); 16 km se Fort St. John 25 Jun 1978-1 (Campbell 1979a); Fort St. John 10 Jun 1962-4 eggs, 7 Jul 1973-3 eggs, 12 Jul 1986-nestlings, 16 Aug 1986-42 in 100 m stretch along Peace Island Park road; 3 km from Simpson Trail jct (Alaska Highway) 13 Aug 1985-1; Ingenika 15 Jul 1977-4 nestlings; Parker Lake (Fort Nelson) 18 Jun 1976-1 (RBCM 15757), 5 Jul 1978-1; Fort Nelson 22 Aug 1985-18 in red-osier dogwoods at airport; 75 km e Fort Nelson 24 Jun 1982-5 eggs, incubation advanced (Campbell and McNall 1982); Raspberry Creek 18 Aug 1953-1; Kwokullie Lake 6 Jun 1982-2; s Kotcho Lake 13 Jun 1982-1, 24 Jun 1982-2 nests, 4 and 5 eggs, respectively (Campbell and McNall 1982); Pine Creek (Atlin) 23 Jun 1975-1 (Swarth 1926); Liard Hot Springs 27 Jul 1983-1 (Grunberg 1983d).

Breeding Bird Surveys: Coastal – Recorded from 3 of 27 routes and on 8% of all surveys. Maxima: Pemberton 22 Jun 1975-7; Chilliwack 25 Jun 1992-4; Seabird 10 Jul 1977-2. **Interior** – Recorded from 45 of 73 routes and on 43% of all surveys. Maxima: Golden 21 Jun 1973-22; Lavington 14 Jun 1975-15; Spillimacheen 22 Jun 1986-9; Pavilion 12 Jun 1977-8; Mabel Lake 19 Jun 1980-8; Wasa 5 Jul 1989-8.

Autumn: Interior – Charlie Lake 2 Sep 1984-1; 24 km s Prince George 8 Sep 1980-2; Wineglass Ranch (Riske Creek) 11 Sep 1992-1 from late nesting; Rock Lake (Riske Creek) 3 Sep 1978-1; Golden 4 Sep 1977-3, 18 Sep 1976-1; Knutsford 8 Sep 1979-1; Swan Lake (Vernon) 11 Sep 1964-1; Summerland 14 Sep 1972-1; Brouse 6 Sep 1982-1; Cranbrook 30 Sep 1940-1 latest seen (Johnstone 1949); Princeton to Merritt 22 Sep 1982-2; Trail 11 Nov 1980-1; Oliver 11 Sep 1971-1; Creston 19 Sep 1982-1. **Coastal** – Sandspit early Sep 1990-1 (Siddle 1991a); Miracle Beach Park 9 Sep 1960-1 (Stirling 1961); Pitt Meadows 31 Oct 1970-1; Jericho Park (Vancouver) 10 Sep 1985-1; Chilliwack 24 Sep 1980-1; Victoria 10 Sep 1985-1.

Winter: No records.

Scissor-tailed Flycatcher STFL

Tyrannus forficatus Gmelin

RANGE: Breeds from eastern New Mexico, southeastern Colorado, and southern Nebraska south to western and eastern Texas and western Louisiana; locally east to Indiana and Mississippi. Winters casually in southern Louisiana and northern Florida but mostly in Mexico and Central America.

STATUS: *Very rare* vagrant on the south coast, including Vancouver Island, and in the southern portions of the interior; *accidental* elsewhere.

CHANGE IN STATUS: Munro and Cowan (1947) do not list the Scissor-tailed Flycatcher among the avifauna of British Columbia. The first record for the province occurred in 1964, but by 1993 it had been recorded on 17 different occasions, with nearly equal frequency on the coast and in the interior.

Figure 108. Scissor-tailed Flycatcher at Long Beach, on the central west coast of Vancouver Island, British Columbia (17 May 1987; Adrian Dorst).

OCCURRENCE: Coastal observations of the Scissor-tailed Flycatcher (Fig. 108) have been limited to the lowlands of both the east and west coasts of Vancouver Island, its neighbouring islands, and the lower Fraser River valley in the vicinity of Matsqui and Hope. The most northern coastal record is from Port Neville. In the interior, it has appeared at Grand Forks and Wynndel in the south and as far north as Quesnel. There is 1 occurrence from the Liard River valley in the far north.

Most records (12 of 17) are of vagrants reported in May and June, which agrees with the species' wandering habits along the west coast of North America. In California, Roberson (1980) noted a peak in frequency from late May to early June, corresponding with appearances of other vagrant "eastern" flycatchers in the state. The remaining records in British Columbia are in summer and autumn. All records but 1 involve single birds. The longest stay at any locality was 3 days on the coast and 11 days in the interior.

This flycatcher was found in or adjacent to open country, including airports, cultivated land, forest edges near roads and highways, dry grasslands, and beaches.

The Scissor-tailed Flycatcher has been recorded on the coast from 16 May to 10 October and in the interior from 13 May to 10 October.

In chronological order, documented records include:

(1) Enderby 24 May 1964-1 flew, working its scissor tail, along the edge of a group of mixed deciduous trees.
(2) Northeast of Quesnel 31 August 1966-1 photographed near Cottonwood House on Barkerville Road (BC Photo 508).
(3) Saltspring Island 2 June 1967-1 on north side Booth Bay Canal (Roberts 1967; Crowell and Nehls 1968a; Tatum 1971; Stirling 1972).
(4) Victoria International Airport (Sidney) 8 to 10 October 1967-1 (Stirling 1968; Crowell and Nehls 1968a; BC Photo 100).
(5) 5 km east of Grand Forks 23 September 1968-1 perched on the top of small bushes, great mullein, and fence wires beside main highway (Millican 1969). On 10 October 1968, presumably the same bird was seen about 3 km east of Grand Forks (R. Walker pers. comm.).
(6) Revelstoke 13 May 1973-1 photographed near airport (BC Photo 289).
(7) Port Neville 7 June 1974-1 feeding on insects.
(8) Kamloops 16 to 26 June 1978-1 sighted on 4 occasions in airport fields.
(9) Junction of Liard and Coal rivers 11 July 1978-1 flycatching from tops of spruce trees (Campbell 1978a).
(10) Hope 1 June 1984-1 at airport.
(11) Hope 22 and 23 May 1985-1 at airport.
(12) Wynndel 27 May 1986-1 photographed near Duck Lake (BC Photo 1085).
(13) Long Beach 16 to 18 May 1987-1 photographed (Campbell 1987c; Mattocks and Harrington-Tweit 1987) (BC Photo 1155; Fig. 107).
(14) 5 km north of Clinton 26 June 1989-1 feeding around farmland at Mound valley (Campbell 1989c).
(15) East of Mt. Lehman Road near Matsqui 5 October 1991-1 flew over Highway 1 to open fields and scattered trees.
(16) Port Alberni 18 July 1992-1 (Siddle 1993a).
(17) Iona Island 16 June 1993-1 female (Siddle 1994c).

REMARKS: Roberson (1980) gives 7 October 1967 as the reported date for the bird at Victoria International Airport, but the correct date is 8 October 1967 (Stirling 1968). Also, details cannot be located for a record listed by the same author of a bird seen northwest of Trail, in the west Kootenay, on 20 May 1973.

The Scissor-tailed Flycatcher was formerly placed in the genus *Muscivora* (American Ornithologists' Union 1957, 1983).

Fitch (1950) and Regosin and Pruett-Jones (1995) provide general information on the life history of the Scissor-tailed Flycatcher.

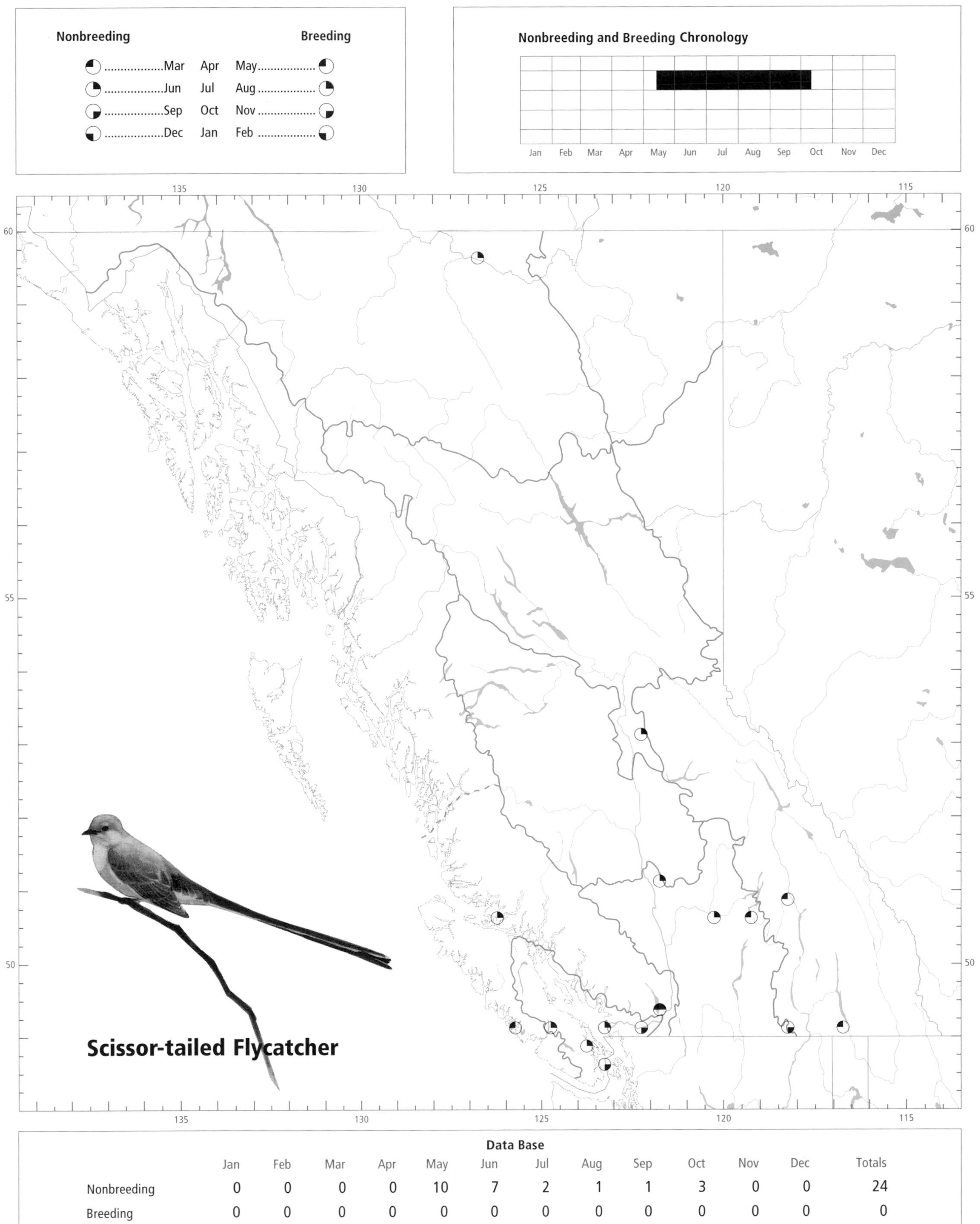

Data Base

	Jan	Feb	Mar	Apr	May	Jun	Jul	Aug	Sep	Oct	Nov	Dec	Totals
Nonbreeding	0	0	0	0	10	7	2	1	1	3	0	0	24
Breeding	0	0	0	0	0	0	0	0	0	0	0	0	0

Sky Lark

SKLA

Alauda arvensis Linnaeus

RANGE: In North America, introduced and established on extreme southeastern Vancouver Island, British Columbia. Small numbers have spread and are now resident on San Juan Island in adjacent Washington state. Naturally breeds from the British Isles, Scandinavia, and northern Siberia south to northwestern Africa, the Mediterranean region, Asia Minor, northern China, Korea, and Japan; winters from the breeding range south to northern Africa, the Persian Gulf, and eastern China. The Sky Lark is established as an introduced species in the Hawaiian Islands, Australia, and New Zealand.

STATUS: Introduced and established as a local resident in the Georgia Depression Ecoprovince only on the Saanich Peninsula of southeastern Vancouver Island; an *uncommon* to *fairly common* resident there; population low but apparently stable. *Casual* elsewhere in the Georgia Depression, including the Gulf Islands and Municipality of Delta; *accidental* on northern Vancouver Island and the Queen Charlotte Islands. Breeds.

Figure 109. Adult female Sky Lark on nest (University of Victoria, 17 July 1965; Tom Sowerby).

CHANGE IN STATUS: In 1903 the British Columbia Natural History Society, with financial support from the provincial government and local residents, imported 100 pairs of Sky Larks (formerly known as Eurasian Skylarks) (Fig. 109) from Great Britain. Two birds died in transit. Half of the consignment was held in Vancouver until the spring of 1903 and released

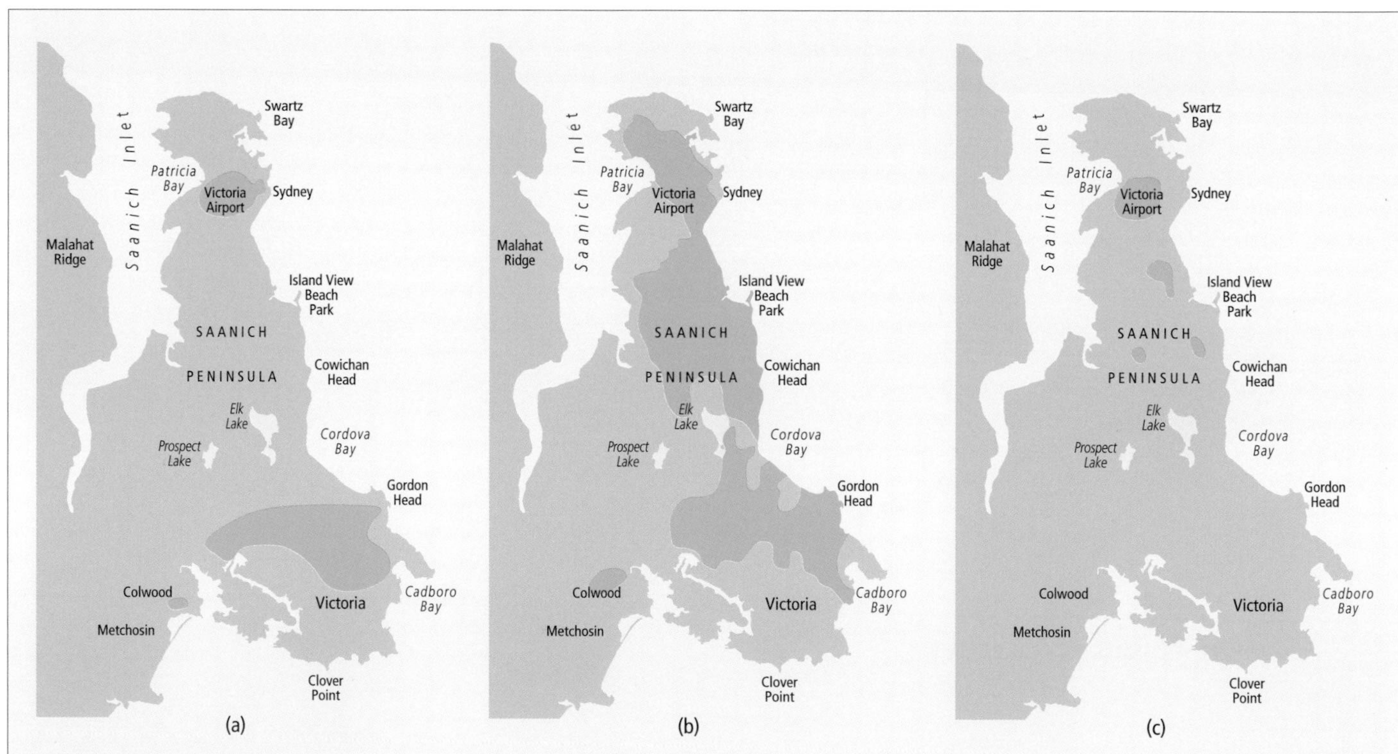

Figure 110. Changes in the distribution of the Sky Lark in Victoria and on the Saanich Peninsula, Vancouver Island, British Columbia, 1903 to 1993. By the mid-1940s (a) populations were established near Sidney and the present site of the Victoria International Airport in North Saanich, locally in Colwood, and in the Greater Victoria area from Gordon Head, Lost Lake, and Burnside Road south and east to the vicinity of Lansdowne Road and the Uplands (modified from Sprot 1937; Meugens 1944). By the early 1960s (b) the population had spread to occupy most of the suitable habitat on the Saanich Peninsula and Victoria area (from Stirling and Edwards 1962). Habitat loss during the 1970s and 1980s resulted in populations being restricted to four locations (c): the vicinity of the Victoria International Airport, the Wallace Drive/Central Saanich Road area, the Island View Road/Lochside area, and between Wallace Drive and Keating Cross Road (adapted from Begg 1990).

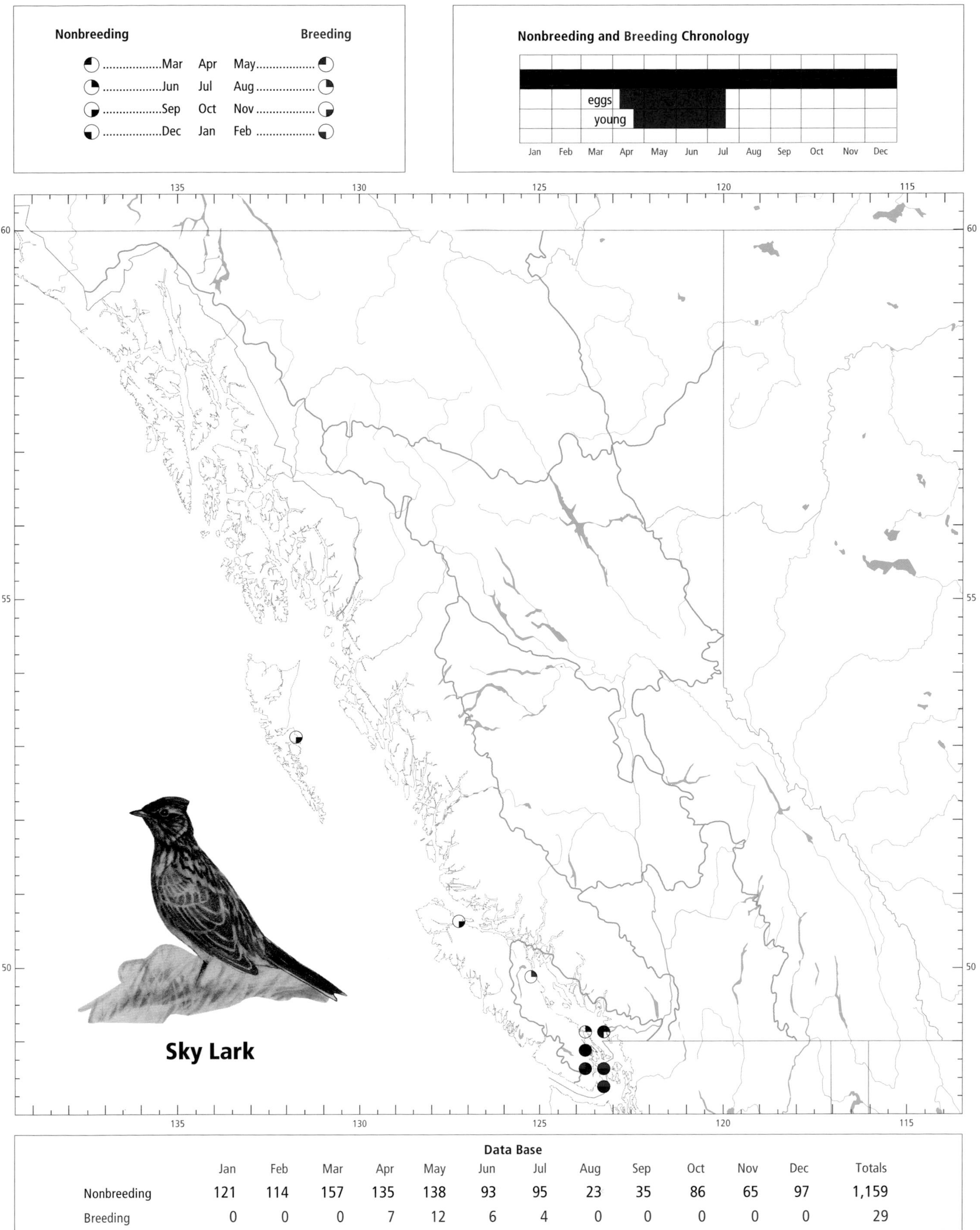

Data Base

	Jan	Feb	Mar	Apr	May	Jun	Jul	Aug	Sep	Oct	Nov	Dec	Totals
Nonbreeding	121	114	157	135	138	93	95	23	35	86	65	97	1,159
Breeding	0	0	0	7	12	6	4	0	0	0	0	0	29

at various points throughout the Lower Mainland. The remainder was held in Beacon Hill Park, Victoria, and released in the autumn of 1903 (Anonymous 1903). Ninety-nine birds were liberated in the vicinity of Duncan, North Saanich, Colwood, and Cedar Hill, and in Victoria in fields near the Jubilee Hospital and at Beacon Hill Park (Grinnell 1936; Scheffer 1935; Sprot 1937; Cooke and Knappen 1941). In April 1913, an additional 49 birds were released on the Saanich Peninsula in the following locations: 34 at Rithet's Farm, 9 at Lansdowne Road, and 6 at Cadboro Bay (Phillips 1928; Sprot 1937). Six more were released privately, at Royal Oak, in 1919 (Tonkin 1971). By the mid-1920s numbers had increased and the species was considered "locally abundant" (Preece 1925). The mainland introductions were unsuccessful, although Brooks and Swarth (1925) mention that the species had "secured a permanent foothold at the mouth of the Fraser River."

A decade later Sky Larks were reported at Sidney near the north end of the Saanich Peninsula, nearly 24 km from the closest release site. The first census, carried out in November and December 1935, found 219 Sky Larks at 9 locations (Sprot 1937).

Consecutive mild winters and the conversion of forests to cultivated land allowed the species to thrive (Meugens 1944; Harwell 1946). By the mid-1940s the Sky Lark had become "firmly established" in Victoria and on the Saanich Peninsula (Munro and Cowan 1947) (Fig. 110).

The increase continued through the 1950s and 1960s, until the Sky Lark appeared to have occupied all available habitat. During the same period it extended its range north to Duncan, west to Saltspring Island (Scheffer 1955; Davidson 1958, 1967; Poynter 1960), south to Sidney Island, and even to San Juan Island in Washington state (Bruce 1961).

In March 1962, another census was completed for Vancouver Island (Stirling 1962; Stirling and Edwards 1962). A total of 694 Sky Larks were counted, mostly in the following areas: Victoria International Airport (129), Saanich Experimental Farm (40), Keating Cross Road (38), Martindale Road (313), Gordon Head (138), and Cedar Hill Cross Road (19), all localities on the Saanich Peninsula. Stirling and Edwards (1962) estimated the provincial population at about 1,000 birds.

The year-end census of 1965 again led to an estimate of just over 1,000 birds (Stirling 1966). This was the highest estimate up to that time and revealed that the Sky Lark had continued to increase despite some loss of habitat. In early 1969, severe weather appeared to threaten the Sky Larks and grain was distributed in an attempt to supplement available feed. On 2 February 1969, 777 birds were counted using the supplemental feed, mostly in the central and north Saanich areas. Only 5 birds were counted in the University–Gordon Head area and 14 elsewhere (Stirling and Beckett 1969).

During the 1970s, urbanization and the accompanying loss of habitat further affected populations of Sky Larks by reducing and fragmenting suitable habitat. On the Saanich Peninsula, they disappeared from the Gordon Head area and the University of Victoria (Tatum 1970a, 1971, 1972, 1973). A few still remained in the Duncan and Cowichan Bay areas north of Victoria (Chambers 1969; Lemon 1970).

The decline in numbers continued during the 1980s and early 1990s. In December 1980, Sky Larks were not found on the Christmas Bird Count (Shepard 1981), for the first time since 1957. In spring 1983, 42 singing males were seen in 12 locations on the Saanich Peninsula (MacLeod 1983), and in the winter of 1984 to 1985 a maximum of 59 birds was estimated for the Martindale and Victoria International Airport areas (Carder 1985). In January 1993, the census total for the entire peninsula was 206 Sky Larks. On 17 March 1993, a similar survey of singing birds resulted in a total of 64 Sky Larks, distributed as follows: Lochside Trail (4 pairs), Maber's Flats (2 pairs), Vantreight's farm (12 pairs), and Victoria International Airport (14 pairs).

The birds now appear to be confined to the above 4 areas within a 9.5 km linear distance, down from 57 km historically

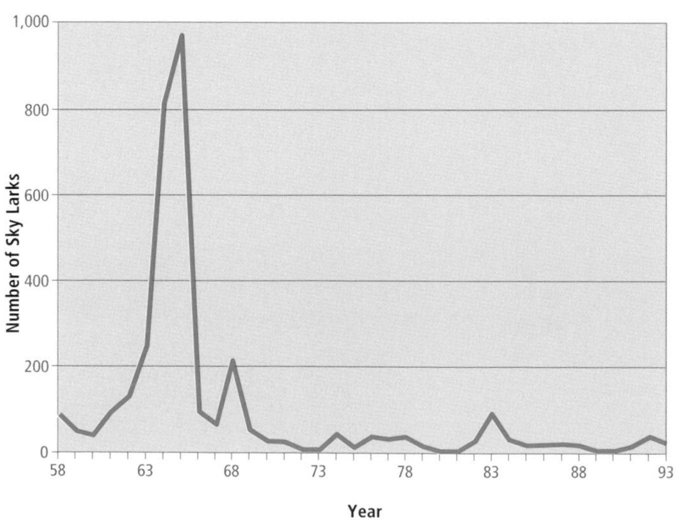

Figure 111. Fluctuation in numbers of Sky Larks recorded on Victoria Christmas Bird Counts between 1957 and 1994.

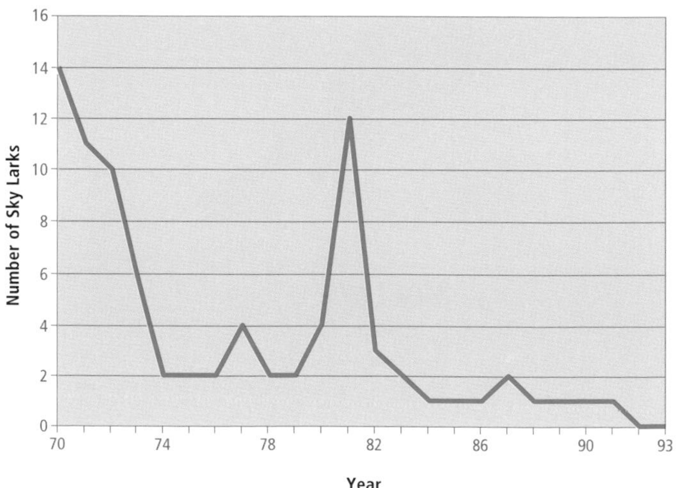

Figure 112. Maximum numbers of Sky Larks recorded annually in the vicinity of Cowichan Bay on Vancouver Island, 1970 to 1993.

(see Fig. 110). Two of these remaining locations are threatened by development (Begg 1990).

In 1970 small numbers of Sky Larks became established in fields surrounding Cowichan Bay, just south of Duncan. Although several pairs nested in the area and birds were occasionally reported into the early 1990s, the population apparently no longer exists (Fig. 112).

In autumn 1953 a few residents of Comox "secretly" released 6 Sky Larks in the vicinity of York Swamp. Adults and young were seen in the summers of 1954 and 1957, after which none was reported.

In early spring 1977, a "few" Sky Larks were released privately near the Vancouver International Airport, south of Vancouver. That introduction failed.

Unless suitable habitat is protected from urban development, it is unlikely that a viable population of the Sky Lark can be maintained on Vancouver Island and the adjacent Gulf Islands.

NONBREEDING: The Sky Lark (Fig. 109) is mainly confined to 4 areas on the Saanich Peninsula of southern Vancouver Island (Figs. 110 and 113), although small numbers of birds can occasionally be found in suitable habitat as far north as Cassidy.

Figure 113. In British Columbia, the highest numbers for the Sky Lark in winter (black) occur in the Georgia Depression Ecoprovince (Saanich Peninsula of southern Vancouver Island); the highest numbers in summer (red) also occur in the Georgia Depression Ecoprovince (North and Central Saanich).

Figure 114. In winter, corn stubble fields are important foraging areas for Sky Larks, especially during snowfalls (near Welch Road, Central Saanich, 31 December 1992; R. Wayne Campbell).

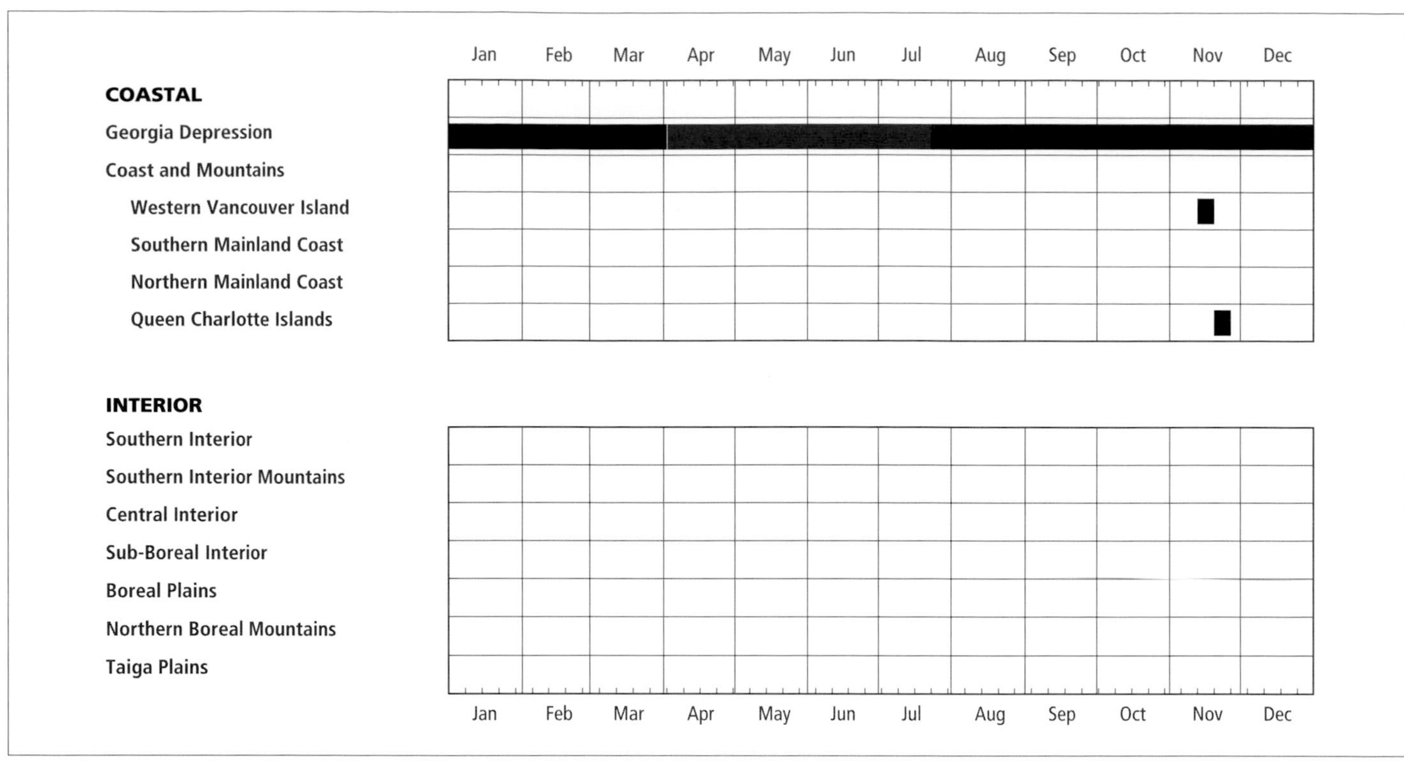

Figure 115. Annual occurrence (black) and breeding chronology (red) for the Sky Lark in ecoprovinces of British Columbia.

Populations at Comox, Cowichan Bay, and Cobble Hill near Duncan probably no longer exist, although individuals are reported irregularly on southeastern Vancouver Island as far north as the Cassidy airport. Vagrants have been seen on Saltspring Island, Pender Island, and Sidney Island in the Strait of Georgia, and on the mainland only on the Fraser River delta (Weber 1975a; Poynter 1977). There are single sight records for northern Vancouver Island and the Queen Charlotte Islands, which are likely vagrants from Asia.

The Sky Lark prefers open country such as beaches, pastures, short-grass fields, cultivated land, golf courses, playing fields, and airfields. As Stirling and Edwards (1962) point out, "the horizon about such places must be low to be suitable, for [the birds] avoid narrow valleys and small fields bordered by trees." The most suitable habitats have low or sparse vegetation, often with a high proportion of bare soil.

In winter, depending on snowfall, the Sky Lark aggregates in grain or corn stubble fields (Fig. 114). In order of preference, the birds can be located in late crops of wheat or oats; light stubble where, on poorer soils, the crops have matured and have been harvested early; and coarse stubble on rich soils that have been cleanly farmed (Sprot 1937). By the early 1960s, the Sky Lark was found in winter at various localities on the Saanich Peninsula where open ground carried weeds, grasses, and fields of vegetable crops (Stirling 1962). As crops changed, the Sky Lark also began to frequent daffodil fields, and occasionally lawns and drainage ditches.

The Sky Lark is resident on southern Vancouver Island, and populations are relatively sedentary. There is no evidence of migratory or dispersal patterns similar to those in Europe. It is suspected, however, that there is a movement in summer, away from wintering areas. Movements on the Saanich Peninsula are local and most noticeable in winter, when flocks shift between snow-covered fields and snow-free fields (Begg 1991).

The Sky Lark is present throughout the year in British Columbia (Fig. 115).

BREEDING: The Sky Lark now breeds only near the Victoria International Airport in North Saanich and in the vicinity of Central Saanich Road and Island View Road in Central Saanich. One or 2 pairs may still nest between Keating Cross Road and Wallace Drive. Formerly it nested locally, near Comox and Cowichan Bay.

It prefers to nest in dry, managed, open habitats such as fields of short to moderately tall grass, and those that have been cultivated before initiation of the breeding season and have a cover of short wild grass and weeds (Tatum 1971) (Fig. 117). The Sky Lark shuns areas with isolated trees, hedges, shrubs, gravel patches, and extensive sandy areas.

The Sky Lark has been recorded breeding in the province from 7 April (calculated) to 18 July (Fig. 115).

Nests: All nests were on the ground in open areas among vegetation such as grasses (Fig. 116) and forbs. Some were placed near paths, airport runways, and ground debris such as sticks and stones. Most nests were in shallow depressions and were constructed of coarse grasses, plant fibres, plant stems, leaves, rootlets, and occasionally flower heads. The nest cup was usually lined with fine grasses.

Figure 116. Incomplete clutch of Sky Lark eggs in nest (University of Victoria, 17 July 1965; Tom Sowerby). The clutch was completed the following day.

Eggs: Dates for 13 clutches (Fig. 116) ranged from 12 April to 18 July, with 60% recorded between 20 April and 5 June. Calculated dates indicate that nests may contain eggs as early as 7 April. Clutch size ranged from 1 to 4 eggs (1E-2, 2E-1, 3E-3, 4E-7), with 7 clutches having 4 eggs. In a sample of 89 clutches from England, Delius (1965) found that 57% had 4 eggs, while clutch size ranged from 3 to 7 eggs. The incubation period in England is 11 days (Delius 1963). In British Columbia, the Sky Lark may lay 2 clutches a year (Bent 1942; Tatum 1970a).

Young: Dates for 10 broods ranged from 20 April to 18 July, with 64% recorded between 5 May and 10 June. Calculated dates indicate that nests may contain young as late as 1 August. Brood size ranged from 1 to 4 young (1Y-2, 2Y-2, 3Y-5, 4Y-1), with 5 nests having 3 young. The nestling period in England is 8 to 10 days and the fledgling period is 18 to 20 days (Delius 1965). Young become independent at about 25 days.

Brown-headed Cowbird Parasitism: In British Columbia, cowbird parasitism was not found in 19 nests recorded with eggs or young. Also, Friedmann (1963), Friedmann et al. (1977), and Friedmann and Kiff (1985) do not report parasitism on the Sky Lark in North America.

Nest Success: Of 2 nests found with eggs and followed to a known fate, 1 produced at least 1 fledgling.

REMARKS: The population of Sky Larks released on Vancouver Island was of the nominate subspecies, *A. a. arvensis.* A Siberian race, *A. a. pekinensis,* is now a regular migrant and casual summer visitant to the islands of western Alaska, with possible breeding on the Pribilof Islands (Byrd et al. 1978; Kessel and Gibson 1978; Gibson 1981). South of Alaska, it is a casual transient on the west coast of North America as far south as California. It is possible that the Sky Larks sighted in Delta (Weber 1975a), Port McNeill, and the Queen Charlotte Islands (Siddle 1992a) represent the latter subspecies (Morlan and Erickson 1983).

The report of Sky Larks at the Vancouver International Airport in late July 1970 is erroneous (Crowell and Nehls 1970d).

In Washington state, the Sky Lark was first found in 1960 on San Juan Island, 18 km east of the Saanich Peninsula on Vancouver Island. It was reported nesting there a decade later (Wahl and Wilson 1971). In August 1973, 63 individuals were counted on the island (Weisbrod and Stevens 1974), and by the late 1980s it was considered a "locally common breeding resident" with its centre of abundance in an area bordered by South Beach, the redoubt, and Pickett's Lane (Lewis and Sharpe 1987).

Attempts to introduce the Sky Lark into North America other than British Columbia have been made off the coast of

Figure 117. Sky Lark nesting area (University of Victoria campus, June 1975; R. Wayne Campbell). By the early 1980s, the species was no longer nesting at this locality because of human disturbance and grass-mowing activities.

Delaware, and in Ohio, New York, California, Oregon, Quebec, Massachusetts, New Jersey, Michigan, and Montana (Cooke and Knappen 1941). All have been unsuccessful (Phillips 1928; Grinnell and Miller 1944; Bull 1974). Twomey (1936) and Garman (1956) suggest that one reason introductions have been successful on southern Vancouver Island but not elsewhere in North America is that there the temperature and rainfall for all the critical months of the year fall within the native range of the subspecies that was imported.

It has also been suggested that cold winters with persistent snow cover have played a part in reducing the Sky Lark population on Vancouver Island (Stirling and Edwards 1962). For example, the winter of 1949 to 1950 was characterized by deep snow; it was suggested that this resulted in a reduction in numbers of Sky Larks in the Victoria area the following December. After that, it took 6 or 7 breeding seasons, and mild winters with little snowfall, to build populations to former levels.

Even though other factors may be contributing, loss of breeding habitat through increasing urban encroachment appears to be the primary reason for the decline in Sky Lark numbers. For example, in the 1950s and 1960s, about 60 pairs nested in the extensive fields surrounding what is now the University of Victoria campus. By early 1970 the population was restricted to only a few pairs (Tatum 1970a) (Fig. 117), and by 1980 the birds had virtually disappeared, partly because of disturbance and grass-mowing activities. Their last stronghold may be in the vicinity of the Victoria International Airport, although some of the Sky Lark habitat there is now also threatened with development.

Conservation efforts that may influence Sky Lark numbers include modifying local mowing and harvesting activities to lessen the impact on breeding Sky Larks (Tatum 1971), and securing commitments from landowners to maintain some fields for breeding and foraging Sky Larks. In addition, providing grain as supplementary winter food during periods of severe weather has been suggested (Stirling and Beckett 1969); however, there is no evidence that this earlier program improved survival.

If the survival of the species in British Columbia is considered important, it is imperative that research be done to determine the ultimate cause of the Sky Lark's steady decline in numbers. Only with reliable information can plans be made to improve its likelihood of survival in Canada.

The Sky Lark was formerly known as the Eurasian Skylark (American Ornithologists' Union 1995).

See Cramp (1988) for additional life-history information on the species in the western Palearctic.

NOTEWORTHY RECORDS

Spring: Coastal – Cattle Point (Victoria) 5 May 1960-1 egg and 3 young in nest; Victoria 30 Apr 1922-1 singing on Uplands golf course, 2 May 1962-3 nestlings; Gordon Head 3 Mar 1962-138; Royal Oak 13 Apr 1958-3 eggs; Royal Oak 20 Apr 1958-3 nestlings; Saanich Apr 1913-49 released throughout area (Sprot 1937); Martindale Flats (Saanich Peninsula) 6 Mar 1960-25 with 10 Horned Larks, 12 Mar 1961-50, 20 Apr 1993-122 with American Pipits in bare soil field; Saanich Peninsula early Mar 1962-694 on census (Stirling 1962), 18 Mar 1974-22 at airport; Central Saanich 3 Mar 1962-172, 12 Apr 1982-55, 22 May 1985-22, 4 Apr 1987-24, 21 Apr 1990-8; North Saanich 4 Apr 1987-36; Sidney 30 Apr 1961-15; Saltspring Island 29 Apr 1959-1 singing; Duncan 9 Mar 1963-6 (Boggs and Boggs 1963c), 10 May 1983-3 singing near Bench Rd, 14 May 1983-5 singing near Old Koksilah Rd; Cowichan Bay 12 Mar 1972-8, 21 Mar 1981-8, 28 Mar 1981-12, 6 May 1971-4 eggs, 7 May 1977-2 singing, 17 May 1980-4, 26 May 1991-1 singing; Richmond 6 May 1977-1 singing at Vancouver International Airport, 29 May 1977-1 singing (Poynter 1977). **Interior** – No records.

Summer: Coastal – Victoria 10 Jun 1942-2 eggs; Ten Mile Point (Victoria) 8 Aug 1935-40 (Grinnell 1936); Cedar Hill Golf Course (Victoria) 10 Jun 1942-1 infertile egg (RBCM 325) and 2 young in nest; University of Victoria 17 Jul 1965-3 eggs (Fig. 116), 18 Jul 1985-4 eggs, 18 Jul 1966-3 young, unable to fly, being fed in grass field; Lakehill 5 Jun 1929-2 eggs (UBC 1106); Mount Newton Aug 1988-1 in seed orchard; Saanich 6 Jun 1993-1 small young being fed by adult in field adjacent to airport, 12 Jun 1958-51 in 2 flocks at airport; Sidney Island 17 Jul 1965-2 (Stirling 1966); Richmond 5 Jun 1977-1 singing at Vancouver International Airport (Poynter 1977); Iona Island 14 Jun 1977-1 singing; Cobble Hill 16 Jul 1975-1; Cowichan Bay 16 Jun 1976-4, 30 Jun 1973-6, 1 Jul 1987-3, 8 Jul 1982-3, 10 Jul 1971-4 singing, 20 Jul 1974-2; Duncan 14 Jun 1983-2 singing over field near Bench Rd, 19 Jun 1983-none seen; Pender Island 19 Jun 1961-1 singing over field along Bedwell Rd; Cassidy 20 Jun to 25 Jul 1990-1 singing over airport field (Siddle 1990c); Comox 16 Jun 1957-2 adults feeding 2 young, 27 Aug 1954-2 adults with 1 juvenile. **Interior** – No records.

Breeding Bird Surveys: Coastal – Recorded from 1 of 26 routes and on 5% of all surveys. Maximum: Victoria 30 June 1974-17. **Interior** – Not recorded.

Autumn: Interior – No records. **Coastal** – Sidney 27 Nov 1976-40 in flight over Galaran Rd and Henry Rd; Saanich Peninsula Sep 1903-99 Eurasian Skylarks released (Sprot 1937; Stirling and Edwards 1962); Central Saanich 30 Nov 1985-108 with 32 Snow Buntings (Hunn and Mattocks 1985); Martindale Flats (Saanich Peninsula) 8 Nov 1973-60, 29 Nov 1981-48, 9 Oct 1982-75; Central Saanich 18 and 19 Oct 1991-19 at Vantreight's daffodil farm (Siddle 1992a); Saanich 18 Nov 1973-60 (Crowell and Nehls 1974a).

Winter: Interior – No records. **Coastal** – Delta 11 Feb 1975-1 near 34th St and 33A Ave (Weber 1975a); Cassidy 20 and 23 Jan 1990-1 singing over airport fields (Siddle 1990c); Quamichan Lake 11 Jan 1970-5; Cowichan Bay 1 Jan 1972-10, 18 Dec 1971-11 (Tatum 1972; Comer 1972); Cobble Hill 27 Feb 1964-2; North Saanich 28 Dec 1976-50 at Henry Ave and Martindale Park Rd; Saanich Peninsula 21 Jan 1969-756 on census (Tatum 1971); Central Saanich 11 Jan 1969-120 at feeding area (Stirling and Beckett 1969), 7 Dec 1980-65 near McHugh Rd and Island View Rd, 31 Dec 1984-90 in fields near McIntyre Rd and w end Mallard Ave; Martindale Flats (Saanich Peninsula) 1 Jan 1965-600, 28 Jan 1965-900 concentrated by snow, 2 Dec 1978-47, 26 Dec 1971-300 (Crowell and Nehls 1972b), 28 Dec 1965-800, 29 Dec 1979-37; Central Saanich 18 Feb 1989-76, 12 Dec 1987-24, 29 Dec 1990-42 all in Vantreight's daffodil farm; Rithet's Bog 7 Feb 1960-20; Mount Tolmie 27 Jan 1925-1 (Preece 1925); Victoria Dec 1935-202 counted in 9 locations (Sprot 1937), 14 Dec 1958-665 near Cedar Hill Crossroad, 1 Feb 1972-27 at university (Tatum 1973); Uplands (Victoria) 23 Feb 1944-18; Tillicum (Victoria) 7 Feb 1960-12; Lost Lake (Victoria) 25 Jan 1952-21 in grass field; Victoria 5 Jan 1991-1 near Dallas Rd, Dec 1935-174 counted in 9 locations (Sprot 1937; Siddle 1991b).

Christmas Bird Counts: Interior – Not recorded. **Coastal** – Recorded from 3 of 33 localities and on 7% of all counts. Maxima: Victoria 2 Jan 1966-969; North Saanich 31 Dec 1960-65; Duncan 18 Dec 1971-11.

Extralimital Records: Coastal – Sandspit 20 and 21 Nov 1991-1 (Siddle 1992a); Port McNeill 14 and 15 Nov 1995-1 in airport (T. Greenfield pers. comm.).

Horned Lark

HOLA

Eremophila alpestris (Linnaeus)

RANGE: Circumpolar. Breeds in North America from western and northern Alaska, the arctic coast of northern Canada, including the southern arctic islands, east to Labrador and Newfoundland, but absent or very local breeder in a broad band of the boreal forest across central Canada; then south to Baja California and southern Mexico, the Gulf coast from Louisiana to Veracruz, and, in the east, North Carolina and Alabama; also in Colombia, Morocco, Asia Minor, and Eurasia. In North America, winters from southernmost Canada south throughout the remainder of its breeding range.

STATUS: On the coast, an *uncommon* to locally *common* migrant and winter visitant to the Georgia Depression Ecoprovince, particularly on the Fraser River delta; *very rare* on Western Vancouver Island and the Queen Charlotte Islands. On the southern and northern portions of the Coast and Mountains Ecoprovince, a locally *uncommon* to *fairly common* migrant and summer visitant above the timberline.

In the interior, a *common* to locally *very common* migrant and summer visitant to the Southern Interior, Southern Interior Mountains, Central Interior, and Northern Boreal Mountains ecoprovinces; *rare* to *uncommon* migrant and summer visitant to the Sub-Boreal Interior Ecoprovince; *very rare* in the Taiga Plains Ecoprovince; *uncommon* spring and autumn migrant in the Boreal Plains Ecoprovince. It is an *uncommon* to locally *very common* migrant and winter visitant to the Okanagan and Thompson river valleys of the Southern Interior; *casual* in winter elsewhere in the interior.

Breeds.

Figure 118. Adult male "Dusky" Horned Lark (*Eremophila alpestris merrilli;* Becher's Prairie, east of Riske Creek in the Cariboo-Chilcotin, 26 June 1978; Richard J. Cannings).

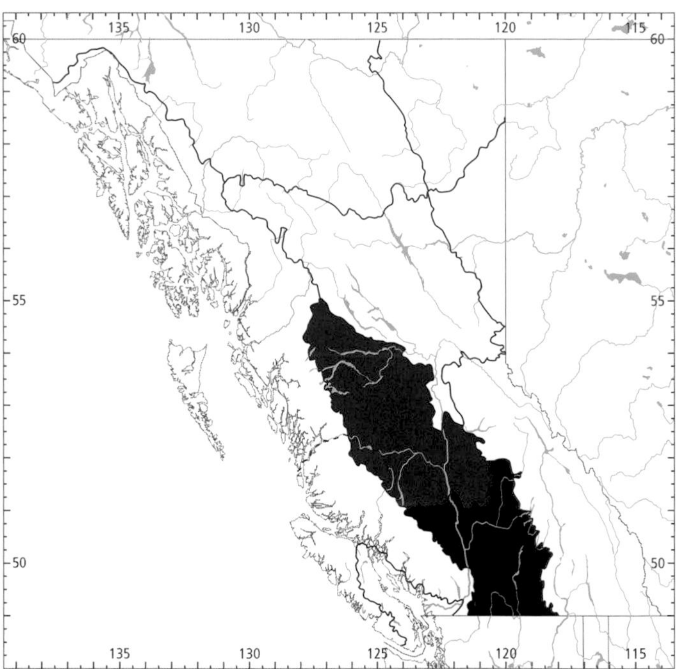

Figure 119. In British Columbia, the highest numbers for the Horned Lark in winter (black) occur in the Southern Interior Ecoprovince; the highest numbers in summer (red) occur in the Central Interior Ecoprovince.

CHANGE IN STATUS: The Horned Lark was formerly resident along southeastern Vancouver Island, from Comox south to Victoria, and in the Fraser Lowland. It has experienced a reduction in numbers to the point of extirpation.

This population, the only population of the subspecies *E. a. strigata* in British Columbia, was first noted on southern Vancouver Island, where it was probably a locally distributed resident in the late 1800s. By the late 1920s, specimens had been collected on the mainland, near Chilliwack, and shortly thereafter it was reported breeding (Behle 1942). Over the next decade or so, additional specimens were collected from local populations adjacent to Vancouver (e.g., University of British Columbia agricultural fields) and at Boundary Bay and Lulu Island.

Small numbers persisted in farmland and prairie areas of the Fraser River delta through the next 30 years. By the mid-1960s, populations on the Fraser River delta were apparently confined to the mowed fields of the Vancouver International Airport on Sea Island. Up to 7 individuals at a time were seen there between 1963 and 1966 (W.M. Hughes pers. comm.). Breeding was suspected there in 1981 (Butler and Campbell 1987), and the latest summer observations occurred in 1987 (W.C. Weber pers. comm.). It was considered a rare resident at the Vancouver International Airport as recently as 1990 (Weber et al. 1990). The destruction of habitat by expanding urbanization has brought this population to the vanishing point (Campbell 1989e), but small numbers may still exist at the airport and in the vicinity of Abbotsford and Chilliwack.

This subspecies has suffered a decline in numbers and restriction of range throughout the coastal plain of Washington and Oregon (Behle 1942; Lewis and Sharpe 1987).

There is no evidence that the other 3 subspecies that occur in the province have experienced long-term changes in distribution or numbers.

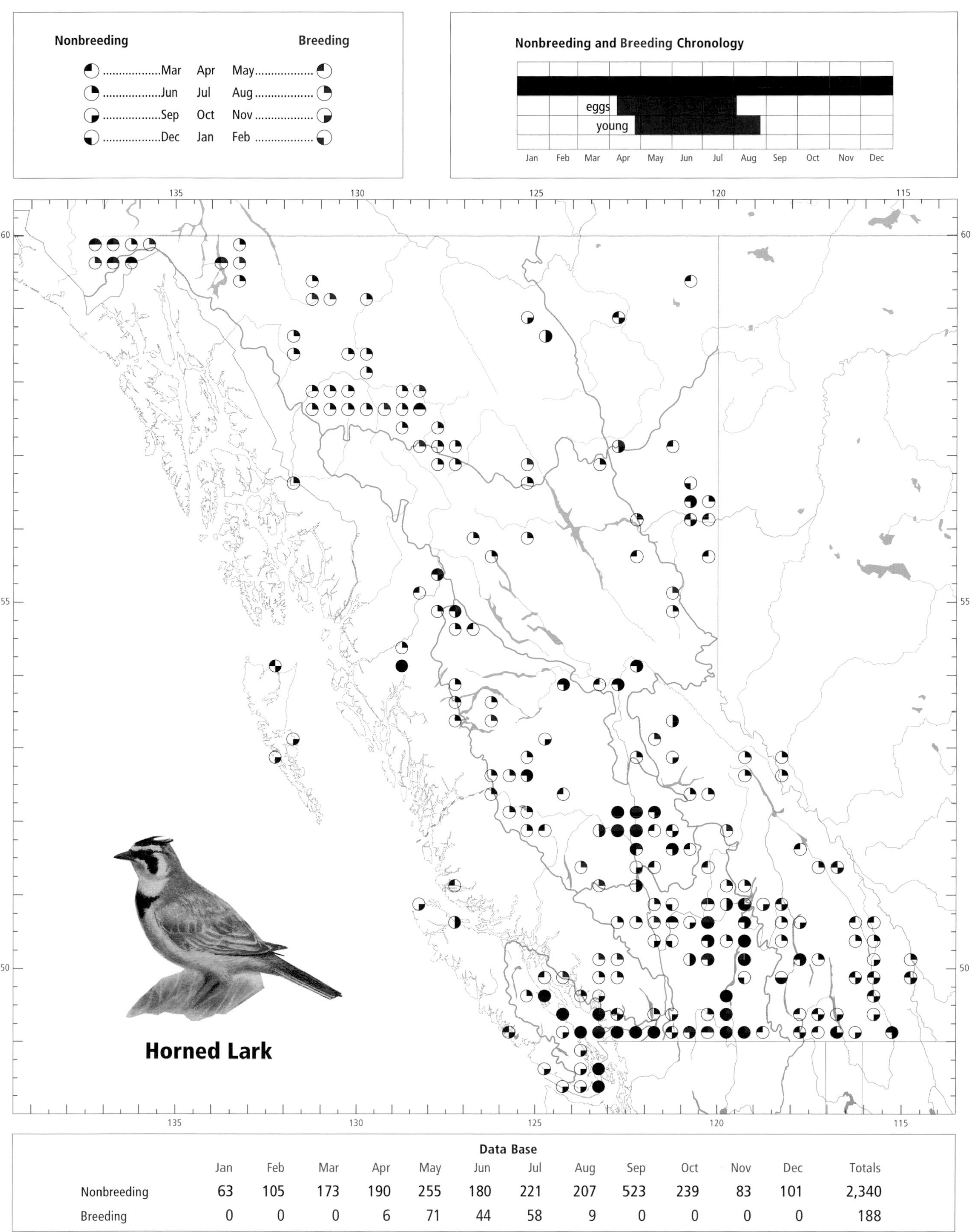

Data Base

	Jan	Feb	Mar	Apr	May	Jun	Jul	Aug	Sep	Oct	Nov	Dec	Totals
Nonbreeding	63	105	173	190	255	180	221	207	523	239	83	101	2,340
Breeding	0	0	0	6	71	44	58	9	0	0	0	0	188

NONBREEDING: The Horned Lark (Fig. 118) is widely distributed in suitable habitat throughout much of British Columbia, including Vancouver Island and the Queen Charlotte Islands. On the mainland coast, it occurs regularly only in the Fraser Lowland. It has not been reported from alpine areas of Vancouver Island or along much of the mainland coast to the north.

The species' elevational distribution changes markedly on a seasonal basis, but during the nonbreeding seasons it frequents lowland valleys and plateaus. During migration and in winter, it has been recorded from near sea level to at least 980 m elevation. In winter it reaches its highest numbers in the valleys of the Southern Interior (Fig. 119).

The Horned Lark is a bird of open, treeless country, and its distribution is closely linked to the presence of such habitat (Fig. 120). In general, Horned Larks in transit or wintering on the coast seek airports, stubble fields, ploughed fields, grassy meadows, road verges, sand dunes, dykes, drier ridges on mudflats exposed at low tide, intertidal zones, and spits of sand and gravel supporting a short growth of salt-hardy plants. In the interior, they also use the wide variety of human-created open habitats mentioned above, as well as short-grass or overgrazed grasslands (Fig. 121), highway verges, dirt and gravel roads, cutlines, and weed patches.

Four geographic races of Horned Lark occur in British Columbia, and representatives of at least 2 of them are to be found in the province throughout the year. At least 3 of the 4 are sufficiently distinct as to be recognizable in the field. Each race differs from the others in its distribution during the nonbreeding and breeding seasons (see Fig. 123), and in the timing and routes of its major migratory movements. To facilitate a better understanding of the distribution and seasonal movements of the Horned Lark in British Columbia, we have included a detailed discussion of the subspecies.

The "Pallid" Horned Lark (*E. a. arcticola*) is the most widespread and abundant subspecies (see Fig. 123), and occurs seasonally throughout the province. It migrates through lower-elevation grassland areas in spring, with the first flocks appearing in southern portions of the province in late March and early April. Numbers build and peak in late April and early May, after which they disperse to alpine nesting areas. In autumn, this race occurs in both high-elevation alpine tundra areas and lowland grasslands. The main movement occurs during the first 3 weeks of September, and by the end of October few birds remain. Coastal migrants are also mainly of this subspecies. Small populations winter across southernmost British Columbia.

The "Dusky" Horned Lark (*E. a. merrilli*) migrates to and breeds at lower elevations in the Okanagan and Thompson valleys of the Southern Interior and in the Cariboo and Chilcotin areas of the Central Interior. There it reaches its

Figure 120. Throughout the year, the Horned Lark is a bird of open, treeless country. In autumn, rolling hills with short grasses attract flocks of migrating Horned Larks (Rose Hill, south of Kamloops, 12 October 1991; R. Wayne Campbell).

Figure 121. In winter, Horned Larks forage in exposed patches of vegetation in overgrazed grasslands created by wind-blown snow on hills with a southern exposure (near Douglas Lake, 9 January 1996; R. Wayne Campbell).

northern distributional limit (Fig. 123). The spring movement is evident by mid-February and persists until mid-April, when flocks disperse over the grasslands to breed. The autumn migration begins in September and continues through October; by early November few birds remain on the nesting grounds. There are few records of this subspecies in British Columbia during December or January.

The "Arctic" Horned Lark (*E. a. hoyti*) is an arctic nesting subspecies. It occurs in the Fort Nelson area as a common spring migrant, less common in autumn (J. Bowling pers. comm.). There is a single record of adults with fledged young at Pink Mountain that may be of this race. West of the Rocky Mountains it has been recorded infrequently, mainly in the Southern Interior and usually in mixed flocks with the "Pallid" Horned Lark during migration to and from its arctic breeding areas. There is a single December record from the Okanagan valley (Munro 1922a).

The "Streaked" Horned Lark (*E. a. strigata*) was formerly a resident in the lower Fraser River valley (Fig. 123) and locally on southeastern Vancouver Island in the vicinity of Comox and Victoria (Munro and Cowan 1947). Any surviving population is restricted to the Vancouver International Airport and perhaps to 2 other sites in the Fraser Lowland.

Generally, the Horned Lark is a very early spring migrant. On the south coast, spring migration is not well defined (Figs. 122 and 124); however, there do appear to be 2 small waves that pass through. The early northbound migrants probably arrive in early February and have passed through the area by the first week of March. A second wave appears in small numbers around the first week of April and continues through the third week of May (Fig. 124).

In the interior, the Horned Lark arrives as soon as bare patches in the snow cover begin to appear. Because of wintering numbers and the differential movements of the subspecies, the spring migration in the Southern Interior is difficult to discern. It appears to occur in 2 main phases. In late winter and early spring (February and March), flocks of the "Dusky" Horned Lark arrive in the Southern Interior and Central Interior; the males begin to establish breeding territories soon after (Behle 1942). Flocks of the "Pallid" Horned Lark begin to arrive in the south as early as March, often in the company of other subspecies (Munro 1958). The migration reaches its peak in the Okanagan valley in April (Fig. 124). During the peak movement, waves of the "Pallid" Horned Lark pass through southern areas when the "Dusky" Horned Lark is already nesting. This differential movement is quite noticeable in the Central Interior, where the first migrants can arrive as early as the end of February. Their numbers build about the first week of March and continue through the end of March. A second movement starts to build in mid-April and continues through the third week of May (Fig. 124). Even in the Southern Interior, birds returning to

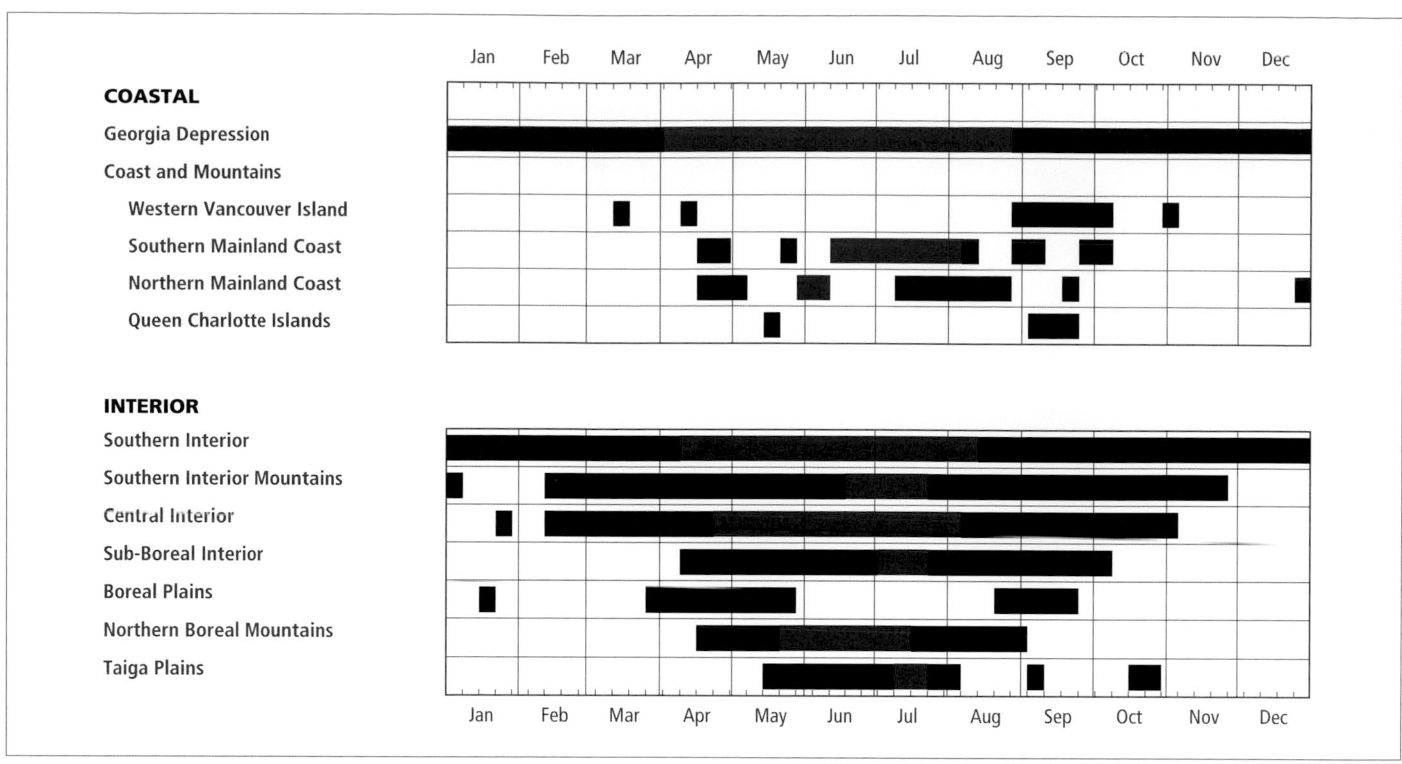

Figure 122. Annual occurrence (black) and breeding chronology (red) for the Horned Lark in ecoprovinces of British Columbia.

alpine areas must wait until late May before their nesting grounds are snow-free.

In the far north, the first migrants may arrive as early as the third week of April; however, the main movement does not begin until the second week of May and continues through the end of the month (Fig. 124). Small numbers pass through the Boreal Plains, arriving the first week of April and reaching a peak in the third week of April; all birds have passed through the region by the third week of May.

The initiation of autumn migration is first apparent in the Northern Boreal Mountains, where nesting birds begin to leave in early August and have left the region by the first week of September (Figs. 122 and 124). In the Taiga Plains, stragglers have been reported in October. In northern British Columbia, most migrants move along alpine areas (Swarth 1936).

The autumn migration is well defined throughout southern portions of the province including the coast. It begins in early August, and a noticeable movement occurs from the third week of August through the third week of September (Fig. 124). Stragglers may occasionally be found in the central regions of the interior until November.

Small wintering populations remain mainly in the Okanagan valley, near Douglas Lake and the Kamloops area of the Southern Interior, and in the Georgia Depression.

In the Georgia Depression and Southern Interior, the Horned Lark has been reported regularly throughout the year. Elsewhere on the coast, it occurs regularly between 15 April and 30 September. In the Southern Interior, it occurs regularly from 25 February through 15 October (Fig. 122).

BREEDING: The Horned Lark breeds over large areas of mainland British Columbia throughout the length of the province. Nesting is suspected, but has not been confirmed, on some coastal and offshore islands, including Vancouver Island and the Queen Charlotte Islands. There is 1 published breeding record from Comox, on eastern Vancouver Island (Flahaut 1953d), which has since been examined and considered to be erroneous.

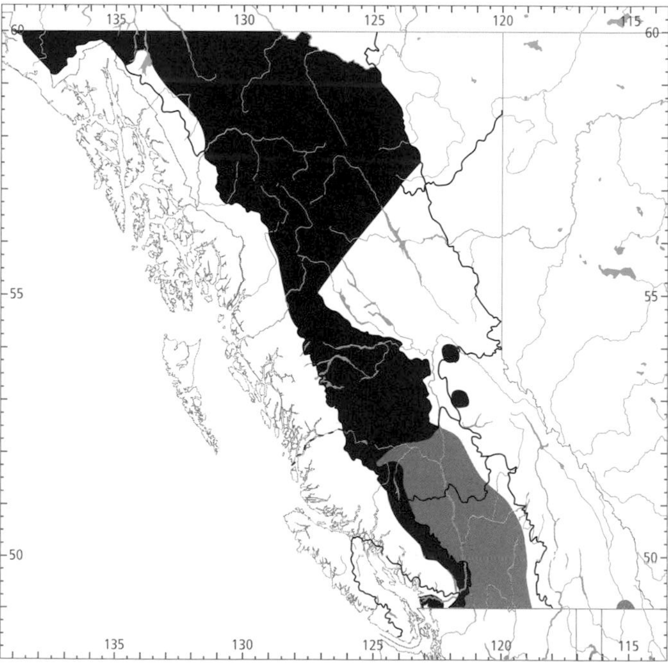

Figure 123. Breeding distribution of 3 subspecies of the Horned Lark (*Eremophila alpestris*) in British Columbia. Red represents the "Pallid" race (*E. a. arcticola*); orange, the "Dusky" race (*E. a. merrilli*); and purple, the "Streaked" race (*E. a. strigata*) on the extreme southwest mainland coat.

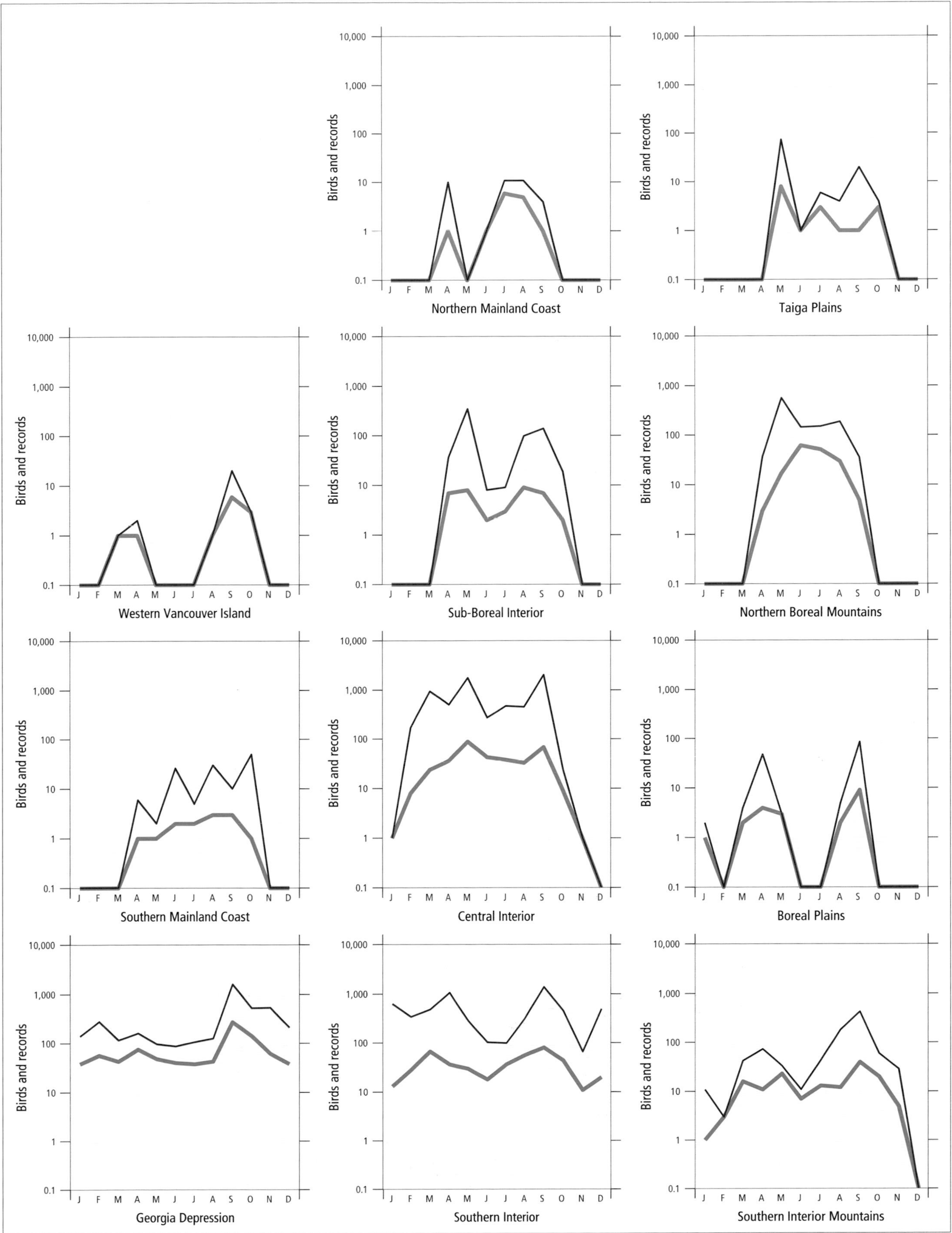

Figure 124. Fluctuations in total number of birds (purple line) and total number of records (green line) for the Horned Lark in ecoprovinces of British Columbia. Christmas Bird Count and nest record data have been excluded.

The Horned Lark reaches its highest numbers in summer in the Chilcotin region of the Central Interior (Fig. 119). Breeding Bird Surveys for both coastal and interior routes for the period 1968 to 1993 contained insufficient data for analysis.

The Horned Lark breeds at elevations from near sea level to at least 2,800 m. The essential feature of its nesting habitat is open space uncluttered with trees, shrubs, or even tall forbs. In British Columbia, this habitat is found above the timberline on mountains (Fig. 125), and on arid prairie land, heavily grazed grasslands (Fig. 126), fallow fields, pastures, airports, and similar open landscapes. The 3 subspecies that breed in British Columbia are each associated with different nesting habitats.

The "Pallid" Horned Lark occupies a variety of windswept alpine habitats above the timberline in mainland mountainous regions over most of the province (Fig. 125), from about 1,800 m to 2,800 m elevation. There are no breeding records on the west-facing slope of the Coast Ranges north of Whistler Mountain. They are also lacking from the British Columbia slope of the Rocky Mountains south of latitude 55°N and from the Selkirk Range (Fig. 123). This subspecies occurs in Jasper National Park right up to the boundary (Cowan 1955), and individuals have been seen on the British Columbia side at Tonquin Pass. It is generally uncommon in the Rocky Mountains on the Alberta side of the summit (Semenchuk 1992).

Its nesting habitats include extensive rocky landscapes with scant vegetation of mat-forming plants; moss-covered slopes; high, dry ridges; fell fields; and alpine tundra (Fig. 125). Swarth (1924) describes the lark's alpine nesting habitat on Nine Mile Mountain in the Skeena River region of the province:

> The country above timberline, covering many miles along the higher ridges, is open and park-like ... White fir and mountain hemlock (*Tsuga mertensiana*) occur, dwarfed and prostrate, forming scattered thickets over ground that otherwise is mostly grass-covered. Snow banks persist throughout the summer, and below the melting snow are occasional little lakes, sometimes an acre or more in extent. On damp slopes grass is replaced by false heather (*Cassiope mertensiana*).

The "Dusky" Horned Lark breeds at elevations from about 250 m to 800 m. It frequents open, rolling grassland areas of the Okanagan and Thompson valleys and the Doug-

Figure 125. Breeding habitat for the "Pallid" Horned Lark (*Eremophila alpestris arcticola*) among short grasses on a slight hump of vegetation (Denain Creek in the Cariboo-Chilcotin, 21 July 1993; R. Wayne Campbell). The nest is located in the lower left corner of the photo.

las Plateau in the Southern Interior, and is also common throughout the southern portions of the Cariboo and Chilcotin areas of the Central Interior (Fig. 126). In the latter region, it does not nest in the lower grasslands, dominated by big sagebrush, but seeks the middle and upper grasslands (A. Roberts pers. comm.). In the Southern Interior Mountains, we have late May and early June specimens, including adults and fledglings, that suggest nesting; these were collected on the Tobacco Plains near Newgate in 1919, 1931, and 1949. No recent nesting has been reported from the east Kootenay, however.

The nesting habitat of the "Dusky" Horned Lark consists mainly of overgrazed grasslands (Fig. 126). The cover requirements of this race of the Horned Lark are fairly precise. In other parts of its breeding range, it has been shown to respond to different grazing regimes (Medin 1986). Many nests are situated in arid grasslands far from water. Although it can obtain adequate moisture from its food, where water is available the Horned Lark uses it readily (Carriger 1899). In some areas of British Columbia the breeding grounds of the "Dusky" and "Pallid" Horned Lark may be separated vertically by only a few hundred metres elevation.

The "Streaked" Horned Lark is known to have nested only in the Georgia Depression, on the Fraser River delta from near sea level to 10 m elevation. Nesting habitat is as described for nonbreeding. Nesting areas are selected where the soil is dry and sandy and the vegetation is predominantly a mixture of short grasses and small forbs.

On the coast, the Horned Lark has been recorded breeding from 5 April (calculated) to 25 August; in the interior, it has been recorded from 15 April (calculated) to 7 August (Fig. 122).

Nests: All nests were situated on the ground in bare, sparsely vegetated, or short-grass areas (Fig. 126). The birds usually scratch a hollow in the earth about 3 inches deep. In Washington, nests have been found in a rut or in the hoofprint of a cow or horse (Bowles 1921). Sites are always well drained. Most nests were built beside a protective feature such as a tuft of grass, rock, clod of earth, plant, small bush or tree, or piece of dead tree, usually on the windward side, a situation reported as common for this species (Dubois 1935; Verbeek 1967).

Nests were small, compact cups, usually deep, with their edges even with the surface of the ground. Nests were composed of dead grasses, plant fibres, plant down, forbs, small pieces of bark strips, dead leaves, sedges, and rootlets. Linings were of fine grasses and plant down.

Eggs: Dates for 64 clutches ranged from 8 April to 4 August, with 54% recorded between 14 May and 22 June.

Figure 126. Breeding habitat for the "Dusky" Horned Lark (*Eremophila alpestris merrilli*) among heavily grazed bluebunch wheatgrass (Becher's Prairie in the Cariboo, 25 May 1994; R. Wayne Campbell). Note the nest with a single young in the lower left corner of the photo.

Calculated dates indicate that eggs can occur as early as 5 April. Sizes of 52 clutches ranged from 2 to 5 eggs (2E-4, 3E-27, 4E-16, 5E-5), with 52% having 3 eggs. The incubation period in British Columbia is 11 or 12 days (*n* = 4). That is consistent with incubation periods reported for Nebraska (Beason 1995), Illinois (Beason and Franks 1974), New York (Pickwell 1931), Wyoming (11 days; Verbeek 1967), and Ontario (11 to 13 days; Peck and James 1987).

The 3 nesting subspecies begin laying eggs at different times in response to differences in their nesting areas. Dates for "Pallid" Horned Lark eggs ranged from 16 May (calculated) to 4 August, those for the "Dusky" Horned Lark from 18 April to 3 July, and those for the "Streaked" Horned Lark from 5 April to 19 July.

Young: Dates for 76 broods (Fig. 126) ranged from 20 April to 26 August, with 53% recorded between 26 May and 20 July. Calculated dates indicate that nests may contain young as early as 20 April. Sizes of 48 broods ranged from 1 to 5 young (1Y-4, 2Y-8, 3Y-24, 4Y-9, 5Y-3), with 50% having 3 young. In the Chilcotin area, young leave the nest in 7 to 9 days and can fly 3 to 5 days later (Cannings 1981).

Dates for "Pallid" Horned Lark young ranged from 27 May to 26 August, those for the "Dusky" Horned Lark from 8 May to 26 July, and those for the "Streaked" Horned Lark from 20 April to 25 August.

Brown-headed Cowbird Parasitism: In British Columbia no instances of cowbird parasitism were reported from 115 nests found with eggs or young. The Horned Lark is an infrequent victim of the Brown-headed Cowbird, with only 3 of the 21 North American races known to have been parasitized by the cowbird (Friedmann 1963); none of these races occurs in British Columbia. Pickwell (*in* Friedmann 1963) suggests that "the early nesting time, the exposed nature of the nest and the habitat, as well as the early termination of the nestling period may mitigate against successful parasitism by the cowbird."

Nesting Success: Of 13 nests found with eggs and followed to a known fate, 6 produced at least 1 fledgling. In the Chilcotin region, Cannings (1981) found an unusually high breeding success rate of 71.4% in a sample of 20 nests that he studied in 1978. The usual success rates are between 30% and 60% (Boyd 1976; Cannings 1977).

REMARKS: Of the world's 76 species of larks, North America is home to only 1 native species: the Horned Lark. It may have the widest natural range of any passerine (MacLean 1970), with the possible exception of the Barn Swallow (Sibley and Monroe 1990).

For additional information on the breeding ecology of the "Dusky" Horned Lark, see Cannings (1977); for the "Pallid" Horned Lark, see Verbeek (1967). Beason (1995) provides details on the life history of the Horned Lark in North America.

NOTEWORTHY RECORDS

Spring: Coastal – *E. a. arcticola:* Victoria Apr 1890-1 (RBCM 646); Sumas Lake 26 Apr 1899-1; Comox 15 Apr 1938-2 (RBCM 15257-8); Masset 15 May 1920-3 (MVZ 102599-601; Brooks and Swarth 1925), 16 May 1946-1 (RBCM 10476). *E. a. strigata:* Victoria Apr 1897-1 (RBCM 1809), 10 Apr 1898-1 (RBCM 645); Sumas Lake 26 Apr 1899-1 (RBCM 643); Sea Island 8 April 1963-3 eggs, incubation started, 25 May 1970-eggs (Campbell et al. 1972a), 28 May 1970-nest with young (BC Photo 3); Lulu Island 2 Apr 1927-1, 20 May 1936-2 (RBCM 7175-6); Chilliwack 22 Apr 1929-1; University of British Columbia (Vancouver) 31 May 1956-nest with 2 young in agronomy fields. *Undetermined subspecies:* Saanich Peninsula 14 Apr 1990-300 (Campbell 1990c); Martindale Flats (Victoria) 6 Mar 1960-10, with 20 Eurasian Skylarks; Ladner 3 Apr 1938-8; Sea Island 2 Mar 1938-20, 1 May 1987-10; Skagit Valley 24 Apr 1971-6; Ross Lake 25 May 1976-2; Chilliwack 19 Mar 1928-1 (RBCM 6571); Stubbs Island 17 Mar 1983-1, 15 Apr 1978-2; Sandspit 12 May 1993-1. **Interior** – *E. a. arcticola:* Manning Park 18 May 1968-3 eggs near microwave station; Beaconsfield Mountain 7 Jul 1968-3 eggs; Okanagan Landing 14 Mar 1922-1; Skookumchuck 6 Apr 1944-1 (RBCM 10965); Knutsford 19 Apr 1933-1; Kamloops 20 Apr 1926-1 (RBCM 7170); 105 Mile House 4 May 1946-5; Becher's Prairie (Riske Creek) 4 Apr 1980-small flocks, 8 Apr 1985-small flocks, 16 May 1991-flocks still passing through; Riske Creek 4 May 1978-110, 14 May 1978-120; Becher's Prairie (Riske Creek) 14 May 1978-4 young about 14 days old (Cannings 1981); near Farwell Canyon 11 May 1990-2 flocks of 200+, still passing through next day; Nulki Lake 3 May 1945-first seen, 6 May 1945-200 (Munro 1949); Tupper Creek 6 May 1938-1 (RBCM 8170; Cowan 1939); Atlin 18 Apr 1933-first seen (Swarth 1936), 12 Apr 1935-arrival. *E. a. hoyti:* Okanagan Landing 12 Mar 1922-1, 6 Mar 1926-1; ne Fort Nelson 15 May 1982-5 (RBCM 17418-22). *E. a. merrilli:* Newgate 26 May 1930-5 adults and 2 immatures (NMC 24656-62); Douglas Lake 29 Apr 1983-3 eggs; Okanagan Landing 4 Mar 1926-1; Coldstream 18 Apr 1892-4 eggs, only lowland nest found in Okanagan valley (Cannings et al. 1987); Lac du Bois (Kamloops) 8 May 1965-3 eggs; Napier Lake 25 Apr 1939-1; Becher's Prairie (Riske Creek) 25 Apr 1969-1 fresh egg in nest; Riske Creek 4 May 1978-15, 14 May 1978-3 eggs. *Undetermined subspecies:* White Lake (Okanagan Falls) 12 May 1977-100; West Bench (Penticton) 4 Apr 1977-200, 12 May 1977-100 (Cannings et al. 1987); Athalmer 4 Mar 1980-2 feeding (Halverson 1981); Celista 17 Apr 1948-350+ in flock; Pete Kitchen Lake 12 May 1981-60; Riske Creek 6 Mar 1983-100, 9 Mar 1979-150 in flock; Rock Lake 14 May 1978-120 in flock (Campbell et al. 1979a); Chezacut 27 Apr 1943-50; Whitesail Lake 25 Apr 1989-10; Willow River 13 Apr 1973-1, earliest spring sighting since 1965; Beverly Lake (Fire Flats) 10 May 1959-200; Quick 19 Apr 1989-90; Peace River 29 Mar 1982-2 (Grunberg 1982c); Baldonnel 20 Apr 1982-20; Nig Creek 24 May 1922-1 (Williams 1933a); Hyland Post 10 May 1982-75; Helmet (s Kwokullie Lake) 15 May 1982-30; Atlin 19 Apr 1981-35; Three Guardsmen Mountain 15 May 1976-1; Chilkat Pass 14 May 1977-150 counted on 30 km survey.

Summer: Coastal – *E. a. arcticola:* Whistler Mountain 1 Aug 1929-young; Prince Rupert Jun 1910-1 (RBCM 15181). *E. a. strigata:* Sea Island 6 Jun 1968-3 eggs, well incubated, 17 Jun 1965-7, 27 Jun 1964-6, 14 Jul 1966-4, 19 Jul 1966-4 eggs, well advanced (W.M. Hughes pers. comm.), 5 Jul 1971-12 (Campbell et al. 1972b), late

Jul 1978-2 fledged young and 11 Jun 1981-adult flushed from field, nesting suspected (Butler and Campbell 1987); North Arm Fraser River 4 Aug 1963-3 eggs; Sumas Prairie 7 Jun 1928-1 (RBCM 6572); Chilliwack 25 Aug 1928-1 fledgling (RBCM 6568). *Undetermined subspecies:* Delta 2 Aug 1975-65; Harrison 2 Aug 1975-25; Comox 25 Aug 1973-15; Port Hardy 31 Aug 1935-1; Tlell River (Queen Charlotte Islands) 26 Aug 1989-15 at river mouth on sand dunes. **Interior** – *E. a. arcticola:* Lakeview Mountain 4 Jul 1965-3 eggs; Ashnola Mountains 21 Jun 1932-1 (RBCM 11422), 22 Jun 1932-eggs; Mount Pearson (near Apex Mountain) 24 Jun 1928-4 eggs, 27 Jul 1932-1 bobtailed fledgling (RBCM 9477), 28 Aug 1919-1 (MVZ 102617); Apex Mountain 20 Jun 1925-3 fresh eggs, 16 Jul 1927-fresh eggs (Cannings et al. 1987); Big Ledge (Monashee Range w Arrow Lake at 2,200 m) 5 Aug 1990-2 adults, 3 fledglings; Mount McLean (Lillooet) 24 Jul 1916-fledgling (RBCM 3137); Rainbow Mountains 20 Jun 1932-14 specimens (Dickinson 1953); Battle Mountain (Wells Gray Park) 1 Jul 1952-young in nest (Edwards and Ritcey 1967); Denain Creek 20 Jul 1993-nest with 3 naked young; Yohetta Mountain 19 Jul 1993-3 broods of flying young being fed by adults; Mount Tatlow 23 Jul 1993-3 recently fledged young being fed by adult; Tonquin Pass 17 Jul 1930-2; unnamed mountain above Barkerville summer 1900-spotted young (Brooks 1903); Vinyards, Cariboo Mountains, 90 km e Prince George 26 Jun 1993-2 adults and 2 young, 24 Jul 1994-4 adults with 5 young; Nulki Lake 28 Aug 1945-50; Nine Mile Mountain 31 Jul 1924-young on wing, 3 adults (MVZ 42206-08); 2 km sw Telkwa 31 Jul 1934-1 (RBCM 1480); Ingenika 27 Jul 1980-1 fledgling (RBCM 17075); Mount Edziza Park 21 Jul 1982-50; between Black Fly and Natcha Hills 4 Jul 1977-nest with 1 egg and 2 nestlings near summit; near Glenora 23 Jul 1919-1 fledgling collected (Swarth 1922); Atlin 9 Jun 1924-young (Swarth 1926), 7 Jun 1958-4 eggs, 2 Aug 1933-1 fledgling (RBCM 5776); Three Guardsmen Mountain 3 Jul 1984-4; Chilkat Pass 15 Jun 1957-5 eggs, 8 Jul 1956-40; Haines Highway 25 Jun 1980-6, including 2 fledglings (RBCM 16810-11); Carmine Plateau, Shini Creek 24 Jun 1983-nestling and 3 adults (RBCM 17750-53). *E. a. hoyti:* Laurier Pass 21 Aug 1912-1. *E. a. merrilli:* Newgate 10 Jun 1949-adult feeding fledgling (RBCM 10971-72); Ashcroft 9 Jul 1892-1 male (ANSP 30902; Rhoads 1893); Lac du Bois (Kamloops) 11 Jun 1941-2 (UBC 285-286); Springhouse 13 Jul 1941-nestlings; Westwick Lake 8 Jul 1955-large numbers of adults and young; Chilko Ranch (Hanceville) 3 Jul 1980-4 eggs; Becher's Prairie 21 Jun 1977-4 well-incubated eggs, 7 Jul 1978-3 young about 8 to 9 days old (Cannings 1981). *Undetermined subspecies:* Cathedral Park 31 Aug 1980-75; Apex Mountain 9 Jun 1980-30; Yoho National Park 9 Jun 1976-20 (Wade 1977); Big Creek 26 Jul 1978-60 in flock; Alkali Lake 17 Jul 1978-17 in flock; Canoe Mountain 25 Aug 1970-100 (Stirling 1971); Willow River 31 Aug 1966-50; Boundary Lake 26 Aug 1985-4; Pink Mountain 21 Jul 1982-12 (BC Photo 820); Stalk Lakes 12 Jul 1976-12; Dawson River 31 Aug 1976-80.

Breeding Bird Surveys: Coastal – Not recorded. **Interior** – Recorded from 7 of 73 routes and on 3% of all surveys. Maxima: Pleasant Valley 15 Jun 1984-24; Meadowlake 3 Jul 1968-13; Chilkat Pass 1 Jul 1976-10.

Autumn: Interior – *E. a. arcticola:* Chilkat Pass 10 Sep 1957-migrating flocks (Weeden 1960); Summit Pass 1 Sep 1943-30 (NMC 29491; Rand 1944); Nulki Lake 3 Sep 1945-200+ (Munro 1949); near Prince George 5 Sep 1994-100 at airport; Lac la Hache 7 Sep 1942-1; 105 Mile House 18 Sep 1942-75; Farwell Canyon 15 Sep 1993-hundreds in large flock above canyon; Horse Lake 17 Sep 1933-1; Sicamous 25 Sep 1893-1 (RBCM 642); Okanagan Landing 1 Sep 1916-30, first seen, 28 Sep 1936-20, latest first arrival, 13 Nov 1930-several small bands flushed from the open range (Munro 1931); Skookumchuck 11 Oct 1939-1 (RBCM 10966); Kokanee Creek 27 Sep 1992-1; Newgate 13 Sept 1939-1 (RBCM 10968). *E. a. merrilli:* Riske Creek 8 Sep 1978-116 (7 flocks), 13 Sep 1978-120 on survey, 23 Sep 1978-125 on survey; 15 km s Riske Creek 30 Oct 1977-2; Buffalo Lake 6 Sep 1953-1 (RBCM 11734). *E. a. hoyti:* Fort Nelson 28 Oct 1986-flock at airport, first autumn sighting (J. Bowling pers. comm.). *Undetermined subspecies:* Fort Nelson 28 Oct 1986-1 (McEwan and Johnston 1987a); Fort St. John 10 Sep 1987-22, 18 Sep 1988-18 (Siddle 1988b); Prince George 2 Oct 1983-15; 50 km e Prince George 17 Sep 1992-15; Chilanko lookout 2 Oct 1992-latest date; Riske Creek 25 Sep 1978-80; Springhouse 3 Nov 1962-1; Williams Lake 16 Oct 1950-1 (RBCM 15707); Revelstoke 19 Nov 1986-1; McGillivray Lake 7 Sep 1970-75 (Stevens et al. 1970); Harmer Ridge 1 Nov 1984-20; Rose Hill (Kamloops) 16 and 17 Sep 1980-1,000+; Douglas Lake to Minnie Lake 20 Oct 1962-72; Quilchena 26 Sep 1963-100; West Bench (Penticton) 1 Sep 1968-50, 24 Sep 1967-50 (Cannings et al. 1987); Vaseux Lake 12 Sep 1960-78, 6 Oct 1973-35 (Cannings et al. 1987); Big Buck Mountain 7 Sep 1966-70; Creston 22 Sep 1984-150. **Coastal** – *E. a. arcticola:* Hazelton 22 Sep 1921-4 (Swarth 1924); Masset 4 Sep 1942-1 (RBCM 10475); Chilliwack 6 Sep 1927-2 (RBCM 6575-76). *E. a. strigata:* Lulu Island 20 Sep 1920-1, juvenal plumage but older than fledgling (RBCM 7174); Victoria 26 Sep 1899-1 (RBCM 1808). *Undetermined subspecies:* near Tlell 14 Sep 1989-15 on beach; Sandspit 8 Sep 1989-1 (Campbell 1990a), 14 Sep 1989-63 feeding in intertidal zone; Takakia Lake 22 Sep 1976-1; Cheam Ridge 19 Sep 1985-40; Maplewood 3 Sep 1983-3 on mudflats; Jericho Park (Vancouver) 22 Sep 1986-10; Tofino 4 Sep 1974-1 (Hatler et al. 1978), 30 Oct 1988-1; Delta 22 Nov 1980-85; Skagit Valley 1 Oct 1971-50; Boundary Bay 30 Oct 1966-14; Central Saanich 23 Oct 1982-30, 22 Nov 1987-25; Anderson Hill 16 Sep 1985-60; Cattle Point 9 Sep 1970-30 (Tatum 1971); Clover Point 17 Sep 1965-55 (Stirling 1966).

Winter: Interior – *E. a. arcticola:* Rose Prairie 21 Jan 1984-1, first winter record (BC Photo 922); Vernon 15 Jan 1980-250, 15 Dec 1984-150, 23 Dec 1960-100; Okanagan Landing 14 Jan 1936-1, 22 Jan 1955-80; Osoyoos 18 Feb 1894-1 (RBCM 644). *E. a. hoyti:* 5 Dec 1918-1 collected from a large flock of *arcticola* (Munro 1922a). *E. a. merrilli:* 15 km s Riske Creek 17 Feb 1984-small flocks, 28 Feb 1981-4 flocks of 12 each scattered over grasslands; near Farwell Canyon 22 Feb 1987-9. *Undetermined subspecies:* Chilcotin River 28 Feb 1979-100 on Junction Range; Riske Creek 16 Feb 1984-4; Williams Lake 22 Jan 1950-1 (NMC 48015); Tranquille 7 Jan 1979-100; Lytton to Kamloops 8 Jan 1978-200 on survey; Kamloops 30 Dec 1970-50; Edgewood 16 Feb 1915-1; Vernon 15 Jan 1980-250; Okanagan Landing 6 Jan 1912-1 (ROM 68218); West Bench (Penticton) 6 Feb 1965-40, first arrival, 6 Mar 1975-75, peak movement (Cannings et al. 1987); Oliver 30 Dec 1990-50 survey between roads No. 9 and 22; Creston 2 Jan 1984-11. **Coastal** – *E. a. arcticola:* Comox 20 Dec 1925-1 (RBCM 6573), 6 Jan 1935-1 (RBCM 9459). *E. a. strigata:* Lulu and Sea islands-6 collected (Turnbull 1929); Lulu Island 31 Jan 1936-1, 17 Feb 1940-6 (RBCM 8844-49); Boundary Bay 5 Jan 1936-1. *Undetermined subspecies:* Kitimat 27 Dec 1986-1; Englishman River 12 Feb 1984-1 on estuary (BC Photo 1140); Chilliwack 14 Feb 1891-1 (AMNH 755660); Lulu Island 7 Dec 1938-100 in 2 flocks; Iona Island 31 Dec 1969-4; Sea Island 17 Feb 1971-1, first spring arrival (Campbell et al. 1972b); Central Saanich 1 Jan 1985-9; Saanich 11 Jan 1969-6 (Tatum 1970).

Christmas Bird Counts: Interior – Recorded from 4 of 27 localities and on 10% of all counts. Maxima: Oliver-Osoyoos 29 Dec 1985-300; Vernon 16 Dec 1984-251; Kamloops 15 Dec 1984-157. **Coastal** – Recorded from 8 of 33 localities and on 6% of all counts. Maxima: Ladner 26 Dec 1964-108; Vancouver 26 Dec 1966-37; Victoria 18 Dec 1982-32.

Purple Martin

Progne subis Linnaeus

PUMA

RANGE: Breeds along the Pacific coast from southwestern British Columbia, western Washington and Oregon, and northern California to northern Arizona, central Utah, and eastern Idaho; and in the interior from central Alberta eastward across the southern parts of the Canadian Prairie provinces and as far east as Nova Scotia; south in the east to southern Texas, the Gulf coast, and southern Florida. Winters in South America from Venezuela and Colombia to Bolivia; casually in Florida.

STATUS: On the coast, a *very rare* to *fairly common* migrant and localized summer visitant to the Georgia Depression Ecoprovince in the Nanaimo Lowland on southeastern Vancouver Island and the Fraser Lowland on the adjacent mainland; *casual* on the Sunshine Coast; *accidental* on Western Vancouver Island.

In the interior, *casual* in the Peace River area of the Boreal Plains Ecoprovince; *accidental* elsewhere.

Breeds.

CHANGE IN STATUS: The Purple Martin (Fig. 127) has never been a common species on the south coast of British Columbia; as with most peripheral species, its numbers fluctuate from year to year. Kermode (1923) noted it was a "common summer resident" in Victoria in the late 1890s, but was quite rare by the early 1920s, with only 2 to 3 pairs reported for several years. He observed that since the arrival of the House Sparrow, a competitor for nesting cavities, martin numbers had "steadily declined." Brooks and Swarth (1925) reported that the Purple Martin was known only from cities on the coast, but did not refer to its abundance. Alford (1928), who reported on 4 years of observations on Vancouver Island between 1912 and 1920, described it as a "summer resident," but not as common as formerly. He did not believe that the House Sparrow was "sufficiently common to menace this or any other species." Cumming (1932), in his review of the birds of Vancouver, found that the species was a scarce summer visitant there, although it was increasing.

By the early 1940s, the Purple Martin had moved north as far as Comox Bay, arriving about 1941 or 1942, attracted to the "holes in pilings on log booming grounds" (Pearse 1946). That range extension suggests an increase in numbers on southeastern Vancouver Island. However, contradictory evidence of such an increase comes from Meugens (1947), who reported the return of the Purple Martin to the Victoria area after an absence of 4 or 5 years. This suggests that in the early 1940s the birds were scarce on southern Vancouver Island.

Numbers declined in the Vancouver area through the 1940s, and by the 1950s reports from the mainland ceased. Between 15 June 1948 and 16 August 1994, when 4 large young were found in a nest box at Maplewood, there was only 1 unsubstantiated report of breeding, in 1966 (Weber 1980).

By the early 1960s, the martin was considered scarce on Vancouver Island, but pairs and small colonies were found in summer along the east coast from Victoria to Campbell River (Stirling 1961). In the 1980s, it was reported from Sooke, Esquimalt Harbour, Cowichan Bay (Fig. 128), Crofton Lake, Chemainus Bay, Ladysmith Harbour, Nanaimo, Saanich, and Campbell River. Numbers have increased slowly since the initiation of nest box programs, and by 1994 D. Copley (pers. comm.) estimated the breeding population for the province at between 35 and 40 pairs.

Figure 127. Pair of adult Purple Martins on pilings at Cowichan Bay, British Columbia (July 1985; Tim Zurowski).

Figure 128. In south coastal British Columbia, the Purple Martin forages and breeds in coastal lowlands in the vicinity of estuaries (Cowichan Bay, 22 July 1992; R. Wayne Campbell).

Historically, the Purple Martin occurred in Vancouver, Lynn Creek, Seymour Creek, Ladner, Lulu Island, New Westminster, Port Coquitlam, Crescent Beach, Surrey, and Mission. Prior to the late 1940s, nesting was confined to the nooks and crannies of the largest buildings in downtown Vancouver. Small colonies and single pairs were reported breeding from the Orpheum Theatre, the Canadian Pacific Railway Station, the Hudson's Bay building, the Bank of Montreal, the old Vancouver Hotel, and the Main Post Office. The last breeding record for downtown Vancouver was in 1948, and the last martin was seen there in 1949. In 1959 a pair nested in a piling near the mouth of the Alouette River, and in 1961, 3 fledged

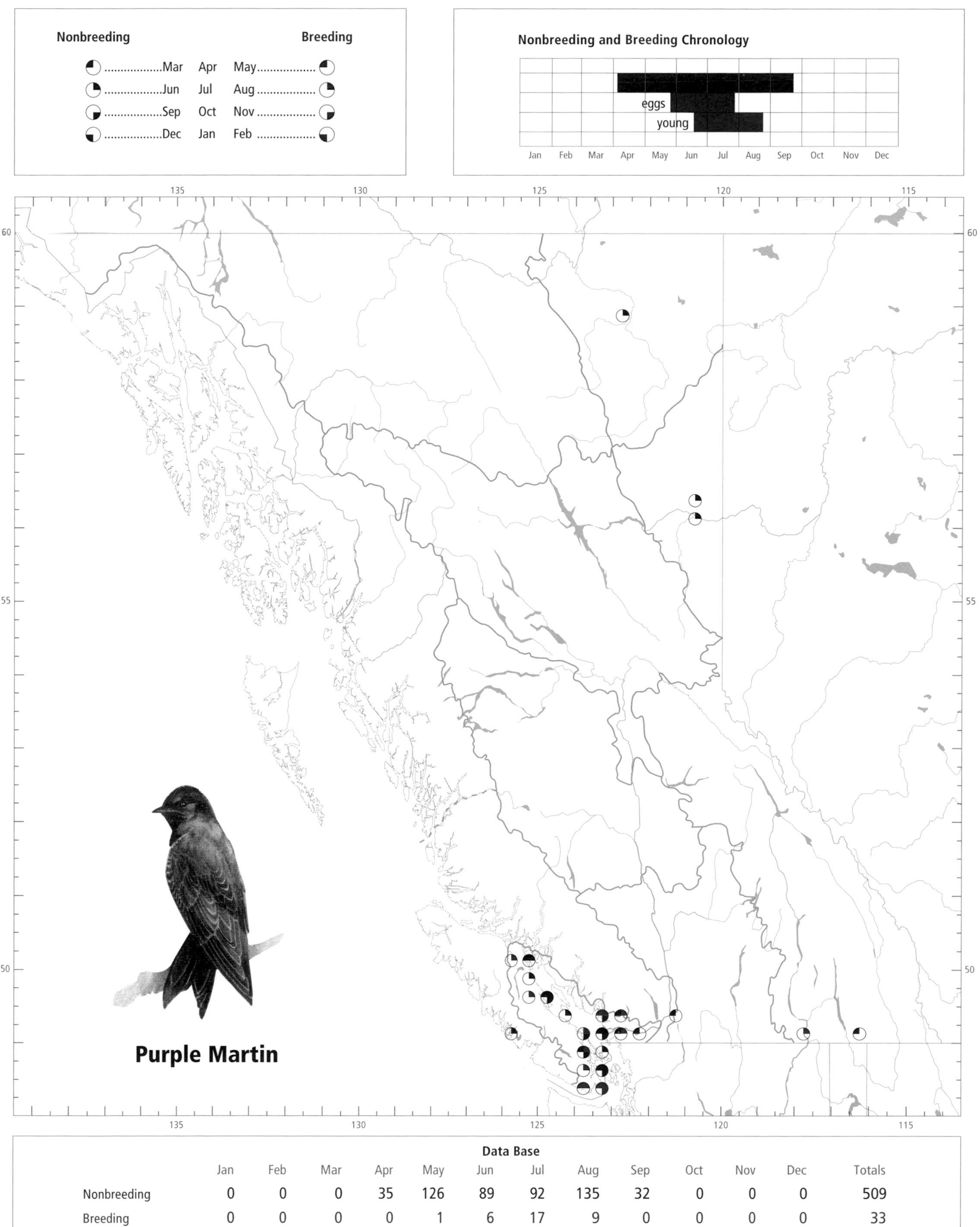

Data Base

	Jan	Feb	Mar	Apr	May	Jun	Jul	Aug	Sep	Oct	Nov	Dec	Totals
Nonbreeding	0	0	0	35	126	89	92	135	32	0	0	0	509
Breeding	0	0	0	0	1	6	17	9	0	0	0	0	33

young were seen being fed by adults at the mouth of the Coquitlam River. No further breeding was reported from the Fraser Lowland until 1994, when a pair nested successfully in a nest box at Maplewood in North Vancouver (Plath 1994). The following year, 3 pairs nested there successfully (T. Plath pers. comm.).

Reports of the Purple Martin breeding in the Peace River region of the Boreal Plains (e.g., American Ornithologists' Union 1983; Godfrey 1986) are apparently based on a record of 2 birds that were collected (1 with a brood patch) from a flock of 8 at Charlie Lake on 12 June 1938 (Cowan 1939). This was the only observation of martins in northeastern British Columbia until Erskine and Davidson (1976) reported a sighting from Fort Nelson in June 1975. Siddle (1982) did not observe the species during a 14-year study period in the Peace Lowland.

NONBREEDING: The Purple Martin (Fig. 127) occurs regularly in British Columbia only on portions of southeastern Vancouver Island and the extreme southwest mainland coast. It has, however, been found in coastal areas of southern Vancouver Island north to Campbell River and west to Tofino, and on the mainland east to Hope (Horvath 1963). It has occurred sporadically elsewhere in the province, including the Peace Lowland (Cowan 1939) and Fort Nelson (Erskine and Davidson 1976).

The Purple Martin occurs from sea level to 700 m elevation. Migrants frequent open areas of the coastal lowland, often foraging over large lakes, harbours, estuaries (Fig. 128), bays, marshes, agricultural fields, and intertidal zones.

Spring migrants reach south coastal areas of the Georgia Depression in early April, with numbers peaking in May (Figs. 129 and 130). First arrivals on Vancouver Island from 1970 to 1994 ranged from 4 April to 27 April (n = 14), with a median arrival date of 22 April. In Oregon and Washington, the Purple Martin usually arrives in mid-April (Mattocks 1986a). In Maryland adult males arrive first, followed later by adult females and subadults (Morton et al. 1990); however, we have no information on a differential migration in British Columbia.

The autumn movement begins in August shortly after the young are on the wing (Fig. 130). At this time, the Purple Martin tends to aggregate in larger premigration foraging and roosting flocks. The largest reported autumn numbers in British Columbia are 40 birds from Florence Lake in 1961 (Stirling 1961), 24 from Prospect Lake in 1968, 21 from Cowichan Bay in 1975, and 20 at Lynn Creek, on the southwest mainland in 1893. In Seattle, just 240 km south of Vancouver, autumn aggregations of nearly 7,000 and 12,500 birds were reported in the early 1940s (Higman 1944; Sharp 1985). Such large numbers may be uncommon today, but they emphasize the peripheral nature of the population in British Columbia.

All birds have left the province by late September (Figs. 129 and 130). Departure dates from Vancouver Island range from 1 to 22 September, with a median date over 14 years of 8 September. In the Vancouver area, the latest date is 7 September. In Seattle the latest departure dates have been reported between 16 September (A. Sprunt *in* Bent 1942) and 16 October (Higman 1944).

The Purple Martin has been reported in British Columbia from 4 April through 22 September (Fig. 129).

BREEDING: The Purple Martin is known to breed only in the vicinity of Maplewood Flats (North Vancouver) on the south-

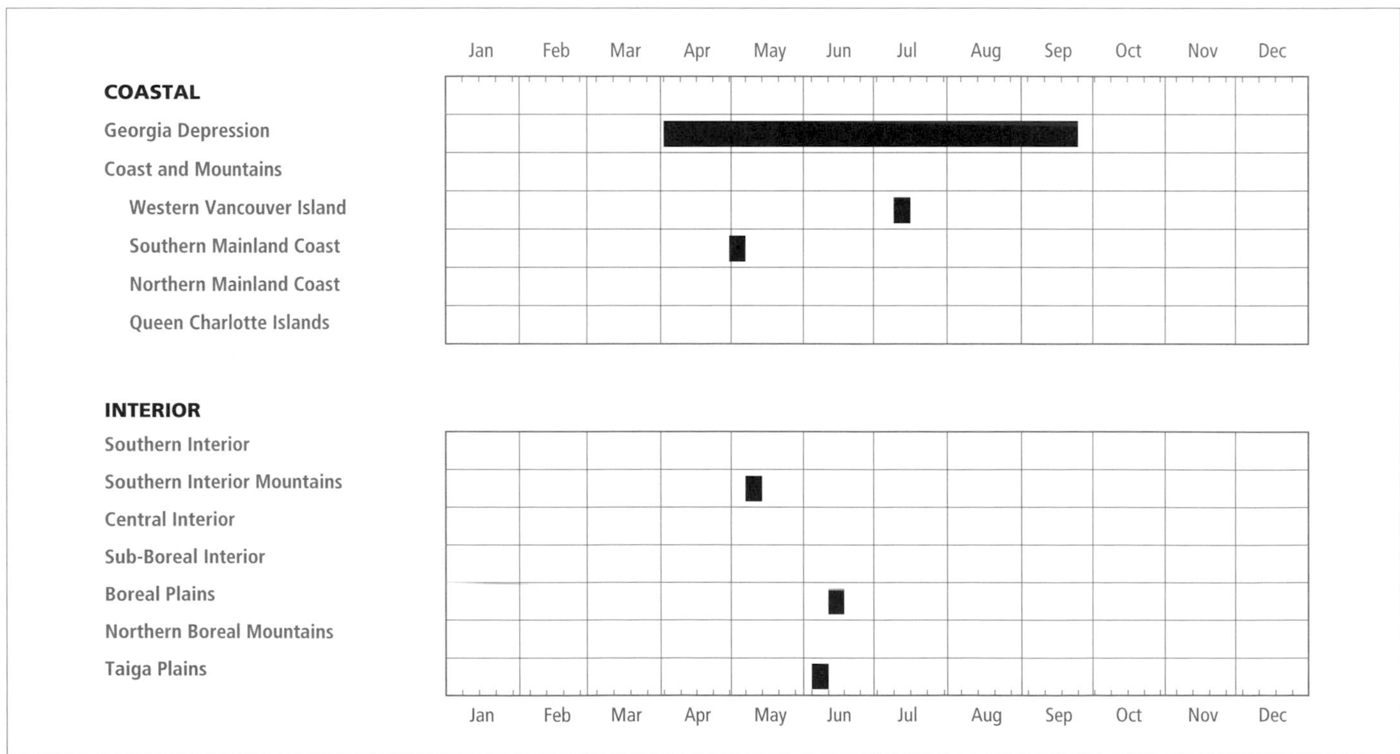

Figure 129. Annual occurrence (black) and breeding chronology (red) for the Purple Martin in ecoprovinces of British Columbia.

west mainland coast, and around Esquimalt Harbour, Cowichan Bay (Fig. 128), and Ladysmith Harbour on southeastern Vancouver Island.

The Purple Martin reaches its highest numbers in the Georgia Depression (Fig. 131). Breeding Bird Surveys for both coastal and interior routes for the period 1968 through 1993 contained insufficient data for analysis. Robbins et al. (1986) note that there were no Breeding Bird Survey strata in North America with significant decreases in Purple Martin numbers over the period from 1965 to 1979. Droege and Sauer (1988), however, show an average annual increase of 2.0% ($P < 0.001$) for the species over the period 1966 through 1987 on North American Breeding Bird Surveys.

The Purple Martin is a bird of British Columbia's lowlands. It frequents sheltered estuaries and harbours, usually where abandoned pilings remain from log-booming activities and serve as nest sites (Fig. 128). It also frequents deciduous second growth near ponds, lakes, mudflats, powerline rights-of-way, and farmland, where insects are plentiful and snags are available for nest sites. Most of the nests reported from British Columbia have been in close association with water, in either a forest or urban setting. Usually these have been in estuaries or on the adjacent shoreline.

The Purple Martin has been recorded nesting in the province from 25 May (calculated) to 26 August (Fig. 129).

Colonies: Throughout North America the Purple Martin is well known for its colonial nesting, although it was once a relatively solitary bird (Turner and Rose 1989). The average colony size is 6 to 8 pairs, but colonies holding over 250 pairs have been reported. In British Columbia, the largest early colony reported was 8 pairs in the preliminary stages of nest building at Fry Lake, near Lower Campbell Lake, in 1959. By 1994, colonies were established at Ladysmith Harbour (8 pairs), Esquimalt Harbour (8 to 10 pairs), Cowichan Bay (8 to 10 pairs), and Victoria Harbour (5 pairs) on southeastern Vancouver Island.

Nests: Most nests (64%; n = 28) were situated in nest boxes (Figs. 132 and 133) or in cavities on wooden pilings over water, followed by natural snags or stumps (14%) and the cornices of tall masonry buildings. At 1 location, pipes in the anchored hull of a large ship were used for nesting.

Nests were constructed of leaves, grass, twigs, feathers, and bark strips. One observer reported the martin's habit of lining the nest cup with fresh, green willow leaves. B. Gates (pers. comm.) observed that adults sometimes covered their eggs with leaves when they left the nest. Hill (1989) discusses a number of theories that have been proposed to explain the adaptive significance of that behaviour.

The heights of 16 nests ranged from 1.7 to 11.0 m, with 11 nests between 2.4 and 5.5 m (Fig. 132).

Egg: Dates for 3 clutches ranged from 28 May to 23 July. Calculated dates indicate that nests could have eggs as early as 25 May and as late as 25 July. Clutch size was 4 or 5 eggs (4E-1, 5E-2). Finlay (1971) gives the incubation period as 15 to 18 days, with a mean of 16.6 days for eastern North America.

Young: Dates for 11 broods ranged from 17 June to 26 August, with 6 recorded between 8 July and 6 August. For most nests the number of young could not be ascertained until

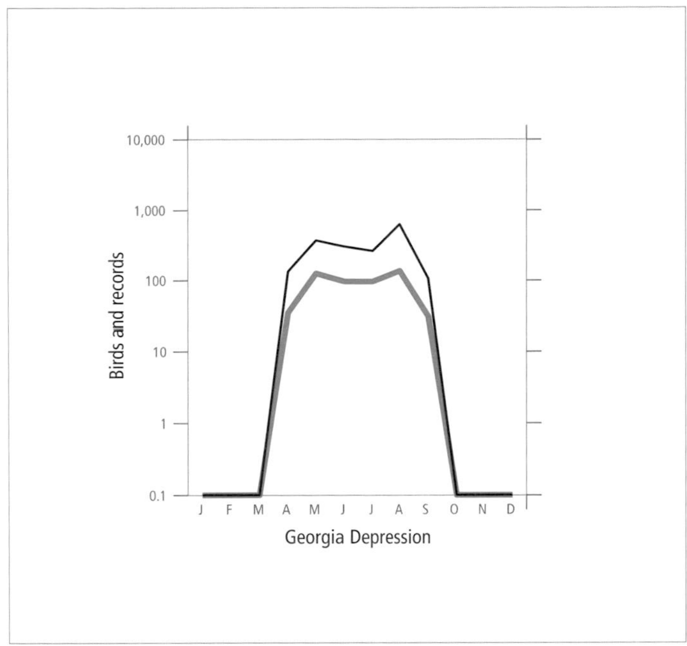

Figure 130. Fluctuations in total number of birds (purple line) and total number of records (green line) for the Purple Martin in the Georgia Depression Ecoprovince of British Columbia. Nest record data have been excluded.

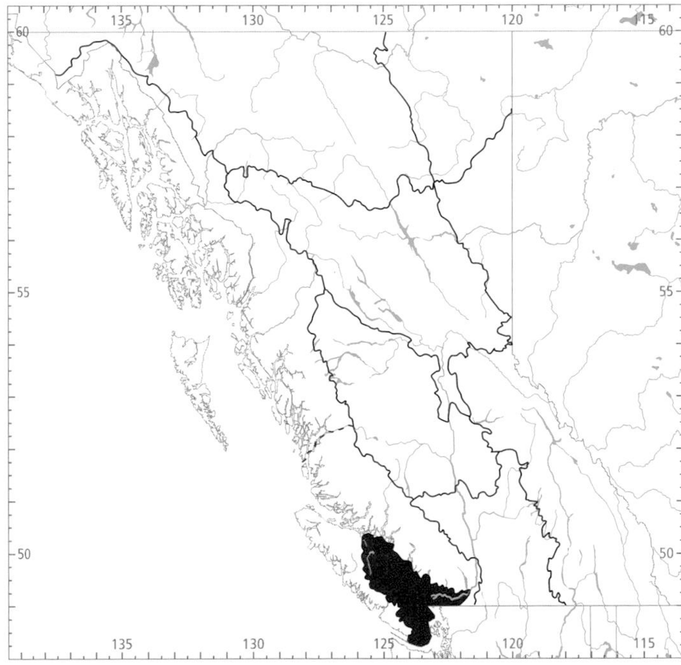

Figure 131. In British Columbia, the highest numbers for the Purple Martin in summer occur in the Georgia Depression Ecoprovince.

they left the cavity. Brood size ranged from 1 to 5 (1Y-3, 2Y-1, 3Y-4, 4Y-2, 5Y-1), with 5 having 2 or 3 young. Nestling period in British Columbia is 31 days. As much as 8 days separated the nest departure of siblings. Finlay (1971) notes the mean nestling period as 27.4 days; it ranged from 26 to 31 days.

Brown-headed Cowbird Parasitism: Cowbird parasitism was not found in British Columbia in 14 nests recorded with eggs or young. Friedmann (1963) notes 1 instance from Detroit, Michigan, but the martins ejected the egg.

Nesting Success: Insufficient data.

REMARKS: Although sparse, the data suggest that there are arbitrary fluctuations in year-to-year numbers of Purple Martins (Fig. 134) that are typical of small or peripheral populations. Any immigration, or the loss of 1 or 2 pairs, is locally significant and causes a visible change in abundance. Such decreases do not necessarily imply the significant changes in population status suggested by some authors (e.g., Weber 1980; Siddle et al. 1991b). There is no doubt that some habitat has been lost, and the birds no longer occur in downtown Vancouver or other areas they regularly inhabited.

Siddle et al. (1991b) estimated that between 300 and 600 pairs of Purple Martins nested annually in southwestern British Columbia from 1900 to 1949. There are only 19 recorded nest sites and, before the recent next box programs, no more than 8 pairs ever nested at any one of those sites. A more conservative estimate of a potential peak historical population would be 152 pairs or, say, 190, giving about 20% for unrecorded nest sites. However, not every nest site was used every year (e.g., nesting at Ladysmith, Chemainus, or Campbell River was not documented until at least the 1950s), and in some years the martins were not reported at all from even the major sites. For instance, Meugens (1947) noted a 4- or 5-year absence from Victoria in the early 1940s, and we have no reports of Purple Martins for 29 of the 50 years from 1900 to 1949. It seems unlikely that the breeding population on the south coast reached 100 pairs over that period.

Figure 132. Cavities in pilings, augmented by nest boxes in the sheltered estuary at Cowichan River, Vancouver Island, provide suitable nesting sites for the Purple Martin in British Columbia (19 April 1993; R. Wayne Campbell). Note the elliptical opening, which is designed to discourage European Starling occupancy.

In 1989 and 1990, 39 and 30 Purple Martins, respectively, were found during autumn censuses on Vancouver Island (Walters et al. 1990; Siddle et al. 1991). In addition, 27 young were known to have fledged that year. The 1990 survey figures did not include all 19 historical sites.

Purple Martin numbers on the west coast of North America began to decline in the late 1940s (Sharp 1985), and that undoubtedly included martin populations in British Columbia. The arrival and subsequent spread of the House Sparrow, and particularly the European Starling, is believed to have had an impact on the Purple Martin population through competition for nest cavities. In British Columbia, the mean number of birds reported each decade fell through the 1950s until the 1980s (Fig. 134). In the first years of the 1990s, their numbers began to rise, and nesting on the mainland was reestablished in 1994. Perhaps this increase in martin numbers has resulted from 2 factors: a decline in European Starling populations and the start of a specialized nest box program on Vancouver Island in 1986.

Some regions, particularly on the Pacific coast, have expressed concern for the future of the Purple Martin (Arbib 1976). It was blue-listed from 1975 to 1981 and has been listed as a species of "special concern" since 1982 (Ehrlich et al. 1992). By 1986 it was reported to be seriously declining in the southern Pacific coast region.

Purple Martins seem particularly sensitive to a scarcity of nest and roosting sites. They appear to respond well to nest box programs (Figs. 132 and 133) and local nesting opportunities, but there have been frustrating setbacks. In 1986, N.K. Dawe observed a pair of second-year birds inspecting cavities in pilings that were situated among islands built during a marsh creation project on the Campbell River estuary; the birds apparently did not nest. In 1988, Dawe again observed Purple Martins (a pair of adults and a pair of second-year birds) inspecting the cavities. Later that same year, a family group of birds appeared on the estuary, suggesting that at least 1 of

Figure 133. Nest box programs, especially in the Cowichan River estuary, have been the main reason Purple Martin populations are stable to increasing in British Columbia (22 July 1992; R. Wayne Campbell).

the pair had nested successfully. Subsequently, however, all the pilings in the estuary were cut down as part of a misguided habitat restoration. It is not clear how the removal of the pilings benefited the estuary, but martins have not been seen there since. A nest box program may encourage the return of the Purple Martin to the estuary.

The Washington Department of Wildlife (1990) provides a number of management recommendations for the Purple Martin that include: retention and protection of old pilings (Fig. 128), retention and creation of snags, use of fire to improve foraging habitat, avoidance of insecticide applications within 12 km of martin nesting colonies, and placement of nest boxes in suitable habitat. Plans for a nest box designed for the Purple Martin are included in the report. Observers throughout coastal areas of the Georgia Depression are encouraged to document sites with pilings and to inform their local wildlife agency. Steps can then be taken to have the sites assessed as potential nesting areas for the martin.

An observation of 1 bird on the Kwinitsa Breeding Bird Survey in 1978 has been excluded from the account for lack of sufficient documentation (see Appendix 3). Records of the Purple Martin are also on file from Kelowna and Erickson (Butler et al. 1986); however, adequate documentation is lacking and they have also been excluded from the account.

General summaries of the natural history of the Purple Martin in North America can be found in Wade (1966), Layton (1988), and Turner and Rose (1989). Coloniality and breeding biology are discussed by Stutchbury (1991), while Finlay (1971) discusses the breeding biology of the species near the northern limit of its range at Edmonton, Alberta. Seasonal foods and the relationship of Purple Martins to mosquito control have been studied by Johnston (1967) and Kale (1968), respectively.

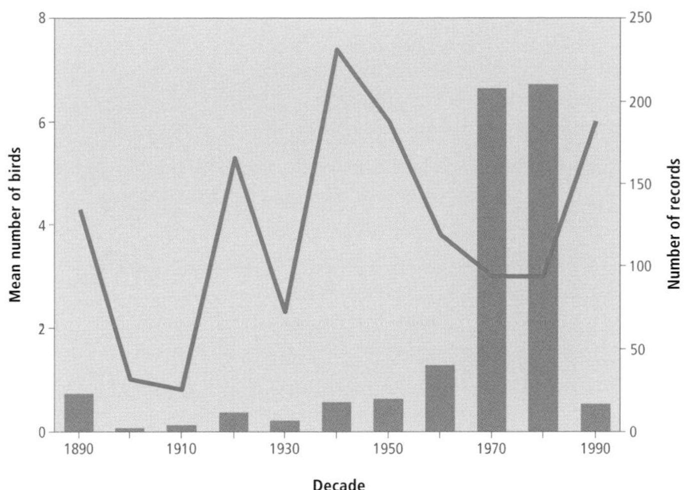

Figure 134. Fluctuations in the mean number of birds per decade (line) and the total number of records per decade (bars) for the Purple Martin in British Columbia. Data for the 1990s include only the years 1990 and 1991. See text for details.

NOTEWORTHY RECORDS

Spring: Coastal – Sooke 28 May 1973-4 eggs; Victoria 4 Apr 1969-1, 19 Apr 1959-6, 16 May 1956-6 flying over inner harbour (Anonymous 1956); Saanichton Spit 28 Apr 1975-15; Cowichan Bay 4 Apr 1983-2, 27 May 1987-8; Lake Cowichan 10 Apr 1978-1 exhausted adult male (Mattocks and Hunn 1978b); Duncan 15 Apr 1961-1; Crofton 8 Apr 1972-1; Crescent Beach 9 Apr 1940-25, 18 May 1947-25; Iona Island 7 Apr 1891-5, 21 Apr 1895-12, 23 Apr 1994-1 adult female, 25 and 26 May 1991-1 pair (Siddle 1991c); Burnaby Lake 5 Apr 1974-2; Port Moody 4 May 1972-1 (Campbell et al. 1974); Hope May 1960-1 (Horvath 1963); Comox 26 and 27 May 1940-1 female (Laing 1942); Campbell River estuary 27 Apr 1974-1. **Interior** – No records.

Summer: Coastal – Sooke basin 24 Aug 1985-2 young at nest hole and 1 on wing, 27 Aug 1985-4 fledglings (Campbell and Gibbard 1986); Victoria 1 Jun 1947-16, 14 Jun 1963-2 eggs in nest, Jul 1967-nesting in pilings in inner harbour (BC Photo 156b), 27 Aug 1956-16; Esquimalt Harbour 2 Aug 1986-12; Cowichan Bay 13 Jun 1987-5, 14 Jun 1986-10, 25 Jun 1987-5 active nests in boxes (B.R. Gates pers. comm.), 22 Jul 1979-1 (Clark 1979), 23 Jul 1985-adult incubating, 11 Aug 1988-17, 16 Aug 1975-21, 26 Aug 1984-3 young (Campbell and Gibbard 1984); Crofton 17 Jun 1971-1 nestling; Ladysmith 21 Jul 1953-2; Nanaimo 1 Aug 1984-3 large young being fed (Campbell and Gibbard 1984); Iona Island (Richmond) 16 Jun 1994-1 female over sewage lagoons (Plath 1994), 27 Aug 1985-1; Tofino 11 Jul 1979-1; Vancouver 15 Jun 1948-last breeding in city, 5 Aug 1929-15; Maplewood 26 Jun 1994-1 adult female at nest box, 16 Aug 1994-4 large young in nest box (first breeding on the mainland coast since 1948; Plath 1994); Coquitlam 28 Jun 1959-5 eggs; Coquitlam River 1 Aug 1961-3 fledged young; Alouette River 28 Jun 1959-5 eggs collected; Lynn Creek 9 Aug 1893-20; Wilson Creek 7 Jul 1980-1 juvenile at pilings on estuary; Port Mellon 31 Jul 1989-1 female; Courtenay 15 Jun 1972-female incubating; Comox 12 Aug 1956-15; Mitlenatch Island 6 and 9 Aug 1964-1 (Campbell 1964); 8 Jul 1986-pair of first spring birds checking cavity in pilings (BC Photo 1118), 27 Jun 1988-1 pair of adults, 1 pair of first spring birds, 11 Jul 1988-6. **Interior** – North Erickson 10 May 1984-1 pair (Butler et al. 1986); Fort St. John 11 Jun 1982-1; Charlie Lake 12 Jun 1938-8 (Cowan 1939); Fort Nelson 4 Jun 1975-3.

Breeding Bird Surveys: Coastal – Recorded from 1 of 27 routes and on less than 1% of all surveys. Maximum: Nanaimo River 4 Jun 1978-1. **Interior** – Not recorded.

Autumn: Interior – No records. **Coastal** – Vancouver 4 Sep 1949-3 flying around private house, 5 to 7 Sep 1929-15 to 20 wheeling around old home; Nanaimo 3 Sep 1991-3 (Siddle 1992a); Cowichan Bay 6 to 19 Sep 1970-3, 22 Sep 1991-1; Mount Richards 12 Sep 1974-5; Oliphant Lake 11 Sep 1976-3; Prevost Hill (Saanich) 17 Sep 1974-3; Swan Lake (Saanich) 7 Sep 1986-7; Beaver Lake (Saanich) 5 Sep 1974-4; Colwood 11 Sep 1972-2 (Tatum 1973); James Bay (Victoria) 2 Sep 1979-1 adult with 2 begging fledglings on Marifield Street; Victoria 12 Sep 1981-2 flying over 1201 Fairfield Road, 12 Sep 1984-3 flying over golf course (Hunn and Mattocks 1985).

Winter: No records.

Tree Swallow

TRSW

Tachycineta bicolor Vieillot

RANGE: Breeds from western Alaska, central Yukon, and the northwestern and southern Mackenzie River Basin east across all Canadian provinces to Newfoundland and Nova Scotia, and south in the west to California and in the east to North Carolina. Winters from southern California and New York (occasionally further north, including British Columbia) south to the Caribbean and the coast of Colombia, Venezuela, and Guyana (Ridgely 1989).

STATUS: Status is complex. On the coast, a *common* to *abundant* (occasionally *very abundant*) spring migrant, *fairly common* to *very common* summer and autumn visitant, and *casual* in winter in the Georgia Depression Ecoprovince except on the Sunshine Coast, where it is *rare; common* to *very common* (occasionally *abundant*) spring migrant, *uncommon* to *fairly common* summer visitant, and *uncommon* in autumn on Western Vancouver Island and the southern mainland portions of the Coast and Mountains Ecoprovince; *fairly common* to *very common* spring migrant, *uncommon* to *fairly common* summer visitant and autumn migrant (occasionally *very common*) to the northern portions of the Coast and Mountains and the Queen Charlotte Islands.

In the interior, a *common* to *abundant* (occasionally *very abundant*) spring migrant, *fairly common* to *very common* summer visitant and autumn migrant (occasionally *abundant*) across most of southern British Columbia north into the Cariboo and Chilcotin areas of the Central Interior Ecoprovince; *uncommon* to *common* spring migrant and *uncommon* to *fairly common* summer visitant and autumn migrant to the Sub-Boreal Interior and the Northern Boreal Mountains ecoprovinces; *common* to *very common* spring migrant, *uncommon* to *fairly common* summer visitant, and *fairly common* to *common* late summer migrant to the Peace Lowland of the Boreal Plains Ecoprovince; *fairly common* to *very common* spring migrant and *uncommon* to *fairly common* summer visitant and late summer migrant to the Taiga Plains Ecoprovince.

Breeds.

Figure 135. Adult male Tree Swallow (Creston, 31 May 1995; R. Wayne Campbell).

NONBREEDING: The Tree Swallow (Fig. 135) has a widespread distribution throughout most of British Columbia, including Vancouver Island and the Queen Charlotte Islands. In the southern half of the province, it is characteristically a bird of passage and a summer visitant from Crowsnest Pass to Tofino and from Osoyoos to Smithers. Along the coast, it is widely distributed throughout lowland areas of southeastern Vancouver Island and the Queen Charlotte Islands. It has a sparse distribution on Western Vancouver Island and along the coast to Alice Arm. North of Smithers, it is infrequent and local except in the Peace Lowland between Tupper Creek and Charlie Lake, where it is fairly widespread. It occurs locally in the Northern Boreal Mountains and Taiga Plains, from Kwokullie Lake in the east to the Tatshenshini River valley in the west. There are no records from the Rocky Mountain Trench between McLeod Lake and the Liard River valley.

The Tree Swallow is a bird of lower altitudes through most of its range, but in portions of the northern interior it may reach 1,500 m elevation. Its distribution is usually associated with wetlands such as beaver ponds, lakes, open

Figure 136. Open streams and marshlands near Pitt Lake, in the lower Fraser River valley, are important staging areas for Tree Swallows during spring migration in British Columbia (26 February 1996; R. Wayne Campbell).

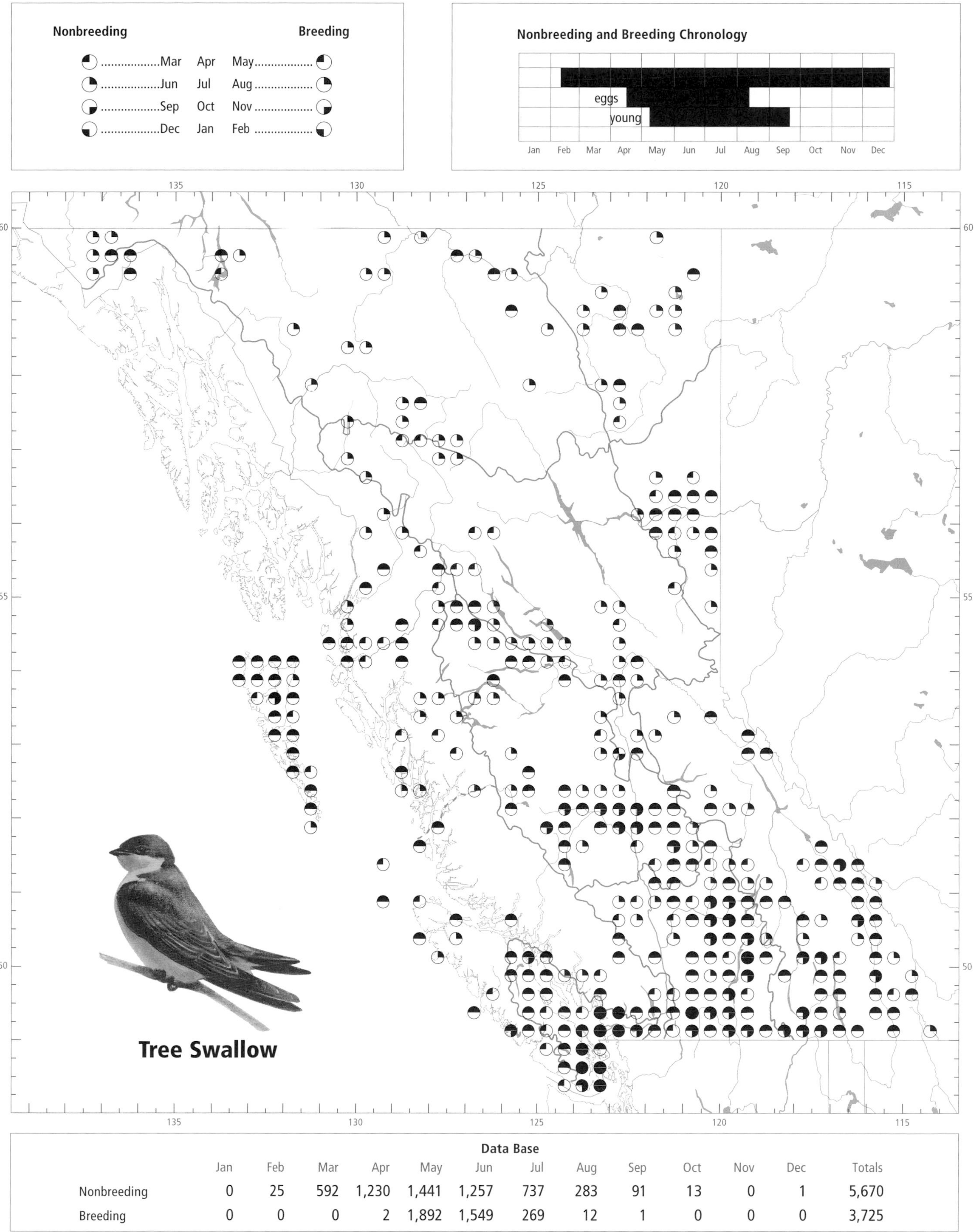

Data Base

	Jan	Feb	Mar	Apr	May	Jun	Jul	Aug	Sep	Oct	Nov	Dec	Totals
Nonbreeding	0	25	592	1,230	1,441	1,257	737	283	91	13	0	1	5,670
Breeding	0	0	0	2	1,892	1,549	269	12	1	0	0	0	3,725

streams and slow-moving rivers (Fig. 136), marshy meadows, reservoirs, sewage lagoons, ditches, and estuaries, all of which generate an abundance of aerial prey. It can also be found over grasslands and farmlands, and in rural and urban residential areas. It is generally absent from subalpine forests.

There are probably 3 discrete migration routes entering the province. These may include a coastal route through the Georgia Depression and along the west coast of Vancouver Island to the Queen Charlotte Islands and the northern Coast and Mountains, a route east of the Cascade Mountains through the valleys of the Southern Interior and Southern Interior Mountains into the middle of the province, and a third route east of the Rocky Mountains (Butler 1988) into the Boreal Plains and the Taiga Plains ecoprovinces.

Spring migrants may enter the Georgia Depression, on the southwest coast, during the second week of February, and by the end of March aggregations of several thousand birds may be seen on the Fraser River delta and near Victoria (Figs. 137 and 138). Migrants spreading inland from the Fraser River delta encounter high mountains that inhibit northward movement. They often congregate at the mouth of the Pitt and Harrison rivers (Fig. 136), where aerial insects are abundant over expanses of water and wetlands. On the west coast of Vancouver Island, the southern and northern mainland of the Coast and Mountains, and the Queen Charlotte Islands, birds arrive in late March or early April and peak in May (Fig. 138).

In the interior, the narrow north-south valleys offer snow-free corridors early in spring migration. Swarms of insects, which emerge as soon as the valley bottom lakes and marshes thaw, permit Tree Swallows and other early insectivorous birds to survive while most of the landscape is still locked in snow and ice. An ability to subsist on a vegetarian diet during cold spells (Robertson et al. 1992) may allow them to survive short periods of severe weather in British Columbia.

A few Tree Swallows occasionally reach the southern Okanagan valley in late February, but most appear in mid-March. Cannings et al. (1987) cite 22 March as the mean of 36 annual first records. By early April, flocks of up to 1,000 birds are moving through. A similar pattern occurs in the Kootenay region of the Southern Interior Mountains, but flock sizes appear to be much smaller, based on average observations of 8 birds per record versus 43 birds per record in the Southern Interior (Fig. 138). Their arrival in the Central Interior has been recorded as early as the second week of March, and numbers there peak in April. Aggregations of 2,500 swallows have been reported from the Central Interior. The migration may reach the Sub-Boreal Interior in late March and the Peace Lowland in early to mid-April; the movement peaks in May (Fig. 138). By the third week of April, the migration has crossed the southern boundary of the Taiga Plains. Further west, in the Northern Boreal Mountains, the Tree Swallow may not arrive until May.

Based on the mean number of birds per record, coastal abundance in spring is generally lower than that reported from the interior, although the largest reported aggregation we have is of 4,000 birds seen in the Fraser Lowland. The mean number of birds per record tends to increase as migration moves north, and peaks in the Central Interior with an average of 58 birds. Mean numbers of birds per record decline further north, although there is an increase in the Peace River area of the Boreal Plains, which further suggests that this population is probably derived from a different migratory corridor than those from west of the Rocky Mountains. The largest aggregation there was 800 birds.

The southward movement of Tree Swallows begins soon after the young fledge. There is no apparent increase in numbers as the young leave the nest, but rather a steady decline in the population over most of the province after July

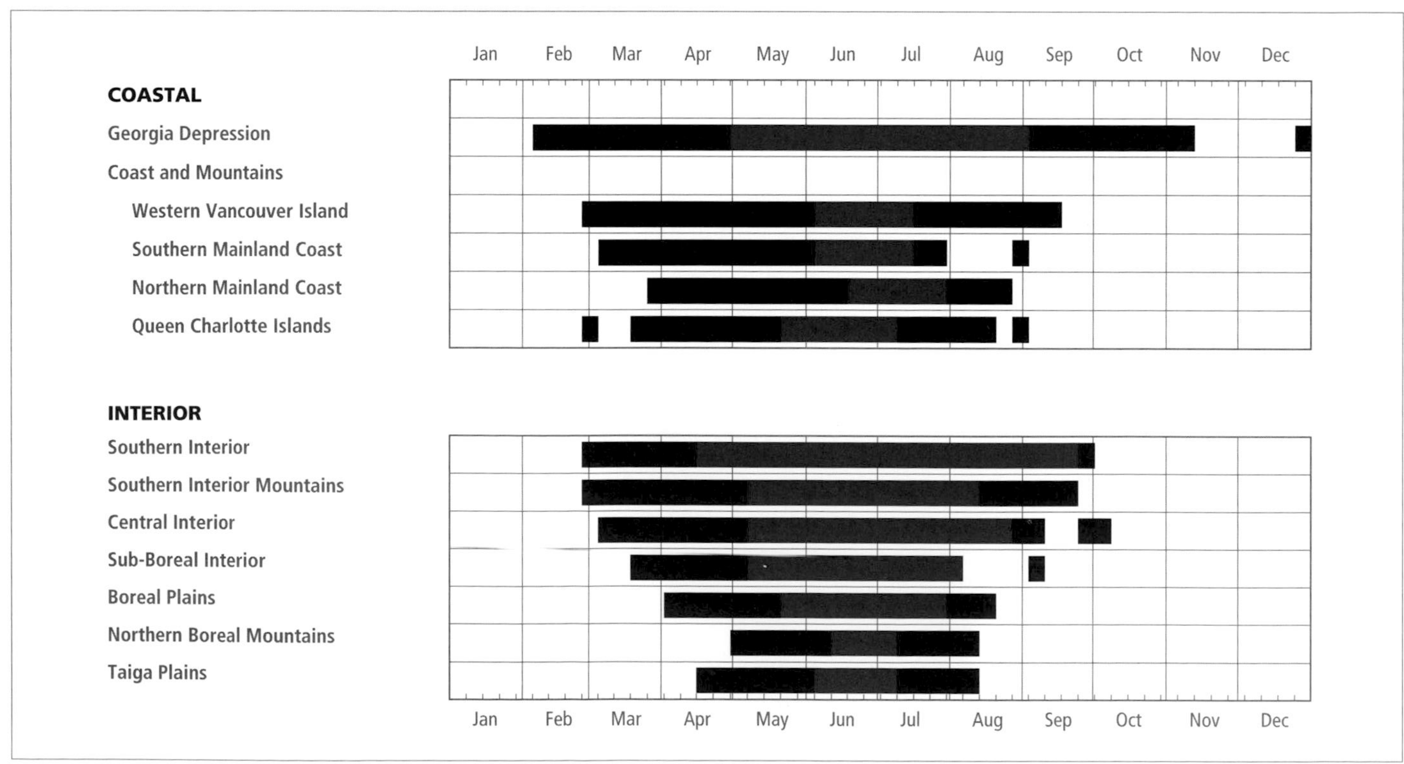

Figure 137. Annual occurrence (black) and breeding chronology (red) for the Tree Swallow in ecoprovinces of British Columbia.

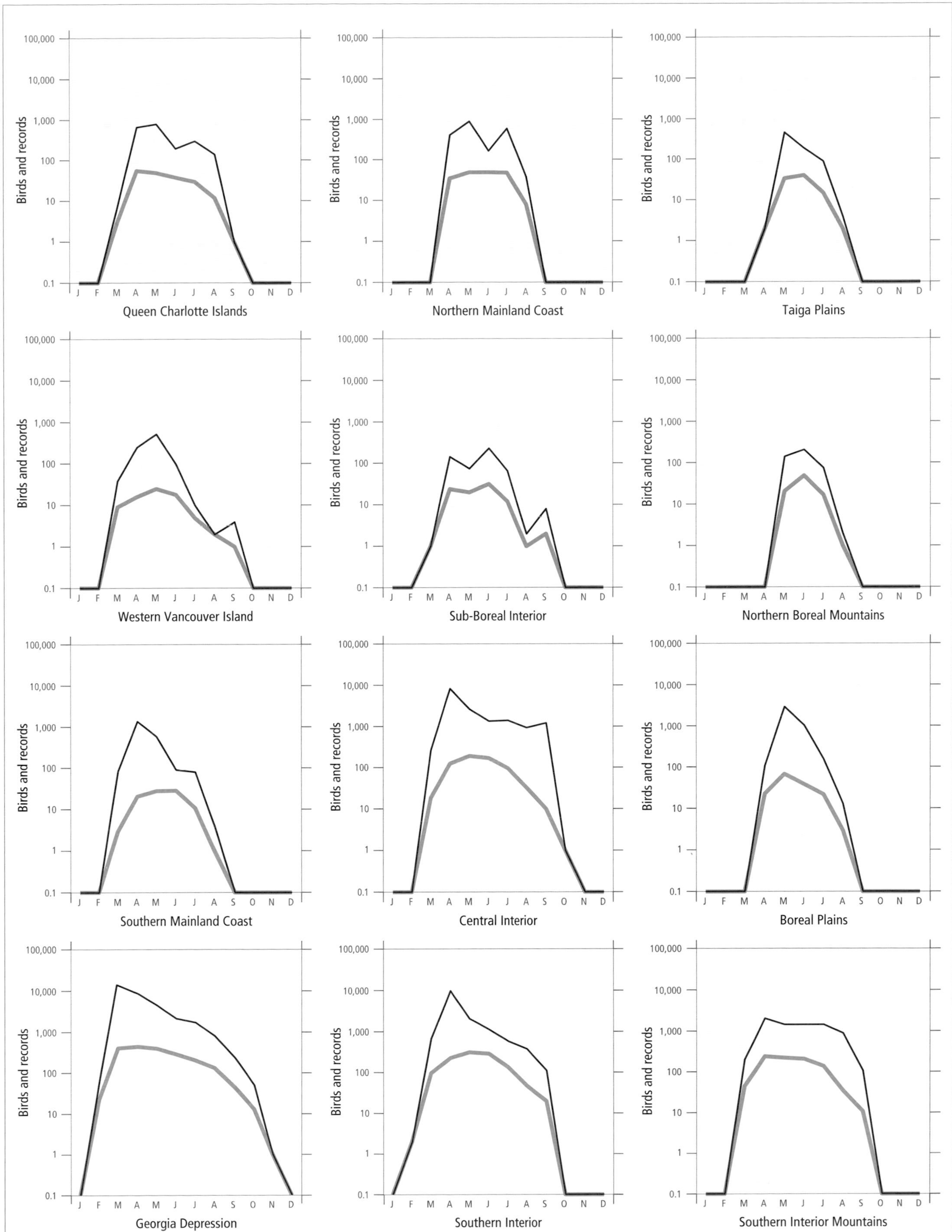

Figure 138. Fluctuations in total number of birds (purple line) and total number of records (green line) for the Tree Swallow in ecoprovinces of British Columbia. Christmas Bird Count and nest record data have been excluded.

(Fig. 138). This observation suggests that, as with the Violet-green Swallow, the young leave the nesting grounds as soon as they are on the wing.

In the northern portions of the province, the Tree Swallow has usually left by the end of July (Fig. 138). Further south, in the Central Interior, Southern Interior, and Southern Interior Mountains, large numbers are still present in August. Most birds, however, have left the interior by early September (Fig. 138). On the Queen Charlotte Islands, the autumn departure occurs in August, while in the Georgia Depression the movement is more gradual and continues into September. A few small flocks may be seen in October, and individuals may remain until November and December.

Throughout the province, most southbound flocks do not exceed 100 birds. The largest interior record is a flock of 500 birds over the marshes at Scout Island, Williams Lake, in early September. There is only 1 coastal record from the southern migration of a flock of 100 birds or more.

On the coast, the Tree Swallow has been recorded fairly regularly from mid February to 11 November, in the central southern interior from 27 February to mid September, and in the far northern areas of the province from 22 April to 12 August. There is 1 early winter record for the Georgia Depression (Fig. 137).

BREEDING: The Tree Swallow normally breeds as single pairs in suitable habitat throughout most of the southern portions of the province. Occasionally, small colonies are found.

On the coast, it breeds on Vancouver Island and the Queen Charlotte Islands and on the adjacent mainland of the Coast and Mountains, north to Prince Rupert and Terrace, although there are few records between the north end of Vancouver Island and the Kitimat area.

In the interior, the Tree Swallow breeds north to Smithers in the Sub-Boreal Interior and in the Peace Lowland of the Boreal Plains. Further north, its distribution becomes spotty. Nesting has been reported from the Telegraph Creek, Atlin, and Tatshenshini River areas, as well as in the Fort Nelson Lowland and near Kotcho Lake.

Figure 140. Wetlands with dead trees for nest sites and open space for foraging provide natural breeding habitat for the Tree Swallow in British Columbia. At least 3 pairs of Tree Swallows nested in cavities in dead trees in this beaver pond 20 km east of Fort Nelson (26 June 1996; R. Wayne Campbell).

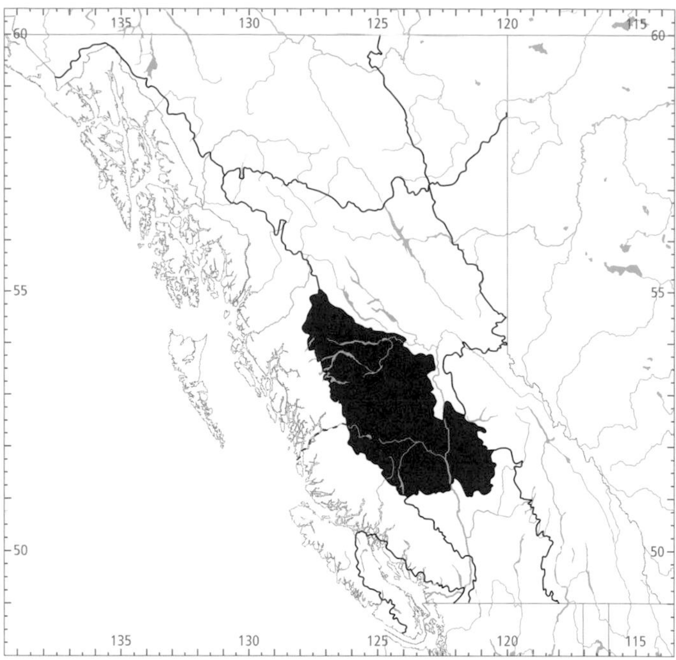

Figure 139. In British Columbia, the highest numbers for the Tree Swallow in summer occur in the Central Interior Ecoprovince.

The Tree Swallow reaches its highest numbers in the Cariboo and eastern Chilcotin areas of the Central Interior (Fig. 139). Analysis of Breeding Bird Surveys for the period 1968 through 1993 could not detect a net change in numbers on either coastal or interior routes.

The Tree Swallow breeds from near sea level to 1,450 m elevation. Most of the breeding records (59%; n = 2,137) were from open, usually human-associated habitats where nest boxes were available. These included cultivated farmlands, rural and suburban areas, rangeland, orchards, golf courses, gardens, and pastures. Within these general habitats, breeding sites were associated with backyards or farmyards (55%; n = 1,042), water (20%), aspen parklands (6%), open fields and meadows, orchards, roadsides, young and mature forests, burns, bottomlands, and campgrounds. Natural habitats include wooded areas with dead trees for nest sites and open space for foraging, such as those associated with swamps, bogs, beaver ponds (Fig. 140), burns, river shorelines, and wet meadows as well as big sage shrublands.

The Tree Swallow has been recorded breeding in the province from 16 April (calculated) to 21 September (Fig. 137).

Nests: Most nests (69%; n = 2,745) were found in nest boxes (Fig. 141), poles, and posts (60%), or buildings (8%), including houses, garages, barns, sheds, water towers, and bridges. Natural nests were found in deciduous and coniferous trees or snags (31%; Fig. 140); 4 nests were found in rock cliffs. Nest boxes of suitable dimensions are readily used, and these made up the bulk of the specific nest locations (87%; n = 3,879), with natural cavities in trees or cavities and crevices in buildings making up the remaining 13%. Nests were reported from exposed pipe ends, an old Barn Swallow nest (Fig. 142), and newspaper boxes; 1 nest was found occupying a small cavity in the side of an abandoned Bonaparte's Gull nest in a living spruce tree (Munro 1945a).

Most nests were constructed with dry grass (92%; *n* = 2,326) and feathers (81%; Fig. 143), but a variety of other materials such as twigs, hair, mud, string, moss, rootlets, and leaves were also reported. Winkler (1993) suggests that the feathers aid chicks directly by preventing hypothermia.

The heights of 2,424 nests in nest boxes ranged from 0.2 to 12.2 m, with 50% between 1.3 and 2.0 m; 564 nests where nest boxes were not used ranged from ground level to 24.4 m, with 56% between 1.5 and 4.6 m.

Unlike the Violet-green Swallow, the Tree Swallow rarely nests on cliffs and, according to Erskine (1979), should be viewed as being an obligate tree-hole nester. The cavities must be in an area where the flight path to the entrance hole is unobstructed (Fig. 141).

Studies in the Cariboo and Chilcotin areas of central British Columbia revealed that the 2 most important characteristics of preferred nest cavities were cavity volume and entrance area (Peterson and Gauthier 1985). Of the 5 species of secondary cavity users studied, including the European Starling and Mountain Bluebird, the Tree Swallow selected the smallest entrance holes (mean entrance area = 23.0 cm^2, mean vertical entrance diameter = 5.5 cm, mean horizontal entrance diameter = 5.2 cm) and the smallest cavity volume (mean = 1,530 cm^3, mean cavity width = 8.9 cm, mean cavity breadth = 11.6 cm, mean cavity depth = 17.5 cm). Peterson and Gauthier (1985) noted that swallows in the Cariboo now use significantly smaller cavities with smaller entrance areas than they did in 1959. However, they could not determine whether the changes resulted from increased competition in the area or from a change in the resources available.

Eggs: Dates for 2,586 clutches ranged from 21 April to 12 August, with 53% recorded between 30 May and 17 June. Calculated dates indicate that nests can contain eggs as early as 16 April and as late as 16 August. Clutch size ranged from 1 to 12 eggs (1E-156, 2E-153, 3E-227, 4E-380, 5E-686, 6E-752, 7E-208, 8E-21, 9E-1, 10E-1, 12E-1), with 70% having 4 to 6 eggs. Clutches larger than 8 eggs are likely the result of more than 1 female. Mean clutch size in the interior was 5 eggs (*n* = 2,496), while on the coast it was 4 eggs (*n* = 54). The incubation period in British Columbia ranged from 14 to 16 days (*n* = 20), with a median of 15 days. Erskine and McLaren (1976) note a period of 14 days. Occasionally second clutches are laid.

Dates of initiation of egg laying are earlier in the southern portions of the province than in the central portions (Fig. 144). In the Southern Interior, the first eggs may be laid in the third week of April; there is a peak by the end of May or early June, and clutch initiation has been as late as the last week of July. In the Cariboo and Chilcotin areas, eggs are not normally laid until the third week of May, although the peak of egg laying was similar to that of the Southern Interior; laying can continue there until mid-July. Erskine and McLaren (1976) found that median dates of clutch initiation were 21 May in a year of warm spring weather and 25 May in a year of cool spring weather. The mean clutch size was also influenced by the weather, with 5.92 eggs in a year with a warm spring and summer and 5.79 in a year with wet weather during the early nesting season.

Young: Dates for 1,789 broods ranged from 8 May to 21 September, with 55% recorded between 17 June and 3 July. Calculated dates suggest that nests can hold young as early as 5 May. Brood size ranged from 1 to 10 young (1Y-61, 2Y-100, 3Y-184, 4Y-407, 5Y-576, 6Y-382, 7Y-74, 8Y-4, 10Y-1), with 55% having 4 to 5 young. The nestling period in British Columbia ranged from 15 to 25 days, with a median of 20 days (*n* = 13). Ehrlich et al. (1988) report a similar period of 20 days (16 to 24 days).

In the east Kootenay, the nesting cycle from beginning of nest construction to fledging was completed in 63 to 64 days (Johnstone 1949).

Nest Success: Of 1,373 nests found with eggs and followed to a known fate, 71% were successful in rearing at least 1 fledgling. Coastal success was 73% (*n* = 26); interior success was 71% (*n* = 1,347). From 125 records where observers noted reasons for mortality, 41% of mortality was due to *Protocalliphora* (blowfly) infestation, while 31% was attributed to poor weather. Often poor weather and *Protocalliphora* infestation combined to cause mortality. Other suspected causes of mor-

Figure 141. The Tree Swallow requires an unobstructed flight path to its nest site, whether in a natural or artificial cavity. This nest box, erected in April 1993, was abandoned in early June because rapid vegetative growth surrounded the nest entrance (Salmon Arm Bay, 24 June 1993; R. Wayne Campbell).

Figure 142. Adult female Tree Swallow incubating in an old Barn Swallow nest (Creston Valley Wildlife Centre, 15 June 1982; Mark Nyhof).

Figure 143. Adult female Tree Swallow with eggs in its feather-lined grass nest (Douglas Lake, 12 June 1993; R. Wayne Campbell).

tality included mite infestation, and squirrel, weasel, House Wren, house cat, mouse, House Sparrow, and American Kestrel predation.

In the Okanagan valley, 57% of the nests fledged at least 1 young; in 75% of the nests that failed to rear young, failure took place in the egg stage, mainly through destruction of eggs by House Sparrows, House Wrens, and European Starlings, or as a consequence of competition for nests with one of those species (Cannings et al. 1987).

In the Cariboo and Chilcotin areas, weather influenced the success rate at which eggs resulted in fledged young. Hatching success was 36% in the good year and 24% in the poor (wet) year. Nest success in producing some young was 48% in the good year and 35% in the poor year (Erskine and McLaren 1976).

A comparison of nest success from nests found in natural cavities versus those found in human-influenced cavities, principally nest boxes, showed that nest success in the latter was higher: 69% (n = 1,263) versus 55% (n = 44), respectively ($\chi^2 = 5.01$; $0.02 < P < 0.01$). Robertson and Rendell (1989), in a similar comparison in another area, found that for 2 breeding seasons, the proportion of nests that fledged young was not different between Tree Swallow populations nesting in nest boxes and those nesting in natural situations.

Brown-headed Cowbird Parasitism: Cowbird parasitism was not found in British Columbia in 3,329 nests recorded with eggs or young. In North America, the Tree Swallow is known only as an infrequent victim of the cowbird; the cowbird has been recorded as a parasite in Wisconsin (Friedmann 1963) and Minnesota (Russell 1969). In Ontario, Peck and James (1987) report only 1 instance of cowbird parasitism in 4,149 nests examined.

Aside from the fact that it is difficult for the cowbird to gain access to most Tree Swallow nests, the swallows appear to be unsuitable cowbird hosts during both incubation and nestling periods. Miles (1988) added cowbird eggs to 15 Tree Swallow nests and found that 1 egg disappeared and 7 failed to hatch, whereas 63 of 66 host eggs hatched. Of the 7 cowbird young that did hatch, all died within 12 days, whereas 6 of the 7 swallow pairs each fledged the cowbird's 4 swallow nestmates.

REMARKS: Butler (1988) developed a life table for the Tree Swallow in which 79.15% of post-fledging young died in their first year and 60.35% died annually thereafter. These yielded a mean life length of 2.7 years and a maximum of 8 years. On the basis of new information, he has refined his life table using a survival estimate of 62.7% for second-year and older birds. This yields a mean life length of 4 years (nearest whole number) and a maximum longevity of 13 years.

Blowflies of the genus *Protocalliphora* are known to parasitize many species of birds, particularly those that nest in cavities or in nests constructed of mud. Thus, the Tree Swallow is often a host for these blood-sucking parasites. Roby et al. (1992) looked at the effects of an instance of blowfly parasitism on Tree Swallow nestlings and found no significant differences in nestling survival or fledging age among blowfly removal, addition, and control treatments. Our data also suggest that blowfly infestations have little effect on the overall nest success of the Tree Swallow: from 76 nests found with blowfly larvae, 58% fledged at least 1 young, which is comparable to nest success from other studies.

In British Columbia, Rendell (1992) studied the influence of old nest material on the numbers of ectoparasites in nest boxes and the breeding ecology of Tree Swallows. His results indicate that the numbers of ectoparasites such as fleas (Ceratophyllidae), blowflies (*Protocalliphora sialia*), and fowl mites (Dermanyssidae) are not necessarily greater in nest boxes containing old nest material, but that other factors are important in determining the parasite numbers. While the swallows preferred empty and clean nest boxes or boxes where the old material had been exposed to microwaves, it could not be determined whether the swallows were avoiding potentially high ectoparasite numbers in nests with old material or whether they simply preferred clean boxes. He also found that nesting phenology, reproductive success, nestling size, and adult feeding effort did not differ between pairs

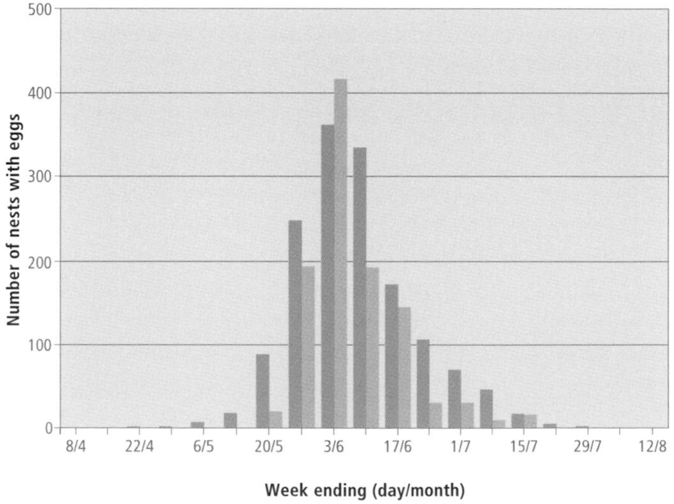

Figure 144. Weeks when nests were found containing Tree Swallow eggs in the Southern Interior Ecoprovince (dark bars) and the Central Interior Ecoprovince (light bars). In southern portions of the province, nests can contain eggs as much as 4 weeks earlier and up to 2 weeks later than nests in the central portions of the province, although the peak of egg laying appears to occur about the same time in both regions. The figure is based on the day the nest was found with eggs.

of swallows using clean nest boxes or boxes containing old material. (See also REMARKS in **Barn Swallow**.)

In a study of organochlorine levels in eggs of several passerine bird species in orchard and non-orchard areas of the Okanagan valley, Elliott et al. (1994) found that Tree Swallows nesting in orchards had, on average, 7.7 times more DDT and 2.8 times more DDE in their eggs than did their non-orchard nesting conspecifics. They suggest that the birds were acquiring DDT and DDE burdens locally, and not on their Central American wintering grounds, as has been hypothesized. There was no evidence of lower reproductive success between the 2 groups, so the levels of DDT and DDE in their eggs "were probably not deleterious." Nevertheless, the persistence of these chemicals, which had not been used since the 1970s, is certainly demonstrated, and a harmful cumulative effect on animals higher in the food chain, such as birds of prey, is suggested.

Muldal et al. (1983), in their experiment on preferred nest spacing in the Tree Swallow, found that the swallows did not show any spacing preferences when their nests were at least 36 m apart. However, when the nest boxes were less than 19 m apart, the swallows nested as far as possible from each other and attempted to prevent conspecifics from nesting nearby. Even so, pairs have been reported breeding 1 to 3 m apart on occasion (Harris 1979; Boone 1982).

Burt and Tuttle (1983) found that female Tree Swallows banded early in the nesting cycle abandoned their clutches significantly more often than females banded late in incubation. They suggest that Tree Swallows be banded as late as possible into the nesting cycle, when the probability of desertion is low, and at night, when capture is most efficient for the bander.

A Tree Swallow banded at Creston on 23 June 1987 was recovered near Yuma, Arizona, in September of that year.

For further details on the biology of this swallow, see Robertson et al. (1992), Turner and Rose (1989), or Bent (1942).

NOTEWORTHY RECORDS

Spring: Coastal – William Head 9 Mar 1981-1; Brentwood 12 Mar 1985-2; Cowichan Bay 11 Mar 1940-first seen; Kennedy Lake 17 Mar 1981-1; Surrey 5 May 1961-2 eggs; Buttertubs Marsh (Nanaimo) 3 May 1986-750; New Westminster 18 May 1971-5 young; Pitt Polder 14 Mar 1976-4,000, 25 Apr 1976-1,200; 24 Apr 1976-1,200; Qualicum Beach 20 Mar 1975-1 (Dawe 1976); Morton Lake 28 Apr 1975-50; Port Hardy 4 Mar 1941-10, 14 Apr 1940-200; East Greenville 24 Apr 1981-200; Port Clements 22 Mar 1972-5; Masset 1 Mar 1942-1 (RBCM 10246), 27 May 1910-5 eggs; Delkatla Slough 15 Apr 1972-200; Kitimat 1 Apr 1981-4; Metlakatla 11 Apr 1905-1; Terrace 29 Apr 1968-1. **Interior** – Grand Forks 18 May 1979-4 eggs; Creston 1 Mar 1981-2; Erie Lake 30 Mar 1983-35; Keremeos 9 Apr 1976-1,000; Vaseux Lake 29 Mar 1986-100; Old Hedley Road 8 Apr 1970-2,000; Taghum 24 May 1978-4 nestlings; Cranbrook 5 Apr 1942-1; Coldstream (Vernon) 27 Feb 1932-1 (RBCM 9535); Box Lake 21 Apr 1984-200; Enderby 21 Apr 1977-6 eggs, 8 May 1977-5 nestlings; Lac la Hache 27 Mar 1944-1 (Munro 1945a); s end Westwick Lake 15 May 1958-1 egg; San Jose River 9 Apr 1961-2,000, 23 Apr 1959-2,500; Williams Lake 10 Mar 1981-1, 19 Mar 1978-5, 29 Mar 1981-60; Chezacut 20 May 1975-nestlings; Vanderhoof 1 and 3 Apr 1994-300; Toms Lake 24 May 1981-3 eggs; Ross Lake (Hazelton) 23 May 1983-adults feeding nestlings; Willow River 3 Apr 1986-4; Lynx Creek 30 Apr 1983-20; Tonis Lake 4 Apr 1975-2; Charlie Lake 15 May 1984-800; Fort Nelson 22 Apr 1977-1; Parker Lake 19 May 1975-100; Atlin 2 May 1981-1.

Summer: Coastal – Duncan 15 Jun 1975-25; Sumas Prairie 4 Aug 1923-200; Aldergrove 12 Aug 1980-5 eggs, 18 Aug 1980-5 nestlings; Ladner 16 Jul 1975-175; Westham Island 20 Jun 1956-200; Agassiz 6 Jun 1971-35; Burnaby Lake 10 Jul 1978-200; Hope 13 Jul 1961-2 nestlings; Qualicum Beach 21 Jul 1974-10; Egmont 24 Jun 1961-2 nestlings; Mitlenatch Island 24 Aug 1973-2; Quadra Island 16 Jun 1973-4; Estevan Point 9 Jun 1960-6 eggs, 14 Jul 1960-2 nestlings; Port Hardy 16 Jun 1940-50; Tasu 18 Jun 1991-4 eggs; Masset 6 Jul 1975-4 nestlings; Delkatla 4 Jun 1985-40, 16 Aug 1974-80; Terrace 18 Jun 1990-4 eggs, 23 Jul 1991-4 nestlings, 25 Jul 1975-150; Kitsault 17 Jun 1980-28. **Interior** – Grand Forks 2 Aug 1977-4 nestlings; Manning Park 11 Aug 1974-30; Balfour 8 Jun 1983-224 on census; Skookumchuck 1 Aug 1976-400; Sorrento 3 Jun 1970-70; Revelstoke 19 Jul 1969-1 egg; Riske Creek 31 Jul 1981-2 eggs; Scout Island (Williams Lake) 17 Jun 1981-200, 24 Aug 1968-200; Nimpo Lake 5 Aug 1990-adults feeding nestlings; Quesnel 13 Jul 1977-5 eggs; Prince George 1 Jul 1969-19; Nulki Lake 29 Jul 1964-450; Moberly Lake 9 Jul 1974-nestlings heard; Fort St. John 17 Aug 1984-1 over sewage lagoons; Charlie Lake 5 Jun 1984-600; Sawmill Lake (Telegraph Creek) 11 Jun 1922-3 eggs; Haworth Lake 6 Aug 1976-2; Fort Nelson 2 Jul 1982-16, 3 Jul 1974-nestlings; Parker Lake 12 Aug 1979-2; Kotcho Lake 26 Jun 1982-5 nestlings; Pine Creek (Atlin) 11 Jun 1975-40; Atlin 4 Jul 1980-5 nestlings, 27 Jul 1979-20.

Breeding Bird Surveys: Coastal – Recorded from 25 of 27 routes and on 64% of all surveys. Maxima: Chilliwack 26 Jun 1976-47; Masset 27 May 1989-44; Albion 20 Jun 1974-42. **Interior** – Recorded from 65 of 73 routes and on 82% of all surveys. Maxima: Prince George 26 Jun 1976-80; Grand Forks 1 Jul 1991-67; Bromley 3 Jul 1992-64.

Autumn: Interior – Smithers 1 Oct 1968-1; Puntchesakut Lake 3 Sep 1944-5; Scout Island (Williams Lake) 9 Sep 1979-500; Kleena Kleene 30 Sep 1962-3; Invermere 19 Sep 1966-6; Swan Lake (Vernon) 27 Sep 1965-2; Penticton 21 Sep 1982-3 new fledglings; Erie Lake 2 Sep 1983-34; Manning Park 8 Sep 1977-64. **Coastal** – Green Island 21 Sep 1977-2; Iona Island 4 Sep 1974-70; Tofino 12 Sep 1983-4; Ladner 3 Oct 1973-10.

Winter: Interior – Coldstream (Vernon) 27 Feb 1932-1 (RBCM 9535); Okanagan Landing 29 Feb 1926-1 (USNM 424889). **Coastal** – Buttertubs Marsh (Nanaimo) 26 Feb 1983-5; Sturgeon Slough 28 Feb 1987-2; Serpentine Fen (Surrey) 20 Feb 1987-2; Reifel Island 10 Feb 1987-5, earliest date ever; Cowichan Bay 26 Feb 1983-4; Quick's Pond (Saanich) 25 Feb 1975-6.

Christmas Bird Counts: Interior – Not recorded. **Coastal** – Recorded from 1 of 33 localities and on less than 1% of all counts. Maximum: Vancouver 27 Dec 1970-1 (Campbell et al. 1972a).

Violet-green Swallow

VGSW

Tachycineta thalassina Swainson

RANGE: Breeds from central Alaska and Yukon, southwestern Mackenzie, northern British Columbia, southwestern Alberta, and Saskatchewan south through the western United States to Baja California and southern Mexico. Winters from central coastal California to southern Mexico and northern Central America.

STATUS: Status is complex. On the coast, *common* to *very abundant* spring migrant on southeastern Vancouver Island and the entire lower Fraser River valley in the Georgia Depression Ecoprovince. *Fairly common* to *very common* summer visitant and autumn migrant there; occasionally *abundant* in autumn and *very rare* in early winter. *Uncommon* to locally *common* migrant and summer visitant to Western Vancouver Island and the southern and northern portions of the Coast and Mountains Ecoprovince; *very rare* transient on the Queen Charlotte Islands.

In the interior, *fairly common* to *very common* (occasionally *abundant*) spring migrant and summer visitant to the south-central portions of the province, including the Southern Interior and Southern Interior Mountains ecoprovinces; *common* to *abundant* (occasionally *very abundant*) autumn migrant; *casual* in winter there. Northward, throughout the rest of the province, *uncommon* to locally *common* migrant and summer visitant except in the extreme northeastern corner, where it is a *casual* spring transient in the Taiga Plains Ecoprovince.

Breeds.

Figure 145. Adult Violet-green Swallow at Victoria, British Columbia (June 1988; Tim Zurowski).

NONBREEDING: The Violet-green Swallow (Fig. 145) is widely distributed across southern British Columbia except for the west coast of Vancouver Island and the central mainland coast, where it is localized and scarce in some years. It is less numerous through the Cariboo and Chilcotin areas, and its distribution is further reduced in the northern parts of the province. Unlike the Tree Swallow, the Violet-green Swallow seldom appears on the Queen Charlotte Islands or in the Taiga Plains, and then only during migration.

The Violet-green Swallow occurs from near sea level to 1,400 m elevation. Migrating flocks closely follow rivers, lakes,

Figure 146. Small, shallow freshwater lakes are important foraging areas for swallows during migration. On 25 April 1993, an estimated 1,100 Violet-green Swallows were actively feeding over Magic Lake on Pender Island (R. Wayne Campbell).

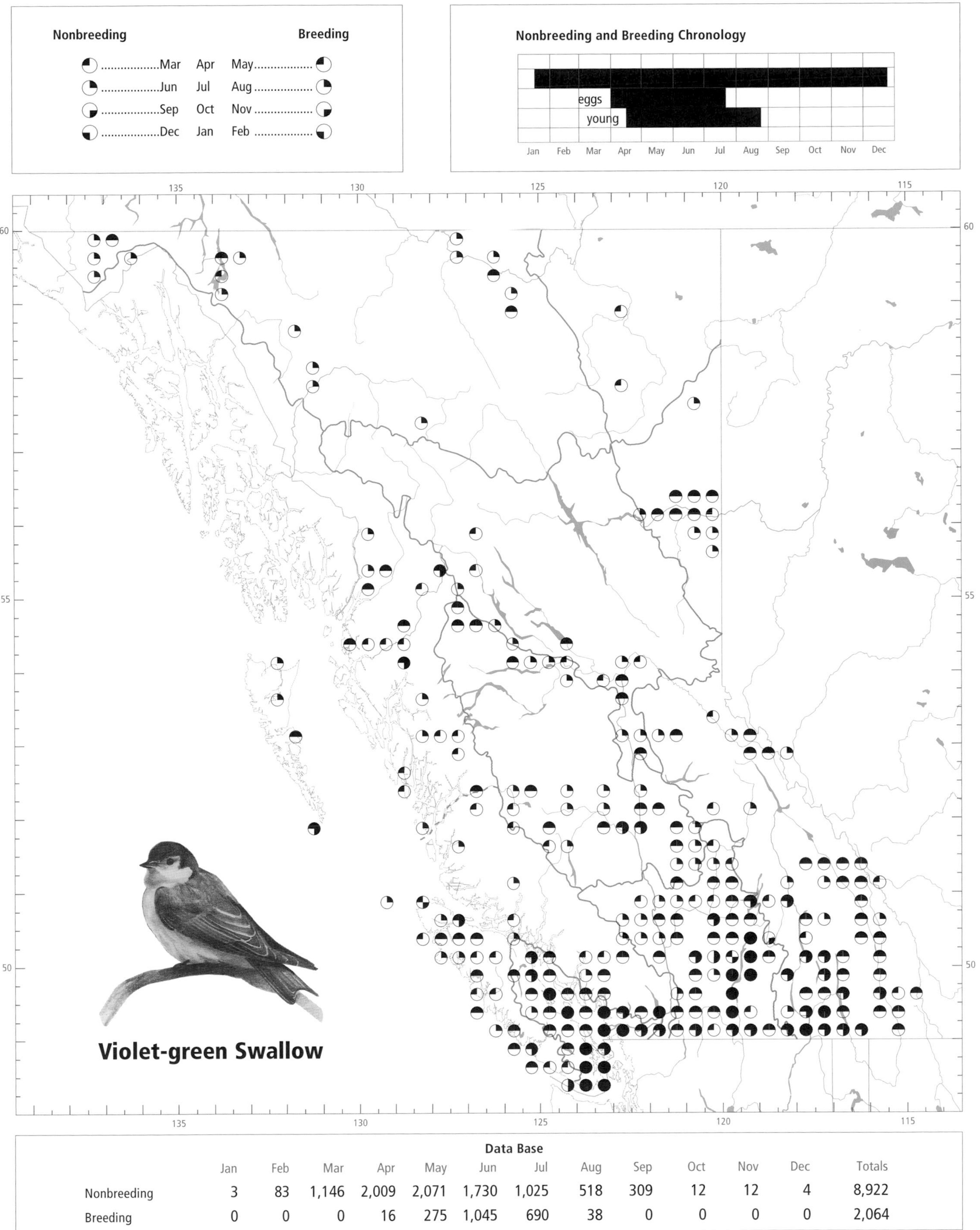

Data Base

	Jan	Feb	Mar	Apr	May	Jun	Jul	Aug	Sep	Oct	Nov	Dec	Totals
Nonbreeding	3	83	1,146	2,009	2,071	1,730	1,025	518	309	12	12	4	8,922
Breeding	0	0	0	16	275	1,045	690	38	0	0	0	0	2,064

and valley bottoms as they move northward during the spring migration before dispersing into breeding areas. Both on the coast and in the interior, it seems to favour lowland areas with open terrain and concentrations of flying insects, on which it feeds. It is usually found over freshwater habitats, including large and small lakes (Fig. 146), rivers, marshes, bogs, wet meadows, and beaver ponds; adjacent forest is also used. This swallow forages from low over water or ground surfaces to heights of 100 m or more above a forest canopy. The Violet-green Swallow occurs less frequently along the coast except over estuaries, where it can be found in brackish meadows, mudflats, fields, and grassy beaches. Rural and urban areas are also used, particularly agricultural pastures and well-treed suburbs. In the early spring, sunny cliff faces may also provide important foraging sites.

The timing of the spring migration varies widely between years. Violet-green Swallows are among the earliest-returning swallows and are particularly dependent on weather conditions, which determine the timing of insect hatches. In exceptionally mild winters, the occasional bird can arrive at the south coast in early February, and in some years the first flocks arrive in late February (Figs. 147 and 148). Normally, however, large numbers of Violet-green Swallows appear suddenly on the Fraser River delta and southeastern Vancouver Island in March, and increase through April. These are birds in passage. Their numbers decline abruptly in May and drop further in June and July. A similar trend is evident from other areas along the coast (Fig. 148). The coastal movement peaks about 2 weeks ahead of migration through the southern interior.

In southern portions of the interior, birds may arrive in late February and early March; their numbers peak in April and then decline. Spring migration through the Southern Interior and Southern Interior Mountains is smaller and less spectacular than the coastal movement. In the Cariboo and Chilcotin areas of the Central Interior, the peak movement occurs in the second and third weeks of April. Few birds move through the Sub-Boreal Interior. In the northern interior, early arrivals occur in mid to late April, becoming most numerous in early May (Fig. 148).

In spring, the Violet-green Swallow usually migrates in flocks of 20 to 200 birds, although flocks of thousands can occur. Of 86 aggregations with more than 200 birds, 84% occurred in the Georgia Depression, 2% in the Southern Interior, 13% in the Southern Interior Mountains, and 1% elsewhere. During peak spring movements, large numbers may accumulate in good foraging areas. For example, all records of more than 1,000 birds are from lakes, all in the extreme south. The 5 maximum counts (2,500 to 9,000 birds) were all recorded in April or the first week of May, at Beaver and Elk lakes, north of Victoria. The highest total count of Violet-green Swallows in the province is 11,000 birds, tallied near Victoria at Elk, Beaver, and Prospect lakes on 15 April 1970 (Tatum 1971). Southeastern Vancouver Island has the largest spring movements in the province.

Post-breeding flocks congregate over larger lakes, rivers, and marshes in July and August before moving southward. During midsummer, mixed flocks of swallows, numbering hundreds of birds, are common sights over interior lakes.

The autumn migration is more protracted than the spring movement and involves smaller flocks of Violet-green Swallows. Peaks vary between years, but generally their movements begin earlier than those of other swallows. Numbers recorded on the southern migration also suggest different behaviour than during the spring movement. On the coast, where there is such a dramatic northern movement in spring,

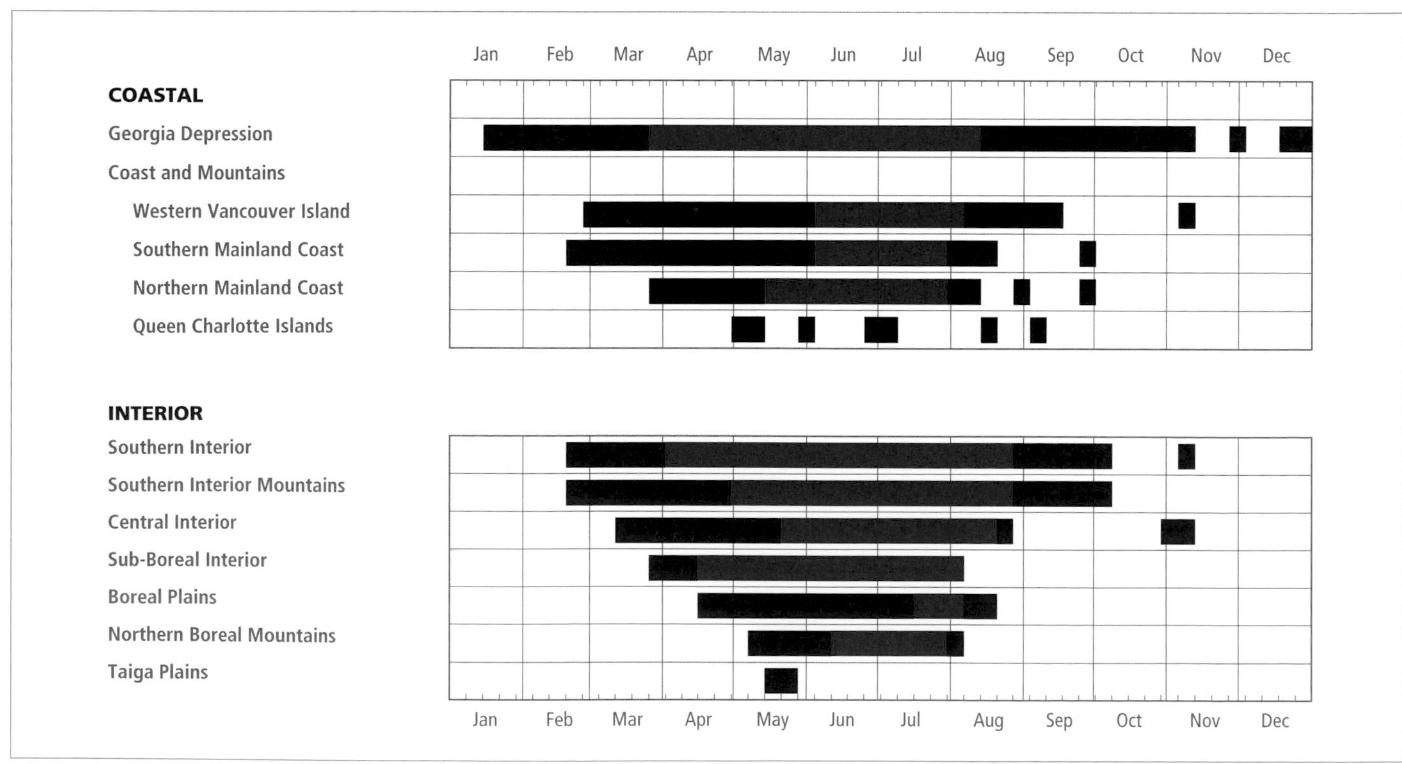

Figure 147. Annual occurrence (black) and breeding chronology (red) for the Violet-green Swallow in ecoprovinces of British Columbia.

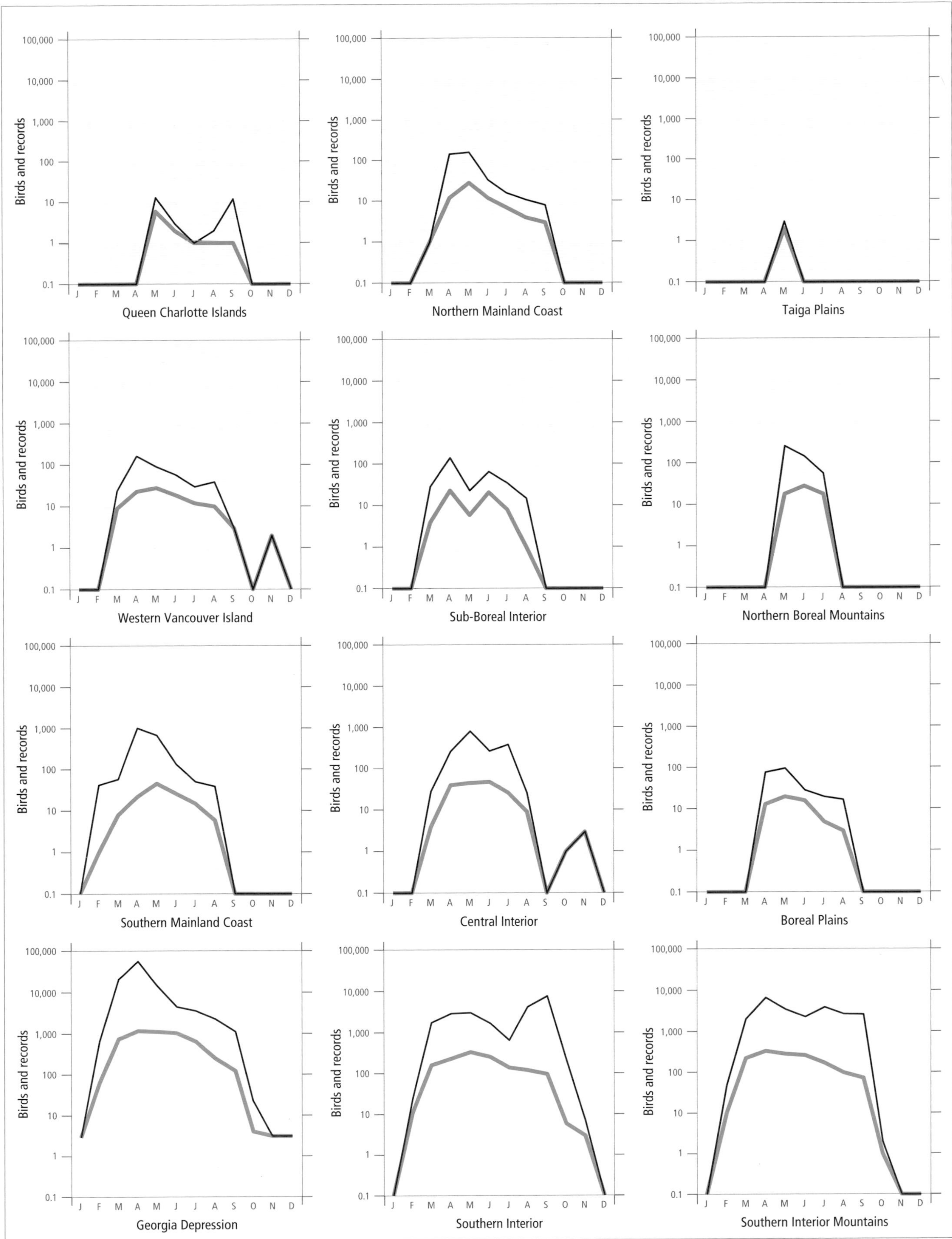

Figure 148. Fluctuations in total number of birds (purple line) and total number of records (green line) for the Violet-green Swallow in ecoprovinces of British Columbia. Christmas Bird Count and nest record data have been excluded.

Figure 149. During autumn migration, flocks of Violet-green Swallows perch and roost on natural and human-made structures, such as telephone lines, throughout the province (Cawston, 4 August 1993; R. Wayne Campbell).

swallows depart almost unnoticed in autumn (Fig. 148), and most have left by the first week of October. In the southern portions of the interior, however, almost the reverse is true. Of 24 autumn flocks of 200 birds or more, only 12% were from the south coast while 88% were from the southern portions of the interior.

In the northern and central regions, migration begins shortly after nesting ends. The highest numbers occur in July, and most birds have gone by early to mid-August (Fig. 148). In the Okanagan, Kootenay, and Columbia valleys, the highest numbers occur in August and September, and most birds have gone by the first week of October. In the Okanagan valley, the autumn migration peaks in early September (Cannings et al. 1987). Almost all Violet-green Swallows have left the province by the end of September.

The pattern of records suggests different routes for spring and autumn migration. In spring, most Violet-green Swallows move northward along the coast, taking advantage of the milder weather and available food. The small number of birds that migrate in spring up the valleys of the southern portions of the interior suggests that the nesting population in the Central Interior and beyond is derived from the large coastal concentrations that move into the interior along major west-east river valleys. Following the breeding season, coastal birds leave shortly after their young fledge and no noticeable concentrations occur. For birds in the interior, there is no advantage in returning to the coast. Swarms of late-summer chironomids and other flying insects create an abundant supply of food over lakes and ponds. Observations of large numbers indicate that the swallows leave through the Okanagan, Creston, and other southern valleys. Mountain passes and montane areas (e.g., Manning Park) may also be important autumn migration corridors, but their use has not been well documented.

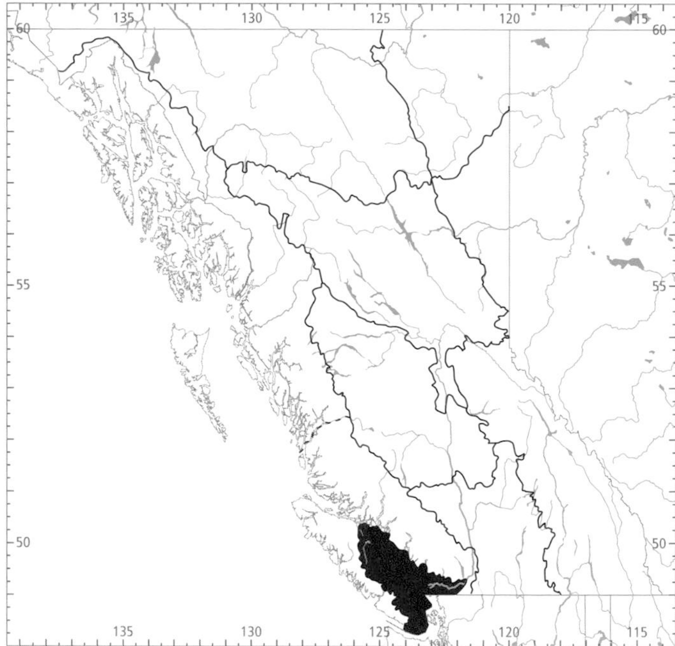

Figure 150. In British Columbia, the highest numbers for the Violet-green Swallow in summer occur in the Georgia Depression Ecoprovince.

Although migrating swallows spend much of their time on the wing, perching and roosting sites are important habitat components. Frequently, they use snags, leafless trees or branches, and roots protruding from eroding river banks. In rural and urban areas, telephone lines (Fig. 149), buildings, wire fences, and bridges also provide perching sites.

On the coast, the Violet-green Swallow has been reported every month of the year, although it occurs in substantial numbers only from late February to late September. In the interior, it has been reported from 19 February to 8 November (Fig. 147).

BREEDING: The Violet-green Swallow breeds throughout most of its range in the province. It is scarce in the Peace Lowland and in the Northern Boreal Mountains. The northernmost breeding records are from the Tatshenshini River, Atlin, Fireside, and Sulphur Creek, along the northern boundary of the province. There is a vast area west of the Rocky Mountains and north of the latitude of Fort St. John where there have been few observers and thus few nest records. For example, although Swarth (1922) found them abundant at Telegraph Creek, there are no other nesting records in the Stikine River drainage. There are just 3 nesting records on the coast north of Vancouver Island: Owikeno Lake, Kitsumkalum Lake, and Alice Arm. The centre of breeding abundance for the Violet-green Swallow is in the Georgia Depression, Southern Interior, and Southern Interior Mountains, where it has adapted well to human-influenced habitats. Breeding has not been reported from the Queen Charlotte Islands or the Taiga Plains.

This species reaches its highest numbers on southeastern Vancouver Island and in the lower Fraser River valley of the Georgia Depression (Fig. 150). An analysis of Breeding Bird Surveys for the period 1968 through 1993 could not detect a net change in numbers on either coastal or interior routes. Both Robbins et al. (1986) for North America and Erskine et al. (1992) for southern British Columbia also note stable Violet-green Swallow populations.

On the coast, the Violet-green Swallow has been found breeding from near sea level to 390 m elevation; in the interior, it breeds from 300 to 1,400 m elevation. It seems to thrive in close association with humans (Weber 1972; Allen et al. 1977). Over 75% of 847 nests were from human-associated habitats. Most nests (62%; *n* = 805) were found in backyard gardens and residential buildings, roadside habitats (21%; Fig. 151), meadows, or shorelines (lakes, streams, marshes, seashores). Only 6% were reported from cliffs (Fig. 152), which suggests a bias due to observer distribution. Other habitats included orchards, pasture and other farmland, parks, golf courses, and campgrounds.

The Violet-green Swallow has been recorded breeding in the province from 1 April (calculated) to 24 August (Fig. 147).

Colonies: Although the Violet-green Swallow nests solitarily, it also nests in substantial colonies, particularly on cliffs (Table 1; Fig. 152). Colony size ranged from 3 to 40 pairs (*n* = 16), with 8 between 5 and 15 pairs. Thirteen of the colonies were found in the interior.

Nests: The Violet-green Swallow has adapted well to nesting in human-made habitats such as nest boxes, the crannies of houses, garages, sheds, barns, and other buildings, and in posts and poles, stone and brick walls, and bridges. Natural nest sites include rock cliffs, snags, and cavities in living deciduous and coniferous trees. Along the Fraser River canyon, near Williams Lake, it nests all along the cliffs, but has never used any of the more than 1,000 boxes on the adjacent plateau (A. Roberts pers. comm.). Violet-green Swallows seem to prefer boxes on buildings rather than those on poles and posts. In suburban environments, the House Sparrow frequently displaces the Violet-green Swallow from nest sites (Erskine 1979).

In the Cariboo and Chilcotin areas, about half of all nests were found in natural situations. Further north, in the Sub-Boreal Interior and Boreal Plains, natural sites such as crevices in cliff faces (Fig. 152) and cavities in a variety of tree species were the preferred nesting sites.

The adaptability of the Violet-green Swallow is indicated by the diversity of nest sites reported; any darkened cavity with an adequate entrance hole or crevice seems suitable. The species has been reported nesting inside the walls of buildings and on fiberglass bats in attics where they have gained access through knotholes or ventilation louvres, under tiles on the ridges of house roofs, between logs in log buildings, under roof shingles and flashing, in chimneys, in a trailer hitch of a semi-trailer truck, under a plywood sheet in a lumberyard, in plumbing vent pipes of buildings, and in the blasting holes of roadside rock-cuts. Sometimes a human-made site can prove to be a death trap for the birds. In the mid-1970s, dismantling of a wooden shed on the Alaksen National Wildlife Area, south of Vancouver, revealed nearly 100 swallow skeletons between the studs of a wall (R. McKelvey pers. comm.). The knothole through which they had entered was about 1.2 m

Figure 151. Small numbers of Violet-green Swallows nest in cliffs exposed by road development throughout the province. This site, near Oyama, supported several pairs of nesting Violet-green and Northern Rough-winged swallows (13 June 1993; R. Wayne Campbell).

Figure 152. At least 15 pairs of Violet-green Swallows nested in crevices in rock cliff faces adjacent to Highway 20, 26.6 km east of Redstone (13 June 1996; Neil K. Dawe).

Table 1. Location and size of major colonies (10 or more nests or pairs) of Violet-green Swallow in British Columbia.

Location	Date	Nests or crevices	Pairs	Nest location
COASTAL				
Non-natural sites:				
Vancouver	15 Jun 1968	11		Building
INTERIOR				
Natural Sites:				
Valemount (18.9 km n)	10 Jul 1991		11	Cliff
Bear Flat	16 May 1977		20	Cliff
Bull Canyon (26.6 km e Redstone)	15 Jul 1992		15	Cliff
Kelowna (Canyon and Scenic Gardens)	4 Jul 1967		10	Cliff
Kleena Kleene (on Maclinchy River 1 mile down)	15 Jun 1961	10	19	Cliff
Kleena Kleene (on Maclinchy River 2 miles down)	15 Jun 1961	4	12	Cliff
Vaseux Lake (Okanagan Falls)	7 Jun 1969	20	20	Cliff
Whipsaw Creek	19 May 1963	20	40	Cliff
Williams Lake (Serpentine Cliff 6 miles e)	10 Jun 1978	10	10	Cliff
Non-natural sites:				
Bridesville	6 Jul 1969	10	10	House
Nelson	14 Jun 1984	20		Pole

Source: All data are from the British Columbia Nest Records Scheme unless otherwise noted.
All data are estimates.

up the wall, and the remains filled the space between the studs to a depth of about 0.8 m. After having gained access, the swallows, for some reason, could not return to the hole to exit.

Reported nests were loosely constructed, the amount of material governed by the size of the cavity occupied. The nest was usually a collection of grasses with a small cup formed in the centre or in 1 corner of the cavity; the cup was formed with finer grasses and lined generously with feathers. White feathers seemed to be preferred. Most nests (94%; n = 515) were made of grasses and were feather-lined (80%). One nest was made entirely of feathers. Other materials included plant fibres, rootlets or twigs (6%), mud (5%), hair (3%), conifer needles, wood chips, plant down, and human-made materials such as fiberglass insulation, string, rope, and paper. Nest sites are reused in successive years.

The heights of 460 nests in nest boxes ranged from 0.3 to 21 m, with 54% between 2.4 and 4.2 m; 344 nests where nest boxes were not used ranged from ground level to 78 m, with 62% between 2.4 and 6.0 m. Many of the known cliff sites have not been included in the samples.

Eggs: Dates for 302 clutches ranged from 7 April to 21 July, with 51% between 29 May and 16 June. Calculated dates indicate that eggs can occur as early as 1 April and as late as 1 August. Clutch size ranged from 1 to 8 eggs (1E-13, 2E-18, 3E-20, 4E-92, 5E-117, 6E-37, 7E-4, 8E-1), with 69% having 4 or 5 eggs. The 8-egg clutch may have been the product of 2 females. The incubation period in British Columbia ranged from 14 to 17 days (n = 16), with a median of 15 days, similar to the 14 or 15 days reported by Turner and Rose (1989) and Brown et al. (1992).

Young: Dates for 416 broods ranged from 15 April to 24 August, with 53% between 20 June and 9 July. Sizes of 353 broods ranged from 1 to 7 young (1Y-20, 2Y-50, 3Y-7, 4Y-142, 5Y-105, 6Y-26, 7Y-3), with 59% having 4 or 5 young and 76% with 3 to 5 young. The nestling period in British Columbia ranged from 23 to 27 days (n = 14), with a median of 25 days. Edson (1943) found a nestling period of 23 days.

Brown-headed Cowbird Parasitism: In British Columbia, cowbird parasitism was not found in 800 nests recorded with eggs or young; the Violet-green Swallow has not been found as a cowbird host elsewhere in North America either (Friedmann et al. 1977; Friedmann and Kiff 1985).

Nest Success: Of 188 nests found with eggs and followed to a known fate, 139 produced at least 1 fledgling, for a success rate of 74%. Coastal nest success was 76% (n = 82); interior success was 74% (n = 106). Weather and parasitism by the larvae of *Protocalliphora* sp. can have significant negative impact on the breeding success of the Violet-green Swallow (see also **Tree Swallow** and **Barn Swallow**). Other causes of egg or young mortality in British Columbia included predation by House Sparrows, chipmunks, squirrels, crows, deer mice, Cooper's Hawks, and American Kestrels; and nest usurpation by European Starlings, Crested Mynas, and House Sparrows. There are 2 reports of birds laying second clutches after the first brood was successfully raised; there are a number of reports of renesting following the loss of eggs.

REMARKS: Godfrey (1986) shows the breeding distribution of the Violet-green Swallow to include the Queen Charlotte Islands and the Taiga Plains Ecoprovince in extreme northeastern British Columbia; however, we could find no evidence that this swallow breeds in those areas.

British Columbia is the centre of Violet-green Swallow breeding abundance in Canada. Yet, as Brown et al. (1992) note, "Despite an extensive distribution, less is known about the Violet-green Swallow than nearly any other North American Swallow." Their review could find no recent studies of the species' breeding biology or behaviour and was based primarily on reports from the 1940s and 1950s. Interested researchers in British Columbia have the opportunity to add to our knowledge of this species.

Erskine (1979) notes that the Violet-green Swallow's ability to nest in remote cliffs and tree cavities and, at the same time, adapt well to human habitation may have saved it from harmful human impact. It readily takes to nest boxes, and once established will return year after year; an adult female banded at a nest box on 30 June 1983 at the Qualicum National Wildlife Area, Vancouver Island, was recaptured in successive years in the same or a nearby nest box until 18 June 1989.

Since the Violet-green Swallow is a secondary cavity-nester, its nesting density may be limited by a scarcity of suitable nest sites in regions lacking rock cliffs (Brawn and Balda 1988).

See Brown et al. (1992) for a summary of the life history of the Violet-green Swallow in North America.

NOTEWORTHY RECORDS

Spring: Coastal – Florence Lake (Victoria) 27 Mar 1982-750; Blenkinsop Lake 13 Mar 1986-240; Quick's Bottom 18 Mar 1986-270, 29 Mar 1986-500; Elk and Beaver lakes (Victoria) 17 Mar 1973-500, 20 Mar 1971-300, 1 Apr 1972-1,000 (Tatum 1973), 2 Apr 1982-3,000, 19 Apr 1975-4,600, 30 Apr 1972-9,000 (Tatum 1973), 2 May 1976-2,500; Magic Lake (Pender Island) 25 Apr 1993-1,100 (Fig. 146); Cowichan Bay 26 Mar 1974-300; Carmanah Point 3 Mar 1991-1; Westham Island 15 Mar 1970-275; Iona Island 1 Mar 1973-40; Pitt Meadows 5 Apr 1964-2,000, 5 Apr 1973-404 (Jerema 1973); Seabird Island 15 Apr 1984-350, 27 Apr 1992-300+; Harrison Hot Springs 6 Mar 1984-1; Qualicum 10 Mar 1975-7 (Dawe 1976), 7 Apr 1979-260 (Dawe 1980); Courtenay 9 Apr 1971-1 female incubating; Mitlenatch Island 15 Apr 1974-3 young; Port Hardy 3 Mar 1942-3; Cape St. James 31 May 1982-1; Sandspit 11 May 1986-2 (Campbell 1986f), 10 to 14 May 1993-5; Terrace 28 Mar 1977-1 first spring arrival, 31 May 1978-feeding young in nest. **Interior** – Yahk 16 Mar 1941-2 (Johnstone 1949); Creston 11 Apr 1983-570, 28 Apr 1981-1,500; Castlegar 1 Mar 1970-2, 29 Mar 1976-250; Naramata 7 Apr 1977-3 eggs; Ta Ta Creek 22 Mar 1939-1 (Johnstone 1949); Okanagan Landing 13 Mar 1965-600 first spring arrivals; Box Lake 20 Apr 1980-300; Enderby 15 Apr 1977-5 eggs; Williams Lake 14 Mar 1992-1 with flock Tree Swallows on wire, 30 Mar 1979-2 first of year; Nechako River 30 Mar 1986-25; Willow River 29 Mar 1972-2; Fort St. James 4 May 1969-feeding young; North Pine 20 Apr 1986-1 spring arrival; Bear Flat (Fort St. John) 16 May 1977-nesting colony; Hudson's Hope 20 Apr 1985-20, 28 May 1974-15; Fort Nelson 19 May 1975-2, 27 May 1974-1; Atlin 21 Apr 1934-present (Swarth 1936), 8 May 1981-2, 16 May 1981-34, 21 May 1981-42 (Campbell 1981); Chilkat Pass 8 May 1957-1 (Weeden 1960).

Summer: Coastal – Tzartus Island 4 Jun 1970-12; Ucluelet 28 Jun 1909-4 eggs; New Westminster 7 Jul 1945-4 eggs; Burnaby Lake 9 Jun 1976-70; Burnaby 12 Aug 1972-1 last nestling; Hope 10 Jun 1963-6 eggs, 28 Jul 1961-nestling heard; Estevan Point 23 Jul 1960-2 nestlings; Garibaldi 15 Aug 1975-3; Discovery Passage (Campbell River) 3 Jul 1977-100; Port Hardy area 6 Jun 1909-4 eggs; Pemberton 4 Jun 1978-25; Port Clements 19 Aug 1957-2; Kitimat River 10 Aug 1975-1; Greenville 16 Jun 1981-10; Alice Arm 26 Jul 1958-1 feeding nestlings; Tatshenshini River 3 Jun 1983-30 (Campbell et al. 1983). **Interior** – Grand Forks 1 Aug 1982-250; Manning Park 5 Aug 1993-23 migrating flock; Rossland 20 Aug 1972-1 about to fledge; Trail 25 Aug 1984-300; Creston 29 Aug 1993-250 adults with flying young; Tadanac 1 Jul 1971-2,000; Jaffray 27 Jul 1943-500 (Johnstone 1949); South Slocan 7 Aug 1994-750 on wires; Naramata 2 Aug 1984-2 nestlings; Fort Steele 13 Jul 1984-2 nestlings; Kaslo 17 Jun 1978-4 nestlings; Salmon Arm 24 Aug 1973-600 (Sirk et al. 1973), 25 Aug 1977-900; Emerald Lake 31 Jul 1962-3 nestlings; Loon Lake (Clinton) 9 Jun 1970-300; Kleena Kleene 24 Jul 1952-200 (Paul 1959), 13 Aug 1966-nestlings; Blue River 10 Jul 1961-5 eggs; Williams Lake 9 Jun 1960-nestlings; 24 km s Prince George 31 Jul 1983-feeding nestlings; Division Lake 11 Aug 1956-1; Charlie Lake 1 Aug 1988-nestlings; Rose Prairie 18 Aug 1975-6; Muncho Lake 28 Jul 1980-5 nestlings; Atlin 24 Jun 1914-nestlings; Smith River 31 Jul 1968-3 (Erskine and Davidson 1976).

Breeding Bird Surveys: Coastal – Recorded from 22 of 27 routes and on 72% of all surveys. Maxima: Victoria 4 Jul 1979-75; Pitt Meadows 9 Jul 1977-68; Albion 26 Jun 1977-50. **Interior** – Recorded from 50 of 73 routes and on 63% of all surveys. Maxima: Grand Forks 1 Jul 1991-79; Salmo 14 Jun 1975-68; Syringa Creek 22 Jun 1991-66.

Autumn: Interior – Atlin 1 Sep 1924-last departure (Swarth 1926); Riske Creek 31 Oct and 8 Nov 1985-1 at Wineglass Ranch, last departure; Okanagan Landing 21 Sep 1929-300, 5 Oct 1905-1 (Cannings et al. 1987), 5 Nov 1929-3 (Cannings et al. 1987), 7 Nov 1927-1 (Munro 1930a); Kalamalka Lake 20 Sep 1970-300; Summerland 4 Oct 1971-100; Penticton 21 Sep 1959-2,000; Taghum 4 Sep 1971-500; Vaseux Lake 6 Sep 1971-300, 7 Sep 1972-300; White Lake (Oliver) 7 Sep 1971-200, 1 Sep 1995-2,000 (pure flock) on telephone wires; Castlegar 29 Sep 1971-10; Salmo 1 Oct 1978-2; Trail 17 Sep 1983-300; Oliver-Richter Pass 15 Sep 1963-450. **Coastal** – Hazelton 26 Sep 1921-2 (Swarth 1924); Cape St. James 3 Sep 1977-12; Cape Scott 22 Sep 1939-1; Port Hardy 7 Nov 1940-1, first snow; Chilliwack Lake 8 Sep 1962-20; Mission 26 Sep 1965-300; Pitt Meadows 13 Sep 1970-15 latest departure (Campbell et al. 1972a); Reifel Island 3 Oct 1965-15, 31 Oct 1979-1; Westham Island 3 Oct 1965-15; Sooke 3 Nov 1972-1 (RBCM 11961), 27 Nov 1972-1 (Tatum 1973).

Winter: Interior – Vernon 19 Feb 1985-3; Kelowna 19 Feb 1988-1; Penticton 22 Feb 1980-first spring arrivals, 23 Feb 1963-3 first spring arrivals; Beaver Creek (Trail) 26 Feb 1983-10. **Coastal** – Weaver Creek (Agassiz) 24 Feb 1977-42; Pitt Meadows 5 Feb 1984-3; Langley 21 Feb 1981-2; Surrey 21 Feb 1977-3; Iona Island 28 Feb 1970-1 earliest (Campbell et al. 1972a); Cowichan Bay 21 Feb 1977-21; Swan Lake (Victoria) 18 and 24 Dec 1982-1, 16 Jan 1983-1; Quick's Bottom (Saanich) 5 Feb 1984-1; Glen Lake (Victoria) 25 Feb 1961-6 (Davidson 1961); Victoria 29 Jan 1987-1, 28 Feb 1970-150.

Christmas Bird Counts: Interior – Not recorded. **Coastal** – Recorded from 1 of 33 localities and on less than 1% of all counts. Maxima: Victoria 18 December 1982-**1**, Canadian high count (Anderson 1983).

Northern Rough-winged Swallow NRWS

Stelgidopteryx serripennis (Audubon)

RANGE: Breeds from north-central British Columbia, southern Alberta, Saskatchewan, and Manitoba to southwestern Quebec and central Maine; south on the Pacific coast to southern Baja California, across the United States to and including most of Florida, and south through Mexico to Costa Rica. Winters from the southern United States to Panama (Ridgely 1989).

STATUS: On the coast, an *uncommon* to *common* migrant and summer visitant (*very common* locally) on eastern Vancouver Island and the southwest mainland coast in the Georgia Depression Ecoprovince. On Western Vancouver Island and the southern and northern portions of the Coast and Mountains Ecoprovince, it is an *uncommon* to *fairly common* migrant and summer visitant; *casual* on the Queen Charlotte Islands.

In the interior, an *uncommon* to *common* migrant and summer visitant (*very common* locally) in the Southern Interior and Southern Interior Mountains ecoprovinces; *uncommon* to *fairly common* in the Central Interior Ecoprovince; *rare* in the Sub-Boreal Interior Ecoprovince; *very rare* in the Boreal Plains and Northern Boreal Mountains ecoprovinces. Absent from the Taiga Plains Ecoprovince.

Breeds.

NONBREEDING: The Northern Rough-winged Swallow (Fig. 153) has a widespread distribution across southern British Columbia, from Vancouver Island to the east Kootenay and north to the Cariboo and Chilcotin areas. On Western Vancouver Island and north along the mainland Coast and Mountains, as well as in the Sub-Boreal Interior, this swallow has a localized distribution. Further north, it is sparsely distributed. There are 2 reports of this species from the Queen Charlotte Islands.

Figure 153. Adult Northern Rough-winged Swallow at a Belted Kingfisher burrow (Long Beach, May 1986; Adrian Dorst).

The Northern Rough-winged Swallow has been recorded from near sea level to 1,700 m elevation. In British Columbia, it frequents areas near water, especially shallow ponds, lakes (Fig. 154), open creeks, slow-moving rivers, sewage lagoons, estuaries, and the sea coast. Open ploughed fields, meadows, and cliff faces are used for foraging when aerial insects are

Figure 154. In spring, shallow sloughs and lakes with meltwater and flying insects are important foraging areas for migrating swallows. Here, a mixed flock of 3 species of swallows, including 4 Northern Rough-winged Swallows, were foraging over open water during a snowstorm (unnamed lake, 8 km north of 70 Mile House, 24 April 1990; R. Wayne Campbell).

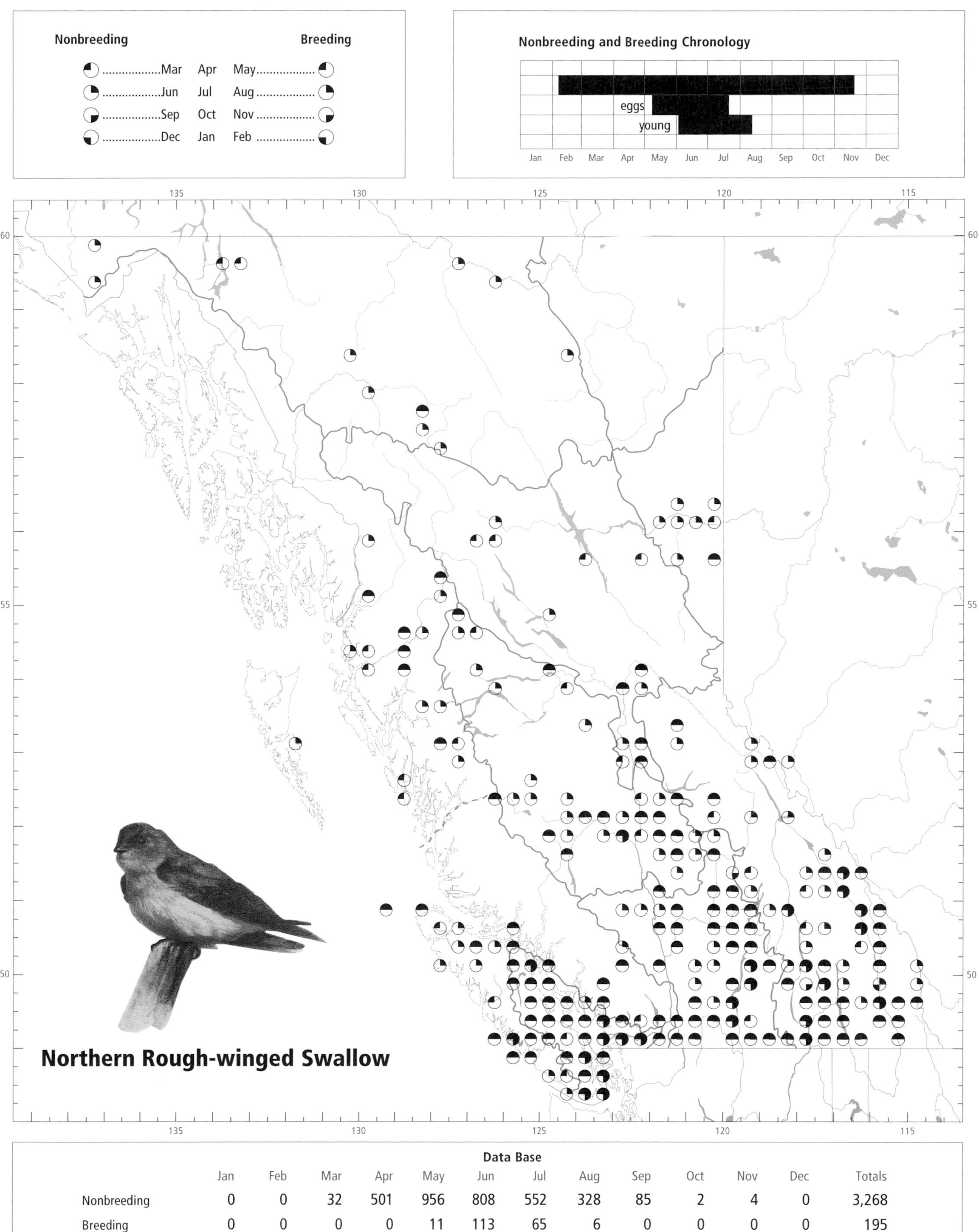

Data Base

	Jan	Feb	Mar	Apr	May	Jun	Jul	Aug	Sep	Oct	Nov	Dec	Totals
Nonbreeding	0	0	32	501	956	808	552	328	85	2	4	0	3,268
Breeding	0	0	0	0	11	113	65	6	0	0	0	0	195

abundant. It pursues insects over these areas in rapid flight and with less frequent changes of direction than the Tree and Violet-green swallows, which share these habitats. It is mainly a swallow of natural habitats, but can be seen around suburban gardens and farmsteads.

Small numbers of this swallow arrive on southern Vancouver Island and in the Fraser River delta as early as March (Figs. 155 and 156). The main movement reaches the south coast, the Thompson Basin, the Okanagan valley, and the east and west Kootenays during April. Numbers reach a peak in these areas in May. In the Okanagan valley, where first arrival dates have been recorded for 33 consecutive years, the dates range from 3 April to 14 May, with a median date of 20 April (Cannings et al. 1987).

The migration front moves northward rapidly and reaches the northern ecoprovinces between late April and mid-May, about the time the largest numbers are present in the southern areas (Fig. 156).

This swallow generally travels in small numbers. Forty-two birds made up the largest migrating flock on record (Cannings et al. 1987). Flocks of 30 birds or so have been seen on Vancouver Island, the Fraser River estuary, and the southern Coast and Mountains, along the lower Skeena River, and across the southern parts of the province, both in spring and late summer. Most migrating flocks contain between 10 and 20 birds.

The Northern Rough-winged Swallow appears to fly by day only. It roosts at night on open perches such as dead trees or shrubs, on roots projecting from cutbanks, and in similar sites. On 11 May 1978, 25 birds were observed roosting on wild rose branches overhanging the waters of a tidal channel in the Little Qualicum River estuary (Dawe 1980).

There is no evident increase in numbers towards the end of the nesting season. This suggests that once the young are on the wing, the swallows begin moving out of the province (Fig. 155). On the southern portions of the coast, numbers remain relatively constant through the summer months, and here also there is no apparent late-summer increase that would likely appear if the young were to remain in the area until autumn departure, such as occurs with the Barn Swallow.

The number of reports and flock sizes in the northern ecoprovinces begin to decline in July, and there appears to be a rapid southward migration as soon as the young have fledged. Most Northern Roughed-winged Swallows have left the far northern parts of the province by the end of July. Even in the Cariboo and Chilcotin areas, there are few records for August. In the Okanagan valley, Cannings et al. (1987) give 13 August and 28 September as the range in dates for latest sightings. Most birds have left the province by early September, although stragglers have been reported through the end of September and, occasionally, into October and November (Figs. 155 and 156).

During the autumn migration, as in the spring, these swallows frequently travel with larger aggregations of other species of swallows. They forage over ponds, streams, or meadows where there is a late-summer abundance of flying insects.

The Northern Rough-winged Swallow has been recorded on the coast from 10 March to 19 November, and in the interior from 3 April to 7 November (Fig. 155).

BREEDING: The Northern Rough-winged Swallow has a widespread breeding distribution across southern portions of the province, including Vancouver Island.

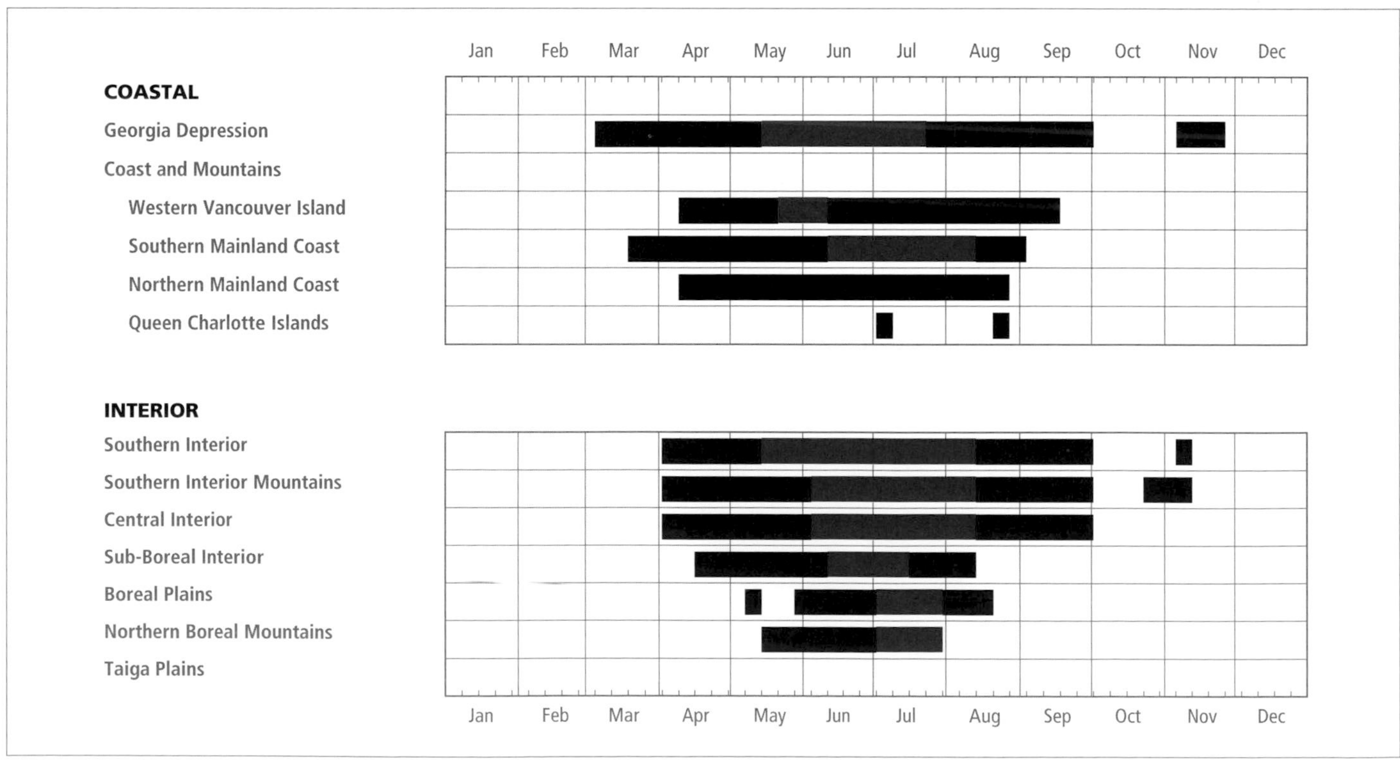

Figure 155. Annual occurrence (black) and breeding chronology (red) for the Northern Rough-winged Swallow in ecoprovinces of British Columbia.

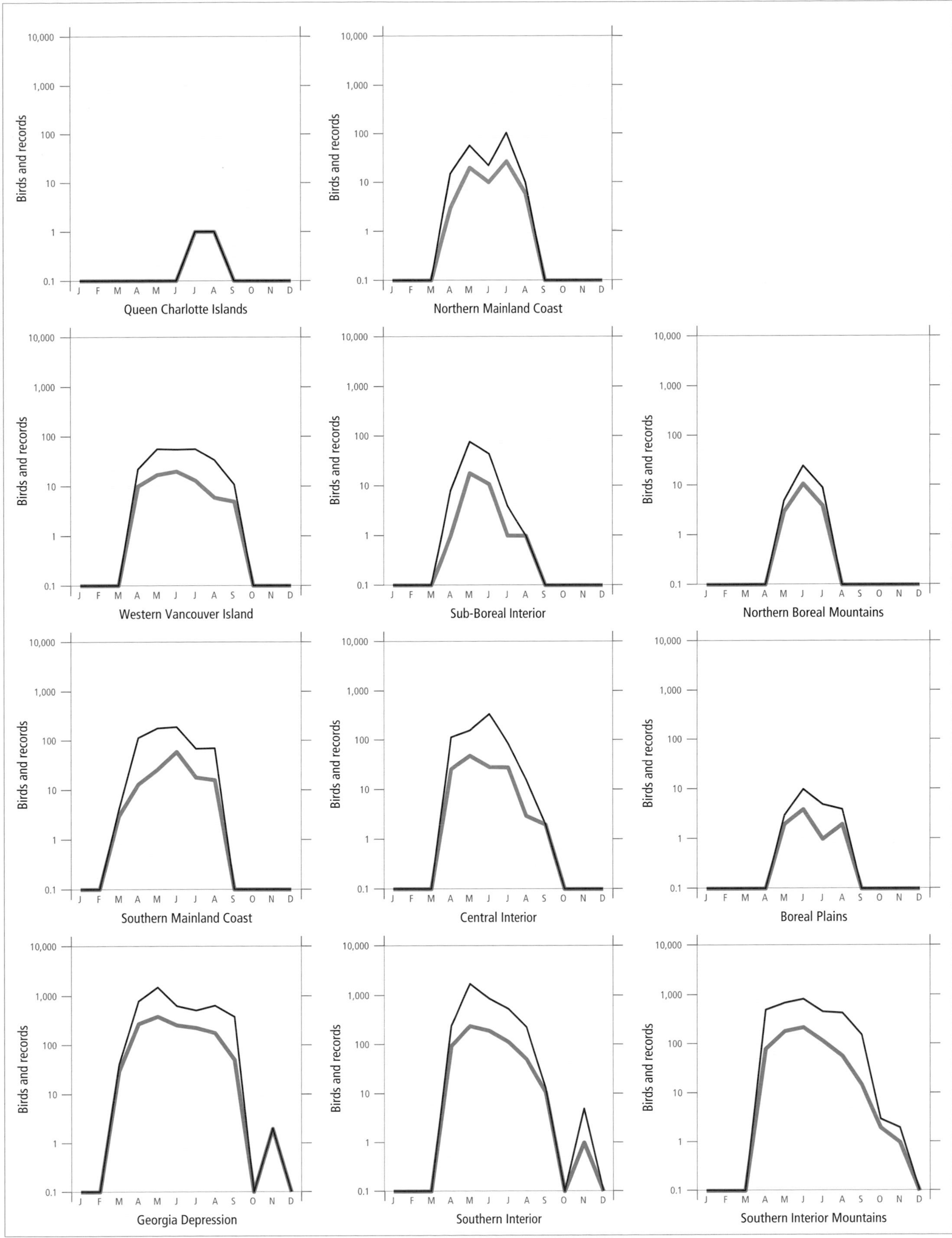

Figure 156. Fluctuations in total number of birds (purple line) and total number of records (green line) for the Northern Rough-winged Swallow in ecoprovinces of British Columbia. Nest record data have been excluded.

There are few breeding records for this species on the mainland Coast and Mountains north of Rivers Inlet. It is not known to breed on the Queen Charlotte Islands. In the interior, it regularly breeds north to at least the vicinity of Smithers and northeast to the vicinity of Fort St. John. Further north it is scarce and of local occurrence; the only confirmed northern nesting site is from Fireside, near the Kechika and Coal rivers.

The Northern Rough-winged Swallow reaches its highest numbers in the Cariboo and Chilcotin areas of the Central Interior (Fig. 157). An analysis of Breeding Bird Surveys for the period 1968 through 1993 could not detect a net change in numbers on interior routes. Data are insufficient to detect trends on coastal routes.

This species breeds from near sea level to 1,700 m elevation. During the nesting season it is seldom seen far from the earthen cutbanks (Figs. 158 and 159) and cliffs in which it nests. The presence of a pre-existing burrow appears to be more important than other ecological circumstances in the choice of breeding habitat.

In Michigan, Lunk (1962) found that when Northern Rough-winged Swallows first arrived in an area, fresh diggings of any description attracted them. In 8 cases the Northern Rough-winged Swallow circled close, "considerably excited," as Lunk installed some artificial nest tubes. In 1 instance, a pair perched at the entrance of a burrow and began calling 2 minutes after he had left the site.

Most nests (54%; n = 224) were found in human-influenced habitats. This indicates a change from the situation 14 years earlier, when Erskine (1979) found that only 25% of the British Columbia colonies were using human-made nesting sites. The difference may arise from a difference in definition of human influence.

More than one-third (37%) of the nest sites were in cutbanks (Fig. 158) or gravel pits (Fig. 159) created by road or railway building; 29% were in stream cutbanks or sea cliffs and 10% were in banks created by industrial activity. A few nests were found in sawdust piles, hog fuel, and landfills.

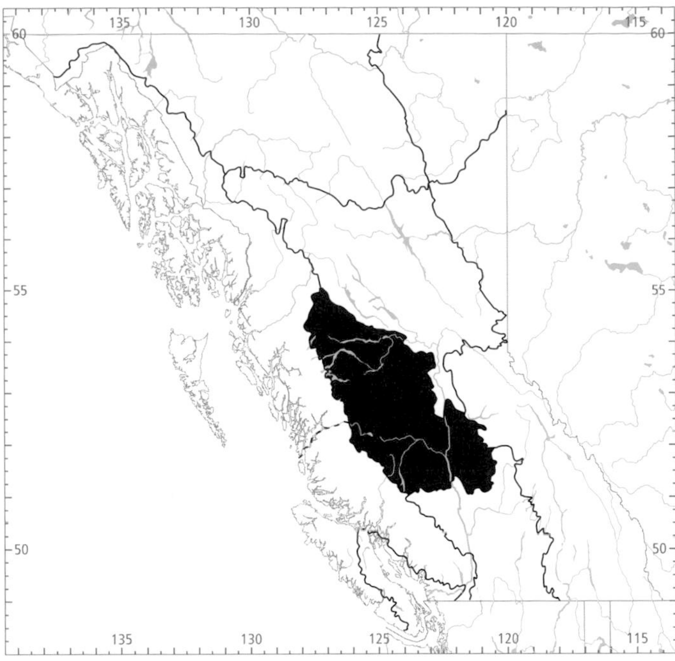

Figure 157. In British Columbia, the highest numbers for the Northern Rough-winged Swallow in summer occur in the Central Interior Ecoprovince.

Figure 158. Biologist Andrew C. Stewart at a Northern Rough-winged Swallow colony in a low cutbank site (Dog Lake Road, south of Alkali Lake, 28 May 1993; R. Wayne Campbell).

Bridge abutments (Fig. 160), rock walls (Fig. 161), and buildings were also used. Overall, 35% of the sites were reported near water. Nest site choices appear to vary among ecoprovinces, with stream and shore banks predominating in the Georgia Depression while cutbanks along transportation corridors were the preferred category elsewhere. However, this difference may be more a reflection of habitat availability than preference.

The Northern Rough-winged Swallow has been recorded breeding in British Columbia from 11 May (calculated) to 12 August (Fig. 155).

Colonies: The Northern Rough-winged Swallow is commonly regarded as a solitary nester (Snapp 1976), although it will nest in groups and should be considered "loosely colonial" (Table 2). Where the birds nest in groups, active burrows are often metres apart, although they can nest closer when sites are scarce. The frequent association of Northern Rough-winged and Bank swallows usually involves only a single pair, or at most a few pairs, of Northern Rough-winged Swallows.

Colonies of 3 to 65 nests account for 45 of the 329 nest records. The average colony size was 16 burrows or, more accurately, 10 pairs, considering that not all nest sites are active at once. Erskine's (1979) calculation of an average of 15 nests per colony was based on a subset of these data. The major breeding colonies in the province are shown in Table 2. All but 2 of these larger colonies are in the interior.

In west Creston, a large Bank Swallow colony, active in 1992, was completely usurped in 1993 by Northern Rough-winged Swallows and some House Sparrows. At least 75 pairs of Northern Rough-winged Swallows were nesting among the 356 burrows, while other sites in the Creston valley that had traditionally been used by this species were abandoned. Bank Swallows reoccupied the site in 1994.

Nests: The Northern Rough-winged Swallow rarely excavates its own burrow (Lunk 1962). Most nests (89%, n = 245) were found in burrows excavated by Bank Swallows or Belted Kingfishers (Fig. 153). Nests were also found in artificial burrows, drainage tiles in retaining walls (Fig. 161), pipes projecting from buildings, or blasting holes drilled into rock faces. A few nests (11%) were in crevices in cliffs. One nest was found in a nest box. This swallow is a versatile and adaptable species, and a wide variety of nest sites have been described elsewhere (see Bent 1942), but the nesting behaviour in British Columbia seems generally similar to that found by Lunk (1962) in Michigan.

Nests and their contents were described from 91 burrows. All were loose structures consisting of grass (88%; Fig. 162), rootlets (31%), twigs (22%), or feathers (17%). Conifer needles, leaves, and bird down were also reported. The high incidence of feathers reported as nesting material may be unusual (Lunk 1962), but could result from this species using old Bank Swallow nests. Typically, fine grasses lined the nests.

The heights of 169 nest entrances ranged from 0.3 to 30.0 m, with 53% between 2.2 and 4.5 m. Burrow length (n = 44) ranged from 15 to 123 cm, with an average length of 60 cm. Burrow height (n = 32) ranged from 4 to 15 cm, with an average height of 5 cm. Burrow width (n = 31) ranged from 6 to 11 cm, with an average width of 8 cm.

Figure 160. Three pairs of Northern Rough-winged Swallows were feeding young in nests in these drainpipes of a bridge abutment (Cache Creek, 18 July 1992; R. Wayne Campbell).

Figure 159. About 30 pairs of Northern Rough-winged Swallows nested in the humus layer among the tree roots in this clay bank above a large gravel pit (Nechako River, east of Cottonwood Island in Prince George, 14 July 1990; R. Wayne Campbell).

Figure 161. Four pairs of Northern Rough-winged Swallows nested in cavities in this concrete retaining wall (north of Merritt, 12 June 1993; R. Wayne Campbell).

Eggs: Dates for 85 clutches ranged from 15 May to 4 July, with 64% recorded between 15 and 20 June. Calculated dates indicate that nests can have eggs as early as 11 May and as late as 23 July. Clutch size ranged from 1 to 7 eggs (1E-3, 2E-7, 3E-7, 4E-12, 5E-33, 6E-19, 7E-4), with 61% having 5 or 6 eggs (Fig. 162). Eggs are normally laid on each successive morning until the clutch is complete (Lunk 1962). Incubation period is between 12 and 16 days, as reported by different observers (Bent 1942; Ehrlich et al. 1988; Turner and Rose 1989).

Young: Dates for 33 broods ranged from 2 June to 12 August, with 55% recorded between 30 June and 22 July. Brood size ranged from 1 to 8 young (1Y-2, 2Y-4, 3Y-8, 4Y-8, 5Y-6, 6Y-4, 8Y-1), with 67% having from 3 to 5 young. In Michigan, the nestling period ranged from 18 to 21 days (Lunk 1962).

Brown-headed Cowbird Parasitism: Cowbird parasitism was not found in 117 nests recorded with eggs or young in British Columbia, nor has it been reported from other areas of North America (Friedmann 1963; Friedman and Kiff 1985).

Nest Success: Insufficient data.

REMARKS: The Northern Rough-winged Swallow (*Stelgidopteryx serripennis*) was formerly known as the Rough-winged Swallow (*S. ruficollis*). The latter was considered a monotypic species, and in North America was separated into 2 races, namely, *S. r. serripennis* and *S. r. psammochrous* (American Ornithologists' Union (1957). Based on morphological and distributional information, 2 species are now recognized: the Northern Rough-winged Swallow (*S. serripennis*) and the Southern Rough-winged Swallow (*S. ruficollis*) (Stiles 1981). Stiles (1981) notes that "geographical variation within the genus *Stelgidopteryx* as a whole must be reassessed."

Except for the Purple Martin, the Northern Rough-

Table 2. Location and size of major colonies (10 or more nests or pairs) of Northern Rough-winged Swallows in British Columbia.

Burrow location	Date	Burrows	Pairs	Location	Source[1]
COASTAL					
Georgia Depression					
Burnaby Lake	11 Jun 1942		12	Bank	
Southern Mainland Coast					
Hope, 40 km e	17 Jun 1969	16	16		
INTERIOR					
Southern Interior					
Heffley Creek	1 Jul 1985	15	15		
Kamloops	26 Jun 1983	25	25	Clay bank	
Oliver	3 Jun 1968	18	10	Clay bank	
Penticton	25 Jun 1913		20	Bank	1
Shingle Creek	4 Jun 1967	20	20	Clay bank	
Six Mile Lake	29 Jun 1962	11	11	Bank	
Stump Lake	22 Jul 1968	30	5	Bank	
Southern Interior Mountains					
Bowron Lake Park	7 Jun 1971	4	10	Bank	2
Bull River (near Trout Hatchery)	20 Jun 1982	50		Cliff	
Kimberley (Tata Creek Hill)	18 Jul 1976	22		Sand bank	
Lazy Lake	2 Jun 1977		10	Sand bank	3
Sparwood	11 Jun 1981	30	60	Sand bank	
West Creston	1993	75			
Wynndel (Duck Lake Road)	27 Jun 1981	12		Bank	
Central Interior					
Alexis Creek (19 km w)	7 May 1977	42	18	Clay bank	
Alexis Creek	29 May 1970	7	15	Sand bank	
Bridge Creek	13 Aug 1982	55		Clay bank	
Chilcotin River	29 May 1970	100	25	Bank	
Fraser Lake	5 Jun 1977	42	15	Cliff	
Lac la Hache	16 Jun 1963	30	50	Bank	
Nimpo Lake	16 Jul 1972		15	Bank	
100 Mile House (7 km se)	26 Jul 1984	65	65	Clay bank	

All data are estimates.

[1] All data are from the British Columbia Nest Records Scheme unless otherwise noted. (1, Anderson 1914; 2, Runyan 1971; 3, Fitz-Gibbon 1977)

winged Swallow is the least numerous of the 7 species of swallows occurring in the province.

Observers are cautioned that, as with many colonial, burrow-nesting species, the size of a colony cannot be estimated by counting nest burrows. Many of the burrows may have survived from previous years and may not be in use. For example, in 1990 a colony in sedimentary cliffs bordering the Chilcotin River had about 100 burrows, but only an estimated 25 pairs of swallows appeared to be nesting there (Table 2). Reaching an estimate of colony size requires determining the number of occupied burrows. A crude estimate of pairs may be reached by counting the number of birds and dividing by 2. This also ignores the presence of incubating or brooding birds as well as the possibility that an unknown number of unpaired birds may be present. This caution is particularly applicable to the Northern Rough-winged Swallow, which often nests in small numbers on the edges of Bank Swallow colonies.

Figure 162. Nest and eggs of the Northern Rough-winged Swallow removed from a burrow at Still Creek near Burnaby Lake (31 May 1962; R. Wayne Campbell). Most of the nest material was dry grasses.

NOTEWORTHY RECORDS

Spring: Coastal – Quicks Bottom 10 Mar 1974-1, 4 May 1904-1 (RBCM 1015; Munro and Cowan 1947); near Ucluelet 9 Apr 1974-2; Mayne Island 22 Apr 1976-2, 25 Apr 1971-6; Chilliwack 9 April 1891-2 (MVZ 102784; MCZ 245520); Iona Island 31 Mar 1990-3; Pitt Meadows 26 May 1963-100; Harrison 25 Mar 1962-1; Burnet Creek (Burnaby) 27 May 1942-12 pairs near dam, 1 nest with 5 eggs collected; Seabird Island 25 Apr 1984-76; Comox 6 May 1931-2 (RBCM 13278 and 13280); McGillivray Creek 17 Apr 1972-35; Guise Bay 18 May 1974-2 incubating; Gold River 25 May 1974-10; Kimsquit River 13 Apr 1985-10; Kitlope Lake 3 May 1991-10. **Interior** – Richter Pass 15 May 1957-5 eggs (Meugens and Cooper 1962); Ward Lake (Grand Forks) 27 Apr 1984-35; Princeton 11 May 1975-200; Trout Creek (Summerland) 3 Apr 1948-1; Summit Creek (Creston) 22 May 1981-building nests; Spences Bridge 23 May 1964-150; Sorrento 27 Apr 1971-14; Revelstoke 7 Apr 1986-1; Bridge Creek (100 Mile House) 10 Apr 1986-3; 100 Mile House 20 Apr 1985-25; Box Lake 20 Apr 1980-among mixed flock of 300 swallows, 4 May 1978-100; Williams Lake 6 May 1984-20; Quesnel 27 May 1993-2 pairs entering burrows; Prince George 23 Apr 1977-8; Noel Creek (s Dawson Creek) 17 May 1992-2 adults; Brassey Creek (sw Dawson Creek) 11 May 1993-2 adults; Peace River 31 May 1973-2 (Penner 1976); Manson River 14 to 23 May 1993-pair trying to nest, driven away by a Belted Kingfisher (Price 1993); Surprise Lake (Atlin) 18 May 1981-2 (Campbell 1981).

Summer: Coastal – Lost Lake (Saanich) 2 Aug 1958-50; Mayne Island 4 Aug 1985-2 nesting; Tofino 16 Jun 1992-adult with noisy young in old Belted Kingfisher burrow; Deroche 16 Jul 1969-nestlings; Burnaby Lake 11 Jun 1942-24; Harrison Hot Springs 31 Aug 1975-1; Kidney Lake 9 Jul 1981-20; Comox 20 Jul 1922-1 (NMC 18063); Mount Currie 4 Jun 1967-12; Keogh River 1 Jun 1951-1 (UBC 5513); 2 km e Gold Bridge 16 Jul 1972-nestling; Minette Bay 23 Aug 1975-1; 64 km e Bella Coola 12 Aug 1979-3 nestlings; Sandspit 7 Jun 1987-1, 25 Aug 1993-1; Kitlope River 9 Jun 1994-pair at burrow; Kitimat 7 Jul 1975-8; Terrace 15 Jun 1975-adults gave alarm calls near nest; Lakelse Lake 10 Jul 1974-32; near Stewart 17 Jun 1975-3. **Interior** – Manning Park 1 Jul 1966-30, 11 Aug 1972-large nestling; Vaseux Lake 6 Jun 1974-32; Syringa Creek 1 Aug 1979-50; Jaffray 27 Jul 1943-200 (Johnstone 1949); Grohman Creek Park (Nelson) 12 Jun 1984-20; Cranbrook 19 Jun 1975-6 eggs; Summit Lake (Nakusp) 13 Jun 1976-50; Kleena Kleene 12 Jun 1962-eggs, 16 Jul 1962-nestling; Williams Lake 1 Jun 1963-200; Bouchie Lake 7 Jun 1944-20 (Munro 1947a); Bowron Lake 10 Aug 1971-2 nestlings; Prince George 20 Jun 1977-7 eggs, 7 Aug 1963-1 (UBC 11996); Red Pass 5 Aug 1972-6 nestlings; Telkwa 14 Jul 1971-4 nestlings; Prophet River (Dawson Creek) 9 Jun 1972-3; Manson Lakes 28 Jun 1993-2; Fort St. John 7 Jun 1976-4, 24 Jul 1977-3 nestlings; Lynx Creek (Peace River) 18 Aug 1986-2 (McEwan and Johnston 1987a); Bear Flats 24 Jul 1977-5; Stalk Lake 12 Jul 1976-6; Klahowya Creek 21 Jul 1976-1; Tuchodi River 4 Jun 1994-1; Donna Creek (56°N, 124°W) 24 Jun 1993-pair carrying nesting material; Dease Lake 14 Jun 1980-3; Kechika River 31 Jul 1980-adults feeding 6 fledglings; Liard Hot Springs 12 Jul 1978-1.

Breeding Bird Surveys: Coastal – Recorded from 19 of 27 routes and on 27% of all surveys. Maxima: Pemberton 21 Jun 1981-12; Seabird 9 Jun 1973-8; Chilliwack 23 Jun 1974-8; Squamish 17 Jun 1984-7. **Interior** – Recorded from 53 of 73 routes and on 65% of all surveys. Maxima: Lac la Hache 30 Jun 1974-72; Horsefly 3 Jul 1988-62; Grand Forks 27 Jun 1993-61.

Autumn: Interior – Riske Creek 18 Sep 1988-1, 28 Sep 1986-1; n Barriere Lake 7 Sep 1974-10; Revelstoke 24 Oct 1985-2, 25 Oct 1985-1; Kootenay Pond 1 Sep 1965-50 (Seel 1965); Edgewater 3 Sep 1937-18; Nakusp 7 Nov 1976-2; Okanagan Landing 8 Sep 1978-2, 28 Sep 1905-1; Okanagan Centre 5 Sep 1931-1 (MVZ 160196); Kelowna 5 Nov 1973-5 on telephone wire in snowstorm (Rogers 1974); Vaseux Lake 3 Sep 1975-2; Glade 15 Sep 1968-3; Fruitvale 1 Sep 1983-50 on wires. **Coastal** – Harrison 1 Sep 1975-8, 5 Sep 1988-5; Sea Island 5 Sep 1971-200 (Crowell and Nehls 1972a); Tofino 14 Sep 1979-1; Langley 30 Sep 1964-50; Ladner 11 Nov 1979-1; Pender Island 9 Sep 1978-50; Swan Lake (Victoria) 19 Nov 1991-1.

Winter: No records.

Bank Swallow

BKSW

Riparia riparia (Linnaeus)

RANGE: Breeds from western and central Alaska, central Yukon, and northwestern Mackenzie across Canada to southern Labrador and Newfoundland, south in the west to California, in the east to South Carolina. Winters from Panama south to central Argentina and Chile. Also occurs over most of Eurasia.

STATUS: On the coast, *rare* migrant and summer visitant in the lower Fraser River valley of the Georgia Depression Ecoprovince; *casual* on the Sunshine Coast and Western Vancouver Island, *rare* migrant to southeast Vancouver Island, and *very rare* north along the mainland portions of the Coast and Mountains Ecoprovince; absent from the Queen Charlotte Islands.

Across the southern interior of the province, it is a *fairly common* to locally *very common* migrant and summer visitant in the Southern Interior, Southern Interior Mountains, and Central Interior ecoprovinces, and the Peace Lowland of the Boreal Plains Ecoprovince, although it may be locally *abundant* near colonies; *uncommon* to *fairly common* (locally *common* or *very common* around colonies) in the Sub-Boreal Interior and Northern Boreal Mountains ecoprovinces, and *uncommon* (locally *abundant* near colonies) in the Taiga Plains Ecoprovince.

Breeds.

CHANGE IN STATUS: The Bank Swallow was not known to nest west of the Coast Mountains when Munro and Cowan (1947) published their review. Brooks (1917), however, suspected a colony within 50 miles of Chilliwack. He notes that this swallow was "tolerably common," but he never found it breeding west of the Cascades. Breeding has since been documented twice from the southwest mainland coast. In 1960 a colony of about 75 pairs was found nesting along Old Yale Road in Surrey. Eight nests were checked and found to contain both eggs and young. Later, in 1964, 8 nests were located in Garibaldi Provincial Park; 1 nest was checked and held 2 eggs. We also have other reports that suggest nesting; however, the burrows were not checked. In May 1964 birds were found at nest sites at Hammond, near Mission. In 1968, 2 birds were seen at a burrow cavity near Glen Valley, and the following year 3 burrows were found near Hope. In 1990 swallows and 12 burrows were also found in the Hope area. Further north, there is an undocumented report of birds nesting in a high gravel bank along the Kimsquit River in 1986.

Figure 163. Swampy areas, like this shallow wetland, are important foraging areas for the Bank Swallow during spring migration in May (south of Pantage Lake, Cariboo, 24 May 1994; R. Wayne Campbell).

NONBREEDING: The Bank Swallow is widely distributed throughout the interior of the province wherever suitable habitat occurs. The highest numbers are found in the southern portions of the province; north of Smithers and Fort St. John it is sparsely distributed. On the coast, the Bank Swallow occurs regularly only in the lower Fraser River valley. It occurs infrequently as a summer visitor to the west coast of Vancouver Island and along the mainland coast, where suitable nesting habitat is almost entirely lacking. There are no records of this swallow from the Queen Charlotte Islands.

The Bank Swallow occurs from sea level to at least 1,500 m. It is usually seen with concentrations of other migrating swallows over ponds, lakes, swampy areas (Fig. 163), muskegs, fields and meadows, roadsides, farmsteads, and wooded areas. In passage, it uses a wide variety of perching sites, including dead trees, fences, telephone and electricity lines, the tips of tall plants, cattails, bulrushes, and a variety of human-made objects.

In British Columbia, it usually prefers the drier, interior ecoprovinces, frequenting grass- and sage-covered rangeland typically bordered by stands of ponderosa pine and interior Douglas-fir. Further north, it occurs in forests dominated by trembling aspen, white spruce, and lodgepole pine along major river valleys. It often forages along ridges and cliff faces. Not only does it nest at such sites if the substrate is suitable but the turbulent air currents bring flying insects up from lower areas, providing an available food source.

On the coast, Bank Swallows may arrive in March, but the main movement begins in April and continues into June, when numbers peak (Figs. 164 and 166). In the interior, birds sometimes arrive as early as the second week of March; however, the main movement begins during the latter part of April and early May (Figs. 164 and 166). The earliest arrivals in the province have been recorded from the Okanagan valley. At Penticton, the mean date over 8 years was 27 April; at Vernon it was 6 May over 11 years (Cannings et al. 1987). Birds entering the Okanagan valley and the east and west Kootenays in April move rapidly northward, taking advantage of the lakes that lose their ice early and offer feeding stops for the migrating birds. Numbers have reached the Cariboo and Chilcotin areas by May and the Nechako and Bulkley river valleys by June. Birds enter the Peace Lowland during May, likely following a migration route east of the Rocky Mountains. In the Northern Boreal Mountains and the Fort Nelson Lowland, the birds arrive in June and peak in July (Figs. 164 and 166).

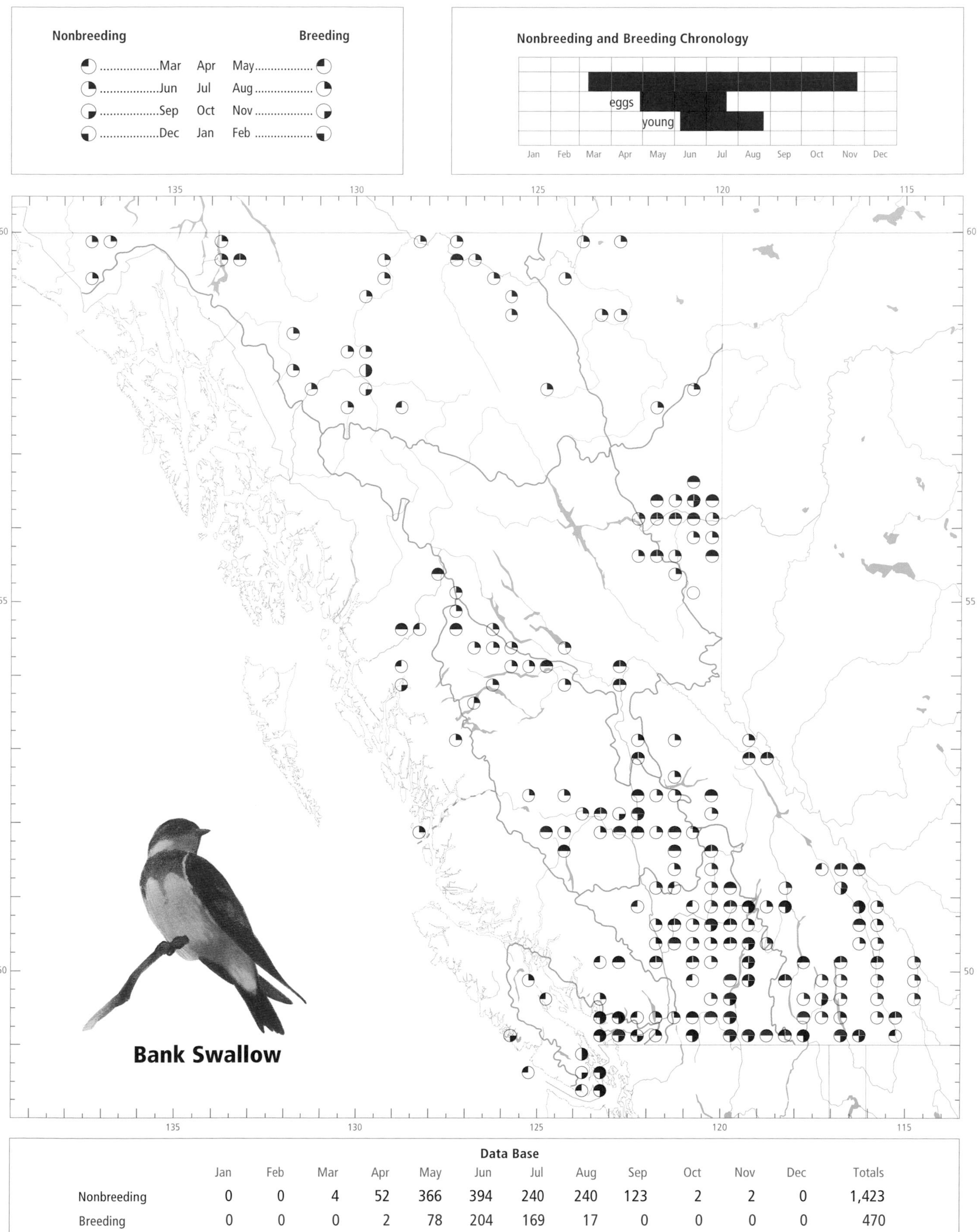

Data Base

	Jan	Feb	Mar	Apr	May	Jun	Jul	Aug	Sep	Oct	Nov	Dec	Totals
Nonbreeding	0	0	4	52	366	394	240	240	123	2	2	0	1,423
Breeding	0	0	0	2	78	204	169	17	0	0	0	0	470

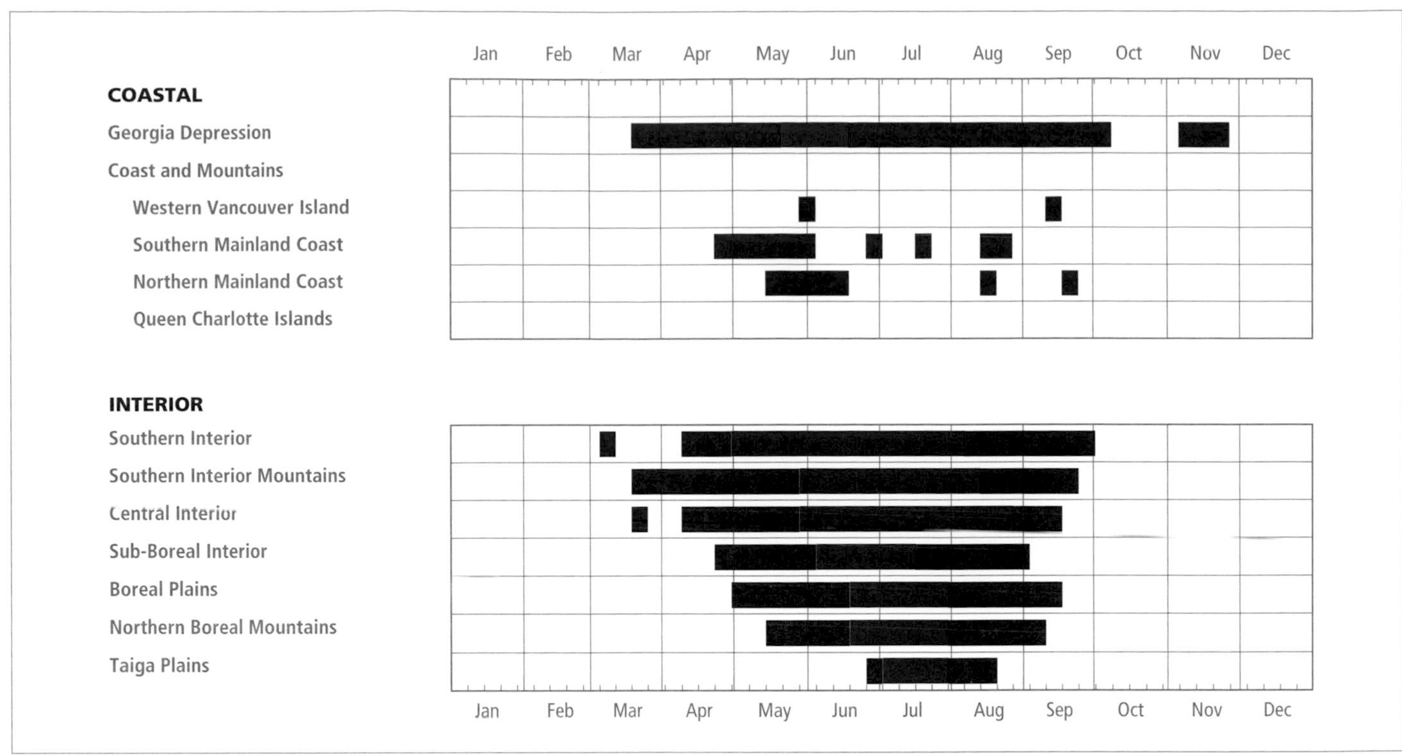

Figure 164. Annual occurrence (black) and breeding chronology (red) for the Bank Swallow in ecoprovinces of British Columbia.

Across the northernmost part of the province, from Hyland River to Fort Nelson, most birds have begun the southern movement by August and have departed by the end of that month. Only rarely do stragglers remain into September. Birds in southern portions of the province, including the Central Interior, Southern Interior, and Southern Interior Mountains, begin their southward movement soon after the young have left the nest. Most birds there have also gone by the end of August, although some remain into September (Figs. 164 and 166). An apparent peak for the Southern Interior in September (Fig. 166) is an artifact created by 1 record of 1,000 birds, the largest flock recorded in the province.

Like many other species, the Bank Swallow stays longer in the lower Fraser River valley than elsewhere in the province. While most of the southbound migrants have left by the end of August, there are a number of sightings in the lower Fraser River valley well into September, and stragglers have been reported into October and November.

Unlike the Tree and Violet-Green swallows, which move south in small groups, large aggregations of Bank Swallows have been seen in the autumn migration. In other parts of North America, this swallow is known to gather in enormous premigratory communal roosts (Ehrlich et al. 1988). In British Columbia, they have been found during passage roosting in the dense stands of rushes and cattails that border some of the most productive lakes along this swallow's routes (Fig. 165). Information about these communal roosts is scarce, however, and observers are encouraged to note the details and locations of any roosts they discover.

The Bank Swallow has been recorded in British Columbia from 9 March to 23 November (Fig. 164).

BREEDING: The Bank Swallow breeds throughout much of British Columbia east of the Coast Mountains; breeding has been reported only twice from west of the Cascades. It is widespread across the southern interior to the east Kootenay and north to the latitude of Fort St. James. Smaller numbers breed in the Peace Lowland and the Fort Nelson Lowland, and west across the northern portions of the Northern Boreal Mountains. Northernmost nesting reports are from the Tatshenshini River valley, Atlin, Coal River, Fireside, and Pine Tree Lake.

The Bank Swallow appears to be absent as a breeding bird from much of the Sub-Boreal Interior, including the Rocky Mountain Trench north of Summit Lake north to the 59th parallel, and adjacent areas of the Northern Boreal Mountains.

In British Columbia, the Bank Swallow reaches its highest numbers in the Southern Interior Mountains (Fig. 167). There, and in other valleys of south and central British Columbia, post-glacial lakes have left deep deposits of glaciolacustrine silt (Matthews 1986). This fine material is ideal

Figure 165. Dense stands of cattails (*Typha latifolia*) bordering lakes provide roosting sites for Bank Swallows in autumn migration (Nicola Lake, May 1994; R. Wayne Campbell).

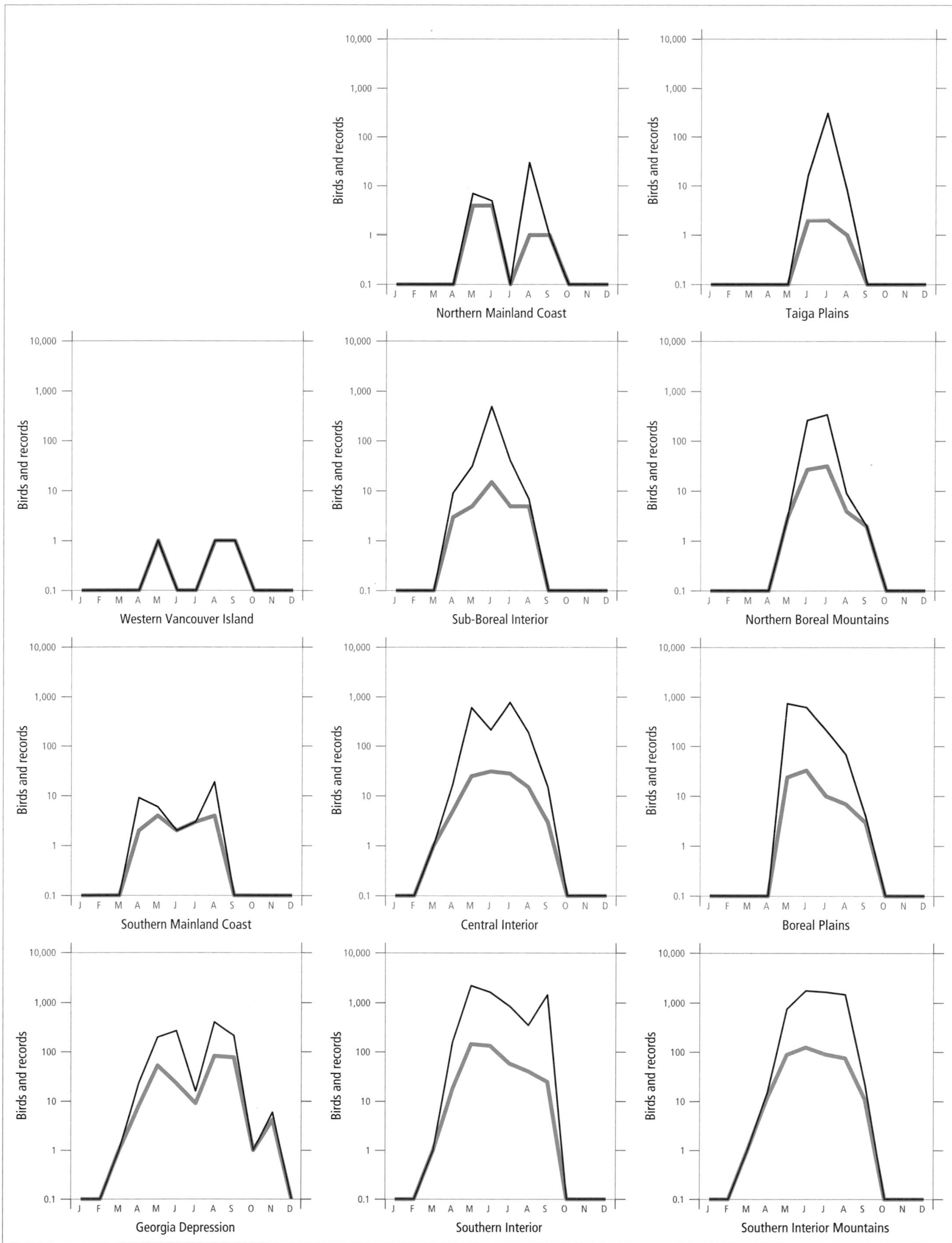

Figure 166. Fluctuations in total number of birds (purple line) and total number of records (green line) for the Bank Swallow in ecoprovinces of British Columbia. Nest record data have been excluded.

for the excavation of nesting burrows because it is unconsolidated and soft but cohesive enough to maintain its form. Rivers and rain have eroded the deposits into steep faces and complex spires such as the hoodoos west of Fairmont. Roads and rail lines have also cut through the same deposits, creating additional nesting opportunities for the Bank Swallow.

An analysis of Breeding Bird Surveys for the period 1968 through 1993 could not detect a net change in numbers on interior routes; insufficient data were available to determine trends on coastal routes for the same period.

The Bank Swallow has been reported nesting from a few metres above sea level to 900 m. Nesting areas occur in habitats similar to those described for the nonbreeding period, but breeding birds are seldom found far from the immediate vicinity of their nesting sites (Fig. 168). They are usually associated with the presence of open grasslands, meadows, cultivated fields, lakes and ponds, or rivers with adjacent riparian growth.

Provincewide, the general habitat used by the nesting birds includes forested lands and woodlands (33%; *n* = 123), cultivated farmlands (36%), sagebrush and grasslands (22%), and the margins of lakes, rivers, and wetlands. Most of that habitat is human-influenced (85%; *n* = 170). Erskine (1979) found that 87% of Bank Swallow nests in British Columbia were in human-made sites.

The Bank Swallow has been recorded nesting in the province from 27 April (calculated) to 24 August (Fig. 164).

Colonies: The Bank Swallow is a colonial nester. There are few instances in British Columbia of a single pair of birds occupying a nesting site. The 491 records of nest colonies in the province for which data are available provide a range of 3 to 3,035 burrows (Fig. 168b), with most colonies (52%) having between 15 and 75 burrows. Erskine (1979) gives the average number of nests per colony in Canada as 42, and Turner and Rose (1989) give 38 as the average colony size in Britain. (See REMARKS in **Northern Rough-winged Swallow** for a caution regarding the determination of colony size.)

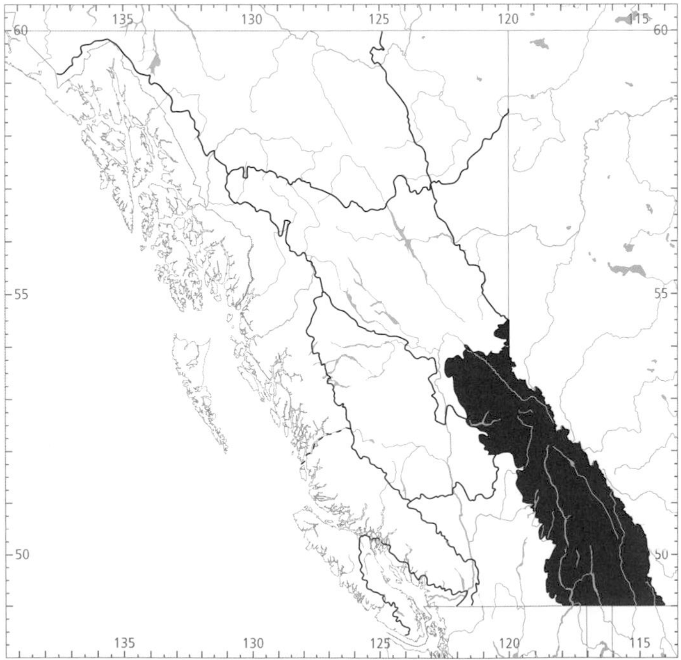

Figure 167. In British Columbia, the highest numbers for the Bank Swallow in summer occur in the Southern Interior Mountains Ecoprovince.

In a Saskatchewan study, as many as 40% of the colonies in 1 area consisted of 1 to 3 pairs (Hjertaas et al. 1988), while in the Sacramento valley of California 60 colonies had a range of 12 to 1,784 pairs and a mean of 269 (Garrison et al. 1987). Thus, it would seem that colonies of the Bank Swallow in the mountains of western North America may be larger than in other parts of its northern range.

The major Bank Swallow breeding colonies in the province are shown in Table 3. In the Okanagan valley and the Kootenay River valley, clusters of nests, occupying several adjacent banks, are regarded as a single colony. The largest, along the Norbury Lake Road 2 km north of Wardner, in the east Kootenay (Fig. 168b), extends almost 1.0 km along a cutbank. In 1991, 3,035 nest burrows were counted there (Campbell and Dawe 1991). Similarly, a large colony in the Okanagan valley has occupied a series of large and small erosion banks on the eastern shore of Skaha Lake for many years.

The southern part of the Rocky Mountain Trench, traversed by the Kootenay River, supports the largest number of colonies containing 200 or more burrows, as well as the 3 largest known colonies in the province.

Many colonies are relatively short-lived because the banks they occupy erode continuously, changing their suitability as nest sites. However, some have persisted for many years. Records of the occupancy of a colony near Castlegar extend from 1975 to 1993. The Norbury Lake Road colony has been in use since at least 1970 despite sometimes heavy wind erosion. During a 1991 visit, the wind was blowing enough sand to blind observers and fill many of the burrow entrances.

Colonies may not be used every year, and in the interim other species may use the site. In 1993 a known Bank Swallow colony along the West Creston Road was occupied entirely by Northern Rough-winged Swallows and House Sparrows. Bank Swallows reoccupied it in 1994.

Nests: Banks created by roadcuts held the greatest number of colonies (27%; *n* = 220; Fig. 168b), followed by lakeshore banks and cliffs (17%), gravel pits (8%; Fig. 168c), and river banks (7%; Fig. 168a). There is little doubt that the number of suitable nesting sites in British Columbia has been greatly increased by human road-building activities. Most nest burrows were reportedly excavated in silt, clay, or sand. There is 1 record of a Bank Swallow nesting in a natural crevice, and 1 large colony was located in a large dump of consolidated sawdust (Fig. 168d). The heights of 49 nests ranged from ground level to 38 m, with 59% between 3 and 8 m. In 39 nests, burrow length ranged from 42 to 180 cm, with a mean of 90 cm. Entrances of 28 burrows averaged 6 cm high (range: 4 to 10 cm) by 7 cm wide (range: 6 to 10 cm).

Studies elsewhere in Canada have revealed that the Bank Swallow selects relatively specific conditions. Almost all nests are in cliffs or banks of fine-textured silt, sand, or soil that have a vertical face above the colluvial apron of at least 3 m

Figure 168. Nesting colonies of Bank Swallows have been found in a wide variety of habitats in British Columbia, including: (a) a colony of 262 burrows in cliffs carved by the Nechako River opposite Prince George, 14 July 1990; (b) 3,035 burrows in the bank along Norbury Lake Road, 2 km north of Wardner, 6 July 1991; (c) 78 burrows in a gravel pit, Lone Butte along Eagle Island Resort Road, 16 July 1990; and (d) 104 burrows in piles of compressed sawdust at the south end of Duck Lake, Creston, 1 June 1968 (all R. Wayne Campbell).

and in which the penetrability involves a force of 6 to 12 kg applied to a pointed probe (John 1991). These characteristics seemed to provide the soil stability required by the birds. The height above the colluvial apron offers some protection from terrestrial predators.

Only 12 nests were reported in detail. Nests were scanty, flat platforms composed of grasses and other materials such as feathers (42%) and twigs (17%), as well as straw, rootlets, plant stalks, or leaves. Most nests were lined with white feathers.

Eggs: Dates for 67 clutches ranged from 16 May to 19 July, with 55% recorded between 14 June and 28 June. Calculated dates indicate that nests could have eggs as early as 27 April. Clutch size ranged from 1 to 7 eggs (1E-8, 2E-11, 3E-13, 4E-16, 5E-13, 6E-5, 7E-1), with 63% having 3 to 5 eggs. The incubation period ranges from 12 to 16 days, with an average of 14 days (Stoner 1936; Turner and Rose 1989).

Young: Dates for 246 broods ranged from 3 June to 24 August, with 78% recorded between 8 July and 21 July. Calculated dates indicate that nests could have young as early as 12 May. Brood size ranged from 1 to 6 young (1Y-57, 2Y-45, 3Y-80, 4Y-57, 5Y-6, 6Y-1), with 74 % having 2 to 4 young. The nestling period ranges from 18 to 24 days, with an average of 22 days (Bent 1942; Turner and Bryant 1979).

Table 3. Location and size of major colonies (more than 200 nests or pairs) of Bank Swallows in British Columbia.

Burrow location	Date	Burrows[1]	Pairs[1]	Location
INTERIOR				
Southern Interior				
Chase, Turtle Valley Road	25 Jul 1974	461	23	Road cut
Grand Forks, 3 km e	8 Jul 1982	750	650	Clay bank
Kamloops	22 Jul 1968	220	60	Sand bank
Naramata	25 Jun 1968	203	225	Gravel bank
Okanagan Falls, nr campground	14 Jul 1961	200	20	Gravel bank
Penticton, 8 km sw	25 Jun 1960	200	150	Sand bank
Penticton, Westbench nr Mac's Lake Road	10 Jul 1988	220	70	Sand bank
Skaha Lake, above w side	21 Jun 1969	300	200	Clay bank
Skaha Lake, e side	8 Jul 1969	587[3]	365	Clay bank
Winfield, 3.7 km w	22 May 1993	391		Sand bank
Southern Interior Mountains				
Athalmer, above Columbia River bridge	13 Jul 1963	250	250	Clay bank
Athalmer, Columbia River marshes	30 Jul 1970	1,500		Clay bank
Athalmer, e of Columbia River	30 Jul 1970	300		Clay bank
Canal Flats, 9 km n	27 Jul 1970	300		Clay bank
Castlegar, old Kinnaird dump	14 Jun 1980	200	50	
Castlegar, on road to golf course	5 Jul 1979	200	200	
Columbia Lake	15 Jul 1970	300		Clay bank
Cranbrook, 40 km se Bull River Road	15 Jun 1975	250	250	
Invermere, gypsum plant loading	21 Jul 1976	275	100	Silt bank
Nelson, w on Kootenay River	20 Jun 1984	200	200	Clay & sand bank
Wardner, ne	13 Jul 1990	500	300	
Wardner, near Bull River	9 Aug 1970	2,000	1,500	Clay bank
Wardner, 2 km n, along Norbury Lake Rd	6 Jul 1991	3,035[3]		Eroding sand
Wasa, 5 km s	7 Aug 1970	200		Clay bank
Wilmer, nr National Wildlife Area	7 Jul 1991	342		Sandstone
Wynndel	22 Jun 1993	233		Sawdust
Central Interior				
Alexis Creek, 5 km e	15 Jun 1967	500		
Anahim Lake Indian Reserve	12 Jun 1975	200	50	
Sub-Boreal Interior				
Peace River, nr WAC Bennett Dam	2 Jun 1969	300		Clay bank
Prince George	16 Jul 1963	295		
Prince George, Salmon River	3 Jul 1972	225		
Northern Boreal Mountains				
Atlin, mile 547.5 Alaska Hwy	28 May 1968	100	300	Bank

Source: All data are from the British Columbia Nest Records Scheme.
[1] All data are estimates unless otherwise noted.
[2] Details are from observer's records.
[3] Totals include 2 or more groups of nests on closely adjacent banks.

Brown-headed Cowbird Parasitism: Cowbird parasitism was not found in British Columbia in 313 nests recorded with eggs or young. Friedmann (1963) reports 1 instance of cowbird parasitism of the Bank Swallow in North America.

Nesting Success: Insufficient data.

REMARKS: There is some historical evidence regarding the response of this species to the increased availability of nest sites provided by human activity, especially road building. In the Kootenay and Columbia river valleys, in the southern Rocky Mountain Trench, Johnstone (1949) studied the bird fauna intensively between 1937 and 1949. He noted just 2 nesting colonies of Bank Swallows, one of 5 pairs and another of 140 nesting burrows. Today in the same region 9 colonies with more than 200 burrows each have been reported (Table 3), along with 18 smaller colonies. Included are the 2 largest known colonies in the province. The impression of long-time observers is that similar but less dramatic changes have occurred in the Okanagan valley and in the Cariboo and Chilcotin areas.

There are additional records of nesting from southern Vancouver Island and the lower Fraser River valley, including reports on Breeding Bird Surveys; however, sufficient data to confirm that the birds in question were indeed Bank Swallows were lacking, and they have been excluded from the account. The descriptions offered were more indicative of the nests of Northern Rough-winged Swallows (i.e., no feathers in the nest material and clutch sizes of 7 and 8 eggs). Observers are encouraged to familiarize themselves with the typical range of a species and to carefully document observations of species in unusual areas. Verifying that eggs or young are present in the burrows, when possible, is an important part of the documentation.

Some colonies are being heavily disturbed and the future of others is in jeopardy because of the collection of sand and gravel for private use. One such colony is in the Okanagan valley, along the Okanagan Centre Road.

The Bank Swallow is known as the Sand Martin in Eurasia, Africa, India, and South America.

NOTEWORTHY RECORDS

Spring: Coastal – Pachena Point 29 May 1975-1; Glen Valley 15 May 1968-2 at burrow in gravel pit; Hammond (Mission) 27 May 1964-30 birds at nests in sand bank, not checked; Iona Island 14 Apr 1986-2; Garibaldi Park (Alouette River) 27 May 1964-2 eggs in 1 of 8 nests; Beaver Lake (Vancouver) 20 Mar 1971-1; Green Lake 25 Apr 1970-7; Kitimat Mission 30 May 1975-2; Terrace 18 May 1987-1. **Interior** – Okanagan Falls 17 May 1962-150; Skaha Lake 16 May 1913-several clutches in advanced incubation plus several newly hatched broods; Penticton 11 Apr 1962-4 at picnic site; Summerland 9 Mar 1959-1; Robson 25 Mar 1971-1; Lillooet 12 Apr 1968-35; Revelstoke 13 Apr 1972-2; Clearwater 29 Mar 1971-2; 100 Mile House 30 May 1984-eggs and shells below burrows; Kleena Kleene 19 May 1961-8, first seen (Paul 1962); Scout Island (Williams Lake) 19 May 1978-380; Anahim Lake 22 Mar 1977-1; Quesnel 29 May 1963-12; Shelley 29 Apr 1981-6; Wright Creek 18 May 1981-1; Farrell Creek 5 May 1984-1, first arrival; Charlie Lake 30 May 1984-300; Klappan River 21 May 1981, colony (Blood et al. 1981).

Summer: Coastal – Surrey (Old Yale Road) 5 Jun 1960-140 birds near 75 burrows containing mixed eggs and young, 12 Northern Rough-winged Swallows in same area, 30 Jun 1960-80 birds still around nest holes feeding young; Sea Island 23 Aug 1940-40; Chilliwack 10 Aug 1888-1; Harrison 23 Aug 1975-1; Brunswick Point 17 Aug 1986-15; Sechelt 29 Aug 1995-1 at airport; Porpoise Bay Park 19 Aug 1994-1; Mitlenatch Island Park 24 Aug 1965-3 (Campbell 1965a); Tahtsa Reach 26 Jun 1976-8; Goose Group 21 Jul 1948-1 (UBC 1934); Kimsquit River 15 Aug 1986-30 assumed nesting in high gravel bank, nests not checked; junction Alsek and Tatshenshini rivers 7 Jun 1983-1 (RBCM 17819). **Interior** – Osoyoos 14 Aug 1974-100 (Cannings et al. 1987); Nick's Island (Creston) 5 Aug 1981-500; Skaha Lake 17 Jul 1961-nestlings near full-grown, 16 Jun 1967-250; Balfour 8 Jun 1983-377; Natal Ridge 24 Jul 1984-300; Wasa 23 Aug 1971-300; Lytton 21 Jun 1986-2; Windermere Lake 12 Aug 1976-nestling; Salmon Arm 12 Jul 1973-200; Chase 28 Jul 1974-young; Adams Lake 1 Jul 1963-2 clutches of 5 eggs; Golden 27 Jul 1982-nestling; 9.5 km n Clinton 30 Jun 1962-2 clutches of 4 eggs; 100 Mile House 10 Jul 1989-3 nestlings seen, others heard (latest nestlings); 123 Mile House 13 Jun 1972-4 fresh eggs; Chilcotin River 24 Aug 1987-nestlings being fed; Riske Creek 26 Aug 1978-70; Kleena Kleene 29 Jun 1961-eggs, 24 Aug 1967-last nestling fledged; Red Pass 19 Jul 1972-2 nests with 5 eggs (latest date) and 3 nestlings; Nulki Lake 29 Jul 1964-350; Sinkut River 20 Jun 1945-136 nests (Munro 1949); Chilako River 27 Aug 1963-1; Prince George 7 Jun 1889-2 eggs, 2 Jul 1974-25, 12 Jul 1990-nestling being fed; Hudson's Hope 24 Jul 1964-nestling; Beatton River 15 Jun 1987-eggs; Fort St. John 25 Jul 1985-50; Boundary Lake 6 Jun 1982-250; Fort Nelson 7 Jul 1978-300, 28 Jul 1978-nestlings; Otter Creek 10 Jun 1975-100; Liard River 14 Aug 1943-8; Atlin 20 Jun 1975-eggs, 5 Jul 1980-4 nestlings (eyes just opened), 26 Jul 1988-brood near fledging; Hyland River 29 Jul 1981-150; Petitot River Bridge 26 Jun 1985-15; Tatshenshini River 17 Jun 1993-15 near colony.

Breeding Bird Surveys: Coastal – Recorded from 5 of 27 routes and on 2% of all surveys. Maxima: Pemberton 11 Jun 1983-6; Kispiox 20 Jun 1993-6. **Interior** – Recorded from 41 of 73 routes and on 34% of all surveys. Maxima: Tokay 29 Jun 1973-230; Golden 21 Jun 1973-137; Oliver 28 Jun 1992-112.

Autumn: Interior – Stikine River 4 Sep 1977-1; Eddontenajon Lake 5 Sep 1977-1; Charlie Lake 8 Sep 1984-2; Fort St. John 16 Sep 1982-1; Nulki Lake 1 Sep 1945-10, latest seen (Munro 1949); Quesnel 11 Sep 1978-3; Riske Creek 5 Sep 1978-10; Scout Island (Williams Lake) 12 Sep 1978-4; Kamloops 1 Sep 1953-1 (PMNH 71849); Nicola Lake 2 Sep 1988-150 roosting in cattails; Kootenay National Park 23 Sep 1982-4; Okanagan Landing 27 Sep 1915-1 (FMNH 141881); Wood Lake 27 Sep 1915-1; Kelowna 2 Sep 1957-1 (NMC 48026); Richter Pass 6 Sep 1984-1,000 at w end; Blackwell Camp (Manning Park) 9 Sep 1984-1. **Coastal** – Kitimat Arm 17 Sep 1974-1 (Hay 1976); Long Beach 11 Sep 1982-1; Sea Island 1 Sep 1970-1 (BC Photo 72); Reifel Island 8 Sep 1989-27, 3 Oct 1985-1, 11 Nov 1965-1, 23 Nov 1965-1 (Campbell 1966); Ladner 2 Sep 1965-3 at sewage pond (Poynter 1965).

Winter: No records.

Cliff Swallow

CLSW

Hirundo pyrrhonota Vieillot

RANGE: Breeds from western and central Alaska, central Yukon, northwestern and south-central Mackenzie, northern Saskatchewan, Manitoba, and Ontario through central Quebec and southern Labrador, east to Nova Scotia and southwestern Newfoundland, and south through the United States to central Mexico. Absent from Florida and the eastern Gulf coast. Winters in central Panama and in western and central South America to northern Argentina.

STATUS: On the coast, *fairly common* to locally *very common* (occasionally *abundant*) migrant and summer visitant to southeastern Vancouver Island and the Fraser River delta of the Georgia Depression Ecoprovince; *fairly common* to *common* autumn migrant. *Rare* on the Sunshine Coast and Western Vancouver Island; *uncommon* along the southern and northern mainland portions of the Coast and Mountains Ecoprovince; *very rare* on the Queen Charlotte Islands.

In the interior, *common* to locally *very common* migrant and summer visitant across much of the Southern Interior, Southern Interior Mountains, and Central Interior ecoprovinces, where it is sometimes *abundant* to *very abundant* during migration periods; generally, a *fairly common* to *common* autumn migrant in those areas. *Uncommon* to locally *very common* migrant and summer visitant in the Sub-Boreal Interior, Boreal Plains, Northern Boreal Mountains, and Taiga Plains ecoprovinces. In the Peace Lowland of the Boreal Plains Ecoprovince, it can be locally *abundant* near large colonies.

Breeds.

Figure 169. Adult Cliff Swallow gathering nesting material (Courtland Flats, Victoria, British Columbia, May 1986; Tim Zurowski).

NONBREEDING: The Cliff Swallow (Fig. 169) is widely distributed in suitable habitat throughout much of British Columbia, particularly in the southern half of the province. On Vancouver Island, it occurs along the east coast from Victoria to Campbell River, but is an infrequent visitor to the west coast and to the northern portions of the island. It is not common on the mainland coast north of Vancouver Island. On the Queen Charlotte Islands, it is known only from isolated occurrences at Delkatla Inlet, Rose Spit, Queen Charlotte City, Sandspit, and Cape St. James. This species has not been

Figure 170. In some years, during spring migration large concentrations of up to 20,000 Cliff Swallows forage over Swan Lake (Vernon), a shallow body of water in the Okanagan valley (14 May 1990; R. Wayne Campbell).

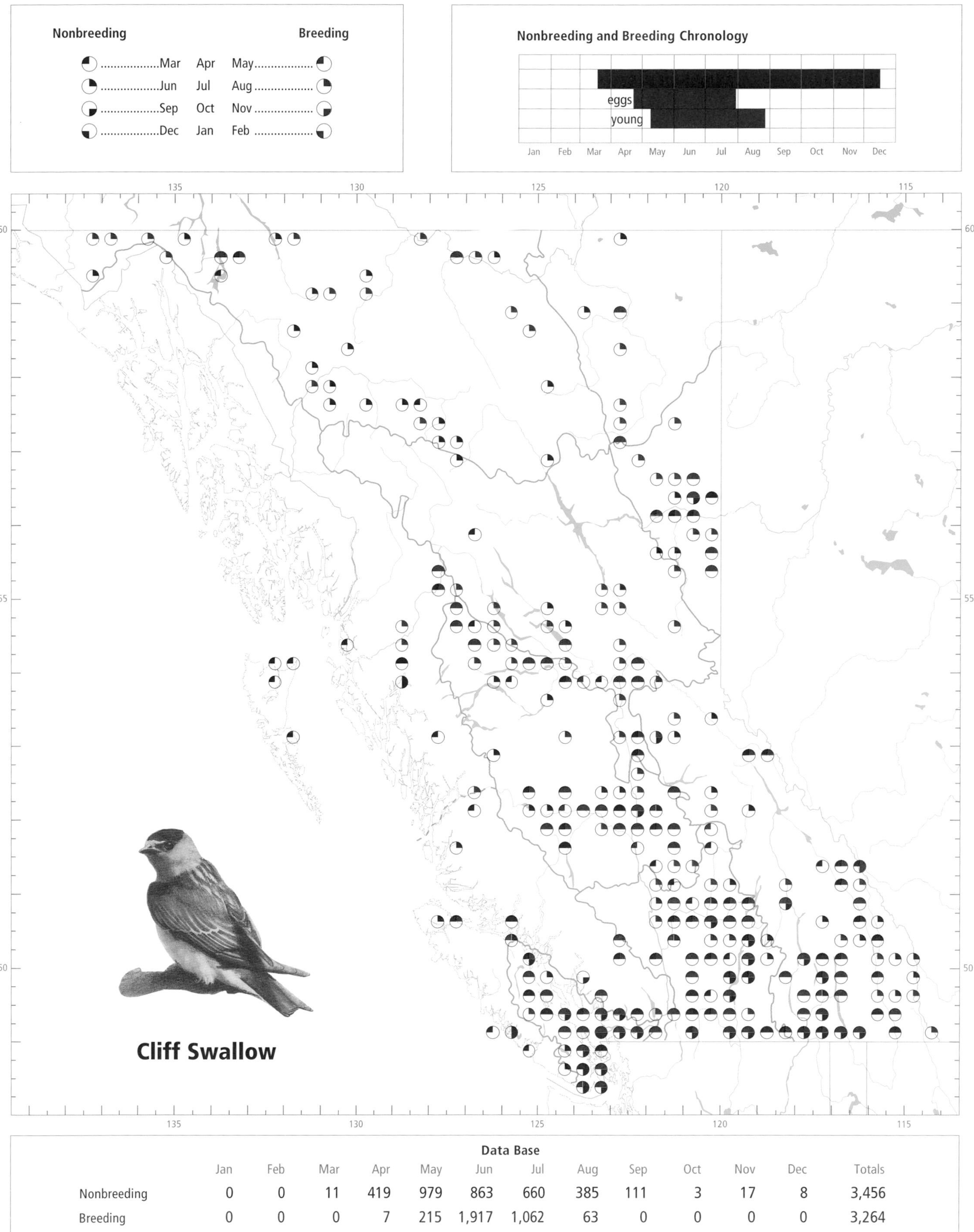

Data Base

	Jan	Feb	Mar	Apr	May	Jun	Jul	Aug	Sep	Oct	Nov	Dec	Totals
Nonbreeding	0	0	11	419	979	863	660	385	111	3	17	8	3,456
Breeding	0	0	0	7	215	1,917	1,062	63	0	0	0	0	3,264

reported from most of the Coast Ranges or from large parts of the Rocky, Columbia, and Selkirk mountains, nor has it been reported from the Rocky Mountain Trench north of Mackenzie. In British Columbia, northernmost records are from the Tatshenshini River area, Tagish Lake, Atlin, Emile Creek, and Petitot River, all near the Yukon boundary.

The Cliff Swallow has been reported from near sea level to 2,200 m elevation. It is a bird of open valleys, rangeland, and parkland. During spring migration, it makes extensive use of lakes (Fig. 170), ponds, sloughs, and their associated marshes, which are important sources of flying insects. It also hawks insects over cliff faces, fields, sewage lagoons, rivers, aspen groves and other deciduous vegetation, farmsteads, beaches, and sand dunes.

The timing of spring migration varies across the province. Along eastern Vancouver Island, in the Fraser River delta, in the Okanagan, Thompson, and Creston valleys, and in the Cariboo and Chilcotin areas, early migrants arrive by late March in some years, but most arrive in April (Figs. 171 and 172). The peak of the movement occurs in early May. The most conspicuous migratory movement occurs in the Okanagan valley, where Swan Lake (Vernon; Fig. 170) and Ellison Lake have produced aggregations of many thousands of Cliff Swallows in early May. In the Columbia and Kootenay river valleys, the migration generally begins in April and does not reach its maximum until June. Part of this stream likely joins other migrants moving into the Cariboo and Chilcotin areas. A smaller number continue into the Sub-Boreal Interior between Quesnel and Prince George, then move northward into the Northern Boreal Mountains and other northern areas, where the migration begins in May and peaks in June.

On the south coast, numbers in June are slightly lower than those in May, probably because some birds continue northward. In July, numbers are up again when fledglings join the population (Fig. 172). The pattern is the same in the Southern Interior Mountains and the Central Interior. In the Southern Interior, however, there is a sharp decline from May to June as migrants move north, leaving only a relatively small breeding population in June and July.

In the Central Interior and northern areas of the province, autumn migration begins soon after the young are on the wing, and most Cliff Swallows have left by the end of August (Figs. 171 and 172). A similar trend occurs in the Southern Interior Mountains. In the Southern Interior, however, August sees a dramatic increase as northern migrants arrive. The migration passes through rapidly, with just a few stragglers remaining into September or later. In the Southern Interior Mountains and on the south coast, most birds migrate south in late July or August, with few remaining into September.

The Cliff Swallow has been reported in winter only from the vicinity of the Fraser River delta. It has been recorded in British Columbia from 18 March to 17 December (Fig. 171).

BREEDING: The Cliff Swallow has a widespread breeding distribution across much of British Columbia wherever suitable habitat occurs. It is a breeding species from the east coast of Vancouver Island and the lower Fraser River valley east to the Flathead valley and much of the southern Rocky Mountain Trench, northward to include the Thompson-Okanagan Plateau and the Chilcotin and Cariboo areas, and north through much of the Sub-Boreal Interior, including the Nechako and Skeena river valleys. Further north, nesting is known from the Peace Lowland north along the Alaska Highway into the Fort Nelson Lowland, to the Petitot River, and west through the Northern Boreal Mountains to Lower Post.

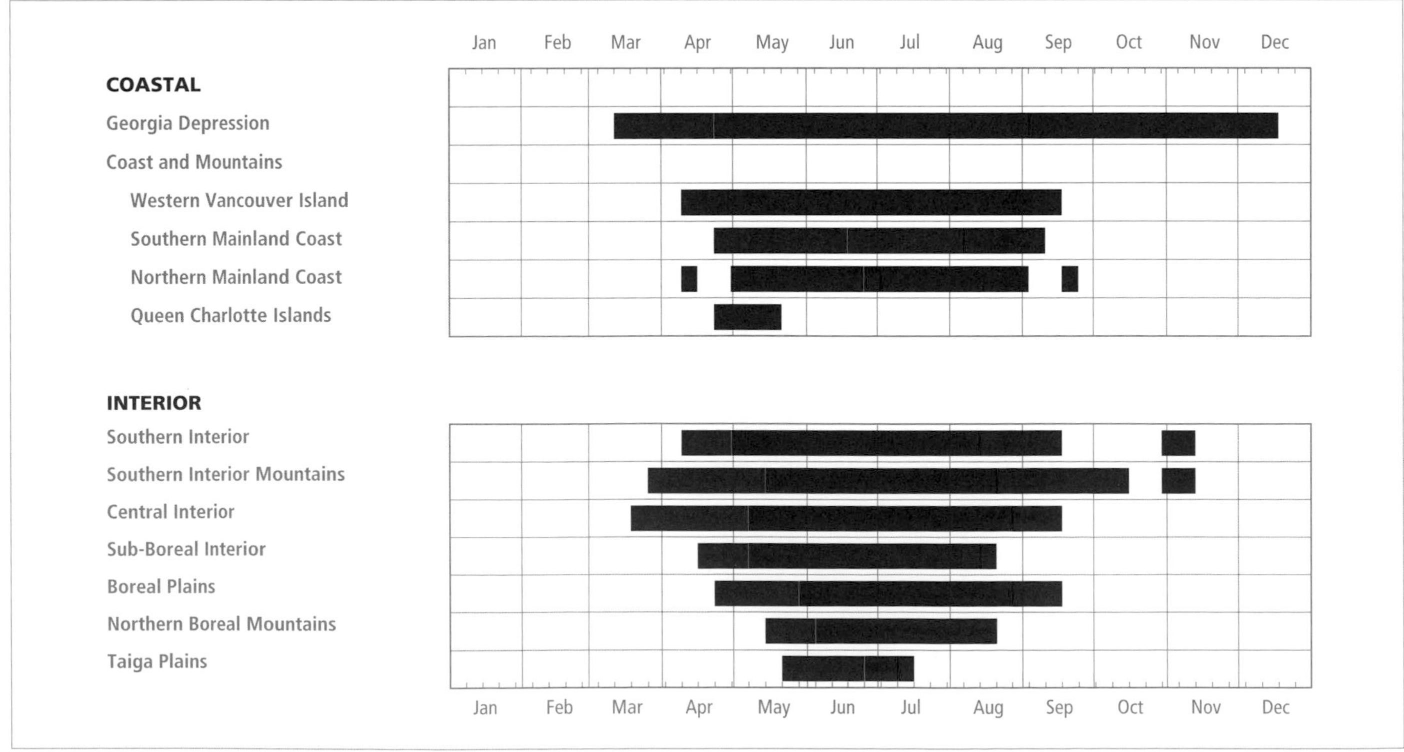

Figure 171. Annual occurrence (black) and breeding chronology (red) for the Cliff Swallow in ecoprovinces of British Columbia.

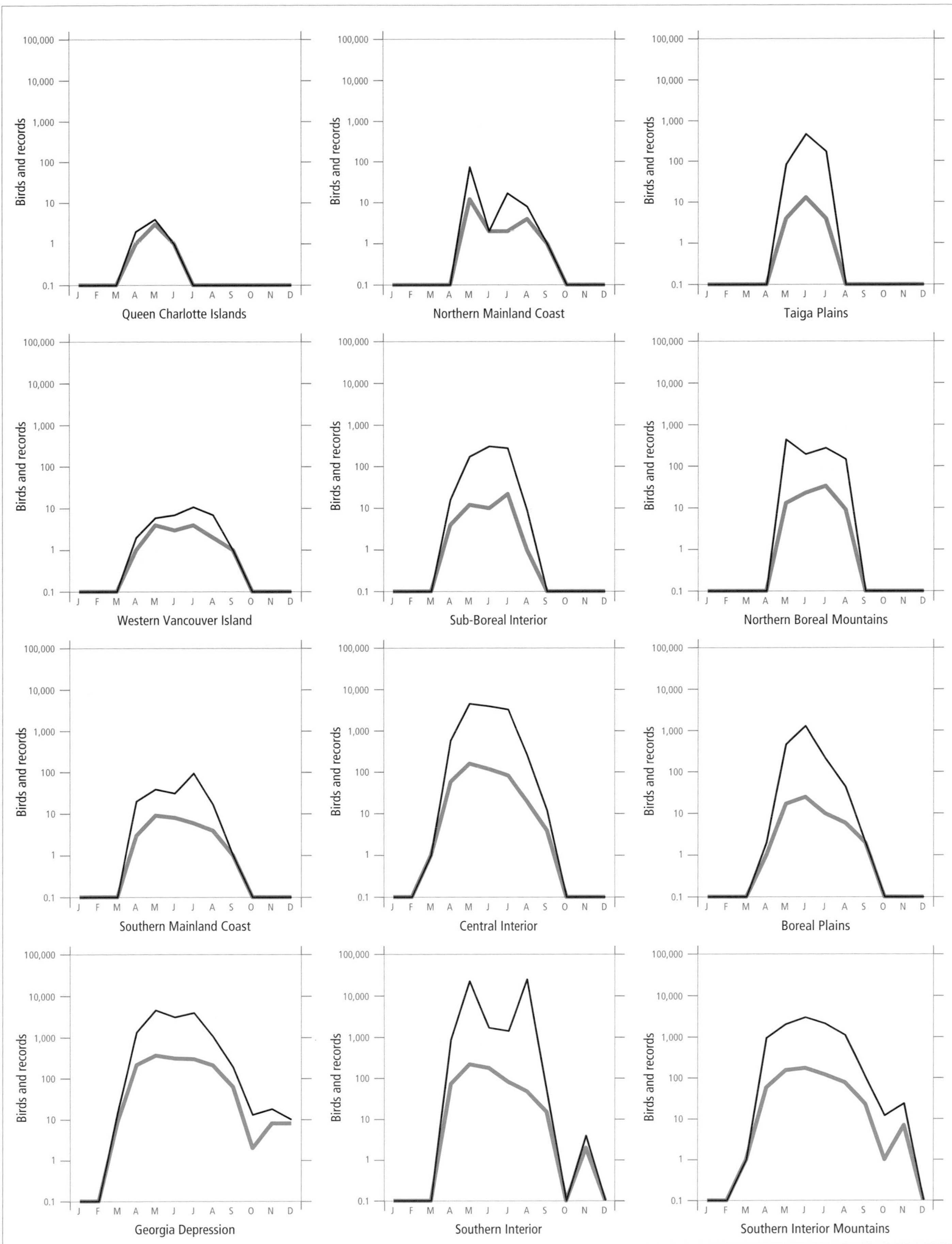

Figure 172. Fluctuations in total number of birds (purple line) and total number of records (green line) for the Cliff Swallow in ecoprovinces of British Columbia. Christmas Bird Count and nest record data have been excluded.

Breeding has also been reported from the Atlin area, the Tatshenshini River valley, Kelsall Lake, and the Chilkat Pass.

There is a large area in the Northern Boreal Mountains, from just east of the Rocky Mountains west to the Alaska panhandle, in which the Cliff Swallow occurs as a summer visitant; however, nesting has been reported from only 2 locations: Tuaton Lake (Osmond-Jones et al. 1977) and Telegraph Creek (Swarth 1922). Bella Coola is the only reported nesting locale for the Cliff Swallow in the Coast and Mountains over the long stretch of coastline and islands between the Squamish River valley and the Skeena River. It is not known to breed on the Queen Charlotte Islands.

The Cliff Swallow reaches its highest numbers in the Central Interior (Fig. 173). An analysis of Breeding Bird Surveys for the period 1968 through 1993 could not detect a net change in numbers on both coastal and interior routes. Robbins et al. (1986) and Erskine et al. (1992) note no significant changes in Cliff Swallow numbers in their reviews of Breeding Bird Surveys for North America and Canada, respectively.

The Cliff Swallow has been reported breeding from near sea level to 1,750 m elevation. Most breeding sites (over 80%; n = 1,299) were found in human-influenced situations, including urban and suburban areas (40%; Fig. 174), rural areas (26%), cultivated farmland or ranchlands (15%), reservoirs, industrial sites, and recreation areas. The presence of water is emphasized as a habitat component of other sites (13%), including rivers, streams, lakes, marshes, sloughs, and estuaries. An essential feature of the nesting habitat includes open foraging areas. Emlen (1954) found that foraging ranges of these birds can be over 6 km from the nest site.

The Cliff Swallow has been recorded breeding in British Columbia from 23 April to 27 August (Fig. 171).

Colonies: There are reports of at least 920 Cliff Swallow colonies in British Columbia. Some observers have followed individual colonies for several consecutive years (Table 4). Colonies ranged in size from 3 to 1,375 nests. Over 60% of the colonies contained 25 nests or fewer, with a further 33% containing between 26 and 100 nests. There are only 6 known colonies of 500 nests or more.

Although the Cliff Swallow is common in the Southern Interior, there are no known large colonies there. Only 4 colonies with more than 200 nests were reported from the province south of Williams Lake and east of the Fraser River delta. The largest colonies were found in the northern third of the province, in the Sub-Boreal Interior, Boreal Plains, and Northern Boreal Mountains ecoprovinces.

Nests: Most nests (85%; n = 1,454) were found attached to buildings (Figs. 174 and 175), including active and abandoned tall, concrete buildings, barns, haylofts, and other farm buildings; houses, sheds, and garages. Other structures (9%) included bridges, water towers, a dam, and old Barn Swallow nests (Fig. 177). The 3 largest colonies reported in British Columbia built their nests on bridges. Natural sites (5%) included rock cliff faces and consolidated sand or clay banks (Fig. 176). All successful nesting locations require an overhang that provides shelter from rain (Taverner 1928; Erskine 1979). If the nests become moistened, their mud structure will fail.

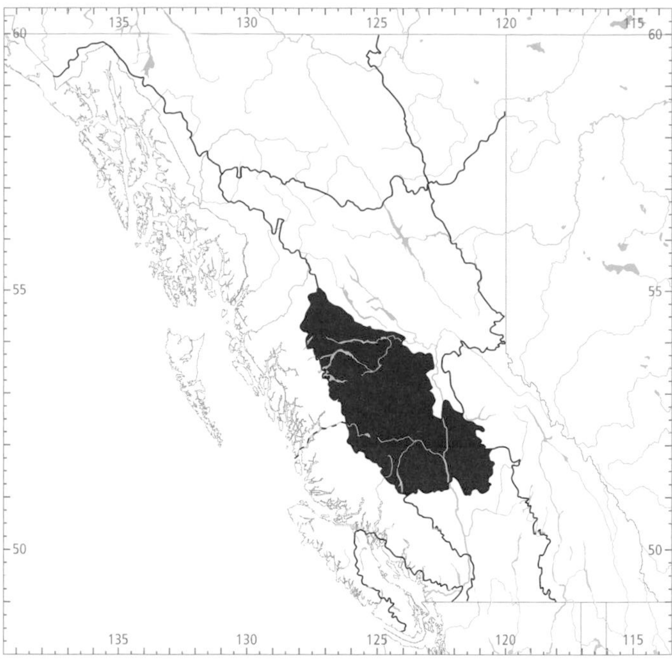

Figure 173. In British Columbia, the highest numbers for the Cliff Swallow in summer occur in the Central Interior Ecoprovince.

Figure 174. In the early 1970s, over 150 pairs of Cliff Swallows built their nests at the top of windows in the Dentistry Building at the University of British Columbia in Vancouver (23 May 1970; R. Wayne Campbell). Note the whitewash on the windows.

Nests were attached to walls of buildings, under eaves or some other overhang (Fig. 175), to rafters of buildings (Fig. 177), or to girders and concrete faces of bridges. Samuel (1971) found that Cliff Swallows nesting inside a building required a relatively large entrance to the site. Almost all colonies he studied were in barns with an open door measuring at least 2.5 m × 2.5 m, and if the door was even partially closed, the birds deserted. There is 1 report in British Columbia of this swallow using a nest box.

Nests are mud bottles consisting of a globe with a short neck at the entrance (Fig. 177). The availability of mud near the nest is another important factor in the swallow's selection of nesting sites. Storer (1927) noted birds from 1 colony travelling more than a mile to obtain mud for their nests. Nest

Table 4. Location and size of major colonies (50 or more nests or pairs) of Cliff Swallows in British Columbia.

Location	Date	Nests[1]	Pairs[1]	Nest location	Source[2]
COASTAL					
Georgia Depression					
Aldergrove (26250 Fraser Hwy)	11 May 1976	97	97	Barn	
Barnston Island (Robson Farm)	11 Jun 1961	68	80	Barn	
Barnston Island (Evans Farm)	22 May 1961	31	50		
Burnaby Curling Club	24 Jun 1961	80	80	Building	
Burnaby (Simon Fraser University gymnasium)	21 Jun 1972	68		Building	
Burnaby (Simon Fraser University quadrangle)	20 Jun 1972	99		Building	
Chilliwack	1 Jul 1960	360			1
Duncan	23 Jul 1972	119		Barn	
Maple Ridge	2 Jun 1973	51	102	Building	
Mission	8 May 1988	300		Bridge	
Mission (bridge over Fraser River)	22 Jul 1986	70	70	Bridge	
New Westminster (B.C. Penitentiary)	21 Aug 1963	90		Building	
Pitt Meadows (Neaves Road)	1 Jul 1958	59		Barn	
Pitt Meadows	11 Jun 1960	516			1
Pitt Lake (Homilk'um)	3 Jul 1988	50	50	Water tower	
Pitt Meadows (Biezefeld pumphouse)	14 May 1961	53	70	Pumphouse	
Port Mann Bridge	8 Jun 1972	300		Bridge	
Reifel Island (Alaksan)	15 Jun 1994	177		Tower/barn	
Ruskin (Stave River dam)	11 Jul 1967	22	50	Dam	
Ruskin	24 May 1972	112		Dam	
Courtenay (3 miles n at Sandwick)	15 Jun 1962		105	Building	
Surrey	10 Jul 1960	192			1
Surrey (Art Knapp Nursery)	2 Jul 1982	59	59	House	
Vancouver International Airport (Arthur Laing Bridge)	13 May 1980	126		Bridge	
Vancouver (s end Oak St Bridge)	5 Jun 1961	70	75	Bridge	
Vancouver (UBC Dentistry Bldg.)	8 May 1972	155		Building	
Vedder Crossing (Vedder Canal)	27 May 1961	200	200	Bridge	
Victoria (Rithet Farm)	15 Jun 1956	70			2
Victoria (Patricia Bay Airport)	15 Jun 1956	60			2
South Coast					
Hope (electrical station)	25 Jun 1969	93	100	Building	
INTERIOR					
Southern Interior					
Aspen Grove (near highway)	21 Jun 1962	90	200	Barn	
Kamloops (near)	28 May 1979	107		Barn	
Kamloops (near)	28 May 1979	112		Barn	
Oliver	20 Jun 1968	75	150	Bridge	
Quilchena Hotel	23 Jun 1974	132	132	Building	
Yellow Lake (56 km w Penticton)	18 May 1969	75	200	Cliff	
Southern Interior Mountains					
Canal Flats (at Columbia River)	27 Jul 1970	150	150	Bridge	
Crescent Valley (on Slocan River)	28 Jun 1982	100			
Creston (16 km w of Interpretive Centre)	5 May 1980	209	418	Building	
Duck Lake (near Sirdar)	15 Jun 1982	50	100	House	
Government Ranch (at bridge)	28 Jun 1967	112		Bridge	
Harrop (s side Kootenay Lake)	23 Jul 1972	119		Building	
Hemp Creek (Wells Gray Park)	15 Jun 1952		100		
Kootenay River Channel (bridge)	6 May 1981	127		Bridge	
Kootenay Crossing	21 May 1982	242	60	Bridge	
Leanchoil (on bridge)	2 Jul 1967	134		Bridge	3
Nelson (16 km w Hydro power plant)	6 Jun 1984	180	180		
Proctor (on Kootenay River)	2 Jul 1981	109	109		
Red Pass (Mount Robson Park)	21 Jun 1972	63	126	Building	
Summit Creek (campground w Creston)	3 Jun 1981	250	250	Bridge	
Wilko (5 km e)	30 May 1972	50	100		

(Continued on next page)

Table 4. (Continued)

Location	Date	Nests	Pairs	Nest location	Source[1]
Central Interior					
Alkali Lake	30 Jun 1961	100	100	Barn	
Chezacut	21 Jun 1973	50	100	Bridge	
Chezacut (Knoll's Ranch)	15 May 1968	80	160	Building	
Chezacut (3 km e)	30 May 1977	104	208	Building	
Fort Fraser (Nechako River)	23 May 1978	319		Bridge	
Foxy Creek Canyon	17 Jun 1981	100		Cliff	
Kleena Kleene	17 Jun 1967	104	104	Building	
Lac la Hache (Chevron station)	15 Jun 1962	153	153	Building	
100 Mile House (6 km e)	13 Jun 1976	200	75	Cliff	
100 Mile House	7 Jul 1987		100	Building	
100 Mile House	6 May 1991	130	65	Building	
108 Mile Ranch (airport)	10 Jul 1988	114		Building	
Owen Lake (Cannon Ranch)	13 Jun 1969	100		Building	
Tatla Lake	25 Jun 1987	184	184	Building	
Westwick Lake (Springhouse)	14 Jun 1978	100		Garage	
Sub-Boreal Interior					
Bowron River (53 km e Prince George)	8 Jun 1990	400	195	Bridge	
McLeod Lake (Hart Hwy)	8 Aug 1968	145	145	Building	
Parsnip River (ne Mcleod Lake)	22 Jun 1978	1,375		Bridge	
Prince George	18 Jun 1972	294		Building	
Prince George (45 km e on Hwy 16)	24 May 1978	140		Bridge	
Tache River (Grand Rapids)	7 Jun 1977	60	100	House	
Willow River (e Prince George)	7 Jul 1991	341	350	Bridge	
Boreal Plains					
Beatton River (bridge)	8 Jun 1956	250		Bridge	
Charlie Lake (cliffs e side)	31 May 1975	200	200	Cliff	
Charlie Lake (nr Fort St. John)	24 Jun 1978	122		Cliff	
Fort St. John (new arena)	4 Jun 1980	151		Building	
Gundy (near Tupper Creek)	9 Jun 1976	118	118	Cliff	
Kiskatinaw River (w Dawson Creek)	26 Jun 1993	1,166	1,100	Bridge	
Peace River (0.5 km e Lynx Creek)	26 Jul 1985	100	100	Cliff	
Peace River (18 km ne Hudson's Hope)	23 Jun 1978	625		Cliff	
Pouce Coupe River (e Dawson Creek)	24 Jun 1993	642	642	Bridge	
Northern Boreal Mountains					
Atlin	6 Jun 1978	141	100		
Atlin	27 Jul 1979	128			
Atlin (*S.S. Tarahne*)	8 Jul 1980	128	128		
Coal River (Mile 536 Alaska Hwy)	28 May 1968	150	200	Building	
Taiga Plains					
Mile 147 Alaska Hwy	13 Jun 1966	1,300		Bridge	
Mile 175 Alaska Hwy (Buckinghorse River)	7 Jun 1975	100		Bridge	

All data are estimates.

[1] All data are from the British Columbia Nest Records Scheme unless otherwise noted. (1, Boggs and Boggs 1960d; 2, Anonymous 1956; 3, Wade 1977)

chambers are lined with an assortment of soft materials, including grass, feathers, rootlets, and (rarely) hair and twigs. Cliff Swallows will also add a funnel entrance to adapt old Barn Swallow nests (Fig. 177). Nests that last over winter are often reused.

The heights of 762 nests ranged from ground level to 75.0 m, with 61% between 4.5 and 6.0 m.

Eggs: Dates for 724 clutches ranged from 28 April to 29 July, with 54% recorded between 8 and 28 June. Calculated dates indicate that eggs may occur as early as 23 April. Clutch size ranged from 1 to 8 eggs (1E-76, 2E-92, 3E-242, 4E-228, 5E-75, 6E-9, 7E-1, 8E-1), with 65% having 3 or 4 eggs. There is some evidence to suggest that coastal clutches tend to be smaller on average than interior clutches (see "Young" below). In some of the southern ecoprovinces, nests may have eggs at least 1 month earlier than nests in the northern regions of the province (Fig. 171). The few clutches begun in late June are probably second attempts after the loss of a first

Figure 175. Adult Cliff Swallows building nests under the eaves of a house (Wonowon, 14 June 1990; R. Wayne Campbell).

clutch or brood. Myres (1957), in his examination of Cliff Swallow colonies at Sorenson Lake, British Columbia, showed that 67% of the clutches laid in the first few days of the nesting season were of 5 eggs, and that the proportion of 4- and 3-egg clutches increased as the season progressed until 3-egg clutches predominated towards the end of the season. He suggests that birds breeding for the first time tend to nest later and lay smaller clutches than older birds.

Incubation periods determined for 15 nests from British Columbia ranged from 14 to 15 days, with a median of 15 days. This is similar to the incubation period noted by Samuel (1971).

Young: Dates for 473 broods ranged from 9 May to 27 August, with 53% recorded between 27 June and 14 July. Brood size ranged from 1 to 6 young (1Y-70, 2Y-151, 3Y-155, 4Y-80, 5Y-15, 6Y-2), with 65% having 2 or 3 young. Nestling periods determined from 6 nests in British Columbia ranged from 18 to 24 days; the nestling period given by Ehrlich et al. (1988) is 21 to 24 days. There is some evidence that coastal broods are significantly ($P < 0.05$) smaller than interior broods. Central Interior broods had an average of 3.2 young ($n = 100$), while broods from the Georgia Depression had an average of 2.4 young ($n = 205$).

Brown-headed Cowbird Parasitism: Only 1 instance of cowbird parasitism was found in British Columbia in 799 nests recorded with eggs or young. Friedmann (1963) notes 8 occurrences of cowbird parasitism of the Cliff Swallow in North America, all but 1 being from Pennsylvania.

Nest Success: Of 276 nests found with eggs and followed to a known fate, 72 produced at least 1 young, for a nest success rate of 26%. Coastal nest success was 30% ($n = 218$); interior nest success was 10% ($n = 58$).

In some cases, low success rates may be attributed to the effect of observers checking nests. The success of colonies also depends on the choice of site: texture of the substrate to which nests are attached, suitability of available construction material, availability of food, and weather. In the western Chilcotin, Paul (1963) reported that 40% of nests in a colony collapsed and were not rebuilt. Many colonies are destroyed by people who object to the mess Cliff Swallows make on buildings.

In Wisconsin, a study by Auman and Emlen (1959) showed that House Sparrows (*Passer domesticus*) commonly evicted Cliff Swallows from nests with eggs or young. In a survey of Cliff Swallow colonies in the Greater Vancouver area, Shepard (1972) suggested that nest usurpation by House Sparrows was restricted to agricultural regions, where it has had an adverse effect on the growth of colonies. For example, at the Gilford Farm on Barnston Island in 1959, 16% of 172 nests were occupied by House Sparrows. In 1960 the colony decreased by 53%, to 80 nests.

In some parts of North America, House Sparrow occupation has contributed to an increase in the numbers of arthropod nest parasites in nests reused by the swallows in following years (Buss 1942). New swallow nests were virtually free of parasites.

REMARKS: Two subspecies of the Cliff Swallow occur in British Columbia (Oberholser 1919; Behle 1976). *Hirundo pyrrhonota pyrrhonota* is frequent in the extreme southwestern portions of the province; *H. p. hypopolia* occurs throughout the northern and central regions of the province and in the southeast, west at least to the Okanagan valley (American Ornithologists' Union 1957). However, Browning (1992) suggests that only 1 subspecies occurs in northern North America: the nominate *H. p. pyrrhonota*.

The Cliff Swallow is subject to large-scale failure of its reproductive effort through the attack of parasites in the nest (Buss 1942; Emlen 1986). In natural cliff sites, the nests are frequently destroyed by weather over the winter, so that each spring the swallows have a clean site and build new nests (Storer 1927). Now that most Cliff Swallow colonies are attached to human-made structures, where protection from winter weather may lead to the survival of one year's nests for use the next year, the swallows have become prone to this form of nest failure. Old nests are heavily infested with swallow bugs (Cimicidae), ticks, and fleas to the extent that birds attempting to reuse the old nests frequently desert, even with young in the nest. Evidence suggests that where the old nests are cleaned off each winter, colonies of Cliff Swallows can be successful in rearing young each year. Where this does not happen, colonies are frequently occupied successfully in alternate years.

Munro (1949) refers to an instance at Nulki Lake where a colony with 39 nests raised no young. He did not establish the cause but suggested an infestation of the blowfly *Protocalliphora*, which was infecting the nestlings. This parasite frequently attacks Cliff and Barn swallows and at least 1 species of the genus is known to kill heavily parasitized nestlings. There was an outbreak of this fly infesting the nestlings of other species in the Nulki Lake area the same year.

Brown and Brown (1988) found that Cliff Swallows in southwestern Nebraska commonly laid their eggs in neighbouring nests in the colony. From 22% to 43% of the nests contained 1 or more parasitic eggs. Spreading eggs over several nests increases the chance that at least 1 of the biological

Figure 176. A Cliff Swallow colony at a natural site on an island in the Peace River south of Hudson's Hope (16 June 1990; R. Wayne Campbell).

parent's offspring will fledge and reduces the risk of reproductive failure in uncertain environments.

Recently, Cliff Swallows have been found to parasitize the nests of other individuals by actually carrying their eggs in their beaks to the host nest, a previously unknown method of brood parasitism (Brown and Brown 1988).

The Cliff Swallow was blue-listed in North America by the National Audubon Society in 1976, 1977, and 1981 (Arbib 1975, 1979; Tate 1981), became a species of "special concern" in 1982 (Tate and Tate 1982), and was of "local concern" in 1986 (Tate 1986). Most of that concern was reported from the northeastern United States, although in 1982 declines were also noted in the southwest and in the midwestern prairie regions of the United States.

See Brown (1986), Shields (1990), and Withers (1977) for additional information on the biology of this species.

Figure 177. An active Cliff Swallow nest built on top of an unused Barn Swallow nest (near Creston, 2 August 1993; R. Wayne Campbell).

NOTEWORTHY RECORDS

Spring: Coastal – Oak Bay 23 May 1953-2 nests complete on stucco house on Musgrave Street (Davidson 1953); Blenkinsop Lake 18 Mar 1989-1; Sidney 12 Apr 1961-16; Genoa Bay 17 May 1972-250; Ladner 28 Apr 1973-eggs; Reifel Refuge 16 Apr 1973-100, 8 May 1973-300, 31 Mar 1983-5; Clayoquot Sound 2 May 1966-1; Cheam Slough 12 Apr 1988-50, 10 May 1984-102; Jericho Park (Vancouver) 18 Mar 1983-5; Pitt Meadows 1 Apr 1974-2; Courtenay 20 May 1940-nests; Port Hardy 10 Apr 1940-2, arrival; Queen Charlotte City 9 May 1987-2; Delkatla Inlet 11 May 1989-1; Kitimat River 2 May 1980-1 over estuary; Rose Spit 29 Apr 1979-2. **Interior** – Creston 17 May 1980-418; West Creston 15 Apr 1982-150 flying over Kootenay River; White Lake (Oliver) 10 Apr 1969-10 at cliff face; Okanagan Falls 30 Apr 1913-incubating eggs; Bull River 20 May 1979-100; Princeton 8 May 1979-1,000, migrating flock; Balfour 16 May 1981-197 on census; Nelson 13 May 1983-adults bringing food to nest; Okanagan Landing 10 Apr 1954-1; Brouse 8 Apr 1984-1; Nakusp 16 May 1978-incubating eggs; Swan Lake (Vernon) 20 Apr 1980-375, 5 May 1968-20,000 (Cannings 1987); Revelstoke 30 Mar 1981-1; Lac la Hache 19 Mar 1950-1 (ROM 78369); Kleena Kleene 30 May 1962-building 40 nests; Sorenson Lake 19 May 1956-5 eggs; Chezacut 12 May 1965-a few nests with eggs; Quesnel 10 Apr 1968-30; Nechako River 23 May 1978-600; 24 km w Prince George 17 Apr 1983-5; Fort St. James 27 May 1973-incubating; Tetana Lake 3 May 1941-4; Hyland Post 16 May 1976-1; Hudson's Hope 24 Apr 1982-2; Charlie Lake 30 May 1984-300; Fort Nelson 27 May 1977-3; Atlin 16 May 1981-6, 20 May 1977-140.

Summer: Coastal – Elk Lake (Victoria) 1 Jun 1971-100; Genoa Bay 17 Jun 1972-250; Nanaimo 10 Jun 1986-2 nestlings; Sea Island (Richmond) 7 Jul 1964-300, 21 Aug 1948-200; Burnaby 27 Aug 1962-2 nestlings; Hope 2 Aug 1962-1 nestling; Pemberton 14 Jun 1971-adults feeding nestlings; Coal Harbour (Vancouver Island) Jul 1980-6; Bella Coola 3 Jul 1967-nestlings; Kitimat 30 Aug 1975-4; Hazelton 1 Jul 1961-adult feeding nestlings. **Interior** – Manning Park 1 Jul 1965-10, 10 Aug 1967-adult feeding nestling; Creston 22 Jul 1980-250, 16 Aug 1982-120; Crescent Valley 28 Jun 1983-200; Quilchena 26 Jun 1974-350; Douglas Lake 2 Jul 1982-500; Swan Lake (Vernon) 27 Aug 1962-20,000; Duck Lake (Vernon) 16 Aug 1982-120, 27 Aug 1962-5,000; Columbia Lake 22 Jul 1979-150; 1 Aug 1977-5; Invermere 10 Aug 1981-3 fledglings; Field 24 Aug 1976-100; Celista 17 Aug 1960-450; Revelstoke (17 km s) 27 Aug 1967-adult feeding nestling; Lac la Hache 21 Jul 1943-2,000 (Munro and Cowan 1947); Kleena Kleene 18 Jun 1962-62 nests occupied, 25 Jun 1962-first eggs hatched, 10 Aug 1962-last fledglings left their nests (Paul 1963); Williams Lake 1 Jun 1963-200, 22 Jul 1978-100; Burns Lake 17 Aug 1956-9; Buffalo Lake 29 Jun 1968-100; Chetwynd 8 Jul 1963-100; Fort St. James 17 Jul 1889-2 eggs (USNM 238470); Hudson's Hope 23 Jun 1978-400; Beatton River 3 Jul 1978-125; Martin Creek 4 Jul 1978-100; Fern Lake 12 Aug 1983-75, 18 Aug 1979-25 (Cooper and Adams 1979); Little Tahltan River 18 Jul 1987-50; Fort Nelson 4 Jun 1987-150, 30 Jun 1985-eggs, 7 Jul 1974-eggs, 14 Jul 1978-14; Fort Nelson River 16 Jun 1922-200 nesting near mouth (Williams 1933a); Atlin 6 Jun 1980-eggs (Anderson 1915); 18 Aug 1977-6; Spruce Creek (Atlin) 17 Jun 1980-most nestlings very large, some fresh eggs; Lower Post 17 Aug 1943-3 nests with well-grown young; Takhini River 27 Jun 1993-43 nests with young, 30 Jul 1980-adult feeding nestlings.

Breeding Bird Surveys: Coastal – Recorded from 15 of 27 routes and on 32% of all surveys. Maxima: Albion 4 Jul 1981-70; Kispiox 20 Jun 1993-40; Chilliwack 27 Jun 1994-26. **Interior** – Recorded from 52 of 73 routes and on 56% of all surveys. Maxima: Pleasant Valley 7 Jul 1988-242; Nicola 17 Jun 1973-234; Grand Forks 26 Jun 1994-141.

Autumn: Interior – Fort St. John 16 Sep 1986-1; Williams Lake 12 Sep 1978-3; Kamloops Lake 9 Nov 1991-2 (Siddle 1992a); Okanagan Landing 1 Nov 1911-2; Nakusp 8 Nov 1981-3, 4 to 7 Nov 1982-up to 8 birds; Salmo 12 Oct 1983-12; Osoyoos Lake 17 Sep 1990-1. **Coastal** – Kitimat Arm 17 Sep 1974-1 (Hay 1976); Egmont 5 Sep 1977-1; Tofino 12 Sep 1984-1; Long Beach (Tofino) 6 Sep 1967 (Campbell 1967), 2 Sep 1968 (Campbell 1968); Iona and Sea islands 31 Oct 1982-10; Sea Island 16 Nov 1983-1; Reifel Island 30 Sep 1986-5; Swan Lake (Victoria) 19 Nov 1991-1 (Siddle 1992a).

Winter: Interior – No records. **Coastal** – Musqueam Park (Vancouver) 16 and 17 Dec 1991-2 (Siddle 1992b); Reifel Island 7 to 10 Dec 1989-1 (Siddle 1990a).

Christmas Bird Counts: Interior – Not recorded. **Coastal** – Recorded from 1 of 33 localities and on less than 1% of all counts. Maximum: Vancouver 15 Dec 1991-**2**, all-time Canadian high count (Monroe 1992).

Barn Swallow

Hirundo rustica Linnaeus

BASW

RANGE: Breeds in North America from south-coastal and southeastern Alaska, southern Yukon, and western Mackenzie across Canada to southern Newfoundland; south into Baja California and the Mexican mainland. Winters from southern Florida south through South America to Tierra del Fuego. Also breeds in Eurasia, from Ireland to Siberia, and south to Japan, China, and Taiwan.

STATUS: *Common* to locally *abundant* migrant and summer visitant to the southern two-thirds of the province, including the Georgia Depression, Southern Interior, Southern Interior Mountains, and the Central Interior ecoprovinces. Exceptions are the west coast of Vancouver Island and coastal areas north of the Georgia Depression, including the Queen Charlotte Islands, where it is an *uncommon* to locally *common* (occasionally *very common*) migrant and summer visitant. It is *very rare* in winter in the Fraser River delta. In the northern third of the province, it is a *rare* to locally *common* (occasionally *very common*) migrant and summer visitant.

Breeds.

NONBREEDING: The Barn Swallow (Fig. 178) has a widespread distribution throughout the province except in areas of high mountains, open ocean, and dense forests such as those found in the Coast and Mountains Ecoprovince. In addition, there are few records of this swallow from the northern portions of the Sub-Boreal Interior Ecoprovince and from the northern Rocky Mountain Trench north of Williston Lake. Apart from those areas, it has a widespread summer distribution throughout the southern two-thirds of the province from southern Vancouver Island east to the Flathead valley, north through the Southern Interior and Southern Interior Mountains ecoprovinces to the Cariboo and Chilcotin areas of the Central Interior, and north to the latitude of Hazelton (55°15'N). In the northern third of the province, its distribution is fairly widespread throughout the Boreal Plains Ecoprovince, becoming more scattered throughout the Northern Boreal Mountains and the Taiga Plains ecoprovinces.

Figure 179. In spring migration, ice-free sloughs and lakes with abundant insect life are important foraging areas for a variety of swallows, including Tree, Violet-green, Northern Rough-winged, and Barn Swallow (unnamed roadside slough 40 km north of Williams Lake, 25 April 1990; R. Wayne Campbell).

Figure 178. The deeply forked tail identifies this adult Barn Swallow (Vaseux Lake, 4 August 1993; R. Wayne Campbell).

The Barn Swallow has been recorded from near sea level to at least 2,400 m. It is a common sight in suburban areas of cities and in towns and villages, where it can be found feeding over garden areas, parks, fields, shopping malls, airports, and similar open spaces in the cityscape. Out of the city, it forages over coastal bays, lagoons, estuaries, beaches and harbours, powerline rights-of-way, forest and woodland glades, river courses, sloughs, marshes, fields, orchards, farmyards, feed lots, and almost anywhere that flying insects are abundant. During migration it uses the same lakes, ponds, and sloughs as other migrating swallows (Fig. 179).

The spring migration enters British Columbia along the coastal flatlands of Washington that terminate in the delta of the Fraser River. The migration also enters the province through interior valleys, such as the Similkameen, Okanagan, and Creston valleys, and the Rocky Mountain Trench.

On the Fraser River delta, the Barn Swallow has been reported throughout the year, although it does not appear in consistent numbers until at least the last 2 weeks of March and early April (Figs. 180 and 181). Numbers continue to increase, with the greatest prebreeding concentrations occurring in May (Figs. 180 and 181). This pattern is similar, though with much smaller numbers, on Western Vancouver Island, the northern mainland of the Coast and Mountains, and the Queen Charlotte Islands.

In the Okanagan valley of the Southern Interior, the west and east Kootenays of the Southern Interior Mountains, and the Cariboo and eastern Chilcotin areas of the Central Interior, the first Barn Swallows are not seen until early April (Figs. 180 and 181). Numbers rise quickly in May to reach a prebreeding maximum. Further north, from Quesnel to Prince George, and across southern portions of the Sub-Boreal

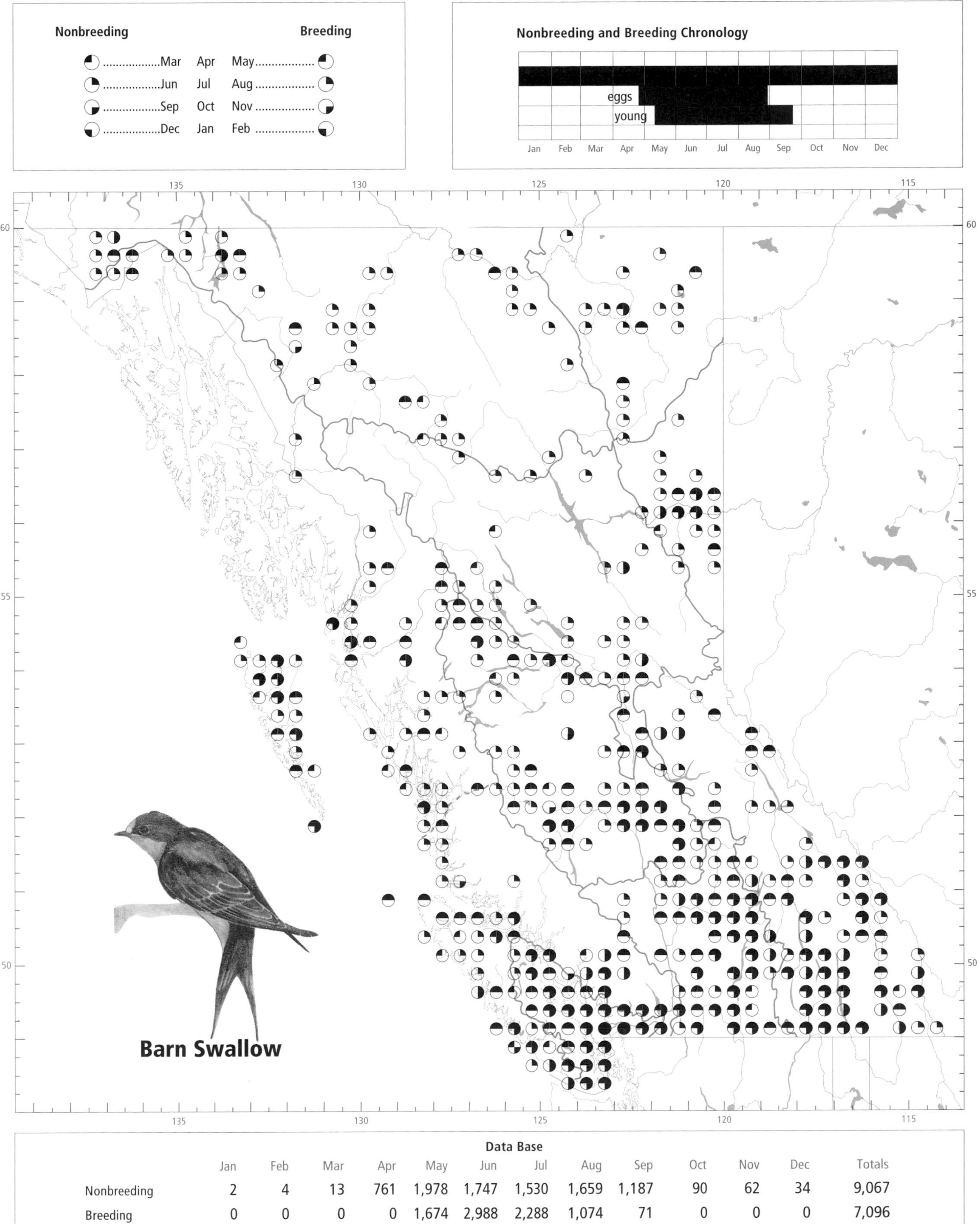

Data Base

	Jan	Feb	Mar	Apr	May	Jun	Jul	Aug	Sep	Oct	Nov	Dec	Totals
Nonbreeding	2	4	13	761	1,978	1,747	1,530	1,659	1,187	90	62	34	9,067
Breeding	0	0	0	0	1,674	2,988	2,288	1,074	71	0	0	0	7,096

Interior, migrants arrive in April in some years, and build slowly to a maximum in June. The front of the migration arrives in May in the Peace Lowland of the Boreal Plains, in the vast boreal plateaus and mountains that make up the Northern Boreal Mountains, and in the Taiga Plains. In most ecoprovinces, there is a decline in numbers after the spring movement followed by an apparent increase in numbers in August and September as young of the year and adults gather in premigration flocks. The Southern Interior Mountains and the Central Interior have a slightly different pattern, featuring a relatively stable population from May until August and September, with only a slight increase in numbers in either of the latter 2 months. The 2 northern ecoprovinces show no autumn peak, simply a decline in numbers from June through August and September.

In all the ecoprovinces for which a clear pattern is evident, the southward movement begins in early August and peaks in late August or early September. In most ecoprovinces, the greatest abundance of Barn Swallows was recorded in August. A conspicuous exception may occur in the Central Interior, where autumn numbers appear to peak in September (Fig. 181). Most birds have left the province by the end of September.

During their southward migration, Barn Swallows are highly aggregative, gathering into flocks sometimes as large as 2,000 birds. Migrating flocks in the autumn often roost for the night in the dense vegetation of cattail or bulrush marshes. Most Barn Swallow flocks of over 200 birds (94%; $n = 96$) were associated with the southward movement; only 6 other large flocks were reported: 3 in May and 3 in June. Reports of large flocks have come from Pitt Lake, Iona Island, Sea Island, Cultus Lake, and Salmon Arm, all in August, but flocks of many hundreds can be seen from late July to early September. They occur predominantly in the southwestern corner of the province, although large aggregations have been recorded from Williams Lake, the Okanagan valley, and the Creston valley. Large aggregations of Barn Swallows have not been reported from the northern regions of the province.

In winter, the Barn Swallow has been reported only from the Fraser River delta in the Georgia Depression.

On the coast, the Barn Swallow has been recorded throughout the year, but regularly only from early April to late September. In the interior, it has been recorded from 1 April to 28 October (Fig. 180).

BREEDING: The Barn Swallow likely breeds throughout most of its range in British Columbia, both in close association with humans and in natural situations.

There are many known nesting localities on southeastern Vancouver Island, but just 5 locations have been recorded on northern and western regions of the island. Nesting is known across the province from the Fraser River delta to the British Columbia–Alberta border and north through the interior of the province to the latitude of Hazelton. Along the mainland coast there are only 2 reported nesting localities in the 400 km between Toba Inlet in the south and the Skeena River valley in the north. North of the latitude of Hazelton, the distribution of the Barn Swallow is sparse, but it extends across the northern extremity of the province from the Tatshenshini Basin in the northwest to Kwokullie Lake in the east.

The Barn Swallow reaches its highest numbers in the Fraser Lowland of the Georgia Depression (Fig. 182). An analysis of Breeding Bird Surveys for the period 1968 through 1993 suggests that the number of birds on coastal routes has decreased at an average annual rate of 4% (Fig. 183); in the interior, no net change in numbers could be detected.

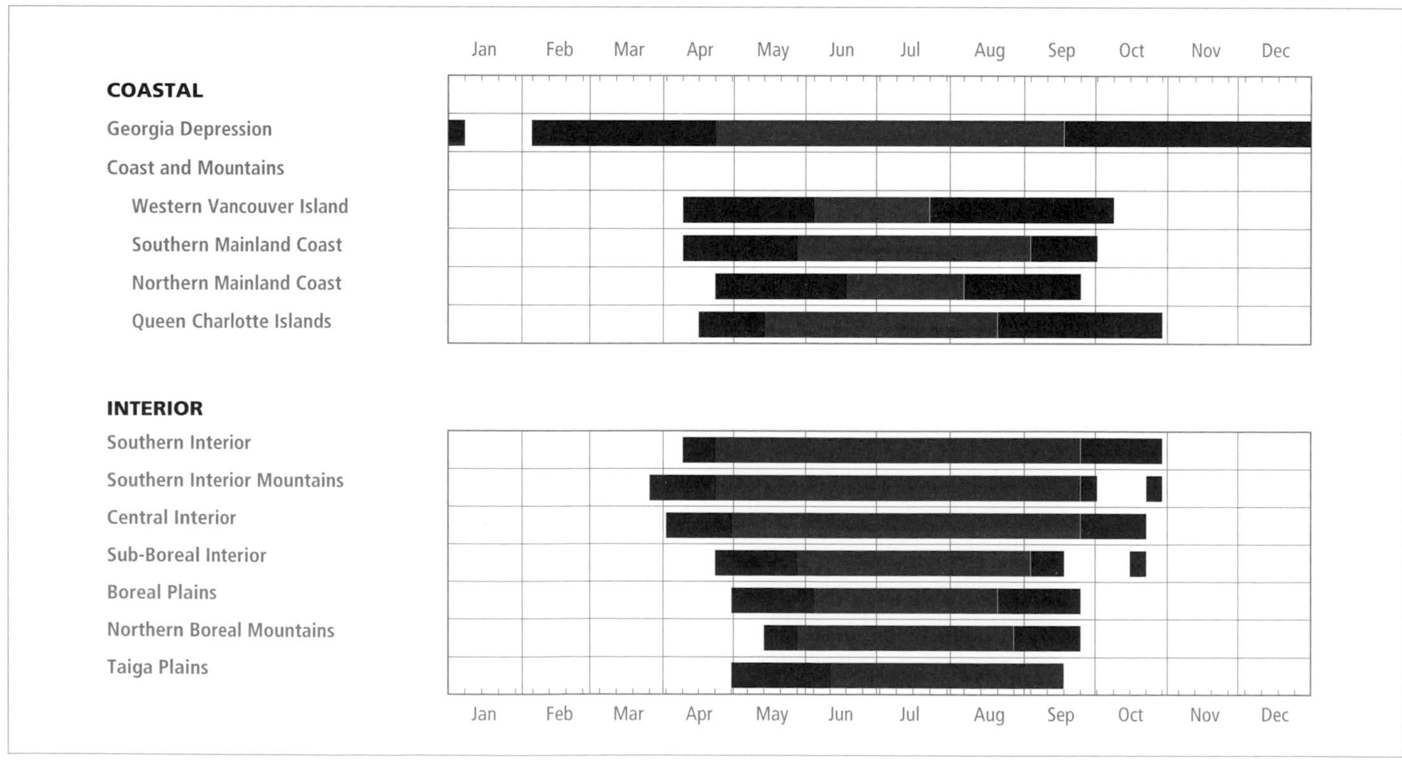

Figure 180. Annual occurrence (black) and breeding chronology (red) for the Barn Swallow in ecoprovinces of British Columbia.

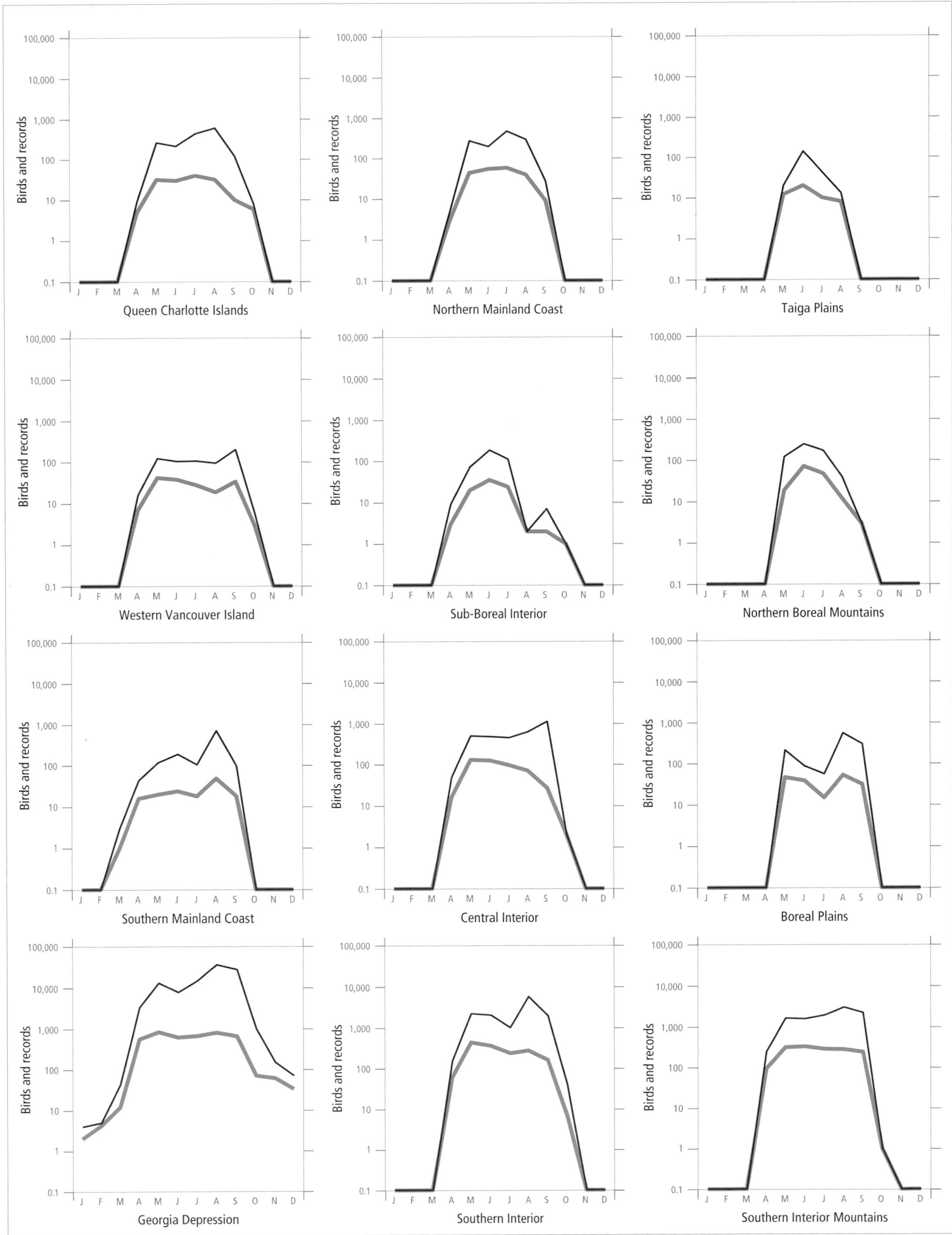

Figure 181. Fluctuations in total number of birds (purple line) and total number of records (green line) for the Barn Swallow in ecoprovinces of British Columbia. Christmas Bird Count and nest record data have been excluded.

Barn Swallow numbers have declined locally in British Columbia. On the coast, this decline has been obvious at Qualicum Beach, where, in the Qualicum National Wildlife Area, numbers of active nests and renests per year have dropped from 24 in 1977 to 4 in 1990 (N.K. Dawe, unpubl. data). In the interior, numbers nesting in barns on Road No. 22 south of Oliver declined from 27 in 1982 to 6 in 1988 (Campbell and Harris 1990). Numbers have also declined in central and western Alaska since the mid-1920s (Turner and Rose 1989), and the species seldom breeds there now (Kessel and Gibson 1978).

Robbins et al. (1986), in their North American summary of Breeding Bird Surveys over the period 1965 through 1979, found that no stratum in North America showed a decrease in Barn Swallow numbers, whereas 28 strata recorded significant increases, more than were found for any other species except Killdeer. That, however, may now be changing. Erskine et al. (1992) note a negative annual long-term trend (–1.1%) in Barn Swallow numbers for southern British Columbia over the period 1966 to 1991; similar trends for the same period were found for the Maritimes (–1.2%), central Ontario and central Quebec (–3.2%), southern Ontario and southern Quebec (–1.0%), and the southern Prairie provinces (–0.3%). With the exception of the central Ontario and central Quebec change, the negative annual percentage changes for those areas were not statistically significant; however, the fact that most show a trend towards a negative mean annual percentage change over so long a period indicates that the Barn Swallow may be having problems, at least in the northern part of its North American range.

This swallow has also recently declined in numbers in the Netherlands, the former West Germany, Denmark, Czechoslovakia, Romania, the Baltic States, Britain, and parts of Israel (Turner and Rose 1989).

The Barn Swallow breeds from near sea level to 2,160 m elevation. Most nests for which details are available (80%; n = 3,060) were reported from human-influenced habitats. Thus, rural areas (34%), cultivated farmlands (21%), and suburban and urban areas (20%) dominate as the settings for this swallow's nesting territories. Associated foraging areas included rural and urban backyards or farmyards, farmlands, roadsides, weedy fields; water-associated habitats such as lakes, marshes, streams, sloughs, and estuaries; meadows, open rangelands and parklands, big sagebrush areas, orchards, and vineyards. In every ecoprovince except the Northern Boreal Mountains, the immediate environment of the nesting territory was human-made.

The Barn Swallow has been recorded breeding in British Columbia from 25 April (calculated) to 22 September (Fig. 180).

Colonies: The Barn Swallow has been described by Snapp (1976) as a facultative colony nester in which neither the number of pairs in a colony nor the degree of nesting synchrony within a colony results in improved reproductive success. She concludes that the colonies should be regarded merely as passive aggregations and not as active colonies. Apparently, the main advantage gained by this swallow in

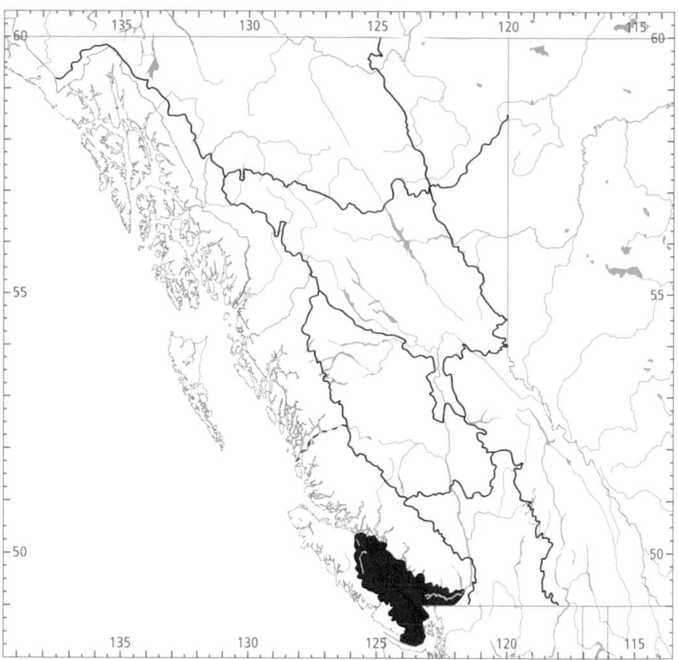

Figure 182. In British Columbia, the highest numbers for the Barn Swallow in summer occur in the Georgia Depression Ecoprovince.

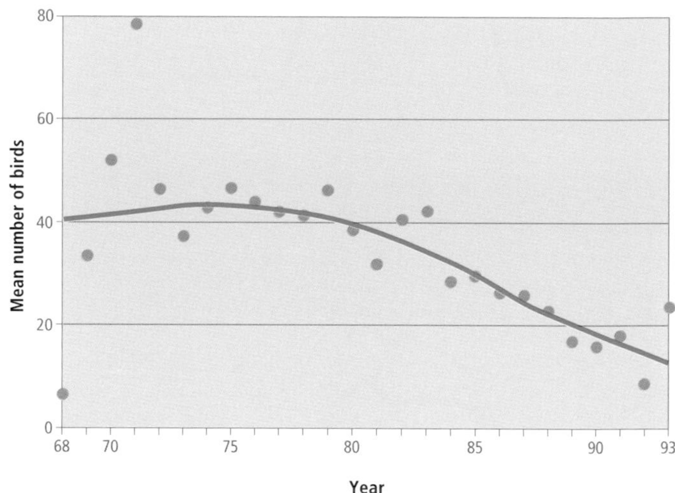

Figure 183. An analysis of Breeding Bird Surveys for the Barn Swallow in British Columbia shows that the mean number of birds on coastal routes decreased at an average annual rate of 4% over the period 1968 through 1993 ($P < 0.05$).

nesting in a colony is allowing more pairs to use a limited supply of nest sites. Similar results were also found by Shields and Crook (1987). In New York state, Snapp (1976) found that 35% used solitary nests and defended a nesting territory. Data do not allow us to calculate a comparable figure for British Columbia.

Table 5 describes the details of nesting colonies of 15 nests or more reported in British Columbia. About half of them, including the 2 largest colonies, are in the Southern Interior. From 135 colonies reported in British Columbia, colony size ranged from 3 to 83 nests, with 56% between 5 and 14 nests. Barn Swallow colonies usually hold fewer than 5 nests, although large colonies of 50 or more nests have been reported from other areas in North America and Europe (Turner and Rose 1989).

N.K. Dawe (unpubl. data), in his study of a colony at Qualicum Beach, found that the distance from one active nest to the next ranged from 1.2 to 2.3 m, with a mean of 1.7 m ($n = 7$).

Nests: Most nests (92%; n = 2,896) were found in or on buildings, including barns (Fig. 184a), sheds (Fig. 184b), garages, and houses. Other structures, such as bridges and wharfs, made up 4% of the general nest locations. Snapp (1976) suggests that an important factor in nest site selection was the size of the entrance to the structure. That is, the smaller the opening, the less likely the structure would be used by Barn Swallows. Natural nesting sites for the Barn Swallow are cliff faces (Fig. 184c) and natural caves where the nests can be placed in situations protected from rain. In British

Table 5. Location and size of major colonies (more than 15 nests or pairs) of Barn Swallows in British Columbia.

Location	Date	Nests[1]	Pairs[1]	Nest location
COASTAL				
Georgia Depression				
Burnaby Lake	24 May 1970	30	70	Barn
Halfmoon Bay	21 Jun 1964	15	15	Shed
Iona Island	18 Jun 1970	48	50	Barn
North Coast				
Kemano River estuary	17 Jun 1979	45	45	Barn
Kitsault, Alice Arm	16 Jun 1980	50	25	Shed
Queen Charlotte Islands				
Eden Lake	15 Jul 1976	15	20	Building
INTERIOR				
Southern Interior				
Douglas Lake, Glimpse Lake Lodge	26 Jul 1980	1	50	Building
Eagle Bay	1 Jun 1969		50	Barn
Kilpoola Lake	11 Jun 1961	15	15	Building
Nicola valley	20 May 1963	45	70	Barn
Okanagan Falls	2 Jun 1968	20	20	Bridge
Oliver, Road 22	29 Jun 1992	27		Barn
Peachland	22 Jun 1969	60	30	Shed
Penticton, Paradise Ranch	14 Jun 1967	15	15	Shed
Penticton	24 May 1969	25	25	Bridge
Quilchena, Quilchena Hotel	23 Jun 1974	17	17	Building
Rock Creek, 6 km e	18 Jul 1980	75	75	Bridge
Tullameen Lake	15 Jun 1967	83		Barn
Tullameen Road nr Voyt Valley Road	20 Jul 1975	19	40	Building
Wilson's Landing, Government Wharf	21 Jul 1968		15	Structure
Southern Interior Mountains				
Creston	11 Jul 1980	55	110	Building
Creston, Wildlife Centre	21 May 1981	20	20	Building
Gray Creek	11 Jul 1974	12	24	Building
La Farms Creek, 23 km n Hwy 23	13 Jul 1989	24	24	Bridge
Central Interior				
Francois Lake, north shore	9 Aug 1972	35	35	Building
Fraser Lake, Piper's Glen Resort	5 Jun 1977	18	18	Building
Kleena Kleene	11 Jul 1973	16	16	Shed
Sub-Boreal Interior				
Parsnip River nr Missinka River	11 Jun 1982	40	40	Bridge

Source: All data are from the British Columbia Nest Records Scheme.
[1] All data are estimates unless otherwise noted.

Figure 184. The Barn Swallow is an adaptable species that builds its nests on human-made structures with nearby foraging areas and large, open flight entrances to nests in: (a) barns (Haynes Ecological Reserve north of Oliver, 14 active nests in May 1994; R. Wayne Campbell) and (b) sheds (4.6 km south of Shelter Bay, West Kootenay, 21 nests, of which 10 were active on 3 July 1991 [Campbell and Dawe 1991]; R. Wayne Campbell), as well as in natural situations including (c) cavities in trees (near Mill Bay, Vancouver Island, 17 July 1988; Grant Keddie) and (d) cliffs (Tatshenshini River, 13 June 1993; John M. Cooper).

Columbia, 1% of reported nests were found in caves, cliffs, or banks of sand or rock. This frequency of use apparently prevails across Canada, where about 1% of nests (n = 5,000) were also in natural sites (Erskine 1979). Speich et al. (1986) review the use of natural nest sites by the Barn Swallow in North America.

Most nests (46%; n = 2,537) were attached to rafters or beams of buildings (Fig. 185). Others were built under eaves (18%); on shelves, ledges, or projections (11%); attached to walls (9%); or over light fixtures (5%; Fig. 186). In the few instances where a natural site was used, the nests were built in a crevice or natural cavity (Fig. 184d). One nest was built on a train that travelled over a 3 km portage between Tagish and Atlin lakes; that particular site was used for at least 20 years by a succession of swallow pairs (Swarth 1935a). There are several instances of Barn Swallows nesting on ferries plying the waters between Vancouver Island and the southern mainland coast.

The nest (Figs. 185 and 186) is constructed primarily of mud and is lined with feathers, grass, and hair. Only occasionally are other materials used, including rootlets, twigs, pine needles, plant down, string, and seaweed. N.K. Dawe (unpubl. data) found that nest building from the start of a new nest to the laying of the first egg took about 15 days (n = 10), as follows: construction of mud bowl, 7 days; grass lining, 3 days; feather lining, 4 days; and completion of feather lining to first egg, 1 day.

The heights of 2,563 nests ranged from 0.3 to 30 m, with 57% between 2.4 and 3.5 m.

Eggs: Dates for 1,705 clutches ranged from 30 April to 29 August, with 52% recorded between 5 June and 11 July. Calculated dates indicate that eggs can occur as early as 25 April. Clutch size ranged from 1 to 10 eggs (1E-107, 2E-105, 3E-237, 4E-548, 5E-643, 6E-59, 7E-4, 8E-1, 10E-1), with 84% having 3 to 5 eggs. There is no significant difference in clutch size either between northern and southern nesting birds or between coastal and interior birds.

N.K. Dawe (*in* Goley 1986) found that about 37% of the birds studied over a 10-year period laid a second clutch (n = 135); there was an interval of about 51 days (n = 25) between the initiation of the first clutch of a successful nesting pair and the second clutch of that pair (Fig. 187). First clutches were significantly larger than second clutches.

Incubation periods from 40 nests in British Columbia ranged between 12 and 17 days, with 58% between 15 and 16 days. In 42 nests studied at Qualicum Beach, the mean incubation period was 14.6 days (N.K. Dawe *in* Goley 1986). Goley (1986) found that both males and females share in the incubation duties during the day, with the female incubating through the night. Turner and Rose (1989) note an incubation period of 14 to 16 days, occasionally 11 to 19 days.

Young: Dates for 1,613 broods (Fig. 188) ranged from 10 May to 22 September, with 51% recorded between 26 June and 30 July. Brood size ranged from 1 to 6 young (1Y-44, 2Y-116, 3Y-346, 4Y-701, 5Y-371, 6Y-35), with 89% having 3 to 5 young (Fig. 188). Nestling periods from 36 nests in British Columbia ranged from 19 to 24 days, with 64% between 21 and 23 days. N.K. Dawe (*in* Goley 1986) found a mean nestling period in 12 nests of 19.5 days, and Turner and Rose (1989) report a nestling period of 18 to 23 days. The nestling period can be as short as 17 days if young leave the nest prematurely when the nest is checked. Direct inspection of the

Figure 185. Most Barn Swallow nests in British Columbia were attached to the face of rafters (Creston, June 1993; R. Wayne Campbell).

Figure 186. Occasionally, Barn Swallows in British Columbia build their nests on top of light fixtures (Burnaby, July 1976; R. Wayne Campbell).

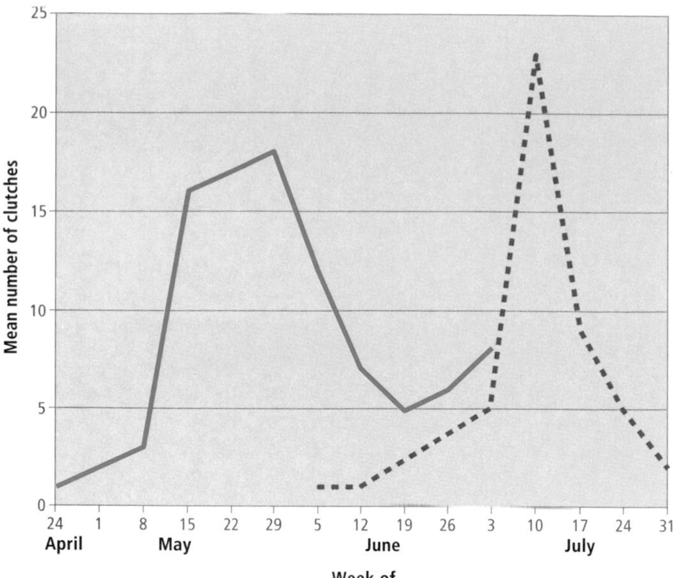

Figure 187. Frequency distribution of Barn Swallow initiation dates for first (solid line) and second (dashed line) clutches at Qualicum Beach, British Columbia, 1975 to 1984 (N.K. Dawe, unpubl. data).

nest should be made with caution after day 15 of the nestling stage.

Brown-headed Cowbird Parasitism: Cowbird parasitism was not found in British Columbia in 2,426 nests found with eggs or young. Friedmann (1963) reports 5 instances of cowbird parasitism in North America and notes that the Barn Swallow is an infrequent victim.

Nest Success: In 609 nests found with eggs and followed to a known fate, 70% produced at least 1 fledgling. Coastal success was 75% (n = 299), while interior success was 65% (n = 310). A number of records reported that young had died in the nest, that young had fallen or were pushed from the nest, or that young were found dead below the nest (Fig. 189). In all likelihood, the majority of these deaths resulted from nest infestation with the larvae of the parasitic blowfly *Protocalliphora,* which often results in the young falling from the nest or the death of the young in the nest. Shields and Crook (1987) found that nests in larger colonies were infested more by *Protocalliphora hirundo* larvae than were nests in smaller colonies. As a result, breeding success was less for birds in large colonies than for those in smaller colonies or in single nests.

Other causes of nest failure included extreme weather, the nest falling or being knocked down, young held in the nest by horse hair or other fibres wrapped around their legs, and predators, including American Kestrel, Sharp-shinned Hawk, Common Raven, American and Northwestern crows, Black-billed Magpie, House Wren, European Starling, domestic cat, rat, red squirrel, and deer mouse. At Qualicum Beach, N.K. Dawe (*in* Goley 1986) found that 65% of first nesting attempts and 90% of second nesting attempts were successful. The leading cause of nestling mortality in his study was *Protocalliphora* infestation. Fewer nestlings there died when parasitized by the larvae of *Protocalliphora hirundo,* an external feeder, than when parasitized by the larvae of an undescribed *Protocalliphora,* a subcutaneous feeder.

REMARKS: Three subspecies of Barn Swallow have been recorded in North America. *H. r. erythrogaster* is widespread throughout the continent. *H. r. rustica* and *H. r. gutturalis,* native to Eurasia, have been found casually in northern and western Alaska (American Ornithologists' Union 1957). An individual of the latter race came aboard a vessel about 150 km west of Tasu Sound, Queen Charlotte Islands, on 15 July 1960 (Cowan and Cowan 1961).

The Barn Swallow frequently returns each year to the same area and, in some cases, to the same nest site. N.K. Dawe (unpubl. data), at Qualicum Beach, banded 26 adult Barn Swallows between 1977 and 1981, and noted the following rates of return to the same barn in successive years: year 1,

Figure 188. Broods of 3 to 5 young were found in nearly 90% of all nests reported in British Columbia (Victoria, June 1991; Tim Zurowski).

Figure 189. Checking nests too early, or disturbing nests that are poorly attached to smooth, flat surfaces, may result in nestling mortality, as with this brood (Kootenay Pass Summit, 3 August 1993; R. Wayne Campbell).

0.46; year 2, 0.39; year 3, 0.70; year 4, 0.20. One mated pair returned to the same barn for at least 5 consecutive years. Most Barn Swallows, however, live fewer than 4 years (Turner and Rose 1989). Davis (1965), in Europe, concludes that adult birds return to their previous breeding place with great fidelity; he could find no satisfactory evidence that they fail to do so if they survive. Møller (1983) looked at breeding habitat selection in the European Barn Swallow (*H. r. rustica*) in Denmark and found that the main feature of farms with larger colonies was the presence of domestic animals. He notes that as soon as the animals are sold or otherwise removed from the farm, the swallows disappear. Tate (1986) notes that nestlings are far more subject to heat-induced mortality in modern metal-roofed barns than in older barns with wooden roofs. Modern farm buildings and procedures may be one reason that numbers of this swallow are declining in some regions.

A Barn Swallow banded in Costa Rica on 11 April 1980 was recovered at Fort St. John in 1981. Another swallow, banded on 20 May 1985 about 11 miles southeast of Startup, Washington, was recovered near Masset, Queen Charlotte Islands, on 28 June 1985 (Campbell 1988b). And a swallow banded near Kaslo on 6 July 1963 was recovered near Ogden, Utah, in May 1964.

See Grzybowski (1979), Medvin et al. (1987), Samuel (1971), and Shields (1984) for additional information on the breeding biology of the Barn Swallow in North America.

NOTEWORTHY RECORDS

Spring: Coastal – Victoria 12 Apr 1971-1; Elk Lake 2 May 1976-500; Shawnigan Lake 2 May 1959-3 eggs; Serpentine River 14 Apr 1975-170; Surrey to Hope 19 Mar 1961-30; Agassiz 28 Mar 1986-1; Flood to Hope 10 Apr 1975-10; Pitt Meadows 1 Apr 1973-10; Stubbs Island 13 Apr 1980-1; Cape Scott 28 Apr 1976-6; Kitimat 25 Apr 1975-2; Juskatla 29 May 1968-feeding young in nest; Sandspit 20 May 1989-2 (Campbell 1989c); Masset Inlet 18 Apr 1977-1, 15 May 1972-150; Khyex River 26 May 1977-42; Three Guardsmen Pass 28 May 1979-4; Atlin 21 May 1981-26. **Interior** – Vaseux Lake 19 Apr 1979-10, 30 May 1974-86; 6 km w Salmo 10 May 1988-feeding young in nest; Okanagan Landing 11 Apr 1941-3; Vernon 11 Apr 1965-1; Kamloops 10 May 1973-feeding young in nest; Revelstoke 4 Apr 1988-1; Leanchoil 4 Apr 1977-1; Emerald Lake 1 May 1979-3; near Satah Mountain 1 Apr 1988-4, 7 Apr 1989-4; Kleena Kleene 16 May 1968-young in nest; Williams Lake 8 Apr 1980-5; Quesnel 29 May 1963-12; Prince George 28 Apr 1982-4; Eaglet Lake 21 Apr 1990-2; North Pine 5 May 1982-4; Charlie Lake 4 May 1984-50; Helmet Lake 16 May 1982-3; Fort Nelson 2 May 1978-1; Hatin Lake 14 May 1979-4; Chilkat Pass 14 May 1977-5.

Summer: Coastal – Bamfield 22 Jun 1970-4 eggs; Cultus Lake 28 Aug 1967-1,176; Tofino 26 Aug 1981-24 at airport; Port Alberni 9 Jun 1969-16; Vancouver 1 Aug 1969-119; Harrison Hot Springs 29 Jun 1976-100; Gibsons 9 Jun 1971-adult feeding young in transit on *M.V. Langdale Queen* (Campbell 1971); Pitt Lake 18 Aug 1982-2,000 apparently roosting in cattails; Garibaldi Park 1 Jun 1963-1 egg; Campbell River 12 Aug 1977-1,000; Alta Lake 27 Aug 1945-4 young (Racey 1948); Port Neville 30 Jun 1985-2 very small young; Namu 11 Jun 1981-5 eggs; 150 km w Tasu Sound 15 Jul 1960-1 (UBC 9888); Queen Charlotte City 17 Aug 1961-3 downy young; Lawnhill 21 Jun 1960-3 eggs; Port Clements 21 Jul 1977-80; Prince Rupert 1 Jun 1970-4, 29 Jun 1982-1 egg; Terrace 13 Jul 1984-5 young, 21 Jul 1975-100; Haines Highway 9 Jun 1975-12; Haines Triangle 12 Aug 1977-5. **Interior** – w Creston 5 Aug 1981-500 along Nick's Island Road; Balfour to Waneta 8 Jun 1980-217 on census; Penticton 22 Aug 1975-eggs; Sparwood 13 Jul 1984-4 nestlings; Wood Lake 3 Jun 1976-140; Salmon Arm 24 Aug 1973-1,230; Riske Creek 11 Aug 1992-1, earliest departure in 10 years; Fort St. James 20 Jul 1964-40; Baldonnel 5 Jun 1985-eggs; Cecil Lake 21 Aug 1980-100; Wonowon 19 Aug 1975-6 young; Fort Nelson 12 Jun 1982-3 eggs; Kledo Creek 28 Aug 1986-4; Parker Lake 1 Jun 1980-80; Kwokullie Lake 6 Jun 1982-nest building; Atlin 6 Jun 1914-3 eggs (RBCM 343), 22 Aug 1957-4 young just out of nest (Weeden 1960), 23 Aug 1974-10; Mile 79.8 Haines Highway 12 Jun 1980-1 egg in nest under bridge at Great Creek.

Breeding Bird Surveys: Coastal – Recorded from 23 of 27 routes and on 82% of all surveys. Maxima: Albion 10 Jul 1982-281; Victoria 30 Jun 1973-154; Pitt Meadows 5 Jul 1981-107. **Interior** – Recorded from 64 of 73 routes and on 86% of all surveys. Maxima: Fraser Lake 20 Jun 1970-122; Grand Forks 3 Jul 1985-114; Ferndale 1 Jul 1980-106; Golden 19 Jun 1976-106.

Autumn: Interior – Atlin 1 Sep 1924-last seen; 7 Sep 1931 (Swarth 1936); Egnell Creek 18 Sep 1986-1; Fort Nelson 16 Sep 1985-3 young fully feathered in nest, 20 Sep 1985-adult feeding 3 young out of nest; Beatton Park 18 Sep 1986-6; North Pine 18 Sep 1981-1; Willow River 16 Oct 1968-1; 32 km se Prince George 1 Sep 1990-2 young in nest; Quesnel 14 Sep 1966-3; Williams Lake 9 Sep 1979-500, 15 Oct 1978-1; Riske Creek 28 Sep 1986-1; Chilcotin River (Wineglass Ranch) 22 Sep 1986-3 young left nest today; Glacier National Park 28 Oct 1981-1; Nakusp 27 Sep 1985-1; Swan Lake (Vernon) 16 Sep 1984-3, 16 Oct 1971-1; Penticton 17 Sep 1988-2 young at nest edge; Castlegar 18 Sep 1976-2 young of 4 still in nest; Osoyoos Lake 23 Oct 1989-1. **Coastal** – Masset 14 Oct 1945-1 (RBCM 10521); Masset Inlet 16 Sep 1984-1, 16 Oct 1971-1; Kitimat River 21 Sep 1974-4; Sandspit 27 Oct 1989-1 (Campbell 1990a); Hemlock Valley 24 Sep 1972-12; Harrison 18 Sep 1975-1 young still being fed in nest, 22 Sep 1975-young fledged; Maple Ridge 24 Sep 1969-15 (Crowell and Nehls 1970a); Vancouver 15 Sep 1987-700 at airport; Iona Island 6 Sep 1970-1,000; Langley 10 Sep 1975-4 young ready to leave nest; Tofino 26 Sep 1980-7; Reifel Island 30 Sep 1986-60, 28 Nov 1969-15; Delta 14 Sep 1974-500; Westham Island 9 Oct 1974-300; Central Saanich 2 Sep 1991-1 adult on nest with 4 to 5 naked young; Port Renfrew 5 Oct 1974-2; Saanich 5 Nov 1973-1 (Tatum 1973); Victoria 1 Oct 1971-1 (Tatum 1972).

Winter: Interior – No records. **Coastal** – Iona and Sea islands 9 Feb 1994-1, 26 Feb 1978-1; Surrey 25 Feb 1970-3; Reifel Island 2 Dec 1972-8 (Crowell and Nehls 1973a), 5 Dec 1968-6, 6 Dec 1971-1 feeding on insects on ice (BC Photo 201), 20 Dec 1966-1; Westham Island 3 Jan to 5 Feb 1994-1, 6 Dec 1969-7 (Campbell 1970a); Ladner 1 Jan 1970-3, 22 Dec 1962-2 feeding.

Christmas Bird Counts: Interior – Not recorded. **Coastal** – Recorded from 1 of 33 localities and on less than 1% of all counts. Maxima: Ladner 26 Dec 1969-4, all-time Canadian high count (Anderson 1976b), 14 Dec 1974-2 (Dawe 1975).

Gray Jay

Perisoreus canadensis (Linnaeus)

GRJA

RANGE: Resident from the timberline in western and central Alaska, northern Yukon, and northern Mackenzie across Canada to northern Labrador and Newfoundland, south in the boreal forest to southern Alaska and British Columbia, including Vancouver Island. Also resident in southwestern Alberta, western Montana, and south along the Rocky Mountains, Cascade Mountains, and coastal ranges into Wyoming, New Mexico, Oregon, and northern California; locally in Arizona and South Dakota.

STATUS: On the coast, *uncommon* to locally *fairly common* resident in the Georgia Depression Ecoprovince, including Vancouver Island, and on the southern mainland of the Coast and Mountains Ecoprovince; *uncommon* on Western Vancouver Island and the northern mainland of the Coast and Mountains. Absent from the Queen Charlotte Islands and other islands along the central and northern mainland coast. Throughout the interior, an *uncommon* to locally *fairly common* resident, mainly at higher elevations.

Breeds.

NONBREEDING: The Gray Jay (Fig. 190) is widely distributed throughout forested portions of British Columbia, occurring, at least seasonally, in every ecoprovince but especially in those with extensive boreal and subalpine forests. It is absent from the Queen Charlotte Islands and from the extensive fringe of islands, many of them mountainous, that characterize the north and central coasts. There are few reports of this jay from the west side of the Coast Mountains north of Squamish. Records indicate that the highest numbers in winter occur in the Southern Interior Mountains Ecoprovince (Fig. 191).

On the coast, the Gray Jay occurs mainly at elevations from 400 m up to the timberline. In the interior, it occurs mainly between 700 and 2,300 m elevation. It usually occupies middle to higher elevations, and ranges well up into the subalpine zone, where it may be relatively abundant in the partially wooded alpine meadows (Carl et al. 1952). The Gray Jay is truly a bird of the cold subalpine and boreal forests. It occurs in montane and subalpine coniferous and mixed coniferous-deciduous forests (Fig. 192) where the winters are long and cold and the summers relatively short. Few other species thrive on a year-round basis in such harsh habitat.

Preferred habitat on the coast is the subalpine association of amabilis fir, alpine fir, mountain hemlock, and yellow cedar forests, but out of the breeding season its habitat can extend downslope to include the mixed forest of western hemlock, western redcedar, grand fir, and Sitka spruce. In the interior, the Gray Jay is usually associated with mature Engelmann spruce forests, but occurs in a wide variety of forested habitats. In the southern interior, it occupies the higher montane forests of Douglas-fir and Engelmann spruce, and tends to avoid the valley bottoms. In the Okanagan valley, for example, Cannings et al. (1987) found the Gray Jay common above 1,000 m, but there were only 3 records from below 600 m elevation. In the northern interior, the Gray Jay occurs in both lowland and higher-elevation forests, where white and black spruce predominate, although Rand (1944) found them rare in the Liard River valley bottom, where the climate is drier and cooler (DeLong et al. 1991). Erskine and Davidson (1976) found this jay throughout the Fort Nelson Lowland, but less abundant in forests with large amounts of trembling aspen and birch. The Gray Jay also frequents subalpine parklands, wooded swamps, bogs, regenerating burns, brushy areas along rivers, and the edges of clearings.

The Gray Jay is readily attracted to the artificial environments provided by ski resorts, campgrounds, bush camps, traplines, logging roads, and garbage dumps. If human food waste is plentiful, as many as 10 or 12 Gray Jays may congregate. It also uses backyard feeders in many high-elevation and high-latitude communities in the province.

(a)

(b)

Figure 190. Four subspecies of Gray Jay are recognized as breeding in British Columbia, although the range of each is not adequately described for the province. There is considerable variation in plumage coloration and degree of intergradation between the subspecies. On Meadow Mountain, in the southwestern Southern Interior Mountains Ecoprovince, the adult, likely *Perisoreus canadensis bicolor* (a), has a white head with dusky nape and gray underparts (21 September 1994; R. Wayne Campbell). In western portions of Manning Park, the adult, *P. c. obscurus* (b), has a dusky crown, nape, and auricular patch, and light gray underparts (3 October 1993; R. Wayne Campbell).

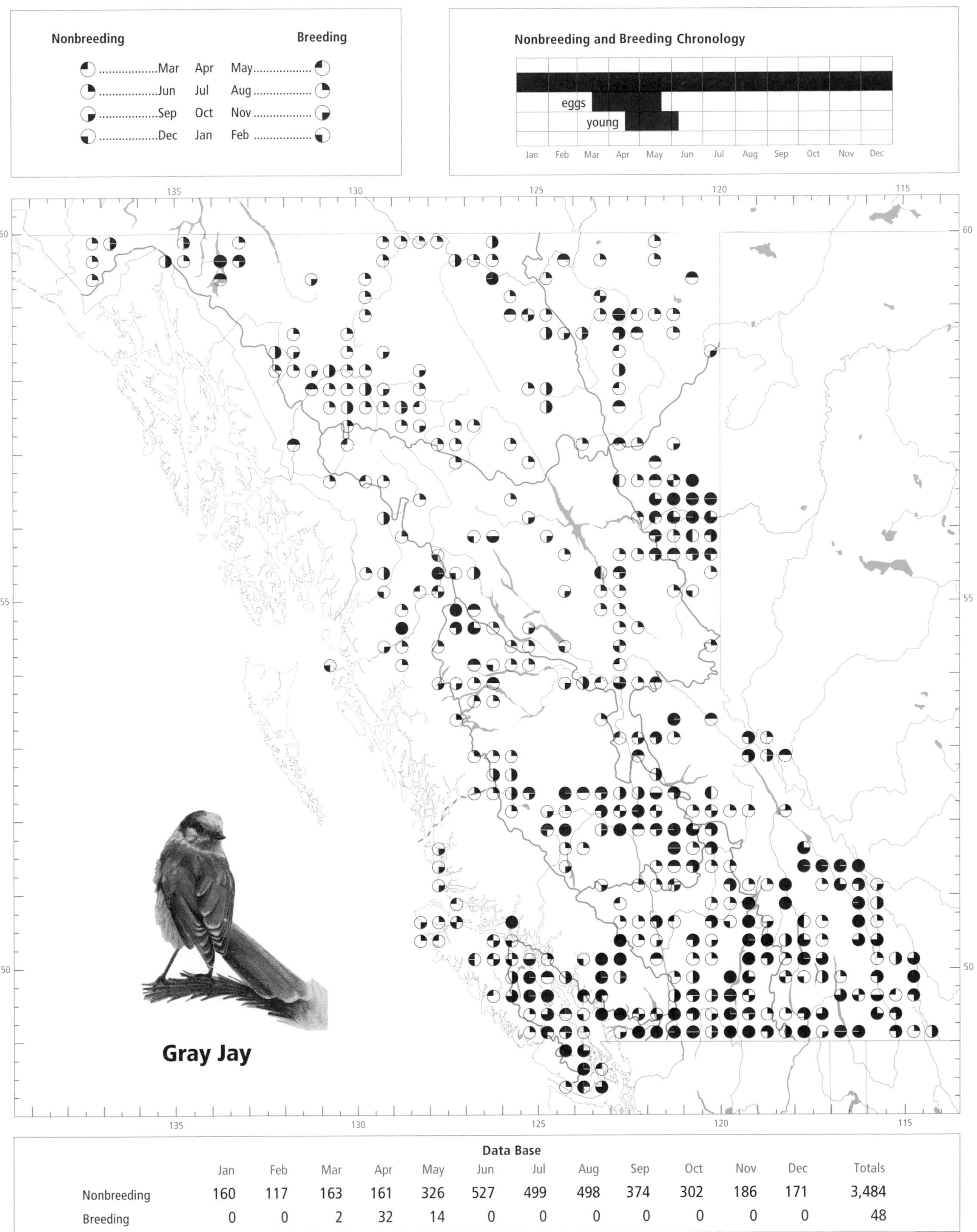

Data Base

	Jan	Feb	Mar	Apr	May	Jun	Jul	Aug	Sep	Oct	Nov	Dec	Totals
Nonbreeding	160	117	163	161	326	527	499	498	374	302	186	171	3,484
Breeding	0	0	2	32	14	0	0	0	0	0	0	0	48

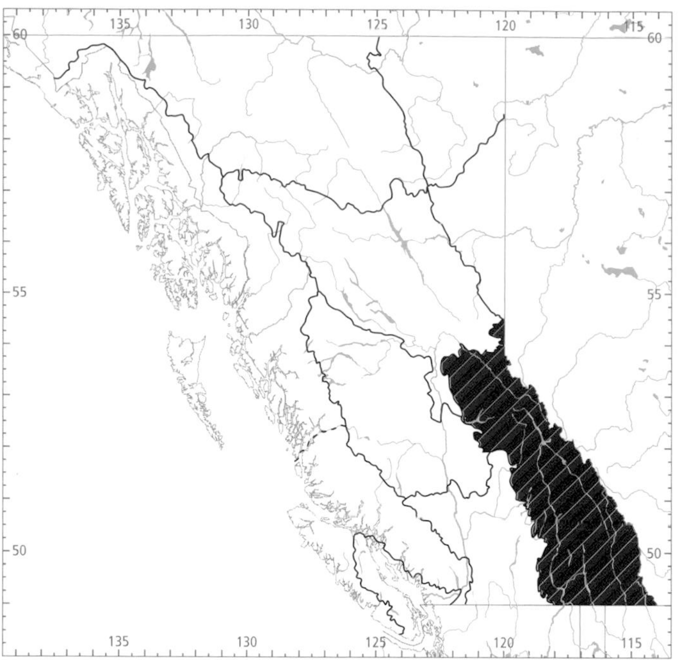

Figure 191. In British Columbia, the highest numbers for the Gray Jay in winter (black) and in summer (red) occur in the Southern Interior Mountains Ecoprovince.

Adult Gray Jay pairs are usually sedentary and territorial, rarely straying from a small home territory (Rutter 1969). There has been no study of the demography and movements of marked populations of the Gray Jay in mountainous western Canada. However, in Ontario and Quebec established adult pairs are resident on their relatively small territories through the entire year. Territories reported in Algonquin Park were about 146 ha and in Réserve de la Vérendrye about 69 ha (Rutter 1969; Strickland 1991). Established pairs do not migrate or even participate in local altitudinal movements. Research on the behaviour of fledglings has revealed that dominant juveniles, more often males, generally expel their siblings from the family group in June (Strickland and Ouellet 1993). The dominant birds stay with their parents through the winter and then disperse to vacant local territories before the next breeding season (Ha and Lehner 1990). The birds ejected from the family, if they survive at all, will often wander from the habitat occupied by the nesting pairs and appear at lower altitudes.

In British Columbia, it is characteristic for small groups of Gray Jays to appear in late summer and winter at lower elevations, where breeding does not occur. Such groups frequently include 2 to 6 birds, but may include as many as 12 birds. In similar situations in Ontario and Quebec, all individually identifiable birds in these wandering groups were birds of the year displaced from their natal territory, or mature birds without a home territory (Strickland and Ouellet 1993). These movements are post-breeding dispersals of nonterritorial individuals, both young and adult birds. To the extent that some of these jays wander to lower elevations, there is some altitudinal movement (Ehrlich et al. 1988). Such movements do not occur in all areas or every year in any area. For example, they did not occur at a study site at 2,000 m elevation in Colorado (Ha and Lehner 1990). However, in Glacier National Park, British Columbia, the Gray Jay became most numerous in winter along the highway corridor, but left that area by early May (Van Tighem and Gyug 1983).

The greatest numbers of the Gray Jay are recorded during summer, when many observers are in the mountains and the jays with their fledglings are highly noticeable (see Fig. 196). Aggregations of more than a dozen birds usually

Figure 192. In portions of the southern interior of British Columbia, Gray Jay winter habitat consists of mixed coniferous-deciduous forests (near Riske Creek, 25 March 1996; R. Wayne Campbell).

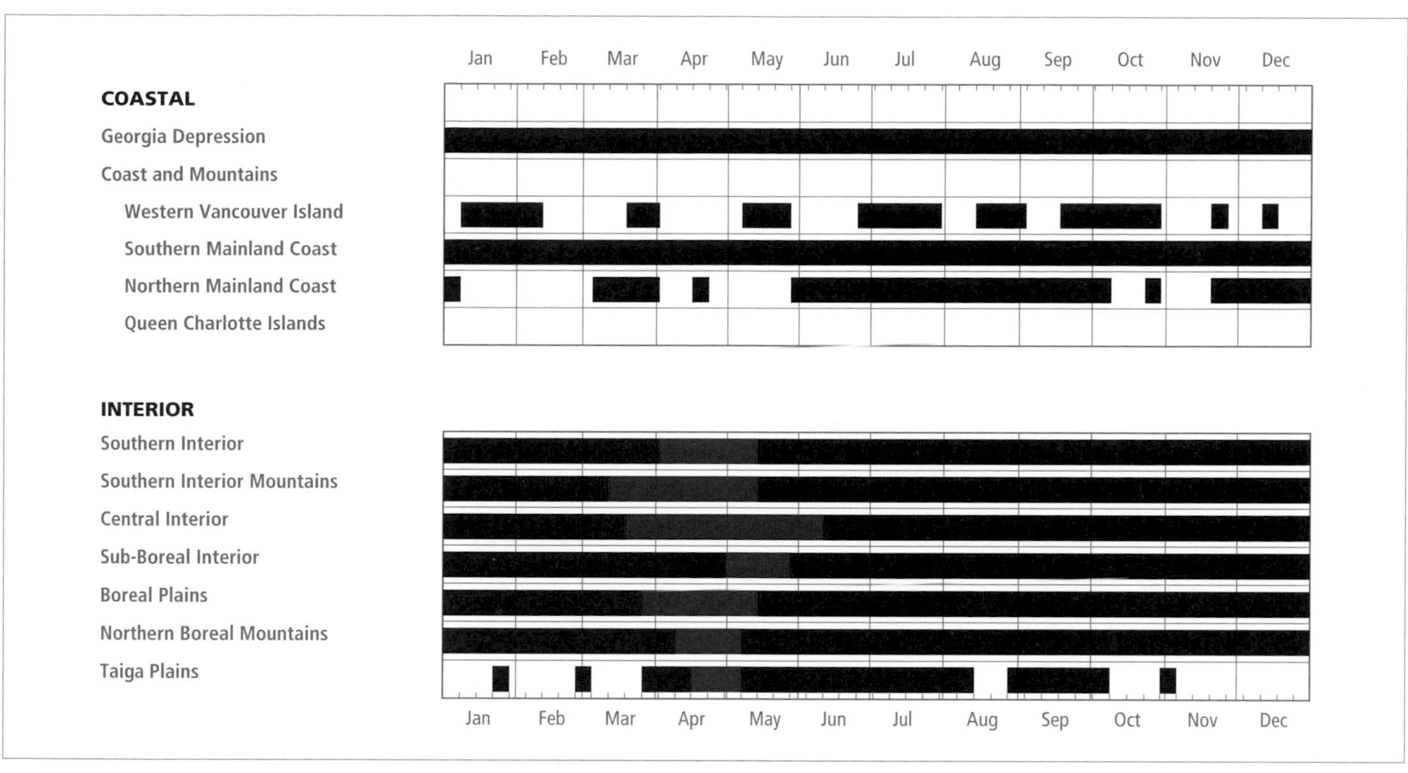

Figure 193. Annual occurrence (black) and breeding chronology (red) for the Gray Jay in ecoprovinces of British Columbia.

occur only at human-influenced sites such as bush camps, picnic grounds, and garbage dumps.

The Gray Jay occurs in British Columbia throughout the year (Fig. 193).

BREEDING: This jay probably breeds in suitable habitats throughout interior British Columbia, as well as at higher elevations on Vancouver Island and in the mountains backing the mainland coast. Despite this extensive probable breeding range, only 19 nests with eggs or young have been reported in the province. The young of the year carry a distinctive plumage until the moult in midsummer (see Fig. 196), and this leads to many so-called breeding records. These records refer to older fledglings with their parents, or groups

Figure 194. Although few Gray Jay nests have been found in British Columbia, breeding habitat is often associated with high-altitude open spruce forests bordering wet meadows (Kootenay Pass Summit, 15 September 1994; Linda M. Van Damme).

composed of nonterritorial adults along with young of the year displaced from the nesting territory by their dominant siblings (Strickland 1991). Neither situation meets our criteria for a breeding record.

Nests have been found the length of the province from Summerland north to Atlin. The Gray Jay reaches its highest summer numbers in the Southern Interior Mountains Ecoprovince (Fig. 191). Breeding Bird Surveys for both coastal and interior routes for the period 1968 through 1993 contain insufficient data for analysis.

Breeding has been documented at elevations between 850 and 2,100 m. The species appears to avoid nesting in dense forests and shows a preference for open areas such as the edges of damp or dry meadows (Fig. 194), spruce bogs, edges of swamps, forest edges and clearings, marshes, islands, and beaver meadows. This is one of the characteristic species of the breeding bird community in Engelmann spruce–subalpine fir forest (Poll and Porter 1984).

The Gray Jay has been reported breeding in the province between 20 March and 25 May (Fig. 193).

Nests: Most nests (Fig. 195) were situated in mixed or coniferous forests; a few were reported from rural situations. They were built mainly in living coniferous trees, including Douglas-fir, western hemlock, lodgepole pine, spruce, and subalpine fir; in the north, they were also found in aspen and willow. Nests were usually placed on a horizontal branch close to the trunk (Fig. 195). The heights of 17 nests ranged from ground level to 31 m, with 12 nests between 1 and 6 m.

In British Columbia, nests were usually well concealed and consisted of a bulky foundation composed of large, coarse twigs, lichen, and bark strips, often including caterpillar cocoons. The nest cup was fashioned of beard-lichen (*Usnea* sp. or *Alectoria* sp.), leaves, grasses, bark strips, or plant stalks, and was lined with feathers, hair or fur, and in 1 case toilet paper. The main body of a nest from near 100 Mile House was about 18 cm across and 13 cm deep; the nest cup was about 6 cm in breadth and depth. It was made of spruce and pine twigs with long blades of grass intertwined. A few feathers and aspen leaves were included (Cresco 1960).

Figure 195. In British Columbia, most Gray Jay nests have been found in living coniferous trees. Note the bulky, coarse twig nest on the horizontal branch next to the trunk of a subalpine fir in the upper right corner of the photograph (Kootenay Pass Summit, 15 September 1994; R. Wayne Campbell).

Figure 196. The plumage of the juvenile Gray Jay is uniformly dark gray; most show a distinctive white whisker mark (Manning Park, 5 August 1993; R. Wayne Campbell).

Eggs: Dates for 11 clutches ranged from 20 March (Paul 1959) to 22 May, with 52% recorded between 6 and 20 April. Clutch size in British Columbia ranged from 1 to 5 eggs (1E-3, 2E-3, 3E-3, 4E-1, 5E-1), with 6 nests having 2 or 3 eggs. Nest failures during the incubation period may be followed by renesting if they occur before mid-April (Strickland and Ouellet 1993). The Gray Jay is one of the earliest species to nest. For example, the mean date of first eggs in Algonquin Park, Ontario (n = 244), was 26.3 March ± 9.4 days (range: 26 February to 17 April); in Réserve de la Vérendrye, Quebec (n = 32), it was 23.1 March ± 5.7 days (range: 14 March to 2 April) (Strickland and Ouellet 1993).

The incubation period in 1 nest from British Columbia was 18 days. In other parts of the Gray Jay's range, Strickland and Ouellet (1993) give the incubation period as 18.5 days but state that the female sits on the nest (apparently without incubating) during the egg laying period. Hatching is thus synchronous and, in a 3-egg clutch, occurs 20 days after the laying of the first egg.

Young: Dates for 9 broods ranged from 16 April to 25 May, with 59% recorded between 23 April and 7 May. Sizes of 8 broods ranged from 3 to 5 young (3Y-6, 4Y-1, 5Y-1). The nestling period from 1 nest in British Columbia was 23 days; in undisturbed nests, young normally fledge in 22 to 24 days (n = 7) (Strickland and Ouellet 1993) (Fig. 196).

Brown-headed Cowbird Parasitism: Cowbird parasitism was not found in British Columbia in 17 nests recorded with eggs or young, nor has it been reported elsewhere in North America (Friedmann et al. 1977; Friedmann and Kiff 1985). The Gray Jay is probably immune to cowbird parasitism because it nests before cowbirds return in the spring and because female jays do not leave their nests unattended during the egg-laying period.

Nest Success: Of 3 nests found with eggs and followed to a known fate, 2 produced at least 1 fledgling.

REMARKS: Four subspecies are recognized by Strickland and Ouellet (1993) as breeding in British Columbia: *P. c. pacificus* in the northwest of the province; *P. c. albescens* in the northeast, east of the Rocky Mountains; *P. c. bicolor* in the Southern Interior Mountains, Southern Interior, and north into the Cariboo and Chilcotin areas (Fig. 190a); and *P. c. obscurus* in the coast forest, including Vancouver Island (Fig. 190b). The American Ornithologists' Union (1957) excludes *P. c. obscurus* from British Columbia, referring to *P. c. griseus* as the coastal subspecies north to Kimsquit; they also refer to *P. c. arcus* as occurring in the Rainbow Range north of Bella Coola. The Gray Jay and the Palearctic species *P. infaustus* may form a superspecies (American Ornithologists' Union 1983; Strickland and Ouellet 1993).

Although the Gray Jay is one of our most widely recognized birds, there has been little research into its biology in British Columbia. Data on breeding ecology are generally difficult to obtain because of logistical problems; Gray Jays nest during late winter and early spring, when heavy snow cover occurs in the subalpine forests.

Experimental studies have shown that there is a potential for serious impact on local populations of the Gray Jay because of their vulnerability to traps set for fur-bearing animals (de Vos et al. 1959). However, after more than a century and a half of fur trapping throughout the mountains of British Columbia, the species remains widely distributed and fairly common.

The Gray Jay was formerly known as Canada Jay and is often referred to by a variety of colloquial names, including whiskey jack, moose bird, and camp robber.

Strickland and Ouellet (1993) provide a recent review of information on the Gray Jay.

NOTEWORTHY RECORDS

Spring: Coastal – Tugwell Creek 1 Mar 1984-4; Jordan Meadows 29 Apr 1979-30; Hollyburn Mountain 3 Mar 1976-10; Cypress Park 23 May 1983-5; Whistler 27 Apr 1968-12; Alta Lake 1 Mar 1968-9. **Interior** – Lightning Lakes 24 Mar 1979-18; Goat Mountain (Kitchener) 8 May 1928-7 (Mailliard 1932); 2 km e Penticton 29 May 1987-3 fledglings; Kootenay Crossing 23 Apr 1982-nest; Arrow Mountain Mar 1926-5 eggs; Golden 11 May 1937-2 adults and 4 fledglings; Lewis Creek 2 Apr 1973-4 eggs; Clearwater 2 Apr 1962-first egg, 5 Apr 1962-3 eggs, 23 Apr 1962-3 nestlings; Wells Gray Park 27 Apr 1960-2 eggs; Bridge Lake 22 May 1963-2 eggs; Horse Lake 27 Apr 1960-2 eggs; Big Lake 17 Mar 1975-nest building, 25 Apr 1975-4 nestlings; Goldbridge 21 May 1986-5; Lac la Hache 10 Apr 1960-1 egg; Kleena Kleene 26 Apr 1965-3 nestlings; Brisco 20 Mar 1977-2 at dump; Stum Lake 25 May 1978-3 nestlings; Dragon Lake 5 May 1985-5 eggs; Burns Lake 10 Apr 1983-3 nestlings; upper Iskut River 31 May 1978-5; Flood Glacier 6 Apr 1919-1; Bear Flat 28 Mar 1982-4; Hyland Post 30 May 1976-8; 35 km s Dawson Creek 16 Apr 1979-3 nestlings; n Brassey Creek 25 Apr 1993-3 eggs, 3 May 1993-3 nestlings; Progress 5 May 1994-2 small nestlings with eyes closed; Manson Creek 17 Apr 1921-2 eggs (RBCM 895); Atlin 15 Apr 1915-3 eggs (RBCM 894).

Summer: Coastal – Sooke Hills 5 Jul 1985-1 young; Mount Arrowsmith 24 Jun 1993-2 adults with 2 fledglings; Paradise Meadows 18 Jul 1976-20; Mount Washington 30 Jul 1983-3 fledglings with adults; Port Neville Inlet 6 Aug 1976-2; Browning Inlet 21 Jul 1968-2 (Richardson 1971); Kimsquit 17 Jun 1939-1 (NMC 29036); Kitsault Lake 27 Aug 1980-11; Bernard Lake 23 Aug 1977-10. **Interior** – Lightning Lakes 11 Jun 1975-12, 6 Aug 1979-20; Blackwall 13 Jun 1965-2 fledglings; Amiskwi River 14 Jun 1975-11 (Wade 1977); Bridge Lake 5 Jun 1976-6 on survey; Titetown Lake Jul 1983-11; Taylor 22 Aug 1975-16; Stoddart Creek (Fort St. John) 19 Apr 1985-2 fledglings; Fern Lake 18 Aug 1983-4 (Cooper and Cooper 1983); Fort Nelson 19 Jun 1976-16 on survey; Steamboat 14 Jun 1976-13; Cabin Lake 20 Jun 1982-2 young; Survey Lake 26 Jun 1980-12.

Breeding Bird Surveys: Coastal – Recorded from 9 of 27 routes and on 7% of all surveys. Maxima: Pemberton 23 Jun 1991-6; Seabird 5 Jul 1992-2; Comox Lake 17 Jun 1973-2; Campbell River 22 Jun 1974-2; Kitsumkalum 22 Jun 1975-2; Port Hardy 1 Jul 1983-1; Meziadin 29 Jun 1977-1; Chilliwack 26 Jun 1993-1; Kispiox 20 Jun 1993-1. **Interior** – Recorded from 50 of 73 routes and on 28% of all surveys. Maxima: Telkwa High Road 23 Jun 1974-22; Princeton 1 Jul 1992-20; Fort Nelson 19 Jun 1976-16.

Autumn: Interior – Chilkat Pass 2 Oct 1981-10; White Pass 11 Sep 1984-1 (MVZ 102838); Andy Bailey Lake 29 Sep 1985-4; Telkwa River 14 Sep 1974-12; Little Big Bar Lake 21 Sep 1986-numerous; Yoho National Park 6 Nov 1976-12; Sorrento 18 Oct 1971-15; Fording River 12 Sep 1979-43; Cathedral Lakes Park 5 Sep 1979-20; Manning Park 26 Sep 1970-19 at various localities in the park, 11 Nov 1982-11. **Coastal** – Kitsault Lake 23 Sep 1980-24; Calvert Island 18 Sep 1934-4 (MCZ 283248-51); Port Neville Inlet 16 Oct 1976-8; Strathcona Park 6 Sep 1986-14; Black Tusk 11 Oct 1971-25; Hollyburn Mountain 5 Oct 1969-30; Chilliwack Lake 30 Nov 1974; Mount Lazar 28 Nov 1975-15; Victoria 22 Oct 1983-1.

Winter: Interior – Liard Hot Springs 10 Dec 1974-4 (Reid 1975), 4 Jan 1975-3; Fort Nelson 25 Jan 1985-1; Farrell Creek 3 Jan 1983-4, 29 Dec 1982-2; Upper Cache Creek Road (Fort St. John) 12 Feb 1983-10; Tumbler Ridge 7 Dec 1994-6; Bulkley River valley 1 Jan 1912-1; Tetana Lake 24 Dec 1937 (Stanwell-Fletcher and Stanwell-Fletcher 1943); Williams Lake 1 Jan 1979-1, 30 Dec 1979-10; Celista 31 Jan 1948-4; Yoho National Park 5 Feb 1976-9 (Wade 1977); Kootenay Crossing 31 Dec 1965-2 (Seel 1965); Apex Mountain 2 Jan 1963-2; nr Vaseux Lake 31 Dec 1977-8 at 1,400 m; Creston area 1 Jan 1981-5 on count; Manning Park 19 Feb 1983-30 over 6 hours. **Coastal** – Terrace 21 Dec 2 1969-1, 2 Jan 1965-2 on Christmas Bird Count; Port Hardy 15 Dec 1958-1; Woss Lake 12 Jan 1984-1; Alta Lake 1 Jan 1936-1; Garibaldi Park 1 Jan 1980-20 on survey, 3 Jan 1981-26 on survey; Mount Washington 12 Dec 1981-14; Mount Seymour Park 1 Jan 1981-10, 31 Dec 1980-10; Mount Arrowsmith 2 Jan 1988-14; Mount Prevost 24 Feb 1973-5.

Christmas Bird Counts: Interior – Recorded from 26 of 27 localities and on 79% of all counts. Maxima: Smithers 29 Dec 1991-42; Yoho National Park 20 Dec 1975-38; North Pine 16 Dec 1990-34. **Coastal** – Recorded from 16 of 33 localities and on 10% of all counts. Maxima: Squamish 1 Jan 1985-35; Whistler 18 Dec 1993-10; Pender Harbour 21 Dec 1993-8.

Steller's Jay

Cyanocitta stelleri (Gmelin)

STJA

RANGE: Resident from southeastern Alaska, northwestern and central British Columbia, southwestern Alberta, western Montana, Wyoming, western Colorado, and New Mexico south to southern California, Arizona, southwestern Texas, and the Middle American highlands to Nicaragua.

STATUS: On the coast, *uncommon* to locally *fairly common* resident, including Vancouver Island and the Queen Charlotte Islands.

In the interior, *uncommon* to locally *fairly common* resident across the Southern Interior and Southern Interior Mountains ecoprovinces; *rare* to *fairly common* in the Cariboo and Chilcotin areas of the Central Interior Ecoprovince; *rare* to *uncommon* in the Sub-Boreal Interior Ecoprovince; *very rare* in the Peace Lowland of the Boreal Plains Ecoprovince and in the Northern Boreal Mountains Ecoprovince. Absent from the Taiga Plains Ecoprovince. Locally, *common* to *very common* during irruptive movements in autumn and winter, across the southern portions of the province.

Breeds.

CHANGE IN STATUS: In 1947 the Steller's Jay (Fig. 197) was unknown in the interior of the province north of latitude 56°N or in the Peace River drainage basin (Munro and Cowan 1947). The first evidence of its entry into the Peace River region is a specimen taken on the banks of East Pine River, 64 km west of Dawson Creek (Jobin 1955). While it has not become a regular occupant of the region, it has recently been observed in the area between Chetwynd and Fort St. John and north as far as Rose Prairie, areas formerly occupied only by the Blue Jay. More recently it has extended its range into the Northern Boreal Mountains, where in 1959 it was recorded at Marion Creek in the Spatsizi Plateau. In 1972 it was discovered at Cassiar and along the Haines Highway, between Miles 52 and 54.

Figure 197. Adult Steller's Jay, coastal race *Cyanocitta stelleri stelleri* (Long Beach, west coast of Vancouver Island, May 1992; Mark Nyhof). Note the absence of the white "eyebrow."

Figure 198. In British Columbia, the highest numbers for the Steller's Jay in winter (black) and summer (red) occur both in the Georgia Depression Ecoprovince and on Western Vancouver Island in the Coast and Mountains Ecoprovince.

NONBREEDING: The Steller's Jay is widely distributed throughout south coastal British Columbia, including Vancouver Island, but is less abundant on the northern mainland of the Coast and Mountains Ecoprovince, the Queen Charlotte Islands, and other offshore islands. In the interior, it has a widespread distribution through the Southern Interior and Southern Interior Mountains, but becomes less numerous in the Central Interior, the Sub-Boreal Interior to the Cassiar Mountains, and northwest to the Tatshenshini Basin. East of the Rocky Mountains, it occurs in the Peace Lowland of the province. It has not yet been reported from the Taiga Plains in the far northeastern corner of the province.

The highest numbers in winter occur on southeastern Vancouver Island and the Fraser River delta regions of the Georgia Depression, and on Western Vancouver Island (Fig. 198).

The Steller's Jay occurs at elevations from near sea level to 1,500 m on the coast, and from the valley bottoms around 300 m up to 2,150 m in the interior. It frequents a wide range of coniferous and mixed coniferous-deciduous forest communities (Fig. 199), from the Coastal Douglas-fir and Mountain Hemlock zones on the coast to the Ponderosa Pine and Engelmann Spruce–Subalpine Fir zones of the interior. In the far north and northeast, it inhabits the Boreal White and Black Spruce Zone.

Although the Steller's Jay can be sympatric with the Gray Jay, especially on winter ranges, it prefers lower elevations than the Gray Jay. For example, in the Southern Interior Mountains from September to March, the Steller's Jay is primarily

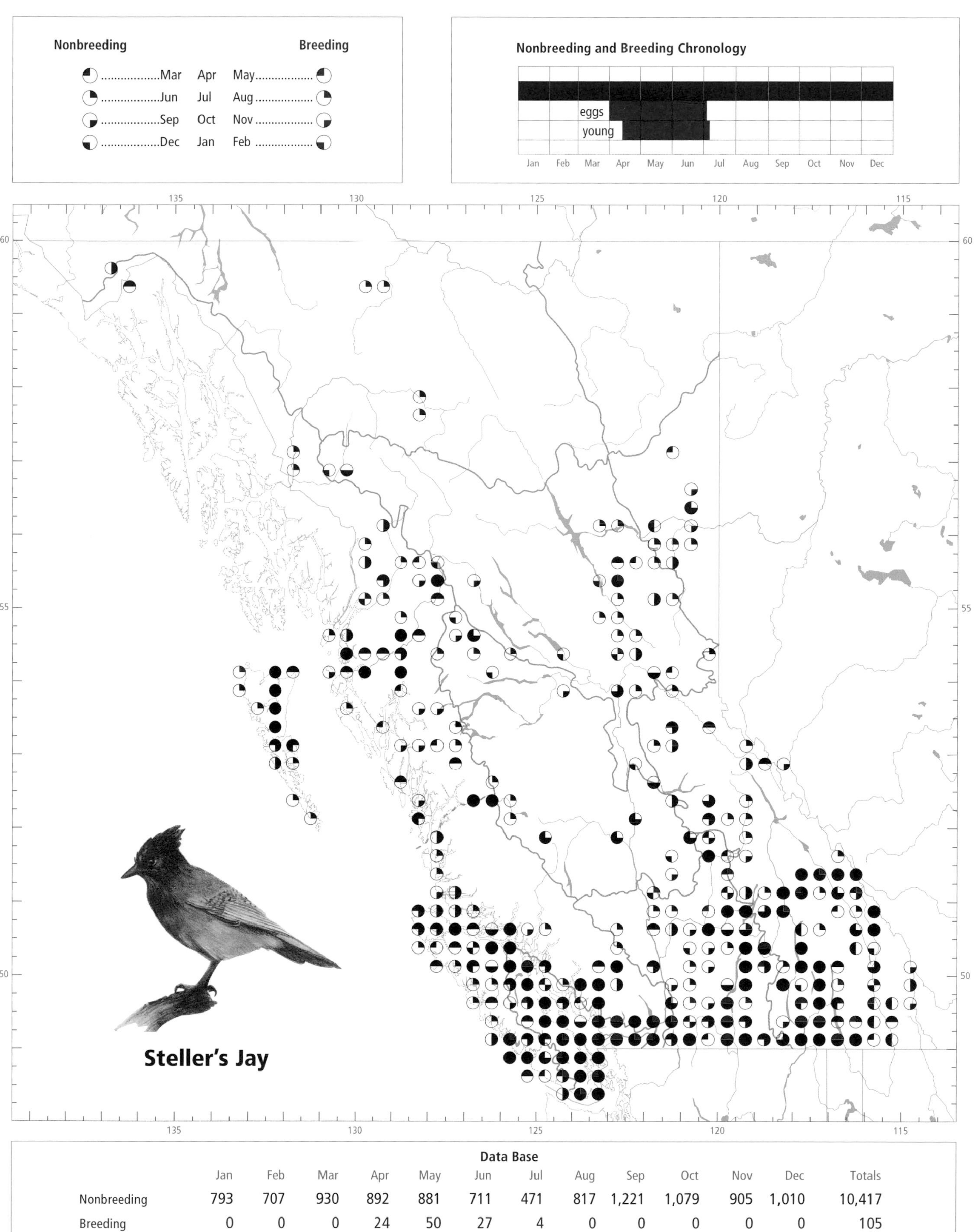

Data Base

	Jan	Feb	Mar	Apr	May	Jun	Jul	Aug	Sep	Oct	Nov	Dec	Totals
Nonbreeding	793	707	930	892	881	711	471	817	1,221	1,079	905	1,010	10,417
Breeding	0	0	0	24	50	27	4	0	0	0	0	0	105

confined to the interior cedar-hemlock forests at a mean elevation of 884 m (n = 221), while the Gray Jay is found in the Engelmann spruce–subalpine fir forests at a mean elevation of 1,280 m (n = 47) (J.G. Woods pers. comm.). The Steller's Jay usually frequents open woodlands, edges of clearings, transmission line rights-of-way, breaks in the forest, and riparian growth along waterways. Other habitats include swamps, bogs, second-growth forests (Fig. 199), and brushy clearcuts. In dense forests, it occurs only along the edges. The Steller's Jay also readily uses human-made habitats such as well-treed residential neighbourhoods, parks, golf courses, nut-orchards, cemeteries, campgrounds, picnic grounds, gardens, road rights-of-way, and garbage dumps. It is often seen on the gravelled shoulders of highways through forested regions.

Although the Steller's Jay is considered a resident throughout its range (Fig. 200), it appears to be an altitudinal migrant. This seasonal movement is more regular in the interior than on the coast, but even there it can be erratic, sometimes starting as early as July. It usually begins in August and reaches a peak in September or October (Fig. 201). In the Cariboo and Chilcotin areas, and in the Okanagan (Cannings et al. 1987), this jay may appear at valley bottom bird feeders in late September and leave by early April. In the east Kootenay, the movement to the valley bottom is noticeable in September; by March the birds begin returning to their nesting elevations. The same behaviour can be found on the south coast. Based on banding returns, individual Steller's Jays are also known to wander erratically over considerable distances (Table 6).

There is also evidence for at least a limited north-south migration. In the autumn of 1927, Steller's Jays were captured and banded at Indianpoint Lake (McCabe and McCabe 1928). The banding revealed that what was believed to be a small

Figure 199. Winter habitat for the Steller's Jay in the Central Interior Ecoprovince may include mixed forests of lodgepole pine, birch, western redcedar, and Douglas-fir (Canim Lake, 26 March 1996; R. Wayne Campbell).

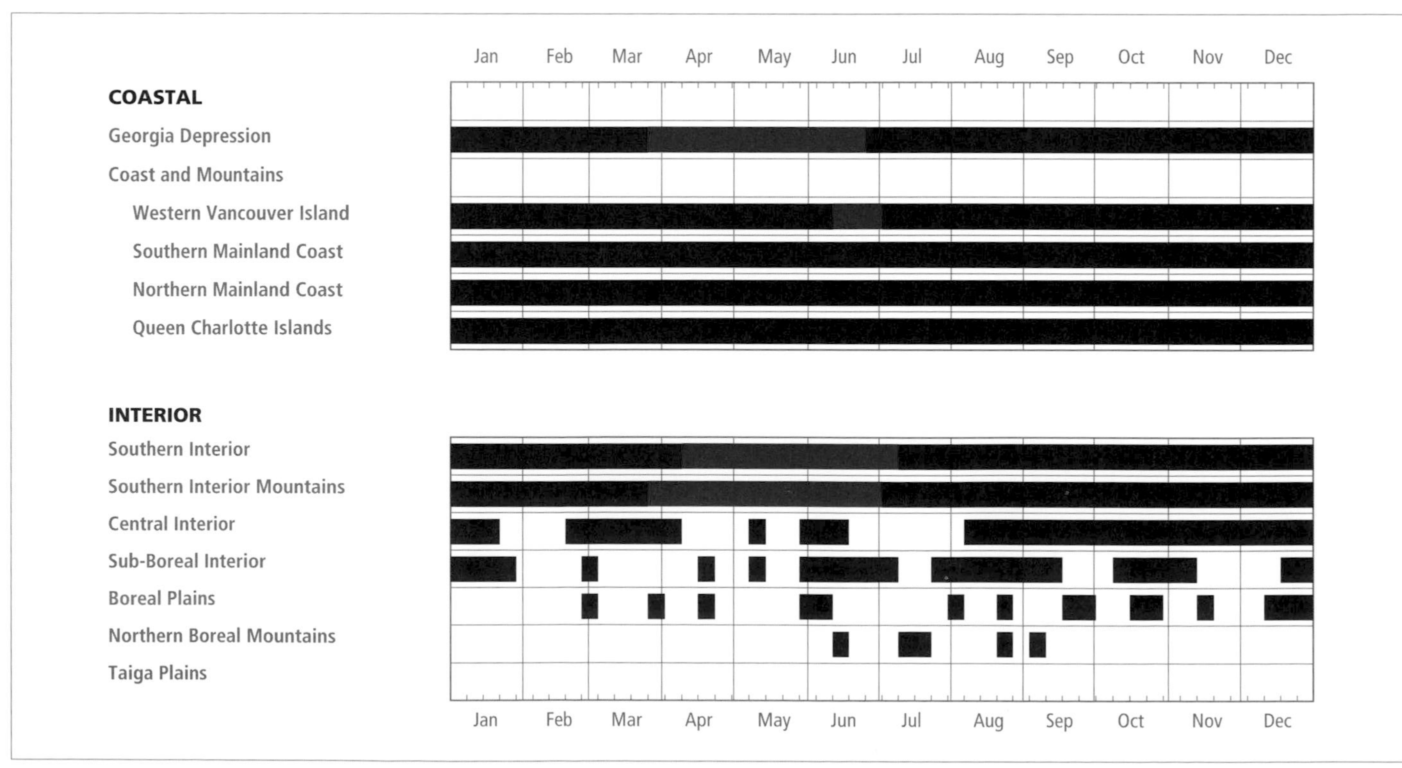

Figure 200. Annual occurrence (black) and breeding chronology (red) for the Steller's Jay in ecoprovinces of British Columbia.

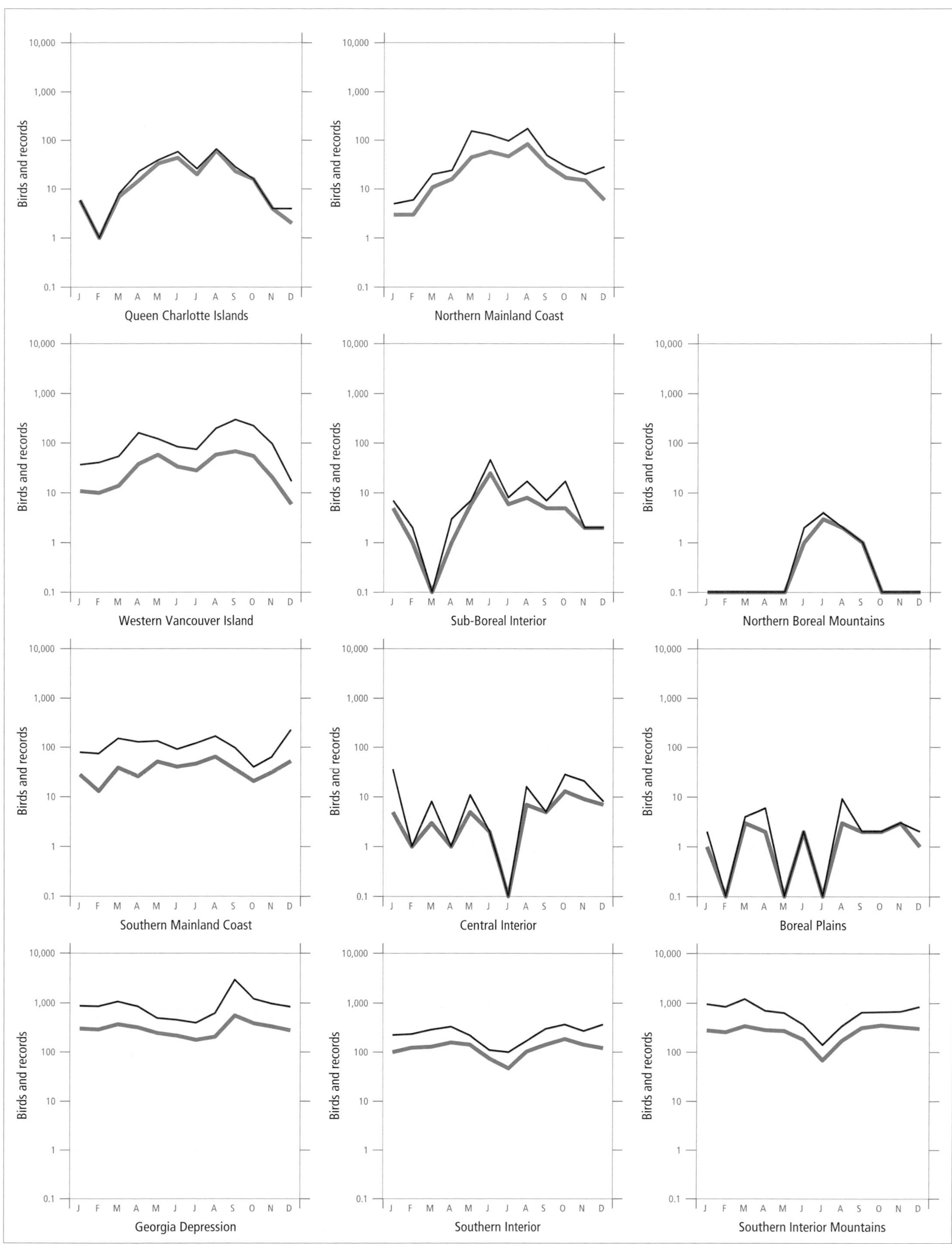

Figure 201. Fluctuations in total number of birds (purple line) and total number of records (green line) for the Steller's Jay in ecoprovinces of British Columbia. Christmas Bird Count and nest record data have been excluded.

resident group involved a steady turnover of individuals as part of a latitudinal movement. Each jay was present at the banding site for an average of 3 days before moving on. One banded bird was recovered 2 weeks later 200 km south, at Vavenby. In another instance, an immature bird banded near Barkerville was recovered 150 km to the south, near Canim Lake (Table 6).

Birds in the northern ecoprovinces appear to move out of those regions, while in the southern ecoprovinces numbers appear relatively constant or build through the winter (Fig. 201). The southern areas could be receiving birds from the north as well as birds moving into the valley bottoms from their higher-elevation breeding areas.

Evidence of a latitudinal movement is strongest on the coast. There the number of records and the total birds seen reach their highest levels in August or September and decline into the winter (Fig. 201). This suggests that the birds are not just arriving on a winter range but are passing through. Further evidence to support this movement comes in the form of banding returns. A bird banded near Richmond was found nearly 170 km to the southeast at Darrington, Washington (Jewett et al. 1953; Table 6). Another bird, banded near Port Hardy, was found near Richmond (Table 6).

The situation at Victoria differs in that when the influx of "migrants" reaches the shoreline at the southern tip of Vancouver Island, it is confronted by a substantial barrier in the form of a sea crossing of at least 11 km to reach the San Juan Islands or 24 km to the Olympic Peninsula. Observation and banding studies undertaken near Victoria during the winter of 1992 revealed that the jays collect in flocks in shoreline trees, the flocks at times consisting of over 90 birds. These birds indicate an urge to continue southward; however, after testing the distance several times, they appear to give up and disperse onto winter areas in and around Victoria. This dispersal of the "migrating" groups may be easily misinterpreted as a departure on continued southward migration. One flock of 12 was seen to embark on the crossing of Haro Strait to San Juan Island; however, there is no evidence that this is a characteristic event (Stewart and Shepard 1994). Banding confirmed that most of the birds pass the winter near Victoria as relatively sedentary groups.

The autumn populations of the Steller's Jay differ significantly from year to year, and from time to time the movement to lowland areas of Vancouver Island results in large concentrations (Munro and Cowan 1947). For example, unusually large numbers of jays occurred in the vicinity of Victoria in the autumns of the following years: 1913 (J.A. Munro *in* Bent 1946), 1919 (Anderson 1920), 1922-23 (J.A. Munro *in* Bent 1946), 1940 (Pearse 1946), 1957 (Schultz 1958a), and at least 5 times over the period 1958 to 1993. These latter irruptions are noticeable on the Victoria Christmas Bird Counts (Fig. 202). Similar, although less dramatic, concentrations can be seen on other Christmas Bird Counts in the province.

In winter, the Steller's Jay is more abundant in lowland habitats, both natural and human-influenced, than during other seasons. For example, Cannings et al. (1987) note a distinct movement in winter to riparian thickets and residential

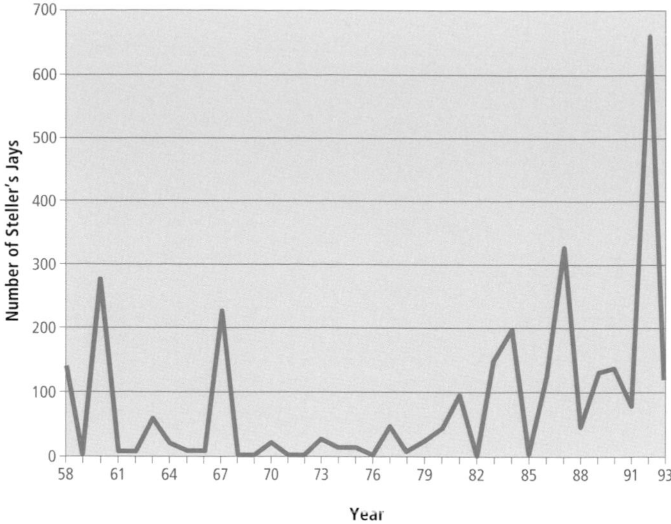

Figure 202. Fluctuations in Steller's Jay numbers on Christmas Bird Counts for Victoria, British Columbia, showing the irruptive nature of the species on Southern Vancouver Island.

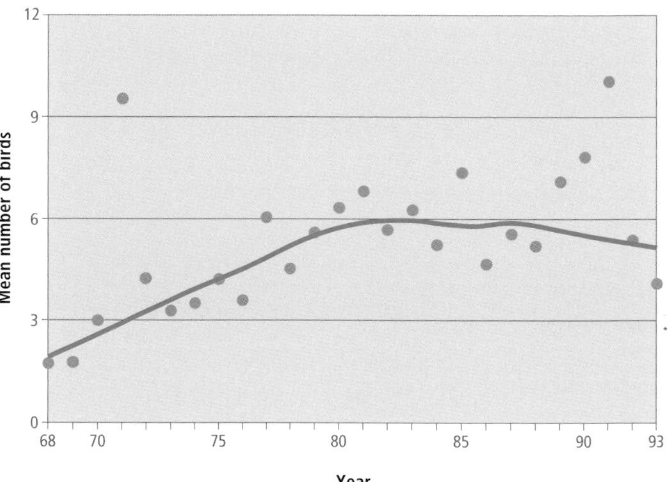

Figure 203. An analysis of Breeding Bird Surveys for the Steller's Jay in British Columbia shows that the mean number of birds on coastal routes increased at an average annual rate of 3% over the period 1968 through 1993 ($P < 0.001$).

areas in the Okanagan valley, and the same is true in the vicinity of Victoria, Vancouver, and the many other towns on eastern Vancouver Island and in the Fraser Lowland. They rarely stay to nest. On the coast, these jays disperse to breeding areas in March or April, and in the interior, in April or May.

The Steller's Jay occurs every month of the year in all but 3 ecoprovinces (Fig. 200). In the Sub-Boreal Interior there are no March records, in the Boreal Plains records are lacking for May and July, and in the Northern Boreal Mountains the few records are between 13 June and 8 September.

BREEDING: The known nesting distribution of the Steller's Jay is concentrated along the southern regions of the province from the International Boundary north to about latitude

51°30'N, although it likely breeds throughout most of its range in the province. Its breeding status in the Peace Lowland of the Boreal Plains and in the Northern Boreal Mountains remains uncertain. There is only 1 nesting record north of the latitude of Rogers Pass, although we have records of fledged young from Barney Creek and Pine Pass in the Sub-Boreal Interior, and from Langara Island and Graham Island on the Queen Charlotte Islands.

The Steller's Jay reaches its highest numbers in summer in the Georgia Depression and on Western Vancouver Island (Fig. 198). An analysis of Breeding Bird Surveys for the period 1968 through 1993 shows that the number of birds on coastal routes has increased at an average annual rate of 3% (Fig. 203). Surveys for interior routes contain insufficient data for analysis. North American populations appear stable (Robbins et al. 1986).

The Steller's Jay nests from near sea level to 1,830 m elevation. Most nests (66%; n = 70) were found in human-influenced coniferous, mixed forest, or woodland habitat. One-third of the nests were found in undisturbed forest (Fig. 205). This sample is probably heavily biased by the greater likelihood of nests close to human habitation being discovered, and does not reflect the normal distribution of nesting habitat. Forest types included Interior Douglas-fir, Interior Western Hemlock/Redcedar, and Subalpine Fir/White Spruce. Specific nest habitat included backyard, farm, and recreational area environments (88%; n = 48), and both young and mature forest, including forest edges (12%). In the Okanagan and Kootenay valleys, most nests were found in treed gardens.

The Steller's Jay has been recorded breeding on the coast from 1 April (calculated) to 28 June; in the interior, it has been recorded from 3 April (calculated) to 6 July (Fig. 200).

Nests: Most nests (85%; n = 41) were situated in coniferous trees. These were usually small second-growth trees, almost equally divided between Douglas-fir, western hemlock, western redcedar, and spruce. Other categories of nests sites included deciduous vegetation (9%), mostly garden shrubs, and buildings (6%; Grant 1949). The majority of nests were set on horizontal branches close to the trunk and near the top of the tree. When nesting near human habitation, the jays frequently placed the nest within a few feet of a window, verandah, or well-used pathway, suggesting that some birds can tolerate a fair amount of disturbance.

The heights of 65 nests ranged from ground level to 9 m, with 68% between 2 and 5 m.

The base of the nests consisted of coarse twigs (69%; n = 55) or branches (28%), with dry grass present in 35% of nests, and leaves, moss, string, plant stems, and other materials in lower frequency. Several nests had a mud cup (24%); the most frequent lining material of the deep cup consisted of fine rootlets, sometimes with fine grass, moss, and paper.

Eggs: Dates for 35 clutches ranged from 4 April to 3 July, with 60% recorded between 19 April and 24 May. Calculated dates indicate that nests can have eggs as early as 1 April. Sizes of 28 clutches ranged from 3 to 5 eggs (3E-1, 4E-26, 5E-1), with 93% having 4 eggs. The incubation period is about 16 days (Ehrlich et al. 1988).

Young: Dates for 40 broods ranged from 13 April to 6 July, with 51% recorded between 14 May and 9 June. Sizes of 26 broods ranged from 2 to 5 young (2Y-6, 3Y-6, 4Y-12, 5Y-2), with 69% having 3 or 4 young. The nestling period is about 20 days (Goodwin 1976).

Brown-headed Cowbird Parasitism: Cowbird parasitism was not found in British Columbia in 66 nests recorded with eggs or young. There are no instances of parasitism reported for North America (Friedmann et al. 1977; Friedmann and Kiff 1985).

Nest Success: Of 8 nests found with eggs and followed to a known fate, 3 produced at least 1 fledgling. This low success rate may well be an artifact of the proximity of many nests to human disturbance.

REMARKS: The American Ornithologists' Union (1957) recognizes 6 subspecies of the Steller's Jay in North America, 3 of

Table 6. Returns for Steller's Jays banded in British Columbia. Recovery locations are in British Columbia unless otherwise noted.

Banding		Recovery		Age[1] of bird	Distance (km)	Direction
Location	Date	Location	Date			
Barkerville	22 Sep 1928	Canim Lake	1 Oct 1928	HY	150	South
Summerland	6 Oct 1932	Revelstoke	? Feb 1934	AHY	190	Northeast
Terrace	12 May 1935	Alice Arm	10 Oct 1935	AHY	120	Northwest
Terrace	13 Oct 1935	Alice Arm	29 Oct 1935	AHY	120	Northwest
Terrace	29 Oct 1935	Alice Arm	13 Feb 1936	AHY	120	Northwest
Richmond	5 Nov 1936	Darrington[2]	27 Aug 1937	AHY	170	Southeast
Kaslo	7 Oct 1960	Nelson	19 Nov 1960	HY	50	South
Kaslo	16 Apr 1962	Fernie	Nov 1962	HY	150	East
Port Hardy	9 Dec 1940	Richmond	12 Aug 1941	AHY	360	Southeast
Cloverdale	8 Jun 1989	Horse Lake	11 Sep 1989	HY	250	North

[1] AHY = after hatch year (adult); HY = hatch year (immature)
[2] Washington state

Figure 204. Nest and eggs of the Steller's Jay (Langley, 27 April 1996; R. Wayne Campbell).

Figure 205. In parts of British Columbia, the Steller's Jay nests in undisturbed mixed coniferous forests (Manning Park, 21 April 1994; R. Wayne Campbell).

which occur in British Columbia. *C. s. stelleri* is generally resident on the coast from southeast Alaska south to Oregon; *C. s. annectens* is resident in the interior east of the Coast Mountains (Brooks 1927); *C. s. carlottae* is an endemic subspecies, resident on the Queen Charlotte Islands. *C.s. carlottae* has the most restricted distribution of the 6 subspecies, and probably a small total population. For details of the identifying features of the subspecies, see Stevenson (1934).

Infrequent hybridization between the Steller's Jay and the Blue Jay is thought by some authors to indicate that the two constitute a superspecies (American Ornithologists' Union 1983).

In 1987 the Steller's Jay was declared the official provincial bird of British Columbia. Over 80,000 residents participated in a vote to select a provincial bird to commemorate the centennial of wildlife conservation in Canada.

NOTEWORTHY RECORDS

Spring: Coastal – Saanich 4 May 1981-incubating; North Saanich 4 Apr 1961-59, 17 Apr 1984-35; Tofino Inlet 3 May 1974-12 in flock; North Surrey 4 Apr 1966-4 eggs; Harrison Lake 13 Apr 1988-26 in flock; North Vancouver 12 Apr 1930-4 eggs; Courtenay 3 May 1973-2 nestlings; Egmont 11 May 1979-22 on survey; Bella Coola 13 Mar 1985-15; Tasu 25 May 1990-1 (Campbell 1990d); Terrace 26 May 1977-12. **Interior** – Creston 4 Apr 1979-15; Oliver 14 Apr 1971-nest building; Okanagan Falls 29 Apr 1913-eggs (RBCM 898); Apex Mountain 23 May 1934-4 eggs; Cranbrook 6 Apr 1915-4 eggs (RBCM 897); Penticton (West Bench) 2 Mar 1980-10, 13 May 1987-5 nestlings; Balfour to Waneta 16 May 1981-36 on survey; Slocan City 14 Apr 1980-4 eggs; 4 May 1990-incubating; Kamloops 11 May 1973-3 fledged young; Celista 10 Apr 1968-10; Mount Revelstoke 13 Apr 1983-3 young; Rogers Pass 25 Mar 1983-103 on survey (Woods 1983); Hudson's Hope 1 Mar 1984-1; Pine Pass 11 May 1982-1.

Summer: Coastal – Goldstream River 24 Jul 1985-16; North Saanich 31 Aug 1967-83 in flock, 27 Aug 1994-121, including 60 in a flock; Port Renfrew 28 Jun 1974-2 young being fed by adults; Surrey 21 Jun 1960-4 nestlings ready to leave nest; Langley 28 Jun 1979-4 nestlings; Rose Harbour 10 Aug 1946-2; Moresby Camp 26 Aug 1979-4; Nesto Inlet 12 Jul 1946-fledgling; Langara Island 29 Jul 1930-1 immature male (ANS 101377); Lakelse Lake 21 Aug 1977-14; Stewart 28 Aug 1977-3. **Interior** – Penticton (West Bench) 20 Aug 1980-12 in flock; Nelson 25 Jun 1976-4 nestlings; Summerland 6 Jun 1966-4 eggs, 6 Jul 1966-2 young on nest; Balfour-Waneta 8 Jun 1983-33 on survey; Winlaw 8 Jun 1985-1 egg; Bolean Lake 3 July 1946-3 nestlings; Glacier National Park 30 Jul 1982-15; Rogers Pass 29 Jun 1977-2 nestlings, 1 egg (Merilees 1977); Quesnel Lake 28 Aug 1963-8; Marion Creek 21 July 1959-1; Hominca River valley 2 Jul 1976-1; Bijoux Falls 1 Jun 1992-9; s Pine Pass 6 Jun 1976-6, 7 Jul 1975-2 fledglings being fed by parents; East Pine 9 Jun 1954-1 (NMC 480401; Jobin 1955); Peace Canyon 25 Aug 1986-1; Hyland Post 22 Aug 1976-1; s Cassiar 13 Jun 1972-2.

Breeding Bird Surveys: Coastal – Recorded from 22 of 27 routes and on 71% of all surveys. Maxima: Port Renfrew 27 Jun 1973-42; Port Hardy 2 Jul 1989-31; Kwinitsa 29 Jun 1980-23. **Interior** – Recorded from 37 of 73 routes and on 18% of all surveys. Maxima: Gosnell 29 Jun 1974-10; Illecillewaet 17 Jun 1982-10; Lavington 15 Jun 1981-8; Pavilion 6 Jun 1993-8; Salmo 14 Jun 1975-5; Gerrard 24 Jun 1979-5.

Autumn: Interior – n Rainy Hollow 8 Sep 1972-1; Fort St. John 28 Sep to 12 Nov 1983-1 (BC Photo 872; Campbell 1984b); Eagle Lake 7 Sep 1992-1; Chilcotin River 24 Nov 1987-10; Wells Gray Park 18 Sep 1959-15; Mount Averil 15 Oct 1983-10; Golden 31 Sep 1982-10; Nelson 20 Nov 1978-25 at feeder. **Coastal** – Bob Quinn Lake 11 Oct 1983-3; Miles 52 to 54 Haines Highway 8 Sep 1972-1 heard; Quadra Island Oct 1977-42; Mission 11 Oct 1965-50; Maple Ridge 24 Sep 1967-51; Fort Langley 25 Sep 1967-43; West Vancouver 20 Sep 1986-150 on survey, 30 Oct 1969-1 full albino (Campbell 1970a); Long Beach 12 Sep 1979-50 on survey; North Saanich 5 Sep 1960-190, 5 Sep 1986-45 in flock, 13 Sep 1967-163; Central Saanich 30 Oct 1983-100; Ten Mile Point (Saanich) 5 Sep 1992-218; Weir Point 25 Sep 1983-36; Rocky Point (Metchosin) 28 Sep 1986-40; Sooke 15 Sep 1977-40 in flock.

Winter: Interior – Bob Quinn Lake 19 Jan 1981-1; Pine Pass 1 Jan 1983-2; Kispiox River 24 Jan 1979-2; Terrace 2 Jan 1972-4 on Christmas Bird Count (Frank 1972), 8 Dec 1986-18; Wells 28 Feb 1979-2; Williams Lake 2 Jan 1995-27; Kleena Kleene 31 Dec 1950-1 (Paul 1959); Dewey 29 Dec 1948-1 (UBC 2187); Revelstoke 20 Dec 1984-32; Salmo 1 Jan 1980-7; Nelson 4 Feb 1979-39 at feeder; Kinnaird 31 Dec 1968-1; Penticton (West Bench) 1 Jan 1970-2; Vaseux Lake 31 Dec 1962-1. **Coastal** – Graham Island 1 Jan 1918-1 (ROM 50255); Skidegate 30 Dec 1982-3; Port Neville 1 Jan 1976-1; Lower Adam River 19 Dec 1983-3; Alta Lake 30 Dec 1931-1 (UBC 9344); Little Qualicum River mouth 24 Feb 1976-56; Harrison Lake 19 Jan 1987-12, 1 Feb 1975-24; Agassiz-Yale 11 Dec 1977-40 on survey along highway; Ucluelet 3 Jan 1976-1; Pachena Bay 19 Jan 1976-7; Sooke 1 Jan 1976-1; Saanich 7 Dec 1995-4 with 1 Blue Jay; Rocky Point (Victoria) 31 Dec 1978-2.

Christmas Bird Counts: Interior – Recorded from 25 of 27 localities and on 85% of all counts. Maxima: Penticton 27 Dec 1992-127; Revelstoke 18 Dec 1985-119; Nakusp 3 Jan 1986-79; Vaseux Lake 2 Jan 1993-79. **Coastal** – Recorded from 31 of 33 localities and on 86% of all counts. Maxima: Victoria 19 Dec 1992-**659**, all-time Canadian high count (Monroe 1993); Vancouver 20 Dec 1992-317; Duncan 2 Jan 1993-293.

Blue Jay

BLJA

Cyanocitta cristata (Linnaeus)

RANGE: Resident from extreme northeastern and southeastern British Columbia, southeastern Alberta, central Saskatchewan, central Manitoba, southern Ontario, southern Quebec, and the Maritime provinces south in the United States to Texas and the Gulf coast and west to the eastern edge of the Rocky Mountains.

STATUS: *Uncommon* resident in the Peace Lowland of the Boreal Plains Ecoprovince; *rare* resident in the east and west Kootenay regions of the Southern Interior Mountains Ecoprovince; *very rare* migrant and winter visitant across the southern areas of the province from the southwest mainland coast to the Okanagan valley; *casual* elsewhere.

Breeds.

CHANGE IN STATUS: The Blue Jay (Fig. 206) is a recent immigrant to British Columbia. It was not found during extensive field work in the Peace River area of the Boreal Plains in 1938 (Cowan 1939), nor was it included in the bird fauna of the province by Munro and Cowan (1947). The first confirmed observation of the Blue Jay in British Columbia, west of the Rocky Mountains, occurred on 26 December 1948, when 3 were seen during the Vancouver Christmas Bird Count. The second occurrence was in April 1951, when a single bird was seen at Clo-oose, on the west coast of Vancouver Island. These were followed by a sighting at Powell River in November 1953 and another at Victoria in 1963. A single bird appeared at Sechelt in December 1970, and 2 were present at Victoria from October 1972 to April 1973.

These early appearances of the Blue Jay in British Columbia were followed by intermittent records of jays at localities across the province west of the Rocky Mountains (Fig. 207). More occurred in the Southern Interior Mountains than in any other region of the province. From the first arrivals until 1990, more than three-quarters of all records were in autumn and winter, and only 2% were in the summer months. There was no indication of nesting in southern British Columbia until 1991.

While these events were proceeding in southern British Columbia, a separate movement of the Blue Jay, westward from its earlier range, occurred into the Boreal Plains. This region is contiguous with an area of Alberta into which the jays had been expanding since the 1950s (Salt and Wilk 1958). The first record in the Peace River area of British Columbia was of 5 birds at Flatrock on 30 October 1951. Twenty-two years passed before the next documented occurrence of the Blue Jay in the Peace River region. Then in 1973 and 1974 there were records of single birds along the Peace River, followed by 3 at Tupper in 1975, 3 separate localities in 1976, and just 1 or 2 records a year until 1982 (Fig. 207). As in the south, these early, sporadic appearances were followed by an increase in records of the Blue Jay starting in 1982. This apparent increase is synchronous with the arrival in the

Figure 206. Since its arrival in British Columbia in 1948, the Blue Jay has become established locally as a breeding species and continues to expand its range in the province (Nakusp, April 1991; Gary S. Davidson).

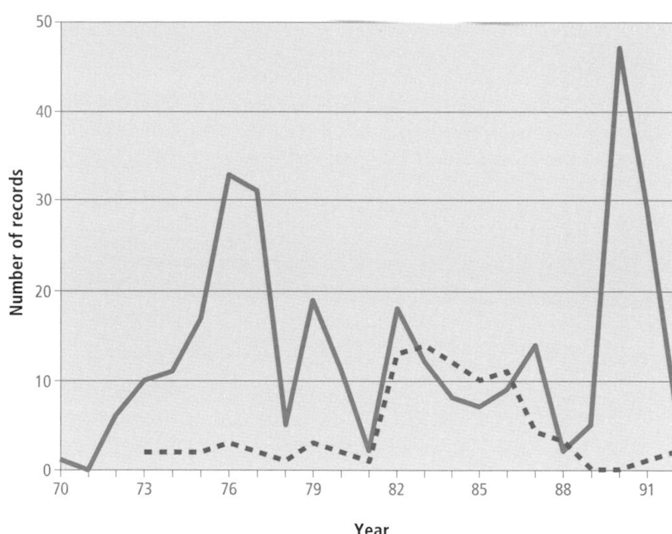

Figure 207. Fluctuations in the numbers of Blue Jay records reported in British Columbia from west of the Rocky Mountains (solid line) and from the Boreal Plains Ecoprovince (dashed line), 1970 to 1992.

region of an ardent student of birds, who searched intensively and documented many records annually between then and his departure in 1991. Thus it is uncertain whether the apparent increase in 1982 was real (Fig. 207).

In the Peace Lowland, 2 nests have been found and others strongly indicated. Evidence suggests that the Blue Jay has established itself as a breeding bird there, whereas this cannot yet be said of the southern invasion, where only 1 nest has been reported.

Data suggest that the 2 invasions were separate in timing. In the north, the jays quite rapidly occupied the potentially suitable habitat and nested successfully, whereas in the south, the area of occurrence is continuing to expand but has failed to result in a new breeding population. Perhaps the

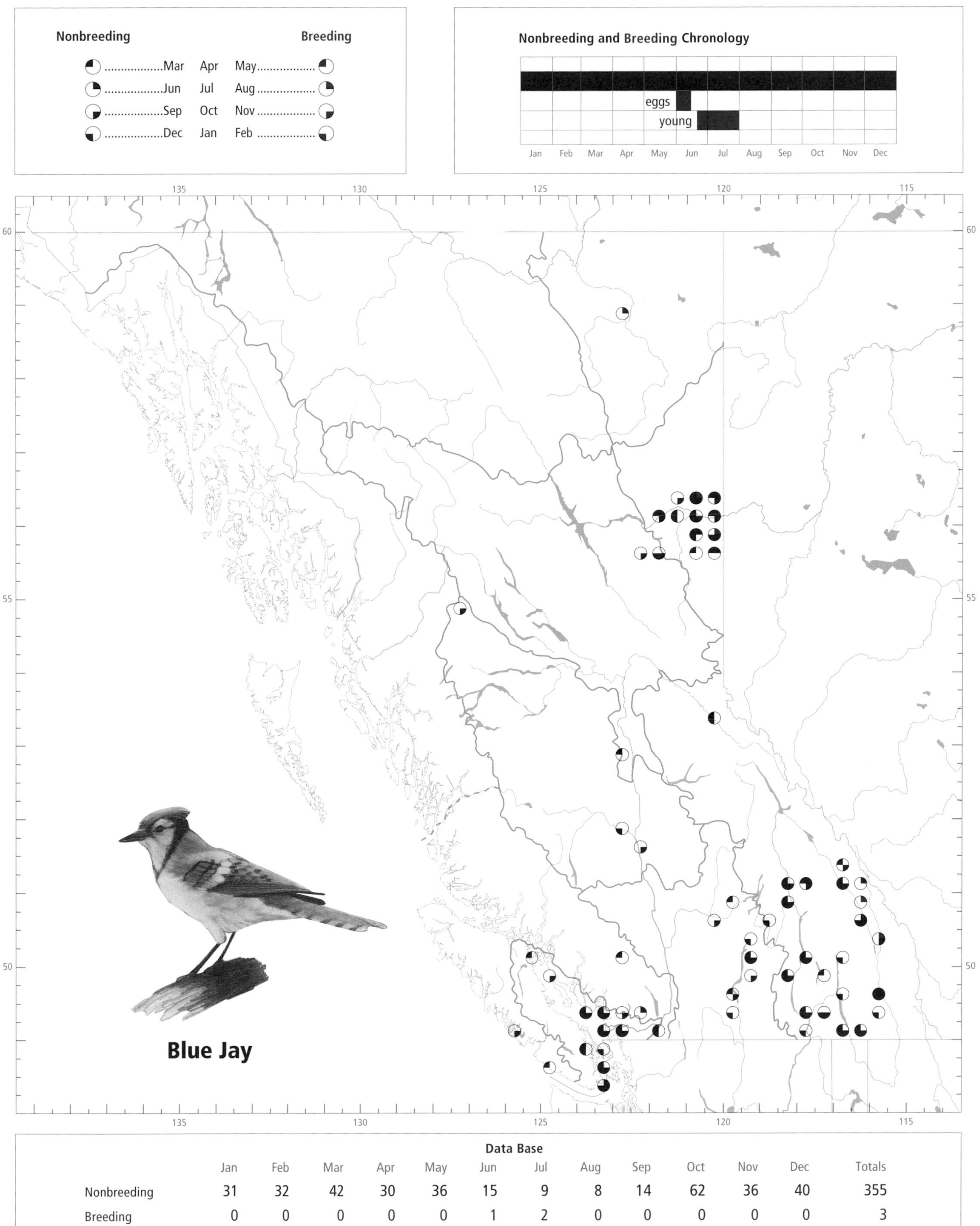

Data Base

	Jan	Feb	Mar	Apr	May	Jun	Jul	Aug	Sep	Oct	Nov	Dec	Totals
Nonbreeding	31	32	42	30	36	15	9	8	14	62	36	40	355
Breeding	0	0	0	0	0	1	2	0	0	0	0	0	3

presence of the Steller's Jay in the south has been a factor in preventing the establishment of the Blue Jay as a breeding species.

NONBREEDING: The Blue Jay is a sparsely distributed resident in the Peace Lowland of the Boreal Plains and possibly in the Rocky Mountain Trench of the east Kootenay. In the Boreal Plains, it occurs from Tumbler Ridge north to Rose Prairie, and from the Alberta border west to Hudson's Hope, Pine Pass, and the Chetwynd area of the Sub-Boreal Interior Ecoprovince. In the northern Kootenays, this jay is restricted to corridors along highways and in towns and villages between Radium and Golden, and west to Revelstoke.

In southernmost British Columbia, the Blue Jay is a scarce winter visitant in the southern east Kootenay, west Kootenay, and Thompson and Okanagan valleys. In extreme southwestern British Columbia, it is mainly an irregular winter visitant, frequenting backyard feeders.

The northernmost occurrence in the province has been at Fort Nelson, while in the west it has been found at Smithers. The centre of abundance in winter is in the Peace Lowland (Fig. 208).

Most observations of the Blue Jay in British Columbia have been from sea level to about 930 m. In natural situations, the Blue Jay frequents low-elevation, relatively open mixed deciduous-coniferous forests, aspen copses, and riparian growth at the edges of thicker coniferous forests or along ponds and rivers. These habitats are somewhat similar to the open, deciduous woodlands dominated by oaks and beeches that they favour in eastern North America (Goodwin 1976) but that are of limited distribution in British Columbia. The

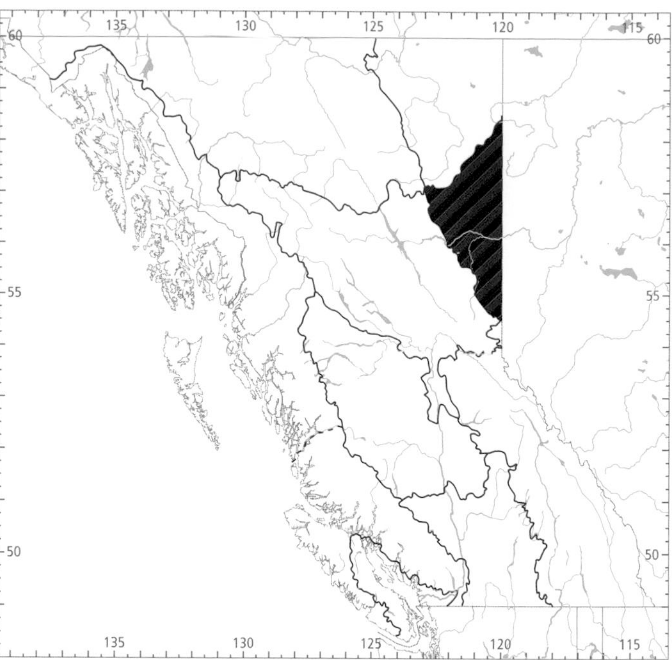

Figure 208. In British Columbia, the highest numbers for the Blue Jay in winter (black) and summer (red) occur in the Boreal Plains Ecoprovince.

Blue Jay does not occur in the more closed-canopy coniferous forests favoured by the Steller's Jay.

In British Columbia, the Blue Jay occupies areas where urban or semi-urban development predominates. Throughout southern areas, most Blue Jays are found near backyard bird feeders, where they usually remain from autumn through spring if a good supply of food is provided. It is likely that

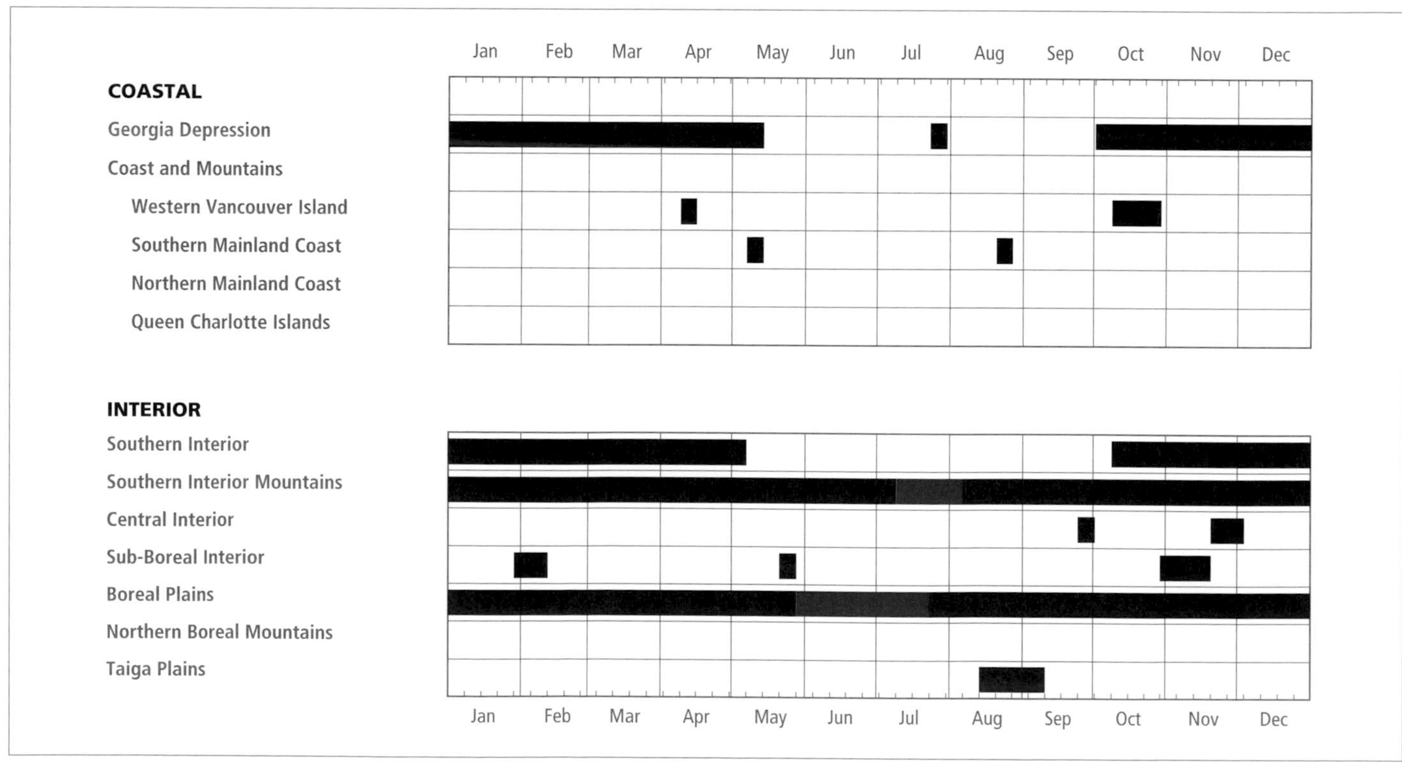

Figure 209. Annual occurrence (black) and breeding chronology (red) for the Blue Jay in ecoprovinces of British Columbia.

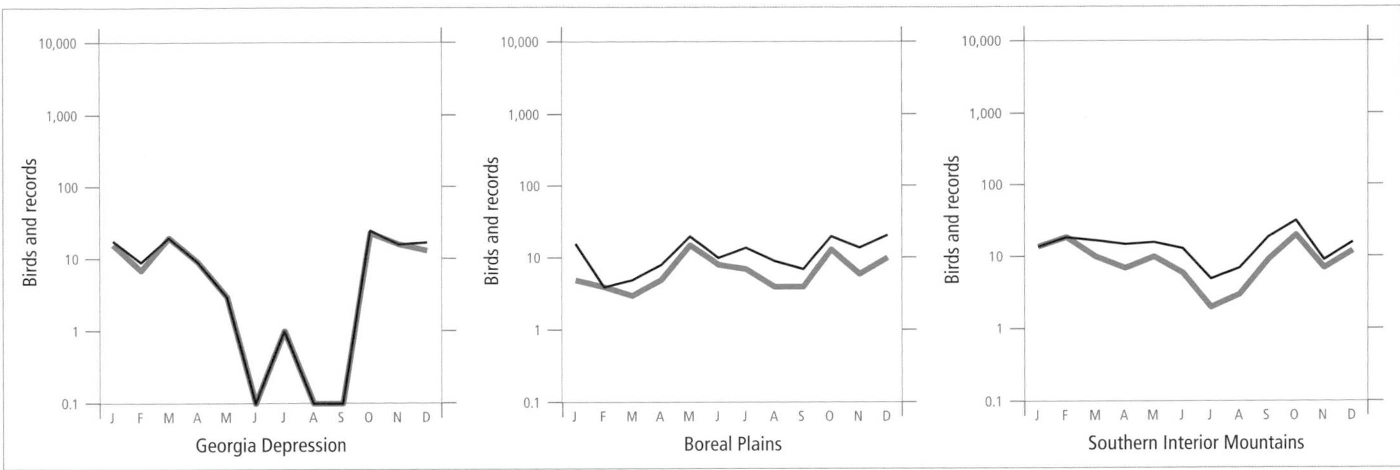

Figure 210. Fluctuations in total number of birds (purple line) and total number of records (green line) for the Blue Jay in ecoprovinces of British Columbia. Christmas Bird Count and nest record data have been excluded.

wintering birds have been encouraged to remain into the nesting season in the Kootenay region because of feeders. In the Peace River region, this jay also winters mainly in the vicinity of farmhouses, barnyards, local parks, and residential neighbourhoods, before dispersing for the summer.

Migration is difficult to discern because of the small number of birds occurring in the province. On the coast, where breeding has not yet been documented, virtually all wintering birds have left by the end of May; summer records are extremely rare on the coast. Wintering birds arrive mainly in October (Figs. 209 and 210).

In the southeastern interior of the province, wintering birds begin dispersing in March, reaching a minimum in July and August. Numbers begin to increase in late August and continue through October, when they again fall (Fig. 210). In southern British Columbia this pattern of autumn arrival, winter presence, and spring departure suggests that the province is a migration corridor and winter range for a small migratory population, but there is no evidence as to its source. What evidence there is suggests that the migration is east-west rather than north-south. An observer who lived in Field between 1975 and 1977 comments that the Blue Jays appeared to be moving westward in the autumn following the Trans-Canada Highway corridor, and returned eastward along this route in the spring (R.R. Howie pers. comm.) (Fig. 210) (see also REMARKS).

In the Boreal Plains, numbers begin to increase in March and continue through May, suggesting an influx of birds from Alberta. Their numbers remain relatively stable throughout the summer, dropping slightly through September and then increasing again in October (Fig. 210). The origin of this autumn influx is unknown.

Local winter populations are well documented by Christmas Bird Counts (Fig. 211), since Blue Jays still make a notable sighting and, as they tend to stay near feeders, nearly all individuals can be counted. Appendix 1 gives a summary of Blue Jay occurrences for all Christmas Bird Counts in British Columbia since 1980; the highest total count of Blue Jays was 55 in 1990.

The Blue Jay has been reported throughout the year in British Columbia (Fig. 209).

BREEDING: The Blue Jay is present in summer in the Peace Lowland from the Alberta border west to Hudson's Hope, and from Tupper north to at least Boundary Lake. Although actual breeding records are few, its nests are probably scattered throughout the area.

In the east Kootenay, breeding has been confirmed only near Windermere, where a family of 3 newly fledged Blue Jays was seen on 31 July and again on 1 August 1991 (Siddle 1992c). This species is present in summer near human-influenced areas along the Columbia River valley from Invermere north to Golden and in the Rogers Pass area.

Breeding Bird Surveys for both coastal and interior routes for the period 1968 through 1993 contain insufficient data for analysis.

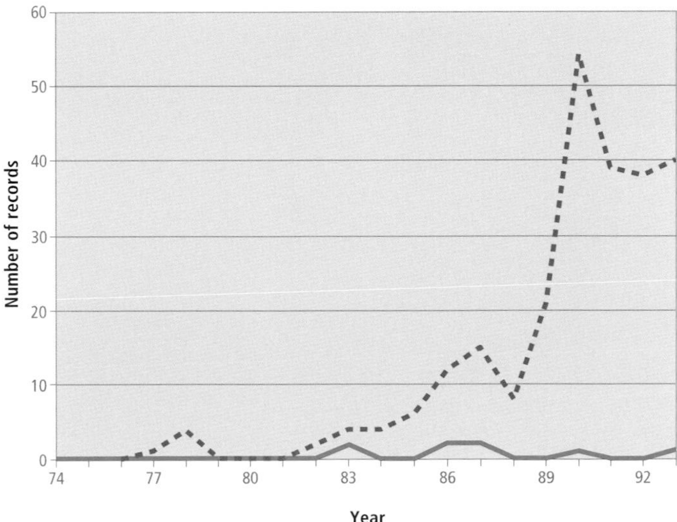

Figure 211. Fluctuations in the number of Blue Jays reported on Christmas Bird Counts in British Columbia, 1974 to 1993. Coastal counts (solid line) and interior counts (dashed line) are shown.

In summer, the Blue Jay frequents the edges of pure and mixed deciduous and deciduous-coniferous forests. It seems to prefer relatively open (Fig. 212), as opposed to dense, stands of trees. It also visits agricultural areas, especially those with patches of tall brush. Occasionally it is a summer visitor to feeders in residential areas.

The Blue Jay has been recorded breeding in British Columbia from 30 May (calculated) to 31 July (Fig. 209).

Nests: Only 1 nest has been found (North Pine, in 1989). It was described as a "bulky stick nest located in the crotch of a spruce tree about 5.6 m above the ground."

Eggs: A single nest containing 5 "well incubated" eggs was found on 14 June. Calculated dates indicate that eggs were in the nest as early as 30 May, although observations of adults gathering nesting material on 8 May, near Pouce Coupe, suggest that eggs may be present even earlier in British Columbia. In southern Ontario, egg dates ranged between 15 April and 8 July, with the main period between 15 and 29 May (Peck and James 1987).

In southern Ontario, the Blue Jay lays from 1 to 6 eggs, with most clutches containing 4 or 5 eggs. The incubation period is 13 to 18 days, and evidence suggests that incubation begins before the last eggs are laid (Peck and James 1987).

Young: Recently fledged young, still with adults, have been found in the Peace River region from 6 to 16 July, and in the east Kootenay on 31 July. Calculated dates indicate that young can occur as early as 20 June.

The nestling period is stated to be from 17 to 21 days or 15 to 18 days (Bent 1946, quoting separate sources).

Brown-headed Cowbird Parasitism: Cowbird parasitism was not found in British Columbia. The Blue Jay is a rejector species, and is therefore rarely reported as a victim of the cowbird (Friedmann et al. 1977).

REMARKS: On the fragile basis of a single apparent hybrid between *Cyanocitta cristata* and *C. stelleri*, found in Colorado, some authors have suggested regarding these 2 species as a superspecies (American Ornithologists' Union 1983). A jay photographed at Osoyoos, British Columbia, in 1990 has some characteristics that might be derived from hybridization between these 2 species, but those who have studied the videotaped record consider the bird to be a leucistic Steller's Jay.

The original range of the Blue Jay was the forested regions of North America, primarily east of the Great Plains, but occurrences west and north of this range were noted in the late 1800s and early 1900s (Bent 1946). Urbanization,

Figure 212. The Blue Jay breeds near the edge of open, rather than dense, trembling aspen woodlands. Understorey vegetation may include willow, rose, high-bush cranberry, and fireweed. The first breeding record for British Columbia was from this site near Cecil Lake, east of Fort St. John, in 1986 (R. Wayne Campbell).

especially the growth of "urban forests" and provision of winter food at bird feeders, is postulated as the explanation for the dispersal of Blue Jays across the Great Plains (Bock and Lepthien 1976). It was first noted in Washington in 1951 (Hudson 1951) and Oregon in 1973 (Gashwiler and Gashwiler 1974). By the late 1960s it was a casual visitor, but increasingly so, in eastern Washington (Alcorn 1971). Sightings became more regular in Washington in the early 1970s (Fitzner and Woodley 1976), leading up to a major influx of Blue Jays into the Pacific Northwest in the autumn and winter of 1976-77 (Van Horn and Toweill 1977; Weber 1977).

In British Columbia, Blue Jays occurring in the Boreal Plains are part of the range extension of this species in Alberta. They were present in central Alberta in 1946 (Bent 1946) and in the Calgary and Edmonton areas in the early 1960s (Sadler and Myres 1976), and reached the Boreal Plains on the Alberta side of the interprovincial boundary at the same time that they entered the contiguous area of British Columbia (Salt and Wilk 1958). There is no biological boundary to inhibit free movement of the Blue Jay between the 2 provinces.

The evidence leaves little doubt that in British Columbia the Blue Jay does cross the Rocky Mountains, apparently following the corridors of transportation with their attendant communities. It is significant that the influx into the Pacific Northwest of the United States was synchronous with that into southern British Columbia. The process was probably the same: westward movement through the Rocky Mountains and other western ranges.

Low food supplies have been correlated with mass movements of Blue Jays in the eastern United States (e.g., mass movements in 1939 and 1962 associated with mast crop failures [Broun 1941]), and this may explain periodic incursions into the Pacific Northwest (see Christmas Bird Count data).

NOTEWORTHY RECORDS

Spring: Coastal – Saanich 22 Apr 1973-1, did not stay (Crowell and Nehls 1973c); Clo-oose 10 Apr 1951-1 (Irving 1953); Duncan 3 Mar to 2 Apr 1977-1; Ladner 3 to 11 Mar 1990-1 (Campbell 1990b), 22 Apr 1990-1, did not stay; Chilliwack 27 Apr 1980-1; Vancouver 4 Apr 1975-1 (BC Photo 402); West Vancouver 8 May 1984-1, 10 and 11 May 1973-1; Green Lake (Whistler) 11 and 12 May 1975-1 (BC Photo 500). **Interior** – Creston 24 Mar 1974-1 (BC Photo 553); Summerland 6 Mar 1979-1; Nakusp 31 Mar to 27 Apr 1991-3 (Siddle 1991c), 4 May 1983-5; Coldstream 3 May 1978-1; Revelstoke 18 Mar 1977-3; Mount Revelstoke 27 Apr 1983-4, 4 May 1983-5; Newlands (50 km ne Prince George) 1 Apr 1992-1 (Bowling 1992); Peace River 14 and 30 May 1973-1 (Penner 1976); Fort St. John 25 Mar 1984-3, 12 May 1979-4 (Siddle 1982); Bissette Creek (w Pouce Coupe) 8 May 1993-2 adults gathering nesting material; Montney 24 Apr 1982-4.

Summer: Coastal – Victoria 23 Jul 1973-1; Alouette Lake 26 Aug 1978-2. **Interior** – Windermere 31 Jul and 1 Aug 1991-3 newly fledged young; Mount Revelstoke 25 Aug to 30 Nov 1982-5 maximum; 4 km w Tupper 4 Aug 1975-3; Dawson Creek 8 Jul 1993-2 adults with 3 young of year; Clayhurst 22 Jun 1976-1 (RBCM 15329); Cecil Lake 16 Jul 1986-2 adults with 3 flying young and another with pin feathers on aspen branch nearby (Campbell and Petrar 1986); North Pine 14 Jun 1989-nest with 5 eggs collected; Stoddart Creek 31 Aug 1985-4; Fort St. John 6 Jul 1988-first fledglings at feeder (Siddle 1988d); Fort Nelson 17 Aug to 12 Sep 1994-1 at feeder.

Breeding Bird Surveys: Coastal – Not recorded. **Interior** – Recorded from 2 of 73 routes and on less than 1% of all surveys. Maxima: Tupper 20 Jun 1994-3; Fort St. John 29 Jun 1987-1.

Autumn: Interior – Flatrock 30 Oct 1951-5; Fort St. John 28 Nov 1982-5 (Siddle 1982); Hudson's Hope 8 Sep 1979-4; 29 km e Pine Pass 29 Oct 1978-1; Smithers 29 Sep to 2 Oct 1991-1 (Siddle 1992a); Riske Creek 22 Nov to 1 Dec 1984-1, finally killed by cat; Mount Revelstoke 29 Sep 1982-4, 14 Oct 1981-5; Revelstoke 24 Oct 1976 to 18 Mar 1977-up to 3 birds seen regularly in downtown area; Nakusp 28 Oct 1990-3; Invermere 12 Nov 1989-1 photographed at feeder (Campbell 1990b); Windermere 17 Sep 1990-1; Vernon Commonage 8 Oct 1985-1; Kelowna 20 Nov to 4 Dec 1987-1 at feeder (Campbell 1988a), 2 Nov 1991-2 (Siddle 1992a); Kalamalka Lake 30 Nov 1977-1 (BC Photo 532); Flathead River 1 Oct 1993-4; Creston 9 to 18 Nov 1987-1 (Campbell 1988a). **Coastal** – Powell River 30 Nov 1953-1 at feeder; Tofino 14 Oct 1987-1; Long Beach 20 and 22 Oct 1977-1 at Incinerator Point (BC Photo 486); Gibsons 15 Nov 1985-1 (Campbell 1986a); Burnaby 6 Oct 1983-1; North Saanich 20 Nov 1975-1 (BC Photo 450); Saanich 4 Oct 1994-2; Victoria 13 Oct 1972-2, 6 Nov 1963-1; Ten Mile Point (Saanich) 5 Sep 1992-1 (Siddle 1993a).

Winter: Interior – Fort St. John 23 Dec 1985-4, 1 and 29 Jan 1982-4 (Siddle 1982); Stoddart Creek 1 Jan 1984-3, 23 Jan 1986-4; North Pine 16 Dec 1989-20; Kiskatinaw River 9 Feb 1986-1; Riske Creek 1 Dec 1984-1; 10 km n Golden 14 Oct to 1 Nov 1995-1; Golden 27 Dec 1990-5 (E. Zimmerman pers. comm. 1995); Invermere 3 Jan 1985-1 (BC Photo, 995); Nakusp 2 Feb 1991-1; Summerland 14 Dec 1979-1; Penticton 26 Dec 1979-1; Nelson 14 Jan 1982-1 (BC Photo 1034); 31 Dec 1981-1; Argenta 17 Jan 1978-1 (BC Photo 517); Grand Forks 11 Dec 1990-1 (Campbell 1990b). **Coastal** – Sechelt 17 Dec 1070-1 in garden; Gibsons 29 Jan to 15 Mar 1986-1 (Campbell 1986b); Chilliwack 20 Dec 1986-2; Vancouver 1 Jan 1988-1; Delta 5 Jan 1984-1, 26 Dec 1948-1 on Christmas Bird Count (Middleton 1949), 27 Dec 1983-2; Victoria 10 Feb 1973-2 (BC Photo 281).

Christmas Bird Counts: Interior – Recorded from 11 of 27 localities and on 10% of all counts. Maxima: North Pine 16 Dec 1990-41; Creston 30 Dec 1990-9; Fort St. John 24 Dec 1978-4. **Coastal** – Recorded from 6 of 33 localities and on 1% of all counts. Maxima: Ladner 27 Dec 1983-2; Chilliwack 20 Dec 1986-2; Vancouver 20 Dec 1987-2; Victoria 21 Dec 1963-1; Sunshine Coast 15 Dec 1990-1; Parksville–Qualicum Beach 2 Jan 1994-1.

Clark's Nutcracker

CLNU

Nucifraga columbiana (Wilson)

RANGE: Resident from central British Columbia, southwestern Alberta, western and central Montana, and western and southeastern Wyoming south through the mountains of the western United States to northern Baja California, east-central Arizona, and southern New Mexico.

STATUS: On the coast, an *uncommon* resident locally on the southern mainland of the Coast and Mountains Ecoprovince. *Rare* and erratic in the Georgia Depression Ecoprovince; *very rare* autumn and winter vagrant to Western Vancouver Island and the northern mainland of the Coast and Mountains; *casual* on the Queen Charlotte Islands.

In the interior, an *uncommon* to locally *fairly common* (occasionally *common*) resident across the southern portions of the province, including the Southern Interior, Southern Interior Mountains, and Central Interior ecoprovinces; *very rare* in the Sub-Boreal Interior Ecoprovince; *casual* in the Northern Boreal Mountains Ecoprovince. *Accidental* in the Boreal Plains and Taiga Plains ecoprovinces of northeastern British Columbia.

Breeds.

Figure 213. Clark's Nutcracker (Manning Park, 24 September 1994; R. Wayne Campbell).

NONBREEDING: The Clark's Nutcracker (Fig. 213) is widely distributed throughout southern and much of central British Columbia. It is mainly a bird of the interior, where it is resident from the southeastern slopes of the Coast Mountains east to the British Columbia–Alberta boundary and north through much of the Central Interior and Cariboo Mountains. Numbers decline rapidly north of Quesnel (latitude 53°N). There are few records from the Sub-Boreal Interior, and wandering birds appear only occasionally in the extreme northern portions of the province.

The western boundary of the normal distribution of the Clark's Nutcracker lies along the eastern slope of the Coast Ranges from Liumchen Mountain (southern Coast and Mountains) and Manning Park (Southern Interior) north to the mountains adjacent to Kleena Kleene and the Rainbow

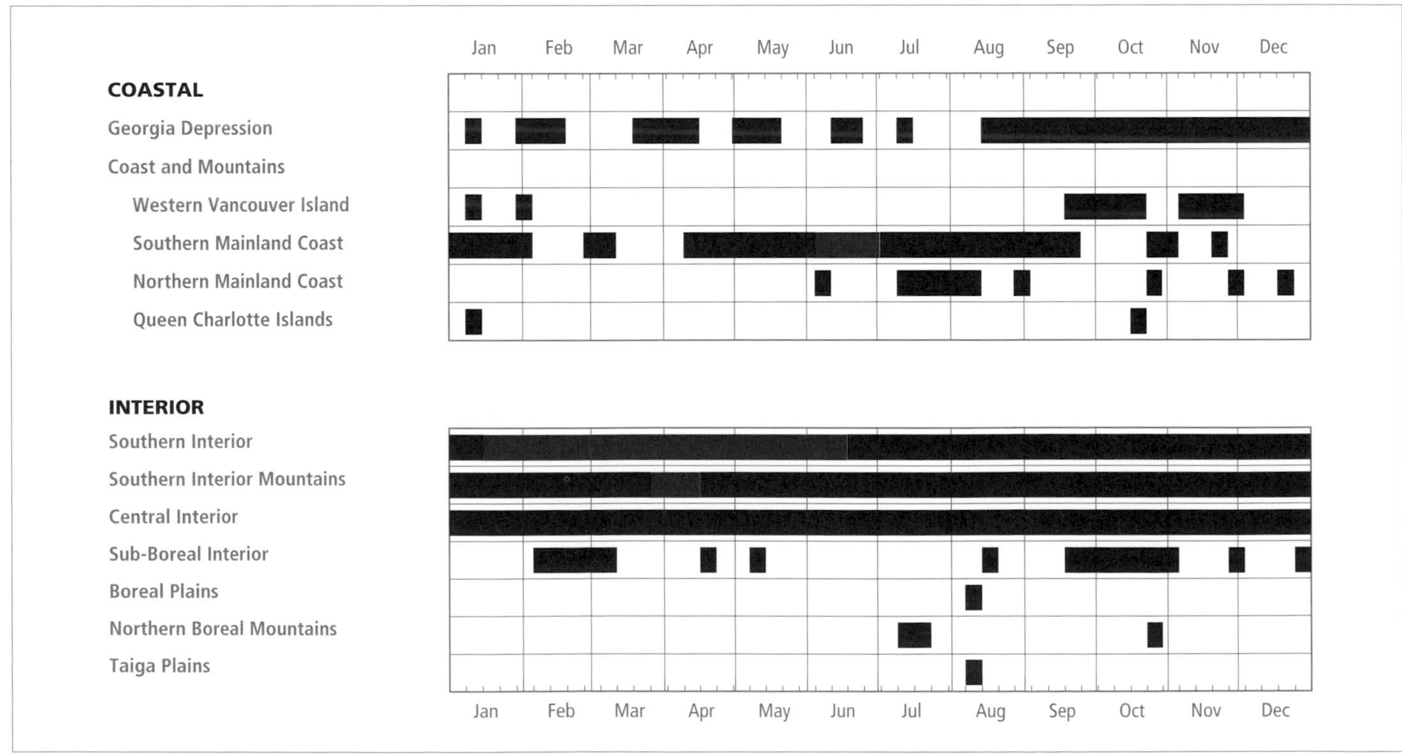

Figure 214. Annual occurrence (black) and breeding chronology (red) for the Clark's Nutcracker in ecoprovinces of British Columbia.

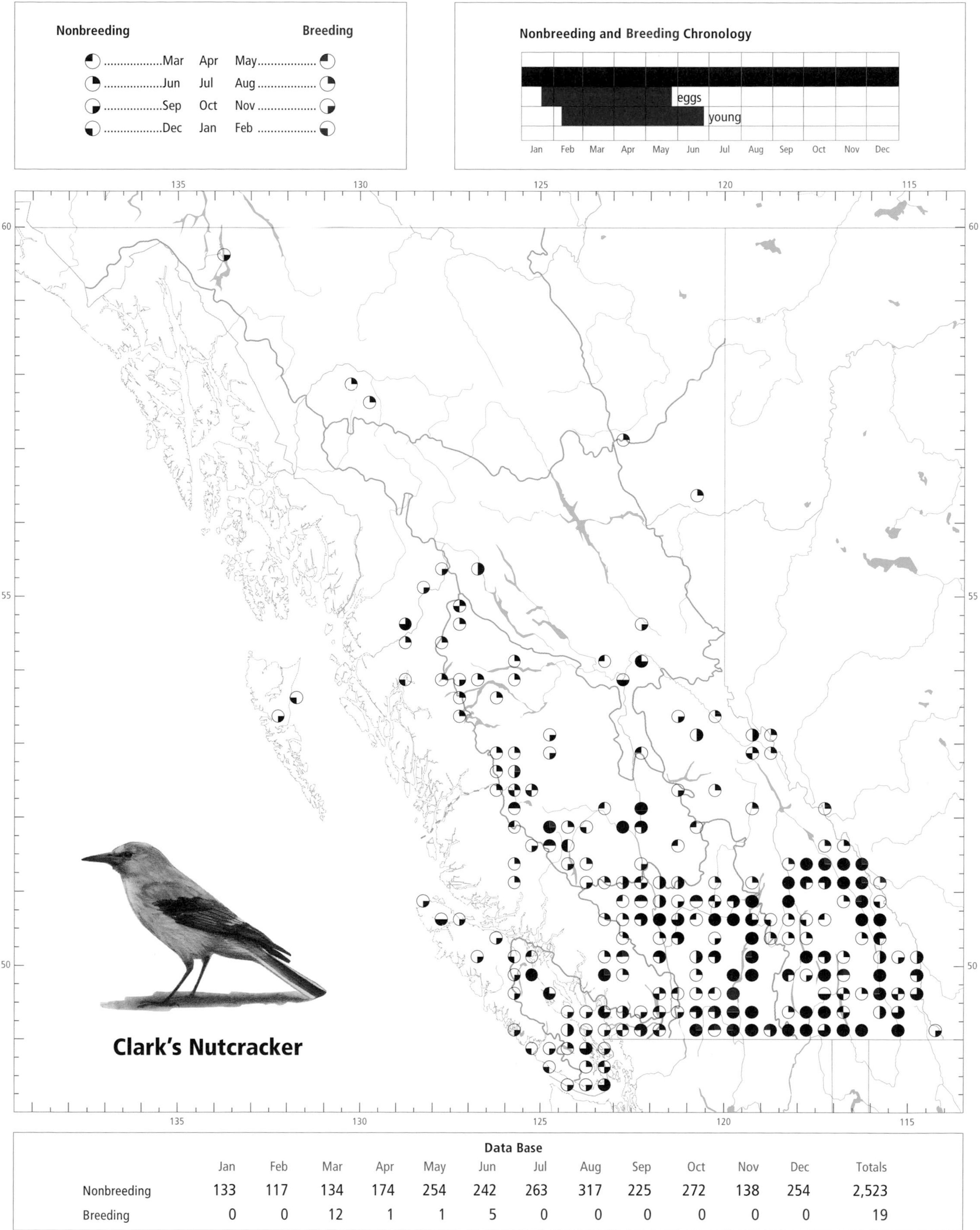

Data Base

	Jan	Feb	Mar	Apr	May	Jun	Jul	Aug	Sep	Oct	Nov	Dec	Totals
Nonbreeding	133	117	134	174	254	242	263	317	225	272	138	254	2,523
Breeding	0	0	12	1	1	5	0	0	0	0	0	0	19

Mountains of the western Chilcotin in the Central Interior. The nutcracker occurs irregularly and seasonally elsewhere in the interior (Fig. 214).

Although there are many records of small numbers of birds from the Georgia Depression and Western Vancouver Island, and a few from the northern Coast and Mountains and the Queen Charlotte Islands, the Clark's Nutcracker is only an erratic vagrant to the maritime environment. Over 70% of coastal records (n = 177) are of single birds.

The highest numbers in winter have been recorded from the Southern Interior (Fig. 215), primarily the Okanagan valley; however, few bird observers have been studying the upper subalpine areas of the province in winter.

The Clark's Nutcracker has been recorded at elevations from near sea level on the coast to 2,600 m in the interior. It occurs mainly in mountainous regions of the province with mature, relatively open coniferous forests. Distribution throughout its range in North America is very closely linked to that of the large-seeded pines since nutcrackers feed mainly on their seeds (Tomback 1978a; see REMARKS).

In interior British Columbia, 2 tree species appear to influence the distribution of the Clark's Nutcracker: whitebark pine (Fig. 216) and ponderosa pine. The whitebark pine occurs the furthest north of the large-seeded pines and is the most important tree species in determining the distributional limits of the nutcracker in British Columbia. This pine occurs in the drier regions of the Engelmann Spruce–Subalpine Fir Zone between 900 m and 2,200 m (Anonymous 1956), and most of the nutcracker population occurs there during much

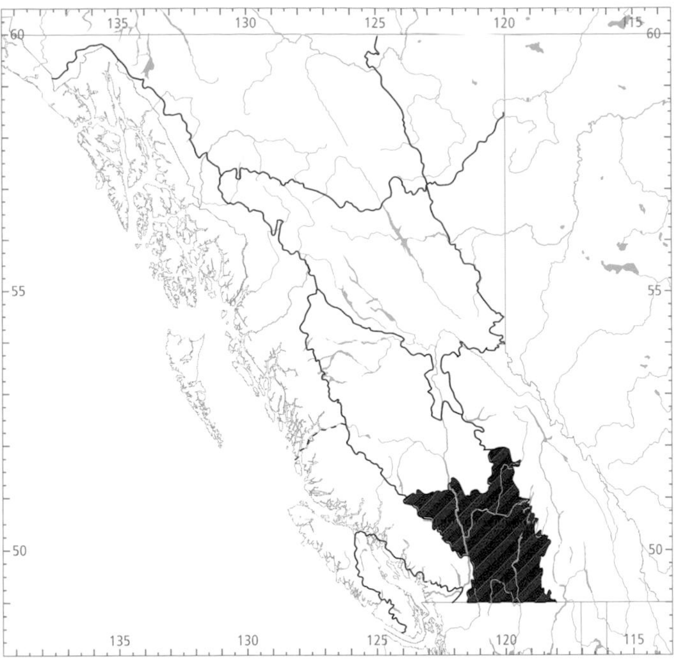

Figure 215. In British Columbia, the highest numbers for the Clark's Nutcracker in winter (black) and summer (red) occur in the Southern Interior Ecoprovince.

of the year. Ponderosa pine is locally important as a food source in parts of the Kootenays and in the Okanagan and Similkameen valleys. The numbers of Clark's Nutcrackers found in a particular area of the province appear to be influenced by the seed-crop production of the whitebark and

Figure 216. In the interior of British Columbia, the distribution of the Clark's Nutcracker is closely linked to that of the whitebark pine (Yohetta Mountain Pass in the Central Interior Ecoprovince, 19 July 1993; R. Wayne Campbell).

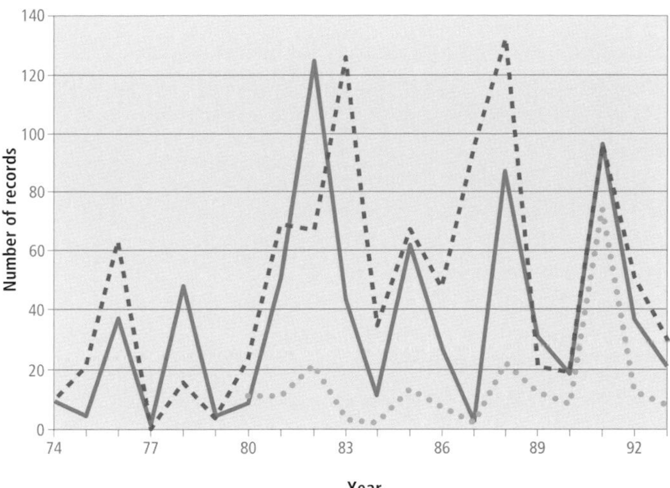

Figure 217. Fluctuations in Clark's Nutcracker numbers on Christmas Bird Counts in the Okanagan valley (Penticton, dashed line; Vaseux, solid line; Oliver-Osoyoos, dotted line) between 1974 and 1993. Irruptive movements can be seen in 1976, 1978, 1982, 1983, 1985, 1988, and 1991.

ponderosa pines. This means that a locality may have many nutcrackers one year and few or none the next.

The Clark's Nutcracker also uses human-influenced habitats such as campgrounds, picnic grounds, ski resorts, garbage dumps, agricultural fields, and residential areas where bird feeders occur. These habitats are used more frequently in autumn and winter, and during years of conifer seed-crop failures. On the mountain slopes, the nutcracker supplements its diet of conifer seeds by searching big-game winter ranges and scavenging carcasses (Edwards 1956).

Much of the nutcracker population frequents subalpine areas from early spring to autumn (Fig. 216). There, in the autumn months, they harvest the pine seeds, sometimes flying long distances to cache the seed in the ground on open slopes such as subalpine meadows, ridges, rocky outcrops, avalanche paths, and other areas that have limited winter snow accumulation and experience early spring melt-off. These caches provide a food supply through the winter and through the nesting season of the following year, including the post-fledging period (Vander Wall and Balda 1977; Tomback 1980; Hutchins 1989; Tomback et al. 1990).

Although a few birds remain near the timberline in at least some winters (e.g., 2,150 m; Cannings et al. 1987), many of the nutcrackers move to other habitats. These include mid-elevation, often steep-sloped, open coniferous forests, snag-filled burns, or rocky slopes, where old-growth stands of ponderosa pine interspersed with Douglas-fir provide most of the food requirements of this bird.

The Clark's Nutcracker is not migratory in the normal sense, although in some years individuals wander widely during the nonbreeding seasons. Regular elevational movements occur as birds move to higher elevations in spring and back to lower elevations during autumn. These seasonal movements are highly variable, and are determined by food availability. Fig. 217 shows fluctuations in the numbers of the Clark's Nutcracker reported on Christmas Bird Counts in the Okanagan valley; irruptive movements to the lowlands can be seen.

Within its range in the United States, the Clark's Nutcracker is known to exhibit mass migration at irregular intervals. These movements of entire populations to lower elevations or other areas outside the normal range of the species are relatively infrequent. Fisher and Myres (1979) suggest that the movements are associated with periods of seed-crop failure in the important pine species, which may, in turn, be influenced by weather systems.

In British Columbia, there have been no instances of mass emigrations such as those reported in Utah (Vander Wall et al. 1981). There, loose flocks of more than 2,000 birds were seen in apparent migration. The largest aggregation of the Clark's Nutcracker reported in British Columbia was 250 birds seen at Scout Island, Williams Lake, in early September 1979. This occurrence, close to the edge of the normal range of the species, may have been an emigration in process. Another apparent emigration was observed north of Lac la Hache in the Central Interior on 10 September 1979, when 162 birds were observed over a 45-minute period as they fed in the tops of Douglas-firs.

There are no winter records for the far northern ecoprovinces and only occasional ones from the Georgia Depression. The Clark's Nutcracker is a wintering species across the southern interior of the province in the Central Interior, Southern Interior Mountains, and Southern Interior ecoprovinces.

It occurs throughout the year in British Columbia (Fig. 214).

BREEDING: The Clark's Nutcracker is likely a widespread breeder in mountainous areas throughout its interior range in the province. In the interior, however, it is known to breed only in the mountains flanking the Okanagan valley. There are a number of reports of family groups in midsummer from Manning Park and west and east Kootenay regions, and from as far north as the Rainbow Range in the Central Interior. There are also reports of nest building at Brisco in the east Kootenay, and from near Bluff Lake in the western Chilcotin.

This species is not known to breed in coastal areas with strong marine influences. However, there are 2 breeding records from the Alta Lake area of the Coast and Mountains, just south and west of the limits of the Engelmann Spruce–Subalpine Fir and Interior Douglas-fir zones. Whitebark pine does occur locally in the Mountain Hemlock Zone on the western slope of the Coast Mountains.

The Clark's Nutcracker reaches its highest known numbers in summer in the Southern Interior (Fig. 215). An analysis of Breeding Bird Surveys in British Columbia for the period 1968 through 1993 could not detect a net change in numbers on interior routes.

The Clark's Nutcracker normally breeds at high elevations in open to semi-open subalpine coniferous forests, in mixed forests of Douglas-fir and ponderosa pine on the lower mountain slopes (Fig. 218), and in ponderosa pine parklands of the benchlands. As with its nonbreeding distribution, the breeding distribution of this species in British Columbia is strongly tied to the presence of whitebark pine and ponderosa pine, the 2 pines that provide most of the food used by the birds during the nesting period. Nesting will occur only at sites where a suitable cone crop exists (Munro 1919).

Because these birds nest early in the year, when much of their subalpine breeding habitat is still snow-covered, few nests with eggs or nestlings have been reported. Those that have been observed were mainly at lower elevations. Known nesting sites in the interior ranged from 450 m to 1,890 m elevation, but the birds likely nest up to the timberline. The single coastal nesting site was at an elevation of 1,830 m.

The Clark's Nutcracker is a definitive species of the breeding bird community in the subalpine meadows of the interior Engelmann Spruce–Subalpine Fir Zone, where islands of conifers are interspersed with open areas (Poll and Porter 1984). It has been recorded breeding in the province from 18 January (calculated) to 25 June (Fig. 214).

Nests: All nests (n = 10) were associated with open coniferous forests or subalpine forests on steep slopes. Most were from lower elevations in the Okanagan valley and were built in ponderosa pine (66%) and Douglas-fir (33%) trees. The few reported from the alpine areas have been found in alpine larch, whitebark pine, and spruce. Nests were placed on horizontal branches usually well out from the trunk and at heights of between 2 and 15 m. They consisted of a bulky platform of dead branches, grass, and twigs. The nest cup in each of 2 nests had an outer wall of rotted wood pulp, and was lined with fine grass and shredded bark of conifers (Fig. 219). The thick walls of the nest cup provide insulation from the cold often encountered during the early nesting period.

Eggs: Dates for 11 clutches ranged from 9 March to 25 May, with 7 recorded between 19 and 27 March. Calculated dates indicate that eggs may be present as early as 18 January (Cannings et al. 1987). Sizes for 10 clutches ranged from 1 (Fig. 219) to 3 eggs (1E-1, 2E-5, 3E-4). However, 2 broods of 4 young have been reported (see below), indicating that in British Columbia some Clark's Nutcrackers lay at least 4-egg clutches. Bent (1946) states that the Clark's Nutcracker, while usually laying 2 or 3 eggs, may lay as many as 6.

Figure 218. In the southern Okanagan valley, the Clark's Nutcracker breeds in mixed forests of Douglas-fir and ponderosa pine on the lower slopes of mountainous terrain (near Kilpoola Lake, 8 August 1996; R. Wayne Campbell).

The incubation period is reported variously as 16 or 17 days (Bent 1946), 18 days (n = 8) (Mewaldt 1956), and 22 days (Skinner 1916).

Young: Dates for 7 broods ranged from 25 February to 25 June. Calculated dates indicate that nests may contain young as early as the first week in February. Sizes for 5 broods ranged from 1 to 4 young (1Y-2, 2Y-1, 4Y-2). The nestling period is between 18 and 21 days (Bent 1946; Mewaldt 1956).

Brown-headed Cowbird Parasitism: Cowbird parasitism was not found in British Columbia in 14 nests recorded with eggs or young, nor has it been reported elsewhere in North America (Friedmann 1963; Friedmann et al. 1977; Friedmann and Kiff 1985).

Nest Success: Insufficient data.

REMARKS: Most research on the Clark's Nutcracker has been conducted in the Rocky Mountain states, where this species has been revealed to have one of the most specialized diets among the corvids. Its annual cycle is dependent on the harvesting and storing of conifer seeds (Vander Wall and Balda 1981), mainly whitebark and limber pine, which have been found to comprise between 85% and 95% of their diet; insects, small mammals, small birds, carrion, and berries make up the rest (Bent 1946; Giuntoli and Mewaldt 1978). In the Okanagan valley, the nutcracker has been recorded preying upon the eggs and nestlings of several species of passerines, including Horned Lark, American Pipit, Hermit Thrush, Cassin's Finch, Pine Siskin, and American Robin (Munro 1919; L.A. Gibbard pers. comm.).

Although this nutcracker will use a variety of other conifer seeds, including Douglas-fir (Giuntoli and Mewaldt 1978) and ponderosa pine, it has apparently coevolved with the whitebark pine (Vander Wall and Balda 1977; Tomback 1978; Tomback et al. 1990).

In the Rocky Mountain states, young in the nest are fed exclusively on pine seeds cached from the previous autumn. After fledging, they learn to find caches themselves by detecting germinating seeds (Vander Wall and Hutchins 1983). Both adults and young are dependent on caches from late winter until the next seed crop ripens the following August. Caches are retrieved from clumped storage areas and are uncovered as soon as the snowpack melts. Nutcrackers follow the receding snowline throughout early summer to uncover caches, reaching alpine storage areas by July.

In the reverse side of what is an apparent dependency relationship, the whitebark and limber pines have evolved large wingless seeds and rely upon nutcrackers for seed dispersal. Tomback et al. (1990) found that in open subalpine habitats, the Clark's Nutcracker harvested much of the pine seed "crop" and "planted" it for later use. A single nutcracker was found to store from 32,000 to 98,000 seeds in a season. The seeds were buried as far as 22 km from the pine source, and the nutcrackers have thus facilitated the rapid recolonization of clearcuts or fire-damaged forests of the large-seeded pines.

The Clark's Nutcracker may roost communally during the nonbreeding seasons, although little work has been done on that aspect of their behaviour. One population under observation during the summer, after spending the late afternoon on west-facing mountain slopes, moved in the early evening to roost in dense coniferous forests on east-facing slopes (Tomback 1978b). This behaviour is thought to maximize daily exposure to warm temperatures.

For further information on behaviour and life history, see Bent (1946), Mewaldt (1956), Tomback (1978a), and Tomback et al. (1990). The latter has a useful bibliography on the interdependence between the large-seeded pines of North America and the Clark's Nutcracker.

Figure 219. Clark's Nutcracker nest, with a single pale gray spotted egg, found abandoned (Richter Pass, June 1993; R. Wayne Campbell).

NOTEWORTHY RECORDS

Spring: Coastal – West Vancouver 20 Mar 1956-1; Whistler Mountain 1 Mar 1976-7, Green Lake (Whistler) 9 Apr 1982-1 at feeder (Campbell 1982c); Campbell River (Discovery Passage) 15 Apr 1971-1; Wowo Lake 5 May 1980-1. **Interior** – Midway 28 May 1975-4 young; Blackwall Mountain 26 May 1962-50; Hedley 15 Apr 1971-200; Vaseux Lake 10 Apr 1977-100; Nahun (Okanagan Lake) 9 Mar 1912-2 eggs; Penticton 27 Mar 1909-3 eggs, 25 Apr 1962-2 nestlings; Okanagan Landing 27 Mar 1912-1 nestling; St. Mary's River 16 May 1937-3 young; Brisco 27 Mar 1962-2 eggs; Yalakom River 13 Apr 1968-70; Botanie Creek 7 May 1966-100; Mount Revelstoke 26 May 1971-12; Field 10 Mar 1977-20; Tatlayoko Valley 17 Mar 1991-4; Cherry Creek w Bluff Lake 21 Mar 1990-pair nest building at 1,889 m elevation; Riske Creek 22 May 1988-51; Willow River 19 Apr 1968-15; Fort St. James 13 May 1889-1 collected (MacFarlane and Mair 1908).

Summer: Coastal – Haley Lake 2 Jun 1931-3; Maple Ridge 17 Jun 1971-5, at research forest (Campbell et al. 1974); Forbidden Plateau 19 Jun 1935-1 (Laing 1942); Mount Buttle 22 Jun 1931-3; Whistler Mountain 25 Jun 1924-2 nests with nestlings. **Interior** – Strike Lake 12 Jul 1984-40; Copper Mountain (Princeton) 26 Aug 1974-24; Vermilion Crossing 16 Jun 1943-fledglings; Anderson Lake (D'Arcy) 22 Jun 1968-30; Pavilion 29 Aug 1968-150; Kappan Lookout 24 Jul 1992-10; Gang Ranch (Tosh Creek) 29 Aug 1978-8; Perkins Peak 10 Aug 1992-7 at 2,819 m elevation; Kleena Kleene 25 Jul 1965-fledglings; Mount Robson Park 3 Aug 1972-30; Nilkitkwa Lake 15 Aug 1986-12; Todagin Lake 13 Jul 1963; Cecil Lake 10 Aug 1977-1; Pink Mountain 11 Aug 1977-1.

Breeding Bird Surveys: Coastal – Recorded from 1 of 27 routes and on less than 1% of all surveys. Maximum: Pemberton 30 Jun 1974-2. **Interior** – Recorded from 22 of 73 routes and on 15% of all surveys. Maxima: Summerland 29 Jun 1980-13; Canford 5 Jul 1990-12; Pleasant Valley 7 Jul 1988-10.

Autumn: Interior – Atlin 28 Oct 1931-1 (RBCM 5901); Prince George 4 Nov 1986-1; Williams Lake 6 Sep 1979-250; Tchaikazan River 16 Sep 1982-25; Jesmond 21 Sep 1986-10; Scotch Creek 10 Sep 1962-30; Nicola Lake 7 Sep 1961-46; Merritt 9 Oct 1982-16; Wasa 18 Oct 1982-20, 20 Oct 1982-100 in flock; Robson 27 Nov 1983-40; Vaseux Lake 22 Oct 1994-45 at mud puddle. **Coastal** – Port Neville 6 Sep 1986-1; Hoomak Lake 11 Nov 1975-1; Snakehead Lake 11 Nov 1980-5; Strathcona Park 6 Sep 1986-1; Stories Beach 15 Oct 1973-1 (BC Photo 359); Harrison Hot Springs 19 Nov 1975-1; Grouse Mountain 29 Oct 1972-1 (Campbell et al. 1974; BC Photo 293); Tofino 14 Oct 1989-1; Mount Prevost 13 Sep 1971-1 (BC Photo 215), 14 Oct 1972-3; Jordan River 11 Nov 1979-1; Victoria 8 Oct 1986-1.

Winter: Interior – Prince George 30 Dec 1972-3 (Rogers 1973b); Towdystan 6 Feb 1976-3; Tatlayoko Valley 21 Jan 1991-7; Williams Lake 27 Dec 1970-20 at feeder; Kootenay National Park 7 Jan 1961-50 (Seel 1965); Invermere 3 Dec 1983-30; Kelowna 22 Dec 1973-45; Naramata 19 Feb 1944-175, 25 Feb 1965-young old enough to fly; Kimberley 11 Dec 1980-19. **Coastal** – Masset Nov 1955-1 (RBCM 10449); Quatsino 1 Dec 1935-1 (UBC 7444); Mount Washington 8 Jan 1984-4; Comox 18 Feb 1904-1 (Brooks 1904); North Vancouver Jan 1956-1 banded (Hughes 1956); Victoria 4 Feb 1914-1.

Christmas Bird Counts: Interior – Recorded from 18 of 27 localities and on 54% of all counts. Maxima: Penticton 27 Dec 1988-**132**, all-time Canadian high count (Monroe 1989a); Vaseux Lake 23 Dec 1982-89; Oliver-Osoyoos 29 Dec 1991-74; Princeton 18 Dec 1993-74. **Coastal** – Recorded from 2 of 33 localities and on less than 1% of all counts. Maxima: Kitimat 17 Dec 1988-1; Chilliwack 28 Dec 1974-1.

Black-billed Magpie BBMA
Pica pica (Linnaeus)

RANGE: Holarctic. Resident in North America from south-coastal and southern Alaska, southern Yukon, northwestern British Columbia, northern Alberta, central Saskatchewan, central Manitoba and western Ontario south, away from the coast, to central California, Nevada, northern Arizona, and western Kansas. Wanders irregularly to the Pacific coast and eastern North America. Also resident in much of Eurasia, Arabia, and northwestern Africa.

STATUS: On the coast, a *rare* vagrant to the Georgia Depression Ecoprovince and the southern mainland of the Coast and Mountains Ecoprovince; *casual* on the northern mainland Coast and Mountains; *accidental* on Western Vancouver Island and the Queen Charlotte Islands.

In the interior, a *fairly common* to *common* resident in the Southern Interior Ecoprovince (*very common* locally in winter); *uncommon* to locally *fairly common* resident in the Southern Interior Mountains and Central Interior ecoprovinces, and the Peace Lowland of the Boreal Plains Ecoprovince; *rare* to *uncommon* in the southern portions of the Sub-Boreal Interior Ecoprovince. *Uncommon* migrant and summer visitant to the Northern Boreal Mountains Ecoprovince, particularly in the extreme northwest; *uncommon* local resident in the Atlin area. *Rare* in the Taiga Plains Ecoprovince.

Breeds.

CHANGE IN STATUS: Since the 1940s, the Black-billed Magpie (Fig. 220) has increased its distribution and abundance in British Columbia, primarily in 2 ecoprovinces: the Southern Interior Mountains and the Boreal Plains.

In the west Kootenay, Munro (1950) listed the Black-billed Magpie at Creston, but only in winter. The first spring occurrence there was not recorded until 1956 (Munro 1958). The magpie now occurs in small numbers in the Creston valley throughout the year (Butler et al. 1986). In the east Kootenay, Johnstone (1949) notes the magpie as a common resident, but the only nesting records he reports are from the extreme southern part of the region, between Roosville and Newgate, near the Montana border. For the period up to the 1940s, Munro and Cowan (1944) report magpies as occasional in spring in Kootenay National Park, and probably less abundant there than they had been in the past. Today, the Black-billed Magpie is present in the east Kootenay throughout the year and breeds in the Rocky Mountain Trench as far north as Radium.

The species has also increased in distribution and in numbers in the Boreal Plains since the late 1940s. Cowan (1939) did not encounter the species but was informed that 1 or 2 individuals had wintered at Charlie Lake. There had been no change in the recorded distribution by 1947, at which time it was regarded as scarce in the Peace River region (Munro and Cowan 1947). By the 1960s, they were known to nest at Fort St. John (A.C. Stewart pers. comm.), and there is 1 occurrence record for Pouce Coupe in 1965 (Erskine 1977). By the mid-1970s, 1 to 4 magpies could be seen daily in the Fort St. John

Figure 220. Adult Black-billed Magpie at nest site in spruce tree (near Mosquito Flats along the Haines Highway, July 1980; R. Wayne Campbell).

area (C. Siddle pers. comm.). It is now a resident there and breeds throughout most of the parkland area. The magpie has recently moved into the Taiga Plains, where it is present in small numbers throughout the late spring and summer in the vicinity of Fort Nelson.

The Black-billed Magpie has also extended its range in the Sub-Boreal Interior, and now occasionally breeds there. That range extension followed a dramatic increase in farmland in the Prince George and Vanderhoof areas over the past 20 years (J. Bowling pers. comm.).

NONBREEDING: The Black-billed Magpie (Fig. 220) has a widespread distribution throughout the southern third of the

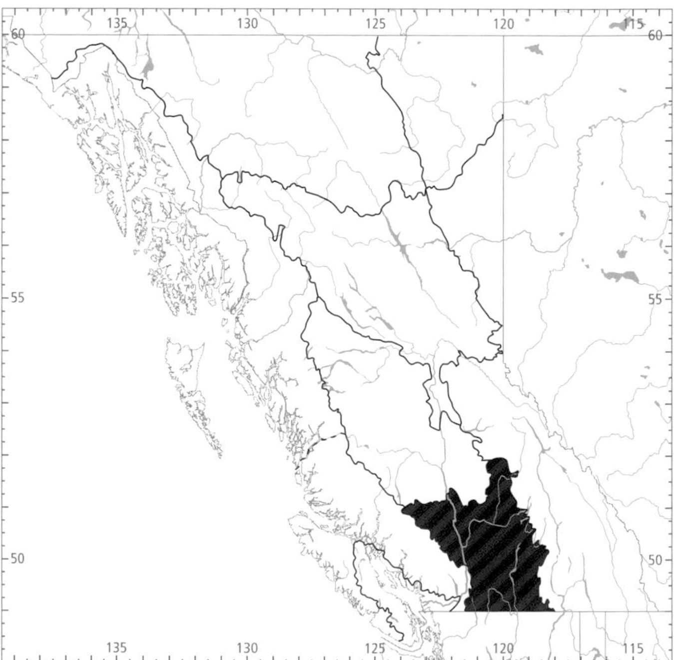

Figure 221. In British Columbia, the highest numbers for the Black-billed Magpie in winter (black) and summer (red) occur in the Southern Interior Ecoprovince.

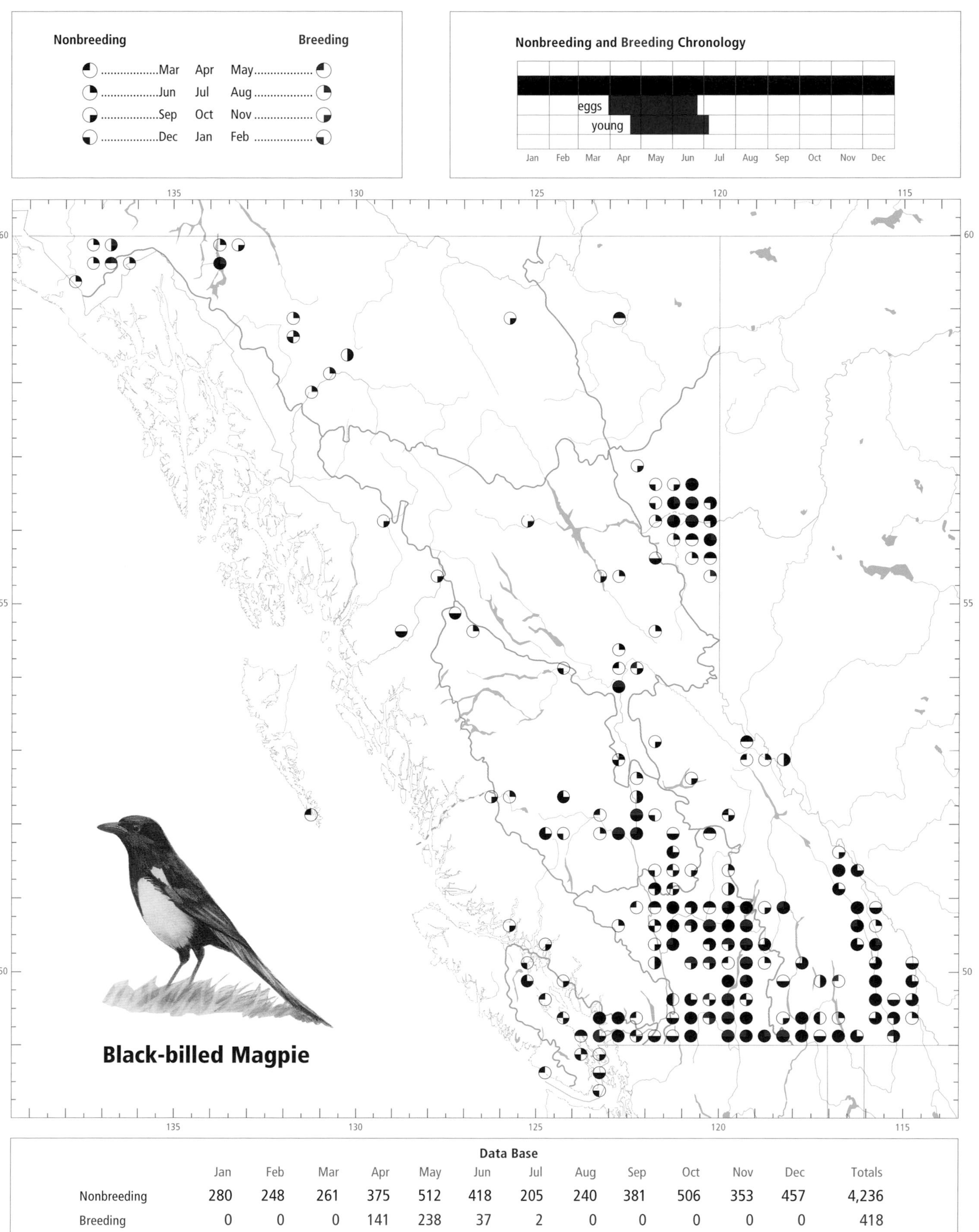

Data Base

	Jan	Feb	Mar	Apr	May	Jun	Jul	Aug	Sep	Oct	Nov	Dec	Totals
Nonbreeding	280	248	261	375	512	418	205	240	381	506	353	457	4,236
Breeding	0	0	0	141	238	37	2	0	0	0	0	0	418

interior of the province, except in the west Kootenay and in the Chilcotin area west of Riske Creek. It has a scattered and irregular distribution in the southern portions of the Sub-Boreal Interior between Quesnel (Munro 1947), Prince George, Vanderhoof, and Nulki Lake (Munro 1949), and occurs only casually in the Skeena River valley. A population has recently occupied the Boreal Plains and is now distributed from Tupper north to Prespatou, and west to the Rocky Mountain foothills. Since 1992, this population has started to colonize the Taiga Plains to the north, where 1 or 2 birds are reported infrequently each late spring and summer.

Further north and west, the Black-billed Magpie has a sparse and scattered distribution. Swarth (1936) refers to it as a breeding species just north of the Yukon boundary at Carcross, but at Atlin he found it common in the autumn and occasional through the winter and spring. It is now a local resident in the Atlin area. Brooks and Swarth (1925) report it as common at Bennett. Its status remains to be determined for most of the Northern Boreal Mountains, but seems to be unchanged over the past 70 years.

On the south coast, the Black-billed Magpie is an irregular visitor to the lower Fraser River valley, the east coast of Vancouver Island, and the southern mainland of the Coast and Mountains Ecoprovince. Records from exposed outer coasts are infrequent, but there is a single occurrence from the west coast of Vancouver Island and another from the Queen Charlotte Islands. Small numbers of Black-billed Magpies have escaped from captivity (Merilees 1992) or have been intentionally released on southeastern Vancouver Island, but to date the species has not become established there.

The centre of winter abundance for the Black-billed Magpie in British Columbia is in the Thompson, Okanagan, and Similkameen river valleys of the Southern Interior (Fig. 221).

The magpie has been reported from sea level to 1,250 m elevation. During the nonbreeding season, it uses a wide variety of habitats, but is most numerous in the vicinity of environments influenced by people. In winter, large numbers of magpies congregate in the vicinity of livestock feedlots, barnyards, garbage dumps, sewage lagoons, railway loading yards, and grain elevators, for the steady supply of food they provide to supplement wild food sources (Reebs and Boag 1987). In the north, such facilities undoubtedly help local populations remain through the winter. At Atlin, for example, all winter records are from the local garbage dump. Increased farming activities may also have aided the extension of magpies into the Peace River area. There the gleaning from agriculture, road-killed animals, and garbage dumps may provide the only dependable supply of winter food.

It has become a common bird of rural and suburban parks and gardens, even city gardens. It follows roads and railroads to scavenge the casualties of passing traffic. When hunters are afield, the Black-billed Magpie is usually one of the first birds to find the offal.

In natural habitats it seeks riparian thickets, wet meadows, open forests (especially those of ponderosa pine and interior Douglas-fir), shrub-choked gullies in grassland areas, lakeshores, and beaver ponds. During the summer months, nonbreeding birds may disperse to higher elevations, where they frequent open areas such as clearcuts, subalpine forest edges, and alpine meadows. A universal characteristic of magpie habitat in natural sites is an association of thickets and riparian areas for roosting, with varied open landscapes such as meadows, grassland, or sagebrush areas for foraging (Bock and Lepthien 1975).

From midsummer through autumn and winter, magpies join together each evening in communal roosts, and a large percentage of each local population may occupy 1 site. Roosts are located in dense thickets of shrubs and trees (Fig. 222) near major food sources. Reebs (1987) found that magpies roosted in deciduous thickets in the autumn and remained in them until the arrival of colder temperatures (< 5°C) and snow cover, when they moved to conifers (Fig. 223). They sometimes use old magpie nests as night roosts (Erpino 1968).

The Black-billed Magpie migration is mainly an elevational movement, although some north-south movement does occur. After the breeding season, individuals may wander widely. By early autumn, they begin to congregate in the valley bottoms where they will spend the winter. In the extreme northwest, where winters are severe, all magpies leave the high country. Some that regularly winter along the

Figure 222. An autumn roost site for the Black-billed Magpie in dense brush bordering a mixed deciduous-coniferous forest (north of Dawson Creek, 24 October 1991; R. Wayne Campbell).

Figure 223. A winter conifer roost site of 11 Black-billed Magpies (Pennask Lake Road, 9 December 1992; R. Wayne Campbell).

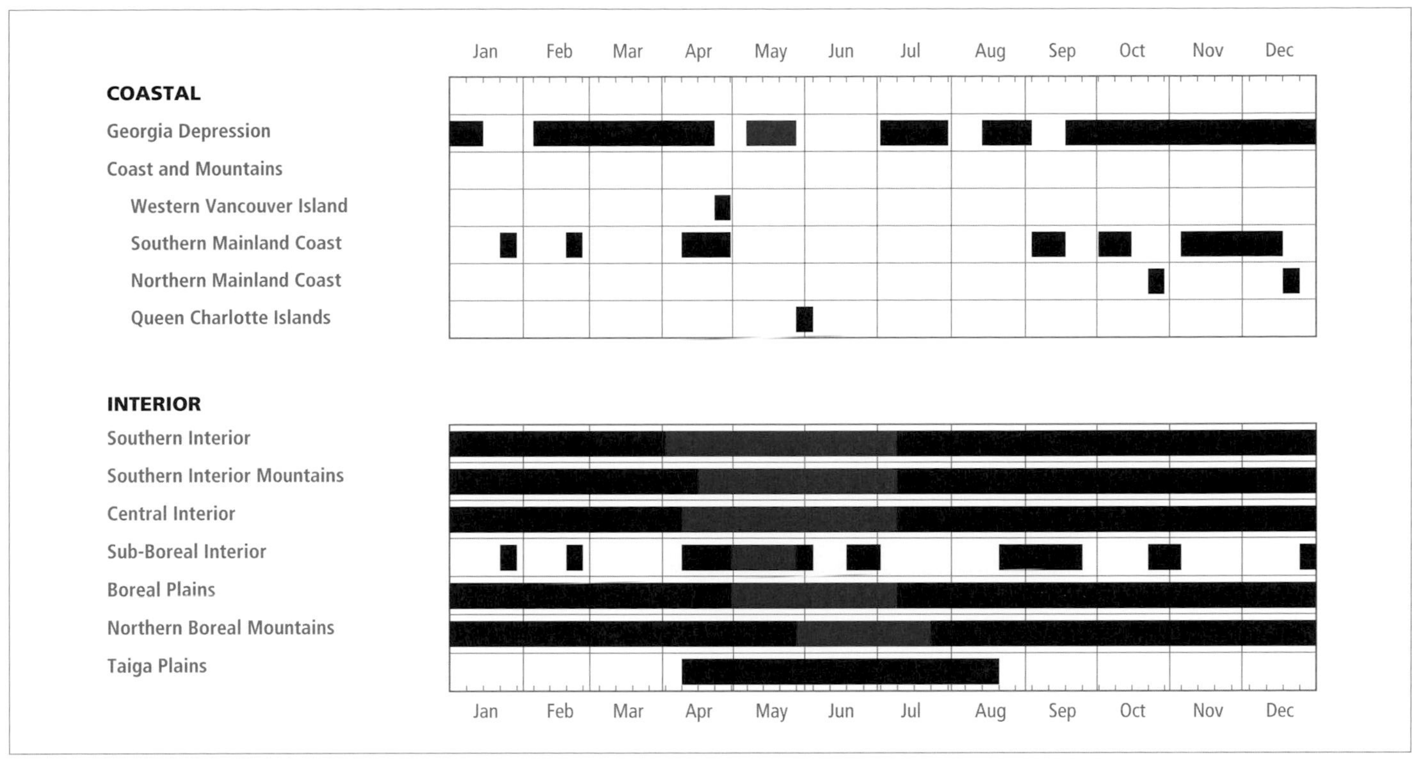

Figure 224. Annual occurrence (black) and breeding chronology (red) for the Black-billed Magpie in ecoprovinces of British Columbia.

coast of southeastern Alaska (Swarth 1922; Gabrielson and Lincoln 1959) probably originate in British Columbia. Much the same movements can be seen in the Peace River region, where there is little doubt there would be few wintering birds were it not for sources of food generated by humans. Further south in the interior of the province, birds move to the valley bottoms for the winter, but continue to forage widely. In some years, small numbers disperse to the southwest coast in autumn. Winter roosts begin to break up in March in the Okanagan (Cannings et al. 1987), and about a month later in the north, with birds dispersing to breeding areas.

The Black-billed Magpie occurs in the interior of British Columbia throughout the year; it appears on the coast primarily between October and March (Fig. 224).

BREEDING: The Black-billed Magpie has a widespread breeding distribution over much of the Southern Interior, including the Okanagan, Thompson, Nicola, and Similkameen valleys. It also nests commonly throughout the grasslands of the Cariboo and Chilcotin regions of the Central Interior (A. Roberts pers. comm.). The small population of magpies that breeds in the Peace Lowland of the Boreal Plains is apparently an extension from the widespread population of the northern Great Plains. Elsewhere in the province, there are scattered records of nesting in both the east and west Kootenays and in the Sub-Boreal Interior. An apparently disjunct breeding population occurs in the Tatshenshini Basin of the Northern Boreal Mountains, and in the nearby Chilkat Pass and Atlin areas. It is probably part of the population that breeds in southern Yukon and southern Alaska, much of which is largely coastal in its distribution.

In the late 1950s and early 1960s, 2 pairs of Black-billed Magpies apparently nested in the Ladner and Delta areas of the Fraser River estuary. One of the pairs was collected and the nest destroyed. The other pair was not seen after the mid-1960s. Information regarding clutch or brood size was not reported.

Godfrey (1986), when he shows the magpie range extending through much of the northern third of the interior, including the Fort Nelson Lowland and the northern mountains west to the Alaska panhandle, implies a wider breeding distribution than was known at the time he published his work. Except for the areas mentioned above, the breeding distribution of the magpie for most of the Northern Boreal Mountains remains to be determined.

The Black-billed Magpie reaches its highest numbers in summer in the Southern Interior Ecoprovince (Fig. 221). An analysis of Breeding Bird Surveys for the period 1968 through 1993 could not detect a net change in numbers on interior routes.

In southern regions and in the Boreal Plains, the Black-billed Magpie breeds in lowland areas, while in the northwestern mountains it breeds in subalpine areas. Breeding has been reported between 270 m and 1,140 m elevation. In the Okanagan, the species is most abundant below 1,000 m elevation (Cannings et al. 1987).

Most nesting territories (41%; n = 149) were associated with cultivated farmland or ranchlands, followed by open forest and woodlands (28%), riparian situations (14%), suburban areas (9%), and open stands of sagebrush. Within these broad habitat classes, the birds used a variety of specific habitats, including grasslands with scattered trees and shrubs, hedgerows in agricultural lands, brushy hillside draws, edges of aspen copses, suburban plantings of trees and shrubs, and the shelter belts around farms and ranches. Loose colonies may form in scrub-filled draws and ravines, where the shrub cover is relatively dense. Such habitat is about equally distributed between wildland and that influenced by humans. The presence of water is common to about a third of the nesting habitats reported.

The Black-billed Magpie has been recorded breeding in the province from 31 March (calculated) to 5 July (Fig. 224).

Nests: Most Black-billed Magpie nests in British Columbia were found in deciduous trees (41%; *n* = 283), with lesser numbers in conifers (24%) and shrubs (19%). A few were placed in barns and other deserted buildings. The trees most frequently used as nest locations were black hawthorn (19%; Fig. 225) and ponderosa pine (19%), followed by willow (9%). The most commonly reported shrub was saskatoon (3%). Another 15 species of trees and shrubs were used less frequently (Fig. 226). In northern areas, most nests were found in spruce and willows. The species of tree or shrub is less important than the form of the vegetation used. Most nests were placed among the dense growth of twigs and branches, frequently towards the top of the shrub or tree. The heights of 280 nests ranged from ground level to 18.0 m, with 59% recorded between 2 and 5 m.

Nests were large accumulations of branches, twigs, mud, grass, rootlets, bark strips, vines, plant fibres, needles, and similar materials, with branches and twigs constituting the base and framework of the nest and the nest cup itself lined with fine rootlets, grass, plant fibres, and other soft material. Most nests had a mud "anchor" that was placed early in the nest-building process and served as the base for the structure. The nest was almost always provided with a hood or dome of loosely assembled twigs and branches, and usually had 1 or more entrances from the sides.

Magpie nests can remain intact for many years, and though they are rarely reused (Erpino 1968), in British Columbia they provide useful nest sites for other species such as the Merlin, Long-eared Owl, and American Kestrel.

Eggs: Dates for 215 clutches ranged from 4 April to 24 June, with 52% recorded between 25 April and 9 May. Calculated dates indicate that nests may contain eggs as early as 31 March. Sizes of 179 clutches ranged from 1 to 9 eggs (1E-7, 2E-6, 3E-11, 4E-23, 5E-32, 6E-43, 7E-46, 8E-9, 9E-2), with 68% having 5 to 7 eggs. Bent (1946) states that 7 is the usual number of eggs in a clutch, with 8 or 9 not uncommon and as many as 13 recorded in 1 nest. In 5 North American studies, the mean clutch sizes had a spread of from 6.12 to 6.50, with completed clutches having between 3 and 10 eggs (Birkhead 1991). The British Columbia sample cannot be compared, because of the unknown component of incomplete clutches, but no significant difference is apparent.

The incubation period has been variously reported as 14 to 21 days (Evenden 1947), slightly less than 18 days (Jones 1960), and 16 to 20 days (Erpino 1968). The Black-billed Magpie begins its incubation part way through the laying of the clutch, so marked eggs are required to determine the incubation period of each egg. Tatner (1982) notes a period of 24 days from the laying of the first egg to its hatching. However, if there is a delay in initiating full incubation, that is not necessarily the incubation period. Birkhead (1991) believes the true incubation period probably lies between 20 and 21 days.

Young: Dates for 110 broods ranged from 20 April to 5 July, with 57% recorded between 10 and 28 May. Sizes of 85 broods ranged from 1 to 9 young (1Y-3, 2Y-8, 3Y-11, 4Y-22, 5Y-19, 6Y-15, 7Y-3, 8Y-3, 9Y-1), with 66% having 4 to 6 young. The nestling period in a nest from British Columbia was 19 days; Erpino (1968) found that the nestling period in 12 broods was 18.3 days and ranged from 16 to 21 days.

Nest Success: Of 23 nests found with eggs and followed to a known fate, 10 produced at least 1 fledgling, for a nest success rate of 43%. The comparable rate for an Alberta population (*n* = 51) was 57% (Dunn and Hannon 1989). The same authors note a mean success rate of 54% in 14 populations in various parts of North America. Birkhead (1991) provides success rates derived from 20 studies, ranging from 32% to 82%. Only 2 of the 20 were as low as those in British Columbia.

REMARKS: The Black-billed Magpie and the Yellow-billed Magpie (*P. nuttallii*) are considered to be a superspecies (American Ornithologists' Union 1983).

Figure 225. A bulky, domed stick nest of the Black-billed Magpie in a black hawthorne (near Kilpoola Lake, Okanagan valley, 1 May 1994; R. Wayne Campbell).

Figure 226. A well-concealed Black-billed Magpie nest in the lower branches of a Douglas-fir (Wineglass Ranch near Riske Creek, 27 April 1990; R. Wayne Campbell).

Historical accounts indicate that the distribution of the Black-billed Magpie in central North America may have been linked to the distribution of the plains bison. As the bison herds of the Great Plains were eliminated in the mid to late 19th century, magpies almost disappeared (Houston 1977). Then as settlers arrived and began to farm and raise cattle on the Great Plains, the magpie recovered. From their centre on the Canadian Prairies they dispersed slowly northward and westward. The magpie was apparently rare in Alberta's Red Deer valley at the turn of the century (Salt and Wilk 1958) and only penetrated into the Peace River region sometime in the 1940s. In the southern interior of British Columbia, it was never associated with bison, as the mammal did not occur west of the Rocky Mountains. The Black-billed Magpie has been abundant in the Okanagan Valley since early settlement, and remains so to this day.

Bock and Lepthien (1975) and Root (1988) suggest that the distribution of the Black-billed Magpie in North America is limited by climatic factors and that it is best adapted to the environments of cool, arid areas.

Like the crow and raven, the magpie was the victim of systematic destruction in British Columbia during the first half of this century; it was considered detrimental to gamebird populations and domestic stock. Bounties were in place for many years, and large numbers of the birds were shot. In 1933 alone, 1,033 Black-billed Magpies were killed in the Okanagan valley by 2 teams of bounty hunters (Bent 1946). In nearby south-central Washington, an estimated 5,000 were killed in the winter of 1920-21 (Kalmbach 1927). The Black-billed Magpie is not protected by legislation in Canada, but today there appears to be a more tolerant attitude towards all corvids.

For additional information on the life history of the Black-billed Magpie, see Linsdale (1937), Bent (1946), Reese and Kadlec (1985), and Birkhead (1991).

NOTEWORTHY RECORDS

Spring: Coastal – Carmanah Point 28 Apr 1952-1; Lulu Island 5 Apr 1980-1; Mud Bay (Ladner) 27 May 1962-active nest with 4 young later destroyed, both adults present; Nanaimo 23 Apr 1991-1 escaped cage bird (Merilees 1992), 12 May 1987-1; Qualicum 28 Mar 1980-1; Oyster River 7 May 1981-1; Anthony Island 29 May 1985-1. **Interior** – Osoyoos 15 May 1968-40; Richter Lake 4 Apr 1961-5 eggs; White Lake (Okanagan Falls) 3 Mar 1968-30; Mirror Lake 29 Apr 1924-4 eggs; Knutsford to Merritt 15 Apr 1968-105 on survey; Alexandria 26 Apr 1962-12; Chilcotin River nr Wineglass Ranch 27 Apr 1988-2 eggs; Newlands 22 May 1969-6 fledglings; Prince George 30 May 1973-newly built nest; Eaglet Lake 22 May 1969-6 young; Fort St. John 7 Apr 1984-25 on survey, 11 Apr 1976-2 pairs at 2 nests, 1 May 1977-5 eggs; Fort Nelson 13 Apr 1987-1; Mile 308, Atlin 15 April 1976-1; Alaska Highway 13 Apr 1987-1; Chilkat Pass 16 May 1977-6.

Summer: Coastal – Vancouver 17 Aug 1979-2; Port Coquitlam 29 Aug 1980-1; Mount Seymour 14 Jul 1973-3; Nanaimo 26 Jul 1990-3. **Interior** – Manning Park 8 Jun 1980-12; Richter Pass 20 Apr 1942-7 eggs (Campbell and Meugens 1971); Oliver 28 Jun 1972-3 nestlings about the size of Gray Jays; Grand Forks 1 Aug 1982-75, roosting flock; Vaseux Lake 25 Jun 1976-29; West Bench 30 Jul 1976-150; 16 km n Penticton 18 Jul 1966-5 young; Bummers Flats 22 Jun 1976-12; Wasa 25 Jul 1971-21 in flock; Otter Lake (Armstrong) 19 May 1982-17; Columbia Lake 31 Aug 1977-14; Windermere 5 Jul 1965-5 nestlings almost fledged; Chilcotin River near Riske Creek 9 Jun 1989-5 fledglings; Williams Lake 11 Jun 1975-nestlings; Hanceville 18 Aug 1978-7, 18 Jun 1962-1; One Island Lake 28 Jun 1978-1 nestling; McQueen's Slough 30 Jun 1978-2 nestlings; Bear Flats 27 Jul 1983-30; Taylor 3 Jul 1976-40 to 60 at garbage dump; Fort St. John 20 Jun 1977-4 nestlings; 14 Jul 1984-50 in flock, 23 Aug 1985-40 (roosting flock); Telegraph Creek 21 Jul 1962-2 (NMC 49949-50), 22 Jul 1910-2 (Brooks and Swarth 1925); Fort Nelson 15 June 1994-2 around town; Tahltan 1 Jul 1975-4; Atlin Lake 18 Jun 1978-young being fed by parents in nest on Number One Island, 26 Jun 1975-5; Mount Mansfield 19 Aug 1977-2; Mile 75 Haines Highway 14 June 1957-6 eggs, 30 Jun 1957-1 young; Mile 74.5 Haines Highway 18 June 1980-4 nestlings, 13 Jul 1980-4 fledglings near nest; Tatshenshini River 20 Jun 1993-family group of 5.

Breeding Bird Surveys: Coastal – Not recorded. **Interior** – Recorded from 30 of 73 routes and on 40% of all surveys. Maxima: Lavington 20 Jun 1976-43; Summerland 18 Jun 1977-33; Prince George 24 Jun 1979-30.

Autumn: Interior – Atlin late Oct to early Apr 1978 to 1981-4 to 6 at garbage dump; Beryl Prairie (nr Hudson's Hope) 12 Nov 1978-3; Montney 27 Nov 1983-10; Cecil Lake 23 Oct 1982-40; Taylor 20 Nov 1983-15; Chetwynd 1 Nov 1980-1; w Williams Lake 2 Sep 1978-25 in flock; Riske Creek 21 Nov 1983-8; Sorrento-Chase 6 Sep 1970-50 on survey; Summerland 28 Nov 1976-25; Lytton 7 Sep 1968-15; Roosville 3 Sep 1948-80 (Johnstone 1949). **Coastal** – Meziadin Lake 12 Oct 1974-1; Port Neville Inlet 10 Oct 1975-1; Pitt Meadows 24 Sep 1964-1, Oct through Dec 1972-2 (Campbell et al. 1974); Maple Ridge 6 Oct 1977-3; Langley Sep 1889-2 (RBCM 1812 and 1813); Skagit River valley 23 Nov 1974-2; Reifel Island 28 Oct 1983-1; Saanich 29 Nov 1980-1.

Winter: Interior– Atlin 20 Dec 1980-26 in flock at garbage dump; Dudidontu River 23 Feb 1982-1; Rose Prairie 2 Jan 1984-1; Fort St. John 7 Dec 1994-10, 26 Jan 1986-11; North Pine 27 Dec 1986-48; Chetwynd 23 Jan 1983-3; Vanderhoof 2 Jan 1991-1; Prince George airport 20 Feb 1982-6; Nulki Lake 26 Dec 1945-2 (Munro 1949); Tatlayoko Valley 8 Feb 1978-1; Chezacut 19 Feb 1943-20; Williams Lake 30 Dec 1979-3, 2 Jan 1985-3; Alkali Lake 25 Feb 1976-20; Golden 1 Jan 1978-1; Kamloops 30 Jan 1990-27 in pre-roost tree; Cache Creek 11 Dec 1987-10; Invermere 8 Feb 1977-9; 4.8 km e Vernon 30 Dec 1905-75 (Cannings et al. 1987); Edgewood 30 Dec 1983-1; Wasa 15 Jan 1977-25; Summerland 2 Feb 1973-30; s end Skaha Lake 31 Dec 1976-61 on count; Creston 26 Dec 1994-7; Osoyoos 1 Jan 1986-9. **Coastal** – Terrace 19 Dec 1969-1 (Vance 1970); Quadra Island 3 Jan 1976-2; Chilliwack 19 Dec 1926-1 (RBCM 6314); Vancouver 23 Dec 1934-1 (RBCM 14766); Burnaby 13 Jan 1980-1; Reifel Island 29 Feb 1984-1; White Rock 30 Dec 1973-1 (Schouten 1974).

Christmas Bird Counts: Interior – Recorded from 23 of 27 localities and on 69% of all counts. Maxima: Kelowna 18 Dec 1993-469; Penticton 28 Dec 1991-448; Vernon 18 Dec 1983-393. **Coastal** – Recorded from 6 of 33 localities and on 1% of all counts. Maxima: Campbell River 3 Jan 1976-2; Vancouver 26 Dec 1960-2; Comox 22 Dec 1985-1; Pitt Meadows 26 Dec 1972-1; Ladner 27 Dec 1983-1; White Rock 30 Dec 1973-1.

American Crow

Corvus brachyrhynchos Brehm

AMCR

RANGE: Breeds from north-central British Columbia and southwestern Mackenzie across central and southern Canada to southern Newfoundland, and south (except along the Pacific coast) to northern Baja California, Arizona, New Mexico, and the Gulf coast. Winters throughout most of its breeding range except for northernmost populations.

STATUS: The status of this species from the Fraser Lowland of the Georgia Depression Ecoprovince to Hazelton and Stewart on the northern mainland of the Coast and Mountains Ecoprovince is uncertain.

In the interior, *common* to *very common* migrant and *fairly common* summer visitant throughout the interior south of the extreme southern portions of the Sub-Boreal Interior Ecoprovince; *uncommon* to locally *very common* or *abundant* winter visitant in the Southern Interior and Southern Interior Mountains ecoprovinces; *uncommon* to locally *very common* in winter in the Central Interior Ecoprovince. Throughout the southern half of the province, it is *very common* locally in late summer and autumn in the vicinity of nocturnal roosts. *Uncommon* to *fairly common* migrant and summer visitant to the remainder of the Sub-Boreal Interior Ecoprovince as well

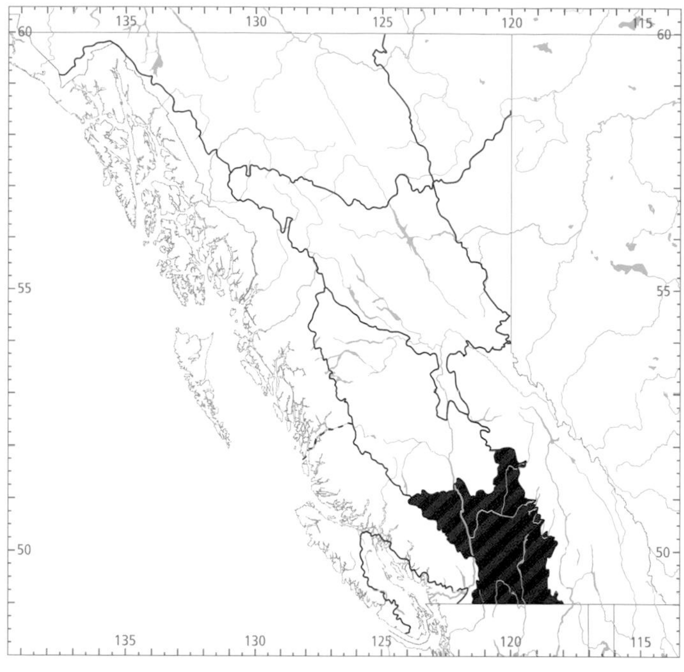

Figure 227. In British Columbia, the highest numbers for the American Crow in winter (black) and summer (red) occur in the Southern Interior Ecoprovince.

Figure 228. In winter, the American Crow often associates with and forages near livestock feeding stations with adjacent stands of black cottonwood trees (near Midway, 25 February 1993; R. Wayne Campbell).

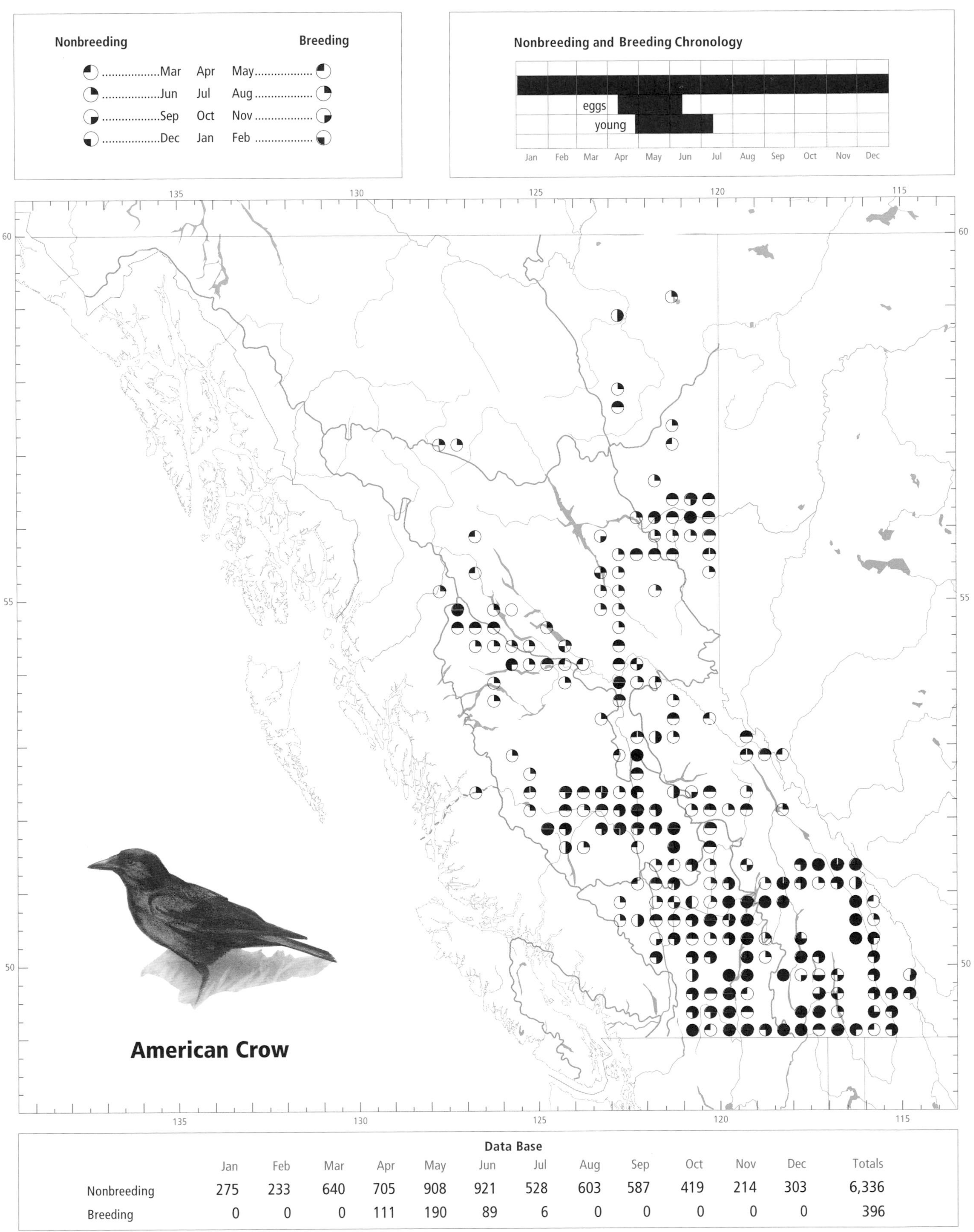

Data Base

	Jan	Feb	Mar	Apr	May	Jun	Jul	Aug	Sep	Oct	Nov	Dec	Totals
Nonbreeding	275	233	640	705	908	921	528	603	587	419	214	303	6,336
Breeding	0	0	0	111	190	89	6	0	0	0	0	0	396

as the Boreal Plains Ecoprovince; *casual* there in winter. *Very rare* migrant and summer visitant to the Taiga Plains Ecoprovince; *casual* in the Northern Boreal Mountains Ecoprovince.

Breeds.

NONBREEDING: The American Crow has a widespread distribution throughout most of the southern and central interior of the province. It occurs from Manning Park east to the Rocky Mountains, north through the valleys of the Southern Interior and Southern Interior Mountains, through much of the Central Interior, to the extreme southern portions of the Sub-Boreal Interior, including the Nechako River valley and the upper Skeena River valley. It also occurs in the agricultural regions of the Boreal Plains, east of the Rocky Mountains. Further north, the American Crow has a sparse distribution that extends to Fort Nelson and Kotcho Lake in the Fort Nelson Lowland of the Taiga Plains. There are only 2 records from the Northern Boreal Mountains, both from Kitchener Lake.

The distribution of this crow in coastal areas, including the lower reaches of the major east-west river valleys that cut through the coastal mountains, is uncertain. There the range of the American Crow meets the area inhabited by the Northwestern Crow. In these areas of overlap, only specimen records that have been recently verified by measurement have been included in the data (see REMARKS).

Although the American Crow occurs at elevations from 300 to 1,800 m, it is most numerous at low to moderate elevations. In the east Kootenay, for example, it is common in the Columbia River valley but infrequent at higher elevations in Kootenay National Park (Poll et al. 1984). This is also true for the Okanagan valley (Cannings et al. 1987). The highest numbers in winter occur in the Southern Interior Ecoprovince (Fig. 227).

This species frequents a wide range of natural and human-influenced habitats. Preferred natural habitat is open to semi-open coniferous and deciduous forests and adjacent grasslands, or riparian woodlands along rivers, valley bottoms, and lakeshores. Other habitats include marshes, sloughs, burns, and clearings. An essential habitat component is mature trees, either coniferous or deciduous (usually cottonwood), free from disturbance, for nocturnal roosting. Roosts are located in stands of tall trees, mainly along rivers.

The American Crow is most abundant and ubiquitous where human activity has produced agricultural fields among woodland, and in the vicinity of communities where such artificial habitat features as garbage dumps, barnyards, parks, residential neighbourhoods, orchards, railways, and road corridors prevail. It readily uses intensely urbanized areas.

Where this crow formerly depended substantially on slaughterhouses for winter food, wintering birds now congregate mainly in the vicinity of landfills, livestock feedlots and feeding stations (Fig. 228), and farmlands (e.g., Vernon) (Munro *in* Bent 1946). Godfrey (1986) notes that garbage dumps have replaced local slaughterhouses as determinants of the winter distribution of this crow in many parts of Canada. In far northern British Columbia, the few birds wintering there are dependent on human refuse at garbage dumps or within town centres. Large areas of coniferous forests appear to act as a barrier to the species.

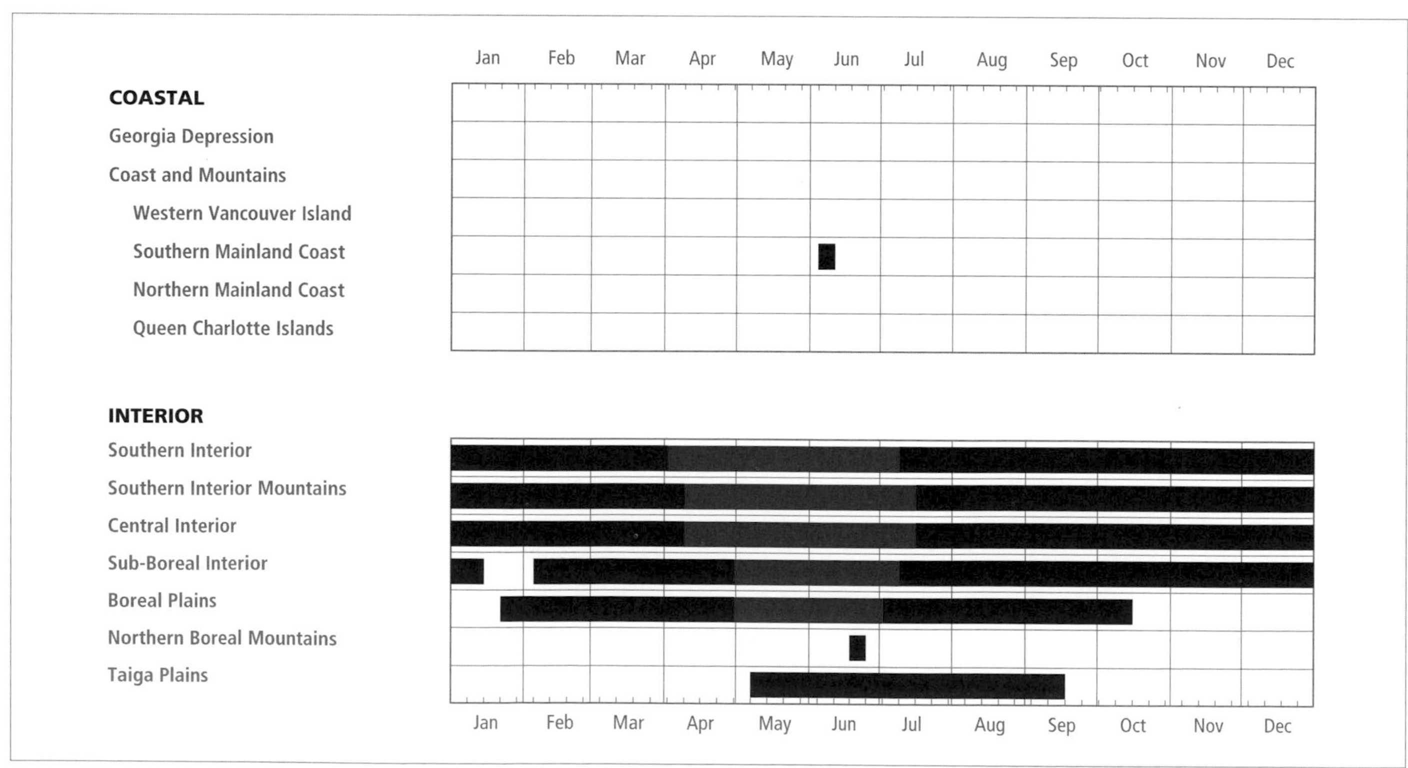

Figure 229. Annual occurrence (black) and breeding chronology (red) of the American Crow in ecoprovinces of British Columbia.

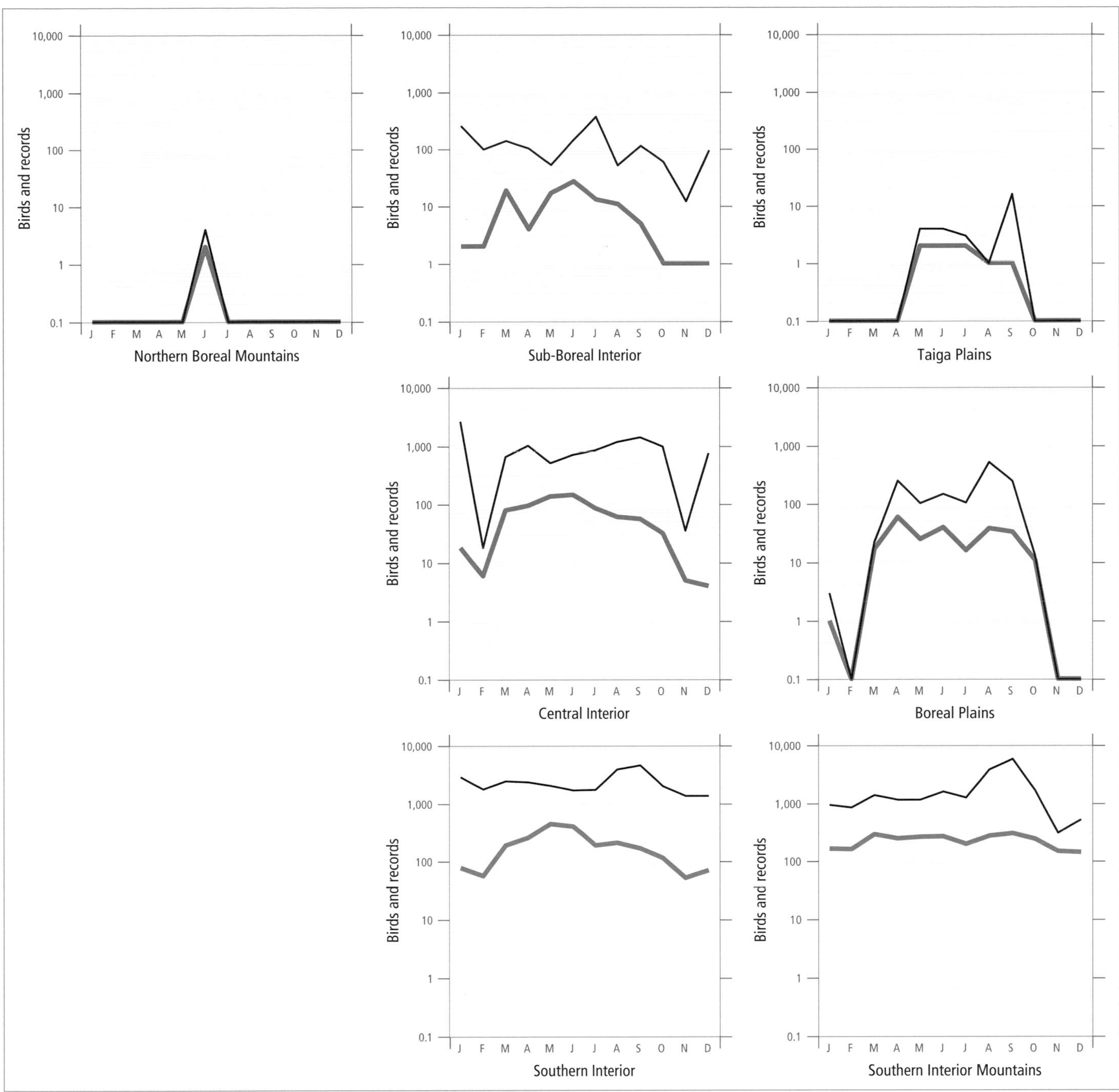

Figure 230. Fluctuations in total number of birds (purple line) and total number of records (green line) for the American Crow in ecoprovinces of British Columbia. Christmas Bird Count and nest record data have been excluded.

The American Crow is migratory over most of British Columbia (Figs. 229 and 230). Spring migration is difficult to discern in the southern portions of the interior of the province, but there is a slight increase in the number of birds reported in March. In the valleys of the Kootenay and Columbia rivers (Johnstone (1949), and in the Okanagan valley (Cannings et al. 1987), migrants begin to move north through the valley bottoms in mid to late February and early March. Peak movements in those areas occur in late March and then decline as transients move northward. In all areas, arrival dates at higher elevations are later than in the valley bottoms. A few birds appear in the Central Interior by the second week of March and numbers build through March, peaking in the first or second week of April. Much the same holds true for the Peace Lowland of the Boreal Plains. The few crows recorded in the Fort Nelson Lowland of the Taiga Plains arrive in late April.

The autumn migration in the Peace Lowland begins in August, and numbers build through the end of that month. Most birds have left the region by the end of September. Further south, in the Sub-Boreal Interior, premigratory flocking begins in July, and only a few stragglers are left after

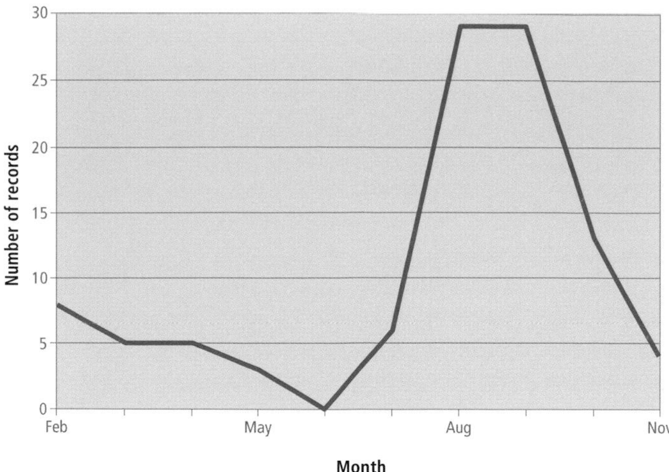

Figure 231. Monthly distribution of the number of records of American Crows with 100 or more birds in British Columbia (n = 102). Counts at roosts and surveys have been excluded.

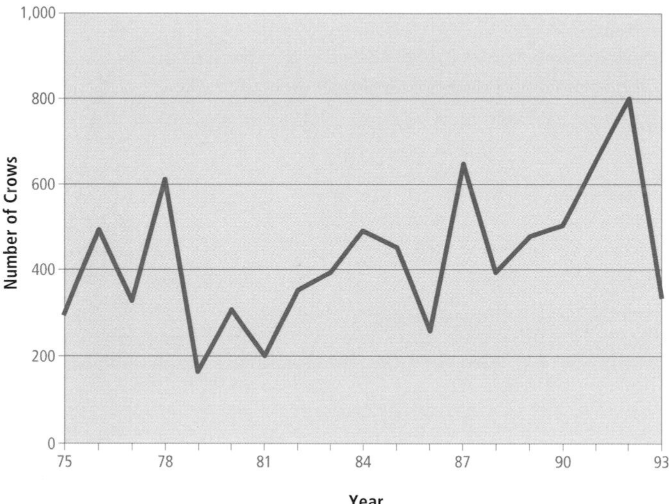

Figure 232. Fluctuation in American Crow numbers summed from Christmas Bird Counts in the Southern Interior Ecoprovince, British Columbia: Shuswap Lake Park, Vernon, Penticton, Vaseux Lake, and Oliver-Osoyoos.

September. In the Central Interior, flocks begin to form in mid-July (e.g., Nulki Lake area; Munro 1949) and numbers increase through August to reach a peak near the end of August and in early September. Most birds have left the region by the end of October, with only a few remaining into November and December. An increase in numbers during August and September can also be seen in the Southern Interior and the Southern Interior Mountains, with the main movement culminating around the last 2 weeks of September (Fig. 230). Most migrants have left the region by the end of October. In the southern ecoprovinces, many American Crows remain through the winter.

The American Crow migrates in flocks rather than as single birds, but flocks are smaller in the spring migration than in the autumn. Among 102 reports of 100 or more birds from February through November, 70% are from August through October (Fig. 231). In the east Kootenay, Johnstone (1949) also noted similar results. There group size in spring averaged 7 birds and in the autumn 104 birds.

It is normal behaviour for this crow, once the young are on the wing, to gather into large flocks to roost. Crows move to the roosts in the early evening and disperse to foraging areas at dawn. Roost sites are used for many years if they remain undisturbed. In British Columbia, there are few records of American Crow roosts; for the most part this aspect of their life history has not been well documented. It is easy to confuse evening flights to roosts with migration flights.

Wintering populations of the American Crow occur at lower elevations in the Southern Interior and in the east and west Kootenay, but small numbers also winter as far north as Prince George. The highest numbers in winter occur in the Okanagan valley (Vernon, Kelowna, Penticton). Other areas reporting high numbers include Smithers, Prince George, Williams Lake, Revelstoke, Sorrento, Salmon Arm, Grindrod, Nelson, and Creston. The numbers of wintering birds vary widely between years (Fig. 232). In severe winters, few American Crows winter in British Columbia. In the east Kootenay, Johnstone (1949) found no evidence of wintering before 1948, while small numbers now do so. Much of the Canadian population winters in the central United States (Bent 1946).

The American Crow occurs in British Columbia throughout the year (Fig. 229).

BREEDING: The American Crow probably breeds throughout most of its range in the interior of British Columbia. Its known breeding distribution extends from the Princeton area east to the Rocky Mountains, north through the valleys of the Southern Interior and Southern Interior Mountains, across the Central Interior to Smithers, and to Fort St. John in the Boreal Plains. Godfrey (1986) shows the breeding distribution of the American Crow to include portions of the Northern Boreal Mountains and the Taiga Plains, but we know of no supporting data from those areas. Johnstone (1949) notes that the American Crow is an abundant nesting species in the east Kootenay; he reports that during a crow shoot he examined 23 nests in 1 day. Eight nests contained 4 young, 12 had 3 young, and 3 had 2 nestlings. He also says that nests were in ponderosa pine trees on a bench above the Kootenay River.

The American Crow reaches its highest numbers in summer in the Southern Interior (Fig. 227). An analysis of Breeding Bird Surveys for the period 1968 through 1993 could not detect a net change in numbers on interior routes.

In British Columbia, the American Crow has been reported nesting between 240 m and 1,200 m elevation. General habitats selected by nesting pairs were much the same as those occupied in the nonbreeding season. Nests were reported both from human-influenced areas and from natural areas. Most nest sites were associated with cultivated farmlands or pasture and rangelands (49%; n = 164), followed by riparian sites (12%); deciduous forest (11%); urban, suburban, and industrial lands (10%); and coniferous forest (9%). Within these general habitats, nests were found near water or in bottomland (56%; n = 62), forest edges (20%), and occupied lands such as gardens, groves, orchards, parks, cemeteries, campgrounds, or farms.

The American Crow has been found breeding in British Columbia from 9 April (calculated) to 11 July (Fig. 229).

Nests: Most nests (59%; *n* = 245) were found in deciduous trees, followed by coniferous trees (23%) and shrubs (4%); all but 3 of the trees or shrubs were living. One nest was found on the ground. The main deciduous nest-trees were birch (24%), willow (20%), black cottonwood (6%), and mountain alder (4%); conifers included lodgepole pine (8%), Douglas-fir (7%), ponderosa pine, and spruce. The heights of 251 nests ranged from ground level to 30 m, with 61% between 4.5 and 9 m.

Where the nest location was reported, most of the nests (98%; *n* = 87) were attached to a branch or forked branch, or were among the intertwined branches, of a tree or shrub. The base of the nests and most of their structure consisted of strong twigs or branches; the cup was lined with bark strips (willow, cottonwood, cedar), grass, moss, fine rootlets, animal hair (cattle, goat, horse), and a variety of human-made materials.

Eggs: Dates for 189 clutches ranged from 9 April to 11 June, with 51% recorded between 24 April and 12 May. Calculated dates indicate that eggs can occur as early as 8 April. Sizes for 141 clutches ranged from 1 to 7 eggs (1E-8, 2E-9, 3E-12, 4E-40, 5E-51, 6E-19, 7E-2), with 65% having 4 or 5 eggs (Fig. 233). One clutch in British Columbia had an incubation period of 18 days; in Saskatchewan, Ignatiuk and Clark (1991) calculated a mean incubation period of 17.7 days from 74 nests.

Young: Dates for 103 broods ranged from 26 April to 11 July, with 54% recorded between 21 May and 7 June. Sizes of 85 broods ranged from 1 to 5 young (1Y-9, 2Y-23, 3Y-18, 4Y-22, 5Y-13), with 74% having 2 to 4 young. The nestling period of the American Crow has been reported as 28 and 35 days (Good 1952) and 31.5 ± 1.6 days (*n* = 71) (Ignatiuk and Clark 1991).

Nest Success: Of 16 nests found with eggs and followed to a known fate, 13 produced at least 1 fledgling, for a nest success rate of 81%. In Saskatchewan, Ignatiuk and Clark (1991) found a success rate of between 45% and 72% in various localities.

REMARKS: The American Crow was formerly known as the Common Crow.

Discussion still continues as to whether the evidence supports species status for the American and Northwestern crows or whether subspecies status better represents their relationship. Johnston (1961) studied the anatomical relationship and

Figure 233. American Crow nest containing five well-incubated eggs (north of Oliver, 22 April 1994; R. Wayne Campbell).

Figure 234. Abandoned American Crow nests are important nest sites for the Long-eared Owl in British Columbia (near Douglas Lake, 19 May 1994; R. Wayne Campbell).

believed that his evidence favoured the subspecies alternative, but the American Ornithologists' Union (1983) concluded that species status for both the American and Northwestern crows should be maintained until more convincing evidence was available. Sibley and Monroe (1990) have followed the same course.

There are a number of coastal sight reports of this species, as well as reports from coastal Breeding Bird Surveys (e.g., Chilliwack, Kispiox, Kitsumkalum, Kwinitsa, Nanaimo River, Pemberton, and Saltery Bay) and Christmas Bird Counts (Chilliwack, Terrace, Victoria, and Whistler). However, these areas lie within the range of the more abundant Northwestern Crow. It is generally agreed that separation of the 2 species in areas of contact can be achieved with certainty only by examining birds in the hand or specimens. Accordingly, field identification of crows from areas of probable contact between the 2 species have been accepted only when supported by recently confirmed specimen evidence. Differences in the bill size and shape and in other dimensions, as well as in certain features of internal anatomy, permit the separation of these 2 species (Johnston 1961; Pyle et al. 1987).

Laing (1935) reported collecting 2 specimens near Courtenay from a large flock that he identified by call notes as being American Crows; however, we have not been able to locate these specimens. Later, Laing (1942) reported the American Crow from Rivers Inlet, Kimsquit (NMC 29044, 29061, and 29063) and Hagensborg (NMC 29062) in the Bella Coola valley, all localities along the central coast where deep fiords give passage to large rivers from the interior. All these specimens are June or July adult females. Our examination of the specimens collected by Laing revealed that the Kimsquit specimens are Northwestern Crows while the single specimen from Hagensborg is an American Crow. Measurements provided by Johnston (1961) of specimens from Bella Coola suggest that they are Northwestern Crows, though he regards them as possibly intermediate between the 2 species. We have not seen any specimens of the American Crow from the Fraser River delta or Vancouver Island.

Small numbers of American Crows were released in the Richmond area of the Fraser River delta in 1968 and 1969, when 2 separate families of 3 and 4 young from the Penticton area were raised and set free there.

The American Crow is reported as sometimes nesting in small colonies (Goodwin 1976; Kilham 1984). Chamberlain-Auger et al. (1990) report that all crows on their study site at Cape Cod, Massachusetts, bred cooperatively in groups of up to 10 birds. They observed shared incubation and brooding by breeders and helpers. We have no reports of this behaviour in British Columbia.

American Crows have long been persecuted by humans because of their depredation of crops, livestock, and the eggs and young of other birds. Perhaps no other North American bird has been persecuted so persistently and so violently. Government-sponsored killing programs were formerly regular events, particularly in the midwestern and southern United States, where bombing of roosts with dynamite was a favoured technique. Although organized programs of destruction are no longer acceptable, recreational shooting of crows is a still a popular activity in parts of the United States; in some years a million crow hunter–days are reported (United States Department of the Interior and United States Department of Commerce 1982).

For many years the American Crow was also systematically destroyed in British Columbia with the encouragement of a bounty or other inducements. The official total of the number of crows killed in the province in predator-control activities between 1900 and 1925 is 6,500 birds. This crow is not protected by legislation anywhere in Canada and is exposed to a year-round open season. Despite that, it continues to flourish.

Abandoned nests of the American Crow are used by several species of birds as nest sites. In British Columbia, the Long-eared Owl nests almost exclusively in such nests (92%; n = 75; Fig. 234). Other species using crow nests include the Northern Goshawk, Great Horned Owl, Merlin, American Kestrel, Barrow's Goldeneye, Canada Goose, and Mallard.

An American Crow banded near Norman, Oklahoma, on 5 February 1936 was recovered at Alexis Creek on 11 April 1940; another banded near Alexis Creek on 17 June 1950 was recovered in the Puget Sound area in November 1950.

For additional life-history information, see Kilham (1989).

NOTEWORTHY RECORDS

Spring: Coastal – No records. **Interior** – Osoyoos 9 Apr 1985-eggs; Nelson 12 Apr 1977-eggs; White Lake (Okanagan Falls) 19 May 1982-57 (flock); Enderby 15 Apr 1941-4 eggs; Balfour to Waneta 16 May 1981-198 on survey; Lillooet 20 May 1968-52; Yoho National Park 26 Mar 1977-31; Moose Lake marsh 28 May 1972-nestlings; Chilcotin River (Riske Creek) 16 Apr 1989-1 egg; n Williams Lake 23 Mar 1977-62; 10 km s Prince George 22 Apr 1984-70, part of a second migration wave; Prince George 5 May 1974-6 eggs; Fort St. James 2 May 1889-4 eggs; 2 km w Dawson Creek 1 May 1992-1 egg; Hudson's Hope 5 May 1979-50; Halfway River (Peace River) 27 Mar 1983-3; Cecil Lake 31 Mar 1984-2; Hyland Post 8 May 1984-2.

Summer: Coastal – Hagensborg 6 Jun 1938-1 (NMC 29062). **Interior** – Manning Park 11 Aug 1979-20; Creston 20 Jun 1981-60; Nelson 10 Jul 1977-2 young; Sparwood 16 Jul 1983-60 (Fraser 1984); Penticton 8 Jul 1987-3 nestlings; Sorrento 4 Aug 1970-200; Salmon Arm 24 Aug 1965-400 (flock); Sorenson Lake 11 Jul 1955-3 young; Yoho National Park 26 Aug 1976-110; Stum Lake 4 Aug 1973-200 (Ryder 1973); Dragon Lake (Quesnel) 16 Jul 1948-300 in flock; Ten Mile Lake (Quesnel) 18 Jun 1990-3 young; Mount Robson Park 24 Aug 1970-125 (Stirling 1971); Sinkut Lake 19 Jul 1945-90 (Munro 1949); Surel Lake 23 Jul 1975-1; Telkwa 30 Aug 1977-150; Bulkley Lake 28 Jul 1982-2; Fort St. John 21 Jun 1984-18 in flock, 1 Jul 1983-30; Montney 19 Aug 1975-152; Beatton River area 23 Jun 1976-2 nestlings fully feathered; Kitchener Lake 20 Jun 1976-3, 22 Jun 1976-1; Fort Nelson 5 Jul 1978-2; Kotcho Lake 26 Jun 1982-3 (Campbell and McNall 1982).

Breeding Bird Surveys: Coastal – See REMARKS and Appendix 2. **Interior** – Recorded from 63 of 73 routes and on 88% of all surveys. Maxima: Prince George 30 Jun 1978-146; Salmon Arm 13 Jun 1982-142; Grand Forks 26 Jun 1982-136.

Autumn: Interior– Fort Nelson 16 Sep 1986-16; Fort St. John 19 Sep 1979-20 in flock, 3 Oct 1982-2, 10 Oct 1982-1, last autumn departure; Smithers 27 Nov 1987-12; 20 km s Prince George 13 Oct 1982-60 in flock; Riske Creek 14 Sep 1978-185 in flock, 22 Sep 1978-200; Kleena Kleene 15 Sep 1956-150 (Paul 1959); 150 Mile House 29 Nov 1965-30; Kootenay National Park 7 Sep 1965-215 in flock (Seel 1965); Kamloops 5 Sep 1978-200; Sorrento–Salmon Arm 23 Sep 1970-1,030 on survey; Wasa 23 Sep 1965-1,200 in flock; Kimberley 11 Oct 1970-225 in flock; Summerland 18 Sep 1970-200; Penticton 16 Oct 1963-200 (Cannings et al. 1987); Princeton 7 Sep 1977-300; Lightning Lakes 11 Nov 1988-12 in flock; Waldo 20 Sep 1944-220 (Johnstone 1949); Creston Sep 1978-500 in flock (Butler et al. 1986). **Coastal** – No records.

Winter: Interior – Fort St. John 24 Jan 1981-3; Prince George 22 Dec 1968-95; Quesnel 7 Feb 1979-90; Kleena Kleene 1 Jan 1950-1; Wineglass Ranch, Chilcotin River 31 Dec 1987-1; Williams Lake 8 Jan 1984-200, 30 Dec 1990-746 on survey; Revelstoke 31 Dec 1985-2; Sorrento 16 Feb 1972-300; Grindrod 26 Jan 1977-294 in flock; Swan Lake (Vernon) 31 Dec 1978-100; Coldstream 2 Jan 1933-200, 13 Feb 1973-100 (Cannings et al. 1987); Penticton 1 Jan 1978-2; Thrums 31 Dec 1979-50; Nelson 18 Dec 1978-50, 4 Feb 1979-150; Hardy Mountain 30 Nov 1979-50; Creston area 1 Jan 1981-4; Manning Park 28 Dec 1969-40. **Coastal** – No records.

Christmas Bird Counts: Interior – Recorded from 24 of 27 localities and on 81% of all counts. Maxima: Vernon 20 Dec 1992-723; Smithers 30 Dec 1990-597; Salmon Arm, 17 Dec 1988-523. **Coastal** – See REMARKS and Appendix 3.

Northwestern Crow

Corvus caurinus Baird

NOCR

RANGE: Resident along the Pacific coast from south-coastal and southeastern Alaska south through British Columbia west of the Coast Mountains and Cascade Mountains, to northwestern Washington, including the Queen Charlotte Islands, Vancouver Island, and all other coastal islands.

STATUS: *Fairly common* to *common* summer resident along the coast. Throughout the rest of the year a *common* to *very common* resident in the Georgia Depression Ecoprovince, including the Sunshine Coast, Gulf Islands, and southeastern Vancouver Island; *fairly common* to *common* resident on Western Vancouver Island, the Southern and Northern Mainland Coast, and the Queen Charlotte Islands, in the Coast and Mountains Ecoprovince. Locally *abundant* to *very abundant* from late summer through early spring in the vicinity of nocturnal roosts, particularly in the Georgia Depression.

Breeds.

NONBREEDING: The Northwestern Crow occurs along the length of the British Columbia coastline, including Vancouver Island and the Queen Charlotte Islands. Numbers dwindle rapidly with increasing elevation and distance from the sea.

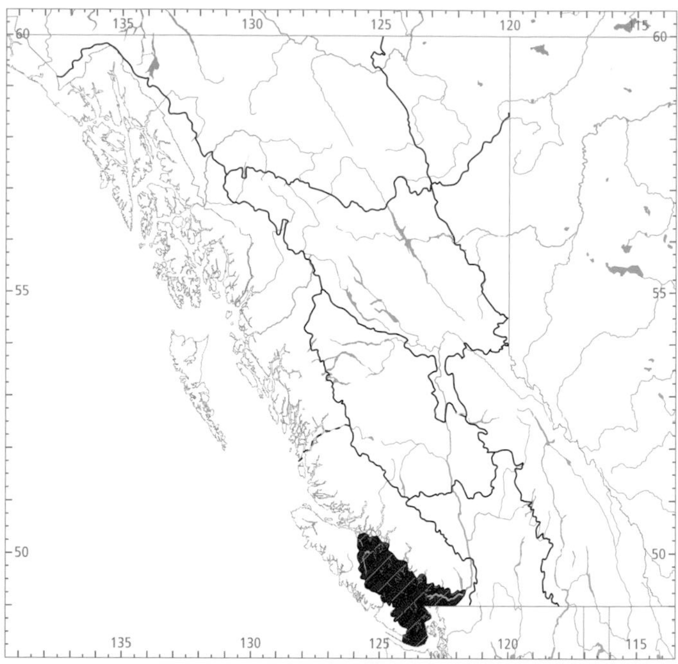

Figure 235. In British Columbia, the highest numbers for the Northwestern Crow in winter (black) and summer (red) occur in the Georgia Depression Ecoprovince.

Figure 236. Shoreline habitat in British Columbia provides important intertidal foraging areas for the Northwestern Crow throughout the year (Langara Island, Queen Charlotte Islands, 7 June 1988; R. Wayne Campbell).

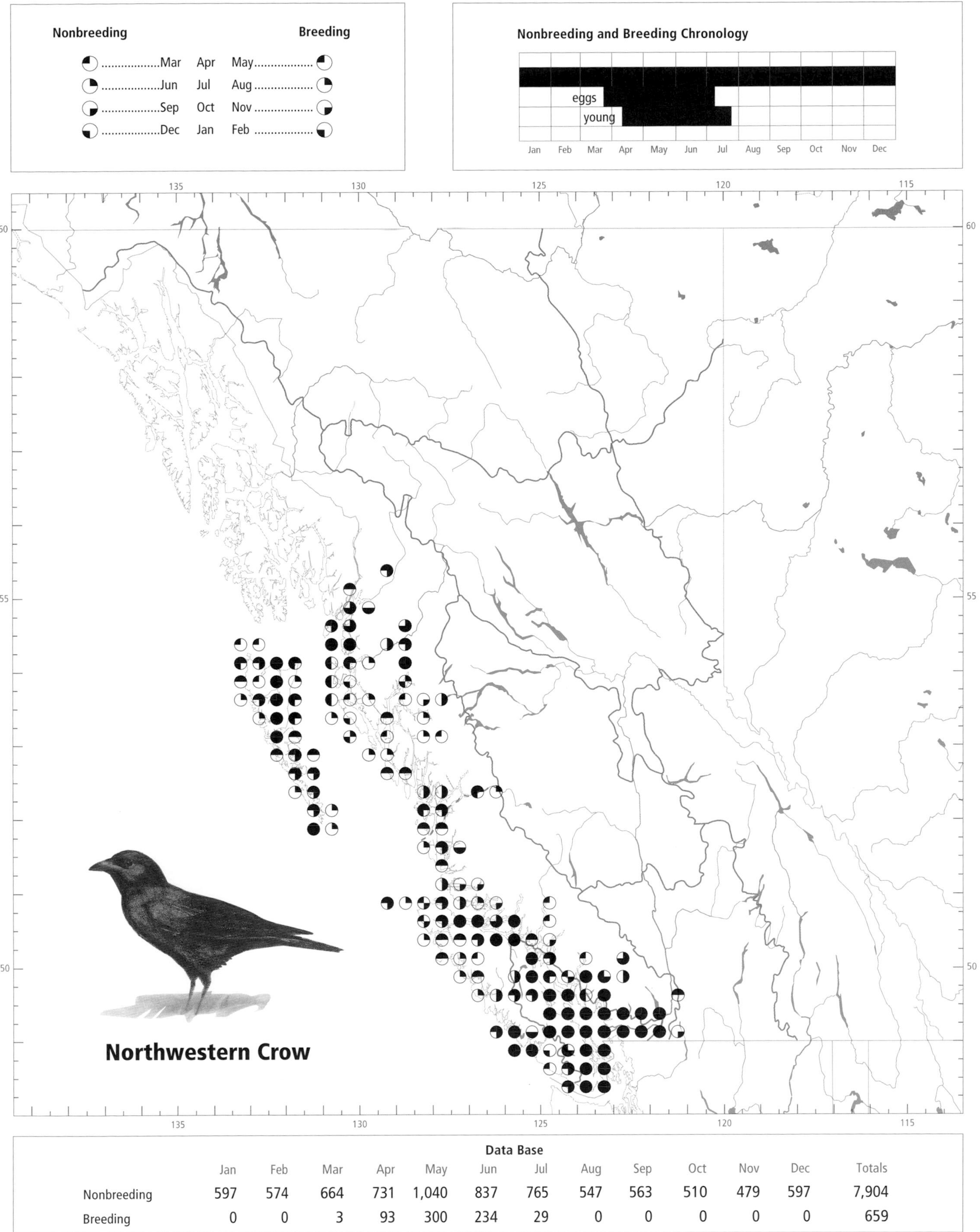

Data Base

	Jan	Feb	Mar	Apr	May	Jun	Jul	Aug	Sep	Oct	Nov	Dec	Totals
Nonbreeding	597	574	664	731	1,040	837	765	547	563	510	479	597	7,904
Breeding	0	0	3	93	300	234	29	0	0	0	0	0	659

The absence of records from the northern Boundary Ranges of British Columbia confirms evidence from the Alaskan southeast (Swarth 1922; Gabrielson and Lincoln 1959) that this crow, at the northern end of its range in regions with intact natural habitats, does not normally occur more than a few kilometres from tidewater. A notable exception to this coastwise concentration occurs when spawning runs of fish enter the large rivers and the Northwestern Crow follows the food source as much as 120 km upriver (W. Prescott pers. comm.). This is especially notable on the Fraser, Skeena, and Nass rivers.

Although the Northwestern Crow is widespread along the coast, its centre of abundance is in the Georgia Depression, including the southeast coast of Vancouver Island and the lower Fraser River valley (Fig. 235).

The Northwestern Crow occurs at elevations from sea level to 1,700 m, but it is most numerous at lower elevations. Although it frequents a wide range of natural and human-influenced habitats, it is closely linked to marine shorelines, where it forages much of the time. Thus, shallow marine shores with interspersed rocky headlands (Fig. 236), boulder beaches, and sand, gravel, or mudflats are the primary feeding grounds of this crow. It occurs infrequently or is absent from the sheer rock margins of the many fiords that characterize the mainland coast. There, the steep, rocky shorelines plunge directly into deep water, providing few foraging opportunities for this bird. Moreover, the surface water of these inlets is low in salinity, and the shore fauna of invertebrates, on which the crow feeds, lacks variety. Tidal estuaries at the heads of the fiords normally support a small population of crows.

The availability of exposed intertidal areas to crows varies with the season. During the winter months, the lowest tides, and therefore the richest feeding habitats, occur at night; during the spring and summer, low tides occur during daylight hours. This crow responds by changing foraging practices with the seasons and concentrating more on human-altered habitats when the beaches are less available. At such

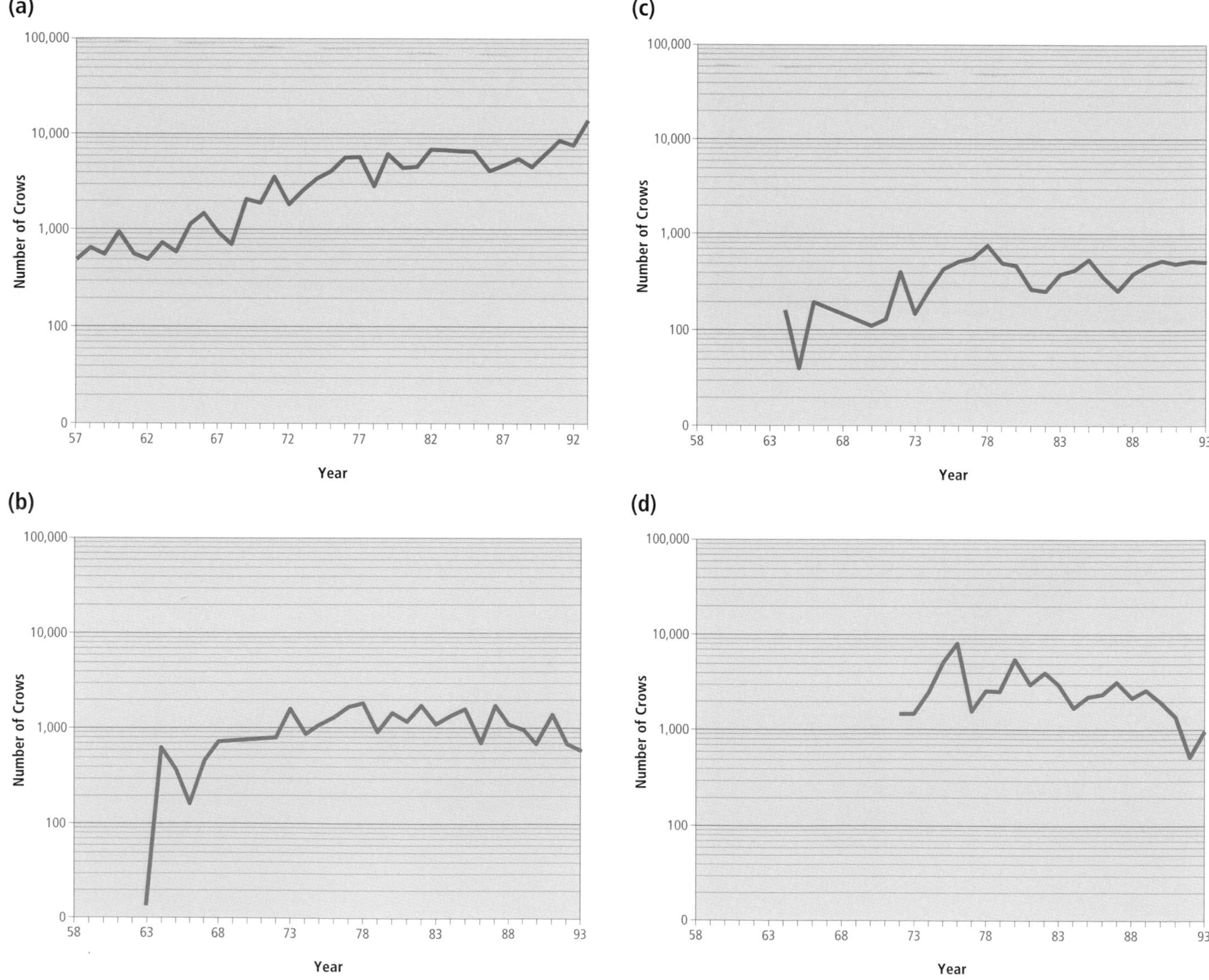

Figure 237. Fluctuations in Northwestern Crow numbers in winter in select areas of southwestern British Columbia, comparing: (a) a large city, Vancouver; (b) a small city, Nanaimo; (c) a rural environment, Pender Island; and (d) an agricultural area, Pitt Meadows. Based on Christmas Bird Counts from 1957 to 1993.

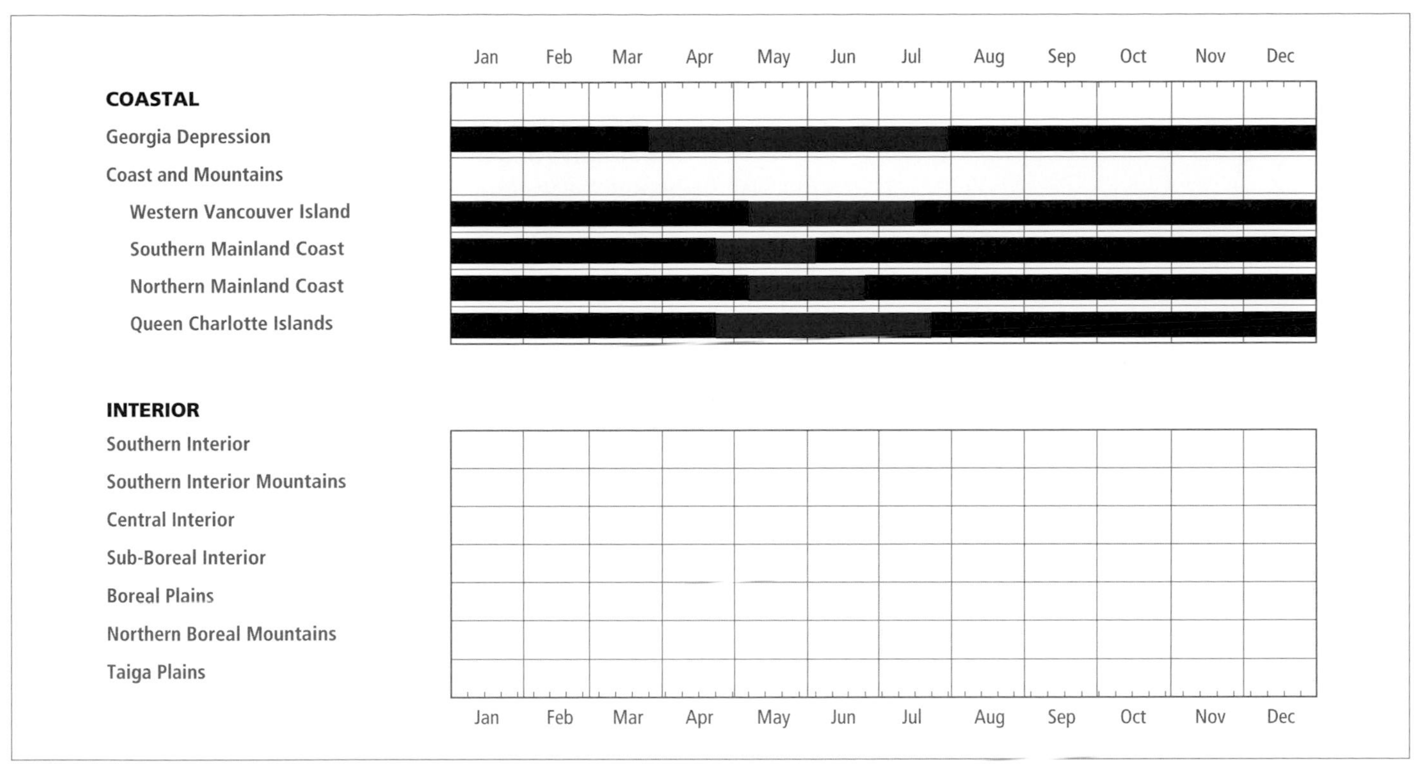

Figure 238. Annual occurrence (black) and breeding chronology (red) for the Northwestern Crow in ecoprovinces of British Columbia.

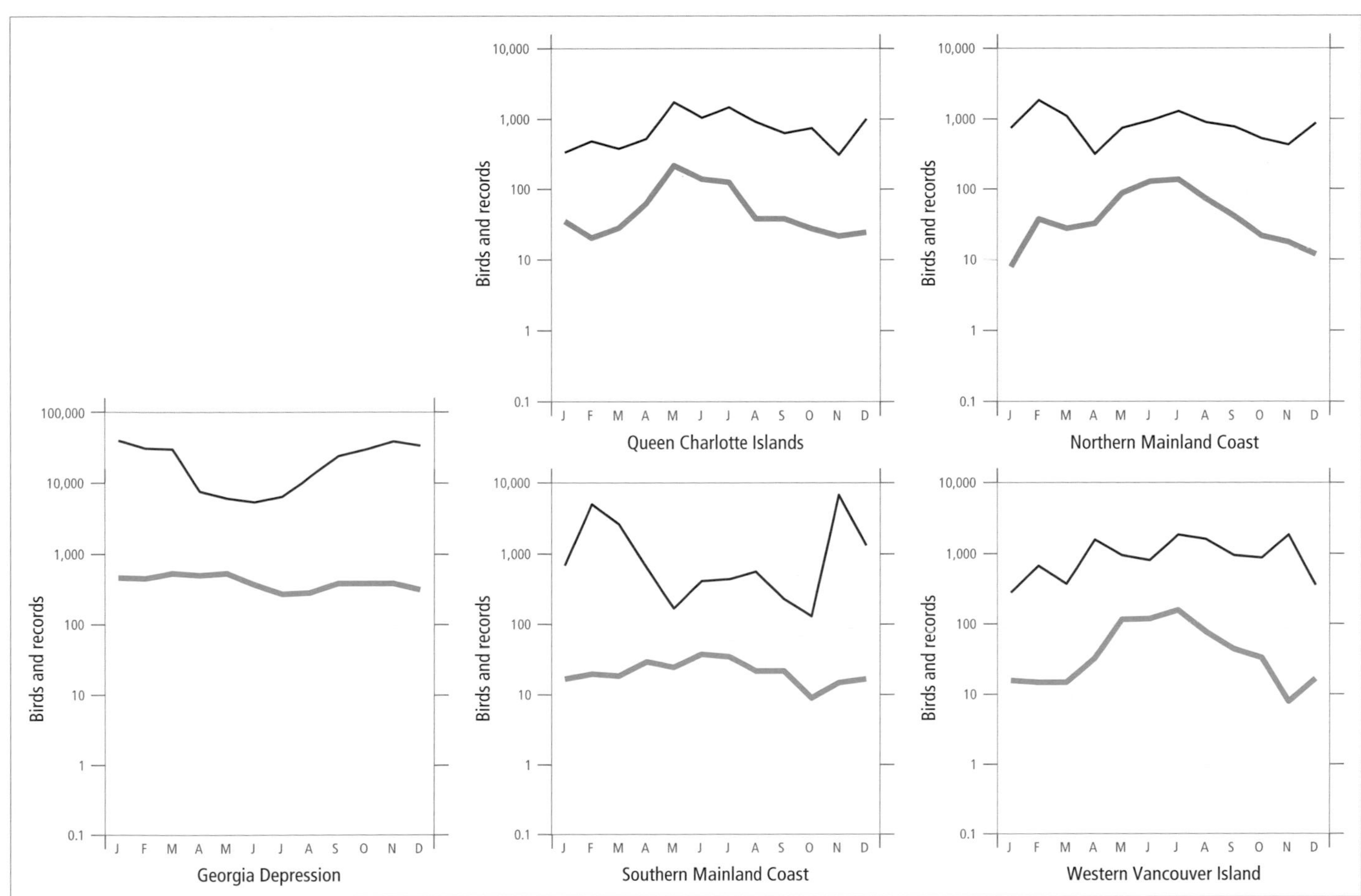

Figure 239. Fluctuations in total number of birds (purple line) and total number of records (green line) for the Northwestern Crow in ecoprovinces of British Columbia. Christmas Bird Count and nest record data have been excluded.

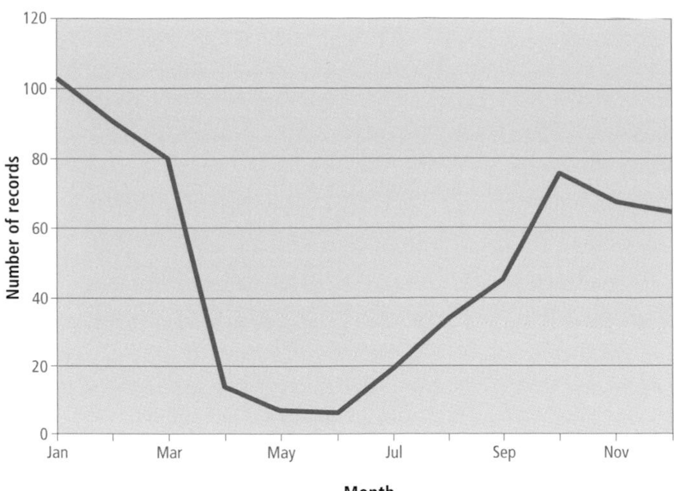

Figure 240. Monthly distribution of the number of records of Northwestern Crows with 100 or more birds in British Columbia (*n* = 441). Counts at roosts and surveys have been excluded.

times, the tameness of the Northwestern Crow becomes notable as it forages on lawns, roadsides, parking lots, ferry wharfs, marinas, refuse piles, garbage landfills, and similar areas.

It adapts easily to the environmental changes imposed by people and, during the nonbreeding parts of the year, is now more abundant around towns and villages than in its natural environment. Deforestation has helped increase crow populations in some areas. Habitats now include farm fields, pastures, playgrounds, residential neighbourhoods, golf courses, cemeteries, and bush camps. Farm fields sown with corn and grains are favoured after the harvest. Garbage landfills have been particularly important in providing a winter food source. This close association with humans and the attendant ease of acquiring food, particularly through the winter months, has likely contributed to the increase in the populations of the Northwestern Crow in urban areas. An analysis of Christmas Bird Counts for the period 1975 to 1994 in select areas of southwestern coastal British Columbia indicates that the larger cities attract more crows in winter than do smaller cities and rural sites, and that agricultural areas are also important foraging sites (Fig. 237), where this crow uses seasonally available food sources such as fruit and grain.

The Northwestern Crow is not normally considered migratory in British Columbia (Fig. 238), but local movements appear to be widespread. Birds move from exposed coasts to more protected areas in autumn (Fig. 239). With the coming of spring, winter flocks disperse, and areas of coastline that were almost without crows during the winter receive their courting pairs. Total numbers of crows reported from the Georgia Depression illustrate this point. They average about 30,000 birds from September through March, then drop to an average of about 7,000 birds from April through August (Fig. 239). This difference can be only partly attributed to flocking behaviour during the nonbreeding season. Pearse (*in* Bent 1946) reports evidence of movements to farmlands on eastern Vancouver Island from the lightly inhabited mainland coast.

Behaviour important to the Northwestern Crow during the nonbreeding season is its congregation into large flocks that move between feeding areas and roosting sites. Of 608 records consisting of 100 or more birds (excluding Christmas Bird Counts or other such surveys), 87% were reported from September through March (Fig. 240). In most areas, local movements occur twice daily: once at dawn, when the entire population of the roost leaves for foraging areas, and again in late afternoon, as the crows muster in noisy flocks to assemble at the roost sites for the night.

On the south coast, some of these movements are spectacular, with birds streaming by in flocks of hundreds or, occasionally, thousands (Fig. 241). For example, in March 1988, 2,071 crows were counted arriving at a roost on Sidney Island between 1810 and 1859 hours. Twenty-one flocks were counted. They ranged in size from 6 to 460 birds, with an average of 99 birds per flock; average time between flocks was 2.4 minutes (R.W. Butler pers. comm.).

Some of the largest known roosts are on Chatham Island, James Island, Grouse Mountain, Douglas Island (Port Coquitlam), Burns Bog, and Burnaby, but there are many local roosts occupied by smaller numbers of crows. The roosts are usually situated in undisturbed areas of large, old conifers, sometimes interspersed with arbutus, Garry oak, and bigleaf maple. Frequently they are on islands away from human activity. Little is known about the characteristics of these traditional roosts sites, or of their role in the biology of the Northwestern Crow. Northern populations have not been reported flocking in such numbers.

The Northwestern Crow is present throughout the year in British Columbia (Fig. 238).

BREEDING: The Northwestern Crow has a widespread breeding distribution along the British Columbia coast from southern Vancouver Island north to at least the Queen Charlotte Islands and Portland Inlet. Its eastward breeding distribution is imperfectly known but reaches at least to Chilliwack in the Georgia Depression and to Kitimat on the Northern Mainland Coast. Most birds nest within a few kilometres of the sea.

Figure 241. Part of a large evening roost flight of Northwestern Crows passing over Cadboro Bay village, Victoria (1942 hours, 18 August 1992; R. Wayne Campbell).

In summer, as in winter, its highest numbers are in the Georgia Depression (Fig. 235). An analysis of Breeding Bird Surveys for the Northwestern Crow in British Columbia shows that the mean number of birds on coastal routes increased at an average rate of 2% over the period 1968 through 1993 (Fig. 242).

This species has been reported breeding from near sea level to 510 m elevation. Much of the coastline of the province is margined by coniferous forests with a varied understorey of berry-producing shrubs. Where the coastal strip of forest has not been removed by clearcut logging, it provides nesting and roosting sites for this crow close to the foraging area of the beach (Fig. 243). Nesting sites are usually near a forest edge that supports many smaller trees and tall shrubs. Other nesting habitats include estuaries, river valleys, lakeshores, offshore islands, marshes, and sloughs.

The population of crows nesting within cities is increasing, but often goes unnoticed except perhaps for the early-morning prelaying courtship-begging of females. As nesting proceeds, however, some crows begin to harass passers-by, occasionally making contact with them, usually about the head. Once the young are off the nest and begin their raucous begging for food throughout the day, they seldom go unnoticed. Autumn leaf-fall exposes the crow nests, which can be seen even in boulevard trees no more than 3 m high.

In selecting its nesting territory, the Northwestern Crow tends to avoid extensive areas of dense forest far removed from the ocean. Nesting habitat includes the hundreds of small rocky islets scattered along the coast. Many of these support nesting colonies of marine birds and feature a lush vegetation of forbs, bulbous plants, grasses, stunted crab apple, elderberry, salmonberry, snowberry, spirea, Pacific ninebark, and even a few wind-wracked conifers. On these islands and islets, the Northwestern Crow frequently nests in low shrubs or on the ground (Fig. 244).

The Northwestern Crow has been recorded breeding in British Columbia from 23 March (calculated) to 24 July (Fig. 238).

Colonies: Colonial nesting has been reported for this species (Bent 1946). In British Columbia, groups of 2 to 19 nesting pairs occurring on small islands and along hedgerows and boulevard trees have been referred to as colonies. However, we do not consider this species to be a colonial nester in British Columbia.

Figure 243. At Bella Coola, the Northwestern Crow breeds in mixed forests close to foraging areas in estuaries and beaches (10 July 1992; R. Wayne Campbell).

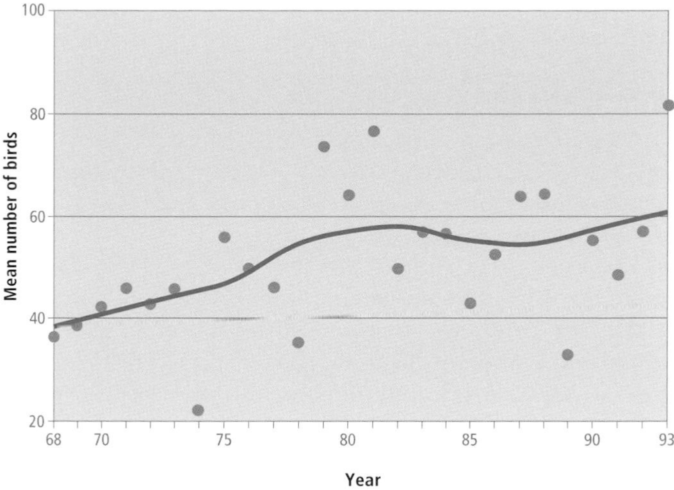

Figure 242. An analysis of Breeding Bird Surveys for the Northwestern Crow in British Columbia show that the number of birds on coastal routes increased at an average annual rate of 2% over the period 1968 through 1993 ($P < 0.07$).

On Mitlenatch and Mandarte islands, each pair defends a territory of about 0.5 ha in which the nest is situated (R.W. Butler pers. comm.); 22% of the breeding pairs are assisted by yearlings in defending the territory (Verbeek and Butler 1981).

Nests: Most nests (64%; n = 391) were found in trees, both coniferous (31%) and deciduous (18%; Fig. 245). The most often used conifers were Douglas-fir (11%), Sitka spruce (10%), and lodgepole pine (4%), followed by western redcedar, hemlock, and grand fir. Pacific crab apple (5%) and bitter cherry (4%) were the most often used deciduous trees, followed by willow, red alder, black cottonwood, arbutus, maple, Pacific dogwood, and Garry oak. On rocky island sites, where trees were scarce or absent, nesting in shrubs (16%) or on the ground (14%; Fig. 244) was commonplace, a behaviour first reported by Darcus (1930). Shrubs included saskatoon (4%), Pacific ninebark (2%), willow, Nootka rose, thimbleberry, and elderberry. In addition, virtually every other local species of native tree and shrub was reported to have harboured a nest.

Where the nest situation was described, most nests (63%; n = 269) were found among the branches of trees or shrubs. Nests in conifers were often saddled on the branches, sometimes close to the trunk. In deciduous trees and shrubs the nests were almost always placed among the branches well away from the trunk. Ground nests were found under or among ground cover (10%), beneath a log or rock, on rocky ledges, in a rock crevice, or among the lower branches of shrubs.

The bulky nests were composed primarily of twigs, sticks, and small branches (Fig. 245). Nests were lined mainly with bark strips, followed by grasses, moss, feathers, fine rootlets, plant fibres, and sometimes hair or human-made items such as paper, rope, fiberglass insulation, and mattress filling. On

Mitlenatch Island, all nests were lined with cedar bark strips even though no cedars grew on the island (Butler 1980).

The heights of 372 nests ranged from ground level to 45 m, with 59% between 1.2 and 8 m.

Eggs: Dates for 270 clutches ranged from 27 March to 8 July, with 52% recorded between 2 and 26 May. Calculated dates indicate that eggs can occur as early as 23 March. Sizes of 241 clutches ranged from 1 to 6 eggs (1E-10, 2E-19, 3E-46, 4E-98, 5E-63, 6E-5), with 67% having 4 or 5 eggs (Fig. 244). In a sample of 187 clutches from Mandarte and Mitlenatch islands, 54% contained 4 eggs (Richardson et al. 1985). On Mitlenatch Island, the mean incubation period from the day the last egg was laid until all were hatched was 18.3 ± 0.85 days (n = 19) (Butler et al. 1984).

Young: Dates for 174 broods (Fig. 245) ranged from 10 April to 24 July, with 53% recorded between 25 May and 20 June. Sizes of 148 broods ranged from 1 to 6 young (1Y-29, 2Y-49, 3Y-43, 4Y-23, 5Y-3, 6Y-1), with 62% having 2 or 3 young. The nestling period was reported as 32 ± 2.5 days (n = 20) on Mandarte Island and 26 ± 3.4 days (n = 55) on Mitlenatch Island; the earlier departure in the latter case may have been induced by observer disturbance (Butler et al. 1984).

Nest Success: Of 28 nests found with eggs and followed to a known fate, 16 produced at least 1 fledgling, for a success rate of 57%. On Mandarte and Mitlenatch islands, 79% of nests fledged at least 1 young (Butler et al. 1984). On the same islands, Richardson et al. (1985) found that the proportion of young that survived from hatching to fledging was 67% and 37%, respectively. Clutches of 3, 4, and 5 eggs fledged an average of 0.98, 1.22, and 1.20 young, respectively; 51% of the nestlings died within 7 days of hatching, and starvation appeared to be the predominant cause of death.

REMARKS: Opposing views have been expressed regarding the systematic status of the 2 crows in British Columbia. The American Ornithologists' Union (1983) notes that the Northwestern Crow and American Crow are closely related and are considered conspecific by some authors (Johnston 1961; American Ornithologists' Union 1983; Sibley and Monroe 1990).

Difficulties in field identification of the 2 species make it impossible to determine, under field conditions, the precise areas of overlap in their distribution (see **American Crow**). On the basis of specimens collected and possible differences in call notes, the Northwestern Crow is thought to occur exclusively west of the coastal mountains, and the American Crow east of the mountains. However, numerous potential zones of contact are available along major river systems such as the Fraser, Skeena, Nass, and Stikine, or along highway corridors. We have deliberately excluded observational reports from these areas, because of the uncertainty of identification. All coastal records of American Crow and all interior records of Northwestern Crow that are not supported by specimens have been excluded. Any interior reports of Northwestern Crow should be treated cautiously, pending further research.

The Northwestern Crow may be highly philopatric. With the exception of 1 report of a crow banded near Comox on 11 June 1979 and recovered north of Victoria on 26 November 1980, all recoveries (n = 38) of the Northwestern Crow were from the banding location, some after a number of years had passed. For example, a bird banded near Campbell River on 22 June 1969 was recovered there in January 1981, nearly 12 years later; another, banded in the same area on 3 June 1977 was recovered there over 7 years later, on 15 November 1984.

Figure 244. Ground nest containing 4 eggs of the Northwestern Crow among branches of a saskatoon (Mitlenatch Island, June 1968; R. Wayne Campbell).

Figure 245. In British Columbia, broods of Northwestern Crows have been recorded between 25 May and 20 June (Mitlenatch Island, 4 June 1969; R. Wayne Campbell).

The Northwestern Crow is a major predator and scavenger on coastal seabird colonies (Fig. 246). On Mandarte Island, for example, it is a major predator upon the eggs of cormorants, the Glaucous-winged Gull, and the Pigeon Guillemot. There it consumes 22% of all eggs laid in the first clutches of Double-crested and Pelagic cormorants (Verbeek 1982), and also preys on the newly hatched young of the cormorants (Butler et al. 1985).

The disturbance caused by people landing among the cormorant nesting colonies results in the parent birds leaving the nests unprotected. The nests are then raided by the crows. Thus the number of fledglings produced by a colony is strongly influenced by the frequency of human disturbance at the colonies. Even a single landing party a week can result in many cormorant nests failing to produce fledglings. On islands with burrow-nesting seabirds, crows not only scavenge food droppings but also prey on the nestlings as they emerge from their burrows (Butler et al. 1985). The Northwestern Crow frequently preys upon the eggs and nestlings of many species of small birds.

Old nests of the Northwestern Crow are used as nest sites by several species of birds. In British Columbia, the Great Horned Owl, Merlin, and Long-eared Owl have been documented nesting in them.

British Columbia probably supports most of the world's population of Northwestern Crows.

Although considered wildlife under the British Columbia Wildlife Act, Northwestern Crows are afforded no protection from shooting for much the same reasons discussed under **American Crow**.

For additional information on the biology and behaviour of the Northwestern Crow, see Butler (1974, 1980), Verbeek and Butler (1981), Butler et al. (1984), and Richardson et al. (1985).

Figure 246. The Northwestern Crow is a major predator on seabird colonies in British Columbia. On Mandarte Island, Verbeek (1982) found that it consumed 22% of all eggs laid in first clutches of Double-crested and Pelagic cormorants (predated Double-crested Cormorant eggs, Mandarte Island, 10 July 1981; R. Wayne Campbell).

NOTEWORTHY RECORDS

Spring: Coastal – Chatham and Discovery islands 16 Mar 1976-1,400; James Island 11 Apr 1984-1,500; Sidney Island 26 Mar 1988-2,071; Cowichan Bay 16 Apr 1977-350; Agassiz 10 Mar 1982-1,000; Vancouver 27 Mar 1972-first day of incubation; 5 km n Agassiz 26 Apr 1975-5 eggs; Harrison Hot Springs 9 Mar 1975-1,000, 10 Apr 1974-200; West Vancouver 5 Mar 1972-2,027; Vedder Crossing 12 Mar 1972-1,000; Egmont 11 May 1979-82; Lennard Island 7 May 1980-1 egg; George Island 27 Apr 1965-4 eggs; Cumshewa Inlet 2 Apr 1979-42; McClinton Bay 22 May 1985-30 (Cooper 1985); Kitimat 10 May 1980-4 eggs; Minette Bay 21 Mar 1975-175 (Hay 1976). **Interior** – No records.

Summer: Coastal – Chatham Islands 31 Jul 1982-700; Mandarte Island Jun 1960-25 pairs (Drent and Guiguet 1961), 24 Jul 1957-3 young; Florencia Island 15 Jul 1969-2 young flew from nest when checked; Gibsons Landing 13 Jul 1967-3 nestlings; Harrison Hot Springs 28 Aug 1975-175; Moos Islet 24 Jun 1975-50; Triangle Island Jul 1984-37, maximum evening roost count (Butler et al. 1984); Storm Islands 7 Jun 1982-26; Darby Channel 8 Aug 1977-150; Bella Coola 17 Jul 1978-75; Tar Islands 13 Jun 1985-44; Skincuttle Inlet 7 Jul 1977-102; Queen Charlotte City 4 Aug 1979-150; Wells Rocks 20 Jun 1980-1 nestling with primaries half emerged; Masset 28 Jun 1919-nestlings; Mace Creek 20 Aug 1983-80; Lawyer Island 16 Jun 1979-1 egg, 2 nestlings; Port Simpson 17 Jul 1967-200; Stewart 17 Jun 1975-6; Bell-Irving River 13 Jul 1978-7. **Interior** – No records.

Breeding Bird Surveys: Coastal – Recorded from 22 of 27 routes and on 79% of all surveys. Maxima: Albion 5 Jul 1979-216; Victoria 30 Jun 1974-206; Point Grey 12 Jun 1988-182. **Interior** – Not recorded.

Autumn: Interior – No records. **Coastal** – Kitimat dump 21 Nov 1974-120 (Hay 1976); Selwyn Inlet 6 Oct 1976-120; Flood 26 Nov 1976-2,500; Alouette Lake 18 Nov 1975-2,300, 19 Nov 1975-3,910; Pitt Meadows 19 Sep 1965-1,657; Tofino Inlet 10 Nov 1961-1,500; Burns Bog 28 Oct 1973-1,500; Boundary Bay 12 Nov 1962-1,500; Saanich 7 Nov 1987-2,000 in cornfield; James Island 16 Oct 1987-3,060; Blenkinsop Lake 17 Nov 1985-1,700.

Winter: Interior – No records. **Coastal** – Prince Rupert 20 Jan 1983-300, 31 Dec 1959-49 (Hammer 1960); Masset 11 Feb 1973-44; Alliford Bay 12 Feb 1973-140; Skidegate Inlet 30 Dec 1982-213; Courtenay 18 Feb 1945-1,500; Grouse Mountain 22 Jan 1983-1,634; Harrison Hot Springs 9 Feb 1987-5,000; mouth of Harrison River 4 Dec 1976-4,911; Langley 28 Jan 1977-3,000; Bamfield 23 Feb 1979-250; Kildonan 31 Dec 1915-2 (NMC 8904-5); Moresby Passage 14 Jan 1986-2,000; Sidney 2 Jan 1992-2,932 passed by in 1.5 hours, 21 Jan 1991-4,261 passed by in 1.5 hours, 5 Feb 1992-4,790 passed by in 1.5 hours; Ucluelet 3 Jan 1976-25; Agassiz 1 Jan 1976-500; Sidney Island 15 Jan 1979-1,000; Cordova Bay 1 Dec 1987-1,000, 26 Dec 1974-2,000; Whiffin Spit (Victoria) 1 Jan 1976-2; Sooke 31 Dec 1983-304.

Christmas Bird Counts: Interior – Not recorded. **Coastal** – Recorded from 32 of 33 localities and on 96% of all counts. Maxima: Vancouver 19 Dec 1993-**13,167**, all-time North American high count (Ortego 1994); Pitt Meadows 27 Dec 1976-7,903; White Rock 30 Dec 1978-7,380; Victoria 1 Jan 1972-7,131.

Common Raven

CORA

Corvus corax Linnaeus

RANGE: Holarctic. In North America, resident from western and northern Alaska, Yukon, and the Northwest Territories (north to southern Ellesmere Island) and east across Canada to Quebec and the Maritime provinces, south throughout Canada (absent from southeastern Alberta, southern Saskatchewan, and southwestern Manitoba) and the United States into Baja California, Mexico, and Nicaragua. In the southern United States, occurs east to the Rocky Mountains and Texas. Palearctic, and south in the Eastern Hemisphere to include northern Africa, the Mediterranean region, the Near and Middle East, the Himalayan ranges, China, Tibet, and Japan.

STATUS: *Uncommon* to *fairly common* resident along coastal and throughout southern British Columbia, becoming locally *uncommon* to *fairly common* from the vicinity of Prince George, in the Sub-Boreal Interior Ecoprovince, northward. Locally *common* to *very common* during the nonbreeding season around concentrated food sources. Breeds.

Figure 247. Adult Common Raven (Fort Nelson, 26 June 1996; R. Wayne Campbell).

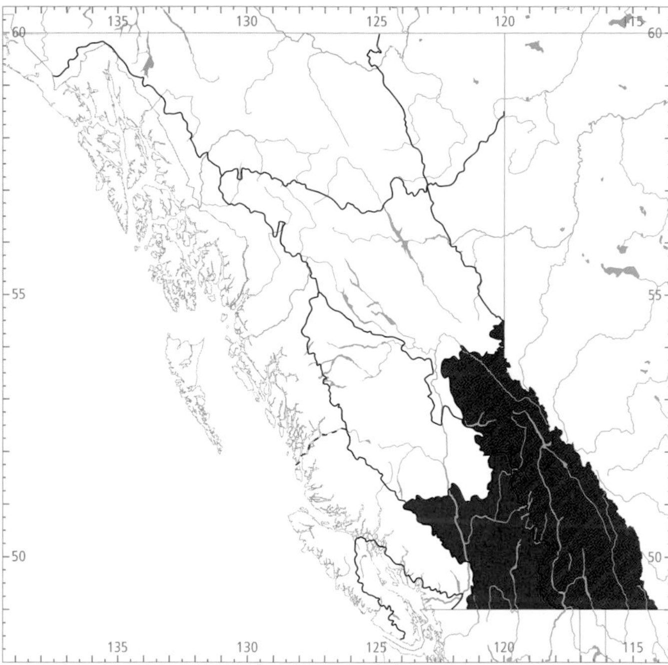

Figure 248. In British Columbia, the highest numbers for the Common Raven in winter (black) occur in the Southern Interior and Southern Interior Mountains ecoprovinces; the highest numbers in summer (red) occur in the Southern Interior Mountains Ecoprovince.

CHANGE IN STATUS: The Common Raven (Fig. 247) has become more widely distributed and more numerous in central and northern British Columbia during the past few decades, but the details of its early status in the far northern parts of the province are unclear. In 1899 Osgood and Bishop explored the Yukon River basin, including the headwater valleys of the river in British Columbia. Bishop (1900) reports that "of all the birds we met the raven occurred most regularly." He encountered a few at White Pass in early June, a flock of "at least 200" at Log Cabin on 20 June, and another of about 50 at Bennett on 22 June (Bishop 1900). This gives a picture of the raven as generally distributed, at least along the travel routes. The Yukon gold rush was at its height at the time and many thousands of people were travelling the rivers and lakes of the Yukon and adjacent British Columbia in search of gold. Their presence led to a local abundance of carrion and garbage attractive to the raven.

In 1914 Kermode and Anderson visited the Atlin region of northwestern British Columbia between late May and late August. They saw no ravens until 20 July and noted that by mid-August the birds were flocking; they commonly saw 40 to 100 of them in a flock (Anderson 1915a). A decade later, Swarth studied the birds of the Atlin region from early May to late September, and reported only a single sighting of the raven at Teslin Lake on 12 September (Swarth 1926). In 1919 Swarth (1922) reported the raven to be "not abundant at any point." He was in the vicinity of Telegraph Creek and Glenora from 23 May to 26 July, and first recorded the raven on 17 June.

Williams (1933a) refers to a record of a raven at Fort St. John in May 1922, and states that the species was occasionally seen on the routes of his travels along the Peace, Athabasca, Liard, and Sikanni Chief rivers in 1921 and 1922. Cowan (1939) records the raven in the Peace River region of the Boreal Plains Ecoprovince only once between 2 May and 30 June. The general impression gained is that in the first quarter of the 1900s, the raven was widely distributed in small numbers through much of northern British Columbia, where it was infrequently encountered during the spring and summer but became locally more obvious during the autumn and winter. The observation by Rand (1944) that the raven became common along the route of the Alaska Highway when the highway construction camps were active confirms the close relationship in the north between the abundance of the raven and access to the food resources that accompany human settlement.

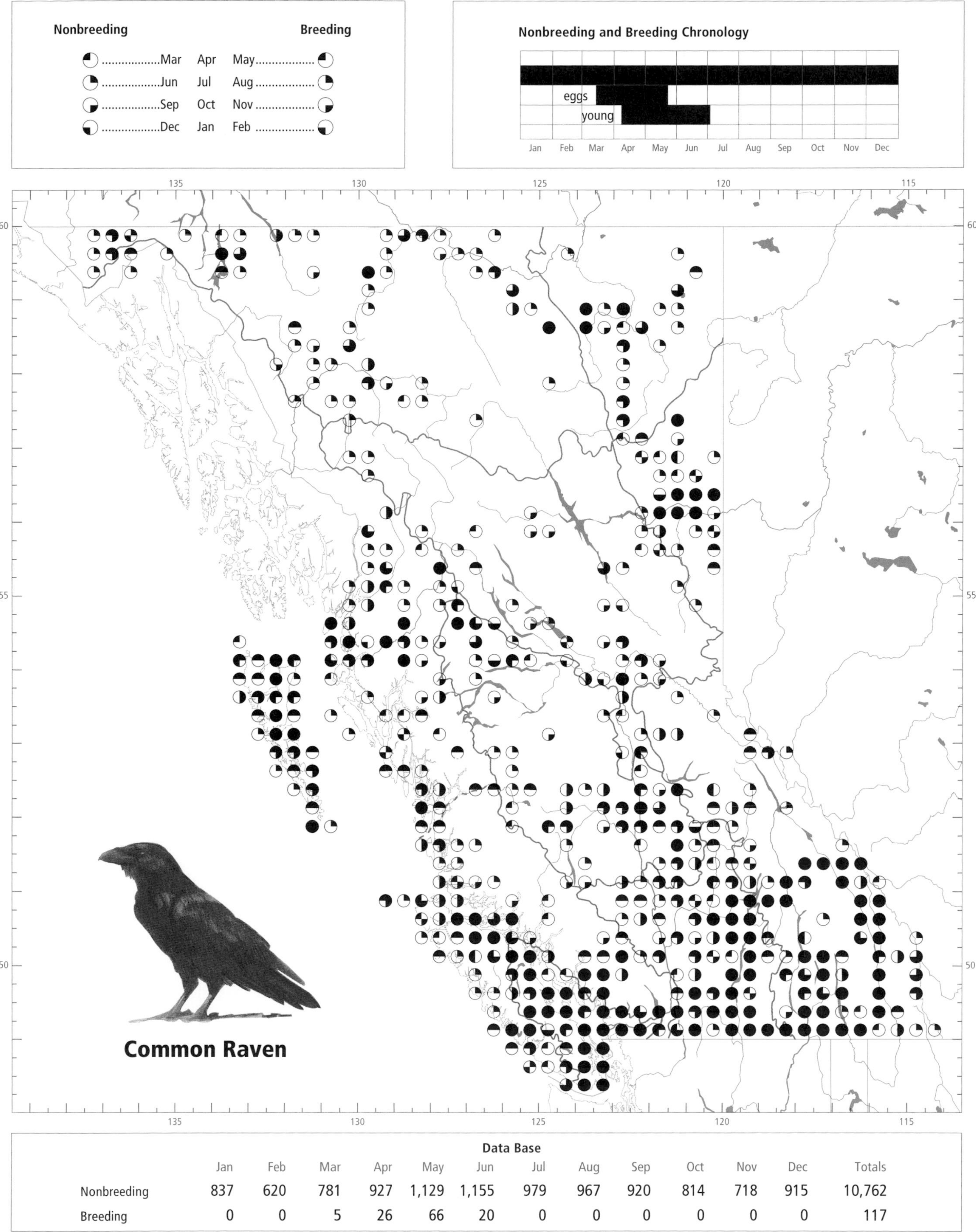

Data Base

	Jan	Feb	Mar	Apr	May	Jun	Jul	Aug	Sep	Oct	Nov	Dec	Totals
Nonbreeding	837	620	781	927	1,129	1,155	979	967	920	814	718	915	10,762
Breeding	0	0	5	26	66	20	0	0	0	0	0	0	117

Further south, in the interior of the province, the raven was not seen in the Vanderhoof region (Sub-Boreal Interior Ecoprovince) in the summers of 1944 and 1945 (Munro 1949), or in the region south of Prince George in 1944 (Munro 1947a). In both areas, small numbers were reported to occur in winter. During field work in Revelstoke National Park (Southern Interior Mountains Ecoprovince) in the summer of 1943, the raven was encountered once, and reference was made to an earlier record of a pair in 1890 (Cowan and Munro 1946). In the same ecoprovince, at Creston, between 1947 and 1949, Munro (1950) saw ravens a few times in the spring of 1948 and in the late summer and early autumn of each year. Munro and Cowan (1947) summarized the evidence as to the status of the species in British Columbia, considering it a local resident most abundant at the coast.

By the late 1940s the raven was a common resident in the east Kootenay, and was noted to be much more numerous than 10 years earlier (Johnstone 1949). The Common Raven is more numerous and widespread today. It now breeds in areas where it was previously known only as a winter visitant, and populations have increased along with increasing human settlement (e.g., Cannings et al. 1987).

NONBREEDING: The Common Raven (Fig. 247) has a widespread distribution throughout British Columbia; it occurs in every ecosection of every ecoprovince, including Vancouver Island and the Queen Charlotte Islands. The highest numbers in winter occur in the Southern Interior and Southern Interior Mountains ecoprovinces (Fig. 248).

This species occurs at elevations from sea level to 2,200 m. It frequents virtually every type of human-influenced and natural habitat available in the province (Figs. 249, 250, and 252). On the coast, it occurs from intertidal ocean beaches and estuaries to alpine peaks, including old-growth coniferous and mixed forests, second-growth coniferous and deciduous forests, clearcuts, mountainous areas, burns, river banks, lakeshores, cliff faces, alpine meadows, marshes, and offshore islands. In the interior, the Common Raven frequents similar inland habitats, but also frequents open grasslands, sage mesas, and rangelands. Mature coniferous trees, particularly snags, are used as roost sites and are important habitat elements.

Although the Common Raven is highly adaptable, it is by no means evenly distributed throughout the province. There are large areas of British Columbia where the food potential is so low that not even as resourceful a bird as the raven can maintain a population. Under natural conditions, over much of northern British Columbia the raven has a commensal relationship with the gray wolf. Every wolf pack that preys upon moose, caribou, elk, or other large mammals has its raven hangers-on during the winter months. The carrion from the hunt is essential to the raven's survival in that inhospitable environment (Fig. 251).

While the Common Raven is generally more wary of humans than other corvids, human-influenced habitats have become important and markedly influence the raven's distribution and behaviour. Garbage dumps are now a major

Figure 249. Snow-covered slag heaps from mine waste frequently contain garbage and provide winter foraging and loafing locations for the Common Raven (Cassiar, 15 January 1989; R. Wayne Campbell).

Figure 250. In wilderness areas of far northern British Columbia, the Common Raven patrols creeks, rivers, hillsides, and mountaintops in search of food, especially the carcasses of mammals (Crehan Creek, west of Fort Nelson, 23 February 1995; R. Wayne Campbell).

source of food for winter populations, especially in the interior, where winters are severe (Fig. 252). Some birds become habitual dump residents. For example, a breeding pair at Masset that foraged intensively in a marshy pasture during the nesting season moved its family each year to the local dump a few days after the young fledged. Almost all of the largest single-locality records in all regions are from dump sites. Other important human-influenced habitats include farm fields, feedlots, highways, logged forests, mine sites, barnyards, public campsites, ski resorts, bush camps, and urban and suburban areas.

Highways are heavily used by ravens, especially during seasonal migration of passerine birds and during the winter, when many large ungulates are on winter ranges close to well-travelled highways. At this time many wildlife species, both birds and mammals, become victims of collisions with vehicles. The ravens search the highways for the road kills (Conner and Adkisson 1976). Even in fairly remote areas of the province, 1 or 2 ravens patrolling a road, highway, or railroad at treetop height is not an uncommon sight. In winter ravens are often observed feeding on road-killed Pine Siskins, crossbills, and Evening Grosbeaks along many highways in the province where these and other finches concentrate on road edges for the salt that accumulates. Ungulates killed along railway tracks are another important winter food source

Figure 251. The Common Raven has a commensal relationship with the gray wolf in northern British Columbia. The carrion from this recent elk kill provided essential food for ravens in an otherwise harsh environment (Crehan Creek, 23 February 1995; R. Wayne Campbell).

Figure 252. Garbage dumps are now a major source of food for the Common Raven throughout the year in British Columbia, but they are especially important in the interior, where winters are severe (Fort Nelson, 24 February 1995; R. Wayne Campbell).

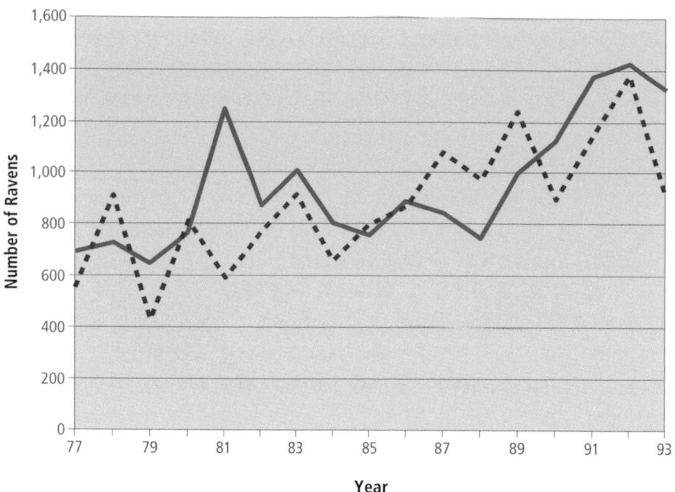

Figure 253. Fluctuations in the number of Common Ravens reported on Christmas Bird Counts on the coast (solid line; $n = 13$) and in the interior (dashed line; $n = 5$) between 1977 and 1993. Included are only counts that have been reporting regularly since 1977. (See Appendix 2 for count locations.)

for ravens (Hatler 1983). Except in the north, urban areas are used less frequently by ravens than by crows, although a few raven pairs are often resident on the fringes of cities, even those the size of Victoria and Vancouver.

Migratory movements are difficult to discern because of the presence of resident birds. The lack of records from the sparsely populated area east of the Coast Mountains and north of latitude 55°N is probably an artifact of the lack of observers. The large number of records of wintering ravens in the Peace Lowland is likely due equally to the presence of enthusiastic observers, to access by the raven population to cattle feedlots and larger garbage dumps, and to the greater number of road kills that tend to occur in areas of higher traffic.

Although generally resident in the southern and coastal parts of the province, the Common Raven wanders widely during the nonbreeding seasons, and congregates near dependable sources of food. In winter, it occurs mainly at lower elevations, although some birds remain throughout the winter at higher altitudes. Anecdotal accounts by T. Pearse (*in* Bent 1946) indicate that there may be some migration along the east coast of Vancouver Island, but he may have mistaken birds flying to roosts as migrants. Movements through Vancouver seem to occur in April and from late August through mid-October (Campbell et al. 1972a). Breeding birds disperse to more widespread nesting areas in early spring, although nonbreeders may remain at dump sites through the summer.

Specific details of major communal roosts, occasionally reported elsewhere (Stiehl 1981), are poorly known in British Columbia.

Analysis of Christmas Bird Count data suggests that winter populations are still increasing provincewide (Fig. 253).

The Common Raven occurs throughout the year in British Columbia (Fig. 254).

BREEDING: The Common Raven has a widespread breeding distribution throughout much of southern British Columbia and along the coast. The main documented nesting regions of the raven include the Queen Charlotte Islands, Western Vancouver Island, and the Georgia Depression, east across the Southern Interior and the east and west Kootenay regions of the Southern Interior Mountains to the Rocky Mountains.

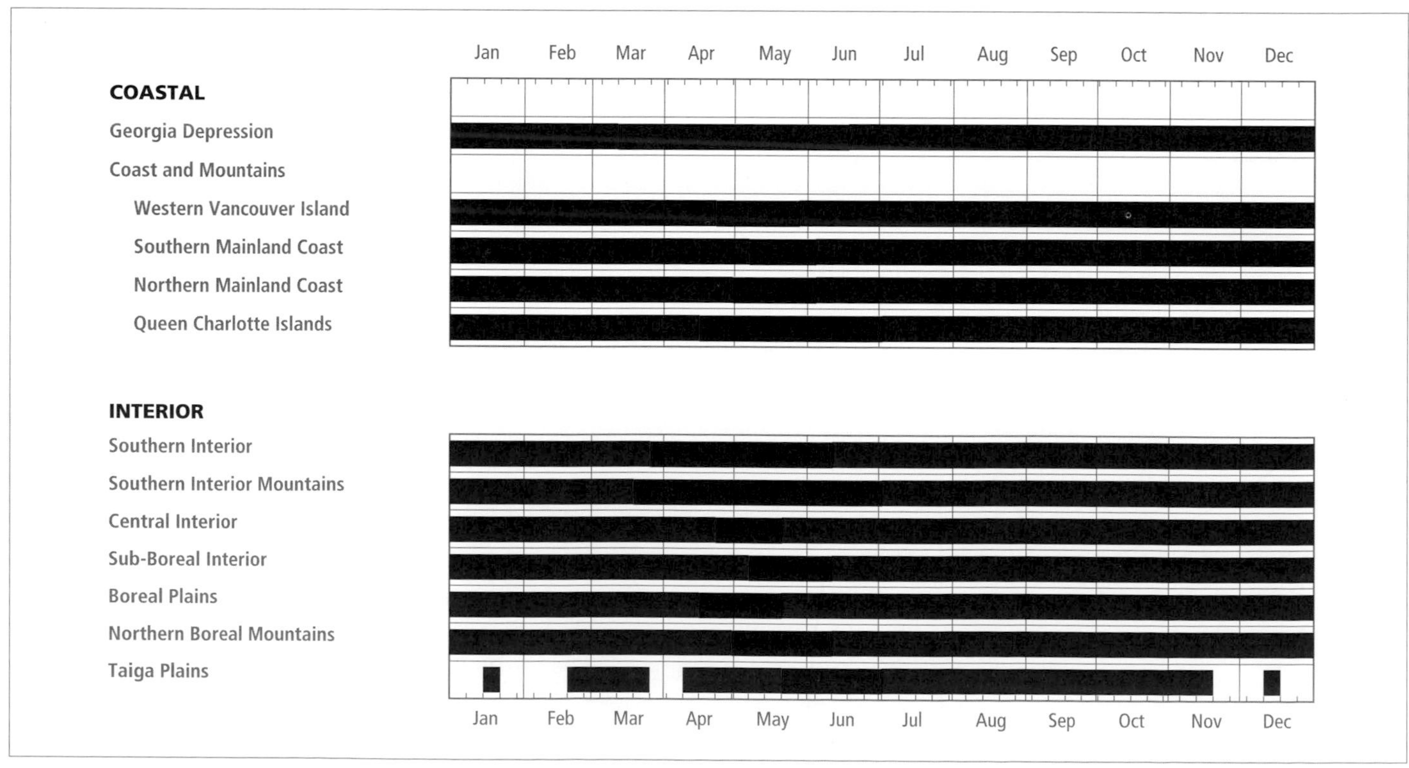

Figure 254. Annual occurrence (black) and breeding chronology (red) for the Common Raven in ecoprovinces of British Columbia.

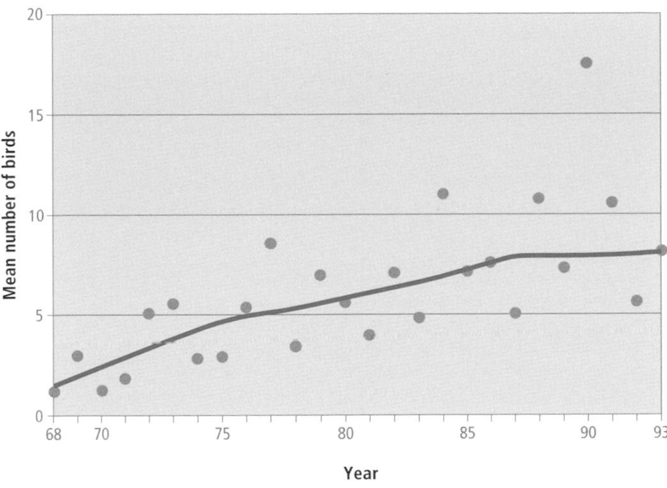

Figure 255. An analysis of Breeding Bird Surveys for the Common Raven in British Columbia shows that the mean number of birds on coastal routes increased at an average annual rate of 5% over the period 1968 through 1993 ($P < 0.001$).

Further north its breeding distribution is not well known, although it likely breeds throughout its range in the province. There are but 10 records of breeding from the central portions of the province in the Central Interior and the Sub-Boreal Interior, including the Cariboo and Chilcotin areas and the Nechako and upper Fraser river valleys. An additional 14 breeding localities have been reported from the Peace Lowland of the Boreal Plains northward.

The Common Raven reaches its highest summer numbers in the Southern Interior Mountains (Fig. 248). An analysis of Breeding Bird Surveys for the Common Raven in British Columbia shows that the mean number of birds on coastal routes has increased at an average annual rate of 5% over the period 1968 through 1993 (Fig. 255); analysis of interior routes over the same period could not detect a net change in numbers. Droege and Sauer (1988) note an annual increase on North American routes of 3.1% over the period 1966 through 1987.

The Common Raven has been reported breeding from near sea level to 600 m elevation on the coast, and from 300 to 1,320 m elevation in the interior. Most nests (56%; n = 96) were found in forested settings (Fig. 256). Fewer occurred in human-altered sites (35%).

Specific breeding habitats reported were forests (40%), with a 3:1 ratio of mature forest to second growth (Fig. 256), cliffs (36%; Fig. 257), industrial sites (7%), residential areas, parks, and burns. The 2 major habitats are not exclusive, as ravens frequently seek out cliff sites within forested habitats. There are a few instances in British Columbia of ravens

Figure 256. In British Columbia most Common Raven nests reported were from forested environments, especially mature forests (Green Lake, near Whistler, June 1968; R. Wayne Campbell).

Figure 257. Cliffs, sometimes within forested habitats, are used as nest sites by Common Ravens in British Columbia. Note the bulky stick nest below the overhang in the top centre of the photograph (Flathead, 30 September 1993; R. Wayne Campbell).

making use of abandoned buildings or other cultural debris, as has been reported in eastern Washington state (Bowles and Decker 1930).

On the coast, the Common Raven has been recorded breeding from 14 March to 15 June; in the interior, it has been recorded from 26 March to 29 June (Fig. 254).

Colonies: There are 2 reports of small nesting colonies of ravens in British Columbia. In 1970, 9 occupied nests were reported from a small area of eroded cliff face at Campbell's Beach on Saturna Island. These nest sites were used for several years by varying numbers of pairs, but by 1993 they had been abandoned. A second colony was reported from Adam's Lake, where 2 to 3 pairs nested for many years on a lakeshore cliff face south of Brennan Creek.

Nests: Nest sites were almost exclusively in 1 of 3 locations: living trees (50%; n = 98), cliff ledges (35%; Fig. 257), and human-influenced sites (14%; Fig. 258). Douglas-fir was the most frequently used tree species (16%), followed by Sitka spruce (5%), lodgepole pine (5%), and 6 other coniferous species totalling 14%; 4% of the nests were in deciduous trees. Even on a cliff location, the actual nest was often placed in a tree. These results differ from those reported in other parts of North America, where nests on cliffs appear to outnumber those in trees by ratios of from 8:1 to 20:1 (Fleming and Speich 1988). Preference was usually for the tallest available tree that had a thickly branched crown. The nest was often built on a branch adjacent to the trunk. On cliffs, the choice is for the darkest, best-sheltered recess, such as a crevice or beneath a rock overhang (Bent 1946). Human-made sites included sheds, water towers, barns, poles, and bridges.

The heights of 88 nests ranged from 2 to 45 m, with 61% between 12 and 24 m.

Nests were bulky platforms, some 90 cm in diameter, of varying depths because of the addition of new material each year. The base of the nest was made almost entirely of sticks, within which the nest cup was constructed of smaller twigs, bark strips, plant fibres, moss, seaweed, and beard-lichen; it often included animal fur.

Eggs: Dates for 20 clutches (Fig. 259) ranged from 14 March to 23 May, with 57% recorded between 2 and 24 April. Sizes of 12 clutches ranged from 1 to 6 eggs (1E-1, 2E-1, 4E-1, 5E-3, 6E-6). Data from 1 nest indicate an incubation period of 17 days. Ehrlich et al. (1988) cites a period of from 18 to 21 days.

Young: Dates for 73 broods ranged from 7 April to 3 July, with 56% recorded between 9 and 29 May. Sizes of 56 broods ranged from 1 to 5 young (1Y-8, 2Y-11, 3Y-19, 4Y-11, 5Y-7), with 73% having 2 to 4 young. The fledgling period is 38 to 44 days (Ehrlich et al. 1988).

Nest Success: Insufficient data.

REMARKS: The Common Raven is also known as the Northern Raven or Holarctic Raven, or, in European literature, simply as the Raven. It is the largest passerine bird species, measuring as much as 67 cm in total length (Godfrey 1986).

Ravens occur in concentrations of up to 500 birds, particularly at landfill sites or other concentrated food sources (see Fig. 252). Small flocks are often seen in flight, soaring high over valleys, mountains, or cliff faces or cruising low over the forest canopy. They are renowned for their acrobatic flight, which appears playful at times. Courtship flights by paired birds begin in February.

The sizes of raven breeding territories in British Columbia have not been studied, although Brenchley (1985) estimates a breeding density on Saltspring Island of 26 pairs per 100 km^2 (3.8 km^2 per pair). In Wales, the Common Raven in 2 separate areas had a mean territory size of between 9.5 and 23.9 km^2 per pair, respectively. In the smaller-size territory, nest sites were plentiful, whereas in the larger area numbers appeared to be limited by a scarcity of secure nest sites (Dare 1986). In Shetland, Ewins et al. (1986) found a mean territory size of 7.4 km^2 per pair.

Old Common Raven nests are used as nest sites by several species of birds. In British Columbia, they have been used by Double-crested and Pelagic cormorants, Red-tailed Hawk, Golden Eagle, Merlin, Peregrine Falcon, Gyrfalcon, Glaucous-winged Gull, and Great Horned Owl.

Along with all other wildlife species, ravens are protected

Figure 258. From 1977 to 1990, a pair of Common Ravens nested on beams of this bridge over the Fraser River north of Williams Lake (7 June 1996; R. Wayne Campbell).

Figure 259. A Common Raven nest containing 5 eggs (Victoria, 15 April 1978; R. Wayne Campbell).

by provincial legislation. However, increasing numbers of ravens have led to requests for control programs in some regions. The Common Raven is the only passerine species included, by name, as a game bird in British Columbia. Daily bag limits of 5 birds are allowed on private lands in the Omineca-Peace and Cariboo wildlife management subregions, and on Saltspring Island in the Vancouver Island subregion. The same bag limit applies to all wild land in the Kootenay region.

For additional life-history information, see Kilham (1989) and Heinrich (1989). A useful bibliography on the Common Raven has been prepared by Knight (1979).

NOTEWORTHY RECORDS

Spring: Coastal – Chain Islets 24 May 1975-4 nestlings, eyes still closed; Chatham Island 7 Apr 1975-4 nestlings, eyes almost open; Saanich 1 Apr 1981-30; Mount Work 10 Mar 1985-60; Saturna Island 23 May 1970-colony of 9 nests; Tofino 15 May 1931-3 nestlings; Burns Bog 16 May 1971-30 (Campbell et al. 1972b); Vancouver 21 Mar 1983-female apparently laying, 20 May 1983-3 nestlings three-quarters grown; New Westminster 8 May 1990-nestlings at Queen's Park; Grouse Mountain 1 Mar 1979-25; Skagit River valley 29 May 1971-3 young; Courtenay 14 Mar 1940-1 egg; Boston Bar 9 May 1982-20; Adam River 20 Mar 1984-27; Klemtu 5 Apr 1976-38; Rachel Islands 29 May 1987-1 young; East Copper Island 24 May 1971-2 to 4 nestlings; Skedans Island 29 May 1971-2 to 4 nestlings near flying; Lihou Island 8 May 1986-nestlings; Queen Charlotte City 12 Mar 1972-40; Kaien Island 11 Mar 1983-50; Tow Hill 24 May 1960-nestlings; Langara Island 28 Apr 1966-20 (Campbell 1969a); Prince Rupert 21 May 1977-100; Three Guardsmen Pass 28 May 1979-2. **Interior** – Creston 1 Apr 1981-58; Okanagan Falls 3 Apr 1904-6 eggs; Osprey Lake Rd (Princeton) 16 Apr 1976-53; near Cranbrook 3 Apr 1985-1 bird incubating; Balfour to Waneta 16 May 1981-133 on survey, 27 May 1982-190 on survey; Winlaw 23 May 1983-nestling; Sparwood 25 May 1984-50 (Fraser 1984); Mara Lake 31 Mar 1953-4 eggs; Sinclair Canyon 26 Mar 1985-first egg laid, 12 Apr 1985-3 eggs hatched, 13 Apr 1985-all 5 eggs now hatched; Kamloops 9 May 1984-4 nestlings; McQueen Lake 22 May 1973-nestlings; Shuswap Lake (Amnesty Arm) 27 Jun 1963-4 nestlings; Revelstoke 29 May 1890-1 (NMC 923); Nicholson 6 Mar 1977-50; Murtle River 2 Apr 1958-6 eggs; Loon Lake (Clinton) 21 Mar 1977-21; Brennan Creek 20 May 1972-3 occupied nests in cliff face; 21 Apr 1959-6 eggs; Vance Creek Ecological Reserve 23 May 1978-nestlings; Alexis Creek 19 May 1984-5 nestlings; Quesnel 29 May 1979-4 nestlings; Fort St. John 23 Mar 1986-31; Noel Creek 12 May 1992-5 young; Pink Mountain 24 Mar 1984-40; Cassiar 27 Mar 1978-100; Lower Post 3 May 1981-eggs; Willow River 29 May 1976-3 nestlings; Atlin 21 May 1981-52 (Campbell 1981).

Summer: Coastal – Qualicum Beach 15 Jun 1975-3 young left nest today, 24 Aug 1977-100; Genesee Creek 19 Aug 1937-belated brood calling; Mitlenatch Island 31 Jul 1922-300; Gold River 1 Jun 1988-50; Yakoun River 5 Aug 1979-30; Masset 4 Jun 1984-20, 5 Jun 1985-12 (Cooper 1985); Skidegate Inlet 2 Jul 1910-1 (NMC 4153); Langara Island 28 Apr 1966-20 (Campbell 1969); Prince Rupert 12 Jun 1970-250; Port Simpson area 24 Aug 1969-70. **Interior** – Kikomun Creek 3 Jun 1983-nestlings; Naramata 3 Jul 1976-5 nestlings; Lytton 11 Jun 1964-1 (NMC 52488); Nakusp 14 Jun 1977-24; Cluculz Lake 9 Jun 1968-young; Line Creek (Sparwood) 18 Aug 1984-40 (Fraser 1984); Shuswap Lake (Amnesty Arm) 27 Jun 1963-4 nestlings; Pine Pass 17 Aug 1975-15; Cassiar 12 Jul 1978-117; Fort Nelson 12 Aug 1979-11; sw Kotcho Lake 29 Jun 1982-2 nestlings, 5 Jul 1982-46 (Campbell and McNall 1982); Atlin 1 Jun 1981-3 nestlings; Lower Post 10 Jun 1981-2 nestlings; Log Cabin, White Pass 20 Jun 1899-1 (FMNH 142193).

Breeding Bird Surveys: Coastal – Recorded from 25 of 27 routes and on 75% of all surveys. Maxima: Masset 21 June 1994-53; Campbell River 24 June 1990-50; Pemberton 17 June 1984-35. **Interior** – Recorded from 66 of 73 routes and on 86% of all surveys. Maxima: Lavington 21 June 1984-47; Bridge Lake 17 June 1979-44; Wingdam 10 June 1970-41.

Autumn: Interior – Muncho Lake 25 Oct 1980-10; Charlie Lake 11 Nov 1982-25; Griscom 3 Oct 1981-12; Bob Quinn Lake 28 Aug 1977-3; Mile 26.5 Clarke Lake Rd 28 Sep 1978-2; Cluculz Lake 20 Oct 1972-15; Francois Lake 3 Oct 1982-32; Natal Ridge 18 Oct 1984-73; Spences Bridge 11 Oct 1982-28; Swan Lake (Vernon) 6 Oct 1943-1 (ROM 83319); Nakusp 30 Nov 1985-30; Munson Lake (Kelowna) 10 Sep 1978-60 in flock. **Coastal** – Stewart 29 Sep 1980-8; Sewell 1 Nov 1975-30; Porcher Island 8 Sep 1920-1 (MVZ 102995); Calvert Island 3 Sep 1934-1 (MCZ 283269); Smythe Island 24 Sep 1935-1 (MCZ 283276); Port Hardy 13 Oct 1950-40; Sayward 3 Sep 1978-35; Cypress Park 18 Oct 1984-22; Harrison Hot Springs 11 Oct 1976-27; Long Beach 15 May 1931-3 nestlings; Fulford Harbour 29 Nov 1980-75; Langford 27 Sep 1984-37; Victoria 23 November 1974-23 in flock flying to evening roost in tall Douglas-firs on Chatham Island.

Winter: Interior – Lower Post 15 Jan 1978-1; Haines Highway 15 Jan 1979-2; Atlin 21 Jan 1981-60; Cassiar 16 Dec 1985-150, 15 Jan 1989-33 on slag heap (Fig. 249); Muncho Lake 15 Dec 1978-4 at garbage container; Fort Nelson 24 Feb 1995-86 at garbage dump (Fig. 252); Dease Lake 15 Dec 1989-15 at garbage dump; North Pine 28 Dec 1979-4; Stoddart Creek 1 Jan 1984-1; Charlie Lake to Halfway River 19 Dec 1981-30; Cache Creek (Peace River) 19 Feb 1983-70 in flock; Chetwynd 23 Jan 1982-5; Crooked River 14 Jan 1990-34; Fort St. James 4 Jan 1992-66; Smithers 2 Jan 1983-264 (Pojar 1983); 20 km s Prince George 17 Nov 1981-2; Golden to Cranbrook 12 Dec 1975-70 (survey); Invermere 26 Dec 1983-67; Kootenay National Park 31 Dec 1965-5 (Seel 1965); Cranbrook 19 Dec 1961-1 (UBC 10635); Nakusp 3 Jan 1982-70; East Trail 1 Jan 1978-18; Armstrong 30 Dec 1978-500 at slaughterhouse; Swan Lake (Vernon) 31 Dec 1978-300; Apex Mountain 31 Dec 1973-25 (Cannings et al. 1987); Osoyoos 31 Dec 1986-1; Keremeos 1 Jan 1974-4; Princeton 21 Jan 1979-75; Manning Park 31 Dec 1983-1. **Coastal** – Terrace 2 Jan 1972-50 (Frank 1972); Triple Island Light Station 31 Dec 1951-2; Kitimat 15 Dec 1974-46 (Hay 1975); Masset 6 Feb 1984-30; Sewell 6 Feb 1984-30; Cape St. James 30 Dec 1981-1; Cape Scott 5 Feb 1976-50; Port Neville 1 Jan 1976-2; Comox 1936-37 winter-133 caught in crow traps (Pearse 1938); Shoemaker Bay (Port Alberni) 16 Feb 1977-40; Tofino 3 Jan 1976-2; Pachena Bay 11 Dec 1974-21; Whiffin Spit 1 Jan 1976-2; Mount Finlayson 31 Dec 1974-3.

Christmas Bird Counts: Interior – Recorded from 27 of 27 localities and on 100% of all counts. Maxima: Kelowna 3 Jan 1987-916; Princeton 30 Dec 1992-734; Revelstoke 20 Dec 1984-573. **Coastal** – Recorded from 32 of 33 localities and on 97% of all counts. Maxima: Campbell River 3 Jan 1982-432; Victoria 14 Dec 1991-382; Terrace 26 Dec 1992-323.

Black-capped Chickadee

Parus atricapillus Linnaeus

BCCH

RANGE: Resident from western and central Alaska, southern Yukon, and southern Mackenzie across forested regions of the continent to Newfoundland, south to northwestern California, northeastern Nevada, central Utah, northern New Mexico, northern Kansas, and central Ohio; and south through the Appalachian Mountains to North Carolina and Tennessee.

STATUS: On the coast, a *common* resident in the Fraser Lowland of the Georgia Depression; *very rare* on the Sunshine Coast. *Fairly common* in the southern Coast and Mountains Ecoprovince, becoming *rare* on the central and northern mainland of that ecoprovince. Absent from Vancouver Island, the Queen Charlotte Islands, and the Gulf Islands.

In the interior, a *common* resident in the southern half of the province, including the Southern Interior, Southern Interior Mountains, and Central Interior ecoprovinces; *fairly common* in the Sub-Boreal Interior, Northern Boreal Mountains, and Boreal Plains ecoprovinces; *uncommon* in the Taiga Plains Ecoprovince.

Breeds.

NONBREEDING: The Black-capped Chickadee (Fig. 260) is the most widely distributed of the 4 species of chickadees that occur in the province. It is resident throughout the interior of the province and on the Fraser Lowland of the southwest mainland coast. It occurs only sporadically between Gibsons and Powell River and at Port Neville on the mainland coast

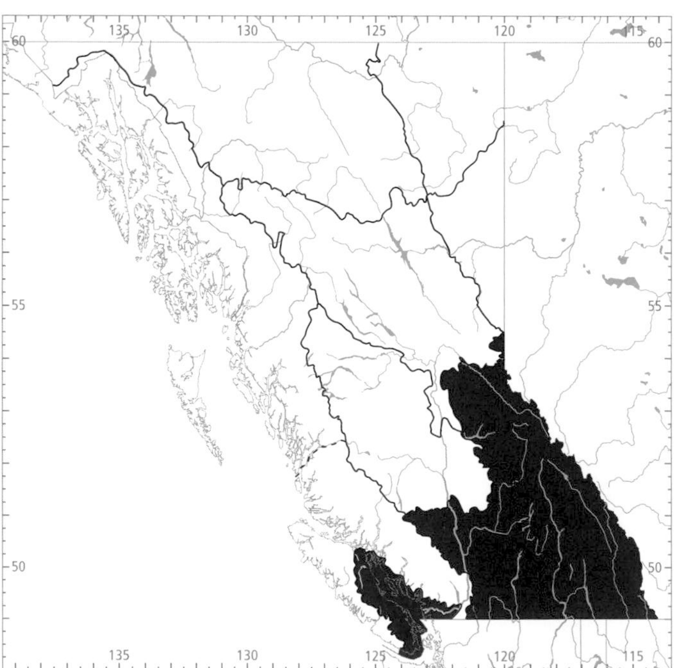

Figure 261. In British Columbia, the highest numbers for the Black-capped Chickadee in winter (black) occur in the Fraser Lowland of the Georgia Depression Ecoprovince; the highest numbers in summer (red) occur in the Southern Interior and Southern Interior Mountains ecoprovinces.

Figure 260. Adult Black-capped Chickadee (near Kelowna, March 1988; Phil Gehlen).

Figure 262. In the interior, cottonwood woodlands bordering streams with thickets of clustered willow provide winter foraging sites for the Black-capped Chickadee (Canim Lake, 26 March 1996; R. Wayne Campbell).

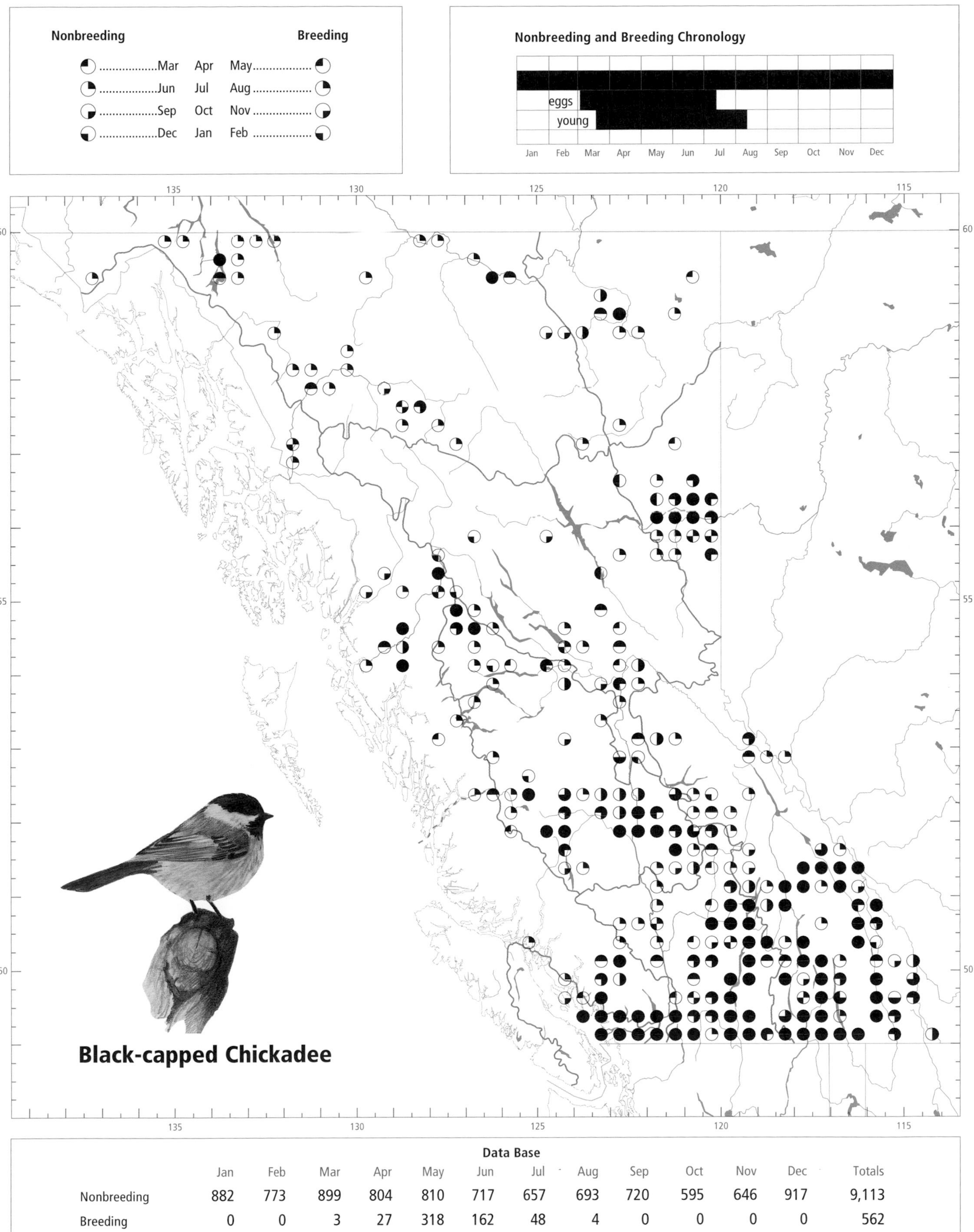

Data Base

	Jan	Feb	Mar	Apr	May	Jun	Jul	Aug	Sep	Oct	Nov	Dec	Totals
Nonbreeding	882	773	899	804	810	717	657	693	720	595	646	917	9,113
Breeding	0	0	3	27	318	162	48	4	0	0	0	0	562

north of Vancouver. On the central and northern mainland coast, north to the Skeena River basin, it occurs locally where major river valleys cut through the coastal mountains. It has not been reliably reported from Vancouver Island, the Queen Charlotte Islands, or the many southern offshore islands.

The Black-capped Chickadee reaches its highest numbers in winter in the Fraser Lowland of the Georgia Depression (Fig. 261). An analysis of Christmas Bird Counts for the period 1944 to 1985 could not detect a change in the provincial population (Brennan and Morrison 1991).

The Black-capped Chickadee is most abundant at lower elevations. On the coast, it occurs from near sea level to the lower mountain slopes; in the interior, it occurs up to about 2,300 m elevation. Favoured habitats include deciduous, mixed deciduous-coniferous, and open coniferous forests; shrub thickets; and riparian woodlands, especially alder, cottonwood (Fig. 262), willow, trembling aspen, and birch stands with a shrub understorey. On the coast, riparian habitats with salmonberry or thimbleberry thickets are heavily used. Thimbleberry is especially attractive to the Black-capped Chickadee, as it is often heavily infested with a gall-producing insect whose larvae are sought as food. In coniferous forests, the Black-capped Chickadee prefers edges and openings along beaver ponds, lakeshores, river banks, bogs, swampy areas, meadows, pastures, regenerating clearcuts, burned forests, orchards, and other human-made clearings where deciduous habitat occurs.

The Black-capped Chickadee is also a familiar visitor to rural and suburban areas, and is readily attracted to backyard bird feeders. It is also found at bush camps, where it comes to glean shreds of meat and fat from animal carcasses and

Figure 263. At least 5 Black-capped Chickadees roosted in a cavity in this Douglas-fir stump (Burnaby Lake, 26 February 1993; R. Wayne Campbell).

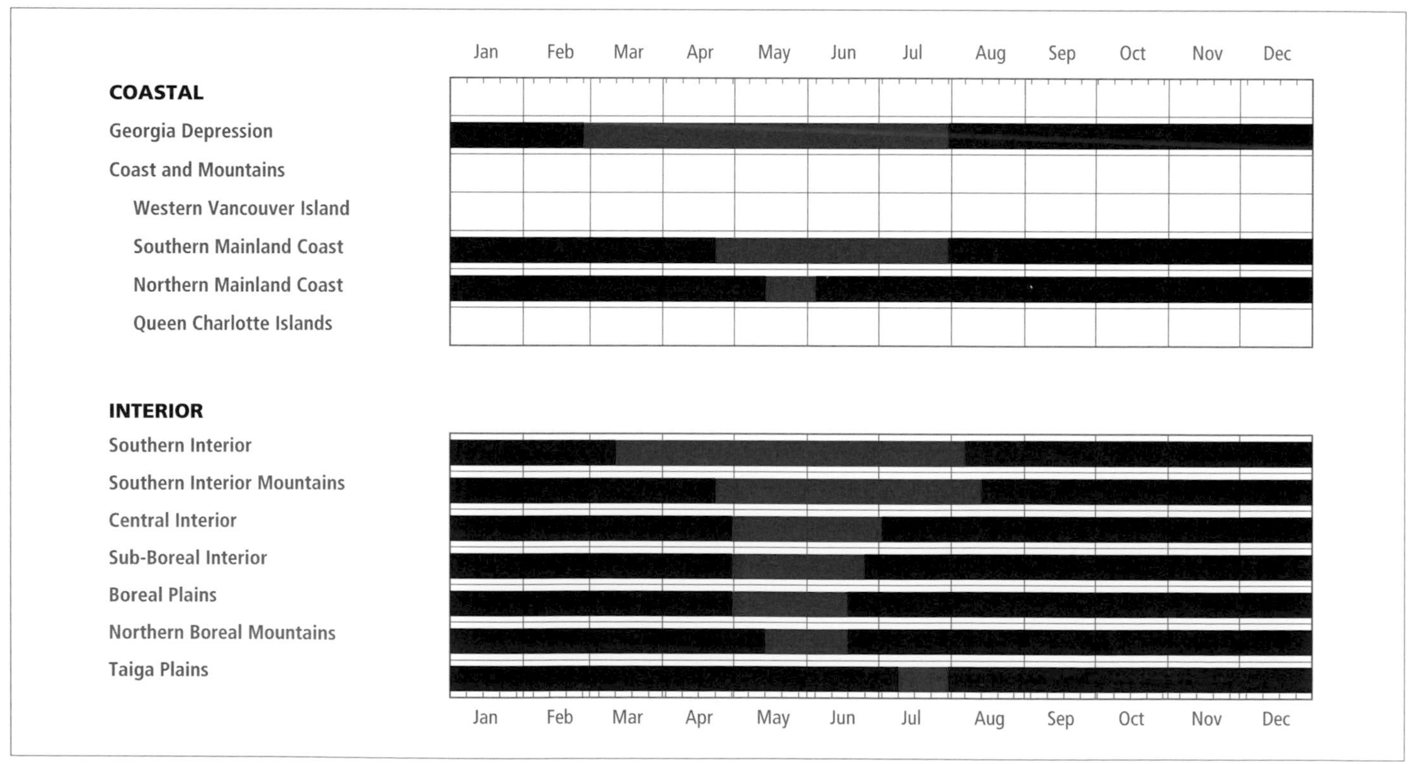

Figure 264. Annual occurrence (black) and breeding chronology (red) for the Black-capped Chickadee in ecoprovinces of British Columbia.

hides. It roosts overnight in cavities in trees (Fig. 263) or in thick tangles of vegetation (Smith 1991).

The range of the Black-capped Chickadee overlaps that of the other chickadee species in much of British Columbia. On the coast, where the Black-capped Chickadee and Chestnut-backed Chickadee coexist, the former favours more deciduous habitats and tends to forage in the shrubby forest understorey, whereas the Chestnut-backed Chickadee forages mainly higher in the canopy and in conifers (Smith 1967; Sturman 1968b; Brennan 1989). In the interior, both the Boreal and Mountain chickadees prefer coniferous forests at higher elevations than does the Black-capped Chickadee. Overlap with the Mountain Chickadee occurs often at lower elevations, but the Boreal Chickadee generally frequents higher elevations than the Black-capped Chickadee.

Aside from seasonal elevational movements and irruptive southward movements during years of low food supply, the Black-capped Chickadee is essentially nonmigratory (Fig. 264). It is more widespread in nonbreeding than in breeding seasons. After the breeding season it may move to higher elevations for the autumn and winter, where it mixes with other chickadees, kinglets, nuthatches, and creepers.

On the coast and in the interior, the Black-capped Chickadee has been recorded year-round (Fig. 264).

BREEDING: The Black-capped Chickadee has a widespread breeding distribution in interior British Columbia. It breeds commonly in the Thompson, Okanagan, and Similkameen river valleys, in the western Columbia River valley north to Revelstoke, in the Rocky Mountain Trench and adjacent valleys north to Brisco and Mount Robson, and into the Central Interior. Breeding populations are less frequently encountered further north, although the species is fairly common along the Nechako and Skeena river valleys and in the Peace Lowland from Tupper Creek to Fort St. John. Other known northern breeding localities are at Telegraph Creek (Swarth 1922), Fort Nelson, and the Tatshenshini River.

On the coast, it breeds from the Fraser River delta and lower slopes of the mountains to the north, east to the Skagit River and north to Squamish and Alta Lake. It also likely breeds in the low-elevation valleys of the southern Coast and Mountains and in the major river valleys (e.g., Bella Coola, Dean, Kitlope, Kitimat, and Skeena rivers) along the central and northern mainland portions of that ecoprovince.

The highest numbers in summer occur in the Southern Interior and Southern Interior Mountains (Fig. 261). An analysis of Breeding Bird Surveys shows that the number of birds on coastal routes decreased at an average annual rate of 2% for the period 1968 through 1993 (Fig. 265); an analysis of interior routes for the same period could not detect a net change in numbers. Breeding densities during a study on the University of British Columbia Endowment Lands were 0.15 to 0.20 pairs/ha (Smith 1967).

The Black-capped Chickadee breeds on the coast from near sea level to 210 m and in the interior from 270 m to about 1,500 m. Its breeding habitat is similar to its nonbreeding habitat. In all habitats it prefers a rich understorey of brush, and tends to nest near the forest edge rather than inside forest stands (Fig. 266). In the interior, its breeding distribution overlaps that of the Mountain Chickadee, which also nests at higher elevations; the 2 species are sometimes sympatric where forests have high structural diversity (Hill and Lein 1989). On the mainland coast and in the west Kootenay, it

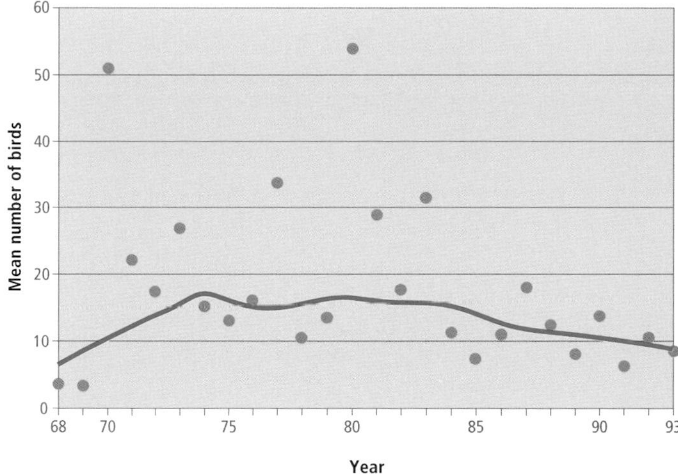

Figure 265. An analysis of Breeding Bird Surveys for the Black-capped Chickadee in British Columbia shows that the number of birds on coastal routes decreased at an average annual rate of 2% over the period 1968 through 1993 ($P < 0.01$).

Figure 266. The Black-capped Chickadee prefers to nest in self-excavated cavities or old woodpecker cavities on the edges of forests (15 km north of Cache Creek, 23 March 1996; R. Wayne Campbell).

overlaps with the Chestnut-backed Chickadee, but the Black-capped Chickadee prefers to nest in open or riparian forests. In the north, the Black-capped Chickadee may nest close to the Boreal Chickadee at lower elevations where thickets of white spruce are interspersed with stands of trembling aspen.

Most of our breeding records were from human-influenced habitats (58%) such as ranches, farms, rural and suburban settings, and clearings. In more natural environments (37%), nests were usually associated with riparian thickets, estuaries, natural forest openings, ponds, or sloughs.

On the coast, the Black-capped Chickadee has been recorded nesting from 6 March (calculated) to 25 July (Fig. 264); in the interior, it been recorded nesting from 12 March (calculated) to 11 August (Fig. 264).

Nests: The Black-capped Chickadee is a primary cavity excavator, but will occasionally use old woodpecker or sapsucker cavities for nesting. Nest cavities were most frequently excavated in dead tree trunks (Fig. 266), dead or diseased branches, and rotten stumps. Deciduous trees were used 8 times as often as conifers, and dead trees were preferred over living trees. The most frequently used tree species were birches in the interior and red alder on the coast. Other nest trees included willow, trembling aspen, black cottonwood, Douglas-fir, ponderosa pine, cascara, Pacific crab apple, and domestic cherry and apple. New nest cavities were built each year. The Black-capped Chickadee readily used nest boxes placed on trees, poles, fence posts, and buildings.

The heights of 109 nests in natural cavities ranged from 0.3 to 23 m, with 62% between 1.2 and 3.6 m. Smith (1991) provides dimensions of 59 nest cavities from various localities in North America: mean depth, 21 cm (range: 10 to 46 cm); nest chamber diameter, 6 to 7 cm. Nests are cups composed mainly of mosses (52%; Fig. 267), animal hair (38%), grasses (27%), feathers (23%), bark strips (12%), and human-made materials (9%).

Eggs: Dates for 128 clutches ranged from 19 March to 12 July, with 52% recorded between 6 May and 30 May. Calculated dates indicate that eggs can be found as early as 2 March. Egg laying normally begins about 1 or 2 weeks earlier in the Georgia Depression than in other ecoprovinces. Sizes of 105 clutches ranged from 1 to 9 eggs (1E-6, 2E-4, 3E-3, 4E-8, 5E-19, 6E-31, 7E-19, 8E-13, 9E-2), with 66% having 5 to 7 eggs (Fig. 267). In a sample from all parts of North America, Smith (1991) found that 80% of clutches held 6 to 8 eggs. The incubation period is 12 to 13 days (Bent 1946; Smith 1991), although Peck and James (1987) list the range as 11 to 17 days in Ontario.

Young: Dates for 220 broods ranged from 18 March to 11 August, with 54% recorded between 23 May and 19 June. Sizes of 134 broods ranged from 1 to 8 young (1Y-6, 2Y-10, 3Y-13, 4Y-25, 5Y-37, 6Y-20, 7Y-15, 8Y-8), with 61% having 4 to 6 young. Bent (1942) and Smith (1991) report the nestling period to be 16 days. Some pairs in southern British Columbia raise 2 broods annually (Kelleher 1963).

Brown-headed Cowbird Parasitism: In British Columbia, there were no cases of cowbird parasitism in 214 nests found with eggs or young. There are only 7 records of parasitism in North America (Friedmann et al. 1977; Friedmann and Kiff 1985).

Nest Success: Of 52 nests found with eggs and followed to a known fate, 36 produced at least 1 fledgling, for a success rate of 69%; interior success was 73% (n = 41), while coastal success was 54% (n = 11). Cannings et al. (1987) suggest that most nest failures in the Okanagan valley may be caused by competition for nest sites with House Wrens and Tree Swallows.

REMARKS: In several areas of British Columbia, all 4 species of chickadee (Black-capped, Boreal, Chestnut-backed, and Mountain) can be found occupying the same area. These areas include Manning Park, the Monashee Mountains, Anahim Lake in the west Chilcotin, the Nelson and Revelstoke regions, and Yoho National Park. This sympatric occurrence offers opportunities for research on resource partitioning and ecological specialization.

Four subspecies are recognized within the province (Smith 1991): *P. a. turneri* in the far northwestern corner; *P. a. septentrionalis* east of the Coast Ranges and northern portions of the interior; *P. a. fortuitus* in the southern portions of the

Figure 267. Two-thirds of all Black-capped Chickadee clutches reported from British Columbia contained 5 to 7 eggs. Note the nest substrate of mosses, feathers, and bark strips (Burnaby Lake, 8 May 1969; R. Wayne Campbell).

Figure 268. This young Black-capped Chickadee fledged from its nest box 16 days after hatching (west of Princeton, 7 June 1994; R. Wayne Campbell).

interior; and *P. a. oregonensis* on the extreme southern mainland coast. At least 3 of these subspecies are readily identifiable by plumage characteristics under field conditions (Smith 1993).

Many people are amazed at how the tiny chickadees can survive the severe winters of interior regions. The Black-capped Chickadee adjusts to colder temperatures by roosting in cavities (Fig. 263) and by dropping its nighttime temperature as much as 10° to 12°C below its daytime temperature in a regulated hypothermia, thus conserving energy (Smith 1991).

Some populations can be enhanced by providing winter food at feeders, by retention of snags along forest edges, by retention of riparian woodlands, and by placement of nest boxes (Smith 1993).

The Black-capped Chickadee has been reported a number of times from Vancouver Island, including the Nanaimo, Pender Island, and Port Alberni Christmas Bird Counts; however, adequate documentation is lacking and these reports have been excluded from the account.

See REMARKS in **Red-breasted Nuthatch** for comments on the relationship between *Armillaria* root disease and the suitability of trees for nesting by weak and strong cavity excavators.

For a complete review of the ecology and biology of the Black-capped Chickadee, see Smith (1991, 1993).

NOTEWORTHY RECORDS

Spring: Coastal – Langley 18 Mar 1992-5 young about 1-day old, 25 Mar 1979-46; Sumas Prairie 25 Apr 1959-7 eggs; Vancouver 10 Apr 1944-eggs (Munro and Cowan 1947); Stanley Park (Vancouver) 5 May 1975-27; North Vancouver 23 Apr 1983-nestlings; Lynn Valley 21 Apr 1964-60; Pitt Lake 26 Apr 1958-7 eggs; Harrison Hot Springs 9 Mar 1980-15; Hope 26 Apr 1961-9 eggs, 5 May 1962-nestlings; Kawkawa Lake 6 May 1961-7 eggs; Squamish 30 Mar 1968-10; Bella Coola 27 May 1979-building nest; Kitlope Lake 3 May 1991-1; Kitimat River 13 Apr 1975-1; Terrace 25 May 1977-nestlings. **Interior** – Taghum 30 Apr 1981-4 nestlings; Penticton 14 Apr 1973-nest building; Nelson 23 Apr 1977-3 eggs, 30 May 1981-nestlings; Balfour to Waneta 16 May 1981-72 on survey; South Slocan 19 Mar 1980-nest building; Merritt 9 May 1981-nestlings; Hidden Creek (Kelowna) 6 Mar 1983-32; Enderby 19 Mar 1977-8 eggs, 1 Apr 1977-nestlings, 20 Apr 1977-fledglings, 21 Apr 1969-7 eggs, 4 May 1969-5 nestlings; Sorrento to Squilax 1 Mar 1971-25 on survey; White Lake (Sorrento) 25 May 1975-20; Celista 12 Apr 1948-145 on survey; Leanchoil to Brisco 10 Apr 1977-12; 100 Mile House 24 Mar 1979-nest building; Williams Lake 13 Mar 1980-20, 11 May 1978-eggs; Chilanko Forks 20 May 1984-8; Wells Gray Park 16 May 1962-8 eggs, 21 May 1962-5 nestlings; Cottonwood 6 Mar 1983-12; Shelley 2 May 1978-eggs; McLeod Lake 25 Apr 1962-2; Terrace 25 May 1977-nestlings; Takla Landing 6 Mar 1938-1 (RBCM 7856); Tupper Creek 6 May 1988-7 eggs; Fort St. John 20 Mar 1984-8; Fort Nelson 2 Mar 1986-4, 6 Apr 1975-7; Liard Hot Springs 22 Apr 1975-2; Atlin 19 May 1981-4.

Summer: Coastal – Vancouver 6 Jul 1969-29; Burnaby 30 Aug 1959-50; Cultus Lake 28 Aug 1967-64; Pitt Meadows 11 Jul 1976-4 eggs; West Vancouver 25 Jul 1958-nestlings; Agassiz 6 Jun 1974-26; Hope 20 Jul 1962-nestlings; Gibsons Landing 23 Jun 1970-1 on Breeding Bird Survey; Pemberton 21 Jun 1981-15 on Breeding Bird Survey; Trail Island (Sechelt) 20 Aug to 18 Sep 1973-1; Bella Coola 5 Jul 1940-1 (UMMC 129598); Hazelton 25 Jun 1917-2 (NMC 11235-6); Tyhee Lake 30 Jul 1976-10; Flood Glacier 28 Jul 1919-1 (MVZ 40257). **Interior** – Vaseux Lake 27 Jun 1974-23 (Cannings 1974); Wasa Lake Park 24 Jul 1977-113 on census; Lytton 12 Jun 1964-2 (WMC 52498); Shuswap Lake Park 19 Aug 1973-27 (Sirk et al. 1973); Rose Hill 24 Jun 1988-2 eggs, 16 Jul 1988-5 young; Adams Lake 12 Jul 1970-incubating eggs; Bridge Lake 8 Jun 1959-8 eggs; Chilcotin River 2 Jul 1989-4 nestlings; Kleena Kleene 1 Jul 1961-4 fledglings; Williams Lake 12 Jun 1979-20, 4 Aug 1979-30; Mount Robson Park 15 Aug 1973-62; Stum Lake 27 Jul 1974-4 fledglings; Prince George 24 Jun 1969-30, 1 Aug 1976-6 fledglings; Fort St. James 4 Jun 1983-4 eggs; Pouce Coupe 28 Jun 1984-4 fledglings; Baldonnel 9 Jun 1984-4 fledglings; Beatton Park 24 Aug 1986-20; Rognaas Creek 26 Jun 1976-10; Laslui Lake 29 Jul 1976-12; Hyland Post 17 Aug 1976-15; Telegraph Creek 14 Jun 1919-9 nestlings (Swarth 1922); Fort Nelson 14 Jun 1976-9, 27 Jul 1986-2 fledglings; Kotcho Lake 3 Jun 1983-2 nestlings; Tatshenshini River 5 Jun 1983-6 fledglings (Campbell et al. 1983).

Breeding Bird Surveys: Coastal – Recorded from 13 of 27 routes and on 49% of all surveys. Maxima: Pitt Meadows 1 Jul 1974-65; Kwinitsa 29 Jun 1973-59; Chilliwack 1 Jul 1990-52. **Interior** – Recorded from 68 of 73 routes and on 87% of all surveys. Maxima: Salmon Arm 8 Jun 1978-62; Creston 18 Jun 1984-48; Mount Morice 16 Jun 1971-40.

Autumn: Interior – Atlin 5 Oct 1980-9; Fort Nelson 26 Oct 1984-6; Cold Fish Lake 21 Sep 1975-6; Taylor 20 Nov 1983-11; 32 km w Prince George 20 Oct 1972-10; Williams Lake 26 Oct 1979-30; Horsefly 12 Nov 1978-6; Golden to Brisco 11 Oct 1976-20; Vaseux Lake 29 Oct 1975-31; Richter Lake 18 Sep 1959-25. **Coastal** – Kitimat River 29 Sep 1974-2; Lakelse Lake 20 Oct 1974-9; Powell River 13 Nov 1979-1 at feeder with Chestnut-backed Chickadees; Gibsons 9 Nov 1985 to 25 May 1986-1; Skagit River 29 Sep 1974-18; Maple Ridge 21 Nov 1964-32; Langley 2 Oct 1967-152.

Winter: Interior – Atlin 3 Dec 1980-1; Liard Hot Springs 18 Feb 1975-1; Fort Nelson 28 Dec 1985-2; Muskwa River 25 Jan 1975-6; Fish Creek (Fort St. John) 6 Feb 1985-7; Bear Mountain (Dawson Creek) 1 Feb 1986-20; Quesnel 26 Feb 1986-11; Williams Lake 3 Jan 1988-256 on Christmas Bird Count; Salmon Arm 3 Jan 1971-101 on Christmas Bird Count; Wilmer 26 Dec 1982-140 on Christmas Bird Count (Jack 1983); Windermere 26 Dec 1986-217 on Christmas Bird Count; Creston 4 Jan 1987-108 on Christmas Bird Count. **Coastal** – Port Neville Inlet 30 Jan 1974-20; Powell River 6 Feb 1976-1 at feeder; Gibsons 20 Feb to 4 April 1992-1; Black Mountain 14 Dec 1963-1 (UBC 11537); Mount Seymour 27 Dec 1963-1 (UBC 11541).

Christmas Bird Counts: Interior – Recorded from 27 of 27 localities and on 100% of all counts. Maxima: Smithers 29 Dec 1991-1,165; Burns Lake–Francois Lake 3 Jan 1993-1,080; Prince George 20 Dec 1992-957. **Coastal** – Recorded from 15 of 33 localities and on 39% of all counts. Maxima: Vancouver 19 Dec 1993-2,613; White Rock 2 Jan 1994-1,402; Pitt Meadows 27 Dec 1980-734.

Mountain Chickadee

MOCH

Parus gambeli Ridgway

RANGE: Resident from northwestern British Columbia south through interior British Columbia, mountainous parts of southwestern Alberta, Montana, and Colorado to northern Baja California, southern Nevada, southeastern Arizona, southern New Mexico, and western Texas. Casual in summer and winter in southeastern Alaska and south-central Yukon.

STATUS: On the coast, *rare* visitant to the southern Coast and Mountains Ecoprovince; *very rare* in the Fraser Lowland of the Georgia Depression Ecoprovince. Absent from Vancouver Island and the Queen Charlotte Islands.

In the interior, *fairly common* to *common* resident across southern portions of the province, including the Southern Interior, Central Interior, and Southern Interior Mountains ecoprovinces; *uncommon* to *fairly common* resident in the Sub-Boreal Interior Ecoprovince. *Rare* resident in the northwest mountains of the Northern Boreal Mountains Ecoprovince.

Breeds.

NONBREEDING: The Mountain Chickadee (Fig. 269) is widely distributed across the southern half of interior British Columbia. It is especially common, and widely distributed, in the Thompson-Okanagan plateaus, the Rocky Mountain Trench south of Golden, and the Cariboo and Chilcotin areas. Small numbers occur in scattered locations east from the Babine Range to the Driftwood Range, and north through mountainous regions to Atlin. On the coast, small numbers occur in the mountains north of North and West Vancouver, along the Squamish valley to Pemberton, in the Pacific Ranges from Jervis Inlet north to Kitimat, Terrace, the Hazelton Mountains, and Hazelton. It is absent from Vancouver Island, the Queen Charlotte Islands, much of the central mainland coast, and, with 2 exceptions, other coastal islands.

Figure 269. Adult Mountain Chickadee incubating eggs in a nest box (Becher's Prairie, west of Williams Lake, June 1983; Anna Roberts).

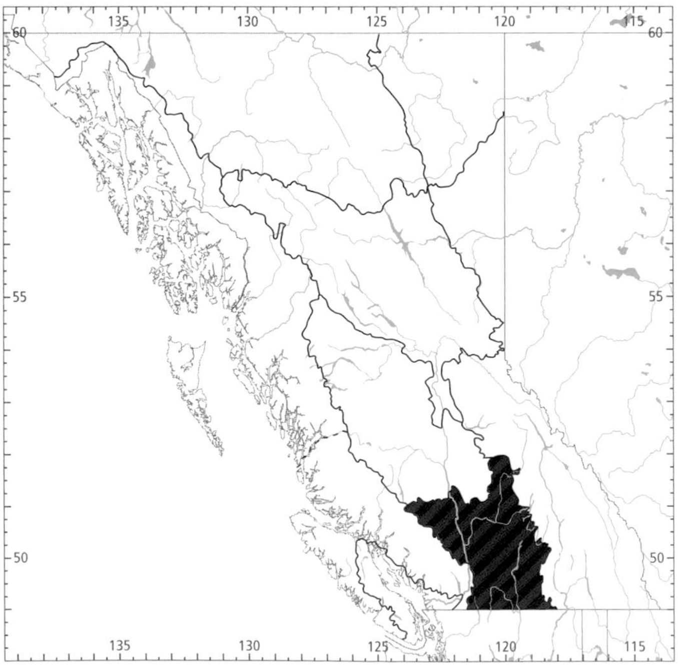

Figure 270. In British Columbia, the highest numbers for the Mountain Chickadee in both winter (black) and summer (red) occur in the Southern Interior Ecoprovince.

The highest numbers in winter occur in the Southern Interior (Fig. 270). An analysis of Christmas Bird Counts in British Columbia for the period 1957 through 1985 shows that the provincial population is stable (Brennan and Morrison 1991).

The Mountain Chickadee occurs at elevations from near sea level on the coast to 2,300 m in the interior. It inhabits mainly open ponderosa pine, Douglas-fir (Fig. 274), lodgepole pine, Engelmann spruce, and subalpine fir forests, but also frequents trembling aspen stands and riparian thickets. It commonly uses backyard bird feeders, especially during winter.

The Mountain Chickadee is essentially sedentary, with most adults remaining near breeding areas. Altitudinal movements to lower elevations occur during winter, and juveniles routinely disperse from natal areas shortly after fledging (Dixon and Gilbert 1964).

During nonbreeding seasons, the Mountain Chickadee is often found in mixed flocks of Black-capped and Chestnut-backed chickadees, Red-breasted Nuthatches, Brown Creepers, Downy and Hairy woodpeckers, or Golden-crowned and Ruby-crowned kinglets.

The Mountain Chickadee occurs throughout the year in the interior (Fig. 272); on the coast, it has been recorded regularly throughout the year in the southern Coast and Mountains and irregularly from mid-July to early May in the Georgia Depression (Fig. 272).

BREEDING: The Mountain Chickadee breeds throughout the interior north to the Nechako River valley and the vicinity of

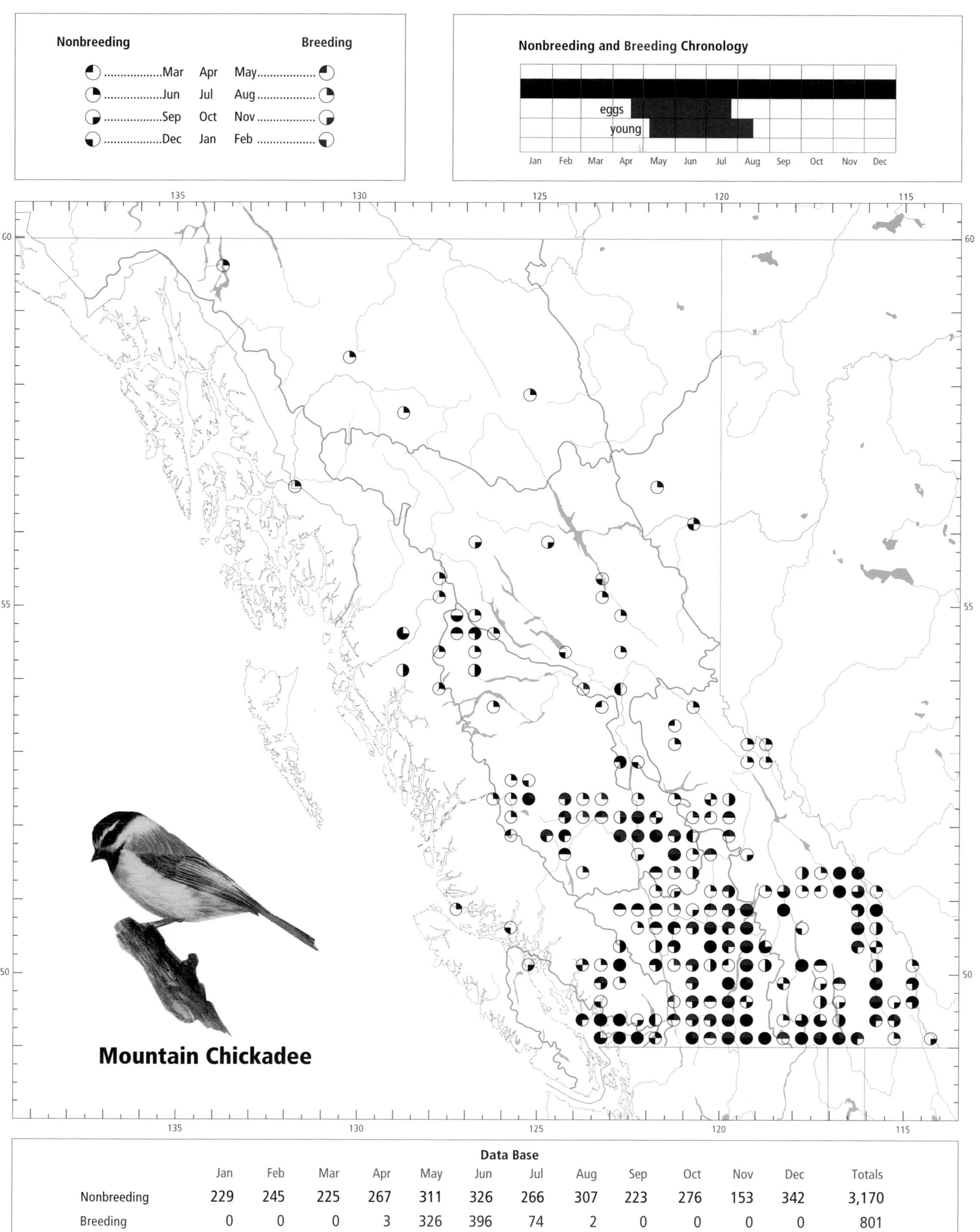

Data Base

	Jan	Feb	Mar	Apr	May	Jun	Jul	Aug	Sep	Oct	Nov	Dec	Totals
Nonbreeding	229	245	225	267	311	326	266	307	223	276	153	342	3,170
Breeding	0	0	0	3	326	396	74	2	0	0	0	0	801

Figure 271. All 4 species of chickadee (Black-capped, Mountain, Boreal, and Chestnut-backed) occur in mixed forests of Douglas-fir, western redcedar, and black cottonwood in Manning Park (Skagit River, 8 August 1996; R. Wayne Campbell).

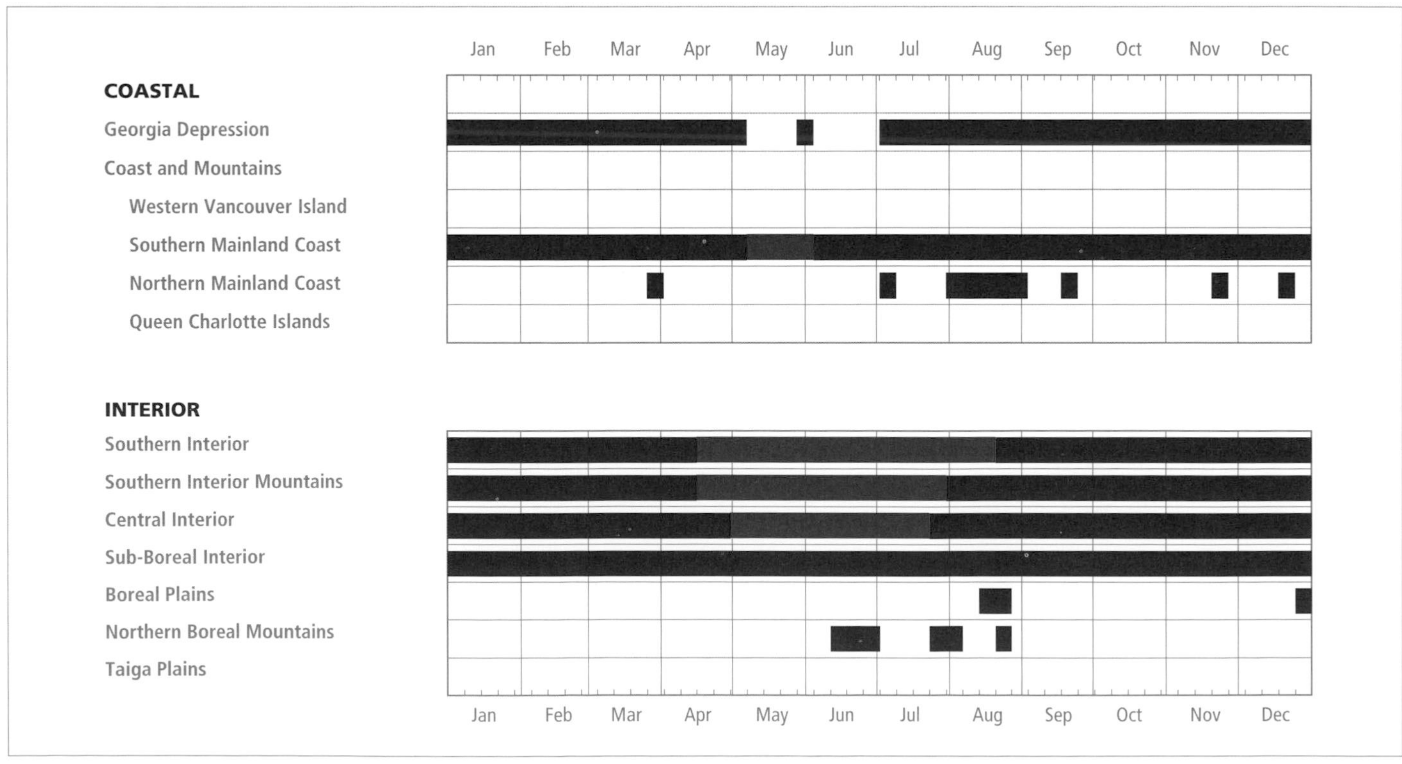

Figure 272. Annual occurrence (black) and breeding chronology (red) for the Mountain Chickadee in ecoprovinces of British Columbia.

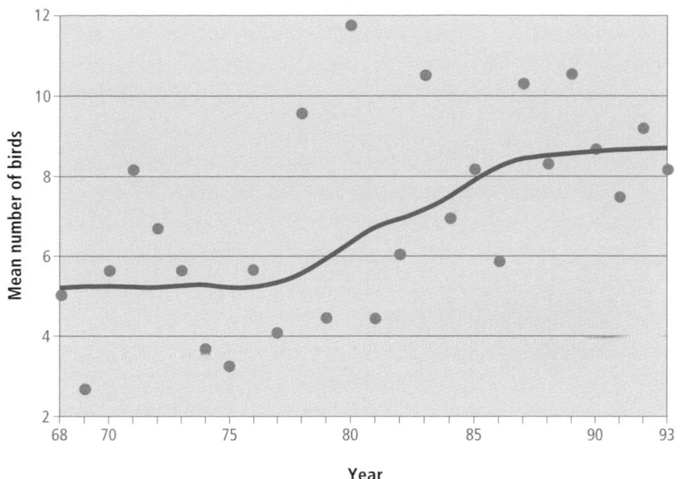

Figure 273. An analysis of Breeding Bird Surveys for the Mountain Chickadee in British Columbia shows that the number of birds on interior routes increased at an average annual rate of 3% over the period 1968 through 1993 ($P < 0.05$).

Smithers. Breeding has not been confirmed in the Babine Range or in the northwestern interior towards Atlin, although it appears to be resident there. On the coast, there is 1 breeding record for the southern Coast and Mountains between Hope and Boston Bar. However, the Mountain Chickadee likely breeds in small numbers in mountainous areas of that ecoprovince; there are occurrence records from the Pemberton valley, Whistler Mountain, and the mountains above North Vancouver.

This chickadee is a characteristic and widespread bird in the interior of southern British Columbia. The highest numbers in summer occur in the Southern Interior (Fig. 270). An analysis of Breeding Bird Surveys shows that the number of birds on interior routes has increased at an average annual rate of 3% for the period 1968 through 1993 (Fig. 273); coastal routes provide insufficient data for analysis.

The Mountain Chickadee breeds at elevations from 320 to 2,100 m. In the southern portions of the interior, it is the most common breeding chickadee and breeds in all forested habitats, from the edge of the grassland zones to near the timberline, including ponderosa pine, interior Douglas-fir (see Figs. 271 and 274), lodgepole pine, subalpine fir, and Engelmann spruce forests. The only coastal breeding area was in a mountain hemlock forest. Breeding habitats included (n = 244) coniferous forest (23%; see Figs. 271 and 274), rural (21%), mixed forest (9%), deciduous forest (9%), farmland (9%), wooded groves in big sagebrush shrublands (7%), and grasslands (6%). Riparian forests along streams, gullies, and lakeshores are also used. In the Okanagan valley, the Mountain Chickadee is most abundant in the open ponderosa pine–interior Douglas-fir forests at elevations from 500 to 1,500 m (Cannings et al. 1987). In the Central Interior, it is most abundant in trembling aspen woodlands.

The Mountain Chickadee has been recorded breeding in the province from 18 April to 15 August (Fig. 272).

Nests: The Mountain Chickadee is a primary cavity excavator that frequently digs its own nest cavity but will also use old sapsucker or chickadee cavities and nest boxes (Fig. 269). Most excavated cavities (n = 139) were in deciduous trees (55%), coniferous trees (30%), stumps (9%; Fig. 274), and posts (2%). Tree species included trembling aspen, Douglas-fir, ponderosa pine, lodgepole pine, birch, willow, white spruce, black cottonwood, subalpine fir, and western larch. Deciduous trees were used more frequently (75%; n = 28) than conifers in the Central Interior, compared with the Southern Interior (52%; n = 87) and Southern Interior Mountains (61%, n = 23). This regional difference may reflect the abundance of sapsucker cavities, which tend to occur mainly in trembling aspen stands, a more common habitat in the Central Interior. Most nests in posts, poles, and buildings were associated with nest boxes and were not included in analyses.

Specific nest sites (n = 411) included nest boxes (64%; Fig. 269) and excavated cavities (33%). The high proportion of nests in nest boxes supports the contention that the Mountain Chickadee is somewhat dependent on existing cavities (Hill and Lein 1988), more so than other chickadees. Unlike the Chestnut-backed and Black-capped chickadees, the

Figure 274. The Mountain Chickadee occurs throughout the year in Douglas-fir forests in parts of the Central Interior Ecoprovince (near Dog Creek, 24 March 1996; R. Wayne Campbell). Note nest cavity near top of stump.

Mountain Chickadee rarely excavated a cavity in the broken end of a tree limb.

The heights of 136 nests in excavated or natural cavities ranged from 0.2 to 18 m, with 52% between 1.2 and 4.5 m (Fig. 274). Nests (*n* = 240) were cups of well-packed material; hair (72%), mosses (35%), grasses (24%), and feathers (23%) were the most frequently reported materials. One nest near Riske Creek was constructed entirely from shredded owl pellets, complete with rodent bones.

Eggs: Dates for 197 clutches ranged from 18 April to 25 July, with 55% recorded between 15 May and 6 June. Sizes of 188 clutches ranged from 1 to 9 eggs (1E-9, 2E-6, 3E-8, 4E-12, 5E-29, 6E-54, 7E-49, 8E-19, 9E-2), with 55% having 6 or 7 eggs.

The incubation period (*n* = 3) in British Columbia is 13 days, which is 1 day less than that reported by Harrison (1979) and Ehrlich et al. (1988).

Young: Dates for 268 broods ranged from 6 May to 15 August, with 53% recorded between 31 May and 20 June. Sizes of 188 broods ranged from 1 to 9 young (1Y-5, 2Y-9, 3Y-14, 4Y-18, 5Y-41, 6Y-47, 7Y-41, 8Y-11, 9Y-2), with 69% having 5 to 7 young (Fig. 275).

Southern populations may raise 2 broods annually. Nests with young in July and August are almost certainly second nestings. In the Okanagan valley, 3 clutches begun between 22 June and 25 June were thought to represent second broods (Cannings et al. 1987).

The average nestling period (*n* = 8) was 18 days, with a range from 14 to 20 days. Cannings et al. (1987) report a mean nestling period of 16 days (*n* = 8) in the Okanagan valley, while in northern California the nestling period was 21 days in 17 of 18 nests (Grundel 1987). A comparison of growth rates of Mountain Chickadee nestlings (Grundel 1987) with those of the Black-capped Chickadee (Ricklefs 1968) indicated that the former grew more slowly than the Black-capped Chickadee nestlings, leading to the longer nestling period observed. The small sample of Mountain Chickadee clutches from southern British Columbia (*n* = 16) (Cannings et al. 1987) and this study do not support the difference observed by Grundel.

Brown-headed Cowbird Parasitism: In British Columbia, there were no cases of cowbird parasitism in 266 nests found with eggs or young. Friedmann et al. (1977) and Friedmann and Kiff (1985) do not list the Mountain Chickadee as being a cowbird host in North America.

Nest Success: Of 117 nests found with eggs and followed to a known fate, 90 produced at least 1 fledgling, for a success rate of 77%. In the Okanagan valley, 15 of 20 nests fledged at least 1 young, and produced 91 fledglings from 127 eggs (Cannings et al. 1987). Causes of nest failure included

Figure 275. Most broods of Mountain Chickadee in British Columbia ranged from 5 to 7 young (west of Princeton, 7 June 1994; R. Wayne Campbell).

predation by squirrels (5 nests), chipmunks (2), and striped skunk (1); loss of an adult (4 nests); and destruction of eggs by a House Wren (1).

REMARKS: Six subspecies are recognized in North America (American Ornithologists' Union 1957); only *P. g. grinnelli* occurs in British Columbia (Behle 1956). See Grinnell (1918), van Rossem (1928), and Behle (1956) for a discussion of the systematics of this chickadee.

The Mountain Chickadee often nests in abandoned sapsucker cavities. Indeed, its local nesting distribution may be strongly linked to the distribution of larger aspens, which are favoured by sapsuckers (Hill and Lein 1988). In British Columbia, it competes for these cavities with House Wrens, Black-capped Chickadees, Red-breasted and Pygmy nuthatches, Tree and Violet-green swallows, and Mountain Bluebirds.

See REMARKS in **Red-breasted Nuthatch** for comments on the relationship between *Armillaria* root disease and the suitability of trees for nesting by weak and strong cavity excavators.

The Mountain Chickadee coexists with Black-capped, Chestnut-backed, and Boreal chickadees. Research in Alberta has shown an ecological separation between the Mountain and Black-capped chickadees: the former nests in areas with more coniferous trees, more dead trees, and larger deciduous trees, and also forages more often in conifers, at greater heights, and in taller trees than the Black-capped Chickadee. Breeding territories rarely overlap because of these ecological differences (Hill and Lein 1988, 1989).

Selective logging of ponderosa pine forests may not adversely affect populations of the Mountain Chickadee (Scott and Oldemeyer 1983).

NOTEWORTHY RECORDS

Spring: Coastal – Beach Grove 27 Mar 1987-1; Stanley Park (Vancouver) 3 Mar 1985-1, 22 Apr 1987-2; Sumas Mountain 20 Mar 1979-6; Pitt Meadows 22 Apr 1987-2; Harrison Hot Springs Mar 1976-1; Hope 21 May 1977-2; Yale 8 Mar 1964-8; Alta Lake 30 Mar 1968-5; Terrace 1 Apr 1977-5. **Interior** – Osoyoos Lake 14 May 1922-1 (NMC 17652); Manning Park 29 Apr 1979-16; White Lake (Okanagan Falls) 28 Mar 1986-30, 30 Apr 1984-7 eggs; Jaffray 26 Apr 1975-7; Princeton 12 May 1989-feeding nestlings; Cranbrook 18 Apr 1938-1 egg, 6 May 1938-6 nestlings (Johnstone 1949); Naramata 5 Apr 1973-nest building, 13 May 1973-7 nestlings; Radium Hot Springs 31 May 1979-nestlings; Celista 12 Apr 1948-62; Golden 6 Mar 1977-6; Clearwater 16 May 1986-nest building; 100 Mile House 3 Apr 1975-10; Chilcotin River 5 May 1989-nestlings; Riske Creek 20 Apr 1978-8, 15 May 1983-8 eggs; 25 km s Prince George 25 Apr 1982-1; Tetana Lake 28 Mar 1938-1 (Stanwell-Fletcher and Stanwell-Fletcher 1943); Bobtail Mountain 2 Mar 1958-1.

Summer: Coastal – Mount Seymour 30 Jul 1967-4; Cypress Park 22 Aug 1985-1; Tetrahedron Peak area 17 Aug 1990-2; Garibaldi Park 3 Jul 1983-2, 29 Jul 1979-5; Kitimat River 7 Jul 1975-1 (Hay 1976); Redslide Mountain 30 Aug 1985-10; Stikine River 14 Jul 1919-1. **Interior** – Cathedral Park 25 Jul 1982-at least 3 eggs, 30 Aug 1980-50; Thompson Mountain (Creston) 25 Jun 1984-8 eggs; Kimberley 21 Jul 1976-4 nestlings; Wasa 15 Jul 1971-2 nestlings; Otter Lake 15 Aug 1982-nestlings; Wilmer 18 Aug 1979-17; Kamloops 30 Jun 1991-2; Spillimacheen 26 Jun 1976-14; Kootenay National Park 29 Jun 1983-1 nestling, 6 Jul 1985-50; Field 4 Jun 1983-pair at nest hole; Alkali Lake 18 Jul 1978-7; Riske Creek 26 Jun 1978-7 eggs; Stum Lake 9 Jun 1973-3 eggs, 1 nestling, 26 Jul 1973-6 nestlings (Ryder 1973); Wells Gray Park 27 Jul 1955-5; Caribou Mountain (Stuie) 16 Jul 1946-1 (FMNH 174619); Anahim Lake 17 Jun 1981-6 eggs; Mount Robson Park 30 Jun 1972-2; Astlais Mountain 2 Jun 1975-1; Sinkut Mountain 23 Jun 1946-several family groups (Munro 1949); Topley 24 Jul 1956-5; Wonowon 19 Aug 1975-8; Cold Fish Lake 26 Jul 1959-1; Haworth Lake 3 Aug 1976-8; Dokdaon Creek 14 Jul 1919-4 (MVZ 40265-68; Swarth 1922); Dease Lake 29 Jul 1962-1; Atlin 12 Jun 1924-1 (MVZ 44960).

Breeding Bird Surveys: Coastal – Recorded from 1 of 27 routes and on 1% of all surveys. Maxima: Pemberton 30 Jun 1974-1. **Interior** – Recorded from 44 of 73 routes and on 58% of all surveys. Maxima: Bridge Lake 12 Jun 1983-43; Beaverdell 22 Jun 1980-39; Summerland 2 Jul 1974-33.

Autumn: Interior – Germansen Landing 21 Oct 1972-4; Tetana Lake 21 Nov 1937-1; Hellroaring Creek 4 Nov 1983-18; Phililloo Lake 24 Oct 1962-10; Bridge Lake 7 Sep 1960-24; Okanagan Landing 5 Sep 1929-25; Nakusp 29 Sep 1975-2; Penticton 26 Oct 1973-37; Fernie 25 Sep 1983-15; Flathead River 21 Oct 1973-1 (RBCM 14489). **Coastal** – Terrace 24 Nov 1986-1; Kitimat River 17 Sep 1974-3 (Hay 1976); Quadra Island 30 Sep 1983-1 photographed with several Chestnut-backed Chickadees at feeder; Marina Island 29 Sep 1977-2; Alta Lake 9 Nov 1944-10; Gibsons 15 Nov 1992-2; Pitt Lake 11 Oct 1982-2; Mount Seymour 5 Sep 1965-9, 13 Sep 1971-5; Stanley Park (Vancouver) 29 Sep 1973-1; Vancouver 5 Nov 1982-2; Reifel Island 21 Sep 1982-3, 3 Oct 1989-1 (Weber and Cannings 1990); White Rock 22 Oct 1986-1.

Winter: Interior – Prince George 21 Jan 1979-1; Quesnel 7 Feb 1979-1; Kleena Kleene 1 Jan 1949-4 (Paul 1959); Williams Lake 2 Jan 1985-144 on Christmas Bird Count; Anahim Lake 4 Feb 1990-10; Golden 18 Jan 1977-12, 30 Dec 1993-25; Invermere 8 Feb 1977-12; Kelowna 1 Jan 1977-53 on Christmas Bird Count; Manning Park 19 Feb 1983-40 in 6 flocks. **Coastal** – Port Neville Inlet 12 Feb 1975-2; Sechelt 22 Feb 1987-1; Harrison Hot Springs 31 Jan 1976-1; Hollyburn Mountain 12 Jan 1975-12; West Vancouver 21 Dec 1986-1; Deroche 12 Dec 1966-1; UBC Research Forest (Maple Ridge) 2 Jan 1988-4; Pitt Meadows 13 Jan 1973-1; Stanley Park (Vancouver) 6 Jan 1973-1 (BC Photo 248), 14 Feb 1985-1; Vancouver 18 Feb 1982-3; Sumas Mountain Park 1 Dec 1979-6; Delta 26 Jan 1986-2.

Christmas Bird Counts: Interior – Recorded from 26 of 27 localities and on 81% of all counts. Maxima: Penticton 27 Dec 1986-**519**, all-time Canadian high count (Cannings 1987a); Vaseux Lake 28 Dec 1989-361; Lake Windermere 30 Dec 1984-334. **Coastal** – Recorded from 8 of 33 localities and on 5% of all counts. Maxima: White Rock 3 Jan 1988-8; Squamish 2 Jan 1983-7; Chilliwack 28 Dec 1974-6.

Boreal Chickadee

Parus hudsonicus Forster

BOCH

RANGE: Resident in the boreal forests south from western and central Alaska, northern Yukon, south-central Mackenzie, and across northern Canada from northern Saskatchewan to Labrador and Newfoundland. South in the west to interior British Columbia, northern Washington, northwestern Montana, and north of the prairie grasslands in southwestern and central Alberta, central Saskatchewan, and southern Manitoba; east through northern Minnesota, northern Michigan, central Ontario, Vermont, and New Hampshire, into New Brunswick and Nova Scotia. Wanders irregularly south as far as West Virginia.

STATUS: On the coast, *casual* in the southern portions of the Coast and Mountains Ecoprovince and on the Fraser Lowland of the Georgia Depression Ecoprovince.

In the interior, *uncommon* but widespread resident in all ecoprovinces, especially throughout northern and central British Columbia and in the Rocky Mountains in the Southern Interior Mountains Ecoprovince; *uncommon* and more locally distributed resident elsewhere in the Southern Interior Mountains and in the Southern Interior Ecoprovince; locally *fairly common* in those ecoprovinces during winter.

Breeds.

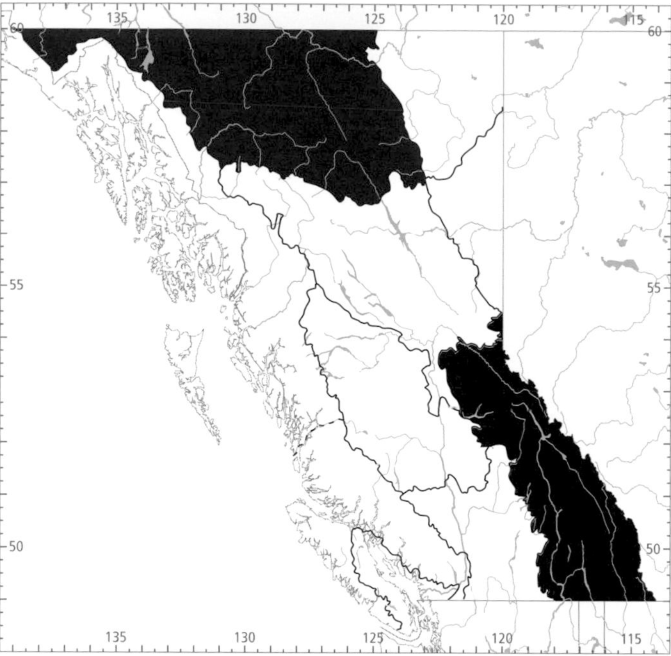

Figure 276. In British Columbia, the highest numbers for the Boreal Chickadee in winter (black) occur in the Southern Interior Mountains Ecoprovince; the highest numbers in summer (red) occur in the Northern Boreal Mountains Ecoprovince.

NONBREEDING: The Boreal Chickadee is mainly a resident of northern British Columbia and high elevations of the central and southern portions of the interior. It appears to be absent from much of the Fraser River Basin and lower elevations of the Central Interior, Southern Interior, and Southern Interior Mountains ecoprovinces. It occurs infrequently in coastal areas and is absent from Vancouver Island, the Queen Charlotte Islands, and all other coastal islands. On the southern mainland of the Coast and Mountains, it appears occasionally on the west slope of the coastal ranges. The highest numbers in winter have been reported from the Southern Interior Mountains (Fig. 276); however, there are few data from northern ecoprovinces with which to compare relative abundances, and the Boreal Chickadee may be as common there as in the Southern Interior Mountains.

The Boreal Chickadee occurs at elevations between 500 and 2,200 m; there is 1 record from North Vancouver at about 100 m elevation. It is widely distributed in coniferous forested habitats of northern and central interior regions, where it occurs at most elevations up to the timberline. In the southern portions of the interior, the Boreal Chickadee occurs mainly at higher elevations in mountainous habitat, rarely in lowland valleys. In the south, it is a species of subalpine parklands with alpine meadows and coniferous forests of Engelmann spruce, subalpine fir, whitebark pine, and alpine larch. In winter, some individuals use lower-elevation lodgepole pine and western larch forests. Cavities in trees are important habitat elements because the Boreal Chickadee roosts in cavities that, during periods of extreme cold, provide critical thermal protection (Smith 1991).

The species is essentially nonmigratory (Fig. 277), even in northern areas. During severe winters or periods of food shortage, a southward movement may occur (Root 1988). In addition, populations that breed at high elevations may move to lower elevations for the winter (Cannings et al. 1987).

This is the quietest of the chickadees; it is a sombre-plumaged little bird that generally travels through the upper and middle forest canopy. It is found alone almost as frequently as in small flocks. From late summer through winter, the Boreal Chickadee may be found in mixed flocks with other small songbirds such as the Black-capped Chickadee, Mountain Chickadee, Chestnut-backed Chickadee, various warblers, Red-breasted Nuthatch, and Golden-crowned Kinglet.

On the coast, the Boreal Chickadee has been recorded in every month but March, April, May, and November; in the interior, it has been recorded throughout the year (Fig. 277).

BREEDING: Although the Boreal Chickadee probably breeds throughout most of the interior, records are lacking from much of its range. It is a relatively widespread breeder throughout much of the northern interior and in the southern Rocky Mountains. In central British Columbia, it is less widespread and breeds at moderate to higher elevations; in the southern portions of the interior, it breeds only at higher elevations. Breeding has not been confirmed on the east slope of the coastal ranges or on the mountains immediately north of Burrard Inlet.

The highest numbers in summer occur in the Northern Boreal Mountains (Fig. 276). Breeding Bird Surveys for both

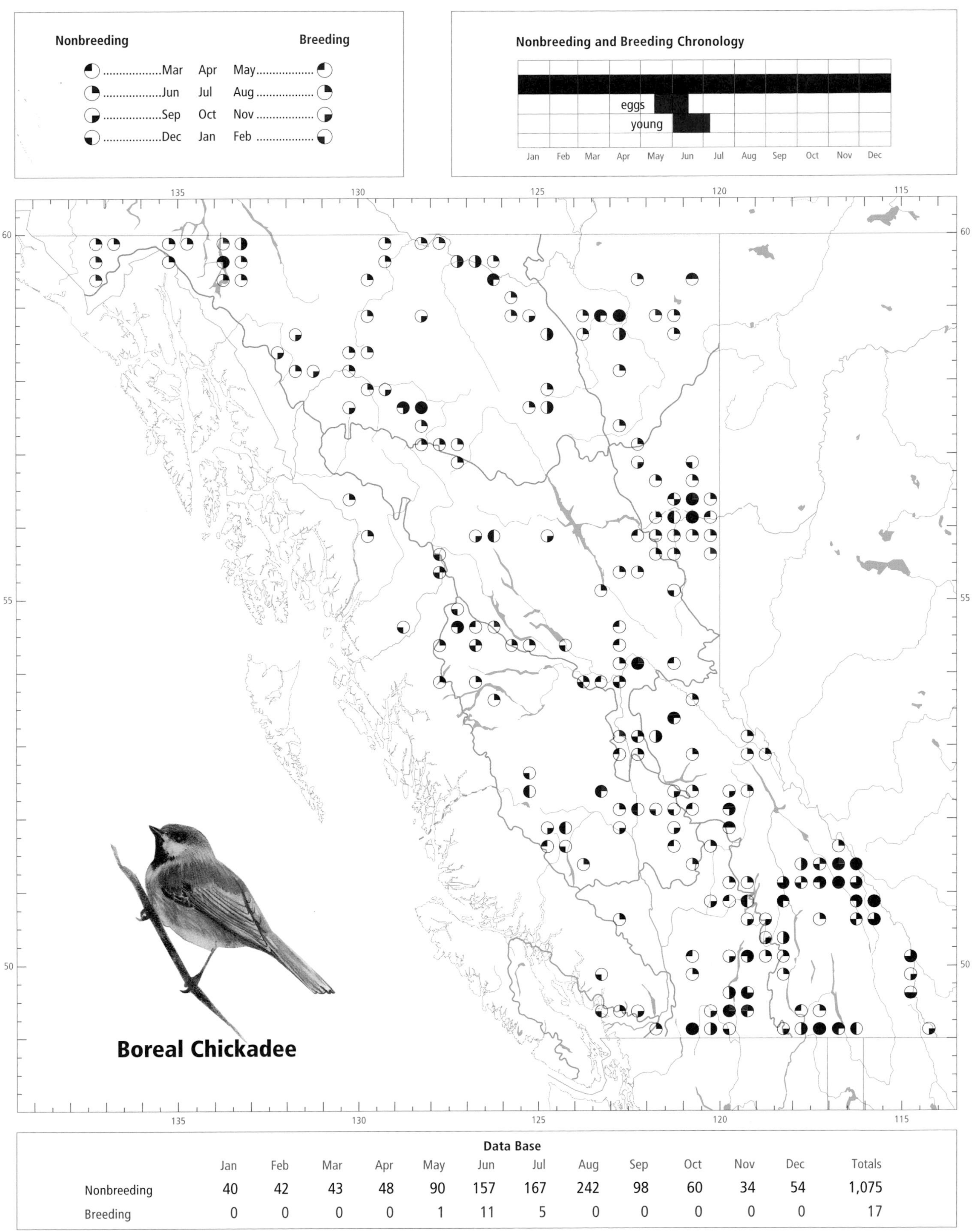

Data Base

	Jan	Feb	Mar	Apr	May	Jun	Jul	Aug	Sep	Oct	Nov	Dec	Totals
Nonbreeding	40	42	43	48	90	157	167	242	98	60	34	54	1,075
Breeding	0	0	0	0	1	11	5	0	0	0	0	0	17

interior and coastal routes for the period 1968 through 1993 contain insufficient data for analysis.

In the north, breeding habitat includes open coniferous, deciduous, and mixed forests (see Fig. 271), or the edges of more closed stands, from valley bottoms up to the timberline (Fig. 278). In the south, it breeds at higher elevations up to the timberline in open coniferous forests of the Montane Spruce and Engelmann Spruce–Subalpine Fir biogeoclimatic zones. Breeding has been documented at elevations from 530 m to about 1,850 m.

In the interior, the Boreal Chickadee has been recorded breeding from 13 May (calculated) to 7 July. Recently fledged young have been recorded as late as 9 August (Fig. 277).

Nests: Few nests have been found in British Columbia. Nest sites included excavated or enlarged natural cavities in decayed stumps (6 nests), snags (3 nests), and live trees (1 nest). Nest trees included spruce, birch, subalpine fir, willow, and trembling aspen. A nest in a living aspen was in a crack in the tree trunk. As with other species of this genus, the Boreal Chickadee is a primary cavity excavator, but it is incapable of preparing nesting cavities in sound wood. Thus it usually uses dead trees softened by fungus or it enlarges natural cavities. It usually excavates its own nest cavity and uses a new cavity each year (McLaren 1975). It may also use cavities excavated by small woodpeckers, other chickadees, or nuthatches.

The heights of 7 nests ranged from 1.0 to 8.0 m, with 5 nests between 1.2 and 5.5 m. One nest cavity was 15 cm in depth. Nest materials included fine grasses, mosses, animal hair, lichens, and wood chips.

Figure 278. The edges of open subalpine fir forests in the Sub-Boreal Interior Ecoprovince support breeding populations of the Boreal Chickadee ("Vineyards," 40 km east of Prince George, 17 July 1992; R. Wayne Campbell).

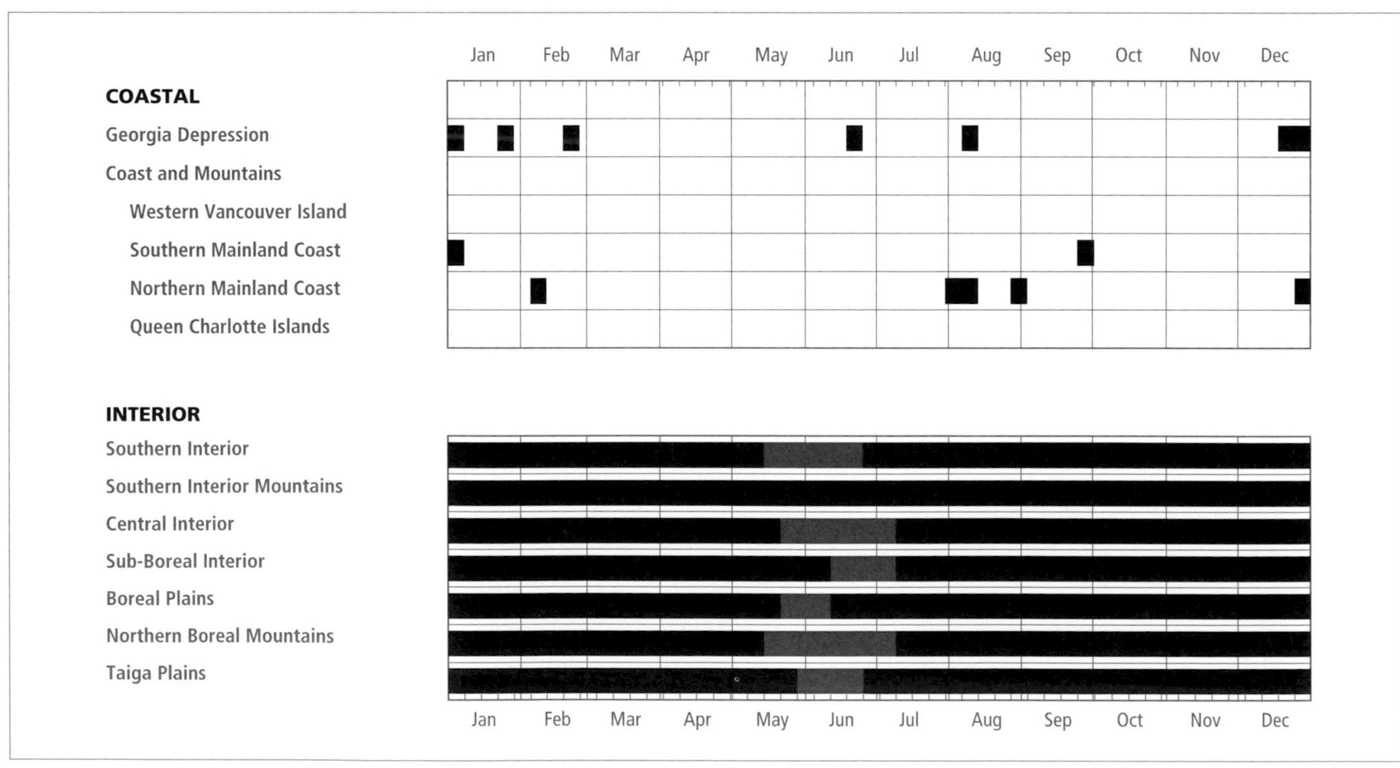

Figure 277. Annual occurrence (black) and breeding chronology (red) for the Boreal Chickadee in ecoprovinces of British Columbia.

Eggs: Dates for 3 clutches ranged from 31 May to 16 June. Calculated dates indicate that eggs can occur as early as 15 May. Clutch size ranged from 2 to 8 eggs (2E-1, 5E-1, 8E-1). In Ontario, clutch size ranged from 4 to 7 eggs (Peck and James 1987). The incubation period ranges from 11 to 15 days (McLaren 1975; Peck and James 1987).

Young: Dates for 13 broods ranged from 1 June to 7 July. Sizes of 9 broods ranged from 2 to 6 young (2Y-3, 4Y-2, 5Y-2, 6Y-2). The nestling period is about 18 days (McLaren 1975). Dates for 12 family groups with newly fledged young ranged from 21 June to 9 August. Those groups contained from 2 to 6 young (2Y-2, 3Y-3, 4Y-3, 5Y-3, 6Y-1). Family groups persist for 2 to 3 weeks before the young disperse (McLaren 1975).

Brown-headed Cowbird Parasitism: In British Columbia, there were no cases of cowbird parasitism in 12 nests found with eggs or young; nor has it been reported for North America (Friedmann et al. 1977; Friedmann and Kiff 1985).

Nest Success: Insufficient data for analysis.

REMARKS: Two of the 5 subspecies recognized in North America occur in British Columbia (Rhoads 1893a; Miller 1943; American Ornithologists' Union 1957). *P. h. columbianus* occurs over most of interior British Columbia, whereas *P. h. cascadensis* is mainly restricted to the Southern Interior.

Concentrations of breeding Boreal Chickadees in British Columbia can reach over 4 pairs/ha. For example, a 40 ha patch of subalpine forest in the Central Interior held 9 breeding pairs (M. Waterhouse pers. comm.).

See REMARKS in **Red-breasted Nuthatch** for comments on the relationship between *Armillaria* root disease and the suitability of trees for nesting by weak and strong cavity excavators.

See Snarski (1969) for a summary of literature on *Parus* spp. in North America.

NOTEWORTHY RECORDS

Spring: Coastal – No records. **Interior** – Strawberry Flats (Manning Park) 27 Mar 1984-4; Venner Meadows 31 Mar 1977-1; Wulf Lake (Salmo) 28 Apr 1983-14; Schoonover Mountain (Okanagan Falls) 31 May 1913-5 eggs (Anderson 1914); Apex Mountain 8 Mar 1978-1; Kootenay Pass 22 Mar 1986-2 at summit; Emerald Lake (Yoho National Park) 1 May 1977-1; Straight Lake 11 May 1983-1 (Graham 1983); Bluff Lake (Tatla Lake) 8 Mar 1981-6; Puntchesakut Lake 18 May 1944-1 (Munro 1947a); Pass Lake (McGregor River) 5 Mar 1983-6; Kerry Lake 8 May 1968-2; mouth of Halfway River 25 Apr 1982-3; Fort St. John 5 May 1993-2; Bear Flat 5 May 1984-3; Hyland Post 30 May 1976-6; Fort Nelson 15 Mar 1975-7 (Erskine and Davidson 1976); Liard Hot Springs 1 Apr 1975-2.

Summer: Coastal – Cheam Peak 8 Aug 1978-4; Mount Seymour Park 18 Jun 1978-1 (Anonymous 1978); Redslide Mountain 30 Aug 1989-2; Stewart 29 Aug 1977-3. **Interior** – Manning Park 4 Aug 1962-2; Frosty Mountain 16 Aug 1974-23; Cathedral Park 20 Aug 1983-20; Wulf Lake (Salmo) 21 Jun 1983-8; Apex Mountain 9 Aug 1969-4 fledglings (Cannings et al. 1987); Wulf Lake (Salmo) 18 Jul 1983-18; Adams Lake 28 Aug 1966-8; Yoho National Park 5 Jun 1975-pair building nest, 16 Jul 1975-28, 4 Aug 1975-25 (Wade 1977); Stum Lake 14 Jun 1973-4 nestlings, 30 Jul 1973-6 fledglings (Ryder 1973); e Likely 9 Jun 1994-2 adults at nest hole in subalpine fir snag; Swiftcurrent Creek 15 Aug 1973-15 (Cannings 1973); Bowron Lake Park 23 Aug 1975-10; Nine Mile Mountain (Hazelton) 30 Jul 1921-1 (MVZ 42577); Topley 7 Jul 1956-5 fledglings, 24 Jul 1956-8; Aleza Lake 3 Jul 1967-5 fledglings; ne Fort St. James 23 Jul 1991-very common in mature forest; Moberly Lake Park 30 Jun 1990-3; Kiskatinaw Park 30 Jun 1981-3 fledglings, 19 Jul 1983-8; Beatton Park 19 Jun 1984-7 new fledglings; St. John Creek 3 Aug 1994-2; Wonowon 19 Aug 1975-10; Stalk Lakes 11 Jul 1976-13; Firesteel River 26 Jun 1976-8; Dease River 25 Jul 1977-40 on survey; Parker Lake (Fort Nelson) 8 Jul 1978-2 (Campbell 1979b); Mile 304, Alaska Highway 17 Jul 1985-12; Bernard Lake 23 Aug 1977-1; ne Kwokullie Lake 3 Jun 1982-8 eggs (RBCM E2021); Atlin 5 Jul 1980-4 fledglings; Tatshenshini River 3 Jun 1983-pair carrying food into nest (Campbell et al. 1983); Chilkat Pass 26 Aug 1974-10; Survey Lake 26 Jun 1980-10; Irons Creek 1 Jun 1980-2 eggs and 2 nestlings.

Breeding Bird Surveys: Coastal – Not recorded. **Interior** – Recorded from 16 of 73 routes and on 4% of all surveys. Maxima: Pennington 13 Jun 1987-7; Kootenay 14 Jun 1983-5; Steamboat 14 Jun 1976-4.

Autumn: Interior – Fort Nelson 1 Nov 1975-12 (Erskine and Davidson 1976); McConachie Creek (Fort Nelson) 27 Sep 1985-12; Little Tahltan River 18 Sep 1986-2; Kwadacha River 1 Sep 1979-4 (Cooper and Adams 1979); Fort St. John 17 Nov 1993-1; Stoddart Creek 18 Oct 1986-8, 21 Nov 1982-12; Telkwa River valley 14 Sep 1974-5; Wells Gray Park 11 Sep 1955-2; Horsefly 10 Nov 1953-1 (NMC 48067); Riske Creek 12 Oct 1989-1; Lac la Hache 4 Oct 1952-6; Wapta Falls 24 Oct 1976-10; Yoho National Park 4 Nov 1976-14; Glacier National Park 12 Sep 1982-10; Apex Mountain 20 Oct 1972-22; Curtis Lake (Salmo) 4 Sep 1983-24; Frosty Mountain 8 Oct 1978-15. **Coastal** – Hemlock Valley 24 Sep 1972-2.

Winter: Interior – Fort Nelson 28 Dec 1985-2, 11 Jan 1975-12; Hyland Post 26 Feb 1982-3; Taylor 19 Jan 1987-5; Prince George 21 Jan 1979-2; Anahim Lake 4 Feb 1990-1; Tatlayoko Lake valley 20 Dec 1991-1; Williams Lake 12 Jan 1988-1; Kootenay National Park 19 Jan 1983-10; Celista 17 Jan 1948-7; Manning Park 19 Feb 1983-20 in 4 flocks. **Coastal** – 24 km e Hazelton 25 Dec 1911-1 (UBC 160); Black Mountain (West Vancouver) 25 Feb 1967-1 (Anderson 1967); Cypress Park 22 Jan 1979-1 (Anonymous 1979); North Vancouver 22 Dec 1985 to 1 Jan 1986-1.

Christmas Bird Counts: Interior – Recorded from 14 of 27 localities and on 22% of all counts. Maxima: Fauquier 2 Jan 1989-20; North Pine 27 Dec 1986-14; Smithers 28 Dec 1986-13. **Coastal** – Recorded from 3 of 33 localities and on 1% of all counts. Maxima: Terrace 26 Dec 1988-13; Squamish 2 Jan 1983-2; Vancouver 22 Dec 1985-1.

Chestnut-backed Chickadee

Parus rufescens Townsend

CBCH

RANGE: Resident from south-central and southeastern Alaska and coastal British Columbia south along the Pacific coast to central California; and in the interior from southern British Columbia and western Alberta through mountainous areas in northern Idaho, northwestern Montana, eastern Washington, and northeastern Oregon.

STATUS: On the coast, a *fairly common* to *common* resident of the islands of the Coast and Mountains Ecoprovince, including the Queen Charlotte Islands and Western Vancouver Island; *fairly common* on the Southern and Northern Mainland coasts of that ecoprovince; *fairly common* to *common* resident in the Nanaimo Lowland and Gulf Islands elements of the Georgia Depression Ecoprovince, where it is *fairly common* on the Fraser Lowland; *rare* in the Boundary Ranges.

In the interior, an *uncommon* resident of the west Kootenay region of the Southern Interior Mountains Ecoprovince and in the eastern Southern Interior Ecoprovince; *very rare* in the southern Rocky Mountains, the Central Interior, Sub-Boreal Interior and western Northern Boreal Mountains ecoprovinces.

Breeds.

NONBREEDING: The Chestnut-backed Chickadee (Fig. 279) is resident and widely distributed along the coast of British Columbia from southern Vancouver Island and the Fraser Lowland to the Nass River and Portland Canal on the north

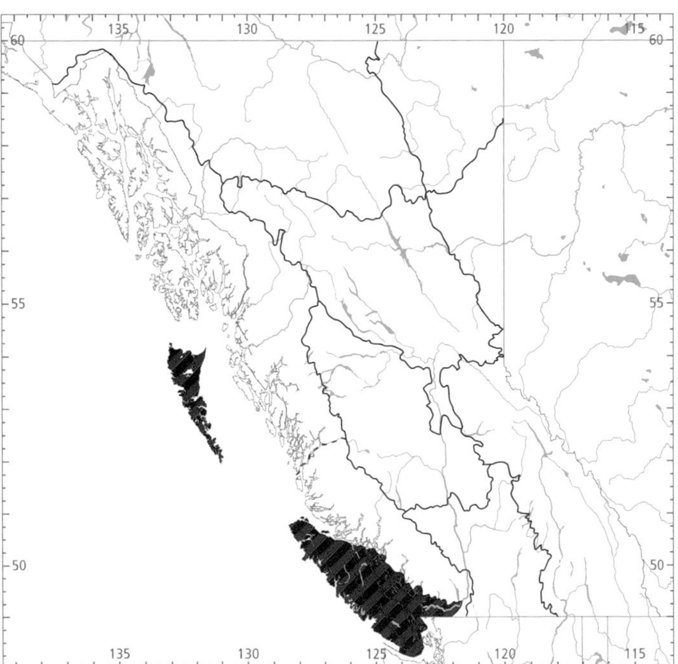

Figure 280. In British Columbia, the highest numbers for the Chestnut-backed Chickadee in both winter (black) and summer (red) occur in the Coast and Mountains Ecoprovince (on Western Vancouver Island and the Queen Charlotte Islands) and in the Georgia Depression Ecoprovince (in the Nanaimo Lowland and Gulf Islands).

Figure 279. Adult Chestnut-backed Chickadee (Victoria, 23 July 1994; R. Wayne Campbell).

Figure 281. On the northeast coast of Vancouver Island, open stands of western redcedar and western hemlock provide foraging and nesting habitat for the Chestnut-backed Chickadee (south of Port Hardy, 16 April 1995; R. Wayne Campbell).

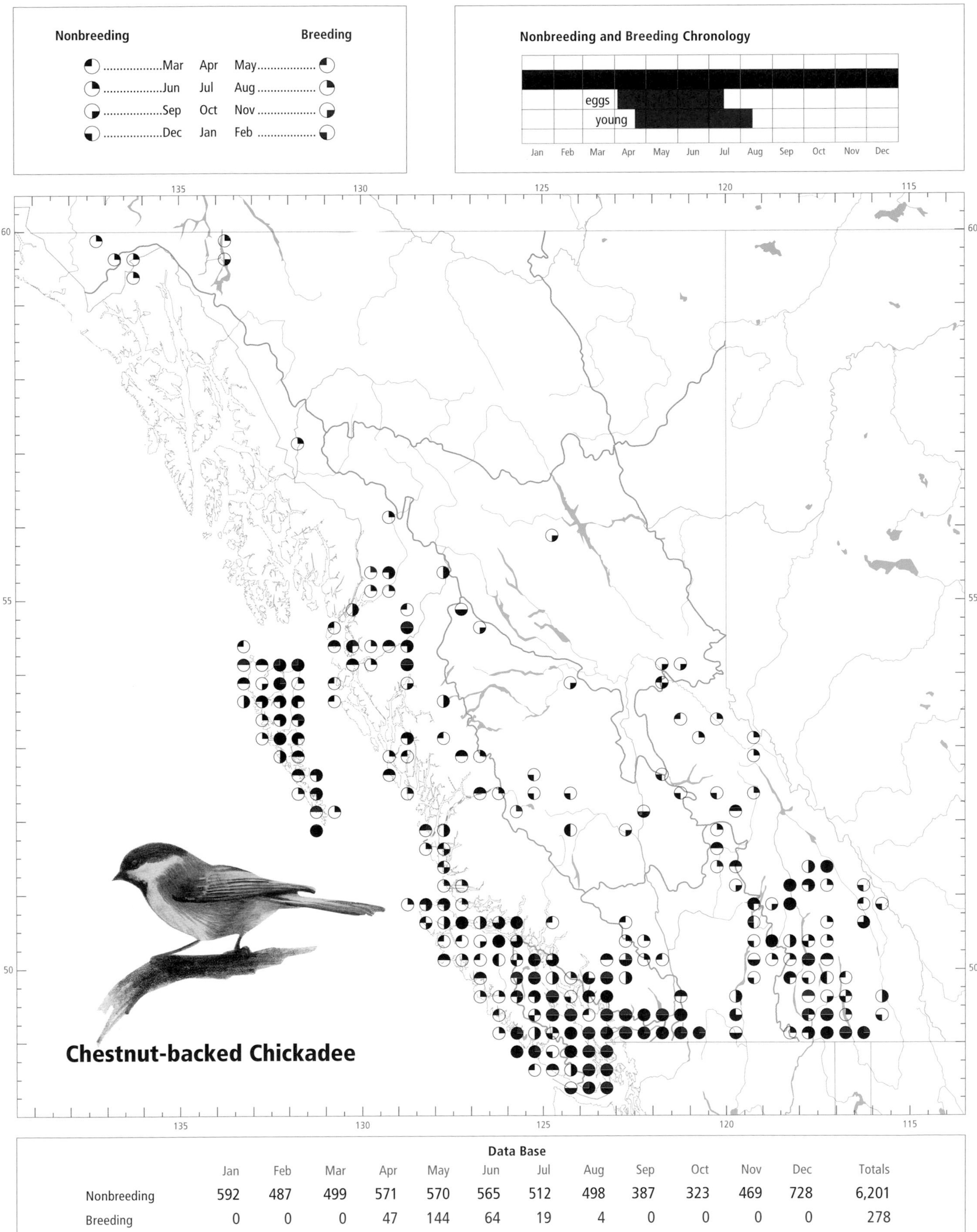

Data Base

	Jan	Feb	Mar	Apr	May	Jun	Jul	Aug	Sep	Oct	Nov	Dec	Totals
Nonbreeding	592	487	499	571	570	565	512	498	387	323	469	728	6,201
Breeding	0	0	0	47	144	64	19	4	0	0	0	0	278

coast. It is most abundant on coastal islands, including Vancouver Island and the Queen Charlotte Islands, where it is the only resident chickadee species. Further north, populations become more locally distributed, with scattered records from Atlin Lake, lower Chilkat Pass, and the Tatshenshini River valley. In the southern portions of the interior, it is resident in the Monashee and Selkirk mountain ranges, and locally in the southern Rocky Mountains and northern Okanagan Highland. Further north it occurs sporadically in the Central Interior and southern Sub-Boreal Interior. The northernmost interior winter records are from along the upper Fraser River and at Germansen Landing. It is absent from the Boreal Plains and Taiga Plains ecoprovinces in northeastern British Columbia.

The highest numbers in winter are found on Western Vancouver Island and the Queen Charlotte Islands of the Coast and Mountains Ecoprovince, and in the Nanaimo Lowland and the Gulf Islands of the Georgia Depression Ecoprovince (Fig. 280). An analysis of Christmas Bird Counts in British Columbia for the period 1957 through 1985 suggests that the provincial population is stable (Brennan and Morrison 1991).

The Chestnut-backed Chickadee occurs at elevations from sea level to at least 2,200 m during nonbreeding seasons. It frequents mainly mature coniferous forests or clumps of larger trees in younger stands, but is most numerous where openings occur (Fig. 281). On eastern Vancouver Island and on the Gulf Islands, where it is the only chickadee species, it is characteristic of the dry Douglas-fir–arbutus forests but also frequents riparian habitats and older lowland stands of bigleaf maple and red alder. On the mainland coast, where its range overlaps that of the Black-capped Chickadee, the Chestnut-backed Chickadee prefers humid, shady areas dominated by mixed western redcedar and western hemlock forests (Fig. 281). It forages mainly in the upper coniferous canopy, while its congener tends to forage in deciduous forest and shrubs (Smith 1967; Sturman 1968a; Brennan 1989). It is a regular visitor to backyard feeders in suburban areas with a significant amount of coniferous forest cover.

The Chestnut-backed Chickadee is essentially nonmigratory (Fig. 282), although some disperse to higher elevations after breeding. In winter, it often occurs in mixed-species flocks that may include any or all of the other chickadees, kinglets, Red-breasted Nuthatches, Bushtits, Downy Woodpeckers, Hutton's Vireos, and Brown Creepers.

The Chestnut-backed Chickadee has been recorded year-round in British Columbia (Fig. 282).

BREEDING: The Chestnut-backed Chickadee is a common and widespread breeder throughout the coast, including all treed coastal islands. It breeds along the entire mainland coast, up the west slopes of the Coast Ranges, and inland along major river valleys. In the southern portions of the interior, breeding occurs locally in Manning Park, the west Kootenay, and the upper Shuswap River region. The northernmost breeding locality in the interior is Lonesome Lake, the only breeding record from the Central Interior. Breeding has not been confirmed in the Okanagan valley (Cannings 1987), east Kootenay, and Quesnel Highland, or in the far northwestern mountains.

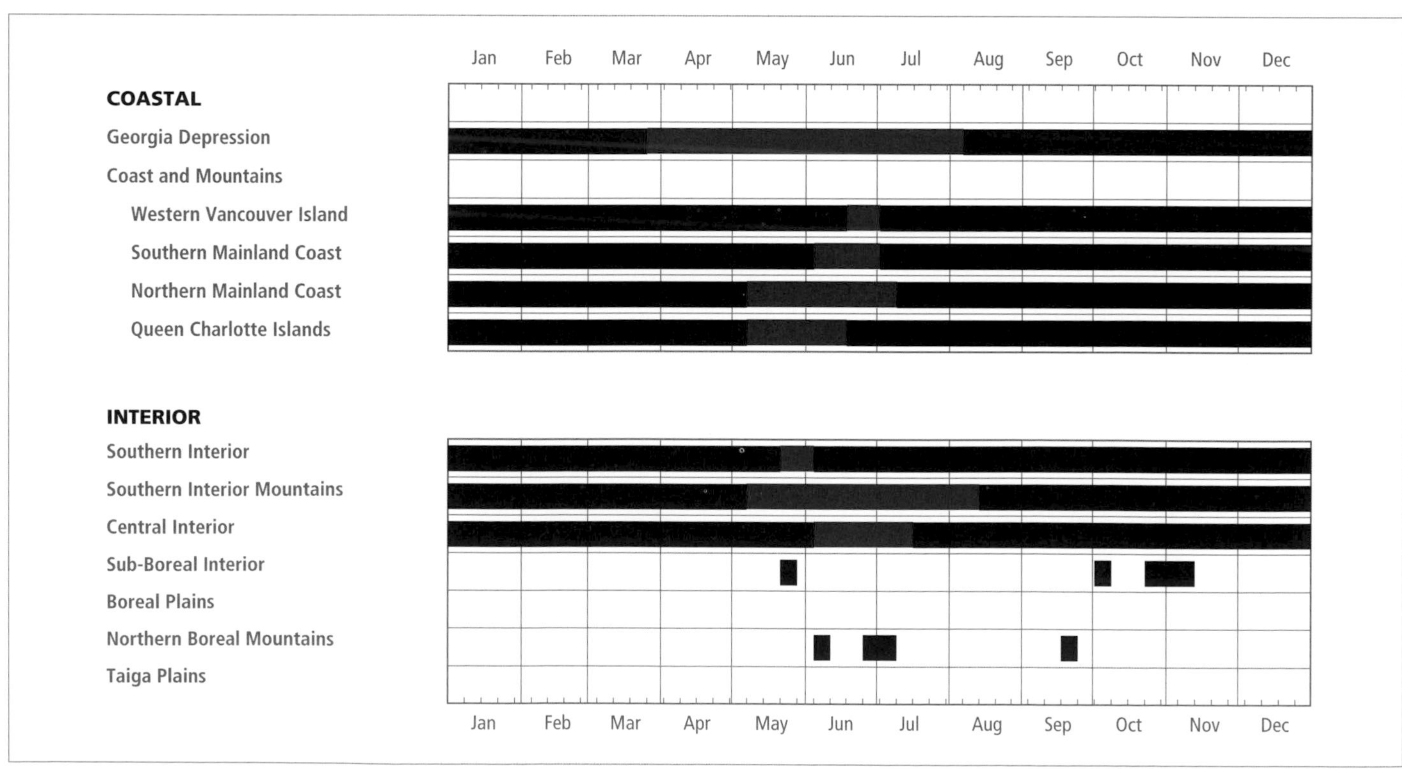

Figure 282. Annual occurrence (black) and breeding chronology (red) for the Chestnut-backed Chickadee in ecoprovinces of British Columbia.

Figure 283. In abandoned aboriginal village sites along the coast of British Columbia, the Chestnut-backed Chickadee often nests in totem poles (Anthony Island, Queen Charlotte Islands, 27 May 1996; R. Wayne Campbell).

The highest numbers in summer occur on Western Vancouver Island and the Queen Charlotte Islands (Fig. 280). An analysis of Breeding Bird Surveys for the period 1968 through 1993 could not detect a net change in numbers on coastal routes; interior routes provided insufficient data for analysis.

The Chestnut-backed Chickadee breeds at elevations from near sea level to 1,350 m on the coast, and from 500 m to 1,450 m in the interior. Along the coast, it is the characteristic breeding chickadee of coniferous forests and breeds mainly in western hemlock, mountain hemlock, Sitka spruce, western redcedar, coastal Douglas-fir, and grand fir forests. On Vancouver Island and the Queen Charlotte Islands, it nests in riparian deciduous forests of red alder, bigleaf maple, black cottonwood, and other species, as well as in drier habitats with stands of open Douglas-fir, Garry oak, and arbutus. On the mainland, where there are other chickadee species, it is restricted to coniferous forests. In the interior, the Chestnut-backed Chickadee breeds mainly in the wet valleys of the Interior Cedar-Hemlock biogeoclimatic zone.

In areas of dense forest, the preferred nesting habitat is along the edges of marine shorelines, ponds, wetlands, meadows, natural clearings, clearcuts, and burns. Well-treed suburbs provide breeding habitat in many areas. Even abandoned aboriginal village sites are used (Fig. 283). In logged areas, the retention or creation of snags and stumps is essential for the long-term maintenance of breeding habitat. In residential areas, it readily uses nest boxes placed on trees or posts.

The Chestnut-backed Chickadee has been recorded breeding in British Columbia from 27 March (calculated) to 12 August (Fig. 282). Breeding begins on the coast about 3 to 4 weeks before it begins in the interior (Fig. 282). In the Georgia Depression, nest building may begin as early as mid-March.

Nests: The Chestnut-backed Chickadee is a weak primary cavity excavator, and requires decayed snags, stumps, or trees with dead limbs with punky heartwood for nest sites. Most excavated nest cavities (n = 94) were found in tree stumps (42%), live trees (36%), dead snags (19%), and poles. Tree species used included Garry oak, Douglas-fir, willows, western redcedar, birches, bigleaf maple, arbutus, Sitka spruce, western hemlock, and black cottonwood. On Anthony Island, in the Queen Charlotte Islands, decaying aboriginal mortuary and house poles have been used.

Specific nest locations (n = 168) included self-excavated cavities (53%), nest boxes (37%), cavities excavated by other species (6%), and natural cavities. In deciduous trees, nest cavities are frequently excavated where a limb has broken away leaving a rotted entry point. Nests in Garry oak,

Figure 284. Chestnut-backed Chickadee nest and eggs (Victoria, 3 June 1996; R. Wayne Campbell).

arbutus, domestic apple, and western flowering dogwood were usually in natural cavities such as rotted knotholes. One unusual nest site on the Queen Charlotte Islands was on the ground in a hummock of *Sphagnum* moss. The heights of 90 nests in natural or excavated cavities ranged from ground level to 26 m, with 56% between 2.1 and 6.0 m. Nest lining (n = 50) included moss (70%), hair (54%), feathers (36%), grass (14%), and human-made fibres (14%).

Eggs: Dates for 62 clutches (Fig. 284) ranged from 3 April to 15 July, with 55% recorded between 24 April and 21 May. Calculated dates indicate that eggs may be present as early as 27 March. Sizes of 44 clutches ranged from 1 to 9 eggs (1E-1, 2E-1, 4E-5, 5E-9, 6E-9, 7E-10, 8E-8, 9E-1), with 64% having 5 to 7 eggs. The incubation period for 1 clutch was 15 days. Data on incubation periods from other regions are not available.

Young: Dates for 107 broods ranged from 20 April to 12 August, with 53% recorded between 17 May and 8 June. Sizes of 54 broods ranged from 1 to 9 young (1Y-1, 2Y-7, 3Y-3, 4Y-6, 5Y-10, 6Y-12, 7Y-9, 8Y-4, 9Y-2), with 57% having 5 to 7 young. The nestling period for 2 broods was 16 and 19 days. Data on nestling periods from other regions are not available.

Brown-headed Cowbird Parasitism: In British Columbia, there were no cases of cowbird parasitism in 81 nests found with eggs or young, nor has the Chestnut-backed Chickadee been recorded as a host species in North America (Friedmann et al. 1977; Friedmann and Kiff 1985).

Nest Success: Of 17 nests found with eggs and followed to a known fate, 12 produced at least 1 fledgling, for a success rate of 71%. Causes of mortality of nests with eggs included predation (2), abandonment (1), being taken over by bees (1), and being destroyed by European Starlings (1).

REMARKS: Three subspecies are recognized in North America; only *P. r. rufescens* occurs in British Columbia (American Ornithologists' Union 1957). See Grinnell (1904) and Burleigh (1959) for a discussion of the systematics of the Chestnut-backed Chickadee.

The Chestnut-backed and Black-capped chickadees occupy the same habitats in some mainland areas, and in the winter months may even travel in mixed flocks. Smith (1967) has shown that the 2 species use foraging habitat differently in British Columbia. The Chestnut-backed Chickadee prefers to forage in coniferous trees, high above the ground (e.g., 14 to 20 m), whereas the Black-capped Chickadee forages mainly in deciduous vegetation close to the ground (0 to 2 m). However, there is some overlap in use of spatial resources (Fig. 285). Additional habitat partitioning results from the

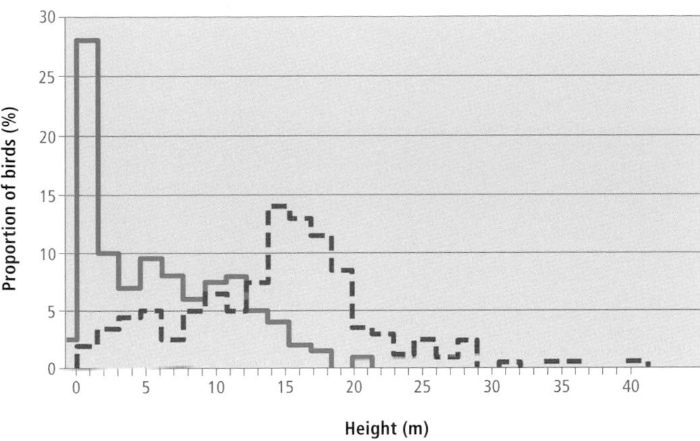

Figure 285. Comparison of feeding heights of Black-capped (solid line) and Chestnut-backed (dashed line) chickadees (adapted from Smith 1967).

Chestnut-backed Chickadee foraging more often than the Black-capped Chickadee on foliage, buds, cones, and the upper surfaces of branches (Sturman 1968b).

The breeding biology of the Chestnut-backed Chickadee is poorly known. Basic data on incubation and nestling periods, numbers of broods raised annually, and breeding territory sizes are not available. This species offers naturalists the opportunity to provide critical data on a common but poorly known Pacific Northwest species.

See REMARKS in **Red-breasted Nuthatch** for comments on the relationship between *Armillaria* root disease and the suitability of trees for nesting by weak and strong cavity excavators.

NOTEWORTHY RECORDS

Spring: Coastal – Oak Bay 4 Apr 1981-7 eggs; Elk Lake (Victoria) 30 Mar 1974-40, 12 Apr 1974-55, 14 May 1984-6 nestlings; North Saanich 16 Apr 1994-7 eggs; Long Beach 25 Apr 1974-10, 25 Jun 1970-5 eggs; Surrey 3 Apr 1960-8 eggs; Langley 23 Apr 1983-nestlings; Skagit River 12 Mar 1971-30; Black Mountain (West Vancouver) 18 May 1974-45; Harrison Hot Springs 26 May 1992-3 fledglings; Courtenay 9 Apr 1971-6 eggs; Hesquiat 7 Mar 1976-8; Klaskish River 7 May 1978-10; Calvert Island 27 Apr 1937-2 (MVZ 281119-20); Hagensborg 11 Apr 1985-17; Alice Lake 5 May 1969-10; Anthony Island 24 May 1989-5 eggs (Campbell 1989f); Kimsquit 11 Apr 1985-15; Tlell 13 Apr 1972-10; Hippa Island 24 Apr 1982-8; Kunga Island 16 May 1977-17; Juskatla 22 Mar 1972-10; Masset 21 May 1909-6 eggs; Yakan Point 12 May 1910-5 eggs; Prince Rupert 5 Mar 1989-3; Terrace 8 May 1977-incubating eggs; Kitsault 13 May 1980-10. **Interior** – Creston 17 May 1979-8 eggs, 29 May 1980-8 nestlings; Erie Creek 8 Mar 1983-7; Sheep Creek (Salmo) 8 Mar 1983-7; Manning Park 29 Apr 1979-8; Slocan 7 May 1980-7 eggs, 14 May 1981-nestlings; Nakusp 12 Apr 1976-building nest; Enderby 24 May 1945-7 eggs (UBC 1442); Celista 16 Mar 1948-12; Mount Revelstoke Park 31 May 1977-1; Clearwater 28 May 1935-1 (MVZ 88731); Mount Purden 27 May 1992-1.

Summer: Coastal – Galiano Island 30 Aug 1986-65; Diana Island 4 Jul 1973-20; lower Sarita River 12 Jun 1978-8; Abbotsford 30 Jul 1990-5 nestlings; Point Atkinson 1 Aug 1968-4 fledglings; Long Beach 25 Jun 1970-5 eggs; Garibaldi Park 30 Jul 1978-family groups; Alta Lake 4 Aug 1941-30; Brooks Peninsula 6 Aug 1981-3; Port Hardy 21 Aug 1935-50; Goose Island (Goose Group) 25 Jun 1948-2 (Guiguet 1953); Namu 4 Jun 1983 20; Aristazabal Island 3 Jun 1936-1 (MVZ 281132); Anthony Island 1 Jul 1977-4; Tlell 8 Jun 1984-2 fledglings; Sewall 14 Jun 1983-4 fledglings; Drizzle Lake 28 Jun 1976-4 fledglings; Tow Hill 2 Aug 1978-30; Lucy Island 27 Jul 1977-46; Kaien Island 9 Jun 1983-adult feeding 2 fledglings; Lakelse Lake 17 Aug 1976-21; Terrace 5 Jul 1977-nestlings; Alice Arm 21 Jun 1956-3 fledglings; Flood Glacier 2 Aug 1919-1 (MVZ 40273); Rainy Hollow 2 Jul 1984-1; Mile 46.5 Haines Highway 21 Jun 1972-6. **Interior** – Creston 9 Aug 1980-2 nestlings; Manning Park 22 Jun 1986-nestlings, 13 Aug 1984-11; Gates Creek 30 Jul 1973-9; Slocan 12 Aug 1980-nestlings; Glacier National Park 24 Aug 1982-20; Lonesome Lake 10 Jul 1956-fledglings just out of nest (Ritcey 1956); Blue Lake (Mount Robson Park) 8 Jun 1977-15; Survey Lake 28 Jun 1980-1; Cliff Lake (Atlin Lake) 5 Jul 1980-2.

Breeding Bird Surveys: Coastal – Recorded from 26 of 27 routes and on 77% of all surveys. Maxima: Chemainus 24 Jun 1973-86; Victoria 5 Jul 1980-76; Courtenay 18 Jun 1992-70. **Interior** – Recorded from 15 of 73 routes and on 8% of all surveys. Maxima: Chilkat Pass 10 Jun 1975-11; Creighton Valley 23 Jun 1983-9; Gerrard 18 Jun 1978-7; Scotch Creek 8 Jul 1990-7.

Autumn: Interior – Williams Creek (Atlin) 18 Sep 1913-2 (Kermode and Anderson 1914); Germansen Landing 23 Oct 1972-8; Smithers 1 Nov 1984-1 (BC Photo 1029); Quick 27 Oct 1984-2; Nulki Lake 26 Sep 1952-1 (ROM 88109); Hungary Creek 31 Oct 1993-8; Riske Creek 22 Sep 1987-1; Glacier National Park 13 Oct 1982-25; Mount Revelstoke 12 Oct 1982-15; Flash Lake 11 Nov 1982-15. **Coastal** – Kitsault 24 Nov 1979-60; Terrace 23 Oct 1985-50; Naden Harbour 21 Oct 1974-20; Delkatla Inlet 11 Nov 1971-15; Kitimat 17 Sep 1974-27; Cape St. James 22 Sep 1981-15; Port Neville Inlet 19 Sep 1976-90; Egmont 7 Oct 1973-25; Forbidden Plateau 23 Sep 1973-2; Hornby Island 6 Sep 1926-100; Qualicum Beach 9 Nov 1974-58; Victoria 20 Oct 1978-116.

Winter: Interior – Smithers 28 Dec 1986-2; Anahim Lake 4 Feb 1990-3; Likely 26 Jan 1953-1 (NMC 48068; Jobin 1953); Williams Lake 18 Jan 1982-1; Horsefly 23 Feb 1953-1 (NMC 48069; Jobin 1953); Revelstoke 10 Feb 1985-6; Grindrod 4 Jan 1948-1 (UBC 7257); Celista 8 Feb 1948-14; Nakusp 31 Dec 1978-5; Burnt Creek (Salmo) 13 Jan 1983-7; Salmo 3 Dec 1981-6; Creston 1 Feb 1982-6. **Coastal** – Terrace 21 Dec 1969-9; Masset 17 Dec 1972-108; Kitimat 1 Feb 1975-6 (Hay 1976); Kumdis Island 20 Feb 1974-14; Port Neville 8 Feb 1977-25; Long Beach 17 Feb 1988-6; lower Sarita River 6 Dec 1978-6.

Christmas Bird Counts: Interior – Recorded from 16 of 27 localities and on 29% of all counts. Maxima: Fauquier 28 Dec 1991-70; Nakusp 3 Jan 1986-52; Revelstoke 16 Dec 1988-24. **Coastal** – Recorded from 30 of 33 localities and on 94% of all counts. Maxima: Victoria 18 Dec 1993-**2,099**, all-time North American high count (Cannings 1994); Duncan 27 Dec 1993-832; Comox 30 Dec 1973-545.

Bushtit

Psaltriparus minimus (Townsend)

BUSH

RANGE: Resident on the Pacific coast from southwestern British Columbia (including southeastern Vancouver Island), western Washington, and Oregon south through California; and in the interior from southwestern Idaho, Utah, and southwestern Colorado through western New Mexico; local in central Texas. Also occurs from Mexico to central Guatemala.

STATUS: On the coast, *common* resident in the Georgia Depression Ecoprovince, including southeastern Vancouver Island, the Fraser Lowland, and most of the Gulf Islands; *rare* on the mainland coast north of Vancouver in the Georgia Lowland; *uncommon* resident in the southernmost part of the Coast and Mountains Ecoprovince; *fairly common*, but very local, on southern portions of Western Vancouver Island.

In the interior, *casual* in the Southern Interior Ecoprovince.

Breeds.

CHANGE OF STATUS: The Bushtit (Fig. 286) has occurred in the lower Fraser River valley since the earliest historical records (Lord 1866; Brooks 1917), but at the turn of the century was considered only an irregular visitor east of the Fraser River delta. Brooks (1917) mentions taking 2 specimens from a large flock on the edge of Sumas Prairie on 25 November 1899, and then seeing a flock at the same place in 23 March 1900. He wrote: "I never saw the species before or since and this must have been only a sporadic northward movement. The Bushtit occurs as a resident some fifty miles nearer the coast at Boundary Bay." It appears that the Bushtit was confined to the Fraser River delta and nearby towns until the mid-1930s.

Munro and Cowan (1947) considered it a fairly common resident in the lower Fraser River valley, and noted that it had recently invaded southeastern Vancouver Island (Hardy 1947). Other authors had also noted increasing populations and distribution limits during the 1920s to 1940s (Meugens 1945; Holdom 1947; Clay 1948a; Thacker 1950), and the Bushtit continued to expand its range over the next several decades.

The first confirmed breeding record for Vancouver Island was at Victoria in 1937 (Butler 1981). Bushtits had spread north to Miracle Beach, just south of Campbell River, by 1953, and to Campbell River by 1971. The species was first recorded on Galiano Island in 1956, on South Pender Island in 1967, and on Quadra Island in 1975. The first record from the west coast of Vancouver Island was in 1965.

Populations have increased in British Columbia during this century (Butler 1981). Besides the increased amount of habitat provided by the clearing of lowland coastal forests, a gradual warming trend on the Pacific coast until about 1940 (Crowe 1963) was correlated with the increased distribution of Bushtits on the southern coast, although over the period of record there have been no significant changes in temperatures measured at Vancouver (Harding and Taylor 1994). Mild winters may also have contributed to large increases in populations in Vancouver and the Fraser River delta during the 1970s (Butler 1981).

Figure 286. Bushtit (Victoria, January 1986; Tim Zurowski).

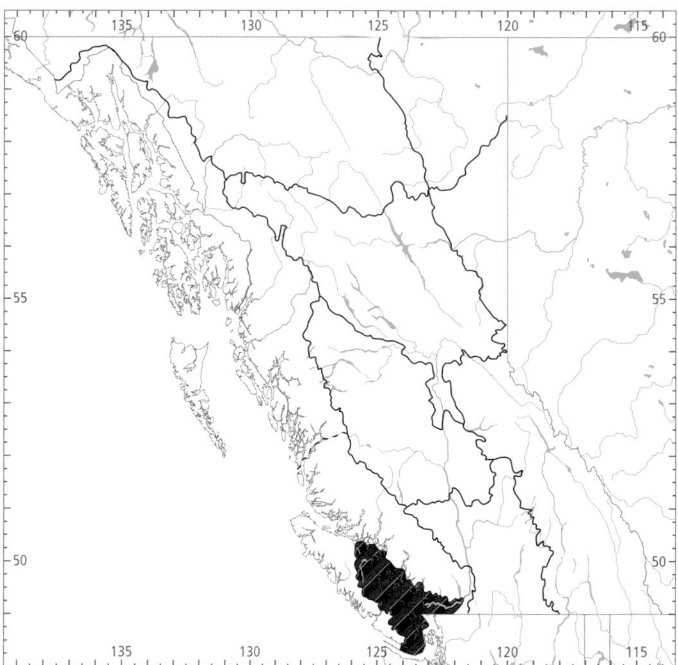

Figure 287. In British Columbia, the highest numbers for the Bushtit in both winter (black) and summer (red) occur in the Georgia Depression Ecoprovince.

NONBREEDING: The Bushtit (Fig. 286) is resident on the southern coast of British Columbia. On the mainland, it is widespread in the Fraser Lowland, where it occurs abundantly from the Fraser River delta east to Matsqui and north to Chilliwack, Harrison Lake, and the suburbs of Port Coquitlam, Port Moody, North Vancouver, and West Vancouver on the lower mountain slopes. It occurs locally on the Sunshine Coast north to Sechelt, in the Squamish River valley, Pemberton, and east of Chilliwack to Hope. On Vancouver Island, it is widespread from Victoria north to Campbell River and west to Sooke, Cowichan Lake, and Port Alberni. It occurs irregularly north to Sayward and west to Jordan River, Bamfield,

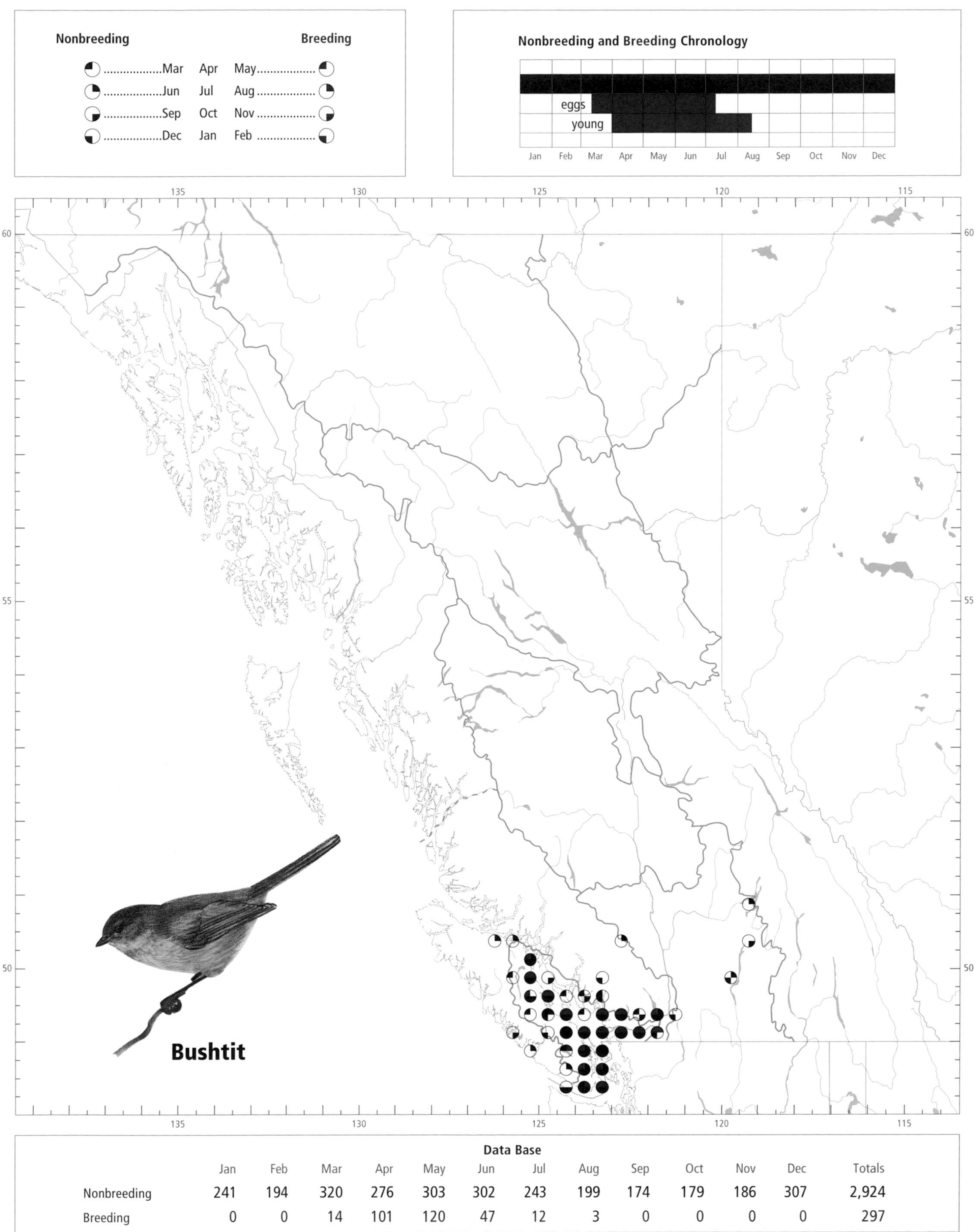

Data Base

	Jan	Feb	Mar	Apr	May	Jun	Jul	Aug	Sep	Oct	Nov	Dec	Totals
Nonbreeding	241	194	320	276	303	302	243	199	174	179	186	307	2,924
Breeding	0	0	14	101	120	47	12	3	0	0	0	0	297

and Pacific Rim National Park. The Bushtit is also a resident on some of the southern Gulf Islands, including Bowen Island, but appears to be absent from James Island and Sidney Island (R.W. Butler pers. comm.). In the interior, it is a casual visitor to the northern Okanagan valley and Shuswap Lake area.

The Bushtit reaches its highest numbers in winter in the Georgia Depression (Fig. 287). The few birds reported from the Southern Interior are thought to have dispersed from coastal British Columbia populations through the Fraser Canyon, rather than from populations in the Yakima valley, Washington (Cannings et al. 1987).

On the coast, the Bushtit occurs at relatively low elevations, from near sea level to 510 m, but most records are from below 200 m; in the interior, it was reported from valley bottoms to about 390 m.

The Bushtit frequents open habitats with a variety of deciduous and coniferous shrubs and trees. It avoids continuous coniferous forests, and its distribution in British Columbia is limited mainly to areas where original forest cover has been removed. Shrubs that are especially attractive for foraging (and nesting) include ocean-spray, common snowberry, thimbleberry, elderberry, low-growing willows, hardhack, blackberries, Scotch broom, choke cherry, cascara, and Nootka rose. These shrubs are common along the edges of open areas. The Bushtit is also attracted to backyard feeders that offer suet. At night, flocks roost in tangles of ivy (Fig. 288), broom, and other dense vegetation.

From July through February the Bushtit is a highly gregarious species that drifts through the shrubbery in tight flocks, each of which appears to be active within a discrete

Figure 288. During the winter of 1992-93, a flock of at least 12 Bushtits roosted here each evening among the tangles of ivy wrapped tightly around black cottonwoods (Haro Woods, Victoria; R. Wayne Campbell).

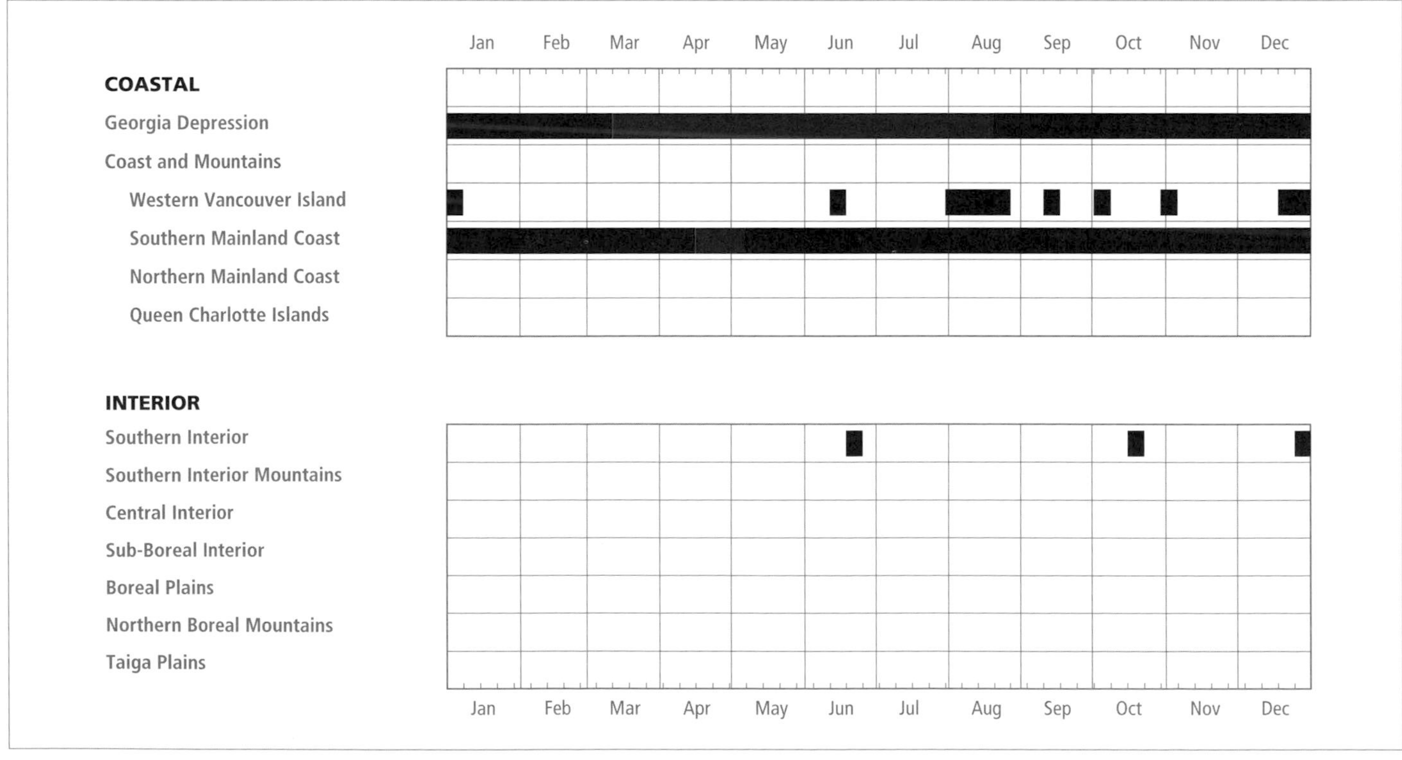

Figure 289. Annual occurrence (black) and breeding chronology (red) for the Bushtit in ecoprovinces of British Columbia.

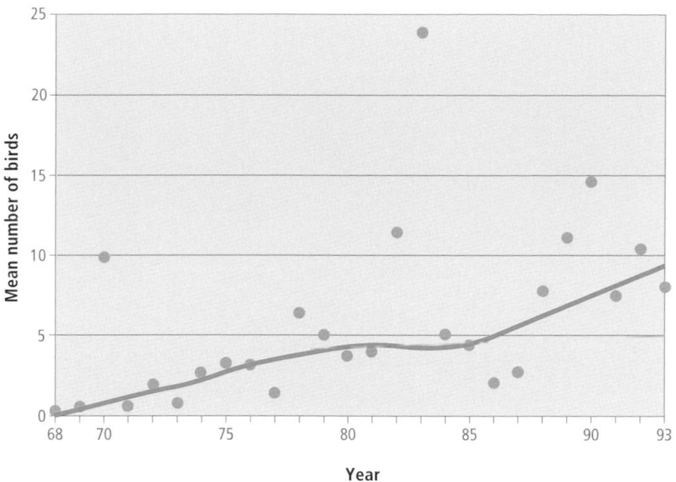

Figure 290. An analysis of Breeding Bird Surveys for the Bushtit in British Columbia shows that the number of birds recorded on coastal routes have increased at an average annual rate of 9% over the period 1968 through 1993 ($P < 0.01$).

home range. Flocks of 10 to 40 birds are most common, and usually consist of 2 or more families. Nonbreeding flocks loosely associate with resident chickadees and Red-breasted Nuthatches, and with migrating warblers. Winter flocks break up in March as breeding birds move to nesting territories.

On the coast, the Bushtit has been recorded throughout the year; in the interior, it has been recorded irregularly in June, October, and December (Fig. 289).

BREEDING: The Bushtit is a widespread breeder throughout its range in the Georgia Depression. On Vancouver Island it breeds along the southeast coast from Victoria north to Campbell River and west to Sooke and Cowichan Lake. There are no records of breeding on the west coast of Vancouver Island, but summer records from Jordan River, Bamfield, and Pacific Rim National Park, and a record of a family group from Sarita River, indicate that breeding probably occurs there. On the mainland, it breeds commonly in the Fraser Lowland from Tsawwassen east to Chilliwack, north to North Vancouver and Harrison Hot Springs, and probably east to Hope and on the Sunshine Coast north to Sechelt. It also breeds on some of the Gulf Islands.

The Bushtit reaches its highest numbers in summer in rural areas of the Fraser and Nanaimo lowlands of the Georgia Depression (Fig. 287). An analysis of Breeding Bird Surveys for the period 1968 through 1993 shows that the number of birds on coastal routes has increased at an average annual rate of 9% (Fig. 290).

The Bushtit breeds at relatively low elevations, and almost all records are from near sea level to 150 m. There is 1 breeding record from Triangle Mountain (Victoria) at 225 m and 1 from Maple Ridge at 510 m. The Bushtit is an edge specialist, and nests are almost always found along hedgerows or the edges of forests and gardens. Preferred habitats include edges of open forests, powerline rights-of-way, rural areas, suburban gardens, brushy areas around clearings, abandoned fields, riparian thickets, boggy areas, early second-growth forest, and the shrubbery of parks (Fig. 291) and golf courses.

On the coast, the Bushtit has been recorded breeding from 11 March (calculated) to 14 August (Fig. 289).

Nests: The Bushtit uses a wide variety of trees and shrubs for its nest site, including Douglas-fir, Scotch broom, willows, hawthorns, Sitka spruce, red alder, vine maple, birch (Fig. 292), western redcedar, ocean-spray, salmonberry, and western flowering dogwood. Most nests (83%; $n = 342$) were in deciduous shrubs or trees; 17% were in conifers (10.5% in Douglas-firs). When building nests in conifers, the Bushtit selects mature trees and usually builds in the larger, lower branches.

Figure 291. The shrubbery of city parks provides nesting habitat for the Bushtit (Beacon Hill Park, Victoria, 25 June 1996; Neil K. Dawe).

Figure 292. An active Bushtit nest (upper right), suspended on a birch branch overhanging a jogging trail (Burnaby Lake, 27 March 1993; R. Wayne Campbell).

Figure 293. A typical Bushtit nest resembles an old sock hanging in a bush or tree. The nest always has a hole near the top, is often constricted in the middle, and is built mostly of interwoven mosses and lichens (Burnaby, 26 April 1968; R. Wayne Campbell).

Nesting begins very early in the year for a passerine species. Early nests are especially visible when built in deciduous shrubs before the shrubs leaf out. When using conifers, bushtits tend to build on a vertically drooped branch that at least partially conceals the nest. Heights for 338 nests ranged from 0.6 to 15 m, with 56% between 2.0 and 3.5 m.

Nests can be reused for second or third broods, and may even be refurbished the next year for another nesting season. Pairs may usurp each other's nests during a nesting season (Ervin 1977).

The Bushtit builds one of the most distinctive nests of all North American birds. The nest is a pendulous "sock," 20 to 25 cm long, often slightly constricted in the middle and with a small circular entrance hole near the top (Fig. 293). Nests are usually attached to several small twigs and hang vertically. They (n = 239) are intricately woven with fine green moss (80%), lichens (53%), spider webs (28%), fine grass (25%), hair (18%), and plant fibres (16%). Paper and string are used occasionally. Nest chambers are lined with soft plant down, insect silk, and feathers (Fig. 294).

Eggs: Dates for 123 clutches ranged from 14 March to 10 July, with 54% recorded between 12 April and 9 May. Calculated dates indicate that eggs could be found as early as 11 March and as late as 30 July. Sizes of 98 clutches ranged from 1 to 12 eggs (1E-6, 2E-6, 3E-6, 4E-12, 5E-7, 6E-12, 7E-37, 8E-9, 10E-1, 12E-2), with 53% having 5 to 8 eggs (Fig. 294). The incubation period is reported to be 12 days (Ehrlich et al. 1988).

Young: Dates for 99 broods ranged from 10 April to 14 August, with 53% recorded between 5 May and 7 June. Calculated dates indicate that young can be found as early as 1 April. Sizes of 22 broods ranged from 2 to 13 young (2Y-2, 3Y-2, 4Y-3, 5Y-4, 6Y-5, 7Y-4, 8Y-1, 13Y-1), with 13 having 5 to 7 young. The nestling period is 14 to 15 days (Ehrlich et al. 1988). Two broods are produced annually, and we have a record of 3 broods from 1 nest, presumably, but not certainly, from the same pair.

Brown-headed Cowbird Parasitism: In British Columbia, there were 2 cases of parasitism by the cowbird in 122 nests recorded with eggs or young. Both were from southern

Vancouver Island (Crowell and Nehls 1973d; Friedmann et al. 1977). One nest, in North Saanich, was found to have an enlarged nest entrance on 13 June 1973, and contained 3 Bushtit eggs and 1 cowbird egg. It was abandoned after the cowbird egg was laid. The other nest was in Victoria.

A total of 8 cases of parasitism of this species have been reported in North America, from California, Washington, and British Columbia (Friedmann et al. 1977; Smith and Atkins 1979; Friedmann and Kiff 1985).

Nest Success: Of 11 nests found with eggs and followed to a known fate, only 1 produced fledglings. Few nests can be easily monitored because nest contents are concealed from view. At least 12% (n = 438) of nests recorded were known to be unsuccessful: 49 nests were destroyed by predators or humans, 3 nests fell to the ground, and 2 nests contained dead nestlings. Predators of eggs and young were mainly the Northwestern Crow, Steller's Jay, and domestic cats.

Figure 294. The chamber of a Bushtit nest is lined with soft plant down, insect silk, and feathers (Burnaby Lake, 10 April 1969; R. Wayne Campbell).

REMARKS: The Bushtit appears to be sensitive to temperatures below 0°C, and populations can be severely reduced during prolonged periods of subzero weather (Larrison and Sonnenberg 1968).

It is puzzling that this species has pioneered Vancouver Island but not all the Gulf Islands, considering that the Gulf Islands lie between mainland and Vancouver Island populations.

There is a published observation of the Bushtit from North Barriere Lake in the Southern Interior (Cannings et al. 1987), but sufficient documentation is lacking and the record has not been considered in this account.

NOTEWORTHY RECORDS

Spring: Coastal – Oak Bay 4 Mar 1984-building nest, 15 Apr 1983-nestlings; Victoria 20 Apr 1979-42, 24 May 1937-nest, first Vancouver Island breeding record (Butler 1981); Colwood 4 Mar 1987-20; South Pender Island May 1967-nesting for first time; North Cowichan 24 Mar 1975-2; Mayne Island 16 Apr 1979-4; Galiano Island 17 May 1986-14; Crescent Beach 6 Apr 1968-15, 10 Apr 1983-nestlings; Sumas Prairie 23 Mar 1900-flock (Brooks 1917); Ladner 22 Mar 1970-6 eggs; Langley 14 Mar 1970-4 eggs; Burnaby Lake 29 Apr 1971-12 eggs (RBCM E1627); Vancouver 17 Apr 1968-12 eggs; Mission 6 Mar 1968-pair building nest; Agassiz 1 May 1970-5; Harrison Hot Springs 13 Mar 1976-30, 18 Apr 1985-7 eggs; West Vancouver 4 Mar 1979-42; Port Alberni 3 Apr 1983-2; Sechelt 14 Apr 1991-2, first occurrence for Sunshine Coast (Siddle 1991c); Squamish 20 Apr 1962-1; Merville 14 Mar 1988-3; Campbell River 4 May 1992-pair and nest with eggs; 13 May 1973-6. **Interior** – No records.

Summer: Coastal – San Juan Point 12 Jun 1983-2; Metchosin 7 Aug 1980-30; Victoria 7 Jun 1981-35, 2 Aug 1988-50; Mount Douglas 30 Jun 1958-50; North Saanich 13 Jun 1973-nest with 3 Bushtit and 1 Brown-headed Cowbird eggs (Friedmann et al. 1977); Sarita River 5 Aug 1965-family group (Hatler et al. 1978); Bamfield 18 Aug 1967-12; White Rock 15 Jul 1973-nestlings; Surrey 14 Jun 1963-13 nestlings, 18 Jun 1963-13 fledglings; Aldergrove 20 Aug 1967-26; Matsqui 14 Aug 1984-feeding nestlings; Nanaimo 3 Jul 1983-20; Mission 18 Jul 1984-24; Pitt Meadows 15 Aug 1976-35; Vancouver 10 Jul 1968-7 eggs, 26 Jul 1981-nestlings (third brood), 26 Aug 1974-151; Parksville 30 Jul 1953-2 (NMC 38686-7); Harrison Hot Springs 16 Aug 1975-20; Royston 22 Aug 1988-20; Pemberton 22 Jun 1975-2; Port Neville 11 Jun 1975-3. **Interior** – 8 km nw Kelowna 21 Jun 1973-5 (Cannings et al. 1987); Shuswap Lake 23 Jun 1970-3 in flock with chickadees and kinglets (Sirk 1970).

Breeding Bird Surveys: Coastal – Recorded from 14 of 27 routes and on 26% of all surveys. Maxima: Albion 5 Jul 1979-38; Victoria 4 Jul 1979-38; Pitt Meadows 6 Jul 1989-21. **Interior** – Not recorded.

Autumn: Interior – Vernon 17 Oct 1977-1 (Cannings et al. 1987). **Coastal** – Quadra Island 27 Sep 1975-20; Comox Oct 1953-30; Halfmoon Bay 14 Sep 1990-15; Harrison Hot Springs 16 Nov 1988-15; Rolley Lake 30 Nov 1974-11; Pacific Rim National Park 2 Oct 1983-10; Pitt Meadows 24 Nov 1975-50; Fort Langley 25 Sep 1967-85; Ladner 10 Nov 1974-150; Huntingdon 5 Sep 1948-45; Swan Lake (Ucluelet) 12 Sep 1973-12 (Hatler et al. 1978); Somenos Lake 23 Oct 1971-60; Cowichan Bay 29 Oct 1984-80; Saltspring Island 22 Nov 1977-100; Blenkinsop Lake 13 Sep 1986-100; Victoria 25 Oct 1985-90; Jordan River 3 Nov 1988-30.

Winter: Interior – Kelowna 28 Dec 1975-2. **Coastal** – Squamish 12 Dec 1990-20, 16 Dec 1989-25 on Christmas Bird Count; Hope 31 Jan 1945-small flock (Thacker 1950); Harrison Hot Springs 7 Dec 1989-30; Chilliwack 17 Dec 1983-330 on Christmas Bird Count; Pacific Rim National Park 28 Dec 1972-6 (Hatler et al. 1978); Sea Island 6 Feb 1978-257; Reifel Island 21 Jan 1979-250; Bamfield 19 Dec 1988-12; Pender Island 16 Dec 1967-32; Victoria 24 Jan 1982-110; Oak Bay 12 Dec 1982-56; Jordan River 2 Jan 1984-30.

Christmas Bird Counts: Interior – Not recorded. **Coastal** – Recorded from 19 of 33 localities and 57% of all counts. Maxima: Victoria 18 Dec 1993-**1,578**, all-time high Canadian count (Cannings 1994); Vancouver 19 Dec 1993-1,392; White Rock 30 Dec 1989-612.

Red-breasted Nuthatch RBNU
Sitta canadensis Linnaeus

RANGE: Breeds from south-coastal and southeastern Alaska, southern Yukon, and southwestern Mackenzie east across the forested areas of Canada to Labrador and Newfoundland. In eastern North America, resident south through the Appalachian Mountains into Tennessee and North Carolina. In western North America, resident along the Sierra Nevada to southern California. Winters throughout the breeding range, except at higher latitudes, south into northern Baja California, southern Arizona, southern New Mexico, southern Texas, the Gulf coast, and central Florida.

STATUS: On the coast, *uncommon* to locally *fairly common* resident throughout the Georgia Depression Ecoprovince and on the southern mainland coast of the Coast and Mountains Ecoprovince; *rare* resident on Western Vancouver Island, the Queen Charlotte Islands, and northern mainland coast of the Coast and Mountains, but also an *uncommon* migrant and summer visitant there.

In the interior, *uncommon* to locally *common* resident in the Southern Interior, Southern Interior Mountains, and Central Interior ecoprovinces. Further north, in the Sub-Boreal Interior, Northern Boreal Mountains, and Taiga Plains ecoprovinces, *very rare* to *uncommon* migrant and summer visitant; *casual* in winter only in the Sub-Boreal Interior. In the Boreal Plains Ecoprovince, *uncommon* migrant and summer visitant, and *very rare* in winter.

Breeds.

Figure 295. Adult Red-breasted Nuthatch (Victoria, 15 June 1994; R. Wayne Campbell).

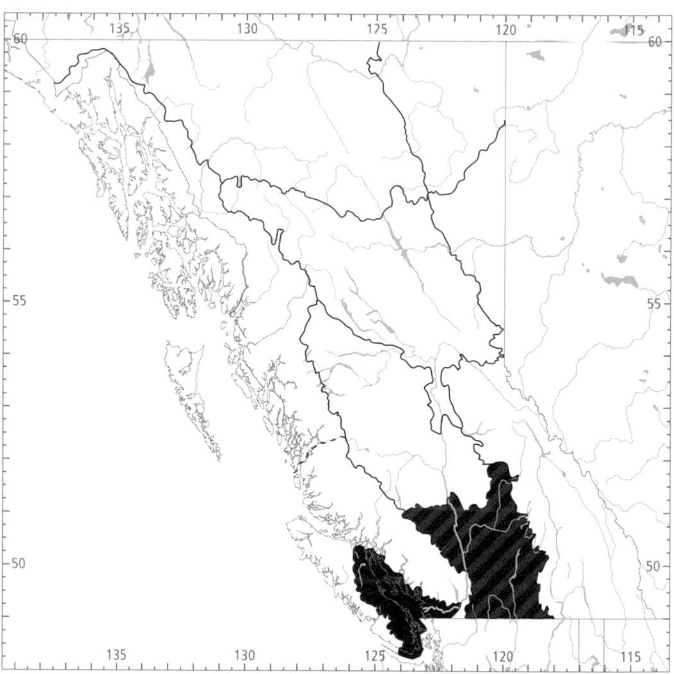

Figure 296. In British Columbia, the highest numbers for the Red-breasted Nuthatch in winter (black) occur in the Southern Interior Ecoprovince (Okanagan valley) and in the Georgia Depression Ecoprovince (southern Vancouver Island); the highest numbers in summer (red) occur in the Southern Interior Ecoprovince.

NONBREEDING: The Red-breasted Nuthatch (Fig. 295) is one of the most widely distributed birds in the southern two-thirds of the province, but is more sparsely distributed in the far north. It is resident along the coast. Populations on the Queen Charlotte Islands and Western Vancouver Island appear to be partially migratory. It is also a scarce resident along the Coast and Mountains north to Portland Canal. In the interior, the Red-breasted Nuthatch is resident from the Central Interior south, but wintering populations are small in northern areas of the Central Interior. The centre of abundance in winter is in the Okanagan valley of the Southern Interior and on southern Vancouver Island in the Georgia Depression (Fig. 296). Root (1988) notes that one of the most consistently concentrated winter populations in North America is in northeastern Washington, which lies adjacent to the southern Okanagan valley of British Columbia.

The Red-breasted Nuthatch has been reported at elevations from sea level to 1,350 m on the coast and up to 2,200 m in the interior. It is a versatile species that occurs in many different forest habitats. It is most abundant in the interior forests of ponderosa pine and Douglas-fir of the Southern Interior and the drier parts of the Southern Interior Mountains (Fig. 297), but only slightly less so in the coastal Douglas-fir forests of the Georgia Depression, the interior forests of western redcedar and western hemlock with some Engelmann spruce, and the altitudinally adjacent Engelmann spruce forest of the moister areas of the Southern Interior Mountains. In the Cariboo and Chilcotin areas, it inhabits mixed forests of Douglas-fir, lodgepole pine, Engelmann spruce, trembling aspen, and balsam poplar. In all areas it prefers old-growth forests or older second growth.

The onset of the spring migration is difficult to detect, as migrants mix with residents that are simultaneously dispersing upslope to summer ranges at higher altitudes (Figs. 298

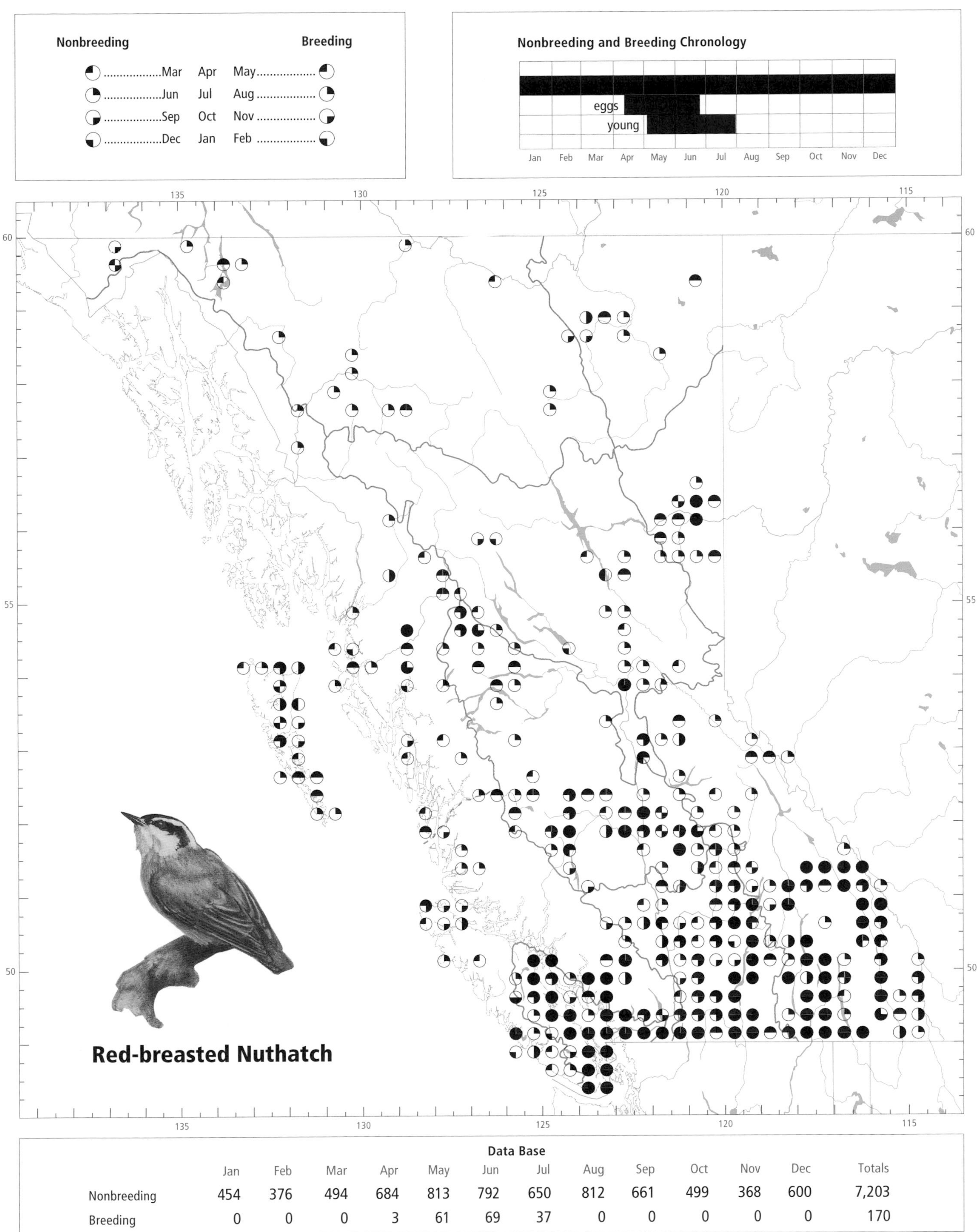

Data Base

	Jan	Feb	Mar	Apr	May	Jun	Jul	Aug	Sep	Oct	Nov	Dec	Totals
Nonbreeding	454	376	494	684	813	792	650	812	661	499	368	600	7,203
Breeding	0	0	0	3	61	69	37	0	0	0	0	0	170

and 299). In the Georgia Depression, including the Fraser River delta and southeastern Vancouver Island, birds begin moving in March and April but the spring migration is subtle. In the Coast and Mountains, the spring movement is somewhat more pronounced. It occurs in March in the southern portions, and reaches the northern mainland region by early April and the Queen Charlotte Islands by mid to late April (Fig. 299).

In the Southern Interior, the arrival of spring migrants peaks in April (Fig. 299). In the Southern Interior Mountains, numbers increase steadily from March into June. Further north, the first spring arrivals can reach the Central Interior in March. Migrants arrive near Prince George in mid to late April and reach the Northern Boreal Mountains and Taiga Plains in early May (Fig. 299). Migrants first arrive in the Peace River region of the Boreal Plains in March and peak there in May.

Autumn migration begins in the northern interior in August, with only stragglers left after September (Figs. 298 and 299). From the Central Interior south, migration occurs from August through October. On the coast, most migrants leave the Queen Charlotte Islands, the northern Coast and Mountains, and the west coast of Vancouver Island in August. In the Georgia Depression and the southern mainland of the Coast and Mountains, the migration is less obvious, but the number of birds reported increases in August and September and declines in October as many migrants continue on their southward passage (Fig. 299).

North of latitude 52°N, the Red-breasted Nuthatch appears to be mostly migratory. In the Taiga Plains and the Northern Boreal Mountains, there are no winter records, and there are only a few records from the Boreal Plains. There is a

Figure 297. In winter, small flocks of Red-breasted Nuthatches forage among cones in ponderosa pine habitat, often in company with Hairy Woodpeckers and Pine Grosbeaks (Wasa Park, 20 February 1993; R. Wayne Campbell).

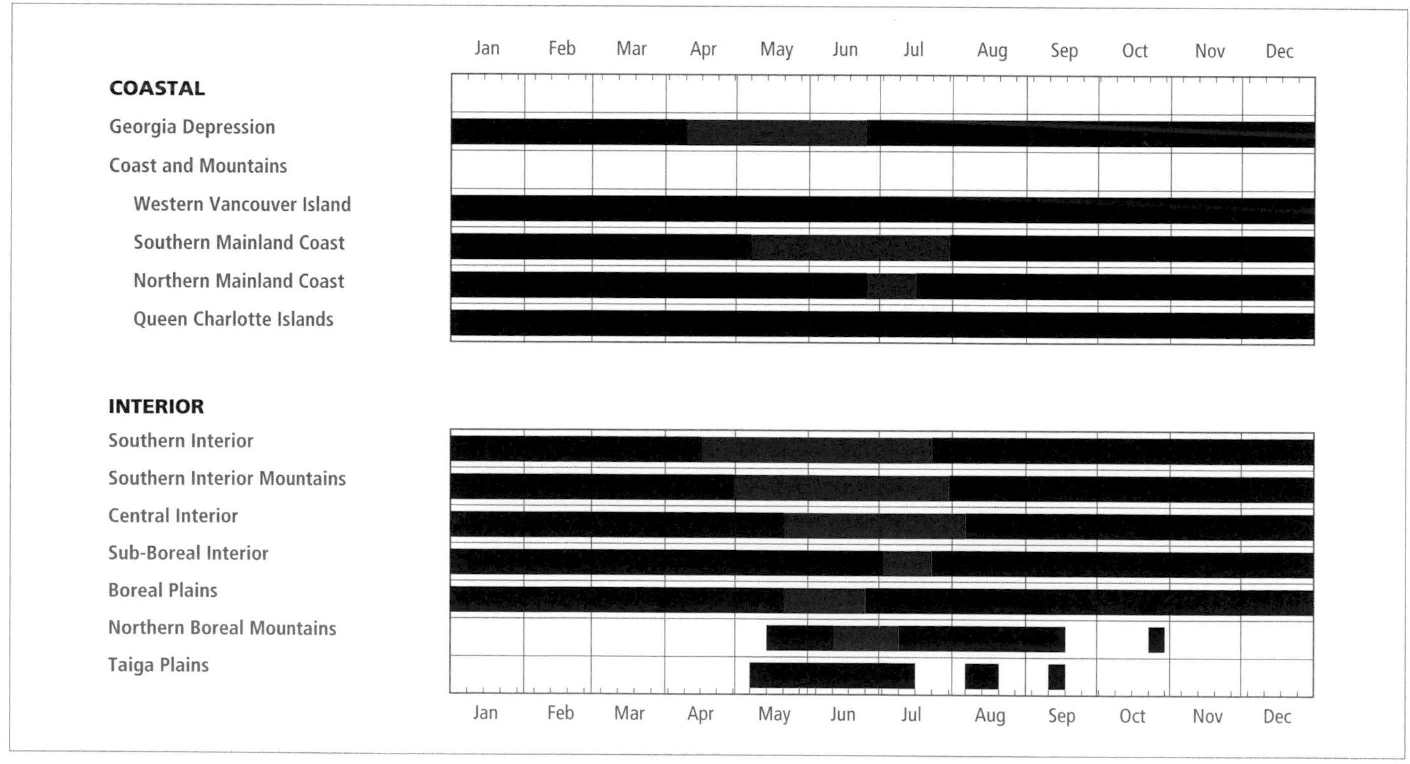

Figure 298. Annual occurrence (black) and breeding chronology (red) for the Red-breasted Nuthatch in ecoprovinces of British Columbia.

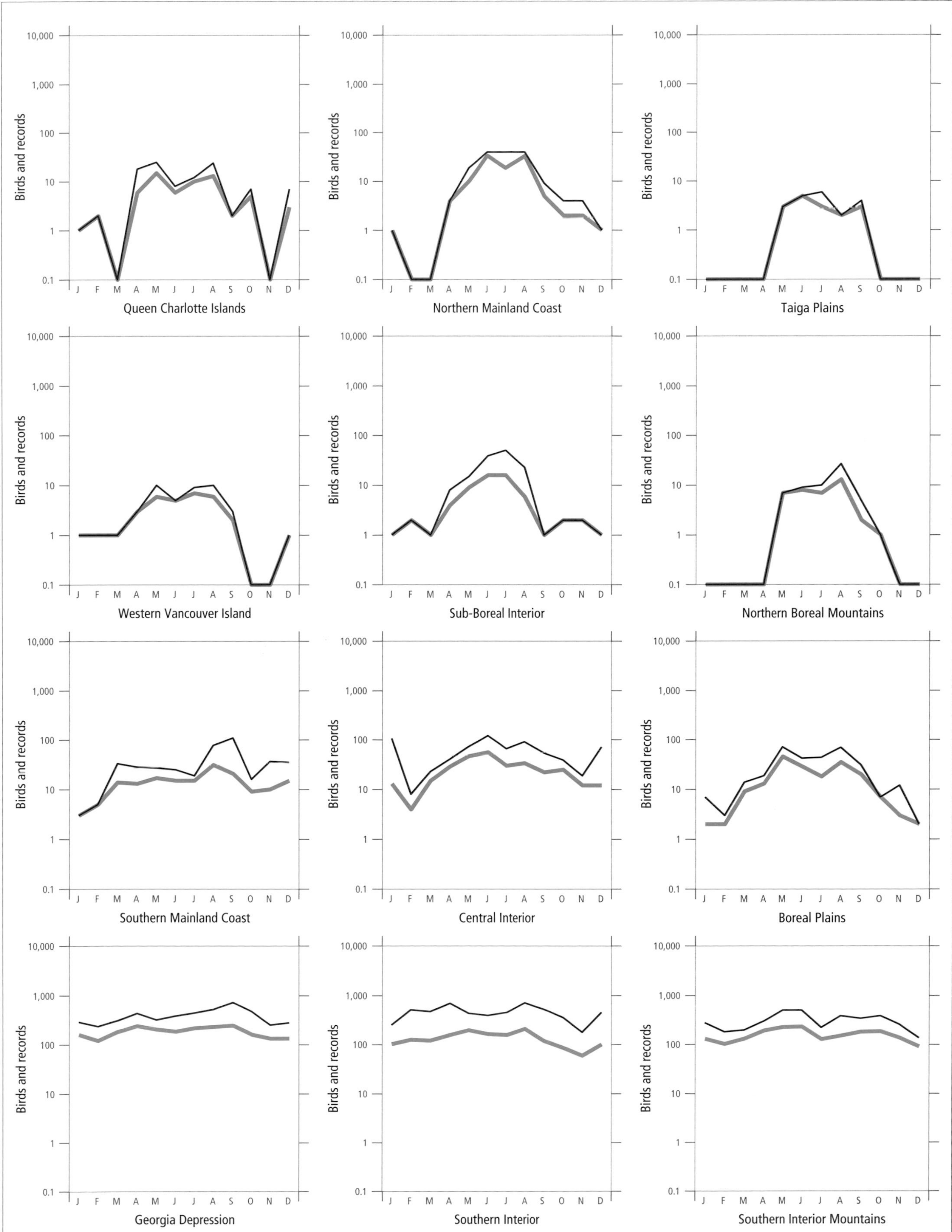

Figure 299. Fluctuations in total number of birds (purple line) and total number of records (green line) for the Red-breasted Nuthatch in ecoprovinces of British Columbia. Christmas Bird Count and nest record data have been excluded.

small and widely distributed wintering population in the Central Interior. At lower elevations in the Southern Interior and Southern Interior Mountains, and in the Georgia Depression, winter populations are not much lower than those of summer. In addition, populations that breed in higher-elevation subalpine forests migrate to valley bottom forests for the winter. In severe winters, this nuthatch may even migrate south from the low-elevation ponderosa pine forests of the Okanagan valley (Cannings et al. 1987).

On the coast, the Red-breasted Nuthatch has been recorded throughout the year. In the interior, it occurs throughout the year in southern and central British Columbia, and from 11 May to 11 September in the Northern Boreal Mountains and Taiga Plains (Fig. 298).

BREEDING: The Red-breasted Nuthatch has a widespread breeding distribution in the southern third of the province south of about latitude 52°N, including the east coast of Vancouver Island. Further north, breeding populations are smaller. Known breeding locations north of Williams Lake include Mount Robson Park, the Skeena and Nechako river valleys, Cox Island on the northwestern tip of the Queen Charlotte Islands, and near Dawson Creek. The northernmost nesting records are from Fort Nelson and Atlin. Although the species is resident throughout the Coast and Mountains, there are very few breeding records. The Red-breasted Nuthatch likely breeds throughout its summer range in the province.

The Red-breasted Nuthatch reaches its highest numbers in summer in the Southern Interior (Fig. 296). An analysis of Breeding Bird Surveys for the period 1968 through 1993 shows that the number of birds on interior routes has increased at an average annual rate of 6% (Fig. 300); an analysis of coastal routes for the same period could not detect a change in numbers.

The Red-breasted Nuthatch has been found breeding at elevations ranging from near sea level to 1,600 m on the coast; in the interior, breeding occurred at elevations from about 230 m to near the timberline at about 2,100 m elevation. Cannings et al. (1987) note that in the Okanagan valley, nesting below 600 m is "generally restricted to moister woods in creek bottoms or north facing slopes." The Red-breasted Nuthatch favours relatively mature forests as nesting habitat. Most nests were found in coniferous (44%; *n* = 102; Fig. 301), mixed (26%), and deciduous (11%) stands. Six percent were found in logged forest. Treed urban, suburban, and rural areas accounted for about 13% of the nesting habitat. The Red-breasted Nuthatch is seldom found during the nesting season in second-growth coniferous forests; 76% (*n* = 38) of forest nesting habitats for which forest age was noted were from mature stands. A variety of other settings were reported, including lakeshores, riparian bottomlands, gardens, and parks.

On the coast, the Red-breasted Nuthatch has been reported breeding from 11 April (calculated) to 27 July; in the interior, it has been recorded from 15 April (calculated) to 30 July (Fig. 298).

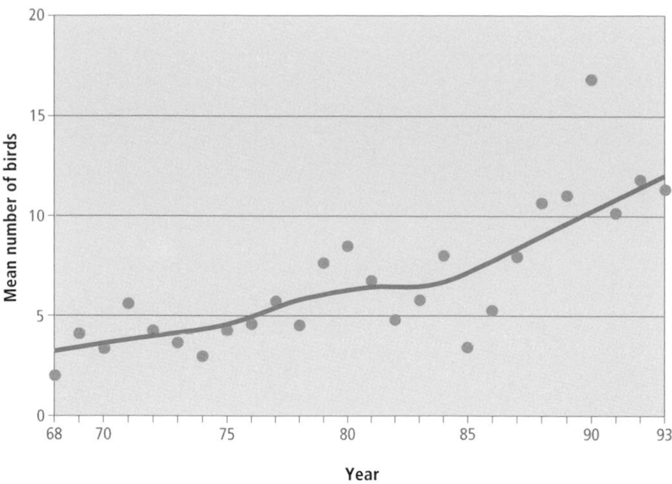

Figure 300. An analysis of Breeding Bird Surveys for the Red-breasted Nuthatch in British Columbia shows that the number of birds on interior routes increased at an average annual rate of 6% over the period 1968 through 1993 ($P < 0.001$).

Figure 301. Mixed coniferous forests of western hemlock and Douglas-fir, with snags and dead trees, provide nesting habitat for the Red-breasted Nuthatch (Victoria, 15 July 1993; R. Wayne Campbell). Note the series of cavities in the dead tree.

Figure 302. Over one-half of all Red-breasted Nuthatch nests in British Columbia were found in dead trees or snags (Hat Creek, 13 July 1996; R. Wayne Campbell).

Figure 303. Red-breasted Nuthatch nest cavity in a western redcedar (Creston, 16 September 1994; R. Wayne Campbell).

Nests: Almost all nests (99%; n = 143) were in cavities; only 2 nests were found in nest boxes. This nuthatch usually excavated its own nest cavities, but abandoned woodpecker and chickadee cavities and naturally occurring crevices and cavities were also used. Nests were found in trunks or branches of unidentified dead trees or snags (54%; n = 171; Fig. 302), deciduous trees or snags (30%), coniferous trees or snags (10%; Fig. 303), and posts or poles. Near Hope, in south coastal British Columbia, most nest excavations were in Douglas-fir (47%; n = 15), followed by bigleaf maple (27%), birch, red alder, and lodgepole pine (Kelleher 1963).

Nests generally included a variety of locally available material, including animal hair, grass, bark strips (often cedar), feathers, moss, lichens, and plant fibres (Fig. 304). The heights for 163 nests ranged from 0.5 to 20.0 m, with 55% between 3.0 and 6.0 m.

The Red-breasted Nuthatch's habit of dabbing pitch around the entrance to its nest hole was described by a number of observers in British Columbia (Fig. 305). The pitch, which is thickest at the bottom of the entrance, is thought to offer protection from competition for the nesting cavity, such as from a chickadee or even a deer mouse. Pitch is added throughout the nesting period. It may pose a problem when entering the cavity, even for the nuthatch. Kilham (1972) reports the death of a female Red-breasted Nuthatch that resulted from her becoming stuck to the pitch at her nest entrance.

Eggs: Dates for 32 clutches ranged from 15 April to 24 June, with 55% recorded between 4 May and 24 May. Calculated dates indicate that eggs can be found as early as 11 April. Sizes of 22 clutches ranged from 4 to 8 eggs (4E-4, 5E-4, 6E-5, 7E-7, 8E-2), with 12 having 6 or 7 eggs. The incubation period is 12 days (Burns 1915).

Young: Dates for 103 broods ranged from 3 May to 31 July, with 53% recorded between 1 June and 6 July. Sizes of 26 broods ranged from 1 to 7 young (1Y-1, 2Y-8, 3Y-5, 4Y-6, 5Y-2, 6Y-3, 7Y-1), with 73% having 2 to 4 young. The nestling period is poorly known and has been variously reported as 18 days (Michael 1934), at least 13 days (Gunderson 1939), and 14 to 21 days (Ehrlich et al. 1988).

Brown-headed Cowbird Parasitism: Cowbird parasitism was not found in British Columbia in 50 nests recorded with

Figure 304. A Red-breasted Nuthatch nest, found in a nest box west of Princeton, British Columbia, was composed mainly of bark strips, grasses, and plant fibres (2 May 1994; R. Wayne Campbell).

eggs or young. Only 2 cases of cowbird parasitism have been documented in North America, 1 each from Saskatchewan and Montana (Friedmann and Kiff 1985).

Nest Success: Of 2 nests found with eggs and followed to a known fate, 1 produced at least 1 fledgling.

REMARKS: Burleigh (1960) divided the Red-breasted Nuthatch into 2 subspecies: *S. c. clariterga* of western North America and *S. c. canadensis* of eastern North America. Later Todd (1963), Phillips et al. (1964), and Banks (1970) all concluded that there was no significant geographic variation in the Red-breasted Nuthatch and considered *clariterga* to be a synonym of *canadensis*.

The 3 species of nuthatches that occur in the province have very different patterns of distribution, density, and population behaviour. The Red-breasted Nuthatch is the most abundant species and occurs in all 9 ecoprovinces. The Pygmy Nuthatch is the second most abundant species and occurs in 5 ecoprovinces. The least abundant is the White-breasted Nuthatch. Although reported more frequently than the Pygmy Nuthatch, it was less numerous; it also occurs in 5 ecoprovinces. The distribution of the 3 species overlaps mainly in the Southern Interior, and there the Pygmy Nuthatch is the most abundant nuthatch.

Steeger and Machmer (in press) studied wildlife trees and their relationship with primary cavity excavators in the Nelson Forest Region. They found that weak excavators (chickadees and nuthatches) nested primarily in conifers (73%; $n = 30$), while strong excavators (woodpeckers) preferred to nest in live or decaying hardwoods (76%; $n = 17$). The nest cavities of the weak excavators were located within the uppermost sections of trees, and most nest trees were "broken off, bare trunks with the nest cavity located directly below the breakage point." Moderately decayed Douglas-firs were the preferred nest tree for the weak excavators.

They also suggest a relationship between *Armillaria* root disease and the suitability of trees for nesting. Once trees (e.g., Douglas-fir or pines) have been infected or killed by *Armillaria*, they are often invaded by the trunk rot, *Fomitopsis* sp. That fungus attacks heartwood and in time creates soft, punky heartwood that provides suitable conditions for the weak excavators. In their study, most Red-breasted Nuthatch nests (78%; $n = 23$) were located in *Armillaria*-killed trees.

Adams and Morrison (1993) found that on the western slope of the Sierra Nevada of northern California, the Red-breasted Nuthatch made greater use of forest stands that were diverse in physical structure and tree species composition. Although naturalists have reported this situation in British Columbia, it is yet to be studied.

The Red-breasted Nuthatch obtains a good part of its food supply from the seeds of conifers (Bent 1948). In years when the cone crop is poor, large numbers move south in winter. Bock and Lepthien (1972, 1976a) examined the relationship between the northern cone crop and the winter invasions of the Red-breasted Nuthatch into the United States. For the period 1956 to 1967, they found a perfect correlation between failure of northern cone crops and the continent-wide movement of the nuthatches far to the south of their normal winter range. Also, this movement usually occurred in alternate years and in company with a cohort of similar "boreal eruptives."

Figure 305. Some nest entrance holes of the Red-breasted Nuthatch in British Columbia are dabbed with pitch. This habit is thought to offer protection from competition for the nesting cavity (Victoria, 3 June 1996; R. Wayne Campbell).

NOTEWORTHY RECORDS

Spring: Coastal – Metchosin 17 Mar 1962-building nest; Victoria 15 Apr 1914-5 eggs (RBCM E391); Thetis Lake 7 Apr 1973-12; Tofino 18 May 1974-5; Meares Island 15 Mar 1987-1; Skagit River 12 Mar 1971-5; West Vancouver 25 May 1979-12; Hope 22 May 1961-nestlings (Kelleher 1963); Fanny Bay 30 Mar 1975-7; Fraser Canyon 26 Apr 1962-6; Egmont 8 Feb 1976-1; Lyell Island 9 Apr 1982-1; Kitlope Lake 3 May 1991-9; Tlell 20 Apr 1972-12; Kitimat 5 Apr 1980-1; Cox Island (Langara Island) 22 May 1981-2 fledglings; Lakelse Lake 13 Apr 1980-1. **Interior** – West Creston 16 May 1984-4 nestlings, 19 May 1981-eggs; Vaseux Lake 27 Apr 1965-6 eggs; White Lake road (Okanagan Falls) 8 May 1990-4 nestlings; Twin Lakes (Keremeos) 3 May 1969-nestlings (earliest brood); Okanagan Falls 21 May 1967-8 eggs; Balfour to Waneta 16 May 1981-43 on census; Pritchard 26 May 1962-14; Celista 19 Mar 1948-16; Kootenay National Park 10 Mar 1981-6; Yoho National Park 30 Apr 1977-7; Moha 13 Apr 1968-8; 100 Mile House 2 Mar 1976-2; Alexis Lake Rd 8 May 1977-7; Anahim Lake 11 Apr 1932-1 (MCZ 281009); Prince George 29 May 1966-6; Highway 29 (Fort St. John) 3 Mar 1984-3; Beatton Park 12 May 1984-5; Cold Fish Lake 21 May 1976-1 (Osmond-Jones et al. 1977); Beaver Lake (Alaska Highway) 11 May 1980-1; ne Kwokullie Lake 11 May 1982-1; Mile 75 Haines Highway (Chilkat Pass) 15 May 1977-1.

Summer: Coastal – Chemainus 25 Jun 1983-31; Galiano Island 30 Aug 1986-15; Chesterman Beach 11 Jun 1981-1; Grouse Mountain 24 Jun 1962-feeding nestlings; Hope 27 Jul 1961-last young left nest (Kelleher 1963); Black Tusk Meadows 20 Aug 1972-20; Cape Scott Park 1 Aug 1981-2; Goose Group 16 Jun 1949-3; Fairfax Inlet 1 Jun 1977-1; Kimsquit 5 Jul 1986-10; Graham Island 1 Aug 1919-4; Kispiox valley 12 Jul 1921-2 nestlings ready to fledge. **Interior** – Anarchist Mountain 24 Jun 1983-6 fresh eggs (latest clutch); Manning Park 26 Jun 1983-31; Apex Mountain 1 Aug 1977-1; Friday Creek 13 Jul 1964-2 nestlings; Crystal Lake (Nelson) 1 Aug 1988-3 nestlings (latest brood); Ta Ta Creek 14 Jul 1979-9; Kaslo 28 Jul 1976-nestlings; Pritchard 19 Jul 1963-10; North Thompson River Park 18 Jun 1984-7 eggs; Bridge Lake 5 Jun 1976-14; Clearwater 9 Jul 1963-6; Kleena Kleene 30 Jul 1962-nestlings; Chezacut 28 Aug 1941-1; Swiftcurrent Creek 15 Aug 1973-54; Prince George 28 Jun 1969-7; Carp Lake 25 Aug 1975-8; Swan Lake (Dawson Creek) 6 Jul 1962-2; Chetwynd 22 Jul 1975-12; Dawson Creek 2 Jun 1992-nestlings, 19 Jun 1992-very small nestlings; Taylor 8 Jul 1984-7; Stoddart Creek 23 Aug 1985-7; Ipec Lake 31 Aug 1979-1 (Cooper and Adams 1979); Spatsizi Plateau 8 Aug 1959-10; Dease Lake 5 Jun 1962-2; Clarke Lake road (Fort Nelson) 9 Jul 1978-1; ne Kwokullie Lake 4 Jun 1968-1; Atlin 1 Aug 1924 (MVZ 103280).

Breeding Bird Surveys: Coastal – Recorded from 23 of 27 routes and on 44% of all surveys. Maxima: Chemainus 25 Jun 1979-12; Port Hardy 19 Jun 1982-11; Victoria 5 Jul 1980-10. **Interior** – Recorded from 64 of 73 routes and on 83% of all surveys. Maxima: Christian Valley 26 Jun 1993-33; Adams Lake 24 Jun 1994-32; Wasa 28 Jun 1992-31.

Autumn: Interior – Mile 52, Haines Highway 4 Oct 1981-1; Steamboat Mountain 11 Sep 1943-2 (Rand 1944); Beatton Park 3 Sep 1986-10, 2 Oct 1982-1; Tetana Lake 21 Sep 1937-1; Glacier Gulch 22 Nov 1981-5; 24 km s Prince George 28 Nov 1983-1; Ten Mile Lake Park 5 Oct 1986-1; Chilcotin River 7 Oct 1987-6; Bridge Lake 7 Sep 1960-15; Yoho National Park 10 Oct 1976-30, 6 Nov 1976-19; Mount Revelstoke National Park 20 Nov 1982-6; Shuswap Lake 6 Sep 1959-50; Penticton 4 Oct 1973-22; Vaseux Lake 3 Nov 1973-8; Manning Park 11 Nov 1982-8; Flathead River 10 Sep 1939-1 at summit (MCZ 281025). **Coastal** – Lawn Point 7 Oct 1971-3; Khutze Inlet 10 Oct 1935-3; Princess Royal Island 1 Sep 1935-2; Garibaldi Park 7 Sep 1969-15; Rebecca Spit 25 Oct 1973-13; Pachena River 1 Sep 1977-2; Victoria 13 Oct 1979-87.

Winter: Interior – Fort St. John 16 Dec 1984-1, 1 Jan 1984-1; Taylor 11 Dec 1983-1; Tetana Lake 5 Feb 1935-1; Prince George 24 Dec 1989-1; Quesnel 28 Jan 1979-1; Williams Lake 30 Dec 1979-55; 100 Mile House 31 Dec 1975-2; Parson 27 Feb 1977-10; Revelstoke 20 Dec 1984-1; Salmon Arm 3 Jan 1971-38 on Christmas Bird Count; Trail 30 Dec 1980-5; Creston 1 Jan 1981-22 on Christmas Bird Count; Manning Park 23 Jan 1971-12. **Coastal** – Kaien Island 4 Dec 1983-1; Prince Rupert 30 Jan 1986-1; Masset Inlet 29 Jan 1983-1; Tlell 5 Dec 1971-3; Queen Charlotte City 26 Dec 1990-1; Cortes Island 21 Jan 1979-1; Alta Lake 31 Dec 1945-4; Comox 21 Dec 1975-22; Tofino 7 Jan 1973-1; Ucluelet 12 Dec 1991-1.

Christmas Bird Counts: Interior – Recorded from 26 of 27 localities and on 87% of all counts. Maxima: Penticton 27 Dec 1983-401; Vaseux Lake 23 Dec 1983-359; Vernon 19 Dec 1993-174. **Coastal** – Recorded from 27 of 33 localities and on 63% of all counts. Maxima: Victoria 18 Dec 1993-433; Sooke 27 Dec 1986-198; Pender Islands 19 Dec 1986-169.

White-breasted Nuthatch

WBNU

Sitta carolinensis Latham

RANGE: Resident from northwestern Washington, the northeastern and southern interior of British Columbia, central Alberta, central Montana, southeastern Saskatchewan, southern Manitoba, southwestern Ontario, Minnesota, Wisconsin, Michigan, southern Ontario, southwestern Quebec, New Brunswick, Prince Edward Island, and Nova Scotia south to southern Baja California, Nevada, and Arizona across the southern United States to northern Florida, and Mexico.

STATUS: On the coast, *very rare* vagrant in the Georgia Depression Ecoprovince; *accidental* on the southern mainland of the Coast and Mountains Ecoprovince.

In the interior, *uncommon* resident in valleys of the Southern Interior Ecoprovince, locally *fairly common* there in autumn and winter; *uncommon* resident in southern valleys of the Southern Interior Mountains Ecoprovince; *rare* resident in the southern Central Interior Ecoprovince and in the Peace Lowland of the Boreal Plains Ecoprovince; *very rare* in the Sub-Boreal Interior Ecoprovince.

Breeds.

Figure 306. Since the 1970s, the White-breasted Nuthatch has been expanding its range in British Columbia. A single bird, present at a feeder in Bella Coola from early December 1995 through 16 January 1996, represents the first coastal occurrence north of Vancouver Island (16 January 1996; Michael Wiggle).

CHANGE IN STATUS: The White-breasted Nuthatch (Fig. 306) was previously restricted to the ponderosa pine forests of the Southern Interior and Southern Interior Mountains, with only 2 records from the coast (Munro and Cowan 1947). Since the 1970s, it has slowly expanded its range. In the west Kootenay of the Southern Interior Mountains, it had been reported only 4 times before 1970, but many times since then, including 2 breeding records. It is now resident in the Central Interior, and small numbers occur regularly in autumn and winter in the Sub-Boreal Interior near Prince George.

In Alberta, Salt and Salt (1976) report a similar increase in numbers of this species and a range extension, first noted in 1950. The northward spread of the White-breasted Nuthatch in Alberta has continued into the 1980s, and the species now occurs in the Peace River region (Siddle 1982; Semenchuk 1992). This suggests that the population in the Peace Lowland of the Boreal Plains originated from east of the Rocky Mountains rather than from the Southern Interior.

Reports of birds in the Central Interior, the Sub-Boreal Interior, and the northern portions of the Southern Interior Mountains follow the first records in the Peace Lowland; thus, those populations may have originated in the Boreal Plains (see REMARKS). The species is also now reported more frequently on the southern coast.

Figure 307. In British Columbia, the highest numbers for the White-breasted Nuthatch in both winter (black) and summer (red) occur in the Southern Interior Ecoprovince.

NONBREEDING: The White-breasted Nuthatch is distributed across southern British Columbia, particularly in the Southern Interior and southern portions of the Southern Interior Mountains. It is a characteristic resident of the Thompson, Similkameen, and Okanagan valleys, where it is more numerous than elsewhere; it is also resident in the Rocky Mountain Trench as far north as Radium. There are few records from the west Kootenay. In the Central Interior, it is distributed widely but sparingly from near Clinton west to Anahim Lake and north to Williams Lake. Small populations also occur near Prince George. The highest numbers during winter are found in the Okanagan valley of the Southern Interior (Fig. 307).

During the nonbreeding seasons, the White-breasted Nuthatch is most abundant at lower elevations from about 300 m to 1,220 m. It is most frequently seen in open ponderosa pine, Douglas-fir, and western larch forests that occur on

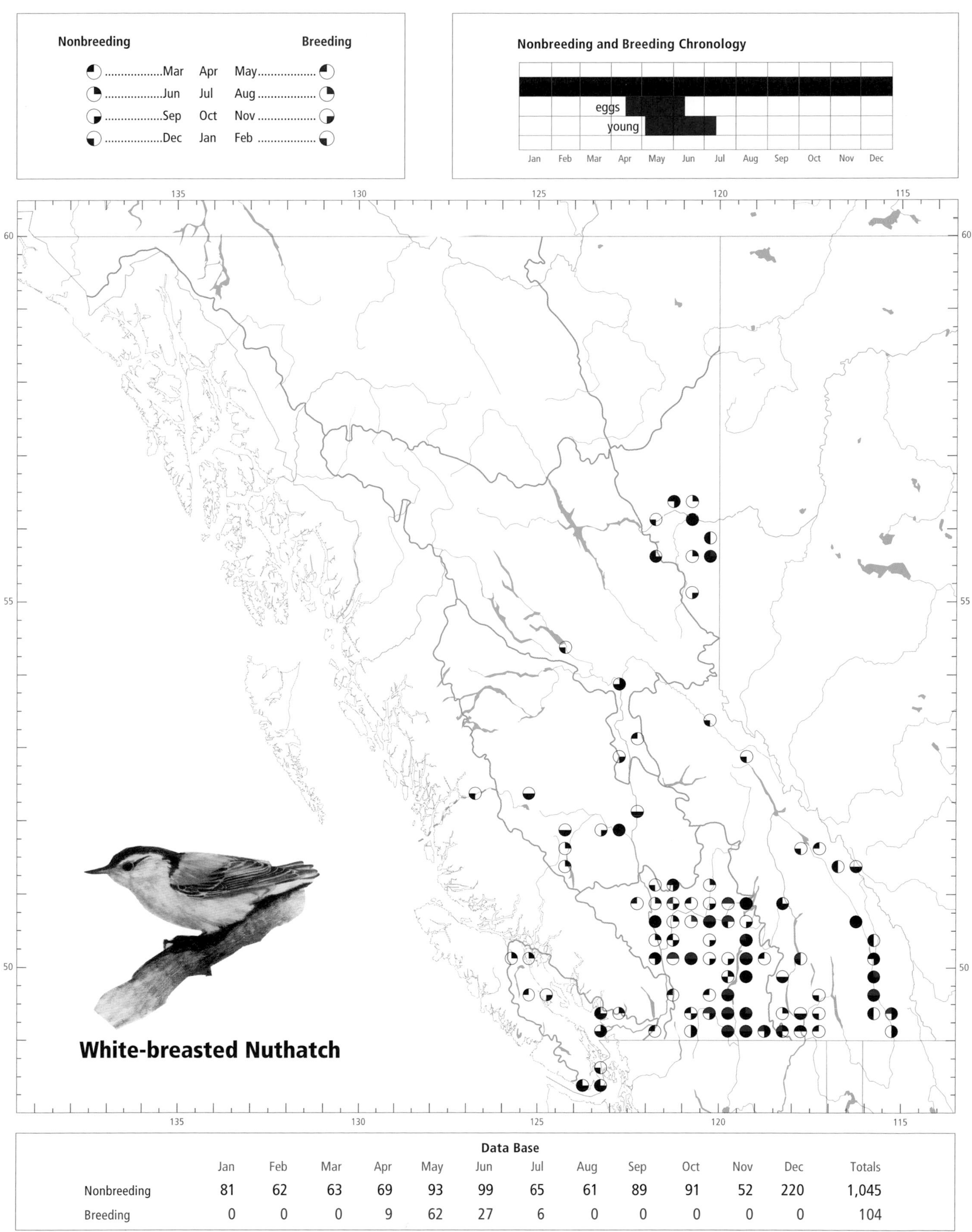

Data Base

	Jan	Feb	Mar	Apr	May	Jun	Jul	Aug	Sep	Oct	Nov	Dec	Totals
Nonbreeding	81	62	63	69	93	99	65	61	89	91	52	220	1,045
Breeding	0	0	0	9	62	27	6	0	0	0	0	0	104

Figure 308. Neil Dawe, Rick Howie, and Chris Siddle observing White-breasted Nuthatches in an open Douglas-fir–ponderosa pine forest (Pruden Pass, north of Kamloops, 30 June 1991; R. Wayne Campbell).

lower slopes across the Southern Interior and Southern Interior Mountains (Fig. 308). At higher elevations it also frequents dry stands of lodgepole pine. Around Shuswap Lake, it has been found foraging in stands of western white pine. In all parts of its range, and particularly in the Boreal Plains, the White-breasted Nuthatch frequents bottomland stands of trembling aspen, cottonwood, or balsam poplar with scattered spruce. It also uses orchards, urban gardens, and bird feeders; all records from the Prince George area have been of birds at backyard feeders. Unlike the Pygmy Nuthatch, it is mainly insectivorous and forages primarily on lower tree trunks and branches in older second-growth to old-growth forest stands.

The White-breasted Nuthatch is resident in the interior of the province, although records from the Cariboo and Chilcotin areas and near Prince George are far more frequent in autumn and winter than during spring and summer (Figs. 309 and 310). There appears to be a small altitudinal movement in the Southern Interior in September (Fig. 310). This movement is likely a post-breeding vertical migration from higher-elevation forests to wintering areas in the valley bottoms (Cannings et al. 1987). The White-breasted Nuthatch has the smallest population of the 3 nuthatch species; their winter populations fluctuate between years. On the coast, higher numbers of winter records than summer records suggest that some birds tend to wander widely after the breeding season (Fig. 310).

Unlike the Pygmy Nuthatch, the White-breasted Nuthatch is not gregarious, although it does join mixed flocks of chickadees, other nuthatches, and Brown Creepers. Breeding pairs tend to remain together on their territories for the winter while juveniles disperse.

On the coast, the White-breasted Nuthatch has been recorded irregularly in all months, especially in spring and summer; in the interior, it occurs throughout the year (Fig. 309).

BREEDING: The White-breasted Nuthatch has a widespread breeding distribution in the Southern Interior as far north as Chase and Hat Creek. It is also known to breed in the Southern Interior Mountains near Christina Lake, near the Arrow Lakes, and east of Kimberley. It undoubtedly breeds at lower elevations throughout the Southern Interior and the southern portions of the Southern Interior Mountains. Although records of nests with eggs or young from the Boreal Plains are lacking, fledglings have been seen in the Peace Lowland near Dawson Creek, which suggests that breeding may occur there. The White-breasted Nuthatch has also been observed during the breeding season in the Central Interior, and could be nesting there as well.

The highest numbers for the White-breasted Nuthatch in summer occur in the Okanagan valley of the Southern Interior (Fig. 307). Breeding Bird Surveys for both coastal and interior routes for the period 1968 through 1993 contain insufficient data for analysis. However, populations are likely increasing, as this nuthatch is extending its range in central and northeastern regions.

Breeding has been documented at elevations from about 300 m to 1,200 m. Breeding habitat includes open but

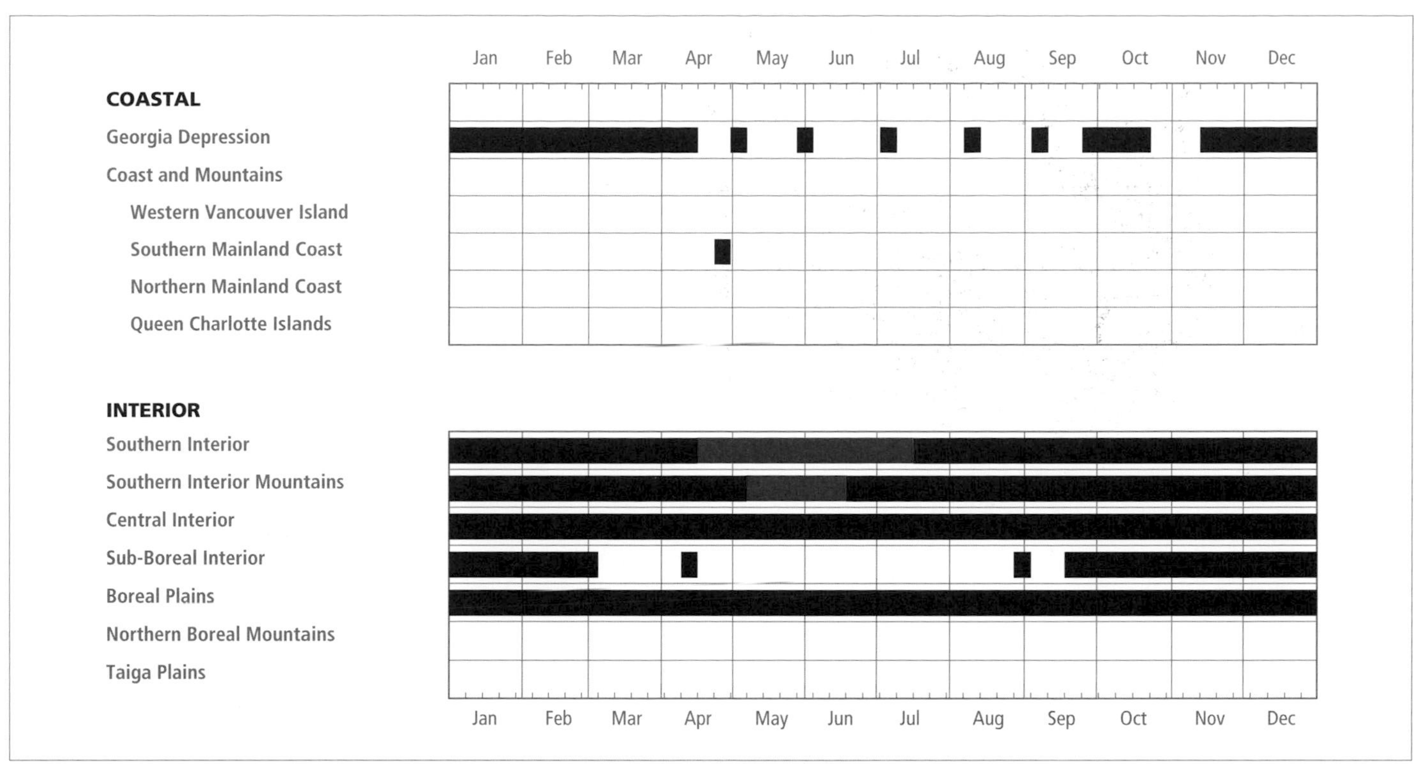

Figure 309. Annual occurrence (black) and breeding chronology (red) for the White-breasted Nuthatch in ecoprovinces of British Columbia.

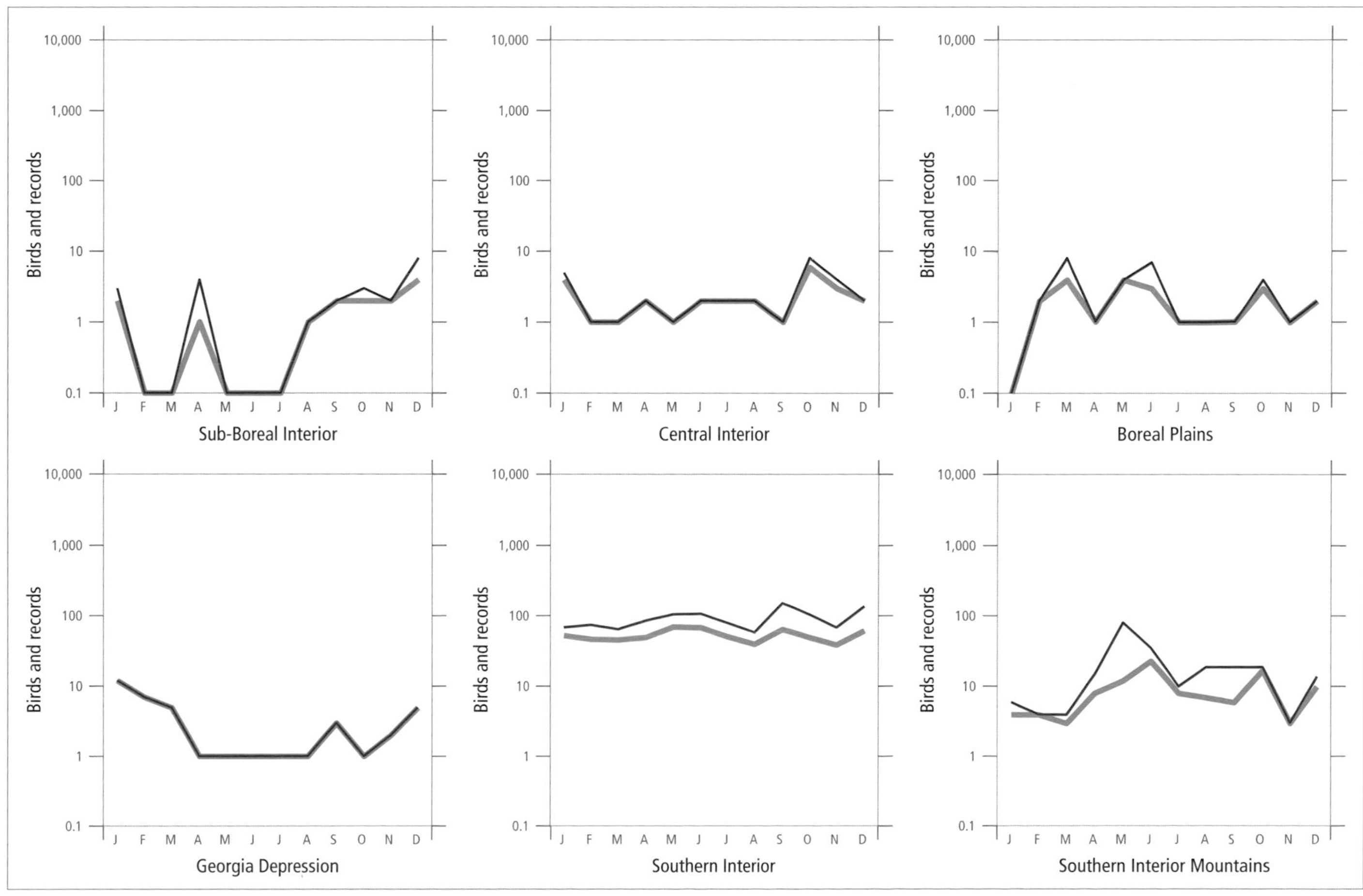

Figure 310. Fluctuations in total number of birds (purple line) and total number of records (green line) for the White-breasted Nuthatch in ecoprovinces of British Columbia. Christmas Bird Count and nest record data have been excluded.

Figure 311. Ponderosa pine forests are the preferred breeding habitat for the White-breasted Nuthatch in British Columbia (Okanagan Valley, 23 March 1996; R. Wayne Campbell).

Figure 312. In 1994, a pair of White-breasted Nuthatches nested in a natural cavity near the top of this ponderosa pine snag (White Lake, near Okanagan Falls, 1 May 1994; R. Wayne Campbell).

relatively continuous forest (78%; $n = 50$), human-influenced forest (20%), and grasslands with scattered stands of trees (2%). More specific habitats within those types include ponderosa pine forest (26%; $n = 38$; Fig. 311), mixed ponderosa pine and Douglas-fir forest (26%; Fig. 308), mixed deciduous and coniferous forest (11%), deciduous bottomland woods (5%), and unspecified rural settings (29%).

In the interior, the White-breasted Nuthatch has been recorded breeding from 13 April (calculated) to 12 July (Fig. 309).

Nests: The White-breasted Nuthatch is a primary cavity excavator but also readily uses existing cavities for nests. Half the nests ($n = 46$) were found in nest boxes and the other half in natural cavities. Where the condition of the nest tree was described in detail, 17 of 18 nests found in natural cavities were located in dead trees (Fig. 312). Tree species included trembling aspen (5 nests), ponderosa pine (4; Fig. 312), Douglas-fir (3), birch (2), western larch (1), and black cottonwood (1). Live ponderosa pines (8 nests) were readily used when nest boxes were provided. Our data contrast with the findings of a study in Colorado where all nests of the White-breasted Nuthatch ($n = 7$) were in cavities in living pines even though many dead ponderosa pines were available at the study site (McEllin 1979). Most of the cavities from McEllin's site had not been excavated by the nuthatches. In addition, in Arizona the White-breasted Nuthatch was the only species of 6 cavity-nesting songbirds that nested more frequently in natural cavities than in nest boxes, even where boxes were provided in excess of need (Brawn and Balda 1988).

In British Columbia, nest cavities were lined with a relatively narrow assortment of materials: feathers, moss, bark strips, and plant fibres. The heights for 21 nests in natural or excavated cavities ranged from 1.5 to 20 m, with 55% between 3.0 and 6.0 m.

Eggs: Dates for 28 clutches ranged from 24 April to 11 June, with 56% recorded between 1 May and 15 May. Calculated dates indicate that eggs can occur as early as 13 April. Sizes of 23 clutches ranged from 3 to 7 eggs (3E-2, 4E-4, 5E-8, 6E-4, 7E-5), with 15 having 5 to 7 eggs. The incubation period is 12 days (Ehrlich et al. 1988).

Young: Dates for 42 broods ranged from 4 May to 12 July, with 55% recorded between 20 May and 12 June. Sizes of 24 broods ranged from 1 to 7 young (1Y-1, 3Y-5, 4Y-5, 5Y-4, 6Y-4,

7Y-5), with 23 having 3 to 7 young. The nestling period is 14 days (Ehrlich et al. 1988).

Brown-headed Cowbird Parasitism: Cowbird parasitism was not found in British Columbia in 34 nests found with eggs or young. In North America, the White-breasted Nuthatch is rarely parasitized by cowbirds (Friedmann et al. 1977).

Nest Success: Of 17 nests found with eggs and followed to a known fate, 11 produced at least 1 fledgling, for a success rate of 65%.

REMARKS: Three subspecies of the White-breasted Nuthatch potentially occur in British Columbia. *S. c. tenuissima* Grinnell is the subspecies occupying southern and central British Columbia east of the Cascade Mountains, including the Rocky Mountain Trench (Godfrey 1955). *S. c. cookei* Oberholser inhabits Alberta, and is probably the subspecies found in the Peace Lowland of northeastern British Columbia. Examination of specimens from that region will indicate whether the population originated in Alberta or southern British Columbia. The habitual use of aspen by birds seen near Prince George and in the Peace Lowland is more characteristic of *S. c. cookei.* The American Ornithologists' Union (1957) suggests that the White-breasted Nuthatch that occurs on the coast is *S. c. aculeata* Cassin, which was shown by Aldrich (1944) to inhabit coastal Oregon and Washington; however, Godfrey (1986) found no specimens that would confirm this.

The published report of 24 White-breasted Nuthatches on the 2 January 1994 Christmas Bird Count for White Rock (MacKenzie 1995) probably should refer to the Red-breasted Nuthatch.

In ponderosa pine forests of Colorado, Stallcup (1968) found that White-breasted and Pygmy nuthatches partitioned foraging habitat. The Pygmy Nuthatch foraged most often in the foliage of living pines, whereas the White-breasted Nuthatch concentrated on their trunks and large branches.

See Tyler (1916), Kilham (1968, 1971, 1972a), and Waite (1987) for additional information on the natural history of the White-breasted Nuthatch.

NOTEWORTHY RECORDS

Spring: Coastal – Metchosin 1 Mar 1986-1; Victoria 28 Mar 1987-1; Stanley Park (Vancouver) 5 Mar 1970-1 (Campbell et al. 1972a); West Vancouver 10 Apr 1985-1; Chilliwack 29 Mar 1891-1; Bowen Island 1 May 1983-1; Fraser Canyon 23 Apr 1987-1. **Interior** – Oliver 11 May 1969-nestlings; Jaffray 20 May 1960-nest; White Lake (Okanagan Falls) 22 Apr 1990-first egg laid (calculated); Midway 25 May 1905-1 (NMC 3216); Anarchist Mountain 18 Apr 1954-1 (UBC 3161); Kimberley 2 Mar 1977-2; Princeton 8 Apr 1975-9, 12 May 1975-nest building; Balfour 16 May 1981-2; ne Wasa May 1977-46 on census; Skookumchuck May 1977-15 on census; Christina Lake 18 May 1980-nestlings; se Penticton 2 Mar 1980-4; Coldstream 4 May 1969-7 nestlings; Nakusp 23 Apr 1977-2 adults in nest, 18 May 1986-2; Spences Bridge 15 Apr 1972-1; Falkland 13 Mar 1977-2; Kamloops 26 Apr 1992-5 eggs; Pritchard 23 Mar 1992-building nest, 14 May 1992-7 nestlings; Lillooet 7 May 1916-1 (RBCM 4446); Hat Creek 26 May 1963-7 nestlings; Loon Lake (Clinton) 5 May 1970-1; Golden 10 Mar 1994-1; Chilcotin River 5 Apr 1987-1; Riske Creek 5 Apr 1987-1; Esker Lake 6 May 1977-2; Chetwynd 15 Apr 1992-4; Swan Lake (Dawson Creek) 9 May 1989-1; Dawson Creek 6 Apr 1987-2; Taylor 12 May 1987-1 (Campbell 1987c); Charlie Lake 2 Mar 1988-2, 9 May 1990-1; Halfway River (Fort St. John) 26 Apr 1981-1 (Siddle 1982).

Summer: Coastal – Surrey 3 Jun 1989-1; Mount Seymour Park 1 Jun 1975-1; Campbell River 4 Jul 1961-1; Sayward 10 Aug 1976-1. **Interior** – Manning Park 3 Jul 1977-1; White Lake Rd (Okanagan Falls) 3 Jul 1990-6 recent fledglings; Gallagher Lake 15 Jun 1974-1; Naramata 8 Jun 1989-fledglings; 12 km e Kimberley 2 Jun 1976-3 nestlings; Wasa 13 Jun 1971-4 young; Vernon 12 Jul 1975-1 nestling; Nicola 24 Jun 1974-3 nestlings; Monck Park 6 Jul 1980-nestlings; Lytton 3 Jun 1964-1 (NMC 52511); Pritchard 4 Jun 1992-young fledged; Chilko Lake 7 Jun 1988-1, 7 Aug 1988-1; Prince George 30 Aug 1994-1; sw Bear Mountain (Dawson Creek) 25 Jun 1987-1 adult with a new fledgling; Taylor 16 Aug 1988-1; Charlie Lake 6 Jun 1987-4.

Breeding Bird Surveys: Coastal – Not recorded. **Interior** – Recorded from 16 of 73 routes and on 10% of all surveys. Maxima: Summerland 21 Jun 1990-7; Wasa 5 Jul 1989-6; Kamloops 26 Jun 1983-5.

Autumn: Interior – Tumbler Ridge 3 to 5 Oct 1991-2; Willow River 27 Nov 1992-1; Prince George 8 Oct 1993-3; Quesnel 25 Sep 1992-1; Anahim Lake 24 Oct 1992-1; Williams Lake 25 to 31 October 1986-1 at feeder (Campbell 1987a); Eagle Lake 15 Oct 1992-1; Field 23 Oct 1977-1; Lytton 28 Sep 1925-2 (UBC 5122-23); Nicola Lake 10 Sep 1984-30; Wasa 19 Sep 1976-5; Penticton 26 Oct 1973-26; Manning Park 7 Oct 1972-1, 8 Nov 1984-1; Osoyoos 14 Oct 1963-9. **Coastal** – Comox 9 Sep 1922-1 (NMC 18278; Taverner 1927); Lighthouse Park (North Vancouver) 25 Sep 1977-1; Vancouver 16 Oct 1977-1; Sumas 10 Oct 1894-1 (Macoun 1900-4); Victoria 16 Nov 1986-1; Metchosin 22 Nov 1985-1 (BC Photo 1080).

Winter: Interior – Gundy 28 Dec 1986-1; Hudson's Hope 30 Dec 1988-1 (Siddle 1989b); Kiskatinaw River near Dawson Creek 18 Jan 1987-1; Chetwynd 18 Jan 1992-2; Giscome 16 Jan 1993-1; Prince George 26 Dec 1984-1; McBride 30 Dec 1991-1 (Siddle 1992b); Valemount 30 Dec 1992-1 (Siddle 1992b); Anahim Lake 28 Jan 1993-1; upper Tatlayoko Lake valley 24 Dec 1993-1; Williams Lake 2 Jan 1985-1; Becher's Prairie (Riske Creek) 14 Feb 1991-1; Golden 22 Dec 1994-1; Nicholson 22 Dec 1977-2; Clinton 11 Dec 1981-1; Kootenay National Park 2 Feb 1964-2; Celista 4 Feb 1948-5; Lillooet 15 Dec 1952-1 (Jobin 1953); Sorrento 2 Jan 1973-5; Kelowna 1 Jan 1977-4; South Slocan 13 Feb 1993-1; Kimberley 14 Jan 1976-1; Cranbrook 13 Jan 1977-1. **Coastal** – Bella Coola early Dec 1995-1 (Fig. 306); Stanley Park (Vancouver) 3 Dec 1972-2 (Campbell et al. 1974); Saanich 20 Dec 1977-1 (Davidson 1978), 1 Jan 1987-1; Victoria 30 Dec 1985-1; Metchosin 1 Jan 1986-1; Sooke 28 Dec 1985-1 (Campbell 1986b).

Christmas Bird Counts: Interior – Recorded from 19 of 27 localities and on 57% of all counts. Maxima: Penticton 30 Dec 1989-64; Vernon 18 Dec 1983-45; Vaseux Lake 2 Jan 1993-42. **Coastal** – Recorded from 2 of 33 localities and on less than 1% of all counts. Maxima: Pender Islands 19 Dec 1988-1; Sooke 28 Dec 1985-1.

Pygmy Nuthatch

Sitta pygmaea Vigors

PYNU

RANGE: Resident from southern interior British Columbia, northern Idaho, western Montana, central Wyoming, and southwestern South Dakota south, generally west of the Rocky Mountains, to Baja California, southern Nevada, southeastern Arizona, and Mexico.

STATUS: On the coast, *casual* in the Georgia Depression Ecoprovince and *accidental* on Western Vancouver Island in the Coast and Mountains Ecoprovince.

In the interior, *fairly common* to *common* resident in the Southern Interior Ecoprovince, particularly in the Okanagan and Similkameen valleys and adjacent plateaus, becoming *uncommon* further north in the Thompson River valley and in the Fraser Canyon between Lillooet and Lytton. In the Southern Interior Mountains Ecoprovince, *rare* resident in the Rocky Mountain Trench south of Radium; *casual* in the west Kootenay. *Accidental* in the Central Interior Ecoprovince.

Breeds.

NONBREEDING: The Pygmy Nuthatch (Fig. 313) has a limited distribution in the province and occurs regularly only in the Southern Interior. It is most numerous in the Okanagan valley, and becomes less so further north and west in the Southern Interior. It also has a limited distribution in the Southern Interior Mountains, mainly in the Rocky Mountain Trench between Newgate and Radium. Only a few occurrences have been reported from the west Kootenay; Christina Lake appears to be as far east as it regularly occurs in that region.

On the coast, there are several records from the Georgia Depression: Comox (several records, including specimens), West Vancouver, and Chilliwack. There is 1 record from China Beach on the southwest coast of Vancouver Island (Campbell 1986b). The highest numbers in winter occur in the Okanagan valley of the Southern Interior (Fig. 314).

In the interior, the Pygmy Nuthatch has been reported from valley bottoms near 300 m to elevations of 1,370 m. Throughout the year it is primarily associated with open forests of ponderosa pine (Fig. 320), mixed ponderosa pine and Douglas-fir, and western larch. The few individuals that have strayed to Vancouver Island were found in open Douglas-fir forests on dry, rocky hillsides. During the winter, most of the Pygmy Nuthatch population remains associated with old-growth or mature ponderosa pine stands, but some flocks enter residential areas along the valleys of the Thompson, Okanagan, Similkameen, and Kootenay rivers, where they forage in ornamental conifers and occasionally visit backyard bird feeders.

Although the Pygmy Nuthatch is a resident species (Fig. 315), a post-breeding dispersal occurs in late summer and autumn (Fig. 316), when individuals may wander outside their normal range. For example, the Pygmy Nuthatch has been reported along the Fraser River near Lillooet, in the eastern Thompson Basin at Adams Lake and Salmon Arm, and as far north as Lac la Hache.

Figure 313. Pygmy Nuthatch foraging in ponderosa pine in the Southern Interior Ecoprovince (Mahoney Lake, 19 September 1995; Linda M. Van Damme).

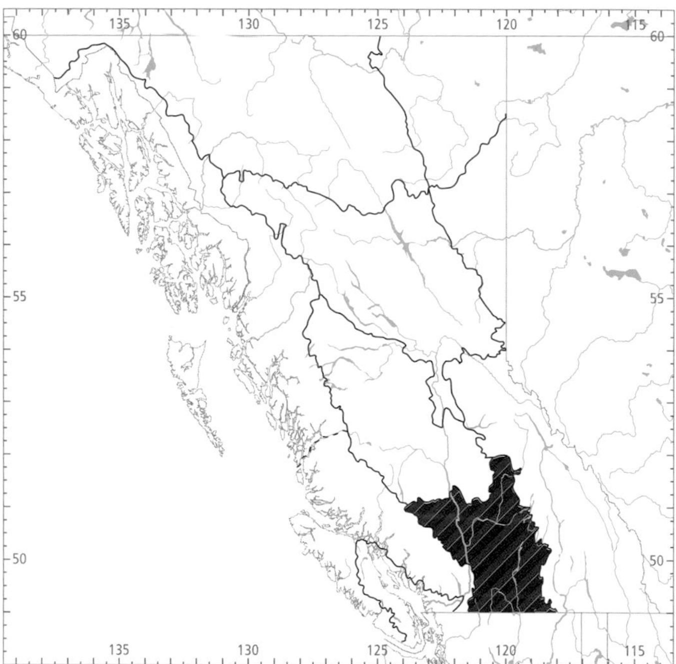

Figure 314. In British Columbia, the highest numbers for the Pygmy Nuthatch in both winter (black) and summer (red) occur in the Southern Interior Ecoprovince.

In the Southern Interior, at least, an altitudinal movement begins in September as birds move to lower elevations (Fig. 316). Root (1988) refers to a downslope movement in winter that is induced by cold temperatures, and notes that a January average minimum temperature of –12°C is too low for this small bird. Nowhere in its range are there reports of a marked latitudinal migration.

The Pygmy Nuthatch is gregarious. Throughout much of the year it often occurs in flocks of 15 to 24 birds, and groups of up to 76 birds have been reported (Fig. 317). Larger flocks are most often seen in the autumn and early winter (Fig. 318).

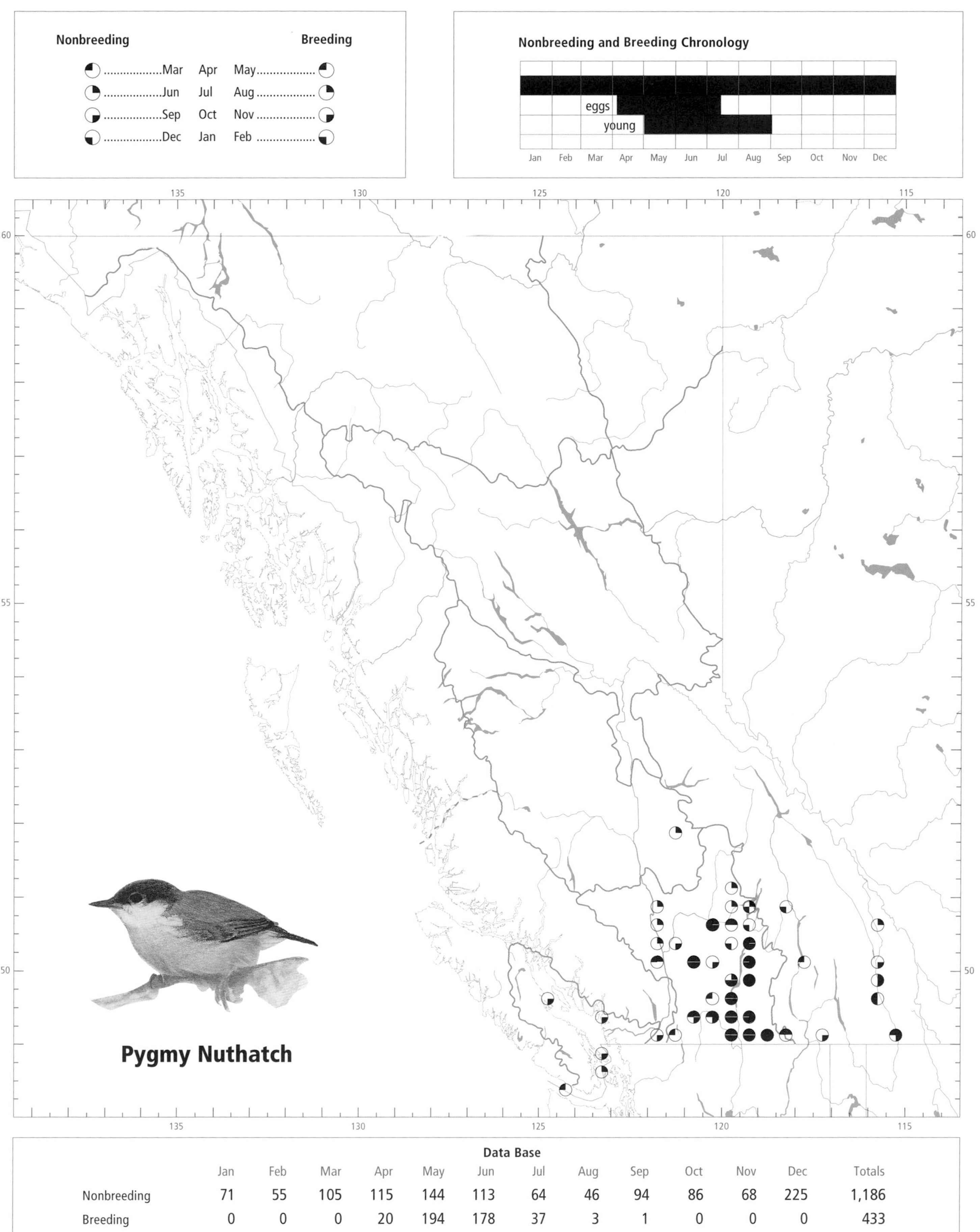

Data Base

	Jan	Feb	Mar	Apr	May	Jun	Jul	Aug	Sep	Oct	Nov	Dec	Totals
Nonbreeding	71	55	105	115	144	113	64	46	94	86	68	225	1,186
Breeding	0	0	0	20	194	178	37	3	1	0	0	0	433

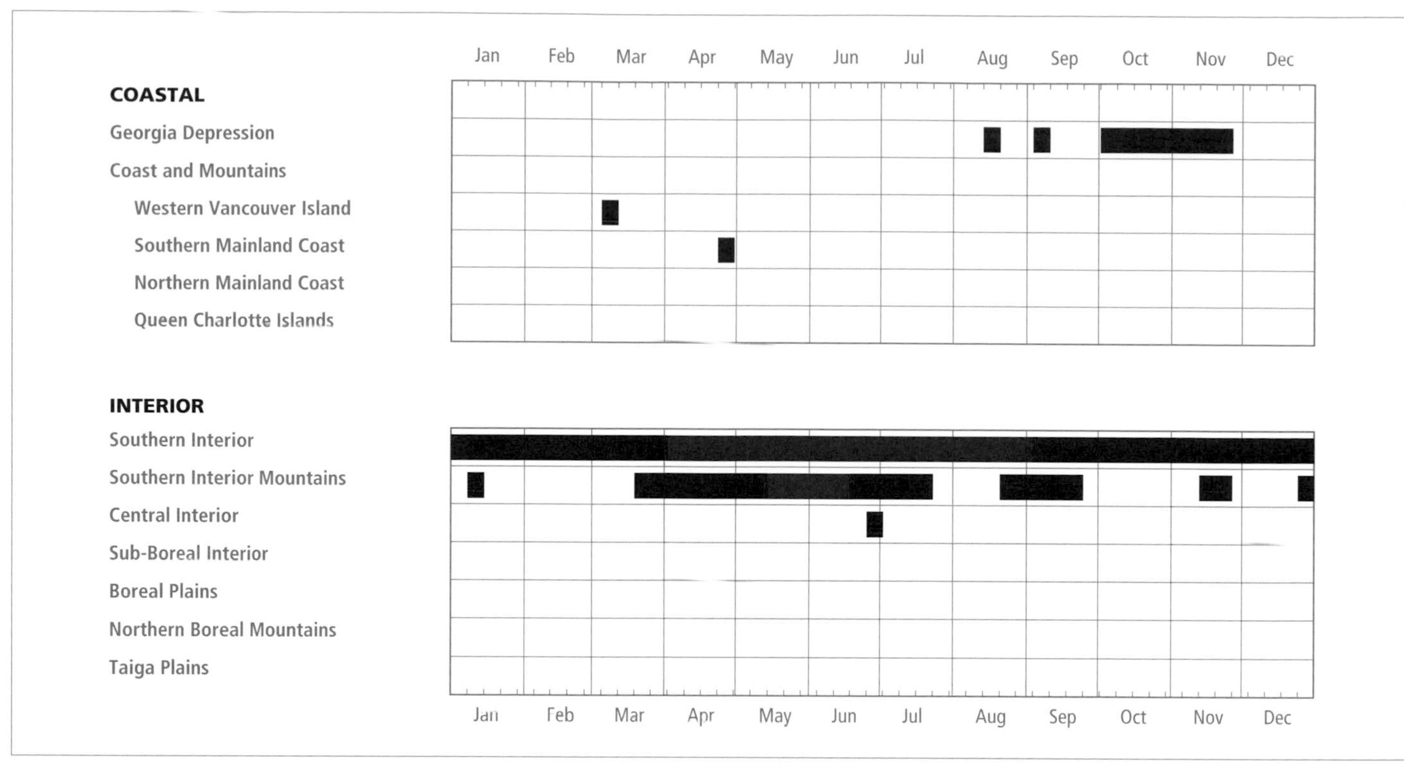

Figure 315. Annual occurrence (black) and breeding chronology (red) for the Pygmy Nuthatch in ecoprovinces of British Columbia.

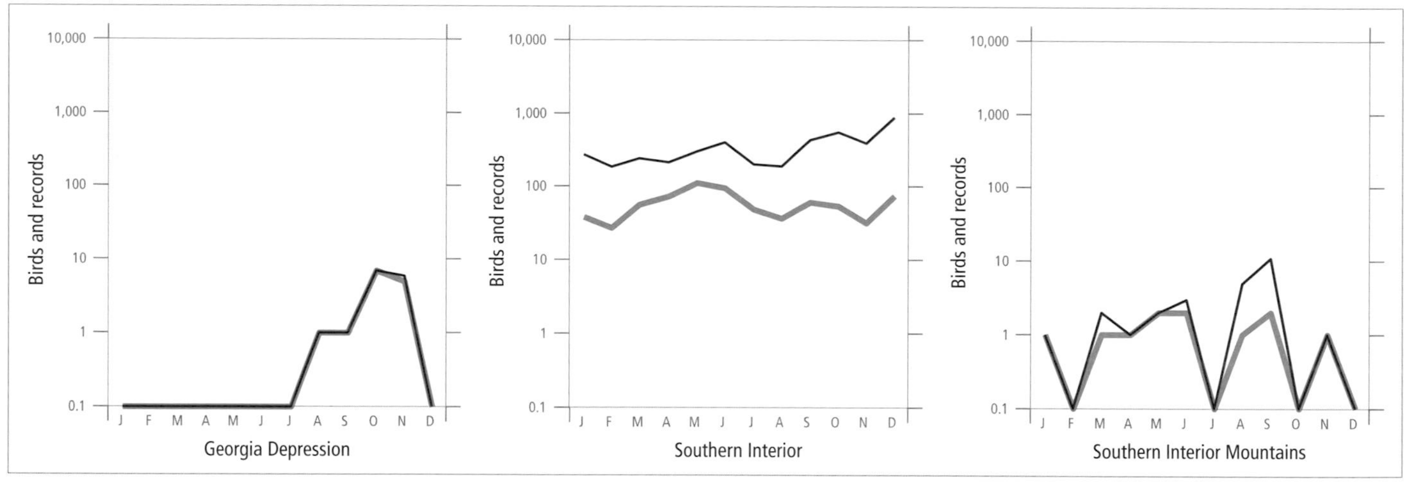

Figure 316. Fluctuations in total number of birds (purple line) and total number of records (green line) for the Pygmy Nuthatch in ecoprovinces of British Columbia. Christmas Bird Count and nest record data have been excluded.

During nonbreeding seasons, it often occurs in mixed flocks with chickadees, other nuthatches, Brown Creepers, and Downy Woodpeckers.

The Pygmy Nuthatch uses cavities as roosting sites and will roost communally. For example, in January, 12 birds were found huddled together within fiberglass insulation 6 m from the ground in the wall of a new house; in December, 7 birds were found huddled together in a cavity 5 m from the ground in a ponderosa pine (Fig. 319); in July, 6 birds were found roosting in a nest box.

Numbers of birds reported from Christmas Bird Counts in the Southern Interior appear to have increased between 1975 and 1993, although an increase in observer effort over the years may be partially responsible.

On the coast, the Pygmy Nuthatch occurs very irregularly; in the interior, it occurs throughout the year (Fig. 315).

BREEDING: The confirmed breeding distribution of the Pygmy Nuthatch is essentially restricted to ponderosa pine forests of the Southern Interior, mainly in the Similkameen and Okanagan valleys (Fig. 320). Breeding also occurs in the Nicola and Thompson valleys. There are 3 breeding records from Christina Lake, on the western edge of the Southern Interior Mountains. Specimens and observations of the Pygmy

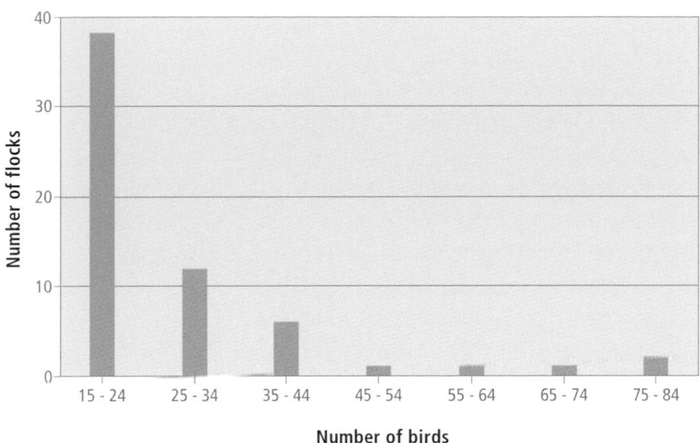

Figure 317. Frequency distribution of records of flock size or groups of flocks of the Pygmy Nuthatch in British Columbia.

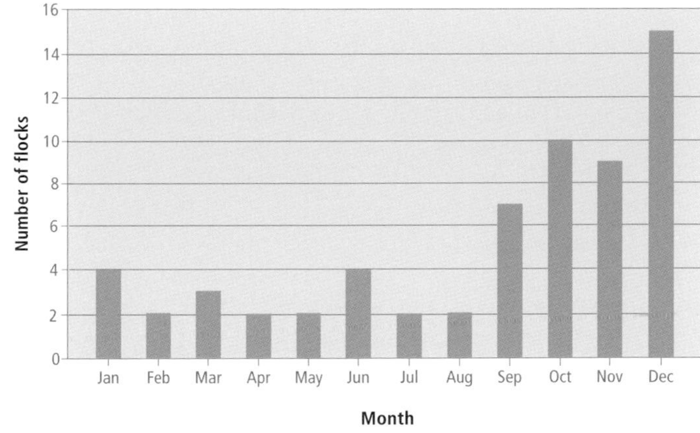

Figure 318. Frequency distribution (by month) of records of flocks or groups of flocks of the Pygmy Nuthatch in British Columbia that contained at least 15 birds.

Nuthatch have been reported from Newgate and Kimberley during the nesting season, and breeding probably occurs locally in the southern Rocky Mountain Trench. The westernmost records of breeding are from Lytton.

The Pygmy Nuthatch reaches its highest numbers in summer in the Okanagan valley (Fig. 314). Breeding Bird Surveys from both interior and coastal routes for the period 1968 through 1993 contain insufficient data for analysis.

The Pygmy Nuthatch has been reported nesting from 285 m to 1,350 m elevation. It frequents much the same habitat during the breeding season as it does throughout the rest of the year. Habitat types described for nesting (n = 183) included unfragmented forest (55%), human-influenced forest (32%), and grassland or big sagebrush and rabbit-brush shrublands with scattered stands of pines (12%; Fig. 320). Burned ponderosa pine forest is heavily used for nesting.

The Pygmy Nuthatch has been recorded breeding in British Columbia from 7 April (calculated) to 1 September (Fig. 315).

Nests: Most nests were found in coniferous trees (69%; n = 211; Fig. 321), posts or poles (11%), and deciduous trees (6%). Specific nest locations (n = 153) included ponderosa pine (74%), fence posts (7%), trembling aspen (6%), Douglas-fir (5%), birch, cottonwood, utility poles, and a pipe. Almost all the tree nests were in stumps (Fig. 321), dead trees, or, occasionally, dead parts of living trees.

The Pygmy Nuthatch typically nests in excavated cavities (73%; n = 213), but will readily use nest boxes (24%). A few nests were in holes in the walls of buildings, and 1 nest was in a crack behind a fireplace chimney. This nuthatch usually excavates its own nest cavity; as a weak excavator, it requires soft, punky wood. Old woodpecker cavities are also used. A study of nest sites used by the Pygmy Nuthatch in Colorado revealed site selection similar to that in British Columbia: all 26 nests were in holes in dead ponderosa pines, and the cavities had been excavated by the nuthatches or other bird species (McEllin 1979).

The heights for 206 nests ranged from 0.3 to 21 m, with 64% between 1.5 and 4.5 m. Nests were jumbled accumulations of feathers (74%; n = 76), hair (67%), moss (32%), and grass (21%). The nest cup was built of finer material, including hair, feathers, fine grass, plant down, and other soft, fibrous material.

Figure 319. Winter roost cavities for 7 Pygmy Nuthatches in a ponderosa pine (White Lake, near Okanagan Falls, 15 December 1992; R. Wayne Campbell).

Eggs: Dates for 100 clutches (Fig. 322) ranged from 21 April to 14 July, with 54% recorded between 6 May and 25

May. In the Southern Interior, records of first clutch initiations peak in late April and early May, whereas second clutch initiations occur mainly in June (Cannings et al. 1987) (Fig. 323). Calculated dates indicate that eggs can occur as early as 7 April. Sizes of 74 clutches ranged from 2 to 12 eggs (2E-2, 3E-1, 4E-1, 5E-8, 6E-16, 7E-30, 8E-10, 9E-4, 10E-1, 12E-1), with 62% having 6 or 7 eggs (Fig. 322). Harrison (1979) notes 10 eggs as the maximum clutch size, and Ehrlich et al. (1988) report a maximum of 9 eggs; thus the clutch of 12 eggs from this study may have been the product of 2 females. The Pygmy Nuthatch is known to often have yearling "helpers" assist a breeding pair in feeding nestlings (Sydeman 1989), but 2 females laying eggs in the same nest has not been reported. The incubation period in British Columbia was most frequently 16 days (L.A. Gibbard pers. comm.). Norris (1958) gives an incubation period of 15.5 to 16 days.

Young: Dates for 156 broods ranged from 1 May to 1 September, with 53% between 27 May and 18 June. The fledging date of nestlings found on 1 September was not determined. Sizes of 66 broods ranged from 2 to 12 young (2Y-8, 3Y-6, 4Y-5, 5Y-3, 6Y-18, 7Y-17, 8Y-5, 9Y-2, 10Y-1, 12Y-1), with 53% having 6 or 7 young. The nestling period from 10 nests ranged from 15 to 26 days, with 6 nests having a nestling period of 17 to 18 days. Ehrlich et al. (1988) cite a period of 20 to 22 days.

Cannings et al. (1987) cite 6 instances that suggest that the Pygmy Nuthatch may raise 2 broods annually (Fig. 323). Although the birds were not banded, evidence pointed to the same adults being involved in the same nest boxes. To our knowledge, the raising of 2 broods in a year has not been reported from other parts of the Pygmy Nuthatch's range.

Brown-headed Cowbird Parasitism: Cowbird parasitism was not found in British Columbia in 99 nests found with eggs or young, nor has it been reported elsewhere in North America (Friedmann 1963; Friedmann et al. 1977; Friedmann and Kiff 1985).

Nest Success: Of 43 nests found with eggs and followed to a known fate, 39 produced at least 1 fledgling, for a success rate of 91%. Most of these nests were in nest boxes.

Figure 320. Second-growth and mature ponderosa pine forests with bluebunch wheatgrass understorey provide habitat for the Pygmy Nuthatch throughout the year in the Southern Interior Ecoprovince (Barnhartvale Road, near Kamloops, 16 June 1996; Neil K. Dawe).

Figure 321. The Pygmy Nuthatch is considered a weak excavator, and therefore requires the soft, punky wood of coniferous stubs in which to nest (White Lake, near Okanagan Falls, 18 June 1993; R. Wayne Campbell).

Figure 322. Nearly two-thirds of all clutches recorded in British Columbia contained 6 or 7 eggs (White Lake, near Okanagan Falls, 10 June 1970; R. Wayne Campbell). The eggs are white and marked with evenly distributed reddish-brown dots and blotches.

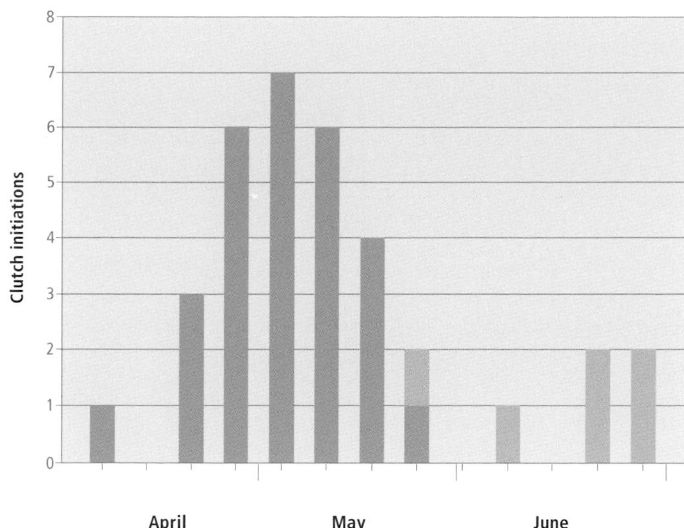

Figure 323. Weekly distribution of first clutch (dark bars) and second clutch (light bars) initiations of the Pygmy Nuthatch in the Okanagan valley (adapted from Cannings et al. 1987).

REMARKS: Four subspecies of Pygmy Nuthatch have been described for North America; only *S. p. melanotis* occurs in British Columbia (van Rossem 1929a; American Ornithologists' Union 1957).

The Pygmy Nuthatch caches ponderosa pine seeds during periods when surpluses are available. Ponderosa pine seeds can make up as much as 65% of its diet (Norris 1958).

See Sydeman (1989) and Sydeman et al. (1988) for additional information on the breeding biology of the Pygmy Nuthatch.

NOTEWORTHY RECORDS

Spring: Coastal – China Beach 8 Mar 1986-1 (Campbell 1986b). **Interior** – Newgate 23 Mar 1941-2, 26 May 1930-1 (NMC 24654); Ponderosa Meadows (Manning Park) 28 Apr 1973-1; Gallagher Lake 10 Apr 1969-adult excavating cavity; Vaseux Lake 8 May 1987-nestlings; Christina Lake 10 May 1980-adult excavating cavity, 23 May 1983-3; Okanagan valley 24 Apr 1906-5 eggs; Kaleden 3 May 1974-18; Whipsaw Creek 22 Apr 1964-1 (UBC 11858); 25 km ne Princeton 17 Apr 1979-16; Naramata 31 Mar 1975-building nest, 21 Apr 1979-10 eggs, 1 May 1979-10 nestlings, 17 May 1979-10 young fledged, 23 May 1975-6 young fledged, 30 May 1969-12 eggs; Summerland 2 Mar 1970-20, 3 Apr 1964-pair excavating cavity; Kimberley 25 Apr 1976-1; Nicola 10 May 1970-eggs; Kalamalka Lake Park 9 May 1990-nestlings; Lytton 23 May 1983-5; Nakusp 17 May 1976-1.

Summer: Coastal – Cordova Ridge (Saanich) 17 Aug 1983-1. **Interior** – Midway 14 Jun 1905-2 (Spreadborough 1905); Anarchist Mountain 3 Jul 1971-3 nestlings; Christina Lake 7 Jun 1981-nestlings; Oliver 22 Jul 1976-20; White Lake (Okanagan Falls) 3 Jun 1970-42; Whipsaw Creek 21 Jun 1983-3 fledglings; Naramata 2 Jun 1973-9 nestlings, 29 Jun 1969-12 young fledged, 9 Jul 1967-7 eggs, 14 Jul 1977-7 eggs, 4 Aug 1967-7 young fledged; Wasa Park 25 Aug 1971-5; Nicola 4 Jul 1970-young fledged; Cosens Bay 26 Jun 1975-66; Lytton Jun 1990-nestlings; 8 km n Lytton 21 Jun 1933-2 (MCZ 280981); Barnhartvale 15 Jun 1990-9 (Campbell 1990c), 21 Jul 1990-9; Kamloops 9 Jun 1981-female building nest; Lee Creek (Chase) 15 Jul 1970-1; Lillooet 21 Aug 1976-1; Kootenay National Park 25 Jun 1982-2; Adams Lake 1 Jul 1963-1; Lac la Hache 21 Jun 1942-1 (Carl 1942).

Breeding Bird Surveys: Coastal – Not recorded. **Interior** – Recorded from 10 of 73 routes and on 8% of all surveys. Maxima: Kamloops 26 Jun 1983-13; Canford 30 Jun 1987-12; Summerland 18 Jun 1977-7; Bromley 3 Jul 1992-7.

Autumn: Interior – Nicola valley 17 Sep 1896-3 (ROM 49629-31); Canal Flats 14 Nov 1939-1 (RBCM 11029); Wasa 19 Sep 1976-8; Summerland 4 Oct 1972-44; Vaseux Lake 1 Sep 1988-nestlings and 2 adults, 14 Sep 1960-15, 29 Oct 1975-76; Salmo 24 Nov 1969-present. **Coastal** – Comox 13 Oct 1931-1 (RBCM 13418); Ambleside Park (West Vancouver) 4 to 12 Nov 1973-1 (Crowell and Nehls 1974a); Ferguson Point (Vancouver) 10 Oct 1983-1; Vancouver 17 Oct 1972-1 (Crowell and Nehls 1973a; Campbell et al. 1974); Chilliwack 3 Oct 1979-1.

Winter: Interior – Revelstoke 30 Dec 1989-1; Salmon Arm 3 Jan 1971-11 on Christmas Bird Count; Okanagan Landing 10 Dec 1905-75 in yard at one time; Merritt 16 Dec 1962-2 (UBC 11592); Kimberley 14 Jan 1976-1; Vaseux Lake 31 Dec 1977-1, 22 Feb 1974-30. **Coastal** – No records.

Christmas Bird Counts: Interior – Recorded from 8 of 27 localities and on 38% of all counts. Maxima: Penticton 30 Dec 1989-**515**, all-time Canadian high count (Monroe 1990a); Vaseux Lake 2 Jan 1993-384; Kelowna 14 Dec 1991-269. **Coastal** – Not recorded.

Brown Creeper

BRCR

Certhia americana Bonaparte

RANGE: Breeds from southwestern, central, and southeastern Alaska; coastal, central, and northeastern British Columbia; the central Prairie provinces, central Ontario, southern Quebec, and Newfoundland south to southern California, Nevada, and Arizona in the west, and Arkansas and Tennessee in the east. During winter, parts of northern populations move to lower altitudes or latitudes. Also occurs in Mexico and Central America as far south as Nicaragua.

STATUS: On the coast, an *uncommon* resident in the Georgia Depression Ecoprovince, especially on southeastern Vancouver Island, and on the southern mainland of the Coast and Mountains Ecoprovince in the mountains north of the Fraser Lowland; *rare* to *uncommon* (sometimes local) resident elsewhere on the Coast and Mountains, including the Queen Charlotte Islands, except *very rare* in winter on the Queen Charlotte Islands and the northern mainland and *rare* to *uncommon* seasonally on Western Vancouver Island.

In the interior, an *uncommon* resident in the Southern Interior and Southern Interior Mountains ecoprovinces; *rare* to *uncommon* in autumn and winter, *very rare* to *rare* in spring and summer in the Central Interior Ecoprovince; locally *very rare* to *rare* in summer in the Sub-Boreal Interior, Northern Boreal Mountains, and the Peace Lowland of the Boreal Plains; *accidental* in the Taiga Plains Ecoprovince.

Breeds.

NONBREEDING: The Brown Creeper has a widespread distribution and is most numerous in coniferous and mixed forests across southern British Columbia south of latitude 52°N. It does not appear to be as widely distributed on the west coast of Vancouver Island and much of the southern mainland coast of the Coast and Mountains, as well as on the plateaus and mountain ranges adjacent to the Thompson, Similkameen, and Okanagan river valleys.

Further north, the Brown Creeper has a scattered and sparse distribution except on the Queen Charlotte Islands, where it occurs regularly throughout the year. In the interior, it is sparsely distributed in the Central Interior and becomes even less abundant in the Sub-Boreal Interior and beyond; there are only a few reports of the Brown Creeper north of latitude 56°N. The northernmost records west of the Rocky Mountains are from Flood Glacier on the Stikine River (Swarth 1922), Dease Lake, and the Haines Highway area, while east of these locations it has been reported in winter as far north as Fort Nelson. The highest numbers in winter occur in the Georgia Depression (Fig. 324).

The Brown Creeper has been reported from sea level to 1,220 m on the coast and from 220 to 2,000 m in the interior. For a description of general habitat, see BREEDING and Figs. 325 and 329. The Brown Creeper forages on the trunks of large and medium-sized trees of all species, but favours those with rough bark, especially conifers. The importance of forests that contain a diversity of species has been demonstrated in the northern Sierra Nevada of California. The abundance and variety of invertebrates on tree trunks of different tree species change with the seasons, and the foraging patterns of the creeper change accordingly, so that more diverse forests provide more stable food supplies (Adams and Morrison 1993). The same principle has been confirmed by Root (1988), who found that the Brown Creeper was most abundant in ecotones between different forest types.

Evidence of a north-south migration for the Brown Creeper in British Columbia is generally inconclusive (Figs. 326 and 327). A flock of 30 creepers seen on the rocks of Cape St. James, at the southernmost tip of the Queen Charlotte

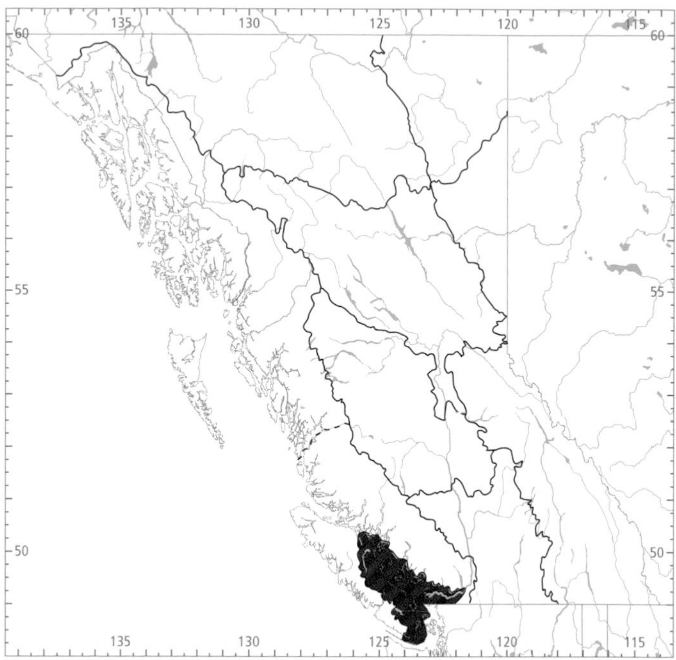

Figure 324. In British Columbia, the highest numbers for the Brown Creeper in both winter (black) and summer (red) occur in the Georgia Depression Ecoprovince.

Figure 325. Mixed western redcedar–western hemlock forests support populations of Brown Creepers in the west Kootenay region of British Columbia (Giant Cedar Trail, 27 km northeast of Revelstoke, 29 June 1993; Mark Nyhof).

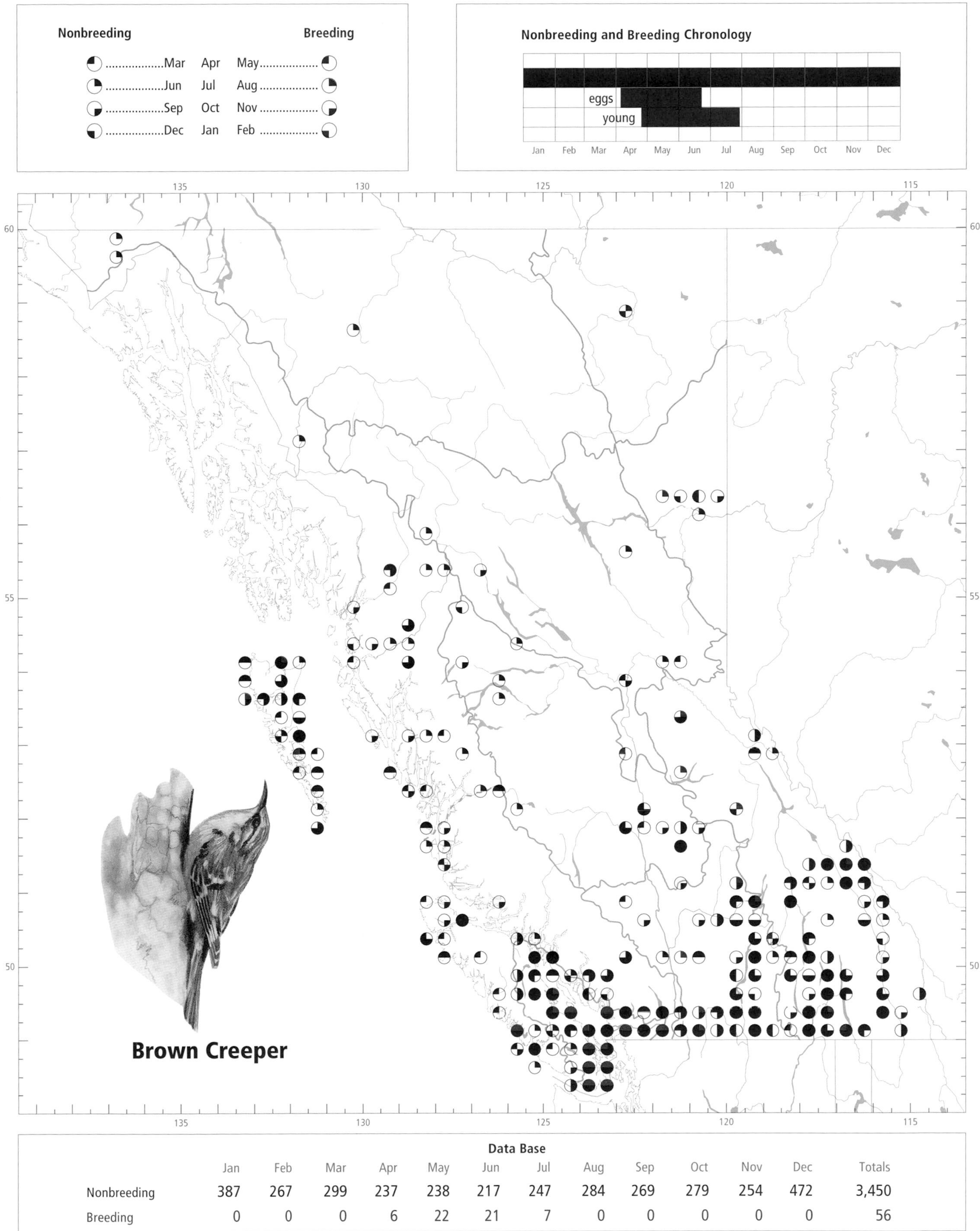

Data Base

	Jan	Feb	Mar	Apr	May	Jun	Jul	Aug	Sep	Oct	Nov	Dec	Totals
Nonbreeding	387	267	299	237	238	217	247	284	269	279	254	472	3,450
Breeding	0	0	0	6	22	21	7	0	0	0	0	0	56

Islands, on 27 December may have been preparing for a southward movement from the islands. On the other hand, the birds may also have simply remained there, behaving much as does the Steller's Jay on southern Vancouver Island (*q.v.*). In the Okanagan valley, Cannings et al. (1987) note that "Creepers begin to appear in late September and early October, and are regularly seen until late March and early April." This suggests at least a movement from higher elevations to lower elevations for the winter, but may also reflect an influx of migrants from northern regions. Similar patterns occur in the Southern Interior Mountains and the Central Interior (Fig. 327). Both latitudinal and altitudinal migration probably occur in British Columbia, but the magnitude of the movements is unknown.

During nonbreeding seasons, 1 or 2 Brown Creepers may normally be found travelling with a mixed flock of small forest songbirds such as kinglets, chickadees, and nuthatches. In the interior, they are nearly always associated with flocks of Golden-crowned Kinglets. On the coast and in the southern and central portions of the interior, the Brown Creeper occurs year-round (Fig. 326).

BREEDING: The Brown Creeper is a widely distributed breeding bird in southern British Columbia, becoming more sparsely distributed further north. The northernmost breeding records are from Masset on the coast and from Quesnel and the Haines Highway area in the interior. Even where the Brown Creeper is numerous, nests are so well concealed that few have been examined. Breeding records are heavily concentrated in the southernmost quarter of the province. Indeed, 94% of nesting records we have examined are from localities south of latitude 51°N, and 68% of these are nests from the Georgia Depression. During March and April, the distinctive songs of the Brown Creeper are a characteristic sound of most old-growth and mature forests of southern British Columbia, which suggests that it breeds in good numbers at least across the extreme southern portions of the interior. Although it likely breeds throughout its summer range, breeding remains to be confirmed along much of the coast and in the central and northern interior.

The highest numbers in summer occur on southeastern Vancouver Island in the Georgia Depression (Fig. 324). An analysis of Breeding Bird Surveys for the period 1968 through 1993 could not detect a net change in numbers on coastal routes; interior routes contain insufficient data for analysis.

On the coast, the Brown Creeper has been recorded nesting at elevations from near sea level to 1,050 m; in the interior, it has been recorded from 340 to 1,200 m. On the coast, it frequents mature and old-growth coniferous forests of Douglas-fir, western hemlock, grand fir, and western redcedar. On the Gulf Islands, it occurs in the Garry oak–arbutus forest, which is characteristic of drier areas (Fig. 325). In the interior it inhabits almost all mature coniferous forests and mixed stands of conifers and trembling aspen, balsam poplar, or birch. There are few records from western and alpine larch forests, or from krummholz habitat near the timberline. It occurs rarely in ponderosa pine forests.

On western Vancouver Island, Bryant et al. (1993) found the Brown Creeper in 55% of 71 old-growth stands (> 200 years old) that were dominated by western redcedar with some Douglas-fir and western hemlock, but in only 19% of 36 younger stands (50 to 60 years old) dominated by Douglas-fir. In southern Washington, Manuwal (1991) found the Brown Creeper most abundant in mature (95 to 190 years old) and old-growth forests (> 200 years old). Also in Washington, Mariani (1987) found that the Brown Creeper occurred mainly

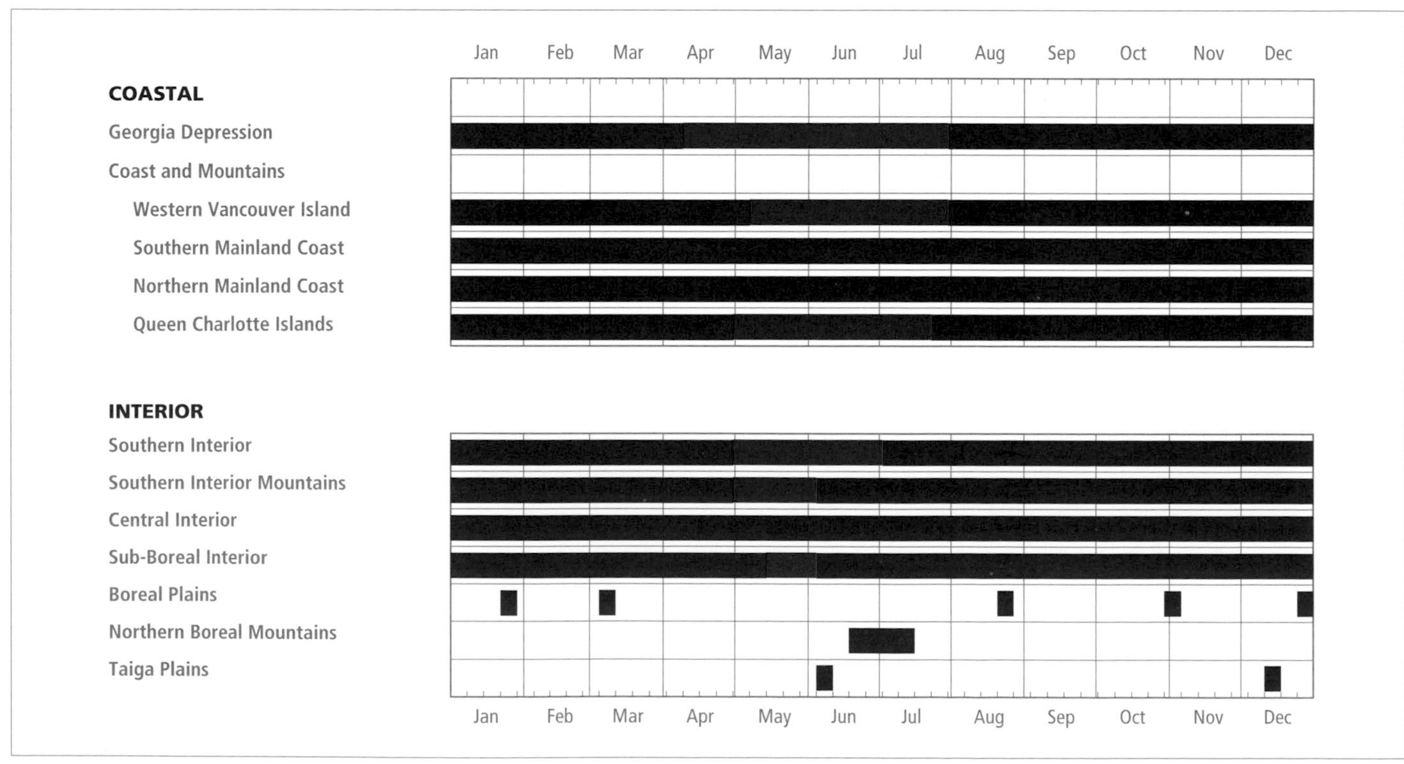

Figure 326. Annual occurrence (black) and breeding chronology (red) for the Brown Creeper in ecoprovinces of British Columbia.

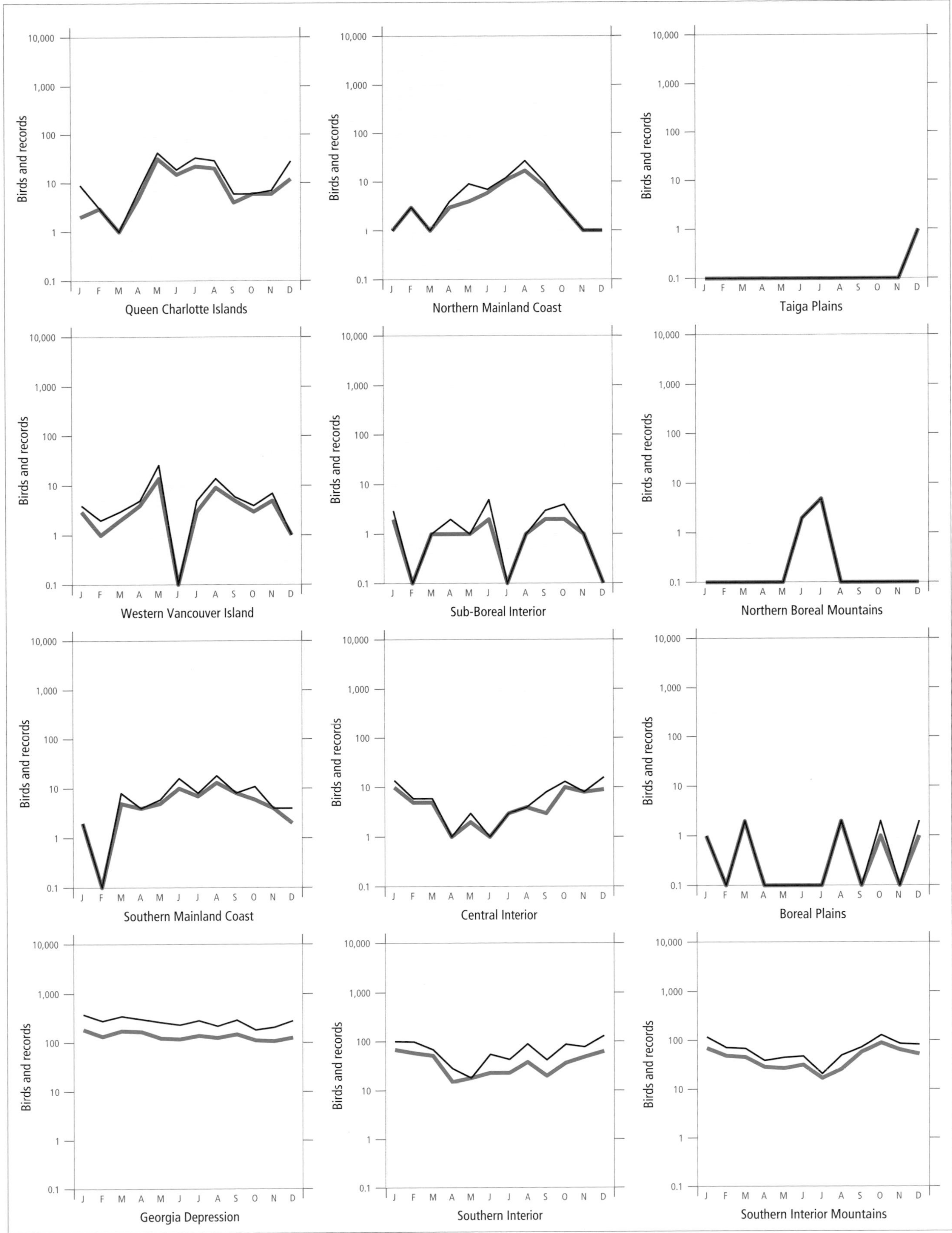

Figure 327. Fluctuations in total number of birds (purple line) and total number of records (green line) for the Brown Creeper in ecoprovinces of British Columbia. Christmas Bird Count and nest record data have been excluded.

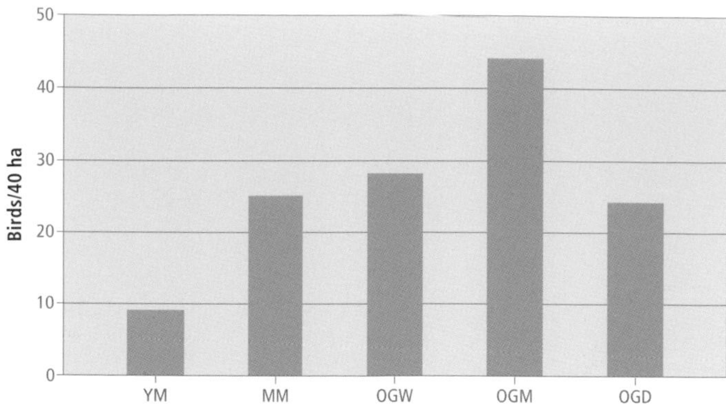

Figure 328. Average densities of the Brown Creeper (birds/40 ha) estimated for 5 age-moisture forest types in the Western Hemlock Zone of the southern Washington and northern Oregon Cascades (from Mariani 1987). Relative abundances differed significantly ($P < 0.001$). YM = young mesic, MM = mature mesic, OGW = old-growth wet, OGM = old-growth mesic, OGD = old-growth dry.

Figure 329. On southeastern Vancouver Island, the Brown Creeper nests in stubs in mixed coniferous forests of western hemlock and Douglas-fir (Haro Woods, Victoria, 21 April 1994; R. Wayne Campbell). Snags and stumps are important elements of the breeding habitat.

in forests of Douglas-fir, western hemlock, western redcedar, western white pine, and grand fir. These studies reveal a clear preference for moderately dry, mature or old-growth forests (Fig. 328).

In British Columbia, most breeding habitats (91%; *n* = 44) were described as natural forest, with only 9% – all in the Georgia Depression – in human-influenced forest habitats. Nineteen of 20 nesting habitats that were described in detail were in coniferous stands, including mixed (6 nests), Douglas-fir (5), western hemlock–western redcedar (2; Fig. 325), ponderosa pine (1), and lodgepole pine (1) stands; only 1 was in a deciduous stand. Nearly twice as many breeding habitats were described as old-growth forest rather than as mature second-growth forest.

Figure 330. Brown Creeper nest box erected on a Douglas-fir by William J. Merilees (Nanaimo, 15 May 1993; R. Wayne Campbell). This nest box design has proven to be successful in attracting nesting Brown Creepers on Vancouver Island.

On the coast, the Brown Creeper has been recorded breeding from 10 April (calculated) to 29 July; in the interior, it has been recorded breeding from 1 May (calculated) to 25 July (calculated) (Fig. 326).

Nests: Most nests reported were found in coniferous trees (70%; *n* = 30), including Douglas-fir (37%; Fig. 329), western redcedar (20%), western hemlock, Sitka spruce, ponderosa pine, and lodgepole pine. Deciduous trees included red alder (10%), bigleaf maple (10%), trembling aspen, black cottonwood, and Garry oak. Living and dead trees were reported about equally.

Most nests (93%; *n* = 42) were built under loose bark; 7% were found in natural cavities. Nests were constructed with twigs (71%), bark strips (26%), moss (19%), grass (19%), feathers (16%), and an assortment of lichens, leaves, hair, and wool. The heights for 43 nests ranged from 0.2 to 15.0 m, with 58% between 2.2 and 6.0 m.

Mariani (1987) studied nesting habitat in a western hemlock forest in Washington; nests were mainly in Douglas-fir (43%), western white pine (32%), and western hemlock (18%). All but 3 nests were in snags in various stages of decay. Recently dead trees were used most often.

Eggs: Dates for 9 clutches ranged from 21 April to 16 June, with 5 recorded between 4 May and 1 June. Calculated dates indicate that eggs can occur as early as 10 April. Clutch size ranged from 3 to 7 eggs (3E-2, 4E-1, 5E-3, 6E-2, 7E-1), with 5 having 5 or 6 eggs. In Michigan, 11 clutches ranged from 4 to 7 eggs (4E-2, 5E-2, 6E-6, 7E-1); incubation periods varied from 15 to 17 days ($n = 5$), with 4 clutches hatching on the 15th day (Davis 1980).

Young: Dates for 31 broods ranged from 25 April to 29 July, with 55% recorded between 20 May and 24 June. Recently fledged young have been noted as late as 12 August. Sizes of 16 broods ranged from 1 to 6 young (1Y-2, 2Y-3, 3Y-5, 4Y-1, 5Y-4, 6Y-1), with 10 having 3 to 5 young. The fledging period is 15 or 16 days (Davis 1980).

Brown-headed Cowbird Parasitism: Cowbird parasitism was not found in British Columbia in 37 nests found with eggs or young. Friedmann and Kiff (1985) report only 3 records of cowbird parasitism for North America.

Nest Success: Of 3 nests found with eggs and followed to a known fate, 2 produced at least 1 fledgling.

REMARKS: Two subspecies of Brown Creeper have been identified in British Columbia: *C. a. occidentalis* Ridgway occurs on the coast, and *C. f. montana* Ridgway occurs east of the Cascade Mountains and Coast Mountains (American Ornithologists' Union 1957).

Merilees (1987) has been able to attract this species to nest boxes in British Columbia (Fig. 330).

There is much yet to be learned about the breeding distribution and ecology of the Brown Creeper, especially at higher latitudes and altitudes in British Columbia. For detailed information on the life history of the Brown Creeper, see Davis (1980) and Franzreb (1985).

NOTEWORTHY RECORDS

Spring: Coastal – Victoria 21 Apr 1968-eggs; Goldstream Park 30 Apr 1963-14 (Edwards 1963); Saanich 16 May 1958-fledglings; Saturna Island 22 May 1961-1 nestling; Crescent Beach 25 Apr 1983-6 eggs; Surrey 3 Mar 1983-15; Florencia Bay 29 Mar 1985-2; Cultus Lake 4 Apr 1965-3; Tofino Inlet 3 May 1974-5; Langley 25 Apr 1970-6 nestlings (Campbell 1972a); Departure Bay 4 May 1986-6 eggs; Hope 8 Mar 1964-2, 21 May 1977-2; Calvert Island 17 May 1933 (NMC 281033); McInnes Island 30 Mar 1964-1; Lyell Island 20 Apr 1982-2; Limestone Islands 2 May 1983-4; Vertical Point 20 May 1983-3 nestlings; Kitlope Lake 3 May 1991-1; Langara Island 18 May 1946-1 (UBC 1184); New Aiyansh 28 Apr 1987-2. **Interior** – West Creston 23 May 1984-5 nestlings about 1 week old; Kelowna 3 Mar 1990-3; Shuswap Falls 30 May 1919-nestlings (Munro and Cowan 1947); Celista 1 Apr 1948-6; Stoddart Creek 8 Mar 1987-1; Pass Lake 16 Apr 1983-2; Parson 6 Mar 1976-5; Ice River 27 Apr 1977-4; Hoodoo Camp Ground (Yoho National Park) 31 May 1975-5 (Wade 1977); 100 Mile House 25 Apr 1985-1; Riske Creek 2 Mar 1990-1; Quesnel 17 May 1987-5 eggs; Nulki Lake 10 Mar 1992-11.

Summer: Coastal – Mount Douglas (Saanich) 26 Aug 1978-10; Jordan Meadows 26 Aug 1947-family group (Hardy 1949); Goldstream Park 15 Jul 1986-20; Towner Bay 25 Jul 1957-5 young fledged; Long Beach 29 Jul 1969-1 nestling fledged; Vancouver 6 Jun 1975-1 nestling; Langley 8 Jun 1985-4 fledglings; Little Qualicum Falls Park 1 Jun 1968-6; Chesterman Beach 12 Jul 1983-2; Brooks Peninsula 5 Aug 1981-3; Alta Lake 4 Aug 1941-4; Woodruff Bay 11 Jun 1986-2; Swanson Bay 18 Jun 1936-7 (MCZ 281040-1); Sandspit 31 Aug 1946-4; Hippa Island 21 Jul 1977-3 nestlings; Frederick Island 22 Jun 1946 (UBC 1185-6); Masset 1 Jun 1910-5 eggs, 4 Jul 1919-4 including 3 fledglings (Patch 1922); Kitsumkalum Lake 31 Aug 1974-1; Hazelton 7 Jul 1917-2 (Spreadborough and Taverner 1917); Alice Arm 2 Aug 1957-3; Flood Glacier 2 Aug 1919-1 (MVZ 40231). **Interior** – Little Muddy Creek 12 Jun 1957-nestlings; Manning Park 26 Jun 1983-12; Lightning Lake 11 Jul 1981-3 fledglings; Princeton 24 Jun 1990-adult feeding nestlings; Slocan River 7 Jun 1980-4; Nelson 21 Jun 1981-3 fledglings; s Lower Nicola 9 Jul 1986-4 eggs; Adams Lake 13 Jun 1962-7 eggs, 16 Jun 1965-5 eggs (RBCM E1826); Shuswap Lake Park 2 Jun 1973-2 fledglings, 24 Jul 1980-5; Amiskwi River 27 Aug 1976-2; Nusatsum River 8 Jul 1938-1 (Laing 1938); Soda Lake 4 Aug 1983-1, only 1 seen all summer (Graham 1983); Spanish Lake (Likely) 15 Jun 1992-3 fledglings; Bowron Lake Park 12 Aug 1975-2 fledglings; 77 km w Chetwynd 16 Jun 1969-1; Division Lake 5 Aug 1956-1; Taylor 22 Aug 1986-1; Hope Springs Ranch (Peace River) Jun 1953-nest.

Breeding Bird Surveys: Coastal – Recorded from 21 of 27 routes and on 39% of all surveys. Maxima: Nanaimo River 3 Jun 1984-14; Pemberton 23 Jun 1991-9; Victoria 5 Jul 1980-8. **Interior** – Recorded from 27 of 73 routes and on 11% of all surveys. Maxima: Princeton 1 Jul 1992-12; Creighton Valley 21 Jun 1980-4; Scotch Creek 8 Jul 1990-4; Adams Lake 24 Jun 1990-4; Osprey Lake 28 Jun 1990-4; Spillimacheen 28 Jun 1980-3; Zincton 14 Jun 1980-3; Illecilewaet 9 Jun 1983-3.

Autumn: Interior – Flatrock 30 Oct 1951-2; Tetana Lake Sep 1937-2 (Stanwell-Fletcher and Stanwell-Fletcher 1943); Nilkitkwa Lake 20 Oct 1979-2; Morice River 29 Sep 1974-1; Prince George 14 Nov 1974-1 (Rogers 1975a), 14 Nov 1992-1; Williams Lake 3 Sep 1987-6; Chilcotin River 13 Oct 1987-3; Glacier National Park 20 Nov 1981-4; D'Arcy 1 Nov 1970-1; Needles 8 Oct 1976-4; Savona 7 Oct 1967-12 (Gornall 1968); Princeton 17 Nov 1979-7; Manning Park 11 Nov 1982-2. **Coastal** – Kwinamass River 14 Sep 1985-2; Delkatla Inlet 10 Nov 1971-2; Hippa Island 17 Sep 1983-3; Campania Island 30 Sep 1938-1 (MCZ 281043); Strathcona Park 6 Sep 1966-2; Egmont 5 Sep 1977-1; Chesterman Beach 26 Nov 1982-1; Vancouver 19 Nov 1981-10; Westham Island 31 Oct 1965-4; Ucluelet 25 Nov 1950-2; Thetis Lake Park 25 Sep 1981-25.

Winter: Interior – Fort Nelson 15 Dec 1985-1; Stoddart Creek 27 Dec 1987-2 (Siddle 1988a); Prince George 29 Dec 1974-1; Williams Lake 30 Dec 1977-6; Revelstoke 21 Dec 1982-10, 26 Feb 1993-2; Yoho National Park 1 Jan 1977-1 (Wade 1977); Celista 8 Feb 1948-5; Nakusp 31 Dec 1976-3; Bull Mountain 1 Jan 1984-5; Cranbrook 30 Jan 1938-1, 18 Feb 1949-1 (Johnstone 1949); Max Lake 1 Jan 1970-2; Vaseux Lake 31 Dec 1975-1; Manning Park 19 Feb 1983-3; Creston 7 Jan 1981-10. **Coastal** – Cape St. James 27 Dec 1981-30, 11 Jan 1982-8; Terrace 21 Dec 1969-2, 30 Dec 1986-2; Kitimat 28 Dec 1985-1, 23 Feb 1980-1; Squamish River estuary 2 Jan 1988-1; Sasquatch Park 6 Dec 1983-3, 2 Jan 1976-3; West Vancouver 3 Feb 1983-35; Tofino 24 Dec 1990-1; Pachena Bay 18 Jan 1976-1; Goldstream Park 1 Jan 1983-1.

Christmas Bird Counts: Interior – Recorded from 24 of 27 localities and on 75% of all counts. Maxima: Penticton 27 Dec 1986-22; Kelowna 20 Dec 1981-16; Oliver-Osoyoos 28 Dec 1981-15; Vernon 23 Dec 1979-15. **Coastal** – Recorded from 29 of 33 localities and on 77% of all counts. Maxima: Victoria 18 Dec 1993-171; White Rock 2 Jan 1993-65; Vancouver 27 Dec 1970-54.

Rock Wren

Salpinctes obsoletus (Say)

ROWR

RANGE: Breeds from southern British Columbia, southern Alberta, southern Saskatchewan, western North and South Dakota south along the east side of the coast mountain ranges of Washington, Oregon, and California to southern Baja California, east to western Nebraska, central southern Texas, and Costa Rica. Winters mainly from northern California, southern Nevada, and southern Utah south through the southern portion of the breeding range; casual in winter north to southern British Columbia, Montana, and Wyoming.

STATUS: On the coast, *very rare* throughout the year on southeastern Vancouver Island and in the Fraser Lowland of the Georgia Depression Ecoprovince.

In the interior, *uncommon* migrant and summer visitant and *very rare* in winter in the Thompson, Okanagan, and Similkameen valleys of the Southern Interior Ecoprovince; *very rare* to *uncommon* migrant and summer visitant in the Southern Interior Mountains Ecoprovince; *uncommon* in the southern regions of the Central Interior Ecoprovince; *casual* in the Taiga Plains Ecoprovince and *accidental* in the Sub-Boreal Interior Ecoprovince.

Breeds.

CHANGE IN STATUS: During the first half of this century, the Rock Wren (Fig. 331) was known only as a summer visitant in the Okanagan and Thompson valleys of the Southern Interior, with occasional records north as far as Hanceville in the Central Interior (Munro and Cowan 1947). At that time, it was considered casual on the Fraser River delta, and there were no nesting records anywhere in the province. A Rock Wren had been collected at Cape Lazo on Vancouver Island in 1932, but the record was not known to Munro and Cowan (1947).

Since then, the Rock Wren has been found throughout the year in the Southern Interior, and breeding has been confirmed there. It has also been found to have a more widespread distribution in the Central Interior, and breeding has recently been reported there as well. In the Southern Interior Mountains, it was first reported from Indianpoint Lake when a specimen was collected in 1931. Subsequently, the Rock Wren was found both north and south of Elko in 1953 (Godfrey 1955). By 1959, it had been reported north to the Wells Gray Park area, and by 1970 it was reported from Mount Robson Park. The Rock Wren now occurs irregularly in the Southern Interior Mountains; however, breeding has yet to be confirmed. The Rock Wren has also recently expanded its range to the Boreal Plains.

On the coast, the Rock Wren now occurs irregularly on southeastern Vancouver Island; it was recorded breeding at Duncan in 1970.

NONBREEDING: The Rock Wren (Fig. 331) has a widespread but very localized distribution in southern British Columbia. It occurs in small numbers in the Okanagan, Thompson, and Similkameen valleys, along the breaks of the Chilcotin and Fraser rivers, and near Williams Lake. Recent observations in the Chilcotin have extended the range of the Rock Wren about 160 km westward, at least to Perkins Peak. It has also been reported, irregularly, in the Southern Interior Mountains from Waneta and Elko in the south to Mount Robson, Indianpoint Lake, Barkerville, and Kinbasket Lake in the north. In recent years, individuals have been observed in the Taiga Plains at Pink Mountain and in the Sub-Boreal Interior near Tumbler Ridge. On the coast, it occurs irregularly in much of the Georgia Depression, particularly on southeastern Vancouver Island. The highest numbers in winter occur in the Southern Interior (Fig. 332).

Figure 331. Adult Rock Wren (Perkins Peak, 250 km west of Williams Lake, 16 July 1992; Ruth E. Travers).

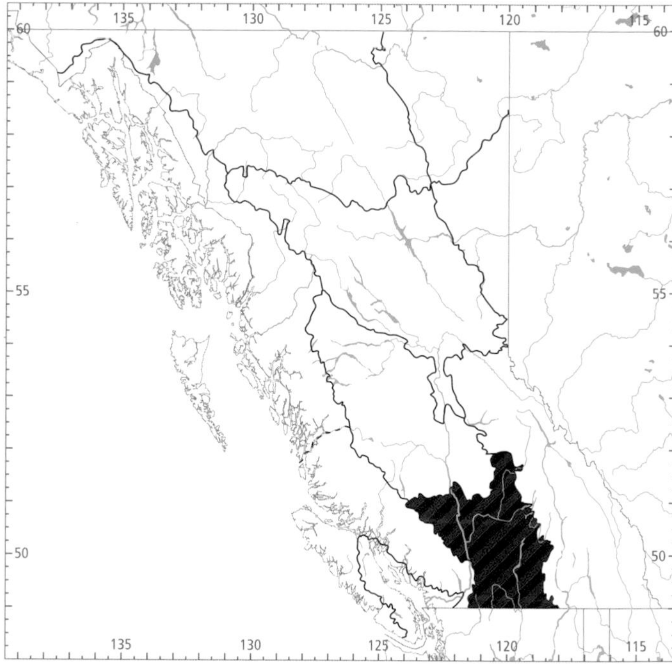

Figure 332. In British Columbia, the highest numbers for the Rock Wren in winter (black) and summer (red) occur in the Southern Interior Ecoprovince.

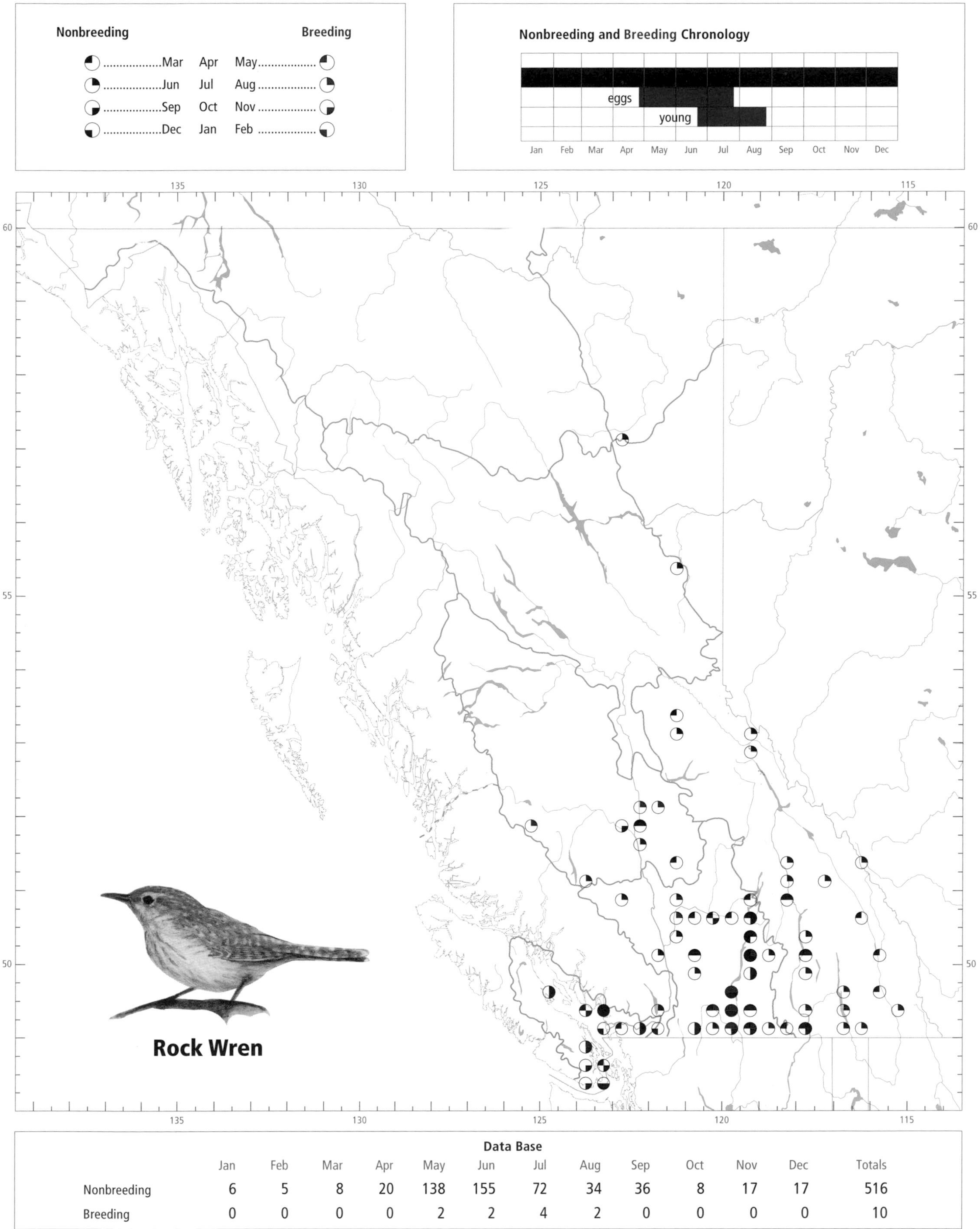

Data Base

	Jan	Feb	Mar	Apr	May	Jun	Jul	Aug	Sep	Oct	Nov	Dec	Totals
Nonbreeding	6	5	8	20	138	155	72	34	36	8	17	17	516
Breeding	0	0	0	0	2	2	4	2	0	0	0	0	10

Figure 333. Rock Wrens often frequent rock bluffs (Signal Hill, Williams Lake, 17 June 1990; R. Wayne Campbell).

Figure 335. Piles of large, loose rocks attract Rock Wrens in alpine habitats (Perkins Peak, 14 July 1992; R. Wayne Campbell).

The Rock Wren has been reported from near sea level on the coast to 2,400 m elevation in the interior. Its habitat is mainly hot and dry rock bluffs (Fig. 333), rock outcrops, cliffs with talus (Fig. 334), piles of loose rocks (Fig. 335), or boulder-strewn slopes on open hillsides with scattered ponderosa pine and Douglas-fir trees. In alpine habitat, it frequents talus slopes or boulder piles (Fig. 335); there is even a record from a glacial moraine in Glacier National Park. This wren has also been recorded in a garden rockery, old dump sites, quarries, bare roadcuts, and orchards. The occurrence near Tumbler Ridge was in a clearcut in subalpine forest, a habitat type that is also used in Oregon (Marshall and Horn 1973). In the United States, Rock Wrens may be found in almost any pile of rocks, even on plains above 2,000 m (Jewett et al. 1953).

Figure 334. Cliffs with talus slopes often support populations of breeding Rock Wrens (Chilcotin River, 8 July 1992; R. Wayne Campbell).

Although this wren occasionally overwinters in the Okanagan valley, most migrate, returning in late April and early May (Figs. 336 and 337). In some years, a few appear as early as late March. In the Columbia valley of the Southern Interior Mountains, the earliest arrival dates are 20 April and 1 May (at Revelstoke). Spring migrants also arrive in Williams Lake and Hanceville by late April and early May, with the largest numbers recorded in June (Fig. 337).

On the southeastern coast of Vancouver Island and the Fraser River delta, the majority of spring arrivals have been recorded in May (7 birds), compared with 2 birds each in February and March (Fig. 337).

In the Okanagan valley, numbers decline sharply through August, and the summer population disappears by the end of September (Fig. 337). Records from the Southern Interior Mountains and the Central Interior, although few, suggest similar trends. On the coast, the Rock Wren occurs too irregularly for a migration pattern to be determined. Some of the birds found on the coast and in the north may be part of a post-breeding dispersal movement.

During nonbreeding seasons the Rock Wren is a solitary species. Eighty percent (n = 527) of all sightings are of single birds, with pairs making up the majority of other records.

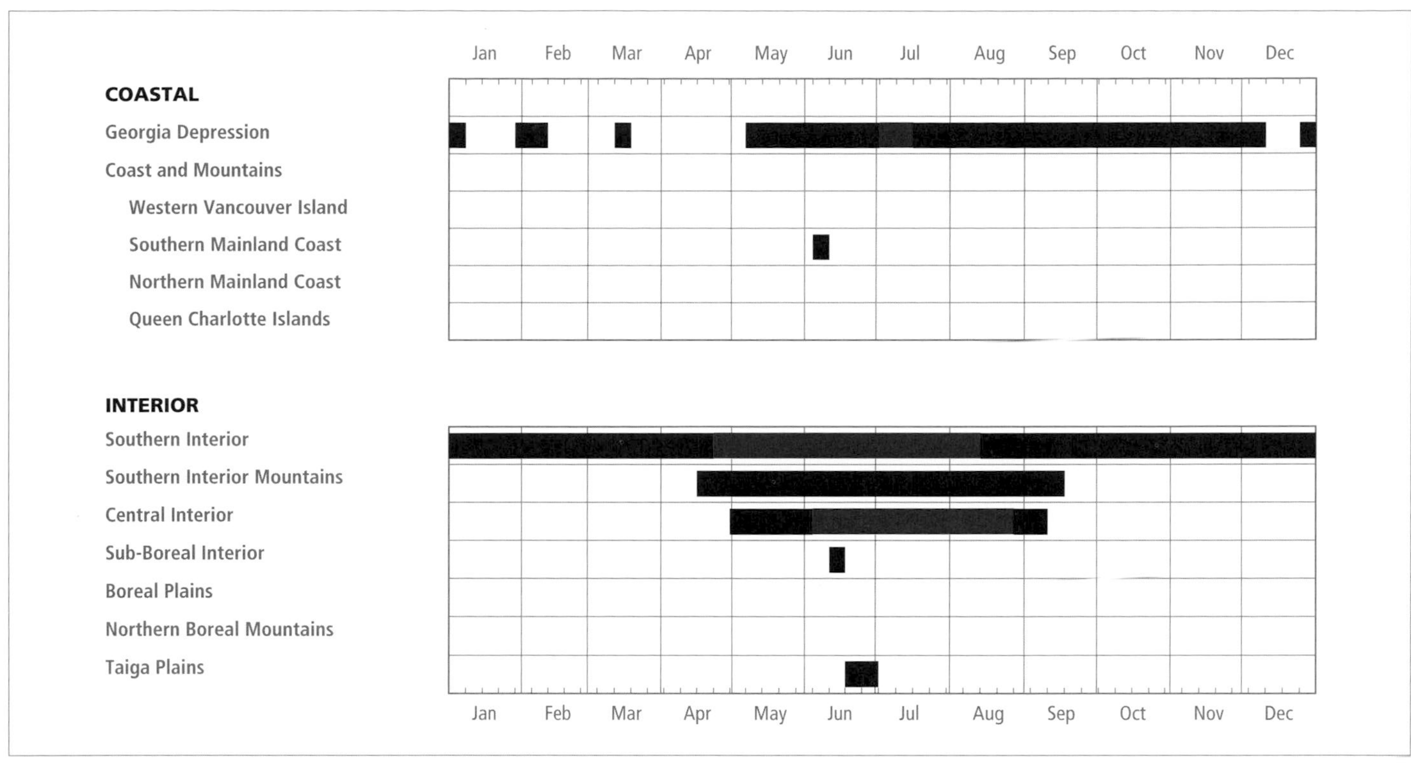

Figure 336. Annual occurrence (black) and breeding chronology (red) for the Rock Wren in ecoprovinces of British Columbia.

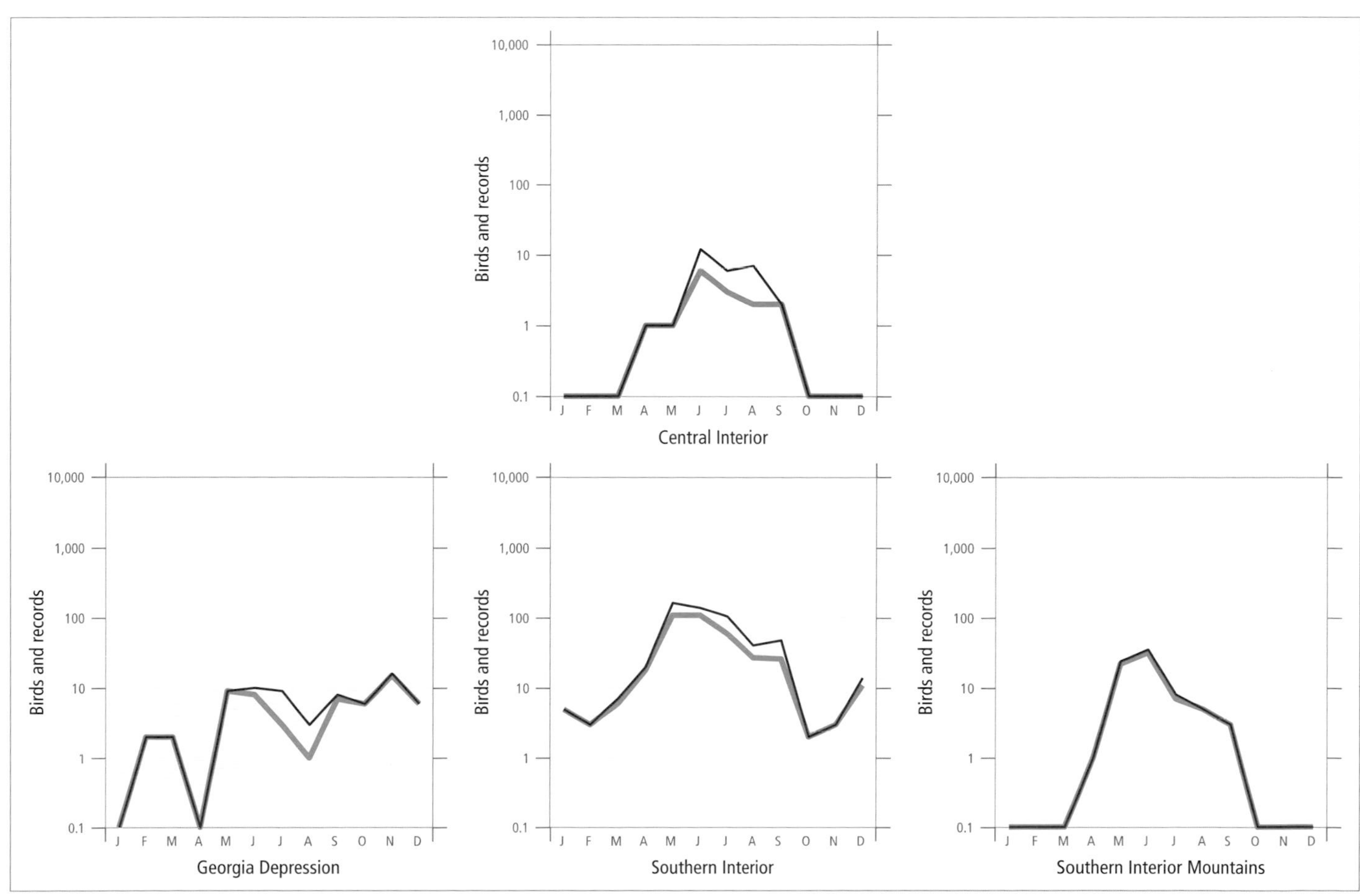

Figure 337. Fluctuations in total number of birds (purple line) and total number of records (green line) for the Rock Wren in ecoprovinces of British Columbia. Christmas Bird Count and nest record data have been excluded.

Figure 338. Site of a nest that contained 3 nearly fledged young, among the boulders at lower left corner (Perkins Peak, 14 July 1992; R. Wayne Campbell).

It does not skulk about in rock crevices and broken cliffs as much as the Canyon Wren. It is more visible, frequently singing from perches on large boulders, rock piles, or bluffs of broken rock, and flies readily between display points within its territory.

There are scattered wintering records from the Okanagan valley, the Fraser Lowland, and southeastern Vancouver Island. Wintering birds have been recorded in the interior from Vernon, Okanagan Landing, Naramata, Vaseux Lake, and Osoyoos Lake, and in coastal areas from Chilliwack, Vancouver, Saanich, Victoria, and Sooke. These areas are all below 600 m elevation.

On the coast, the Rock Wren has been reported irregularly throughout the year; in the interior, it has been recorded throughout the year but regularly between 15 April and 29 September (Fig. 336).

BREEDING: The known breeding distribution of the Rock Wren in British Columbia is concentrated in the Southern Interior, especially the valleys of the Okanagan and Similkameen rivers, where two-thirds of all recorded nesting has occurred. The core breeding area includes the south Okanagan valley between Osoyoos Lake and Summerland and Naramata, the western Kettle River valley, and the Similkameen River valley from Keremeos to just east of Princeton.

Outside this centre of distribution, breeding has been recorded at Ashcroft and in the Central Interior at 150 Mile House, Williams Lake, and Perkins Peak in the western Chilcotin. On the coast, the only breeding locality is on Vancouver Island, near Duncan.

The highest numbers in summer occur in the Southern Interior (Fig. 332). Breeding Bird Surveys for coastal and interior routes for the period 1968 through 1993 contain insufficient data for analysis. However, the Rock Wren appears to have been expanding its range in British Columbia in recent years, and populations may be increasing.

In interior valleys, the Rock Wren has been found nesting at elevations between 320 and 750 m; nests in alpine habitats (Fig. 338) have been found up to 2,100 m. The only coastal nest site was at about 30 m elevation. Breeding and nonbreeding habitat are similar at lower elevations. Breeding habitat features a highly isolated environment of arid grassland and open stands of ponderosa pine and Douglas-fir on steep slopes interspersed with rock outcroppings (Fig. 333), rubble, scree (Fig. 334), and fractured cliffs; coulees choked with deciduous vegetation; or rocky shrub steppes with sagebrush, rabbit-brush, and antelope-brush. Human-influenced habitats such as railway tunnels through rock cliffs, road cutbanks, and railway cutbanks also provide breeding habitat. In alpine areas, the Rock Wren may nest in piles of boulders (Figs. 335 and 338). The Duncan nest site, with its rocky terrain and summer drought, mimicked the dry, hot rocky breeding habitat used in the Southern Interior.

The Rock Wren has been recorded breeding in the province from 29 April (calculated) to 26 August (Fig. 336).

Nests: Seven of 8 nests were found on the ground, 4 nests in a rock cliff, and 3 nests on a rocky hillside among boulders (Fig. 338). One nest was in a building on a railway line along a steep bank. A nest found near Penticton was under a rock at the end of a 30 cm passage (Cannings et al. 1987), which is typical of the Rock Wren (Bent 1948). Nest material (n = 3) consisted of stones, dry grass, and feathers. The Rock Wren often lines its nest with small stones, and may also line a path to the nest with stones (Bent 1948; Merola 1995). The nest in the building was 0.8 m above ground.

Eggs: Only 2 nests with eggs have been found in British Columbia. One, near Oliver, contained 7 eggs (Fig. 339) on 5 May, and the other, in Penticton, held 6 eggs in advanced incubation on 13 May. Egg laying in the Oliver nest would have begun by 29 April at the latest. Calculated dates indicate that eggs can be found as late as 26 July. The incubation period is not well known; 2 clutches in New Mexico hatched after 12 and 14 days (Merola 1995).

Figure 339. Nest and eggs of the Rock Wren. The cover rock was moved to obtain the photograph (Oliver, 5 May 1966; R. Wayne Campbell).

In Washington, the Rock Wren is reported to lay 2 clutches annually, 1 in May and a second in late June or July (Jewett et al. 1953). The same appears to apply in British Columbia (Cannings et al. 1987).

Young: Dates for 7 broods ranged from 21 June to 26 August. Calculated dates indicated that nests could contain young as early as 20 May. Parental care of fledglings from second broods has been found as late as early September (J. Comer pers. comm.). Sizes of 5 broods ranged from 3 to 5 young (3Y-1, 4Y-3, 5Y-1). Sizes of 20 fledged broods ranged from 1 to 7 young (1Y-3, 2Y-4, 4Y-5, 5Y-6, 6Y-1, 7Y-1). In New Mexico, the nestling period was between 14 and 16 days (Merola 1995).

Brown-headed Cowbird Parasitism: Cowbird parasitism was not noted in British Columbia in 7 nests found with eggs or young. The Rock Wren is an infrequent host of the Brown-headed Cowbird (Friedmann 1963).

Nest Success: Insufficient data for analysis.

REMARKS: The Dog Lake (now Skaha Lake) record occurred in 1936, not 1916 as stated in Munro and Cowan (1947).

Godfrey (1986) shows the Rock Wren as breeding throughout much of the Southern Interior Mountains; to our knowledge, however, breeding has not been established in that area.

The Rock Wren reaches the northern limit of its range in British Columbia. Its distribution is very localized, and it appears to be irregularly expanding its range. However, more information is required before the status, distribution, migration, and ecological requirements of the Rock Wren in British Columbia will be adequately understood. Observers are encouraged to watch for this species in areas with suitable habitat. As with the Canyon Wren, little study has been carried out on the Rock Wren. See Wolf et al. (1985) for a discussion of nesting in high-altitude environments, and Merola (1995) for information on nesting behaviour.

NOTEWORTHY RECORDS

Spring: Coastal – Mount Tuam 10 May 1988-1; Vancouver 21 to 29 May 1977-1 in University of British Columbia botanical garden (BC Photo 480); Wilson Creek 10 May 1987-1, first Sunshine Coast record. **Interior** – Richter Pass 18 May 1963-1 (Campbell and Meugens 1971); Oliver 5 May 1966-7 eggs; Vaseux Lake 2 Mar 1988-1, 15 Apr 1968-1, 10 May 1977-6; Hedley 29 May 1928-1 (RBCM 13464); Penticton 13 May 1942-6 eggs; Okanagan Landing 2 Mar 1966-1; Waneta to Balfour 27 May 1982-2 on survey; Nakusp 1 May 1977-1; Columbia Lake 19 May 1969-1 (Wilson et al. 1972); Grindrod 29 Apr 1928-1 (Munro 1953a); Kamloops 15 Mar 1987-2 singing; Tranquille 13 Mar 1994-1; Revelstoke 20 Apr 1986-1; Doc English Gulch 30 Apr 1978-1 singing; 100 Mile House 14 May 1934-1 (MCZ 283413).

Summer: Coastal – Genoa Bay 7 Jun to 31 Aug 1970-nesting pair, 11 Jul 1970-5 young, first breeding record for Vancouver Island (BC Photo 162); Harrison Hot Springs 10 Jun 1984-1. **Interior** – Osoyoos 26 Aug 1979-8; Manning Park 10 Jun 1987-2; Mount Kobau 25 Jun 1993-4 fledglings; Fairview 25 Jun 1984-7 fledglings; Vaseux Lake 10 Jul 1973-10, 23 Jul 1979-building nest, 12 Aug 1982-adult feeding unknown number of young in nest; Kettle River 24 Jul 1985-4 fledglings; Elko area 26 Jun 1953-several (Godfrey 1955); Syringa Creek 13 Jun 1982-2; 1 km w Stirling Creek (Hedley) 24 Aug 1975-1 fledgling; Nickel Plate Mine (Hedley) 19 Jun 1959-4 fledglings; Penticton 29 Jul 1969-5 fledglings; Naramata 12 Jun 1976-5 fledglings; Kalamalka Lake 18 Jul 1993-1 adult with 1 fledgling, 21 Jul 1993-2 fledglings; Kicking Horse River canyon, 16 km e Golden 10 Jun 1994-1 singing; Lytton 2 Aug 1964-1; Spences Bridge 4 Jun 1889-1 (NMC 727); Lillooet 18 Aug 1979-2; Hat Creek 29 Jun 1936-1; Ashcroft Jun 1892-1 (Rhoads 1893), Aug 1938-1 (RBCM 8832); Dog Creek 10 Jun 1952-1 (NMC 48103); n end Kinbasket Lake 17 Aug 1994; Yohetta Creek 19 Jul 1993-1 singing; Perkins Peak 14 Jul 1992-nest with 3 nearly fledged young, and adult with 3 flying young (Campbell and Dawe 1992); 10 km w Alexis Creek 10 Jul 1992-1 singing; Williams Lake 21 Jun 1977-4 nestlings, 26 Aug 1970-adult feeding unknown number of young in nest; Mount Robson Park 19 Jul 1970-2; Cunningham Pass 10 Jun 1994-1; Meikle Creek 11 Jun 1989-1; Pink Mountain 18 Jun 1992-1 (Siddle 1993a), 28 Jun 1994-1.

Breeding Bird Surveys: Coastal – Recorded from 1 of 27 routes and on less than 1% of all surveys. Maxima: Seabird 25 Jun 1994-2. **Interior** – Recorded from 12 of 73 routes and on 5% of all surveys. Maxima: Beaverdell 24 Jun 1984-3; Summerland 23 Jun 1992-3; Grand Forks 24 Jun 1973-2; Syringa Creek 25 Jun 1988-2; Oliver 24 Jun 1990-2; Bromley 1 Jul 1993-2; Brookmere 19 Jun 1974-1.

Autumn: Interior – Hanceville 7 Sep 1946-1 (UBC 1285); Murtle Lake Sep 1959 (Edwards and Ritcey 1967); Kamloops 14 Sep 1893-1 (RBCM 1136); 4 km e Kamloops 10 Sep 1959-3 (UKMU 38379), 14 Sep 1893-1 (RBCM 1136); Cosens Bay 12 Oct 1981-1; Giants Head 29 Sep 1981-1, first on this mountain for several years, 5 Nov 1989-1; Vaseux Lake 1 Sep 1973-9; Trail 12 Sep 1970-1; Manning Park 27 Sep 1972-1. **Coastal** – Cape Lazo 19 Nov 1932-1 (MVZ 103505); Denman Island 19 and 20 Sep 1964 (Hesse and Hesse 1965a); Point Grey (Vancouver) 29 Nov 1988-1 to 18 Mar 1989-1; Chilliwack Nov 1889-1 (Brooks 1900); Genoa Bay 1 to 15 Sep 1970-family group of 7 birds; Victoria 12 Oct 1959-1 (BC Photo 95); Mount Douglas (Victoria) 17 Nov 1975-2; Sooke 1 Nov 1969-1 (NMC 57377; Tatum 1970b); Rocky Point (Metchosin) 20 Sep 1990-1 (Siddle 1991a).

Winter: Interior – Vernon 31 Dec 1966-1, 12 Jan 1967-1; Okanagan Landing 25 Dec 1968-2; Naramata 20 Feb 1977-1 near Chute Creek; Vaseux Lake 12 Jan 1979-1. **Coastal** – Vancouver 4 Feb 1989-1; Genoa Bay 11 Dec 1970-1; Luckakuck Creek 6 Dec 1888-1 (MVZ 103506); Mount Douglas (Victoria) 30 Dec 1975-1; Victoria 1 Jan 1960-1 (Stirling 1960a).

Christmas Bird Counts: Interior – Recorded from 2 of 27 localities and on 2% of all counts. Maxima: Vaseux Lake 28 Dec 1989-**2**, all-time Canadian high count (Monroe 1990a); Vernon 18 Dec 1983-1. **Coastal** – Recorded from 1 of 33 localities and on less than 1% of all counts. Maximum: Victoria 27 Dec 1975-1.

Canyon Wren

Catherpes mexicanus (Swainson)

CAWR

RANGE: Resident from extreme southern interior British Columbia, eastern Washington, west-central Idaho, Wyoming, southeastern Montana, and southwestern South Dakota south (east of the coastal ranges) to southern Baja California and southern Mexico.

STATUS: In the interior, an *uncommon* local resident in the southern Okanagan valley of the Southern Interior Ecoprovince; *very rare* in the southern west Kootenay of the Southern Interior Mountains Ecoprovince; *accidental* or *casual* elsewhere in those ecoprovinces. Breeds.

CHANGE IN STATUS: The Canyon Wren (Fig. 340) reaches the northern limit of its range in southern British Columbia. It was first reported in the province at Vaseux Creek in March 1909 (Brooks 1909a). Munro and Cowan (1947) note it as a resident species in the southern part of the Okanagan valley from Osoyoos to Penticton, although it had also been reported north of Naramata (Plowden-Wardlaw 1944).

During the 1960s, the Canyon Wren was observed at several localities outside its normal range in the Okanagan valley. One or 2 birds were recorded at Westbank (1960), Hedley (1963), Vernon (1966), Kelowna (1968), and Kamloops (1968). None of these occurrences appear to have led to the establishment of populations. The Hedley record is still the only one in the Similkameen River valley north of the United States border.

Following a severe winter in 1968-69, when the mean minimum January 1969 temperature at Oliver was –28.6°C, the Canyon Wren was virtually extirpated from the Southern Interior (Fig. 341). Cannings et al. (1987) state that not one Canyon Wren was reported in its former haunts in the Okanagan valley during 1969, despite many attempts to find the species, although a bird was found dead near Rock Creek on 20 December 1969. Indeed, none was seen in the Okanagan valley until August 1970.

The Canyon Wren has recolonized most of its former range in the Okanagan valley in the years since the "big freeze." In late 1970, it was again reported from Vaseux Lake. It reached Penticton by 1976 and Naramata by 1977. The origin of the new population is unknown, but the quick reappearance of the Canyon Wren north to Naramata suggests that a few birds may have survived the general winter kill. It is also probable that there was immigration from populations in Washington.

Figure 340. Adult Canyon Wren singing (Vaseux Lake; Stephen R. Cannings).

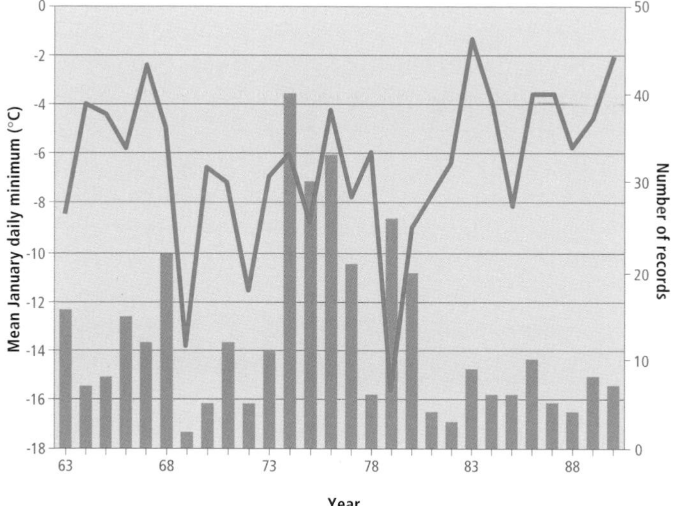

Figure 341. Fluctuations in the number of Canyon Wren records from British Columbia (bars) compared to winter temperatures (line) in the south Okanagan valley from 1963 to 1990. Temperature is the mean January minimum at Oliver (modified from Cannings et al. 1987).

Figure 342. In British Columbia, the highest numbers for the Canyon Wren in both winter (black) and summer (red) occur in the Southern Interior Ecoprovince.

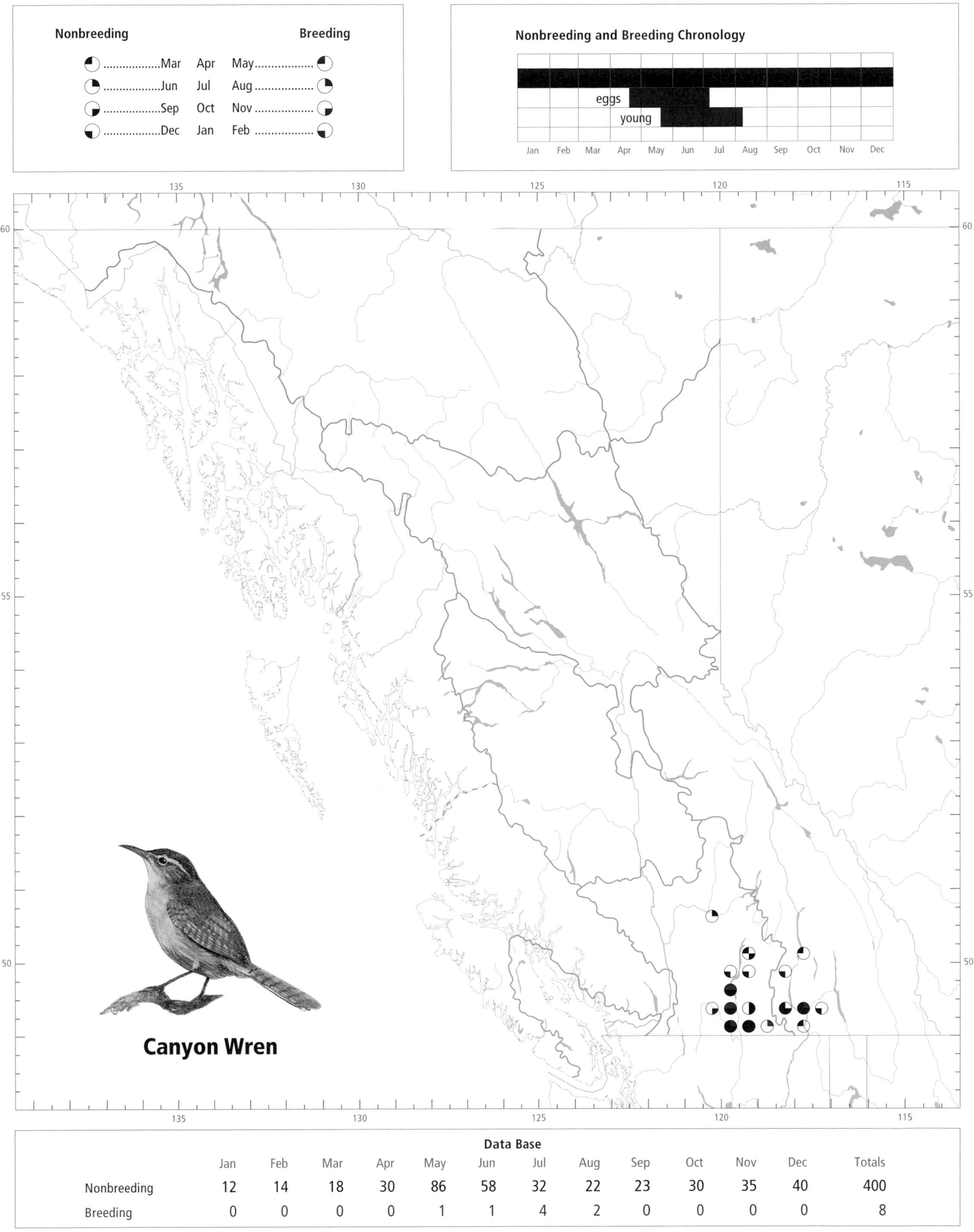

Data Base

	Jan	Feb	Mar	Apr	May	Jun	Jul	Aug	Sep	Oct	Nov	Dec	Totals
Nonbreeding	12	14	18	30	86	58	32	22	23	30	35	40	400
Breeding	0	0	0	0	1	1	4	2	0	0	0	0	8

The Canyon Wren was first reported in the west Kootenay in 1963 and 1966, just north of Christina Lake, and then at Castlegar (1968, 1971), Brilliant (1969), and Nelson (1971). All but 1 of the west Kootenay records from the 1960s were autumn occurrences. None of those birds was found to overwinter. In 1983, 1990, and 1991, Canyon Wrens were found in spring and summer at Deer Park and Syringa Creek, and a pair nested at Syringa Creek in 1992. At Christina Lake, a single bird was monitored monthly from January to December 1990. These events in the west Kootenay, including the tenuous establishment of what may be a small resident population, are the most important changes in status since 1947.

NONBREEDING: The Canyon Wren (Fig. 340) is locally distributed mainly in the Okanagan valley of the Southern Interior between Osoyoos and Naramata, where over 90% (n = 402) of our records are from. North of Naramata and south of Vernon, the Canyon Wren has occurred occasionally, mainly in the autumn, but has not established wintering populations. In the Similkameen River valley, although there is apparently suitable habitat, there are only 2 known localities where the Canyon Wren occurs: Hedley and near the border crossing at Nighthawk, at the south end of Richter Pass. A single record from Kamloops is its northernmost occurrence.

In the Southern Interior Mountains, the Canyon Wren occurs in arid parts of the Columbia River from Deer Park to Castlegar and near Christina Lake. The highest numbers in winter are in the Southern Interior (Fig. 342), especially in the vicinity of Vaseux Lake. Winter populations fluctuate (Fig. 343).

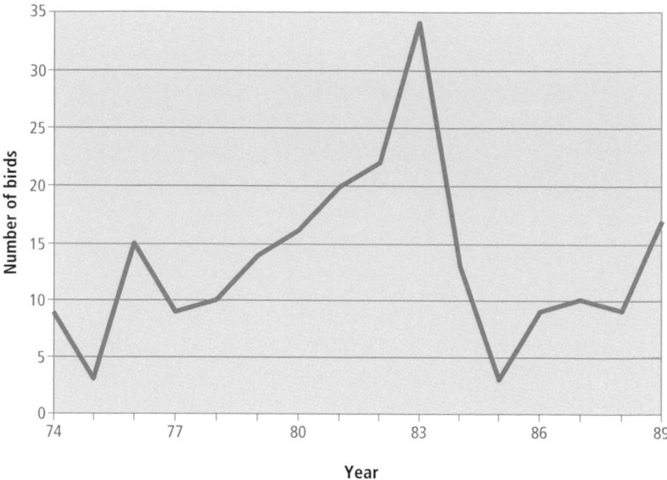

Figure 343. Fluctuations in Canyon Wren numbers on Christmas Bird Counts in the Okanagan valley from 1974 to 1989. The number of birds is the sum of the totals reported on the Oliver-Osoyoos, Vaseux Lake, and Penticton Christmas Bird Counts for each year.

Figure 344. Typical Canyon Wren habitat of rocky bluffs in ponderosa pine parklands (near Vaseux Lake, 20 June 1996; Neil K. Dawe).

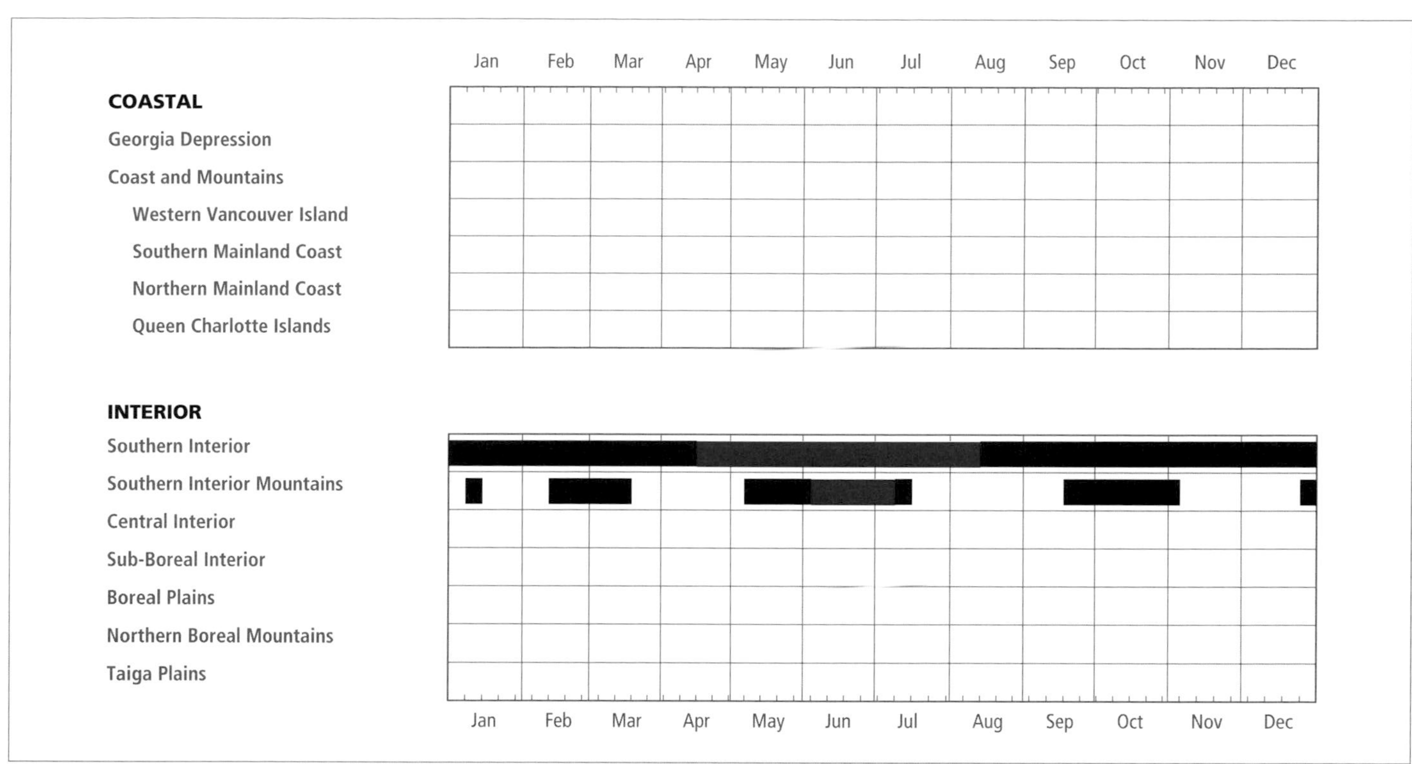

Figure 345. Annual occurrence (black) and breeding chronology (red) for the Canyon Wren in ecoprovinces of British Columbia.

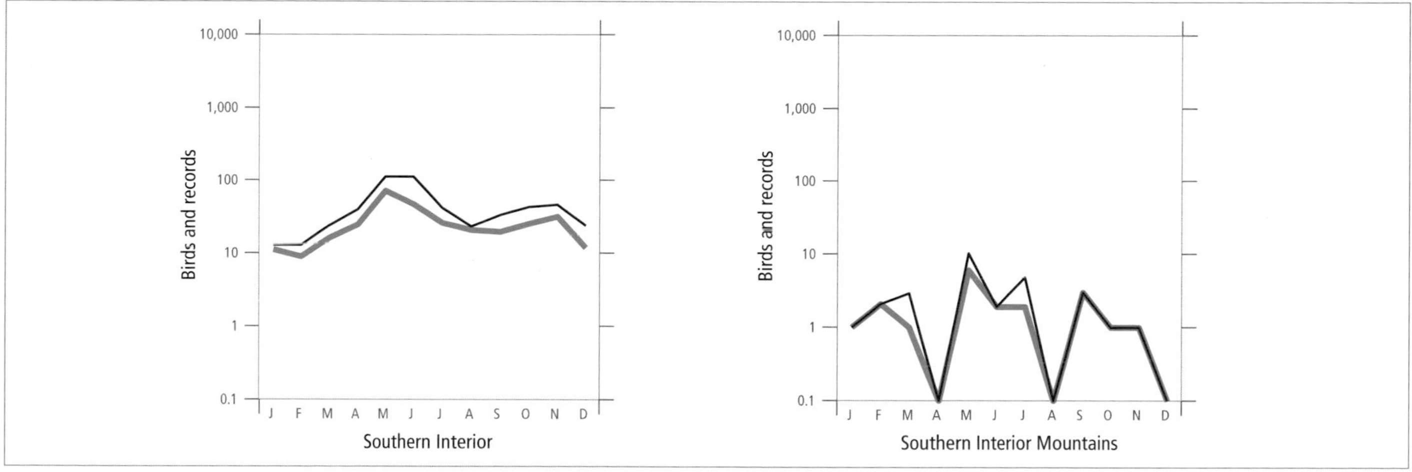

Figure 346. Fluctuations in total number of birds (purple line) and total number of records (green line) for the Canyon Wren in the Southern Interior and Southern Interior Mountains ecoprovinces of British Columbia. Christmas Bird Count and nest record data have been excluded.

For a description of habitat, see BREEDING (Fig. 344). There is no indication of regular migratory movements in British Columbia (Figs. 345 and 346), but post-breeding wanderers occasionally move beyond the Canyon Wren's normal range during late summer and autumn. Two-thirds of all records are of single birds and another 20% are of pairs. Small parties of up to 6 birds recorded in late summer and autumn are probably family groups. Winter flocks are uncommon, but a loose group of 8 birds occupied a cliff on the east shore of Vaseux Lake on 21 December 1977. Small numbers winter regularly in the Okanagan valley from Penticton to Osoyoos, and single birds have been seen in winter near Syringa Creek in the west Kootenay.

The Canyon Wren regularly occurs throughout the year in the Southern Interior; it is found irregularly year-round in the Southern Interior Mountains (Fig. 345).

BREEDING: The Canyon Wren breeds mainly in the southern Okanagan valley. All but 2 (n = 19) confirmed nesting records are from the vicinity of Vaseux Lake (10 records), Naramata (5), and Oliver (2). The other 2 known breeding localities are Rock Creek and north of Castlegar, at Syringa Creek.

The Canyon Wren reaches its highest numbers in summer in the Southern Interior (Fig. 342). Breeding Bird Surveys for interior routes for the period 1968 through 1993 contain insufficient data for analysis.

The Canyon Wren frequents habitat that is used by few other bird species, including fractured cliff faces (Fig. 347), talus slopes, rocky bluffs (Fig. 344), gorges, canyon walls with large boulders (Fig. 348), or rock outcroppings where there is little vegetation. These sheer or steep rocky habitats are interspersed with open, patchy forest of ponderosa pine and Douglas-fir or shrublands featuring big sagebrush, rabbitbrush, and antelope-brush. Suitable habitat is scattered throughout the Okanagan valley, but is concentrated from Kelowna south to Vaseux Lake at elevations between 275 and 435 m.

The Canyon Wren is a secretive bird that frequents shady crannies, caves, and rock crevices, where it forages for insects, spiders, and other invertebrates. It usually remains out of sight. Lower on the slopes, it creeps among the larger talus and appears occasionally to call from a boulder top. Its loud, clear, musical song is one of the most haunting sounds of spring in the Okanagan valley. Although Godfrey (1986) states that there is a preference for canyon walls along water courses, this is not typical of habitat used in British Columbia.

In the interior, the Canyon Wren has been recorded breeding from 19 April (calculated) to 7 August (Fig. 345).

Nests: Only 5 nests have been found and described in British Columbia. One nest was on a ledge below an overhanging cliff, with some large boulders lying in front of the cliff. One each was found in a rock "chimney" on a cliff face, on a narrow ledge on a cliff face, on top of a stone that was wedged into a cleft, and in an abandoned railroad tunnel.

Nests were described as bulky cups of twigs with lichens, moss, and fine grass; some had a base of coarse, dry plant stems. One nest was lined with fiberglass, tissue paper, fabric, string, yarn, deer hair, and a Northern Flicker feather. Elsewhere in its range, the Canyon Wren is known for its versatility as a nest builder. Nest cups often include large amounts of spider webs and wool, and a variety of human-made artifacts are known to have been used (Bent 1948). The heights for 3 nests ranged from 2.8 to 4.5 m above the cliff base.

Nest building can begin significantly earlier than the earliest dates known for eggs. The earliest date recorded was 17 March, when a partially completed nest was found at Vaseux Lake (Cannings et al. 1987).

Eggs: Dates for 3 clutches ranged from 28 June to 6 July. Calculated dates, based on an estimate of 30 days for the incubation and nestling period, indicate that eggs may be present as early as 19 April. Clutch size ranged from 1 to 7 eggs (1E-1, 5E-1, 7E-1). Clutch sizes are poorly documented anywhere, but usually contain 5 or 6 eggs (Jones and Dieni 1995). The incubation period is 12 to 18 days, with a mean of 16 days (Jones and Dieni 1995).

Young: Dates for 4 broods ranged from 19 May to 7 August (brood partially fledged). Recently fledged young have been found as early as 19 and 21 May and as late as 7 August. Calculated dates, based on an estimated nestling period of at least 15 days, indicate that nestlings can occur as early as 4 May. Sizes of 9 broods ranged from 2 to 4 young (2Y-3, 3Y-4, 4Y-2). Elsewhere in its range, the Canyon Wren may raise 2 broods in a season (Bent 1948). Although direct evidence for double-brooding in British Columbia is lacking, records of recently fledged young in all months from May to August suggest that some pairs produce 2 broods annually. The nestling period is 12 to 17 days, with a mean of 15 days (Jones and Dieni 1995).

Brown-headed Cowbird Parasitism: Cowbird parasitism was not found in British Columbia in 4 nests found with eggs

Figure 347. Fractured cliff faces with large rock rubble are good habitat for the Canyon Wren (Syringa Creek, 14 November 1993; R. Wayne Campbell).

Figure 348. Aggregations of large boulders provide year-round cover for the Canyon Wren (Haines Ecological Reserve, north of Oliver, 4 August 1993; R. Wayne Campbell).

or young, nor in 8 fledged broods. There are no cases of parasitism reported for North America (Friedmann 1963; Friedmann and Kiff 1985).

Nest Success: Only 1 nest was found with eggs and followed to a known fate; it was preyed upon.

REMARKS: The British Columbia population of Canyon Wren belongs to the subspecies *C. m. conspersus* (American Ornithologists' Union 1957).

The Canyon Wren population in British Columbia is small; often fewer than 5 breeding pairs are reported in a year. Preston (1990), in a systematic survey of potential habitats in the southern Okanagan valley, counted 43 singing males and another 8 birds at 40 sites between 15 May and 11 July 1990. Cannings (1992), after considering these counts and using knowledge of apparently suitable habitats that had not been included in the survey, estimated a population of about 100 pairs in the Okanagan valley. He also estimated 5 pairs in the Similkameen valley and "perhaps 10 to 15 pairs in the Castlegar area," estimates that we think are too high. In the same paper, he also suggested that the Canadian (and British Columbian) Canyon Wren population probably fluctuates between zero (after very severe winters) and 300 or more birds (in late summer and autumn after a series of milder winters).

The history of the Canyon Wren in British Columbia demonstrates post-breeding dispersal as a behaviour of the population during the periods when it is apparently near capacity in the southern Okanagan valley. This dispersal behaviour is probably a way by which the species probes the available habitat which may permit its breeding range to expand. Thus the area of the province within which the species breeds is surrounded by areas from which there are scattered records of single birds that are almost always seen during the late summer and autumn. Given the cryptic nature of the species, many more of its probings must have been unreported. During some 90 years that this wren has been under observation in the province, it appears to have had little success in discovering new nesting habitats. The only known instance is from the Syringa Creek area in the Columbia valley, where nesting took place in 1992.

Since this wren is highly localized, a search of potential habitats throughout the Okanagan valley, along the lower Kettle River valley, near Christina Lake, and in the Castlegar area may yield better estimates of populations. Given the sensitivity of the species to cold winter temperatures, a survey of all suitable habitats during a single breeding season should be carried out if accurate estimates are to be made.

The available information on the Canyon Wren in British Columbia shows the difficulty of managing a species at the northern limit of its range. Even when suitable habitat exists and a population has been present for many years, unusually extreme winter weather can almost wipe out the population. It is likely that little can be done to enhance populations other than ensuring that suitable habitat is maintained.

As Jones and Dieni (1995) note, the Canyon Wren has been little studied. Naturalists who can record details of the Canyon Wren's life history, behaviour, and habitat requirements in British Columbia could provide important conservation data.

Formerly known as Cañon Wren.

NOTEWORTHY RECORDS

Spring: Coastal – No records. **Interior** – Osoyoos 21 May 1941-1 (RBCM 8407); Anarchist Mountain 2 May 1974-3; Haynes Point Park 18 May 1990-7 (Preston 1990); Christina Lake 21 May 1980-1, 22 May 1966-1; Vaseux Creek Mar 1909-1 (NMC 14185; Brooks 1909a); Vaseux Lake 17 Mar 1979-nest almost complete (Cannings et al. 1987), 6 Apr 1975-4, 15 May 1990-8, 18 May 1977-6, 19 May 1990-4 or 5 fledglings (Siddle 1990b), 31 May 1979-4; Deer Park 13 to 14 Mar 1971-3, 15 Apr 1983-1, 25 May 1991-1; Nelson 8 May 1971-1 singing; near Naramata 21 May 1990-2 fledglings; Summerland 19 Apr 1964-6; Cosens Bay 3 Apr 1966-1, 8 May 1994-1 singing.

Summer: Coastal – No records. **Interior** – Nighthawk (s Richter Pass) 16 Jun 1990-3 on cliffs w of border crossing; Osoyoos 14 Jul 1979-1 (RBCM 8107); Anarchist Mountain 31 Jul 1976-1 at highway lookout; Rock Creek 7 Jul 1963-3 fledglings with 2 adults; 10 miles s Oliver 9 Jul 1963-3; Gallagher Lake 9 Jun 1990-3 fledglings with parent (Preston 1990); McIntyre Bluff 21 Jul 1979-4 fledglings, 23 Jul 1979-4 nestlings left nest, late Jul 1975-nest with 1 egg; Vaseux Lake Jun 1976-1 adult (BC Photo 983), 21 Jun 1979-4 fledglings just out of nest with 1 parent, 4 Jul 1974-5; Syringa Creek 23 Jun 1990-1, 2 Jul 1988-2, 10 Jul 1992-2 fledglings with 2 adults; Penticton 27 Jun 1936-2 (Munro and Cowan 1947); 4 km n Naramata 28 Jun 1988-4 eggs, 5 Jul 1988-5 eggs, 7 Aug 1984-2 nestlings and 2 fledglings; Castlegar 8 Jul 1971-1; Kamloops 1 Jul 1968-1 (BC Photo 701).

Breeding Bird Surveys: Coastal – Not recorded. **Interior** – Recorded from 2 of 73 routes and on less than 1% of all surveys. Maxima: Syringa Creek 24 Jun 1990-3; Needles 18 Jun 1994-2.

Autumn: Interior – Cosens Bay 10 to 13 Oct 1981-1; Deer Park 4 Oct 1970-1; Hedley 16 Oct 1963-2; White Lake (Okanagan Falls) 14 Nov 1964-1; Brilliant 19 Oct 1969-1; Castlegar 26 Sep 1968-1; Vaseux Lake 4 Oct 1974-6, 29 Oct 1930-1 (Munro and Cowan 1947), 15 Nov 1974-3; near Christina Lake 17 Sep 1963-2; Syringa Creek 3 Nov 1987-1. **Coastal** – No records.

Winter: Interior – Adventure Bay 25 Jan 1985-1 (Cannings et al. 1987); Cosens Bay 7 Dec 1984-1 (Cannings et al. 1987); Lambly Creek 21 Dec 1968-1 (Cannings et al. 1987); Westbank Dec 1960-1; Fauquier 31 Dec 1994-1; Skaha Lake 29 Dec 1978-3 on e side; Deer Park 19 Feb 1971-1; Syringa Creek 14 Jan 1990-1, 15 Feb 1983-1 singing (Siddle 1990a); Shuttleworth Creek 20 Jan 1979-1; Vaseux Lake 7 Dec 1979-4, 21 Dec 1978-8, 31 Dec 1962-4; Rock Creek 20 Dec 1969-1 (RBCM 18894). **Coastal** – No records.

Christmas Bird Counts: Interior – Recorded from 3 of 27 localities and on 18% of all counts. Maxima: Vaseux Lake 23 Dec 1983-**21**, all-time Canadian high count (Monroe 1984a); Oliver-Osoyoos 28 Dec 1979-11; Penticton 27 Dec 1983-7. **Coastal** – Not recorded.

Bewick's Wren

Thryomanes bewickii (Audubon)

BEWR

RANGE: Resident from southwestern British Columbia, including southern Vancouver Island, south along the coastal slope to southern California and Baja California. Also breeds across the middle of the continent from southern Wyoming and eastern Nebraska to Michigan, southern Ontario, and southeastern New York south to central Texas, northern Arkansas, Georgia, South Carolina, and the northern portion of the Gulf states. Winters throughout most of its breeding range. Resident in parts of the Mexican highlands.

STATUS: On the coast, a *fairly common* resident in the Georgia Depression Ecoprovince; *uncommon* resident at lower elevations in the southern portions of the Coast and Mountains Ecoprovince; *very rare* on Western Vancouver Island in the Coast and Mountains. Breeds.

NONBREEDING: In British Columbia, the Bewick's Wren (Fig. 349) occurs almost exclusively in the Georgia Depression, where 99% of all observations have been recorded. It has not been reported from the coast north of Cortes Island. There are only 2 records from the interior, both from Manning Park.

On Vancouver Island, the Bewick's Wren occurs regularly from Sooke and Victoria along the east coast, north to Campbell River and Cortes Island. It occurs less regularly west to Port Renfrew, Bamfield, Port Alberni, and Long Beach. On the mainland, it occurs regularly in the lower Fraser River valley from the coast east to Cultus and Harrison lakes. The easternmost records are from the western slope of Manning Park, and the northernmost are from the Sechelt Peninsula, Pemberton, and Emory Creek. The highest numbers in winter occur in the Georgia Depression, specifically near Victoria and on the Fraser River delta from Ladner to White Rock (Fig. 350).

The Bewick's Wren is a bird of low elevations that occurs mainly between sea level and 100 m. About 90% of all records for this wren in the nonbreeding season are within 100 m of sea level. Where roads and subdivisions have created suitable habitat, it can reach higher elevations. There is 1 record from 1,500 m in Garibaldi Park.

The Bewick's Wren is a bird of edge habitats where coniferous forest gives way to a tangled understorey of shrubs (Fig. 351), trailing blackberry, salal, western flowering dogwood, cascara, willow, pink spirea, thickets of Nootka rose, and Himalayan blackberry; and of brush areas along the borders of fields, powerline cuts, around swampy ponds or sloughs, river and creek banks, lakeshores, and road edges. It is also a common visitor to brushy parks, residential gardens, and bird feeders.

During nonbreeding seasons it is mainly a solitary bird, except for resident pairs, which often occur together. Like most secretive, ground-dwelling birds, the Bewick's Wren is more noticeable during the spring, when males are singing.

There is no evidence that this wren is migratory in British Columbia. It winters throughout its range on the Fraser River delta, along the Sunshine Coast, and on southeastern Vancouver Island, the Gulf Islands, and the extreme southern end of the south Coast and Mountains near Harrison Lake.

Figure 349. A Bewick's Wren with food for its young in a natural nest cavity in a living Garry oak tree (Oak Bay, May 1980; Mark Nyhof). Note the characteristic white eyebrow of the adult.

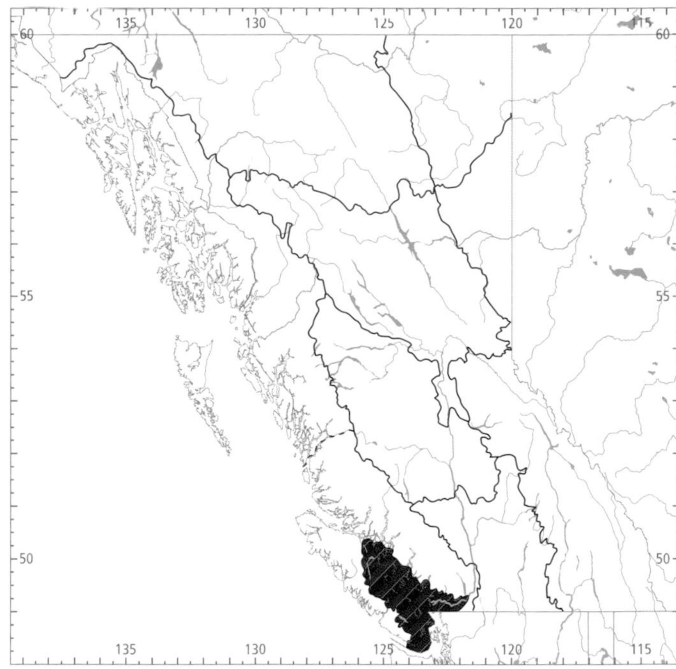

Figure 350. In British Columbia, the highest numbers for the Bewick's Wren in both winter (black) and summer (red) occur in the Georgia Depression Ecoprovince.

On the coast, the Bewick's Wren has been recorded throughout the year (Fig. 352).

BREEDING: The Bewick's Wren breeds throughout its primary range on southeastern Vancouver Island, north to Campbell River and Cortes Island, on the Gulf Islands, and on the Fraser Lowland east to Chilliwack. It probably also breeds in the Fraser Lowland east to Hope, and possibly north to Emory Creek, Squamish, and the Sechelt Peninsula.

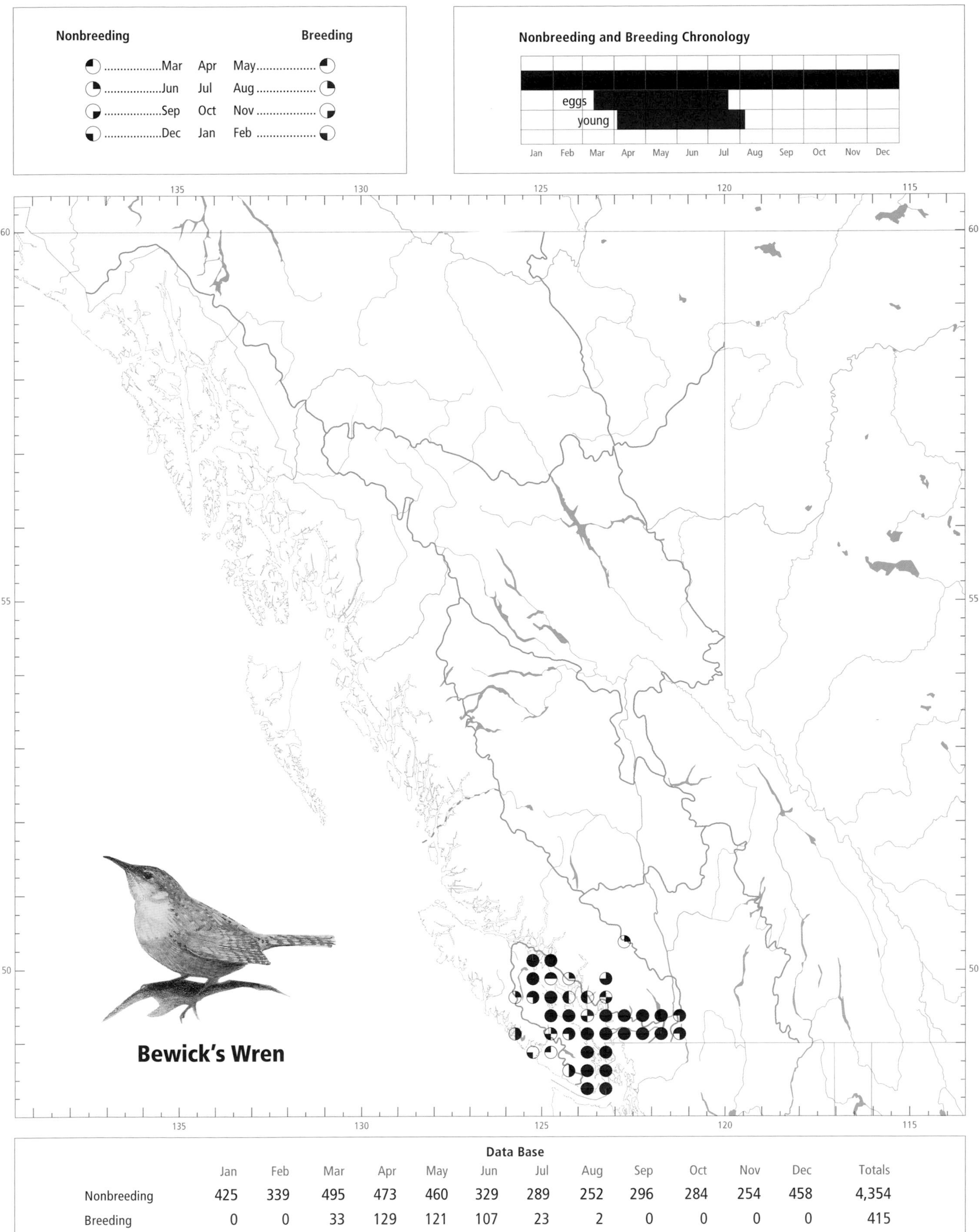

Data Base

	Jan	Feb	Mar	Apr	May	Jun	Jul	Aug	Sep	Oct	Nov	Dec	Totals
Nonbreeding	425	339	495	473	460	329	289	252	296	284	254	458	4,354
Breeding	0	0	33	129	121	107	23	2	0	0	0	0	415

Figure 351. Edges of coniferous forests with tangles and shrubs provide year-round habitat for the Bewick's Wren (Burnaby Lake, 27 March 1993; R. Wayne Campbell).

The highest numbers in summer occur in the Georgia Depression (Fig. 350). An analysis of Breeding Bird Surveys for the period 1968 through 1993 could not detect a net change in numbers on coastal routes.

Nesting occurs mainly below 100 m elevation. The highest elevation reported for breeding is 225 m. The Bewick's Wren is one species of songbird that can survive and breed in habitat that has been heavily altered by humans. Primary breeding habitat includes brushy, open woodlands in suburban and rural areas, or farmlands along the edges of forested areas (90%; n = 99; Fig. 351). Nearly 80% of nests (n = 92) were described as occurring in backyards, on farms, or in gardens. Although the Bewick's Wren remains near the ground most of the time, territorial males may perch at the top of trees as tall as 20 m when singing.

The Bewick's Wren has been recorded breeding in British Columbia between 17 March and 5 August (Fig. 352).

Nests: Most nests were found in buildings (47%; n = 135), including sheds, garages, barns, and living or dead trees (40%), including Douglas-fir, Garry oak (Fig. 349), western redcedar, various birches, red alder, and black cottonwood. Posts and poles (5%), abandoned automobiles (4%), banks, and shrubs were also used. It nested successfully in greenhouses, garages, or sheds that were actively used by humans, which suggests a relatively high tolerance for disturbance. The Bewick's Wren requires cavities such as abandoned woodpecker nests, natural cavities in trees, or nest boxes for nesting. In suburban and rural areas it often uses cavities created by cracks in buildings or openings in abandoned machinery and junk piles.

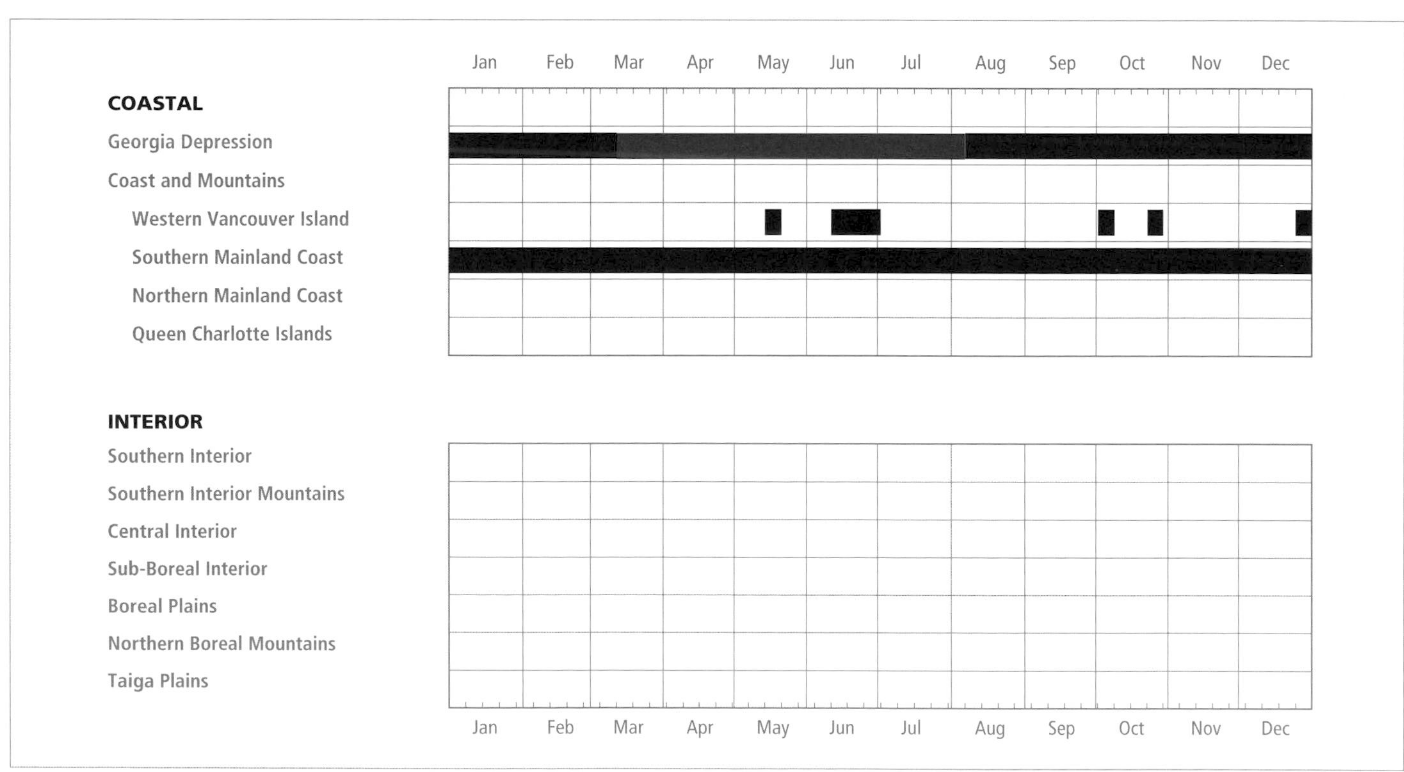

Figure 352. Annual occurrence (black) and breeding chronology (red) for the Bewick's Wren in ecoprovinces of British Columbia.

Specific sites included artificial cavities, containers, or nest boxes (42%; n = 157), natural cavities or crevices (34%), and rafters or other supports in buildings (15%). Some of the more unusual nest sites chosen by this adaptable wren included an enclosed space in a brush pile, the roots of upturned tree stumps, pockets or sleeves of clothing left hanging in abandoned buildings, saddlery and other bits of harness, garage drawers and cupboards left slightly ajar, behind a frying pan hanging on a post, mailboxes, a box of wood scraps under a sundeck, empty flowerpots, old Barn Swallow nests, and a paper bag half filled with nails. The heights for 118 nests ranged from 0.3 to 6.0 m, with 68% between 0.9 and 1.8 m.

In 126 Bewick's Wren nests, the 6 most frequently used materials were twigs (74%), grasses (49%), feathers (38%), moss (32%), leaves (28%), and hair (27%).

Eggs: Dates for 102 clutches ranged from 17 March to 15 July, with 51% recorded between 9 April and 30 May. Clutch size ranged from 1 to 7 eggs (1E-7, 2E-4, 3E-4, 4E-19, 5E-38, 6E-26, 7E-4), with 63% having 5 or 6 eggs. The incubation period in British Columbia is 15 to 16 days (n = 3). In California, the incubation period is reported as 14 days (Miller 1941).

Young: Dates for 66 broods ranged from 5 April to 5 August, with 53% recorded between 28 April and 15 June. Brood size ranged from 1 to 7 young (1Y-2, 2Y-5, 3Y-7, 4Y-23, 5Y-21, 6Y-7, 7Y-1), with 67% having 4 or 5 young. The nestling period in British Columbia is 18 to 22 days (n = 5). Miller (1941) gives the nestling period in California as 14 days. In British Columbia, the Bewick's Wren is often double-brooded and may use the same nest for successive broods (Pearse 1957).

Brown-headed Cowbird Parasitism: In 147 nests recorded with eggs or young in British Columbia, 2 were parasitized by the cowbird. In 1 nest, the adults reared a young cowbird to the fledging stage, but not their own young (Lemon 1969). Friedmann (1963, 1971) and Friedmann et al. (1977) note only 12 records of parasitism by cowbirds, but suggest that the Bewick's Wren may be more frequently parasitized than is known.

Nest Success: Of 36 nests found with eggs and followed to a known fate, 24 produced at least 1 fledgling, for a success rate of 67%.

REMARKS: The subspecies of Bewick's Wren that occurs in British Columbia is *T. b. calophonus* (American Ornithologists' Union 1957). This subspecies ranges from southwestern British Columbia south through western Washington and western Oregon.

The Bewick's Wren has been blue-listed in the eastern and central United States since 1972 (Ehrlich et al. 1988), principally because of declines in the Appalachian, Southern Atlantic Coast, and Central Southern regions. Breeding Bird Surveys show significant declines there for the period 1968 through 1979 (Robbins et al. 1986); those trends, however, did not appear in the west. Declines have been attributed to loss of woodlots and outbuildings and the use of agricultural pesticides on modern farms (Ehrlich et al. 1988). As well, the problem for wrens and many other small insectivorous birds probably lies in the poorly controlled use of domestic pesticides, combined with an increased abundance of feral and domestic cats.

NOTEWORTHY RECORDS

Spring: Coastal – Victoria 17 Mar 1984-1st egg of clutch, 2 May 1964-22 on census, 23 May 1986-12; Mayne Island 14 Apr 1974-4; Carnation Creek 17 May 1981-1; Surrey 23 Mar 1986-5 eggs, 5 Apr 1986-nestlings, 28 Mar 1966-66 on census; Matsqui 28 Apr 1987-4 young fledged; Chilliwack 18 Apr 1972-2 nestlings; Harrison Hot Springs 22 Mar 1980-10, 1 May 1986-4; Emory Creek 12 Apr 1968-4; Port Alberni 6 Apr 1988-1; Campbell River 14 May 1975-5 fledglings, 30 May 1975-10; Cortes Island 21 Apr 1958-4 eggs. **Interior** – No records.

Summer: Coastal – Elk and Beaver lakes (Victoria) 20 Aug 1978-9; White Rock 5 Aug 1978-4 young left nest; Langley 15 Jul 1977-4 eggs; Sumallo River 22 Jun 1986-6; Cultus Lake 3 Jul 1965-5; Stanley Park (Vancouver) 20 Jul 1961-2 eggs, 1 nestling; Protection Island 15 Aug 1976-2; Haney 30 Jul 1974-nestlings; Harrison Hot Springs 19 Jun 1978-5; Grouse Mountain 17 Aug 1979-1; Hope 16 Jun 1962-4 likely nesting (Kelleher 1963); Silver Sands Creek (Sechelt Peninsula) 2 Jul 1959-5 nestlings; Mitlenatch Island 30 Jun 1970-1; Saltery Bay 22 Jun 1975-1; Hernando Island 3 Aug 1970-2; Garibaldi Park 21 Aug 1982-1; Pemberton 22 Jun 1986-3 on Breeding Bird Survey. **Interior** – Manning Park (Skagit River trail) 28 Jun 1984-1, 29 Jun 1973-1.

Breeding Bird Surveys: Coastal – Recorded from 17 of 27 routes and on 46% of all surveys. Maxima: Victoria 30 Jun 1974-57; Chemainus 25 Jun 1979-21; Point Grey 16 Jun 1974-21; Nanaimo River 4 Jun 1983-18. **Interior** – Not recorded.

Autumn: Interior – No records. **Coastal** – Pender Harbour 13 Oct 1968-1; Brackendale 2 Sep 1916-1 (NMC 9563); Squamish 7 Nov 1982-1; Agassiz 20 Nov 1895-1 (Oberholser 1920); Chilliwack 29 Sep 1927-1 (RBCM 6600); Maple Ridge 12 Sep 1979-2; Cultus Lake 9 Oct 1922-1 (USNM 425263); Long Beach 26 Oct 1986-1; Boundary Bay 12 Nov 1962-15; Mayne Island 12 Oct 1974-1; Port Renfrew 5 Oct 1974-3; Jordan River area 30 Nov 1986-1; Victoria 6 Oct 1978-17.

Winter: Interior – No records. **Coastal** – Cortes Island 31 Dec 1975-1; Campbell River 1 Feb 1975-4 on estuary; Comox 26 Dec 1949-2; Squamish 2 Jan 1988-1; Harrison Hot Springs 29 Feb 1980-12, 13 Dec 1980-3; Alouette Lake 16 Jan 1966-5; Chilliwack 4 Dec 1889-1 (MCZ 102159), 15 Jan 1889-1 (ROM 51872); Sumas Mountain 10 Jan 1976-1; Green Timbers (Surrey) 15 Feb 1966-12; Boundary Bay 12 Nov 1962-15; Bamfield 27 Dec 1987-1; Mayne Island 5 Dec 1970-2, 25 Feb 1973-2; Victoria 13 Dec 1988-3.

Christmas Bird Counts: Interior – Not recorded. **Coastal** – Recorded from 21 of 33 localities and on 69% of all counts. Maxima: Victoria 19 Dec 1987-**218**, all-time Canadian high count (Monroe 1988a); Ladner 26 Dec 1977-155; Vancouver 18 Dec 1976-93.

House Wren HOWR

Troglodytes aedon Vieillot

RANGE: Breeds from northeastern British Columbia, northern Alberta, and central Saskatchewan east to New Brunswick, south through southern British Columbia to northern Baja California, and across the southern United States from California to South Carolina. Winters across the United States from California to Florida; occasionally further north. Resident in southern California and much of Arizona.

STATUS: On the coast, *uncommon* to *fairly common* migrant and summer visitant on southeastern Vancouver Island in the Georgia Depression Ecoprovince; *rare* to *uncommon* migrant and summer visitant on the Fraser Lowland of the Georgia Depression, *accidental* there in winter; *casual* on the Southern Mainland Coast and *accidental* on the Northern Mainland Coast of the Coast and Mountains Ecoprovince.

In the interior, *fairly common* migrant and summer visitant in the Southern Interior Ecoprovince; *uncommon* to *fairly common* in the southern portions of the Southern Interior Mountains and Boreal Plains ecoprovinces; *very rare* in the southern Central Interior, the Sub-Boreal Interior, and Taiga Plains ecoprovinces.

Breeds.

NONBREEDING: The House Wren (Fig. 353) has a widespread distribution during spring and summer across extreme southern British Columbia from the south and east coast of Vancouver Island north to Campbell River, east through the lower Fraser River valley to Manning Park and north to the Nicola valley; through the Okanagan valley, north to Adams Lake and Clearwater (Dickinson 1953); and through southern portions of the west and east Kootenay. In the Central Interior, there are fewer than 20 records, all from the area between 100 Mile House, Williams Lake, and Riske Creek. A specimen from the Kispiox valley is the only record from the Northern Mainland Coast of the Coast and Mountains Ecoprovince (Swarth 1924; Brooks and Swarth 1925). In the Sub-Boreal Interior, there are 4 known localities where this wren has occurred: Chetwynd, Prince George (Godfrey 1986), Quesnel, and Smithers. The House Wren also occurs regularly in the Peace Lowland, and locally but fairly regularly in the Fort Nelson area.

On the coast, it occurs mainly at elevations from sea level to about 300 m, but there is 1 record at about 1,400 m. In the interior, it has been found up to 1,200 m.

The House Wren is a bird of drier habitats. In the southern interior, it is most abundant in deciduous groves within ponderosa pine forests (Cannings et al. 1987), but also occurs in relatively open stands of trembling aspen, as it does in the northeast. On Vancouver Island and the Gulf Islands, it is most numerous along the edges of dry forests featuring arbutus, Garry oak, and Douglas-fir (Fig. 355), particularly in brushy areas at the forest edge or along sun-baked rocky bluffs that border the sea. Specific habitats include hedgerows, bramble tangles, thickets of Scotch broom and gorse, open brushy areas along roads and powerline rights-of-way, and thickets around ponds (Fig. 355) and seashore. Throughout its range, it occurs frequently in orchards, gardens, and abandoned homesteads.

Figure 353. Adult male House Wren (Arawana, 19 June 1993; R. Wayne Campbell).

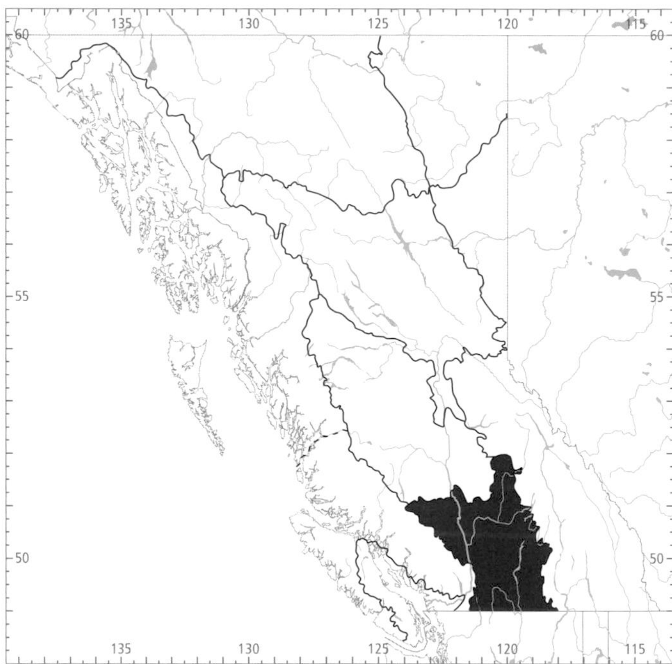

Figure 354. In British Columbia, the highest numbers for the House Wren in summer occur in the Okanagan valley of the Southern Interior Ecoprovince.

On the coast, spring migration begins slowly in early April, rarely in late March, and peaks in May (Figs. 356 and 357). East of the coastal mountain ranges, the House Wren arrives after the first week of April and numbers continue to increase into May and June. In the Peace Lowland, spring migrants arrive in May and numbers peak in late May and early June. The northeastern population is continuous with that of western Alberta (Semenchuk 1992), and reaches that area via a migration route east of the Rocky Mountains. West

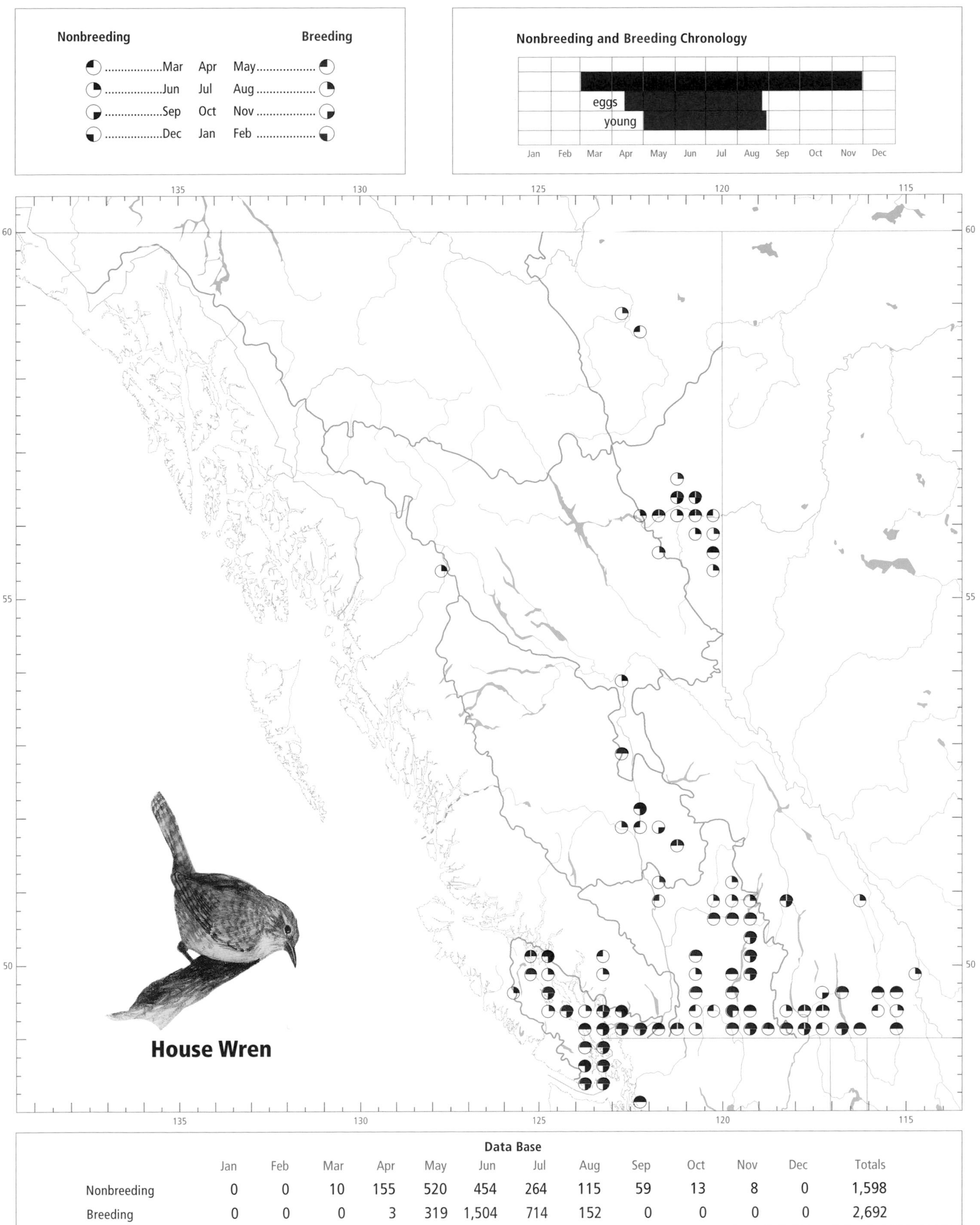

Data Base

	Jan	Feb	Mar	Apr	May	Jun	Jul	Aug	Sep	Oct	Nov	Dec	Totals
Nonbreeding	0	0	10	155	520	454	264	115	59	13	8	0	1,598
Breeding	0	0	0	3	319	1,504	714	152	0	0	0	0	2,692

of the Rocky Mountains, the House Wren migrates through intermountain valleys.

In the Peace Lowland, the southward migration occurs mainly in August, and most birds have gone by mid-September. In the Southern Interior, the autumn movement can be detected as early as late July but most birds leave in August, with few reported in September and only casual occurrences thereafter. On the coast, the autumn migration is more protracted, occurring in August and September, with only a few individuals left in October and November.

There is only 1 convincing late winter record for the House Wren in British Columbia: a specimen from Chilliwack (see also REMARKS). Root (1988) states that the House Wren winters along the Pacific coast mostly where the average January minimum temperature does not drop below –1°C. Our data suggest that even the relatively mild winters on the coast of British Columbia are not suitable for this wren.

On the coast, the House Wren has been recorded regularly from 27 March to 28 October; in the interior, it has been recorded from 13 April to 6 October (Fig. 356).

BREEDING: The House Wren breeds throughout most of its summer range, from southeastern Vancouver Island across the extreme southern portions of the province to the foothills of the Rocky Mountains. A disjunct population breeds in the Peace River region and at Fort Nelson. There are many nesting records from the valleys of the Thompson, Okanagan, and Similkameen rivers; in the Columbia River valley near Revelstoke, and the southern west Kootenay; and near Cranbrook. The northernmost breeding records west of the Rocky Mountains are near 100 Mile House, Williams Lake, Quesnel, and Prince George.

Figure 355. Open Garry oak and Douglas-fir forest, with dense shrub undergrowth surrounding ponds, is typical habitat for the House Wren on southeastern Vancouver Island (Langford, 11 May 1994; R. Wayne Campbell).

The highest numbers in summer occur in the Southern Interior, especially in the Okanagan valley (Fig. 354). House Wrens are also locally common breeders on southeastern Vancouver Island, the Gulf Islands, and the Peace Lowland. On the lower Fraser River delta, within a few kilometres of the sea, there are many records of House Wrens in spring and summer, but few nesting records.

Beyond the centre of its range, the House Wren may breed irregularly. In the Prince George area, this wren seemed to disappear in the 1970s, but had returned to breed by 1993 (J. Bowling pers. comm.). In the Arrow Lakes valley north of Castlegar, House Wrens were not recorded from 1975 to 1990, but several singing males were found in the early 1990s (G.S. Davidson pers. comm.).

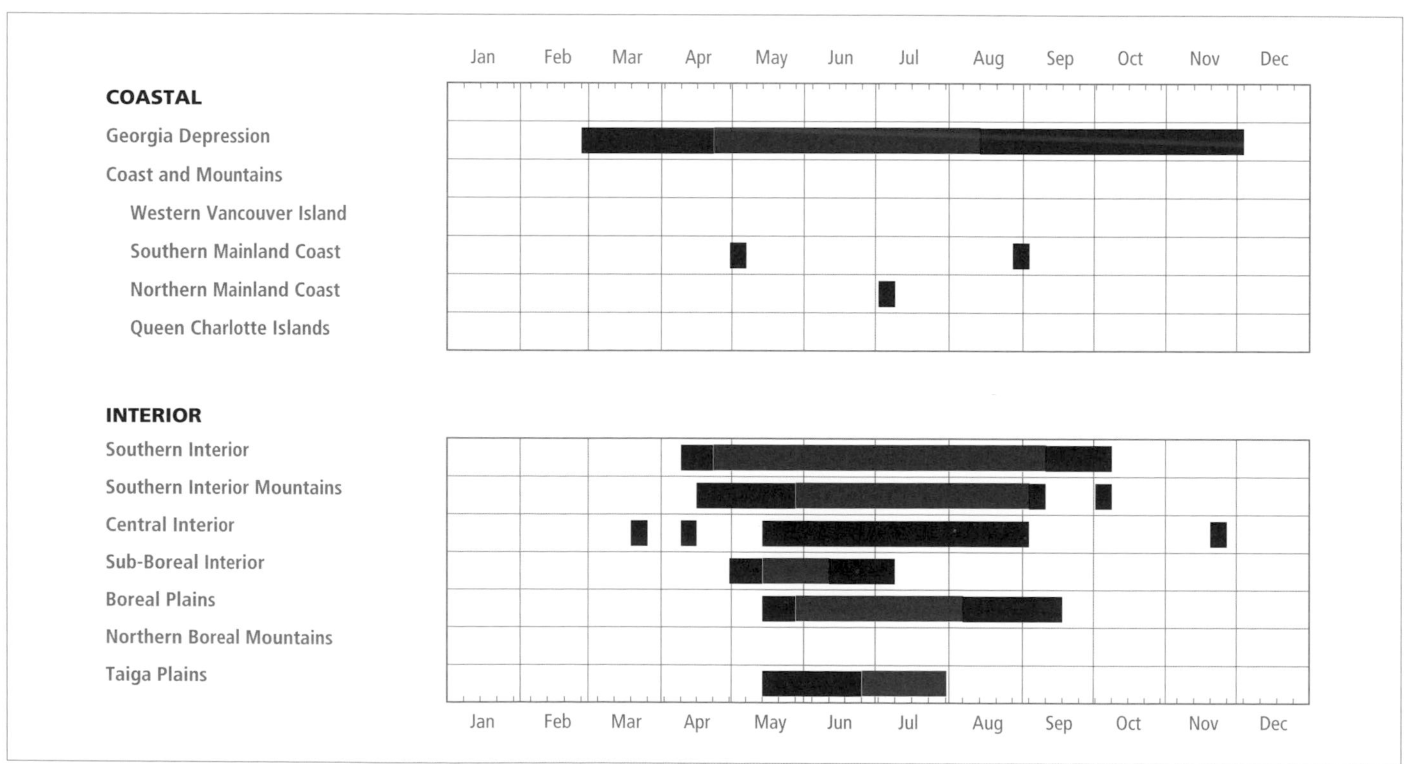

Figure 356. Annual occurrence (black) and breeding chronology (red) for the House Wren in ecoprovinces of British Columbia.

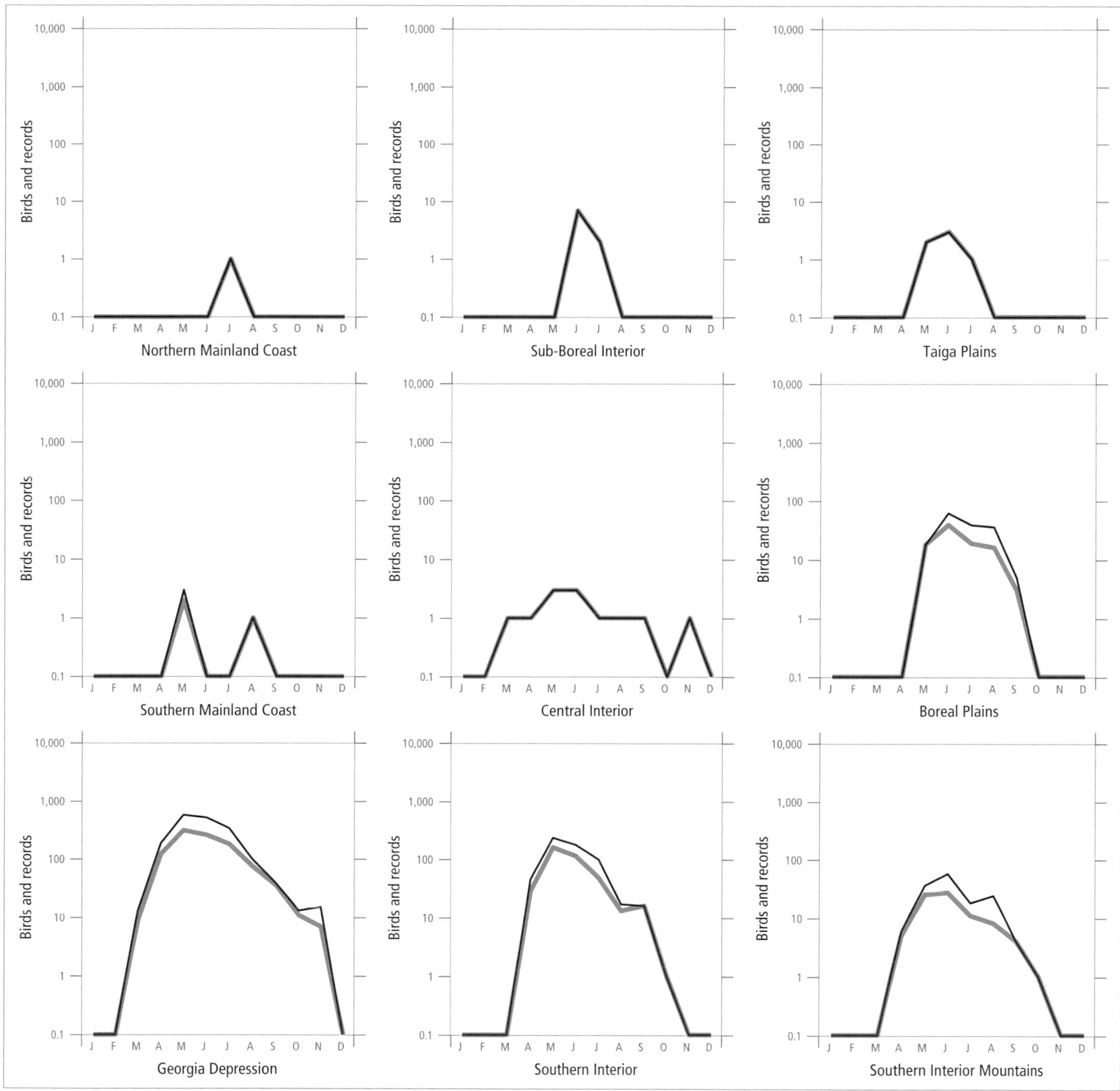

Figure 357. Fluctuations in total number of birds (purple line) and total number of records (green line) for the House Wren in ecoprovinces of British Columbia. Christmas Bird Count and nest record data have been excluded.

An analysis of Breeding Bird Surveys for the period 1968 through 1993 shows that the number of birds on interior routes has increased at an average annual rate of 4.0% (Fig. 358). Data for coastal routes were insufficient for analysis.

In the interior, breeding has been recorded at elevations from 270 m to about 1,150 m; on the coast, from near sea level to 390 m. Most House Wren nests were reported from human-altered habitats (79%; *n* = 832) and forests (12%; Fig. 359), but this summary is biased by the predominance of nests studied at 1 rural residence in the Okanagan valley. Deciduous and mixed woodlands were preferred to coniferous forest.

Specific nesting habitats included brushy backyards or farmyards (59%; *n* = 812), pasture and open rangeland (13%), woodlands (11%; Fig. 359), orchards (8%), and riparian situations (6%; Fig. 355). Other sites included overgrown fields with brambles, fence rows, and railway rights-of-way.

In the interior, the House Wren has been recorded breeding from 26 April (calculated) to 28 August; on the coast, it has been recorded from 25 April (calculated) to 12 August (Fig. 356).

Nests: Most nests were found in nest boxes (84%; *n* = 942; Figs. 360 and 361), cavities such as abandoned woodpecker nests or natural cavities left by the rotting of branches (13%), and crevices (2%). House Wrens often used nest boxes set out for bluebirds or swallows. Specific sites for non–nest box nests

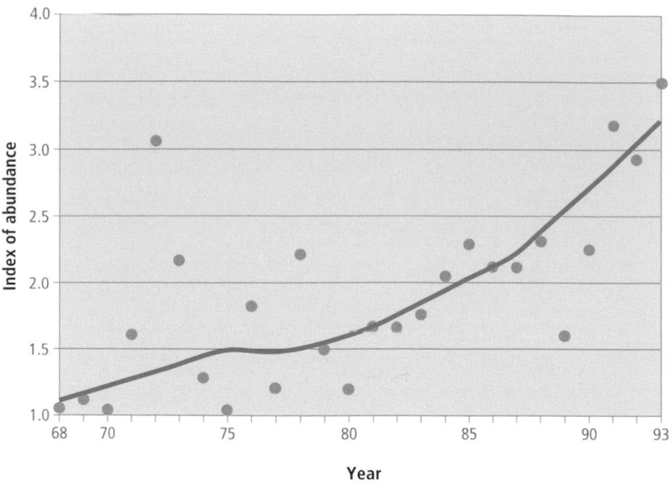

Figure 358. An analysis of Breeding Bird Surveys for the House Wren in British Columbia shows that the mean number of birds on interior routes increased at an average annual rate of 4.0% over the period 1968 through 1993 ($P = 0.10$).

Figure 359. Pockets of trembling aspen in open rangeland provide breeding habitat for the House Wren in the southern portions of the interior of British Columbia (Kilpoola Lake, 1 May 1994; R. Wayne Campbell).

(*n* = 130) included deciduous trees (58%), stumps (16%), coniferous trees (13%), and buildings (12%). The heights for 139 nests in natural cavities ranged from 0.9 to 15 m, with 55% between 1.8 and 4.5 m.

Nest materials included mainly twigs (99%; *n* = 813; Fig. 361), feathers (65%), grass (24%), hair (3%), moss, and occasionally spider webs, rootlets, mud, rope, and leaves. Nests varied in size, as materials usually filled the available space in cavities. The lining of the nest cup usually included fine grass, animal hair, feathers, plant fibres, leaves, plant stalks, pine needles, and other fine materials.

Eggs: Dates for 707 clutches ranged from 28 April to 24 August, with 52% recorded between 1 June and 29 June. Calculated dates suggest that nests can have eggs as early as 25 April. Sizes of 684 clutches ranged from 1 to 9 eggs (1E-17, 2E-12, 3E-25, 4E-53, 5E-123, 6E-167, 7E-220, 8E-59, 9E-8), with 57% having 6 or 7 eggs. In 3 nests, the incubation period was between 14 and 15 days. In Ontario, Peck and James (1987) give a range of 10 to 19 days, with an average of 12 to 14 days.

Cannings et al. (1987) note that clutch size in the Okanagan valley declines through the summer; 146 clutches started before 15 June had a mean clutch size of 6.4 eggs, while 71 clutches initiated later than that had a mean clutch size of 5.2 eggs. About half of the nesting females in the Okanagan valley attempt to produce 2 broods each breeding season.

Young: Dates for 683 broods ranged from 15 May to 28 August, with 51% recorded between 18 June and 17 July. Sizes of 598 broods ranged from 1 to 11 young (1Y-9, 2Y-7, 3Y-25, 4Y-51, 5Y-132, 6Y-153, 7Y-175, 8Y-43, 9Y-3), with 55% having 6 or 7 young. In 13 nests, the nestling period ranged from 14 to 22 days, with a median of 16 days.

Brown-headed Cowbird Parasitism: Cowbird parasitism was not found in British Columbia in 802 nests recorded with eggs or young. In Ohio, Murphy (1984) reported 2 cases of cowbird parasitism in a sample of over 900 House Wren nests. In Ontario, Peck and James (1987) found 4 nests parasitized in a sample of 1,120 nests. Friedmann and Kiff (1985) list only 9 records for North America.

Nest Success: Of 535 nests found with eggs and followed to a known fate, 449 produced at least 1 fledgling, for a success rate of 84%; coastal nest success was 100% (*n* = 2); interior nest success was 84% (*n* = 533).

Finch (1989) showed that the nesting success of House Wrens was greater when nests were placed in open habitats. Belles-Isles and Picman (1986b) demonstrated the same results and suggested that greater nesting success in sparser vegetation resulted from a more effective defence of nests against other marauding House Wrens in such sites.

REMARKS: Three subspecies of House Wren occur in North America; the one occurring in British Columbia is *T. a. parkmanii* (Godfrey 1986).

All winter records of the House Wren must be carefully documented. We have a number of winter reports of this wren on file, including published Christmas Bird Counts from both the coast and the interior – Campbell River, Comox, Ladner, Nanaimo (including the all-time Canadian high count of 4 birds on 30 Dec 1979 [Anderson 1980]), Vancouver, Victoria, and Shuswap Lake Provincial Park (Appendix 2). All lack adequate details, and we have excluded them from this account. In addition, 10 House Wrens recorded on the Telkwa High Road Breeding Bird Survey on 23 June 1974 are also without documentation and have not been considered here.

The House Wren is notorious as a nest predator. Territorial House Wrens are known to destroy nests of other species and other House Wrens, whether or not they are within its nesting territory. The behaviour includes removing the soft nest lining, killing small nestlings, and pecking holes in eggs (Hill 1869; Sherman 1925; Creaser 1925; Kendeigh 1941; Belles-Isles and Picman 1986a). Experiments indicate that this behaviour has the effect of making preferred nesting sites available for their own use (Pribil and Picman 1991).

Kennedy and White (1992) discuss nest box selection by the House Wren, and conclude that boxes with wide, slot-shaped entrances were more likely to contain a nest with eggs, apparently because the male could more readily carry twigs through slots than holes. McCabe (1961, 1965) and Lumsden (1986) provide additional information on the use and selection of nest boxes by House Wrens.

Figure 360. Les and Violet Gibbard inspecting a House Wren nest box (Naramata, 13 June 1993; R. Wayne Campbell). Most of the breeding data for the House Wren in British Columbia have been gathered by the Gibbards over the past 20 years, from a dozen or so pairs of House Wrens nesting in their backyard.

Figure 361. An active House Wren nest (Kilpoola Lake, 1 May 1994; R. Wayne Campbell). Note the nest box crammed with small twigs, which are characteristic of House Wren nests.

NOTEWORTHY RECORDS

Spring: Coastal – Victoria 21 Apr 1987-8, 21 May 1984-7 eggs; Saanich 24 May 1982-15; Galiano Island 7 May 1964-1 (Campbell 1964a); Crescent Beach 27 Mar 1958-1; Pitt Meadows 10 Mar 1963-2; Huntingdon 6 May 1944-nest with eggs; Haney 28 Apr 1967-2 eggs; Deroche 28 Apr 1968-4 eggs; Harrison Hot Springs 1 May 1986-1 singing; Parksville 28 Apr 1910-2 (MVZ 16505-6); Comox 18 May 1925-1 (RBCM 6597); Cortes Island 19 Mar 1981-1. **Interior** – Osoyoos 13 Apr 1985-1; Kilpoola Lake 26 May 1959-7 eggs; Grand Forks 29 May 1985-7 eggs; Erie Lake 25 Apr 1977-2; Twin Lakes (Keremeos) 18 Apr 1978-4; South Slocan 27 May 1979-6; Tulameen 10 May 1987-nest building; Naramata 30 Apr 1975-3 nestlings, 28 May 1978-6; Celista 22 May 1948-7 eggs; 100 Mile House 15 Apr 1978-1; Williams Lake 20 Mar 1965-1, 22 May to late June 1982-pair nesting in box in yard; Becher's Prairie (Riske Creek) 19 May 1975-1; Quesnel 3 May 1985-1; Tupper Creek 20 May 1938-2 (RBCM 8186-87); Peace River 10 May 1975-1 (Penner 1976); Beatton River 17 May 1986-1; Clarke Lake Rd (Fort Nelson) 19 May 1980-1.

Summer: Coastal – Victoria 12 Aug 1969-4 young in nest; Skirt Mountain 6 Jun 1983-15; Saanich 2 Jul 1983-20; Sumas 20 Jun 1940-6 eggs (UBC 1249); Chilliwack 26 Jul 1983-1, 2 Aug 1889-1; Comox 30 Jun 1922-2 (NMC 17909 and 17913); Brackendale 28 Aug 1916-1 (NMC 9564); Miracle Beach 8 Aug 1966-6; Campbell River 16 Jun 1973-nestlings; Strathcona Park 20 Aug 1987-1; Cortes Island 30 Jun 1975-6; Kispiox valley 2 Jul 1921-1 (MVZ 42561; Swarth 1924). **Interior** – Newgate 11 Jun 1953-1 (NMC 38701); Grand Forks 19 Aug 1954-1 (NMC 39027); Rock Creek 1 Jul 1976-2; South Slocan 3 Jun 1979-14; Nelson 4 Jun 1976-7 eggs, 18 Aug 1981-3 nestlings about 1 week old from third nesting attempt of pair; Cranbrook 16 Jun 1945-7 eggs, 11 Aug 1945-young fledged (Johnstone 1949); Turtle Lake (Cranbrook) 14 Jun 1947-6 eggs (Johnstone 1949); Tulameen 20 Jun 1985-nestlings, 10 Aug 1985-nestlings; Naramata 2 Jun 1969-3 eggs; Summerland 1 Jun 1975-2; Kelowna 24 Aug 1974-5 eggs (latest eggs), 28 Aug 1974-4 nestlings from second renest; Vernon 1 Aug 1964-1; Drywash Mountain 9 Aug 1948-12; Black Pines 14 Aug 1942-1 (RBCM 9410); Revelstoke 29 Jun 1970-4 young, 27 Jul 1966-5 young; 8 km se 100 Mile House 17 Aug 1986-fledgling being fed by parent; Becher's Prairie (Riske Creek) 19 May 1975-adults carrying twigs to cavity in trembling aspen; Williams Lake 3 to 20 Jun 1972-1 singing constantly in backyard, nest built but not used, 4 Jun 1971-1; s of Prince George 22 Jun 1969-1; Prince George 12 Jun to 1 Jul 1969-1 singing opposite Jacobs Hotel (W.E. Godfrey pers. comm.); Salmon River (Prince George) Jun 1993-nesting; Tupper 10 Jun 1976-8; Chetwynd 24 Jul 1964-nestlings; Pouce Coupe 7 Jul 1993-nestlings; 40 km sw Dawson Creek 9 Jun 1993-incubating; Dawson Creek 3 Jul 1974-8; McQueen Slough 18 Jun 1993-7 eggs; Beryl Prairie 8 Jul 1979-1; Hudson's Hope 25 Jun 1990-nestlings; Taylor 22 Aug 1975-12; Charlie Lake 3 Jun 1954-1 (RBCM 15740), 21 Jun 1973-5 nestlings, 24 Jun 1973-7 eggs, 2 Aug 1976-3 nestlings; Fort Nelson Jun 1974-pair nesting, 16 June 1975-1, 23 Jul 1978-adult feeding unknown number of young in nest in old building.

Breeding Bird Surveys: Coastal – Recorded from 7 of 27 routes and on 4% of all surveys. Maxima: Victoria 12 Jun 1994-4; Chemainus 24 Jun 1973-3; Nanaimo River 17 Jun 1973-3. **Interior** – Recorded from 28 of 73 routes and on 25% of all surveys. Maxima: Lavington 9 Jun 1979-17; Tupper 20 Jun 1994-14; Oliver 21 Jun 1994-14.

Autumn: Interior – Bear Flat 8 Sep 1986-2; Charlie Lake 12 Sep 1982-2; 100 Mile House 2 Sep 1979-1; Revelstoke 3 Oct 1984-1; Goose Lake (Vernon) 10 Sep 1971-1; Okanagan Landing 6 Oct 1905-1; Kelowna 17 Sep 1975-1; Slocan River 9 Sep 1981-1; Rossland 1 Sep 1971-1; Sirdar 6 Sep 1948-1. **Coastal** – Brackendale 1 Sep 1916-1 (NMC 4788); Marpole late November through April-1 around yard all winter, roosted in shed; Reifel Island 2 Oct 1976-1; Canoe Pass 6 Oct 1963-1 (Boggs and Boggs 1964a); North Cowichan 15 Oct 1976-1; Saltspring Island 22 Nov 1977-6; Prevost Island 8 Nov 1980-4; North Pender Island 28 Oct 1978-1; North Saanich 25 Oct 1983-1; Sooke 19 Oct 1976-1.

Winter: Interior – No records; see REMARKS. **Coastal** – Chilliwack 28 Feb 1891-1 (MCZ 245482).

Christmas Bird Counts: Interior – see REMARKS. **Coastal** – see REMARKS.

Winter Wren

WIWR

Troglodytes troglodytes (Linnaeus)

RANGE: Resident from coastal southern and southeastern Alaska south along the Pacific coast to central California and the southern portions of the interior of British Columbia. Breeds from northern British Columbia through the forested parts of Canada to Newfoundland, south into northwestern Oregon and northern Idaho in the west, and Georgia and Tennessee in the east. Winters within, but mostly south of, the breeding range, including the southern United States. Also occurs in Eurasia and northwestern Africa.

STATUS: On the coast, *fairly common* to *common* resident in southwestern British Columbia, including the Georgia Depression Ecoprovince and Western Vancouver Island, Queen Charlotte Islands, and Southern Mainland Coast of the Coast and Mountains Ecoprovince, becoming *uncommon* along the Northern Mainland Coast of that ecoprovince, including offshore islands other than the Queen Charlotte Islands.

In the interior, *fairly common* resident in the Southern Interior and the Southern Interior Mountains ecoprovinces; *uncommon* migrant and summer visitant, and locally *very rare* throughout the year, in the Central Interior and Sub-Boreal Interior ecoprovinces; *very rare* migrant and summer visitant to the Northern Boreal Mountains and Boreal Plains ecoprovinces; *casual* in the Taiga Plains Ecoprovince.

Breeds.

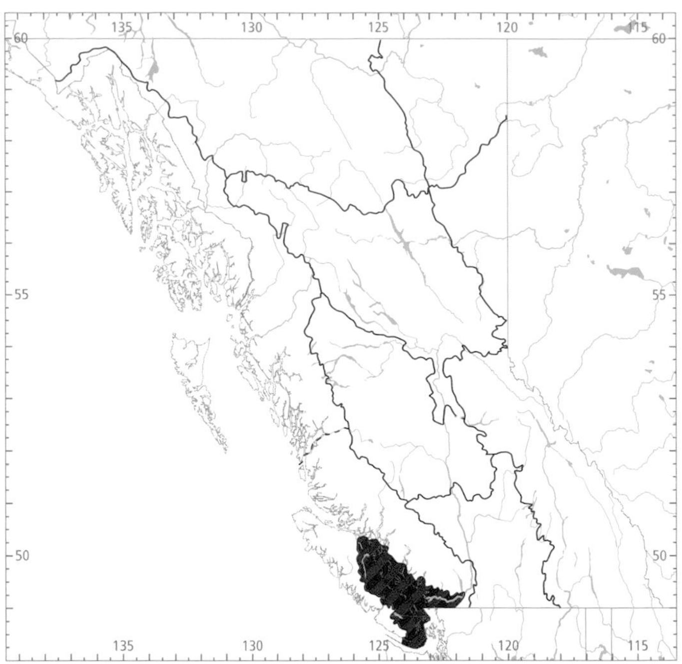

Figure 362. In British Columbia, the highest numbers for the Winter Wren in both winter (black) and summer (red) occur in the Georgia Depression Ecoprovince.

Figure 363. Mixed coastal forests of western redcedar, Douglas-fir, and western hemlock with a dense understorey of salal provide year-round habitat (Pender Island, 24 April 1993; R. Wayne Campbell).

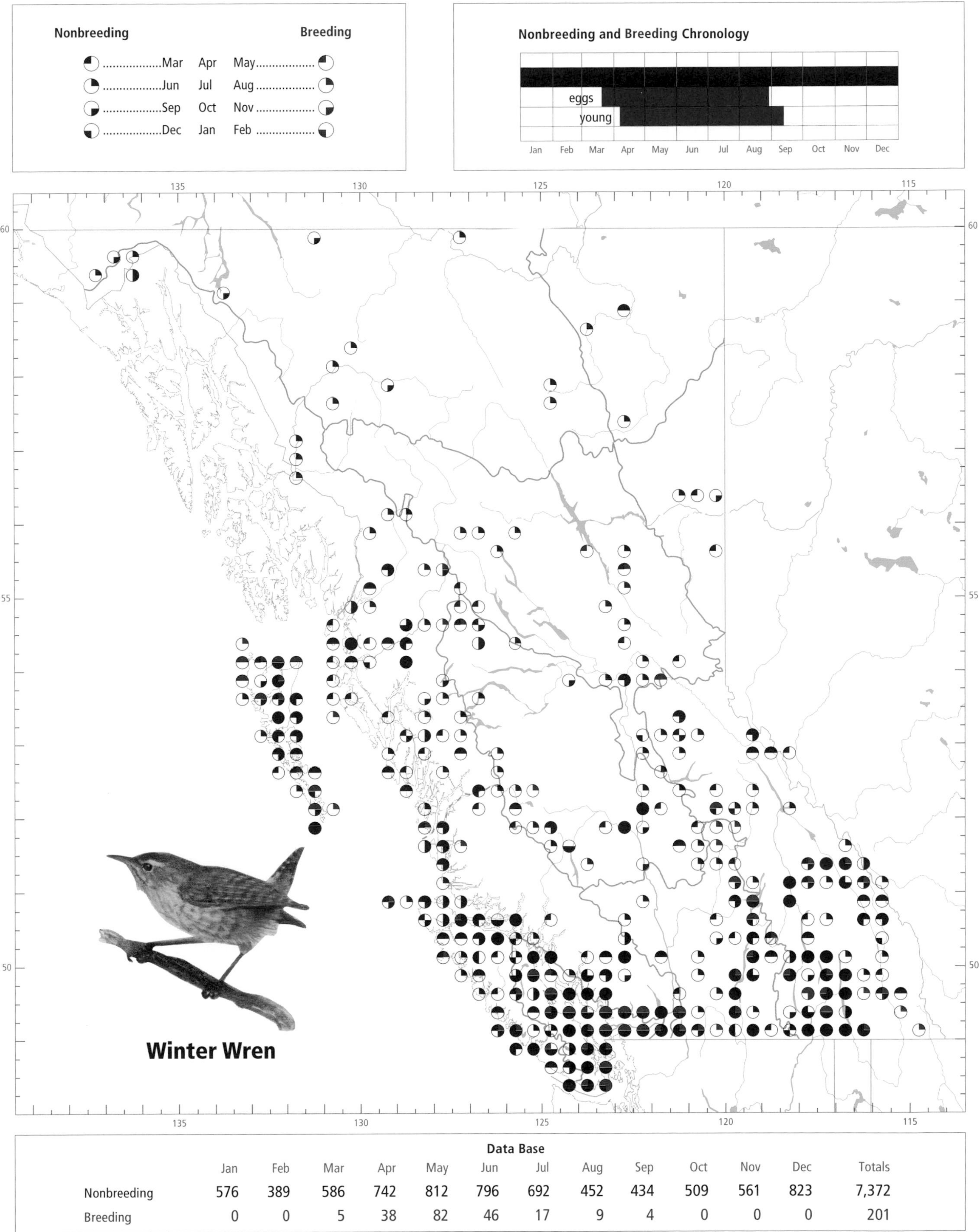

Data Base

	Jan	Feb	Mar	Apr	May	Jun	Jul	Aug	Sep	Oct	Nov	Dec	Totals
Nonbreeding	576	389	586	742	812	796	692	452	434	509	561	823	7,372
Breeding	0	0	5	38	82	46	17	9	4	0	0	0	201

Figure 364. Coastal forest habitats, with ground litter and fallen trees in various stages of decomposition, support resident Winter Wrens (Windy Bay, Queen Charlotte Islands, 29 May 1996; R. Wayne Campbell).

CHANGE IN STATUS: Munro and Cowan (1947) did not include the Cariboo and Chilcotin areas of the Central Interior, or northeastern British Columbia north of the Peace Lowland, within the range of the Winter Wren. It is now known to be a local resident at lower elevations and a fairly common summer visitant in the subalpine forests of the Central Interior, and a scarce summer visitant to the Taiga Plains and eastern Northern Boreal Mountains.

NONBREEDING: The Winter Wren (Fig. 370) is widely distributed in forested habitats throughout much of the province. It is most numerous and occurs throughout the year in the dense coniferous forests along the coast. East of the coastal mountains, it occurs in forested areas in much of the southern and central portions of the interior east to the Rocky Mountains and north in local populations to the Tatshenshini River, Liard River, and Fort Nelson areas. It is sparsely distributed in northern regions.

The highest numbers in winter occur in the Georgia Depression and adjacent mountain slopes (Fig. 362). Root (1988) notes that the Winter Wren winters mainly in areas that receive an annual precipitation of at least 80 cm, which occurs on most of the coast.

On the coast, the Winter Wren has been recorded at elevations from sea level to 1,800 m. In the interior, it has been recorded up to 2,250 m in the Southern Interior Mountains and 1,950 m in the Central Interior. Although it uses a wide range of habitats, it is seldom found more than a few metres above ground. On the coast, the Winter Wren prefers forested habitats where the forest floor is shaded (Fig. 363) and has "old-growth" characteristics such as ground litter and fallen trees in various stages of decomposition, especially where the

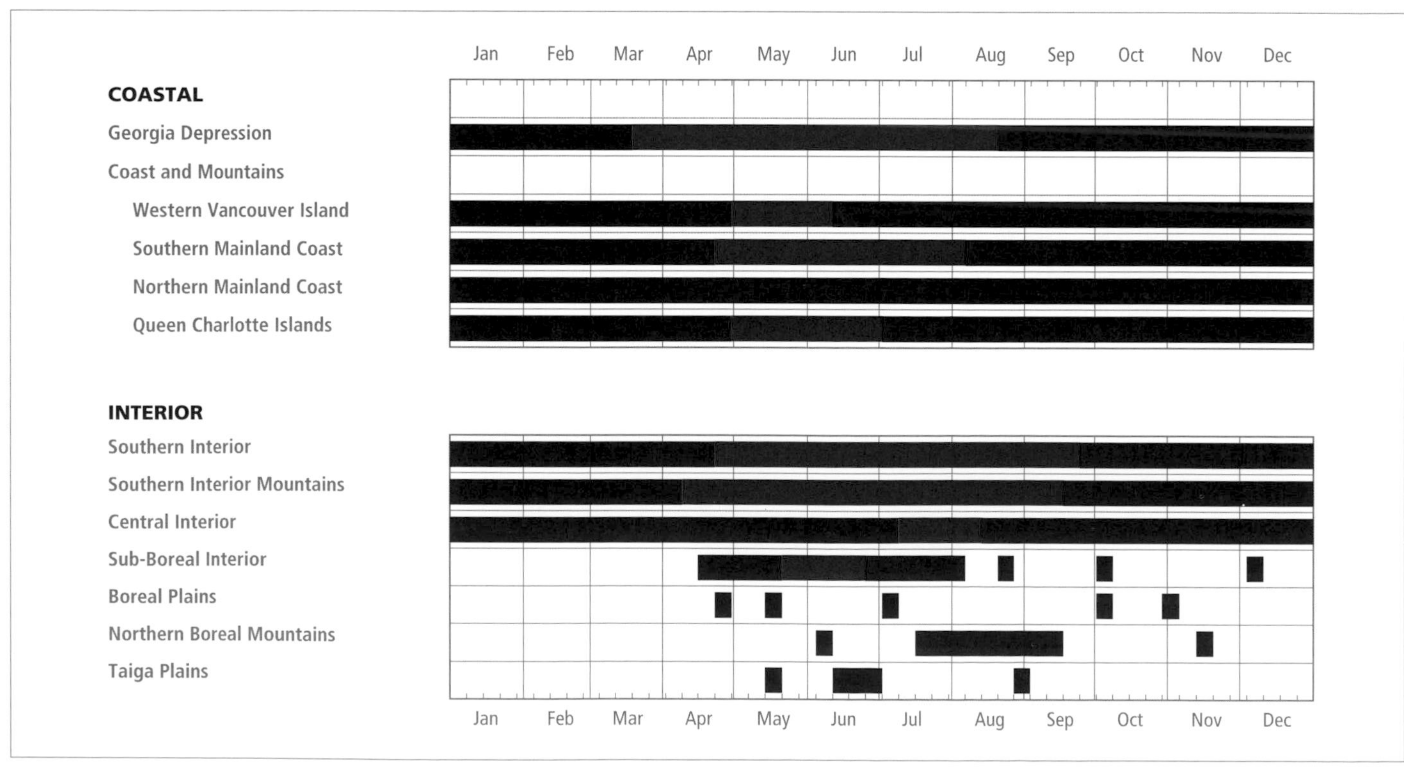

Figure 365. Annual occurrence (black) and breeding chronology (red) for the Winter Wren in ecoprovinces of British Columbia.

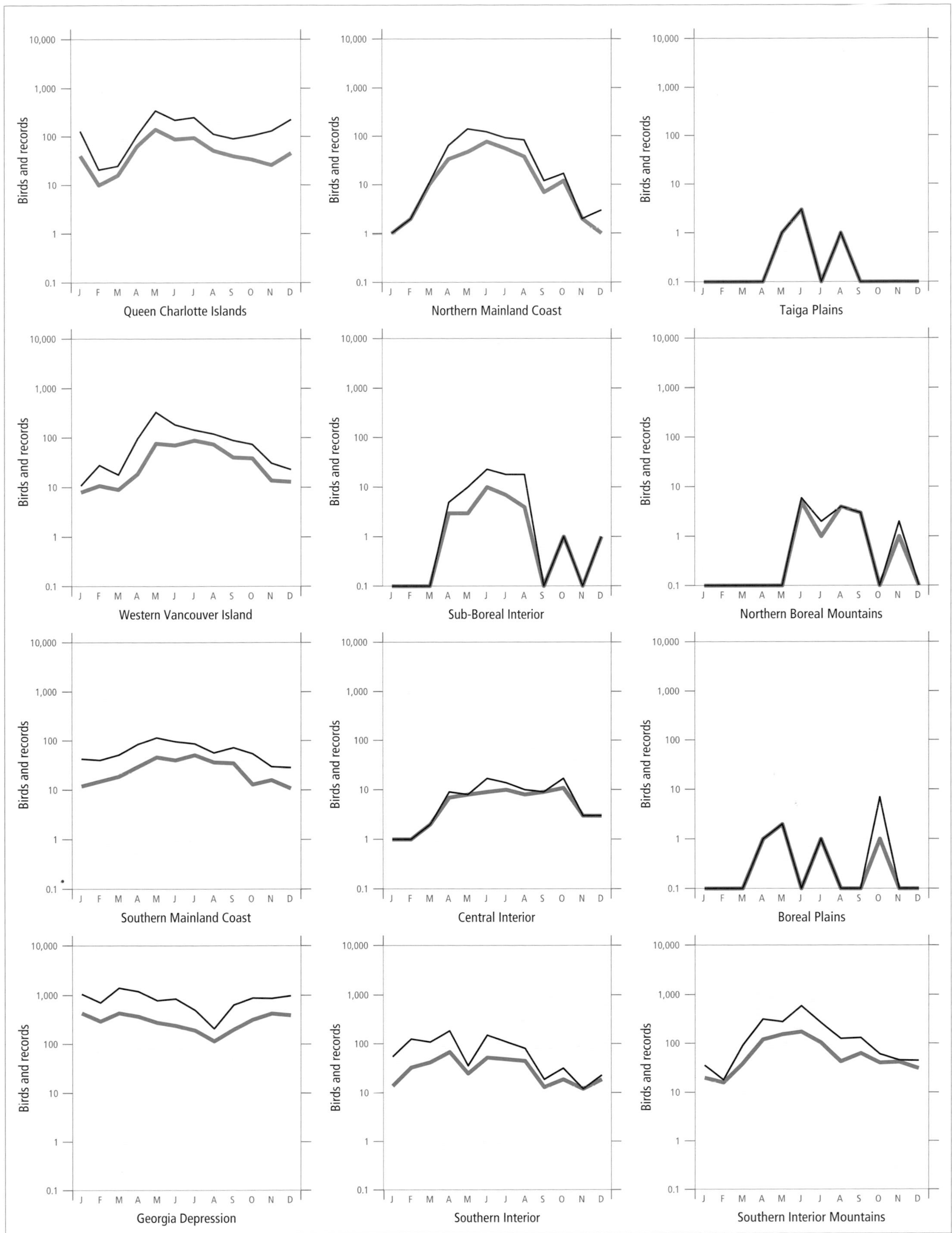

Figure 366. Fluctuations in total number of birds (purple line) and total number of records (green line) for the Winter Wren in ecoprovinces of British Columbia. Christmas Bird Count and nest record data have been excluded.

skeletons of upturned root-masses are a prominent landscape feature, and where mosses are the dominant ground vegetation (Fig. 364). It also occurs in areas of dense brush and shrubs associated with the edges of human-made corridors, in thickets at the edges of wetlands, and in more open forests with a shrubby ground layer. On treeless offshore islands, it frequents dense, wind-pruned, shrubby vegetation. Along coastal beaches, it uses piles of driftwood above the upper tideline as foraging habitat.

On the west coast of Vancouver Island, Bryant et al. (1993) found the Winter Wren present in 96% (n = 71) of old-growth plots, 69% (n = 36) of 50- to 60-year-old forest plots, 64% (n = 36) of 30- to 35-year-old forest plots, and less than 25% of plots in forests that were less than 21 years old. In Pacific Rim National Park, on the other hand, it was common in a 2-year-old clearcut and in a 24-year-old forest replanted after logging (Roe 1974).

In the interior, the Winter Wren occurs in more dispersed populations, especially where old-growth forest occurs. At lower elevations, dense woods in moist ravines or gulleys that have thick underbrush are used. At higher elevations, dense spruce and fir forests are used, including krummholz habitat at the timberline. An exception is the Douglas-fir–ponderosa pine forests of the southern valleys, where the Winter Wren occurs mainly in migration; even there it may be resident in shaded, cooler gullies where western redcedar is present.

The Winter Wren, a versatile little bird, also occurs at the edges of clearings of powerline rights-of-way, in logging slash, and in shrubbery of parks, golf courses, and gardens. During the winter it finds sheltered roosting sites in old woodpecker nest cavities, bird and mammal burrows, abandoned or open outbuildings, and even crevices in occupied homes.

Most of the population along the coast and in the Southern Interior and Southern Interior Mountains is resident, but elsewhere in the province this wren is migratory (Figs. 365 and 366). Spring migration is not discernible in the Georgia Depression and southern regions of the Coast and Mountains, because of the presence of resident birds. On the Northern Mainland Coast, spring migration seems to occur in April. In the Southern Interior there is an increase in observations between January and April, which may reflect increased numbers due to spring migration, but a steep decline in records between April and May certainly reflects a movement of wrens from valley bottom localities to higher-elevation forests (Fig. 366). In the Southern Interior Mountains, the first spring migrants arrive in March, but April probably sees the greatest movement. The small number of records for the northern regions of the province indicates that April is the month of spring arrival there.

Figure 367. On the Queen Charlotte Islands, an unusual nest site of the Winter Wren included a decaying totem pole (Anthony Island, 27 May 1996; R. Wayne Campbell).

Figure 368. Most nest sites of the Winter Wren in British Columbia were found in natural cavities in trees (Victoria, 4 June 1996; R. Wayne Campbell).

Figure 369. Although most nests in British Columbia were found in tree cavities, some were built among patches of concealing mosses in small banks within the coastal forest (Mount Douglas, 31 July 1994; R. Wayne Campbell).

In the interior, the Winter Wren migrates out of regions north of about latitude 52°N in autumn, roughly the latitude of Williams Lake. It also moves away from high-elevation mountain slopes to winter in protected, low-elevation riparian habitats (e.g., Cannings et al. 1987). In the northern half of the province, southward migration begins by the end of July and is essentially complete by the end of August, although late migrants have been found in October in the Peace Lowland and in November near the Yukon border at Redfish Creek. In the Central Interior, the main departure occurs from August to October, although a few birds overwinter. In the Southern Interior Mountains, migrants leave mainly in September and October. Southern populations that breed near the timberline in the interior move to lower elevations by mid-September. It is unknown whether or not some interior populations migrate to the coast for the winter.

Autumn migration in coastal areas is difficult to discern. For example, although there seem to be higher numbers in winter than in summer in the Georgia Depression (Fig. 366), the extent or timing of migration cannot be determined from our data.

Along the coast and in the southern third of the interior, the Winter Wren occurs throughout the year; in the central interior, it has been recorded regularly from 16 April to 23 October; in the northern third of the interior, it has been recorded regularly from 23 April to 10 September, and as late as 18 November (Fig. 365).

BREEDING: The Winter Wren breeds across much of the southern two-thirds of the province, including the Queen Charlotte Islands, Vancouver Island, other offshore islands, and the mainland coast, east across the southern portions of the interior to the Flathead River valley and north to the Skeena and Nechako river valleys. The northernmost breeding records are from Terrace, Smithers, Pine Pass, and Williston Lake. Nests with eggs or young have not been recorded north of latitude 56°N. Although we know of no confirmed breeding records, the Winter Wren probably breeds north to the Yukon border, as suggested by Godfrey (1986).

The Winter Wren reaches its highest numbers in summer in the Georgia Depression and Coast and Mountains (Fig. 362). An analysis of Breeding Bird Surveys for the period 1968 to 1993 could not detect a net change in numbers on either coastal or interior routes.

The Winter Wren has been reported breeding from near sea level to 2,100 m elevation. Most breeding sites were described as forest (78%; $n = 171$) or human-influenced habitats (19%). In forested habitats, coniferous forest was most frequently used (41%; $n = 103$; Fig. 364), followed by mixed forest (22%; Fig. 363) and human-made corridors through forest

(18%). Both mature forests (26%; n = 76) and young forests (26%) were used, as were roadsides (20%) and backyards or farmyards (12%).

The Winter Wren has been recorded breeding in British Columbia from 23 March to 12 September (Fig. 365).

Nests: Nests were found mainly in living and dead coniferous or deciduous trees (52%; n = 122) and banks and cliffs (24%; Fig. 369). A few were found in buildings and sheds. Unusual nest sites included shrubs, a bridge, an abandoned automobile, and standing totem poles (Fig. 367) in an abandoned Indian village.

Most nests were built in a cavity. Specific nest sites included natural cavities (Fig. 368) and those excavated by woodpeckers in living and dead trees or stumps (33%; n = 106), among the roots of overturned trees (26%), under an overhang in a soil bank or cliff or an overhang of a building (26%), attached to rafters, under loose bark or in bark crevices, suspended above ground in vertically oriented branches and sticks beneath clumps of shrubby vegetation, attached to a tree branch, in seabird burrows, and in a pair of snowshoes; 1 nest was found in a nest box.

The globular nest (Fig. 369) is usually well concealed and is composed of moss, twigs, grass, leaves, feathers, hair, plant fibres, rootlets, needles, shreds of rotten wood, and similar soft debris. The heights for 86 nests ranged from ground level to 7.0 m, with 61% between 0.8 and 1.8 m.

Eggs: Dates for 61 clutches ranged from 23 March to 29 August, with 53% recorded between 20 April and 5 June. Sizes of 53 clutches ranged from 1 to 7 eggs (1E-7, 2E-3, 3E-3, 4E-7, 5E-17, 6E-13, 7E-3), with 57% having 5 or 6 eggs. The incubation period is variously stated to be 14 to 16 days (Bent 1948), 14 to 17 days (Harrison 1979), 11? to 16 days (Ehrlich et al. 1988), and 14 to 20 days (Kluijver et al. 1940). Armstrong and Whitehouse (1977) state that incubation periods longer than 16 days are probably the result of delayed or irregular brooding. See further comments on incubation period in REMARKS.

Young: Dates for 30 broods (Fig. 370) ranged from 6 April to 12 September, with 52% recorded between 13 May and 14 June. Brood size ranged from 1 to 7 young (1Y-2, 2Y-6, 3Y-3, 4Y-7, 5Y-9, 7Y-3), with 53% having 4 or 5 young. The nestling period is 15 to 20 days (Harrison 1979).

Ehrlich et al. (1988) suggest that the Winter Wren may produce 2 broods a year, and some populations in British Columbia probably do so. Double-brooding can be established positively only where the nesting birds are banded or otherwise individually identifiable; however, records from the Queen Charlotte Islands suggest that at least part of that island population nests twice in a summer. Records of 2 broods out of the nest on 29 May at Anthony Island and a pair nest building at Masset on 4 April establish the timing of the first broods, while a pair feeding nestlings at Rose Harbour on 26 August and a similar record for Hippa Island on 12 August suggest a second brood. There are similar data for Vancouver Island and southern portions of the interior.

Figure 370. Adult Winter Wren at nest site in a natural tree cavity (Elk Lake, Victoria, May 1981; Mark Nyhof). The nest contained 3 nestlings.

Brown-headed Cowbird Parasitism: Cowbird parasitism was not found in British Columbia in 83 nests recorded with eggs or young. Friedmann (1963), Friedmann et al. (1977), and Friedmann and Kiff (1985) do not list any occurrences for North America.

Nest Success: Of 7 nests found with eggs and followed to a known fate, 2 produced at least 1 fledgling.

REMARKS: Two subspecies of Winter Wren occur in the province: *T. t. pacificus* west of the Rocky Mountains (Munro and Cowan 1947; American Ornithologists' Union 1957) and *T. t. hiemalis* in the Boreal Plains and possibly other boreal regions along the Yukon border. See Oberholser (1902, 1920) for more details on subspecies.

The Winter Wren is considered "exceptional and perhaps unique as a North American passerine which has successfully extended its range from North America into Asia, Europe, and North Africa" (Armstrong and Whitehouse 1977). These authors discuss variations in behaviour that lead to alterations in the apparent incubation period. They cite Kluijver et al. (1940) for the finding that in April and May incubation usually begins on the day the last egg is laid, whereas in June and July it often starts before the clutch is complete, thus leading to variation in incubation period. Some wrens may even begin incubation upon laying the first egg: 1 nest recorded in British Columbia was discovered on 23 May with a full clutch of 7 eggs. The hatching of the first nestling was noted on 26 May and proceeded with 1 egg hatching daily until 31 May. The nest was checked daily until 13 June, when all 7 young were still present.

NOTEWORTHY RECORDS

Spring: Coastal – Beacon Hill Park (Victoria) 22 Mar 1981-31; Oak Bay 9 Apr 1984-5 eggs; Saanich 23 Mar 1984-1 egg; Klanawa River 23 May 1991-30; Carnation Creek 27 Apr 1981-50, 15 May 1981-3 fledglings; Surrey 6 Apr 1967-2 nestlings; Deroche 23 Apr 1970-5 nestlings; Lighthouse Park (West Vancouver) 1 Apr 1967-5 eggs, 14 Apr 1967-4 nestlings; Black Mountain (North Vancouver) 18 May 1974-18; Kawkawa Lake 12 May 1961-2 eggs; Strathcona Park 14 Mar 1988-3; Alice Lake (Whistler) 4 Apr 1970-12; Skookumchuck Narrows Park 15 May 1982-11; Egmont 27 Mar 1975-13; Calvert Island 16 to 18 May 1933-3 (MCZ 283365-67); Cape St. James 21 Mar 1982-6; Anthony Island 29 May 1977-2 nests fledged young; Bolkus Island 17 May 1977-18; Kitlope Lake 3 May 1991-10; Kitimat River 13 Apr 1975-15; Masset 4 Apr 1984-nest building, 5 May 1912-6 eggs; Watson Island (Prince Rupert) 26 Mar 1979-2. **Interior** – Creston 11 May 1928-1 (Mailliard 1932); Sheep Creek (Salmo) 29 Mar 1984-22; Kokanee Creek 19 May 1994-pair with 2 fledglings; Nelson 24 May 1994-1 adult with 5 fledglings; Slocan River 10 Apr 1984-eggs; Eagle River 21 May 1977-6; Pass Lake area 16 Apr 1983-3; Scotch Creek 13 May 1970-4 nestlings; Celista 16 Mar 1948-10; Bridge Creek 18 Apr 1982-3; Kleena Kleene 28 Mar 1958-1; Williams Lake 4 May 1978-1; Prince George 29 May 1966-8 (Grant 1966); Purden Lake 24 May 1978-6 eggs (RBCM E1750); Blackwater Creek (Mackenzie) 26 May 1975-5 eggs; Stoddart Creek 23 Apr 1983-1; Tupper Creek 17 May 1938-1 (RBCM 8188); Fort Nelson 20 May 1974-1 (Erskine and Davidson 1976); Wann River 5 May 1983-1 (Grunberg 1983c).

Summer: Coastal – Goldstream Park 3 Jun 1982-2; Florencia Islet 22 Jul 1968-3 fledglings; Surrey 13 Jun 1963-7 newly hatched nestlings, 4 Jul 1964-25; between Skagit and Chilliwack Lake 27 Jul 1905-3 young; Little Qualicum River estuary 4 Jun 1976-8; Quadra Island 24 Jun 1975-20; Gaultheria Lake 1 Aug 1981-5 (Campbell and Summers, in press); Spout Islet 9 Jul 1975-4; Quatse Lake 9 Jun 1978-16; Lanz Island 16 Jun 1950-seen daily; Pine Island 15 Jun 1976-28; Goose Group 12 Jul 1953-3 fledglings; Rose Harbour 26 Aug 1946-fledglings; Bonilla Island 5 Jun 1970-2 fledglings; Hippa Island 3 Jun 1977-12, 12 Aug 1983 fledglings; Selvesen Island 22 Jul 1977-4 fledglings with natal down on head; Mercer Lake (QCI) 9 Jul 1977-7; Frederick Island 25 Jul 1977-4 fledglings; Lucy Island 10 Jun 1977-15; Alice Arm 1 Aug 1957-5; Stewart 12 Jul 1974-6, 23 Aug 1977-2; Haines Highway 22 Jun 1972-2. **Interior** – Manning Park 26 Jun 1983-44 on survey, 2 Aug 1967-3 young, 29 Aug 1972-2 eggs, a second clutch in same nest; Vallican 15 Aug 1986-nestlings, 20 Aug 1983-nestlings; Kokanee Glacier Park 7 Aug 1974-5 eggs; Bolean Lake 7 Aug 1974-5 eggs; Adams Lake 24 Jun 1990-19 on survey; Marble Canyon (Kootenay National Park) 22 Jul 1965-7 nestlings nearly fledged; Young Creek (Tweedsmuir Park) 10 Jul 1978-4; Tahtsa Reach 25 Jun 1976-6; Division Lake 18 Aug 1956-2; Telkwa 21 Jul 1974-eggs; Monk Lake 4 Aug 1984-17; Pine Pass 18 Jul 1969-nestlings (Webster 1969a); near Takla Lake 20 Jul 1938-12 (Stanwell-Fletcher and Stanwell-Fletcher 1943); near Bear Flats 7 Jul 1976-1; Meziadin Lake 29 Jun 1977-19 on survey; Mason Lake 28 Aug 1982-1; Ipec Lake 30 Aug 1979-1 (Cooper and Adams 1979); Mount Edziza Park 5 Aug 1978-1; near Dease Lake 7 Jun 1962-2; Fort Nelson 15 Jun 1975-1; Steamboat 25 Jun 1974-1; Coal River 10 Jul 1953-2.

Breeding Bird Surveys: Coastal – Recorded from 27 of 27 routes and on 95% of all surveys. Maxima: Queen Charlotte City 25 Jun 1994-112; Masset 21 Jun 1994-88; Port Hardy 2 Jul 1989-81. **Interior** – Recorded from 46 of 73 routes and on 43% of all surveys. Maxima: Gerrard 27 Jun 1983-33; Scotch Creek 8 Jul 1990-33; Gosnell 29 Jun 1971-20.

Autumn: Interior – Redfish Creek 18 Nov 1979-2; Haines Highway 9 Sep 1972-1; Atlin Lake 4 Sep 1975-1; McBride River 10 Sep 1977-1; Flatrock 30 Oct 1951-7; Riske Creek 4 Nov 1987-1; Walcott 23 Oct 1985-4; Kleena Kleene 13 Sep 1958-1 (Paul 1959); Celista 1 Sep 1963-3; Scotch Creek 19 Oct 1964-7; Vallican 6 to 12 Sep 1990-nestlings; Nakusp 3 Nov 1984-2; Penticton 1 Nov 1976-1; Porcupine Creek (Salmo) 3 Sep 1984-21. **Coastal** – Kispiox River 10 Sep 1921-1, last seen (Swarth 1924); Cape St. James 14 Nov 1981-23; Nanika River 4 Oct 1974-2; Port Neville 7 Nov 1975-1; Victory Mountain 3 Sep 1975-4; Squamish 10 Nov 1962-4; Egmont 6 Oct 1973-16; Harrison Hot Springs 27 Oct 1990-10, 18 Nov 1988-6; Boundary Bay 12 Nov 1962-12; Ucluelet 15 Nov 1950-7; Roche Cove 11 Oct 1987-28.

Winter: Interior – 24 km n Quesnel 8 Dec 1986-1, 20 Feb 1987-1; Williams Lake 15 Jan 1990-1, 15 Feb 1990-1; Riske Creek 5 Dec 1989-1; Celista 31 Jan 1948-7, 18 Feb 1948-42 on survey; Penticton 26 Dec 1978-8 on survey; Nelson 2 Dec 1989-1; Salmo 22 Feb 1983-2; Sheep Creek (Salmo) 7 Jan 1984-8. **Coastal** – Prince Rupert 25 Feb 1983-2; Kitimat 10 Dec 1982-3 on survey; Alliford Bay 1 Jan 1942-10; Cape St. James 6 Dec 1981-23; Egmont 22 Feb 1984-19; Harrison Hot Springs 3 Jan 1976-20.

Christmas Bird Counts: Interior – Recorded from 17 of 27 localities and on 48% of all counts. Maxima: Vernon 21 Dec 1986-10; Penticton 27 Dec 1982-9; Shuswap Lake Park 20 Dec 1988-8; Salmon Arm 17 Dec 1989-8; Fauquier 27 Dec 1989-8; Nakusp 30 Dec 1991-8. **Coastal** – Recorded from 31 of 33 localities and on 94% of all counts. Maxima: Victoria 15 Dec 1990-**451**, all-time Canadian high count (Monroe 1991a); Vancouver 20 Dec 1992-444; Duncan 2 Jan 1993-195.

Marsh Wren

MAWR

Cistothorus palustris (Wilson)

RANGE: Breeds from southwestern, central, and northeastern British Columbia, northern Alberta, central Saskatchewan, and across much of southern Canada to New Brunswick, south to northeastern Baja California, the Gulf coast, and Florida. Winters along the coasts throughout its range, and in the interior from southern British Columbia south to the southern United States, Baja California, and mainland Mexico.

STATUS: On the coast, *fairly common* to *common* resident in the Georgia Depression Ecoprovince; *rare* resident on Western Vancouver Island of the Coast and Mountains Ecoprovince; *casual* on the Southern Mainland Coast of that ecoprovince. Absent from the Queen Charlotte Islands.

In the interior, *uncommon* resident and *uncommon* to *fairly common* migrant and summer visitant in the Southern Interior and the Southern Interior Mountains ecoprovinces; *uncommon* to *fairly common* migrant and summer visitant to the Central Interior Ecoprovince; *accidental* in early winter. *Uncommon* migrant and summer visitant in the Peace Lowland of the Boreal Plains Ecoprovince; *very rare* further north, in the Taiga Plains Ecoprovince; *casual* in the Sub-Boreal Interior Ecoprovince; *accidental* in the Northern Boreal Mountains Ecoprovince.

Breeds.

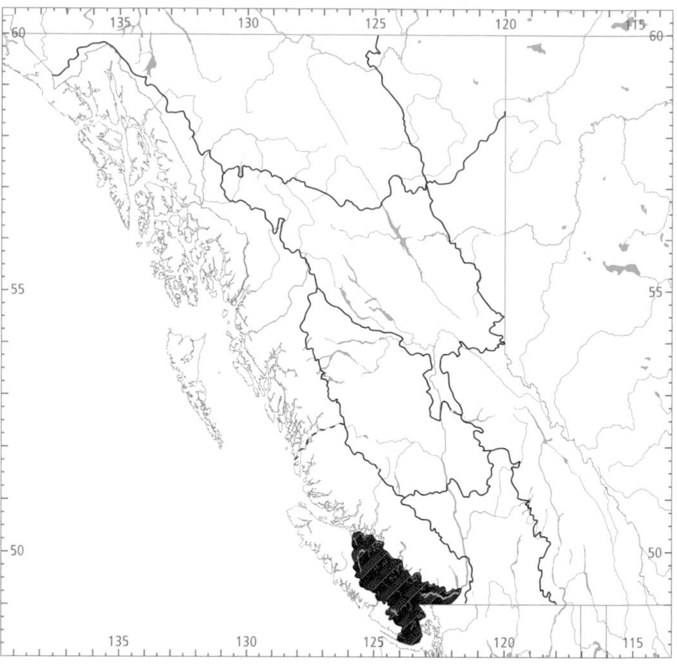

Figure 371. In British Columbia, the highest numbers for the Marsh Wren in both winter (black) and summer (red) occur in the Georgia Depression Ecoprovince, especially in the Fraser River delta and on southeastern Vancouver Island.

Figure 372. Mixed vegetation of cattails, shrubs, and grasses provide suitable habitat for resident Marsh Wrens in southwestern British Columbia (Burnaby Lake, 15 March 1969; R. Wayne Campbell).

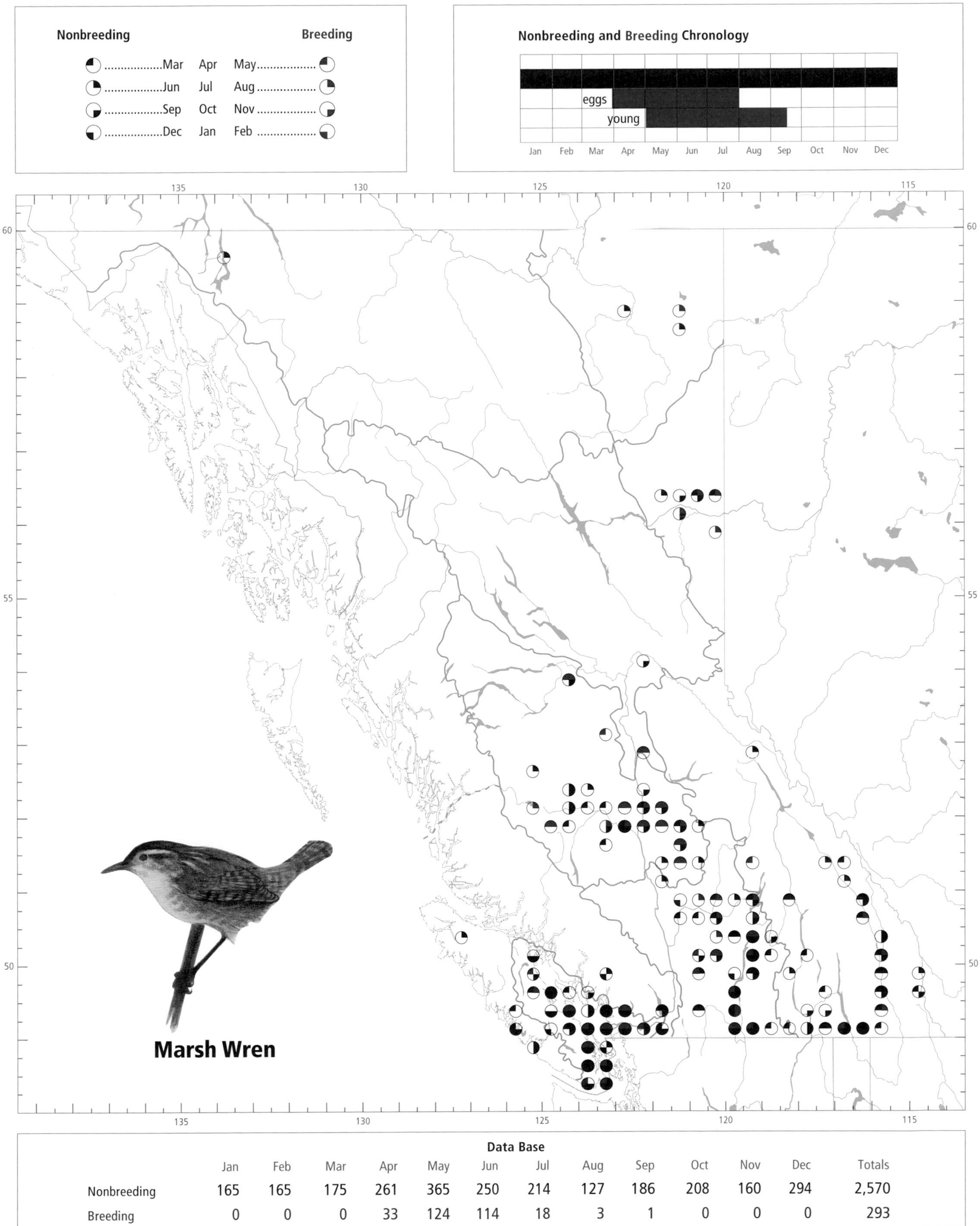

Data Base

	Jan	Feb	Mar	Apr	May	Jun	Jul	Aug	Sep	Oct	Nov	Dec	Totals
Nonbreeding	165	165	175	261	365	250	214	127	186	208	160	294	2,570
Breeding	0	0	0	33	124	114	18	3	1	0	0	0	293

CHANGE IN STATUS: In 1947, the Marsh Wren had not been recorded from areas north of Tachick Lake, near Vanderhoof. On Vancouver Island, it was noted at the same time as "Resident ... formerly at least, on southeastern Vancouver Island," suggesting that the Marsh Wren was not as common there as in the past (Munro and Cowan 1947). This wren is now locally but fairly abundantly distributed on southeastern Vancouver Island, and it now breeds in small numbers in northeastern British Columbia.

NONBREEDING: The Marsh Wren has a local but widespread distribution across southern British Columbia, including southeastern Vancouver Island north to Campbell River, western Vancouver Island north to Tofino, the Fraser Lowland, the Okanagan, Nicola, and Thompson river valleys, the Thompson Plateau, the Creston valley and other southern regions of the west Kootenay, and the east Kootenay from Cranbrook north to Golden. Further north, it occurs locally in the Central Interior north to near Vanderhoof. Small populations occur east of the Rocky Mountains in the Peace Lowland and Fort Nelson Lowland. The Marsh Wren is absent from northern Vancouver Island, the Northern Mainland Coast and Queen Charlotte Islands of the Coast and Mountains, the Northern Boreal Mountains (except for a record from Atlin), and virtually all of the Sub-Boreal Interior (contrary to Godfrey 1986).

The highest numbers in winter occur in the Georgia Depression (Fig. 371), specifically the western edge of the Fraser River delta and the Saanich Peninsula. On Vancouver Island, few birds winter north of Nanaimo. Root (1988) states that areas of high concentrations of Marsh Wrens in winter occur where the minimum January temperature exceeds –4°C. Temperatures along the coastal fringe of the Fraser River delta frequently fall below this point, but only for brief periods.

During nonbreeding seasons, the Marsh Wren occurs at low elevations from sea level to about 1,040 m. During winter on the coast, most birds frequent marshes at or near sea level (Fig. 372). Important wintering sites include the tidal marshes of the Fraser River estuary, including Boundary Bay and the lower reaches of the Nicomekl and Serpentine rivers; the freshwater marshes of southeastern Vancouver Island from Victoria to Campbell River; and meadows and marshes at Tofino Inlet, on the west coast of Vancouver Island. In the Okanagan valley, there is a small wintering population along the valley floor, at about 300 m elevation, from Vernon south to Osoyoos Lake.

The Marsh Wren is seldom found in any habitat except wetlands with emergent vegetation. On the coast, it occurs mainly in brackish marshes. In the interior, it prefers lacustrine and palustrine marshes with heavy stands of cattail. In all areas it can also be found in stands of bulrush, tall sedges, and mixed vegetation consisting of shrubs, forbs, rushes, and grasses along the slightly drier circumference of many marshes (Fig. 372). The Marsh Wren also frequents dense thickets of hardhack and sweet gale, typical littoral vegetation of many freshwater marshes on the coast.

In the Georgia Depression, principally the western parts of the Fraser River delta and near Victoria, the Marsh Wren is fairly common through the winter (Fig. 373). Numbers seem to increase there from February to April (Fig. 374),

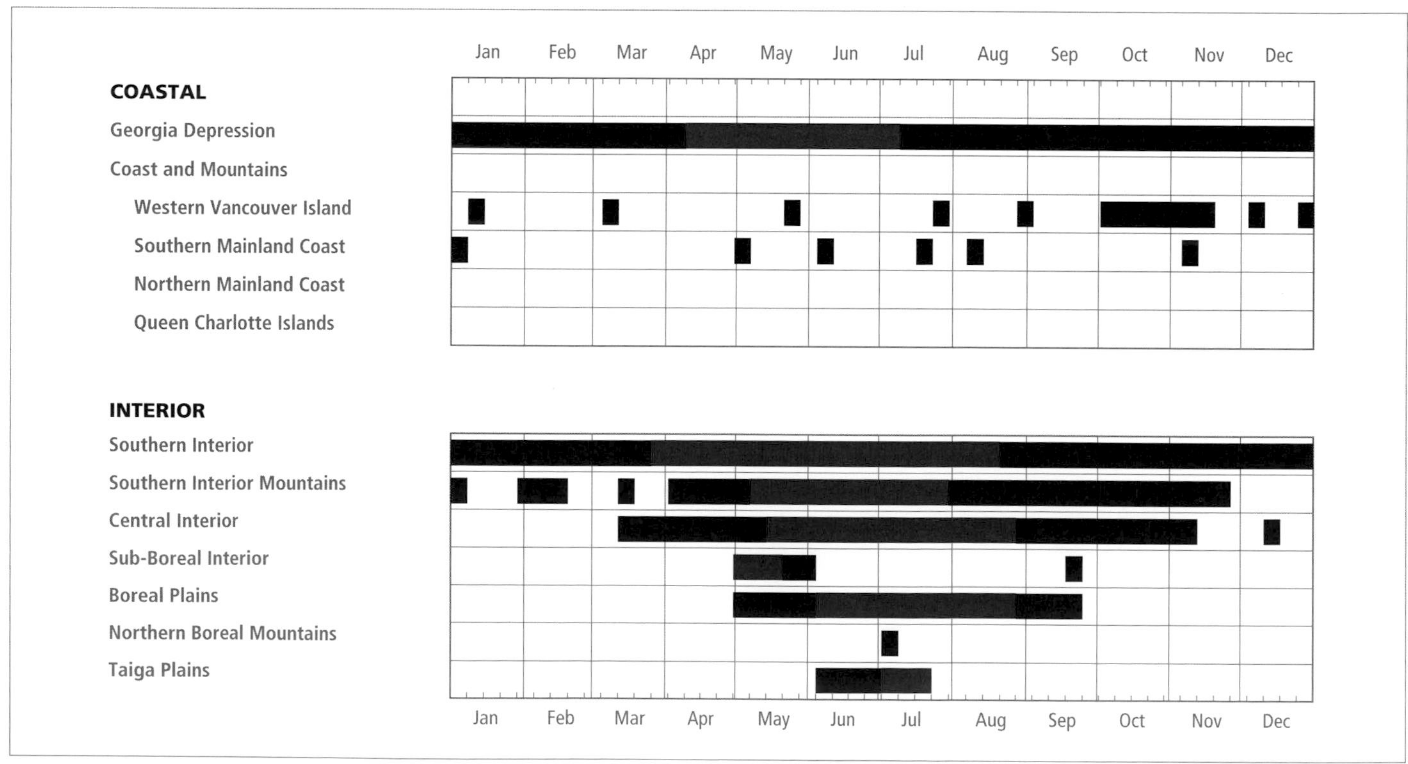

Figure 373. Annual occurrence (black) and breeding chronology (red) for the Marsh Wren in ecoprovinces of British Columbia.

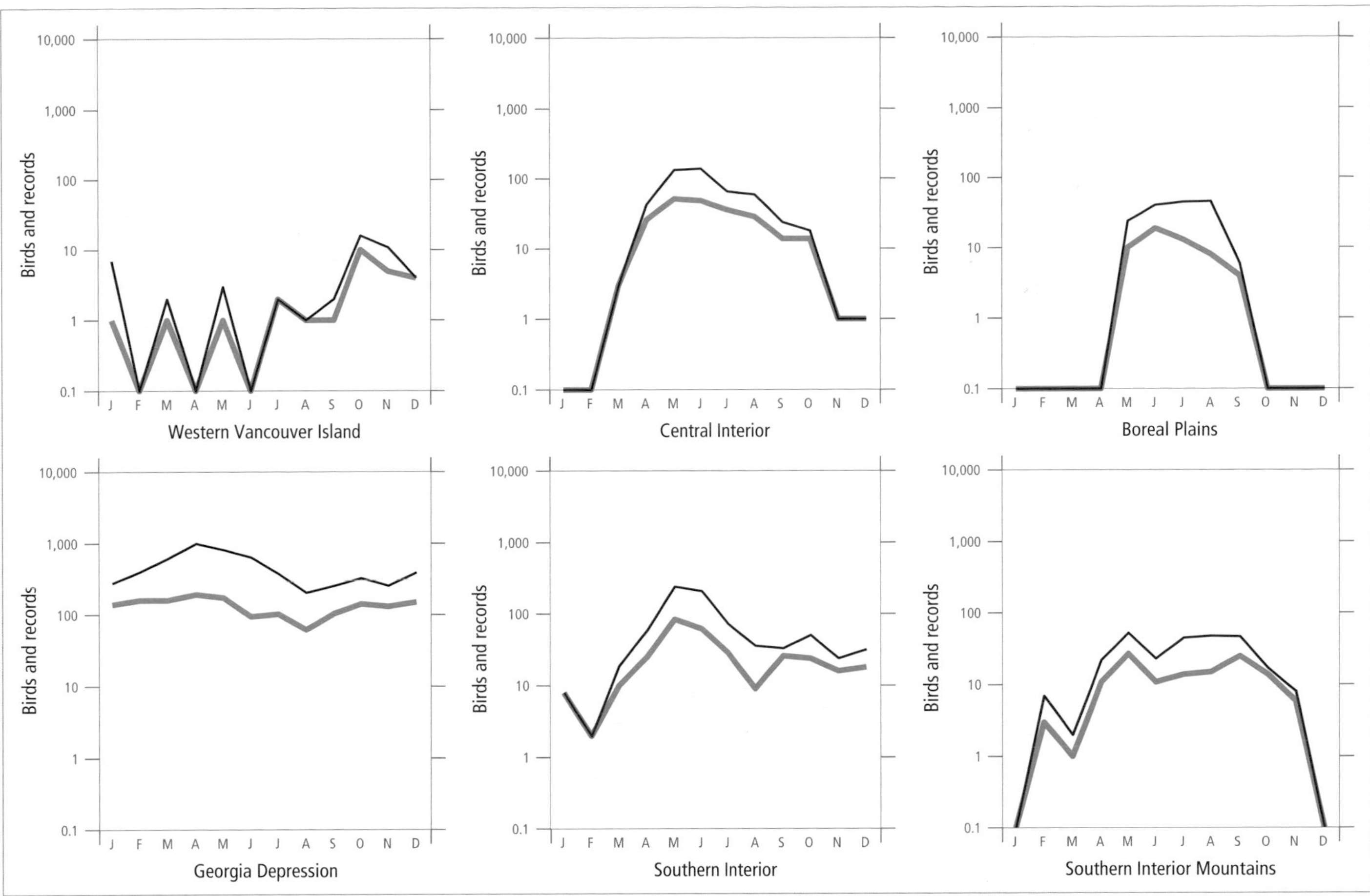

Figure 374. Fluctuations in total number of birds (purple line) and total number of records (green line) for the Marsh Wren in ecoprovinces of British Columbia. Christmas Bird Count and nest record data have been excluded.

which suggests a northward movement into the western Fraser Lowland and southeastern Vancouver Island from the coast of Washington state. Alternatively, this may simply reflect the beginning of the breeding season, when singing by males greatly enhances detectability. In the northern and eastern parts of the Fraser Lowland, there are no recent winter records of the Marsh Wren. This part of the population may move the short distance to the Fraser River delta or migrate out of the province. The winter status of this wren remains to be thoroughly investigated, however, and is difficult to determine because of its secretive nature during winter.

In the interior, the Marsh Wren is mainly migratory except for small wintering populations in the Okanagan valley and possibly the Creston valley (Figs. 373 and 374). In the Okanagan valley, spring migrants return in March and April (Cannings et al. 1987). Arrival in the valleys of the west and east Kootenay occurs mainly in April, and populations peak in May. In the Central Interior, early spring migrants can arrive after the first week of March, but the population peaks in May. In the Boreal Plains, spring migrants arrive in early May.

The autumn migration of populations in the Taiga Plains and Boreal Plains occurs in August, and only a few birds remain as late as mid-September. In the Central Interior, the Marsh Wren moves southward in August and September, but small numbers remain until early October, and occasionally as late as November. Further south in the interior, most of the population leaves during September and October. The migration of the Marsh Wren from the interior of British Columbia apparently occurs entirely east of the coastal mountain ranges. There are no specimens of the interior subspecies from the coastal marshes of the Fraser River delta or Vancouver Island. In the Georgia Depression, an unknown portion of the population migrates southward in August and September, leaving a smaller wintering population.

On the coast, the Marsh Wren has been recorded regularly throughout the year; in the interior, it is present regularly from 10 March to 20 October, except in the Okanagan valley, where it occurs regularly throughout the year (Fig. 373).

BREEDING: The Marsh Wren has a localized breeding distribution throughout much of southern British Columbia. Generally, south of the latitude of Williams Lake, its known breeding distribution is similar to its nonbreeding distribution. However, breeding has not been confirmed on western Vancouver Island, Port Alberni, Campbell River, Squamish, or Golden. North of Williams Lake, there are 8 known nesting locations: Pantage Lake near Quesnel, Tachick Lake near Vanderhoof (Fig. 375), Cranberry Marsh near Valemount, 4 sites from the Boreal Plains (McQueen Slough, Fort St. John,

Figure 375. Islets of dense cattail beds in shallow lakes provide excellent breeding habitat for Marsh Wrens in the interior of British Columbia (Tachick Lake, 24 May 1994; R. Wayne Campbell).

Boundary Lake, and Cecil Lake), and a site near Kotcho Lake in the Taiga Plains (Campbell and McNall 1982). The latter is this wren's northernmost breeding location in the province. In recent years, 1 or 2 adults and empty nests have been found in July at Parker Lake, near Fort Nelson. However, breeding (nests with eggs or young) has not been confirmed.

The Marsh Wren reaches its highest numbers in summer in the Georgia Depression, specifically on the Fraser River delta and southeastern Vancouver Island (Fig. 371). Breeding Bird Surveys for both coastal and interior routes for the period 1968 through 1993 contain insufficient data for analysis.

The Marsh Wren nests in isolated pairs or in loose colonies, from near sea level to 60 m elevation on the coast and from 270 to 1,040 m elevation in the interior. All breeding records for British Columbia are from wetland habitats (Figs. 372 and 375), mainly estuarine, lacustrine, and palustrine marshes. The wren's preferred habitat within these wetlands includes emergent beds of cattail (Fig. 375), bulrush, and sedge, but some nesting territories are established where shrubs such as hardhack, sweet gale, and willow predominate (Fig. 372). Wet ditches with emergent vegetation also provide nesting habitat. Shrub habitats are used for nesting more frequently on the coast than in the interior.

On the coast, the Marsh Wren has been recorded breeding from 11 April to 30 June; in the interior, it has been recorded breeding from 31 March (calculated) to 22 August (Fig. 373).

Nests: Most nests (n = 124; 59% from coastal populations) were built in stands of cattail (51%; Fig. 376), hardhack (23%), sedge (8%), and willow (6%). Other nest substrates included tall grass, red elderberry, sweet gale, salmonberry, and rushes. Nests were typically built over water and were attached to several stems of vegetation. The heights for 213 nests ranged from near water level to 2.4 m above the water, with 61% between 0.6 and 0.9 m above the water.

Nests were elliptical structures constructed of strips of cattail, grasses, sedges, and other plant fibres (Fig. 377), and frequently included large amounts of cattail down (Fig. 378), willow fluff, or other very soft material. Some nests were constructed with large amounts of the water alga, *Spirogyra* sp. (Munro 1943c). The nest lining is predominantly cattail down, but plant fibres, fine grass, feathers (Munro 1942), moss, and hair are often included.

Munro (1945a) states that in the Cariboo, the "roundstem" bulrush was the most commonly used nesting material but that local populations differed in nest materials used. A population at Westwick Lake used duck feathers extensively; at Lac la Hache and Horse Lake, most wrens made their nests of flat sedges woven around willow branches.

During the courtship period, males may build several "dummy" nests within their territories (Fig. 376). These nests are not lined with soft materials; lining is added by the female only to the chosen nest (Verner 1963). Dummy nests have been shown to have adaptive importance, in that active nests built near larger numbers of dummy nests were more successful than those near fewer dummy nests (Leonard and Picman 1987b) (see REMARKS).

At Pitt Meadows, Runyan (1979) found dummy nests as early as 14 March and lined nests ready for use on 13 April, with the first eggs 6 days later. In his study area, simultaneously active nests reached a density of 2.6/ha. Nests were placed about 93 cm above ground or water, in vegetation 157 cm in height. Nests averaged 8.2 m from the water's edge over water 6.3 cm deep. Forty-three percent of the nests were built over land.

Eggs: Dates for 202 clutches (Fig. 378) ranged from 4 April to 31 July, with 53% recorded between 11 May and 13 June. Calculated dates indicate that eggs can be laid as early as 31 March and can hatch as late as 6 August. Sizes of 175 clutches ranged from 1 to 8 eggs (1E-17, 2E-14, 3E-12, 4E-33, 5E-56, 6E-27, 7E-9, 8E-7), with 51% having 4 or 5 eggs. The incubation period is 12 to 16 days (Verner 1965; Ehrlich et al. 1988), and is longest early in the season (Verner 1963). Incubation begins before the clutch is complete, and eggs hatch asynchronously, with a spread of 1 to 3 days between hatching of first and last eggs (Verner 1965).

Young: Dates for 57 broods ranged from 1 May to 22 August, with 52% recorded between 29 May and 20 June. Calculated dates indicate that young may occur as early as 21 April. Sizes of 33 broods ranged from 1 to 7 young (1Y-4, 2Y-4, 3Y-8, 4Y-6, 5Y-7, 6Y-3, 7Y-1), with 64% having 3 to 5 young. The nestling period is 13 to 16 days (Ehrlich et al. 1988). The extent of double-brooding in British Columbia is unknown, but it may occur frequently. In Washington, most pairs raised 2 broods (Verner 1963). For pairs that raised 2 broods, the elapsed time between the beginning of the first clutch and the beginning of the second clutch ranged from 44 to 51 days, with a mean of 47 days. In Ontario, Peck and James (1987) suggest that some pairs raise 3 broods annually.

Brown-headed Cowbird Parasitism: In British Columbia, there were no cases of cowbird parasitism in our sample of 219 nests found with eggs or young. However, Picman (1986) examined 1,200 Marsh Wren nests at Westham Island, Delta, and found 1 nest with 2 broken Marsh Wren eggs and a cold,

Figure 376. Three Marsh Wren nests in cattails. Two were "dummy" nests; the other contained 2 fresh eggs (Pantage Lake, 24 May 1994; R. Wayne Campbell).

Figure 377. Marsh Wren nest. The spherical structure is constructed mainly of narrow strips of cattails and marsh grasses (Duck Lake, Creston, 23 June 1993; R. Wayne Campbell).

Figure 378. Clutch of 7 Marsh Wren eggs on lining of cattail down. The shell of the central egg lacks the normal dark brown pigmentation (Tachick Lake, 24 May 1994; R. Wayne Campbell).

undamaged cowbird egg. This record is the first documented case of attempted cowbird parasitism of the Marsh Wren in North America.

Nest Success: Insufficient data for analysis.

REMARKS: Two subspecies of the Marsh Wren are present in British Columbia (American Ornithologists' Union 1957): *C. p. paludicola* occurs on the coast and *C. p. plesius* occurs in the interior. Although we know of no specimens from northeastern British Columbia, it is probable that populations in the Boreal Plains and Taiga Plains belong to the subspecies *C. p. iliacus,* which occurs in Alberta. In the decade between 1965 and 1975, the breeding range of the Marsh Wren in Alberta expanded northward (Salt and Salt 1966, 1976), and this may account for the fairly recent expansion of the species into northeastern British Columbia.

Marsh Wrens often destroy the eggs of neighbouring wrens or blackbirds. Picman (1977, 1980a, 1980b, 1984) studied the aggressive interaction between Marsh Wrens and between the wrens and Red-winged Blackbirds. He concluded that nest destruction by aggressive Marsh Wrens helps exclude other members of the same species and other small, marsh-nesting songbirds. Leonard and Picman (1986) showed that, in Manitoba, wren nesting territories increased in size when blackbirds were removed. In the western United States, Bump (1986) found that Marsh Wrens destroyed or disrupted 5.3% (n = 189) of nesting attempts by Yellow-headed Blackbirds.

Gutzwiller and Anderson (1987) discuss 4 habitat variables that they use in their model to characterize the suitability of a wetland for supplying cover and reproductive needs for Marsh Wrens. These include growth form of emergent hydrophytes, percent canopy cover of emergent herbaceous vegetation, mean water depth, and percent canopy cover of woody vegetation. In British Columbia, more young were fledged at a site with denser vegetation and deeper water than at a site with less vegetation and shallower water (Leonard and Picman 1987a).

The Marsh Wren was formerly known as the Long-billed Marsh-Wren.

NOTEWORTHY RECORDS

Spring: Coastal – Saanich 31 Mar 1984-3 nests built; Quamichan and Somenos lakes 6 Apr 1974-200 on census, 21 May 1970-79; Ladner 19 Apr 1972-8 eggs; Reifel Island 14 Mar 1987-30, 12 Apr 1968-2 eggs; Sea Island 2 May 1968-7 nestlings; Grice Bay 9 Mar 1981-2; Serpentine River 17 Apr 1984-4 eggs; Pitt Meadows 3 Apr 1966-3, 1 May 1961-3 nestlings; Agassiz 17 Mar 1977-1; Newman Creek 8 Apr 1987-2; Courtenay 13 Mar 1989-1. **Interior** – Haynes Point Park 28 Mar 1976-4; Vaseux Lake 19 Apr 1974-7; Duck Lake (Creston) 20 May 1981-nest with 5 eggs and nest with nestlings; Penticton 4 Apr 1928-5 eggs (RBCM 406), 29 April 1928-6 eggs (Cannings et al. 1987); Wasa 1 May 1977-2; Swan Lake (Vernon) 29 May 1978-25; Golden 3 May 1987-5; near 70 Mile House 17 May 1958-5 eggs; 105 Mile Lake 1 May 1943-earliest arrival date (Munro 1945a); Riske Creek 20 May 1978-10; Rock Lake (Riske Creek) 24 Apr 1979-1; Bond Lake (Williams Lake) 29 Apr 1979-16; Williams Lake 13 Mar 1983-1; Alexis Creek 27 May 1946-1; Dragon Lake (Quesnel) 31 May 1994-1; Boundary Lake (Fort St. John) 11 May 1986-4; North Pine 2 May 1983-1.

Summer: Coastal – Saanich 23 Jul 1977-adult entered nest; Quamichan Lake 1 Jun 1970-106, 11 Aug 1971-34; Sarita River 26 Jul 1977-1; Campbell River Park (Langley) 6 Jul 1979-28; Ladner 25 Jun 1973-4 eggs; Burnaby Lake 30 Jun 1962-4 nestlings; Seabird Island 19 Jul 1992-pair; Courtenay 24 Jun 1964-2; Squamish 8 Jun 1975-1. **Interior** – Osoyoos Lake 27 Jul 1974-10; Vaseux Lake 3 Jun 1968-5 eggs, 22 Jun 1974-18; Duck Lake (Creston) 16 Aug 1982-6; Elizabeth Lake (Cranbrook) 15 Jul 1984-4 eggs early in incubation; Glimpse Lake 20 Jul 1978-2 nestlings; 2 km w Brisco 7 Jul 1991-26; Tatton Lake 15 Jul 1943-fledglings (Munro 1945a); Horse Lake 15 Jul 1943-2 nests; Bald Mountain (Riske Creek) 15 Jul 1980-6 eggs; Riske Creek 7 Jul 1959-6 eggs; Felker Lake 31 Aug 1969-7; Kleena Kleene 17 Jul 1931-1 (MCZ 280958); Westwick Lake 4 Aug 1948-some nests with nestlings; Chilanko Forks 1 Jul 1984-8; 150 Mile House 13 Jun 1970-18; Williams Lake 31 Jul 1982-5 eggs, 22 Aug 1982-5 nestlings ready to fledge; Moose Lake (Mount Robson) 2 Jun 1993-2; Pantage Lake 24 May 1994-2 nests with 5 eggs; Tachick Lake 23 May 1994-5 nests with eggs, 8 Jul 1945-8 nests (Munro 1945a); 8 km n junction St. John and Beatton rivers 6 Jun 1977-7 eggs; Cecil Lake 3 Jul 1978-5 nestlings, 22 Aug 1984-1; Boundary Lake (Fort St. John) 29 Jun 1978-5, 6 Jul 1990-12, 9 Aug 1986-10; Parker Lake (Fort Nelson) 7 Jun 1985-1, 14 Jul 1985-2; Kotcho Lake 30 Jun 1982-1, 6 Jul 1982-nest with 2 fresh eggs (Campbell and McNall 1982); McDonald Lake (Atlin) 5 Jul 1980-1.

Breeding Bird Surveys: Coastal – Recorded from 6 of 27 routes and on 5% of all surveys. Maxima: Albion 13 Jun 1968-4; Nanaimo River 11 Jun 1989-4; Victoria 2 Jul 1990-2; Pitt Meadows 9 Jul 1977-2; Squamish 8 Jun 1975-1; Seabird 10 Jul 1977-1. **Interior** – Recorded from 16 of 73 routes and on 16% of all surveys. Maxima: Wasa 28 Jun 1978-8; Spillimacheen 20 Jun 1982-5; Golden 23 Jun 1991-5; Salmon Arm 30 Jun 1993-4; Williams Lake 15 Jun 1986-4.

Autumn: Interior – 35 km w Fort St. John 18 Sep 1983-1 (latest record); Williams Lake 9 Nov 1980-1; Riske Creek 8 Sep 1978-5, 4 Nov 1991-1; Columbia Lake 19 Sep 1976-4; Quilchena 12 Oct 1975-6; Otter Lake 19 Nov 1950-8; East Trail 10 Oct 1979-1; Creston 23 Nov 1983-1 calling. **Coastal** – Campbell River 15 Sep 1974-1, 11 Nov 1978-1; Courtenay 13 Nov 1987-10; Porpoise Bay Park 24 Oct 1987-1; Seabird Island 6 Nov 1982-1; Cheam Lake 27 Oct 1982-2; Stubbs Island 29 Oct 1991-1; Tofino Inlet 19 Oct 1978-3, 18 Nov 1982-4; Reifel Island 3 Oct 1965-42; Westham Island 29 Sep 1976-40; Pachena River 1 Sep 1977-2.

Winter: Interior – near Riske Creek 15 Dec 1992-1; 10 km e Cache Creek 20 Jan 1985-1; Kamloops 14 Dec 1991-1; Otter Lake (Vernon) 15 Feb 1977-1; Kelowna 1 Jan 1977-2, 6 Jan 1968-1; Vaseux Lake 31 Dec 1975-1; Creston 4 Jan 1987-1, 1 Feb 1982-3, 14 Feb 1993-1; Osoyoos Lake 15 Feb 1974-1. **Coastal** – Campbell River 6 Dec 1985-1; Squamish 2 Jan 1988-1; Comox 23 Dec 1967-1; Qualicum Beach 10 Dec 1975-1, 23 Jan 1976-3; Port Alberni 4 Jan 1976-11; Grice Bay 12 Jan 1979-7 in meadow; Tofino Inlet 26 Dec 1982-1; Chilliwack 9 Feb 1891-1 (MCZ 244647); Pitt Meadows 20 Feb 1976-5; Nanaimo 26 Feb 1983-10, 28 Feb 1987-25; Sea Island 15 Jan 1989-15; Reifel Island 27 Feb 1975-13; Duncan 26 Feb 1983-12; Pender Island 19 Dec 1972-1; Saanich 14 Jan 1984-12; Tugwell Creek 31 Dec 1983-3.

Christmas Bird Counts: Interior – Recorded from 7 of 27 localities and on 23% of all counts. Maxima: Oliver-Osoyoos 28 Dec 1980-19; Vernon 15 Dec 1991-13; Vaseux Lake 23 Dec 1979-10. **Coastal** – Recorded from 16 of 33 localities and on 41% of all counts. Maxima: Ladner 23 Dec 1991-**102**, all-time Canadian high count (Monroe 1992); Vancouver 20 Dec 1981-81; White Rock 2 Jan 1989-56.

American Dipper
Cinclus mexicanus Swainson

AMDI

RANGE: Resident from the Aleutian Islands, western and northeastern Alaska, and north-central Yukon south through mountainous areas of British Columbia, southwestern Alberta, Montana, and western South Dakota; south to California, Nevada, Arizona, and New Mexico; and south through Central America to Costa Rica and Panama.

STATUS: *Uncommon* resident across most of southern British Columbia, including Vancouver Island, becoming *fairly common* locally during autumn and winter in the Georgia Depression, southern Coast and Mountains, Southern Interior, and Southern Interior Mountains ecoprovinces; *rare* to *uncommon* resident locally in the Central Interior, Sub Boreal Interior, and Northern Boreal Mountains ecoprovinces; *rare* throughout the year in the Boreal Plains and Taiga Plains ecoprovinces.

Breeds.

Figure 379. American Dipper with insect larvae. Note its characteristic stout body, slate-gray plumage, and short tail (Goldstream River, Vancouver Island, May 1983; Mark Nyhof).

NONBREEDING: The American Dipper (Fig. 379) has a widespread distribution throughout most of the province from Western Vancouver Island and the Queen Charlotte Islands east to the Rocky Mountains and the Peace Lowland, and north to the Yukon border. However, records are lacking for the north-central parts of the Central Interior in the vicinity of Ootsa and Francois lakes; the central parts of the Sub-Boreal Interior in the vicinity of Stuart, Babine, and Takla lakes; much of the central portion of the Northern Boreal Mountains; and

Figure 380. American Dipper winter habitat in an open, clear-water stream in northwestern British Columbia (near Cassiar, 15 January 1989; R. Wayne Campbell).

Figure 381. In the Taiga Plains Ecoprovince in northeastern British Columbia, the American Dipper aggregates around open patches of fresh, clear water in winter despite temperatures well below zero (Upper Muskwa River, 25 February 1995; R. Wayne Campbell).

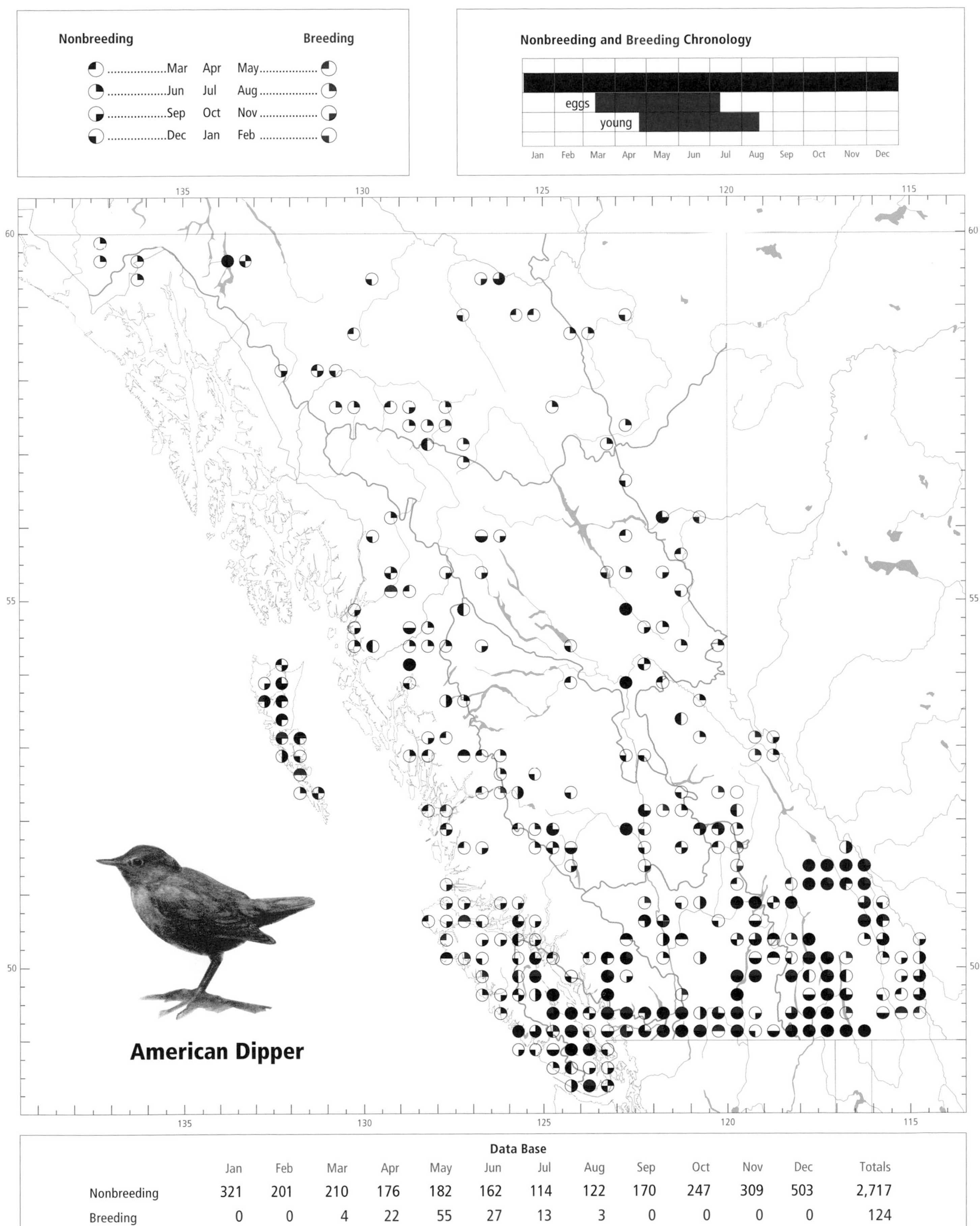

Data Base

	Jan	Feb	Mar	Apr	May	Jun	Jul	Aug	Sep	Oct	Nov	Dec	Totals
Nonbreeding	321	201	210	176	182	162	114	122	170	247	309	503	2,717
Breeding	0	0	4	22	55	27	13	3	0	0	0	0	124

the Boundary Ranges bordering southeastern Alaska. The American Dipper probably occurs throughout these areas where cold, turbulent rivers or streams are present (Figs. 380 and 381).

Small numbers winter in the northern half of British Columbia, but significant wintering populations occur only in southern regions. The highest midwinter counts from single rivers are from the south coast in the Georgia Depression (e.g., Skagit River, Goldstream River, and Weaver Creek), but high numbers also occur in winter in the Southern Interior Mountains (Fig. 382).

On the coast, the American Dipper occurs regularly during nonbreeding seasons at elevations from near sea level to 1,400 m. In the interior, it has been recorded at elevations between 270 and 1,240 m. At higher elevations in winter, it occurs where warm springs or fast-flowing water maintain open water.

The American Dipper frequents mainly the shallow margins of rapids and riffles in clear-water streams (Figs. 380 and 381), the edges of pools at the base of waterfalls, or the rocky edges of lakes. Stanwell-Fletcher and Stanwell-Fletcher (1943) and Johnstone (1949) report wintering American Dippers actively foraging along holes in ice where ambient temperatures were as low as –40°C. It can even overwinter in spots where warm springs keep holes open in otherwise ice-covered lakes. Stretches of waterways with exposed rocks and logs are preferred over those without them.

Along the coast, including many of the larger coastal islands, and in the interior within the watershed of the Fraser River, the American Dipper winters mainly along streams that have runs of salmon and steelhead or contain resident trout. There it finds an abundant food source consisting of aquatic insect larvae and drifting fish eggs in the autumn and winter,

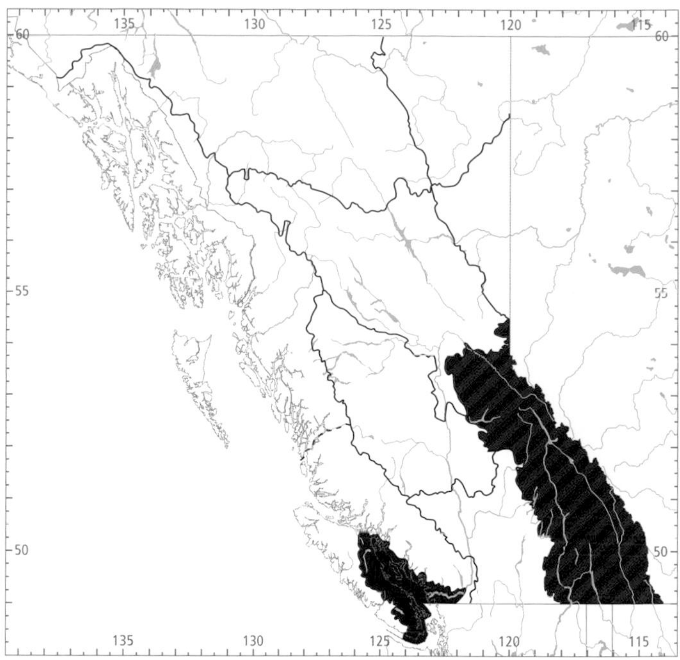

Figure 382. In British Columbia, the highest numbers for the American Dipper in winter (black) occur in the Georgia Depression and Southern Interior Mountains ecoprovinces; the highest numbers in summer (red) occur in the Southern Interior Mountains Ecoprovince.

and salmon fry in spring. Family groups of up to 6 birds often occupy the mouths of small rivers, until winter territoriality causes them to disperse.

Along the Skagit River, the largest numbers of wintering birds are associated with shallower sections of river with fewer conifers on the banks, many logs along the shores, and exposed rocks in the water (King et al. 1973). Notably, the highest concentrations were along the river channel on the

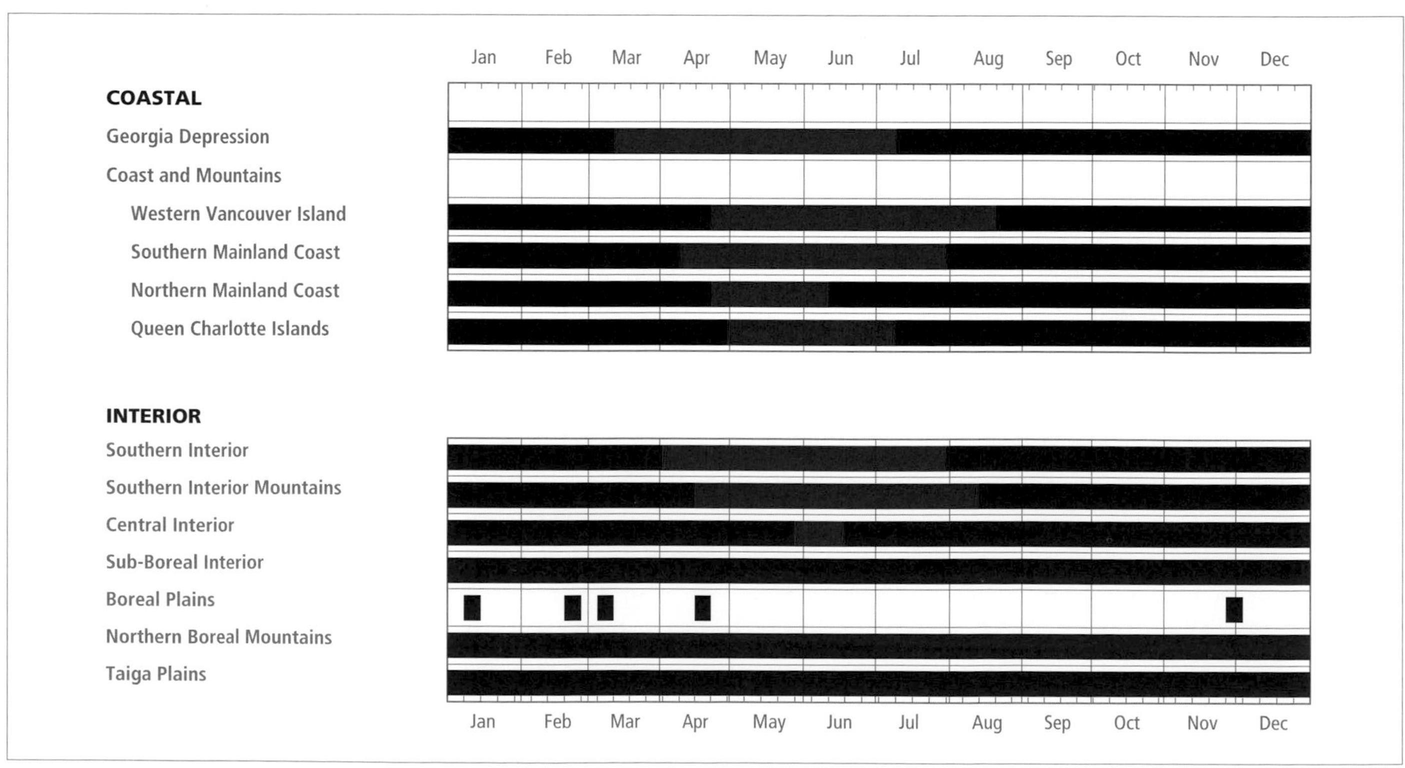

Figure 383. Annual occurrence (black) and breeding chronology (red) for the American Dipper in ecoprovinces of British Columbia.

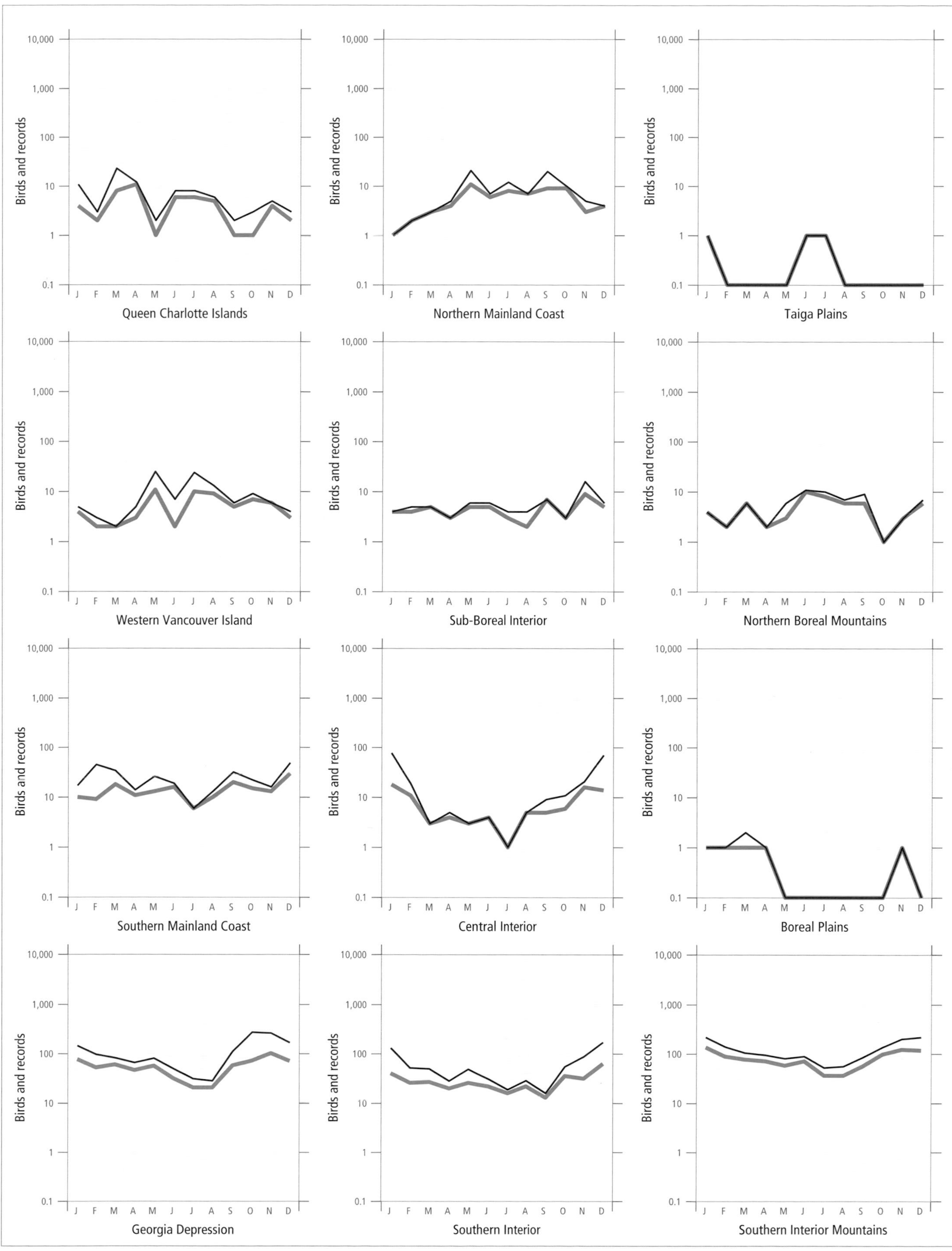

Figure 384. Fluctuations in total number of birds (purple line) and total number of records (green line) for the American Dipper in ecoprovinces of British Columbia. Christmas Bird Count and nest record data have been excluded.

muddy floor of a reservoir exposed by a drawdown of water. In recent years at Tatlayoko Lake, higher water levels in winter have covered rocks along one stretch of lakeshore that was regularly used by small numbers of wintering birds; none have wintered there since the rocks were covered (R. Travers pers. comm.).

While the American Dipper is present during all seasons throughout most of the province (Fig. 383), there is evidence of local seasonal movements and larger-scale migration (Fig. 384). Local movements involve birds leaving their widely scattered breeding areas, which become untenable after freeze-up, for winter ranges with open water and adequate food supply. In Montana and Colorado, the autumn "migration" appears to involve short moves downstream (Bakus 1959; Price and Bock 1983). In British Columbia, the distances travelled may be greater, but it is not known whether this involves significant migration out of the province to wintering areas further south or migration into the province from Alaska or Yukon. It is probable that most of the northern populations are more migratory than southern populations.

The autumn movement occurs mainly in August and September in the northern interior, and in September and October in the central and southern portions of the interior and on the coast (Fig. 384). On the Skagit River, numbers increase steadily in September and October, peak from November to February, then decline to summer levels in March (King et al. 1973). In the Okanagan valley, many birds return to winter on the Okanagan River in October or November, and most leave for breeding areas by mid-March (Cannings et al. 1987). In the Georgia Depression, the number of American Dippers peaks from October through December and declines through January and February (Fig. 384).

Figure 385. In summer, the highest numbers of American Dippers are found in the Southern Interior Mountains Ecoprovince of southeastern British Columbia, where clear, fast-moving creeks and rivers with boulders and deadfall provide foraging and nesting sites (Enterprise Creek, 10 August 1996; R. Wayne Campbell).

Figure 386. Large boulders in shallow, fast-moving rivers provide back eddies for foraging and substrates for nesting for the American Dipper in parts of British Columbia. Note the mossy nest on top of the boulder (Similkameen River, April 1967; Stephen R. Cannings).

Spring dispersal from wintering areas in the Georgia Depression is difficult to detect, but records suggest a steady decline after December (Fig. 384). In general, the American Dipper probably begins to return to breeding streams in February on the coast and in southern areas of the interior, and in March and April in the northern interior.

The American Dipper has been reported throughout the year in British Columbia (Fig. 383).

BREEDING: The American Dipper has a widespread breeding distribution along the coast from the Goldstream River on southern Vancouver Island and the Chilliwack River north to the Queen Charlotte Islands and the Nass River. In the interior, it also has a widespread breeding distribution in mountainous areas from Manning Park east to the Rocky Mountains and north to the latitude of Mount Robson. Although there are no breeding records for the interior north of Mount Robson, it is probable that the American Dipper breeds throughout its range in the province.

The American Dipper reaches its highest numbers in summer in the Southern Interior Mountains Ecoprovince (Fig. 382). Breeding Bird Surveys for both coastal and interior routes for the period 1968 through 1993 contain insufficient data for analysis.

On the coast, the American Dipper frequently nests close to sea level, but has been recorded as high as 1,250 m. In the interior, it has been reported breeding at elevations from 300 to 2,140 m. Although the critical habitat element appears to be clear, fast-moving streams where rills, waterfalls, and rocky bottoms provide a rich aquatic fauna for food (Fig. 385), the presence of suitable nesting sites is an important variable in determining breeding density (Price and Bock 1983).

(a)

(b)

(c)

(d)

Figure 387. Nesting sites for the American Dipper in British Columbia include: (a) ledges behind waterfalls (Anna Inlet, Queen Charlotte Islands, 4 June 1990; R. Wayne Campbell); (b) girders on human-made structures such as bridges (Howell Creek, Flathead, September 1993; Linda M. Van Damme); (c) ledges adjacent to fresh, and sometimes salt, water (Bag Harbour Creek, Queen Charlotte Islands, 15 June 1989; R. Wayne Campbell); and (d) roots of standing and fallen trees (Klaskish River, Vancouver Island, 16 May 1978; Michael S. Rodway).

Along the coast, fast-flowing, gravel-bottomed streams that support runs of salmon, steelhead, and cutthroat trout are used frequently for nesting (Pritchard 1934; Munro 1936). On nesting streams, rocky streambeds provide boulders and back eddies for feeding sites, while cutbanks, large rocks in midstream (Fig. 386), cliffs, waterfalls, and log jams provide nest sites. When nesting along steep, wet cliffs above marine waters, the dipper also forages in intertidal areas. It breeds mainly along streams where human disturbance is minimal. Slow-moving, soft-bottomed rivers are not used for breeding.

On the coast, the American Dipper has been recorded breeding from 16 March (calculated) to 17 August; in the interior, it has been recorded breeding from 8 April to 8 August (Fig. 383).

Nests: Most nests in British Columbia were found on cliffs (51%; *n* = 91) and often behind waterfalls (Fig. 387a); on human-made structures such as bridges or culverts (31%; Fig. 387b); in cavities in streamside logs, cutbanks, or similar sites (14%; Fig. 387c); or among the roots of trees exposed during spring freshets (Fig. 387d). Nests were always adjacent to fresh water.

More specifically, nests were situated on or alongside rocks, rock ledges (Fig. 387a), crevices in canyon walls, or edges of cliffs (47%; *n* = 79); on girders or beams of bridges (22%); or under an overhang (18%). Although the American Dipper readily uses nest boxes in some parts of its range (Hawthorne 1979), no instances of such use have been reported from British Columbia. There is a record of an American Dipper nesting in a cavity excavated by a Northern Flicker.

The nest consists of 2 parts: an outer dome-shaped shell of moss (Fig. 388) and an interior nest bowl composed largely of grass, but also including twigs, rootlets, leaves, and plant fibres (Fig. 389).

The heights for 68 nests ranged from just above water level to 9 m, with 56% between 1.2 and 3 m.

Eggs: Dates for 22 clutches ranged from 18 March to 10 July, with 64% between 2 April and 4 May. Calculated dates indicate that eggs could be found as early as 16 March and as late as 23 July. Sizes of 14 clutches ranged from 3 to 5 eggs (3E-2, 4E-8, 5E-4), with 8 having 4 eggs (Fig. 389). The incubation period has been reported as 13 to 17 days (Ehrlich et al. 1988), or 16 days (Hann 1950).

Young: Dates for 68 broods ranged from 23 April to 17 August, with 52% recorded between 16 May and 22 June. Calculated dates indicate that young could be found as early as 4 April. Sizes of 36 broods ranged from 1 to 5 young (1Y-4, 2Y-7, 3Y-15, 4Y-6, 5Y-4), with 61% having 2 or 3 young.

Hann (1950), in Colorado, found that in 2 nests the nestling period was 24 to 25 days, but if a nest was disturbed the young could leave as early as 18 days. Several observers in British Columbia noted that nestlings left the nest when the nest was checked. In a Colorado population, one-third of pairs had 2 broods annually (Price and Bock 1983). There is no confirmation of double-brooding in British Columbia, but records of nesting from March through August suggest that some pairs may have 2 broods a year.

Feeding of fledglings by parents has been noted as late as 7 October (Burcham 1904).

Brown-headed Cowbird Parasitism: Cowbird parasitism was not found in British Columbia in 56 nests recorded with eggs or young. Friedmann (1968) and Friedmann and Kiff (1985) do not list the American Dipper as a host species for the Brown-headed Cowbird.

Nest Success: Of 2 nests found with eggs and followed to a known fate, 1 produced at least 1 fledgling. In Colorado, nest success varied between 50% and 84% over 3 years (Price and Bock 1983).

REMARKS: British Columbia rivers support large wintering populations of the American Dipper. For example, King et al. (1973) estimated a wintering population of 40 birds along 30 km (1.3 birds/km) of the Skagit River (a tenfold increase from the breeding population along the same stretch of river). More than 50 birds have been counted along a 12 km stretch (4.2 birds/km) of the Squamish and Cheakamus rivers (R.J. Cannings pers. comm.). In the interior, as many as 35 American Dippers have been counted along 1 km of the Okanagan River at Okanagan Falls (Cannings et al. 1987).

Figure 388. Typical moss nest of the American Dipper, showing nest bowl of grasses and rootlets (near Skedans Island, Queen Charlotte Islands, 5 June 1990; R. Wayne Campbell).

Figure 389. In British Columbia, the American Dipper lays its eggs in a nest bowl lined with leaves, plant fibres, and a few mosses (Goldstream River, 7 May 1973; R. Wayne Campbell).

High winter densities reported elsewhere include 2.1 birds/km in Montana (Bakus 1959) and 0.45 to 0.71 birds/km in Colorado (Price and Bock 1983).

In the interior north of about latitude 52°N, and in coastal mountains north of Prince Rupert, overwintering areas are limited to streams kept open by their turbulence, or lakes and streams where thermal springs prevent ice formation (Stanwell-Fletcher and Stanwell-Fletcher 1943; Figs. 380 and 381). In southern areas, there are many wintering localities with concentrations of birds. The importance of these areas for the American Dipper should be considered during any land use planning that may affect the quality of habitat.

Munro (1923c, 1924a) and Thut (1970) provide information on the feeding habits and role of the American Dipper as a predator on the eggs and fry of trout and salmon in British Columbia.

NOTEWORTHY RECORDS

Spring: Coastal – Goldstream River 23 Apr 1990-1 fledged from nest, 25 Apr 1975-3 nestlings; Cowichan River 30 Mar 1984-4 eggs; Aldergrove 18 Mar 1970-3 eggs; Chilliwack River 2 Apr 1958-4 eggs, 3 Apr 1946-5 eggs; Kanaka Creek 24 Mar 1970-4 eggs; Harrison Hot Springs 12 Apr 1986-1; Nicolum River 29 Apr 1962-4 eggs (RBCM 1920), 12 May 1978-5 nestlings; 18 km s Port Alberni May 1991-nesting in culvert (Campbell and Harris 1991); Strathcona Park 14 Mar 1988-1; Elaho River 5 May 1976-5; Klaskish River 16 May 1978-3 nestlings (Fig. 387d); Port Hardy 14 May 1935-3 nestlings about 1 day old; Namu Lake 5 Mar 1983-2; Quatlena River 1 May 1911-4 eggs (RBCM E914); Sunday Inlet 12 May 1986-1 (Campbell 1986f); Kimsquit River 9 May 1985-1; Kitlope Lake 3 May 1991-1; Deena Creek 14 Mar 1975-8; Awun River 20 May 1985-at least 2 nestlings (Cooper 1985); De la Beche Inlet 21 May 1989-4 nestlings (Campbell 1989f); Masset 11 Apr 1939-1; Tseax River 28 May 1978-at least 2 nestlings; Atlin 15 May 1979-1. **Interior** – Ashnola River 13 Apr 1958-4 eggs (RBCM E395), 2 May 1974-2 nestlings; Manning Park 18 May 1963-5; Okanagan Falls 20 May 1973-fledglings (Cannings et al. 1987); near Hedley 8 Apr 1976-eggs; Flathead River 6 May 1990-at least 4 eggs; Bull River 23 Apr 1938-building nest (Johnstone 1949); Nelson 14 Mar 1982-4; Kuskanax Creek bridge 25 Apr 1972-eggs, 14 May 1983-4 nestlings; Sorrento 21 Mar 1972-6; Glacier National Park 1 May 1983-6; Summit Lake (Kicking Horse Pass) 3 Apr 1983-4; Success Creek 19 May 1975-4; Williams Lake River 1 Mar 1968-1; Toboggan Creek 4 Apr 1978-2; Coldstream Creek (Chetwynd) 8 Mar 1978-2; Peace River 22 Apr 1974-1; Kluayaz Creek 26 May 1979-4; Beatty Creek (Stikine River) 2 Mar 1978-1; Toad River 8 Mar 1995-2.

Summer: Coastal – Goldstream River 4 Jul 1978-5; Chilliwack Lake Park 3 Jun 1983-2; Englishman River Falls 6 Jul 1985-nestlings; Silburn Creek 17 Aug 1982-nestlings; Megin River 13 Jul 1985-11; Comox 2 Aug 1940-2; Tahsis 3 Aug 1949-2 fledglings and 1 adult (Flahaut 1949d); Lesser Garibaldi Lake 27 Jul 1955-3 nestlings; Anna Inlet 8 Jun 1987-3; Pallant Creek 2 Jul 1979-nestlings; Mace Creek 20 Aug 1983-2; Kleanza Creek 8 Jun 1976-2; New Aiyansh 7 Jun 1976-nestlings; Alice Arm 2 Aug 1957-1. **Interior** – Flash Lake 10 Jul 1974-2; Penticton Creek 25 Jun 1957-nestlings (Cannings et al. 1987); Kokanee Glacier Park 1 Jul 1978-4; Whatshan River 3 Jun 1984-3; Cariboo Creek 5 Jul 1977-4; Scotch Creek 14 Aug 1963-3; 14 km w Westwold 7 Jul 1991-2 nestlings; Hung-abee Lake 27 Aug 1975-5 (Wade 1977); Lake O'Hara 8 Aug 1982-3 nestlings; Tatlayoko Lake 5 Aug 1984-1; Rainbow Range 10 Jul 1992-2 eggs and 1 nestling; Narraway River 7 Jul 1977-1; Bijoux Creek 25 Jun 1969-2; Pine Pass 17 Aug 1975-3; Buckinghorse Creek 17 Jul 1976-2; Ipec Lake 29 Aug 1979-2 (Cooper and Adams 1979); Tetsa River 15 Jun 1980-2; Atlin 15 Jun 1979-1.

Breeding Bird Surveys: Coastal – Recorded from 7 of 27 routes and on 5% of all surveys. Maxima: Chemainus 25 Jun 1969-1; Alberni 10 Jun 1969-1; Seabird 22 Jun 1972-1; Pemberton 4 Jun 1989-1; Campbell River 18 Jun 1983-1; Squamish 10 Jun 1973-1; Meziadin 29 Jun 1977-1. **Interior** – Recorded from 9 of 73 routes and on 4% of all surveys. Maxima: Creighton Valley 15 Jun 1988-2; Grand Forks 3 Jul 1985-2; Kuskonook 30 Jun 1977-2; Kootenay 15 Jun 1981-2; Zincton 26 Jun 1971-1; Needles 19 Jun 1982-1; Syringa Creek 24 Jun 1990-1; Christian Valley 25 May 1994-1.

Autumn: Interior – Atlin 22 Nov 1931-1 (RBCM 5905); Little Tahltan River 18 Sep 1986-2; Peace River 27 Nov 1983-1; Babine River 15 Nov 1978-6; Hazelton 14 Sep 1938-1; Lightning Creek (Barkerville) 11 Nov 1982-6; Chilcotin River 5 Nov 1986-2; Lone Cabin Creek 25 Sep 1988-5; Revelstoke 30 Sep 1974-4; Lillooet 7 Oct 1904-still feeding fledglings (Burcham 1904); Fitztubbs Creek 1 Oct 1977-4; Deep Creek (Peachland) 29 Oct 1979-7; Okanagan River 10 Nov 1973-25; Similkameen River 12 Oct 1890-2 (MCZ 45584-5); Beaver Creek Park 27 Nov 1982-6. **Coastal** – Bearskin Lake 21 Sep 1986-1; Talahaat Creek 30 Sep 1985-6; Kitimat 19 Nov 1978-2; Naden Harbour 7 Oct 1974-3; Mitchell Inlet 4 Nov 1986-2; Kingcome Inlet 25 Sep 1936-1 (NMC 27425); Fulmore River 30 Sep 1975-5; Tsitika River 6 Oct 1979-2; Little Bear Bay 24 Oct 1974-3; Ralph River 6 Sep 1986-4; Harrison Hot Springs 1 Sep 1975; Mercer Creek 14 Sep 1983-2; Goldstream River 22 Oct 1986-27 on survey.

Winter: Interior – Fourth of July Creek (Atlin) 9 Dec 1980-2; Pine Creek (Atlin) 30 Dec 1980-1; Liard Hot Springs 20 Jan 1975-1 (Reid 1975); Cassiar 15 Jan 1989-1; Muskwa River 25 Feb 1995-11 on helicopter survey; Fort Nelson 10 Jan 1986-1; Tanzilla River 4 Dec 1980-1; Chowade River 1 Jan 1984-1; Hudson's Hope 20 Jan 1975-1; Stewart 18 Jan 1989-1 at warm spring; Tetana Lake 11 Dec 1937-2 (Stanwell-Fletcher and Stanwell-Fletcher 1943); Pine River (w Chetwynd) 24 Feb 1989-1 (Siddle 1989b); Wolverine River (Murray River) 16 Feb 1977-2; Prince George 11 Jan 1989-1; Indianpoint Creek 19 Dec 1929-2 (MCZ 283333-4); Wells Gray Park 7 Feb 1953-6; Williams Lake 30 Dec 1977-41 on Christmas Bird Count; Kleena Kleene 1 Jan 1950-1 (Paul 1959); Field 20 Dec 1975-24 on Christmas Bird Count; Columbia River (Athalmer to Radium) 4 Jan 1981-12; Sorrento 24 Feb 1971-6; Okanagan River (McIntyre Bluff) 31 Dec 1977-35 (Cannings et al. 1987); East Trail 31 Dec 1980-1. **Coastal** – Stewart 18 Jan 1989-1 at warm spring; Yakoun River 31 Dec 1931-2; Pallant Creek 19 Jan 1978-4; Lagoon Inlet 26 Jan 1976-5; Alta Lake 30 Dec 1935-1; Weaver Creek Park 20 Feb 1983-28; Sarita River 6 Dec 1978-2; Harrison Hot Springs 25 Dec 1990-5, 9 Feb 1974-2; Port Alberni 4 Jan 1976-9; Skagit River 1 Jan 1971-2, 12 to 13 Feb 1971-36 (King et al. 1973); Pender Island 4 Dec 1994-1.

Christmas Bird Counts: Interior – Recorded from 25 of 27 localities and on 86% of all counts. Maxima: Vaseux Lake 31 Dec 1977-50; Revelstoke 21 Dec 1991-32; Shuswap Lake Provincial Park 18 Dec 1986-25. **Coastal** – Recorded from 25 of 33 localities and on 55% of all counts. Maxima: Squamish 1 Jan 1985-**69**, all-time Canadian high count (Monroe 1985a); Chilliwack 19 Dec 1981-20; Pitt Meadows 27 Dec 1977-18.

Golden-crowned Kinglet

GCKI

Regulus satrapa Lichtenstein

RANGE: Breeds in coniferous forests from southern Alaska and Yukon, the northern parts of the Prairie provinces, central Ontario, southern Quebec, Prince Edward Island, Nova Scotia, and Newfoundland south along the coastal and interior mountains into parts of California, Nevada, Utah, Arizona, New Mexico, Minnesota, and Michigan, and from New York to western North Carolina. Also occurs in Mexico. Winters from south-coastal Alaska and southern Canada south to Baja California, the Gulf coast, central Florida, and Mexico.

STATUS: On the coast, *uncommon* to *fairly common* in summer in the Georgia Depression Ecoprovince and in the Southern Mainland Coast of the Coast and Mountains Ecoprovince, becoming *fairly common* to locally *common* through the rest of the year; *uncommon* to *fairly common* throughout the year on Western Vancouver Island and the Queen Charlotte Islands in the rest of the Coast and Mountains; *uncommon* migrant and summer visitant in the Northern Mainland Coast of that ecoprovince, becoming *very rare* there in winter.

In the interior, an *uncommon* to locally *common* resident in the south-central and southeastern portions of the province in the Southern Interior and Southern Interior Mountains ecoprovinces; *uncommon* migrant and summer visitant, *rare* in winter, to the Central Interior and the Sub-Boreal Interior ecoprovinces; *rare* to *uncommon* migrant and summer visitant to the Boreal Plains, Northern Boreal Mountains, and Taiga Plains ecoprovinces.

Breeds.

Figure 390. Female Golden-crowned Kinglet (Victoria, January 1991; Tim Zurowski).

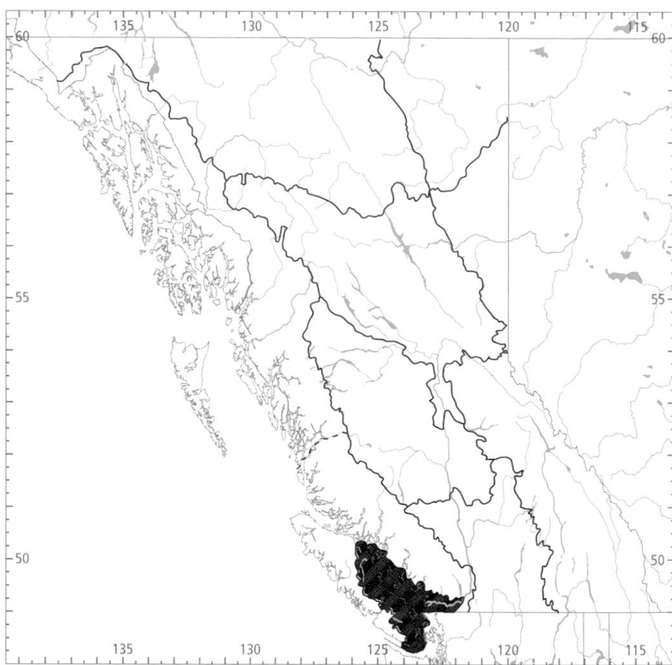

Figure 391. In British Columbia, the highest numbers for the Golden-crowned Kinglet in both winter (black) and summer (red) occur in the Georgia Depression Ecoprovince.

NONBREEDING: The Golden-crowned Kinglet (Fig. 390) is widely distributed along the coast from southern Vancouver Island north to the Queen Charlotte Islands and along the northern mainland coast to Stewart. In the interior, it has a widespread distribution across the southern portions of the province north to the Cariboo and Chilcotin plateaus and to the extreme southern and eastern portions of the Sub-Boreal Interior. Further north, its distribution becomes more scattered and localized. It occurs in the Peace Lowland and the Spatsizi area, along the northern boundary at Fireside on the Liard River, and in the vicinity of Atlin Lake and the Saint Elias Mountains in the extreme northwestern part of the province.

This kinglet occurs in every ecoprovince, but its centre of winter abundance is in the Fraser Lowland and throughout southeastern Vancouver Island and the Sunshine Coast in the Georgia Depression (Fig. 391).

The Golden-crowned Kinglet occurs from sea level to 2,100 m (Mount Scaia) in the nonbreeding seasons. It prefers dense coniferous forests (Fig. 392), but wanders widely; it can frequently be found in deciduous and mixed forests, shrublands, and thickets. It often travels in mixed-species flocks that may include Downy Woodpeckers, Ruby-crowned Kinglets, chickadees, Brown Creepers, Hutton's Vireos, nuthatches, warblers, and Chipping Sparrows.

Across southern areas of the province, the Golden-crowned Kinglet occurs at low elevations and prefers the more humid forest types. In the southern and central ecoprovinces, the species is present throughout the year (Fig. 393), but at lower elevations it appears to be more numerous in winter than in summer. In the Central Interior and the northern ecoprovinces, the Golden-crowned Kinglet is migratory, and is seen only rarely in winter (Figs. 393 and 394).

On eastern Vancouver Island, and in the lower Fraser River valley of the Georgia Depression, the onset of spring sees the start of a steady decline, from March through July, in the numbers of Golden-crowned Kinglets recorded (Fig. 394).

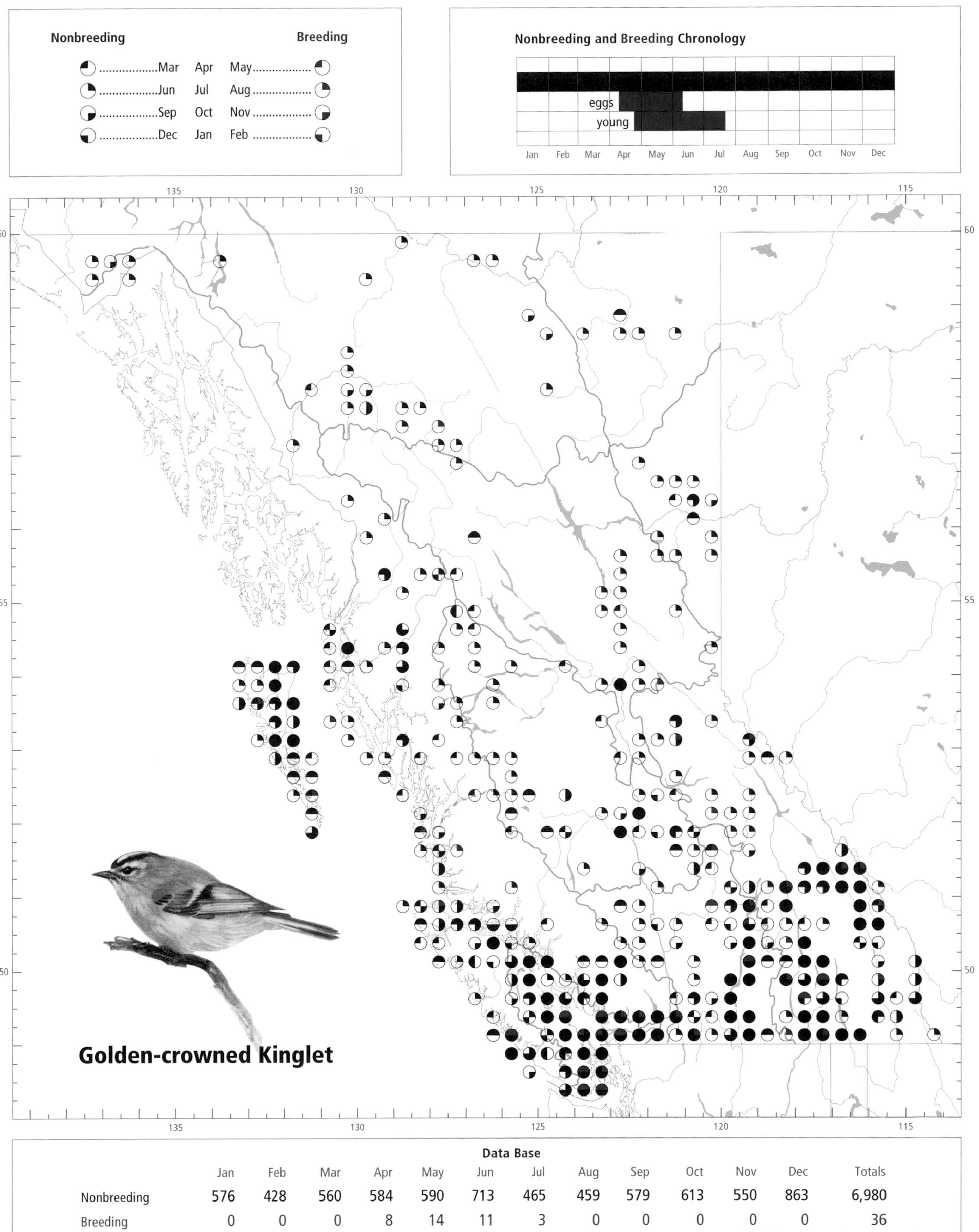

Data Base

	Jan	Feb	Mar	Apr	May	Jun	Jul	Aug	Sep	Oct	Nov	Dec	Totals
Nonbreeding	576	428	560	584	590	713	465	459	579	613	550	863	6,980
Breeding	0	0	0	8	14	11	3	0	0	0	0	0	36

Part of this decline may result from the dispersal of winter flocks as birds pair and establish their nesting territories, and part may reflect emigration. On the southern portions of the Coast and Mountains, after an initial decline in numbers in February and March, numbers in April doubled, suggesting a migration going through. Numbers decline again in May. On the Queen Charlotte Islands, the arrival of the spring migrants is reflected in increased numbers recorded in April and May. On the north coast, arrivals are apparent in May (Fig. 394).

In the Southern Interior and Southern Interior Mountains, the numbers of Golden-crowned Kinglets decline from February through May, as part of the wintering population leaves for nesting areas elsewhere (Fig. 394), but the change in numbers is not as dramatic as in the Georgia Depression. In the Cariboo and Chilcotin plateaus, the returning migrants begin to arrive in March; they reach the Sub-Boreal Interior and the Boreal Plains in April, and the 2 most northerly ecoprovinces, the Northern Boreal Mountains and the Taiga Plains, in May (Figs. 393 and 394).

From the northern boundary south through the Central Interior, the autumn migration begins in August, and most birds have left these regions by the end of September or early October. In the Southern Interior and the Southern Interior Mountains, where there are substantial wintering populations, the highest numbers of Golden-crowned Kinglets are reached in August and September. This probably reflects a migration southward out of British Columbia (Fig. 394).

In the interior, the Golden-crowned Kinglet is not a regular wintering species north of latitude 52°N. The most northerly wintering records are from the Skeena River valley at Smithers (Pojar 1983) and from Prince George, but they are isolated occurrences. Winter records for the Williams Lake area and Riske Creek are only somewhat more frequent. Along the coast, the northernmost winter records are at Prince Rupert and Terrace, but there are few winter records for the Northern Mainland Coast. On the Queen Charlotte Islands, December and January numbers are comparable to those of the summer months. The Golden-crowned Kinglet appears to be less numerous in winter than in summer on Western Vancouver Island, but Hatler et al. (1978) list the species as common throughout the year in Pacific Rim National Park.

Figure 392. Nonbreeding habitat of the Golden-crowned Kinglet includes dense, mixed Douglas-fir and western hemlock forests (Mount Douglas Park, Victoria, 9 September 1994; R. Wayne Campbell).

Lepthien and Bock (1976) found that the Golden-crowned Kinglet preferred to winter in areas with significant levels of winter precipitation, including the moist forests of the Pacific Northwest. Root (1988) noted that the species tended to be limited to areas with minimum January temperatures above –18°C, and that all areas with high winter abundance

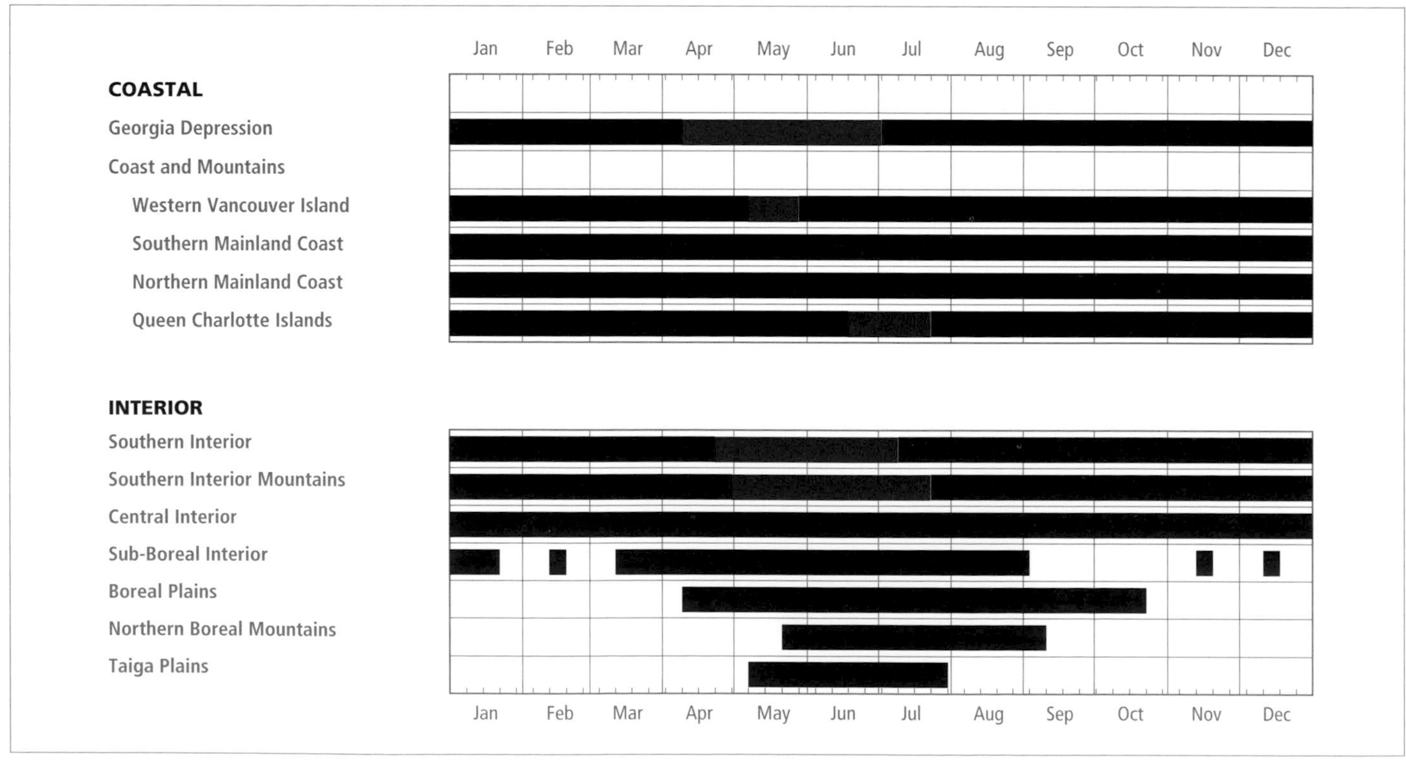

Figure 393. Annual occurrence (black) and breeding chronology (red) for the Golden-crowned Kinglet in ecoprovinces of British Columbia.

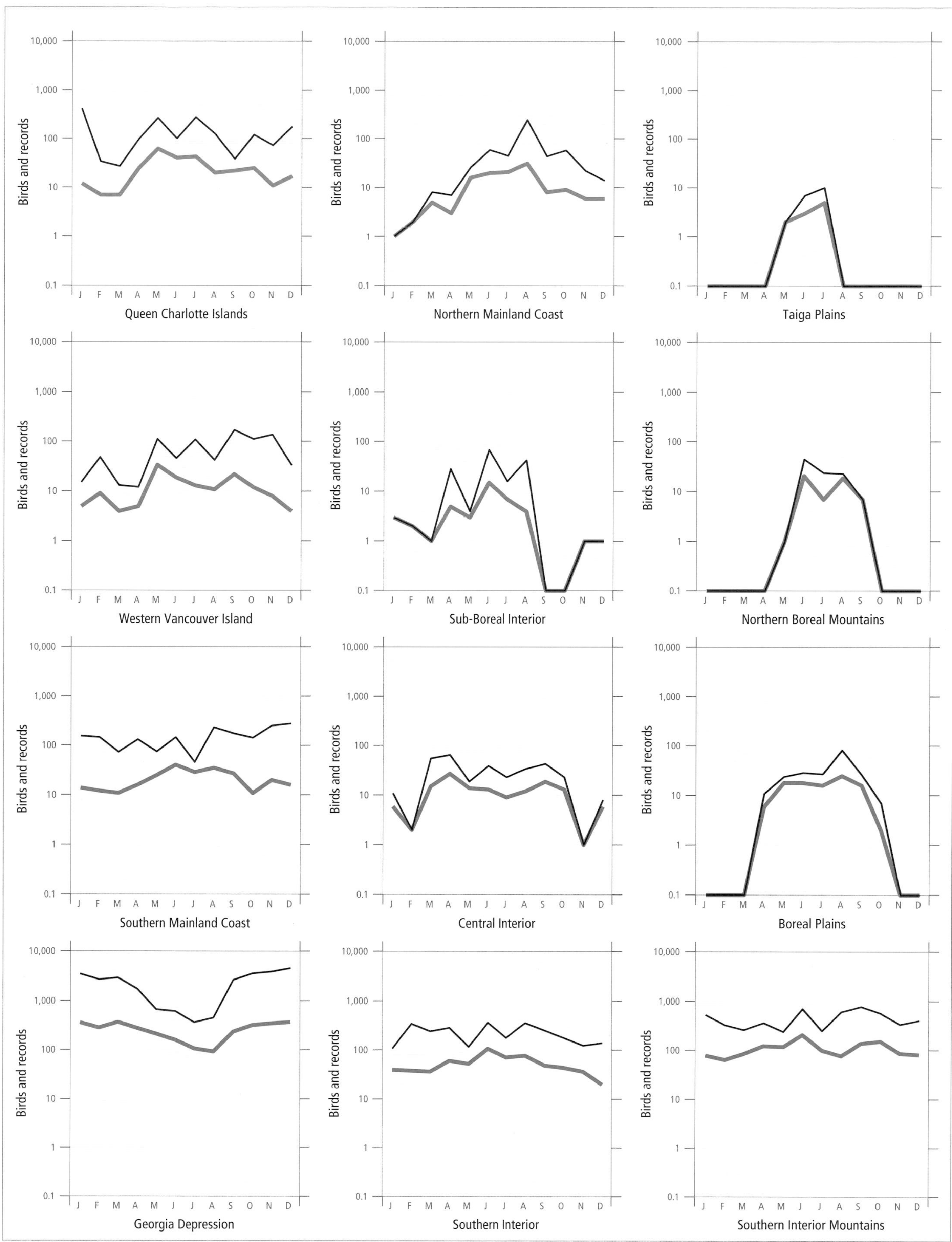

Figure 394. Fluctuations in total numbers of birds (purple line) and total number of records (green line) for the Golden-crowned Kinglet in ecoprovinces of British Columbia. Christmas Bird Count and nest record data have been excluded.

in North America occurred where temperatures seldom fall below freezing.

The Golden-crowned Kinglet has been recorded regularly in the northern half of the interior from 6 April to 31 August; in the southern half it occurs throughout the year (Fig. 393). On the coast, it occurs throughout the year on the Queen Charlotte Islands, the Southern Mainland Coast, the Georgia Depression, and Western Vancouver Island. Although there are records for every month for the Northern Mainland Coast, few are from the winter months.

BREEDING: The known breeding distribution of the Golden-crowned Kinglet is fairly scattered throughout the southern quarter of the province; it likely nests in suitable habitat over its summer range in British Columbia, but the nests are seldom found. On the coast, it is known to nest on Vancouver Island, the southwest mainland coast, and the Queen Charlotte Islands. In the interior, nesting has been documented in the Thompson Basin, the Okanagan valley, and the east and west Kootenays. There are also reports of fledged young from northern parts of the province, including 2 from the Northern Boreal Mountains; these suggest more widespread nesting than is indicated by the current records.

In summer, the highest numbers of Golden-crowned Kinglets occur on southeastern Vancouver Island in the Georgia Depression (Fig. 391). An analysis of Breeding Bird Surveys for the period 1968 through 1993 shows that the number of birds on the coastal routes has decreased at an average annual rate of 2% (Fig. 395); an analysis of interior routes for the same period could not detect a net change in numbers.

The Golden-crowned Kinglet has been reported nesting along the coast from near sea level to 435 m elevation. In the interior, it has been reported from 350 to 2,070 m elevation.

In British Columbia, as elsewhere in North America (Bent 1949; Galati 1991), nesting is almost always associated with forests in which spruce, hemlock, subalpine fir, or Douglas-fir are important elements. The densely vegetated but pendulous limbs of these trees appear to provide the ideal conditions needed by the Golden-crowned Kinglet for its nest site. The forests occupied by nesting pairs are primarily dense old-growth or advanced second growth, more often mixed forests of fairly open structure. Even isolated trees in parks or similar settings may be used. There is just a single record of this kinglet nesting within a major city, even where the stands of trees would appear to be suitable.

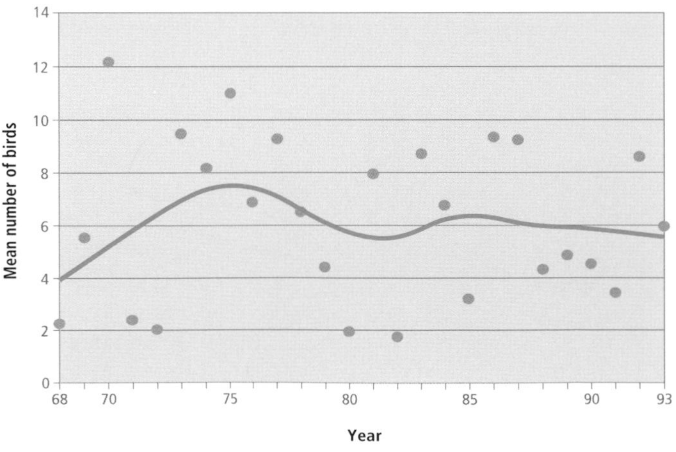

Figure 395. An analysis of Breeding Bird Surveys for the Golden-crowned Kinglet in British Columbia shows that the number of birds on coastal routes decreased at an average annual rate of 2.0% over the period 1968 through 1993 ($P < 0.10$).

The Golden-crowned Kinglet has been recorded breeding in British Columbia from 10 April (calculated) to 21 July (Fig. 393).

Nests: Most nests were in coniferous trees (9 of 15), 2 were in deciduous trees, and 3 were built in unidentified trees. Nests were found mainly near the end of a branch, with the structure attached to adjacent branchlets (Fig. 396). Only a few were attached to a branch close to the trunk. The body of the nest was composed of new and old moss and lichens interwoven with fine plant fibres, bark strips, and grass (Fig. 396). Conifer needles were usually present. The nest cups were lined with soft vegetable fibres, small feathers, or animal fur.

The heights for 17 nests ranged from 4.5 to 15.0 m, with 8 nests between 7.5 and 10.5 m. However, these data are not from a systematic search and are biased in favour of accessible nests. In Minnesota, the mean height ($n = 19$) was 15.3 m, with a range of 8.2 to 19.6 m (Galati and Galati 1985). Nest building in Minnesota began as early as 18 April and required 4 to 6 days (Galati and Galati 1985).

Eggs: Dates for 12 clutches ranged from 18 April to 10 June, with 8 found between 24 April and 30 May. Calculated dates indicate that eggs may be laid as early as 10 April. Clutch size ranged from 4 to 9 eggs (4E-2, 5E-1, 7E-2, 8E-6, 9E-1), with 6 having 8 eggs. In Minnesota, Galati and Galati (1985) found that each of 8 first clutches had 9 eggs; 2 complete second clutches had 8 eggs and 4 had 9. The incubation period in Minnesota was 14 or 15 days.

Young: Dates for 13 broods ranged from 24 April to 21 July, with 9 recorded between 22 May and 6 June. Broods ranged from 1 to 6 young ($n = 6$) (1Y-2, 2Y-3, 6Y-1). There are no observations of the nestling period for British Columbia. In Minnesota, the nestling period ranged from 16 to 19 days (Galati and Galati 1985). There, the Golden-crowned Kinglet was double-brooded, with first clutches initiated between 12 May and 27 May, and second clutches between 22 June and 1 July. Females started building a second nest as early as 8 days after the hatching of the first brood, while the male cared for the nestlings. Dates of clutches in British Columbia suggest that at least part of the population may be double-brooded.

Brown-headed Cowbird Parasitism: Of 22 nests recorded with eggs or young in British Columbia, 3 were parasitized by the cowbird (Fig. 396). Six additional instances were recorded of fledgling cowbirds being fed by Golden-crowned Kinglet foster parents. Friedmann and Kiff (1985) report 5 instances of parasitism in North America.

Nest Success: Insufficient data. In Minnesota, 102 eggs from 12 nests produced 62 fledglings, a success rate of 61% (Galati and Galati 1985).

REMARKS: Two geographic races of Golden-crowned Kinglet occur in British Columbia (American Ornithologists' Union 1957). Cowan (1939) found the subspecies *R. s. satrapa* in the Peace Lowland; Rand (1944) found it along the Alaska Highway at Summit Pass, at the transition between the Taiga Plains and the Northern Boreal Mountains, but it was scarce. The rest of the province, from Atlin south, is occupied by the western subspecies, *R. s. olivaceus* (Swarth 1926). Some ornithologists (Godfrey 1986) recognize a third subspecies, *R. s. amoenus*, as the race inhabiting the interior of British Columbia (van Rossem 1945), and restrict the range of *R. s. olivaceus* to the coast.

See Galati (1991) for a personal account of the breeding ecology of the Golden-crowned Kinglet.

Figure 396. Golden-crowned Kinglet nest with 6 eggs, 1 of which is a Brown-headed Cowbird (Mount Douglas, 14 May 1974; R. Wayne Campbell). In British Columbia, nests are placed on conifer branches where branchlets provide support. The nest is constructed mainly of mosses and plant fibres.

NOTEWORTHY RECORDS

Spring: Coastal – Metchosin 1 May 1982-2; Victoria 1 Mar 1975-1; Mount Douglas 14 May 1974-5 eggs and 1 Brown-headed Cowbird egg (Fig. 396); Langley 18 April 1970-8 eggs; Cultus Lake 7 Apr 1968-50; Tofino Inlet 27 May 1931-6 bob-tailed young fed by parents, could just flutter from branch to branch; Stubbs Island 15 Apr 1978-5; Whytecliffe Park 30 Mar 1980-150; Pitt Lake 21 Apr 1974-65; Black Mountain (North Vancouver) 10 May 1974-20; Lower Adam River 21 Mar 1984-6; Egmont 27 Mar 1975-27; Alice Lake 3 May 1969-10; Kitlope Lake 3 May 1991-8; Tlell 10 Mar 1972-10; Kitkatla Inlet 25 Apr 1976-5; Delkatla Slough 20 Apr 1979-12; Langara Island 22 May 1977-23; Prince Rupert 5 Mar 1983-3. **Interior** – Stirling Creek 14 Apr 1971-20; Riondel 18 Apr 1981-20; Nakusp 20 Mar 1977-16; Enderby 29 Apr 1942-7 eggs (UBC 1418); Celista 16 Mar 1948-30; Kootenay National Park 9 May 1975-8 eggs; Soda Lake 21 May 1964-5; Riske Creek 31 Mar 1984-12; Chilcotin 6 Apr 1984-6; Kleena Kleene 9 Apr 1955-1 (Paul 1959); Prince George 6 Apr 1987-4; Stoddart Creek 11 Apr 1987-4; Telegraph Creek 27 May 1919-1 (MVZ 40279; Swarth 1922); Fort Nelson 13 May 1987-1.

Summer: Coastal – Langley 29 Jun 1975-2 nestlings; Miracle Beach 29 Jul 1974-20; Mitlenatch Island 5 Aug 1965-1 (Campbell 1965); Rugged Point 16 Jul 1985-30; Brooks Peninsula 14 Aug 1981-20; Nahwitti Lake 9 Jun 1978-8; Goose Island 26 Jun 1988-10; Ramsay Island 9 Jul 1977-60; Skincuttle Inlet 7 Jul 1977-2 fledglings; Gray Bay 6 Aug 1979-30; Hippa Island 20 Jul 1977-adult feeding weak flying fledgling; Lakelse Lake 17 Aug 1977-25. **Interior** – Three Brothers Mountain 10 Aug 1974-25; Manning Park 26 Jun 1983-78, 14 Aug 1957-3 fledglings; Slocan 4 Jun 1981-nestlings; Coldstream 2 Jul 1956-2 nestlings; Shuswap Lake Park 5 Jun 1973-nestlings; Stillwater Lake (Tweedsmuir Park) 13 Aug 1982-10; Young Creek 18 Jul 1978-7; Moose River 24 to 29 Jul 1911-common (Riley 1912); Mount Robson 21 Jul 1973-4 recently fledged young; Swiftcurrent Creek 15 Aug 1973-87; Prince George 30 Jun 1969-10; Purden Lake 27 Jun 1979-3 fledglings; Francois Lake 4 Jun 1977-6; Kinuseo Falls 30 Aug 1981-1; Bear Lake 15 Aug 1975-6; Chetwynd 20 Jun to 8 Jul 1969-43; Charlie Lake 23 Jun 1985-5; Beatton Provincial Park 13 Jul 1983-3, 31 Aug 1986-30; Rognaas Creek 26 Jun 1976-6; Pink Mountain (at Alaska Highway) 7 Jun 1994-4; Puti Lake 6 Aug 1976-3; Fort Nelson 19 Jun 1976-2; Parker Lake 6 Jul 1978-3; Clarke Lake 28 Jul 1985-1; junction Coal and Liard rivers 11 Jul 1978-6.

Breeding Bird Surveys: Coastal – Recorded from 26 of 27 routes and on 75% of all surveys. Maxima: Chemainus 24 Jun 1973-48; Gibsons Landing 4 Jul 1973-46; Queen Charlotte City 25 Jun 1994-39. **Interior** – Recorded from 53 of 73 routes and on 59% of all surveys. Maxima: Gerrard 18 Jun 1978-34; Illecilewaet 9 Jun 1983-26; Kootenay 6 Jun 1994-25.

Autumn: Interior – Haines Road 8 Sep 1972-1; Summit Pass (Alaska Highway) 5 Sept 1943 (Rand 1944); Tatogga Lake 3 Sep 1977-1; Beatton Provincial Park 1 Sep 1985-3; Riske Creek 7 Oct 1986-4; Empire Valley 25 Sep 1988-8; Glacier National Park 5 Oct 1982-15; Mount Revelstoke 12 Sep 1963-60; Shuswap Park 23 Sep 1982-30; Stoddart Creek 18 Oct 1986-1; Scotch Creek 7 Oct 1964-14; Nakusp 21 Nov 1976-34; Mount Scaia (nw Edgewood) 8 Oct 1995-flock at 2,100 m; Logan Lake 9 Sep 1990-2; Sparwood 24 Oct 1983-1, 3 Dec 1982-1 (Fraser 1984); Princeton 17 Nov 1979-10. **Coastal** – Hippa Island 17 Sep 1983-10; Masset Sound 20 Nov 1974-20; Minette Bay 29 Sep 1974-20; Namu 23 Oct 1981-20; Port Neville 16 Nov 1965-25; Carrington Lagoon 20 Sep 1975-65; Garibaldi Lake 3 Oct 1965-30; Alice Lake Provincial Park 2 Nov 1986-30; Cypress Bowl Provincial Park 27 Sep 1987-200; Ucluelet 27 Sep 1986-30; Victoria 20 Oct 1978-189; Chatham Island 29 Nov 1957-250.

Winter: Interior – Smithers 2 Jan 1983-1; Prince George 2 Jan 1991-1, first winter record (Siddle 1991b), 17 Feb 1992-1; Williams Lake 6 Jan 1979-6 at –40°C; Riske Creek 11 Dec 1988-2; Nakusp 31 Dec 1978-36; Kelowna 31 Dec 1977-12; 1 Jan 1977-17; Sirdar 1 Jan 1981-2. **Coastal** – Terrace 2 Jan 1972-8; Alliford Bay 1 Jan 1942-25; Skidegate Inlet 30 Dec 1982-18; Cape St. James 17 Jan 1982-301; Kitimat River 6 Feb 1975-1; Port Neville 1 Jan 1976-10; Lower Sarita River 6 Dec 1978-15; Alta Lake 31 Dec 1935-8; Egmont 22 Feb 1974-65; Harrison Hot Springs 12 Dec 1982-52; Addington Marsh, Coquitlam 1 Dec 1989-60; Vedder Crossing 26 Dec 1937-80 (Ricker 1935); Ucluelet 3 Jan 1976-2; Weeks Lake 20 Feb 1983-100; River Jordan 31 Dec 1985-2; Saanich 1 Jan 1976-5.

Christmas Bird Counts: Interior – Recorded from 22 of 27 localities and on 74% of all counts. Maxima: Shuswap Lake Provincial Park 3 Jan 1981-211; Nakusp 30 Dec 1979-108; Salmon Arm 17 Dec 1988-93. **Coastal** – Recorded from 31 of 33 localities and on 95% of all counts. Maxima: Victoria 19 Dec 1992-**3,337**, all-time North American high count (Monroe 1993); Vancouver 20 Dec 1992-1,907; Sooke 31 Dec 1988-1,721.

Ruby-crowned Kinglet
Regulus calendula Linnaeus

RCKI

RANGE: Breeds from northwestern and north-central Alaska, central Yukon, northwestern and southern Mackenzie across the Prairie provinces east to Labrador and Newfoundland; south along the cordillera to southern California and across the north-central United States. Winters from southern British Columbia, Idaho, and southern Ontario south through forested mountains as far as Guatemala and Baja California.

STATUS: On the coast, *uncommon* to locally *very common* during spring migration, *uncommon* in summer, and *uncommon* to *common* in winter in the Georgia Depression Ecoprovince; *uncommon* migrant and summer visitant on Western Vancouver Island, in the southern and northern Coast and Mountains Ecoprovince; *rare* in winter in the southern Coast and Mountains; *casual* in winter on Western Vancouver Island and in the northern Coast and Mountains; *very rare* in spring and summer and *casual* in autumn on the Queen Charlotte Islands.

In the interior, *fairly common* to *very common* migrant and summer visitant to the Southern Interior, Southern Interior Mountains, Central Interior, and Sub-Boreal Interior ecoprovinces, becoming *uncommon* in the Boreal Plains, Taiga Plains, and Northern Boreal Mountains ecoprovinces. In winter, *very rare* in the Southern Interior and Southern Interior Mountains.

Breeds.

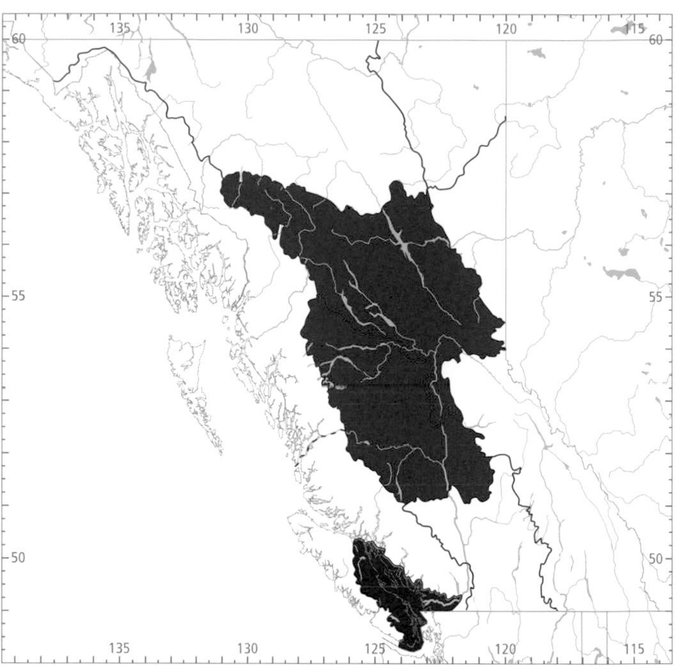

Figure 397. In British Columbia, the highest numbers for the Ruby-crowned Kinglet in winter (black) occur in the Georgia Depression Ecoprovince; the highest numbers in summer (red) occur in the Central Interior and Sub-Boreal Interior ecoprovinces.

Figure 398. Mixed coniferous forests of Douglas-fir and western hemlock provide winter habitat for the Ruby-crowned Kinglet on the southwest mainland coast (Stanley Park, Vancouver, 25 March 1994; R. Wayne Campbell).

NONBREEDING: The Ruby-crowned Kinglet is widely distributed in coniferous and mixed forests throughout the province during spring and autumn migration, although it has not been reported from large areas of the Coast Mountains, the Rocky Mountain Trench north of Donald, and the Muskwa, Cassiar, Kechika, and Tahltan mountains. Its centre of abundance in winter is in the Georgia Depression (Fig. 397); winter records are scarce elsewhere in the province.

It occurs from near sea level to 1,220 m on the coast and from 300 to 2,340 m elevation in the interior. In winter, most records are between sea level and 100 m on the coast and near 390 m in the interior valleys.

The Ruby-crowned Kinglet prefers mixed coniferous forests from mid-elevations to the timberline. Most of its habitats are dominated by Douglas-fir, but it also inhabits Engelmann spruce–subalpine fir, Engelmann spruce–subalpine larch, Engelmann spruce–lodgepole pine, and white spruce–black spruce–tamarack forests. It often uses habitats on the edge of muskegs, lakes, ponds, and alpine meadows. During migration, the Ruby-crowned Kinglet can be found in a wide variety of woodlands and shrublands,

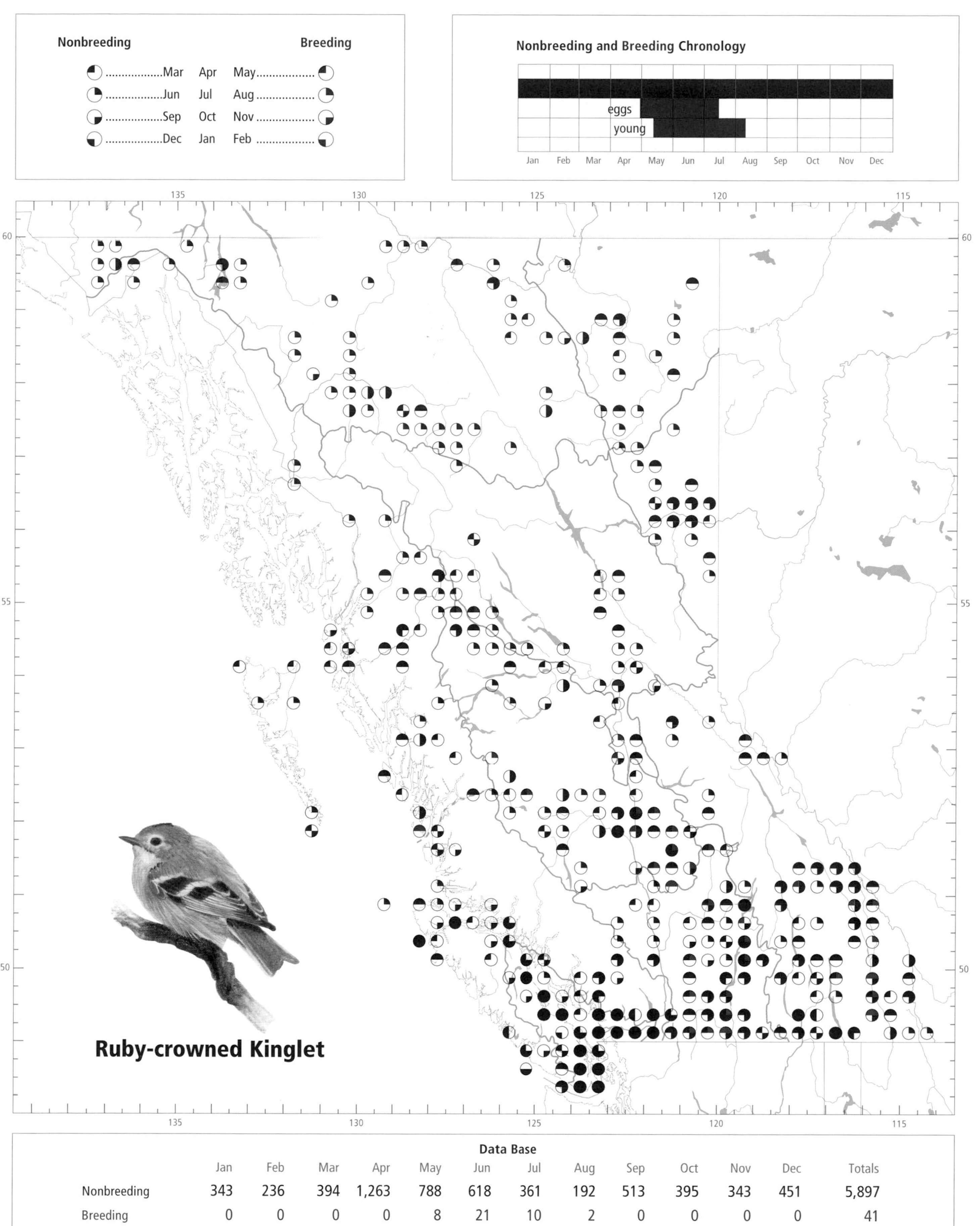

Data Base

	Jan	Feb	Mar	Apr	May	Jun	Jul	Aug	Sep	Oct	Nov	Dec	Totals
Nonbreeding	343	236	394	1,263	788	618	361	192	513	395	343	451	5,897
Breeding	0	0	0	0	8	21	10	2	0	0	0	0	41

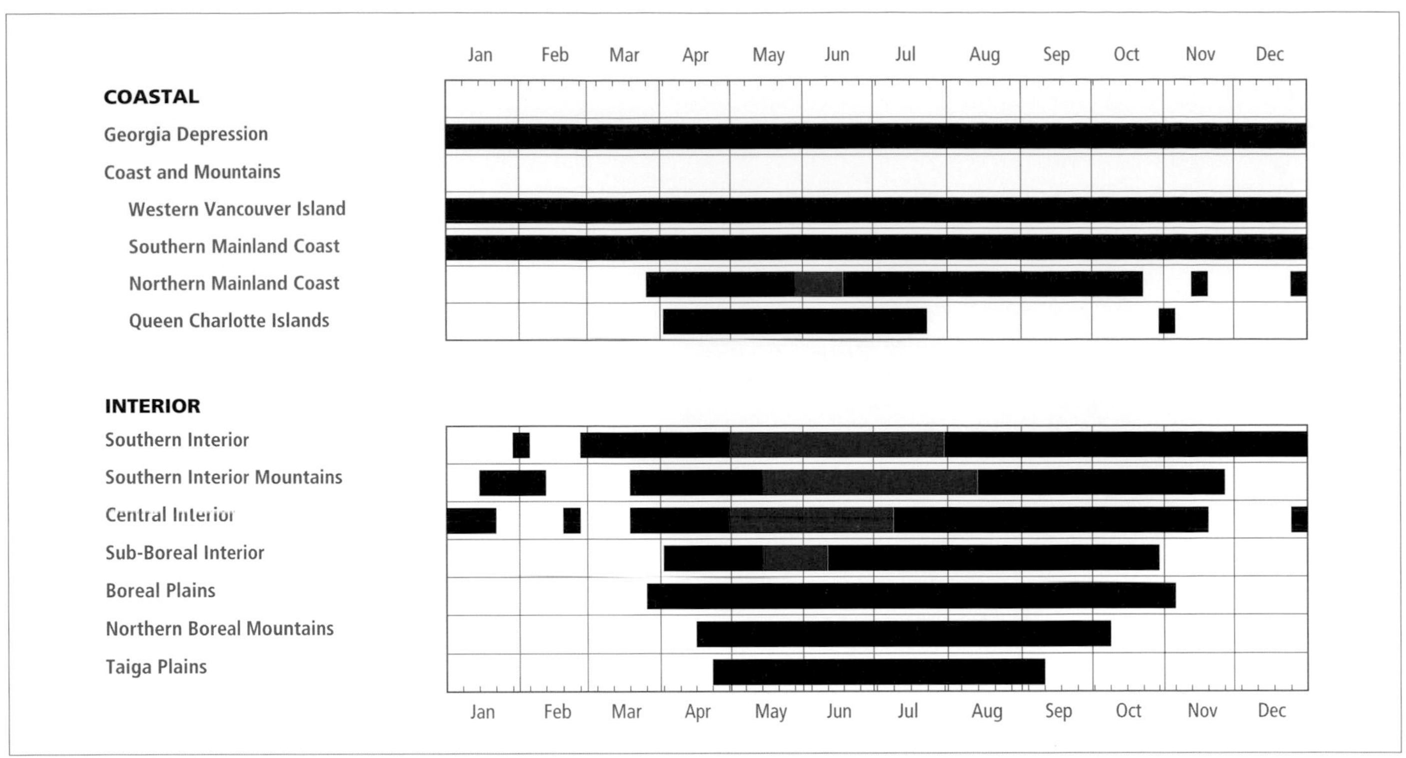

Figure 399. Annual occurrence (black) and breeding chronology (red) for the Ruby-crowned Kinglet in ecoprovinces of British Columbia.

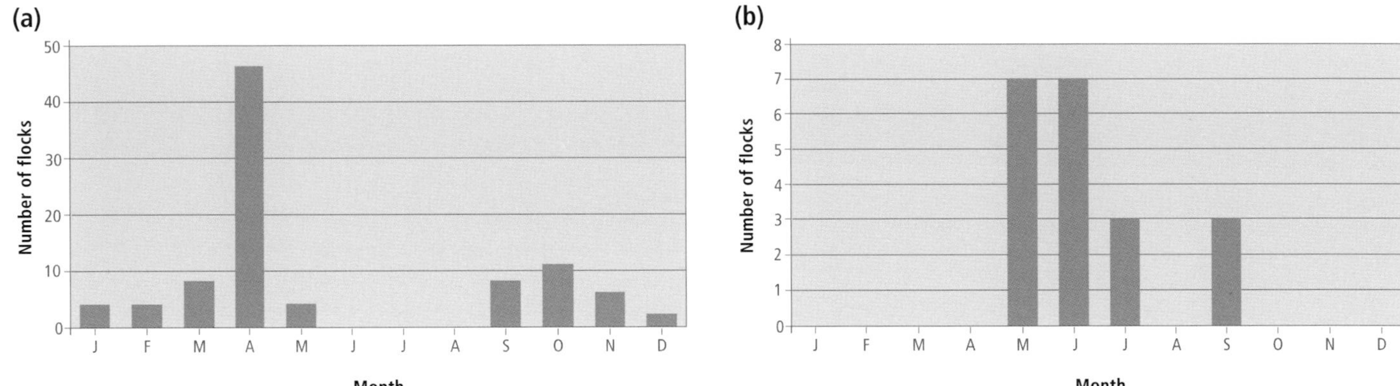

Figure 400. Seasonal distribution of large flocks or groups of flocks (> 14 birds) of the Ruby-crowned Kinglet in the (a) Georgia Depression and (b) Southern Interior ecoprovinces of British Columbia. The figure suggests a fairly substantial April movement in the Georgia Depression Ecoprovince, while the interior movement is more protracted.

and in human-made clearings such as powerline corridors, farms, and gardens. In those habitats it often uses trembling aspen or birch.

In winter, it frequents mixed coniferous (Fig. 398) and mixed coniferous-deciduous forests and second growth that includes Douglas-fir, grand fir, western hemlock, western redcedar, lodgepole pine, willows, red alder, bigleaf maple, vine maple, and tall shrubs. It frequently enters gardens, where it may include crevices of buildings in its foraging.

The Ruby-crowned Kinglet is highly migratory over most of British Columbia (Figs. 399 and 401), and moves both latitudinally and altitudinally to avoid winter cold. Root (1988) notes that this kinglet is fairly common in winter in almost all parts of North America in which January temperatures below freezing are rare.

The arrival of spring migrants in the Georgia Depression is masked by the continued presence of wintering birds, but numbers rise in late March and reach a peak about the third week of April (Fig. 401). At that time, 5 or more flocks, each with up to 12 Ruby-crowned Kinglets, may pass through a single locality in a day (Fig. 400). These observations of large groups suggest that a major movement takes place through the Georgia Depression. Numbers drop sharply through May and June, as migrants move to higher latitudes or altitudes to nest in the coniferous forests of mountain slopes. Similar patterns apply to other coastal regions, although numbers are smaller and the decline in observed numbers is gradual from early summer through August.

The peak movement of Ruby-crowned Kinglets into the interior follows the earliest migrants by at least a month. Forerunners of the migration may appear in the Southern Interior in early March, in the Southern Interior Mountains and the Central Interior in late March, in the Sub-Boreal Interior and Boreal Plains in early April, and in the north of the province

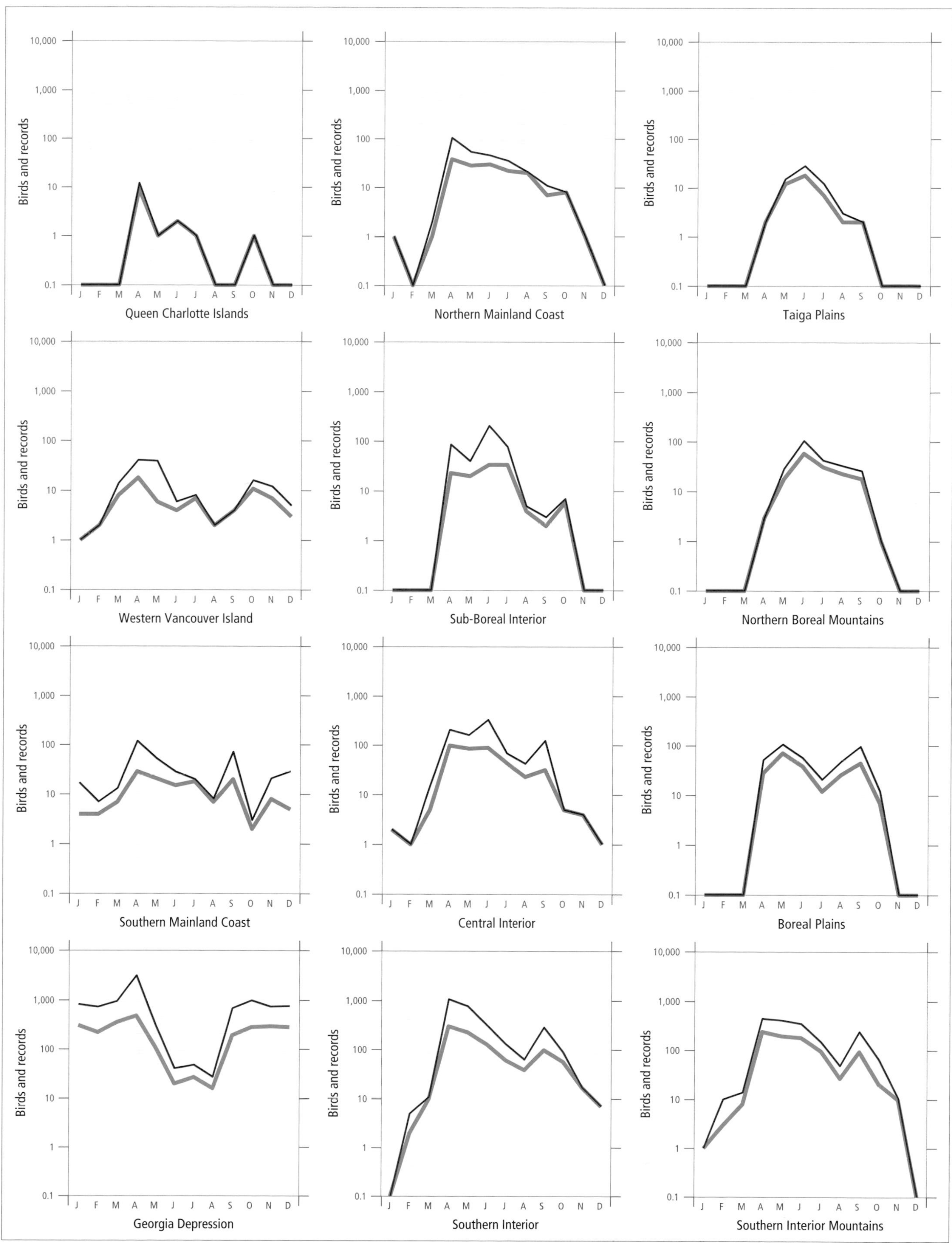

Figure 401. Fluctuations in total number of birds (purple line) and total number of records (green line) for the Ruby-crowned Kinglet in ecoprovinces of British Columbia. Christmas Bird Count and nest record data have been excluded.

near the end of April (Figs. 399 and 401). In the Southern Interior, the number of birds per record is highest in April (Fig. 401). The migration in the Southern Interior and Southern Interior Mountains reaches a peak around the end of April or early May. The main migration reaches the Central Interior and the Prince George region of the Sub-Boreal Interior by the third week of April, the Peace Lowland in the Boreal Plains by the end April or the first week of May, and the far north of the province by late May.

The autumn migration is more protracted than the spring movement (Fig. 401). In some ecoprovinces, it is barely discernible, as declines in numbers coincide with reduced detection (a result of the cessation of songs) and with somewhat reduced observer activity during the summer months. In the Boreal Plains, a small movement is detectable at the end of August and the first week of September, and most birds have left that region by the end of September. In the Vanderhoof region, Munro (1949) found the southward migration at its height between 29 August and 3 September. In the Central Interior, Southern Interior Mountains, and Southern Interior there is an increase in numbers in the last 3 weeks of September, which may reflect the arrival of migrants from the north (Fig. 401). Almost all Ruby-crowned Kinglets have left these regions of the province by the end of October (Figs. 399 and 401). On the coast, autumn migrants move into the Georgia Depression during the last 2 weeks of September and the first week of October (Fig. 401). Numbers appear to be stable through the winter months.

In winter, the Ruby-crowned Kinglet is abundant in the Georgia Depression, and occasionally occurs on the coast as far north as the lower Skeena River valley. In the southern portions of the interior, there are scattered records from the Okanagan valley, from the Columbia valley north as far as Nakusp, and at Williams Lake and Riske Creek.

On the coast, the Ruby-crowned Kinglet is present throughout the year; in the interior, it is regularly present from 21 March through 22 November (Fig. 399).

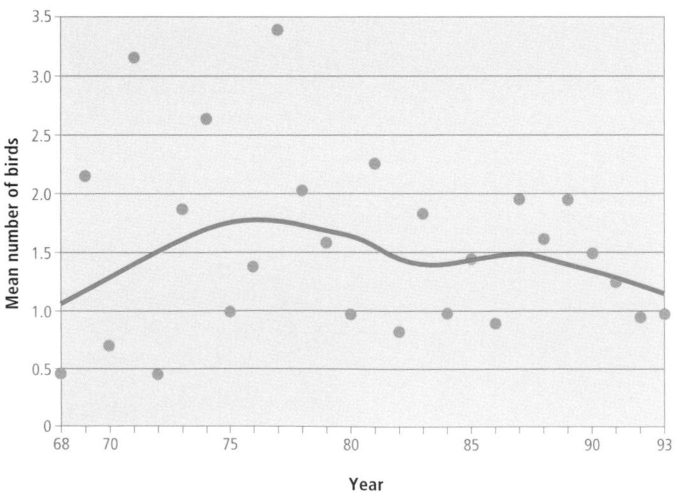

Figure 402. An analysis of Breeding Bird Surveys for the Ruby-crowned Kinglet in British Columbia shows that number of birds on coastal routes decreased at an average annual rate of 1% over the period 1968 through 1993 ($P < 0.05$).

BREEDING: The breeding distribution of the Ruby-crowned Kinglet is not well known, but the species probably breeds throughout its summer range in the province. East of the Coast Mountains, it nests in the Similkameen River valley (Manning Park), the Okanagan valley north to Coldstream and Enderby at elevations mostly above 1,000 m (Cannings et al. 1987), in the Thompson River valley (McQueen Lake), the Rocky Mountain Trench (Cranbrook and Windermere), the Cariboo and Chilcotin plateaus (e.g., Bridge Lake, Loon Lake, 158 Mile House, Stum Lake, and Williams Lake), the Quesnel area, and the Prince George region. All reports of nesting north of Prince George are of adults with fledglings.

The only acceptable record of nesting on the coast is of a nest and egg collected from Porcher Island, near Prince Rupert in the northern Coast and Mountains (Brooks and Swarth 1925). All other coastal reports are without supporting documentation (see REMARKS).

The highest numbers in summer occur in the Central Interior and Sub-Boreal Interior ecoprovinces (Fig. 397). An analysis of Breeding Bird Surveys for the period 1968 through 1993 shows that the number of birds on coastal routes has decreased at an average annual rate of 1% (Fig. 402). Analysis of interior routes for the same period could not detect a net change in numbers.

The Ruby-crowned Kinglet has been reported breeding from near sea level on the north coast to 2,070 m elevation in the interior mountains. It nests in both old-growth and advanced second-growth forests. Suburban and cultivated areas provided 8 of 25 nesting records; however, this is probably biased by observer distribution. Of the remainder that described general habitats, 8 specified coniferous forest, 4 mixed forest (Fig. 403), and 1 deciduous woods as the habitat. In eastern North America, this kinglet has been recorded as a typical bird of the black spruce bogs (Bent 1949). It uses similar habitat during the breeding season in northern Alberta and in the Peace Lowland of British Columbia (Cowan 1939), but nests have not yet been found there (Fig. 404). Black spruce bogs are a dominant feature of the Boreal and Taiga plains.

In her study of the bird communities in interior trembling aspen forests, Pojar (1993) found 1 to 8 Ruby-crowned Kinglets per 10 ha in mature (50 to 60 years) aspen stands; 2 to 17 birds per 10 ha in old (> 100 years) aspen stands; 15 to 31 birds per 10 ha in mixed stands of aspen, spruce, and pine; and the same in stands of conifers (pine). This kinglet was not reported from clearcuts or thickets of aspen saplings, most of them younger than 7 years. Recently, it has been reported at densities of 2.3 birds per 10 ha in old-growth stands of Engelmann spruce–subalpine fir at 1,500 m in the Quesnel Highlands (L. Waterhouse pers. comm.).

The Ruby-crowned Kinglet has been recorded breeding in British Columbia from 30 April (calculated) to 10 August (Fig. 399).

Nests: Most nests have been found in coniferous trees (22 of 25), including 13 Douglas-fir (Fig. 405), 6 spruce, 2 western hemlock, and 1 pine. Nests were attached to a limb, usually near the end (8 of 10). The heights for 21 nests ranged from 0.6 to 15.0 m, with 16 nests between 2 and 9 m. These records were not from systematic nest searches and are likely biased towards nests near the ground. Nests were nearly spherical, pendent structures composed primarily of mosses and plant fibres, with lichens, plant down, grass, feathers, and bark strips. Occasionally twigs, rootlets, and fur were used.

Eggs: Dates for 16 clutches ranged from 16 May to 15 July, with 9 recorded between 27 May and 22 June. Calculated dates indicate that eggs can occur as early as 30 April. Sizes of 15 clutches ranged from 1 to 9 eggs (1E-2, 2E-1, 4E-2, 5E-2, 6E-3, 7E-2, 8E-1, 9E-2), with 9 having 4 to 7 eggs (Fig. 405). The incubation period is about 12 days (Ehrlich et al. 1988).

Young: Dates for 11 broods ranged from 12 May to 10 August. Sizes of 5 broods ranged from 2 to 7 young (2Y-1, 3Y-1, 6Y-2, 7Y-1). The nestling period is about 12 days (n = 1).

Brown-headed Cowbird Parasitism: In British Columbia, 3 of 21 nests found with eggs or young were parasitized by the cowbird. In addition, there were 8 records of Ruby-crowned Kinglets feeding fledged cowbird young. Friedmann and Kiff (1985) list only 2 instances in North America of a Ruby-crowned Kinglet successfully raising a cowbird. One of these records notes a kinglet feeding a fledged cowbird 80 km north of Kamloops on 1 August 1979 (Boxall 1981).

Nest Success: Insufficient data.

REMARKS: Two geographic races of the Ruby-crowned Kinglet occur in British Columbia. Their status was recently reviewed by Browning (1979), who showed that the race *R. c. grinnelli* inhabits the Pacific slope from southern Alaska to southwestern British Columbia; it is characterized by having significantly shorter wings than those elsewhere in the province. The other race, *R. c. calendula*, occurs throughout the rest of North America. Browning (1979) also gives evidence that the coastal population is nonmigratory in a latitudinal sense, as it does not occur in winter along the coastal slope of the continent south of British Columbia. However, data from this study reveal marked reductions in the numbers of this kinglet from October to March, with an attendant increase in numbers in the Fraser Lowland and the southeast coast of Vancouver Island. This suggests that a latitudinal movement does take place, but that it may be confined to coastal British Columbia. The numbers that enter the Georgia Depression to winter may also be the result of an altitudinal movement.

It is possible that the large spring movement noted on the southwest coast in April includes kinglets of the continental population *R. c. calendula* that are using the coastal route to reach the nesting grounds east of the coastal mountain

Figure 403. Mixed forests of white spruce, birch, and willow provide breeding habitat for the Ruby-crowned Kinglet in British Columbia (Pantage Lake, 24 May 1994; R. Wayne Campbell).

Figure 404. In the Peace River area of British Columbia, the Ruby-crowned Kinglet frequents black spruce muskegs in summer (Boot Lake, 11 June 1996; R. Wayne Campbell).

ranges. An analysis of the migrants to determine the subspecies involved has yet to be done.

A number of reports in the literature suggest that the Ruby-crowned Kinglet breeds along the British Columbia coast, but, except for the Porcher Island record, evidence of a nest containing eggs, nestlings, or recently fledged young is lacking. For example, nesting records reported from San Josef and Spider islands (Munro and Cowan 1947) are of fledged young. A record from Goose Island is of a female with an egg in the oviduct (Guiguet 1953). Laing (1942) notes that he found the Ruby-crowned Kinglet breeding in mixed woods of spruce, Douglas-fir, birch, willow, and trembling aspen, at salt water, along the shore of the Dean River flat, as well as in patches of alpine fir near the timberline (1,580 m) at the head of Mosher Creek in the Stuie area. However, detailed supporting information is not mentioned. A report that the species had been found nesting at Cumshewa Inlet on southern Graham Island in the Queen Charlotte Islands was based only upon an adult male taken there in July (Osgood 1901). There is still no evidence that the Ruby-crowned Kinglet nests anywhere on the Queen Charlotte Islands, contrary to the breeding distribution map in Godfrey (1986). Horwood (1992) shows the Ruby-crowned Kinglet as a breeding species in the Kitimat River valley, but this, too, is probably inferred from summer presence.

A Ruby-crowned Kinglet banded near Modesto, California, on 5 December 1983 was recovered in Vancouver on 24 April 1985.

See Laurenzi et al. (1982) for more information on the winter biology of this kinglet.

Figure 405. Nest and eggs of the Ruby-crowned Kinglet (west of Williams Lake, 21 June 1978; R. Wayne Campbell).

NOTEWORTHY RECORDS

Spring: Coastal – Sombrio Beach 8 Apr 1985-6; Wickaninnish 17 Mar 1981-2; Stanley Park (Vancouver) 22 Mar 1970-34, 19 Apr 1985-225 on census; North Vancouver 18 Apr 1984-150; Harrison Hot Springs 29 Apr 1986-26; Golden Ears Park 6 May 1963-10; Klaskish River 19 May 1978-6 at 1,036 m elevation; Namu 22 Mar 1982-2; Cape St. James 2 Apr 1982-1; Kitimat River 13 Apr 1975-12; 10 May 1975-14; Langara Island 8 May 1981-1; Kelsall Lake (Haines Highway) 20 May 1979-3. **Interior** – Cranbrook 16 May 1915-2 eggs; West Bench (Penticton) 8 May 1974-50 (Cannings 1974); Naramata 21 Mar 1972-1; Kelowna 12 Apr 1974-50; Edgewood 29 Mar 1924-6; Kootenay National Park 16 May 1983-20, 27 May 1980-nest completed, pair mating; Tranquille 15 Apr 1989-350, in every bush (Campbell 1989c); Yoho National Park 29 Apr 1977-37; Loon Lake (Clinton) 26 May 1963-4 eggs; Watson Lake 19 May 1964-10; Westwick Lake 12 May 1958-adults feeding nestlings; Scout Island (Williams Lake) 15 Apr 1978-15; Anahim Lake 27 Apr 1932-3 (MCZ 283991-4); Prince George 6 Apr 1982-6, 13 Apr 1985-20, 26 May 1974-7 eggs, 27 May 1978-20; Stoddart Creek 28 Apr 1987-13; Jackfish Creek 25 May 1975-3; Taylor 15 May 1983-6; Beatton River 11 Apr 1985-1; Parker Lake 30 May 1975-1; Fort Nelson 27 Apr 1987-1; Atlin Lake 29 May 1987-7.

Summer: Coastal – Langley 9 Jul 1977-6; Northwest Bay 31 Aug 1962-8; "Cassiope" Lake (Brooks Peninsula) 31 Jul 1981-2 (Campbell and Summers, in press); Port Hardy 10 Jun 1978-2; Goose Island 21 Jun 1948-nesting pair collected (Guiguet 1953); Spider Island 6 Jul 1939-fledgling; Greenville 16 Jun 1981-10; Hippa Island 1 Jun 1977-1; Kemano Bay 3 Jul 1985-5; Tlell 17 Jul 1974-1; Porcher Island 1 Jun 1921-1 egg (USNM 459370); Kitimat Jun 1990-adult feeding 3 nestlings (Horwood 1992); Lakelse Lake Park 23 Aug 1977-2; Telkwa 22 Aug 1919-1 (NMC 13800); Great Glacier 10 Aug 1919-1. **Interior** – Manning Park 7 Jun 1994-7 eggs, 26 Jun 1983-13, 18 Jul 1962-nestlings, 19 Jul 1962-4 fledged; Brookmere 17 Jun 1974-22; Anarchist Mountain 4 Jul 1970-10; Botanie Valley 4 Jul 1964-6; Scotch Creek 21 Aug 1966-8; Ewin Creek 20 Jun 1982-30; Windermere 10 Aug 1984-3 nestlings plus cowbird; McQueen Lake 1 Jun 1973-adults feeding nestlings; Bridge Lake 5 Jun 1976-41; Alexis Creek 22 Jun 1974-8 eggs; Williams Lake 1 Jun 1963-40; Horn Lake 30 Jun 1914-9 eggs; Crowfoot Mountain 7 Jul 1973-8; Mount Robson Park 18 Aug 1973-10; Division Lake 18 Aug 1956-7; Sinkut Creek 11 Jun 1975-50; Prince George 2 Jul 1974-10; Smithers 1 Jul 1974-4; Steamboat 14 Jun 1976-7; Fort St. John 31 Jul 1982-6; Beatton Provincial Park 31 Aug 1982-5; Prophet River 19 Jun 1974-1 (DMNH 40842); Dease Lake 18 Jun 1975-23; Kinaskan Lake 3 Aug 1979-2; Fern Lake area 18 Aug 1983-8; Fort Nelson 15 Jun 1975-3; Muncho Lake 31 Jul 1943-1 (NMC 28524); Kelsall Lake 12 Jun 1988-1; Liard 10 Aug 1977-1; Cormier Creek 25 Jul 1981-5.

Breeding Bird Surveys: Coastal – Recorded from 14 of 27 routes and on 22% of all surveys. Maxima: Kwinitsa 18 Jun 1978-26; Saltery Bay 23 Jun 1974-18; Kitsumkalum 14 Jun 1981-11. **Interior** – Recorded from 63 of 73 routes and on 72% of all surveys. Maxima: Ferndale 7 Jun 1969-63; Punchaw 16 Jun 1974-55; McLeod Lake 14 Jun 1969-51.

Autumn: Interior – Atlin 4 Oct 1931-1 last seen (Swarth 1936); Fort Nelson 4 Sep 1986-1; Pearson Mountain 29 Aug 1919-1; Gladys Creek 20 Sep 1976-1; Stoddart Creek 1 Apr 1983-1; Beatton Park 26 Sep 1982-8; Fort St. John 15 Oct 1983-1; Tetana Lake 13 Sep 1938-1; Prince George 23 Oct 1992-1; Williams Lake 25 Sep 1988-30, 5 Nov 1985-1; Glacier National Park 8 Oct 1981-30; Mount Revelstoke 21 Sep 1982-15; Revelstoke 13 Nov 1988-2; Scotch Creek 12 Sep 1962-10; Summerland 6 Nov 1972-1; Kinnaird 22 Nov 1969-1; Vaseux Lake 8 Oct 1975-7; Manning Park 5 Oct 1974-2. **Coastal** – Kelsall Lake (Haines Highway) 8 Sep 1972-6; Cape St. James 29 Oct 1978-1; Namu 6 Oct 1981-2; Green Island 12 Nov 1977-1; Port Neville 8 Nov 1975-3; Skagit Valley 29 Sep 1974-12; Hope 18 Nov 1968-6; Qualicum Beach 9 Nov 1974-21; East Sooke Park 23 Sep 1983-35; Beacon Hill Park 11 Oct 1981-40.

Winter: Interior – Williams Lake 1 Jan 1986-1; Riske Creek 15 Jan 1989-1, 31 Dec 1988-1 at Wineglass Ranch; 100 Mile House 24 Feb 1985-1; Celista 1 Feb 1948-4; Kinnaird 19 Jan 1971-1, 22 Nov 1969-1; Okanagan Landing 29 Dec 1913-1; Taghum 10 Feb 1974-8. **Coastal** – Squamish 2 Jan 1983-8; Harrison Hot Springs 3 Jan 1976-6, 1 Feb 1983-2; Hope 27 Dec 1952-12 (Houlden 1953); Langley 20 Feb 1974-43; Ladner 2 Jan 1960-54 (Erskine 1960a); Tofino 6 Dec 1986-2; Bamfield 28 Feb 1976-1; Sooke 1 Jan 1976-1.

Christmas Bird Counts: Interior – Recorded from 7 of 27 localities and on 8% of all counts. Maxima: Penticton 26 December 1975-7; Kelowna 14 December 1991-5; Creston 2 Jan 1989-4. **Coastal** – Recorded from 24 of 33 localities and on 75% of all counts. Maxima: Victoria 17 December 1988-**353**, all-time Canadian high count (Monroe 1989a); Vancouver 17 December 1989-254; White Rock 30 December 1978-125.

Western Bluebird

Sialia mexicana Swainson

WEBL

RANGE: Resident in western North America from southern British Columbia, western and south-central Montana, and north-central Colorado south through the mountains to northern Baja California, western and southern Nevada, southern Utah, western and southeastern Arizona, western Texas, and New Mexico south to the Mexican highlands.

STATUS: On the coast, *rare* migrant and summer and winter visitant on southeastern Vancouver Island, *very rare* in the Fraser Lowland and *casual* in the Georgia Lowland of the Georgia Depression Ecoprovince; *casual* on Western Vancouver Island and in the southern portions of the Coast and Mountains Ecoprovince.

In the interior, a *fairly common* to locally *common* migrant and summer visitant, *rare* winter visitant, in the Okanagan valley of the Southern Interior Ecoprovince; *rare* to *uncommon* in the east and west Kootenay regions of the Southern Interior Mountains Ecoprovince; *casual* in the Central Interior Ecoprovince.

Breeds.

Figure 406. Adult male Western Bluebird (R. Wayne Campbell). Although populations appear stable in the southern portions of the interior of British Columbia, numbers have declined in recent years in the Georgia Depression Ecoprovince, especially on southern Vancouver Island.

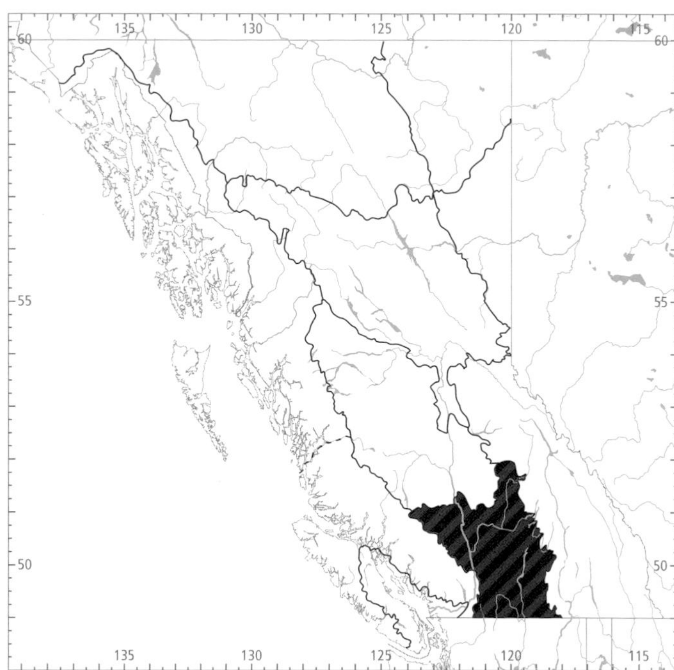

Figure 407. In British Columbia, the highest numbers for the Western Bluebird in winter (black) and summer (red) occur in the Southern Interior Ecoprovince.

CHANGE IN STATUS: Over the past century, populations of Western Bluebirds have declined significantly in the Georgia Depression. Accounts of the bluebird's status in the province in the late 1880s and early 1890s refer to the species as a common summer resident both east and west of the Cascade Mountains, but more numerous to the west (Fannin 1891, 1898; Rhoads 1893; Kermode 1904, 1909). Brooks and Swarth (1925) refer to it as being resident on the lowlands of extreme southwestern British Columbia, and found it in summer on both sides of the Cascades. Later, Cumming (1932), writing on the birds of the Vancouver district, refers to the species as a common visitant with numerous winter records. By the late 1950s a decline was evident. The field notes of Kenneth Racey, one of the most active naturalists in this region over the period 1921 to 1960, noted an increase in the number of nesting Western Bluebirds from 1946 to 1956, but this was followed by a sharp reduction in the numbers recorded on the Fraser River delta. By the 1970s the Western Bluebird had all but disappeared from the southwest mainland coast (Campbell et al. 1972a, 1972b, 1974). Weber et al. (1983) further comment on the decline in numbers, referring to only 2 wintering records of the Western Bluebird in the Boundary Bay area, near Vancouver, in the "past 15 years." As a result of coastal declines, the Western Bluebird was included on the proposed list of threatened and endangered species in British Columbia (Weber 1980).

At the same time, populations on southeastern Vancouver Island appeared stable and, with the initiation of a vigorous nest box program, numbers increased. In fact, from 1961 through 1994, there were 61 reports of nestings. The greatest number of nests reported in 1 year was 16 (1988). However, between 1990 and 1994 not more than 1 or 2 pairs were known to nest in any one year. The nest box program has recently been abandoned as a failure due to loss of habitat, competition with European Starlings and House Wrens, and a decreasing nucleus of breeding birds (Pollock 1990, 1991, 1992, 1993, 1994). Today the Western Bluebird seldom occurs anywhere in the Georgia Depression.

This reduction has not occurred in the Okanagan valley (Cannings et al. 1987) (see REMARKS).

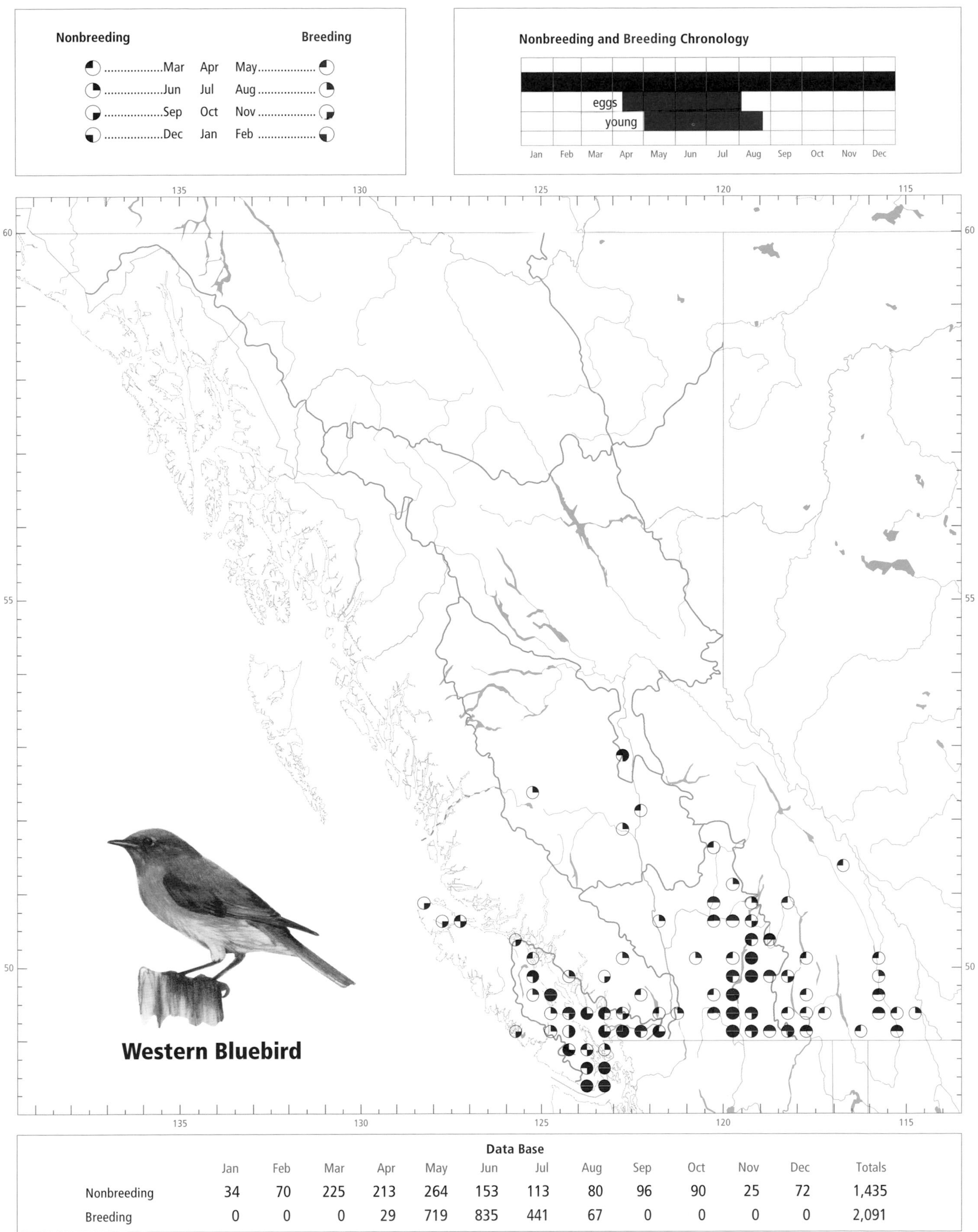

Data Base

	Jan	Feb	Mar	Apr	May	Jun	Jul	Aug	Sep	Oct	Nov	Dec	Totals
Nonbreeding	34	70	225	213	264	153	113	80	96	90	25	72	1,435
Breeding	0	0	0	29	719	835	441	67	0	0	0	0	2,091

NONBREEDING: The Western Bluebird (Fig. 406) is widely distributed across southern portions of the province from Vancouver Island east to the British Columbia–Alberta border, and north through the interior to southern portions of the Central Interior and Sub-Boreal Interior. Formerly it occurred regularly, in small numbers, on southeastern Vancouver Island. There are scattered records for western and northern Vancouver Island and along the mainland coast north to Powell River. In the interior, it is most numerous in the Okanagan valley, but also occurs locally across the province east of the Cascade Mountains from Manning Park to the Rocky Mountain Trench. Small numbers can be found in the Thompson and Nicola valleys and in the east and west Kootenays. It has been reported as far north as Anahim Lake in the Central Interior. The Western Bluebird reaches its highest numbers in winter in the Okanagan valley of the Southern Interior (Fig. 407). It occurs from sea level to about 1,200 m, favouring open plant communities but not requiring the extensive, open habitats sought by the Mountain Bluebird. On southeastern Vancouver Island, as recently as 1990 it occurred more often on the sparsely forested ridges than in the Garry oak woodlands close to sea level.

Habitat includes flower fields; weedy, logged, or burned forests, farms, Garry oak woodlands (Fig. 408), and log-strewn or stony beaches, where the bluebird forages on intertidal and upper-beach invertebrates. During the 1950s, it was common on the Fraser River delta, where mixed farming was practiced and nest boxes were provided.

In the interior, preferred habitat is open-growth forests of ponderosa pine and Douglas-fir, but includes orchards, farm fields, fenced roadsides, suburban parks with plantings of Russian olive, swales with red-osier dogwood thickets, rangeland edges, and stands of big sagebrush.

A study of habitat selection by Western Bluebirds in Washington found that 44% of the birds recorded were in "edge" habitat and 56% in burned forest (Herlugson 1978). The Western Bluebird forages as much from the ground as it does searching for insects or fruit from tree perches.

In British Columbia, most of the population of Western Bluebirds is migratory (Figs. 409 and 410). On the coast, the few spring migrants may reach the Georgia Depression as early as the third week of February, but the main movement is most evident from early March to the first week of April (Fig. 410). Late migrants do not arrive until May.

In the Southern Interior, the small number of wintering birds is augmented by returning migrants about the second week of March, with the number of birds reaching a peak by the end of March (Fig. 410). Although there are few records for the east and west Kootenay districts, the spring migration is likely similar there.

Figure 408. Open country surrounding mixed woodlands of Garry oak, Douglas-fir, and lodgepole pine is one of the preferred foraging and breeding habitats for the Western Bluebird on southern Vancouver Island (Langford, 7 September 1994; R. Wayne Campbell).

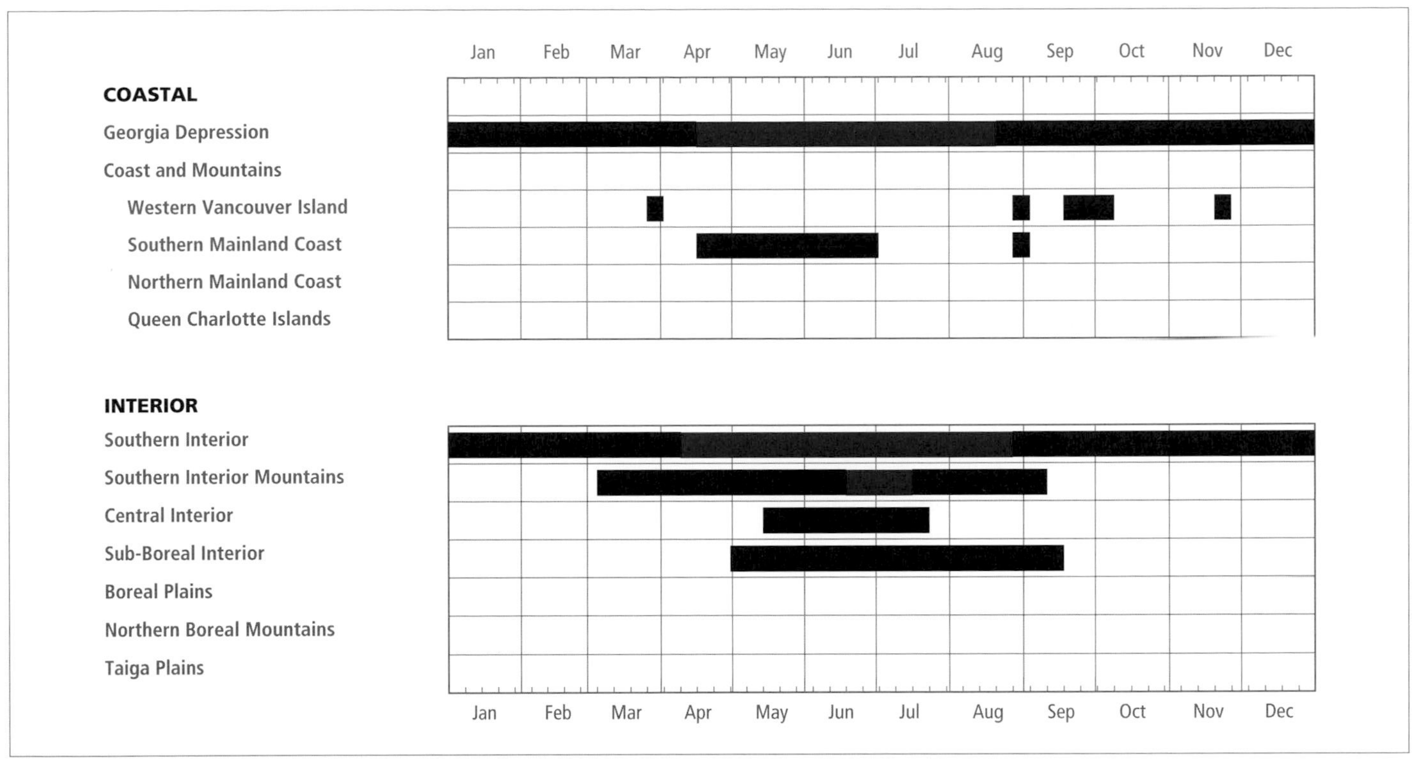

Figure 409. Annual occurrence (black) and breeding chronology (red) for the Western Bluebird in ecoprovinces of British Columbia.

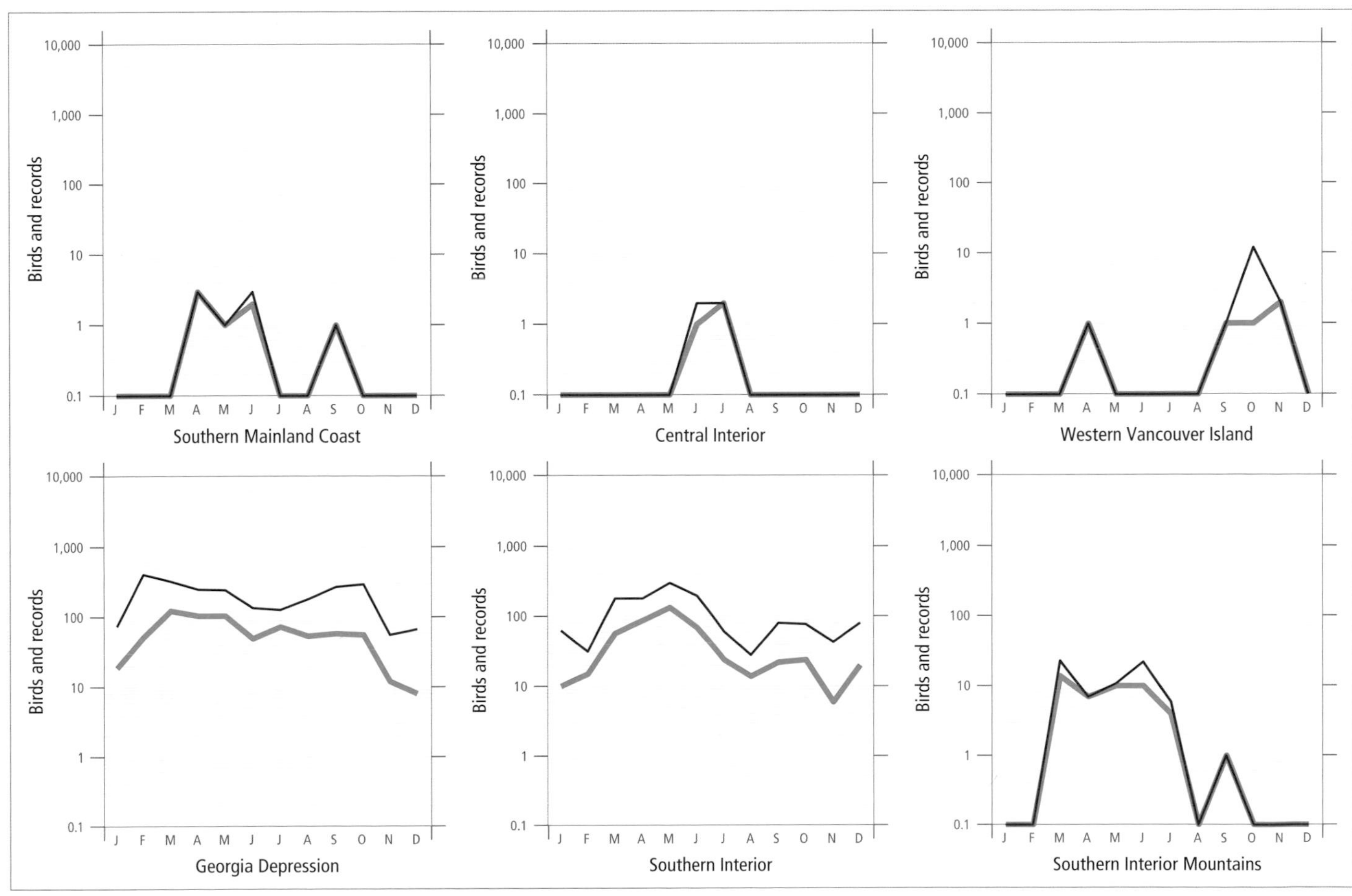

Figure 410. Fluctuations in total number of birds (purple line) and total number of records (green line) for the Western Bluebird in ecoprovinces of British Columbia. Christmas Bird Count and nest record data have been excluded.

Figure 411. Parklike woodlands of trembling aspen, with adjacent big-sagebrush–dominated country, provide breeding and foraging habitat for the Western Bluebird in British Columbia (Kilpoola Lake, Richter Pass, 8 August 1996; R. Wayne Campbell).

Figure 412. Naturalists Margaret Harris and Al Preston banding a nestling bluebird in the Okanagan valley (White Lake, 13 June 1993; R. Wayne Campbell). They have been monitoring and banding Western and Mountain bluebirds as part of the "Bluebird Nestbox Trail" program in British Columbia.

Figure 413. Full clutch of 5 eggs of the Western Bluebird in its typical grass nest (White Lake, 13 June 1993; R. Wayne Campbell).

The autumn migration is more protracted. In the Georgia Depression, when this bluebird was more abundant, migration started in August, continued through October, then declined sharply in November (Fig. 410). There is no obvious peak in the autumn migration in the Okanagan valley, but there is an increase in the number of records from a low point in August to a higher level through September and October (Fig. 410). This is accompanied by an increase in the number of birds per record that probably reflects the Western Bluebird's tendency to form small flocks during migration. Small numbers may linger through the winter on the coast (irregularly) and in the southern portions of the interior. On Vancouver Island, wintering birds have been reported from Comox to Victoria, although only from the latter since the 1980s. In the Fraser Lowland, there are only 2 winter records, both from the delta, that would not be considered early migrants. In the interior, there have been winter occurrences from the Vernon area south to Osoyoos, although since the 1980s reports of wintering birds have come from only as far north as Summerland.

The Western Bluebird has been reported throughout the year in British Columbia (Fig. 409).

BREEDING: The Western Bluebird breeds across southern areas of the province from southeastern Vancouver Island east to Castlegar in the Columbia River valley and Fort Steele in the Rocky Mountain Trench (Johnstone 1949), north in the Okanagan and Thompson valleys to Enderby, Kamloops, and Heffley Creek. On Vancouver Island there are no records of nesting north of Comox. This species reaches its highest numbers in summer in the Southern Interior (Fig. 407). An analysis of Breeding Bird Surveys for the period 1968 through 1993 could not detect a net change in numbers on interior routes; coastal routes contain insufficient data for analysis.

Nesting elevations range from 30 to 600 m on the coast and from the valley floor at 300 m to 1,150 m in the interior (Cannings et al. 1987).

The Western Bluebird nests primarily within human-influenced habitat, including forests, grasslands, and parklike woodlands (Fig. 411). In all habitats it prefers interspersed trees and grassy or forb-clothed openings. On the coast, nesting habitat includes sparsely forested slopes and summits of the many hills that characterize the landscape of extreme southeastern Vancouver Island and the adjacent Gulf Islands (Fig. 408). It also nests on logged lands, burned forest, farms, and pastures. In the interior, cultivated farms, rural and suburban lands, orchards, and vineyards provide 53% of the habitat used (n = 546); open stands of ponderosa pine and Douglas-fir interspersed with areas of big sagebrush and grasses provide another 33%.

The Western Bluebird has been recorded breeding in the province from 10 April to 24 August. On the coast, the Western Bluebird has been recorded breeding from 16 April (calculated) to 14 August; in the interior, it has been recorded breeding from 10 April (calculated) to 24 August (Fig. 409).

Nests: Most nest sites were in the vicinity of farms or rural neighbourhoods and along the edges of open parkland forests, adjacent to pastures and rangelands (Fig. 411). The Western Bluebird is a secondary cavity nester that makes use of sites excavated by woodpeckers, as well as natural crevices and cavities in dead trees, stumps, posts, or poles. Most of the information on nests has been derived from active nest

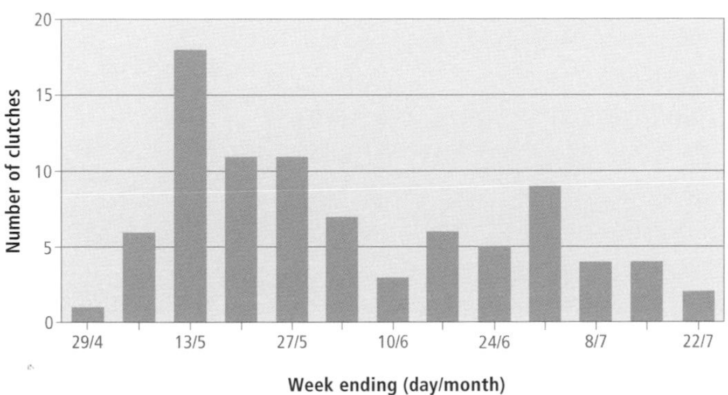

Figure 414. Weekly distribution of first eggs of clutches of the Western Bluebird in the Southern Interior Ecoprovince. Nests were included only if clutches were discovered while the laying phase was in progress. Note the second peak in clutches, indicating that some pairs raise a second brood.

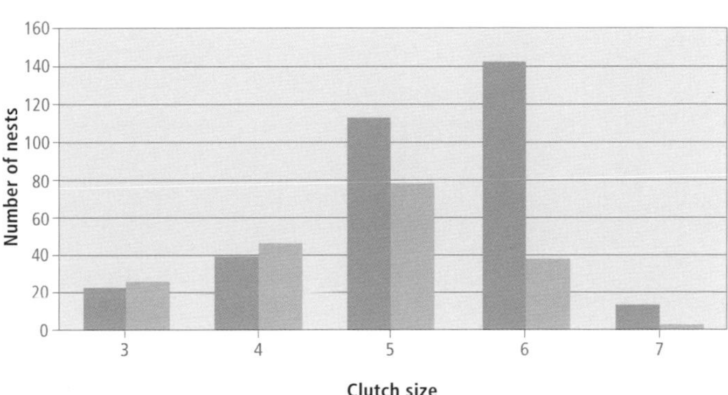

Figure 415. Distribution of the numbers of eggs in early and late clutches of the Western Bluebird. Dark bars indicate clutches begun on or before 7 June; light bars indicate clutches begun after 7 June.

box programs as part of a "Bluebird Nestbox Trail" program in the province (Fig. 412).

Grass was the most frequently used nesting material (93% of nests; n = 523), followed by feathers, conifer needles, plant stems, mosses, mammal hair, fine rootlets, string, and plant down.

The heights for 94 nests not in nest boxes ranged from 0.6 to 16.5 m, with 53% between 2.1 and 6.0 m; 524 nests found in nest boxes ranged from 0.5 to 3.7 m, with 72% between 1.5 and 1.8 m. The Western Bluebird appears to select higher nest cavities than does the Mountain Bluebird, both when nest boxes are used and when they are not.

Eggs: Dates for 565 clutches ranged from 15 April to 3 August, with 52% recorded between 17 May and 27 June. Calculated dates indicate that eggs may be found as early as 10 April. Sizes of 558 clutches ranged from 1 to 10 eggs (1E-22, 2E-16, 3E-49, 4E-85, 5E-191, 6E-179, 7E-15, 10E-1), with 66% having 5 or 6 eggs (Fig. 413). The clutch of 10 eggs was likely the product of 2 females.

In the absence of a population of marked birds, it has not been possible to clearly separate first and second nestings, but it was possible to identify dates at which clutches were initiated, based upon clutches found while laying was in progress. The earliest and latest dates of clutch initiation were 10 April and 6 August, a slightly longer span than reported by Cannings et al. (1987). In 87 nests found in the Southern Interior while egg laying was in progress, there was a concentration of first eggs laid during the second, third, and fourth weeks of May (46% of all clutches; Fig. 414). A second peak of clutch initiation occurred in the second, third, and fourth weeks of June (23% of all clutches). It is probable that these 2 peaks of clutch initiation reflect the concentrations of first and second nestings. Clutch sizes in nests initiated between 10 April and 7 June had a mean of 5.3 eggs (n = 330), while those after 7 June had a mean of 4.7 eggs (n = 190). The difference in mean clutch size between early and late nests results largely from the proportion of 6-egg clutches in the early nests (43%) as opposed to late nests (20%), and the corresponding change in proportions of 4- or 5-egg clutches from 46% in early nests to 65% in late nests (Fig. 415).

The incubation period in British Columbia ranged from 12 to 17 days, with a median of 14 days (n = 17). Cannings et al. (1987) give the incubation period as 11 to 14 days (n = 8), with 2 other nests calculated as 17 and 19 days. They suggest that the females of these nests may have delayed the beginning of incubation after their clutches were complete.

Young: Dates for 601 broods (Fig. 416) ranged from 1 May to 24 August, with 52% recorded between 1 June and 3 July. Sizes of 526 broods ranged from 1 to 7 young (1Y-7, 2Y-27, 3Y-46, 4Y-116, 5Y-194, 6Y-125, 7Y-11), with 61% having 5 or 6

Figure 416. Western Bluebird nest containing 2 eggs and 1 recently hatched young (Castlegar, 3 July 1994; Linda M. Van Damme).

young. The nestling period ranged from 17 to 24 days, with a median of 21.5 days ($n = 24$).

Brown-headed Cowbird Parasitism: Of 767 nests found with eggs or young in British Columbia, 2 contained eggs of the Brown-headed Cowbird. These instances provide evidence of a new host species for the Brown-headed Cowbird in North America (Friedmann 1963; Friedmann et al. 1977; Friedmann and Kiff 1985).

Nest Success: Where nest boxes were not used, of 8 nests found with eggs and followed to a known fate, 6 were successful in fledging at least 1 young. Of 375 nests in nest boxes found with eggs and followed to a known fate, 281 were successful in fledging at least 1 young, for a success rate of 75%. All but 9 nests used in these calculations were from the interior.

REMARKS: The decrease in numbers of Western Bluebirds in the coastal mainland of British Columbia parallels changes in numbers in Washington state, where, in the regions west of the Cascade Mountains, there were only 2 records of the species between 1948 and 1974 (Herlugson 1978). There were several breeding season records for the same area between 1894 and 1947. Breeding Bird Surveys in Washington between 1968 and 1976 detected no consistent change in breeding populations in the state east of the Cascades, but encountered no Western Bluebirds west of that range (Herlugson 1978). This period of decline of Western Bluebird numbers west of the Cascades in both Washington and British Columbia coincided with the arrival of the European Starling and its rapid increase in numbers. This has led to suggestions that the 2 species compete for nest sites. Although competition is probably one cause of the decline, other factors may be involved. There has been no research on the extent of the interaction between the 2 species in British Columbia.

In most of British Columbia, where the habitat is otherwise suitable, the Western Bluebird responds to the provision of nest boxes with an increased nesting population. In 1993, more than 4,600 nest boxes were in place on Bluebird Trails in southern British Columbia, and over 9% of them were occupied by Western Bluebirds (Pollock 1990, 1991, 1992, 1993, 1994). Only on the south coast, including southeastern Vancouver Island, have the bluebirds failed to respond to the provision of nest boxes.

See **Mountain Bluebird** for references that discuss the establishment of Bluebird Trails and various aspects of bluebird management.

For a bibliography of the genus *Sialia*, see Gutzke (1985).

NOTEWORTHY RECORDS

Spring: Coastal – Metchosin 4 Mar 1979-10, 21 April 1983-6 eggs; Sooke 21 Mar 1983-6 eggs; Victoria 24 May 1971-20 in Highlands district; Saanich 30 Apr 1981-1; Cowichan Lake 21 Mar 1939-1 (RBCM 9008); Saltspring Island 31 Mar 1990-10; Huntingdon 9 Apr 1944-nest building, 28 May 1944-fledged; Douglas 24 Mar 1906-1 (NMC 3308); Cultus Lake 4 Apr 1965-2; Sardis 6 May 1904-1 (RBCM 1246); Chilliwack 27 Mar 1891-1 (MCZ 245469); Vancouver 26 Apr 1982-1 female; Roberts Creek 20 Mar 1983-1 male; Courtenay 8 Apr 1972-10; Port Hardy 1 Apr 1951-1. **Interior** – Mount Kobau 4 May 1991-6 eggs; Midway 17 Apr 1905-2 (NMC 3146-7); Grand Forks 30 Apr 1983-7 eggs; Christina Lake 26 May 1972-6; Oliver 20 May 1978-9; Moyie 24 Mar 1938-2 (Johnstone 1949); White Lake 15 Apr 1983-6 eggs; Penticton 21 Mar 1948-25; Summerland 31 May 1971-4 at research station; Edgewood district 31 Mar 1924-1; Okanagan 31 Mar 1930-1 (MVZ 103799); Canal Flats 31 Mar 1939-1 (RBCM 11005); Okanagan Landing 1 Apr 1922-25; Nakusp 10 Mar 1976-1; Clearwater 22 Apr 1935-1 (MCZ 283795).

Summer: Coastal – Metchosin 4 Jun 1984-8, 4 Aug 1984-3 nestlings; Huntingdon 9 Jul 1945-second brood fledged, 16 Aug 1958-35; North Vancouver Jun 1975-male feeding fledglings (BC Photo 984); Courtenay 1 Jul 1924-7; Forbidden Plateau 19 Aug 1959-15; Alta Lake 18 Jun 1924-pair nested (Racey 1948). **Interior** – Newgate 11 Jun 1953-1 (NMC 38717); Grand Forks 8 Jul 1980-2 eggs and 3 nestlings; Pend-d'Oreille valley 30 Jun 1984-2; White Lake 31 Jul 1975-3; Vaseux Lake 5 May 1991-5 eggs, 31 Jul 1991-5 eggs; Okanagan Falls 28 Jul 1991-hatching; Mount Broadwood 4 Jul 1975-2; Naramata 20 Aug 1986-5 nestlings; Balfour to Waneta 8 Jun 1983-8; Summerland 7 Jun 1972-8; Fort Steele 14 Jul 1948-pair feeding 2 fledglings (Johnstone 1949); Lytton 21 Jun 1931-1 (MCZ 283792); Jones Creek (Lillooet District) 13 Jul 1977-1; Shuswap Falls 22 Jul 1918-1; Lac la Hache 30 Jul 1936-2 broods accompanied by parents (Munro 1945a); Chilcotin Valley nr Riske Creek 21 Jul 1931-1; Anahim Lake 3 Jun 1948-2.

Breeding Bird Surveys: Coastal – Not recorded. **Interior** – Recorded from 5 of 73 routes and on 4% of all surveys. Maxima: Oliver 24 Jun 1993-3; Summerland 29 Jun 1981-3; Beaverdell 22 Jun 1991-2; Osprey Lake 29 Jun 1994-2.

Autumn: Interior – Edgewood 5 Sep 1914-1; Okanagan 30 Sep 1989-1 (MVZ 103804); Okanagan Landing 5 Sep 1929-15, 22 Nov 1955-5; Summerland 13 Nov 1988-15; Grand Forks 15 Nov 1985-3. **Coastal** – Cape Scott 20 Sep 1935-1; Sea Otter Cove 2 Oct 1935-12; Port Hardy 24 Nov 1936-1; Courtenay 24 Sep 1933-40; Sechelt 27 Sep 1980-1; Brackendale 2 Sep 1916-1 (NMC 9588); Agassiz 9 Nov 1896-1 (ROM 69776); Chilliwack 24 Oct 1901-1; Iona Island (Richmond) 2 Oct 1971-2 (Campbell et al. 1972b); Cowichan Lake 30 Nov 1923-1 (ROM 85957); Victoria 7 Oct 1953-35; Metchosin 31 Oct 1975-8; Rocky Point (Victoria) 30 Nov 1984-12.

Winter: Interior – Okanagan Landing 16 Jan 1918-12 (Munro and Cowan 1947); Osoyoos 1 Jan 1986-10; Kelowna 2 Jan 1967-8; Penticton 31 Dec 1964-3. **Coastal** – Comox 2 Jan 1925-1 (RBCM 6564), 31 Jan 1926-1 (Cowan 1940); Sechelt 21 Feb 1985-1 male; False Creek (Vancouver) 17 Feb 1896-150; Chilliwack 25 Feb 1928-1; Delta 1 Jan 1983-1; Blenkinsop Lake 1 Jan 1965-5; Victoria 8 Jan 1960-20.

Christmas Bird Counts: Interior – Recorded from 4 of 27 localities and on 10% of all counts. Maxima: Penticton 27 Dec 1988-**73**, all-time Canadian high count (Monroe 1989a); Kelowna 18 Dec 1993-14; Vaseux Lake 27 Dec 1985-13. **Coastal** – Recorded from 1 of 33 localities and on less than 1% of all counts. Maximum: Victoria 27 Dec 1964-59.

Mountain Bluebird

Sialia currucoides (Bechstein)

MOBL

RANGE: Breeds from east-central Alaska, southern Yukon, north-central Alberta, central Saskatchewan, and western Manitoba south in the mountains east of the Coast Mountains to southern California and New Mexico; east to North Dakota, Nebraska, and Oklahoma. Winters from southern British Columbia (casually) to Baja California, and, in the eastern parts of its range, from Montana south to Mexico.

STATUS: On the coast, an *uncommon* spring transient in the Georgia Depression Ecoprovince, *very rare* in summer, and *rare* to *uncommon* the rest of the year; *uncommon* to locally *fairly common* migrant and summer visitant in the southern mainland portions of the Coast and Mountains Ecoprovince, *very rare* in the northern mainland of the Coast and Mountains; *casual* on Western Vancouver Island and the Queen Charlotte Islands.

In the interior, a *common* to *very common* migrant and *fairly common* to *common* summer visitant across southern British Columbia north to the Cariboo and Chilcotin areas of the Central Interior Ecoprovince; *rare* to *uncommon* migrant and summer visitant from Quesnel north through the Sub-Boreal Interior Ecoprovince; *rare* in the Peace Lowland of the Boreal Plains Ecoprovince and across the Northern Boreal Mountains Ecoprovince; *casual* transient in the Taiga Plains Ecoprovince.

Breeds.

Figure 417. Adult male Mountain Bluebird with grasshopper for nestlings (Becher's Prairie, 40 km west of Williams Lake, August 1987; Anna Roberts).

NONBREEDING: The Mountain Bluebird (Fig. 417) is widely distributed across the southern regions of the province. On the coast, it occurs regularly but infrequently on southeastern Vancouver Island, from Victoria to Campbell River, and in the Fraser Lowland on the mainland. It occurs infrequently on the rest of Vancouver Island and north along the coast to the upper Skeena River valley, as well as on the Queen Charlotte Islands. The northernmost coastal records are from Meziadin Lake and Stikine River.

Figure 418. In early autumn, subalpine habitats in the Southern Interior Mountains Ecoprovince attract Mountain Bluebirds, which feed on grasshoppers and in turn are food for migrating Prairie Falcons (Meadow Mountain, near Kaslo, 22 September 1994; R. Wayne Campbell).

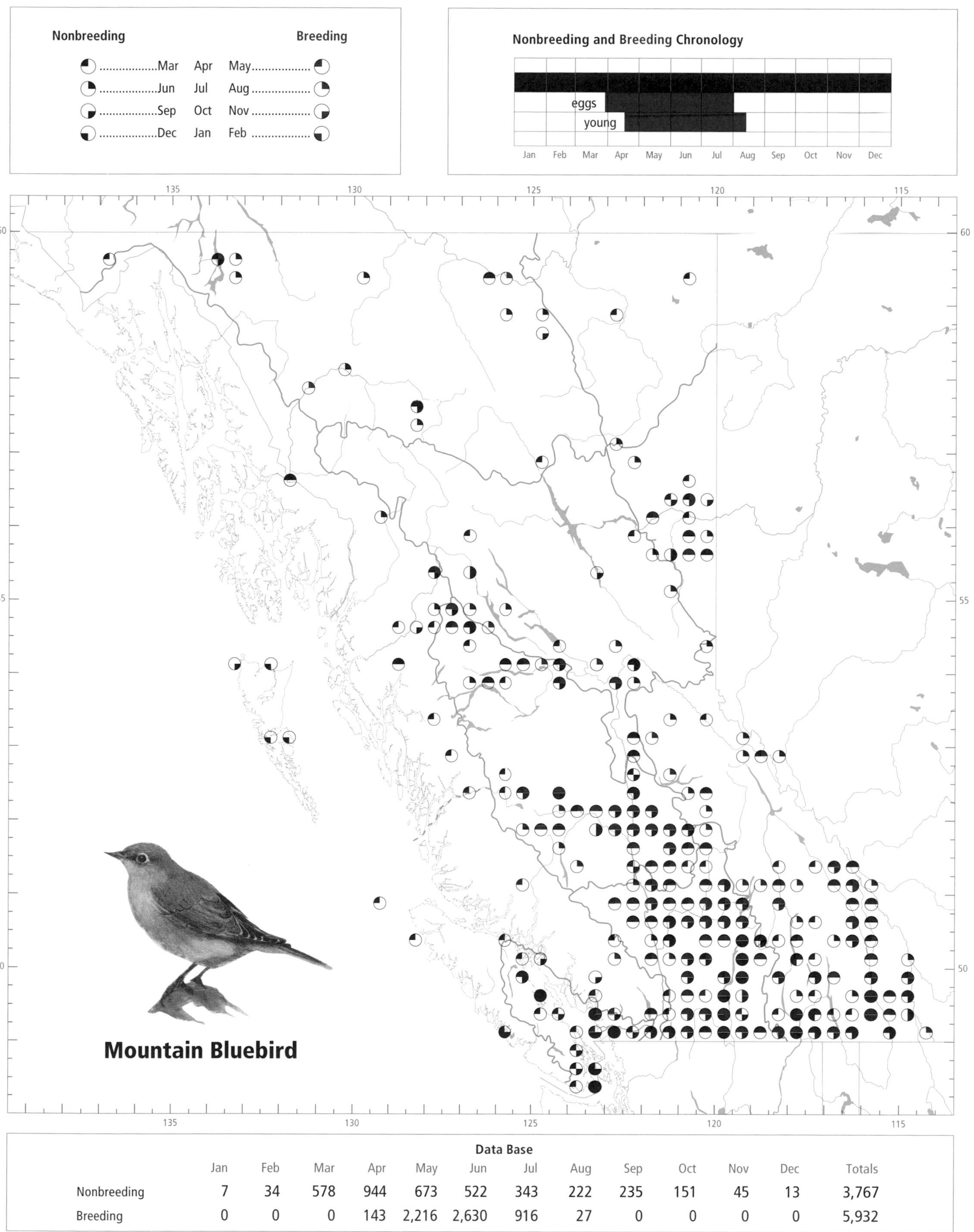

Data Base

	Jan	Feb	Mar	Apr	May	Jun	Jul	Aug	Sep	Oct	Nov	Dec	Totals
Nonbreeding	7	34	578	944	673	522	343	222	235	151	45	13	3,767
Breeding	0	0	0	143	2,216	2,630	916	27	0	0	0	0	5,932

In the interior, the Mountain Bluebird occurs north to the Cariboo and Chilcotin areas and locally in the Nechako and Bulkley river valleys and in the Peace Lowland. Further north, it is sparsely distributed. Specific localities along the Yukon boundary extend from Chilkat Pass, in the far west, to the Kwokullie Lake area in the northeastern corner of the province. The Mountain Bluebird has not been recorded from some large areas of the province, including the Nechako Plateau and Nazko Upland in the Central Interior, the Rocky Mountain Trench and adjacent mountain ranges north of McBride, central Vancouver Island, and the western hemlock forests of the British Columbia coast, including those in the Coast-Cascade ranges between Jervis Inlet and the lower Skeena River valley.

On the coast, the Mountain Bluebird occurs near sea level; in the interior it ranges from valley bottoms at 260 m elevation to the high mountain passes and alpine meadows at over 2,700 m elevation.

In migration, this bluebird uses a variety of habitats, including ponderosa pine forests, rangeland, aspen parkland, areas of burned or cut-over forest with standing snags, the arbutus–Garry oak–Douglas-fir association of eastern Vancouver Island, subalpine meadows (Fig. 418), farmsteads, meadows, and pastures, especially those where fence posts or powerlines offer perching sites. On southeastern Vancouver Island, migrant Mountain Bluebirds also seek the open summits of hills and low mountains.

In the southern portions of the interior, the spring migration can begin as early as the second week of February (Figs. 419 and 420); however, the main movement begins the first week of March and is at its height during the last week of March and the first week of April (Figs. 420 and 421). The spring migration is dramatic, as many of the bluebirds travel in loose flocks that are concentrated in snow-free areas. The migration continues until the end of April, as birds move onto their nesting territories or to regions further north. On the coast, the few migrants pass through from about the second week of March to the third week of April, and decline quickly thereafter.

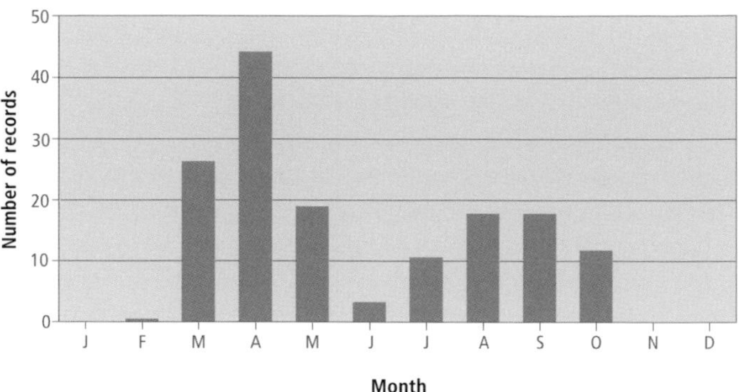

Figure 419. Monthly distribution of records of more than 10 birds each for the Mountain Bluebird in the Southern Interior Ecoprovince of British Columbia. The figure shows the timing of the migration periods.

In northern regions, the autumn migration likely begins in late July and early August. In the central and southern regions of the interior, it is quite evident by mid-August, and peaks from late August to mid-September (Figs. 419 and 421). This migration is more protracted than the one in spring, and, as the birds move over more extensive areas from valley bottoms to alpine elevations, it is not as dramatic. The autumn movement is still noticeable during late September and early October, especially in subalpine areas (Fig. 418). Stragglers

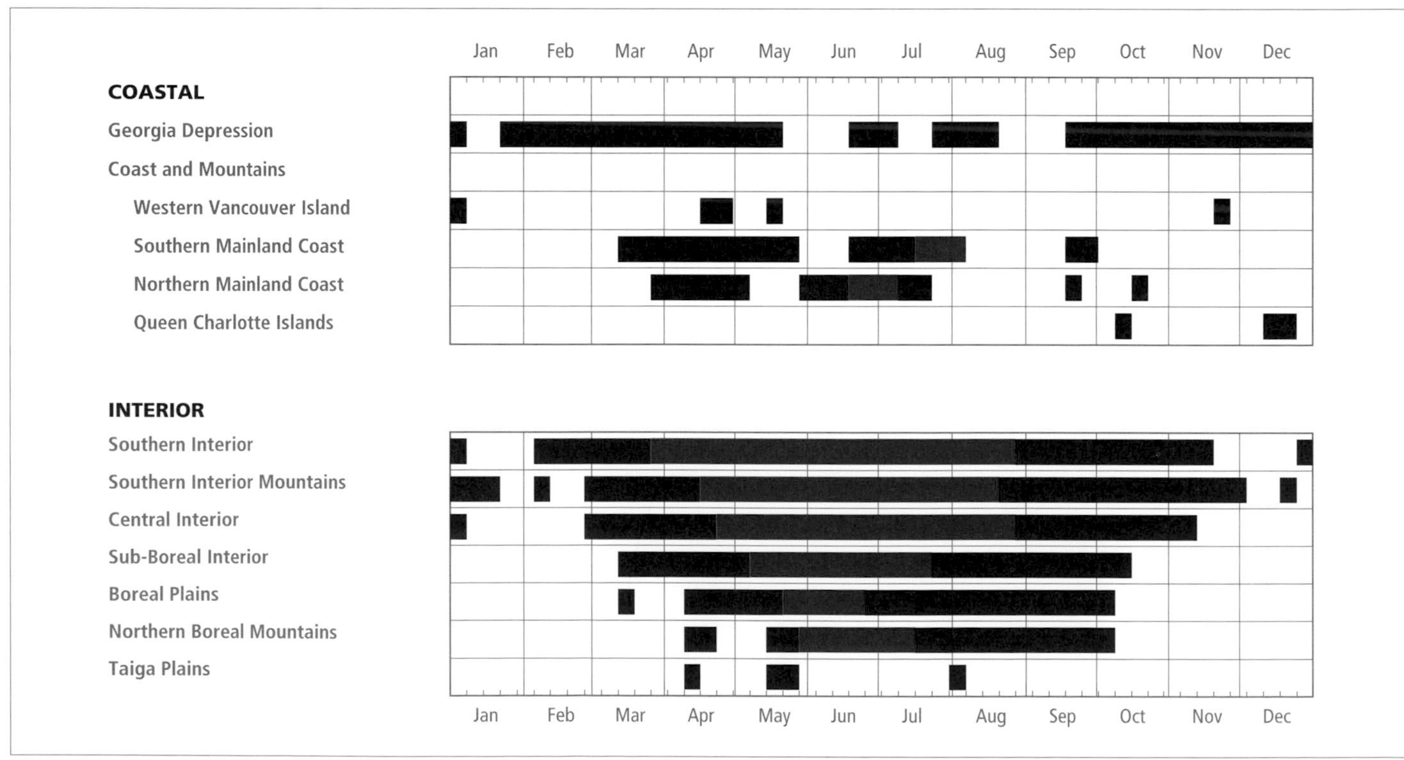

Figure 420. Annual occurrence (black) and breeding chronology (red) for the Mountain Bluebird in ecoprovinces of British Columbia.

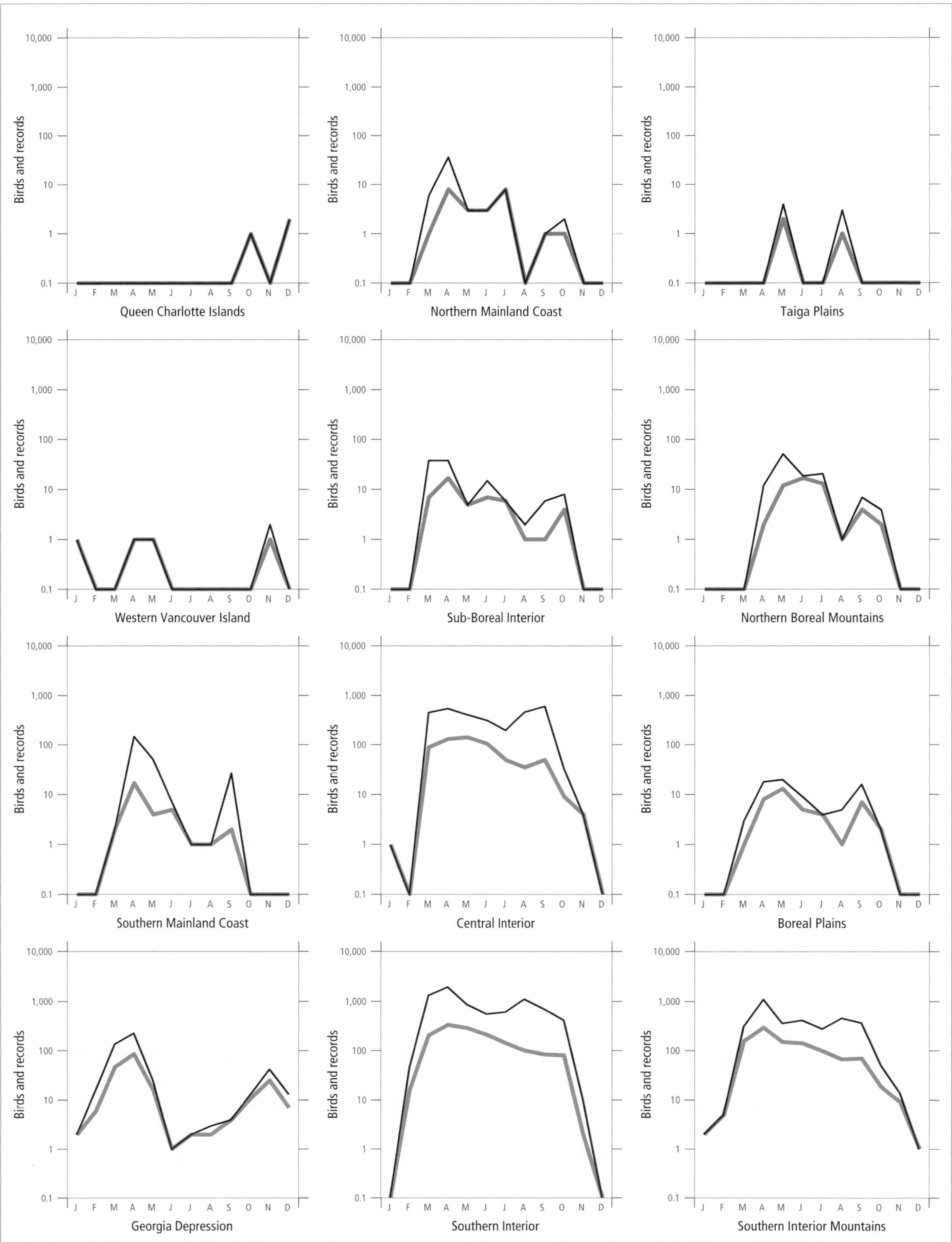

Figure 421. Fluctuations in total number of birds (purple line) and total number of records (green line) for the Mountain Bluebird in ecoprovinces of British Columbia. Christmas Bird Count and nest record data have been excluded.

are reported from early November near Williams Lake and Revelstoke and in the Okanagan valley. There is no obvious autumn migration through the coastal regions, but more Mountain Bluebirds are seen in the Georgia Depression in October and November than from June to September.

In the interior, the northernmost winter record is from Chezacut; other winter observations are from Kamloops, Nakusp, Okanagan Landing, Nelson, and Castlegar, all from the west Kootenay and Okanagan valleys. On the coast, the northernmost winter records are from Terrace on the mainland and from Masset and Sandspit on the Queen Charlotte Islands. Other winter records are from Vancouver Island (Comox and Saanich) and the Fraser River delta (South Vancouver and Boundary Bay).

On the coast and in the interior, the Mountain Bluebird has been recorded every month of the year, but there are few records from either region between late September and early March (Figs. 420 and 421).

BREEDING: The Mountain Bluebird has a widespread breeding distribution across the province east of the Coast Ranges from Manning Park to the Crowsnest Pass and north to the Cariboo and Chilcotin areas. Its known breeding distribution north of latitude 55°N includes only 4 localities (Beryl Prairie, Sawmill Lake [Telegraph Creek], Atlin, and Aline Lake).

This bluebird does not normally nest west of the summit of the Cascade Mountains. On the coast, there are only 2 confirmed nesting occurrences: Whistler (Alta Lake; Racey 1948) and Hazelton (Swarth 1924).

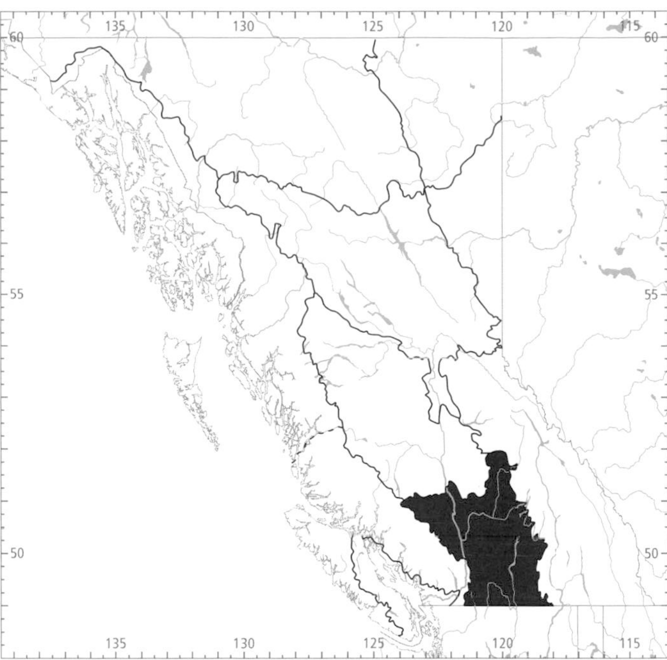

Figure 422. In British Columbia, the highest numbers for the Mountain Bluebird in summer occur in the Southern Interior Ecoprovince.

The Mountain Bluebird reaches its highest numbers in summer in the Okanagan valley in the Southern Interior (Fig. 422). An analysis of Breeding Bird Surveys for the period 1968 through 1993 could not detect a net change in numbers on interior routes; coastal routes for the same period contain insufficient data for analysis.

Figure 423. Snags with woodpecker cavities and natural crevices are used as nesting sites by Mountain Bluebirds in subalpine areas of southern British Columbia (Thompson Mountain, 10 km east of Creston, 16 September 1994; R. Wayne Campbell).

(a)

(b)

(c)

(d)

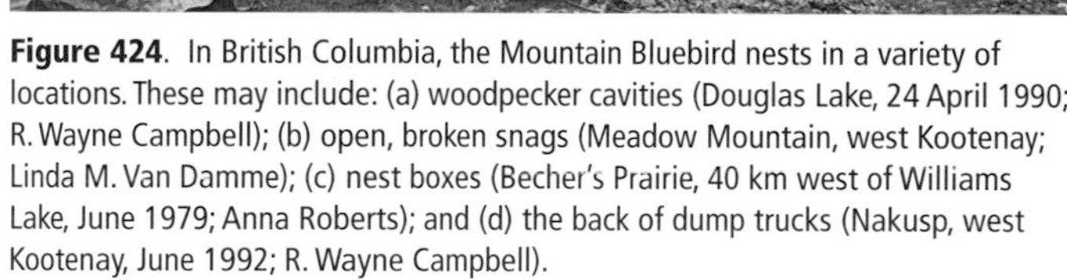

Figure 424. In British Columbia, the Mountain Bluebird nests in a variety of locations. These may include: (a) woodpecker cavities (Douglas Lake, 24 April 1990; R. Wayne Campbell); (b) open, broken snags (Meadow Mountain, west Kootenay; Linda M. Van Damme); (c) nest boxes (Becher's Prairie, 40 km west of Williams Lake, June 1979; Anna Roberts); and (d) the back of dump trucks (Nakusp, west Kootenay, June 1992; R. Wayne Campbell).

The Mountain Bluebird has been reported nesting between 260 and 2,700 m elevation. The characteristic nesting habitat of this bluebird in British Columbia features open space, and is centred upon the rangelands and parklands of the valleys of the Thompson, Okanagan (Fig. 411), and lower Similkameen rivers; the east and west Kootenays; the Rocky Mountain Trench north to Edgewater; and the Cariboo and Chilcotin areas. Throughout these areas most Mountain Bluebird nests (59%; *n* = 2,177) were associated with grassland, including those dominated by big sagebrush (see Fig. 411). Following in importance were areas influenced by human activity (26%; farmlands, rangeland, rural and suburban areas), and open forest stands of ponderosa pine, interior Douglas-fir, and trembling aspen (13%). Subalpine meadows (Fig. 423) were also used, as were burns. In adjacent areas of Washington, a similar close association has been reported between nesting Mountain Bluebirds and fire: 81% of the birds observed were in burned areas, 13% in subalpine areas, and 7% in farmland (Herlugson 1978).

The Mountain Bluebird has been recorded breeding in British Columbia from 29 March (calculated) to 13 August (Fig. 420).

Nests: The Mountain Bluebird is a secondary cavity nester, with 96% of nests (*n* = 2,728) occurring in a variety of cavities (Figs. 424 and 425). Under natural circumstances, abandoned woodpecker nesting excavations were frequently used (Fig. 424a), as were other natural cavities left by accident or decay in dead trees, stumps, or the tops of broken snags (Fig. 424b). Other nesting cavities were found in crevices in rock cliffs, under a cluster of rocks, or in partially complete or abandoned nests of Cliff Swallows, Barn Swallows, American Robins, and Dark-eyed Juncos.

Recently, specific nest locations have been heavily influenced by human activity. Eighty-five percent of all nest locations reported were nest boxes (Figs. 424c and 425); this reflects the number of Bluebird Nestbox Trails that have been established in the province and the ease of checking these nests compared with checking nest sites where nest boxes were not used. Even in the Central Interior and Boreal Plains ecoprovinces, humans have encouraged the presence of the bluebirds by putting out nest boxes. Only on southeastern Vancouver Island and in the lower Fraser River valley has the provision of nest boxes failed to increase the numbers of nesting Mountain Bluebirds (Pollock 1990, 1991, 1992, 1993, 1994). See also REMARKS.

While nest boxes were the most frequently reported nest locations of the Mountain Bluebird in British Columbia, many other artificial sites were used, including a ledge or shelf under a building eave, the beam of a barn, the ledge of a hayloft, a nail keg, a mail box (Campbell and Harris 1990), the gauge cover on a propane tank, an old car and dump truck (Fig. 424d), a broken irrigation pipe, and the pipe of a cattle guard.

The heights for 1,700 nests found in nest boxes ranged from 0.3 to 9.9 m, with 72% between 1.2 and 1.5 m, the height of a fence post. The heights for 390 nests found where nest boxes were not used ranged from ground level to 15 m, with 58% between 1.2 and 3.0 m.

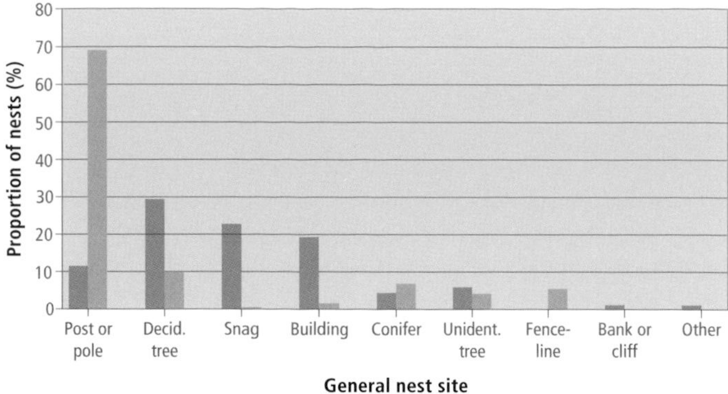

Figure 425. Proportion of Mountain Bluebird nests in British Columbia at various general nest sites, found in nest boxes (light bars; *n* = 1,527) and found at other sites (dark bars; *n* = 428).

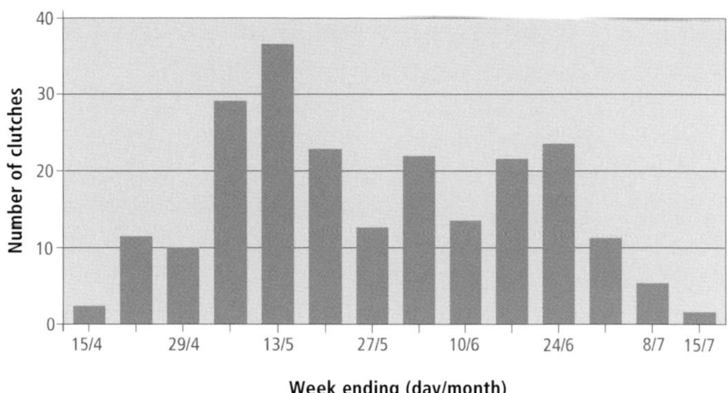

Figure 426. Weekly distribution of first eggs of clutches of the Mountain Bluebird in British Columbia. Nests were included only if clutches were discovered while the laying phase was in progress. Note the peaks in clutches, indicating that some pairs raise a second, and perhaps a third, brood.

Grass was the primary nest material (83% of nests; *n* = 1,827; Fig. 428), followed by feathers, plant fibres, and other fine, soft natural and human-made components.

Eggs: Dates for 1,968 clutches ranged from 7 April to 1 August, with 52% recorded between 18 May and 19 June. Calculated dates suggest that nests can be found with eggs as early as 29 March. Sizes of 1,862 clutches ranged from 1 to 11 eggs (1E-102, 2E-103, 3E-131, 4E-268, 5E-822, 6E-400, 7E-28, 8E-4, 9E-3, 11E-1), with 58% having 4 or 5 eggs (Fig. 428). The incubation period in British Columbia is 12 to 15 days (*n* = 5).

Nests with 8 or more eggs are likely the result of 2 females using the same nest. Seven eggs appears to be the upper clutch size for the Mountain Bluebird in British Columbia. Of the 8 nests with 8 or more eggs, only 22% of the eggs hatched, compared with 76% in the 28 nests with 7 eggs.

Of 222 clutches for which the week of initiation was known, 56% were initiated between 9 April and 27 May, and 44% between 28 May and 15 July (Fig. 426); most of the latter clutch initiations were likely renests or second nestings. Fig. 427 shows the clutch sizes for early and late nests in

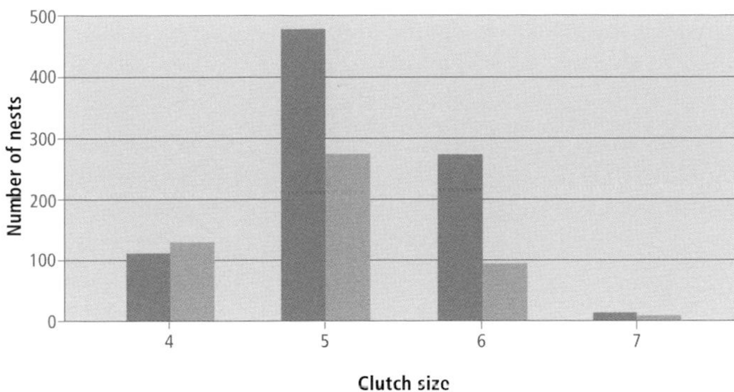

Figure 427. Distribution of the numbers of eggs in early and late clutches for the Mountain Bluebird in British Columbia. Dark bars indicate clutches begun on or before 28 May; light bars indicate clutches begun after 4 June.

British Columbia. Nests initiated before 28 May had a mean clutch size of 5.2 eggs (n = 870), while those initiated after 4 June had a mean of 4.9 eggs (n = 507). The proportions of nests believed to be second nests from Bluebird Nestbox Trails in British Columbia in 1991 and 1992 were 31% (n = 864) and 29% (n = 950), respectively (Pollock 1992, 1993). These figures likely include both renestings and second nests.

Young: Dates for 1,776 broods (Fig. 429) ranged from 16 April to 13 August, with 53% recorded between 31 May and 26 June. Sizes of 1,581 broods ranged from 1 to 7 young (1Y-44, 2Y-78, 3Y-156, 4Y-388, 5Y-635, 6Y-269, 7Y-11), with 64% having 4 or 5 young. The nestling period in British Columbia is 18 to 23 days (n = 7), with a median of 22 days.

Brown-headed Cowbird Parasitism: In British Columbia, there were 2 definite cases and 1 possible case of parasitism in 2,550 nests recorded with eggs or young (Fig. 428). In the latter instance, a cowbird was observed trying unsuccessfully to gain access to a Mountain Bluebird nest; 12 days later, 2 bluebird eggs were found below the nest and the nest had been abandoned.

Only 4 cases of cowbird parasitism have been reported for North America, including the 2 from British Columbia (Friedmann 1963; Friedmann et al. 1977; Friedmann and Kiff 1985).

Nest Success: Where nest boxes were not used, of 58 nests found with eggs and followed to a known fate, 36 were successful in fledging at least 1 young, for a success rate of 62%. A study of a nesting population in a clearcut area in the Beaverfoot River valley yielded lower success rates. There the birds were nesting predominantly in Northern Flicker nest cavities at a mean height of 1.5 m (the height of remaining stumps). Mean predation rates for 1993 and 1994 were 45% (n = 40) (R.F. Holt pers. comm.). Among 1,233 nests in nest boxes found with eggs and followed to a known fate, 884 were successful in fledging at least 1 young, for a success rate of 72%.

A second expression of success is available from the Bluebird Nestbox Trails program in British Columbia. Best estimates of the number of fledglings produced per pair of nesters during the summer of 1992, including second nests, were as follows: Southern Interior, 5.8; Southern Interior Mountains, 7.0; and Central Interior, 4.2.

REMARKS: Bluebird populations across North America are believed to have declined between the 1920s and the 1970s (Zelney 1976; Ehrlich et al. 1987; Grooms and Peterson 1991). As noted above, however, the data for British Columbia do not indicate a significant change over that period. Studies in Washington have also concluded that there has been no consistent change in numbers of the Mountain Bluebird there (Herlugson 1978).

Since 1990, data collected on the Bluebird Nestbox Trails in British Columbia suggest that the placing of nest boxes designed to attract bluebirds has been accompanied by an increased abundance of these birds. However, it is difficult to ascertain how much the apparent increase is an artifact of luring the bluebirds to nest in nest boxes, where they are more visible than in more widely dispersed natural sites. By 1993, more than 4,600 nest boxes were in place, widely distributed over much of the bluebird range in the province from Fernie to Smithers. About 16% of the boxes are occupied by Mountain Bluebirds annually, and this percentage has not declined as the number of boxes has increased. The number of fledglings leaving these boxes was 3,924 in 1992 and 3,885 in 1993 (Pollock 1994).

A Mountain Bluebird banded near Calgary, Alberta, on 13 July 1987 was recovered near Vancouver on 10 August 1987, a longitudinal movement of some 680 km.

A bibliography of the technical literature on *Sialia* has been prepared by Gutzke (1985). For information on establishing Bluebird Nestbox Trails and on aspects of bluebird life history and management, see Zeleny (1976), Campbell and Hosford (1979), Shantz (1986), Scriven (1989), Grooms and Peterson (1991), and Power and Lombardo (1996). For tips on identifying female bluebirds, see Dunn (1981).

Figure 428. Of the 4 instances of known Brown-headed Cowbird parasitism on the Mountain Bluebird in North America, 2 are from British Columbia (near Oliver, 13 June 1993; R. Wayne Campbell).

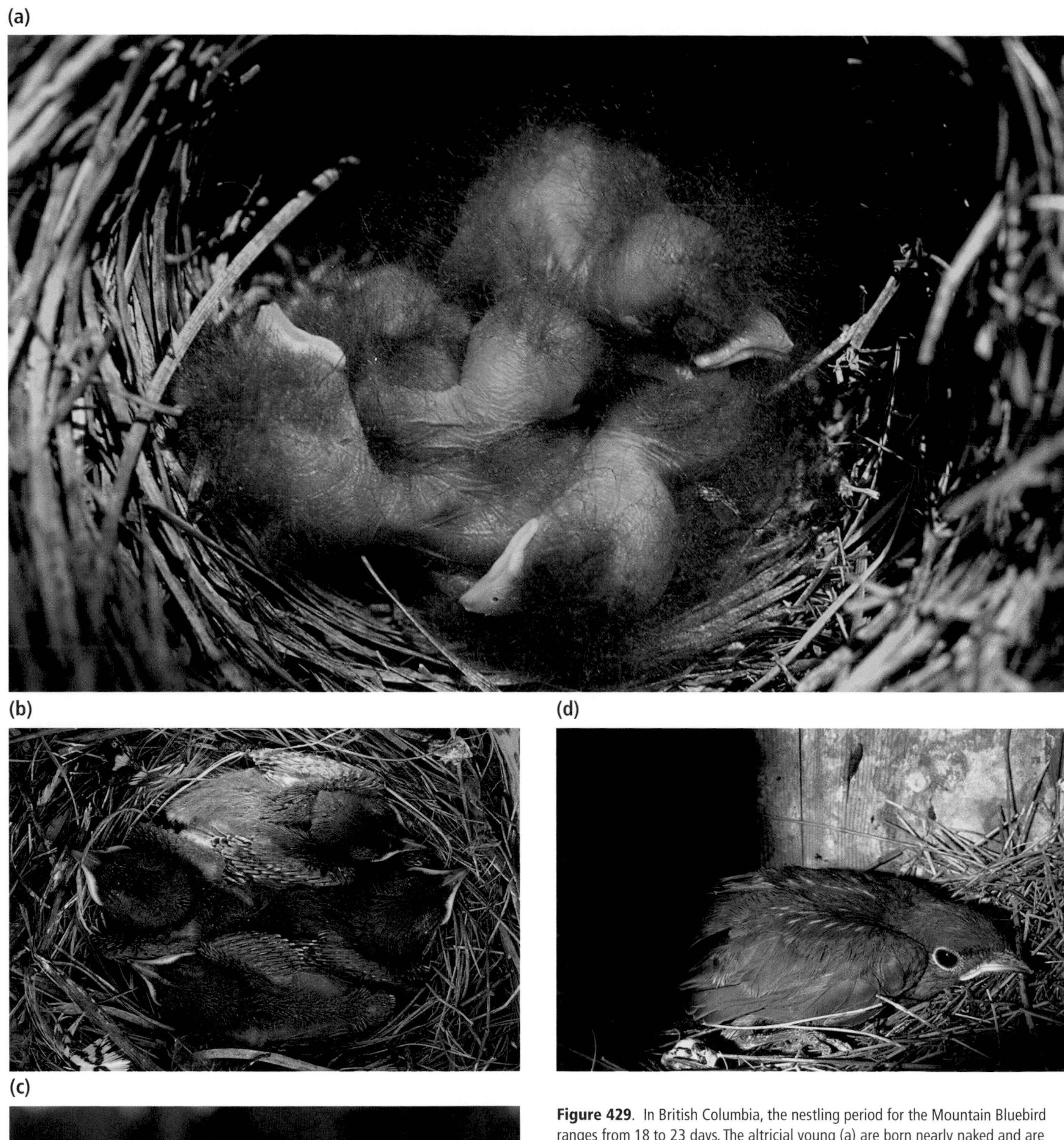

Figure 429. In British Columbia, the nestling period for the Mountain Bluebird ranges from 18 to 23 days. The altricial young (a) are born nearly naked and are unable to open their eyes completely until about 8 days old. The young develop rapidly, and between 9 and 10 days old (b) the pin feathers of the primary feathers begin erupting through the quill. At 14 or 15 days old (c), the young can unexpectedly flop from their nest. Between 18 and 23 days (d), the young have well-developed flight feathers and are ready for their first flight. (a, Rock Lake [Cariboo], 25 May 1994; b, Riske Creek, 25 July 1993; c, Becher's Prairie [Cariboo], 27 July 1993; d, White Lake [Okanagan], 4 August 1993; all R. Wayne Campbell)

NOTEWORTHY RECORDS

Spring: Coastal – Victoria 21 Mar 1986-2; Albert Head 1 Apr 1936-1 (RBCM 5520); Long Beach 27 Apr 1976-1; Skagit Valley 23 Apr 1971-25 at Ponderosa Meadows; Reifel Island 14 May 1987-4; Chilliwack 27 Mar 1891-1 (MVZ 103826); Vancouver 1 Apr 1974-1; Englishman River 21 Mar 1989-3 males and 1 female on estuary (Campbell 1989b); Pitt Meadows 6 Mar 1972-2, 25 Mar 1972-12; Agassiz 7 Mar 1975-1; Oyster River 15 May 1939-pair showed unmistakably that they had a nest, although time did not allow search for the site (Laing 1942); Squamish 20 Mar 1976-1; Terrace 31 Mar 1969-6; Kispiox 15 Apr 1976-12; Telkwa 2 May 1985-1; Stikine 18 May 1983-1. **Interior** – Oliver to Bridesville 20 May 1962-32; West Creston 15 Mar 1979-10; Trail 20 Apr 1985-3 eggs; Vaseux Lake 4 May 1969-6 eggs; Fernie 5 May 1904-2 (NMC 4278-9); Penticton 23 Mar 1977-150, 16 Apr 1981-5 nestlings; Merritt 16 Apr 1968-85; Columbia Lake 3 Apr 1982-90; Enderby 20 Apr 1973-first egg; Tranquille 7 Apr 1990-5 eggs; Celista 11 Apr 1962-12; Revelstoke 1 Apr 1985-1; Yoho National Park 29 May 1977-20; Watson Lake 6 Mar 1983-2; 150 Mile House 14 Mar 1987-3; Riske Creek 5 Mar 1984-60, 1 May 1983-2 eggs; Chezacut 2 Mar 1941-4, 8 May 1943-40; Anahim Lake 23 Apr 1932-1 (MCZ 283810); Prince George 18 Mar 1983-4, 9 May 1983-eggs; Hazelton 29 May 1921-1 (MVZ 42633); Tupper Creek 13 May 1938-1 (RBCM 8147); Dawson Creek 29 May 1993-5 eggs; Baldonnel 22 Apr 1982-6; Charlie Lake 14 May 1987-3; Stikine 18 May 1983-1; Ingenika Point 12 Apr 1985-10; Kwokullie Lake 20 May 1982-2; Atlin 19 May 1981-10.

Summer: Coastal – Victoria 19 Jun 1927-1; New Westminster 10 Aug 1927-1; Courtenay 3 Jul 1939-1; Elma Bay 10 Aug 1938-2; Horseshoe Lake (10 miles inland from mouth of Jervis Inlet) 8 Jul 1936-pair nesting (Laing 1942); Alta Lake 2 Aug 1928-4 nestlings; Anderson Lake 23 Jun 1968-8; Hagensborg 16 Apr 1976-6; Hazelton 12 Jun 1921-1, 4 Jul 1924-first fledgling (Swarth 1924); Meziadin Lake 26 Jun 1991-1. **Interior** – Creston 15 Jun 1978-50; Manning Park 31 Jul 1965-15, 1 Aug 1967-20, 13 Aug 1975-4 nestlings; Princeton 24 Aug 1977-244; Keremeos to Osoyoos 1 Jun 1980-15; Christina Lake 24 Aug 1980-100; Norbury Lake 23 Aug 1984-50; Harmer 23 Jul 1984-30; Westbank 7 Jun 1969-nest; Nicola 5 Jun 1970-eggs; Wilmer 18 Aug 1977-20; Kamloops 4 Jul 1961-40, 29 Jul 1967-1 egg, 4 nestlings; Pavilion 15 Jul 1979-25; Revelstoke 17 May 1969-5 eggs, 5 Jun 1969-5 nestlings; Beaverfoot River 1 Aug 1975-62; Field 29 Aug 1892-5 (ANS 31609-13); Buffalo Lake 12 Jul 1965-20; Moha 28 Aug 1968-120; Clearwater 6 Jun 1971-4 nestlings; 100 Mile House 9 Jul 1986-4 eggs; Westwick Lake 1 Jul 1955-16; Riske Creek 30 Jul 1979-4 nestlings; Chezacut 1 Jun 1943-1, 1 Aug 1941-100; Chilanko Forks 22 Jul 1972-4 fledglings; Nimpo Lake 9 Aug 1991-4 nestlings; Anahim Lake 14 Jun 1964-eggs; Quesnel 10 Jun 1965-6; Wells 3 Aug 1978-4 nestlings; Nulki Lake 27 Jun 1945-nestlings (Munro 1949); Topley 17 Jul 1956-3 nestlings; Moose Lake 25 July 1969-2 nestlings; Dawson Creek 1 Aug 1975-5; 22 km w Hudson's Hope 13 Jun 1990-young in nest; Bullmoose Creek 6 Aug 1976-2; Omineca Mountains 1 Jul 1941-1; Dease Lake 21 Jul 1962-1 (NMC 50031); Mile 130 Alaska Highway 12 Jun 1986-4; Murray River 1 Jul 1978-1; Pink Mountain 1 Sep 1983-3; Hyland Post 14 Jun 1976-2; Sawmill Lake 14 Jun 1919-4 eggs (Swarth 1922); Muncho Lake 20 Jul 1943-3; Aline Lake 22 Jun 1980-nestlings; Surprise Lake 23 Aug 1975-1; Atlin 15 Jun 1981-5 eggs; 16 Jul 1980-4 fledglings.

Breeding Bird Surveys: Coastal – Recorded from 2 of 27 routes and on 4% of all surveys. Maxima: Seabird 30 Jun 1974-2; Chilliwack 23 Jun 1974-1. **Interior** – Recorded from 41 of 73 routes and on 37% of all surveys. Maxima: Riske Creek 2 Jul 1992-23; Brookmere 3 Jul 1992-13; Pavilion 12 Jun 1977-13; Kamloops 2 Jul 1974-10; Princeton 4 Jul 1991-10.

Autumn: Interior – Atlin 4 Oct 1980-2; Montney 4 Sep 1982-6, 1 Oct 1982-1; Mackenzie 3 Oct 1983-1; Lone Prairie 13 Sep 1977-4; Coffin Lake 10 Nov 1978-1; Vanderhoof 14 Oct 1972-1; Willow River 2 Oct 1972-3; Riske Creek 6 Sep 1985-200; Chezacut 1 Oct 1941-10; Wineglass Ranch (Chilcotin River) 6 Sep 1985-200+, 9 Nov 1987-1 male (Campbell 1988a); Williams Lake 5 Nov 1988-1; Pavilion 17 Nov 1986-8; Merritt 8 Sep 1984-25; Ta Ta Creek 1 Oct 1938-2 (RBCM 11002-3); Kimberley 26 Sep 1971-50, 3 Oct 1975-10; Crowsnest Pass 6 Sep 1937-1 (MCZ 38723); Ashnola River 9 Oct 1890-4 (MCZ 45946-9); Manning Park 7 Oct 1978-8; Richter Pass 7 Oct 1978-30; Columbia Gardens 3 Nov 1978-4. **Coastal** – Hazelton 22 Sep 1921-1; Terrace 16 Oct 1969-2 (Crowell and Nehls 1970a); Langara Island 11 Oct 1937-1 (MVZ 103810); Loughborough Inlet 13 Nov 1972-1; Ross Lake 30 Sep 1971-20; Schooner Cove 23 Nov 1972-2 (Hatler et al. 1978); Golden Ears 17 Sep 1986-1; Chilliwack 31 Oct 1888-1 (MVZ 103829); Englishman River 2 Nov 1991-1 on estuary; Boundary Bay 16 Nov 1987-1 (BC Photo 2501); Elma Bay 1 Oct 1933-2; Saratoga 15 Apr 1979-1; Cowichan Bay 16 Nov 1974-1; Victoria 4 Oct 1896-1 (RBCM 1259); Oak Bay 23 Nov 1983-7.

Winter: Interior – Chezacut 4 Jan 1943-1, 1 Jan 1979-1; Kamloops 18 Dec 1983-1; Nakusp 27 Feb 1983-1; Okanagan Landing 9 Feb 1910-16; Nelson 18 Dec 1970-1; Castlegar 27 Feb 1983-1. **Coastal** – Masset 12 Dec 1939-1; Sandspit 20 Dec 1986-1; Comox 4 Jan 1924-1, 22 Jan 1920-1 (MVZ 103824); Boundary Bay 30 Dec 1982-3, 5 Feb 1983-3 (Weber et al. 1983); Long Beach 5 Jan 1976-1 (BC Photo 511); Central Saanich 13 to 17 Dec 1990-1 to 5 (Siddle 1991b).

Christmas Bird Counts: Interior – Recorded from 4 of 27 localities and on 2% of all counts. Maxima: Penticton 27 Dec 1979-**12**, all-time Canadian high count (Anderson 1981); Vaseux Lake 28 Dec 1989-3; Vernon 19 Dec 1993-1. **Coastal** – Recorded from 4 of 33 localities and on 1% of all counts. Maxima: Victoria 15 Dec 1990-3; Ladner 23 Dec 1991-2; Terrace 22 Dec 1985-1; Skidegate 16 Dec 1990-1.

Townsend's Solitaire

Myadestes townsendi (Audubon)

TOSO

RANGE: Breeds from east-central and southern Alaska, southern Yukon, and west-central and southwestern Mackenzie south in mountainous country to southern California, northern Arizona, central New Mexico, and Mexico; east to southwestern Saskatchewan, western Montana, northeastern Wyoming, southwestern South Dakota, and northwestern Nebraska. Winters from southern British Columbia, southern Alberta, Montana, and South Dakota south to northern Baja California and Mexico.

STATUS: On the coast, an *uncommon* migrant and summer visitant in the Georgia Depression Ecoprovince and in the southern mainland portions of the Coast and Mountains Ecoprovince, becoming *rare* in winter; *very rare* migrant along Western Vancouver Island, the Queen Charlotte Islands, and in the northern mainland of the Coast and Mountains.

In the interior, a *fairly common* migrant and summer visitant in the Southern Interior, Southern Interior Mountains, and Central Interior ecoprovinces; in winter, *uncommon* in the Southern Interior and Southern Interior Mountains and *rare* in the Central Interior. Further north, an *uncommon* migrant and summer visitant in the Sub-Boreal Interior, Northern Boreal Mountains, and Boreal Plains ecoprovinces; *casual* in the Taiga Plains Ecoprovince; *accidental* in winter in the Sub-boreal Interior.

Breeds.

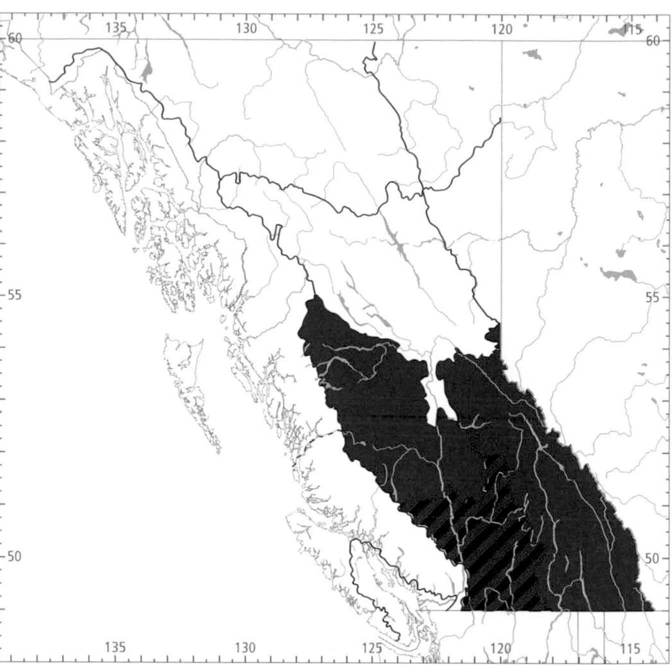

Figure 430. In British Columbia, the highest numbers for the Townsend's Solitaire in winter (black) occur in the Southern Interior Ecoprovince; the highest numbers in summer (red) occur in the Southern Interior, Southern Interior Mountains, and Central Interior ecoprovinces.

Figure 431. During migration, some Townsend's Solitaires forage in open logged areas throughout the province (west of Tatla Lake, 14 July 1992; R. Wayne Campbell).

NONBREEDING: The Townsend's Solitaire is widely distributed across the southern interior of British Columbia, where it is a characteristic species of mountainous areas. In the central portions of the interior, north of the Cariboo and Chilcotin areas and east of the Coast Mountains, there is less terrain suitable for the species and it is infrequently seen. It becomes more numerous again in mountainous regions of the far north. On the coast, the Townsend's Solitaire occurs regularly, during migration and in winter, in the Georgia Depression and in the southern mainland of the Coast and Mountains, but infrequently elsewhere on the coast (Fig. 433). There are large regions devoid of records, including the central mainland coast, much of the Fraser Plateau, the northern portion of the Boreal Plains, and most of the Taiga Plains.

The Townsend's Solitaire reaches its highest numbers in winter in the Southern Interior (Fig. 430).

During spring migration in the interior, the Townsend's Solitaire is a bird of snow-free lower slopes and valley bottoms, where it forages in agricultural areas, open forests, burns, clearcuts, and logged areas (Fig. 431). In the summer and autumn, much of the population is high on the mountainsides. There the solitaire often frequents forests that have been destroyed by fire, with standing spars left above a ground cover of low shrubs. In late autumn and winter in the interior, it also occurs from the valley bottoms to about 1,500 m elevation, where it forages on berries and fruits. On the coast in winter, most birds are found at elevations between sea level and 600 m.

In all seasons, the Townsend's Solitaire is a bird of broken country or open forest, often with sparsely vegetated hillsides (Fig. 432). Low shrubs, such as common juniper, soopolallie, kinnikinnick, or grouseberry, are frequently present. Along the east coast of Vancouver Island, the Gulf Islands, and Sunshine Coast, this species occurs mainly along rocky outcrops, open hilltops, sun-drenched rocky bluffs, and

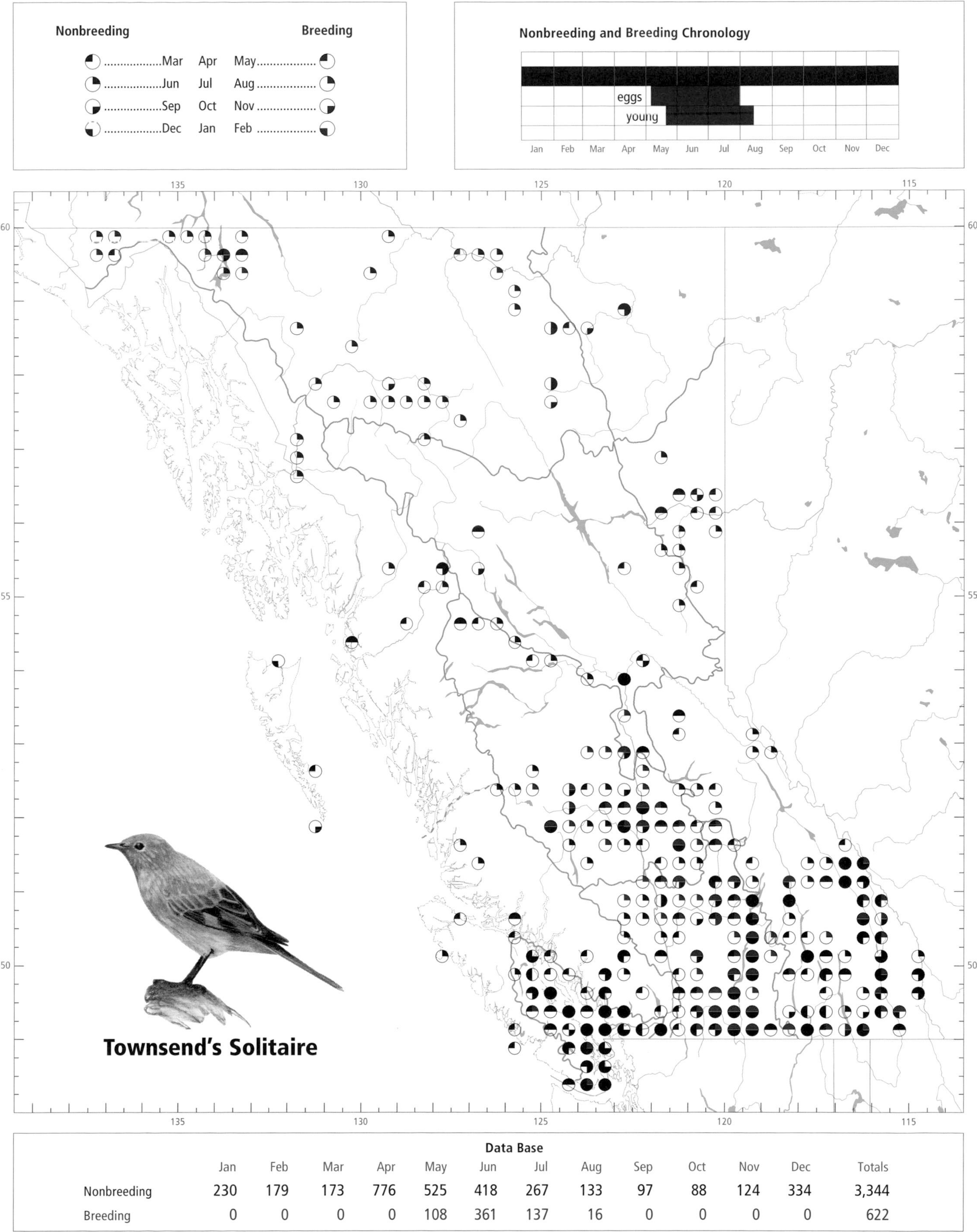

Data Base

	Jan	Feb	Mar	Apr	May	Jun	Jul	Aug	Sep	Oct	Nov	Dec	Totals
Nonbreeding	230	179	173	776	525	418	267	133	97	88	124	334	3,344
Breeding	0	0	0	0	108	361	137	16	0	0	0	0	622

seaward slopes where low rainfall and rapid runoff lead to an open forest of Douglas-fir, arbutus, Garry oak, and mountain juniper along with shrubs such as manzanita, common juniper, and kinnikinnick. The Townsend's Solitaire avoids dense forest both on the coast and in the interior.

Migrants appear to move across a broad front from the coast to the Rocky Mountains. Spring migrants arrive in the south, both on the coast and in the interior, mainly from early April to early May (Figs. 433 and 434), peaking on the coast in the third week of April and in the Okanagan valley as early as the second week of April. Further north, spring migrants arrive mainly in late April and early May. They then disperse to higher-elevation breeding areas, and low-lying areas become almost devoid of solitaires.

Autumn migrants probably move southward mainly at higher elevations along ridges and peaks, but the extent of the high-elevation movement remains to be determined. In northern areas, the autumn migration begins in mid-August, and most birds are gone by mid to late September (Figs. 433 and 434). In central and southern regions, birds appear to begin leaving the area as early as July, with most birds gone by September and early October. In the Southern Interior Mountains and Southern Interior, there is an abrupt increase in October and November, respectively (Fig. 434). Cannings et al. (1987) also report a sharp increase in numbers in the Okanagan valley during November, from the lowest numbers of the year between August and October. Presumably this late autumn increase occurs as populations move to lower elevations for the winter. On the coast, the autumn migration is most noticeable in late September and early October.

The Townsend's Solitaire winters in the Cariboo and Chilcotin areas of the Central Interior; in the Thompson,

Figure 432. In the interior, Douglas-fir forests on open hillsides provide winter foraging habitat for the Townsend's Solitaire (near Westwold, 12 December 1992; R. Wayne Campbell). Snags are important habitat components as perch sites.

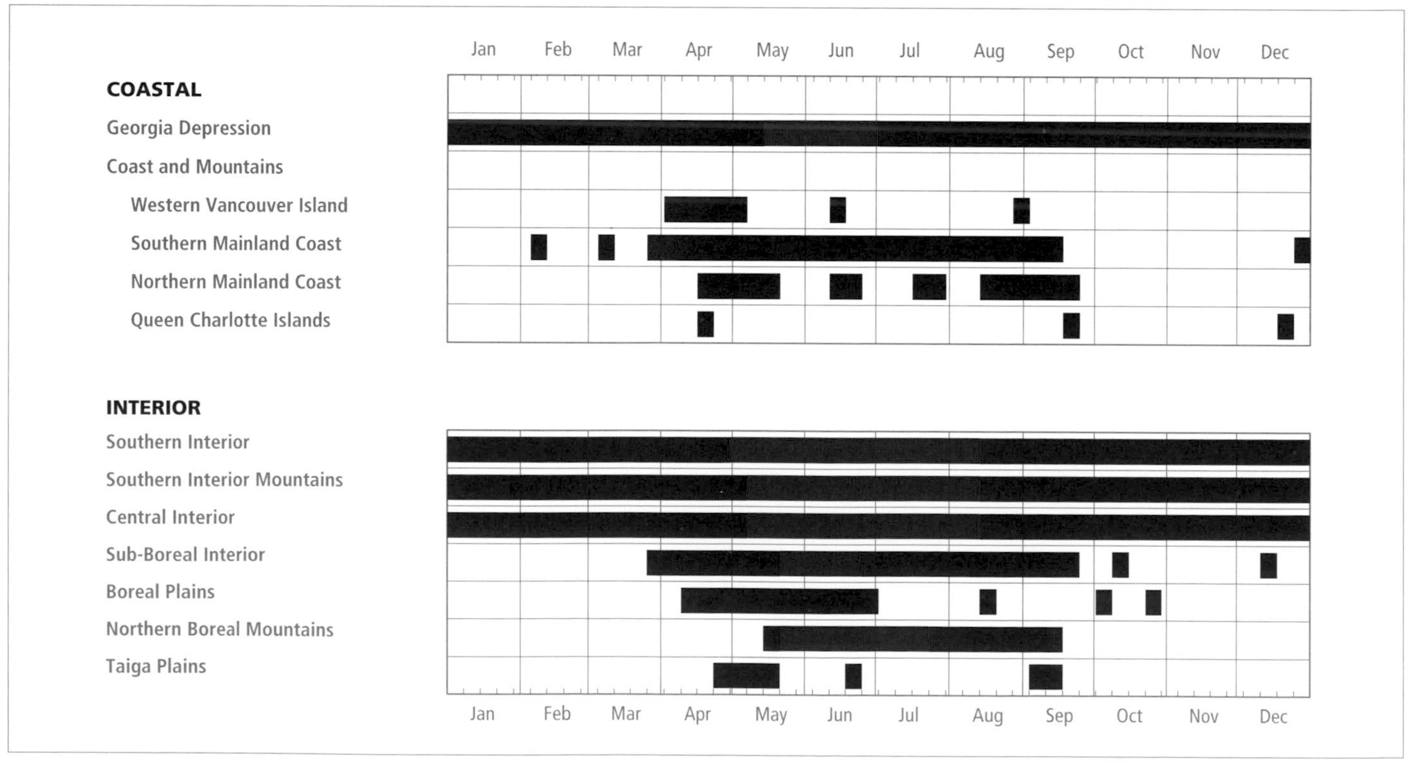

Figure 433. Annual occurrence (black) and breeding chronology (red) for the Townsend's Solitaire in ecoprovinces of British Columbia.

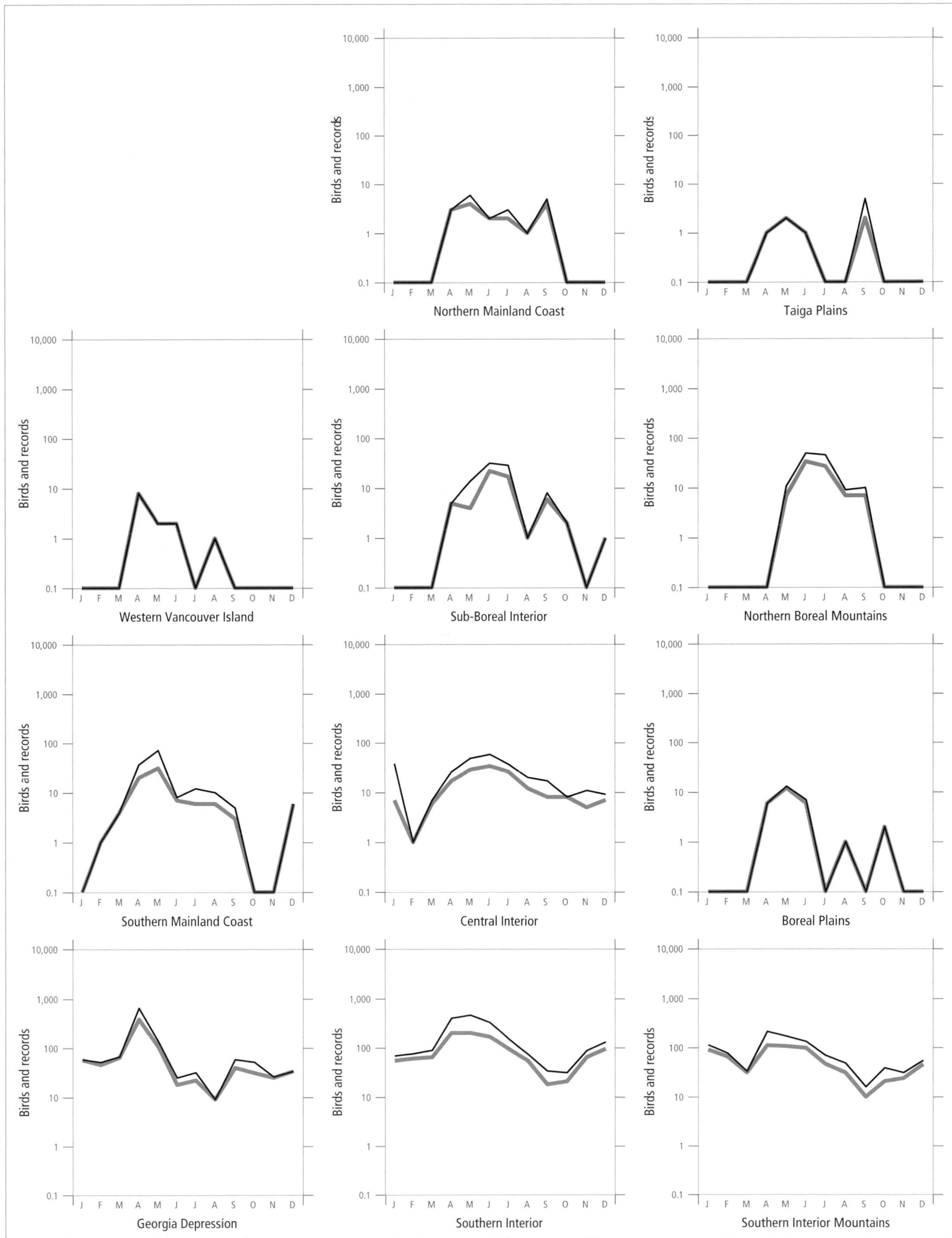

Figure 434. Fluctuations in total number of birds (purple line) and total number of records (green line) for the Townsend's Solitaire in ecoprovinces of British Columbia. Christmas Bird Count and nest record data have been excluded.

Nicola, and Okanagan river valleys in the Southern Interior; and in the Columbia and Kootenay river valleys of the Southern Interior Mountains. On the south coast, there are records for all winter months from the lower Fraser River valley and from southeastern Vancouver Island (Fig. 433).

In the Rocky Mountain states, large flocks of Townsend's Solitaire have been reported on the winter range and during migration (Bent 1949). In British Columbia, however, this species is usually a relatively solitary bird; seldom are more than 2 or 3 found together. Both in the southern portions of the interior and on the coast, individual birds on the winter range often establish feeding territories and defend them from other solitaires, Cedar and Bohemian waxwings, and American Robins. Flocks of 6 to 10 birds have been reported from the coast and southern interior during migration. The largest flock on record consisted of 25 to 30 solitaires in North Vancouver between 5 and 25 April 1953 (Racey 1953).

The Townsend's Solitaire occurs throughout the year in the Georgia Depression, Southern Interior, Southern Interior Mountains, and Central Interior, north to Williams Lake (Fig. 433). Elsewhere, it occurs regularly between early April and late October.

BREEDING: The Townsend's Solitaire breeds primarily throughout the mountainous areas of the Southern Interior, the Southern Interior Mountains, the Central Interior, the southern portion of the Sub-Boreal Interior north to Hixon, as well as in the Northern Boreal Mountains. Breeding has not been confirmed in the Boreal Plains, Taiga Plains, most of the Sub-Boreal Interior, and the southern parts of the Northern Boreal Mountains. On the coast, breeding has been confirmed at only 3 localities on southern Vancouver Island: north of Sooke, Copper Canyon near the Chemainus River, and Doran Lake, north of Sproat Lake. Most breeding records are concentrated in the southern half of the province from the latitude of Princeton and Cranbrook (49°30′N) north into the Cariboo and Chilcotin areas.

Figure 436. In subalpine and alpine areas in the interior of southern British Columbia, the Townsend's Solitaire builds its nest under the protective cover of rocks, cliff faces, and rocky peaks (Mount Tatlow, 23 July 1993; R. Wayne Campbell).

Figure 435. Most Townsend's Solitaire nests reported in British Columbia were found in vertical banks along roads that cross forested slopes (Hat Creek, 13 June 1996; R. Wayne Campbell).

There is not a single breeding record of the Townsend's Solitaire between latitudes 53°N and 59°N, a distance of about 650 km. Near the Yukon border, however, it has nested near the Tatshenshini, Liard, and Beatton rivers, at Tagish and Atlin lakes, and near Fireside.

The Townsend's Solitaire reaches its highest numbers in summer in the Southern Interior, Southern Interior Mountains, and Central Interior ecoprovinces (Fig. 430). An analysis of Breeding Bird Surveys for the period 1968 through 1993 could not detect a net change in numbers on interior routes; coastal route surveys contain insufficient data for analysis.

From late spring through autumn, the Townsend's Solitaire is most often found between 500 and 1,800 m, but occurs up to at least 2,500 m. Breeding has been recorded at elevations between 360 and 2,400 m.

It breeds mainly in forested mountainous areas where vertical banks are available for nesting (Fig. 435). Forest habitat used is primarily lodgepole pine and trembling aspen in the Central Interior, or Douglas-fir and ponderosa pine in the Southern Interior and Southern Interior Mountains ecoprovinces. Natural or selectively logged open forests,

(a)

(b)

(c)

(d)

Figure 437. In British Columbia, the Townsend's Solitaire builds its nest in a variety of locations. These include: (a) overhangs in dirt banks along roads in forested habitats (25 km west of Douglas Lake, 2 June 1994; R. Wayne Campbell); (b) under rocks in open lowland habitats (Hat Creek, June 1989; John M. Cooper); (c) under rocks in alpine habitats (Mount Tatlow, 2,400 m elevation, 23 July 1993; R. Wayne Campbell); and (d) human-made structures such as this snow shovel leaning against a shed (Nimpo Lake, 10 July 1992; R. Wayne Campbell).

burned or cut-over areas, and even denser forests are all used during the breeding season. Nest sites are always in relatively open areas; virtually all breeding records are from alongside roads (Fig. 435). Smaller numbers breed in alpine areas, on steep slopes, cliff faces, rocky peaks, or natural cutbanks, but the relative importance of this habitat is unknown (Fig. 436).

The Townsend's Solitaire has been recorded breeding in British Columbia between 3 May and 13 August (Fig. 433).

Nests: The Townsend's Solitaire usually selects a nest site in a crevice on a vertical bank or cliff. Of 331 nests, 95% were in dirt banks (Fig. 437a) and 4% were on rock cliff faces. One nest was among the exposed roots of a ponderosa pine on a steep hillside (Cannings et al. 1987), 2 were under a rock overhang in lowland and alpine environments (Fig. 437b, c), and another was built on the flat blade of a snow shovel leaning against a shed (Fig. 437d). Almost all nests were found along roads through hilly or mountainous terrain, where road building has created countless kilometres of vertical cutbanks (Fig. 435). Natural nest sites included dirt banks along rivers; natural banks created by slides, slumping ground, or erosion; and banks along game trails. Nests were placed in a hollow in the bank, usually near the top of the bank where overhanging grass, sod (Fig. 437a), tree roots, rocks (Fig. 437b, c), or other debris provided cover.

Although all nests were built on the ground, vertical heights above level ground for 245 nests ranged from 0 to 10 m, with 75% between 0 and 1.5 m.

Nests (n = 334; Fig. 438) were loose cups of grass (77%), twigs (44%), plant stems (13%), conifer needles (12%), and lichens and moss (11%), and were lined with fine grass and rootlets. They often contained bits of fallen dirt. Nests were sometimes used for second broods.

Eggs: Dates for 295 clutches ranged from 5 May to 30 July, with 51% recorded between 30 May and 25 June. Calculated dates suggest that eggs could be found as early as 3 May. Sizes of 276 clutches ranged from 1 to 6 eggs (1E-22, 2E-24, 3E-51, 4E-143, 5E-35, 6E-1), with 51% having 4 eggs (Fig. 437a). The incubation period is unknown. In British Columbia, 2 clutches appeared to hatch after 11 and 13 days.

Young: Dates for 179 broods (Fig. 439) ranged from 20 May to 9 August, with 52% recorded between 14 June and 8 July. Calculated dates suggest that nestlings could be found as late as 13 August (Cannings et al. 1987). Sizes of 160 broods ranged from 1 to 5 young (1Y-14, 2Y-21, 3Y-42, 4Y-69, 5Y-14), with 72% having 3 or 4 young. The nestling period is unknown. In southern regions, some pairs raise 2 broods each year (Paul 1964; Cannings et al. 1987), but the extent of double-brooding is unknown.

Brown-headed Cowbird Parasitism: In British Columbia, 8 of 409 nests found with eggs or young were parasitized by the cowbird, all from the interior (Fig. 438). Seven cases were from the Southern Interior. Elsewhere in the province, parasitism rarely occurs. Friedmann and Kiff (1985) list the Townsend's Solitaire as a cowbird egg host species, but do not include it as a species that will raise cowbird young to fledging. Near Tranquille, north of Kamloops, a nest with 3 Townsend's Solitaire eggs and 2 Brown-headed Cowbird eggs was found on 24 June 1979. On 7 July 1979, the nest contained 2 cowbird nestlings about 10 cm in length and a solitaire nestling about 7 cm in length. Although the final outcome is not known, it is probable that the cowbird nestlings fledged.

Nest Success: Of 59 nests found with eggs in the interior and followed to a known fate, 21 produced at least 1 fledgling, for a nest success of 36%. Many nests are destroyed by

Figure 438. Townsend's Solitaire nest with 3 solitaire eggs and 1 Brown-headed Cowbird egg (near Falkland, 12 June 1993; R. Wayne Campbell).

Figure 439. Townsend's Solitaire nest in a crevice in a vertical bank, containing 2 large young (20 km east of Douglas Lake, 2 Jun 1994; R. Wayne Campbell).

predators (Paul 1964), probably because nests are fairly visible and accessible to mammalian predators. However, there are few data on specific predators. Causes of nest mortality (n = 34) include unknown predator (62%), abandonment (15%), collapsed bank (12%), trampled by cattle (2 nests), human interference (1 nest), and nestlings eaten by a garter snake (1 nest) (Cannings et al. 1987).

REMARKS: Little is known of the breeding biology or ecology of the Townsend's Solitaire. Naturalists in the interior of British Columbia have an opportunity to increase our knowledge of this species because of the large numbers that breed there. For further life-history information, see Bent (1949) and Salomonson and Balda (1977).

NOTEWORTHY RECORDS

Spring: Coastal – Jocelyn Hill 15 Apr 1990-8; Spectacle Lake 4 May 1983-5; Ucluelet 5 Apr 1982-1; Tofino 3 May 1986-1; Florencia Bay 5 Apr 1974-1; Vancouver 26 Mar 1985-2, 9 Apr 1972-9; Pitt Meadows 13 Apr 1971-7; Harrison Hot Springs 11 May 1984-9; Comox 25 Apr 1954-40 in area (Flahaut and Schultz 1954c); Port Neville Inlet 25 May 1976-6; Ramsay Island 19 Apr 1984-1; Kootenay Inlet 14 May 1986-1 (Campbell 1986f); Metlakatla 24 Apr 1908-1 (Keen 1910); Terrace 2 May 1970-1 (Crowell and Nehls 1970c). **Interior** – Creston 11 Apr 1984-10; 8 km w Elko 19 May 1983-3 eggs, 28 May 1983-3 nestlings; McKinney Creek Road 1 May 1974-18; Vaseux Lake 12 May 1979-5 eggs; Apex Mountain 5 May 1914-3 eggs (Cannings et al. 1987); Summerland 12 Mar 1977-4; Naramata 28 May 1967-4 nestlings; Kelowna 20 May 1965-4 nestlings (Cannings et al. 1987); Silverton 23 Apr 1984-7; Coldstream 31 Mar 1954-1; Nakusp 17 Mar 1978-2, 20 Apr 1985-10; Invermere 15 May 1976-3 eggs; Revelstoke 26 Apr 1980-5; Skwaam Bay 20 May 1984-9; Loon Lake (Clinton) 25 May 1963-4 eggs (RBCM E1935); 100 Mile House 25 Mar 1975-2; Straight Lake (100 Mile House) 11 May 1983-eggs, 25 May 1983-nestlings (Graham 1983); Alexis Creek 8 May 1977-7; Kleena Kleene 12 May 1965-building nest; Williams Lake 20 May 1966-4 eggs; 18 km w Williams Lake 22 Mar 1987-5 in vicinity of Doc English Ecological Reserve; Puntchesakut Lake 21 May 1944-eggs (Munro and Cowan 1947); Prince George 9 Apr 1986-1; Willow River 1 Apr 1966-1; Nechako and Chilako rivers 22 May 1960-7 at junction; Telkwa 28 Apr 1977-4; Pine Pass 30 Apr 1965-1; Tetana Lake 8 Apr 1938-1 (Stanwell-Fletcher and Stanwell-Fletcher 1943); Fort St. John 10 Apr 1987-1, 1 May 1985-1; Tetsa River 27 May 1981-1; Fort Nelson 28 Apr 1974-1 (Erskine and Davidson 1976), 12 May 1987-1; Atlin 16 May 1981-1 (Campbell 1981); Coal River 30 May 1980-5 nestlings; Chilkat Pass 16 May 1977-3.

Summer: Coastal – Tugwell Lake 2 Jul 1983-1 fledgling with parents; Mount Finlayson 4 Aug 1985-3; Copper Canyon 21 Jun 1968-1 nestling; Jordan Meadows 27 Aug 1947-1; Quamichan Lake 11 Jun 1970-pair with 2 fledglings; Goldie Lake 16 Aug 1974-1; Doran Lake 29 Jun 1995-3 young; Garibaldi Park 12 Aug 1968-4; Stuie 16 Jul 1938-1 (NMC 28745); San Cristoval Range (above De La Beche Inlet) 11 Jun 1986-1 (Campbell 1986g); Metlakatla 16 Jun 1903-1; Rocher Deboule 22 Jul 1944-2 (Munro 1947a); Hazelton 17 Aug 1938-1 (UMMZ 97904); Doch-Da-On Creek 19 Jul 1919-1 (MVZ 40204). **Interior** – Crater Mountain 14 Jun 1971-4 eggs; Windy Joe Mountain 10 Jul 1966-7; Manning Park 26 Jun 1983-6; 27 Jul 1961-eggs, 22 Aug 1974-3 nestlings (O'Brien 1974); Hedley area 12 Apr 1976-16 along road; Penticton 7 Aug 1964-5; Shingle Creek Rd 22 Jun 1988-23; Okanagan Lake 5 Aug 1965-1 nestling being eaten by garter snake (Cannings et al. 1987); Wasa 6 Aug 1970-3 nestlings; Kelowna 20 Jul 1962-4 nestlings; Lardeau 21 Jun 1981-6; Silverstar Mountain 30 Jul 1962-4 eggs; Gold Bridge 17 Jun 1972-1 nestling; Crowfoot Mountain 16 Jul 1960-8; Field 4 Aug 1975-4; Horse Lake area (100 Mile House) 8 Jul 1983-4; 100 Mile House 9 Aug 1976-3 nestlings; Kleena Kleene 8 Aug 1967-3 nestlings; Mount Tatlow 23 Jul 1993-4 eggs; Chilanko Forks 17 Jun 1961-10; Anahim Lake 9 Jun 1966-5 eggs; Berg Lake 3 Jul 1973-1; 72 km nw Quesnel 21 Jun 1971-5 nestlings; Blackwater River 21 Jun 1967-5 nestlings; Topley 22 Jun 1956-6; East Pine 16 Aug 1930-1 (Williams 1933a); Todagin Lake 22 Jul 1963-12; Hyland Post 7 Jun 1976-3; Fern Lake (Kwadacha Wilderness Park) 13 Aug 1983-3 (Cooper and Cooper 1983); Fort Nelson 23 Jun 1981-1; Muncho Lake 7 Jul 1943-1 (Rand 1944); Atlin 19 Jul 1919-3 eggs (RBCM E970); Coal River 30 May 1980-5 downy nestlings; Samuel Glacier 26 Jun 1983-4 (Campbell et al. 1983); Chilkat Pass 30 Jun 1958-6, 19 Jul 1959-4 eggs (Weeden 1960); Mount Reilly 19 Jul 1959-4 eggs (Weeden 1960); Cloutier Peak 13 Jul 1975-4 newly hatched nestlings.

Breeding Bird Surveys: Coastal – Recorded from 3 of 27 routes and on 3% of all surveys. Maxima: Pemberton 30 Jun 1974-5; Comox Lake 15 Jun 1974-1; Kitsumkalum 11 Jun 1978-1. **Interior** – Recorded from 40 of 73 routes and on 35% of all surveys. Maxima: Summerland 22 Jun 1988-26; Oliver 28 Jun 1992-11; Canford 1 Jun 1975-10.

Autumn: Interior – Atlin 16 Sep 1931-1 (CAS 34140); Steamboat Mountain 11 Sep 1943-3 (Rand 1944); Fort Nelson 8 Sep 1987-2; Tetsa River 11 Sep 1970-1 juvenile; Fern Lake (Kwadacha Wilderness Park) 6 Sep 1979-3 (Cooper and Adams 1979); Beatton Park 6 Oct 1986-1; North Pine 26 Oct 1985-1; Nilkitkwa Lake 15 Sep 1980-1; Nechako and Chilako rivers 7 Sep 1959-2 at junction; Giscome 25 Oct 1992-1; Willow River 13 Oct 1971-1; Williams Creek and Fraser rivers 9 Nov 1974-4 at junction; Riske Creek 14 Oct 1978-1, 8 Nov 1986-3; Horse Lake 18 Sep 1983-8; Pavilion 17 Nov 1986-5; Kamloops 16 Sep 1990-1 fledgling fed by parents; Kootenay National Park 27 Sep 1982-3, 30 Nov 1981-1; Radium Hot Springs 3 Nov 1965-6 (Seel 1965); Harmer Ridge 2 Oct 1983-3 (Fraser 1984); Summerland 31 Oct 1966-1, 30 Nov 1972-1; Vaseux Lake 6 Sep 1971-6; Oliver 10 Oct 1985-6. **Coastal** – Kispiox River 13 Sep 1921-2 (Swarth 1924); Hazelton 17 Sep 1921-1 (Swarth 1924); Cape St. James 17 Sep 1978-1; Barriere Lake 13 Sep 1974-2 (Thomson 1974); Comox 16 Nov 1930-2; Chilliwack 19 Nov 1927-1 (RBCM 9498); Reifel Island 30 Nov 1971-1; Mount Prevost 14 Oct 1972-15 (Tatum 1973).

Winter: Interior – Prince George 15 Dec 1991-1 (Siddle 1992b); Riske Creek 11 Dec 1983-1, 13 Jan 1989-1; Kleena Kleene 1 Feb 1968-1; Horse Lake 13 Dec 1983-1; Athalmer 4 Jan 1981-5; Invermere 8 Feb 1977-5; Fairmont Hot Springs 7 Dec 1975-5; Nakusp 1 Jan 1991-1; New Denver 7 Jan 1978-2; Summerland 10 Jan 1974-4, 1 Feb 1974-2; Keremeos 28 Dec 1964-1 (UBC 11971); East Trail 31 Dec 1981-1; Manning Park 19 Jan 1979-1. **Coastal** – Masset 18 Dec 1982-1; Alta Lake 29 Dec 1945-1 (Racey 1953); Comox 15 Dec 1991-1; Lions Bay 5 Feb 1980-1; Chilliwack 9 Jan 1927-1 (RBCM 13568); Vancouver 26 Dec 1966-8 (Smith 1967); Crescent Beach 29 Dec 1947-1; Maple Bay 6 Jan 1979-2; Mount Finlayson 6 Feb 1974-3; Victoria 1 Jan 1972-2 (Tatum 1972).

Christmas Bird Counts: Interior – Recorded from 19 of 27 localities and on 52% of all counts. Maxima: Lake Windermere 26 Dec 1982-**62**, all-time Canadian high count (Anderson 1983); Penticton 28 Dec 1985-35; Vernon 26 Dec 1975-28. **Coastal** – Recorded from 16 of 33 localities and on 12% of all counts. Maxima: Vancouver 26 Dec 1966-8; Victoria 14 Dec 1991-6; Duncan 14 Dec 1974-3.

Veery

VEER

Catharus fuscescens (Stephens)

RANGE: Breeds from south-central British Columbia, central Alberta, central Saskatchewan, southern Manitoba, southern Ontario, southern Quebec, New Brunswick, central Nova Scotia, and southwestern Newfoundland south to Georgia in the east and central Oregon in the west; also in eastern Arizona. Winters in South America from Colombia and Venezuela to Brazil.

STATUS: On the coast, a *very rare* migrant and local summer visitant to the Fraser River delta and along southeastern Vancouver Island in the Georgia Depression Ecoprovince; *uncommon* to *fairly common* locally in the southern mainland of the Coast and Mountains Ecoprovince; *rare* to *uncommon* locally in the northern mainland of the Coast and Mountains, including the lower Skeena and Nass river valleys. Absent from Western Vancouver Island and the Queen Charlotte Islands.

Figure 440. Because the Veery arrives directly on its breeding grounds in British Columbia, nonbreeding and breeding habitats are similar. In the interior, shady, moist woodlands of black cottonwood with a dense shrub understorey of crab apple, red-osier dogwood, willow, and water birch provide summer habitat for the Veery (Barriere, 8 July 1991; R. Wayne Campbell).

Figure 441. On the coastal mainland, the Veery is suspected of breeding in dense mixed woodlands of mountain-ash, black cottonwood, and red alder scattered with western redcedars and Douglas-firs (Hagensborg, 11 July 1992; R. Wayne Campbell).

In the interior, an *uncommon* to locally *fairly common* migrant and summer visitant to the Okanagan and Thompson valleys of the Southern Interior Ecoprovince, the Kootenay and Columbia valleys of the Southern Interior Mountains Ecoprovince, and the Cariboo and Chilcotin areas of the Central Interior Ecoprovince; *very rare* in the extreme southern Sub-Boreal Interior Ecoprovince; *rare* and local in the Peace Lowland of the Boreal Plains Ecoprovince.

Breeds.

NONBREEDING: The Veery is a characteristic summer bird of the southern third of the province east of the Pacific and Cascade Mountains, from Princeton east to the Crowsnest Pass and north to the Cariboo and Chilcotin areas. Its numbers are highest in the Okanagan and Similkameen valleys, the southern portions of the west and east Kootenays, the Thompson and South Thompson river valleys, and the

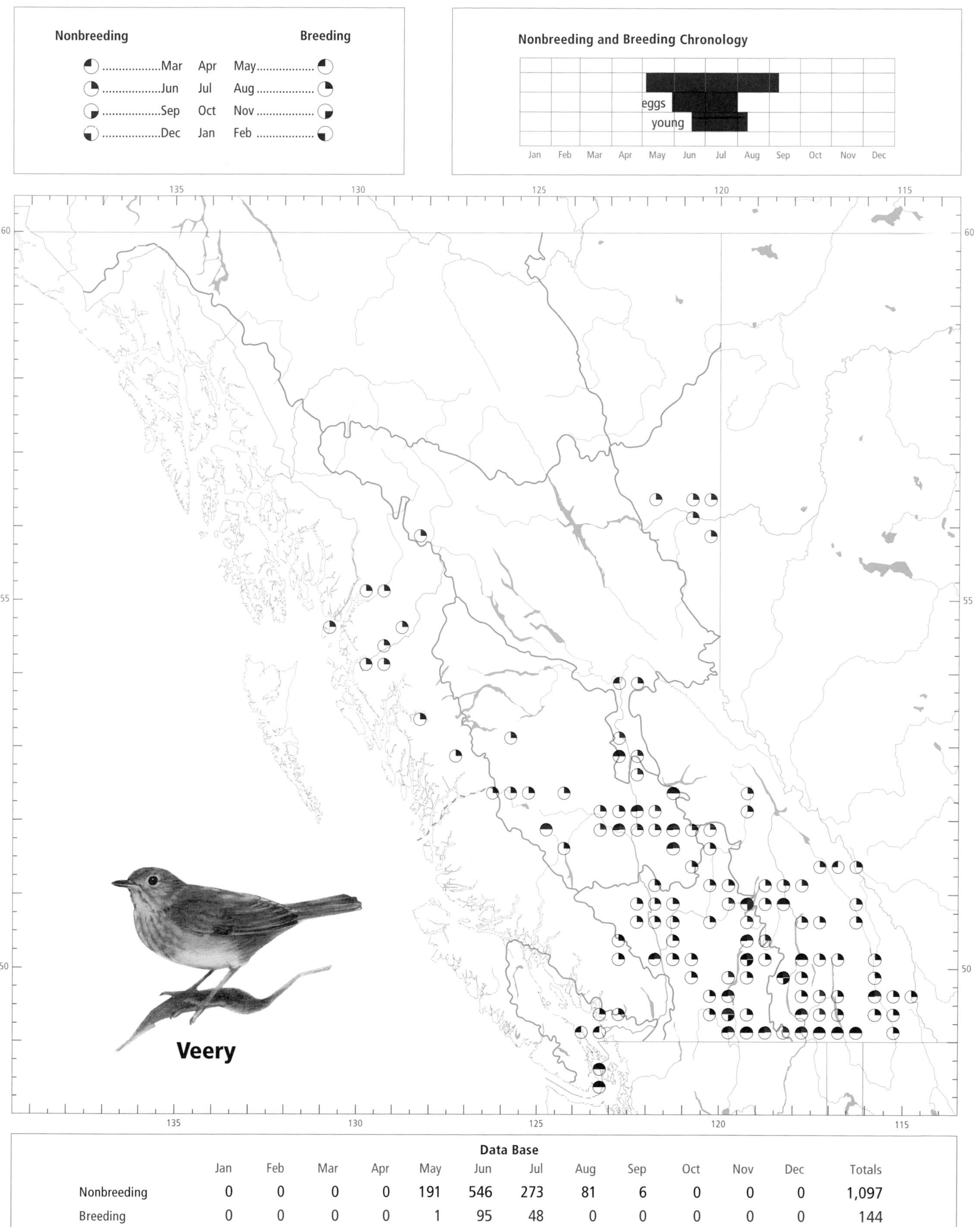

Data Base

	Jan	Feb	Mar	Apr	May	Jun	Jul	Aug	Sep	Oct	Nov	Dec	Totals
Nonbreeding	0	0	0	0	191	546	273	81	6	0	0	0	1,097
Breeding	0	0	0	0	1	95	48	0	0	0	0	0	144

Figure 442. In the Peace Lowland of northeastern British Columbia, the Veery inhabits dense shrubbery bordering beaver ponds (northeast of Boundary Lake, 24 June 1996; R. Wayne Campbell).

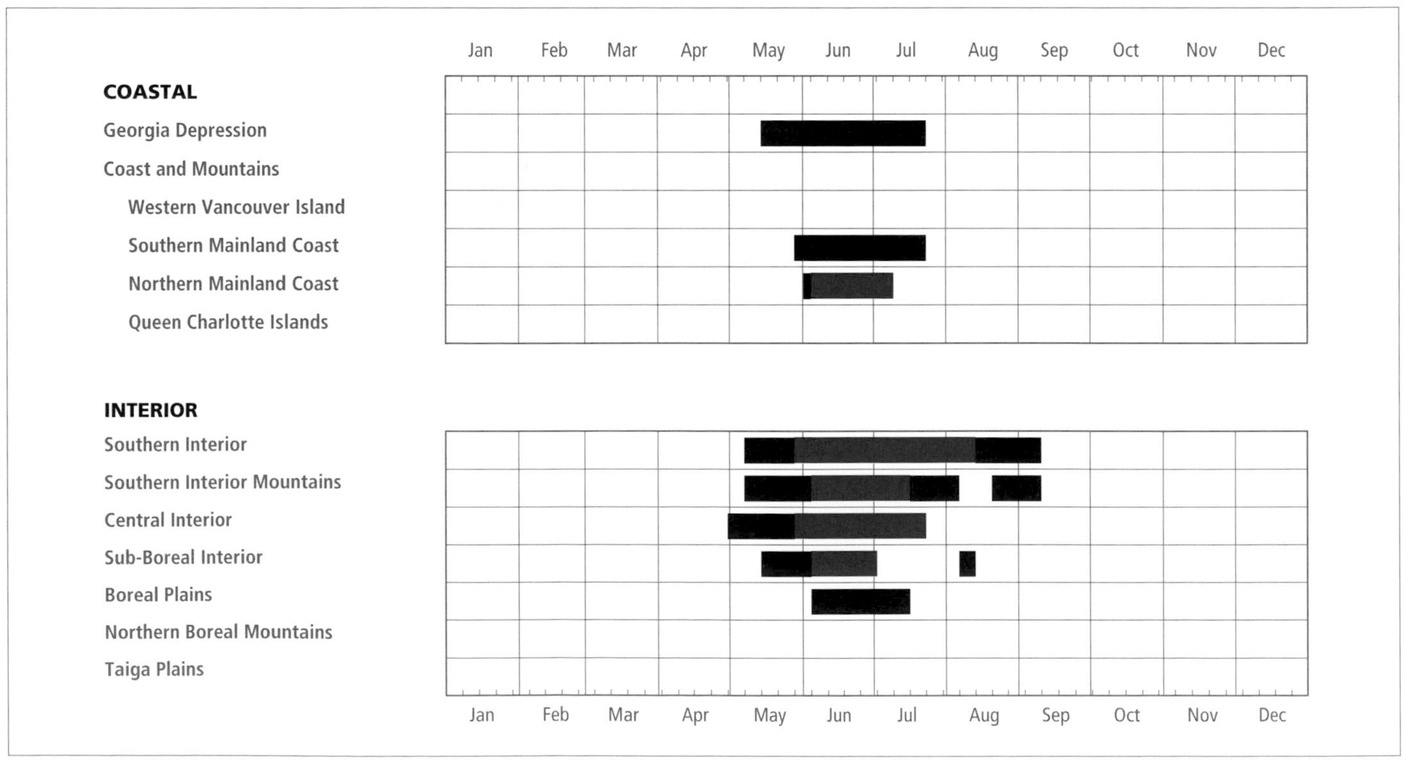

Figure 443. Annual occurrence (black) and breeding chronology (red) for the Veery in ecoprovinces of British Columbia.

southern portions of the Central Interior. Further north its distribution is sparse and scattered. On the coast, it has a localized distribution, and is known to occur only in the east-west aligned valleys such as those of the Nass, Skeena, Bella Coola, and Kemano rivers.

West of the Rocky Mountains, the Veery has been reported from as far north as the Prince George region in the interior, and Stewart at the head of Portland Canal on the coast. East of the Rocky Mountains, it has been recorded only in the Peace Lowland near Cecil and Boundary lakes. It occurs infrequently on the Fraser River delta and along southeastern Vancouver Island.

In migration, the Veery has been reported from the understorey of second-growth deciduous and mixed forests and along forest edges. However, because of its secretive nature during migration, there is little other information on its nonbreeding habitat. The characteristics of its summer habitat are described under BREEDING (Figs. 440, 441, and 442).

The Veery is among the last of the spring migrants to reach British Columbia. It likely arrives in late April but is

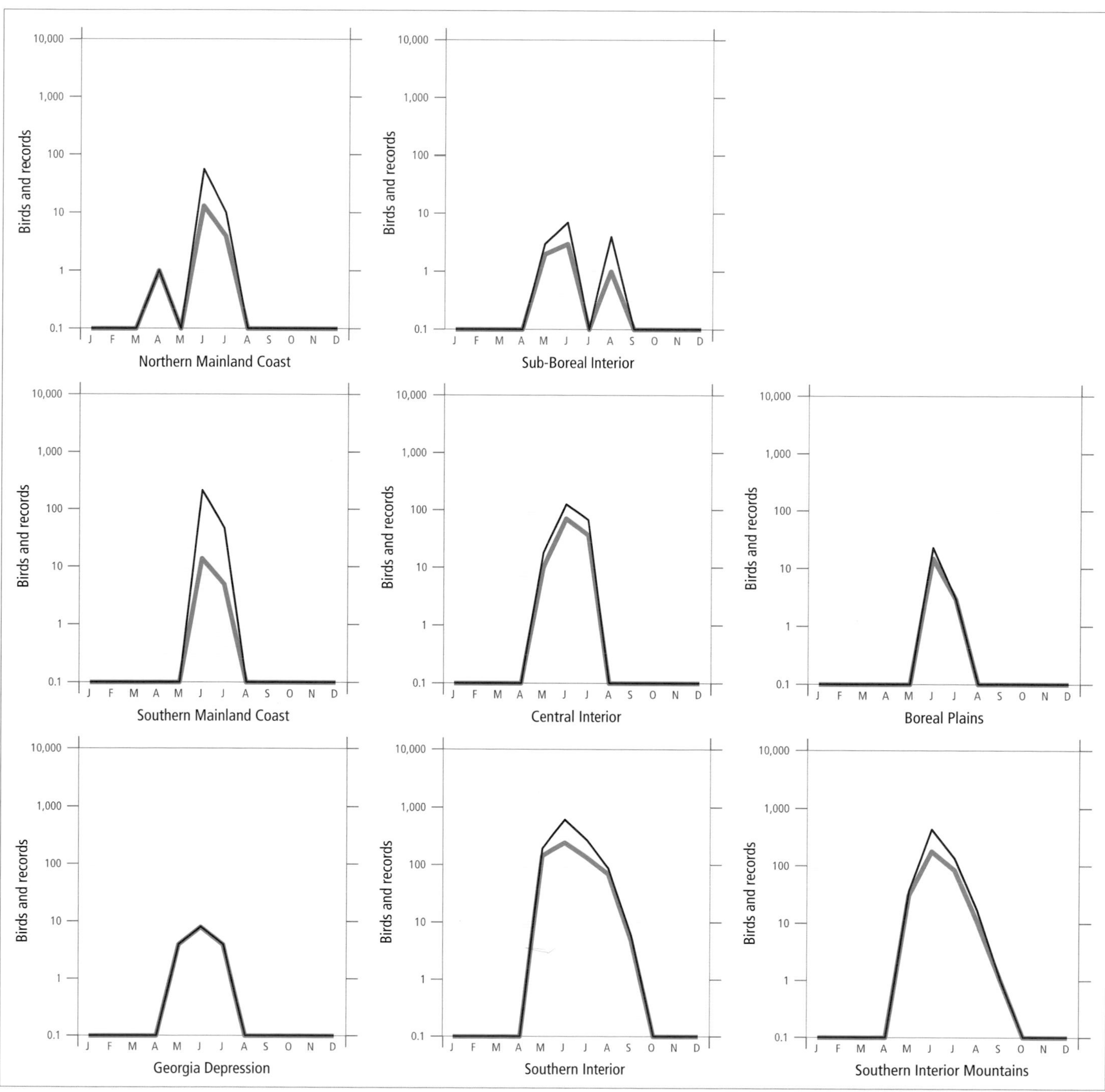

Figure 444. Fluctuations in total number of birds (purple line) and total number of records (green line) for the Veery in ecoprovinces of British Columbia. Nest record data have been excluded.

silent and unnoticed for the first week or two after arrival (Bent 1949). It is first reported in southern British Columbia in early May (Figs. 443 and 444), and migration continues into early June.

In the Southern Interior, and probably also in the Southern Interior Mountains, autumn departure begins in mid-August, and most birds have left by the end of August, although a few stragglers can be found until the second week of September (Figs. 443 and 444). Elsewhere in the province, there are few records past the end of July. In the Central Interior and areas further north, the Veery has not been reported after the end of July. Both spring and autumn migrations are rapid.

This thrush has been reported in the interior between 4 May and 9 September, although only a single September record has been filed since 1944. On the coast, it has been reported between 20 May and 21 July (Fig. 443).

BREEDING: The Veery has a widespread distribution in suitable habitat throughout the southern one-third of the interior. It is most abundant as a nesting species in the Okanagan and Thompson river valleys. It also breeds in the west Kootenay, occasionally in the east Kootenay; in the Cariboo and Chilcotin areas from 100 Mile House, Lac la Hache, and Williams Lake west to Kleena Kleene; and in the extreme southern portions of the Sub-Boreal Interior in the vicinity of Quesnel. It also breeds in the vicinity of Terrace and Ascaphus Creek near the Skeena River on the northern mainland coast. Although no nests have been found, the presence of singing males suggests that they breed as far west as Bella Coola and as far north as the Nass River valley.

In summer, the Veery reaches its highest numbers in the Southern Interior (Fig. 445). An analysis of Breeding Bird Surveys for the period 1968 through 1993 could not detect a net change in numbers on interior routes; coastal routes contain insufficient data for analysis.

The species has been found nesting between 105 and 1,050 m elevation, but over 80% (n = 39) of the reports were from sites below 650 m. Cannings et al. (1987) note that above 500 m the Veery tends to be replaced by the Swainson's Thrush. The Veery inhabits shady, riparian woodlands of black cottonwood or balsam poplar with a dense shrub understorey (Fig. 440), a habitat that develops mainly on river floodplains, the margins of lakes, or creek bottoms. In British Columbia, this thrush also occurs in aspen and birch forests; moist mixed forests (Fig. 441); steep hillside gullies filled with shrubbery that are a feature of the Ponderosa Pine and Interior Douglas-fir zones; margins of clearings or abandoned farmland; borders of beaver ponds (Fig. 442); and other moist woodlands with dense shrubbery (Figs. 440 and 441). The same type of habitat is favoured elsewhere in its breeding range (Dilger 1956; Bertin 1977; Moskoff 1995).

The Veery has been recorded breeding in British Columbia from 29 May to 31 July (Fig. 443). It may frequently attempt 2 broods per summer (Cannings et al. 1987).

Nests: Nests were generally found near or on the ground, among the branches or stems of shrubs or trees (68%; n = 42; Fig. 447), or on the ground in the shelter of dense vegetation or among leaves (32%; Fig. 446). Most (63%; n = 56) were situated in shrub thickets; other locations included dead tree stumps (21%) and deciduous trees (13%). The heights for 47 nests ranged from ground level to 3 m, with 57% between 0.1 and 1.2 m.

Grass, the most frequent structural component, was found in 89% of the nests (n = 48; Fig. 446), followed by leaves (34%), bark strips (21%), plant fibres (19%), moss (13%), and plant stems (11%). Some nests contained conifer needles, hair, twigs, and rootlets.

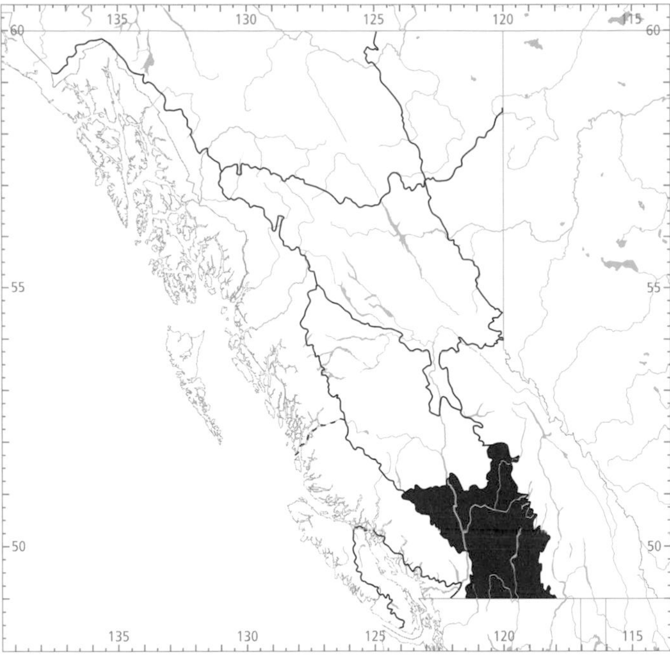

Figure 445. In British Columbia, the highest numbers for the Veery in summer occur in the Southern Interior Ecoprovince.

Figure 446. Veery nest with 3 eggs and 1 Brown-headed Cowbird egg (Creston, 1 June 1995; R. Wayne Campbell). Nearly 25% of nests found with eggs and young in British Columbia were parasitized by the cowbird.

Eggs: Dates for 61 clutches ranged from 29 May to 31 July, with 57% recorded between 9 June and 2 July. Sizes of 50 clutches ranged from 1 to 6 eggs (1E-6, 2E-6, 3E-11, 4E-23, 5E-3, 6E-1), with 68% having 3 or 4 eggs (Fig. 446). In 3 nests in British Columbia, the incubation period was 11, 13, and 13 days; Ehrlich et al. (1988) give a period of 10 to 12 days.

Young: Dates for 22 broods ranged from 17 June to 31 July, with 54% recorded between 25 June and 15 July. Sizes of 20 broods ranged from 1 to 4 young (1Y-3, 2Y-4, 3Y-5, 4Y-8), with 13 having 3 or 4 young. In British Columbia, nestling periods at 3 nests were 12, 13, and 15 days. Harrison (1979) gives the period as 10 to 12 days, while Ehrlich et al. (1988) mention a period of 10 days.

Brown-headed Cowbird Parasitism: The Veery is a frequent cowbird host. In British Columbia, 24% of 74 nests found with eggs or young were parasitized by the cowbird. Friedmann et al. (1977) report a parasitism rate of 20.1% (n = 139) for studies in Ontario, while Cannings et al. (1987) report that 38.7% (n = 31) of nests in the Okanagan valley were parasitized.

Nest Success: Of 11 nests found with eggs and followed to a known fate, 4 produced at least 1 fledgling.

REMARKS: Of the 3 North American races of the Veery, only *C. f. salicicola* occurs in British Columbia (American Ornithologists' Union 1957).

Cannings et al. (1987) suggest that Veery populations in the Okanagan valley have "undoubtedly declined markedly in the past 50 years," due primarily to the cutting and clearing of riparian woodlands. Breeding Bird Surveys in the interior of the province do not show a decline in the population between 1968 and 1993, but in the Okanagan valley the decline noted may have occurred earlier.

See Moskoff (1995) for a summary of the life history and conservation of the Veery in North America.

Figure 447. Typically, the Veery builds its nest close to the ground, usually at the base of a shrub (Kootenay River Trout Hatchery at Bull River, early July 1993; Mark Nyhof).

NOTEWORTHY RECORDS

Spring: Coastal – Victoria 20 May 1893-1 (NMC 1590); Pender Island 20 May 1981-1. **Interior** – Oliver 27 May 1975-4; Salmo River 28 May 1984-2; Lytton 20 May 1970-4; Lavington 8 May 1967-1; Revelstoke 31 May 1898-1 (NMC 930); 100 Mile House 19 May 1985-1; Kleena Kleene 29 May 1964-2 eggs; Williams Lake 31 May 1979-1; Milburn Lake (Quesnel) 4 May 1983-2; Prince George 26 May 1977-1; Summit Lake 31 May 1980-1.

Summer: Coastal – Beacon Hill Park (Victoria) 16 Jun 1979-1; Stanley Park (Vancouver) 21 Jul 1985-1; Mount Seymour Park 20 Jun 1975-1; Golden Ears Park 14 Jul 1985-1; Pemberton Meadows 5 Jun 1983-50 (Mattocks et al. 1983c); 17.4 km e Stuie 13 Jul 1992-1 (Campbell and Dawe 1992); Stuie 18 Jul 1938-1 (NMC 28770); Hagensborg 11 Jul 1992-1 in full song (Campbell and Dawe 1992); Ferry Island (Skeena River) 2 Jul 1983-2; Terrace 12 Jun 1978-4 eggs; 7 Jul 1972-4 nestlings; Ascaphus Creek 31 Jul 1994-2 nestlings; Canyon City to Greenville 29 Jun 1993-39 singing beside the lower Nass River; Greenville 16 Jun 1981-4; Stewart 9 and 10 Jun 1992-3 (Siddle 1992c), 26 Jun 1993-2. **Interior** – Jaynes Lake (Richter Pass) 17 Jun 1960-4 eggs and 1 Brown-headed Cowbird egg (Campbell and Meugens 1971); Grand Forks 10 Jun 1959-3 eggs; Newgate 1 Jul 1953-1 (NMC 38715); Okanagan Falls 4 Jun 1980-3 eggs; Jaffray 22 Jun 1968-26; Nelson 7 Jun 1972-1 egg; 23 Jun 1972-3 hatched; 6 Jul 1972-4 near fledging; Naramata 31 Jul 1935-2 eggs; Kimberley 14 Jun 1976-4 eggs; Wasa Slough 31 Jul 1977-1; Stein River 17 Jun 1979-4 nestlings; Okanagan Landing 7 Jun 1958-2 eggs; Shuswap Falls 15 Jun 1916-3 eggs; Anderson Lake 28 Aug 1968-3; Lillooet 3 Jul 1916-1 (NMC 9817); Salmon Arm 13 Jun 1975-18 (Cannings 1973); Ashcroft 16 Jul 1964-1 (NMC 52578); Sicamous 3 Jul 1903-1 (FMNH 146277); Carpenter Lake 28 Aug 1968-2; Scotch Creek 14 Aug 1963-4; Three Valley Gap 27 Aug 1963-2; Canim Lake 30 Jun 1977-1; Wells Gray Park 4 Jun 1962-1 egg; Lac la Hache 11 Jul 1958-10; Kleena Kleene 16 Jul 1962-1 fledgling (Paul 1964); Williams Lake 24 Jun 1966-8, 29 Jun 1980-4 nestlings; Quesnel 4 Jun 1966-4 eggs; Milburn Lake (Quesnel) 30 Aug 1990-4; Terrace 10 Jun 1978-2 fresh eggs; Boundary Lake 6 Jun 1982-3; Cecil Lake 20 Jun 1982-1.

Breeding Bird Surveys: Coastal – Recorded from 3 of 27 routes and on 10% of all surveys. Maximum: Pemberton 12 Jun 1988-26. **Interior** – Recorded from 40 of 73 routes and on 52% of all surveys. Maxima: Grand Forks 26 Jun 1994-64; Pavilion 12 Jun 1977-50.

Autumn: Interior – Celista 1 Sep 1963-2; Edgewood 9 Sep 1919-1 (RBCM 4824); Okanagan Landing 2 Sep 1944-1, 9 Sep 1943-1 (ROM 87730; Munro and Cowan 1947); Horn Lake (Twin Lakes) 5 Sep 1919-1. **Coastal** – No records.

Winter: No records.

Gray-cheeked Thrush

GCTH

Catharus minimus (Lafresnaye)

RANGE: In North America, breeds from northern Alaska, northern Yukon, and northern Mackenzie across Canada into northern Quebec, Labrador, and Newfoundland; south to southern Alaska, northwestern British Columbia, and the northern parts of the Prairie provinces, and east to New Brunswick, northern Nova Scotia, Vermont, and New York. Winters in South America from Colombia south to Peru and Brazil. Also breeds in eastern Siberia.

STATUS: On the coast, *casual* migrant and summer visitant to the Boundary Ranges adjacent to southeastern Alaska in the Northern Boreal Mountains Ecoprovince.

In the interior, a *rare* to *uncommon* migrant and summer visitant to the Northern Boreal Mountains, becoming locally *fairly common* in the extreme northwestern corner of that ecoprovince; *very rare* to *rare* migrant and summer visitant to the Taiga Plains, Boreal Plains, and Sub-Boreal Interior ecoprovinces; *casual* in the Rainbow Mountains of the Central Interior Ecoprovince.

Breeds.

NONBREEDING: The Gray-cheeked Thrush is sparsely distributed in widely scattered areas throughout the interior of British Columbia. Most records are from higher elevations in the northern third of the province north of Mackenzie (latitude 55°N). The Gray-cheeked Thrush usually occupies inaccessible subalpine habitats. It is a difficult species to identify unless its song is heard or it is seen under exceptional field

Figure 448. Nesting habitat for the Gray-cheeked Thrush in the Northern Boreal Mountains Ecoprovince of northwestern British Columbia includes stands of white spruce with a dense undergrowth of *Vaccinium* and willow (Shini Lakes, July 1983; R. Wayne Campbell).

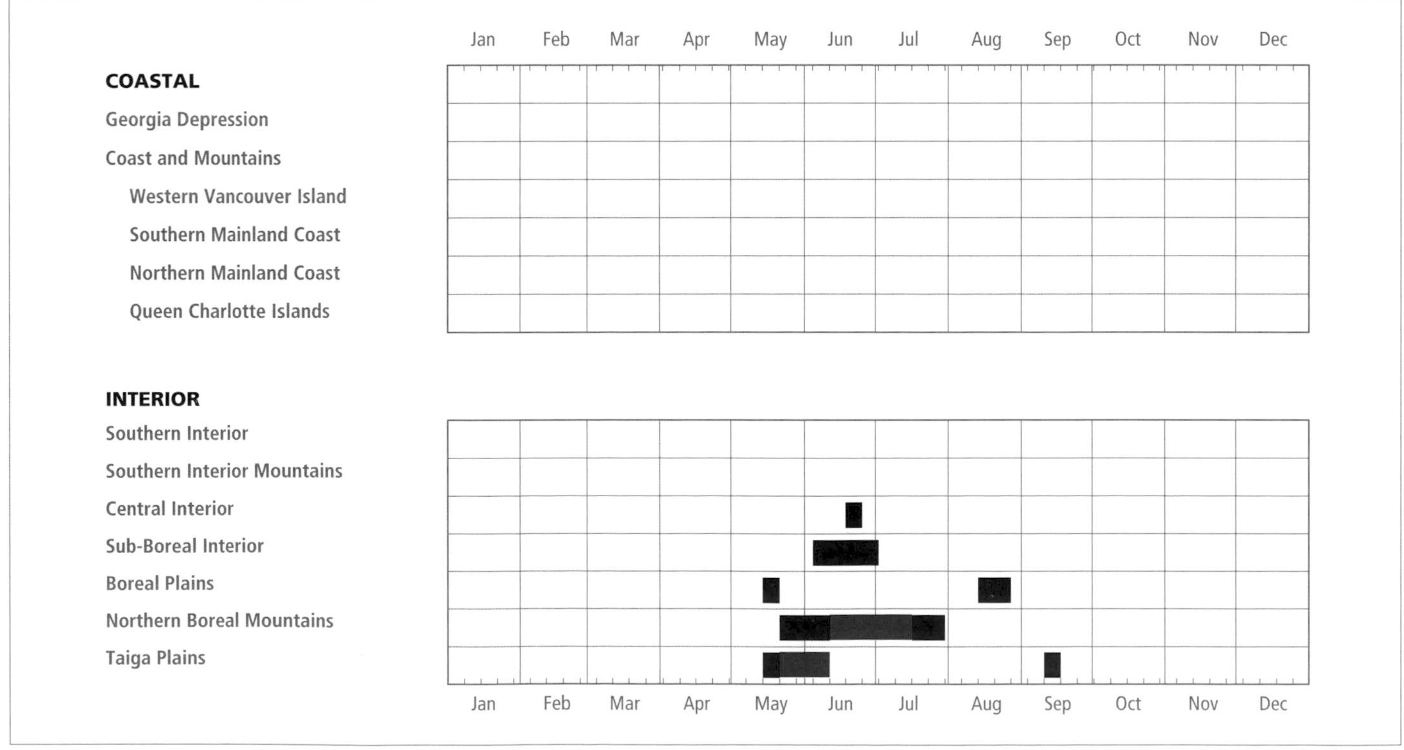

Figure 449. Annual occurrence (black) and breeding chronology (red) for the Gray-cheeked Thrush in ecoprovinces of British Columbia.

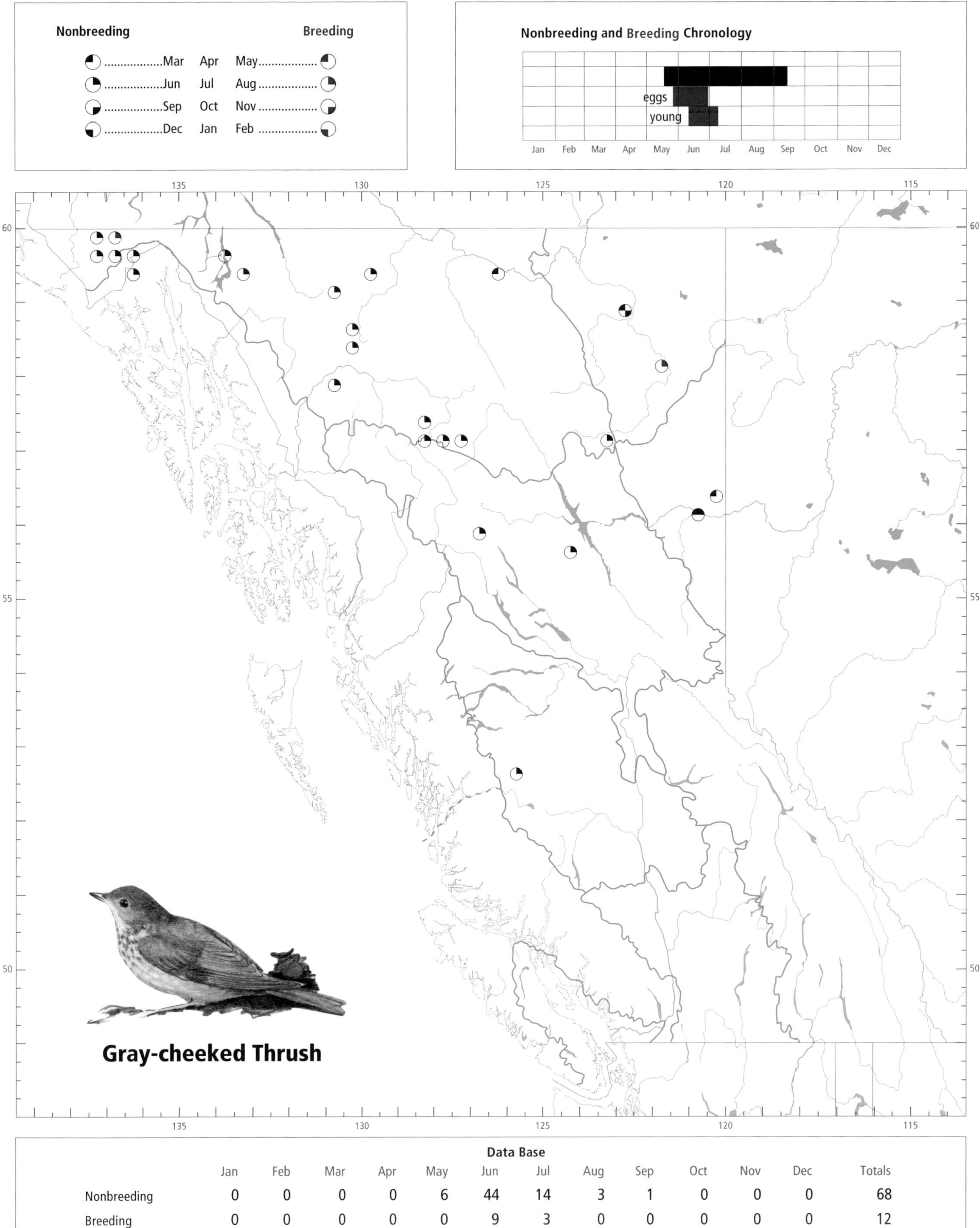

Data Base

	Jan	Feb	Mar	Apr	May	Jun	Jul	Aug	Sep	Oct	Nov	Dec	Totals
Nonbreeding	0	0	0	0	6	44	14	3	1	0	0	0	68
Breeding	0	0	0	0	0	9	3	0	0	0	0	0	12

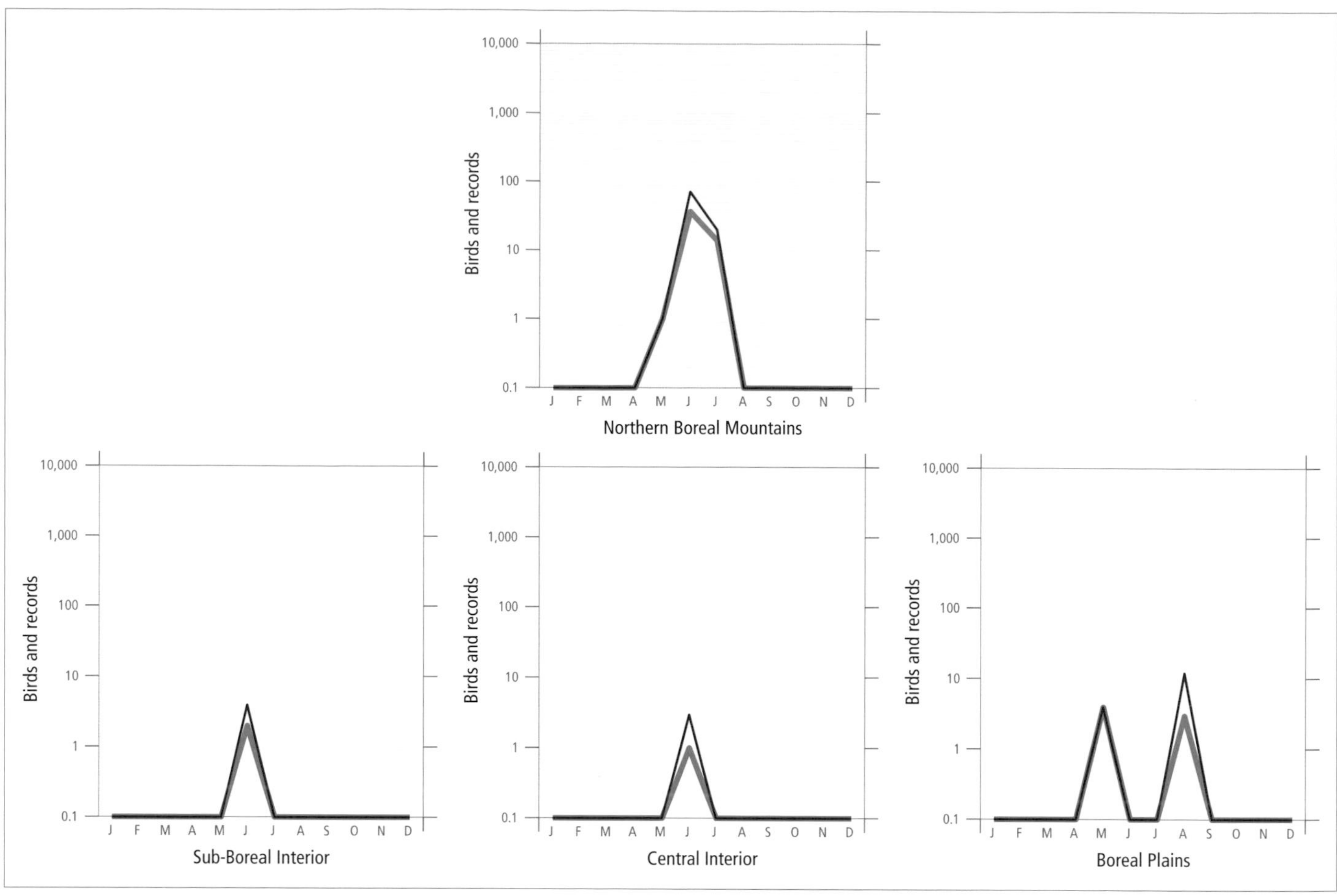

Figure 450. Fluctuations in total number of birds (purple line) and total number of records (green line) for the Gray-cheeked Thrush in ecoprovinces of British Columbia. Nest record data have been excluded.

conditions. The sparse coverage by observers in this thrush's remote habitat probably obscures its true status and distribution during migration periods in British Columbia.

The Gray-cheeked Thrush is an elusive species during migration, when it has been reported only from forest edges (Fig. 448), river bottomland, and willow edges in muskeg.

This species migrates east of the Rocky Mountains and, south of latitude 55°N, it is rarely reported further west. The absence of records in southern British Columbia and Washington state suggests that the Gray-cheeked Thrush enters and leaves the province from the northeast.

This thrush has been reported in northeastern British Columbia as early as the third week of May (Figs. 449 and 450), consistent with arrivals in Alaska (Gabrielson and Lincoln 1959; Kessel and Gibson 1978). Further south, in the mountains bordering the Driftwood River valley, this thrush was among the latest of the spring migrants to arrive (Stanwell-Fletcher and Stanwell-Fletcher 1943).

Autumn migration takes place between mid-August and the second week of September (Figs. 449 and 450). Several birds observed at Taylor in the Boreal Plains on 19 and 22 August 1986 were likely part of the southward migration. The latest dates reported from Atlin and Fort Nelson are 1 September and 13 September, respectively.

Lincoln (1939) refers to this thrush as the champion migrant among our thrushes. He notes that "it does not appear in the United States until the last of April – April 25 near the mouth of the Mississippi and April 30 in northern Florida. A month later, or by the last week of May, it is to be seen in northwestern Alaska, the 4,000 mile trip from Louisiana having been made at an average speed of 130 miles per day."

The Gray-cheeked Thrush has been recorded in British Columbia from 17 May to 13 September (Fig. 449).

BREEDING: The breeding distribution of the Gray-cheeked Thrush in British Columbia is poorly known. There are only 6 nesting records, but it likely nests throughout its summer range in the province. Five of the records are from neighbouring sites in the extreme northwest, along the Haines Highway close to Haines Junction, Yukon. The sixth nesting locality is from the extreme northeast, near the Fort Nelson River (Williams 1933a).

The Gray-cheeked Thrush reaches its highest numbers in summer in the Northern Boreal Mountains (Fig. 451). Breeding Bird Surveys for both coastal and interior routes for the period 1968 through 1993 contain insufficient data for analysis.

The Gray-cheeked Thrush has been reported nesting in British Columbia between 840 and 870 m elevation. Throughout its range, this thrush has the narrowest range of habitat requirements of all our thrushes. It inhabits thickets of stunted subalpine fir and spruce interspersed with willows (Fig. 448), and the neighbouring subalpine parkland coniferous forest, where there is a thick undergrowth of *Vaccinium*, willow, and other shrubs. Its territorial song is frequently given from the highest available perches, but most of its activity is within the dense thickets; 4 of 5 nests were found there. In Alaska, it nests in tall shrubs through a variety of habitats (Kessel and Gibson 1978). In New York, Dilger (1956) found the Gray-cheeked Thrush to be the most habitat-specific of the 5 species of North American forest thrushes; there it was confined to the cloud-drenched, stunted fir and spruce tangles of mountaintops.

The Gray-cheeked Thrush has been reported nesting in British Columbia from 28 May (calculated) to 9 July (Fig. 449).

Nests: Four of 6 nests were placed in a crotch of a willow shrub, 1 was in an unidentified deciduous tree, and another was found under a root. Nest materials include grass, sedge, and mud. The heights for all nests were within a metre of the ground, with 3 nests between 0.3 and 0.8 m.

Eggs: Dates for 3 clutches range from 17 June to 29 June. Calculated dates indicate that eggs can occur as early as 28 May. All 3 clutches had 4 eggs. The incubation period is 13 to 14 days (Ehrlich et al. 1988).

Young: Dates for 5 broods range from 10 June to 9 July, with 3 between 27 June and 4 July. Broods included 4 or 5 young (4Y-4, 5Y-1). The fledging period is 11 to 13 days (Ehrlich et al. 1988).

Brown-headed Cowbird Parasitism: In British Columbia, cowbird parasitism was not found in 8 nests recorded with eggs or young, nor has it been reported elsewhere in North America (Friedmann 1963; Friedmann et al. 1977; Friedmann and Kiff 1985).

Nest Success: Insufficient data.

REMARKS: There are 3 summer occurrences far to the south of what is regarded as the centre of its distribution in British Columbia. In the Driftwood Range, the Stanwell-Fletchers (1943) reported that the Gray-cheeked Thrush was a common breeder at the timberline, and collected a specimen at 1,800 m. They do not report finding any nests. Two others records are from the Rainbow Range, east of Bella Coola. Three birds were collected at the timberline north of the junction of the Atnarko and Hotnarko rivers in June 1932; they were probably nesting there. In addition, Laing (1942), in the same general locality in the summers of 1938, 1939, and 1940, notes that the Gray-cheeked Thrush "shared the breeding ground at Mosher Creek (north of the Atnarko River) at high elevation" with the Hermit Thrush, although complete dates and numbers of birds are not given.

A Gray-cheeked Thrush banded near Beloit, Illinois, on 29 September 1973 was recovered at Fort St. John in May 1978, nearly 5 years later.

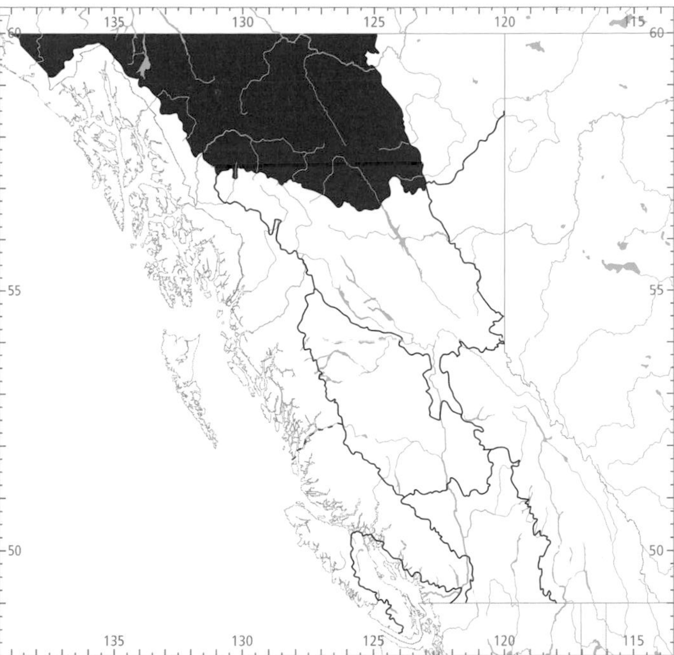

Figure 451. In British Columbia, the highest numbers for the Gray-cheeked Thrush in summer occur in the Northern Boreal Mountains Ecoprovince.

NOTEWORTHY RECORDS

Spring: Coastal – No records. **Interior** – Boundary Lake 17 May 1987-1; Fort Nelson 18 May 1988-1; Liard Hot Springs 27 May 1975-1.

Summer: Coastal – Three Guardsmen Lake 2 Jun 1981-2; Mile 44 Haines Highway 25 Jun 1980-1. **Interior** – Mosher Creek (n of Atnarko River) – shared breeding ground at high elevations with the Hermit Thrush in the summers of 1938, 1939, and 1940 (Laing 1942); Rainbow Range 19 Jun 1932-3 (MCZ 283785-87); w Latham Creek 9 Jun 1981-1 (Blood et al. 1981); Driftwood Range 27 Jun 1941-1 (RBCM 8937; Stanwell-Fletcher and Stanwell-Fletcher 1943); 80 km nw Mackenzie 4 Jun to 14 Jul 1993-up to 3 males singing (Price 1993); Boundary Lake 11 Jul 1976-1; Taylor 22 Aug 1986-1; Spatsizi Park Jul 1976-1 (Osmond-Jones et al. 1977); Buckley Lake 25 Jul 1910-1 (Brooks and Swarth 1925); Skelhorne Creek 17 Jul 1976-1; Fort Nelson River 10 Jun 1922-4 nestlings (Williams 1933a); Goat Creek (Mile 78 Haines Highway) 19 Jun 1980-1 (RBCM 16845); Mile 87.5 Haines Highway 29 Jun 1980-1 (RBCM 16843); Survey Lake 26 Jun 1980-20 (RBCM 16839), 27 Jun 1980-nestlings fed by adult (BC Photo 986); Wilson Creek (Atlin) 13 Jun 1914-1 collected (Brooks and Swarth 1925); Mile 80 Haines Highway 17 Jun 1980-3 eggs, 19 Jun 1981-first egg of a 4-egg clutch, 7 Jul 1949-1 (NMC 35431); Stalk Lake 20 Jun 1976-1 (Osmond-Jones et al. 1977), 11 Jul 1976-1; Tatshenshini River at Alsek River 27 Jun 1983-2; West Nadahini Creek 18 Jul 1979.

Breeding Bird Surveys: Interior – Recorded from 4 of 73 routes and on less than 1% of all surveys. Maxima: Gnat Pass 18 Jun 1975-5; Cassiar 19 Jun 1975-1; Chilkat Pass 10 Jun 1975-1. **Coastal** – Not recorded.

Autumn: Interior – Atlin 1 Sep 1929-1 (CAS 32533; Swarth 1930); Fort Nelson 13 Sep 1986-1. **Coastal** – No records.

Winter: No records.

Swainson's Thrush

SWTH

Catharus ustulatus (Nuttall)

RANGE: Breeds from western and central Alaska, central Yukon, and western and southern Mackenzie across the Prairie provinces north of the grasslands, and east into Labrador and Newfoundland; in the west, south to southern California; in the east, to southern New York, New Hampshire, and West Virginia. Winters primarily from southern Mexico south to Brazil, Peru, and Argentina.

STATUS: On the coast, a *fairly common* to *common* migrant and summer visitant in the Georgia Depression Ecoprovince; *uncommon* to *fairly common* on Western Vancouver Island, the mainland coast, and the Queen Charlotte Islands of the Coast and Mountains Ecoprovince. *Casual* in winter in the Georgia Depression.

In the interior, a *fairly common* to *common* migrant and summer visitant in the Southern Interior and Southern Interior Mountains ecoprovinces, becoming *uncommon* to *fairly common* further north. *Accidental* in winter in the Southern Interior.

Breeds.

NONBREEDING: The Swainson's Thrush is widely distributed throughout the province, including offshore islands. Information is lacking regarding this thrush's occurrence in the mountainous areas of the Southern Mainland Coast, the western Chilcotin Plateau, the Blackwater region, the lower Nechako and McGregor river basins, and the Rocky Mountain Trench northward between Mackenzie and Lower Post. While some of these gaps may reflect the lack of suitable habitat, others suggest the need for further biological exploration.

In migration along the coast, the Swainson's Thrush has been reported from sea level to about 800 m elevation. In the southern portions of the interior, it usually occupies elevations between 450 and 1,200 m; further north the species is rarely found above 800 m.

Since the Swainson's Thrush is a characteristic summer bird in much of the province, there is considerable overlap between the wide variety of habitats it occupies during the nonbreeding and breeding periods. In the Coast and Mountains, it is most numerous along marine and freshwater shorelines where dense thickets of salal, salmonberry, or devil's club produce a supply of fruit close to the abundant invertebrates of the beaches. It frequently forages along the foreshore close to cover (Fig. 452); in riparian forests of alder, willow, or cottonwood; in mixed stands of western redcedar, thimbleberry, and crab apple that surround lakes or beaver ponds; as well as the edges of forest openings, fields, sewage ponds, creeksides, roadsides, and muskeg edges. On the west coast of Vancouver Island, where the Swainson's Thrush is particularly widespread, it prefers 30- to 35- and 50- to 60-year-old forests of western hemlock, amabilis fir, Douglas-fir, western redcedar, and Sitka spruce, with red alder in the wet areas (Bryant et al. 1993).

In wetter parts of the interior, the Swainson's Thrush occurs in the widespread coniferous forests of lodgepole pine, Douglas-fir, Engelmann spruce, white spruce, western larch,

Figure 452. On the Queen Charlotte Islands, the Swainson's Thrush forages along the foreshore close to the cover of Sitka spruce forests (Anthony Island, June 1989; R. Wayne Campbell).

Swainson's Thrush

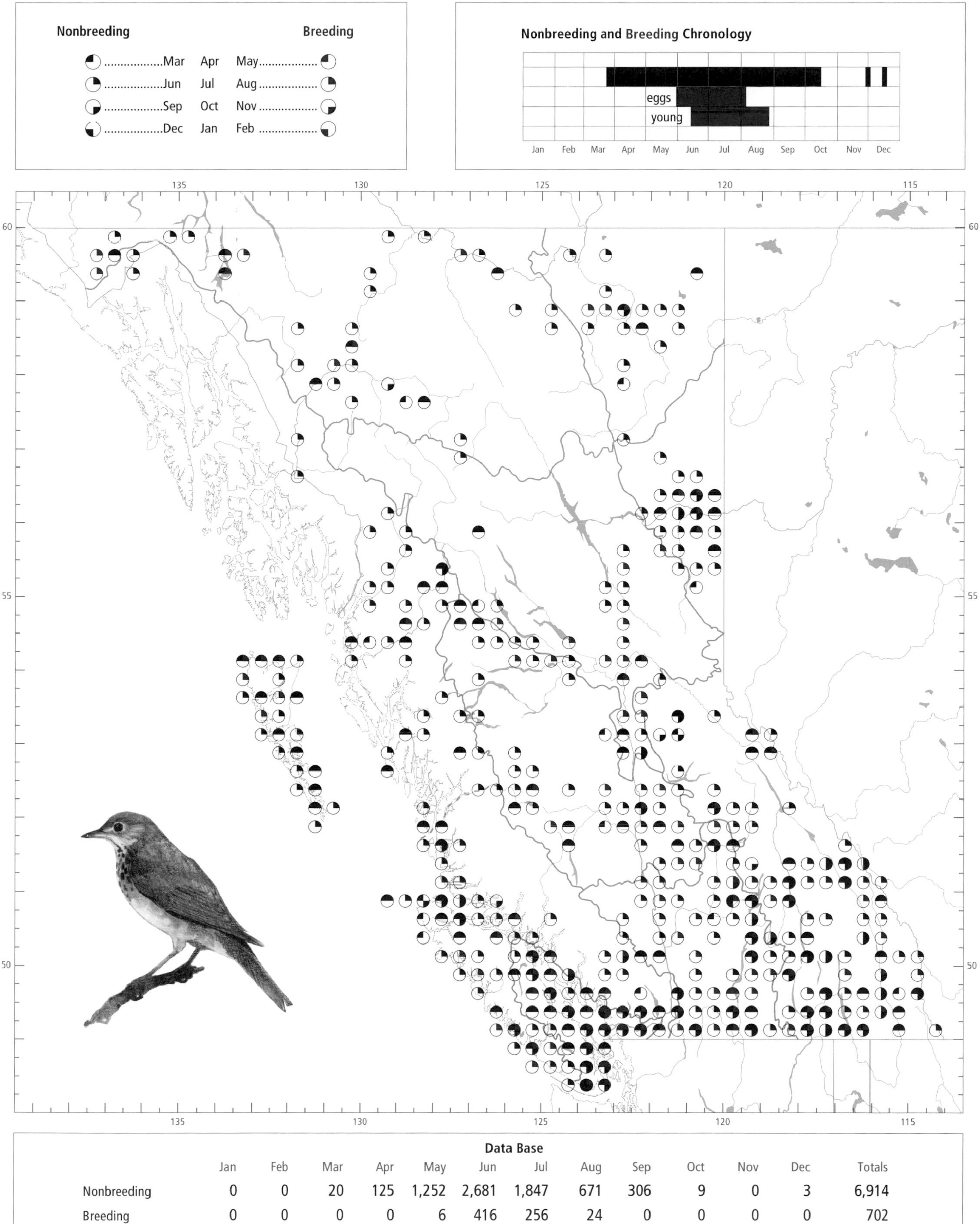

Data Base

	Jan	Feb	Mar	Apr	May	Jun	Jul	Aug	Sep	Oct	Nov	Dec	Totals
Nonbreeding	0	0	20	125	1,252	2,681	1,847	671	306	9	0	3	6,914
Breeding	0	0	0	0	6	416	256	24	0	0	0	0	702

Figure 453. In drier areas of the southern portions of the interior of British Columbia, the Swainson's Thrush inhabits mixed forests of young black cottonwood and trembling aspen with scattered western hemlock and western redcedar (31 km north of Revelstoke, 17 June 1996; Neil K. Dawe).

and subalpine fir. In drier areas, it occupies deciduous woodland with scattered conifers (Fig. 453), as well as thickets of black hawthorn, choke cherry, and saskatoon that occur in the swales and gullies of grassland slopes. It coexists with the Veery in riparian cottonwood stands with a dense shrub understorey. In the north, the Swainson's Thrush occurs in mixed deciduous woodlands (Fig. 454), in forests of white spruce mixed with aspen, birch, or willow, or where black spruce and tamarack predominate, and where berry-producing shrubs are widespread.

The Swainson's Thrush coexists with other closely related thrushes in British Columbia. In the southern interior, where it shares habitat with the Veery, the Swainson's Thrush

Figure 454. In the Peace River region of British Columbia, the Swainson's Thrush inhabits mixed forests of white spruce and trembling aspen (30 km south of Wonowon, 25 June 1996; R. Wayne Campbell).

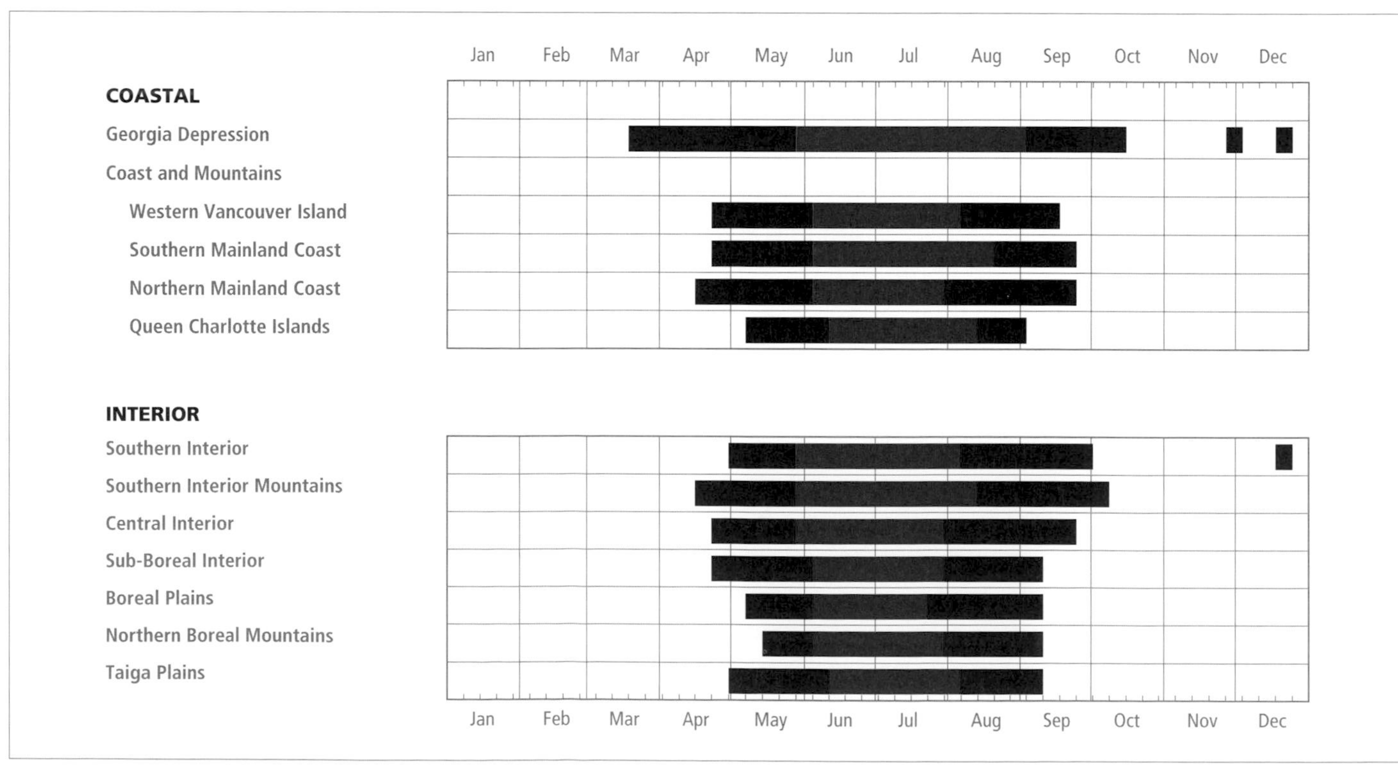

Figure 455. Annual occurrence (black) and breeding chronology (red) for the Swainson's Thrush in ecoprovinces of British Columbia.

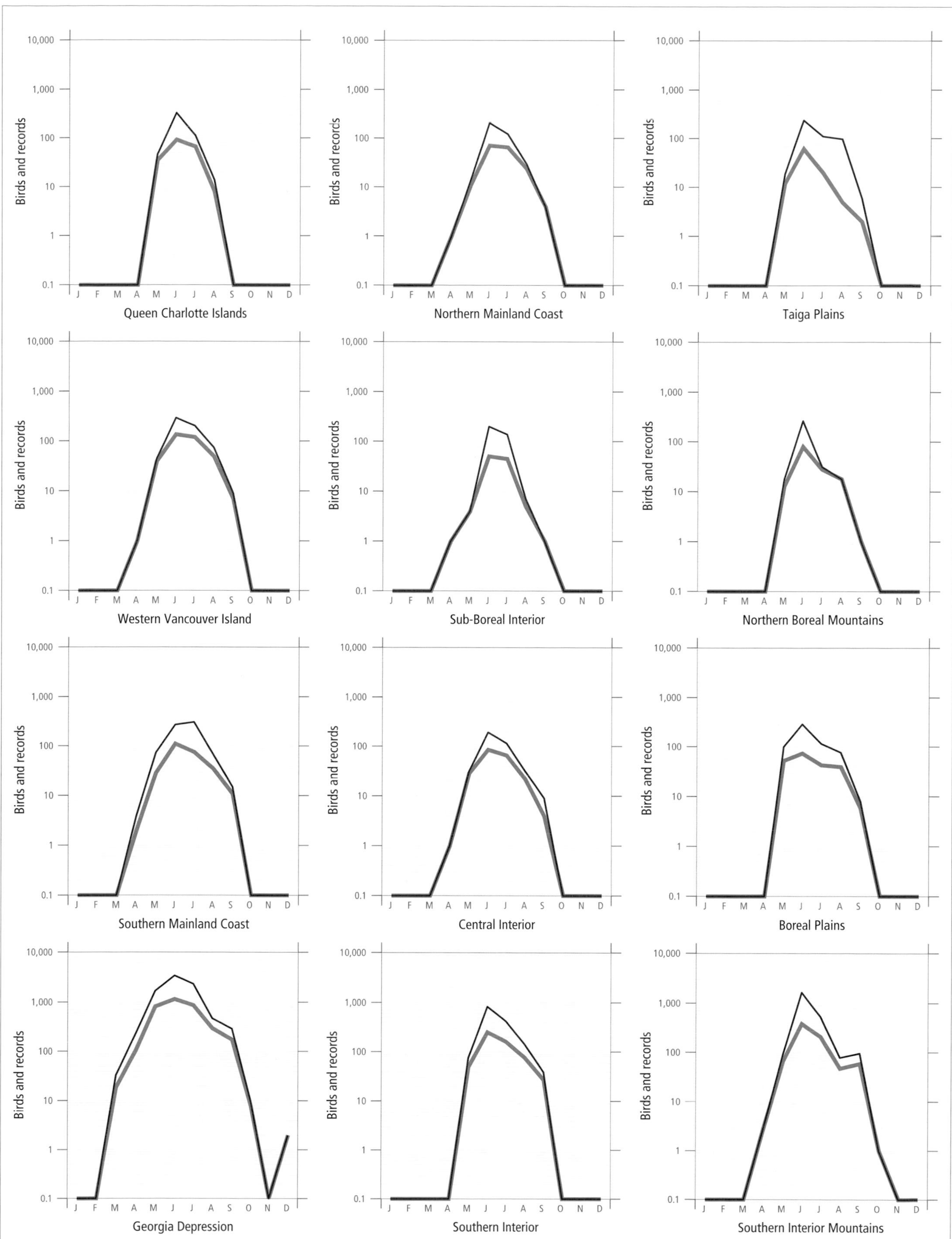

Figure 456. Fluctuations in total number of birds (purple line) and total number of records (green line) for the Swainson's Thrush in ecoprovinces of British Columbia. Christmas Bird Count and nest record data have been excluded.

forages higher above the ground, and prefers habitat with less undergrowth than that preferred by the Veery. On the Queen Charlotte Islands (Fig. 452), where it shares habitat and perhaps the same food sources with the Hermit Thrush, the spring migration of the Swainson's Thrush reaches the islands later than that of the Hermit Thrush. Thus, although the same resource base may be used by the 2 species, the periods of peak demand are separated (Sealy 1974).

On the south coast, the spring migration begins in April, or occasionally in very late March (Figs. 455 and 456). In early May, numbers increase tenfold. Migration in the Georgia Depression reaches a peak towards the end of May. This peak is a little later further north, reaching the northern mainland Coast and Mountains and Queen Charlotte Islands in early June. In the southern portions of the interior, the first birds may arrive as early as the third week of April in the east and west Kootenays, but not until early May in the Okanagan; however, the main movement does not occur until late May or early June. The timetable for spring arrival is similar further north, and indicates the rapidity of the northward movement.

The southward migration is more protracted than the spring movement. In the northern ecoprovinces, south to the Central Interior, an autumn migration is difficult to discern. In the Southern Interior and Southern Interior Mountains, the autumn migration begins in early August, with a notable movement occurring between the third week of August and the second week of September. Most birds have left the north by the end of August and the southern regions by the end of September. On the coast, the autumn departure from the Queen Charlotte Islands and the Northern Mainland Coast is completed about a month before it is on the Southern Mainland Coast and 2 months before the Georgia Depression. Throughout much of the coast, migration begins in late July, is at its height in August, and is completed in September. Exceptions are the Queen Charlotte Islands, where there are no records after the end of August, and the Georgia Depression, where there are a few late autumn records (Figs. 455 and 456).

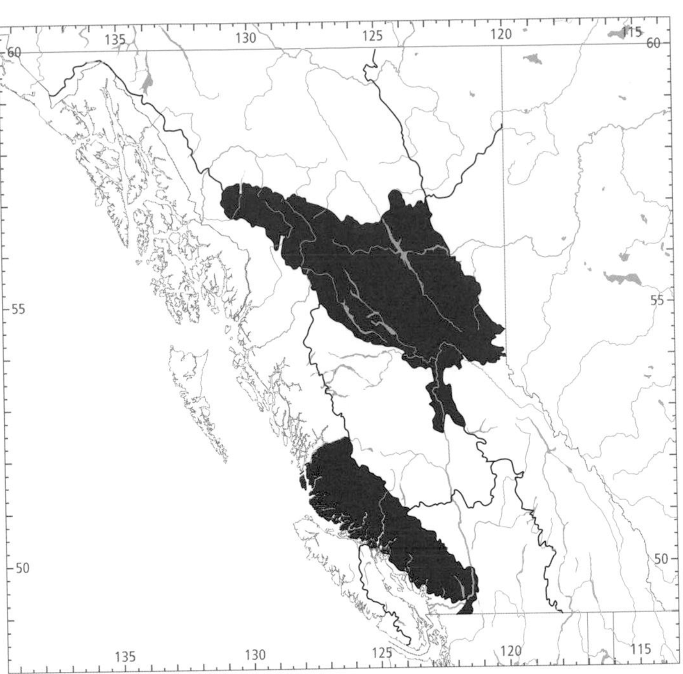

Figure 457. In British Columbia, the highest numbers for the Swainson's Thrush in summer occur in the southern portions of the Coast and Mountains Ecoprovince and throughout the Sub-Boreal Interior Ecoprovince.

Figure 458. The Swainson's Thrush breeds in mixed coniferous forests, often building its nest in small conifers near the edge (Christina Lake, July 1990; Mark Nyhof).

Figure 459. In some areas along the coast, breeding habitat for the Swainson's Thrush includes shrub areas of red elderberry with young red alder, black cottonwood, and lady fern (Nusatsum River, 7.8 km east of Hagensborg, 12 June 1996; Neil K. Dawe).

Rarely are birds recorded in winter.

In the interior, the Swainson's Thrush occurs regularly from 18 April to 1 October; on the coast, it occurs from 23 March to 12 October (Fig. 455).

BREEDING: The Swainson's Thrush undoubtedly breeds throughout its summer range in the province. On the coast, however, there are no records of nesting from northern Vancouver Island, from the south Moresby Island archipelago, or along the entire mainland coast between Queen Charlotte Strait and the mouth of the Skeena River.

In northern British Columbia, nesting has been documented in the Tatshenshini River valley, Atlin, the Rabbit River valley near Coal River, and the vicinity of Fort Nelson. Elsewhere across this vast area, nesting has been recorded only at Junction and Telegraph creeks, near the south end of Dease Lake (Swarth 1922), and at a number of locations in the Peace Lowland. Much remains to be learned about the nesting distribution of the Swainson's Thrush in northern British Columbia.

The highest numbers for the Swainson's Thrush in summer occur in the southern Coast and Mountains and Sub-Boreal Interior (Fig. 457). An analysis of Breeding Bird Surveys for the Swainson's Thrush in British Columbia for the period 1968 through 1993 could not detect a net change in numbers on either coastal or interior routes.

On the coast, the Swainson's Thrush nests from sea level to about 930 m elevation. In summer, Weber (1975) found the Swainson's Thrush to be the most abundant songbird on Mount Seymour, north of Vancouver, between 90 and 370 m, while the Hermit Thrush was the most abundant species between 740 and 1,050 m elevation (Weber 1975). In the interior, the Swainson's Thrush ranges from valley bottoms to 1,850 m, but in the Southern Interior, few Swainson's Thrushes are found below 500 m elevation (Cannings et al. 1987).

Along the coast, this thrush is a characteristic nesting species in second-growth forests of Douglas-fir, Sitka spruce, western redcedar, western hemlock, and grand fir where there is a dense undergrowth of salal, thimbleberry, huckleberry, salmonberry, and other fruiting shrubs. Similar habitats are used in the Southern Interior, where nesting habitat is usually in relatively undisturbed Douglas-fir and lodgepole pine forests mixed with trembling aspen, spruce, and willow. In the northern interior, the Swainson's Thrush frequents mixed spruce, paper birch, and trembling aspen forests (Fig. 454).

Figure 460. Nest with incubating Swainson's Thrush in a shrub tangle (Campbell Valley Park, Langley, 30 July 1989; Glen R. Ryder).

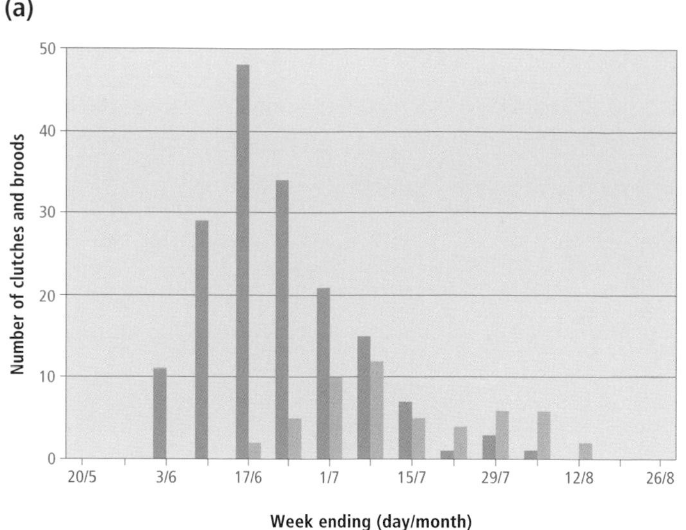

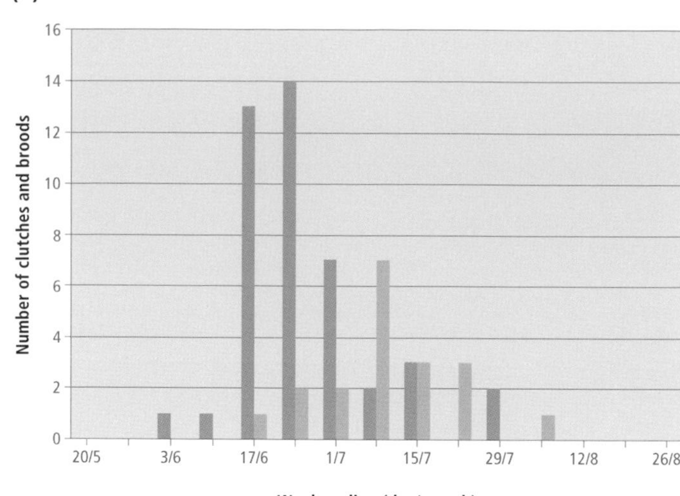

Figure 461. Weekly distribution of clutches (dark bars) and number of broods (light bars) for the Swainson's Thrush in the (a) Georgia Depression and (b) Southern Interior ecoprovinces of British Columbia. The figures are based on the week eggs or young were found in the nest. Note the slight second peak in broods for the last week of July and first week of August in the Georgia Depression. This peak may result from renests after the first nest is lost, although some pairs may have 2 broods.

Mixed forest was the most frequently reported nesting habitat class (39%; *n* = 146; Figs. 453 and 454), followed by deciduous stands (17%), coniferous stands (17%; Fig. 458), rural and suburban areas (12%), and shrublands (9%; Fig. 459). Most nests (84%; *n* = 244) were associated with undisturbed forest, followed by disturbed forest and shrubland (Fig. 459). Nearly 40% (*n* = 106) of nests were reported from riparian situations regardless of the general forest type. In New York, Dilger (1956) found that, of the 5 species of North American forest thrushes studied, the Swainson's Thrush was the most closely associated with undisturbed coniferous growth.

The Swainson's Thrush has been recorded breeding in the interior between 31 May and 6 August; on the coast it has been recorded from 30 May to 27 August.

Nests: Most nests were placed in trees (60%; *n* = 340), including deciduous trees (30%), conifers (24%; Fig. 458), and snags. Shrub and brush tangles (38%; Fig. 460) accounted for most of the other general nest locations. Nests were built among branches (Fig. 460), saddled on a branch, or lodged in the fork or crotch of a branch (96%; *n* = 288). A few nests were found among tree roots or in shrubs, and 1 nest was found on a post. The heights for 329 nests ranged from ground level to 24 m, with 60% between 1 and 2 m.

Nests were cups of grass, moss, leaves, fine twigs, plant fibres, and rootlets (Fig. 460).

Eggs: Dates for 328 clutches ranged from 30 May to 7 August, with 51% recorded between 14 and 30 June (Fig. 461). Sizes of 313 clutches ranged from 1 to 5 eggs (1E-30, 2E-25, 3E-72, 4E-172, 5E-14), with 54% having 4 eggs. The incubation period in British Columbia ranged from 11 to 14 days (*n* = 4). Harrison (1979) gives the incubation period as 10 to 13 days.

Young: Dates for 115 broods ranged from 13 June to 27 August, with 52% recorded between 29 June and 15 July (Fig. 461). Sizes of 74 broods ranged from 1 to 5 young (1Y-9, 2Y-23, 3Y-18, 4Y-23, 5Y-1), with 86% having 2 to 4 young. The nestling period in British Columbia is 11 to 14 days (*n* = 4). Populations breeding in the Georgia Depression may have 2 broods a year (Fig. 461). Ehrlich et al. (1988) state that the number of broods produced each year is uncertain.

Brown-headed Cowbird Parasitism: In British Columbia, 8% of 395 nests found with eggs or young were parasitized by the cowbird. There was an additional record of adults feeding a fledged cowbird. The parasitism rate both on the coast (*n* = 250) and in the interior (*n* = 145) was 8%. Friedmann et al. (1977) report this species as an infrequent host for the cowbird.

Nest Success: Of 31 nests found with eggs and followed to a known fate, 5 produced at least 1 fledgling, for a nest success rate of 16%.

REMARKS: Two subspecies of Swainson's Thrush are found in British Columbia. The "Olive-backed" Thrush (*C. u. almae*) occurs east of the Coast Mountains and Cascade Mountains,

Figure 462. Swainson's Thrush nest with 3 thrush eggs and 1 Brown-headed Cowbird egg (Victoria, 16 June 1973; R. Wayne Campbell). Seven percent of all Swainson's Thrush nests (*n* = 368) found in British Columbia with eggs or young were parasitized by the cowbird.

while the "Russet-backed" Thrush (*C. u. ustulatus*) inhabits coastal forests (Bond 1963).

There are 2 banding reports of the Swainson's Thrush from British Columbia: a bird banded at Madison, Wisconsin, on 11 September 1961 was recovered at Okanagan Landing on 8 April 1963; 1 banded near Hays, Kansas, on 24 May 1973 was recovered near Grand Forks on 21 August 1973.

There are a number of reports of the Swainson's Thrush before 23 March and after 12 October, including sightings from Christmas Bird Count locations where the very similar Hermit Thrush is known to winter. These include Vancouver 27 December 1955-1 (Weber and Weber 1975) and Pender Islands 28 December 1965-5 (Stevens and Stevens 1966). See also the following Christmas Bird Counts: Victoria 23 Dec 1961-1 (Stirling 1962a; Anderson 1976b), 1 Jan 1969-1 (Tatum 1970; Anderson 1976b); Pender Islands 23 Dec 1978-1 (Anderson 1979; McLardy 1979); Duncan 15 Dec 1979-1 (Comer 1980); White Rock 28 Dec 1975-1 (Anderson 1976a; Schouten 1974); Vancouver 26 Dec 1975-1 (Anderson 1976a; Kautesk 1976); Nanaimo 31 Dec 1977-1 (van Kerkoerle 1978). All but 3 records, however, lack convincing details and have been excluded from the account.

NOTEWORTHY RECORDS

Spring: Coastal – Mt. Metchosin 23 Mar 1983-1; Beaver Lake (Victoria) 30 Apr 1982-1; Carnation Creek 29 Apr 1981-1; Surrey 31 Mar 1966-4; Reifel Island 30 May 1973-2 eggs; Vancouver 23 Mar 1983-1; Tofino 21 May 1980-3; Halfmoon Bay 12 May 1979-24; Comox 20 Apr 1934-2 (CMNH 115849-50); Port Neville Inlet 23 Apr 1975-3 to 4 birds; Pine Island 20 May 1932-1 (RBCM 5298); Calvert Island 21 May 1937-1 (MCZ 283781); Kimsquit River 27 Apr 1982-1; Terrace 30 May 1977-1; Masset 12 May 1920-1 (FMNH 145981); Langara Island 10 May 1927-arrival, 29 May 1946-4; Metlakatla 19 Apr 1910-earliest arrival between 1900 and 1910, 29 May 1904-latest arrival date (Keen 1910); Quick 4 May 1977-1. **Interior** – Fruitvale 18 Apr 1978-1; Elko 26 May 1926-2; Summerland 6 May 1965-1 (Cannings et al. 1987); Okanagan Landing 9 May 1917-earliest arrival 8 years; Shuswap Lake Park 8 May 1977-2; Heinz Mountain 18 Apr 1978-1; Yoho National Park 21 May 1975-2 (Wade 1977); 108 Mile House 28 Apr 1983-1; Williams Lake 24 May 1988-3; 24 km s Prince George 29 Apr 1976-1; Taylor 24 May 1986-7; Boundary Lake 11 May 1986-1; North Pine 11 May 1986-2; Telegraph Creek 15 May 1919-1 (MVZ 40289); Parker Lake Rd 5 May 1980-1; Komie Creek (lat. 59°23'N, long. 120°47'W) 5 May 1982-1; Liard Hot Springs 28 May 1981-3; Chilkat Pass 15 May 1977-1.

Summer: Coastal – Sooke River 4 Aug 1931-3 eggs; Saanich 2 Jul 1980-35; Duncan 3 Aug 1974-15; Tofino 11 Jul 1925-4 eggs, 1 Aug 1971-4 nestlings; Surrey 27 Aug 1960-4 young following female thrush in woods; Manson's Landing 7 Aug 1977-2 nestlings; Deroche 7 Aug 1967-3 large nestlings; Qualicum 15 Jun 1987-3 new nestlings; Hope 7 Jun 1964-1 egg, 20 Jul 1964-4 nestlings; Alta Lake 3 Aug 1942-3 eggs; Quadra Island 24 Jun 1975-106; Port Neville 4 Aug 1975-4; Storm Islands 12 Jun 1976-22; Rivers Inlet 16 Aug 1937-1; Kitimat 10 Aug 1975-3; Marble Island 20 Jun 1977-111; Ramsay Island 21 Jul 1961-10; Bruin Bay 10 Jun 1914-4 eggs; Mayer Lake 8 Aug 1985-1 fledgling; Masset 8 Jul 1920-2 eggs, 21 Aug 1920-1 (MVZ 103719); Langara Island 17 Jun 1947-3 eggs; Prince Rupert 2 Jul 1976-10; Terrace 9 Jun 1978-3 eggs; 29 Jul 1974-1 nestling; Kitwanga 26 Jun 1975-57. **Interior** – Castlegar 31 Jul 1982-2 nestlings hatched; Manning Park 26 Jun 1983-100 on survey, 1 Aug 1956-4 nestlings; Horn Lake (Twin Lakes) 1 Jun 1914-4 eggs; Nelson 9 Jun 1968-4 eggs; Naramata 28 Jun 1988-4 eggs; Brookmere 19 Jun 1974-39; Slocan Valley 6 Aug 1983-1 nestling; Trinity Valley 12 Jun 1965-4 eggs; Creighton Valley 2 Jul 1979-34; Lumby 12 Jun 1965-4 eggs; Spillimacheen 26 Jul 1976-67; Celista 22 Aug 1960-10; McLure Lake 2 Jun 1978-2 eggs; Amiskwi River 1 Jul 1976-32; Monk Lake 4 Aug 1984-12; Lac la Hache 26 Jul 1959-3 nestlings; Williams Lake 27 Aug 1973-8; Prince George 23 Jun 1969-28; Nulki Lake 20 Aug 1945-1 (USNM 426261); Tabor Lake 12 Jun 1966-4 eggs; 8 km s Shelley 8 Jul 1966-4 nestlings; Morice River 29 Jun 1975-50; Division Lake 10 Aug 1956-2; Topley 3 Jul 1956-12; Telkwa 4 Jul 1975-3 newly hatched nestlings; Smithers 1 Jul 1974-25; South Pine 24 Jul 1929-1 nestling; Taylor 8 Jul 1984-14; North Pine 22 Aug 1986-8; Charlie Lake 8 Jun 1986-3 thrush eggs and 1 cowbird egg, 7 Jul 1974-3 eggs; Flood Glacier 3 Aug 1919-1 (MVZ 40305); Doch-Da-On Creek 21 Jul 1919-1 (MVZ 40301); Steamboat 14 Jun 1976-58 on Breeding Bird Survey; Fort Nelson 19 Jun 1976-25; 22 Aug 1985-25; 1.5 km e Fort Nelson 9 Jul 1986-1 nestling; near Kledo Creek 25 Jul 1967-1 egg; Cassiar 19 Jun 1975-56; Liard River 11 Jul 1978-4; Towagh Creek 5 Jun 1983-9, 7 Jun 1983-egg shells in recently destroyed nest (Campbell et al. 1983); Rabbit River valley near Coal River 6 Jul 1981-2 fledglings; Mile 316.5 Alaska Highway 17 Jun 1976-4 eggs; Lower Liard Crossing 4 Aug 1943-1; Atlin 13 Jun 1975-3 eggs, 13 Jul 1914-3 eggs (RBCM 947), 30 Aug 1934-1 (CAS 42160).

Breeding Bird Surveys: Coastal – Recorded from 27 of 27 routes and on 100% of all surveys. Maxima: Port Renfrew 26 Jun 1974-195; Squamish 21 Jun 1986-115; Queen Charlotte City 25 Jun 1994-109. **Interior** – Recorded from 71 of 73 routes and on 97% of all surveys. Maxima: McLeod Lake 3 Jul 1993-146; Mount Morice 23 Jun 1969-140; Scotch Creek 1 Jul 1992-121.

Autumn: Interior – Fort Nelson River 2 Sep 1983-2; Fort Nelson 8 Sep 1985-3; Halfway River 1 Sep 1979-3; St. John Creek 7 Sep 1985-1; Stikine River (McBride) 8 Sep 1977-1; Hazelton 21 Sep 1921-1 (MVZ 283781); Barkerville 6 Sep 1932-1; Quesnel 6 Sep 1929-1 (RBCM 14833); Williams Lake 21 Sep 1986-6; Emerald Lake 12 Sep 1975-2 (Wade 1977); Scotch Creek 29 Sep 1963-1; Celista 1 Sep 1963-4; Cranbrook 1 Oct 1941-1 (RBCM 10957). **Coastal** – Hazelton 21 Sep 1921-1, last one seen this year; Calvert Island 11 Sep 1934-3 (MCZ 283779); Cape Scott 15 Sep 1935-1 (NMC 26158); Pachena River 1 Sep 1977-2; Hope 19 Sep 1946-1 (UBC 4516); Bowen Island 1 Oct 1933-1 (RBCM 14837); Surrey 31 Oct 1965-2; Boundary Bay 12 Nov 1962-1; West Saanich 7 Oct 1987-1; Victoria 28 Sep 1992-1, 10 Oct 1974-1 (RBCM 14891).

Winter: Interior – See Christmas Bird Counts. **Coastal** – North Vancouver 18 to 23 Dec 1970-1 with American Robins and Varied Thrushes feeding on fallen apples (Weber and Weber 1975); Victoria 1 and 2 Dec 1976-1 with American Robins feeding on *Cotoneaster* and mountain-ash berries. See REMARKS.

Christmas Bird Counts: Interior – Recorded from 1 of 27 localities and on less than 1% of all counts. Maximum: Shuswap Lake Park 20 Dec 1988-**1**, all-time Canadian high count (Munroe 1989a). **Coastal** – See REMARKS.

Hermit Thrush

Catharus guttatus (Pallas)

HETH

RANGE: Breeds from western and central Alaska and southern Yukon through the boreal forest of northern Canada to Labrador and Newfoundland, south in the Rocky Mountains, and to the west as far as California, Nevada, Arizona, New Mexico, and western Texas; east of the Rocky Mountains, from central Alberta east to central Ontario and southern Quebec, and south to western Virginia. Winters from southern British Columbia and southern Ontario to southern Baja California and through Mexico to Guatemala.

STATUS: On the coast, an *uncommon* to *common* migrant and summer visitant to the Georgia Depression and Coast and Mountains ecoprovinces, including Western Vancouver Island and the Queen Charlotte Islands; *rare* to *uncommon* in winter on the south coast, including the Georgia Depression and Western Vancouver Island; *casual* in winter on the Queen Charlotte Islands.

In the interior, an *uncommon* to *common* migrant and summer visitant to the Southern Interior and Southern Interior Mountains ecoprovinces, becoming *uncommon* to *fairly common* from the Central Interior Ecoprovince north throughout the rest of the province; *casual* in winter in the Southern Interior.

Breeds.

NONBREEDING: The Hermit Thrush is widely distributed throughout the province, including offshore islands, and is one of the few species that has been recorded in all ecoprovinces. However, as in the case of the Swainson's Thrush, large parts of British Columbia have no reports of this thrush.

The Hermit Thrush reaches its highest numbers in winter on southeastern Vancouver Island and in the Fraser Lowland of the Georgia Depression (Fig. 463).

Outside the breeding season, the Hermit Thrush has been reported from sea level to 1,350 m, although rarely below 500 m in the interior. In winter it frequents the dense mature and second-growth coastal Douglas-fir, western redcedar, and western hemlock forests of the Georgia Depression (Fig. 464), often associated with moist areas such as lakeshores, floodplains, and pond or marsh edges. It is also found in Garry oak and arbutus associations, blackberry and Nootka rose thickets, and occasionally in suburban gardens.

Spring migration begins on the south coast in mid to late March (Fig. 466) with the arrival of a small number of thrushes on the Fraser River delta and surrounding hills. The movement increases through April, reaching its height in the last week of April and first week of May. Numbers decline through the rest of May and June. On Western Vancouver Island, the Hermit Thrush arrives in April and reaches maximum numbers about the third week of May, before declining to a much lower number of summer residents (Fig. 466).

Migrants arrive on the Queen Charlotte Islands and the north coast during the third week of April. Sealy (1974) notes

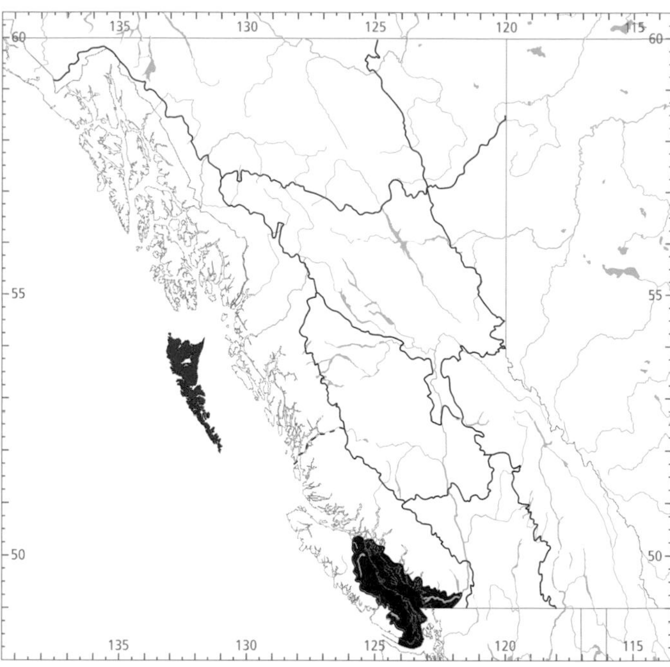

Figure 463. In British Columbia, the highest numbers for the Hermit Thrush in winter (black) occur on southeastern Vancouver Island and in the Fraser Lowland of the Georgia Depression Ecoprovince; the highest numbers in summer (red) occur on the Queen Charlotte Islands in the Coast and Mountains Ecoprovince.

Figure 464. Winter habitat for the Hermit Thrush in the Georgia Depression Ecoprovince of British Columbia includes second-growth forests of Douglas-fir, western redcedar, and western hemlock with a dense understorey (Stanley Park, Vancouver, 26 February 1994; R. Wayne Campbell).

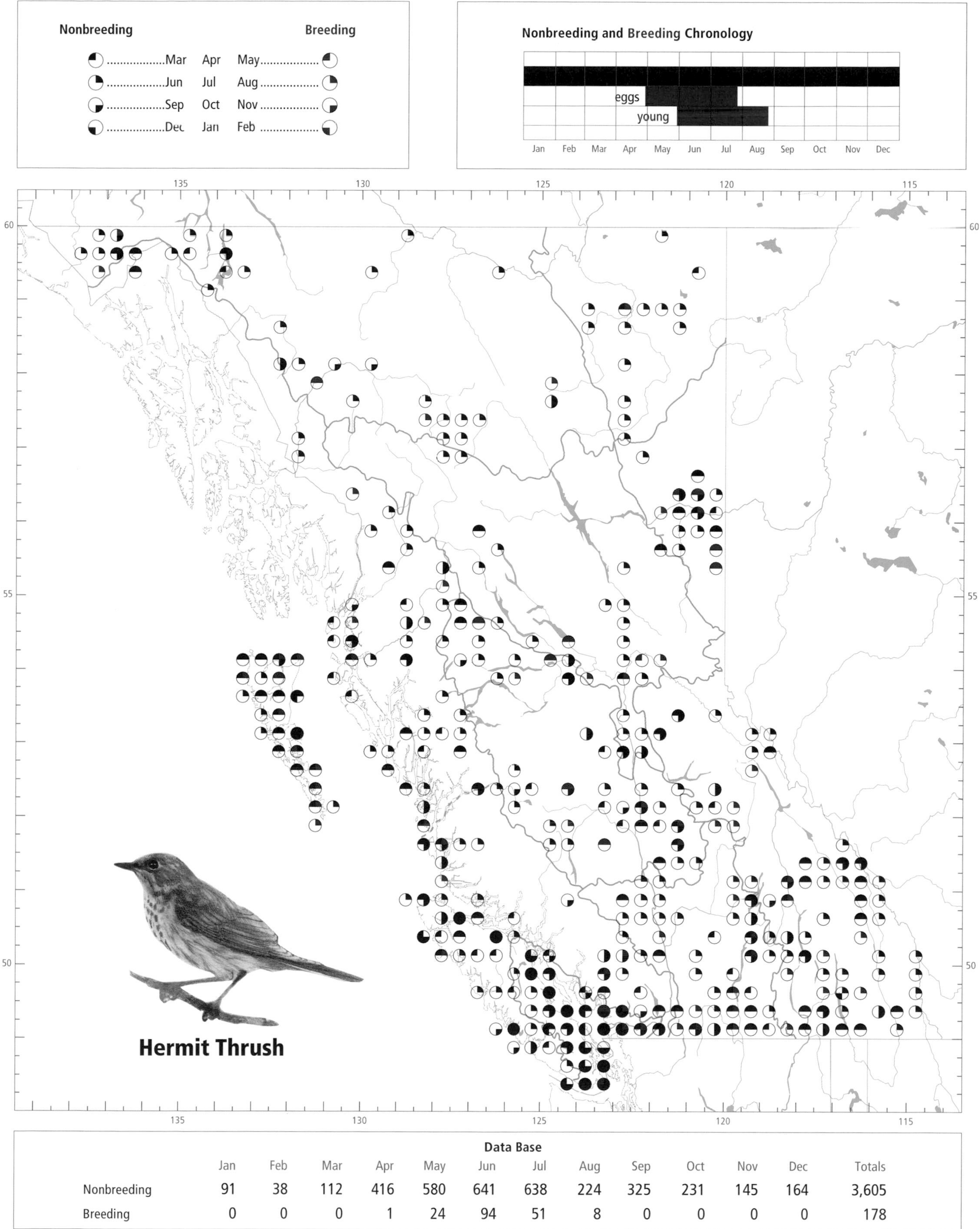

Data Base

	Jan	Feb	Mar	Apr	May	Jun	Jul	Aug	Sep	Oct	Nov	Dec	Totals
Nonbreeding	91	38	112	416	580	641	638	224	325	231	145	164	3,605
Breeding	0	0	0	1	24	94	51	8	0	0	0	0	178

the first arrival of the Hermit Thrush on Langara Island on 16 April. Numbers appear to peak in the northern Coast and Mountains about the second and third week of May, and on the Queen Charlotte Islands in mid-June, well into the egg-laying period (see "Eggs"). In the Okanagan valley, the first arrivals may appear as early as the first week of April, but elsewhere in the interior they may arrive in early May, with the peak very shortly thereafter (Fig. 466). In the Peace Lowland, migrants probably follow routes east of the Rocky Mountains rather than through southern British Columbia.

Unlike the Veery and the Swainson's Thrush, which move northward along the valleys, the Hermit Thrush tends to avoid the arid ponderosa pine forests and grasslands of the Okanagan and Thompson valleys. In June and July, it is abundant above 1,100 m in the mountains bordering these valleys (Cannings et al. 1987) and the more humid valleys of the Columbia and Kootenay river basins.

In the Fort Nelson Lowland, the southward migration appears to begin in July. In the central and southern portions of the interior, the autumn migration takes place rapidly in August, with only a few birds remaining through September and occasionally into October.

Similarly, on the coast there is a dramatic decline in numbers recorded between July and August, and only occasional occurrences after that. It has not been determined how much of this midsummer decline in birds reported is a consequence of the end of the song period and how much is due to the migration of the birds. There is no indication of an autumn migration along Western Vancouver Island.

The southern migration in the Georgia Depression is unlike that in other parts of the province. The lowest numbers for the year occur in August, as local birds leave for wintering grounds before northern migrants reach the south coast. Numbers build in early September and peak during the last week of September and the first 2 weeks of October. At that time, migration brings the populations that nested in the mountains, along the north coast and perhaps in parts of the interior, down onto the coastal lowlands. Later there is a gradual decline in numbers as many of these birds move further south, leaving only those that will spend the winter. Root (1988), referring to United States occurrences, notes that the Hermit Thrush selects warm areas in which to winter and is common only in forests that rarely drop below –1°C in January.

On the coast, the Hermit Thrush has been reported throughout the year only in the Georgia Depression, although it has been reported twice in winter from the Queen Charlotte Islands (Fig. 465). In the interior, it has been reported regularly from 2 April through 14 October.

BREEDING: The Hermit Thrush breeds throughout much of the province. In the southern half, it has been reported nesting in most areas south of Prince George and Hazelton. There is only 1 nesting record from Vancouver Island: a recently fledged young was caught in a snap trap on the Brooks Peninsula (Campbell and Summers, in press). There are no nesting records from the Coast Mountains north of Garibaldi Provincial Park, and the few from the mainland coast north of Rivers Inlet are from smaller offshore islands. There are also few breeding records from the southern Rocky Mountains, where the species is abundant in summer. In the northern half of the province east of the Rocky Mountains, the species has been found nesting at Tupper Creek, Charlie Lake (Cowan 1939), and Fort Nelson. West of the Rocky Mountains, there are records from Kwadacha Wilderness Park (Cooper et al. 1979), Spatsizi Park (Osmond-Jones et al. 1977), Telegraph

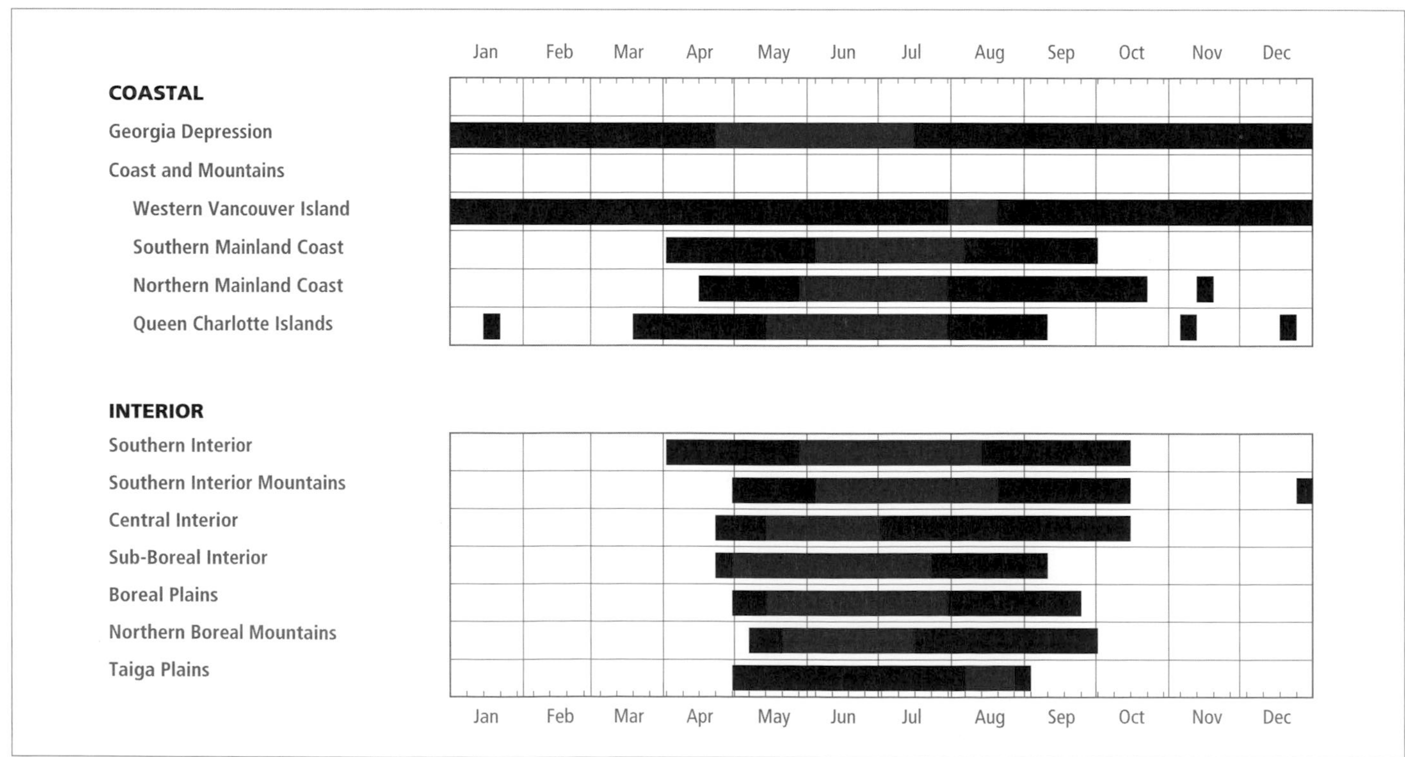

Figure 465. Annual occurrence (black) and breeding chronology (red) for the Hermit Thrush in ecoprovinces of British Columbia.

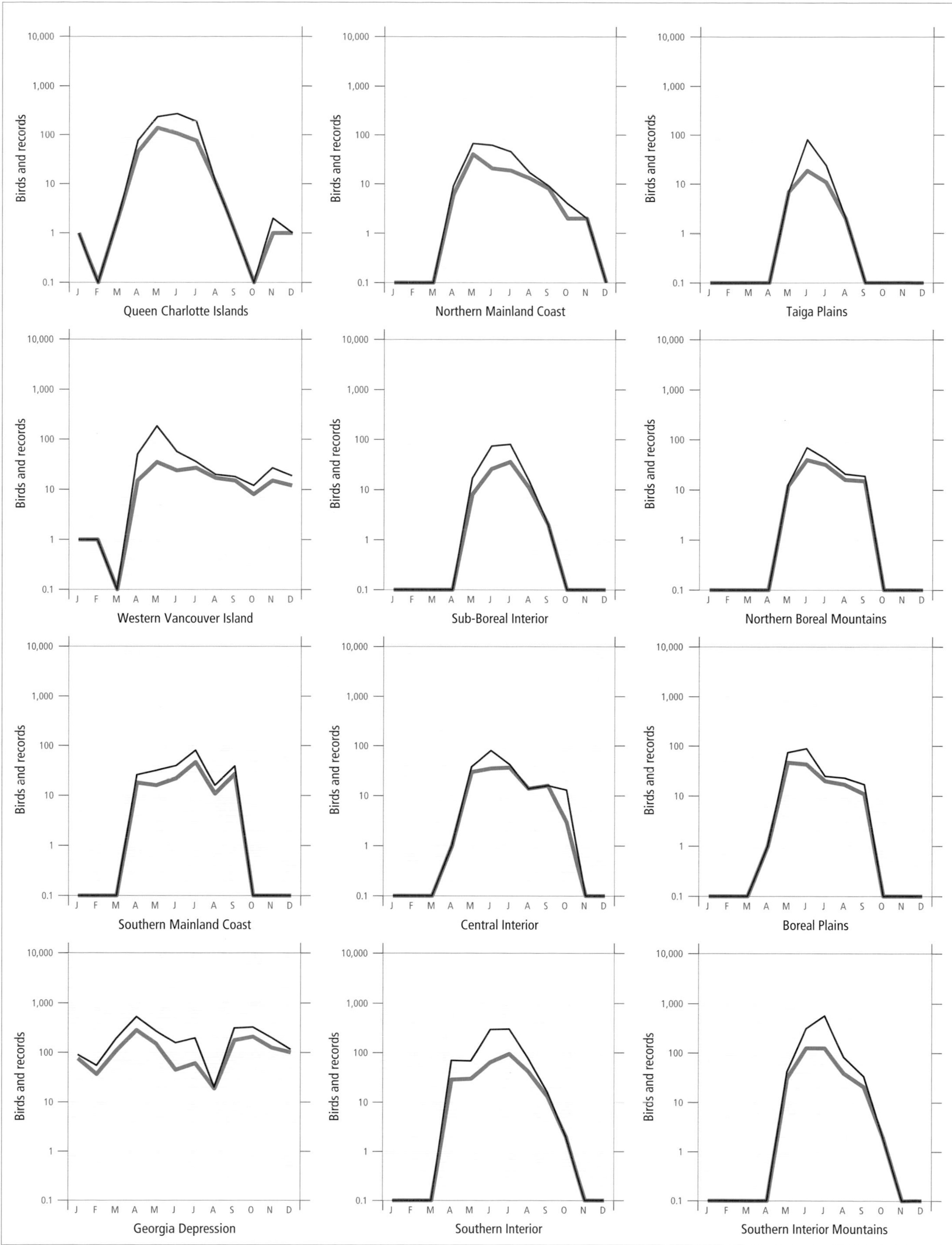

Figure 466. Fluctuations in total number of birds (purple line) and total number of records (green line) for the Hermit Thrush in ecoprovinces of British Columbia. Christmas Bird Count and nest record data have been excluded.

Creek (Swarth 1922), Atlin (Swarth 1926), and the Tatshenshini River valley (Campbell et al. 1983).

The Hermit Thrush reaches its highest numbers in summer on the Queen Charlotte Islands (Fig. 463). An analysis of Breeding Bird Surveys for the period 1968 through 1993 could not detect a net change in numbers on interior routes; coastal routes for the same period contained insufficient data for analysis.

The Hermit Thrush breeds from near sea level to 1,500 m elevation on the coast and from 435 to 2,200 m in the interior.

During the breeding season this thrush occupies a wide variety of forest types, but prefers areas with a dense shrub understorey (Fig. 467). On the Southern Mainland Coast, it nests in the upper elevations of the Coastal Western Hemlock and in the Mountain Hemlock biogeoclimatic zones. There it is the most abundant species between 750 and 1,050 m elevation, whereas the Swainson's Thrush is most abundant between 90 and 400 m (Weber 1975). Along the Northern Mainland Coast, the Hermit Thrush is a characteristic nesting species from near sea level through the western hemlock, Sitka spruce (Fig. 468), and mountain hemlock–yellow cedar forests. Only in these northern portions of the Coast and Mountains does this thrush nest at low elevations. There it is most abundant in forest edge communities where berry-bearing shrubs are plentiful. Away from the coast, the Hermit Thrush nests in subalpine spruce, alpine fir, and krummholz forests near the timberline (Figs. 469 and 470). It is a species of the upper edge of the forest, where openings and shrub patches, including several species of blueberries, provide late summer and autumn food. In the interior, it also breeds at lower elevations in interior Douglas-fir, Engelmann spruce, and lodgepole pine forests (Southern Interior and Southern Interior Mountains); in trembling aspen forests (Central Interior); pine-aspen, spruce-tamarack, and trembling aspen woodlands (Boreal Plains); or black spruce muskegs (Taiga Plains). In all areas, it favours moist sites (Fig. 471).

On the Queen Charlotte Islands (Sealy 1974) and other coastal islands north of Rivers Inlet (Dickinson 1953), both the Hermit and Swainson's thrushes can be heard in song close to the beaches, and there appears to be no significant difference in their habitat preference. There does seem to be a temporal separation in their habitat use (see REMARKS).

Figure 467. Mixed coniferous forests with a dense shrub understorey are preferred nesting habitat for the Hermit Thrush in British Columbia (Revelstoke, July 1993; Mark Nyhof).

Figure 468. The Hermit Thrush breeds in the Sitka spruce forests of the Queen Charlotte Islands (Woodruff Bay, 26 May 1996; R. Wayne Campbell).

In much of the interior, however, nesting habitats of the Hermit Thrush and Swainson's Thrush are usually separated altitudinally and by cover type. For example, in the Telegraph Creek area, Swarth (1926) found the Hermit Thrush to be a species mainly of the spruce woods at elevations of 1,900 m, while the Swainson's Thrush was most numerous in the willow and poplar thickets at an elevation 300 m lower.

On the coast, the Hermit Thrush has been recorded nesting from 28 April to 13 August; in the interior, it has been recorded between 11 May and 25 August (Fig. 465).

Nests: Most nests (64%; *n* = 89) were found in trees (Fig. 471), including snags, or among tree roots. Coniferous trees (42%) were preferred over deciduous trees (8%). Nests were placed among tree branches (71%; *n* = 76), either saddled on the branch or in a fork or crotch (45%), against the tree trunk (20%), or near the top of the tree (6%). A few nests (6%) were found in shrubs. Many (57%) were on or close to the ground, at the base of trees, on banks, in tree or rock crevices, or under fallen logs.

Nests were neatly constructed cups with a lining of fine grass, other soft plant material, or hair, resting on a pad in which moss was usually prominent along with grass, twigs, and fine rootlets. Grass, moss, twigs and stems, rootlets, and hair were the most frequently used nest materials, with about half the nests having each of the first 2 items. The heights for 84 nests ranged from ground level to 6 m, with 57% between 0.3 and 1.5 m.

Figure 469. In portions of the central interior of British Columbia, the Hermit Thrush breeds in the upper edge of lodgepole pine and white and Engelmann spruce forests adjacent to openings where sedges and willows are the dominant plant cover (west of Tatla Lake, 14 July 1992; R. Wayne Campbell).

Eggs: Dates for 88 clutches ranged from 28 April to 26 July, with 55% recorded between 4 June and 26 June. Sizes of 87 clutches ranged from 1 to 6 eggs (1E-1, 2E-8, 3E-27, 4E-48, 5E-2, 6E-1), with 55% having 4 eggs (Fig. 472). In 1 nest in British Columbia, incubation from the laying of the last egg to the hatching of the first egg was 11 days. Elsewhere, the incubation period is reported as 12 to 13 days (Ehrlich et al. 1988).

Young: Dates for 39 broods ranged from 14 May to 25 August, with 56% recorded between 20 June and 29 July. Eggs laid by 28 April could produce young by 10 May. There is a report by Racey (1948) of a nest with eggs found on 24 August, with no other particulars. Because of the late date and inadequate details, we have excluded the record from the account; however, if this was an active nest, it would mean that nestlings could be found as late as 4 September. Sizes of 37 broods ranged from 1 to 5 young (1Y-8, 2Y-7, 3Y-9, 4Y-12, 5Y-1), with 57% having 3 or 4 young. The nestling period is 12 days (Ehrlich et al. 1988).

Brown-headed Cowbird Parasitism: In British Columbia, where most of the Hermit Thrush populations nest in habitats inhospitable to the cowbird, only 6% of 110 nests found with eggs or young were parasitized; cowbird parasitism rates for both the coast and interior were equal. There is an additional record of an adult thrush feeding a fledged cowbird young. One nest in the Fort St. John area contained 5 cowbird eggs. Friedmann and Kiff (1985) suggest that the Hermit Thrush is a "regular, if not favored, host choice." In a northern Michigan study, 22% (*n* = 60) of nests found over a 24-year period were parasitized (Southern and Southern 1980).

Nest Success: Of 16 nests found with eggs and followed to a known fate, 5 produced at least 1 fledgling.

REMARKS: Five subspecies of the Hermit Thrush are known to occur in British Columbia. Aldrich (1968) identifies the paler subspecies that breeds on the coastal islands and mainland of southern Alaska and northern British Columbia as the "Alaska" Hermit Thrush (*C. g. guttatus*); a smaller and darker thrush on the Queen Charlotte Islands and the outer islands of southeastern Alaska as the "Dwarf" Hermit Thrush (*C. g. nanus*); birds breeding on Vancouver Island and in a small mountainous area north of Vancouver as the "Vancouver" Hermit Thrush (*C. g. vaccinius*); birds that breed in the Kootenays and adjacent regions of Alberta, Idaho, and Washington as "Audubon's" Hermit Thrush (*C. g. auduboni*); and birds breeding in south-central British Columbia, between the Cascade Mountains and the Kootenays, as the "Cascade" Hermit Thrush (*C. g. oromelus*).

On the Queen Charlotte Islands, Sealy (1974) found that different nesting chronologies reduced competition between sympatric populations of the Hermit and Swainson's thrushes. The Hermit Thrush returned to the islands 6 weeks

Figure 470. In Tweedsmuir Park, British Columbia, the Hermit Thrush breeds in alpine fir krummholz forests near timberline (Rainbow Range, 12 July 1992; R. Wayne Campbell).

Figure 471. In British Columbia, the Hermit Thrush prefers moist sites to build its nest. Note adult on nest in lower centre of photograph (Tatshenshini River, June 1993; John M. Cooper).

Figure 472. Nest and eggs of the Hermit Thrush (Tatshenshini River, 10 June 1983; R. Wayne Campbell).

earlier and had fledged young by the time the Swainson's Thrush arrived. The earliest dates for Hermit Thrush eggs or young on the Queen Charlotte Islands are 1 month earlier than those for the Swainson's Thrush. Similar observations were made on nearby Forrester Island, Alaska, by Willett (1915) and Bailey (1927).

The records of eggs at Cheakamus on 24 August 1929 (Munro and Cowan 1947; Racey 1948) were probably an abandoned clutch and have not been included in the account.

A Hermit Thrush banded near Fort St. James on 2 August 1981 was recovered near Beloit, Illinois, on 18 May 1984.

NOTEWORTHY RECORDS

Spring: Coastal – Victoria 1 Mar 1980-5; Bamfield 29 Apr 1981-37; Langley 19 Mar 1979-5; Vancouver 25 Apr 1982-16; North Vancouver 28 Apr 1974-3 eggs; Port Mellon 12 May 1985-12; Halfmoon Bay 12 May 1979-15; Klaskish River 20 May 1978-18; Calvert Island 16 Apr 1937-1 (MCZ 283679); Aristazabal Island 1 May 1936-1 (MCZ 283704); Bella Coola 5 Apr 1933-4 (MCZ 283656-9); Kimsquit River 18 Apr 1985-2; Port Clements 22 Mar 1972-1; Drizzle Lake 17 May 1978-3 eggs; Delkatla Inlet 27 Apr 1979-6; Langara Island 16 Apr 1971-first arrival (Sealy 1974); Lord Bight 21 May 1977-7. **Interior** – Manning Park 29 Apr 1979-7; Penticton 2-Apr 1986-1; Nelson 25 May 1991-5; Okanagan Landing 5 Apr 1907-1; Celista 16 Apr 1948-5; Mount Revelstoke 1 May 1986-1; Coffin Lake 1 May 1983-1; Puntchesakut Lake 8 May 1944-10; Prince George 10 May 1975-1; Fraser Lake 28 May 1960-4 eggs; Stuart Lake 11 May 1989-6 eggs; 10 km s North Pine 30 Apr 1977-1; Alcock Lake 22 May 1993-3 eggs; Tupper Creek 25 May 1938-3 eggs and 1 cowbird egg (Cowan 1939); Fort St. John 2 May 1985-1; Bear Flat Hill 5 May 1984-6; Telegraph Creek 23 May 1922-3 eggs; Parker Lake (Fort Nelson) 6 May 1977-1; Atlin 10 May 1932-1; Chilkat Pass 15 May 1977-1; Three Guardsmen Pass 28 May 1975-2; Km 66 Haines Highway 28 May 1975-5.

Summer: Coastal – Mount Becker 5 Jun 1978-15; Mt. Klitsa (w Sproat Lake) 27 Jul 1986-nestlings; North Vancouver 13 Jul 1931-1 fledged female (FMNH 175465); Cypress Park 15 Jul 1990-20; Grouse Mountain 17 Aug 1969-2; Brooks Peninsula 13 Aug 1981-1 recent fledgling; Quatse Lake 9 Jun 1978-11; Cape Scott Park 1 Aug 1981-3; Goose Group 21 Jun 1948-1 nestling; Spider Island 4 Jul 1939-fledglings; Kimsquit 17 Jun 1980-23 around upland lakes; Langara Island 13 Jun 1946-32, 20 Jun 1971-first fledglings (Sealy 1974); Lawyer Islands 17 Jul 1979-3 nestlings; Prince Rupert 20 Jun 1979-4 eggs; Port Simpson 6 Jun 1903-4 eggs; Hazelton 13 July 1927-4 eggs. **Interior** – Manning Park 26 Jun 1983-120 on count; Rossland 4 Jul 1929-1; Keremeos 1 Jun 1948-3 eggs; Apex Mountain 22 Jul 1974-20 at summit; Kimberley 4 Aug 1977-2 young; Edgewood 19 Jul 1924-4 eggs; Curtis Lake 7 Jun 1983-15; Summit Lake (15 km se Nakusp) 15 Jul 1978-22; Monashee Pass 24 Jun 1937-2 (RBCM 4147-48); Phyllis Lake 7 Jun 1977-4 eggs; Mount Revelstoke National Park 14 Aug 1988-2 nestlings; Allan Creek 14 Jun 1961-4 eggs; Bridge Lake 5 Jun 1976-22; Horsefly 1 Jul 1975-3; Tatla Lake 11 Jun 1970-4 eggs; Chezacut Lake 22 Jun 1959-4 nestlings; Prince George 24 Jun 1969-10; Nulki Lake 8 Jul 1945-3 eggs (Munro 1949); Fraser Lake 28 May 1960-4 eggs; Mount Mussen 30 Aug 1975-1; Sinkut Mountain 8 Jul 1949-3 eggs; Topley 3 Jul 1956-9; Tuaton Lake Jul 1970-nestling (Osmond-Jones et al. 1977); Fort St. John 23 June 1973-4 eggs, of which 2 were Brown-headed Cowbird; Dawson Creek 3 Jul 1974-2; Hudson's Hope 7 Jun 1976-10; Lynx Creek 25 Jul 1986-3 nestlings; Hotlesklwa Lake 3 Aug 1976-3; Telegraph Creek 4 Jun 1919-2 (MVZ 40311-12); Fern Lake (Kwadacha Wilderness Park) 25 Aug 1979-feeding fledglings (Cooper and Adams 1979); Moosehorn Lake 18 Jul 1987-7; White Birch 22 Jun 1982-30; Fort Nelson Airport 31 Aug 1985-1; Atlin 13 Jun 1924-eggs, 12 Jul 1924-4 fresh eggs (Swarth 1926); Cassiar Junction 13 Jun 1972-26; Tatshenshini River valley 7 Jun 1983-3 eggs (Campbell et al. 1983).

Breeding Bird Surveys: Coastal – Recorded from 13 of 27 routes and on 26% of all surveys. Maxima: Queen Charlotte City 25 Jun 1994-85; Masset 21 Jun 1994-38; Kitsumkalum 11 Jun 1978-36. **Interior** – Recorded from 57 of 73 routes and on 46% of all surveys. Maxima: Pennington 30 Jun 1991-41; Bridge Lake 28 Jun 1987-33; Chilkat Pass 1 Jul 1976-32.

Autumn: Interior – Atlin 30 Sep 1930-1 (RBCM 5853); Mile 52 Haines Highway 8 Sep 1972-4; St. John Creek 7 Sep 1985-4; Taylor 23 Sep 1984-1; Vanderhoof 2 Sep 1934-1 (MVZ 103662); Nanika River 4 Oct 1974-11; Indianpoint Lake 10 Oct 1928-1; Quesnel 6 Sep 1900-1 (MVZ 103686); Lac la Hache 6 Oct 1946-1; Buffalo Lake 1 Sep 1932-1 (RBCM 11452); Hellroaring Creek 18 Sep 1983-4; Grindrod 9 Oct 1948-1; Scotch Creek (Shuswap) 12 Sep 1962-2; Kamloops 14 to 30 Nov 1995-1; Vernon 10 Oct 1935-1; Okanagan Landing 7 Oct 1921-1 (MVZ 42672); Cranbrook 1 Oct 1941-1 (Johnstone 1949). **Coastal** – Terrace 12 Nov 1977-1; Prince Rupert 18 Oct 1956-3; Masset 4 Sep 1937-1 (RBCM 10473); Queen Charlotte City 5 Nov 1986-2; Calvert Island 4 Sep 1934-2 (MCZ 283668-9); Cheam Lake 27 Oct 1982-2; Cape Scott 16 Sep 1935-2 (NMC 26064-5); Port Neville 21 Oct 1971-2; Snake Island 14 Oct 1979-1; Alta Lake 15 Sep 1941-7; Cypress Park 27 Sep 1987-18; Diana Island 15 Nov 1985-4; Victoria 17 Nov 1985-9; Metchosin 15 Oct 1987-15.

Winter: Interior – Kamloops 1 to 25 Jan 1995-1 at feeder; see Christmas Bird Counts. **Coastal** – Masset 18 Dec 1993-1; Tlell River 19 Jan 1952-1 (ROM 87823); Skidegate Inlet 20 Dec 1986-1, 19 Jan 1952-1; Koin's Island (Forward Inlet) 1 Feb 1936-1 (UBC 7542); Vargas Island 24 Jan 1969-12 (Hatler et al. 1973); Tofino 26 Dec 1972-2; Saanich 31 Dec 1977-1, 2 Feb 1985-5; Jordan River 3 Dec 1983-2, 7 Jan 1969-1; Goldstream Park 31 Jan 1980-1; Metchosin 1 Jan 1984-1.

Christmas Bird Counts: Interior – Recorded from 1 of 27 localities and on less than 1% of all counts. Maximum: Oliver-Osoyoos 28 Dec 1987-1. **Coastal** – Recorded from 20 of 33 localities and on 27% of all counts. Maxima: Victoria 2 Jan 1966-**34**, all-time Canadian high count (Anderson 1976b); Sooke 2 Jan 1988-17; Campbell River 17 Dec 1972-15.

American Robin
Turdus migratorius Linnaeus

AMRO

RANGE: Breeds from western and northern Alaska, northern Yukon, northern Mackenzie, and southern Keewatin east to Labrador and Newfoundland and south through most of North America, except interior and southeastern desert regions. Winters from southern Alaska (casually) but mainly from southern British Columbia and the northern United States south through the rest of the United States. Also winters in Mexico, Guatemala, Bermuda, and, at least irregularly, in western Cuba.

STATUS: On the coast, *fairly common* to *very common* resident in southwestern British Columbia, including southeastern Vancouver Island, the Gulf Islands, the lower Fraser River valley, and the Sunshine Coast in the Georgia Depression Ecoprovince, becoming locally *abundant* to *very abundant* in winter, notably in agricultural areas and at roosts; *common* to *abundant* spring and autumn migrant there; *fairly common* to locally *very common* (occasionally *abundant*) migrant and summer visitant to Western Vancouver Island and to the southern and northern mainland of the Coast and Mountains Ecoprovince; *uncommon* to locally *common* on the Queen Charlotte Islands, becoming *rare* to *fairly common* in winter.

In the southern portions of the interior, a *fairly common* to *common* resident in the Southern Interior and Southern Interior Mountains ecoprovinces, becoming a *very common* to locally *abundant* spring migrant, and locally *very common* autumn migrant; *fairly common* to locally *very common* migrant and summer visitant to the Central Interior Ecoprovince; *uncommon* to locally *fairly common* in the Sub-Boreal Interior ecoprovince; *casual* to *very rare* in winter. *Fairly common* to locally *very common* migrant and summer visitant to the Boreal Plains Ecoprovince; *uncommon* to locally *fairly common* migrant and summer visitant to the Northern Boreal Mountains and Taiga Plains ecoprovinces; *casual* in winter in the Taiga Plains.

Breeds.

Figure 473. In southern British Columbia, the American Robin is almost entirely dependent on berries and small fruit in winter (Williams Lake, 24 March 1996; R. Wayne Campbell).

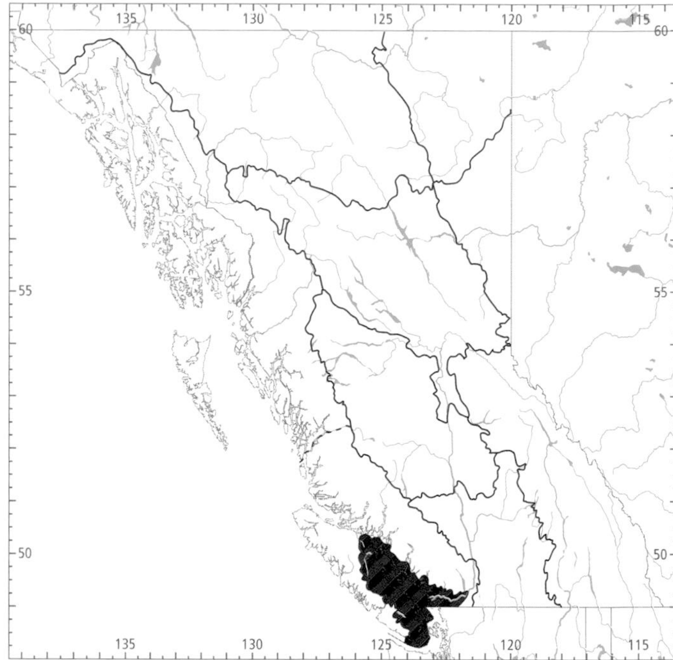

Figure 474. In British Columbia, the highest numbers for the American Robin in both winter (black) and summer (red) occur in the Georgia Depression Ecoprovince.

NONBREEDING: The American Robin (Fig. 473) is widely distributed throughout the province from spring through autumn, but is absent from most of the northern and central interior during winter. It reaches its highest numbers in winter in the Georgia Depression (Fig. 474).

The robin is found mainly at lower elevations during the nonbreeding seasons, except during the post-breeding period in late summer, when much of the population apparently moves up to the timberline to feed on ripening berry crops. It occurs at elevations from sea level on the coast to 2,425 m in the interior (e.g., Ashnola Mountains and Mt. Robson Park) (Cannings 1974).

It is mainly a bird of fields, open woodlands, edges where forests and clearings meet, and rural and suburban habitats. During spring migration, flocks occur in almost any open grassy or bare-earth area that provides foraging opportunity. These include pastures, fields, park lawns, golf courses, open swamps, meadows, school fields, and beaches. On the coast, migrants and wintering birds are most numerous in agricultural areas, but also occur in suburbs, logged areas, and burns, along open river banks, on estuaries, and on intertidal sand beaches and mudflats. In unbroken coastal forests, the robin occurs mainly along beaches and estuaries, where it forages on intertidal mudflats at low tide. In the interior, it occurs mainly in agricultural and suburban areas, grasslands, and open forests, and along lakeshores and marsh edges. In mountainous areas, post-breeding birds remain at or move to higher

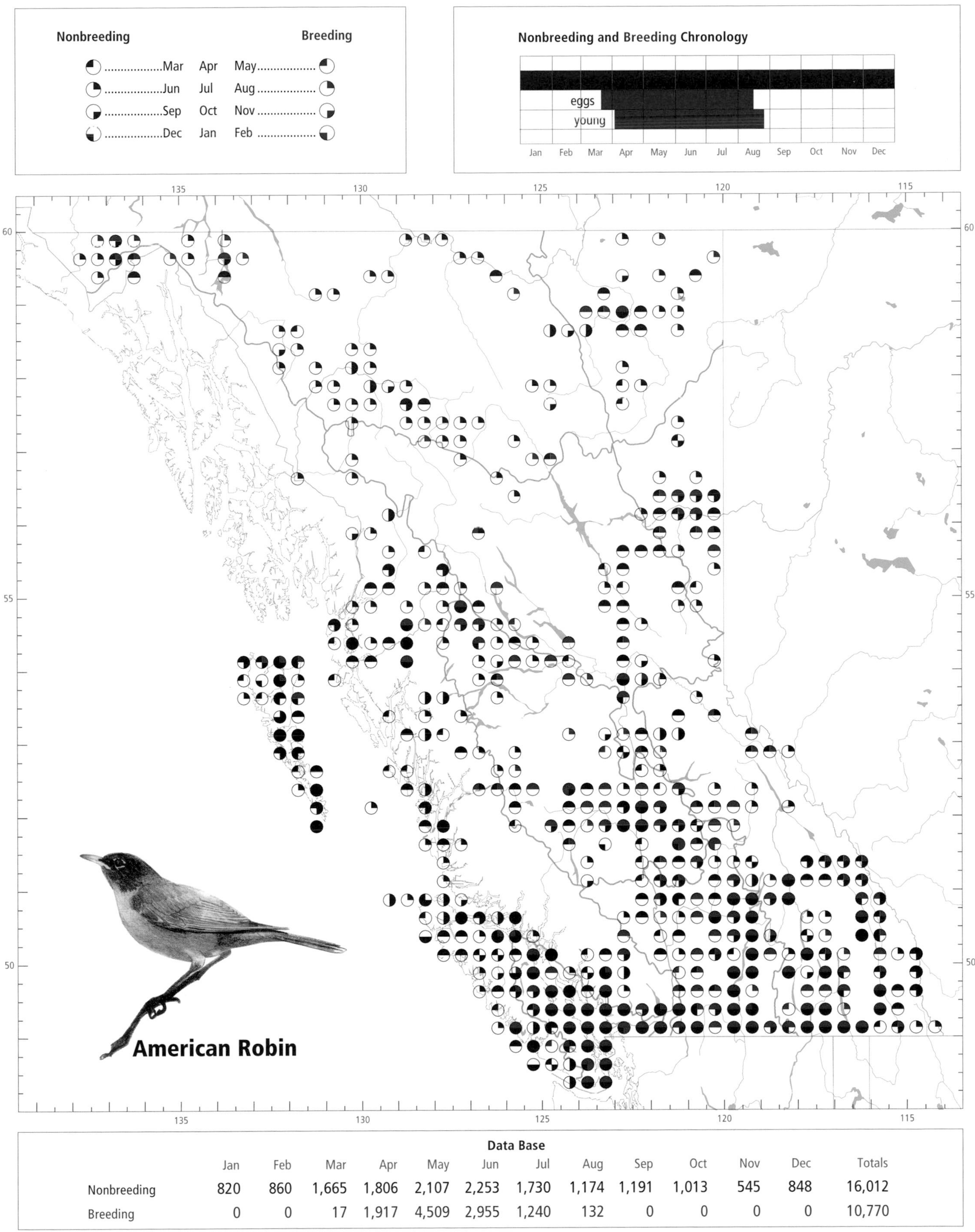

Data Base

	Jan	Feb	Mar	Apr	May	Jun	Jul	Aug	Sep	Oct	Nov	Dec	Totals
Nonbreeding	820	860	1,665	1,806	2,107	2,253	1,730	1,174	1,191	1,013	545	848	16,012
Breeding	0	0	17	1,917	4,509	2,955	1,240	132	0	0	0	0	10,770

elevations for a period of time before migrating south. From July through early September, robins frequent subalpine meadows (Fig. 479), shrubby avalanche chutes, stream floodplains, and forest berry patches. During the autumn movement, the American Robin uses environments similar to those used in the spring, but migrates at higher elevations. It congregates in areas with berry- and fruit-bearing trees and shrubs such as mountain-ash, holly, arbutus, Pacific crab apple, blueberries, and ornamental shrubs. Commercial blueberry farms, vineyards, and orchards often attract large numbers of summering and autumn migrant birds. During periods of subfreezing temperatures, the American Robin either migrates further south or remains in the province and becomes entirely dependent on berries and small fruits.

The onset of spring migration is difficult to determine in southern and coastal regions of the province because of the presence of overwintering or resident birds (Fig. 475). On the coast, substantial numbers overwinter, but large flocks of migrants appear in the Georgia Depression throughout March and into mid-April (Fig. 476). Many of these birds come from wintering areas in California, Oregon, and Washington (Fig. 477). The lower Fraser River valley sees the largest of the spring flocks. Smaller numbers migrate along the outer coast, peaking in mid-March on the west coast of Vancouver Island and early April on the Queen Charlotte Islands and Northern Mainland Coast. In the interior of North America, migrant robins apparently follow the 3°C isotherm during the northward spring movement (Bent 1949). This isotherm usually passes through the southern portions of the interior of British Columbia during late February or early March (Farley 1979). In the Okanagan valley in the Southern Interior, early migrants usually arrive in late February (Cannings et al. 1987), with the movement peaking around the end of March and first 2 weeks of April (Fig. 476); in the Southern Interior Mountains, the peak occurs during the last 2 weeks of March and the first week of April. Spring migrants appear in the Chilcotin-Cariboo Basin of the Central Interior about mid-March, with the movement peaking quickly in late March and the first week of April. Further north, migrants reach Prince George in the Sub-Boreal Interior in late February or early March, the Peace Lowland of the Boreal Plains by late March, and the Fort Nelson Lowland of the Taiga Plains and Atlin in the Northern Boreal Mountains by late April (Fig. 476).

The autumn movement begins in early to mid-August in the far northern interior, peaking in early September in both the Northern Boreal Mountains and the Taiga Plains; in mid-September in the Peace Lowland of the Boreal Plains, the Sub-Boreal Interior, and the Central Interior; and during late September and early October in the Southern Interior and Southern Interior Mountains (Fig. 476). On the coast, the autumn buildup occurs from late August to early October. The increase includes flocks descending from the subalpine berry patches as well as migrants from the north and interior. Numbers on the Fraser Lowland and on the east coast of Vancouver Island continue to increase through November into December (Fig. 476).

Regular wintering populations occur mainly on the coast, in the valleys of the Southern Interior, and in the southern valleys of the Southern Interior Mountains. During mild winters, large numbers remain on the Nanaimo and Fraser lowlands. Even there, numbers appear to decline during periods of subfreezing temperatures. In part, this reflects a movement closer to the sea, but there may be some movement into Washington and areas further south (see Fig. 477). The largest winter concentrations in the province occur on the

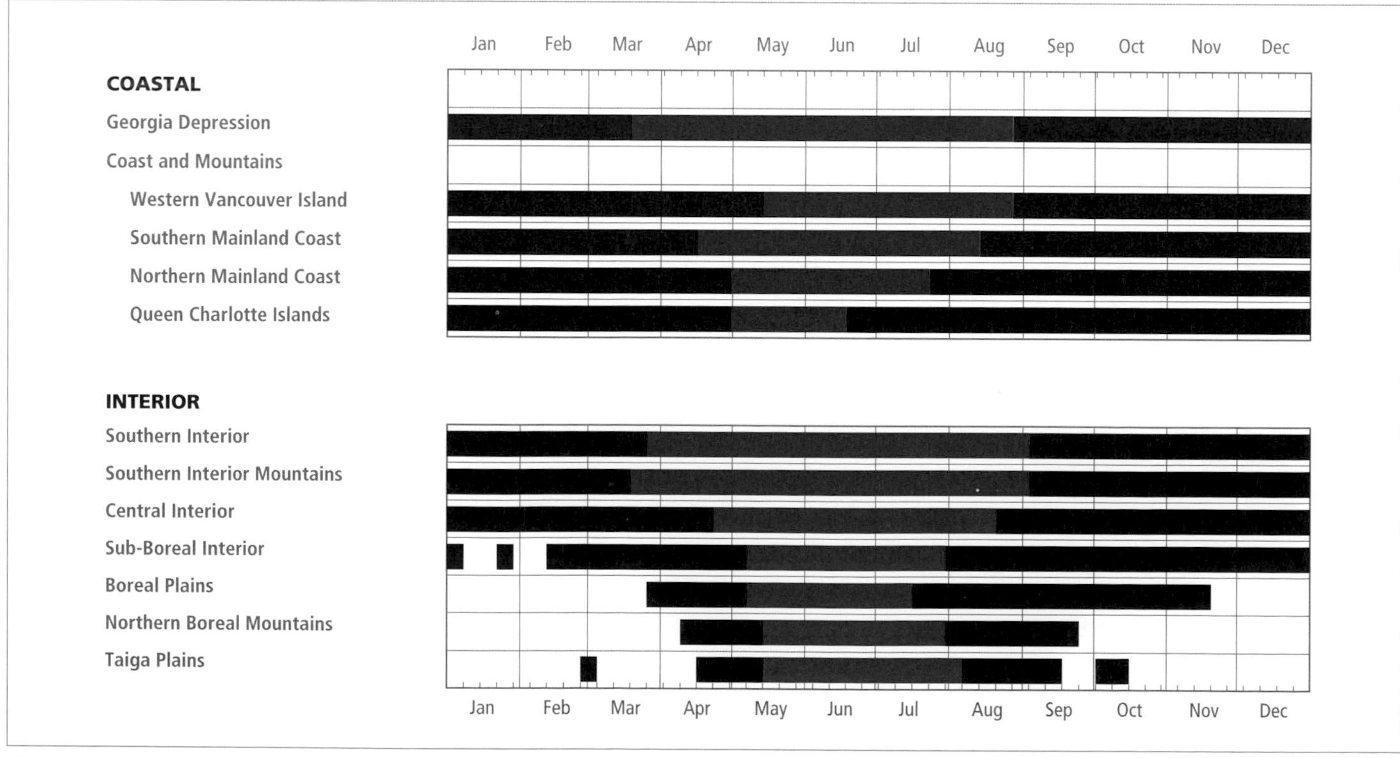

Figure 475. Annual occurrence (black) and breeding chronology (red) for the American Robin in ecoprovinces of British Columbia.

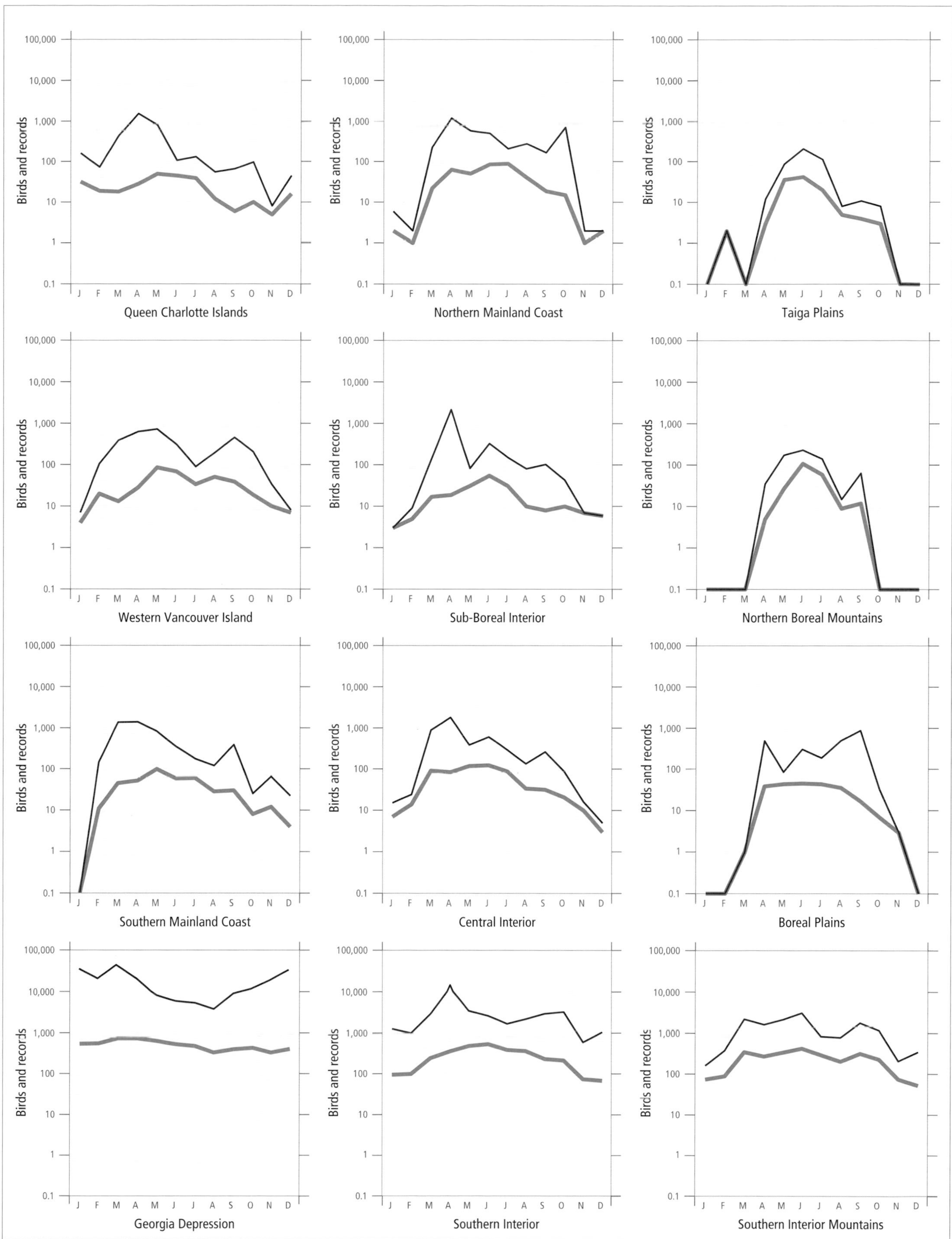

Figure 476. Fluctuations in total number of birds (purple line) and total number of records (green line) for the American Robin in ecoprovinces of British Columbia. Christmas Bird Count and nest record data have been excluded.

Saanich Peninsula, near Victoria. Although there are scattered winter records from Williams Lake, Smithers, Prince George, and Fort Nelson, the American Robin is not a regular wintering species in the interior north of the Okanagan valley. Wintering populations in the Georgia Depression and the Southern Interior fluctuate annually depending on local climatic conditions (Fig. 478).

The American Robin forms communal, nocturnal roosts throughout the year (Eiserer 1976). Roosts may number from a few individuals to several thousands of birds. The largest assemblages of roosting birds occur on the south coast during the nonbreeding season, particularly in late autumn and winter. For example, there are 2 reports of American Robin roosts exceeding 5,000 birds: 7,000 at Beacon Hill Park, Victoria, and 10,000 at the University of Victoria. Daily movements to and from roosting sites occur around sunset and sunrise, respectively, a time when flocks are often observed passing overhead. The characteristics and locations for most of these important habitat components for the American Robin in British Columbia are lacking.

The American Robin tends to flock during the daylight hours of the nonbreeding seasons and during migration. The largest nonroosting flocks reported range from a few hundred to 2,240 birds. This flocking behaviour breaks down gradually in early spring as resident birds become territorial (Kemper 1971).

On the coast and in the southern portions of the interior, the American Robin has been recorded regularly throughout the year; in the Central Interior and northern regions, it has been recorded regularly between 21 February and 7 October (Fig. 475).

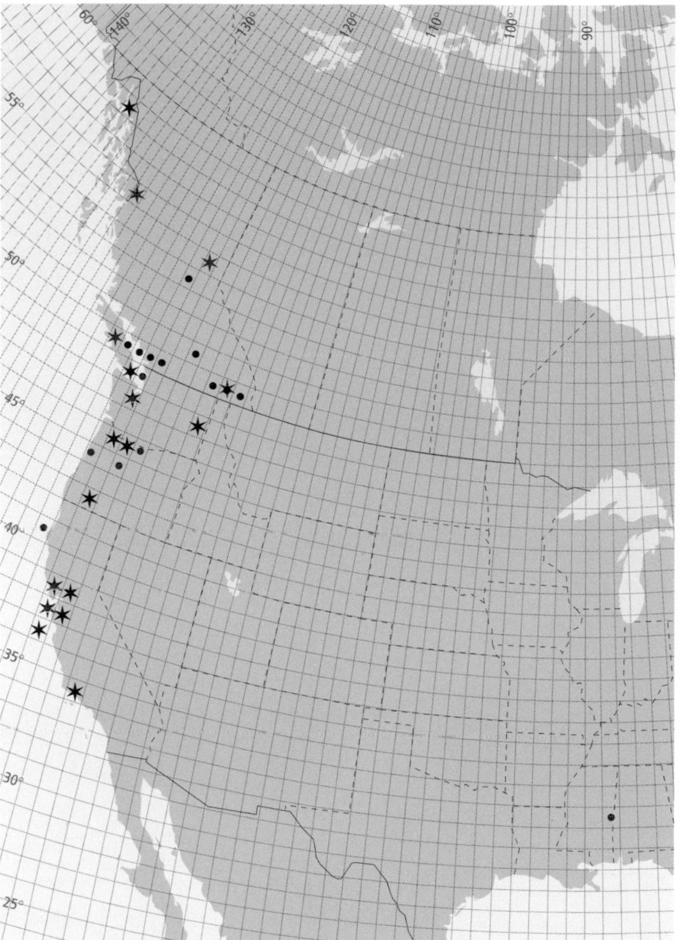

Figure 477. Banding locations (stars) and recovery sites (circles) of American Robins associated with British Columbia. Red indicates birds banded in British Columbia; black indicates birds banded elsewhere.

BREEDING: The American Robin breeds throughout the province, including large and small coastal islands. It is the most widely distributed breeding songbird in British Columbia. It is a common breeder in the Peace Lowland of the Boreal Plains, but becomes much less abundant away from human-influenced habitat in the lowland spruce forests of the Taiga Plains and the mountainous boreal forests of northwestern British Columbia. Exceptions are the valley bottoms of the major northwestern rivers, and timberline krummholz forests, where it is a fairly common breeder. Although

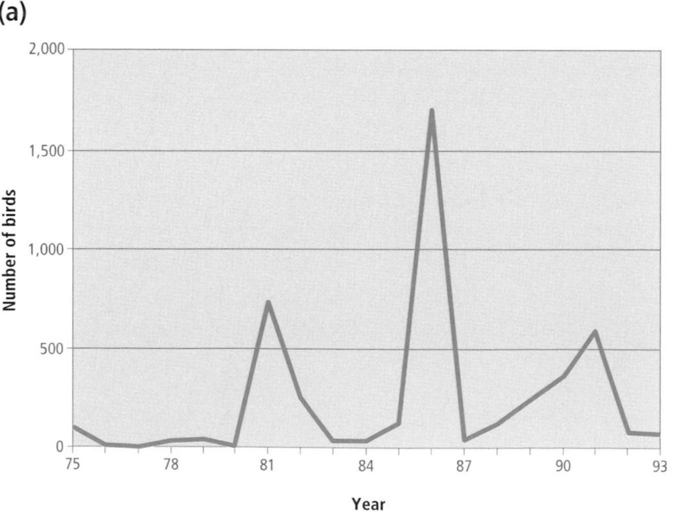

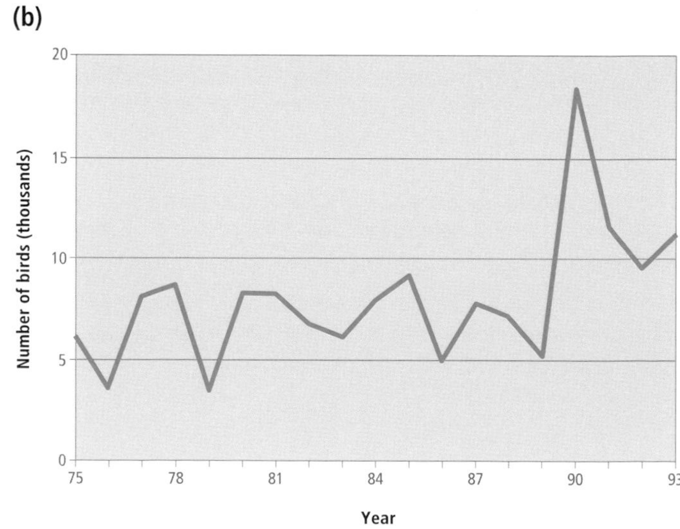

Figure 478. Fluctuations in winter numbers of the American Robin (1975 to 1993) in (a) the Southern Interior (Vernon, Penticton, Vaseux Lake, Oliver-Osoyoos) and (b) the Georgia Depression (Vancouver, Nanaimo, Ladner, Duncan, Victoria) ecoprovinces, based on Christmas Bird Counts.

Figure 479. In subalpine areas, the American Robin prefers patches of subalpine firs in parkland basins and cirques over steep mountain slopes (Buck Mountain, 20 July 1993; R. Wayne Campbell).

documentation of breeding is lacking along much of the central and northern mainland coast, small numbers are known to nest near the sandy bays scattered along the exposed coastlines from northern Vancouver Island north to Portland Canal.

This thrush reaches its highest numbers in the Georgia Depression (Fig. 474). An analysis of Breeding Bird Surveys in British Columbia for the period 1968 through 1993 could not detect a net change in numbers on either coastal or interior routes.

The American Robin breeds from near sea level on the coast to at least 2,200 m in the interior. In all regions, it is more abundant at lower elevations in the floodplains, valleys, plateaus, and lower slopes. Small numbers breed in higher-elevation forests and in the subalpine shrub zone near the timberline, where the American Robin is more numerous in subalpine parkland basins and cirques than on steep, forested mountain slopes (Fig. 479).

It breeds in the widest variety of habitats of all our songbirds. Unlike other thrushes, the robin thrives in human-influenced environments, and the vast majority of our nest records are from suburban gardens, backyards, and agricultural areas. In city and suburb, the robin breeds in residential neighbourhoods, parks, cemeteries, golf courses, and almost anywhere ornamental trees, hedges, and shrubs occur near grass lawns, where they can forage for earthworms. In rural and agricultural areas it nests in similar habitats, but also along hedgerows; in orchards, vineyards, and plant nurseries; in patches of woods within pastures; and in thickets along roadsides. In more natural situations, it breeds mainly along

Figure 480. The American Robin occasionally nests in cattail marshes (Sea Island, 14 May 1971; R. Wayne Campbell).

Figure 481. American Robin nests were most often found in the forks of the mainstems of deciduous trees (Kruger Mountain Road, Richter Pass, 18 June 1993; Linda M. Van Damme).

the edges of mature coniferous, deciduous, and mixed forests; in riparian strips along rivers and lakes; in regenerating burns and young second-growth forests; in the forested borders of ocean beaches; along the brushy edges of beaver ponds, marshes, and swamps; in copses of trees in rangelands; in subalpine parklands; and in the krummholz fringe at the timberline. In many environments it tends to nest near streams, marshes (Fig. 480), or ponds. In an extensive logged and burned area near Comox that had a 10-year-old regenerating Douglas-fir forest and an abundance of water, the density of breeding birds was about 1 pair per 5.5 ha; nests were an average of about 100 m apart in denser habitat and 75 m apart in more open habitat (Martin 1973).

The species seldom breeds in dense, mature forests. In mixed-age coastal forests, it breeds along the edges of clearings, clearcuts, and roads; in natural openings in the forest;

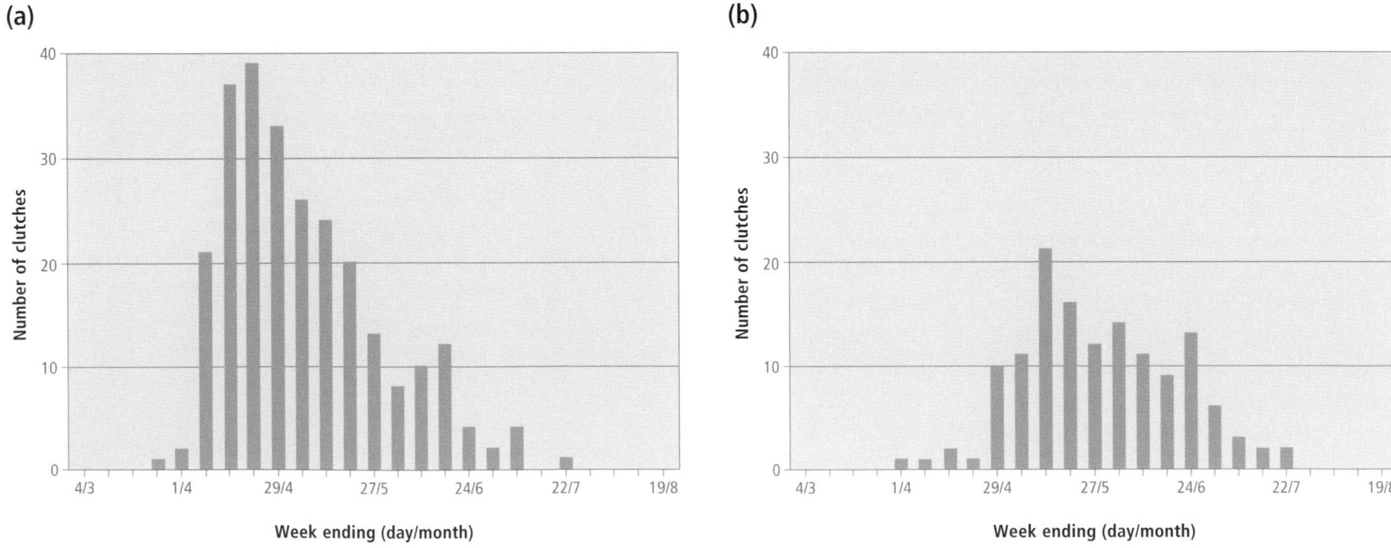

Figure 482. Weekly distribution of first eggs in American Robin clutches in the (a) Georgia Depression and (b) Southern Interior ecoprovinces. Nests were included only if clutches were discovered while laying was in progress. Note the peaks in clutches, indicating that some pairs raise a second brood.

(a)

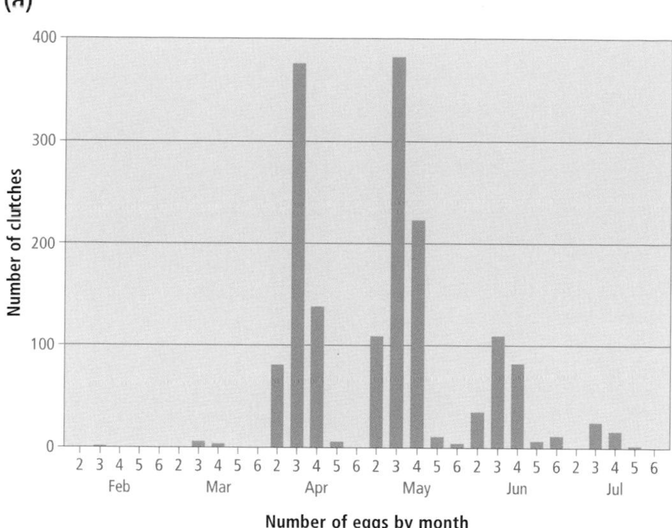

(b)

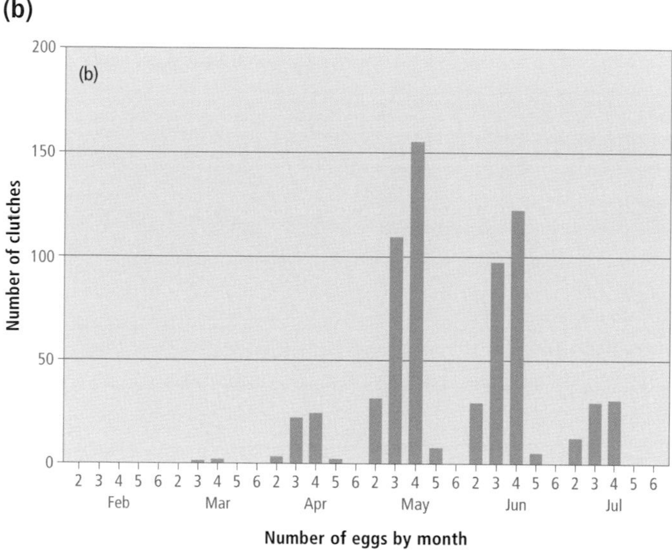

(c)

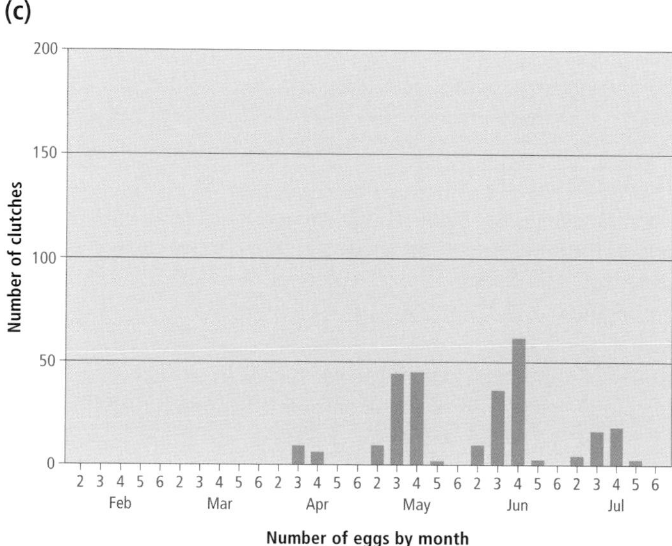

Figure 483. Distribution of clutch sizes of the American Robin between February and July for the (a) Georgia Depression, (b) Southern Interior, and (c) Southern Interior Mountains ecoprovinces.

or in forest that has a diverse canopy structure and little shrub layer (Horvath 1963). Bryant et al. (1993) found the species common and widespread in all age classes of forest they studied on the west coast of Vancouver Island, where it was the most abundant bird encountered during their study. The highest number of robins there was found in 50- to 60-year-old forests, followed by forests 30 to 35 years of age. Robin abundance was positively correlated with crown closure and negatively correlated with distance to the coast and woody debris. These authors note that the location of their sampling points, along roads, may have contributed to the apparent abundance of this "edge" species.

In interior forests, the American Robin breeds in similar habitats, but also within mature ponderosa pine and interior Douglas-fir forests, which are relatively open, and in denser forests that have been selectively logged, where removal of large trees creates sufficient open space. It also breeds in sagebrush and antelope-brush shrublands of the Southern Interior, and in the willow- and alder-choked avalanche chutes of steep mountain slopes.

On the coast, the American Robin has been recorded breeding from 21 March (calculated) to 21 August; in the interior, it has been recorded from 22 March (calculated) to 25 August (Fig. 475).

Nests: Nests were situated in a diverse array of sites. Most (60%; *n* = 4,926) were placed in trees. Deciduous trees had the highest frequency of use (34%; see Fig. 481) followed by coniferous trees (26%), buildings (13%), and shrubs (13%). Nest trees included at least 19 species of native deciduous trees, 14 species of native conifers, and more than 30 species of ornamental trees. At least 25 species of deciduous shrubs, 9 species of evergreen shrubs, and 1 fern were also reported. Most nest trees were living, but a small percentage of nests were found in dead trees. In burned and logged areas, nests were often found in broken tops or side cavities of tree stumps (1%). Along mountain roads, nests were placed in hollows in road banks or were supported by tree roots protruding from the bank (1%). A few nests were located in cattail marshes (Fig. 480) and in clumps of big sagebrush. In addition to buildings, other man-made structures used as nest sites included ornamental vines growing on trellises (1%), posts and poles (1%), bridges, abandoned equipment, underground parking garages (Cooper 1982), tunnels, and billboards.

There were regional, seasonal, and elevational differences in specific nest sites chosen. Deciduous trees were used most often in all ecoprovinces except the Central Interior, where coniferous trees predominated. Shrubs were used frequently in the Georgia Depression (17%) and the Southern Interior (10%), but were seldom used in the central and northern interior (2%). During the early part of the nesting season, coniferous trees, with their additional concealment and thermal cover, were used more often than deciduous trees, which were not yet in leaf. In the Georgia Depression, 46% of 61 nests found in March were built in coniferous trees; only 21% were in deciduous trees, 16% in dense or evergreen shrubs, and 15% on buildings. Later in the season, when deciduous trees began to leaf out, nests were placed in deciduous trees more

Figure 484. One of only 6 American Robin nests in British Columbia that were found parasitized by the Brown-headed Cowbird (Victoria, 20 May 1973; R. Wayne Campbell).

often. In the southern portions of the interior, nests found at elevations above 1,000 m (n = 68) were, as expected, placed more often in coniferous trees (40%) than in deciduous trees (31%).

The heights for 4,761 nests ranged from ground level to 24 m, with 65% between 2 and 4 m. Most nests (28%; n = 3,197) were placed in the crotch of a branch (Fig. 481), saddled on a branch next to the trunk (17.3%), or saddled on a branch away from the trunk (13.2%). Many nests adjacent to the tree trunk were supported by growths of suckers. Other natural sites included the crowns of small conifers, among many small branches in shrubs, in the roots of fallen trees, in crevices of rock walls and caves, on fallen tree trunks leaning against another tree, in witches' broom, on the ground under an overhanging stump, on willows pollarded by moose browsing, in a Pileated Woodpecker cavity, and in old nests of American Robin, Steller's Jay, Red-winged Blackbird, and Western Kingbird. Other studies in British Columbia have also noted that nests are often reused for second and third clutches (Kemper 1971; Cannings, et al. 1987). Nests in woodland habitats were usually situated just inside the "edge" where vegetated and open areas meet.

On human-made structures, specific nest locations included almost any flat spot imaginable, such as on the tops of beams and rafters (9%); on the tops of window and other ledges (8%); on drainpipes and light fixtures; on wires or brackets under overhanging eaves; or on and in nest boxes. Other sites included the steps of ladders leaning against buildings; poles of rail fences; boards in lumber piles; hanging flower baskets and bird feeders; outside building vents; sawed-off tops of trees; hanging tires; old cars, trucks, and machinery; and between the antlers of a mounted deer head.

Nests were bulky cups of grass and mud (Figs. 480 and 481), with small amounts of twigs, moss, plant stems, leaves, and rootlets. Those near human settlements occasionally included string or rope (206 nests) and paper (37 nests). Nests were lined almost exclusively with fine grasses; a few contained small amounts of mammal hair. Nests reused during the same or following years were relined with new material.

Eggs: Dates for 3,086 clutches (Fig. 484) ranged from 23 March to 15 August, with 51% recorded between 30 April and 7 June. Calculated dates indicate that nests can hold eggs by 21 March. There were significant latitudinal differences in the onset of breeding by the robin in British Columbia (Fig. 475). Dates of first eggs from the Georgia Depression were about the same as those in southern portions of the interior (Fig. 482), which were some 6 weeks earlier than in the Northern Mainland Coast, Queen Charlotte Islands, Central Interior, and southern end of the Sub-Boreal Interior. There was a further 1-week delay in the onset of nesting in the Boreal

Plains, and a 2-week delay in the Northern Boreal Mountains and Taiga Plains. In total, egg laying began in the southern regions of the province 8 weeks before it did along the northern boundary (Fig. 475). Although egg laying may begin by late March in the Georgia Depression and the southern portions of the interior, there are only 11 reports from March. Egg laying in the Georgia Depression peaked in the third and fourth weeks of April (Fig. 482); Kemper and Taylor (1981) found similar results in the Vancouver area. In the Southern Interior, egg laying peaked about 4 weeks later than in the Georgia Depression.

Clutch size in 3,086 nests ranged from 1 to 8 eggs (1E-290, 2E-364, 3E-1,362, 4E-1,022, 5E-41, 6E-3, 7E-3, 8E-1), with 77% having 3 or 4 eggs. There were significant seasonal and regional differences in clutch size. Clutches on the coast were more likely to contain 3 eggs than 4 eggs (Fig. 483). In the southern portions of the interior, clutches were more likely to contain 4 eggs than 3 eggs. Previously reported mean clutch sizes for the province include 2.98 and 3.07 eggs for 2 years in logged habitat on eastern Vancouver Island (Martin 1973), 3.04 eggs at the University of British Columbia in Vancouver (Kemper and Taylor 1981), and 3.63 eggs in the Okanagan valley (Cannings et al. 1987). At least 54 two-egg clutches were known to contain a maximum number of eggs. This unusually low definitive clutch size probably represents mainly replacement clutches after depredation or other nest loss.

The incubation period in southern British Columbia ranged from 11 to 16 days, with a median of 13 days (n = 53). Mean incubation periods in the Comox Burn were 13.1 days (n = 31) in 1971 and 12.7 days (n = 36) in 1972 (Martin 1973). An 11-day incubation period is abnormal, but has been reported (Young 1955; Brackbill 1977). There was no difference in incubation period between coastal and southern interior regions. There are no data on the incubation period for northern British Columbia.

Unlike most songbirds, the American Robin lays its eggs well past dawn, during the late morning or afternoon rather than during the early morning (Weatherhead et al. 1991; Scott 1993).

Young: Dates for 2,252 broods ranged from 3 March to 25 August, with 51% recorded between 14 May and 24 June. Brood size ranged from 1 to 6 young (1Y-161, 2Y-390, 3Y-1,028, 4Y-649, 5Y-23, 6Y-1), with 74% having 3 or 4 young. The nest with a brood of 6 young eventually produced 5 fledglings. The nestling period of the American Robin in southern British Columbia ranged from 11 to 21 days, with a median of 15 days (n = 55). Nestling periods are often a few days longer early in the season or during periods of inclement weather. Mean nestling periods for the 2-year study near Comox were 13.2 and 13.5 days (Martin 1973); a period of 14 days (n = 22) was reported for the Okanagan valley (Cannings et al. 1987). There are no data on nestling periods for northern British Columbia.

In Vancouver, most pairs raised 2 broods, and some 3 broods, in a year (e.g., Kemper 1971). On the Queen Charlotte Islands, 2 broods were produced annually. During 5 years of observations at Delkatla Inlet, large numbers of fledglings appeared at foraging areas between 31 May and 5 June, with another wave of fledglings in late July. Two broods were usually raised in the southern portions of the interior. In the British Columbia–Yukon boundary area, 1 brood was the norm (J.N.M. Smith pers. comm.). Fledged young still being fed by parents may be found into early September in southern British Columbia.

Brown-headed Cowbird Parasitism: In British Columbia, 0.1% of 4,522 nests found with eggs or young were parasitized by the cowbird (Fig. 484). On the coast, 2 nests were parasitized (0.1%; n = 2,541); in the interior, 4 nests were parasitized (0.2%; n = 1,981). Throughout North America, the robin is an infrequent host of the cowbird (Rothstein 1971; Friedmann and Kiff 1985). It usually rejects cowbird eggs, but in 1 instance a robin is known to have fledged a young cowbird (Lowther 1981).

Nest Success: Of 1,408 nests found with eggs and followed to a known fate, 698 produced at least 1 fledgling, for a nest success rate of 50%. Nest success on the coast was 42% (n = 777); in the interior, it was 58% (n = 631).

In a study conducted between 1967 and 1969 on the University of British Columbia campus in Vancouver, 53 nests were discovered in the building stage. The number reaching the egg-laying stage was not stated, but eggs hatched in 27 nests, or 51%. Eighty-two eggs were laid, and these resulted in 72 nestlings and 71 fledglings (Kemper 1971). A later study at the same location, by which time the campus was a more populous place, found that 47% of nests (n = 122) produced at least 1 fledgling (McLean et al. 1986).

Another high success rate in British Columbia was encountered in a study of birds and pesticide use in orchards in the south Okanagan and Similkameen areas. Here about two-thirds of robin eggs in unmanipulated nests (n = 23) produced fledged young (Sinclair and Elliott 1993).

A contrast is provided by the Comox study of the American Robin nesting in a young fir forest. There, only 30% of 120 nests where incubation had begun produced at least 1 fledged young. Sources of nest loss were predation (49%) and abandonment (21%) (Martin 1973). This low success rate was attributed mainly to depredation by marten.

In our provincewide sample, 364 nests were reported depredated, destroyed, or interfered with sufficiently to cause loss of eggs or young. At least 27 different causes were identified or suspected. The 6 most important nest predators were American and Northwestern crows, squirrel, Steller's Jay, domestic cat, and Clark's Nutcracker (Fig. 485).

REMARKS: Three subspecies of the American Robin occur in British Columbia: *T. m. caurinus* occurs on the coast, *T. m. propinquus* in the Southern Interior north to the Cariboo and Chilcotin plateaus, and *T. m. migratorius* in the northern interior south to the Nechako River (Munro and Cowan 1947; American Ornithologists' Union 1957). There are subtle but significant variations in the morphology and plumage of the American Robin throughout North America (Aldrich and James 1991), resulting in at least 6 recognized subspecies.

The American Robin is one of the most ubiquitous birds in the province. It seems equally at home catching sand-hoppers (amphipods) on sandy coastal beaches, pulling earthworms from suburban lawns, or harvesting blueberries in alpine meadows.

The species rapidly responds to habitat changes that open up forested areas; because it requires open areas within its breeding territory, it often increases in numbers after logging and fires (Bock and Lynch 1970; Kilgore 1971). Numbers remain low on clearcuts and recently burned areas until regrowth of the forest begins. Numbers increase as tree heights exceed 1 m, before declining again as tree heights exceed 10 m and overstorey density increases (Martin 1973).

Adult robins tend to return to breed in or near their previous year's territory while juveniles disperse. In the Comox study, 10 of 14 banded adults returned the following year, while none of 35 juveniles banded returned (Martin 1973). A bird banded on Vancouver Island on 28 July 1959 was recovered at Healdsburg, California, in February 1962 (Stoner 1969; see also Fig. 477).

Elliott et al. (1994) found that eggs of the robin in the Okanagan valley in 1990-91 contained high levels of both DDE (up to 103 mg/kg) and DDT (up to 26 mg/kg), despite the fact that these chemicals had not been used in the valley since the 1970s. The levels of both organochlorines were significantly higher in orchard habitats than in non-orchard habitats. A suggested explanation for the high levels of DDT-related compounds in robins' eggs from the Okanagan orchards compared with other orchard sites (such as the Niagara region of Ontario) has to do with differences in earthworm species. Two species of earthworms dominant in the Okanagan valley feed and remain near the soil surface, while the principal earthworm species in the Niagara valley transports its fecal material as deep as 1 m, thus diluting the DDT-related compounds in the soil profile. The data of Elliott et al. (1994) suggest that robins were acquiring DDT and DDE locally, on the breeding grounds, and not on their wintering grounds.

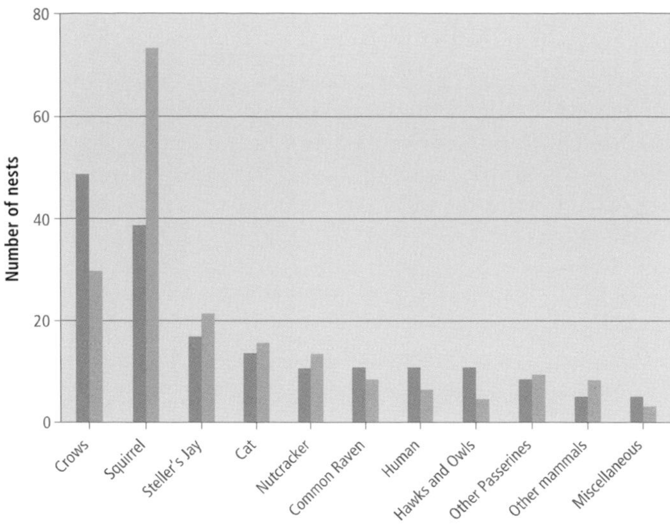

Figure 485. This summary of observed (dark bars; $n = 174$) and suspected (light bars; $n = 190$) causes of nest loss for the American Robin in British Columbia provides evidence of the diversity of predators but not necessarily the relative frequency with which they attack robins' nests. "Miscellaneous" includes snake, blowfly, weather, and poison spray.

The American Robin is one of the most often recorded wildlife species killed by highway traffic in southern British Columbia (Campbell 1984f).

For further information on the American Robin in British Columbia, see Kemper (1971), Martin (1973), and McLean et al. (1986). Additional references on life history include Bent (1947), Eiserer (1976), and Howell (1942).

NOTEWORTHY RECORDS

Spring: Coastal – Victoria 23 March 1984-3 eggs; North Saanich 18 Apr 1992-1 fledged young; Carmanah Point 2 Mar 1991-84; Sumas 1 Mar 1977-2,240, 4 Mar 1965-800; Aldergrove 2 Mar 1968-400; Richmond 3 Mar 1979-800; Tofino 3 May 1974-200; Sea Island 20 Mar 1975-500; Rosedale 30 Mar 1974-1,200; Pitt Meadows 25 Mar 1979-800, 28 Mar 1976-1,500; Vancouver 7 Feb 1955-nest with 3 eggs, earliest record for province; Stanley Park (Vancouver) 12 Mar 1980-1,000; Harrison 30 Mar 1985-681; Hope 24 Apr 1964-3 eggs; Sayward 6 Mar 1977-100; Cape Scott 5 Mar 1975-100; Bella Coola 3 Mar 1990-1, first spring arrival; Sandspit 19 Apr 1982-87; Queen Charlotte City 26 Mar 1972-300; Tlell 17 May 1935-young in nest (Munro and Cowan 1947); Port Clements 10 Apr 1972-300; Delkatla Inlet 17 Apr 1972-250, 15 May 1972-250; Kitimat 4 Apr 1974-100 (Hay 1976), 4 Mar 1975-4 (Hay 1976); Prince Rupert 26 Mar 1984-20; Terrace 21 Apr 1968-100, 7 May 1977-4 eggs. **Interior** – Grand Forks 1 Mar 1981-25, 24 Mar 1980-200; Manning Park 8 Apr 1976-250, 19 Apr 1979-1,000; Hedley 15 Apr 1976-1,500; Princeton 3 Apr 1982-250, 8 April 1979-2,000; Cranbrook 26 Apr 1940-1 egg (Johnstone 1949); Summerland 28 Mar 1970-2 eggs (Cannings et al. 1987); Naramata 25 Mar 1984-building nest; Lytton 23 May 1964-75; Sorrento 27 Mar 1972-202; Erie Lake 21 Mar 1970-100; Slocan 7 Apr 1981-3 nestlings; New Denver 7 Mar 1983-40; Nakusp 6 Apr 1979-400; Radium 3 Mar 1965-1 (Seel 1965); Clearwater 29 Apr 1989-2 eggs; 100 Mile House 28 May 1975-3 young in nest; Dog Creek (Cariboo) 17 Mar 1991-4; Horse Lake 3 Apr 1976-100; Williams Lake 10 Mar 1985-60, 11 Mar 1979-20, 20 Mar 1980-194, 19 May 1984-3 young in nest; Horsefly 2 May 1973-4 eggs; Quesnel 10 Mar 1979-1; 28 km s Prince George 10 May 1983-4 eggs; Prince George 20 Mar 1983-100, 10 Apr 1985-2,000, 11 Apr 1990-1,000 (Siddle 1990b), 14 May 1971-3 eggs; Vanderhoof 14 Mar 1946-first spring arrival (Munro and Cowan 1947); Fraser Lake 19 Apr 1981-50; Quick-Telkwa 23 Mar 1990-3, 5 Apr 1976-750; Tyhee Lake 6 Apr 1978-200; Walcott 19 May 1990-4 young in nest; Dawson Creek 7 May 1993-4 eggs; Baldonnel 5 Apr 1984-1, first spring arrival; Bear Flat 27 Mar 1983-1, first spring arrival; Taylor 30 Apr 1986-150; Fort St. John 21 Apr 1988-100, 13 May 1981-4 eggs; Ingenika Point 12 Apr 1985-1; Fort Nelson 23 Apr 1977-9, 29 Apr 1975-1, first spring arrival, 3 May 1975-30, 21 May 1975-4 eggs; Liard Hot Springs 30 Apr 1975-15 (Reid 1975); Atlin 20 Apr 1934-first spring arrival (Munro and Cowan 1947), 23 Apr 1981-13; Chilkat Pass 14 May 1977-40, 27 May 1979-4 eggs.

Summer: Coastal – Quick's Bottom 22 Aug 1981-100; Central Saanich 6 Jun 1982-250; Qualicum 6 Aug 1982-5 eggs; Deer Lake (Burnaby) 21 Jul 1983-165; Vancouver 18 Aug 1958-1 young in nest; Stanley Park (Vancouver) 16 Aug 1958-150; Cheam Slough 6 Jul 1975-200; Port Neville 7 Aug 1967-2 eggs; Alta Lake 21 Aug 1944-young in nest; Stuie (Bella Coola) 11 Aug 1982 4 nestlings; Greenville 16 Jun 1981-130; Lakelse Lake 29 Aug 1976-71. **Interior** – Creston 4 Aug 1980-42 in roost; Castlegar 15 Aug 1977-4 eggs; Cranbrook 14 Aug 1942-young in nest (Johnstone 1949); Naramata 20 Aug 1974-young fledged (Cannings et al. 1987), 25 Aug 1991-3 nestlings; Balfour to Waneta 8 Jun 1983-705 (count); Needles 25 Jun 1983-59; Kamloops 29 Aug 1960-95; Glacier National Park 24 Aug 1982-2 nestlings; Hemp Creek 19 Jul 1956-3 eggs incubated, 17 Aug 1959-1 young in nest; Chilcotin River (Wineglass Ranch) 4 Aug 1990-2 nestlings; Stum Lake 18 Jul 1973-33 in roost (Ryder 1973); Prince George 27 Aug 1984-50; Houston 20 Jul 1986-1 egg and 3 young; Dawson Creek 2 Jun 1993-1 fledged young; Stoddart Creek 25 Aug 1986-100; Fort St. John 31 Aug 1975-15; Charlie Lake 6 Jun 1986-fledged young; Fort Nelson 5 Aug 1978-2 young in nest; Atlin 17 Jul 1914-4 nestlings; Tatshenshini River 7 Jun 1983-16 (Campbell et al. 1983).

Breeding Bird Surveys: Coastal – Recorded from 27 of 27 routes and on 100% of all surveys. Maxima: Chilliwack 26 Jun 1976-267; Port Renfrew 26 Jun 1974-220; Albion 4 Jul 1970-183. **Interior** – Recorded from 72 of 73 routes and on 99% of all surveys. Maxima: Telkwa High Road 27 Jun 1990-156; Grand Forks 2 Jul 1989-148; Kuskonook 18 Jun 1968-132; Zincton 17 Jun 1968-132.

Autumn: Interior – Kelsall Lake 5 Sep 1977-10; Chilkat Pass early Aug-common migrant (Weeden 1960); Atlin Lake 15 Sep 1972-15, 4 Oct 1931-late departure (Swarth 1936); Fort Nelson 2 Sep 1985-8, 7 Oct 1986-5, late departure, 9 Oct 1985-1, late departure; Samotua River 18 Sep 1986-6; Gladys Lake (Spatsizi) 19 Sep 1976-3; Kwadacha River 1 Sep 1979-2 (Cooper and Adams 1979); Ingenika Point 24 Sep 1990-3; Fort St. John 14 Sep 1985-140, 10 Nov 1985-1, late departure, 13 Nov 1982-1; Taylor 23 Sep 1984-350; Goodlow 2 Oct 1982-20; Quick 12 Sep 1982-100, 3 Oct 1985-25; Prince George 7 Sep 1985-30, 29 Sep 1981-1, 16 Nov 1992-1; Riske Creek 28 Nov 1986-7; Loon Lake (Clinton) 4 Oct 1985-12; Glacier National Park 2 Oct 1982-50; Revelstoke 18 Sep 1985-100; Radium 19 Nov 1977-1; Cranbrook 2 Nov 1939-late departure (Johnstone 1949); West Bench 21 Sep 1975-140 (Cannings et al. 1987), 24 Oct 1966-100; Trail 13 Oct 1980-100; Osoyoos 5 Nov 1976-80. **Coastal** – Kitsault 23 Sep 1980-110; Kitimat 10 Oct 1974-600, 1 Nov 1974-2 (Hay 1976); Delkatla Inlet 24 Oct 1982-50; Drizzle Lake 17 Oct 1976-15; Kemano 1 Oct 1990-50; Sewell 22 Sep 1975-50; Port Neville 19 Sep 1976-90; Grouse Mountain 10 Oct 1971-300; Tofino 2 Oct 1971-50, 13 Oct 1980-60, 11 Nov 1982-17; Vancouver 12 Nov 1973-1,300; Reifel Island 14 Nov 1973-500; Mill Bay 26 Oct 1976-300; Central Saanich 23 Oct 1975-800; Port Renfrew 5 Oct 1974-50; Victoria 16 Oct 1977-300, 12 Nov 1972-1,500; Witty's Lagoon 22 Sep 1989-450.

Winter: Interior – Fort Nelson 27 Feb 1987-1; Smithers 1 Dec 1987-1, 22 Dec 1990-1; Prince George 31 Dec 1993-3, 12 Feb 1984-1, 27 Feb 1986-4; Williams Lake 3 Jan 1980-1, 7 Jan 1991-6; Sheep Creek (Williams Lake) 30 Jan 1988-3; Williams Lake 26 Dec 1943-1, 3 Jan 1982-2, 7 Jan 1991-6; Riske Creek 15 Jan 1988-1; 100 Mile House 6 Feb 1984-1, 21 Feb 1985-6; Big Creek 20 Dec 1981-3; Revelstoke 23 Dec 1988-16, 1 Jan 1989-2; Vernon 4 Jan 1967-100; Nakusp 31 Jan 1978-3; Invermere 12 Jan 1982-1; Wilmer 22 Feb 1980-first spring arrival; West Bench 11 Dec 1974-100; Nelson 17 Dec 1939-100, 1 Jan 1979-3, 15 Jan 1972-10, 27 Feb 1979-50; Summerland 25 Dec 1951-140, 1 Jan 1931-500 (Tait 1931), 24 Feb 1952-200, 18 Feb 1970-50 (Cannings et al. 1987); Castlegar 8 Dec 1975-1; Trail 17 Dec 1979-100, 1 Jan 1983-7, 24 Feb 1983-16; Sirdar 30 Dec 1981-20; Salmo 10 Jan 1976-5; Creston 1 Feb 1984-33. **Coastal** – Terrace 4 Jan 1982-4; Haida 20 Jan 1972-12; Kitimat 12 Dec 1982-31, 4 Jan 1975-2, 15 Feb 1975-2 (Hay 1976); Tlell 29 Feb 1972-1; Queen Charlotte City 24 Feb 1972-18; Sandspit 24 Feb 1972-16; Cape St. James 1 Jan 1982-17; Namu 26 Feb 1981-20; Port Hardy 14 Jan 1979-1; Port Neville 27 Dec 1975-1; Langley 11 Dec 1976-650, 12 Feb 1968-500; Aldergrove 21 Feb 1974-600; Bamfield 2 Feb 1976-1, 27 Feb 1988-10; Cowichan Bay 22 Jan 1977-500; Central Saanich 26 Jan 1985-400, 2 Feb 1985-800; Victoria 9 Dec 1958-7,000 roosting with 8,500 starlings, 11 Dec 1976-1,000, 28 Jan 1963-10,000 in roost.

Christmas Bird Counts: Interior – Recorded from 18 of 27 localities and on 52% of all counts. Maxima: Penticton 27 Dec 1987-797; Vernon 20 Dec 1987-700; Kelowna 19 Dec 1987-330. **Coastal** – Recorded from 30 of 33 localities and on 89% of all counts. Maxima: Victoria 27 Dec 1966-**14,847**, all-time Canadian high count (Anderson 1976b); Vancouver 26 Dec 1957-3,911; Duncan 16 Dec 1978-3,471.

Varied Thrush VATH

Ixoreus naevius (Gmelin)

RANGE: Breeds from western and northern Alaska, northern Yukon, and northwestern and western Mackenzie south to northern California and east through Oregon, Washington, and Idaho into northwestern Montana. Winters along the coast from southern Alaska and British Columbia south as far as Baja California.

STATUS: On the coast, resident as well as *fairly common* to locally *common* migrant and summer visitant to the Georgia Depression and Coast and Mountains ecoprovinces, including Western Vancouver Island and the Queen Charlotte Islands; *fairly common* to locally *very common* in the Georgia Depression during the winter and *fairly common* winter visitant elsewhere.

In the interior, *fairly common* to locally *common* migrant and summer visitant in the Southern Interior and Southern Interior Mountains ecoprovinces; *uncommon* to locally *fairly common* winter visitant there; *uncommon* to locally *fairly common* migrant and summer visitant to the Central Interior Ecoprovince; *rare* to *uncommon* there in winter; *rare* to *uncommon* migrant and summer visitant to the Sub-Boreal Interior, Boreal Plains, Taiga Plains, and Northern Boreal Mountains ecoprovinces, where it may be *rare* or absent in winter.

Breeds.

NONBREEDING: The Varied Thrush (Fig. 486) has a widespread distribution across much of the province, including Vancouver Island and the Queen Charlotte Islands. On the coast, it

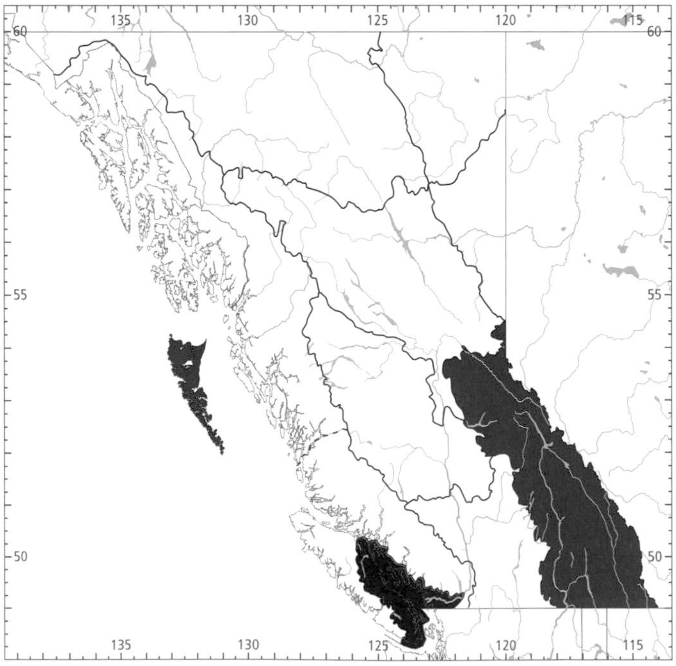

Figure 487. In British Columbia, the highest numbers for the Varied Thrush in summer (red) occur on the Queen Charlotte Islands of the Coast and Mountains Ecoprovince and in the Southern Interior Mountains Ecoprovince; the highest numbers in winter (black) occur in the Georgia Depression Ecoprovince.

Figure 486. During severe winters in southern British Columbia, the Varied Thrush is a familiar bird of parks and gardens where berries and fruits are plentiful (Victoria, 15 January 1982; Mark Nyhof).

Figure 488. Some of the highest numbers of Varied Thrushes in British Columbia can be found in the moss-laden Sitka spruce forests on the Queen Charlotte Islands (Windy Bay, 29 May 1996; R. Wayne Campbell).

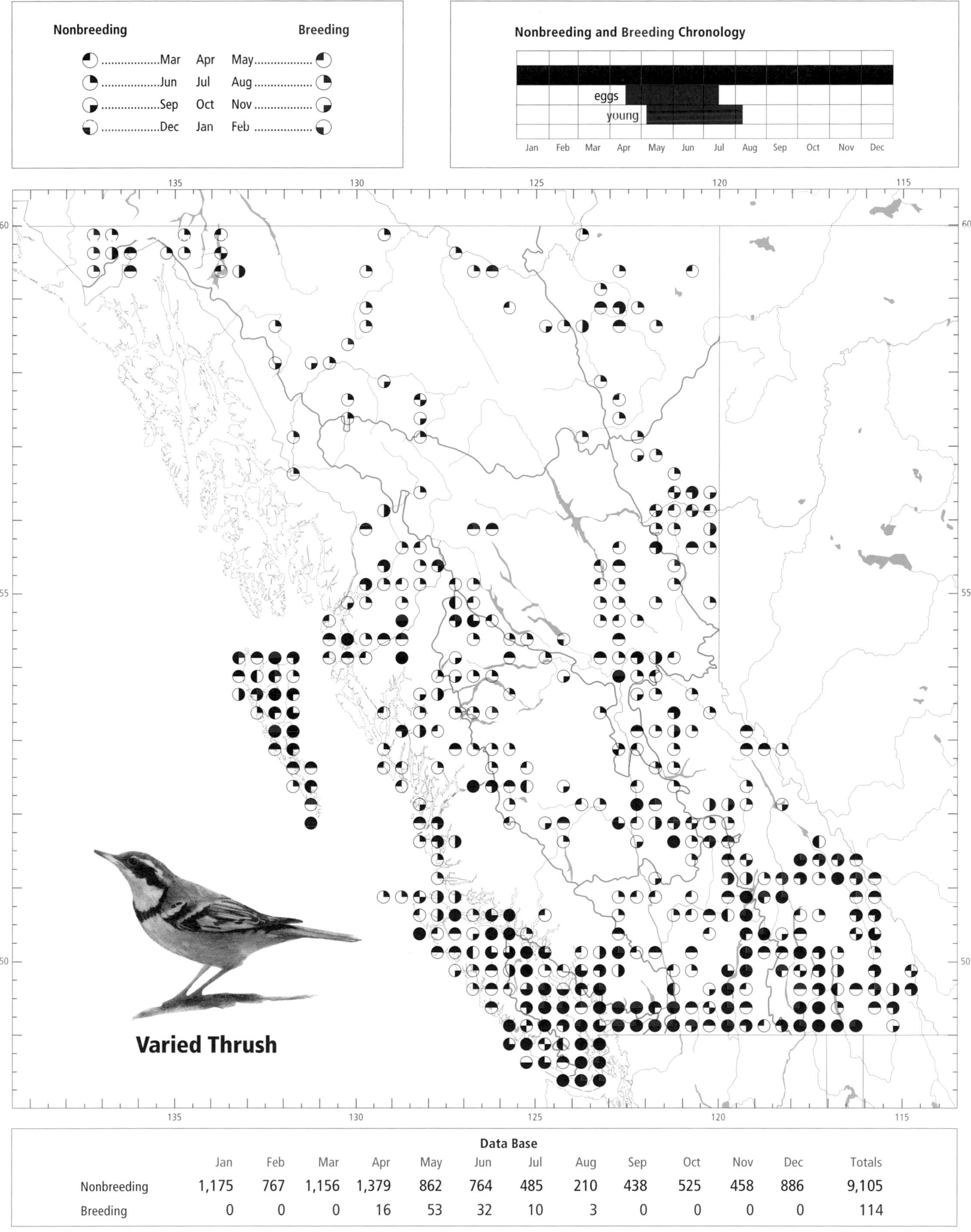

Data Base

	Jan	Feb	Mar	Apr	May	Jun	Jul	Aug	Sep	Oct	Nov	Dec	Totals
Nonbreeding	1,175	767	1,156	1,379	862	764	485	210	438	525	458	886	9,105
Breeding	0	0	0	16	53	32	10	3	0	0	0	0	114

has been reported throughout the year from Victoria to the lower reaches of the Skeena River. In the interior, it can be found throughout the year across southern portions of the province from the Okanagan, Kootenay, and Columbia valleys north to the latitude of Quesnel (53°N). Further north, the Varied Thrush is widespread but sparsely distributed from spring through autumn, occurring in the northwest to the Tatshenshini River valley and in the northeast to Kwokullie Lake.

It reaches its highest numbers in winter in the Georgia Depression (Fig. 487), where over 1,000 birds have been reported several times on Christmas Bird Counts.

In mountainous areas, the Varied Thrush occurs from sea level to near 2,000 m elevation in the south and 1,200 m in the north. It is a species of the moist coniferous forest. Along the coast, it inhabits shady old-growth forests of Douglas-fir, western hemlock, western redcedar, and Sitka spruce (Fig. 488), where moss clothes the ground beneath an understorey of devil's club, false-azalea, red huckleberry, and other shade-tolerant plants. It forages along the edges of streams, estuaries, muskegs, beaver ponds, and other forest openings where increased light encourages berry-bearing plants. On the west coast of Vancouver Island, this thrush uses second-growth forest, but significantly more Varied Thrushes are found in old-growth than in either 30- to 35-year or 50- to 60-year stands (Bryant et al. 1993).

The Varied Thrush also uses other habitats, including red alder groves on logged or burned sites, where dogwood and cascara produce fruit and where leaf litter encourages the presence of terrestrial invertebrates that can be excavated easily. Salal thickets, which often dominate the ground cover along the edges of coastal forests, provide fruit and a more protected habitat in otherwise exposed areas. This thrush will also forage on sand and gravel beaches and along the upper tideline for small invertebrates (Weisbrod 1974; Egger 1979).

The dry shorelines in much of the southern Georgia Depression support stands of arbutus trees that regularly produce a heavy crop of fruit. This is a favoured food source for the Varied Thrush, and supports large overwintering populations in the ecoprovince.

During the winter months, dwindling food supplies often force the Varied Thrush into parks and gardens to feed on the berries of cotoneaster, pyracantha, mountain-ash, and other garden trees and shrubs. When heavy snowfalls occur in the Georgia Depression, the Varied Thrush becomes a frequent visitor to bird feeders.

Birds begin to leave the Georgia Depression in the last 2 weeks of April, and most of the large winter numbers have left the area by the second week of May as they move to nesting areas in the north or to the adjacent mountains (Fig. 490). A noticeable movement begins in the southern Coast and Mountains in the third week of March, and numbers peak near the end of April. In the northern Coast and Mountains, including the Queen Charlotte Islands, early migrants arrive in the first or second week of April, with the peak movement from mid-April into the first week of May (Fig. 490).

In the Southern Interior, the west and east Kootenay of the Southern Interior Mountains, and the Cariboo and Chilcotin portion of the Central Interior, the small number of overwintering birds increases slightly from February through early March as the first northbound migrants enter the province. The main movement arrives about the third week of

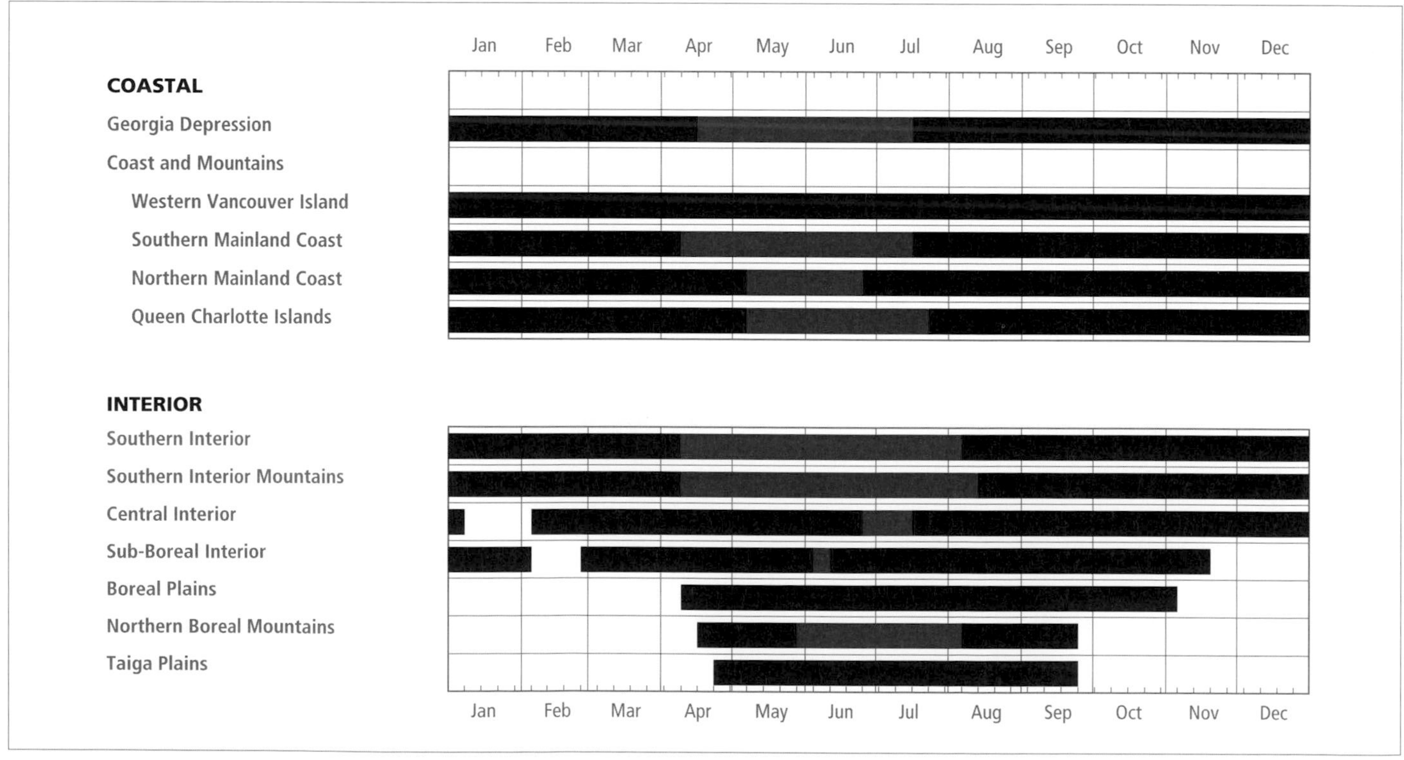

Figure 489. Annual occurrence (black) and breeding chronology (red) for the Varied Thrush in ecoprovinces of British Columbia. Records are shown for the week in which they occurred.

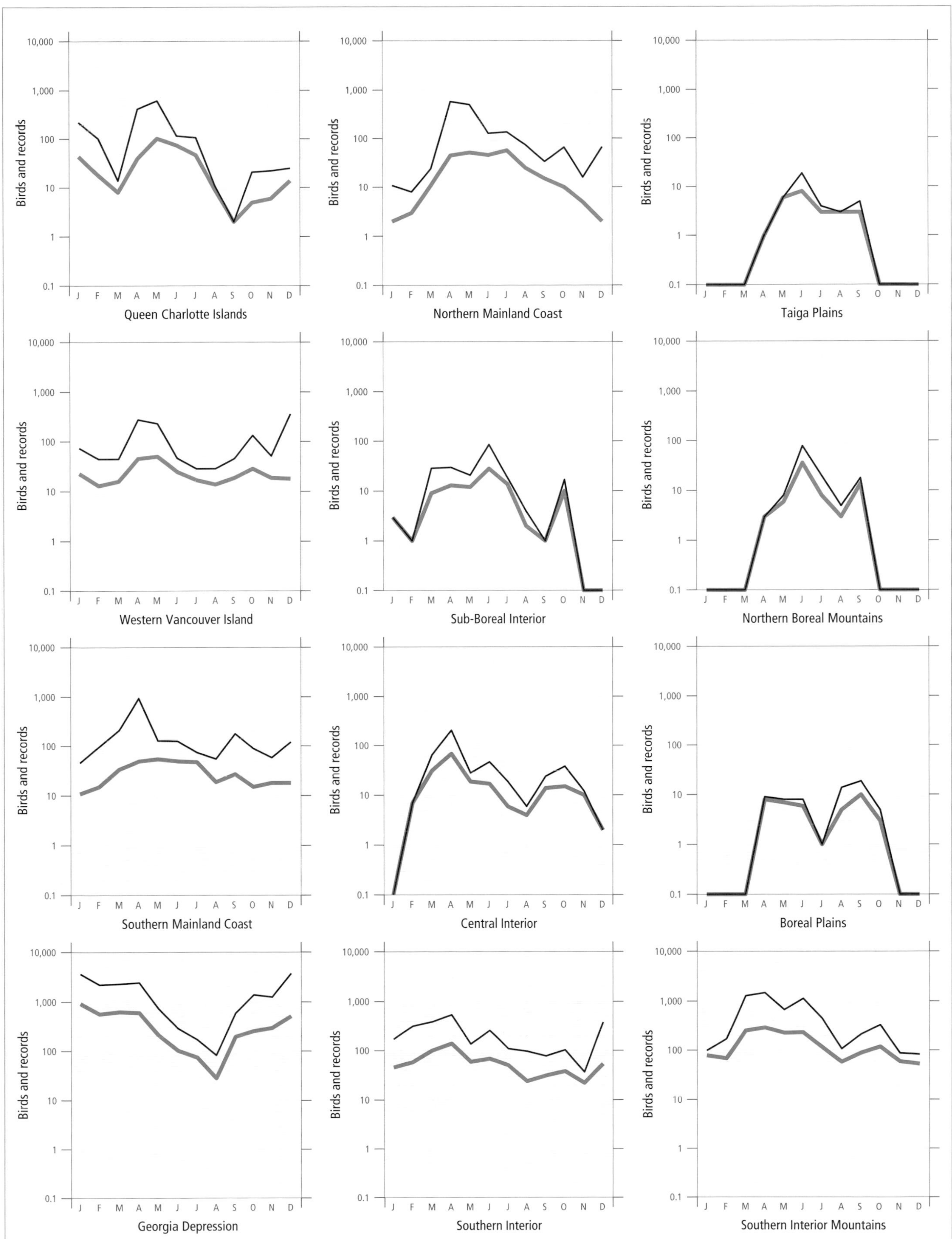

Figure 490. Fluctuations in total number of birds (purple line) and total number of records (green line) for the Varied Thrush in ecoprovinces of British Columbia. Christmas Bird Count and nest record data have been excluded.

March and peaks in mid to late April. Most birds have moved through the valley bottoms to higher elevations or northern areas by the first week of May (Fig. 490).

In the Sub-Boreal Interior, the earliest arrivals reach the Quesnel and Prince George areas in March. In the Boreal Plains and other far northern ecoprovinces, the first spring arrivals are reported in April (Figs. 489 and 490).

In the northern interior, the autumn movement is weak, with small numbers moving through in late August and September; most have left by mid-September. In the Central Interior, a larger autumn movement begins in late August and peaks in early October, with most birds gone by the end of October. In the Southern Interior, numbers are stable from July to October (Fig. 490), and the southward passage is not readily apparent. In the Southern Interior Mountains, migration is more obvious, with numbers peaking about the second week of October.

On the north coast, birds begin leaving in August and continue through early October. Most birds have left by November (Fig. 490). In the Georgia Depression, numbers of Varied Thrush begin to increase about mid-September and grow through December as migrants arrive from northern regions and higher elevations. The interior subspecies has not been reported on the coast (see REMARKS), which suggests that there is little or no migration from the interior to the coast.

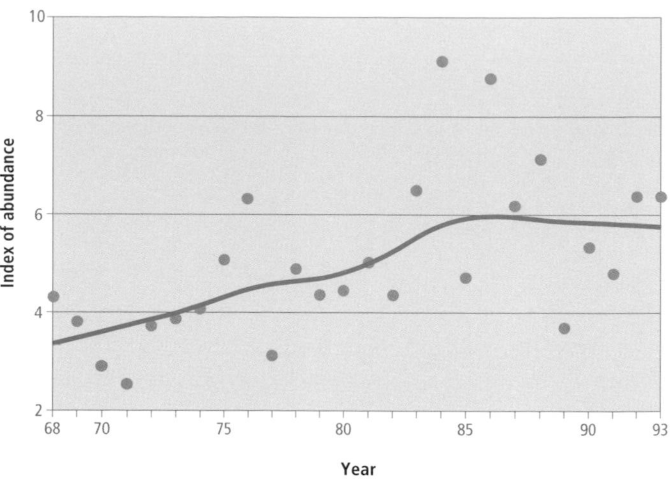

Figure 491. An analysis of Breeding Bird Surveys for the Varied Thrush in British Columbia shows that the number of birds on interior routes has increased at an average annual rate of 4% over the period 1968 through 1993 ($P < 0.1$).

Figure 492. Subalpine breeding habitat for the Varied Thrush in British Columbia consists of dense stands of spruce and subalpine fir (east of Prince George, 17 July 1992; R. Wayne Campbell).

The Varied Thrush has been reported throughout the year from both the coast and southern portions of the interior (Fig. 489).

BREEDING: The Varied Thrush has a widespread nesting distribution throughout southern portions of the province, from the west coast of Vancouver Island east to the Rocky Mountains and north, in the interior at higher elevations, to the latitude of Quesnel (53°N). Further north, nesting reports are scattered and sparse. In northern coastal areas, nesting has been reported from the Queen Charlotte Islands and the Skeena River valley east to Hazelton. It is likely that the Varied Thrush nests in suitable habitat throughout its summer range.

It reaches its highest numbers in summer on the Queen Charlotte Islands and in the Southern Interior Mountains (Fig. 487). An analysis of Breeding Bird Surveys for the period 1968 through 1993 could not detect a net change in numbers on coastal routes; the mean number of birds on interior routes increased at an average annual rate of 4% over the same period (Fig. 491).

On the coast, the Varied Thrush nests from sea level to at least 1,500 m elevation; in the interior, it nests from 420 to 1,860 m elevation (Fig. 492).

Most habitats in which nesting occurred were forested, including coniferous forest (78%; n = 36), mixed forest (11%), and deciduous forest (6%). Two nests were found in suburban locations. Age classes of the forests used for nesting included mature forest (38%; n = 21) and second growth (34%). Brooks (1905) described nesting habitat as dense coniferous forest where moss blanketed the ground and draped branches of the small conifers chosen for the nest sites.

The Varied Thrush has been recorded nesting on the coast from 15 April to 20 July, and in the interior from 17 April to 7 August (Fig. 489).

Nests: Most nests (87%; n = 69) were in trees, including coniferous (46%) and deciduous (35%) trees, followed by shrubs or vines (7%; Fig. 494). One nest was found in a building, another on the ground. In interior Alaska, most nests were placed on the ground among scrubby Sitka alders (Kyllingstad *in* Bent 1949). Nests in trees and shrubs were placed on branches (91%; n = 45), typically close to the trunk or near the top of the nest tree. The heights for 66 nests ranged from ground level to 11.0 m, with 60% between 2 and 4.0 m.

The nest is a bulky cup of grass, moss, and rootlets, on a base of twigs (Fig. 494). It is usually lined with fine grass, although leaves and shredded redcedar bark are also frequent components.

Figure 493. In extreme northwestern British Columbia, the Varied Thrush breeds in dense coniferous forests (Haines Highway, 1 July 1996; R. Wayne Campbell).

Figure 494. Some Varied Thrush nests in southwestern British Columbia are built among vines that hug the trunks of large deciduous trees (Haro Woods, Victoria, 7 May 1993; R. Wayne Campbell).

Eggs: Dates for 58 clutches ranged from 15 April to 15 July, with 52% recorded between 3 May and 5 June. Sizes of 57 clutches ranged from 1 to 5 eggs (1E-4, 2E-3, 3E-35, 4E-14, 5E-1), with 61% having 3 eggs (Fig. 494). The incubation period for 1 nest in British Columbia was 12 days. Ehrlich et al. (1988) cite a tentative incubation period of 14 days. Some late June and July clutches suggest occasional attempts at second broods.

Young: Dates for 37 broods ranged from 5 May to 7 August, with 57% recorded between 11 May and 25 June. Sizes of 28 broods ranged from 1 to 4 young (1Y-8, 2Y-10, 3Y-8, 4Y-2), with 64% having 2 or 3 young. The nestling period for 1 brood in British Columbia was 13 to 15 days.

Brown-headed Cowbird Parasitism: Of 78 nests recorded with eggs or young in British Columbia, 1 was parasitized by the cowbird. This is the first report of parasitism for this species in North America (Friedmann 1963; Friedmann et al. 1977; Friedmann and Kiff 1985).

Nest Success: Of 7 nests found with eggs and followed to a known fate, 2 produced at least 1 fledgling. This species is prone to deserting the nest when disturbed by humans.

REMARKS: Two subspecies are known from British Columbia: *I. n. naevius*, a buff-toned race, occupies the coastal regions; *I. n. meruloides*, a grayer subspecies, occupies the interior (Munro and Cowan 1947; American Ornithologists' Union 1957).

A Varied Thrush banded at Modesto, California, on 4 January 1978 was recovered at Creston in March 1978.

NOTEWORTHY RECORDS

Spring: Coastal – Victoria 12 May 1972-2 nestlings; Klanawa River 23 May 1991-25; Cowichan Bay 17 Mar 1985-150; Bamfield 4 Mar 1976-7; North Surrey 21 Apr 1968-3 eggs; Vedder River near Chilliwack Lake 15 Apr 1895-2 eggs; Pitt Meadows 5 May 1968-50; Coquitlam River 27 Apr 1961-3 eggs; Como Lake 17 May 1977-2; North Vancouver 21 Apr 1974-70; Nicolum River 23 Apr 1962-3 eggs; Comox 8 May 1933-2 new fledglings; Alice Lake 14 May 1961-12; Garibaldi Park 26 Apr 1976-200, 25 May 1963-3 eggs; Port Hardy 10 Apr 1951-30; near Kimsquit 11 Apr 1985-25, 19 Apr 1985-80; Cumshewa Inlet 15 Mar 1976-4; Kitlope Lake 3 May 1991-52; Delkatla Slough 17 Apr 1972-120; 13 May 1972-200; Masset 15 May 1914-3 eggs; Terrace 19 May-3 eggs; Km 80 Haines Highway (Kelsall Lake) 14 May 1979-4. **Interior** – Skaist River (Manning Park) 30 Apr 1978-1 egg; Creston 17 Apr 1988-4 eggs; Crescent Bay (Nakusp) 20 Mar 1983-50; Silver Star (Vernon) 9 May 1981-20; Lumby 18 May 1978-4 eggs; Enderby 17 Apr 1944-3 eggs; Sorrento 27 Mar 1972-51; Celista 7 Apr 1962-15; Adams Lake 19 May 1966-3 eggs; Erie Creek 6 May 1984-40; Williams Lake 14 Mar 1978-25; Quesnel (Fraser River) 14 March 1985-6; Smithers 30 Apr 1985-50; Prince George 1 Mar 1987-10; Beatton Park 9 Apr 1983-1; mouth of Halfway River 25 Apr 1982-1; Fort Nelson 28 Apr 1987-1; Mile 65 Alaska Highway (Pleasant Camp) 14 May 1977-30; Jackfish Creek 12 May 1975-1; Atlin 20 Apr 1981-1, 22 Apr 1934-1 (CAS 42167); Mile 437 Alaska Highway 18 May 1975-1.

Summer: Coastal – Rock Bay 14 Aug 1973-10; Bear Creek Reservoir 3 Jun 1984-25; Harrison 20 Jul 1987-1; Mount Seymour 2 Jul 1980-15; Crescent Lake (Hope) 15 Jul 1964-2 nestlings; w Fanny Bay 15 Jul 1987-3 nestlings at 1,000 to 1,200 m; Adam/Eve River 12 Jul 1983-5; Nootka Sound 2 Aug 1910-8; Whistler Mountain 24 Jun 1924-many nesting; Alta Lake 21 Jul 1949-6; Pine Island 15 Jun 1976-4; Rivers Inlet 8 Jul 1937-1 (NMC 28755); Goose Island 7 Jun 1948-2 (UBC 1937-38); 1 Kemano Bay 3 Jul 1975-12; Hippa Island 3 Jun 1977-6; Wiah Point, 18 Jun 1947-3 eggs; Kieta Island 31 Jul 1977-13; Langara Island 20 Jul 1972-1 new fledgling; Lakelse Lake 16 Aug 1976-21; Terrace 15 May 1977-adult on nest, 19 May 1978-3 eggs, 18 Jun 1978-2 nestlings; Kitsault 17 Jul 1980-27 around upland lakes; Kispiox 22 Jun 1924-new fledglings (Swarth 1924); Anthony Island 1 Jul 1977-3 eggs; Flood Glacier 1 Aug 1919-4 (MVZ 40328-40331); Mile 87.5 Haines Highway (Kelsall Lake) 24 Jun 1980-nestlings; Towagh Creek 7 Jul 1983-11 (Campbell et al. 1983). **Interior** – Manning Park 26 Jun 1983-70, 22 Jul 1977-2 fledged this date, 1 Aug 1985-2 near fledging; Rossland 8 Jul 1971-nestlings; McIntyre Bluff, Vaseux Lake 20 Jun 1966-incubating; Barrett Lake 2 Jun 1983-40; St. Mary River 21 Nov 1945-5 (Johnstone 1949); Creighton Valley 28 Jun 1981-19; Bridge River valley 15 Aug 1985-5; Celista 16 Jul 1950-9; Adams Lake Plateau 20 Aug 1966-12; Balsam Lake 7 Aug 1943-brood ¾ grown; Rogers Pass 16 Jun 1977-2 fledglings; Amiskwi River 1 Jul 1976-30; Emerald Lake 16 May 1975-5 (Wade 1977); Williams Lake 9 Jul 1980-2; Wells Gray Park 8 Jun 1985-4 eggs; e Likely 5 Aug 1991-male feeding nestlings; Quesnel (Mount Milburn) 17 Jun 1987-4 eggs; Prince George 13 Jun 1965-11; Morice River 26 Jun 1975-8; Walcott 15 Jul 1989-2 nestlings; Smokehouse Creek (w Gwillim Lake) 6 Jun 1976-3 eggs; Pine Pass 25 Jun 1963-1 (UBC 11691); Toms Lake 4 Aug 1975-6; Dawson Creek 3 Aug 1975-3 fledglings; Mile 65 Alaska Highway (Pleasant Camp) 19 Jul 1979-2; Cliff Creek 25 Jun 1985-4 eggs; Nuttlude Lake 21 Jul 1963-5; Dease Lake 5 Jun 1962-3 eggs; Steamboat 14 Jun 1976-6; Liard River 6 Aug 1943-1 (NMC 29515), 5 Aug 1944-new fledgling; Liard Hot Springs 27 Jun 1985-5; Clear Creek 5 Jun 1962-3 eggs.

Breeding Bird Surveys: Coastal – Recorded from 25 of 27 routes and on 66% of all surveys. Maxima: Queen Charlotte City 25 Jun 1994-90; Masset 25 Jun 1991-80; Port Renfrew 13 Jun 1992-66. **Interior** – Recorded from 53 of 73 routes and on 48% of all surveys. Maxima: Wingdam 10 Jun 1970-113; Illecilewaet 9 Jun 1983-62; Chilkat Pass 1 Jul 1976-49.

Autumn: Interior – Atlin 9 Sep 1943-1; Fort Nelson 2 Sep 1985-1; Bearskin Lake 21 Sep 1986-3; Cameron River (37 km w Mile 122 along Alaska Highway) 17 Oct 1978-1; Beatton Park 17 Sep 1983-4, 2 Oct 1983-3; Sukunka River 23 Sep 1978-1; Prince George 25 Nov 1991-1; Nazko valley 2 Nov 1986-2; Williams Lake 16 Oct 1971-12; Riske Creek 1 Nov 1984-2; Kleena Kleene 29 Oct 1950-3 (Paul 1959); Glacier National Park 9 Oct 1981-25; Mount Revelstoke 3 Nov 1982-8; Celista 4 Oct 1964-16. **Coastal** – Terrace 24 Nov 1986-6; Masset 6 Oct 1921-8; Kemano 1 Oct 1990-30; Port Neville 16 Oct 1975-15; Bamfield 16 Nov 1979-15; Mount Seymour 3 Oct 1970-200.

Winter: Interior – Smithers 2 Jan 1983-1 (Pojar 1983); Quesnel 22 Dec 1992-11, Jan 1993-1; Anahim Lake 13 Dec 1989-1; Williams Lake 27 Dec 1988-1; Riske Creek 9 Feb 1987-1; Tatlayoko valley 9 Feb 1991-1; Celista 1 Jan 1948-8; Eagle Pass 16 Dec 1983-9; Revelstoke 1 Jan 1986-1; Okanagan Landing 13 Feb 1936-100; Castlegar 2 Jan 1975-4; Creston 22 Feb 1971-77. **Coastal** – Terrace 30 Dec 1986-25; Masset 17 Dec 1982-26, 11 Feb 1985-25; Cape St. James 1 Jan 1982-4; Kitimat 18 Dec 1983-64; Kaien Island 3 Jan 1982-7; Siwash Island 1 Feb 1972-9; Squamish River estuary 3 Jan 1980-3; Harrison 9 Jan 1991-32; Stubbs Island 18 Dec 1983-125; North Vancouver 31 Dec 1978-10; Vancouver 1 Jan 1975-1; Tofino 1 Jan 1987-10; Lennard Island 28 Dec 1979-2; Bamfield 25 Feb 1979-12; Jordan River 13 Jan 1985-12; Sooke-Jordan River 23 Feb 1986-300.

Christmas Bird Counts: Interior – Recorded from 17 of 27 localities and on 37% of all counts. Maxima: Revelstoke 21 Dec 1987-120; Nakusp 27 Dec 1987-58; Penticton 27 Dec 1983-29. **Coastal** – Recorded from 31 of 33 localities and on 89% of all counts. Maxima: Victoria 27 Dec 1966-**1,243**, all-time Canadian high count (Anderson 1976b); Vancouver 18 Dec 1983-1,212; Duncan 2 Jan 1993-581.

Gray Catbird

GRCA

Dumetella carolinensis (Linnaeus)

RANGE: Breeds from the southern mainland of British Columbia, the southern half of the Canadian Prairie provinces, southern Ontario, southwestern Quebec, New Brunswick, Prince Edward Island, and Nova Scotia south through Washington, eastern Oregon, Utah, and eastern Texas, east to the Atlantic coast. Winters mainly on the coastal fringe of the Gulf coast and eastern Mexico, south to Panama and the Caribbean islands.

STATUS: On the coast, *rare* to *uncommon* migrant and summer visitant locally to the lower Fraser River valley, Squamish River valley, Pemberton valley, and Bella Coola River valley; *casual* elsewhere on the coast.

In the interior, an *uncommon* migrant and summer visitant to the Southern Interior, Southern Interior Mountains, and Central Interior ecoprovinces, except locally *fairly common* in the Okanagan valley and in the vicinity of Williams Lake; further north, a *casual* summer visitant in the Sub-Boreal Interior and the Boreal Plains ecoprovinces; *casual* in winter in the Southern Interior, *accidental* in the Southern Interior Mountains.

Breeds.

Figure 495. Adult Gray Catbirds feeding insects to 3 nestlings (near Osoyoos, June 1981; Mark Nyhof).

NONBREEDING: The Gray Catbird (Fig. 495) is sparsely but widely distributed across the interior of southern British Columbia north to Quesnel and Golden, and from the southern Rocky Mountains west to Riske Creek in the Cariboo and Chilcotin areas and casually in the Fraser Lowland on the south coast. In addition, there are several occurrence records from outside the known breeding range, including Sechelt, Comox, Bella Coola, Kitimat, Hazelton, Kleena Kleene, Bowron Lake, and Fort St. John.

The Gray Catbird has been recorded from sea level to 150 m on the coast and between 270 and 2,050 m in the interior. It is most abundant in riparian thickets (Fig. 496) or in

Figure 496. Gray Catbird habitat in the west Kootenay region consists of riparian thickets of willow, red-osier dogwood, saskatoon, and wild rose (Penny Sidings Road, Slocan Valley, 14 June 1996; Neil K. Dawe).

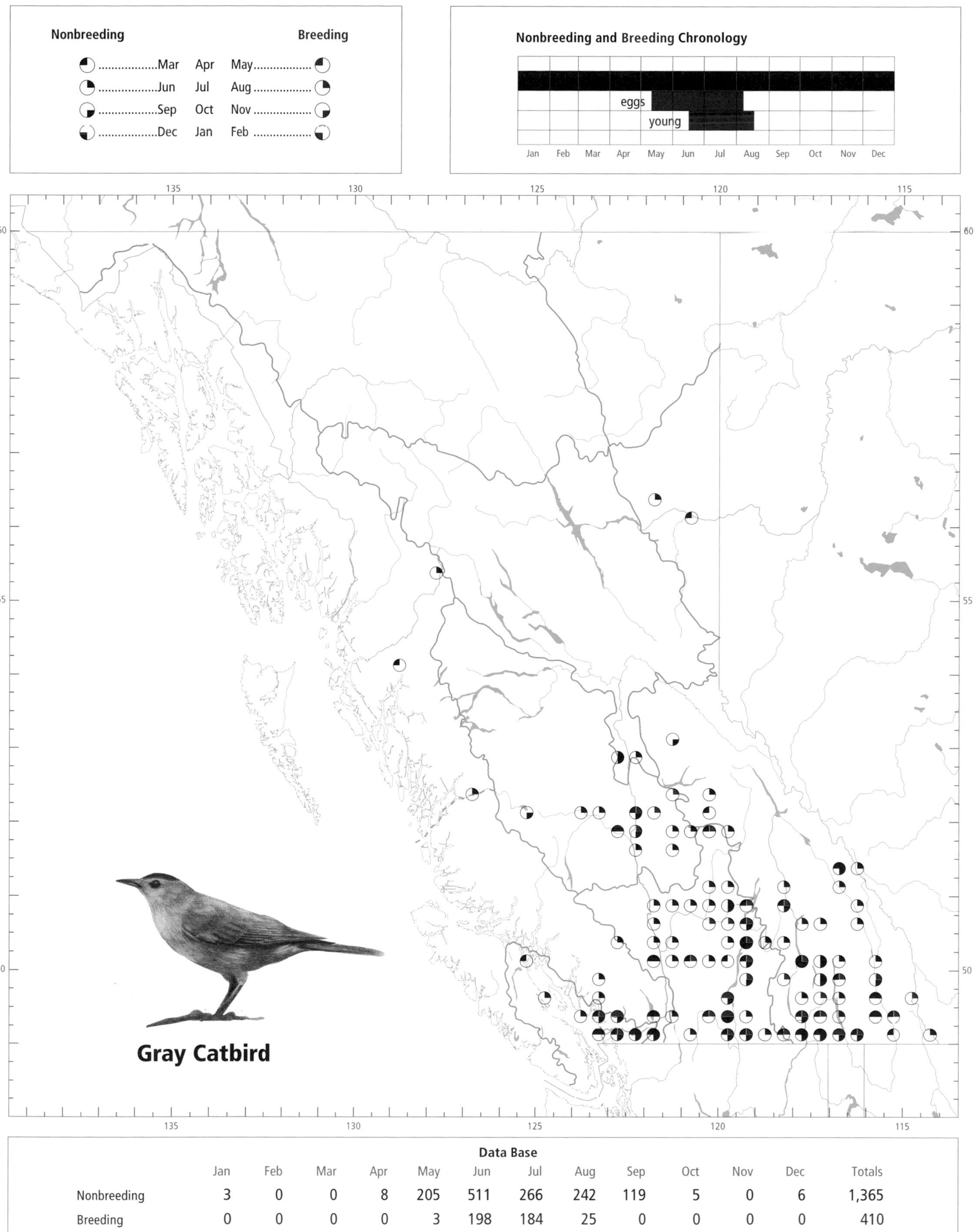

Data Base

	Jan	Feb	Mar	Apr	May	Jun	Jul	Aug	Sep	Oct	Nov	Dec	Totals
Nonbreeding	3	0	0	8	205	511	266	242	119	5	0	6	1,365
Breeding	0	0	0	0	3	198	184	25	0	0	0	0	410

open deciduous forests with an understorey of dense deciduous shrubs, small trees, and scattered conifers. Preferred habitats are thickets along the edges of streams, lakes, and ponds, fencelines, roadsides, ditches and woodland edges, and shrub-choked gullies on sagebrush-grassland slopes. Closed forest habitats are avoided. Vegetation in preferred habitat often includes clustered wild rose, red-osier dogwood, saskatoon, willows, water birch, black hawthorn, choke cherry, trembling aspen, and Oregon-grape. Human-influenced habitats include shrubby gardens and vine-tangled hedgerows along orchards and farm fields, often with Himalayan blackberry on the coast. Gray Catbirds rarely venture far from dense cover. They are heard more often than seen, and are one of the few songbirds in the province that frequently sing at night.

Detection of nonsinging birds is difficult; thus, birds may be present, but silent, before territorial activities are detected. Apparent spring migration begins relatively late, normally from mid to late May, although in some years early migrants arrive in mid to late April (Figs. 497 and 498). In the Okanagan valley, 28 of 37 first-of-year records occurred between 21 May and 1 June, with a mean and median date of 23 May (Cannings et al. 1987). In the south, migration peaks from late May to early June and ends abruptly (Fig. 498). In the Williams Lake area, at the northern edge of its range, the earliest Gray Catbird arrivals are in mid-May, but few are seen until June. The autumn migration occurs mainly in late August and early September. Information on autumn departures is scarce, but by mid-September few catbirds are reported (Fig. 498). In some years, small numbers may remain into early October in southern areas. Migratory movements are inconspicuous, as catbirds are secretive. From evidence of large numbers of catbirds killed by colliding with tall structures, such as lighthouses and tall buildings, Root (1988) concluded that it is a night migrant.

The only winter occurrences are individuals at Nakusp, Irish Creek near Vernon, and Vaseux Lake.

The Gray Catbird has been recorded regularly in British Columbia from 10 April to 15 October (Fig. 497).

BREEDING: The Gray Catbird is a distinctive member of the summer avifauna in the warm southern valleys of the interior of British Columbia. Its highest breeding densities are in the Okanagan, Similkameen, and South Thompson river valleys; it is slightly less abundant in the valleys of the Kootenay and Columbia rivers. In the interior, nesting has been confirmed north to Wasa, Revelstoke, Horsefly Lake, and Williams Lake, and west to the Chilcotin River in the vicinity of Riske Creek. The Gray Catbird also regularly breeds in the riparian thickets along the Fraser River between Lytton and Williams Lake, and in the valley of the San Jose River south of Williams Lake. It is scarce in summer elsewhere in the central portions of the interior, where breeding likely occurs north to Quesnel.

On the coast, small numbers breed in the Fraser Lowland, mainly in the Pitt Meadows area. The Gray Catbird was reported as common in summer at Stuie in the Bella Coola River valley (Laing 1942), although breeding there has not been confirmed.

The species reaches its highest numbers in summer in the Okanagan valley of the Southern Interior (Fig. 499). Breeding Bird Surveys for both coastal and interior routes contain insufficient data for analysis.

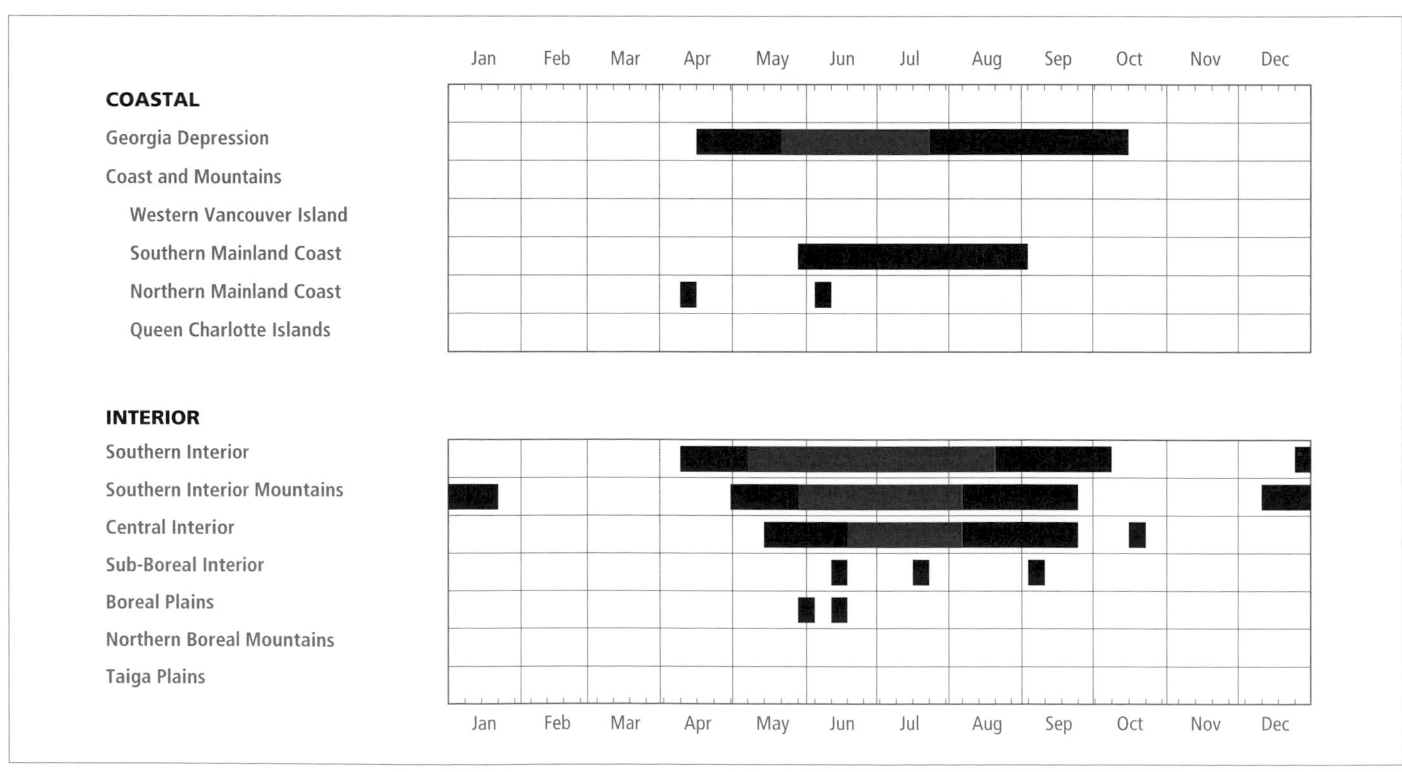

Figure 497. Annual occurrence (black) and breeding chronology (red) for the Gray Catbird in ecoprovinces of British Columbia. Records are shown for the week in which they occurred.

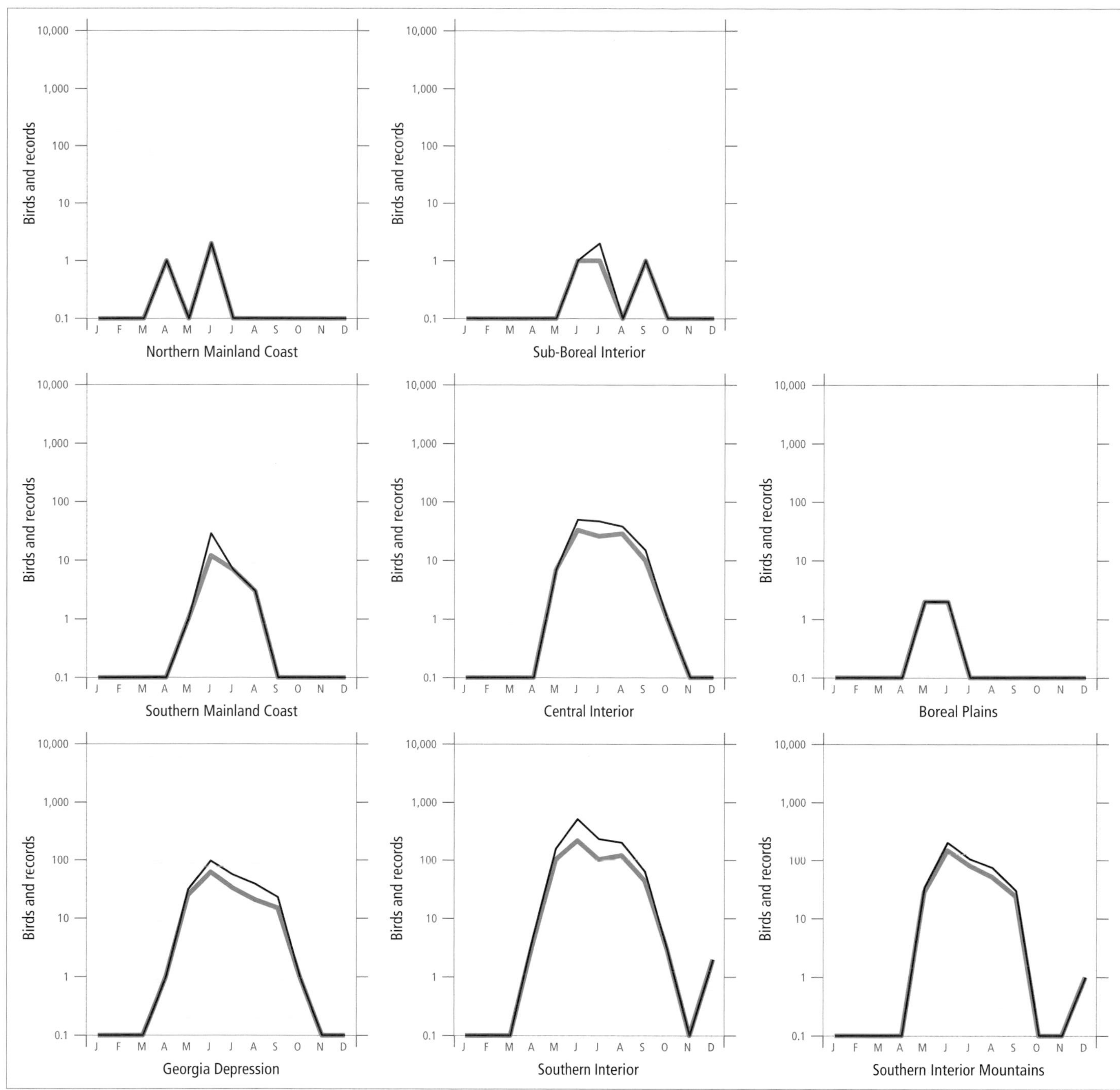

Figure 498. Fluctuations in total number of birds (purple line) and total number of records (green line) for the Gray Catbird in ecoprovinces of British Columbia. Christmas Bird Count and nest record data have been excluded.

Breeding habitat is similar to nonbreeding habitat, but thickets including wild rose are preferred. General habitats for reported nests were natural shrub thickets (80%; Fig. 496) and human-influenced habitats (20%), mainly gardens, orchards, vineyards, hedgerows, and uncultivated land associated with farms. In the Okanagan valley, natural sites are frequently in moist draws and riparian areas. Cannings et al. (1987) state that only 2 of 93 nests were not in deciduous shrubs or small trees; they were in dense garden firs.

The Gray Catbird breeds at elevations below 100 m in the lower Fraser River valley and up to 1,100 m in the interior.

Populations are small and local throughout British Columbia, with pairs widely scattered in suitable habitat. Surveys have revealed relatively dense concentrations (5 to 10 pairs) at the north end of Okanagan Lake, the west side of Vaseux Lake, the north end of Osoyoos Lake, and the east end of Williams Lake.

The Gray Catbird has been recorded breeding in British Columbia between 9 May and 16 August (Fig. 497).

Nests: Nests are usually built in dense tangles of vegetation (Figs. 495 and 500). Specific nest sites (n = 139) included wild rose (47%), lilac bushes (25%), tall Oregon-grape (7%), red-osier dogwood (4%), willows (4%), and 20 other species of shrubs and small trees. The nest is a bulky assemblage of coarse material. The nest cup is lined with fine grass, plant fibres, and other soft items (Fig. 501). The most consistently used nest materials were grass (84%), twigs (44%), plant fibres (29%), rootlets (21%), and leaves (9%); a wide assortment of other materials appeared in fewer nests. The heights above ground for 156 nests ranged from 0.2 to 8 m, with 53% between 1 and 2 m.

Eggs: Dates for 137 clutches ranged from 10 May to 6 August, with 55% recorded between 12 June and 2 July. Calculated dates indicate that nests may contain eggs on 9 May. Egg laying begins about a month later in the Cariboo and Chilcotin areas than in more southern areas (Fig. 497). Some pairs, in the Okanagan valley at least, are double-brooded. The laying of first clutches peaks in early June, second clutches about 9 July (Cannings et al. 1987). It is not known what proportion of second clutches followed the loss of a first clutch rather than a successful first brood. In an Ontario study, only 3 of 22 pairs that successfully raised 1 brood attempted a second, but pairs that lost the first clutch routinely attempted a second nest (Scott et al. 1988). The same study revealed a gradual decline in mean clutch size as the season progressed.

Sizes of 122 clutches ranged from 1 to 6 eggs (1E-3, 2E-7, 3E-22, 4E-57, 5E-32, 6E-1), with 73% having 4 or 5 eggs (Fig. 501). The mean clutch size in British Columbia was 3.8 eggs (n = 124 clutches visited 2 or more times), which was almost identical to the average number of 3.9 eggs per clutch laid in Ontario between 13 May and 2 June (n = 53) (Scott et al. 1988). Both Canadian studies revealed larger clutch sizes than those recorded from Michigan, which averaged 3.3 eggs (n = 22 clutches of 2 or more). The primary difference was in

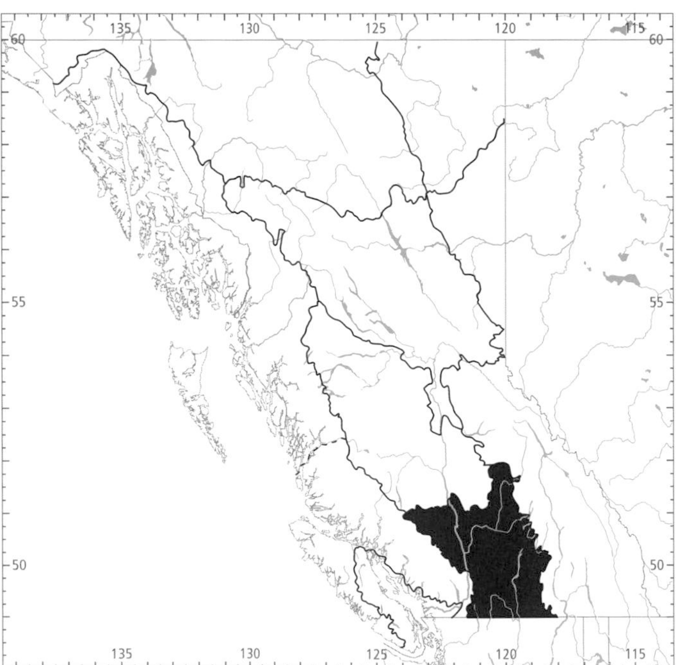

Figure 499. In British Columbia, the highest numbers for the Gray Catbird in summer occur in the Southern Interior Ecoprovince.

the large component of 5-egg clutches in British Columbia (25%), whereas no clutches over 4 eggs were found in Michigan (Zimmerman 1963).

The incubation period in British Columbia was 12 to 15 days (n = 11).

Young: Dates for 108 broods ranged from 15 June to 16 August, with 54% recorded between 1 and 20 July. Sizes of 94 broods ranged from 1 to 5 young (1Y-4, 2Y-12, 3Y-18, 4Y-45, 5Y-15), with 67% having 3 or 4 young. The nestling period in British Columbia was 12 to 15 days (n = 10). Family groups remained together for a short time after the young left the nest.

Brown-headed Cowbird Parasitism: In British Columbia, 5% of 183 nests found with eggs or young were parasitized by the cowbird. None of 8 coastal nests was parasitized. Most of the parasitism occurred in the Southern Interior, but the overall rate of 6% was higher than the 4% reported for the Okanagan valley (Cannings et al. 1987). In southern Ontario, Peck and James (1987) reported a parasitism rate of 1.5% (n = 1,193), although they were careful to mention that the percentage of attempted parasitism was undoubtedly higher.

Three parasitized nests were closely observed in British Columbia. In each the cowbird egg(s) disappeared from the nest within 2 days. A catbird from 1 nest that had received 2 cowbird eggs was seen carrying a cowbird egg in its beak, and both eggs were missing from the nest a day after they were deposited. The Gray Catbird is known to routinely eject cowbird eggs within a day after they are laid (Nickell 1958; Friedmann 1963; Rothstein 1975; Scott 1977). The latter author suggested that the attentiveness of the Gray Catbird, and its vigorous defence of the nest, made it difficult for cowbirds to lay their eggs and remove 1 or more eggs of the host. In British Columbia, there is a record of a cowbird nestling together with a catbird nestling, but there are no records in British Columbia of the catbird fledging a cowbird. Elsewhere there are a few cases of successful parasitism (Woodward 1976; Lowther 1980).

Nest Success: Of 59 nests found with eggs and followed to a known fate, 37 produced at least 1 young, for a success rate of 63%. The success rate in Michigan was 61% (n = 23) (Zimmerman 1963). In British Columbia, predation on eggs or young was suspected in 17 cases, although confirmation was lacking. Reported predators included squirrel or chipmunk (11 nests), domestic cat (3 nests), shrew, gopher snake, and American Crow (1 nest each). Heavy rain was thought to have caused the death of 1 brood.

REMARKS: The Gray Catbird was formerly known as Catbird.

Belles-Isles and Picman (1986c) report that the Gray Catbird destroyed a variety of eggs of other species placed near the nest.

For additional information on breeding biology and on distinguishing age and sex, see Bent (1948), Nickell (1965), Johnson and Best (1980), and Suthers and Suthers (1990).

Figure 500. The Gray Catbird builds its nest in dense tangles of vegetation (Pitt Meadows, August 1995; R. Wayne Campbell). Note nest with eggs in lower left centre.

Figure 501. Nearly three-quarters of all Gray Catbird clutches found in British Columbia had 4 or 5 eggs (Pitt Meadows, 31 May 1959; R. Wayne Campbell).

NOTEWORTHY RECORDS

Spring: Coastal – Sardis 4 May 1904-1 (RBCM 1952); Queen Elizabeth Park (Vancouver) 8 to 14 May 1985-1; Pitt Meadows 8 May 1964-1, 31 May 1959-nest with 4 eggs (Fig. 501); Harrison Hot Springs 31 May 1986-1; Quadra Island 17 Apr 1977-1; Kitimat April 1958-1 (Hay 1976). **Interior** – Grand Forks 28 Apr 1984-1; Creston 5 May 1928-1 (CAS 32073); Horn Creek (Okanagan) 10 May 1915-2 eggs, earliest date; Wolfe Lake (Hedley) 10 Apr 1974-2; McCuddy Creek 29 Apr 1962-1 (Cannings et al. 1987); Penticton 15 May 1898-4 eggs; Naramata 26 May 1969-building nest; Lytton 4 May 1964-1 (NMC 52527); Vernon 21 Apr 1963-1 (Cannings et al. 1987); Tappen 11 May 1971-2; Revelstoke 18 May 1971-1; Williams Lake 18 May 1965-1, 18 May 1985-1, first spring arrival; Fort St. John 31 May 1990-1.

Summer: Coastal – Vancouver 19 Jun 1945-nest with 4 eggs; Coquitlam River 12 Jun 1933-5 eggs (UBC 1256; Munro and Cowan 1947); Pitt Meadows 1 Jun 1989-1, 6 Jun 1971-7, high count (Campbell et al. 1972b), 6 Jul 1972-4 nestlings, 22 Jul 1972-4 young, 15 Aug 1970-3 young with 2 adults (Campbell et al. 1972a); Pitt River 5 Jul 1972-4 fledglings; Harrison Hot Springs 19 Jun 1978-2; Comox 20 Jun to 3 Jul 1947-1 (Pearse 1948); Squamish 22 Jun 1975-1; Brackendale 19 Jun 1916-1 (NMC 1962); Pemberton Meadows 9 Jun 1985-8, 30 Aug 1984-1; Bella Coola 4 Jul 1940-1 (UMMZ 161390). **Interior** – Spotted Lake (Richter Pass) 23 Jun 1957-3 eggs plus 1 Brown-headed Cowbird egg (Campbell and Meugens 1971); Bridesville 29 Jun 1973-5; Sirdar 3 Jun 1981-5 eggs, 17 Jun 1981-5 nestlings; Vaseux Lake 4 Jun 1976-15 on survey, 6 Jun 1974-14 on survey, 20 Aug 1994-10 (banding); White Lake (Okanagan Falls) 12 Jun 1963-5; Hedley 14 Jun 1950-5 eggs (UBC 1255); Jaffray 22 Jun 1968-6; Penticton 15 Jun 1969-2 nestlings, 16 Aug 1969-3 young in nest; Nelson 18 Jun 1991-5 eggs, 8 Jul 1972-4; Summerland 1 Jun 1928-5 eggs, 10 Jun 1969-6 eggs, 21 Jul 1974-6 (Cannings 1974); Naramata 20 Jun 1970-3 nestlings, 16 Aug 1969-4 nestlings; Bummers Flats 7 Jun 1991-1 carrying food; Ta Ta Creek 26 Jul 1970-3 nestlings; Sparwood 22 Jul 1983-3 nestlings; Apex Mountain 6 Jun 1983-4; Skookumchuck 6 Jul 1976-4 eggs; Edgewood 11 Jun 1977-5; Balfour to Waneta 8 Jun 1983-11 on survey; Fauquier 29 Jul 1988-3 young; Lytton 12 Jul 1963-9; n end Okanagan Lake 2 Jul 1979-18 on survey; Brouse 4 Jun 1982-4; Wilmer 18 Aug 1977-5; Chase 17 Jun 1973-4; Golden 2 Jul 1976-5 on survey; Hemp Creek 14 Aug 1959-1 (RBCM 11009); Robson 7 Jul 1960-4 eggs; 150 Mile House 4 Jul 1959-4 eggs; Chilcotin River 14 Aug 1991-adults feeding young; s end Williams Lake 21 Jun 1970-4 eggs, 8 Jul 1949-12; Williams Lake 5 Jul 1972-7; Horsefly Lake 22 Jun 1970-4 eggs (RBCM 1892); Dragon Lake (Quesnel) 17 Jul 1948-2, 13 Jun 1986-2; Hazelton 10 Jun 1921-1 (MVZ 42560); Hope Springs Ranch (Upper Cache Creek, Peace River district) 14 Jun 1959-1 (Symons 1967).

Breeding Bird Surveys: Coastal – Recorded from 5 of 27 routes and on 4% of all surveys. Maxima: Pemberton 2 Jul 1976-7; Albion 4 Jul 1970-2; Pitt Meadows 6 Jul 1976-1; Point Grey 16 Jun 1974-1; Seabird 6 Jun 1971-1. **Interior** – Recorded from 32 of 73 routes and on 30% of all surveys. Maxima: Golden 23 Jun 1974-12; Grand Forks 24 Jun 1973-10; Salmon Arm 12 Jun 1975-8.

Autumn: Interior – Pass Creek (Robson) 8 Sep 1968-2; Bowron Lake Park 1 Sep 1971-1; Dragon Lake 6 Sep 1986-1; Williams Lake 3 Sep 1980-4, 19 Sep 1980-2; Chilcotin River (Wineglass Ranch) 5 Sep 1993; Hooch Lake 15 Oct 1987-1; Golden 19 Sep 1976-1; Enderby 16 Sep 1941-1 (UBC 7124); Lavington 23 Sep 1979-1; Okanagan Landing 25 Sep 1906-1, last autumn departure (Cannings et al. 1987); Brouse 2 Sep 1990-3, 20 Sep 1981-1; Okanagan Falls 5 Oct 1962-1; Vaseux Lake 4 Oct 1977-1, last autumn departure (Cannings et al. 1987). **Coastal** – Pemberton-Meadows 30 Aug 1984-1; Agassiz 6 Sep 1980-1; Pitt Meadows 20 Sep 1971-1, last autumn departure (Campbell et al. 1972b); Vancouver 10 Oct 1926-1 (RBCM 7179); Sumas 2 Sep 1923-6, 25 Sep 1921-1; Huntingdon 3 Sep 1947-1 (UBC 7429).

Winter: Interior – Nakusp 16 Dec 1981 to 15 Jan 1982-1; Irish Creek (Vernon) 26 Dec 1987-1; McIntyre Bluff (Okanagan River) 28 Dec 1987-1. **Coastal** – No records.

Christmas Bird Counts: Interior – Recorded from 2 of 27 localities and on less than 1% of all counts. Maxima: Nakusp 3 Jan 1982-1; Vaseux Lake 29 Dec 1987-1. **Coastal** – Not recorded.

Northern Mockingbird

NOMO

Mimus polyglottos (Linnaeus)

RANGE: Occurs in Canada sporadically throughout the year, from southern British Columbia east across extreme southern Canada to Nova Scotia. Resident in the United States primarily from southeastern Oregon and northern California east to southern New England and New York, and south throughout the contiguous United States. Northern populations are partially migratory. Also occurs in Mexico, the Bahamas, and Greater Antilles.

STATUS: On the coast, *rare* visitant locally throughout the year in the Georgia Depression Ecoprovince; *very rare* on Western Vancouver Island, *casual* on the southern and northern mainland of the Coast and Mountains Ecoprovince; *accidental* on the Queen Charlotte Islands.

In the interior, *rare* summer visitant locally, mainly in valleys of the Southern Interior and Southern Interior Mountains ecoprovinces, *very rare* there at other times of the year; further north, *casual* in summer and *accidental* in autumn in the vicinity of Fort St. John in the Boreal Plains Ecoprovince.

Breeds.

CHANGE IN STATUS: In the early 1900s, the Northern Mockingbird (Fig. 502) was abundant in southern California, but its range did not extend north of the San Joaquin Valley. By the 1930s it was slowly moving northward along the coast (Grinnell and Miller 1944). Ingles (1939) provided some details on the range expansion, and noted that the mockingbird first appeared in a region in winter before becoming established as a breeding species.

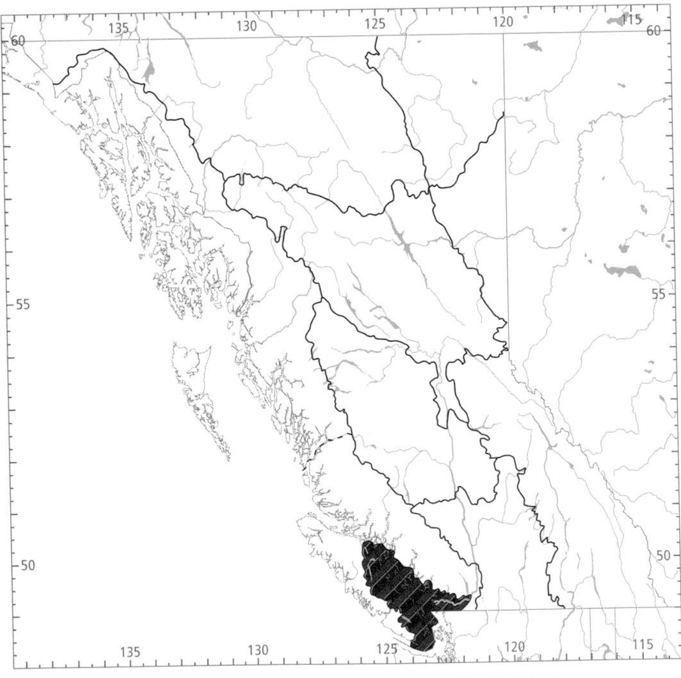

Figure 503. In British Columbia, the highest numbers for the Northern Mockingbird in winter (black) and summer (red) occur in the Georgia Depression Ecoprovince.

Figure 502. Since its first appearance in 1931, the Northern Mockingbird has become widely distributed in British Columbia and now breeds (Tim Zurowski).

The Northern Mockingbird was first observed in British Columbia at Port Alberni in 1931 (Racey 1933). A year later it was found nesting at Port Townsend, Washington (Jewett et al. 1953). The first 5 occurrences in British Columbia were coastal, and all were from southern Vancouver Island (Racey 1933; Brooks 1942a; Irving 1953).

Between 1941 and 1960, 8 occurrences of single birds were reported in British Columbia. Beginning in 1964, its frequency of occurrence increased. There have been 1 or more records every year since then, with 51 separate occurrences from 1964 to 1979, increasing to 128 records of apparently different individuals through 1990.

During the 65 years since its first appearance in the province, nearly 70% of all reports have been from along the coast. They range from Vancouver Island and the adjacent southwest mainland coast east to Langley and north to McInnes Island and the Queen Charlotte Islands.

Since its first record in the interior, in 1951 (Jobin 1952), the Northern Mockingbird has become more widely distributed and has been recorded there every month of the year.

The first nesting attempt was in 1967, and the first successful nesting was recorded in 1993. The species is now a fairly regular but scarce member of the provincial avifauna.

NONBREEDING: The Northern Mockingbird has a widespread, though spotty, distribution in British Columbia. It has been observed along the coast, including Vancouver Island and the Queen Charlotte Islands, and eastward across the southern portions of the interior to Fernie in the east Kootenay, north to Dease Lake in the northwest and Fort St. John in the northeast. It has been recorded from 68 different localities in every ecoprovince except the Taiga Plains. About 57% of all reports are from southeastern Vancouver Island, the Gulf Islands, and the adjacent mainland coast. Another 11% are from the coast north of Vancouver Island, 13% are from the Southern Interior, 9% are from the east and west Kootenays, and 5% are from other regions of the province.

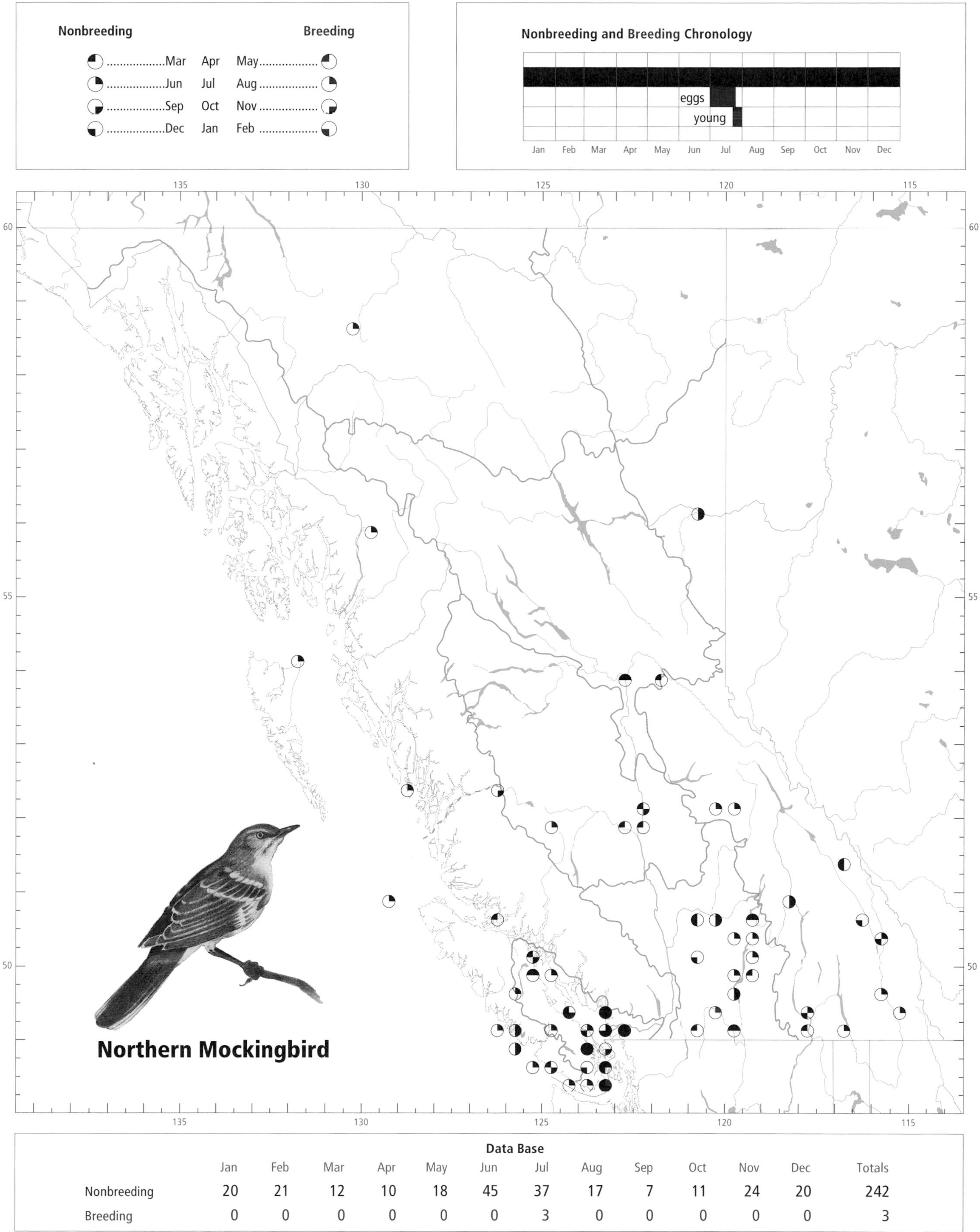

Data Base

	Jan	Feb	Mar	Apr	May	Jun	Jul	Aug	Sep	Oct	Nov	Dec	Totals
Nonbreeding	20	21	12	10	18	45	37	17	7	11	24	20	242
Breeding	0	0	0	0	0	0	3	0	0	0	0	0	3

Figure 504. On southeastern Vancouver Island, the Northern Mockingbird has been found in brushy and shrubby areas bordering woodlands of Garry oak (Langford, 11 May 1994; R. Wayne Campbell).

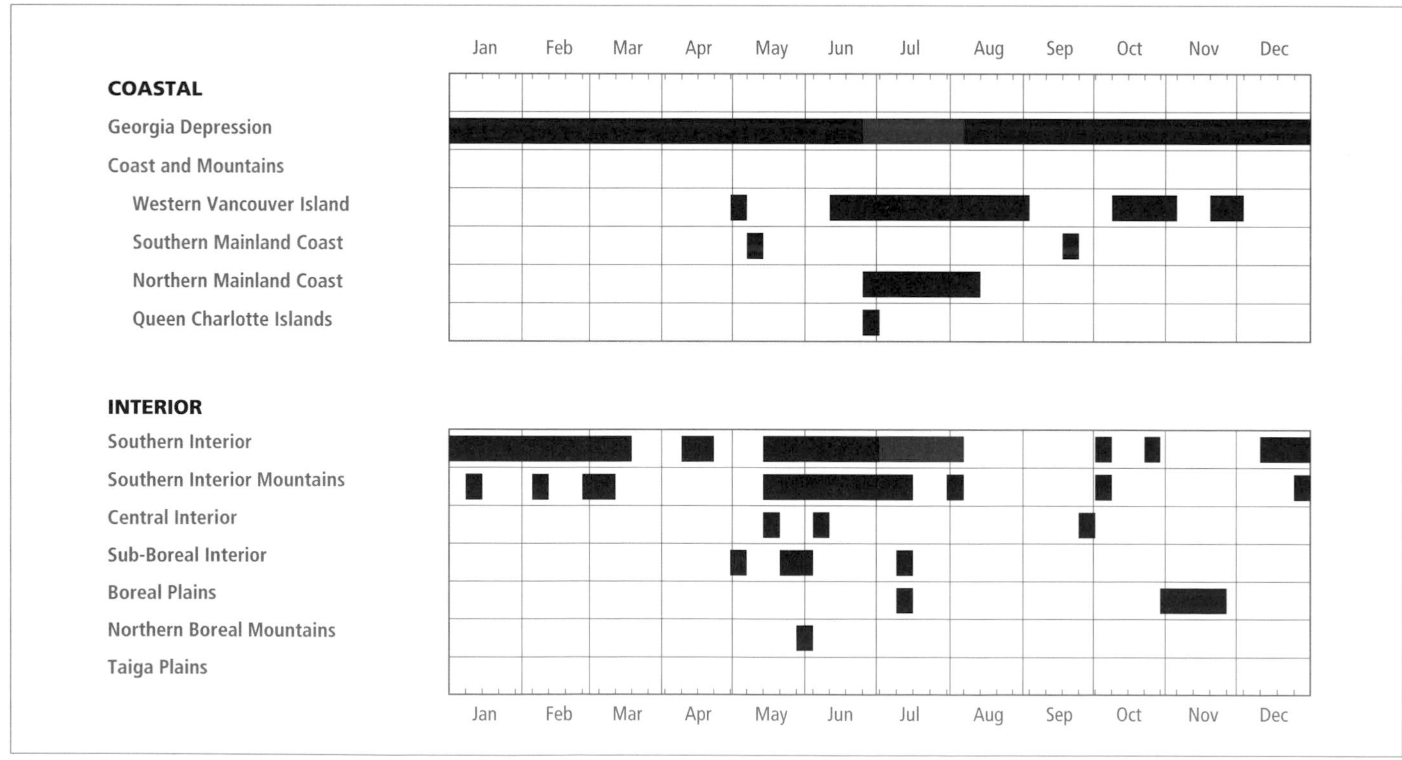

Figure 505. Annual occurrence (black) and breeding chronology (red) for the Northern Mockingbird in ecoprovinces of British Columbia. Records are shown for the week in which they occurred.

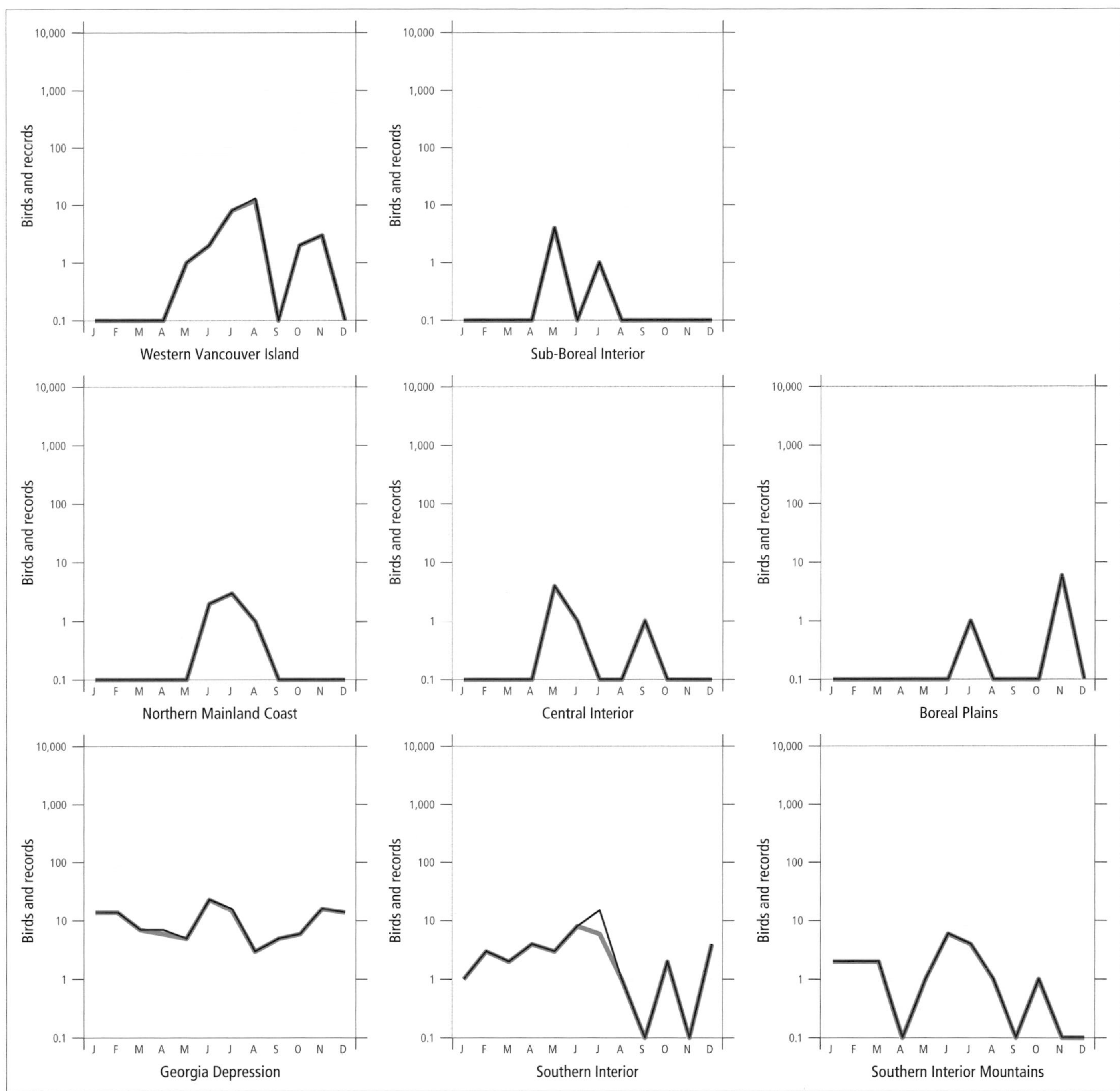

Figure 506. Fluctuations in total number of birds (purple line) and total number of records (green line) for the Northern Mockingbird in ecoprovinces of British Columbia. Christmas Bird Count and nest record data have been excluded.

The Northern Mockingbird reaches its highest numbers in winter in the Fraser Lowland and Nanaimo Lowland of the Georgia Depression (Fig. 503). Root (1988) notes that it winters in areas that have an average minimum January temperature no lower than 20°F (–7°C). In the interior of British Columbia, the few wintering birds have occurred at low elevations in the Okanagan valley. There the mean January temperature has ranged from –2°C to –7°C (Cannings et al. 1987), and the mean minimum temperatures are substantially colder.

Habitats used by the Northern Mockingbird in British Columbia are generally associated with cultivated lands or with suburban gardens and parks with shrubs, trees, and open spaces (Fig. 504). It also occurs in the riparian willow thickets of swampy areas, clearings and edges in forested areas, logging slash, coastal beaches, and garbage dumps.

Seasonal occurrence differs between the coast and the interior (Figs. 505 and 506). Along the coast, 36% of occurrences have been from May through August and 30% from December through February. In the interior, 68% of occurrences have been from May through August and only 12% from December through February.

Migration is difficult to detect in a species of erratic occurrence. On the coast, the number of observations declines from January to April, but increases sharply in May and continues to increase in June before declining in July (Fig. 506). In the interior, a spring migration in May is suggested by a large increase in the number of observations in that month. In the Cariboo and Chilcotin areas and in the vicinity of Prince George, the only spring records are in May (Figs. 505 and 506). In the interior of the province, there are many records of the Northern Mockingbird in July but only a few for August and September; this suggests that many of the summering birds leave the province in late summer and early autumn. On the coast, the same decline in numbers occurs in August, but then numbers rise steadily up to December and January (Fig. 506). Numbers then decline steadily to reach a low point from March through May. It is not possible to say whether this decline indicates a steady mortality among the wintering Northern Mockingbirds or a departure for other areas.

The Northern Mockingbird has been recorded in every month of the year both on the coast and in the interior (Fig. 505).

Figure 507. Breeding site at Ten Mile Point, Victoria, British Columbia, where the Northern Mockingbird nested unsuccessfully in 1967 (R. Wayne Campbell).

BREEDING: The Northern Mockingbird has nested at only 2 locations in the province. The first nesting occurred on southern Vancouver Island, at Ten Mile Point (Fig. 507). A mockingbird was observed there between 27 June and 27 July 1967. On 7 July, a bird was found on a nest incubating 4 eggs, and was photographed (Lemon 1967, 1968) (BC Photos 80 and 98; Fig. 508). The single bird present was presumed to be the female (see Derrickson and Breitwisch 1992). The nest was watched for 18 days, and on 25 July the eggs were "candled" and found to be infertile; both the nest and clutch were collected. A second adult was not seen at any time.

The nest was among shrubs in a suburban garden, well treed with the remnants of the original forest of Garry oak, arbutus, Douglas-fir, and lodgepole pine, and a mixed shrubbery dominated by ocean-spray and mock-orange (Fig. 507).

The second nesting was discovered beside Highway 5A about 3 km north of Princeton (MacKenzie et al. 1995). An empty nest was found there on 31 July 1993 with 3 recently fledged young nearby (Fig. 509). It was situated in dry ranchland habitat with scattered ponderosa and lodgepole pines, saskatoon thickets, grassland, and some lawn.

The Northern Mockingbird reaches its highest numbers in summer in the Georgia Depression Ecoprovince (Fig. 503).

The 2 breeding records in British Columbia span the period 28 June (calculated) to 31 July (Fig. 505).

Nests: The Victoria nest was a relatively small structure and weighed only 60.8 g (dry weight). It was composed almost entirely of small pieces of dead vegetation. Small twigs of Douglas-fir made up most of the nest, along with bits of cotoneaster, ocean-spray, horsetail, cudweed, grass, and a lining of grass, moss, and bark (Campbell 1973). The nest was placed 1.5 m from the ground in a cotoneaster shrub.

Eggs: A nest with 4 eggs found on 7 July is the only record of a clutch size for British Columbia. It is not known when the clutch was initiated, but from the behaviour of the adult female (see Derrickson and Breitwisch 1992), it is likely that

Figure 508. Incubating Northern Mockingbird (Ten Mile Point, 25 July 1967; Ralph Fryer). This was the first nesting attempt for this species in British Columbia.

eggs were laid during the last week of June. The incubation period is 12 to 13 days (Graber et al. 1970).

Young: Three recently fledged but flightless young, found on 31 July, represent the only record of successful nesting in British Columbia (MacKenzie et al. 1995) (Fig. 509). The nestling period is 11 to 13 days (Derrickson and Breitwisch 1992).

REMARKS: Two subspecies of the Northern Mockingbird are described for Canada (American Ornithologists' Union 1957): *M. p. polyglottos* inhabits the eastern region from Nova Scotia west to Manitoba, and *M. p. leucopterus* occurs from Saskatchewan to British Columbia. The first specimen taken in British Columbia, at Duncan in 1940, was the eastern subspecies (Brooks 1942a). To our knowledge, no other British Columbia specimens have been examined for subspecific identity, but they are presumed to be of the western race.

There are reports of Northern Mockingbirds either being intentionally released in the province (e.g., 2 released in Esquimalt, Vancouver Island, in late April 1955 [Campbell 1973]) or having escaped captivity. When known, these records have been excluded from the account.

Figure 509. This Northern Mockingbird fledgling, still in the saskatoon containing its nest, represents the first successful nesting of this species in British Columbia (3 km north of Princeton, 1 August 1993; R. Jerry Herzig).

See Laskey (1962), Adkisson (1966), Safina and Utter (1989), and Derrickson and Breitwisch (1992) for additional information on the breeding biology and natural history of the Northern Mockingbird in North America.

NOTEWORTHY RECORDS

Spring: Coastal – Colwood 2 May 1975-1; Carmanah Point 5 May 1950-1 (Irving 1953); Cleland Island 22 to 30 May 1969-1 (Hatler et al. 1978); Point Grey (Vancouver) 2 Mar 1968-1; Parksville 30 Mar 1974-1; Campbell River 4 Apr 1973-2; Port Neville 13 May 1987-1. **Interior** – Manning Park 27 May 1981-1 at lodge (BC Photo 774); between Tadanac and Trail 11 Mar 1975-1; Oliver 15 May 1993-1; Mission Creek 15 Mar 1989-1; Kelowna 21 Apr 1966-1 (Rogers 1966c); Tranquille 1 Mar to 10 Apr 1983-1; Salmon Arm 28 May to 21 Jun 1972-1 (Anderson 1972); Yoho National Park 16 May 1974-1; Springhouse 18 May 1951-1 (NMC 48108; Jobin 1952); Riske Creek (Wineglass Ranch) 16 May 1992-1, 14 May 1986-1; Prince George 6 May 1986-1; Bowron River (s Aleza Lake) 26 May 1968-1 (Rogers 1968c; BC Photo 206).

Summer: Coastal – Sooke 8 Jul 1955-2, believed to be birds that escaped their cage 6 weeks previously (Campbell 1973); River Jordan 29 Jun 1983-1; Victoria 27 Jun to 27 Jul 1967-1 adult, 25 Jul 1967-nest with 4 eggs (RBCM E992; Lemon 1967, 1968; BC Photo 98; Fig. 508); Saanich 2 Jun 1985-1; Carmanah Point Aug 1948-1 (Irving 1953); Ucluelet 17 Jun 1986-1; Westham Island 27 Sep 1977-1; Cleland Island 1 Jul 1977-1 (BC Photo 502); Sea Island 30 Aug 1980-1; Port Alberni 7 Jun 1931-1 (Racey 1933), 25 Jul 1992-1 (Siddle 1993a); Miracle Beach Park 8 Jun 1959 (Stirling 1960b); Mitlenatch Island 1 Aug 1976-1 (Butler and Butler 1976); McInnes Island 28 Jun to 11 Aug 1964-1; Triangle Island 19 and 20 Jul 1986-1 (BC Photo 1117), 22 Aug 1976-2 (Vermeer et al. 1976); North Beach (Masset) 25 Jun 1992-1 (Siddle 1993d). **Interior** – Chopaka 17 Jun 1989-1 calling; Creston 5 Jun 1994-1; Oliver 4 Jul 1991-1; 1.5 km n Princeton 31 Jul 1993-3 fledglings and nest; Fernie 4 and 5 Jul 1991-1 calling all night (Campbell and Dawe 1991); Summerland 2 to 15 Jul 1958-1 (BC Photo 321); Naramata 26 Jun 1966-1; Kimberley 3 Aug 1991-1 (Siddle 1992a); Monte Lake 18 Jul 1977-1; Kamloops 24 Jun 1990-1 (Siddle 1990c), 15 to 20 Jun 1983-1; Salmon Arm 1 Jun 1972-1 (BC Photo 329); Revelstoke 9 to 10 Jun 1990-1 (Siddle 1990c); Clearwater Lake 4 Jun 1983-1; Murtle Lake 9 Jul 1959-1 (NMC 48894; Edwards and Ritcey 1967); Goat River 30 Jun 1974-1; Kleena Kleene 9 Jun 1964-1 (Paul 1964a); 80 km e Prince George 15 Jul 1968-1 (Rogers 1968d); Baldonnel to Two Rivers 13 Jul 1985-1; Dease Lake 1 to 5 Jun 1990-1 photographed.

Breeding Bird Surveys: Coastal – Not recorded. **Interior** – Not recorded.

Autumn: Interior – Fort St. John 1 Nov 1982-1 (Grunberg 1983a); Taylor 23 Nov 1985-1 (BC Photo 1076); Williams Lake 24 Sep 1990-1; Revelstoke 4 Oct 1992-1 (Siddle 1993a), 4 Nov 1982-1; near Knutsford 4 to 6 Oct 1992-1; Summerland 24 Oct 1992-1 (Siddle 1993a). **Coastal** – 30 km e Bella Coola 17 Sep 1976-1; Tofino Airport 11 Oct 1984-1, 28 Nov 1985-1; Pacific Rim National Park 20 Oct 1992-1 (Siddle 1993a); Vancouver 12 and 15 Nov 1969-1 (Crowell and Nehls 1970a; Campbell 1970a); Richmond 1 Sep 1979-1; Duncan 30 Nov 1975-1; South Pender Island 9 Sep 1971-1 (Tatum 1972); Carmanah Point 2 Nov 1950-1 (Irving 1953).

Winter: Interior – Golden 10 Feb 1993-1 (Siddle 1993b); Lake Windermere 8 Jan 1983-1 (Campbell 1983; BC Photo 877), 27 Feb 1983-1 (Rogers 1983); Westsyde 16 Dec 1982 to 19 Mar 1983-1; Tranquille 16 Dec 1982 to 28 Feb 1983-1, 27 Feb 1989-1; Kamloops 5 Dec 1990-1; Quilchena 29 Dec 1981-1, 29 Dec 1986 to 5 Jan 1987-1. **Coastal** – Nanaimo 27 Dec 1981-1 (BC Photo 746), 26 Jan 1982-1; West Vancouver 1 Jan 1987-1; Vancouver 26 Feb to 22 Mar 1968-2 (Campbell and Anderson 1968; BC Photo 6); Sea Island 1 Jan 1978-1; Fry's Corner (Surrey) 31 Dec 1976-1; Duncan 20 Jan 1940 (MVZ 103512; Ashby 1940); Cowichan Lake Jan 1940-1; Esquimalt Lagoon 3 Dec 1975-1; Victoria 12 Dec 1972 (Tatum 1973); Metchosin 5 Feb 1975-1.

Christmas Bird Counts: Interior – Not recorded. **Coastal** – Recorded from 4 of 33 localities and on less than 1% of all counts. Maxima: Vancouver 21 December 1986-1; Ladner 29 December 1979-1; Nanaimo 27 December 1981-1; Duncan 16 December 1972-1.

Sage Thrasher

Oreoscoptes montanus (Townsend)

SATH

RANGE: Breeds from south-central British Columbia, central Idaho, south-central Montana, southwestern Alberta, southwestern Saskatchewan, northern and southeastern Wyoming, and Colorado south through eastern Washington, eastern Oregon, east-central California, southern Nevada, southern Utah, northeastern Arizona, northern New Mexico, northern Texas, western Oklahoma, and southwestern Kansas. Winters from central California, southern Nevada, northern Arizona, New Mexico, and central Texas south to southern Baja California. Also occurs in mainland Mexico.

Figure 510. In British Columbia, the Sage Thrasher breeds only in the southern portion of the Southern Interior Ecoprovince (Chopaka, Similkameen valley, 1 July 1981; Richard J. Cannings).

STATUS: On the coast, *casual* in the Georgia Depression Ecoprovince; *accidental* on Western Vancouver Island in the Coast and Mountains Ecoprovince.

In the interior, locally *rare* in sagebrush communities in the southern Similkameen and Okanagan valleys in the Southern Interior Ecoprovince; *casual* elsewhere in the Southern Interior Mountains and Central Interior ecoprovinces.

Breeds.

CHANGE IN STATUS: In the early 1900s, the Sage Thrasher (Fig. 510) was considered a scarce local breeder in the Osoyoos district. The earliest record for the province was of an adult male and nest with 4 eggs collected in the summer of 1909 in the Osoyoos district by C. de B. Green (Brooks 1909b). This locality reference was repeated by Brooks and Swarth (1925), and shortened to Osoyoos by Munro and Cowan (1947).

The bird's status did not change through the mid-1920s, and the distribution is stated as Osoyoos, White Lake, and Keremeos in the Okanagan and Similkameen valleys (Brooks and Swarth 1925). Cannings (1992a) suggests that the pre-1920 spring populations in British Columbia "may have been as high as 30 pairs." By the mid-1940s the thrasher was considered an "irregular and local summer visitant ... in the Osoyoos-Arid Biotic Area; casual in the Dry Forest Biotic Area" (Munro and Cowan 1947). All occurrences were still from the Okanagan and Similkameen valleys. Even acknowledging the erratic fluctuations in populations of the Sage Thrasher (Reynolds 1981), numbers in British Columbia appear to have decreased since 1947, and it is now regarded as

Figure 511. Sage Thrasher breeding habitat in British Columbia includes open country with big sagebrush (*Artemisia tridentata*), antelope-brush (*Purshia tridentata*), and rabbit-brush (*Chrysothamnus nauseosus*), with forbs and a ground cover of perennial grasses (White Lake, Okanagan valley, 18 June 1993; R. Wayne Campbell).

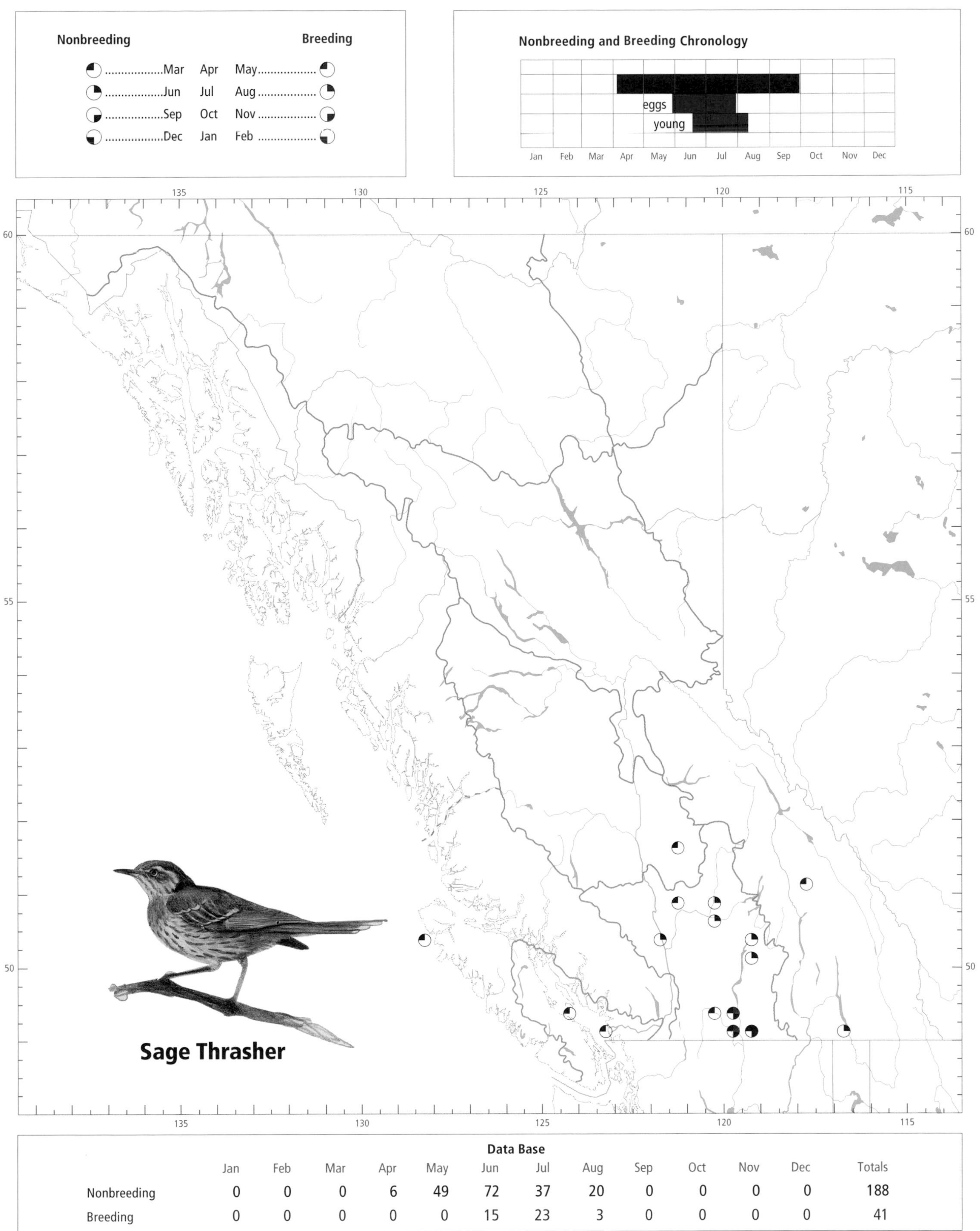

Data Base

	Jan	Feb	Mar	Apr	May	Jun	Jul	Aug	Sep	Oct	Nov	Dec	Totals
Nonbreeding	0	0	0	6	49	72	37	20	0	0	0	0	188
Breeding	0	0	0	0	0	15	23	3	0	0	0	0	41

an endangered species. No more than 5 to 10 pairs are estimated to have bred in the province in any one year since 1980 (Cannings 1992a), and some of the areas formerly occupied have not been used for many years. For example, Cannings (1992a) states that the "highest count at White Lake in the past 30 years was five pairs in 1969, and apparently none has nested there since 1980, although a male was present for some time in 1991." In 1990, the Chopaka, Richter Pass, Kilpoola Lake, and White Lake areas were thoroughly searched for 6 days but only 4 birds were found (Preston 1990). A year later D.H. Harvey (pers. comm.) found 11 Sage Thrashers singing in the same areas. However, the presence of a singing male early in the breeding season, even in traditional nesting habitat, cannot be taken as evidence of the presence of a nesting pair (see REMARKS).

NONBREEDING: In British Columbia, the Sage Thrasher is at the northern limit of its range. Numbers fluctuate from year to year, and in some years it may not be present at all. Since its discovery in 1909, the Sage Thrasher has never had a wide distribution in British Columbia. It occurs regularly only in the Okanagan and Similkameen valleys of the Southern Interior. Historically, records have extended from Chopaka and Osoyoos, near the international boundary with Washington, north to Vernon at the northern end of the Okanagan valley. Most recent records in the province are from the White Lake area (Fig. 511) and near Chopaka.

Outside the Southern Interior, there are 6 records of vagrant Sage Thrashers, all between 1969 and 1990. These are from Iona Island (Richmond), Little Qualicum River, the northwest coast of Vancouver Island, Buffalo Creek near 100 Mile House (the most northerly occurrence), Duck Lake northwest of Creston, and Mount Revelstoke National Park. Five of these records are from spring, and were probably deviant migrants. In 1982, the year of the occurrence in the Fraser River delta, there were spring observations of Sage Thrashers on the coast of Oregon at Yaquina Bay and the coast of Washington at Cape Flattery (Mattocks and Hunn 1982b).

The Sage Thrasher has been recorded from near sea level to 950 m elevation. It is a species of the shrub-steppe habitats characteristic of the Great Basin of western North America. This habitat (Fig. 511), with its often extensive stands of big sagebrush, antelope-brush, and rabbit-brush, extends north up the Similkameen and Okanagan valleys of southern British Columbia. Braun et al. (1976) consider the Sage Thrasher as a sagebrush obligate, being almost entirely dependent on mature big sagebrush environments. In British Columbia, this bird is seldom seen in any other habitat. Individuals that have strayed far from their normal environment have been seen feeding on lawns with American Robins, or hunting sand-hoppers (amphipods) along the tideline of a marine beach.

Migrants arrive in the Okanagan valley as early as the first week of April (Fig. 512), but more usually in May or even early June. Few records for the province are earlier than 1 June. In autumn most leave the province in late August and early September. All birds have gone by the end of September (Fig. 512).

On the coast, the Sage Thrasher has been recorded sporadically from 8 to 18 May; in the interior, it has been found fairly regularly from 4 April to 29 September (Fig. 512).

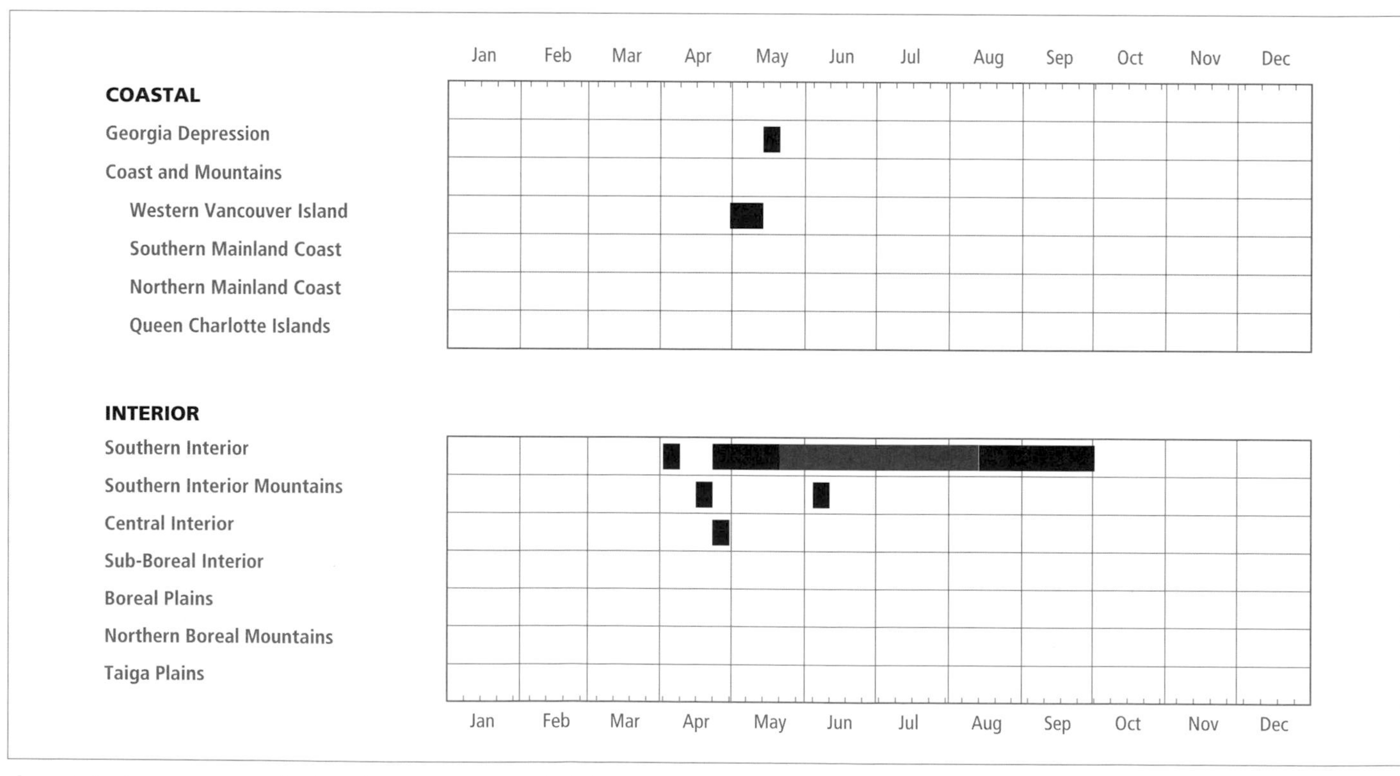

Figure 512. Annual occurrence (black) and breeding chronology (red) for the Sage Thrasher in ecoprovinces of British Columbia. Records are shown for the week in which they occurred.

BREEDING: All confirmed nesting records for British Columbia are from the arid sagebrush basins of the southern Okanagan and Similkameen valleys, from the vicinity of Chopaka, Osoyoos, Kilpoola Lake, Richter Pass, and White Lake (Fig. 511). Darcus (1932) reported nesting at Penticton and Oliver between 1928 and 1931, and Brooks and Swarth (1925) recorded breeding at Keremeos, but specific details are lacking. The northernmost nesting report in the province, from Okanagan Landing (Brooks 1925), was based on the observation of nesting behaviour; the nest was not found, nor were young subsequently seen. The only other potential nesting area is west of Cache Creek, where an old nest was found in big sagebrush in July 1990 (see Fig. 515).

During the nesting season, the Sage Thrasher reaches its highest numbers in the extreme southern portion of the Southern Interior (Fig. 513). Most recent nests have been found in the vicinity of Chopaka, near the British Columbia–Washington border (Fig. 514), and Kilpoola Lake in Richter Pass. Breeding Bird Surveys for both coastal and interior routes contain insufficient data for analysis.

This thrasher breeds between 300 and 500 m elevation. Breeding habitat is generally similar to nonbreeding habitat, and is located in areas of rangeland characterized by low annual rainfall and a long history of overgrazing. Large and small areas dominated by the big sagebrush community are preferred. Habitats selected for territory placement include very large sagebrush bushes with high vegetation density. Although the big sagebrush near nest sites may be less clumped than at other sites, the total percentage of cover is highest near the nest. The mean density of the canopy is estimated to be 70% (R. Millikin pers. comm.).

The Sage Thrasher has been recorded breeding in British Columbia from 28 May (calculated) to 11 August (Fig. 512).

Figure 514. Large and full big sagebrush bushes with high vegetation density provide optimum nesting sites for the Sage Thrasher in British Columbia (Chopaka, 18 June 1993; R. Wayne Campbell).

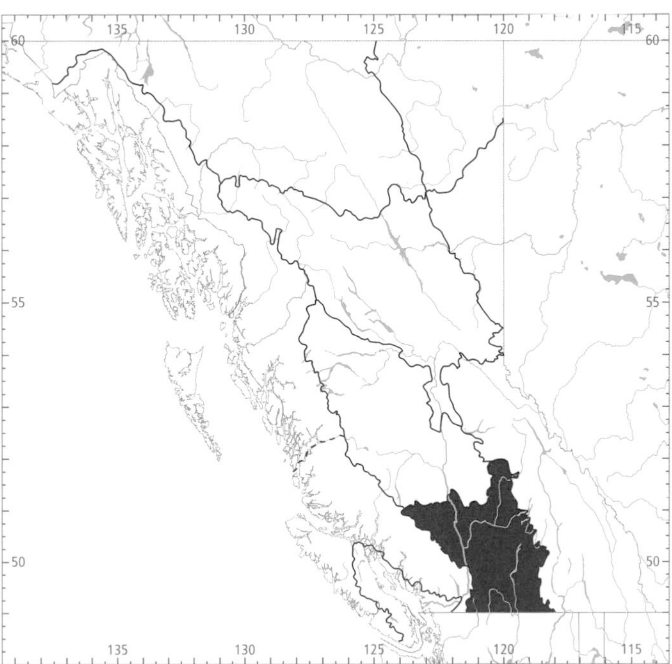

Figure 513. In British Columbia, the highest numbers for the Sage Thrasher in summer occur in the Southern Interior Ecoprovince.

Nests: All but 3 nests (n = 24) were found in a big sagebrush shrub. The 3 nests were in a peach tree, a saskatoon bush, and a hawthorn (Fig. 516).

Sagebrush selected by the thrashers for nest sites is larger than the surrounding shrubs and larger than randomly selected bushes on census transects (R. Millikin pers. comm.). Big sagebrush used as nest sites measured as follows: total height, 132 ± 32 cm; height of crown, 113.6 ± 37 cm; width of crown, 168 ± 58 cm.

The heights for 19 nests ranged from 8 to 154 cm, with 55% at 36 cm. None of the nests reported in British Columbia was on the ground. In 1993 and 1994, nests found in the vicinity of Chopaka and White Lake were at a minimum of 26 cm from the ground and a minimum of 53 cm from the top of the crown. The minimum horizontal distance from the nest to the perimeter of the sage bush was 7 cm, but the average was 27.3 cm. The average distance of the nest from the main stem of the nest bush was 37.7 cm (R. Millikin pers. comm.).

Studies of nest habitat in Idaho reveal no significant difference from the habitat selected in British Columbia

(Reynolds and Rich 1978; Peterson and Best 1991). There, 72% of nests were in big sagebrush shrubs over 70 cm tall, whereas this height class represented only 7% of the shrubs in the habitat. The amount of vegetation above, shading the nest, appeared to be important, as the distance between the top of the nest and the top of the bush was much the same (66 ± 16 cm) whether the nest was on the ground or elevated (Reynolds and Rich 1978).

Nests were bulky structures of coarse twigs, rootlets, bark strips, and plant stems; they were lined with fine grasses, horse hair, and fine rootlets (Fig. 517).

Eggs: Dates for 21 clutches ranged from 1 June to 28 July, with 65% between 15 and 22 June. Calculated dates indicate that nests could have eggs as early as 29 May. Clutch size ranged from 1 to 5 eggs (1E-1, 2E-3, 3E-2, 4E-10, 5E-5), with 71% having 4 or 5 eggs (Fig. 517). In a British Columbia nest, the minimum incubation period was 14 days (R. Millikin pers. comm.). Incubation period was 14 to 17 days, with a mean of 15 days (n = 4) (Reynolds and Rich 1978; Reynolds 1981). In Washington, the earliest nest with a complete clutch of eggs was 4 April, and the mean egg-laying date was 2 June (Bowles 1921).

Young: Dates for 15 nests with young ranged from 18 June to 11 August, with 56% between 7 and 15 July. Brood size ranged from 1 to 5 young (1Y-1, 2Y-3, 3Y-7, 4Y-3, 5Y-1), with 87% having 2 to 4 young. A British Columbia nest yielded a nestling period of 10 days (R. Millikin pers. comm.). The nestling period ranges from 11 to 14 days (Ehrlich et al. 1987). Cannings et al. (1987) suggest that the Sage Thrasher can probably raise 2 broods per season. In southern Idaho, Reynolds (1981) reports that the nesting season was divided into 2 periods: 18 first nests hatched between 4 May and 22 May, while 4 second nests hatched between 12 June and 14 July. None hatched between 22 May and 12 June, a period of about 3 weeks. However, the interval between the hatching of the earliest nests in each group was 39 days and between the latest of each was 53 days. No distinction was made between renesting after a failed first attempt and renesting after the successful raising of a brood.

Brown-headed Cowbird Parasitism: Cowbird parasitism was not found in British Columbia in 24 nests recorded with eggs or young, nor has it been reported elsewhere in North America (Friedmann et al. 1977; Friedmann and Kiff 1985).

Nest Success: There are not enough data for British Columbia. In Idaho, the loss of nestlings was significantly higher than the loss of eggs. Nest success was 43%, and the mean number of young fledged per successful nest was 2.6 (n = 30) (Reynolds and Rich 1978).

REMARKS: Since it was first recorded in the province, the Sage Thrasher has been erratic in its presence, absent from suitable nesting territory for several years at a time and then reappearing. Since 1970 it has been apparent that its numbers and distribution are being reduced because of habitat destruction (Cannings 1992a). Residential and agricultural developments and range management practices such as removal of big sagebrush and burning, herbicide, and pesticide treatments of shrublands have reduced the quality and quantity of Sage Thrasher habitat throughout its range. In British Columbia, the most important cause of habitat destruction has been agriculture. This author concluded that "suitable habitat for the Sage Thrasher no longer exists near Oliver and Penticton, where Darcus (1932) found them nesting in the 1920s." He also concluded that clearing for alfalfa production near Chopaka had destroyed optimal habitat, and that housing developments now threaten good habitat in the Richter Pass area. There is serious concern that the Sage Thrasher may become extirpated in British Columbia.

The future of the Sage Thrasher in British Columbia depends on protection of its habitat. Sizeable tracts of sagebrush are found in the Thompson valley from Pritchard west to Cache Creek and Spences Bridge, and in the Fraser River canyon from Lillooet to the Chilcotin River. For some reason, the Sage Thrasher has not colonized these areas, even though

Figure 515. This Sage Thrasher nest is supported by secondary stems and placed in the dense protective foliage in the crown of a tall sagebrush (west of Cache Creek, July 1990; R. Wayne Campbell).

Figure 516. In the early 1960s, a pair of Sage Thrashers nested successfully for several years with their nest built low in this red hawthorn tree (White Lake, 18 June 1993; R. Wayne Campbell). In the mid-1960s, Black-billed Magpies usurped the nest site.

much of the habitat seems suitable and single birds have been seen there. Thus the remaining sizeable tracts of big sagebrush in the Okanagan and Similkameen valleys appear to be the only habitat in British Columbia suitable for the Sage Thrasher.

At present, there are only 2 such areas, although 1 can be divided into 3 separate units that together comprise about 700 ha of habitat potentially suitable for Sage Thrashers. Richter Pass–Kilpoola Lake–Chopaka is the largest tract of sagebrush in the south Okanagan–Similkameen. High-quality sagebrush habitat can be found there around Spotted Lake in Richter Pass (about 100 ha), in the Lone Pine Creek area south of Kilpoola Lake (about 300 ha), and between the Chopaka border crossing and Highway 3A (about 150 ha). The other area is White Lake: this large basin contains about 150 ha of sagebrush habitat suitable for thrashers (Cannings 1992a) (Fig. 511).

As with many songbirds, the presence of the Sage Thrasher can be determined most easily during the song period, and it is tempting to use singing males as evidence of the presence of a breeding location. However, intensive field studies of the Sage Thrasher conducted from 1992 to 1994 by the Canadian Wildlife Service have produced several instances in which the presence of singing males in areas where nesting had taken place in previous years did not lead to nesting. For example, at Chopaka in 1993, 3 singing males were first heard on 23 May and were gone a few days later. On 9 June, 3 males were singing in the same area; all were colour-banded but only 2 later defended territories. In the same area in 1994, 2 singing males were present in May but did not establish territories. At White Lake a singing male was present from 23 May to early June 1993, but was gone by the second week of June (R. Millikin pers. comm.). The 1 record from Kamloops was of a singing male in apparently suitable nesting habitat. It persisted for more than a week before disappearing (R.W. Ritcey pers. comm.). Such observations suggest that in British Columbia, where the thrasher is at the northern limit of its distribution, males are "pioneering" and a nesting attempt depends upon the subsequent arrival of females.

Figure 517. The Sage Thrasher nest is a bulky structure composed of coarse twigs, rootlets, and bark strips and lined with grass. In British Columbia, nearly three-quarters of all nests found contained 4 or 5 eggs (7 km west of Osoyoos, 10 June 1978; R. Wayne Campbell).

In Alberta, the Sage Thrasher is observed even less frequently, and was not confirmed as a nesting species until 1988 (O'Shea 1988; Semenchuk 1992).

For additional information on the ecology, behaviour, conservation, and life history of the Sage Thrasher, see Braun et al. (1976), Castrale (1982), Kantrud and Kologiski (1982), Reynolds and Trost (1981), Rich (1980), and Wiens and Rotenberry (1981).

NOTEWORTHY RECORDS

Spring: Coastal – Iona Island 16 to 18 May 1982-1 (Mattocks and Hunn 1982b); Little Qualicum River 18 May 1980-1 on estuary; Grant Bay 8 May 1969-1 (RBCM 11611; Richardson 1971). **Interior** – Chopaka 31 May 1982-4 along road; Spotted Lake 4 May 1974-2 (Cannings 1974); White Lake (Okanagan) 27 May 1936-8, 4 Apr 1974-2; between Princeton and Hedley 28 May 1964-1; near Savona 21 May 1966-1; Mount Revelstoke National Park 17 Apr 1981-1 at Lauretta picnic area (Weber 1991a); 40 km e Cache Creek 23 Apr 1973-1 (McNicholl 1973); Buffalo Creek (15 km ne 100 Mile House) 27 and 28 Apr 1987-1.

Summer: Coastal – No records. **Interior** – Chopaka 15 Jun 1989-1 (Preston and Harris 1989), 4 and 13 July 1990-2 (Preston 1990), 9 Jun 1992-7 singing males (C. Antoniazzi pers. comm.), 12 Jun 1993-5 eggs, same nest, 4 nestlings hatched 26 Jun and fledged 6 Jul (R. Millikin pers. comm.); near Nighthawk 30 Jul 1976-5 along road; Osoyoos 21 Jun 1909-nest with 4 eggs, first record of nesting in the province (Brooks 1909b), 21 Aug 1919-3 (MVZ 82576, 103578-9); Kilpoola Lake 19 Jun 1975-3 eggs (BC Photo 987), 13 Jul 1990-2 (Preston 1990), 6 Jul 1993-nest with 2 eggs; Richter Pass Jul 1975-nest with 3 eggs (BC Photo 987); Oliver 9 Jun 1992-2 singing males along Camp McKinney Road; Duck Lake (Creston) 7 to 10 Jun 1990-1 (Weber 1991a); Vaseux Lake 11 Aug 1986-1 in birch; White Lake (Okanagan) 1 Jun 1910-4 eggs, earliest nesting date for British Columbia (BCPM E925), 13 Jun 1912-5 eggs, 11 Jun 1931-nest with 2 eggs, 5 Jul 1954-nest with 4 eggs, 12 Jul 1969-19, 28 Jun 1974-nest with 5 eggs, 11 Aug 1963-3 young just out of nest, 26 Aug 1932-young out of nest (Munro and Cowan 1947), 28 Aug 1931-1 (ROM 83903); Okanagan Landing 7 Jun 1925-apparent breeding pair (Brooks 1925), 28 Aug 1931-1 (ROM 83903); 15 km n Lytton 10 Jun 1990-1 (Weber 1991a); Bachelor Hill (Kamloops) 28 and 29 Jun 1983-1 (BC Photo 885); Ashcroft 14 Jun 1991-1.

Breeding Bird Surveys: Coastal – Not recorded. **Interior** – Not recorded.

Autumn: Interior – White Lake 4 September 1919-1 (Cannings et al. 1987); Chopaka 29 Sep 1991-5 (Cannings 1992a). **Coastal** – No records.

Winter: No records.

Brown Thrasher

BRTH

Toxostoma rufum (Linnaeus)

RANGE: Breeds from southeastern Alberta south through the Rocky Mountain states to New Mexico and east across southern Canada, including southern New Brunswick, to the New England states, and south to southern Florida, the Gulf coast, and northern and eastern Texas. Winters regularly from eastern New Mexico, Kentucky, and southern Maryland south to southeastern Texas, the Gulf coast and southern Florida. Casual on the Pacific coast.

STATUS: On the coast, *casual* in the Georgia Depression Ecoprovince and *accidental* in the northern Coast and Mountains Ecoprovince.

In the interior, *casual* in the Southern Interior and Southern Interior Mountains ecoprovinces and *accidental* in the Boreal Plains Ecoprovince.

Figure 518. Brown Thrasher (Fruitvale, 18 December 1993; G.S. Davidson). The Brown Thrasher was reported 14 times in British Columbia between 1968 and 1995.

CHANGE IN STATUS: The Brown Thrasher (Fig. 518) is considered a rare bird anywhere on the Pacific coast, and its appearance in British Columbia is best understood in the context of its occurrence in the Pacific coast states. Between 1870 and 1944, there was a total of 4 reported occurrences of the Brown Thrasher in California, all in the southern half of the state (Grinnell and Miller 1944). Three of the 4 California records were reported between 1932 and 1941. By 1980 the number of records had risen to 84, still mostly in southern

Figure 519. In British Columbia, the Brown Thrasher has most often been reported in brushy and shrubby habitats (Riverview, Creston, 11 August 1996; R. Wayne Campbell).

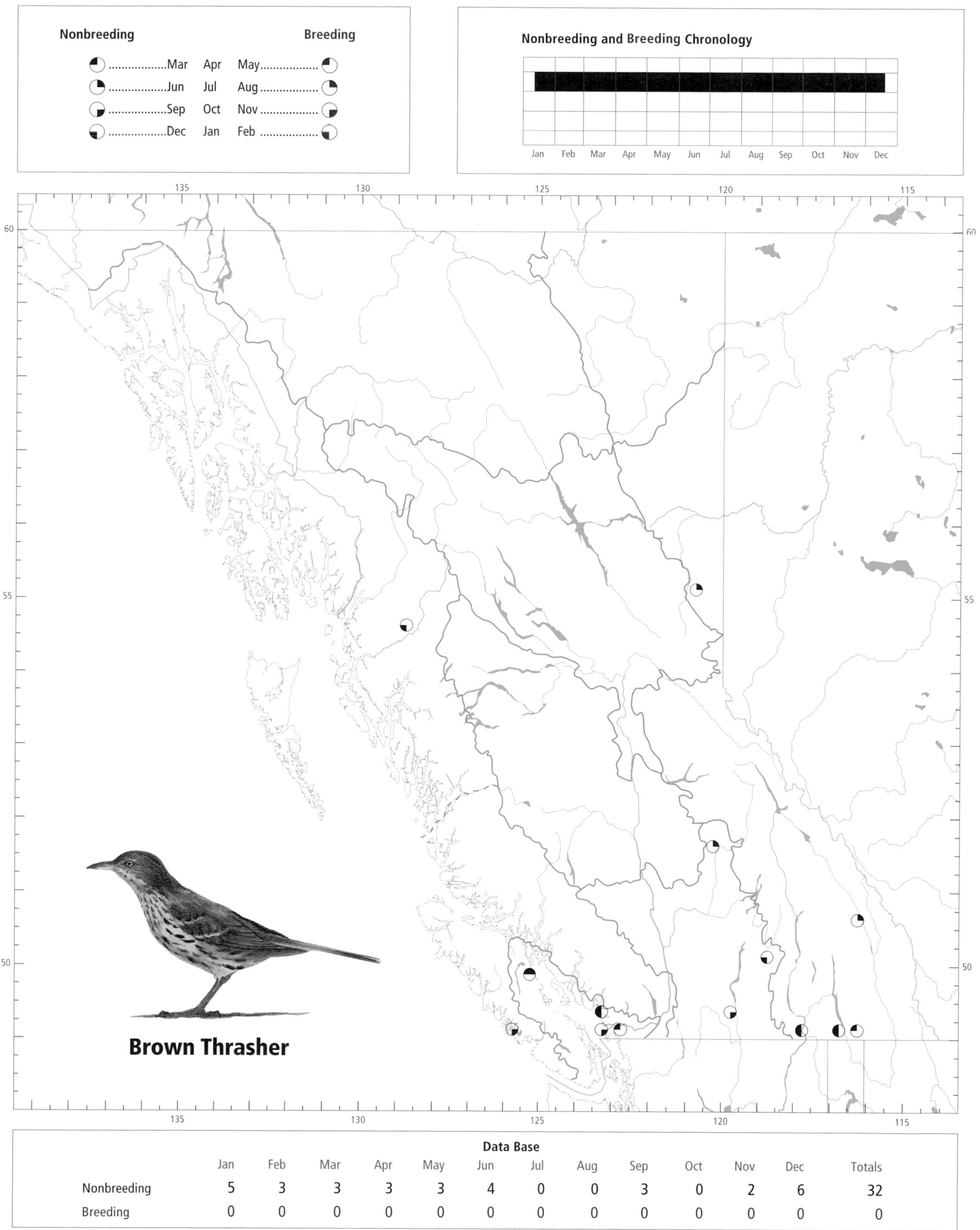

Data Base

	Jan	Feb	Mar	Apr	May	Jun	Jul	Aug	Sep	Oct	Nov	Dec	Totals
Nonbreeding	5	3	3	3	3	4	0	0	3	0	2	6	32
Breeding	0	0	0	0	0	0	0	0	0	0	0	0	0

Figure 520. The Brown Thrasher was found in these dense patches of lodgepole pine on Mitlenatch Island (27 August 1969; R. Wayne Campbell).

California but a few north of San Francisco (Roberson 1980). The species was first taken in Oregon in 1940 (Bagg 1941), and over the next 5 decades was found there on at least 11 occasions, mainly from late May to late August and occasionally in winter (Schmidt 1989). In Washington, it was first reported in 1972 (Manuwal 1973). Since then, it has been seen infrequently throughout coastal and interior portions of the state. In southern California it is considered of regular occurrence (about 6 records a year), mainly in autumn or spring (Small 1974). There is a single record from Alaska, in September 1974 (Roberson 1980).

In British Columbia, the first Brown Thrasher was seen in the summer of 1968 at Clearwater, in the Southern Interior. Since then it has been reliably reported 13 times. The entry of the Brown Thrasher into British Columbia has been part of its slow spread westward to, and northward along, the Pacific coast over a period of about 60 years.

A recent occurrence in the Peace Lowland was probably of a bird that originated in Alberta; the Brown Thrasher is relatively common in the southern portions of that province (Semenchuk 1992).

OCCURRENCE: The Brown Thrasher has been reported at widely scattered localities from the west coast of Vancouver Island and Terrace on the Northern Mainland Coast east to the Columbia River valley in the Rocky Mountain Trench. There is a single record east of the Rocky Mountains, in the Peace Lowland.

The Brown Thrasher frequents brushy and shrubby areas in suburban and rural backyards, seashores, and lakeshores, the edges of deciduous forests, clearings (Figs. 519 and 520), fields, and roads; it comes readily to feeders.

The Brown Thrasher has been reported from sea level to 790 m elevation. It occurs as a vagrant, with records from every month of the year but July, August, and October (Fig. 521). Four birds have been reported from the winter months; 2 apparently wintered successfully at bird feeders. All records are of single birds.

Before 1970, occurrences of the Brown Thrasher in western North America were concentrated in autumn and winter. Between 1970 and 1980, there was a noticeable change to spring and summer occurrences (James and Richardson 1982). In Oregon and California, reports of this species are primarily from spring and autumn (Schmidt 1989; Small 1994). In British Columbia, however, most birds have been reported from winter and spring.

In chronological order, British Columbia records, updated from Van Damme (1993), are as follows:

(1) Clearwater 9 June 1968-1 (correspondence from John and Hettie Miller to Charles J. Guiguet; Royal British Columbia Museum files).
(2) Penticton (West Bench) 21 and 22 September 1970-1 in ornamental shrubbery and frequenting a pool in a rural garden (Campbell and Stirling 1971; Cannings 1972; BC Photo 109).

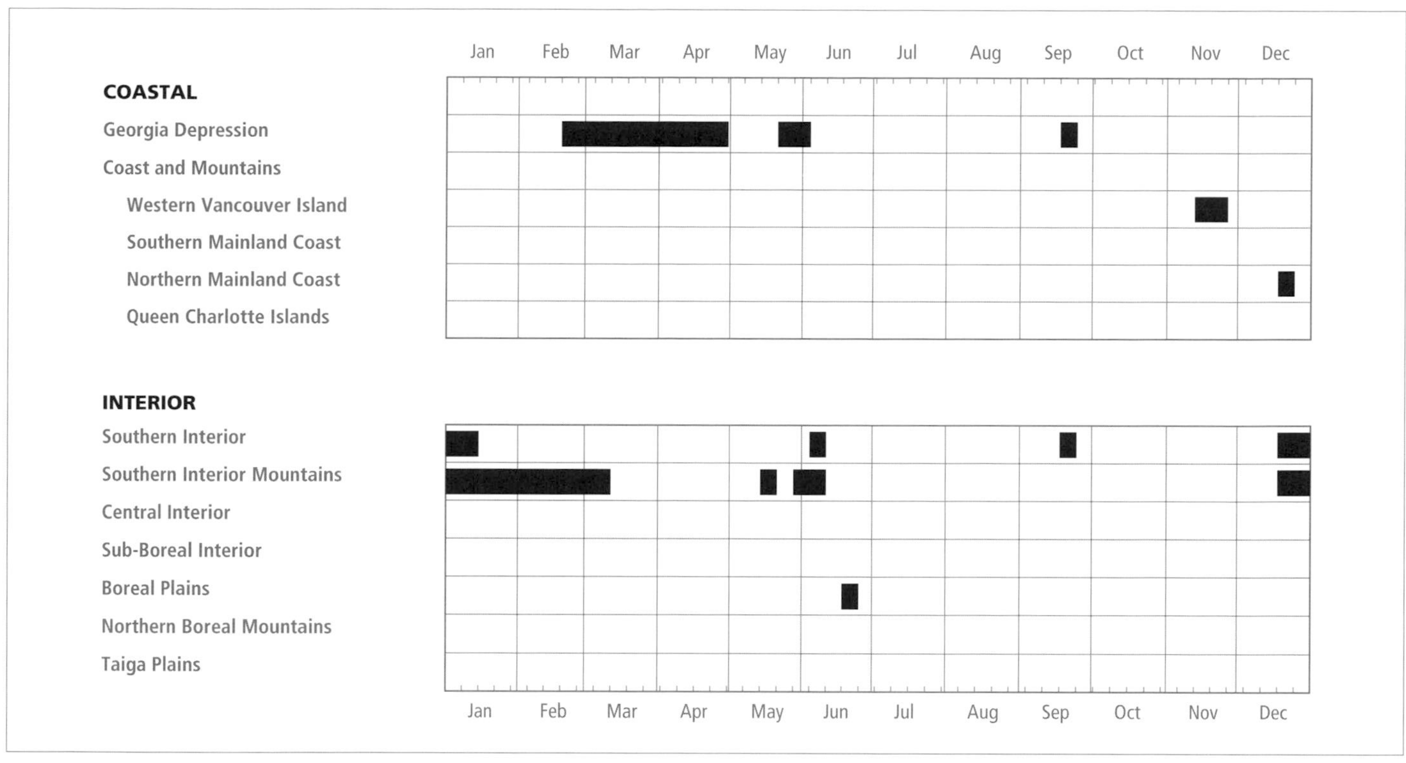

Figure 521. Annual occurrence for the Brown Thrasher in ecoprovinces of British Columbia. Records are shown for the week in which they occurred.

(3) West Vancouver 23 February to 29 April 1972-1 at feeding station on Eagle Harbour Road nearly daily for 66 days (Crowell and Nehls 1972b [with photograph], 1972c; Campbell 1974; BC Photo 200). When not at a feeder, it used blackberry brambles and dense shrub growth.
(4) Sandhill Creek (Long Beach, Vancouver Island) 17 and 21 November 1973-1 near Comber's Beach (Campbell 1974; Hatler et al. 1978).
(5) Terrace 19 December 1976-1 photographed on Christmas Bird Count (Weismiller and Weismiller 1977).
(6) Mitlenatch Island 30 May to 2 June 1979-1 in clump of lodgepole pine (James and Richardson 1982; BC Photo 525; Figs. 520 and 522).
(7) Monte Creek (Duck Meadows) 12 June 1980-1 observed singing for 4 minutes (David J. Low pers. comm.).
(8) Riverview (Creston) 18 May 1983-1 (Butler et al. 1986).
(9) Beaver Falls (between Trail and Fruitvale) 18 December 1983 to 6 March 1984-1 at feeder, often with Varied Thrushes (BC Photo 907; Fig. 518).
(10) Cherryville 17 December 1985 to 11 January 1986-1 taped singing and frequently seen at feeder at Silver Hills Ranch (Rogers 1986).
(11) Langley 22 April 1988-1 male singing in trees and shrubs (Campbell 1988b).
(12) Delta 18 September 1990-1.
(13) Wilmer National Wildlife Area (4 km north of Wilmer) 5 June 1993-1 in willow thicket (Kimpton 1993).
(14) Tumbler Ridge (southwest of Dawson Creek) 20 June 1994-1 singing from shrubby hedgerow along Border Highway on Breeding Bird Survey.

REMARKS: See Fischer (1981) for details on the wintering ecology of this thrasher.

Figure 522. The sixth record of the Brown Thrasher in British Columbia occurred on Mitlenatch Island between 30 May and 2 June 1979 (Paul C. James; BC Photo 525).

American Pipit

AMPI

Anthus rubescens (Tunstall)

RANGE: Breeds throughout Alaska, Yukon, the Canadian low Arctic, and Newfoundland; also on isolated mountaintops in New Brunswick, Maine, and New Hampshire; also south in the west through central, southeastern, and southwestern British Columbia, and locally in mountainous country to California, Arizona, and New Mexico. Winters from south coastal British Columbia, southern Idaho, Colorado, and New York south to the southern United States, Mexico, and El Salvador. Also breeds in eastern Siberia and western Greenland; winters in southern Asia.

STATUS: On the coast, *very common,* occasionally *abundant,* migrant in the Georgia Depression Ecoprovince and *fairly common* migrant in the Coast and Mountains Ecoprovince; *uncommon* summer visitant to alpine areas in the Southern Pacific Ranges of the southern Coast and Mountains and *rare* in alpine areas elsewhere on the coast; *uncommon* in winter in the Georgia Depression; *casual* elsewhere on the coast.

In the interior, *very common,* occasionally *abundant* migrant except *uncommon* in the Taiga Plains Ecoprovince; *fairly common* summer visitant to alpine areas throughout the interior, becoming *very rare* at lower elevations; *casual* in winter in the Southern Interior Ecoprovince.

Breeds.

Figure 523. The American Pipit is one of the most widely distributed migrant birds in British Columbia (Victoria, November 1993; Tim Zurowski).

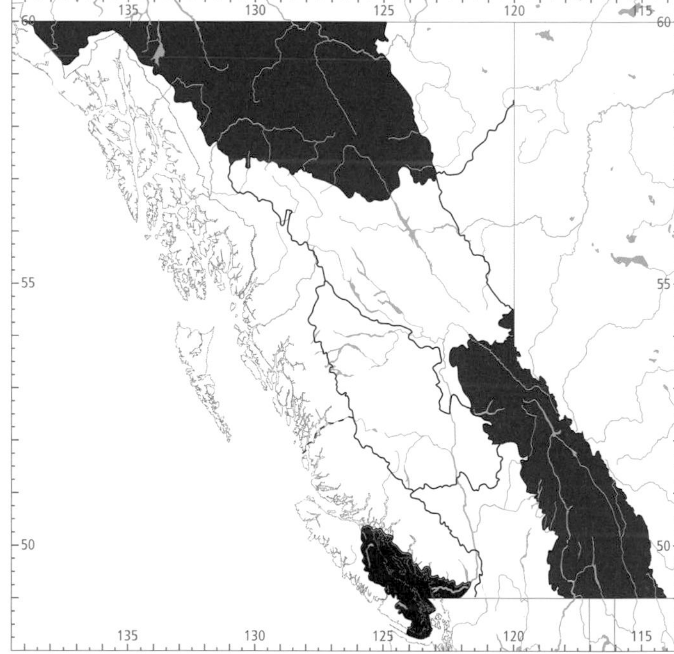

Figure 524. In British Columbia, the highest numbers for the American Pipit in winter (black) occur in the Fraser Lowland of the Georgia Depression Ecoprovince; the highest numbers in summer (red) occur in the Southern Interior Mountains and Northern Boreal Mountains ecoprovinces.

NONBREEDING: The American Pipit (Fig. 523) is one of the most widely distributed migrant birds in British Columbia. During some parts of the year, it occurs from offshore islands along the coast east through the Coast Ranges and the interior to the Rocky Mountains, and north through the Boreal Plains, Taiga Plains, and Northern Boreal Mountains ecoprovinces. The highest numbers in winter occur in the Fraser Lowland of the Georgia Depression (Fig. 524).

In British Columbia, the American Pipit has been recorded during migration and winter at elevations between sea level and 2,819 m. During the spring migration, loose flocks of a few to several hundred birds can be observed sweeping over snow-free open habitats, such as estuaries, intertidal mudflats, agricultural fields (Fig. 525), stubble fields, pastures, and rangeland. During the autumn migration, it uses the same habitats as when it is northbound, but also uses the mountaintop routes with their alpine tundra, scree slopes, sparsely vegetated fell-fields, and seepage basins below melting snow. In forested regions, migrants may pause along highway rights-of-way, river gravel bars, airports, and playgrounds. In winter, pipits are found mainly on open areas near the coast, including sand dunes, sea beaches, rocky shorelines, grassy marshes, cultivated fields, golf courses, pastures, sewage lagoons, airports, playing fields, tidal flats, and river estuaries.

Spring migration of the American Pipit along the coast becomes apparent as a sharp increase in the numbers of birds occurs between late March and April on the Fraser Lowland and the Nanaimo Lowland (Fig. 527). During peak movements, flocks can pass by one after another all day, following low-elevation routes along coastal shores and interior valleys and plateaus. This spring movement is remarkably synchronous throughout the southern regions of the province, from Western Vancouver Island to the Rocky Mountain Trench, and north to include the northern Coast and Mountains and the Central Interior. The speed of advance is shown by the dates of first arrivals (see Appendix 1). Only about 10 days separate the first arrivals in the Okanagan from those at Prince George, and the time of peak movements in southern and northern interior localities are almost identical. A few nonbreeding birds occasionally remain at lower elevations

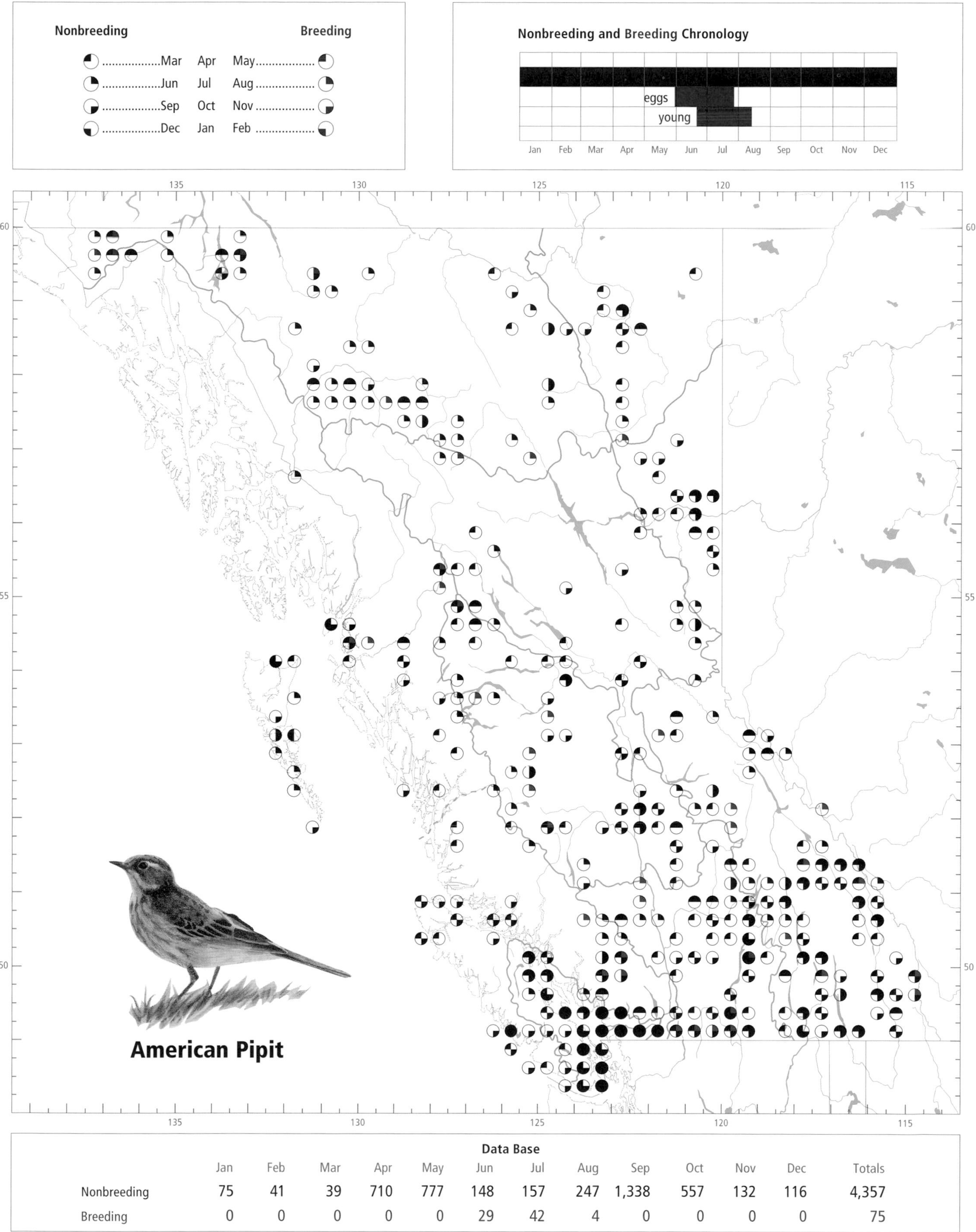

Data Base

	Jan	Feb	Mar	Apr	May	Jun	Jul	Aug	Sep	Oct	Nov	Dec	Totals
Nonbreeding	75	41	39	710	777	148	157	247	1,338	557	132	116	4,357
Breeding	0	0	0	0	0	29	42	4	0	0	0	0	75

during the summer, but the American Pipit is quite rare outside alpine areas in June and July.

The autumn migration is as dramatic as that in spring (Fig. 527). After nesting is completed, juveniles and adults join loose flocks in alpine areas and then move southward, mainly following mountain ranges. Flocks remain in alpine areas until the weather deteriorates, then move to lower elevations (Verbeek 1970). Peak movements are evident in early September in the north and mid-September to early October in the south (Fig. 527). By the end of October, the migration is essentially over, except for a few November stragglers in the south (Figs. 526 and 527).

A comparison between the accumulated totals of American Pipits recorded on northbound and southbound migrations at coastal and interior localities in the southern third of the province reveals important differences. The total number of birds recorded during the spring migration in the Georgia Depression is close to the combined total of the Southern Interior and Southern Interior Mountains. During the autumn migration, however, the total on the coastal route is about 8 times that of the 2 southern interior ecoprovinces combined. Furthermore, the total of all autumn records in the southern regions is roughly 3 times the total recorded during the spring migration (Fig. 527). The difference between the apparent numbers moving south and those returning in spring suggests heavy losses on the winter range or in transit (see REMARKS).

In winter, the few American Pipits remaining in British Columbia are concentrated in the lower Fraser River valley and on southeastern Vancouver Island, but there are a few December records for the Queen Charlotte Islands, Western Vancouver Island, and the Okanagan valley (Fig. 526).

Figure 525. Open agricultural fields with short vegetation provide good foraging habitats for the American Pipit during spring migration (Martindale Flats on the Saanich Peninsula, Vancouver Island, 5 March 1994; R. Wayne Campbell).

On the coast, the American Pipit occurs throughout the year; in the interior, it has been recorded regularly from 10 March to early November (Fig. 526).

BREEDING: In British Columbia, the American Pipit nests in alpine habitats throughout mountainous areas of the interior, and to a lesser extent in coastal alpine habitats. Nesting has been confirmed from the coastal ranges east to the Rocky Mountains and north to the Yukon boundary. On the coast, nesting has been recorded in the Eastern Pacific and Southern Pacific ranges of the Coast and Mountains from Whistler Mountain south to Golden Ears Park northeast of Vancouver. Although it has been recorded in summer on Vancouver Island, in seemingly suitable alpine habitat, there is no evidence of nesting. It is not known to breed on the Queen Charlotte Islands.

It is relatively numerous throughout its breeding range.

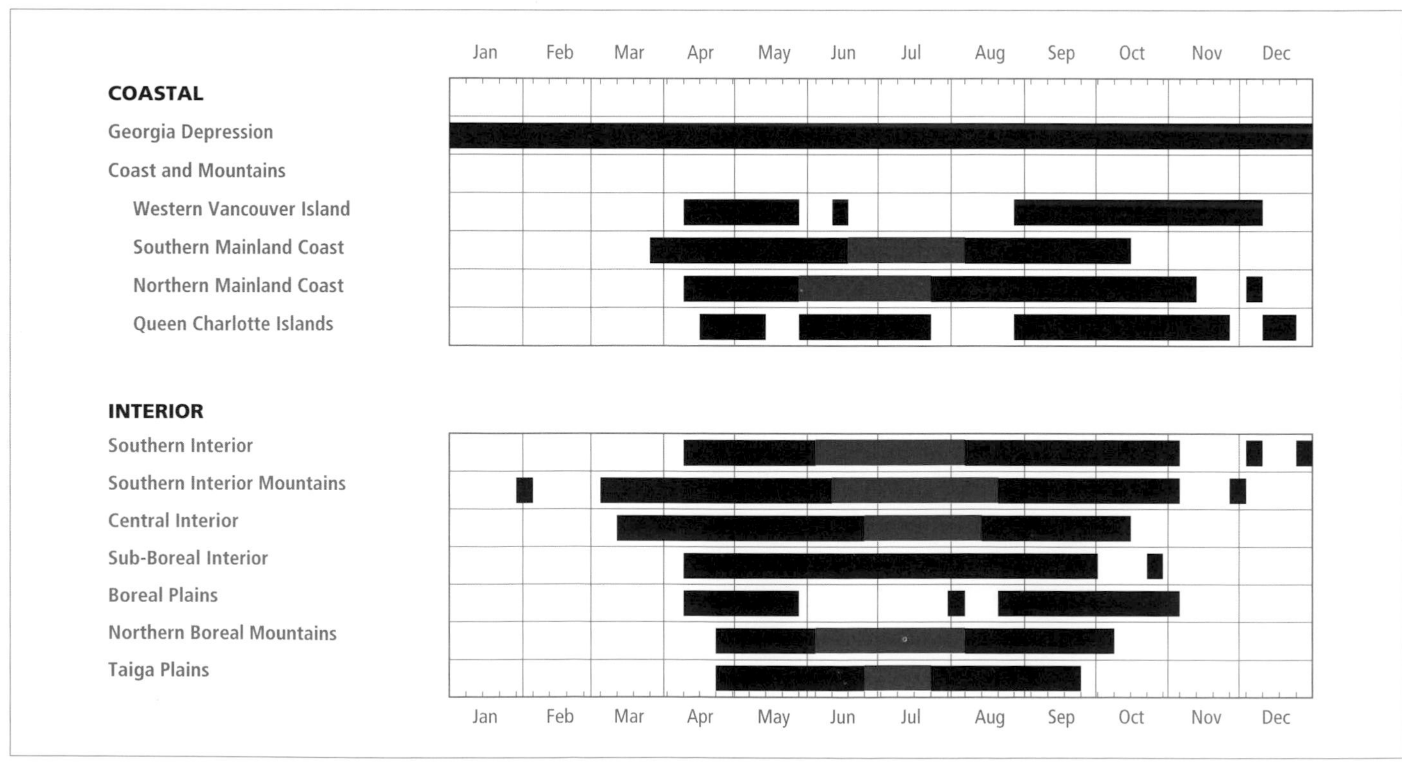

Figure 526. Annual occurrence (black) and breeding chronology (red) for the American Pipit in ecoprovinces of British Columbia. Records are shown for the week in which they occurred.

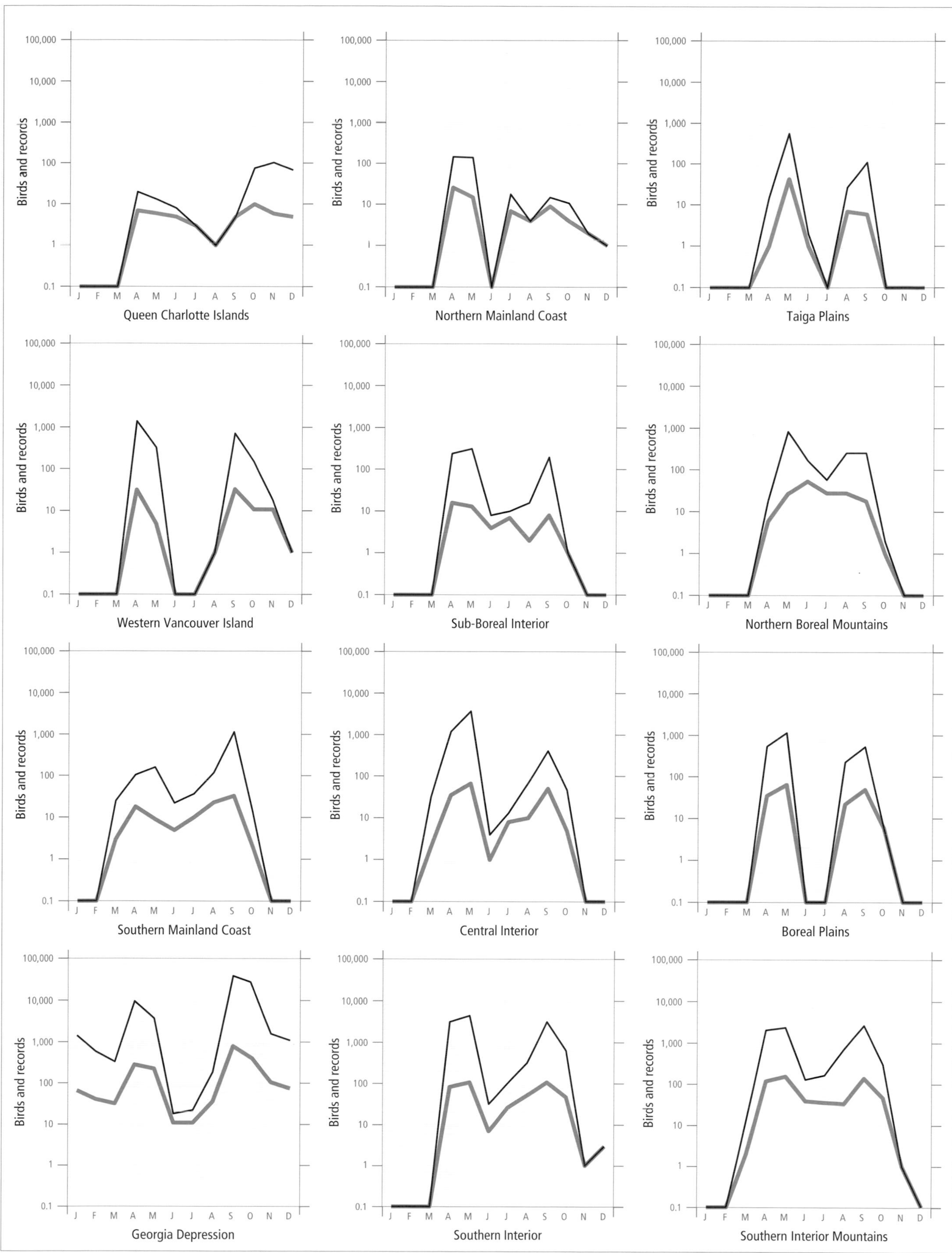

Figure 527. Fluctuations in total number of birds (purple line) and total number of records (green line) for the American Pipit in ecoprovinces of British Columbia. Christmas Bird Count and nest record data have been excluded.

Figure 528. Sparsely vegetated alpine areas are breeding habitat for the American Pipit in British Columbia (Taseko Mountain, 20 July 1993; R. Wayne Campbell).

The highest numbers in summer have been recorded in the Southern Interior Mountains and Northern Boreal Mountains ecoprovinces (Fig. 524). Breeding Bird Surveys for both coastal and interior routes for the period 1968 to 1993 contain insufficient data for analysis.

Nests have been found at elevations between 600 and 2,400 m, but breeding always occurs in alpine habitats (Fig. 528). With the latitudinal change of timberline, nesting in the south is at higher elevations. Habitat occupied during the breeding season includes sparsely vegetated alpine tundra (Fig. 528), heath meadows, more richly vegetated alpine meadows, boulder fields, scree slopes with scattered vegetation between the rocks, cliff ledges, and gravelly stream beds. The pipit often forages for insects on snow patches remaining during the summer.

Nesting habitat is mainly in well-vegetated sites, from just above the timberline to where rock and scree become predominant. Nest sites are usually situated on sloping hillsides where solifluction and erosion leave overhanging clumps of vegetation, sod, and stones as cover for the nest (Fig. 529).

The American Pipit initiates nesting as soon as there are patches of bare ground on the alpine slopes (Verbeek 1970). In some parts of Alaska, where alpine conditions occur at lower elevations and snowfall is often lighter than in many more southern mountain ranges, bare ground can be exposed and the pipit can begin its nesting cycle before populations of southern, higher-elevation alplands (e.g., Wyoming) can find suitable conditions (Verbeek and Hendricks 1994). However, the relatively small samples of nesting data from southern and northern alplands of British Columbia do not show this effect in the province.

In British Columbia, the American Pipit breeds between 31 May and 13 August (Fig. 526).

Nests: All nests (n = 47) for which details are available were on the ground (Fig. 529). Nests were small cups placed in shallow depressions under tufts of vegetation. Nests (n = 46) were composed mainly of grasses (85%), sedges (17%), mosses (9%), plant stems, rootlets, lichens, and animal hair, and were lined with finer grass.

Eggs: Dates for 47 clutches ranged from 5 June to 27 July, with 60% recorded between 20 June and 16 July. Calculated dates indicate that eggs may occur as early as 31 May. Late clutches were probably replacements of lost first clutches; only 1 brood is raised annually (Verbeek 1970). Sizes of 39 clutches ranged from 2 to 7 eggs (2E-1, 3E-4, 4E-11, 5E-11, 6E-10, 7E-2), with 54% having 5 or 6 eggs. Clutch sizes are known to increase with increasing latitude (Verbeek and Hendricks 1994). The incubation period is about 14 days (Verbeek 1970). In western Alaska, earliest egg laying is calculated to begin about 11 to 12 May, but most clutches are begun in the first half of June (Kessel 1989).

Young: Dates for 26 broods ranged from 20 June to 13 August, with 61% recorded between 3 July and 28 July. Calculated dates indicate that nestlings can be present as late as 13 August. Sizes of 15 broods ranged from 2 to 5 young (2Y-1, 3Y-1, 4Y-12, 5Y-1), with 12 of the broods having 4 young. The nestling period is about 14 days (Verbeek 1970). If disturbed, nestlings may leave the nest before they can fly. Young remain with parents for about 2 weeks after leaving the nest, before joining loose flocks (Verbeek and Hendricks 1994).

Brown-headed Cowbird Parasitism: Cowbird parasitism was not found in British Columbia in 69 nests recorded with eggs or young, nor are there cases of parasitism elsewhere in its range in North America (Friedmann et al. 1977; Friedmann and Kiff 1985; Verbeek and Hendricks 1994).

Nest Success: Of 2 nests found with eggs and followed to a known fate, 1 produced at least 1 young. Hendricks and Norment (1992) discuss the impacts of a midsummer snowstorm on nesting success of 2 populations (alpine and subalpine) in Wyoming. During the storm, the alpine population (3,200 m) had a nestling mortality rate of 79%, while the subalpine population (2,900 m) experienced 7% nestling mortality.

REMARKS: Two subspecies occur in British Columbia: *A. r. pacificus,* which occurs in North America west of the Rocky

Figure 529. A nest site of the American Pipit in a well-vegetated site on a sloping hillside above the timberline (Yohetta Mountain Pass, Cariboo and Chilcotin areas, 19 July 1993; R. Wayne Campbell). The actual nest is located beneath the clump of heather in the bottom centre of the photograph.

Mountains, and *A. r. alticola*, which is thought to breed in the Rocky Mountains from southern British Columbia south into New Mexico, Arizona, and the Sierra Nevada of California (Miller and Green 1987). There are no data on the boundaries between these 2 subspecies in British Columbia.

The American Pipit was formerly known as the Alpine Pipit (Munro and Cowan 1947), and was believed to be a subspecies of the Water Pipit (*A. spinoletta*). Recently it was determined that the American Pipit is specifically distinct from the Water Pipit of Eurasia, and the subspecies name of the North American population (*rubescens*) became its species name (American Ornithologists' Union 1989).

A report of 100 "Alpine" Pipits on the Vernon Christmas Bird Count on 30 December 1951 is questionable (Grant 1952).

Although data from British Columbia are lacking, some American Pipits are known to reuse nesting territories in subsequent years; for example, 10% of females in Wyoming nested within 40 m of their previous year's nest (Hendricks 1991).

For additional information on the life history of the American Pipit, see Bent (1950), Verbeek (1965), Hendricks (1993), and Verbeek and Hendricks (1994). King (1981) and Parkes (1982) provide information on the field identification of North American pipits.

NOTEWORTHY RECORDS

Spring: Coastal – Victoria 2 Mar 1953-6; Central Saanich 27 Mar 1988-53; Sea Island 19 Apr 1966-400, 27 Apr 1965-500, 3 May 1985-200; Stubbs Island 12 Apr 1979-2; Hope 20 May 1986-100; Boston Bar 24 Apr 1959-25; Pemberton 30 Mar 1968-12; Port Hardy 14 Apr 1941-500, 2 May 1939-200; Bella Coola 9 May 1933-7; Kitimat 22 Apr 1975-30 (Hay 1976); Kootenay Inlet (Moresby Island) 18 May 1988-3 (Campbell 1988e); Masset 19 Apr 1979-18, 4 May 1979-4; Prince Rupert 17 Apr 1983-22; Kaien Island 6 May 1979-36; Metlakatla 10 Apr 1907-1, first spring arrival (Keen 1910); Three Guardsmen Mountain 19 May 1978-1 (Blood and Chutter 1978). **Interior** – Anarchist Mountain 1 May 1974-400 (Shepard 1975a); Richter Lake 12 Apr 1969-200; Creston 20 Apr 1986-100; Duck Lake (Creston) 10 Mar 1981-12; Castlegar 10 May 1970-200; White Lake (Okanagan Falls) 20 Apr 1976-755; Penticton 10 Apr 1977-30; Brouse 17 Apr 1977-3; Nakusp 20 Apr 1978-30; Tunkwa Lake 12 May 1968-500; Tatla Lake 3 May 1991-40, earliest spring sighting; Riske Creek 10 May 1978-300; Prince George 23 Apr 1984-200; Quick 24 Apr 1980-100; Baldonnel 24 Apr 1980-200; Fort St. John 13 May 1986-121, 24 May 1986-60; Charlie Lake 14 Apr 1984-2; North Pine 20 Apr 1984-10, first flock; Jack Fish Creek (Fort Nelson) 13 May 1982-50; Fort Nelson 29 Apr 1987-15; Atlin 24 Apr 1934-1, first arrival (CAS 42152); Liard Hot Springs 28 Apr 1975-20 (Reid 1975); Chilkat Pass 14 May 1977-150, 24 May 1979-100.

Summer: Coastal – Esquimalt Lagoon 31 Aug 1980-17; Chesterman Beach (Tofino) 31 Aug 1974-1; Cheam Peak 6 Aug 1981-100; Golden Ears Park 1 Jul 1975-3; Mount Albert Edward 24 Jul 1976-6; Mitlenatch Island 28 Jun 1970-7; Garibaldi Park 6 Jul 1982-7 eggs, 2 Aug 1968-adult feeding fledgling; Whistler Mountain 26 Jun 1924-5 eggs (Munro and Cowan 1947); Taseko Mountain 19 Jul 1993-4 eggs; Rainbow Mountains 30 Jun 1932-1 (Dickinson 1953); Bigsby Inlet 7 Jun 1986-4; Tasu 26 Jun 1961-1 (UBC 10457); Peel Inlet 17 Jul 1960-1; Mount La Pérouse 29 Aug 1961-1; Sibola Mountains 17 Jul 1976-7; Nine Mile Mountain 5 Jun 1958-5 eggs (earliest eggs). **Interior** – Okanagan 10 Jun 1915-6 eggs (RBCM 976); Ashnola River 20 Jun 1932-6 eggs found in mountains; Cathedral Park 20 Jul 1968-5 nestlings, 20 Aug 1982-100 (many adults feeding young); Apex Mountain 13 Jul 1984-4 eggs; Michel Ridge 15 Jun 1982-5 eggs; Penticton 27 Jun 1972-10; China Head Mountain 20 Jul 1982-adult feeding 2 fledglings; Niut Mountain 4 Aug 1992-4 adults feeding fledglings; Emerald Lake 5 Jun 1976-40 (Wade 1977); Perkins Peak 31 Aug 1991-2; Wells Gray Park 15 Jul 1976-4 nestlings; Tsitsutl Peak 22 Jun 1979-4; Battle Mountain (Clearwater) 27 Jul 1977-5 eggs (latest eggs); Canoe Mountain 20 Aug 1971-200 (Rogers 1972d); Blackwater River 25 Jul 1979-4 eggs found near headwaters; McBride (summit Little Bell Mountain) 10 Jul 1991-20; Smithers 30 Jun 1990-5 eggs; Kinuseo River 30 Aug 1981-75; Nation Mountain (Spatsizi Park) 9 Jul 1959-14; Tatlatui Lake 27 Jul 1974-1 nestling; Valhalla Mountain 13 Aug 1975-many fledglings; Ipec Lake 29 Aug 1979-100 (Cooper and Adams 1979); Mount Edziza 20 Jul 1977-2; Tuaton Lake 10 Jun 1979-4; Pink Mountain 5 Jul 1982-6 eggs; Fort Nelson 12 Jun 1982-2, 31 Aug 1985-7; Boulder Creek (Atlin) 11 Jun 1958-7 eggs; Carmine Mountain 24 Jun 1983-16 (Campbell et al. 1983); Chilkat Pass 20 Jun 1960-4 nestlings (Weeden 1960), 9 Jul 1956-40.

Breeding Bird Surveys: Interior – Recorded from 1 of 73 routes and on less than 1% of all surveys. Maximum: Chilkat Pass 1 Jul 1976-9. **Coastal** – Not recorded.

Autumn: Interior – Feather Creek (Atlin) 3 Oct 1980-2; Muncho Lake 6 Sep 1974-25; Fort Nelson 20 Sep 1988-1; Mile 379 Alaska Highway 6 Sep 1974-50; Charlie Lake 30 Oct 1986-1; Fort St. John 1 Oct 1983-1; Nation River 27 Oct 1972-1; Perkins Peak 7 Oct 1992-latest record; Riske Creek 11 Oct 1959-25; Yoho National Park 7 Sep 1976-115 (Wade 1977); Golden 8 Oct 1977-50; Scotch Creek 11 Sep 1962-500; Sorrento 1 Oct 1970-100; West Bench 1 Nov 1964-1; Vaseux Lake 9 May 1974-30; Trail 31 Oct 1982-1, 26 Nov 1953-1; Manning Park 4 Oct 1974-30; Osoyoos 27 Oct 1976-20, latest flock. **Coastal** – Portland Inlet 22 Sep 1966-6; Masset 25 Oct 1971-25, 25 Nov 1971-25; Kemano 1 Oct 1990-4; McInnes Island 6 Nov 1963-1; Cape Scott 22 Sep 1935-18; Port Hardy 26 Oct 1971-1; Whistler Mountain 2 Sep 1944-150; Sechelt 13 Sep 1985-200; Tofino 30 Nov 1981-1; Pacific Rim National Park 24 Sep 1974-200, 16 Nov 1981-6; Sea Island 15 Sep 1966-1,000, 8 Oct 1965-5,000, 28 Oct 1965-2,700; Serpentine Fen (Surrey) 11 Sep 1978-500, 25 Nov 1976-235; Reifel Island 29 Oct 1986-300; Victoria 10 Sep 1985-200.

Winter: Interior – Vernon 9 Dec 1979-1; Penticton (West Bench) 9 Dec 1976-1. **Coastal** – Green Island 3 Dec 1977-1; Masset 11 Dec 1972-3; Sandspit 20 Dec 1986-21; Chesterman Beach (Tofino) 4 Dec 1985-1; Brunswick Point 28 Feb 1981-40; Chilliwack 11 Jan 1929-1; Delta 28 Feb 1990-50; Cowichan River estuary 17 Jan 1987-13; Blenkinsop Lake 1 Jan 1978-16; Central Saanich 31 Dec 1960-3, 7 Jan 1969-400, 8 Feb 1987-103.

Christmas Bird Counts: Interior – Recorded from 2 of 27 localities and on 1% of all counts. Maxima: Oliver-Osoyoos 28 Dec 1993-8, Kelowna 19 Dec 1992-1. **Coastal** – Recorded from 20 of 33 localities and on 21% of all counts. Maxima: Ladner 22 Dec 1962-**317**, all-time Canadian high count (Anderson 1976b); Victoria 18 Dec 1993-126; Vancouver 20 Dec 1981-40.

Sprague's Pipit

SPPI

Anthus spragueii (Audubon)

RANGE: Breeds on the Great Plains from south-central British Columbia (locally), south-central Alberta, central Saskatchewan, and west-central Manitoba south to eastern Montana, western South Dakota, and northwestern Minnesota. Winters from south-central and southeastern Arizona, southern New Mexico, central and eastern Texas, Arkansas, northwestern Mississippi, and southern Louisiana south through Mexico.

STATUS: In the interior, *casual* and local migrant and summer visitant to the southeastern portion of the Central Interior Ecoprovince in the vicinity of Becher's Prairie near Riske Creek. Breeds.

CHANGE IN STATUS: In North America, the Sprague's Pipit (Fig. 530) is primarily a breeding bird of the Great Plains, but it has recently been recorded on the Pacific coast in autumn and winter. In California, it was first recorded in 1974 (McCaskie 1975), and is now reported annually and considered a "very rare fall transient and even scarcer winter visitor from late September to about mid-March" (Small 1994).

It was first recorded and found breeding in British Columbia in the summer of 1991 (McConnell et al. 1993). It has been seen occasionally in subsequent years, but not found breeding (Hooper 1996). The nearest point in its normal breeding range is about 540 km to the southeast, in the vicinity of Calgary in south-central Alberta (Semenchuk 1992).

Figure 530. The Sprague's Pipit is a very local breeder in the southeastern region of the Central Interior Ecoprovince of British Columbia (Becher's Prairie near Riske Creek, 19 June 1991; Anna Roberts).

NONBREEDING: The Sprague's Pipit has been found in only a small portion of the Chilcotin near Riske Creek, north of Highway 20 between Fish Lake Road and Meldrum Creek Road. The area includes the grasslands of Becher's Prairie. (See BREEDING for a description of habitat.)

The Sprague's Pipit migrates primarily through the eastern Great Plains, but a few birds in autumn migration pass through southwestern Alberta; they may include individuals

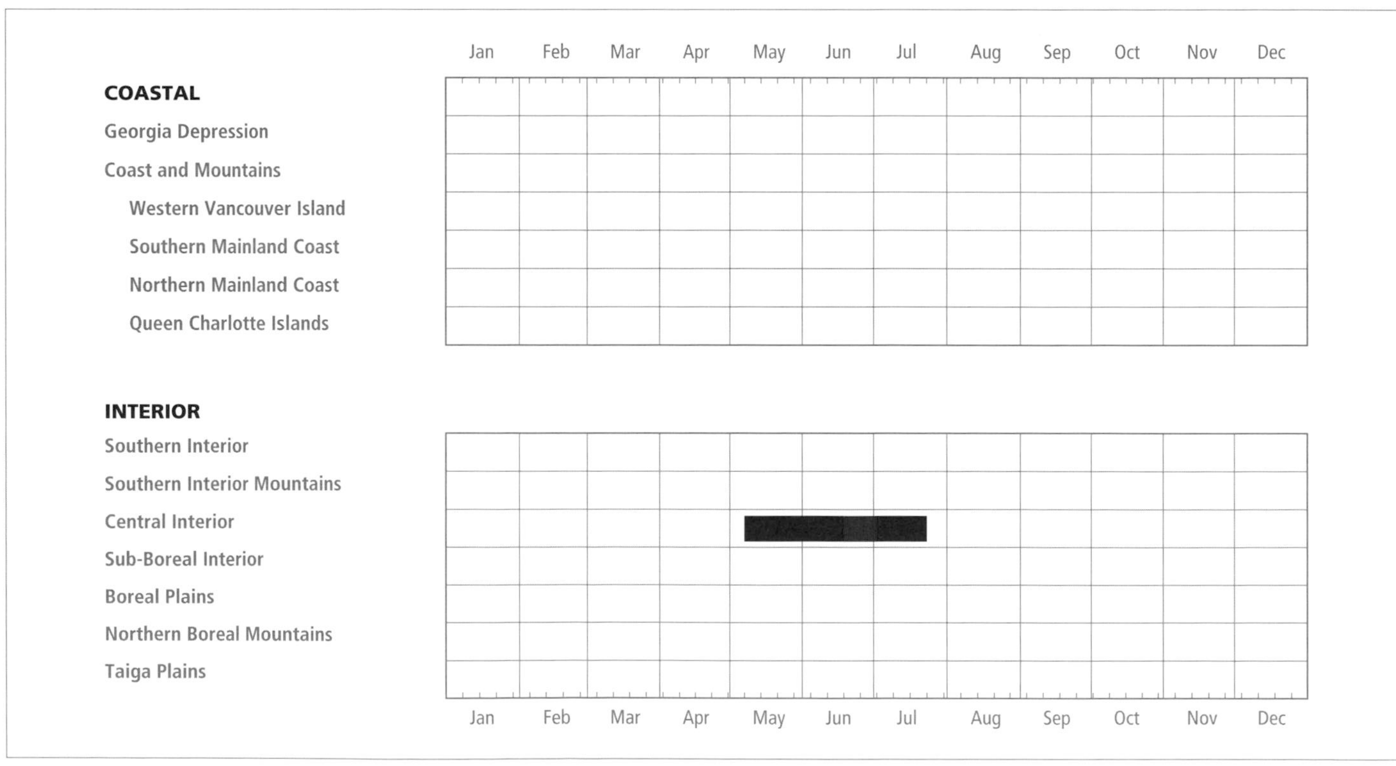

Figure 531. Annual occurrence (black) and breeding chronology (red) of the Sprague's Pipit in ecoprovinces of British Columbia. Records are shown for the week in which they occurred.

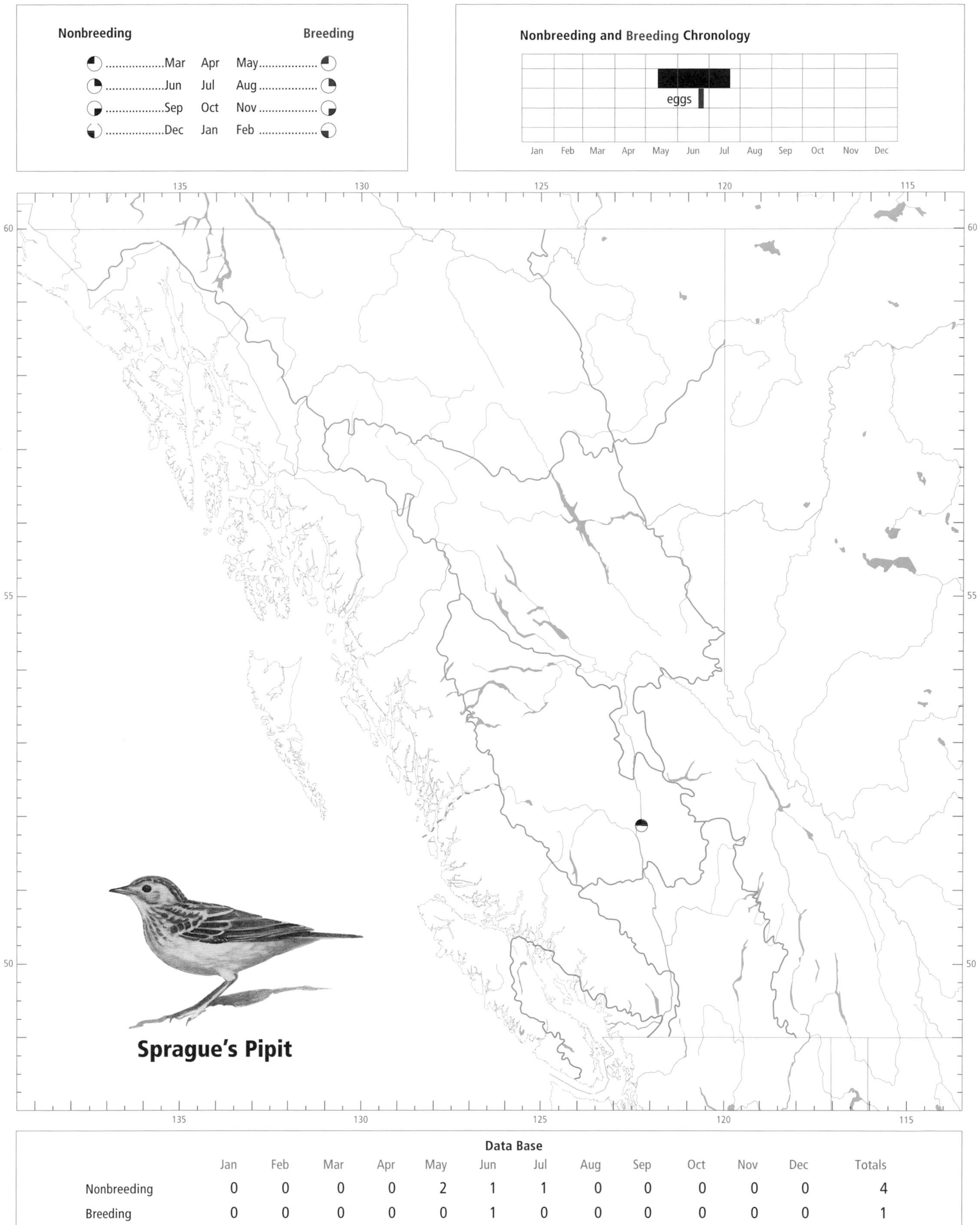

Data Base

	Jan	Feb	Mar	Apr	May	Jun	Jul	Aug	Sep	Oct	Nov	Dec	Totals
Nonbreeding	0	0	0	0	2	1	1	0	0	0	0	0	4
Breeding	0	0	0	0	0	1	0	0	0	0	0	0	1

(a)

(b)

Figure 532. In British Columbia, the Sprague's Pipit breeds in disturbed grasslands with high grass cover and little bare ground (a and b) (Becher's Prairie near Riske Creek, 8 July 1992; R. Wayne Campbell). Dominant vegetation was bluebunch wheatgrass (*Agropyron spicatum*), porcupinegrass (*Stipa spartea*), spreading needlegrass (*Stipa richardsonii*), and Rocky Mountain fescue (*Festuca saximontana*).

that find their way to California. There is no information on spring and autumn migration periods in British Columbia.

In 1992, a singing male Sprague's Pipit was seen many times in the same area on Becher's Prairie between 12 May and 26 June. On 30 June and 8 July, another individual was discovered at a different location in the same area. Evidence of nesting was not found in 1992, nor were any sightings reported in 1993, 1994, or 1995.

In the interior, the Sprague's Pipit has been recorded from 12 May to 21 July (Fig. 531).

BREEDING: The Sprague's Pipit is known to nest only in the Southern Interior, on Becher's Prairie in the Fraser Plateau near Riske Creek, about 45 km southwest of Williams Lake (latitude 51°52′N, longitude 122°21′W).

Its general nesting habitat lies between 585 and 700 m elevation in the grassland and shrub-steppe within the Bunchgrass and Interior Douglas-fir biogeoclimatic zones (Fig. 532a). The dominant vegetation includes mainly bluebunch wheatgrass, porcupinegrass, spreading needlegrass, and Rocky Mountain fescue; pasture sage, junegrass,

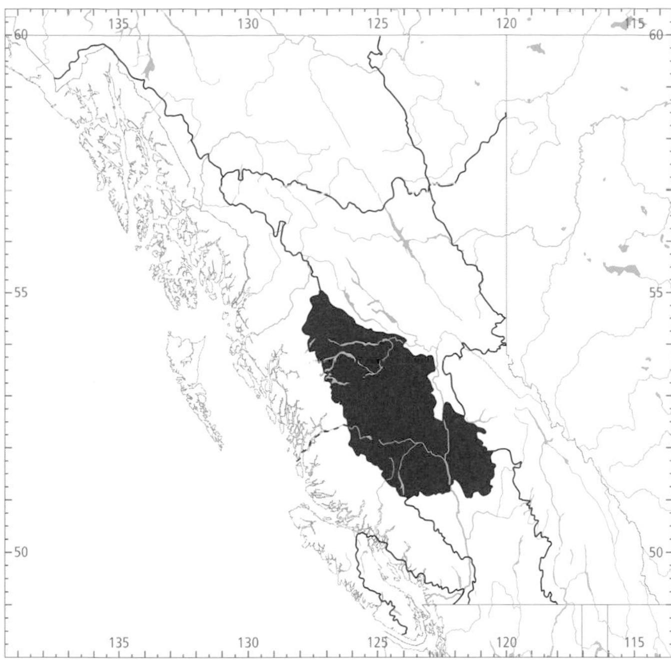

Figure 533. In British Columbia, the highest numbers for the Sprague's Pipit in summer occur in the Central Interior Ecoprovince.

Figure 534. The only nest of the Sprague's Pipit discovered in British Columbia (Becher's Prairie, 25 June 1991; Ruth van den Driessche).

Kentucky bluegrass, and woolly cinquefoil are also common (Hooper and Pitt 1996) (Fig. 532b).

In British Columbia, the Sprague's Pipit has occurred only in the Central Interior (Fig. 533).

The only nest was found on a gently sloping, north-facing site (Fig. 534). The site included a grass cover of 62%, bare ground 27%, and a litter cover of 12%. It had been subject to heavy spring grazing but no summer or autumn grazing (Hooper and Pitt 1996). Elsewhere in its range, this pipit is associated with large areas dominated by grasses of medium height (Johnsgard 1979) where grazing has been light or absent (Maher 1973; Owens and Myres 1973; Kantrud and Kologiski 1983) (Fig. 532).

On 25 June 1991, a nest containing 6 eggs was found on Becher's Prairie (McConnell et al. 1993) (Fig. 534). It was described as "a small depression lined with coarse and fine grass woven into a cup. Almost three-quarters of the nest was covered with a domed roof, formed from long grass adjacent to the nest with additional loose grass interwoven ... The vegetation surrounding the nest was mainly spreading needle grass *Stipa richardsonii*."

On 5 July, the nest was found to have been destroyed by a predator. By 21 July, a male was singing and displaying in the immediate vicinity, suggesting another breeding attempt.

The incubation period for the Sprague's Pipit is unknown, but the fledgling period is 10 to 11+ days (Bent 1950). In British Columbia, calculated dates indicate that breeding occurs from at least 19 June to 11 July (Fig. 531).

REMARKS: In 1959, 2 nests with eggs of the Sprague's Pipit were reportedly discovered in dry grassy fields and followed to fledging at the Kimberley airport, about 18 km northeast of Kimberley in the east Kootenay. The nests were built in clumps of dried grass, and both included domes of protective grasses. In addition, males were watched singing their characteristic flight songs. While this record may be valid, we have not included it in our analysis because it lacks adequate documentation.

Hooper and Pitt (1996) describe the communities and habitat associations of breeding birds in the grasslands of the Chilcotin region of British Columbia, and Johnsgard (1979) provides a general account of grassland birds of the Great Plains.

Dobkin (1992) states that the Sprague's Pipit is declining significantly over its entire range in North America, and attributes the trend to the loss of native prairie habitat and to habitat degradation caused by livestock grazing.

NOTEWORTHY RECORDS

Spring: Coastal – No records. **Interior** – Becher's Prairie 12 May 1992-1 male singing, 17 May 1991-1 male singing (Siddle 1991c; McConnell et al. 1993).

Summer: Coastal – No records. **Interior** – Becher's Prairie 7 Jun 1991-1 male singing (McConnell et al. 1993), 19 Jun 1991-1 (Siddle 1992a), 21 Jul 1991-1 male singing (Siddle 1993b), 25 Jun 1991-female flushed from nest with 6 eggs (McConnell et al. 1993), 3 Jul 1991-nest empty and predation suspected (McConnell et al. 1993), 5 Jul 1991-1 male in flight display (McConnell et al. 1993), 11, 12, and 21 Jul 1991-1 male singing, 30 Jun to 8 Jul 1992-1.

Breeding Bird Surveys: Not recorded.

Autumn: No records.

Winter: No records.

Bohemian Waxwing

Bombycilla garrulus (Linnaeus)

BOWA

RANGE: Breeds from western and northern Alaska, central Yukon, northwestern and southern Mackenzie, northern Saskatchewan, and northern Manitoba south to south coastal Alaska, through the interior of British Columbia and southwestern Alberta to central Washington, northern Idaho, and northwestern Montana. Winters from central Alaska, northern British Columbia, southwestern Mackenzie, and Alberta east to southern Quebec and the Maritime provinces, and south irregularly to Mexico. Also occurs in Eurasia.

STATUS: On the coast, a *very rare* summer visitant in the Georgia Depression Ecoprovince, but locally *common* in autumn and winter during some years; *rare* resident on slopes of the mainland Coast and Mountains Ecoprovince, becoming a *very rare* migrant and winter visitant on the outer coast and *casual* autumn and winter visitant to the Queen Charlotte Islands.

In the interior, an *uncommon* to *fairly common* migrant and summer visitant in the Sub-Boreal Interior, Northern Boreal Mountains, and Boreal Plains ecoprovinces, where it becomes *very rare* in winter; *uncommon* summer visitant and *common* to locally *abundant* migrant and winter visitant in the Southern Interior and Southern Interior Mountains ecoprovinces; *uncommon* to *fairly common* summer visitant and *very common* to *abundant* migrant and winter visitant in the Central Interior Ecoprovince; *uncommon* migrant and summer visitant in the Taiga Plains Ecoprovince, where it is absent in winter.

Breeds.

NONBREEDING: The Bohemian Waxwing (Fig. 535) occurs regularly in spring and autumn migration and during the winter throughout the southern and central portions of the interior. Smaller numbers may be found in winter in the far north. It occurs locally and irregularly on the inner coast during autumn, winter, and spring. It is a highly nomadic species during the nonbreeding seasons. Because the Bohemian Waxwing depends mainly on fruits and berries during winter, and because these vary in abundance, its local distribution varies from year to year. It may occur in large numbers one year and be absent or present in only small numbers the next.

The vast majority of birds that winter in British Columbia do so in the interior from the Chilcotin-Cariboo Basin south through the Southern Interior and Southern Interior Mountains. Substantial numbers also winter in the Peace Lowland. Few remain in winter north of latitude 56°N. During some winters, a few large flocks appear in the Fraser Lowland, but occurrences on the coast are erratic. Few individuals wander to the outer coasts during any season. On the Queen Charlotte Islands and the west coast of Vancouver Island, it occurs mainly as a winter vagrant. The highest numbers in winter occur in the Southern Interior (Fig. 536).

The Bohemian Waxwing occurs from near sea level on the coast to at least 2,400 m in interior mountains. After breeding, it remains at moderate to higher elevations to forage on ripening berry crops and flying insects. During late summer, it occurs in wet coniferous forests, with white or black spruce, tamarack, subalpine fir, and pines growing close to lakeshores, muskegs, swampy areas, river banks, and old burns. These habitats are often interspersed with birches, mountain alder, willows, and shrubs such as red-osier dogwood and highbush-cranberry. Beginning in mid-autumn and during the winter, most birds move to lower elevations, where habitats include open forests, orchards, vineyards, parks, roadside stands of ornamental trees, riparian strips along rivers

Figure 535. The Bohemian Waxwing is a highly nomadic species during the nonbreeding seasons in British Columbia (Nelson, 3 February 1996; Linda M. Van Damme). Numbers vary locally from year to year, depending on the abundance of berries in winter.

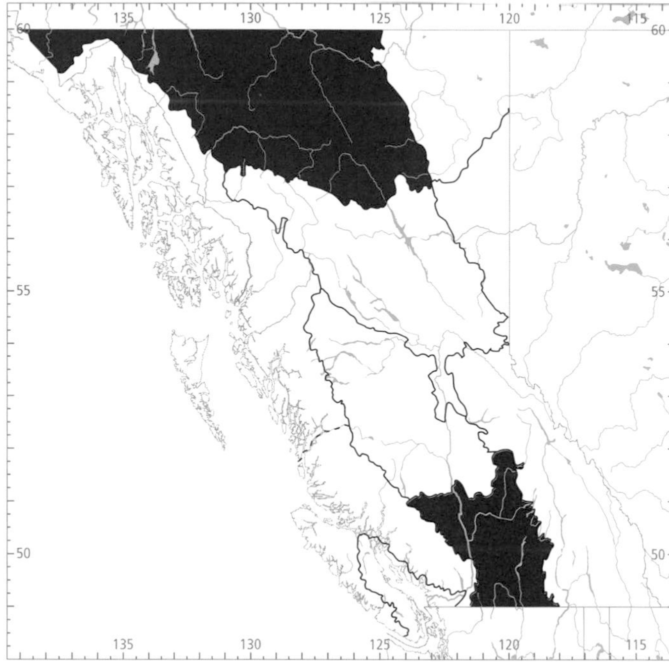

Figure 536. In British Columbia, the highest numbers for the Bohemian Waxwing in winter (black) occur in the Southern Interior Ecoprovince; the highest numbers in summer (red) occur in the Northern Boreal Mountains Ecoprovince.

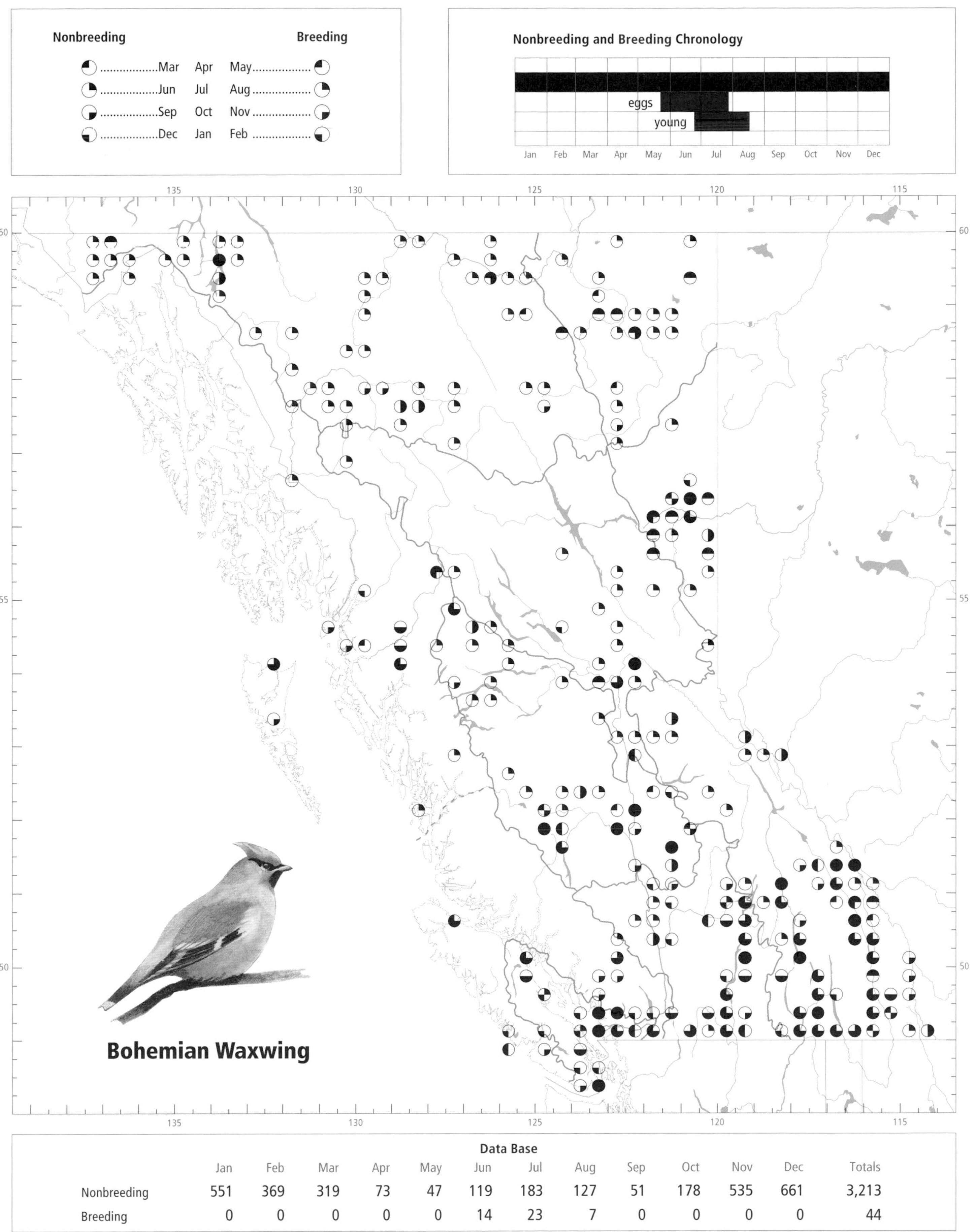

Data Base

	Jan	Feb	Mar	Apr	May	Jun	Jul	Aug	Sep	Oct	Nov	Dec	Totals
Nonbreeding	551	369	319	73	47	119	183	127	51	178	535	661	3,213
Breeding	0	0	0	0	0	14	23	7	0	0	0	0	44

and lakes, and trembling aspen and black cottonwood stands. From mid-autumn through spring, the Bohemian Waxwing forages mainly on berries of mountain-ash, kinnikinnick, cranberries, juniper, and western flowering dogwood, as well as wild rose hips and buds of trembling aspen or black cottonwood. Wintering populations in the Southern Interior take advantage of a dependable supply of food in fruit orchards and vineyards.

Flocks begin to leave valleys in the southern parts of the interior in late February through March, and most wintering birds have gone by April (Fig. 538). Throughout settled parts of southern and central British Columbia, the Bohemian Waxwing all but disappears during late April and May as it disperses to mountainous breeding areas or migrates northward. In the north, the spring movement peaks from late March to mid-April, with smaller nonbreeding flocks still occurring in May and early June. Spring migration for this waxwing occurs about 6 to 8 weeks earlier than for the Cedar Waxwing.

The autumn movement begins in the northern interior in August and peaks in October (Fig. 538); in southern parts of the interior and on the mainland coast, it begins in September and peaks in late November. Southbound migrants have been observed in September moving through alpine and subalpine areas of Yoho National Park (R.R. Howie pers. comm.) and near Cayoosh Creek on the eastern slope of the Eastern Pacific Ranges, but the extent of this alpine movement is unknown. During mid to late autumn, most flocks move to the lower elevations of major valley bottoms, where they occasionally mix with smaller numbers of Cedar Waxwings. Winter populations in the southern parts of the interior build from November through December before peaking in January (Fig. 538).

The Bohemian Waxwing is a gregarious bird during nonbreeding seasons. Most winter flocks in the southern and central portions of the interior and on the mainland coast contain between 50 and 300 birds, although flocks of up to 3,000 birds have been recorded in the Okanagan valley (Cannings et al. 1987). This waxwing is most noticeable when tightly knit flocks cascade down from roost trees to swarm over a fruit-laden tree or when they fly off at high speed, skimming over the treetops.

In the interior, the Bohemian Waxwing has been recorded regularly throughout the year; on the coast, it has been recorded regularly from 13 June to 21 April (Fig. 537).

BREEDING: The Bohemian Waxwing breeds throughout most of the northern and central interior from the east slope of the Rocky Mountains west to the east slope of the Coast Mountains and south, at high elevations, to the Flathead River along the Rocky Mountains and to near Whistler on the south coast. Breeding remains unconfirmed over much of its summer range, including the Taiga Plains and much of the Northern Boreal Mountains and Boreal Plains. Breeding records are scarce in the southern portions of the interior, but small populations likely breed locally in subalpine areas along the major mountain ranges. There are only 3 breeding records from the Southern Interior Mountains, 1 near Kootenay Crossing (Munro and Cowan 1944) and 2 near the Flathead River. There are no confirmed breeding records from the Southern Interior, but breeding is suspected in the Okanagan Highland (Cannings et al. 1987). The only coastal breeding record is from Alta Lake, near Whistler Mountain, in the Eastern Pacific Ranges.

The highest numbers in summer occur in the Northern Boreal Mountains (Fig. 536). Breeding Bird Surveys for

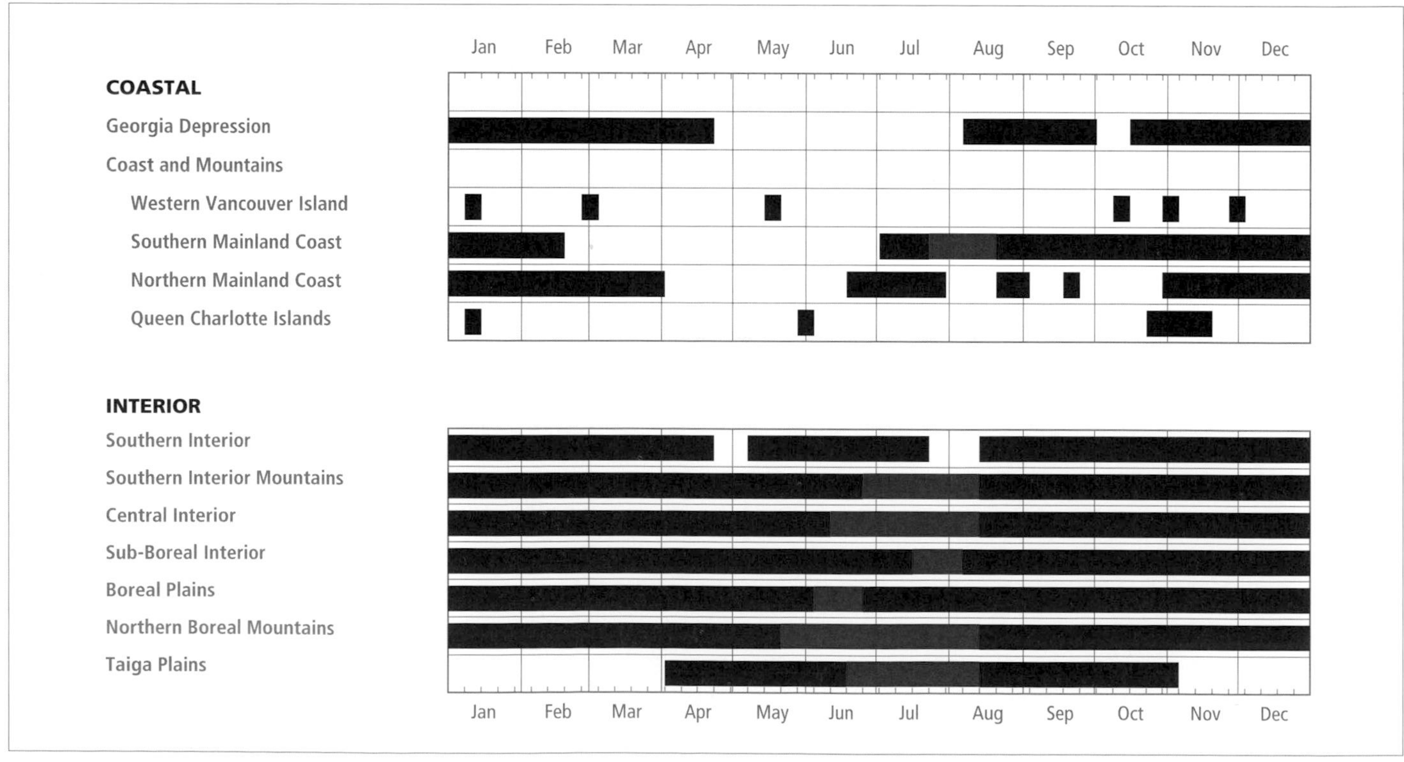

Figure 537. Annual occurrence (black) and breeding chronology (red) for the Bohemian Waxwing in ecoprovinces of British Columbia. Records are shown for the week in which they occurred.

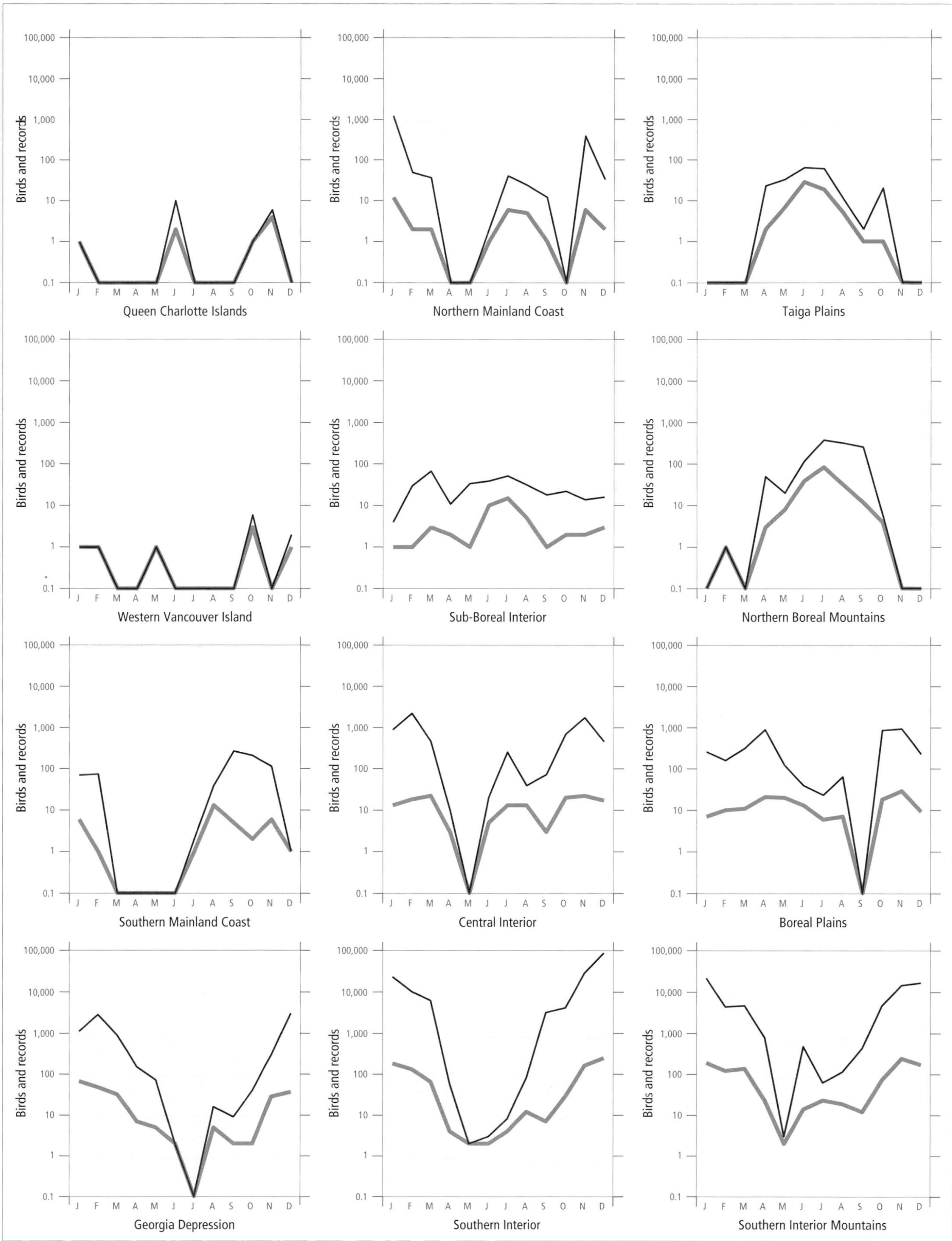

Figure 538. Fluctuations in total number of birds (purple line) and total number of records (green line) for the Bohemian Waxwing in ecoprovinces of British Columbia. Christmas Bird Count and nest record data have been excluded.

both coastal and interior routes contain insufficient data for analysis.

The Bohemian Waxwing has been found breeding at elevations between 900 and 1,550 m in southern and central British Columbia, and between 530 and 880 m in the northern interior. It breeds in conifer-dominated forest habitats with access to an abundance of berries and with damp open areas or bodies of water nearby. Beaver ponds, swamps, muskegs (Fig. 539), and old burns provide good breeding habitat. Such habitat usually includes open mature coniferous (mainly spruce) forest edges with aspens, alders, and willows interspersed and numerous drowned or dead trees. The edges of ponds, swamps, or bogs provide open foraging habitat, and dead trees are used for perch sites when birds are hawking flying insects. In the north, this habitat is common in the Boreal White and Black Spruce and lower Spruce-Willow-Birch biogeoclimatic zones, whereas in the south it occurs in the Engelmann Spruce–Subalpine Fir biogeoclimatic zone.

The Bohemian Waxwing is gregarious even during breeding seasons, and often nests in loose aggregations near good foraging habitat. For example, several pairs nested on a small island in Atlin Lake (Anderson 1915b), 5 pairs were found nesting at a site near 158 Mile House (Brooks 1903), and 3 nests were found along a 30 m section of forest edge at Liard Hot Springs. In addition, there are several records of flocks of 10 to 30 birds foraging together during the nesting season; although confirmation was lacking, observers have speculated that a number of pairs were nesting near each other.

The Bohemian Waxwing breeds relatively late compared with other forest songbirds, with most recorded nestings in late June and July. However, extraordinary records of fledged young in late June near Quesnel and the Liard River suggest that much earlier nesting may occur in some years. The Bohemian Waxwing has been recorded breeding in the province from 22 May (calculated) to 16 August (Fig. 537).

Nests: Most nests ($n = 26$) were in coniferous trees, mainly spruce (9 nests), lodgepole pine (6 nests), and subalpine fir (3 nests). Two nests found in southeastern British Columbia were in a western larch and a mountain alder (Fig. 540), respectively. Two nests near the Yukon border were in deciduous trees – a trembling aspen and a willow. Seven of 10 nests were saddled on horizontal branches near or against the trunk, while 3 were near the tip of the branch. The heights for 19 nests ranged from 1.8 to 9 m, with 12 nests recorded between 3.0 and 7.6 m. Nests were bulky cups of grasses, twigs, and lichens (Fig. 540). Nest lining included feathers, fine grasses, and plant down.

Eggs: Dates for 28 clutches ranged from 11 June to 27 July, with 63% recorded between 26 June and 14 July. Calculated dates indicate that eggs can occur as early as 22 May. Sizes of 25 clutches ranged from 2 to 6 eggs (2E-6, 3E-1, 4E-9, 5E-8, 6E-1), with 69% having 4 or 5 eggs. The incubation period for 1 clutch was 13 days. The incubation period has been reported to be about 14 days (Ehrlich et al. 1988).

Young: Dates for 10 broods ranged from 23 June to 16 August. Calculated dates indicate that young could be found as early as 8 June. Fledged young still dependent on their parents can be found until the end of August. Sizes of 8 broods ranged from 2 to 6 young (2Y-1, 3Y-1, 4Y-5, 6Y-1). The nestling period is 13 to 16 days (Ehrlich et al. 1988).

Brown-headed Cowbird Parasitism: In 34 nests of the Bohemian Waxwing in British Columbia, there were no confirmed instances of cowbird parasitism. The same is true elsewhere in North America (Friedmann 1963; Friedmann and Kiff 1985). A single suggested instance of parasitism involved a deserted waxwing nest containing 3 eggs of the host and 2 of the Brown-headed Cowbird, found at Grand Forks on 27 June 1957. However, because adequate documentation was lacking, and because of the habitat and elevation where the nest was found and the possibility of confusion with the more widespread and common Cedar Waxwing, this record is questionable and has therefore been excluded from the account.

Nest Success: Insufficient data.

REMARKS: There are 2 subspecies of the Bohemian Waxwing, but only *B. g. pallidiceps* is known to occur in North America (American Ornithologists' Union 1957).

The Bohemian Waxwing is a species noted for irruptive movements of large numbers in autumn and winter during some years. At such times they roam more widely than usual,

Figure 539. In northern British Columbia, breeding habitat for the Bohemian Waxwing includes beaver ponds and swamps within coniferous forests. Shorelines usually have numerous dead trees and branches, which provide the necessary perching habitat (Kudwat Creek, Tatshenshini River, 11 June 1993; John M. Cooper).

Figure 540. Bohemian Waxwing nest in the crotch of a mountain alder branch (near Sage Creek, Flathead River, 17 July 1994; R. Wayne Campbell). Nest material included grasses, twigs, plant down, and tree lichens.

and appear in larger flocks. Periods of irruption in British Columbia include the winters of 1947-48, 1966-67, 1976-77, and 1978-79. Kelso (1926) reported "immense numbers" at Revelstoke in the winter of 1917-18. Bent (1950) provides additional dates for irruptions into northern Washington (1916-17, 1919-20, 1931-32), which likely apply to British Columbia as well. These irruptions probably occur after summers of high breeding success during years with deficient food on the normal winter ranges.

Godfrey (1986) includes all of interior British Columbia within the breeding range of the Bohemian Waxwing. In southern portions of the province, it is likely that breeding occurs only locally at high elevations and is much less widespread than suggested.

This species may be involved in east-west movements at times: a bird banded at Penticton on 15 February 1933 was recovered near Rapid City, South Dakota, on 20 March 1934, and a bird banded in Saskatchewan on 19 February 1968 was recovered 5 years later at Okanagan Landing.

The best general reference on life history is still Bent (1950). The breeding biology of the Bohemian Waxwing remains largely unknown.

NOTEWORTHY RECORDS

Spring: Coastal – Victoria 21 Apr 1978-4; Vancouver 4 Mar 1975-150, 16 Mar 1982-34, 9 Apr 1975-10; Chilliwack 29 Mar 1980-33; Agassiz 21 Mar 1976-22; Campbell River 24 Mar 1976-17; Port Hardy 14 May 1936-1; Kitimat 11 Mar 1975-1 (Hay 1976); 60 km e Prince Rupert 28 Mar 1977-36. **Interior** – Summerland 10 Mar 1967-1,000; Ta Ta Creek 9 Apr 1948-180 (Johnstone 1949); Okanagan Landing 12 May 1922-1 (UMMZ 16259); Nakusp 17 Apr 1976-1; Salmon River (Falkland) 19 Apr 1977-23; Brisco 20 Mar 1977-250; Celista 19 Apr 1948-18; Yoho National Park 29 Mar 1977-1,000 (Wade 1977); Kleena Kleene 24 Mar 1956-60 (Paul 1959); Williams Lake 15 Mar 1973-100, 2 Apr 1977-6; Quesnel 3 Mar 1979-64; Nechako River 14 May 1960-34; Chetwynd 18 Apr 1980-10; Moberly River 24 Mar 1985-150; Cache Creek (Peace River) 20 May 1979-20; Hudson's Hope 14 Apr 1979-150, 5 May 1979-30; Halfway River (Peace River) 14 Apr 1984-205; Fort Nelson 4 Apr 1987-20; Helmet (ne Kwokullie Lake) 15 May 1982-12; Liard Hot Springs 27 Apr 1975-40 (Reid 1975).

Summer: Coastal – Victoria 7 Aug 1892-1; Mount Seymour 15 Aug 1985-6; Widgeon Lake 31 Jul 1970-1 adult, 3 fledglings (Campbell et al. 1972a); Berkey Creek 26 Aug 1978-10; Alta Lake 14 Aug 1941-nestlings (Munro and Cowan 1947); Shearwater 29 Jul 1986-3; Kimsquit River 12 Jul 1986-30; Masset 2 Jun 1946-5; Kispiox River 22 Aug 1921-20 (Swarth 1924); Haines Highway 19 Jul 1956-20 at Mile 48. **Interior** – Manning Park 10 Jul 1985-2, 16 Aug 1965-12; Couldrey Creek 7 Aug 1957-nestlings; Mara Meadows 1 Jul 1971-3 (Rogers 1971d); Bridge River 15 Aug 1985-40; Scout Mountain 21 Aug 1983-7; Kootenay Crossing 19 Jun 1943-pair nesting (Munro and Cowan 1944); Kootenay National Park 23 Jul 1976-1 fledgling; 158 Mile House 11 Jun 1901-10, 15 Jun 1901-2 nests with 4 eggs (Brooks 1903); Nimpo Lake 7 Aug 1985-4 nestlings (Campbell and Gibbard 1986); Berg Lake 22 Aug 1978-20; Indianpoint Lake 1 Jul 1941-4 eggs, 25 Jul 1934-6 nestlings, 26 Jul 1937-4 nestlings just hatched; Whitesail River 21 Jul 1936-200 over several miles (Munro and Cowan 1947); Nechako River 22 Jul 1960-20; Narraway River 25 Aug 1977-10; Burnie Lakes 23 Aug 1975-fledglings; Mount Milburn 22 Jun 1991-3 fledglings; Pine Pass 24 Aug 1975-13; Sudeten Park 9 Jun 1976-11; Sukunka River 7 Jul 1976-3; Manson River 6 Jul 1967-building nest; Chetwynd 22 Jul 1975-10; Dawson Creek 27 Jul 1975-8; Jackfish Lake 27 Jul 1930-2 fledglings; Fort St. John 23 Aug 1985-20; Hyland River 6 Jul 1978-50; Dokdaon Creek 15 Jul 1919-eggs (Swarth 1922); Kwadacha Wilderness Park 19 Aug 1979-100 (Cooper and Adams 1979); Telegraph Creek 26 Jun 1919-eggs (Swarth 1922), 5 Jul 1919-6 (MVZ 40132-37); Firesteel River 22 Jun 1976-14; Liard Hot Springs 15 Jun 1976-3 pairs building nests; Fishing Creek 29 Jun 1981-4 fledglings; Atlin 13 Jun 1924-6 eggs (UBC 1186), 4 Jul 1980-4 nestlings, 8 Jul 1914-5 eggs (RBCM E450; Anderson 1915), 14 Jul 1931-2 nests with eggs (Hanna 1931); Mile 48 Haines Highway 19 Jul 1956-20; Tatshenshini River 12 Jun 1993-25; Survey Lake 27 Jun 1980-5 eggs.

Breeding Bird Surveys: Coastal – Recorded from 1 of 27 routes and on less than 1% of all surveys. Maximum: Pemberton 2 Jul 1977-2. **Interior** – Recorded from 13 of 73 routes and on 4% of all surveys. Maxima: Punchaw 3 Jul 1976-16; Steamboat 8 Jul 1979-8; Prince George 16 Jun 1986-8; Chilkat Pass 1 Jul 1976-6.

Autumn: Interior – Liard River 23 Oct 1980-3; Clarke Lake (Fort Nelson) 7 Sep 1978-2; Hyland Post 10 Sep 1976-100; Buckinghorse River 29 Oct 1980-20; North Pine 26 Oct 1985-100; Fort St. John 29 Oct 1986-500, 12 Nov 1983-75; Willow River 8 Sep 1982-18, 10 Nov 1966-8; Black Dome Mountain 26 Sep 1988-50; Walcott 27 Oct 1985-200; Williams Lake 16 Oct 1958-60, 25 Nov 1982-300; Kleena Kleene 22 Oct 1992-28; Golden 30 Oct 1975-1,000; Emerald Lake 3 Oct 1973-1,000 (Wade 1977); Nakusp 21 Oct 1984-26; Vernon 24 Oct 1968-2,000; Coldstream 20 Nov 1956-2,000; Nelson 30 Nov 1978-1,000; Michel Creek 20 Oct 1984-300; Penticton 7 Oct 1977-2 (Cannings et al. 1987), 31 Oct 1980-500, 21 Nov 1979-1,900. **Coastal** – Terrace 24 Nov 1966-300; Masset 7 Nov 1945-3 (Munro and Cowan 1947); Queen Charlotte City 27 Oct 1971-1; Garibaldi Park 23 Sep 1980-200; Squamish 25 Nov 1979-40; Pitt Lake 18 Nov 1975-51; West Vancouver 25 Sep 1932-large numbers (Turnbull 1933); North Vancouver 30 Oct 1971-30 (Campbell et al. 1972b); Stanley Park (Vancouver) 25 Nov 1973-80; Granite Creek (Nitinat Lake) 10 Oct 1980-4.

Winter: Interior – Atlin 24 Feb 1946-1 (UBC 4604); Fort St. John 19 Dec 1981-50, 23 Jan 1987-70, 2 Feb 1987-30; Topley 21 Jan 1979-4; Prince George 20 Dec 1975-10; Quesnel 24 Feb 1979-50; Williams Lake 3 Dec 1984-300, 2 Jan 1989-1,140; Chilcotin River 9 Feb 1986-1,250; Tatlayoko Lake 2 Jan 1989-200; Westbank 30 Dec 1978-3,000+ (Cannings et al. 1987); Summerland 1 Jan 1979-2,600, 8 Jan 1968-1,500; Cranbrook 26 Dec 1991-2,759; Nelson 30 Dec 1978-2,000, 26 Jan 1986-1,000; Princeton 30 Jan 1977-375; Creston area 1 Feb 1982-425. **Coastal** – Terrace 25 Jan 1979-300; Masset 13 Jan 1948-1 (RBCM 10529); Kitimat 5 Jan 1975-450 (Hay 1976); Malcolm Island 8 Jan 1976-1; Alta Lake 2 Jan 1937-3 (UBC 4607-09); Harrison Hot Springs 16 Feb 1976-75; Agassiz 29 Dec 1981-25; Mission 26 Jan 1985-65; Vancouver 19 Dec 1947-1,000, 2 Feb 1948-900; Nanaimo 8 Jan 1972-264; Tofino 1 Dec 1985-2; Ucluelet 28 Feb 1975-1 (BC Photo 418).

Christmas Bird Counts: Interior – Recorded from 25 of 27 localities and on 80% of all counts. Maxima: Vernon 16 Dec 1990-11,133; Penticton 28 Dec 1985-6,893; Nelson 5 Jan 1986-5,600. **Coastal** – Recorded from 16 of 33 localities and on 17% of all counts. Maxima: Terrace 22 Dec 1985-3,037; Vancouver 26 Dec 1971-436; Kitimat 18 Dec 1978-259.

Cedar Waxwing
Bombycilla cedrorum Vieillot

CEWA

RANGE: Breeds from southeastern Alaska, central British Columbia, northern Alberta, Saskatchewan, Manitoba, Ontario, central Quebec, New Brunswick, Prince Edward Island, Nova Scotia, and Newfoundland south through most of the United States to northern California and Georgia. Winters from southern British Columbia, southern Ontario, southern Quebec, and the Maritime provinces south through much of the United States and Central America to Panama.

STATUS: *Fairly common* to *common* summer visitant across southern British Columbia from southeastern Vancouver Island to the Rocky Mountains; becoming a locally *abundant* autumn migrant and *uncommon* winter visitant in the lower Fraser River valley, southeastern Vancouver Island, and the Okanagan valley, and locally in the west Kootenay. *Fairly common* migrant and summer visitant in the Southern Interior Mountains, Central Interior, and Sub-Boreal Interior ecoprovinces north to Prince George, and in the Peace Lowland of the Boreal Plains Ecoprovince; *very rare* in winter in the southern Sub-Boreal Interior and northern Southern Interior Mountains; *rare* summer visitant in the Northern Boreal Mountains Ecoprovince; *very rare* on the Queen Charlotte Islands and exposed outer coasts of the Coast and Mountains Ecoprovince; *fairly common* summer visitant in the Taiga Plains Ecoprovince.

Breeds.

Figure 541. The Cedar Waxwing is widely distributed throughout British Columbia in summer, but in winter the abundance of food in the form of berries and fruit determines its distribution across the southern parts of the province (Richter Pass, 9 August 1996; R. Wayne Campbell).

Figure 542. In British Columbia, the highest numbers for the Cedar Waxwing in winter (black) and summer (red) occur in the Georgia Depression Ecoprovince.

NONBREEDING: The Cedar Waxwing (Fig. 541) occurs regularly throughout south and central British Columbia south of the Nechako and Skeena rivers, including the mainland coast and southeastern Vancouver Island. Small numbers occur irregularly on the Queen Charlotte Islands and the west coast of Vancouver Island. In the Boreal Plains and Taiga Plains in northeastern British Columbia, it is widely distributed in suitable habitat. Elsewhere in the far north it is scarce.

During winter, the Cedar Waxwing is usually absent north of the inner south coast, the Okanagan valley, and the Kootenay region. A few birds winter along the Skeena River. As in the case of the Bohemian Waxwing, winter numbers are highly variable because of the Cedar Waxwing's dependence on foods that vary in abundance from year to year. The highest numbers in winter occur in the Georgia Depression Ecoprovince, particularly on southeastern Vancouver Island (Fig. 542).

The Cedar Waxwing occurs at elevations from sea level on the coast to at least 2,250 m in the interior, but it is mainly a species of lower elevations. It tends to frequent valley bottoms and lower mountain slopes, and is generally segregated from the Bohemian Waxwing, which prefers higher elevations, except during winter, when both species may occur together in lowland areas.

During migration and the post-breeding period, the Cedar Waxwing frequents many open forest habitats, but is most numerous in habitats resulting from human activities, such as roadside thickets, orchards, urban parks, brushy gardens, private yards with berry trees (Fig. 543), hedgerows, and shrubbery along powerlines and sewage lagoons. In natural situations, it frequents second-growth Douglas-fir forest; mixed-aged forests of western hemlock, western redcedar, Engelmann spruce–subalpine fir, ponderosa, and lodgepole pine, the forested margins of tidal flats, creekside thickets of willow, red-osier dogwood, and elderberry (Fig. 544); riparian stands of black cottonwood, groves of Rocky Mountain juniper, beaver ponds, burns, and spruce and larch muskegs.

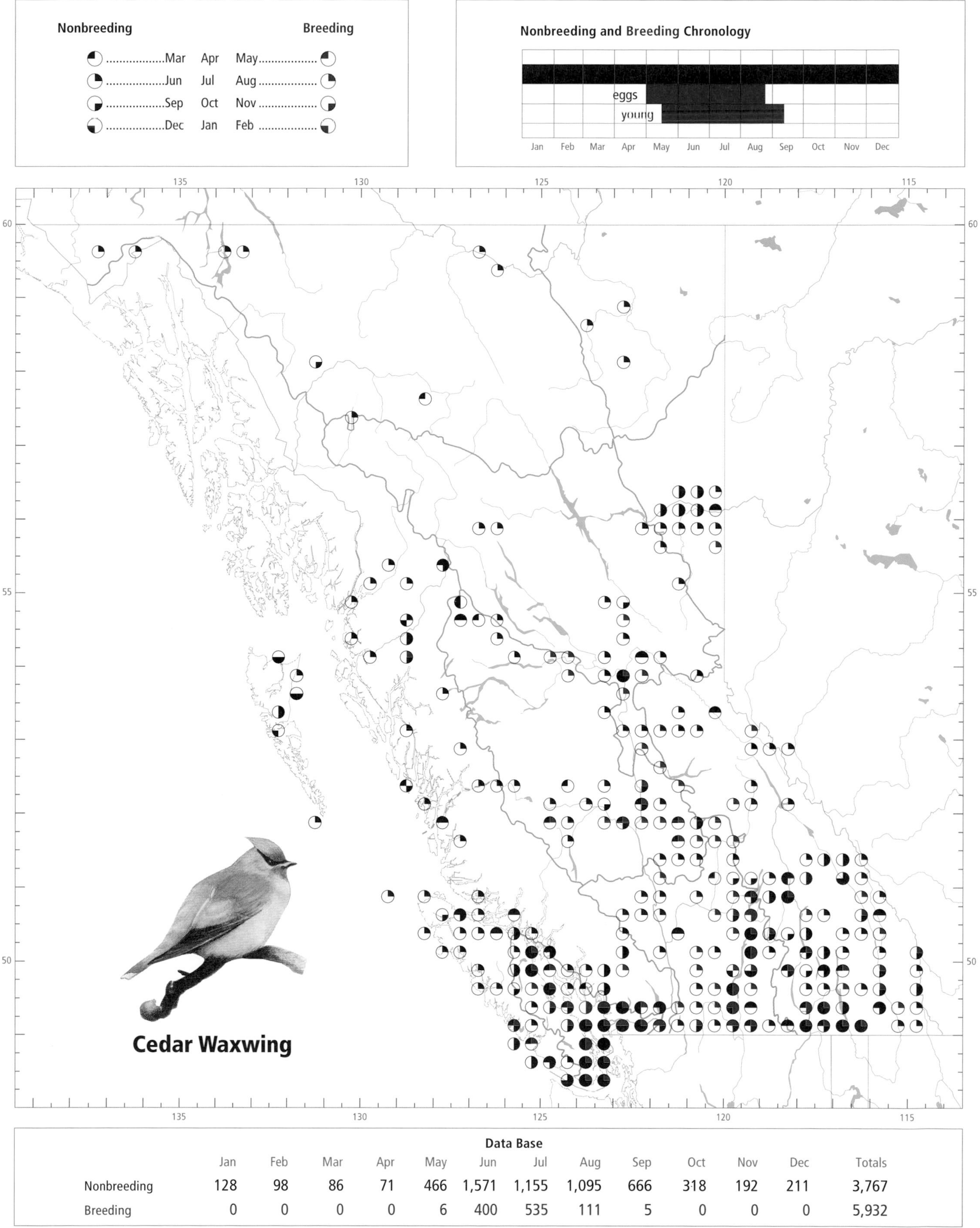

Data Base

	Jan	Feb	Mar	Apr	May	Jun	Jul	Aug	Sep	Oct	Nov	Dec	Totals
Nonbreeding	128	98	86	71	466	1,571	1,155	1,095	666	318	192	211	3,767
Breeding	0	0	0	0	6	400	535	111	5	0	0	0	5,932

Figure 543. In autumn and winter, mountain-ash trees laden with berries attract flocks of Cedar and Bohemian waxwings (Revelstoke, 14 October 1992; R. Wayne Campbell).

Like the Bohemian Waxwing, the Cedar Waxwing is attracted by abundant crops of berries and small fruits (McPherson 1987) (Fig. 543). During autumn and winter, fruits of salal, mountain ash, arbutus, hawthorns, and many ornamental shrubs are eaten.

Spring migration may begin in April, but most birds arrive in the last part of May across the southern and central portions of the interior and on the south coast (Figs. 545 and 546). The spring movement peaks in early June from southern areas north through the Chilcotin-Cariboo Basin to Prince George. In the Boreal Plains, northern interior, and Queen Charlotte Islands, early migrants appear in June (Fig. 546).

Post-breeding flocks of juveniles and adults begin to gather in late July, although many birds are still nesting. Throughout its range in the province, populations grow through August as flocks form and the autumn movement begins. All birds have departed by the end of September in the northern interior, south to the Skeena and Omineca mountains, and by mid-November in the Central Interior. Large numbers occur regularly in the southern portions of the interior and in the Georgia Depression from late August through November. Migrants and wintering birds usually occur in flocks of between 10 and 50 birds, but flocks of 100 to 200 may occur at good foraging sites. In the interior, flocks often mix with larger numbers of Bohemian Waxwings during winter.

Regular wintering populations occur in southern British Columbia in the Georgia Depression, the Okanagan valley, the west Kootenay from Revelstoke south through the Arrow Lakes and Columbia River valley to Trail, Golden, and the Creston valley (Fig. 545). In the north, a few birds winter regularly near Terrace and recently have been found near Prince George. In the south, large numbers are present in winter only during some years. In Vancouver and Victoria, all the highest winter counts occurred between 1956 and 1973. There are several winter records for the Queen Charlotte Islands.

On the south coast and in the southern interior, the Cedar Waxwing occurs regularly throughout the year; north of Williams Lake, it occurs regularly only from 18 May to 28 September (Fig. 545).

BREEDING: The Cedar Waxwing breeds across southern British Columbia from Western Vancouver Island to the Rocky Mountain Trench and north to the upper Fraser River and Nechako River valleys, and in northeastern portions of the province. Breeding occurs mainly in the lowland valleys of the Kootenay, Columbia, Okanagan, Similkameen, Thompson, and lower Fraser rivers; the Cariboo and Chilcotin areas; and southeastern Vancouver Island. West of the Rocky Mountains, the northernmost nesting locality is Kerry Lake, about 70 km north of Prince George. It is likely that breeding occurs in northern British Columbia more often than is known at present, particularly in the Peace Lowland and along the central and northern mainland coast. Along the hundreds of kilometres of coast north of Powell River, breeding has been confirmed only at Kitimat.

The Cedar Waxwing reaches its highest numbers in summer in the Georgia Depression (Fig. 542). An analysis of Breeding Bird Surveys for the period 1968 through 1993 could not detect a net change in numbers on coastal or interior routes.

Figure 544. Wetlands with red alder, sedges, willows, and red-osier dogwood provide late summer and early autumn foraging areas for the Cedar Waxwing in southern British Columbia (Vaseux Lake, 9 August 1996; R. Wayne Campbell).

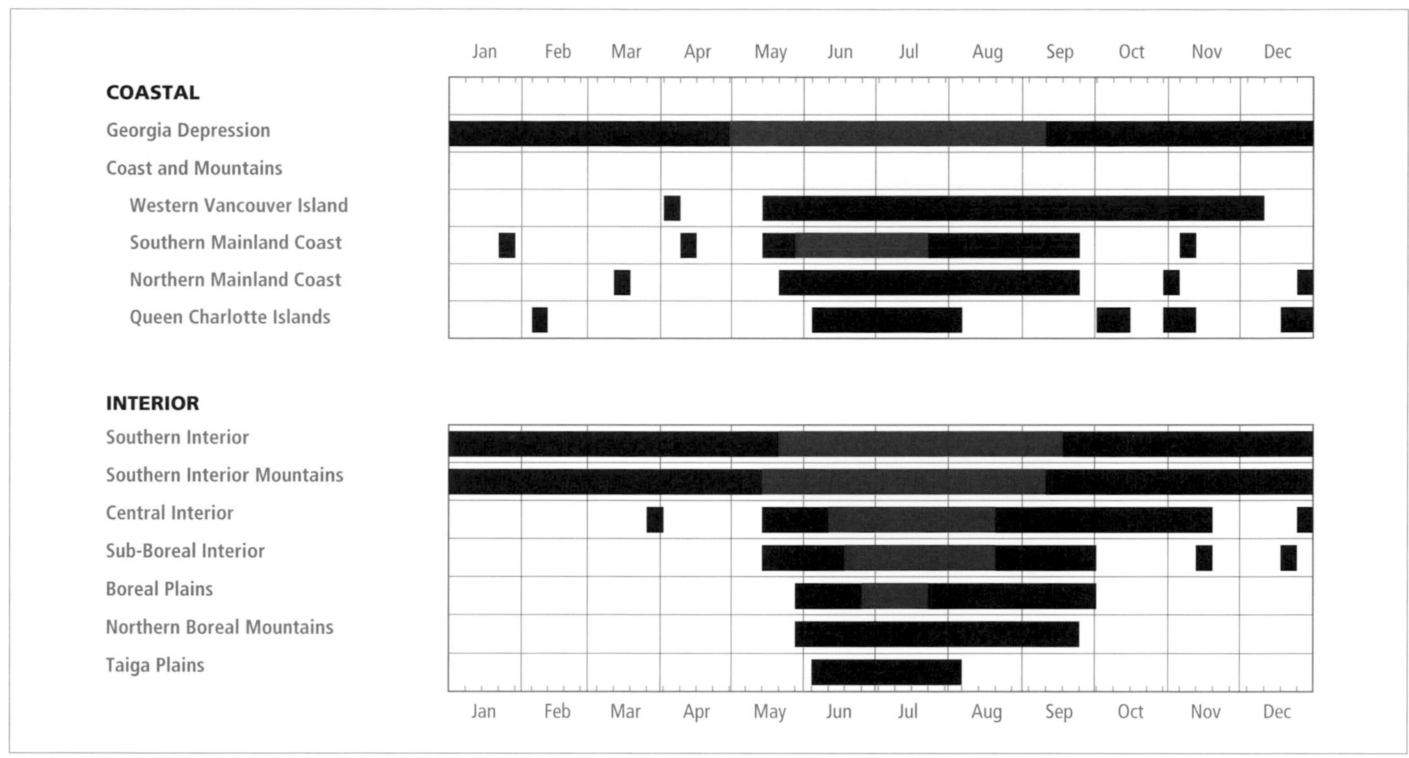

Figure 545. Annual occurrence (black) and breeding chronology (red) for the Cedar Waxwing in ecoprovinces of British Columbia. Records are shown for the week in which they occurred.

Breeding occurs at elevations from just above sea level on the coast to at least 1,530 m in the interior. In general, the Cedar Waxwing breeds at much lower elevations than does the Bohemian Waxwing. The Cedar Waxwing prefers the edges of mixed woodlands, often near water, where there are large shrubs and open spaces. Most nests (n = 391) were reported from habitats created by humans, such as suburban gardens (42.7%) and rural areas (18.7%), where low trees and shrubs are abundant (Fig. 547), water is available, and there are good supplies of insects, berries, and other soft fruit. Orchards are also extensively used. Natural breeding habitats include riparian thickets and forests along rivers, edges of lakes and beaver ponds (Fig. 548), tamarack muskegs, the edges of second-growth pine, spruce, or Douglas-fir forests, and heavily burned areas regenerating to willow, vine maple, red huckleberry, and trailing blackberry. During the breeding season, fruits of red elderberry, soopolallie, red-osier dogwood, saskatoon, Pacific crab apple, salmonberry, red huckleberry, blueberry, cranberry, and cherries are consumed. The Cedar Waxwing also captures flying insects during spring and summer, and has been observed taking insect larvae from floating lily pads (Thacker 1938).

On the coast, the Cedar Waxwing has been recorded breeding between 1 May (calculated) and 3 September; in the interior, it has been recorded breeding between 6 May (calculated) and 7 September (Fig. 545). In southern areas, the breeding season extends for about 4 months, whereas in central regions of the interior it extends for about 10 weeks.

Nests: Most nests (n = 474; Fig. 549) were built in trees (87.6%) or shrubs (11.6%). Other nest sites included in vines (2 nests), on a cliff (1), and on a building (1). Deciduous trees were used about 3 times as often as coniferous trees, except in the Southern Interior, where they were used about twice as often. Twenty-nine species of deciduous trees were used for nest sites, mainly willows, vine maple, red alder, and domestic fruit trees; 9 species of conifers were used, mainly spruce and Douglas-fir, and 15 species of shrubs.

Nests (n = 222) were mainly saddled on branches (58.1%), in the crotch of a branch (23.0%), or on a branch next to the trunk (11.3%). Heights for 492 nests ranged from ground level (a cliff nest) to 12 m, with 56% between 2.0 and 4.0 m. All nests were bulky cups of grass and twigs, moss, plant stems, rootlets, and lichens, and often contained human-made fibres. In the Georgia Depression and the Coast and Mountains, moss was used most frequently (41%; n = 126 nests), whereas in the interior the use of lichens predominated. Nest linings included finer grasses, plant down, hair, and feathers.

Eggs: Dates for 381 clutches ranged from 12 May to 24 August, with 53% recorded between 19 June and 12 July. Calculated dates indicate that eggs can occur as early as 1 May. Sizes of 334 clutches ranged from 1 to 7 eggs (1E-11, 2E-36, 3E-52, 4E-101, 5E-120, 6E-13, 7E-1), with 66% having 4 or 5 eggs. Similar clutch sizes were reported for a large sample, mainly from eastern North America (Leck and Cantor 1979). The incubation period in British Columbia is 12 to 13 days (n = 7). Egg laying generally occurs 3 to 4 weeks earlier in the south than in central parts of the province, and about a week earlier on the south coast than in the southern portions of the interior (Fig. 545).

Young: Dates for 238 broods ranged from 15 May to 11 September, with 50% recorded between 7 July and 28 July.

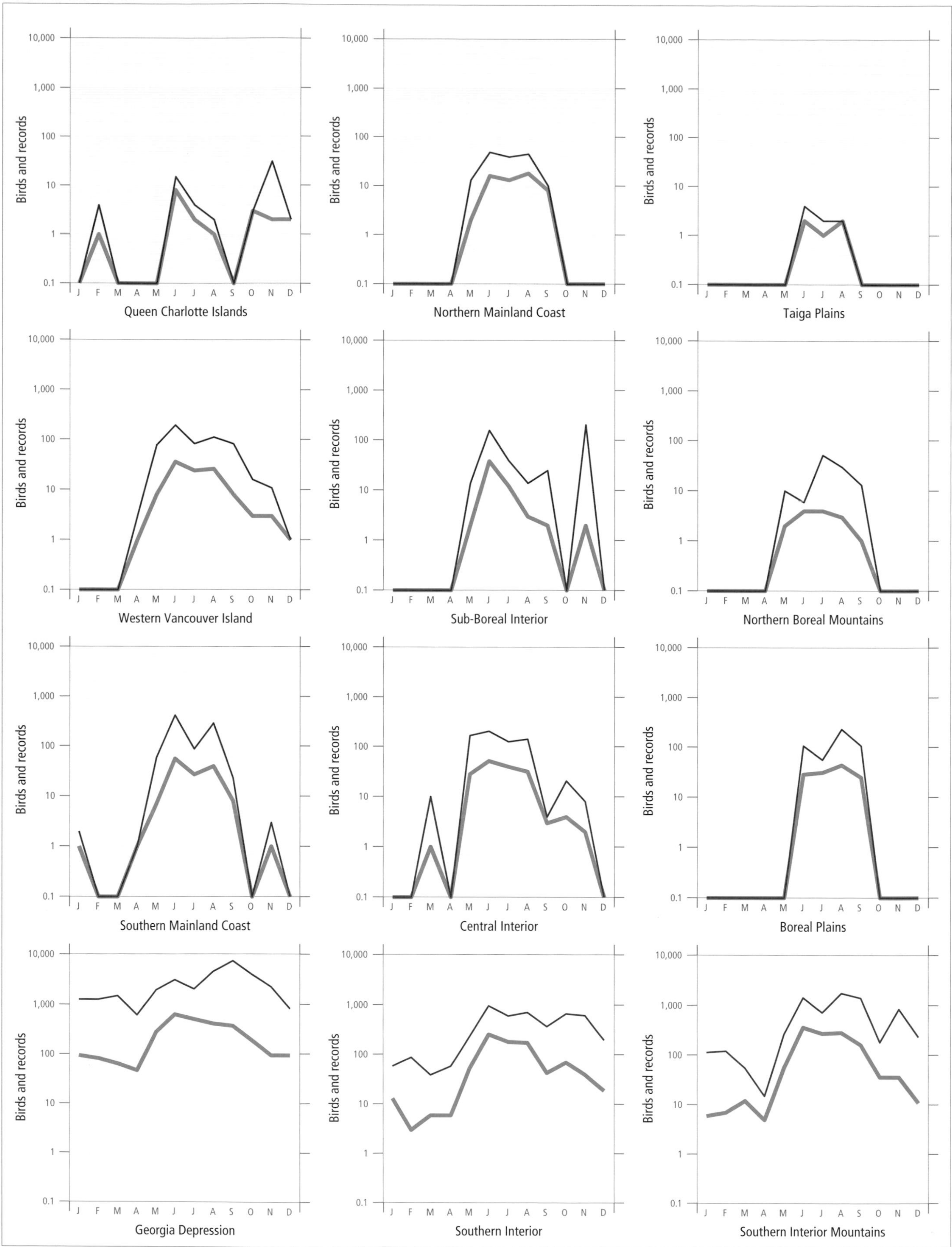

Figure 546. Fluctuations in total number of birds (purple line) and total number of records (green line) for the Cedar Waxwing in ecoprovinces of British Columbia. Christmas Bird Count and nest record data have been excluded.

Figure 547. Cedar Waxwing breeding habitat in the west Kootenay region of British Columbia may include open rural fields with bushes for nesting (Castlegar, 19 June 1994; Linda M. Van Damme).

Figure 548. Riparian thickets surrounding beaver ponds and sloughs provide breeding habitat for the Cedar Waxwing in the Peace River region of British Columbia (Bear Flat, 16 June 1990; R. Wayne Campbell).

Sizes of 218 broods ranged from 1 to 6 young (1Y-5, 2Y-20, 3Y-60, 4Y-79, 5Y-50, 6Y-4), with 64% having 3 or 4 young. The nestling period in British Columbia is 15 to 18 days (n = 6). There is no convincing evidence that the Cedar Waxwing is double-brooded in British Columbia, but the unusually long breeding season suggests that some pairs may raise 2 broods annually.

Brown-headed Cowbird Parasitism: In British Columbia, 4% of 445 nests recorded with eggs or young were parasitized by the cowbird (Fig. 549). All cases of parasitism were in the interior (7%; n = 290); on the coast, none of 155 nests was parasitized. Of the 20 nests found with cowbird eggs or young, 18 were in the Southern Interior (Fig. 549), 1 was near Castlegar, and 1 was near Prince George. Of 13 nests with cowbird eggs and followed to hatching, 6 were deserted before eggs hatched, 4 hatched waxwing young only, 2 hatched waxwing and cowbird young, and 1 was destroyed. Only 4 of the 6 parasitized nests containing nestlings were followed to completion, and each produced only waxwing fledglings. However, there were 3 records of a Cedar Waxwing feeding a cowbird fledgling. Other studies showed variable but low or no rates of parasitism: 5.6% to 9.8% in Michigan (Rothstein 1976a), 1.2% in southern Quebec, and 0% in Ohio (Putnam 1949). The Cedar Waxwing is one of the smaller host species that routinely ejects cowbird eggs, but also has a relatively high rate of desertion when cowbird eggs are present. The normally prompt ejection of cowbird eggs by the Cedar Waxwing probably results in lower estimates of attempted parasitism than actually occurs (Rothstein 1976b).

Nest Success: Of 195 nests found with eggs and followed to a known fate, 104 produced at least 1 fledgling, for a success rate of 53%. Nest success at the coast was 40% (n = 40); in the interior, it was 57% (n = 155). Causes of nest losses (n = 63) included predation (68%), weather (11%), young dead in nest of unknown causes (11%), and desertion after cowbird parasitism (10%). Eleven predators were implicated in nest losses: squirrels (15 nests), American Crow (4 nests), Steller's Jay (3 nests), chipmunks (2 nests), domestic cats (2 nests), House Wren, American Kestrel, mink, snake, mouse, and hawk.

REMARKS: There are 20 records of birds banded in the Modesto and Alameda regions of California that were recovered on the southwest coast of British Columbia (Stoner 1969). One bird banded in California was recovered near Grand Forks. There is also a record of a bird banded in Iowa on 1 November 1969 and recovered near Salmo in June 1970.

For additional life-history information, see Putnam (1949) and Bent (1950).

Figure 549. Nest with 4 Cedar Waxwing eggs and 1 egg (far right) of the Brown-headed Cowbird (Castlegar, 24 June 1994; Linda M. Van Damme). About 4% of Cedar Waxwing nests found in British Columbia were parasitized by the cowbird.

NOTEWORTHY RECORDS

Spring: Coastal – Victoria 9 Mar 1983-200, 20 May 1986-225; Saanich 1 Mar 1983-92, 9 Apr 1983-75; Tofino 3 Apr 1983-3; West Vancouver 15 May 1968-3 nestlings; Chilliwack 29 May 1984-nestlings; Harrison Hot Springs 10 Apr 1980-1; Garibaldi Park 14 May 1963-6; Box Island 29 May 1983-40; Nootka Sound 8 May 1910-1 (MVZ 16316); Port Neville 25 May 1978-14; Namu 26 May 1983-12. **Interior** – Oliver 30 May 1956-7 eggs plus 1 cowbird egg; Vaseux Lake 26 May 1974-eggs; Summerland 21 Apr 1967-30, 9 May 1967-30; South Slocan 27 May 1979-26; Vernon 13 Mar 1971-22, 12 May 1936-first arrival (Cannings et al. 1987); Nakusp 22 Apr 1985-8; Mount Revelstoke 21 Mar 1971-20; Barriere 27 May 1984-8; Williams Lake 18 May 1988-35; s Prince George 20 May 1987-10; Willow River 26 May 1986-4; Beatton River 28 May 1922-first arrival (Williams 1933a); Hyland Post 30 May 1976-5.

Summer: Coastal – Oak Bay 20 Aug 1985-225; Saanich 30 Aug 1980-175; Carmanah Point 7 Jul 1979-8; Bamfield 13 Aug 1983-2 adults with 2 young; Fraser Mills 2 Jun 1960-100; Pitt Meadows 1 Jul 1959-50, 15 Aug 1976-4 eggs; nw Pitt Meadows (UBC Research Forest) 6 Jun 1960-5 eggs, 13 Jul 1960-4 nestlings; Harrison Hot Springs 3 Jun 1984-35; Courtenay 23 Aug 1972-4 nestlings; Alta Lake 17 Aug 1937-66 in 4 flocks; Port Neville 15 Jun 1975-16; Cape Scott 1 Aug 1981-20; Triangle Island (Scott Islands) 23 Jun 1974-2 (Vermeer et al. 1976); Bella Bella 1 Jun 1968-8; Cape St. James 5 Jun 1982-1; Hotspring Island 16 Jun 1986-1 (Campbell 1986g); Queen Charlotte City 6 Jul 1982-2; Kimsquit River 8 Jul 1985-20; Naikoon Park 2 Aug 1974-2; Prince Rupert 11 Jun 1982-17; Rainy Hollow 2 Jul 1984-2. **Interior** – Osoyoos 1 Jul 1969-20; Oliver 11 Jun 1970-3 fledglings; Salmo 28 Aug 1975-60; Flathead River 31 Jul 1956-1 (NMC 40591); Balfour to Waneta 8 Jun 1983-71 on survey; Nelson 4 Jun 1983-4 nestlings; Naramata 6 Jun 1969-fledglings (Cannings et al. 1987); Kelowna 5 Jun 1966-84; Kaslo 29 Aug 1975-nestlings; Zincton 17 Jun 1990-33 on Breeding Bird Survey; Kalamalka Lake 5 Jun 1975-5 nestlings; Nakusp 1 Jun 1984-30; Enderby 8 Jul 1978-4 eggs plus 2 cowbird eggs; Lavington 22 Aug 1964-3 eggs; Botanie Lake 18 Jul 1964-15; North Barriere Lake 4 Jul 1974-20; Yoho National Park 27 Aug 1976-160 (Wade 1977); Williams Lake 14 Jun 1970-4 eggs, 29 Aug 1979-35; Stum Lake 14 Jun 1973-15 (Ryder 1973); Quesnel 15 Aug 1976-4 nestlings; Prince George 16 Jun 1969-20, 2 Jul 1982-4 fledglings; Fraser Lake 4 Aug 1959-4 nestlings; Tetana Lake 23 Jun 1938-1 (RBCM 8263); Wolverine River 26 Jul 1976-6; Tupper area 2 Jul 1993-5 eggs; Bear Flat 2 Jun 1985-1; Fort St. John 10 Jul 1982-6, 27 Aug 1986-33; Charlie Lake 17 Jun 1938-12 (Cowan 1939); Beatton Park 19 Aug 1994-2; s of Eddontenajon 13 Jul 1978-40; Sikanni Chief 5 Jun 1980-1; Fort Nelson area 5 Jul 1974-1 (Erskine and Davidson 1976); Parker Lake (Fort Nelson) 3 Aug 1968-1 (Erskine and Davidson 1976); Tats Creek 28 Jun 1993-1; Liard Hot Springs 16 Jul 1982-2; Coal River 11 Jul 1978-8; Fish Lake (Atlin) 17 Aug 1977-18.

Breeding Bird Surveys: Coastal – Recorded from 23 of 27 routes and on 73% of all surveys. Maxima: Albion 4 Jul 1970-51; Pemberton 21 Jun 1981-45; Chilliwack 26 Jun 1976-41. **Interior** – Recorded from 58 of 73 routes and on 65% of all surveys. Maxima: Succour Creek 17 Jun 1993-55; Kuskonook 18 Jun 1968-39; Golden 19 Jun 1994-37.

Autumn: Interior – Clarke Lake 28 Sep 1985-1; Middle Creek (Stikine River) 18 Sep 1986-13; Charlie Lake 12 Sep 1982-25; Fort St. John 18 Sep 1986-4; Hudson's Hope 8 Sep 1979-adults feeding fledged young; Firth Lake (Prince George) 29 Sep 1993-20, 18 Nov 1993-3; Williams Lake 30 Oct 1982-17, 11 Nov 1985-5; upper Tatlayoko valley 8 Oct 1992-8; Riske Creek 16 Nov 1986-3; Lavington 7 Sep 1964-3 young fledged; Nakusp 15 Sep 1984-47; Revelstoke 8 Nov 1985-500; Sparwood 24 Sep 1984-50; Naramata 5 Sep 1974-3 nestlings; Summerland 15 Oct 1979-48; Nelson 1 Sep 1976-4 nestlings, 8 Sep 1976-4 fledglings, 13 Nov 1978-100; Penticton 10 Nov 1978-70; East Trail 12 Oct 1980-25; Manning Park 8 Sep 1984-25. **Coastal** – Lakelse Park 19 Sep 1977-2; Kitimat 10 Sep 1975-2 (Hay 1976); Tlell 3 Nov 1942-30 (Cook 1947); Queen Charlotte City 1 Oct 1971-1; Alta Lake 13 Sep 1941-1; Sechelt 3 Sep 1959-5 nestlings; Harrison Hot Springs 7 Nov 1979-3; Vancouver 24 Nov 1946-550; Tofino 4 Sep 1979-45, 22 Nov 1983-3; Cowichan Bay 12 Sep 1972-800; Island View Beach (Saanich) 22 Oct 1975-250; Beacon Hill Park (Victoria) 24 Sep 1973-300, 9 Oct 1977-90, 25 Oct 1985-120.

Winter: Interior – Prince George 19 Dec 1993-28; Golden 27 Dec 1990-40; Revelstoke 21 Dec 1982-34, 22 Jan 1983-20, 28 Feb 1983-15; Vernon 31 Dec 1966-31, 5 Feb 1971-14; Summerland 19 Jan 1967-15, 6 Feb 1968-50; Nelson 23 Jan 1993-37; East Trail 27 Dec 1982-100; Taghum 9 Jan 1979-67, 20 Feb 1979-55. **Coastal** – Masset 5 Feb 1994-4, 26 Dec 1994-1; Tlell 24 Dec 1971-1; North Vancouver 29 Feb 1980-30; Mission 26 Jan 1985-2 in flock of 65 Bohemian Waxwings; Vancouver 14 Feb 1951-61; Burnaby 19 Feb 1985-74; Nanaimo 27 Feb 1983-50; North Saanich 2 Jan 1983-53; Victoria 2 Feb 1966-154 (Stirling 1966), 17 Feb 1983-150, 25 Feb 1983-80; Jordan River 8 Dec 1983-1.

Christmas Bird Counts: Interior – Recorded from 13 of 27 localities and on 25% of all counts. Maxima: Revelstoke 21 Dec 1987-328; Lake Windermere 27 Dec 1981-120; Kelowna 20 Dec 1981-114. **Coastal** – Recorded from 17 of 33 localities and on 30% of all counts. Maxima: Vancouver 26 Dec 1958-737; Victoria 27 Dec 1970-602; Pender Islands 23 Dec 1978-108.

Northern Shrike

Lanius excubitor Linnaeus

NOSH

RANGE: Breeds in northern North America from western and central Alaska, central and southern Yukon, northwestern British Columbia, western and southern Mackenzie, extreme northern Alberta, Saskatchewan, and Manitoba across north-central Quebec to central Labrador. Winters in North America from central Alaska and the southern parts of the breeding range in Canada south to central California in the west and to Missouri in the east. Also occurs across north-central Eurasia.

STATUS: On the coast, an *uncommon* to *fairly common* migrant and winter visitant in the Georgia Depression Ecoprovince, especially on the southwest mainland coast; *very rare* transient on Western Vancouver Island and the southern portions of the Coast and Mountains Ecoprovince, the Queen Charlotte Islands, and the mainland regions of the northern Coast and Mountains.

In the interior, an *uncommon* to *fairly common* migrant and winter visitant in the Southern Interior Ecoprovince; *uncommon* migrant and winter visitant in the Southern Interior Mountains and Central Interior ecoprovinces; *uncommon* spring and autumn transient, and *rare* in winter, in the Sub-Boreal Interior, Boreal Plains, and Taiga Plains ecoprovinces; *accidental* in summer in the Boreal Plains; *uncommon* spring and autumn transient, and probably *rare* throughout the year in the Northern Boreal Mountains Ecoprovince.

Breeds.

Figure 550. The Northern Shrike is primarily a spring and autumn migrant and winter visitant in British Columbia (Riske Creek, 25 March 1996; R. Wayne Campbell).

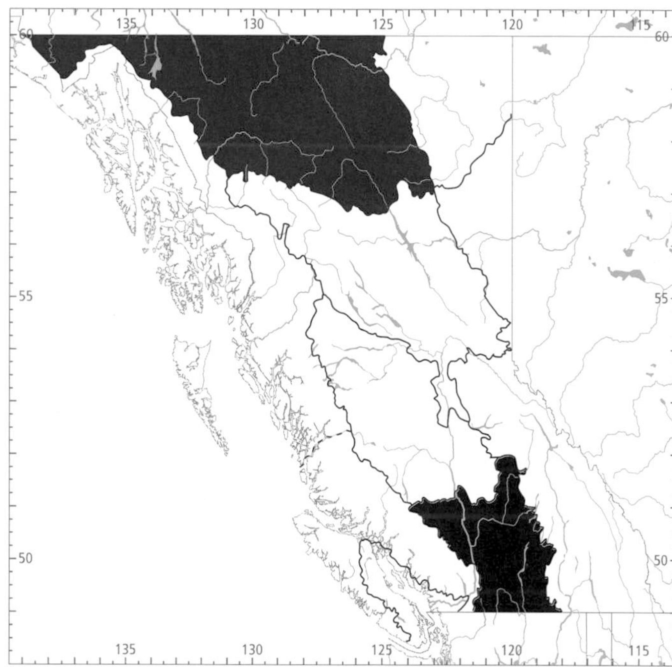

Figure 551. In British Columbia, the highest numbers for the Northern Shrike in winter (black) occur in the Southern Interior Ecoprovince; the highest numbers in summer (red) occur in the Northern Boreal Mountains Ecoprovince.

NONBREEDING: The Northern Shrike (Fig. 550) occurs widely across the southern half of the province. It is a scarce species on the Queen Charlotte Islands, in the northern Coast and Mountains, and in the interior generally in the northern half of the province north of latitude 55°N. It has been recorded in the Peace Lowland in northeastern British Columbia in all seasons. Over the 12-year period from 1982 to 1993, the highest numbers in winter have occurred in the Okanagan valley and adjacent areas of the Southern Interior (Fig. 551).

During the winter months, the Northern Shrike occurs at lower elevations, from sea level to 300 m on the coast, and from 270 to 900 m in the interior, generally below the heavily forested slopes. There it occupies grasslands, windbreaks (Fig. 552), sparse forests, and other open landscapes, including the floodplain of the lower Fraser River. On Vancouver Island, it seeks open beach areas bordered by thickets of crab apple and other shrubs. Deforested landscapes are sometimes used. On the Fraser River delta it is seen most frequently along field borders, dykes, roadsides, margins of wetlands, and river banks, where observation perches are provided by groves of red alder, willow, and other deciduous trees, fence posts, and utility poles carrying wires.

In the southern portions of the interior, grassland; sagebrush shrublands with bordering fencelines; streams, ponds, and lakes rimmed by aspens and shrubs (Fig. 553), ranchland, farmland, orchards, and roadsides provide an abundance of open but vegetated foraging areas and convenient perches (Figs. 550 and 552). Although the requirement of open terrain is fulfilled at the timberline in the mountains that characterize much of the provincial landscape, the Northern Shrike seldom occurs there. All records for alpine habitat come from the Atlin region and other parts of the Northern Boreal Mountains, where this shrike has been seen as high as 900 m.

The Northern Shrike occurs in small numbers both in migration and in winter. The onset of spring migration on the south coast is revealed by a decline in the number of records through March and April as the wintering birds leave

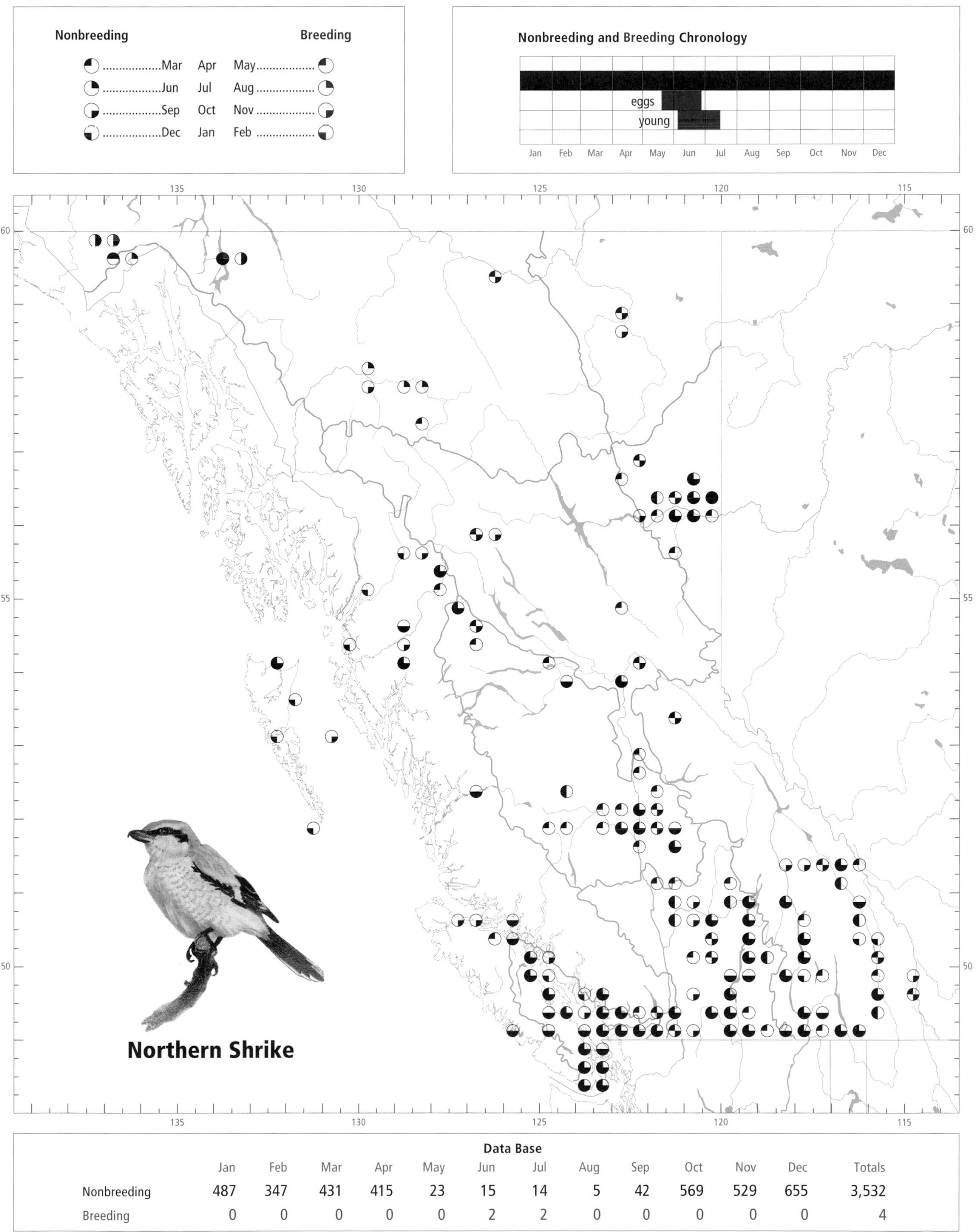

Data Base

	Jan	Feb	Mar	Apr	May	Jun	Jul	Aug	Sep	Oct	Nov	Dec	Totals
Nonbreeding	487	347	431	415	23	15	14	5	42	569	529	655	3,532
Breeding	0	0	0	0	0	2	2	0	0	0	0	0	4

Figure 552. In winter, the Northern Shrike selects hunting perches with unobstructed vantage points, such as the tops of trembling aspens in this windbreak in the east Kootenay region of British Columbia (Parsons, 23 February 1993; R. Wayne Campbell). Note perched bird.

Figure 553. In spring, the Northern Shrike hunts small migrating passerines along the shrubby borders of small streams (north of Cache Creek, 23 March 1996; R. Wayne Campbell).

for the north (Figs. 554 and 555). Very few birds remain into May. In the Southern Interior, this decline in numbers takes place abruptly in April. In the Okanagan valley, a noticeable increase in Northern Shrikes during the last week of March and the first week of April is assumed to represent an influx of migrants from the south. The local wintering shrikes are caught up in the migration, and the numbers recorded drop rapidly during the second week of April. In most years all shrikes are gone by mid-April (Cannings et al. 1987). In the valleys of the east and west Kootenay, and to a lesser extent in the Cariboo and Chilcotin areas, there is a gradual increase in numbers from February to a high in April, when the migration passes through and out of the region (Fig. 555). In the Peace Lowland of the Boreal Plains, the spring migration is remarkably abrupt. Here, the highest total numbers of Northern Shrikes occur in April and there are no records for May and June (Figs. 554 and 555). In the Prince George area of the

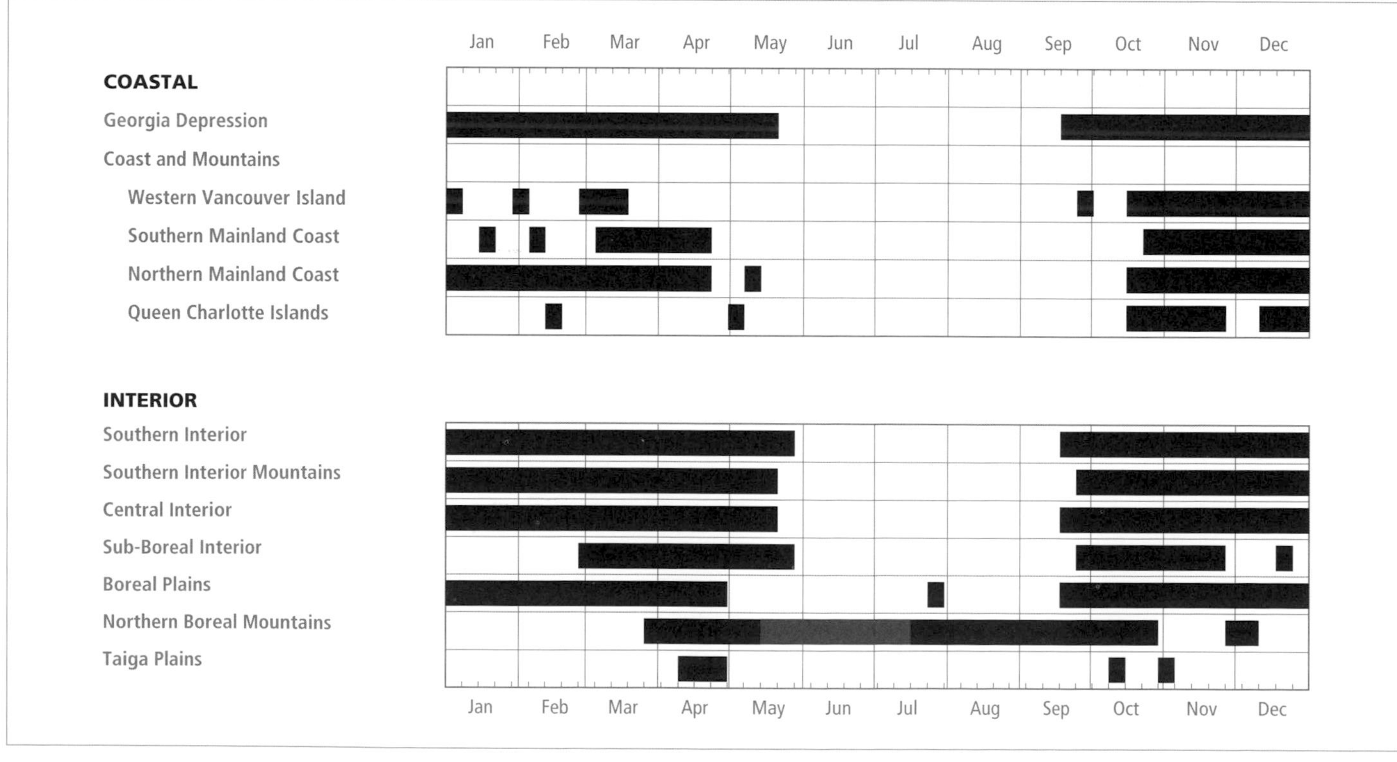

Figure 554. Annual occurrence (black) and breeding chronology (red) for the Northern Shrike in ecoprovinces of British Columbia. Records are shown for the week in which they occurred.

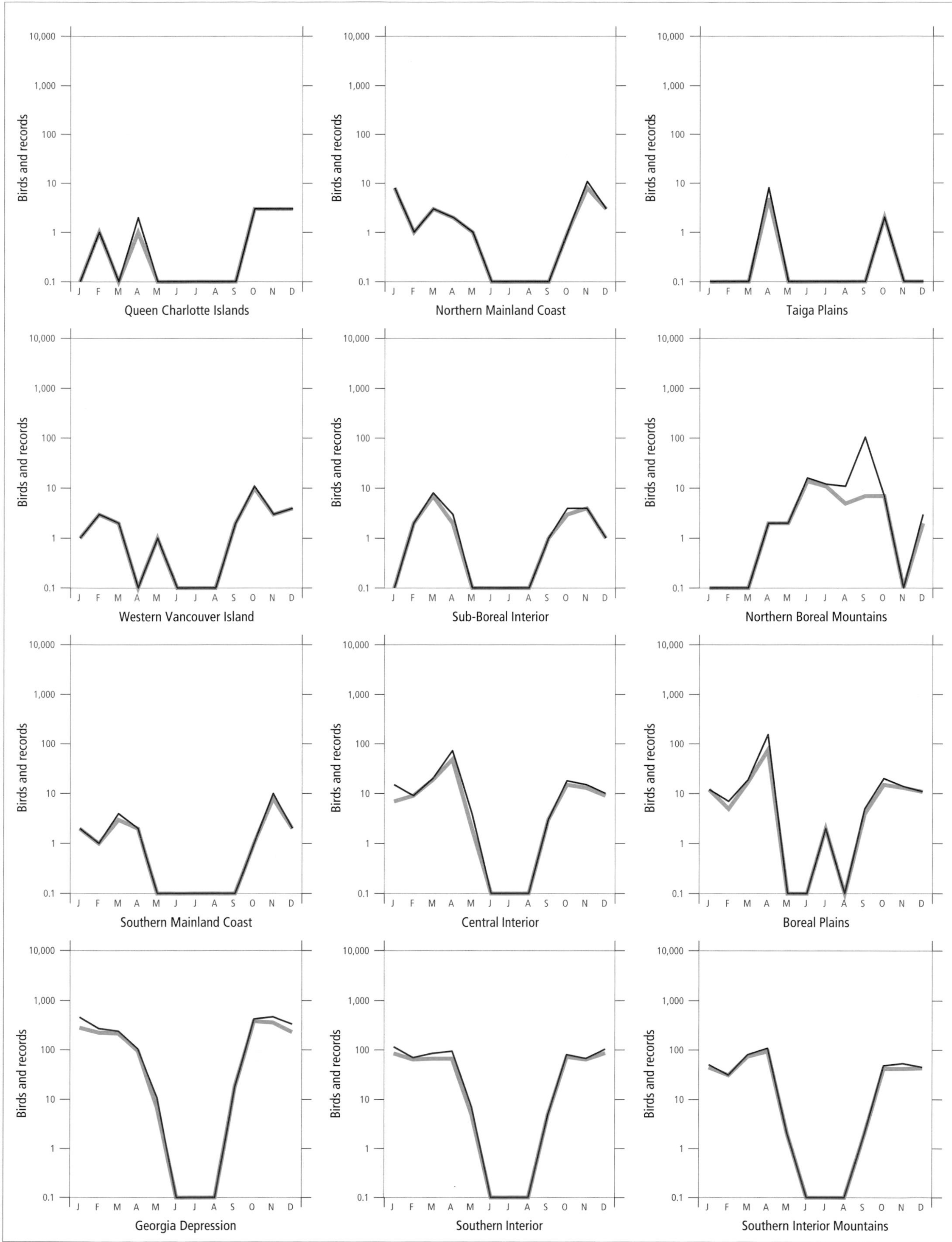

Figure 555. Fluctuations in total number of birds (purple line) and total number of records (green line) for the Northern Shrike in ecoprovinces of British Columbia. Christmas Bird Count data and nest record data have been excluded.

Sub-Boreal Interior, the earliest arrivals appear in late February and early March. The Northern Shrike is probably resident year-round in the Northern Boreal Mountains, but is never abundant. Spring migration there is not obvious, but numbers are highest in the summer months.

The autumn migration can be seen first in the Northern Boreal Mountains as numbers decline in September (Fig. 555). Across the Central Interior, especially the Cariboo and Chilcotin areas, the southward migration is indicated by early arrivals from the north in late August (Fig. 554) and the increase in birds through October and November. Elsewhere in the southern portions of the interior, autumn migrants arrive in September and numbers reach wintering levels in October (Cannings et al. 1987). The timing of autumn arrivals on the Fraser and Nanaimo lowlands is similar. The scarcity or absence of records of the Northern Shrike along the coast between latitudes 50°N and 54°N suggests that the Fraser River valley serves as a natural corridor through the Cascades and the coastal ranges for Northern Shrikes migrating to and from their wintering grounds on the Fraser River delta and southern Vancouver Island.

Root (1988) states that the Northern Shrike winters where the average minimum January temperature is below –7°C, except for the wintering area in the Georgia Depression, where the minimum January temperature is higher. She also refers to periodic irruptions of Northern Shrikes out of Canada into the northern United States occurring in 1965, 1966, 1970, 1971, and 1972.

On the coast, the Northern Shrike has been recorded from 23 September to 23 May; in the interior, south of the Sub-Boreal Interior, it has been found from 20 September to 22 May (Fig. 554).

BREEDING: All but 1 of the summer occurrences for the province are from the Northern Boreal Mountains Ecoprovince. There are 4 breeding records. Swarth (1926) collected 6 fledglings, just out of the nest, at the head of Canon Creek, near Atlin on 30 June 1924. The other 3 records are from along the Haines Highway. On 23 July 1944, young were found at Mile 80 on this highway, but specific details of the number and age of the birds are lacking. On 24 June 1980, a nest containing 5 half-grown nestlings was found at Mile 89.5. The following year, 2 adult shrikes with 4 young fledglings were found at Mile 90 of the Haines Highway on 15 July (C. Siddle pers. comm.). Further field work will undoubtedly establish that the Northern Shrike nests regularly in that part of the province.

While studying the fauna of Spatsizi Plateau Wilderness Park, Osmond-Jones (1977) found 2 adult Northern Shrikes with 4 young at Tuaton Lake. This suggests that the species may breed over a larger area of northern British Columbia, extending some 3° of latitude south of the presently established nesting areas.

The highest numbers in summer occur in the Northern Boreal Mountains Ecoprovince (Fig. 551). Breeding Bird Surveys for interior routes for the period 1968 through 1993 contain insufficient data for analysis.

Elevations occupied by the Northern Shrike during the breeding season range from 800 to 1,250 m. The brood of fledglings discovered by Swarth (1924) was at the timberline, at an altitude of 1,250 m. There the forest was a krummholz mix of stunted spruce and alpine fir scattered across alpine slopes and meadows. The broods adjacent to the Haines Highway occurred in similar terrain and habitat (Fig. 556).

Calculated dates indicate that the Northern Shrike breeds in the province from 19 May to 15 July (Fig. 554).

Nests: The single nest examined in British Columbia was placed on a branch next to the trunk of a dense spruce tree. The nest consisted of a bulky base of twigs and was lined with finer material and an abundance of feathers. Nest height was 3.6 m. In Alaska and Alberta, the Northern Shrike nests in deciduous trees and shrubs at least as often as in conifers (Kessel 1989; Semenchuk 1992).

Eggs: No clutches have been reported in British Columbia. Calculated dates from British Columbia broods indicate that eggs could be present as early as 19 May. This is consistent with dates of 11 May to 20 June reported for Alaska and the Northwest Territories (Kessel 1989; Bent 1950). Incubation period is at least 15 days (Ehrlich et al. 1988).

Young: Dates for 3 broods ranged from 3 June to 15 July. Brood sizes were 4, 5, and 6 young. Nestling period is 20 days (Ehrlich et al. 1988).

Brown-headed Cowbird Parasitism: Cowbird parasitism was not found in British Columbia, nor are there North American reports of parasitism (Friedmann et al. 1977; Friedmann and Kiff 1985).

Nest Success: Insufficient data.

REMARKS: Two geographic races of the Northern Shrike are recognized in North America. The race found in British Columbia is *L. e. invictus*.

Many late spring, summer, and early autumn reports of the Northern Shrike in British Columbia have been without the details that would indicate with certainty the species of bird observed. These have been omitted from the account.

Figure 556. In northwestern British Columbia, the Northern Shrike breeds in spruce forests, often near wetlands (Haines Highway, 2 July 1980; R. Wayne Campbell).

We encourage observers to document adequately such occurrences in the future.

Many naturalists in the province have observed the Northern Shrike capturing or impaling food (Fig. 557). Following is a list of prey items identified: insects and very young ptarmigan (Swarth 1926), Downy Woodpecker, Horned Lark, Gray Jay (immature), Black-capped Chickadee, Chestnut-backed Chickadee, Golden-crowned Kinglet, American Robin, Varied Thrush, Bohemian Waxwing, Yellow-rumped Warbler, Savannah Sparrow (Fig. 557), Song Sparrow, White-crowned Sparrow, Golden-crowned Sparrow, Fox Sparrow, Dark-eyed Junco, Common Redpoll, House Finch, Red Crossbill, Evening Grosbeak, Snow Bunting, Lapland Longspur, House Sparrow (Campbell 1988b), Townsend's vole (*Microtus townsendii*), red-backed vole (*Clethrionomys gapperi*), meadow vole (*Microtus pennsylvanicus*), and shrew (*Sorex* sp.).

See Cade (1967) for a discussion of the foraging ecology of the Northern Shrike.

Figure 557. Shrikes often impale their prey on pointed objects, eating them immediately or later. Here a Northern Shrike has impaled a Savannah Sparrow on a barbed-wire fence (Reifel Island, November 1974; Neil K. Dawe).

NOTEWORTHY RECORDS

Spring: Coastal – Blenkinsop Lake 12 Mar 1985-4; Quicks Bottom 23 May 1985-1; Westham Island 21 Apr 1968-3; Sea Island 9 May 1971-2, 20 May 1971-1 (Campbell et al. 1972b); Adam River 12 Mar 1982-1 in lower watershed; Floods 19 Mar 1961-2; Agassiz 1 Apr 1897-2; Delkatla Slough 30 Apr 1979-2; 8 miles n Kitimat 7 May 1975-1; Kispiox 15 Apr 1976-1. **Interior** – Merritt 15 Apr 1968-4; Douglas Lake 28 Mar 1981-4; Duck Lake (Okanagan) 20 May 1978-1; Sorrento 12 May 1970-3; Revelstoke 27 Mar 1977-2; 9 km e Cache Creek 22 May 1985-1; Spokin Lake 6 Mar 1983-2; Williams Lake 2 Apr 1977-6; Quesnel 13 Mar 1896-1 (RBCM 1063); Fraser Lake 1 May 1981-2; Prince George 9 Apr 1985-2; 6 km e Houston 14 May 1977-2; w side Peace River 29 Mar 1981-2; North Pine 18 Apr 1982-31 (C. Siddle pers. comm.); ne Fort St. John 29 Apr 1985-2, last migrants; Spatsizi Park 1 to 11 Apr 1990-1; Fort Nelson 12 Apr 1975-1 (Erskine and Davidson 1976); Liard Hot Springs 19 May 1975-1; Atlin 18 Apr 1934-1 earliest arrival (Swarth 1936).

Summer: Coastal – No records. **Interior** – Boundary Lake (Peace River) 23 Jul 1988-1 juvenile; Tuaton Lake 26 Jul 1976-6 in family group (Osmond-Jones 1977); Spatsizi Plateau 12 Aug 1976-6 (2 adults and 4 young); Gnat Lake 19 Jun 1952 3 (NMC 50067-69); 40 km n Skagway (Alaska) 24 Jun 1979-5; Canon Creek (Atlin) 20 Jun 1924-6 young just able to fly, huddled together in a spruce thicket (CAS 44897-44899 and 3 specimens in MVZ); Mile 80, Haines Highway 23 Jul 1944-2 young (ROM 71071-72); Mile 89.5, Haines Highway 9 Jul 1980-5 nestlings; Mile 90, Haines Highway (68.5 km sse Haines Junction) 15 Jul 1981-2 adults and 4 fledglings (Siddle 1995 pers. comm.); Chilkat Pass 14 Aug 1984-1 (CAS 83513); Bennett 9 Jun 1903-1 (Swarth 1926).

Breeding Bird Surveys: Not recorded.

Autumn: Interior – Atlin 6 Oct 1931-1 (CAS 34116); Griffith Creek 6 Sep 1976-2; Mile 278, Alaska Highway 10 Oct 1985-1; Fort Nelson 29 Oct 1987-1; Charlie Lake 19 Sep 1985-2 at south end; Fort St. John 17 Sep 1987-1, 13 Nov 1985-2; Driftwood Valley 28 Sep 1937-1 (RBCM 7838); Beryl Prairie 14 Oct 1979-1; Willow River 2 Sep 1966-1; Indianpoint Lake 26 Sep 1929-1 (MVZ 284041); Williams Lake 6 Nov 1983-2; 153 Mile House 23 Sep 1950-1 (NMC 48156); Lac la Hache 10 Oct 1959-2; Golden 10 Nov 1899-3; Tranquille 22 Oct 1927-1 (RBCM 4849); Okanagan Landing 20 Sep 1908-1; Vaseux Lake 6 Oct 1973-1 (Cannings et al. 1987); Trail 20 Oct 1966-1; Creston 14 Oct 1988-1, first arrival (Campbell 1989a). **Coastal** – Kispiox Valley 21 Oct 1980-1; Terrace 21 Nov 1974-2; Masset 17 Oct 1971-1, 11 Nov 1943-1; Kitimat 17 Sep 1974-11; Bella Coola 23 Oct 1942-1 (RBCM 10020), 21 Nov 1988-1 (Campbell 1989a); Keogh River 19 Oct 1950-1 (UBC 4583); Jericho Beach (Vancouver) 23 Sep 1985-1; Vancouver 16 Oct 1971-8; Tofino 27 Sep 1984-1; Lennard Island 27 Oct 1983-2; Reifel Island 29 Oct 1966-1 (Campbell 1967b); Island View Beach (Saanich) 23 Sep 1973-1.

Winter: Interior – Atlin 5 Dec 1931-2 (RBCM 5903); Montney 3 Dec 1981-1; Fort St. John 6 Jan 1985-1, 30 Dec 1986-1; North Pine 21 Feb 1987-3; 24 km s Prince George 27 Feb 1981-1, 20 Dec 1983-1; Nulki Lake Feb 1946-1 (Munro 1949); Williams Lake 2 Jan 1977-7, 30 Dec 1979-2; 111 Mile House 4 Feb 1983-1; Golden 1 Jan 1977-1; Revelstoke 31 Dec 1986-1; Toby Creek 31 Jan 1982-1; Skaha Lake 1 Jan 1978-1; Penticton (West Bench) 6 Jan 1974-10; Oliver to Vaseux Lake 10 Feb 1974-3; Creston 6 Jan 1982-6. **Coastal** – Terrace 1 Jan 1964-1, 27 Dec 1982-1; Masset 12 Dec 1937-1 (RBCM 10517); Tlell 12 Feb 1973-1; Cape St. James 28 Dec 1978-1; 12 km n Kitimat 15 Dec 1974-1; Mitlenatch Island 20 Dec 1965-1 chasing Song Sparrow (Campbell 1965); Squamish River 16 Dec 1989-1 on estuary; Campbell River 1 Jan 1976-1; Jones Creek 19 Jan 1979-1; Nanaimo 2 Jan 1965-1; Vancouver 13 Feb 1971-13 in area (Campbell et al. 1972b); Chilliwack 7 Jan 1888-1 (MCZ 244653), 22 Dec 1977-4; Iona and Sea islands 31 Dec 1977-3; Long Beach (Tofino) 6 Jan 1987-1, 1 Feb 1978-1, 18 Dec 1987-1; George C. Reifel Refuge (Reifel Island) 1 Jan 1970-5; Saanich 29 Jan 1984-3.

Christmas Bird Counts: Interior – Recorded from 23 of 27 localities and on 70% of all counts. Maxima: Vernon 18 Dec 1983-24; Kelowna 17 Dec 1988-21; Oliver-Osoyoos 28 Dec 1988-26. **Coastal** – Recorded from 25 of 33 localities and on 58% of all counts. Maxima: Ladner 22 Dec 1973-30; Vancouver 21 Dec 1969-18; White Rock 22 Dec 1974-14.

Loggerhead Shrike

Lanius ludovicianus Linnaeus

LOSH

RANGE: Breeds from eastern Washington, eastern Oregon, most of California, central Alberta, central Saskatchewan, southern Manitoba, southern Ontario, southwestern Quebec, central New York, and Pennsylvania south to southern Florida and the Gulf coast, southern Baja California, and throughout Mexico south to Oaxaca and Veracruz. Winters from central Washington, eastern Oregon, California, southern Nevada, northern Arizona, northern New Mexico, and, east of the Rocky Mountains, the southern half of the breeding range south to the southern limits of the breeding range.

STATUS: On the coast, *very rare* in spring, autumn, and winter in the Fraser Lowland portion of the Georgia Depression Ecoprovince. In the interior, *very rare* in spring and winter and *casual* in summer and autumn in the Southern Interior Ecoprovince, especially the Okanagan valley; *casual* in the Southern Interior Mountains Ecoprovince; *accidental* elsewhere.

CHANGE IN STATUS: Munro and Cowan (1947) list 4 specimen records of the Loggerhead Shrike for the province before the mid-1940s. At the time, 2 other specimens were unknown to them. Of the 6 specimens taken before 1943, 1 was from the Fraser River delta, 4 from the Okanagan valley, and 1 from the Southern Interior Mountains. Today, 50 years later, the evidence of another specimen is available, along with 3 clearly identifiable photographs taken between 1977 and 1985, and the demonstrated range has been extended eastward to Golden. However, with an army of birdwatchers scouring the province, sightings are being reported almost annually in the southern portions of the interior and about every 4 years on the south coast, adding more detail to the distribution of the Loggerhead Shrike in British Columbia. The evidence does not suggest a change in numerical status over the past 50 years. This shrike still appears to occur infrequently in the province, but it is a species that may be overlooked by many field observers.

Figure 558. The Loggerhead Shrike is seen infrequently in British Columbia, with most occurrences recorded between March and May (Chopaka, Richter Pass, 3 May 1987; Richard J. Cannings).

OCCURRENCE: The Loggerhead Shrike (Fig. 558) has been reported from widely scattered locations across the southern mainland of British Columbia, from the Fraser River delta and Vancouver in the Georgia Depression east to Parson and Golden in the Southern Interior Mountains and north to Riske Creek in the Central Interior. Although published records exist (e.g., Flahaut and Schultz 1956), there are no confirmed occurrences for any of the coastal islands, including Vancouver Island.

The Loggerhead Shrike has been recorded from near sea level to 800 m. Along the coast, it is most often seen in open country, including beaches (Fig. 559), brushy edges of fields, dykes, lakeshores with scattered bushes, and agricultural lands. In the interior, especially in the Okanagan valley, it

Figure 559. On the south coast of British Columbia, the Loggerhead Shrike frequents open country, including beaches (Iona Island, Richmond, 7 August 1996; R. Wayne Campbell).

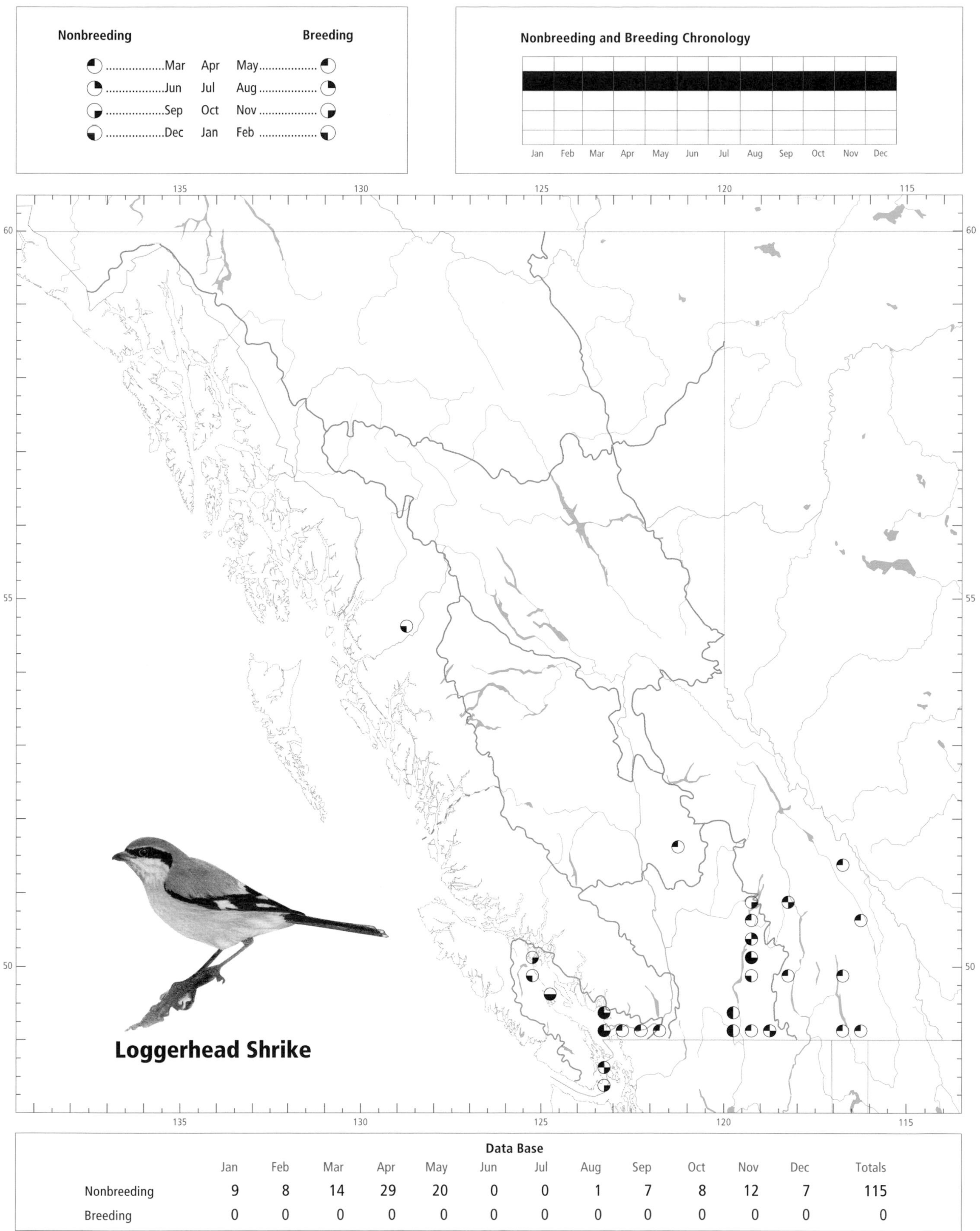

Data Base

	Jan	Feb	Mar	Apr	May	Jun	Jul	Aug	Sep	Oct	Nov	Dec	Totals
Nonbreeding	9	8	14	29	20	0	0	1	7	8	12	7	115
Breeding	0	0	0	0	0	0	0	0	0	0	0	0	0

Figure 560. In the Southern Interior Ecoprovince of British Columbia, the Loggerhead Shrike inhabits mixed woodlands in open country with adjacent patches of shrubs and grasslands (Similkameen River, Richter Pass, 3 October 1994; R. Wayne Campbell).

inhabits the mixed forest that features open stands and coppices of ponderosa pine and Douglas-fir interspersed with grassland slopes and sagebrush plains where gullies and wetlands support aspen, serviceberry, hawthorn, choke cherry, and mock-orange (Fig. 560). Thorny species such as the hawthorn are used by the shrike to impale prey while eating it. The shrike prefers areas with a short ground cover, which makes it easier to capture its prey, mainly large insects. The Loggerhead Shrike frequently forages from fence posts, telephone poles, or overhead wires.

Most occurrences in British Columbia (40%; $n = 59$) have been recorded between 9 March and 28 May (Fig. 561), a period including the known migration for the species in Oregon and Washington. Also, most of the spring observations in British Columbia occur in the arid regions of the southern portions of the interior, in environments equivalent to those occupied by the shrikes in Washington. There are only 2 summer records for the province, early June and August. Most of the autumn observations are from October and November (Fig. 561).

Although the Loggerhead Shrike is present throughout the year in neighbouring Washington state, part of the population migrates south for the winter, returns to the Washington breeding areas during the last 2 weeks of March, and migrates south again in late August and early September (Wahl and Paulson 1991). In Alberta, it arrives during the last week of April and usually starts moving southward before the end of August (Semenchuk 1992).

Winter records in British Columbia are from 14 December to 28 February (Fig. 561). Root (1988) states that the northern boundary of the winter range corresponds roughly to latitude 40°N. The British Columbia experience reveals that in some years the Loggerhead Shrike may winter well north of its normal range.

On the coast, the Loggerhead Shrike has been recorded from 2 January to 14 May, in early September, and from early October through December; in the interior, it has been found sporadically from 8 January to 1 June, in August, and from 15 October to 28 December (Fig. 561).

REMARKS: The Loggerhead Shrike is the only member of its family with a range restricted to North America (Howard and Moore 1994).

There are no known nesting records for this shrike in British Columbia, and published references that include the province within the breeding range of the species (e.g., Hands et al. 1989; Telfer 1993) are inaccurate. The historical nesting range of this shrike includes a wide area of the arid regions of southern Alberta, Saskatchewan, Manitoba (Godfrey 1986), and central Washington; there are summer occurrences at many points just south of the International Boundary (Jewett et al. 1953). Given the similarity between the habitat in which this shrike has nested in Washington and the parts of British Columbia in which it has occurred, it would not be surprising to find the species nesting in the province.

In the United States and Canada, the Loggerhead Shrike has been declining in numbers throughout its range (Cadman 1990, 1991) and has been on the National Audubon Society's "Blue List" since 1972 (Tate 1986). Its decline in its eastern Canadian range in southern Ontario has led to its inclusion in the Canadian list of endangered species (Cadman 1991). The decline in the northern breeding range includes Alberta, Manitoba, and, to a lesser degree, Saskatchewan, where it is considered a threatened species (Telfer et al. 1989). Johns et

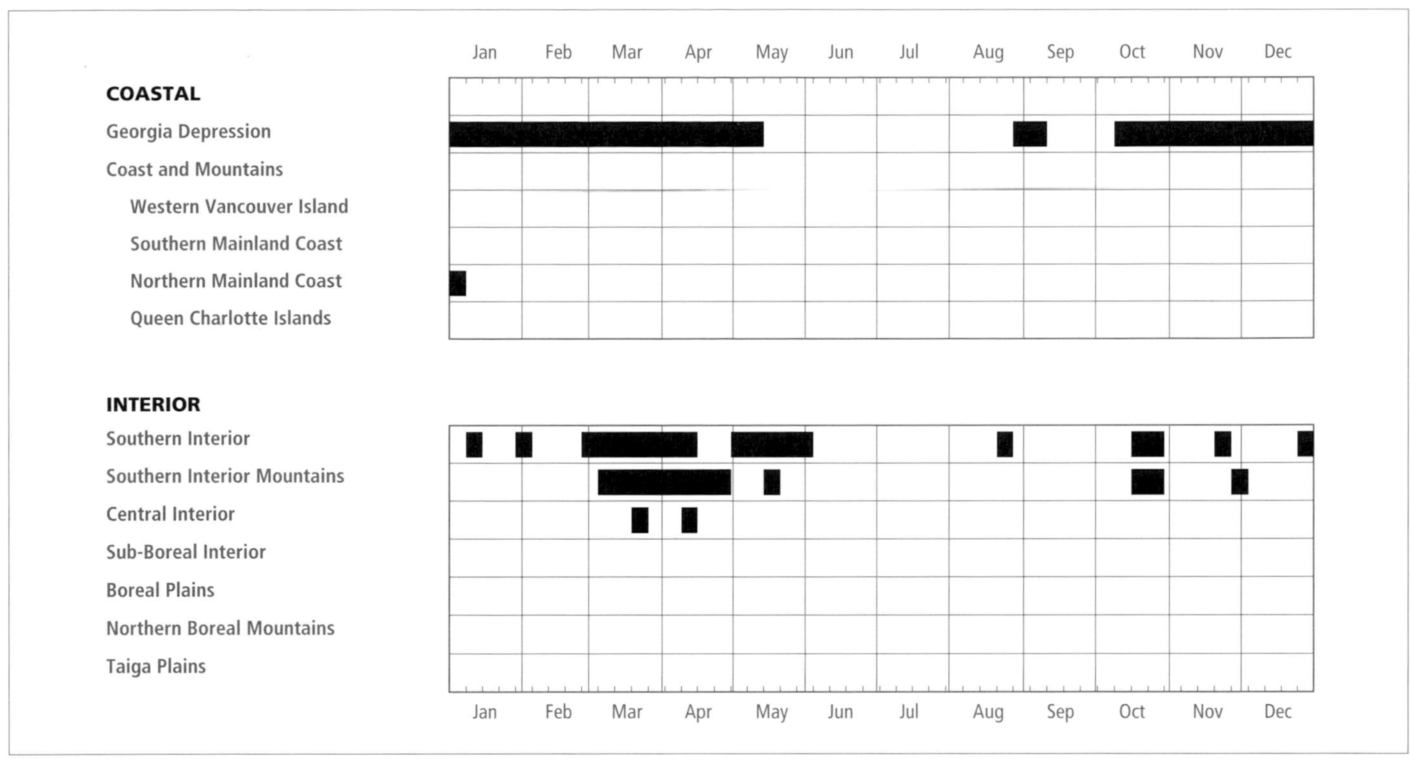

Figure 561. Annual occurrence for the Loggerhead Shrike in ecoprovinces of British Columbia. Records are shown for the week in which they occurred.

al. (1994) describe a national recovery plan for Canada. Fraser and Luukkonen (1986) provide additional information on the status of the Loggerhead Shrike in the United States.

At times it is difficult to differentiate the Loggerhead Shrike from the more common Northern Shrike, which is a regular migrant and wintering species in British Columbia, and we have been cautious in accepting field sightings of the former species from the province. All accepted records have accompanying field notes that provide details of the sighting. Many other field observations and published records without convincing details have not been included in this account (e.g., Victoria 1 November 1955-1 [Flahaut and Schultz 1956]; Port Neville 2 September 1987-1 [Campbell 1987d]; Osoyoos March 1909-1 [Brooks 1909]; Vermilion Pass June 1973-1 [Scott 1973]; Kitimat 2 January 1966-1, winter 1960-1 [Vance 1966; Hay 1976]; Terrace 2 Jan 1966-2 on Christmas Bird Count; White Rock 2 Jan 1984-1 on Christmas Bird Count, 24 June 1974-1 on the Ashnola Breeding Bird Survey).

NOTEWORTHY RECORDS

Spring: Coastal – Delta 6 and 7 May 1985-1 along River Road (Campbell 1985c; BC Photo 1045); Chilliwack 9 Apr 1888-1 (MVZ 244657; Fannin 1891; Brewster 1893); Blackie Spit (Delta) 22 Mar 1981-1 (Weber 1982); Iona Island (Richmond) 26 Apr 1970-1 (Campbell et al. 1972a), 14 May 1976-1; Jericho Park (Vancouver) 8 to 12 April 1989-1. **Interior** – Midway 6 Apr 1915-1 (Macoun and Macoun 1909; Kelso 1926); Osoyoos Lake 9 Mar 1908-1 (MVZ 104025; Brooks 1909); Osoyoos 5 Apr 1984-1 along oxbows; Chopaka (Richter Pass) 3 May 1987-1 (BC Photo 1179); Oliver 18 May 1959-1 (Cannings et al. 1987); White Lake 28 May 1963-1 (Cannings et al. 1987); Swan Lake (Princeton) 7 and 8 Apr 1993-1; Vaseux Lake 15 May 1976-1 at north end; Penticton 9 Apr 1969-1 (Cannings et al. 1987); Okanagan Landing 28 Mar 1927-1 (MVZ 104024); Edgewood 16 May 1917-1 (Brooks and Swarth 1925; Kelso 1926); Grindrod 23 May 1950-1 (UBC 7174; Munro 1953a); Kamloops 20 Apr 1984-1; Golden 29 Apr 1977-1 (BC Photo 2010); Riske Creek 7 and 10 Apr 1985-1.

Summer: Coastal – No records. **Interior** – Swan Lake (Vernon) 21 Aug 1976-1 (Cannings et al. 1987); Tranquille 1 Jun 1985-1.

Breeding Bird Surveys: Not recorded (see REMARKS).

Autumn: Interior – Sorrento 23 Oct 1970-1 (Schnider et al. 1971); Okanagan Landing 15 Oct 1928-1 (Munro and Cowan 1947). **Coastal** – Jericho Park (Vancouver) 2 to 10 Sep 1987-1; Beach Grove (Delta) 2 to 5 November 1980-1; Martindale (Victoria) 5 to 7 Sep 1995-1.

Winter: Interior – 6 km n Parson 23 Feb 1993-1; Kamloops 14 Dec 1991-1 along Lac Le Jeune Rd (Siddle 1992b); Vernon 1 Feb 1981-1, 26 Dec 1978-1; Okanagan Landing 28 Feb 1943-1 (MVZ 88341); Kelowna 8 Jan 1978-1 (Cannings et al. 1987); Penticton 26 Dec 1977-1 (Cannings et al. (1987); Oliver 9 Jan 1981-1 (Cannings et al. 1987). **Coastal** – West Vancouver 21 Dec 1969-1 (Crowell and Nehls 1970b); Coquitlam 27 Dec 1991-1 at Colony Farm (Siddle 1992b); Sea Island (Richmond) 7 Jan to 3 Mar 1979-1; Delta 3 to 6 Feb 1989-1; Mud Bay (Delta) 2 Jan 1984-1.

Christmas Bird Counts: Interior – Recorded from 3 of 27 localities and on 2% of all counts. Maxima: Oliver-Osoyoos 28 Dec 1986-1; Vernon 26 Dec 1978-1; Kamloops 14 Dec 1991-1. **Coastal** – Recorded from 4 of 33 localities and on less than 1% of all counts. Maxima: Vancouver 17 Dec 1978-1; Ladner 22 Dec 1973-1 (see REMARKS).

European Starling
Sturnus vulgaris Linnaeus

EUST

RANGE: Native to Europe, western and southern Asia, and North Africa. Introduced into North America (New York) in 1890, and has spread to inhabit a breeding range from the Anderson River Delta, southern Alaska, southern Yukon, southern Mackenzie, and northern British Columbia east across Canada to Labrador; south to Baja California and across the southern United States to Florida. Winters throughout most of the breeding range, south to central Mexico, the Bahamas, and Cuba.

STATUS: On the coast, a *very common* to *very abundant* resident in the Georgia Depression Ecoprovince; *fairly common* to locally *common* on Western Vancouver Island and in the southern and northern mainland portions of the Coast and Mountains Ecoprovince; *uncommon* on the Queen Charlotte Islands; *very rare* in the latter 3 regions in winter.

In the interior, a *common* to locally *very common* resident in the Southern Interior Ecoprovince; *uncommon* to *fairly common* in the Southern Interior Mountains and Central Interior ecoprovinces, becoming rare in winter in the Central Interior; *uncommon* to locally *common* migrant, and summer visitant, in the Sub-Boreal Interior and Boreal Plains ecoprovinces; *very rare* to *rare* in the Northern Boreal Mountains Ecoprovince, and *very rare* in the Taiga Plains Ecoprovince.

Breeds.

CHANGE IN STATUS: The European Starling (Fig. 562) was first reported in British Columbia in 1945 (Munro 1947b; Jobin 1952a), and the first specimen was taken in 1947 (Munro and Cowan 1947). The first sighting in British Columbia occurred just 2 years after the first starlings were seen in Washington (Wing 1943) and only a year after the first sighting in California (Grinnell and Miller 1944). Thus, British Columbia was caught up in the extraordinary dynamics of the spread of this introduced species into western North America.

Figure 562. The European Starling was first recorded in British Columbia in 1945. Fifty years later, it breeds throughout the province. The bird shown is in nonbreeding plumage and lacks the yellow bill (Vancouver, 22 November 1992; Richard J. Cannings).

The European Starling was present at 3 localities in 1947: Oliver (Munro 1947b), Bella Coola (Godfrey 1949), and Wistaria (Myres 1958). In 1948, a specimen was collected at 150 Mile House, and a female was seen near there, at Onward Ranch, carrying food into an abandoned woodpecker nesting cavity. This was the first recorded nesting in the province; the second recorded nesting in that region occurred in 1951, at Williams Lake (Jobin 1952a).

Table 7. Numbers of European Starling records for British Columbia by decade and ecoprovince. Each record can include from 1 to many birds. Note the steady spread across the ecoprovinces from south to north, and the increases in recorded observations.

Ecoprovince	1940s	1950s	1960s	1970s
COASTAL				
Georgia Depression	2	153	1,468	3,021
Coast and Mountains				
Western Vancouver Island	–	6	24	106
Southern Mainland Coast	7	–	23	85
Northern Mainland Coast	1	5	24	157
Queen Charlotte Islands	–	5	1	77
INTERIOR				
Southern Interior	9	136	232	1,300
Southern Interior Mountains	1	37	34	459
Central Interior	13	65	34	326
Sub-Boreal Interior	–	8	18	29
Boreal Plains	–	1	11	66
Northern Boreal Mountains	–	–	–	14
Taiga Plains	–	–	3	10

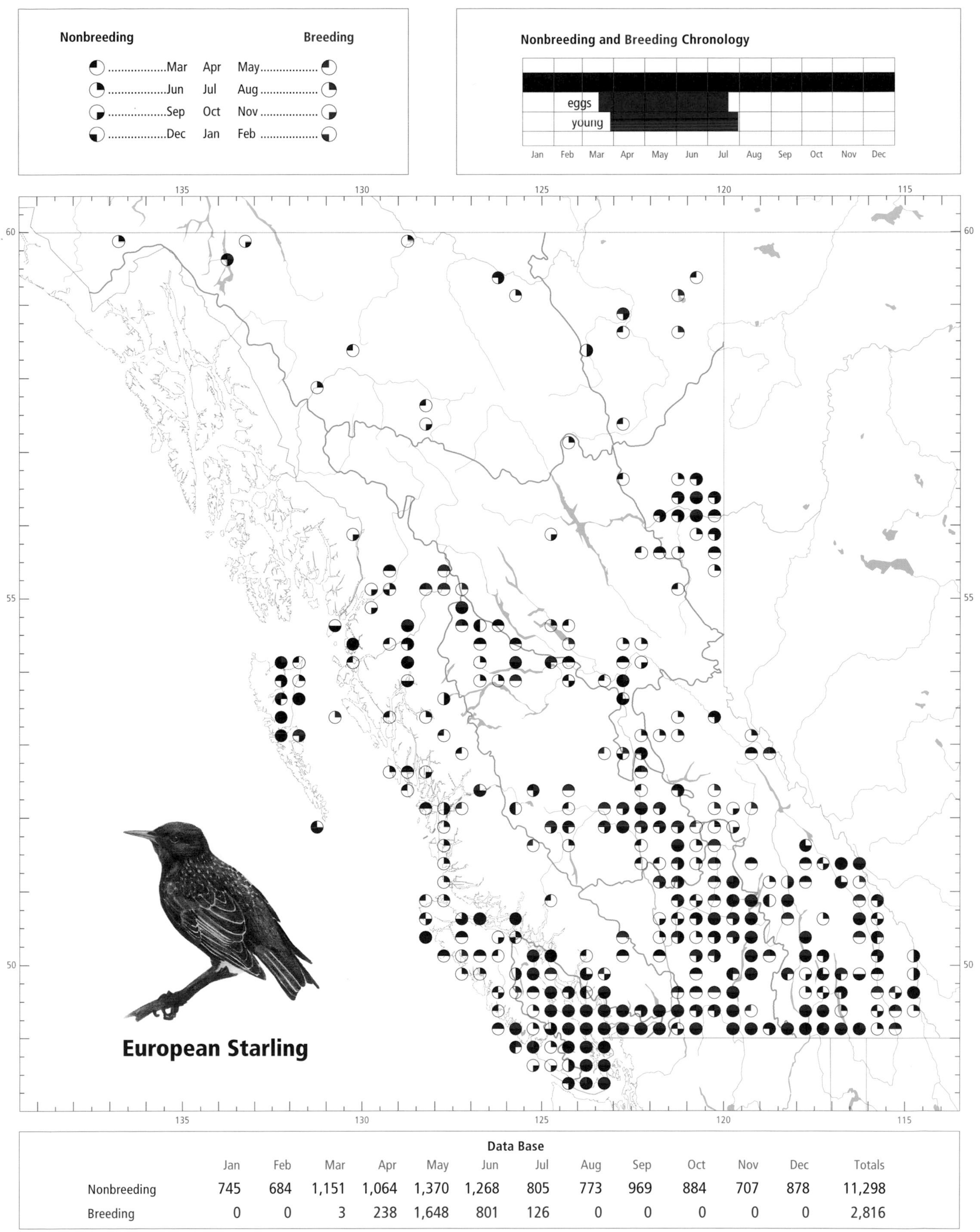

Data Base

	Jan	Feb	Mar	Apr	May	Jun	Jul	Aug	Sep	Oct	Nov	Dec	Totals
Nonbreeding	745	684	1,151	1,064	1,370	1,268	805	773	969	884	707	878	11,298
Breeding	0	0	3	238	1,648	801	126	0	0	0	0	0	2,816

In 1949, the species appeared in the lower Fraser River valley at Cloverdale (Myres 1958); in the Southern Interior at Okanagan Landing and Vernon (Munro 1953a); and in the Central Interior at Alkali Lake and Burns Lake. In the winter of 1949-50, it was recorded on the Fraser River delta at Surrey in December, and on Lulu Island and in Vancouver in January (Racey 1950). Except in the Okanagan, all these early reports refer to just 1 or 2 birds.

The first observation of flocks of European Starlings in British Columbia occurred at Burns Lake and Wistaria in the Nechako River valley between October and December 1949 (Racey 1950). A flock of over 30 was seen at Wistaria. In the east Kootenay, the European Starling was first reported in the spring and summer of 1950 at Brisco, Canal Flats, and Cranbrook (Myres 1958). That year it nested at Brisco and Cranbrook.

The spread in distribution continued through the 1950s. It was first reported from Victoria in September 1951 (Guiguet 1952a) and Comox in December (Pearse 1953). That same year, it occurred in the western Chilcotin at Kleena Kleene in October (NMC 48161; Jobin 1952a). In 1952 it was recorded 10 km north of Dawson Creek in the Boreal Plains and on the Queen Charlotte Islands, at Queen Charlotte City (Guiguet 1952b). In 1953, Munro (1956) found it at Vanderhoof in the Sub-Boreal Interior. A specimen was taken at Port Hardy in 1954 (PMNH 72000), and it was reported at Alexandria in 1955. The last ecoprovinces to be reached by pioneering starlings were the Taiga Plains (Fort Nelson in 1967) and the Northern Boreal Mountains (Liard Hot Springs in 1975) (Reid 1975).

Table 7 lists the numbers of early European Starling records from the different ecoprovinces. The numbers of records probably roughly chart the course of spread and increase during the period of rapid expansion, when starlings were a novelty in many regions. It is apparent that in every ecoprovince the pioneers were a small number of birds that could be regarded as casuals, birds that had overflown or dispersed from the main inhabited areas. It was usually several years before nesting was documented in a newly occupied area.

However, some of the invasions involved substantial numbers of birds spread over a broad front. For example, there was intensive ornithological field work in the Skeena and Nechako river valleys in the summer of 1958, but European Starlings were not seen by, or reported to, the field scientists. A year later, in August 1959, there was a flock of 25 to 30 starlings at the Prince George airport, and in the summer of 1960, starlings were found nesting at 5 widely dispersed localities along the Skeena River valley. The increase was fairly rapid in regions with sizeable human populations, but more gradual in thinly settled regions. In each ecoprovince, the starling reached peak density in the third or fourth decade after its arrival. Only in the Southern Interior Mountains did the increase extend into the fifth decade.

The European Starling appears to have entered the province from the northeast via the Peace River region and from the south through Washington and Idaho (Racey 1950; Munro 1956; Johnson and Cowan 1974). The strongest argument for an eastern source of the earliest birds invading northern British Columbia is the large numbers involved at a time when the species was not yet abundant in southern British Columbia. For example, the first noted invasion of the Vanderhoof district in 1953 involved about 250 birds that arrived in September, nearly 320 km north of the nearest British Columbia population. There was also possibly an influx along a fairly broad front through the passes along the boundary with Alberta. Starlings were present in 1948 close to the Crowsnest Pass (Coleman), and also in Banff and Jasper national parks (Myres 1958).

Figure 563. In British Columbia, the highest numbers for the European Starling in both winter (black) and summer (red) occur in the Georgia Depression Ecoprovince.

NONBREEDING: The European Starling is distributed across much of the province, usually in close association with humans. In the southern half of the province, it occurs from southern Vancouver Island east to the Rocky Mountains and north to the Skeena River valley. Further north, its distribution is spotty except in the Peace Lowland. The most northerly occurrences have been from the Kelsall River valley in the west and Kwokullie Lake in the east. The highest numbers in winter occur in the Georgia Depression in the vicinity of Vancouver (Fig. 563).

The European Starling has been reported from sea level on the coast to 1,100 m elevation in the interior. It avoids heavily timbered regions and muskegs, and is absent above the timberline. It is most abundant around human habitations. In cities, it frequently roosts under large bridges, on buildings, or in urban and suburban trees, usually conifers in winter months but also deciduous species when in leaf, especially horse-chestnut trees. During daylight hours it moves from its roosts to nearby cultivated areas (Fig. 564), parks, gardens, vegetated playgrounds, fallow fields, airports, farms, cattle yards, rangelands, orchards, berry plantations (Fig. 565), and natural locations such as sloughs, marshes, estuarine wetlands, tidal flats, and debris-rich beaches.

Throughout its winter range in the province, the European Starling is heavily dependent on human-made habitats for both foraging and roosting. Towards the northern parts of its range, it seeks the shelter and food resources of feedlots, barnyards, poultry runs, garbage dumps, and sewage lagoons, and may roost in barns and other buildings. It frequently makes use of artificial heat sources such as chimneys, light fixtures, incinerators, and manure piles.

The European Starling is a resident of most inhabited areas of the province, except in parts of the north and northeast, but seasonal changes in numbers reflect regional movements or migration (Fig. 567). Over the southern half of the province, the spring migration is abrupt and takes place largely in late February and March. In the Georgia Depression, wintering numbers have dropped sixfold by the second week of April as the large flocks break up and birds move to their breeding areas within the ecoprovince as well as to areas further north. This decline continues into May.

On Western Vancouver Island, few birds remain through the winter, but the breeding population returns in April and May. The fact that numbers do not decline into May and June suggests that Western Vancouver Island is not a migration

Figure 564. In the Fraser River delta, where most European Starlings in the province occur, farmlands and cultivated fields attract the largest foraging flocks (Delta, 2 April 1970; R. Wayne Campbell).

Figure 565. In autumn, large flocks of European Starlings forage on grapes in vineyards in the Okanagan valley (5 km west of Osoyoos, 13 October 1992; R. Wayne Campbell).

route. There is a small movement near the end of March through the Southern and Northern Mainland Coast, but numbers do not reach the Queen Charlotte Islands until May (Fig. 567).

In the interior, data do not indicate a major migration, nor is there mention by Cannings et al. (1987) of a spring migration through the Okanagan valley. However, evidence from the returns of banded birds does reveal a spring movement through the Okanagan valley (Johnson and Cowan 1974). In the east and west Kootenays, there is little to suggest that the valleys are used as a migration route to the north.

In the Cariboo and Chilcotin areas, spring migration is dramatic. Numbers begin to build about the second week of March and peak in the last week of March. The main movement occurs rapidly and is completed by the second week of April (Fig. 567), after which the numbers remain relatively constant.

In the southern Sub-Boreal Interior, there is a rapid increase through March and April as the migrating birds arrive to nest or move to other nesting grounds (Fig. 567). This peak is followed by a decline in May, to summer levels in June. Spring migration in the Peace Lowland is marked by a large increase in numbers beginning the second week of March. The main movement continues to the end of the month. A few birds reach the Northern Boreal Mountains and Taiga Plains between the middle and the end of March, and numbers increase into May.

In the far northern regions, the European Starling withdraws for the winter. In the Taiga Plains and Northern Boreal Mountains, it is absent from the end of October until March.

In the Peace Lowland, the migration begins about mid-August, peaks near the end of August, and tapers off through September (Fig. 567). The rise in numbers from July to September appears too great to be accounted for solely by the input of young of the year raised locally; migration from areas to the north or east may be involved. Departure is abrupt, with most birds leaving before the end of October.

In the Sub-Boreal Interior, most starlings have left by the end of October. Throughout the Central Interior and southern portions of the interior, numbers decline during August, and the migration is at its height through September; most individuals have left the regions by mid-October. In the Okanagan valley, where there is a substantial wintering population, winter peaks are reached in November.

On the Queen Charlotte Islands and in the Northern Mainland Coast, migration is difficult to discern. After increasing slightly between September and November, numbers decline slightly to winter levels. On Western Vancouver Island, numbers decline rapidly between the end of September and December, and few birds remain for the winter.

In the Fraser Lowland, and in the Nanaimo Lowland from Victoria north to Campbell River, numbers begin to build through June as the young of the year flock together and increase the summer population. During August, populations increase as southbound migrants from regions inland and to the north arrive. Numbers continue to increase through mid-October and then slowly drop, presumably as some birds pass through on their way to wintering areas further south.

Movement out of the Fraser Lowland is supported by the observations of large numbers of the European Starling leaving Point Roberts, Washington, just south of Vancouver, in the early morning hours during October, and flying over the sea towards the mainland of Washington (Campbell et al. 1972b). The absence of any observed return flights suggests

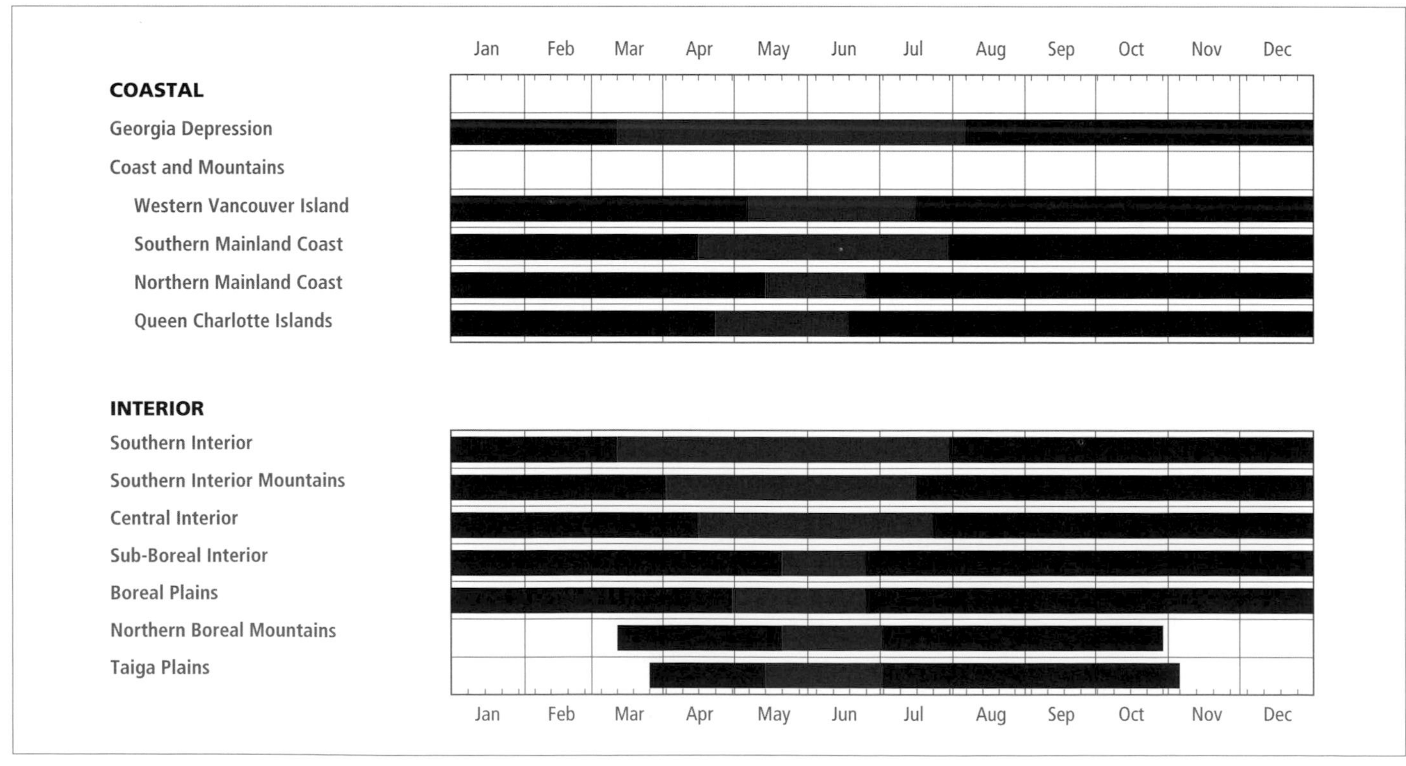

Figure 566. Annual occurrence (black) and breeding chronology (red) for the European Starling in ecoprovinces of British Columbia. Records are shown for the week in which they occurred.

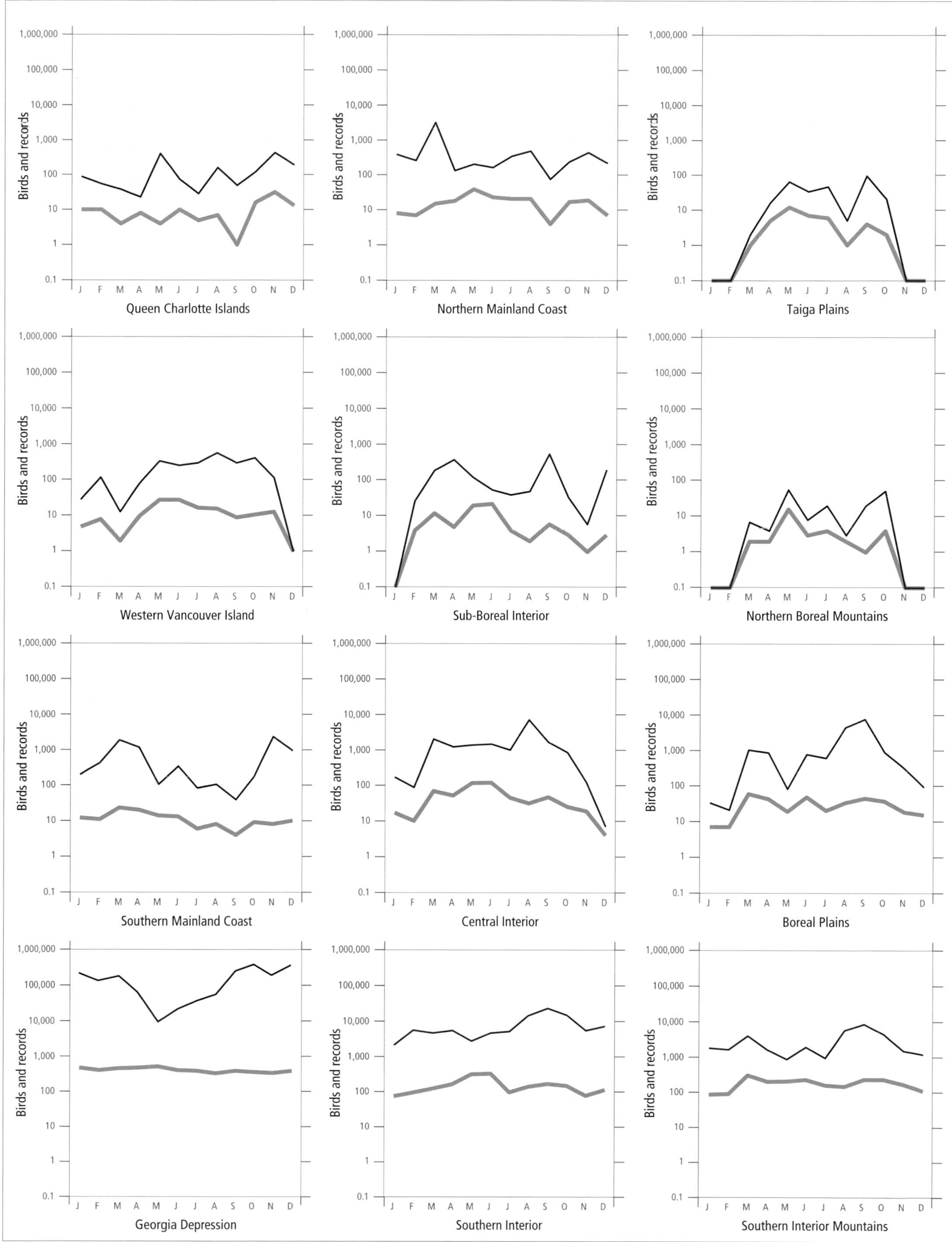

Figure 567. Fluctuations in total number of birds (purple line) and total number of records (green line) for the European Starling in ecoprovinces of British Columbia. Christmas Bird Count and nest record data have been excluded.

that the flocks seen leaving are not merely dispersing from night roosts to daytime feeding grounds. As much as 25% of the European Starling population that is in the Fraser River delta area in September may leave by the end of November.

Evidence of migratory movements and destinations of European Starlings banded in British Columbia are shown in Figure 568. Recoveries of birds banded in Washington, Oregon, and California were about equally divided between the Georgia Depression and the southern portions of the interior of the province. Most of the recoveries were in the spring.

Whatever the origin of starlings in the Boreal Plains, there is now a migratory exchange between the Peace River summer range and a California winter range: 2 birds banded in California were recovered in the Peace River region (Johnson 1974).

Analysis of the Christmas Bird Counts for Vancouver and Ladner between 1957 and 1993 shows that the wintering population of the European Starling reached maximum numbers in 1966 (Fig. 569), roughly 20 years after its arrival in the area. This was followed by a sharp decline to a low point in 1975, when winter numbers were less than a tenth of the earlier maximum, followed by a slight rebound in numbers to 1980. Since then, with some fluctuations, the general trend through the 1980s to 1993 has been towards a gradual decline. The reason for this is not known, but changes in numbers have also been noticed in Seattle, Portland, Sacramento, and San Francisco (Searing et al. 1990a).

The European Starling is present in coastal areas and southern portions of the interior throughout the year (Fig. 566).

BREEDING: The European Starling is widely distributed throughout the province during the nesting season. However, there are few known nesting occurrences north of the Skeena, Bulkley, and Nechako river valleys and the Peace Lowland. There are also few nesting records along the coast north of Howe Sound and in the vast area north of latitude 56°N and west of the Rocky Mountains. In the far northwest, the European Starling has been reported nesting only in the Atlin region. All breeding records in the Cariboo and Chilcotin areas are adjacent to the highway between Williams Lake and Bella Coola, and there is the same association with major transportation corridors throughout the northern half of the province. This is probably a response of the starling to the new nesting and foraging opportunities provided by the human communities that the highways attract.

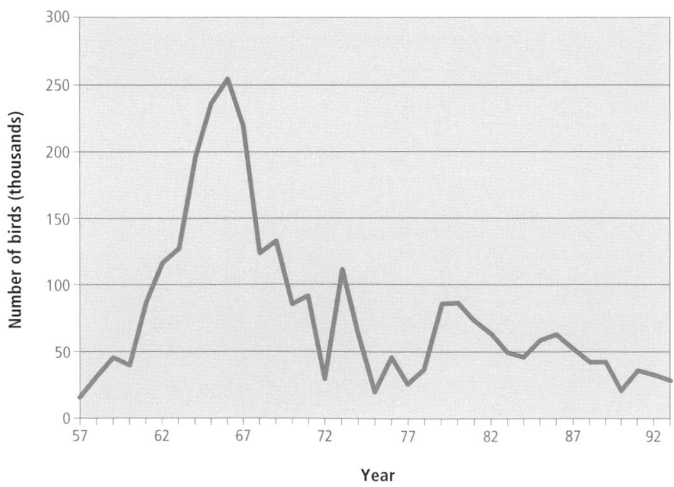

Figure 569. Summed data from the Vancouver and Ladner Christmas Bird Counts showing fluctuations in European Starling numbers over the period 1957 through 1993. The peak in starling numbers occurred in 1966, about 20 years after they were first reported in the area.

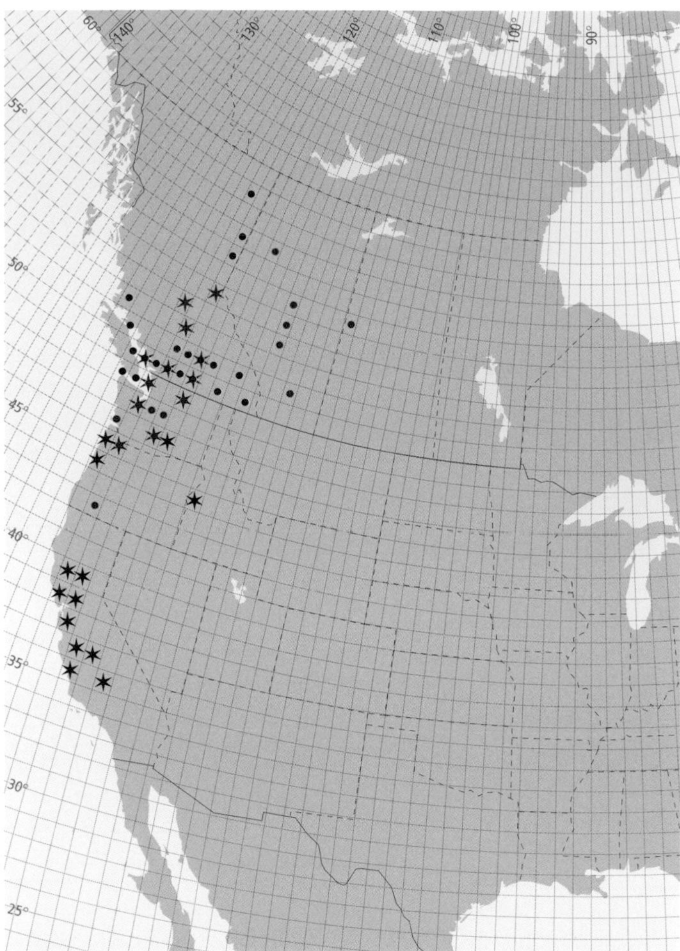

Figure 568. Banding locations (stars) and recovery sites (circles) of European Starlings associated with British Columbia. Red indicates birds banded in British Columbia; black indicates birds banded elsewhere.

The European Starling reaches its highest numbers in summer on the Fraser River delta and southeastern Vancouver Island in the Georgia Depression (Fig. 563). An analysis of Breeding Bird Surveys for the period 1968 through 1993 shows that the number of birds on coastal routes decreased at an average annual rate of 4%; the number of birds on interior routes decreased at an average annual rate of 2% (Fig. 570).

During the nesting season, the European Starling has been reported from sea level on the coast to 1,450 m elevation in the interior.

The most frequently selected nesting habitats were associated with human settlement (50%; n = 1,521), followed by forest (38%; Fig. 571) and grasslands. In each habitat, water (pond, lake, or stream) was frequently part of the setting (Fig. 572). Within these broad habitat types, most nests were found in rural, suburban, or urban settings (54%; n = 1,395), followed by woodlands – including deciduous (18%; Fig. 571),

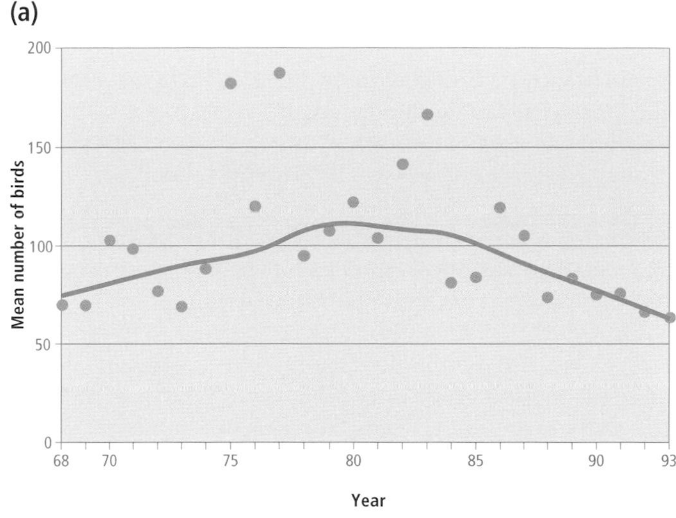

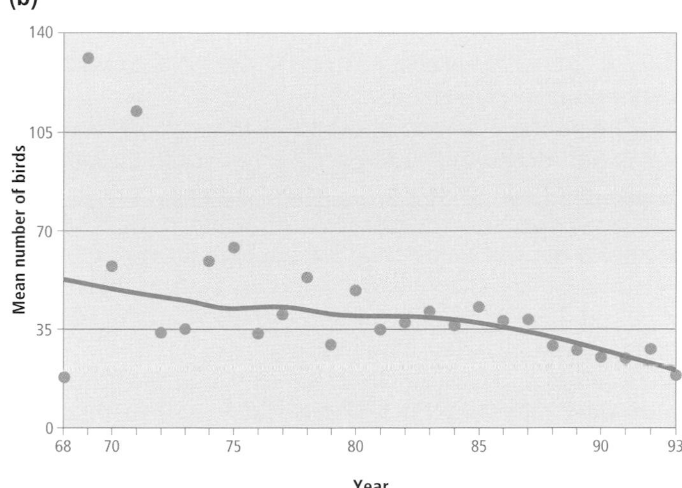

Figure 570. An analysis of Breeding Bird Surveys for the European Starling shows that (a) the number of birds on coastal routes decreased at an average annual rate of 4% ($P < 0.05$) over the period 1968 through 1993, and (b) the number of birds on interior routes decreased at an average annual rate of 2% ($P < 0.001$) over the same period.

riparian (6%), mixed (6%), and coniferous (2%) – and grasslands.

The European Starling has been recorded breeding in the province from 15 March (calculated) to 27 July (Fig. 566).

Figure 571. Although most breeding habitats in British Columbia are associated with human settlements, the European Starling regularly nests in black cottonwood forests (Cottonwood Island, Prince George, 14 July 1990; R. Wayne Campbell).

Nests: Nest sites were in trees (60%; *n* = 1,736) (deciduous, 35% [Fig. 573]; snags or stubs, 16%; and conifers, 5%); human-made structures (28%) such as buildings, houses, barns, sheds, bridges, other structures, chimneys, silos or grain elevators, posts or poles (Fig. 574) and rock or clay cliffs or banks. Most nests were located in cavities or crevices (69%; *n* = 1,726), including about 17% in cavities excavated by other species, mainly the Northern Flicker and Lewis' Woodpecker, in living or dead trees, including cottonwood, trembling aspen, red alder, bigleaf maple, arbutus, Garry oak, western redcedar, and Douglas-fir. A further 21% of the starling nests were in nest boxes.

Nests were bulky aggregations of grass, feathers, plant fibres, forbs, bark strips, leaves, twigs, rope, human-made trash, and other debris. The nest cup was woven mainly of grass, feathers, plant fibres, and other soft materials. The heights for 1,317 nests found in sites other than nest boxes ranged from ground level to 75 m, with 79% between 2.1 and 6.0 m. Occupied nest boxes were as low as 1.2 m from the ground.

Eggs: Dates for 682 clutches ranged from 28 March to 21 July, with 52% recorded between 2 May and 24 May. Calculated dates suggest that nests can contain eggs as early as

Figure 572. On the coast, the European Starling breeds in natural crevices and woodpecker cavities in wooden pilings, often over water (Cowichan River estuary, 22 July 1992; R. Wayne Campbell).

16 March. Sizes of 609 clutches ranged from 1 to 8 eggs (1E-41, 2E-41, 3E-63, 4E-142, 5E-187, 6E-117, 7E-16, 8E-2), with 54% having 4 or 5 eggs (Fig. 575).

The timing of breeding events at the coast differs from that in the interior, depending on the season and the latitude. In the Georgia Depression, the largest number of nests with eggs occurs in the last week of April, while in the Southern Interior it is a week later (Fig. 576). In both regions, the largest number of nests with nestlings occurs in the third week of May, and there is evidence of birds, in the Southern Interior, attempting a second nesting in June.

Johnson and Cowan (1974) found that in the vicinity of Vancouver, mean size for first clutches was 5.45 ± 0.12 eggs (n = 76), which is almost identical to that given by Kessel (1957) for New York. For second clutches, excluding replacement clutches, it was 4.73 ± 0.30 eggs (n = 32), which exceeds those found elsewhere.

The incubation period in British Columbia is 10 to 13 days, with 11 and 12 days being the most frequent (Johnson and Cowan 1974).

Young: Dates for 1,247 broods ranged from 28 March to 27 July, with 50% recorded between 16 May and 10 June. Sizes of 464 broods ranged from 1 to 8 young (1Y-28, 2Y-60, 3Y-67, 4Y-140, 5Y-118, 6Y-42, 7Y-8, 8Y-1), with 56% having 4 or 5 young. The nestling period is reported as 18 to 21 days (Ehrlich et al. 1988), although Feare (1984) notes that chicks usually fledge on the 21st day. Those that leave early stand little chance of surviving.

In southern British Columbia, the European Starling is double-brooded. Nesting may begin as early as mid-February (e.g., a starling carrying nesting material was observed in Victoria on 14 February 1985). In Vancouver the first clutch of eggs was present between 5 April and 10 May, followed by the eggs of the second clutch between 21 May and 18 June. Hatching success was 84% for first clutches (76 nests) and 69% for second clutches (32 nests). Thirty-eight percent of females nested successfully twice (Johnson and Cowan 1974). Hatler et al. (1978) refer to a pair of starlings successfully raising 2 broods and attempting a third.

In a study of 193 marked female starlings in Ithaca, New York, Kessel (1953b) showed that only 21% of the females attempted a second brood after success with the first, and only 5% were successful in raising 2 broods in a summer.

Cannings et al. (1987) give the mean date of second clutch initiation in the Okanagan as 28 May.

Brown-headed Cowbird Parasitism: Cowbird parasitism was not found in British Columbia in 1,727 nests recorded with eggs or young. Only rarely is the European Starling a cowbird victim, and there are few records of the species as a successful foster parent. Friedmann et al. (1977) suggest that the "aggressive and pugnacious nature of the starling, coupled with its habit of nesting in holes, apparently keeps it largely free of cowbird molestation."

Nest Success: Of 49 nests from sites other than nest boxes found with eggs and followed to a known fate, 31 produced at least 1 fledgling, for a success rate of 63%; success was 50% on the coast and 65% in the interior. Of 59 nests from nest boxes found with eggs and followed to a known fate, 30 produced at least 1 fledgling, for a success rate of 51%; success was 36% on the coast and 54% in the interior. In a British

Figure 573. Over one-third of all European Starling nests found in British Columbia were located in crevices and cavities in deciduous trees such as this Garry oak (Mount Tolmie, Victoria, 8 May 1991; R. Wayne Campbell).

Figure 574. The European Starling is an adaptable species that builds its nest almost anywhere a hole, crevice, cranny, or cavity can be found, including concrete power posts (Durrance Lake, 22 May 1995; R. Wayne Campbell).

Figure 575. Over half of all European Starling nests located in British Columbia contained 4 or 5 eggs (Uplands Park, Victoria, 25 April 1995; R. Wayne Campbell).

Columbia study (Johnson and Cowan 1974), nesting success, defined as the percentage of eggs that produced fledged young, was 76% for first clutches and 71% for second clutches.

REMARKS: The history of the introduction of the European Starling into North America is summarized by Kalmbach and Gabrielson (1921). After many attempts from 1872 to 1891, the introductions of 1890 and 1891 in Central Park, New York, resulted in the permanent establishment of the species. No one involved in this introduction could have anticipated that within a century the starlings would number in the millions in North America and be resident from coast to coast and from Alaska to Mexico.

The European Starling is well known for its flocking behaviour during the nonbreeding season. From late summer until the onset of the nesting season in the spring, the starling feeds in flocks of a few dozen to a few hundreds. Beginning as early as late June, large numbers gather to roost. The most spectacular aggregations occur at communal roosts during the winter months. Only 8 years after the expansion of the starling into the Vancouver area, a major roost had become established. In the winter of 1957-58, about 15,500 starlings occupied a group of conifers in a Vancouver residential area (Weber 1967). In 1966, 7 major roosts were identified in Vancouver. During the winter of 1966-67, 3 of these roosts were estimated to contain the following numbers: Cambie Bridge, 190,000; Second Narrows Bridge, 54,000; and Lapointe Pier, 47,000 – a total of about 290,000 birds (Weber 1967). At approximately the same time, a roost in Beacon Hill Park, Victoria, was used by 7,500 European Starlings and 7,000 American Robins (Poynter 1959).

While the large roosting aggregations of the summer and autumn attract a lot of attention, they are not the pattern over much of British Columbia. More frequently, a few hundred to a few thousand starlings will establish roosts in early summer, choosing either deciduous or coniferous trees with dense foliage. As autumn progresses, numbers increase, and the colonies in deciduous trees are abandoned in favour of conifers or human-made structures. The winter roosts are apparently chosen primarily for their microclimate, especially shelter from cold winds (Kalmbach and Gabrielson 1921; Kelty and Lustick 1977; Yom-Tov et al. 1977; Walsberg and King 1986).

Many changes have occurred in the nearly 30 years since the first inventory of roosts in and around Vancouver. In 1989, there were 10 major winter roosts in the Vancouver area, not one of which was known to have been in use 20 years earlier (Searing et al. 1990a, 1990b). Anti-starling campaigns in the city had removed or altered most of the previously used major roost sites.

The European Starling is widely recognized as a problem species. In cities it becomes a nuisance when it roosts by the thousands on buildings, bridges, and other structures, or in treed boulevards – excrement accumulates on the structures (Fig. 577a) and is showered onto pedestrians passing under the roosts (Fig. 577b). There may be health hazards, especially for children, when large amounts of fecal material accumulate on sidewalks or other areas (Dodge 1965; Tosh et al. 1970). In addition, the noise made by large numbers of roosting starlings may be distressing to those living or working nearby.

(a)

Nests with eggs or young

80
60
40
20
0

1/4 15/4 29/4 13/5 27/5 10/6 24/6 8/7 22/7 5/8 19/8

Week ending (day/month)

(b)

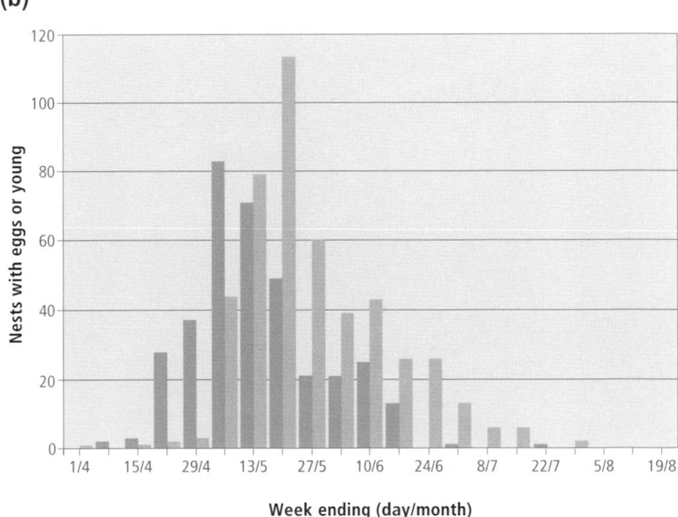

Figure 576. Weekly number of European Starling nests with eggs (dark bars) and young (light bars) for the (a) Georgia Depression and (b) Southern Interior ecoprovinces. Note the slight peaks in both eggs and young after May, which indicates that many pairs raise 2 broods.

There is extensive literature on attempts to control or otherwise mitigate the problems created by the European Starling throughout its range in Europe and North America. Techniques involve various methods to reduce their numbers (shooting, netting, poisoning, and trapping) or to scare them from areas where the nuisance is extreme (gunshots, noisemakers, helium-filled balloons, recorded starling alarm calls, and harassment by trained falcons). The general conclusion has been that killing is futile because of the high recruitment rate (Kalmbach and Gabrielson 1921; Feare 1984). Scaring techniques can provide local relief, but merely move the problem from one place to another. Within a city, the most effective long-term solution is to render buildings and bridges unattractive to starlings by design and construction changes, by retrofitting where new construction is uneconomical, and by removing problem trees.

Several cities in British Columbia, especially Vancouver and Victoria, have struggled with the nuisance problems and are still trying to find satisfactory and economically viable solutions. Vancouver's health department recently commissioned a review of the nuisance problems and potential solutions; the reports of the consultants (Searing et al. 1990a, 1990b) provide a useful review of these. Since 1990, starlings have been successfully controlled at all problem roosts in Vancouver on an annual basis. Control is nonlethal, highly effective, and economical. It includes habitat alteration, enclosures and coverings, noise bombs and explosions, pyrotechnics, distress calls, nonlethal chemicals, and balloons.

In suburban and rural areas, feeding flocks of starlings are a major hazard to small-fruit growers (Feare 1980). In the United States, grapes, blueberries, and tart cherries have suffered the most serious damage (Dolbeer and Stehn 1979); however, there has been no analysis of starling damage to agricultural crops in British Columbia.

There has been concern that the introduction of large numbers of European Starlings to the ecosystems of North America has caused the numbers of some native species to decline. The starling is a cavity-nesting species. In British Columbia, when away from an urban setting, it frequently nests in cavities excavated by the larger woodpeckers. These cavities are also used by such native species as Mountain and Western bluebirds, Tree and Violet-green swallows, Purple Martin, House Wren, American Kestrel, Northern Pygmy-Owl, Northern Saw-whet Owl, Western Screech-Owl, Bufflehead, Barrow's and Common goldeneyes, Wood Duck, and woodpeckers. A study of nest cavities previously used by the Northern Flicker, Mountain Bluebird, and Tree Swallow at Westwick Lake (near Williams Lake), found that the proportion usurped by the starling increased from 40% to 80% over 3 years (Myres 1957). In the same region, European Starlings usurped nest cavities previously occupied by Buffleheads (Erskine 1972). However, there has been no detailed study to determine the impact of the starling in British Columbia as a major competitor of the native species for scarce nest sites.

An anecdotal account from California (Weitzel 1988) describes the experience in a small area (0.35 ha) on which 14 pairs of native species had nested before the European Starling invasion. The native species included American Kestrel, Mountain Bluebird, Northern Flicker, Olive-sided Flycatcher, Mourning Dove, Tree Swallow, House Wren, House Finch, and House Sparrow. The European Starling displaced all of them by nest takeover or harassment, and none nested on the property for 5 years. After the starlings were removed, the native species returned.

Because it is very abundant during most of the year, the European Starling is an important prey species. It is the principal prey taken by the small breeding population of Peregrine Falcons on the islands in the Strait of Georgia, and is taken throughout the year by an urban population of nesting Cooper's Hawks on southern Vancouver Island.

For further information on the biology of the European Starling, see Feare (1984).

(a)

(b)

Figure 577. The European Starling has become a nuisance where its excrement from large pre-roosts and evening roosts (a) mar signs and (b) soil public walkways (Tsawwassen ferry terminal, 10 September 1995; R. Wayne Campbell).

NOTEWORTHY RECORDS

Spring: Coastal – Victoria 12 Apr 1983-1 egg in nest; Mount Douglas 8 Apr 1975-23,000; Sidney Island 31 Mar 1974-50; Turtle Island 21 May 1972-large young (Hatler et al. 1978); Crescent Beach 23 Mar 1969-25,000; Tofino 3 May 1974-200; Stubbs Island 11 Apr 1979-20; New Westminster 28 Mar 1969-eggs; Iona and Sea islands 10 May 1975-600; Marpole 10 Apr 1960-2 pairs nest building in old pilings (Guiguet 1961); Pitt River 18 May 1958-fledgling with parents, first nesting record (Guiguet 1961); Langley 30 Mar 1983-adults feeding young; Agassiz 8 Mar 1975-1,000; Hesquiat 7 Mar 1976-12 in village; Squamish 16 Apr 1981-eggs in nest; Campbell River 17 May 1960-first nesting (Guiguet 1961); Bella Coola Mar 1947-2 (Godfrey 1949); Kwatna River 18 May 1995-nesting, shells of hatched eggs on ground; Klemtu 5 Apr 1976-6; Sandspit 28 Apr 1977-eggs; Queen Charlotte City 26 Mar 1972-25; Port Clements 15 May 1976-4 eggs; Pallant Creek estuary 2 Apr 1979-6; Delkatla 12 May 1988-5; Prince Rupert 28 Mar 1976-3,000, 29 May 1975-15; Terrace 12 May 1979-5 eggs; Kispiox 30 May 1974-2 young. **Interior** – Manning Park 31 Mar 1972-3; Balfour to Waneta 16 May 1981-201; Old Hedley Road 8 Apr 1976-2,500; Creston 28 May 1956-brood (Munro 1958); South Slocan 17 Apr 1984-6 eggs; Cranbrook 20 Mar 1951-6 (Myres 1958); Swan Lake (Vernon) 13 May 1968-75; Canal Flats 8 Mar 1950-1 (RBCM 11114); Nakusp 3 Apr 1977-eggs; Invermere 14 May 1952-3 (Myres 1958); Tranquille (Kamloops) 16 Mar 1979-eggs; Lac du Bois 14 May 1954-1 (UBC 4479); Brisco 14 May 1950-nesting (Myres 1958); Sorrento 22 Mar 1972-505; Louis Creek 7 Apr 1961-5 eggs; Golden to Edgewater 10 Apr 1977-300; Parson 26 Mar 1977-500; Empire Valley 3 May 1955-2 (Myres 1958); 100 Mile House 29 Mar 1975-200; Elliot Lake 3 May 1958-5 eggs; 150 Mile House 30 Apr 1948-1 (NMC 48159); Westwick Lake 5 May 1955-2, 30 May 1955-2 broods of nestlings (Myres 1958); Wineglass Ranch (Riske Creek) 28 April 1988-nestlings; Riske Creek 4 May 1978-600; 16 km se Prince George 10 Apr 1985-300; Prince George 7 May 1968-30; Vanderhoof 24 Mar 1977-122; Moricetown 5 Jun 1960-nestlings; Attachie 9 May 1968-30; Fort St. John 29 Mar 1987-112, 8 May 1968-5 eggs; Bear Flat 13 Apr 1985-100; Hyland Post 29 May 1976-6; Dease Lake 18 Mar 1987-1; Fort Nelson 28 Mar 1977-2, 15 Apr 1987-5, 6 May 1975-30, 19 May 1975-eggs; Kotcho Lake 26 Jun 1982-4 young; Kwokullie Lake 18 May 1982-2; Liard Hot Springs 9 and 23 Apr 1975-3 (Reid 1975); Atlin 25 Mar 1981-6, 17 May 1981-9 (Campbell 1981).

Summer: Coastal – Oak Bay 8 Jun 1982-3,000; Cadboro Bay 5 Jun 1953-1 (RBCM 10073); Dixon Island 21 Aug 1976-200; Florencia Island 15 Jul 1969-4 nestlings; Wickaninnish Provincial Park 20 Jul 1967-4 nestlings; Langley 23 Jul 1978-3 nestlings; Iona Island 31 Jul 1979-2,000; Sea Island 7 Aug 1971-5,000 (Poynter 1972); Cleland Island 31 Aug 1976-2; Cheam Slough 21 Aug 1975-750; Agassiz 6 Jun 1971-164; Kent 6 Jul 1975-250; Hope 27 Jul 1962-nestlings; Miracle Beach 26 Jun 1960-first nesting (Guiguet 1961); Bella Coola 1 Jun 1968-6; Sandspit 11 Jun 1974-2 young; Port Clements 31 Jul 1971-12; Delkatla 11 Jun 1988-40; Masset 1 Aug 1988-80; Kitimat 4 Jun 1975-35; Terrace 21 Jun 1978-young in nest, 23 Aug 1977-150; Kitwanga 2 Jun 1960-2 pairs nesting in cavities in totem poles (Guiguet 1961); Hazelton 4 Jun 1960-adults carrying food to nest (Guiguet 1961); Kispiox 3 Jun 1960-3 pairs nesting (Guiguet 1961). **Interior** – s Creston valley 14 Jul 1984-2 nestlings; near Oliver 14 Jul 1971-2 eggs; Balfour to Waneta 8 Jun 1983-509; Wasa 15 Aug 1976-850; Penticton June 1950-5 first summer record (Myres 1958); 23 Aug 1977-2,000; Rutland 26 Jul 1958-4 nestlings; Kaslo 15 Jul 1975-nestlings; Kamloops-Chase 13 Jun 1982-300; Salmon Arm 30 Jul 1973-2,000 (Cannings 1973); Golden 2 Jul 1976-130; Southbanks 18 Jul 1983-4 nestlings; Riske Creek 15 Jun 1978-200; Prince George 24 Jun 1972-4 young, late Aug 1959-flock 25-30 (Guiguet 1961), 2 Jul 1974-30; Burns Lake 7 Jun 1959-nesting (Guiguet 1961); Fort St. James 7 Jun 1977-9; Smithers 15 Aug 1967-500; Moricetown 5 Jun 1960-1 adult feeding fledgling; Chetwynd 22 Jun 1978-4; Dawson Creek 29 Jul 1952-2 (Myres 1958); Fort St. John 12 Jul 1982-120, 28 Aug 1986-600; Baldonnel 23 Jun 1984-nestlings; Telegraph Creek 8 Jun 1976-3; 27 Aug 1977-1; Muncho Lake 27 Jul 1980-16; Fort Nelson 7 Jun 1982-nestlings, 27 Jul 1967-5 fledglings (Erskine and Davidson 1976), 7 Jun 1974-nestlings, 2 Jul 1982-27; Kotcho Lake 26 Jun 1982-4 nestlings; Atlin 5 Jun 1981-nestlings, 28 Jun 1981-latest nestlings.

Breeding Bird Surveys: Coastal – Recorded from 24 of 27 routes and on 76% of all surveys. Maxima: Albion 1 Jul 1983-3,726; Chilliwack 22 Jun 1975-703; Victoria 30 Jun 1973-488. **Interior** – Recorded from 56 of 73 routes and on 84% of all surveys. Maxima: Prince George 23 Jun 1973-539; Lavington 19 Jun 1974-307; Salmon Arm 26 Jun 1972-266.

Autumn: Interior – Atlin 11 Oct 1980-30; Lower Liard River 25 Oct 1980-3; Fort Nelson 27 Sep 1985-34, 29 Oct 1986-16; Taylor 20 Nov 1983-30; Fort St. John 2 Sep 1984-2,500, 9 Oct 1962-130; Germansen Landing 21 Oct 1972-30; Spatsizi 21 Oct 1980-8; Willow River 3 Sep 1977-300; Smithers 27 Nov 1967-22; Vanderhoof 30 Sep 1953-250 estimated (Munro 1956); Nulki Lake 21 Sep 1953-4 (ROM 83106-09); Ootsa Lake Nov 1949 (UBC 2227); Williams Lake 8 Sep 1951-1 (RBCM 15785); 5 Oct 1979-80; Riske Creek 14 Sep 1978-230; Alkali Lake 20 Nov 1949-4 (Racey 1950); Kleena Kleene 22 Oct 1951-1 (NMC 48161); Field 15 Nov 1975-2 (Wade 1977); Armstrong (Otter Lake) 12 Nov 1950-20 (Myres 1958); Nakusp 3 Oct 1976-350, 5 Nov 1978-200; Okanagan Landing 12 Nov 1949-21 (Munro 1953a); Sparwood 31 Oct 1984-10; Penticton 13 Sep 1980-3,000; Manning Park lodge 28 Oct 1978-13. **Coastal** – Kitimat 21 Oct 1974-70, 6 Nov 1974-62; Sandspit 17 Sep 1979-50; Cape St. James 21 Oct 1981-21; Port Hardy 28 Oct 1954-1 (PMNH 72000), 7 Oct 1959-200; Alert Bay 2 Oct 1968-4; Hope 26 Nov 1976-2,000; Vancouver 18 Sep 1966-15,000, 25 Nov 1961-50,000; Agassiz 12 Oct 1975-1,500; Tofino 11 Nov 1957-10; Tofino Inlet 26 Sep 1970-12 (Campbell and Shepard 1971); Huntingdon 17 Aug 1959-1 (PMNH 71999); Port Renfrew 21 Sep 1974-120; Victoria 6 Sep 1951-2 (Guiguet 1952a), 28 Oct 1975-30,000 (Boggs and Boggs 1962a).

Winter: Interior – Fort St. John 29 Dec 1985-1, 1 Jan 1983-4, 14 Feb 1986-9; Smithers 2 Jan 1983-84; Burns Lake Oct-Dec 1949-many (Racey 1950); Prince George 30 Dec 1972-51, 27 Feb 1983-20; 24 km s Prince George 21 Feb 1987-1; Wistaria winter 1949-50-flock of 30 (Racey 1950); Quesnel 9 Dec 1965-150; Williams Lake Dec 1945-1 (Godfrey 1949), 2 Jan 1984-25, 30 Dec 1979-3; 100 Mile House 28 Feb 1978-50; Vernon 15 Dec 1950-5 (Myres 1958); Nakusp 31 Dec 1981-30; Kelowna 1 Jan 1977-984, 25 Feb 1968-1,200, 31 Dec 1977-241; Nelson 13 Dec 1969-150, 21 Jan 1977-319, 5 Feb 1977-500; Summerland-Penticton 27 Dec 1976-1,892; Creston 1 Jan 1981-70; Oliver 15 Jan 1947-8 (Munro 1947b). **Coastal** – Terrace 1 Jan 1964-21; Prince Rupert 31 Dec 1959-25, 22 Feb 1975-145; Kitimat 2 Jan 1975-160; Masset 1 Jan 1972-20, 15 Feb 1983-12; Queen Charlotte City 28 Feb 1952-2 (RBCM 10028, 10367); Skidegate Inlet 31 Dec 1982-140; Bella Coola 1 Dec 1948-1 (Godfrey 1949), 27 Dec 1989-2; Egmont 3 Jan 1975-1; Comox 21 Dec 1951-2 (Pearse 1953); Vancouver 25 Jan 1950-2 (Racey 1950); Agassiz 31 Jan 1983-1,000; Tofino 3 Jan 1976-4; Cloverdale 4 Dec 1956-1 (Myres 1958), 10 Feb 1968-10,000; Leonard Island 31 Dec 1978-1; Surrey mid-Dec 1950-4 (Racey 1950); Boundary Bay 18 Feb 1979-1,200; Sarita River 11 Feb 1977-60; Sooke 31 Dec 1983-83.

Christmas Bird Counts: Interior – Recorded from 25 of 27 localities and on 86% of all counts. Maxima: Kelowna 3 Jan 1987-5,783; Penticton 30 Dec 1989-3,558; Vernon 18 Dec 1988-2,755. **Coastal** – Recorded from 32 of 33 localities and on 97% of all counts. Maxima: Vancouver 26 Dec 1966-**254,068**, all-time Canadian high count (Anderson 1976a); Ladner 26 Dec 1969-56,019; Pitt Meadows 27 Dec 1980-36,148.

Crested Myna

Acridotheres cristatellus (Linnaeus)

CRMY

RANGE: In North America, introduced and presently established in Greater Vancouver, British Columbia. Native and resident in central and southern China, Taiwan, Hong Kong, Macao, Vietnam, central Laos, and eastern Burma (American Ornithologists' Union 1983; De Schauensee 1984).

STATUS: On the coast, a *rare* to *uncommon* local resident in the Georgia Depression Ecoprovince, where it is now confined to small areas in Vancouver, Burnaby, and New Westminster, and Sea Island and northern Lulu Island in Richmond. Breeds.

CHANGE IN STATUS: Before 1897, 1 or 2 pairs of Crested Mynas (Fig. 578) that originated from Hong Kong to Macao escaped or were intentionally released in Vancouver (Kermode 1921). The first specimen was collected in downtown Vancouver in 1904. By 1921, the species had established a communal winter roost in downtown Vancouver in some elaborately corniced buildings on the corner of Carrall and Cordova streets.

Figure 578. The Crested Myna was introduced into southwestern British Columbia in the late 1800s. Its population reached peak abundance in the 1920s and 1930s, and by 1959 had declined to less than 3,000 birds. In the mid-1990s, there are probably less than 100 birds in the Greater Vancouver area (Mark Wynja drawing).

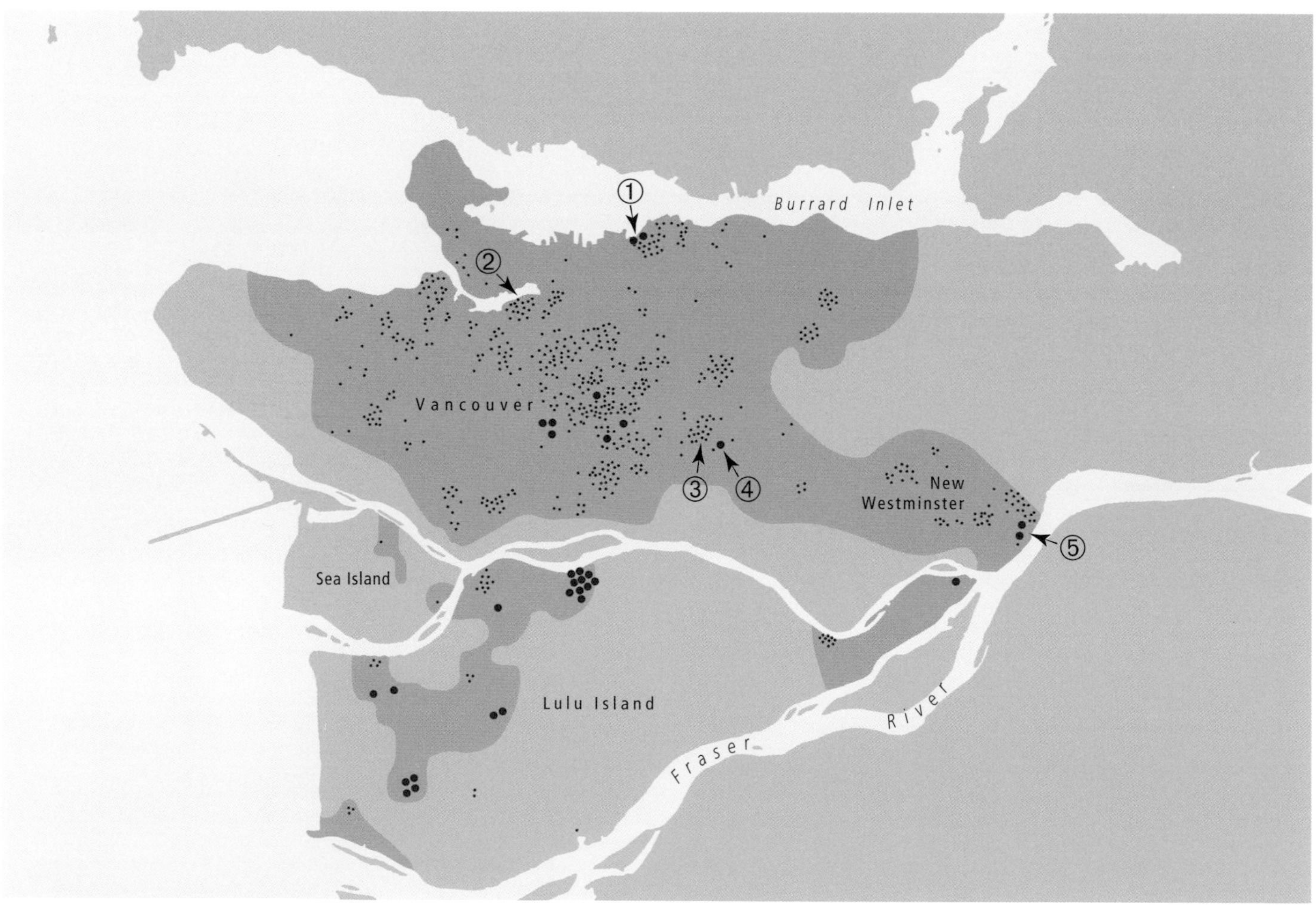

Figure 579. Distribution of Crested Mynas in the Greater Vancouver area, 1959-1960. The 1959 spring distribution is shown by the large dots and solid small circles, representing 2 and 50 birds, respectively. The locations of winter roosts in 1960 include (1) Lapointe Pier, (2) Connaught bridge, (3) Sir Guy Carleton School, and (4) Collingwood United Church, all in Vancouver; and (5) Russell Hotel in New Westminster. The approximate boundary of the built-up areas is shown by the shaded portion (after Mackay and Hughes 1963).

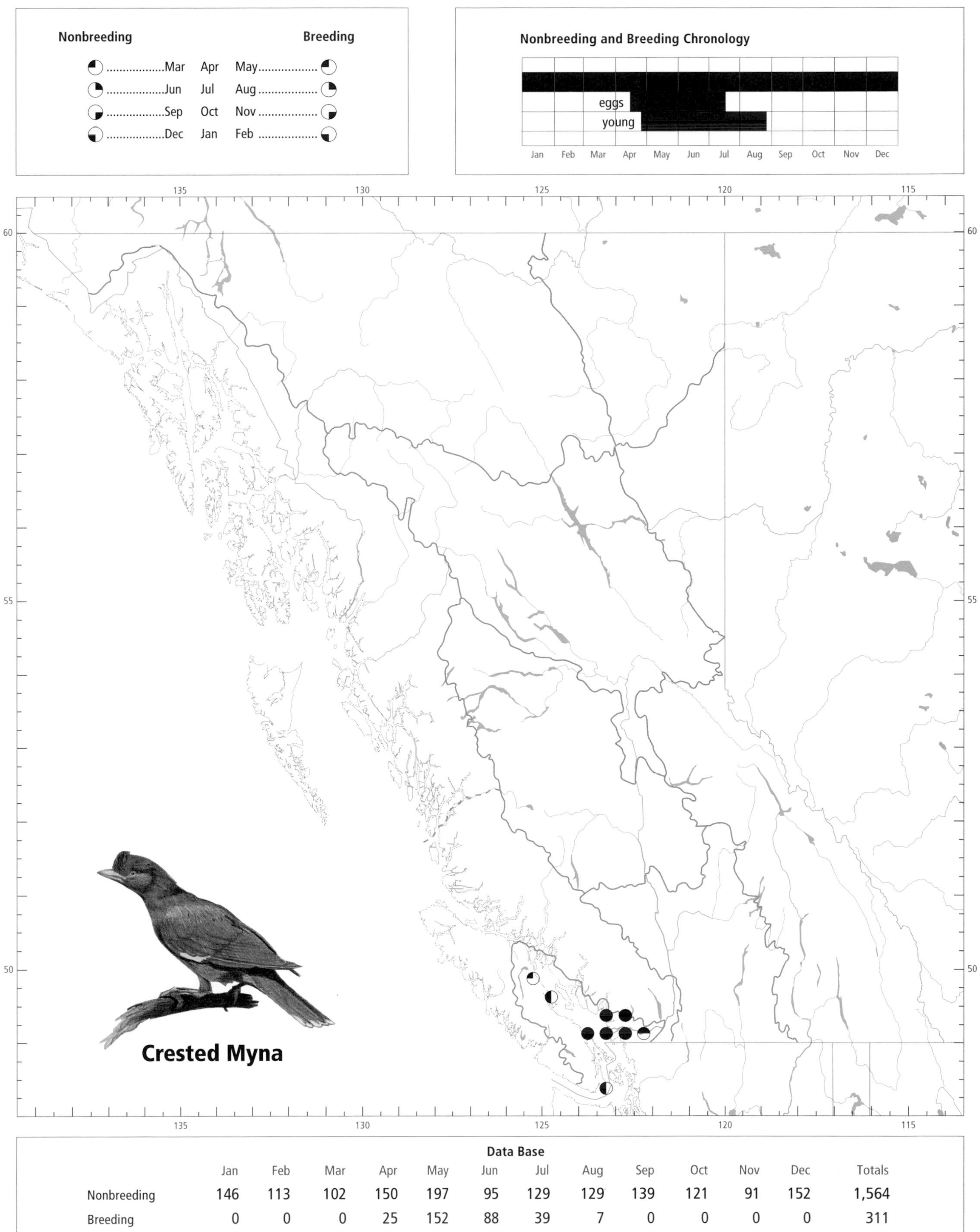

Data Base

	Jan	Feb	Mar	Apr	May	Jun	Jul	Aug	Sep	Oct	Nov	Dec	Totals
Nonbreeding	146	113	102	150	197	95	129	129	139	121	91	152	1,564
Breeding	0	0	0	25	152	88	39	7	0	0	0	0	311

(a)

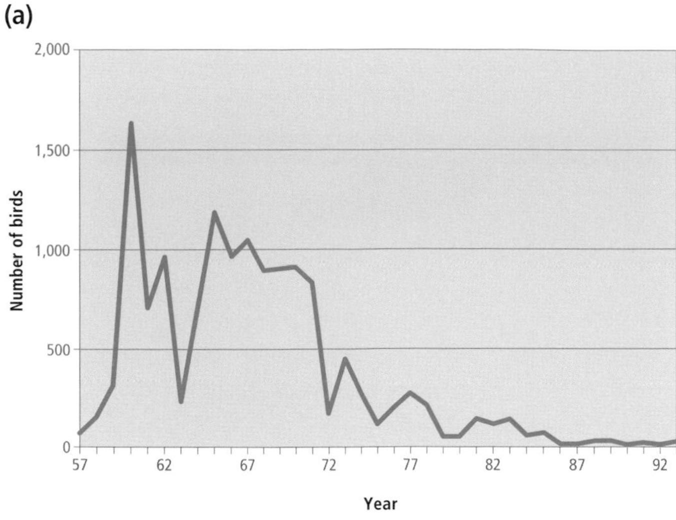

(b)

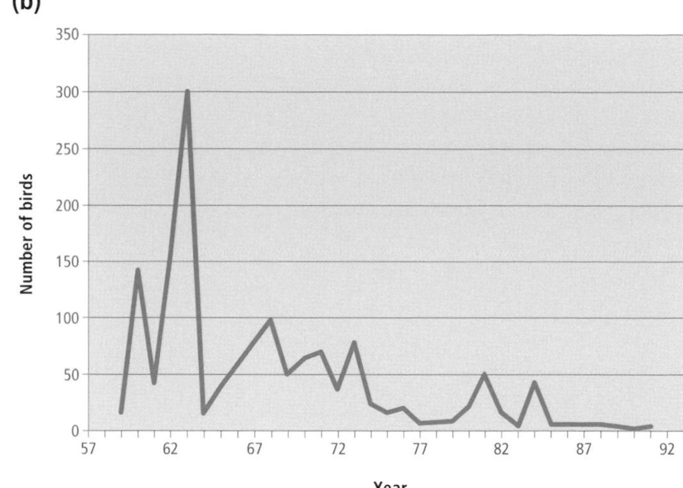

Figure 580. Fluctuations in Crested Myna numbers from Christmas Bird Counts for 1957 to 1993 for the (a) Vancouver and (b) Ladner count areas, showing the decline in numbers from their highs in the 1960s to their lows in the 1990s.

Kermode (1921) estimated that as many as 1,200 mynas used this roost in midwinter.

The winter concentration was in a heavily industrialized area featuring streets lined with 4- to 6-storey buildings, railroad yards, and waterfront industries of a busy seaport handling a large tonnage of grain, agricultural produce, and general cargo, in an era when most of these products were shipped in bulk. Horses were the main source of traction. Food for the mynas, including waste grains and produce, was close at hand and abundant even in this densely built-up urban centre.

Butler and Toochin (1984) suggest that the mynas also sought the warmth generated by the heated but uninsulated buildings.

The species continued to increase in numbers. In 1925, there were an estimated 6,000 to 7,000 birds (Cumming 1925), and in 1927, Kelly (1927) suggested that the population in downtown Vancouver may have reached 20,000 birds occupying roosts in about 4 square city blocks. Kelly (1927) also refers to smaller bands roosting at sites in the west end and east end of Vancouver and in New Westminster. Both the Cumming and Kelly figures were informal estimates, but they were quoted by Scheffer and Cottam (1935) as if they were accurate estimates and have since been repeated by others. However, at least a few thousand birds were certainly present.

Also in the mid-1920s, Crested Mynas were being reported from southern Vancouver Island and southward as far as Blaine, Bellingham, and Seattle in Washington state, and Portland in Oregon.

No attempts were made to estimate populations by direct count or to determine the distribution (Fig. 579) of mynas until Mackay and Hughes (1963) completed their study. Their surveys revealed a population of 2,000 to 3,000 birds in the Greater Vancouver area. The population behaviour between the late 1920s and 1960 is unknown. There are records of slow expansion of range, but the species was apparently not an aggressive colonizer. Scheffer and Cottam (1935) suggest that the Crested Myna had already spread as far as North Vancouver in the north and Ladner in the south, but there are no data to support this. The earliest available records of regional expansion are from Chilliwack (Munro 1930b), Ladner (1941), Crescent Beach (1942), Pitt Meadows (1959), Port Moody (1960), Squamish (1961 and 1962), west Mission (1964), Cultus Lake (1964), and West Vancouver (1968). Except for Ladner, however, there is no evidence that the myna was able to establish itself in these peripheral areas.

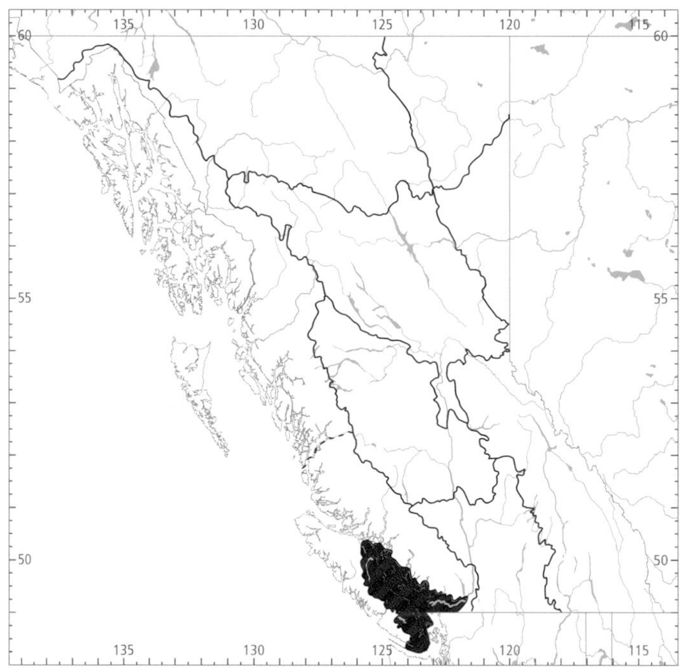

Figure 581. In British Columbia, the highest numbers for the Crested Myna in winter (black) and summer (red) occur in the Georgia Depression Ecoprovince.

Scheffer and Cottam (1935) refer to 12 Crested Mynas at the head of Lake Washington, Washington, in 1929. A small group numbering at least 7 birds reached Vancouver Island in 1937, and was reported from Union Bay, Merville, Courtenay, and Comox in 1937 and 1938 (Pearse 1938). Another myna reached Comox in 1940; a pair attempted to nest in Victoria in 1946 (Clay 1946), but was collected in an

Figure 582. Foraging habitat for the Crested Myna in Vancouver includes open spaces adjacent to boulevards with trees, city streets, and edges of abandoned railroad rights-of-way (Vancouver, West Boulevard and 49th Street, 7 August 1996; R. Wayne Campbell).

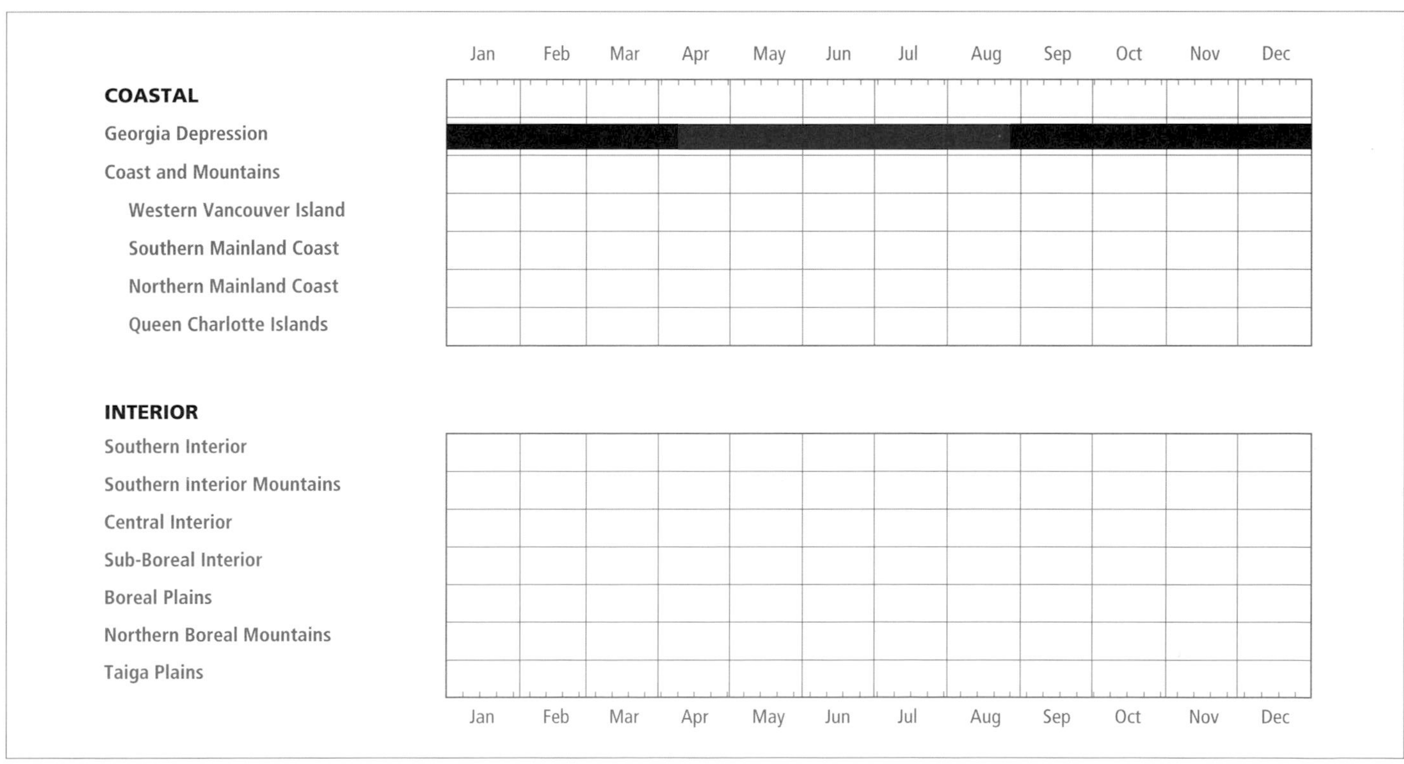

Figure 583. Annual occurrence (black) and breeding chronology (red) for the Crested Myna in ecoprovinces of British Columbia. Records are shown for the week in which they occurred.

attempt to prevent the spread of the species to southern Vancouver Island. It was again recorded in Victoria in 1950 and 1961. The Crested Myna arrived in Nanaimo in 1952, and established a breeding colony that survived for some 16 years (Merilees 1985; Johnson and Campbell 1995).

A survey of the Crested Myna in Greater Vancouver in late February 1980 found 650 birds and revealed that the myna was absent from downtown Vancouver, where the large roosts had been found in 1960 (Butler and Toochin 1984). From the 1980 survey, the population was estimated to be 2,537 birds, which is similar to the 1960 estimate by Mackay and Hughes (1963), although the mynas had changed their distribution and winter roost habits.

However, the Christmas Bird Count data for the same region indicate a major decline in numbers between 1957 and 1993 (Fig. 580). Certainly the recent Christmas Bird Counts suggest that the present population consists of fewer than a hundred birds. The discrepancy between the 2 sources of evidence suggests that at least 1 of the 1960 and 1980 estimates is inaccurate or that the Christmas Bird Counts have represented only a changing proportion of the actual population.

Figure 584. Over one-third of all Crested Myna nest sites in British Columbia were reported in buildings. This nest site, located in the upper left corner of the brick building, was active from 1992 through 1995 (2nd Avenue and Yukon Street, Vancouver, 7 August 1996; R. Wayne Campbell).

Based on our analysis of available data, we believe that the present population of Crested Mynas in the Greater Vancouver area is probably fewer than 100 birds, and that there are no other populations in British Columbia. At its present rate of decline, it is likely to disappear from North America within a few decades (Johnson and Campbell 1995).

NONBREEDING: The Crested Myna occurs locally in small numbers in several places in urban, suburban, and rural environments around Vancouver, mainly south of Burrard Inlet. A few birds may still occur at the south end of Lonsdale Avenue in North Vancouver. The southernmost populations are in Ladner, and the easternmost in New Westminster. All occurrences have been between sea level and about 300 m elevation.

The Crested Myna is resident year-round and is confined to the extreme southwest mainland coast in the Georgia Depression (Fig. 581), where it is closely associated with human environments. Habitats occupied include farms with low trees, bramble tangles, meadows, grain fields, orchards, small-fruit plantations, farmlands with manure heaps, parks, gardens, and campuses featuring a mixture of open meadow or lawn with low trees, tall shrubs, and ivy or other dense, climbing vines. Mynas also use parking lots, drive-in restaurants, alleys with garbage dumpsters, electricity service lines with boulevard trees and open grassy edges, and abandoned railroad rights-of-way (Fig. 582). In 1960, many of the mynas were using collective winter roosts on bridges, schools, and other buildings (Fig. 579).

Although nonmigratory in British Columbia, the Crested Myna changed its habits with the seasons during the first 60 or more years of its presence here. In the autumn it aggregated and used communal roosts, which, when the species was abundant, at times accommodated from 5,000 to 7,000 birds. Large communal roosts of that size have not been reported in the past 25 years. Unlike the European Starling in the same habitat, the myna arrived at and left the roosts in small flocks; it usually foraged in pairs or small groups.

The Crested Myna is a year-round resident in British Columbia.

BREEDING: The geographic distribution of the Crested Myna during the breeding season is almost the same as that for the rest of the year, but within that area there is a tendency, first noted by Scheffer (1931), for the species to favour the more rural parts of its range as the nesting season approaches. In spring and summer, the birds are most frequently seen around barns, farms, suburban gardens, and parks. Today the mynas are most often seen in southern areas of Vancouver close to the north arm of the Fraser River and on adjacent islands forming the outer delta.

Breeding Bird Surveys for the period 1968 through 1993 contain insufficient data for analysis.

The Crested Myna reaches its highest numbers in summer in the extreme southwest mainland coast of the Georgia

Figure 585. The nest of the Crested Myna in British Columbia consists mainly of dried grasses and leaves, often intermingled with feathers and other debris such as peach stones, bones, and paper (near Iona Island, 8 May 1995; R. Wayne Campbell).

Depression (Fig. 581). Elevations occupied during the nesting season are similar to those of the nonbreeding season. All nests were found in human-influenced habitats (Fig. 582), including suburban (51%), urban (39%), and rural (9%) sites.

The Crested Myna has been recorded breeding in British Columbia from 11 April (calculated) to 26 August (Fig. 583).

Nests: The Crested Myna nests primarily in human-made sites, including nest boxes, cavities in power poles, drainpipes, pilings, semi-enclosed spaces among the eaves and decorations of buildings (Fig. 584), in holes in tall tree stumps or in similar natural cavities or cavities excavated by woodpeckers, and sometimes in dark recesses within densely growing climbing vines. Analysis of nest sites ($n = 71$) shows that 34% of nests were in buildings, 29% in poles, 13% in houses, 10% in trees, 3% in barns, 1% in silos, and 10% in various other structures (Johnson and Campbell 1985). Specific nest placements included nest boxes (44%), crevices (20%), cavities (17%), cavities excavated by other species (6%), under the eaves of buildings (11%), and on beams (1%).

In southwestern British Columbia, the Northern Flicker and Lewis' Woodpecker were formerly the most important excavators of cavities used by mynas. Although the Northern Flicker is still present, the Lewis' Woodpecker has virtually disappeared from southwestern British Columbia, and most of the tall Douglas-fir snags they used have been removed during land clearing.

The first indication of nesting activity is the occurrence in the nest cavity of fresh, green leaves and colourful flowers such as dandelion (Johnson and Campbell 1995). The base of the nest is usually composed of a loose mass of dried grasses and leaves, intermingled with feathers, roots, snakeskins, mummified bird and mammal carcasses (e.g., American Goldfinch, Barn Swallow, and house mouse), and a variety of cultural detritus (manufactured material occurs in 62% of nests). The nest cup is lined with finer plant fibres, feathers, and other fine material (Fig. 585). Nest building takes about 14 days to complete (Bent 1950). The heights for 47 nests ranged from ground level to 12 m, with 64% between 3 and 8 m.

Eggs: Dates for 71 clutches ranged from 12 April (Mackay and Hughes 1963) to 17 July, with 51% between 4 and 30 May. Calculated dates indicate that eggs can occur as early as 11 April. Sizes of 67 clutches ranged from 1 to 6 eggs (1E-7, 2E-1, 3E-7, 4E-26, 5E-21, 6E-5), with 70% having 4 or 5 eggs (Fig. 585). The mean clutch size for first clutches was 4.9 ($n = 31$); for second clutches it was 3.8 ($n = 5$) (Johnson 1972; Johnson and Cowan 1974). Second clutches were produced by 9% of pairs, from late May to early August. The

Table 8. Numbers of Crested Mynas and European Starlings tallied on Christmas Bird Counts in Nanaimo, Vancouver Island, between 1963 and 1968 (from Merilees 1985).

Year	Mynas	Starlings
1963	63[1]	102
1964	26	236
1965	20	267
1966	6	187
1967	6	2,495
1968	4	805

[1] W.J. Merilees, who is familiar with the myna population in Nanaimo, believes there were never more than 30 birds present.

incubation period was 12 to 15 days and averaged 14 days (Johnson and Cowan 1974).

Young: Dates for 62 broods ranged from 25 April to 26 August (Johnson 1972), with 51% between 23 May and 28 June. Sizes of 56 broods ranged from 1 to 5 young (1Y-9, 2Y-9, 3Y-14, 4Y-20, 5Y-4), with 61% having 3 or 4 young. First broods occurred from about 23 May to 27 June, while second broods were scattered from early June through mid-August (Johnson 1972). The nestling period in British Columbia was 22 days (Johnson 1972; Johnson and Cowan 1974, 1975).

Brown-headed Cowbird Parasitism: Cowbird parasitism was not found in British Columbia in 121 nests recorded with eggs or young, nor has it been recorded elsewhere in North America (Friedmann and Kiff 1985).

Nest Success: Of 18 nests found with eggs and followed to a known fate, 9 produced at least 1 fledgling for a nesting success of 50%. Hatching success for first broods was 61% (n = 32); for second broods it was 58% (n = 5). About 46% of eggs from first clutches and 35% of those from second clutches produced fledglings.

REMARKS: The reasons for the decline of the Crested Myna in British Columbia have not been identified. Despite its early success, there is evidence that the species finds the ambient temperature marginal. The myna is a less attentive brooder than the European Starling, which nests in the same environment (Johnson and Cowan 1974). This results in greater fluctuations in egg temperature, and apparently leads to the poor success rate from egg to fledgling. Estimates show that starlings fledged 547 young per 100 pairs per year, whereas the myna fledges 238 young per 100 pairs (Johnson and Cowan 1974; Johnson and Campbell 1995).

A further indication of the myna's unsuitability to winter temperatures in Vancouver comes from the finding that the birds are concentrated in 8 of 11 "urban heat islands." The observation that the mynas are selecting warmer microclimates during colder weather, and that they have not adapted their warm-climate incubation behaviour to the cooler conditions of British Columbia, led to the suggestion (Butler and Toochin 1984) that the early success of the species may have been associated with the artificial warmth of poorly insulated buildings during the winter and an opportunity to select better-insulated nest sites during the nesting season. Both conditions have changed in the last few decades.

Probably the most important factor has been the complete alteration of the ecological situation in which the myna flourished. Horses have vanished from Vancouver, and with them the tons of manure that were deposited throughout the area, serving as a source of food for the mynas. The grain-handling technology of the Port of Vancouver has become more efficient, with less spillage. The green space of interspersed fields, gardens, and city that provided foraging and nesting habitat for the mynas has been replaced by densely packed housing, paved roads, and parking lots.

The arrival of the European Starling in southwestern British Columbia brought an obvious and dramatic new element into the environment of the Crested Myna, introducing an aggressive competitor for both food and nest sites. The European Starling began its invasion of the province in 1945 (Myres 1958), and its increase in numbers in the region occupied by the myna has paralleled the decline of the latter.

The Nanaimo colony of mynas, on southeastern Vancouver Island, offers an opportunity to examine the short history of a myna population that established itself in an already developed and mechanized city just before starlings arrived there (Merilees 1985). In 1952 the first mynas were reported in Nanaimo. Their numbers increased gradually, reaching a high point in 1963, when, on the basis of Christmas Bird Count data, the colony was estimated to number about 63 birds. Numbers subsequently declined to 4 birds in 1968 and none some years later. The decline coincided with a rapid increase in the numbers of newly arrived European Starlings (Table 8). Although there are no data to support or refute a cause-and-effect relationship between the advent of the European Starling and the decline of the Crested Myna, either at Nanaimo or throughout its range in the province, it is likely that the two are linked. For whatever reasons, the Crested Myna is approaching extinction in North America.

See Johnson and Campbell (1995) for a summary of the natural history of the Crested Myna in North America.

NOTEWORTHY RECORDS

Spring: Coastal – Victoria 29 Apr 1946-1 (RBCM 9162), 7 May 1946-1 (RBCM 9163); Tsawwassen 11 May 1963-2; Richmond 25 Apr 1970-1 egg and 4 nestlings; 1 km e Sumas 4 May 1984-1; New Westminster 16 May 1949-1 (CMNH E8668); Vancouver 7 Mar 1979-30, 27 Mar 1990-16 foraging near Oak Street and Marine Drive, 1 Apr 1976-30, 13 Apr 1970-1 egg; Port Moody 15 May 1960-6; North Vancouver 8 May 1985-1; Nanaimo 1 May 1963-8; Union Bay 26 Mar 1937-3; Squamish (Alice Lake) 14 May 1961-6; Squamish 20 Apr 1962-1; Merville 24 May 1937-1 (RBCM 14375); Courtenay 27 Apr 1937-4 (Pearse 1938). **Interior** – No records.

Summer: Coastal – Cultus Lake 18 Jun 1964-1; Lulu Island 27 Jun 1991-5 adults with 3 juveniles near 7100 Granville Road; Iona Island 8 May 1995-4 eggs (Fig. 585); Richmond 28 Jun 1993-4 adults with 2 young at Railway and Steveston roads; Sea Island 22 Aug 1964-32; Mission 1 Jul 1964-1; Dinsmore Island 3 Jun 1939-4 eggs; South Vancouver 26 Aug 1942-3 nestlings, 24 Jul 1993-4 adults with 2 young at 64th and Granville streets; Vancouver 17 July 1968-6 eggs, 29 Jul 1979-20; North Vancouver 6 Jun 1979-16; West Vancouver 21 July 1968-1; Pitt Meadows 1 Jul 1959-1, 29 Jun 1975-1. **Interior** – No records.

Breeding Bird Surveys: Coastal – Recorded from 1 of 27 routes and on less than 1% of all surveys. Maxima: Point Grey 22 Jun 1975-1. **Interior** – Not recorded.

Autumn: Interior – No records. **Coastal** – North Vancouver 1 Sep 1989-2; Vancouver 22 Oct 1949-25, 25 Nov 1904-1 (RBCM 737); New Westminster 6 Sep 1992-5 along waterfront; Sea and Iona islands 11 Nov 1973-37.

Winter: Interior – No records. **Coastal** – Courtenay 7 Feb 1938-1 (RBCM 13668); Nanaimo 19 Jan 1952-1 (Guiguet 1952); Stanley Park to Point Atkinson 1 Jan 1961-62; Vancouver 19 Feb 1921-125, 26 Jan 1944-160, 9 Jan 1971-200 at roost on Thompson School at 55th Avenue and Commercial Street (Campbell et al. 1972b), 19 Jan 1972-60 at Shaughnessy (Campbell et al. 1974), 7 Feb 1992-9 at n end Oak Street Bridge; Canoe Pass 14 Feb 1961-50; Victoria 11 Dec 1961-1.

Christmas Bird Counts: Interior – Not recorded. **Coastal** – Recorded from 4 of 33 localities and on 14% of all counts. Maxima: Vancouver 26 Dec 1960-**1,632**, all-time North American high count (Anderson 1976b); Ladner 21 Dec 1963-300; Nanaimo 28 Dec 1963-63.

Solitary Vireo
Vireo solitarius (Wilson)

SOVI

RANGE: Breeds from northern British Columbia, southeastern Yukon, and southwestern Mackenzie across Canada to southern Quebec, southern Newfoundland, and the Maritime provinces; south through mountainous and forested regions to Baja California, New Mexico, Texas, Tennessee, Georgia, and Guatemala. Winters mainly from southern California and the Gulf coast south through Central America to Costa Rica.

STATUS: On the coast, *uncommon* to *fairly common* migrant and summer visitant in the Georgia Depression Ecoprovince; *rare* to *uncommon* on the southern mainland and *very rare* on the northern mainland of the Coast and Mountains Ecoprovince; *casual* on Western Vancouver Island; absent from the Queen Charlotte Islands.

In the interior, *uncommon* to *fairly common* migrant and summer visitant in the Southern Interior, Southern Interior Mountains, and Central Interior ecoprovinces; *rare* to *uncommon* in the Sub-Boreal Interior and Boreal Plains ecoprovinces; *rare* in the Taiga Plains Ecoprovince and *casual* in the Northern Boreal Mountains Ecoprovince.

Breeds.

Figure 586. In British Columbia, the Solitary Vireo has been found breeding from 20 April to 29 July (Victoria, June 1993; Tim Zurowski).

NONBREEDING: The Solitary Vireo (Fig. 586) has a widespread distribution across most of southern and central British Columbia. It occurs regularly from southeastern Vancouver Island east to the Rocky Mountain Trench and north to the Cariboo and eastern Chilcotin. North of Williams Lake, the Solitary Vireo is more sparsely distributed, although it is relatively common in the east-central interior north to the Peace Lowland. It is sparsely distributed in the western Chilcotin and Taiga Plains, and very scarce in the Northern Boreal Mountains. The northernmost records are from near Kwokullie Lake, Coal River, and Dease River. On the central and northern mainland coast, it occurs only in small numbers from the Bella Coola valley north to Stewart, on estuaries and along major river valleys that cut through the mountains from the interior. It is absent from the Queen Charlotte Islands and nearly all other mainland coastal islands north of Vancouver Island.

Figure 587. On southeastern Vancouver Island, the Solitary Vireo inhabits mixed forests of Douglas-fir, Garry oak, and arbutus on dry, rocky sites (Nanoose Hill, 18 April 1992; R. Wayne Campbell).

The Solitary Vireo has been recorded at elevations from near sea level to 500 m on the coast, and from 300 to 1,200 m in the interior. It occurs mainly in valley bottoms and on plateaus and lower mountain slopes (Fig. 587). For a detailed discussion of habitat, see BREEDING.

Early migrants may arrive in the Georgia Depression as early as the last week of March; however, birds normally arrive around the first week of April, with the main movement between the third week of April and early May (Figs. 588 and 589). In the Southern Interior and Southern Interior Mountains, early migrants may arrive in early April, but the main movement begins in late April and peaks quickly in early to mid-May. In the Central Interior and southern Sub-Boreal Interior, early spring migrants may arrive in late April and the movement peaks in mid-May. In the Boreal Plains and Taiga Plains, spring migrants arrive through Alberta

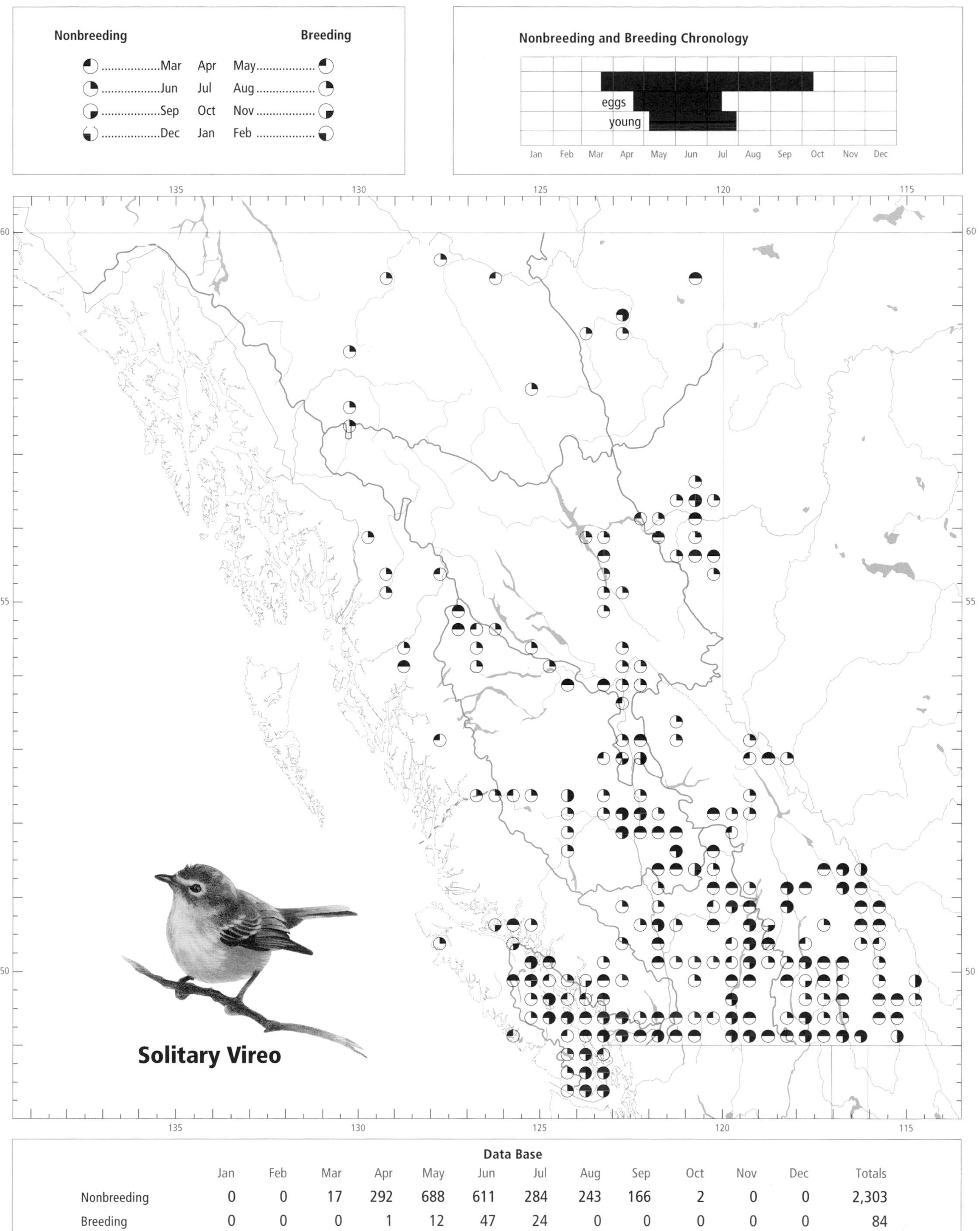

Data Base

	Jan	Feb	Mar	Apr	May	Jun	Jul	Aug	Sep	Oct	Nov	Dec	Totals
Nonbreeding	0	0	17	292	688	611	284	243	166	2	0	0	2,303
Breeding	0	0	0	1	12	47	24	0	0	0	0	0	84

beginning in mid-May; the migration peaks quickly in late May and finishes in early June.

In the northeast, autumn migration appears to begin as soon as the young are independent, probably around late July. The main movement occurs during the last 2 weeks of August, with a few stragglers remaining into early September. In the interior south of Prince George, the autumn migration is remarkably synchronous, beginning in August and ending by mid-September. Only in the Southern Interior is there a record of a straggler as late as the second week of October (Figs. 588 and 589). On the coast, the main autumn movement begins in late August and peaks in the first 2 weeks of September; most birds have gone by the third week of September.

On the coast, the Solitary Vireo has been recorded from 23 March to 5 October; in the interior, it has been recorded from 25 April to 27 September (Fig. 588).

BREEDING: The Solitary Vireo breeds throughout much of southern British Columbia from southeastern Vancouver Island east to the Rocky Mountains and north to the Cariboo and Chilcotin areas. Further north, breeding records are scarce, but nesting has been confirmed in the Sub-Boreal Interior, in the Peace Lowland, and near the Yukon border. The Solitary Vireo is likely a more widespread breeder in the northern interior than is currently suggested by the data. On the coast, breeding has not been confirmed on Western Vancouver Island or on the north coast.

The Solitary Vireo reaches its highest numbers in summer in the Southern Interior (Fig. 590). An analysis of Breeding Bird Surveys for the period 1968 through 1993 shows that the mean number of birds on interior routes has increased at an average annual rate of 5% (Fig. 591); analysis of coastal routes for the same period could not detect a net change in numbers.

On the coast, the Solitary Vireo has been recorded breeding at elevations from near sea level to 500 m; in the interior, it breeds at elevations from 340 to 1,200 m. This vireo inhabits a variety of open forest types. On southeastern Vancouver Island and the Gulf Islands, the Solitary Vireo is most numerous in mixed forests on dry, rocky sites with open stands of Douglas-fir, Garry oak, and arbutus (Fig. 587). Elsewhere on the coast, it is found along the edges of mixed woods where bigleaf maple, red alder, black cottonwood, and western flowering dogwood occur with scattered Douglas-fir and western redcedar. It is also found along powerline rights-of-way, field margins, rural roads, and similar clearings, as well as in middle-aged, regenerating forests on dry sites.

In the southern portions of the interior, the Solitary Vireo, unlike the Warbling and Red-eyed vireos, occurs frequently in both coniferous and deciduous woodlands. It is a characteristic summer bird of western larch, Douglas-fir, ponderosa pine, and lodgepole pine forests on lower mountain slopes. In the Central Interior, Sub-Boreal Interior, and Peace Lowland, it occurs mainly in mixed forests of trembling aspen, willow, lodgepole pine, and spruce. In northern coniferous forests, it occurs more frequently in forest stands near water.

Most breeding habitats (n = 30) have been described as mixed (47%), coniferous (30%), or deciduous (10%) forest. Rural, suburban, and other human-influenced habitats were also used (13%). More specifically, the Solitary Vireo has been found nesting in younger second-growth forest (33%; n = 21), mature forest (24%), and riparian forest (19%); adjacent to roadsides; and in backyard gardens and suburban parks.

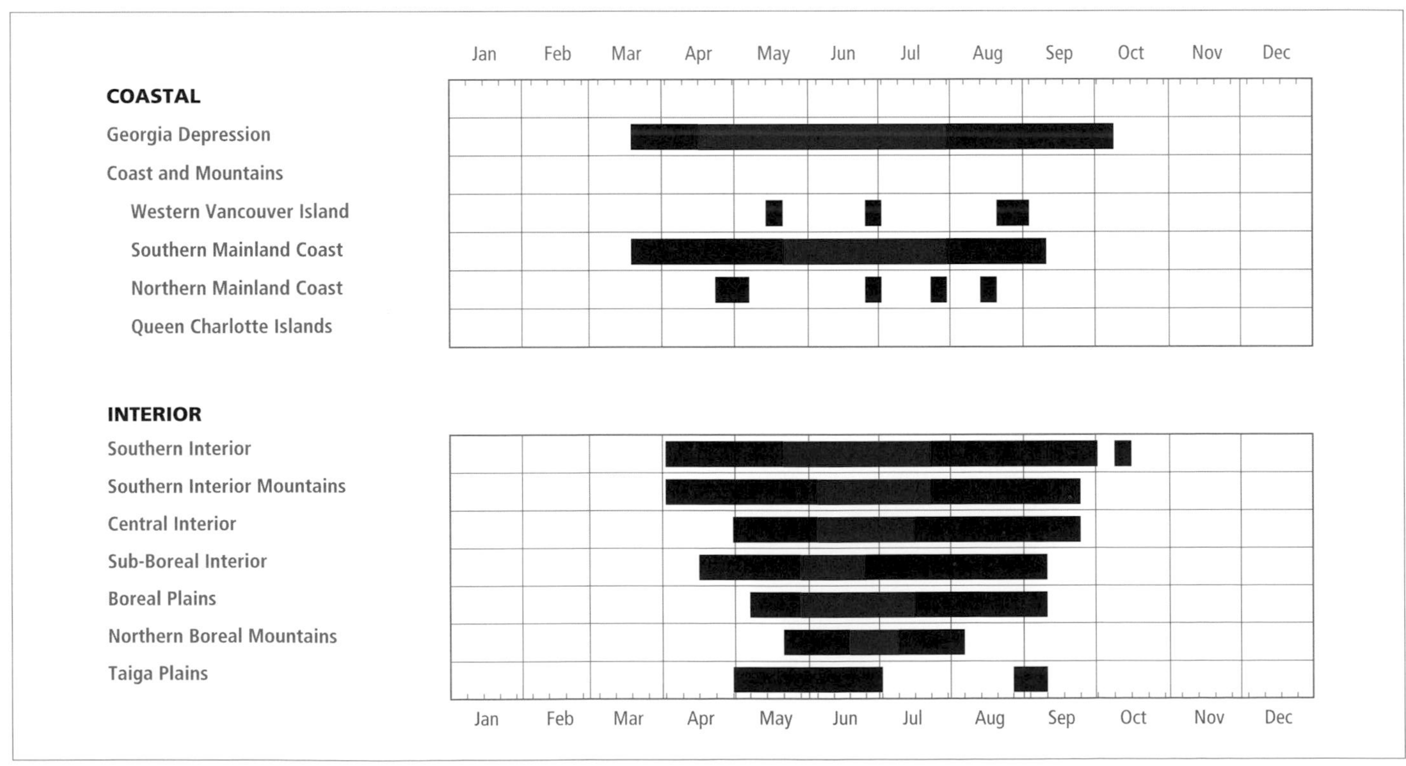

Figure 588. Annual occurrence (black) and breeding chronology (red) for the Solitary Vireo in ecoprovinces of British Columbia. Records are shown for the week in which they occurred.

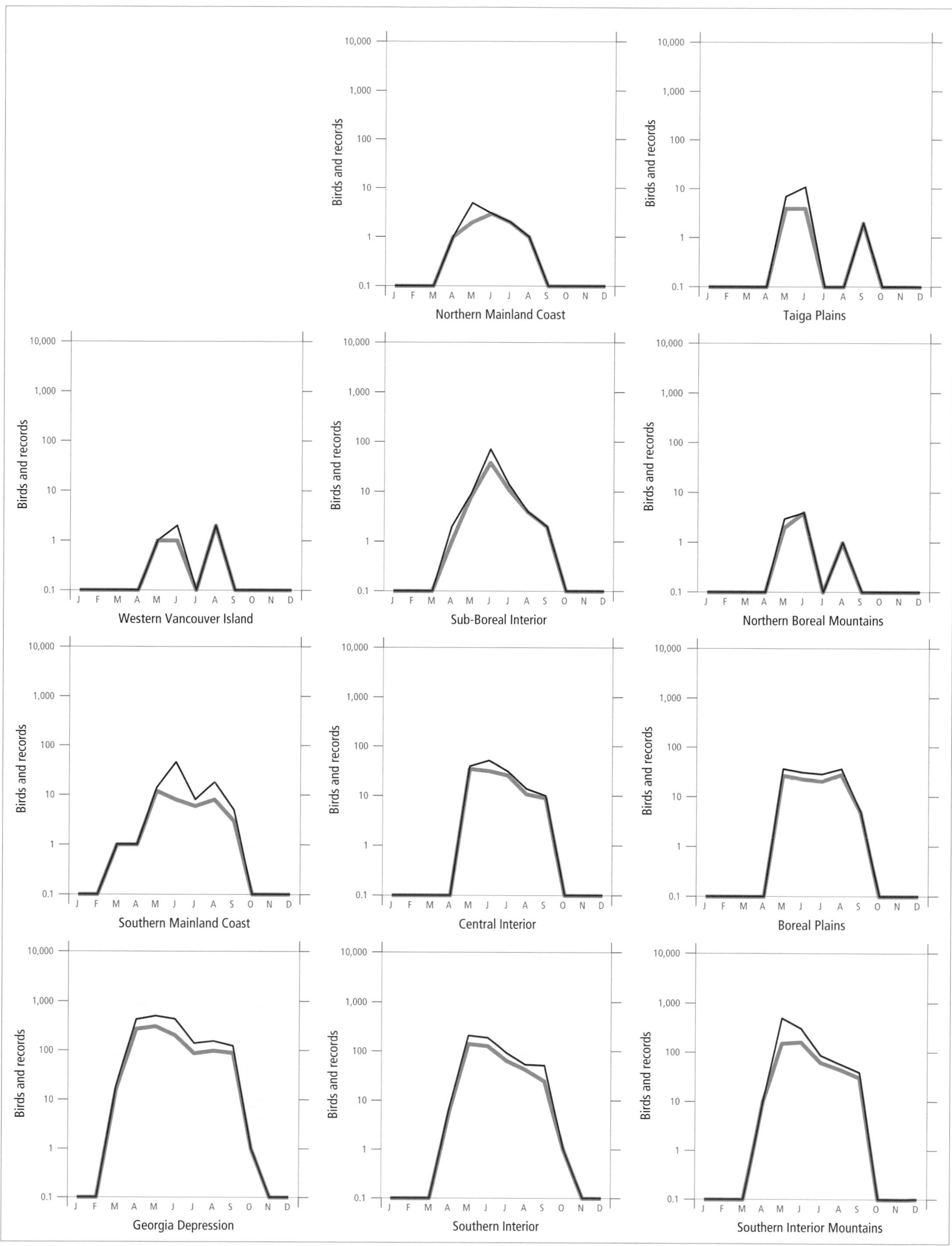

Figure 589. Fluctuations in total number of birds (purple line) and total number of records (green line) for the Solitary Vireo in ecoprovinces of British Columbia. Nest record data have been excluded.

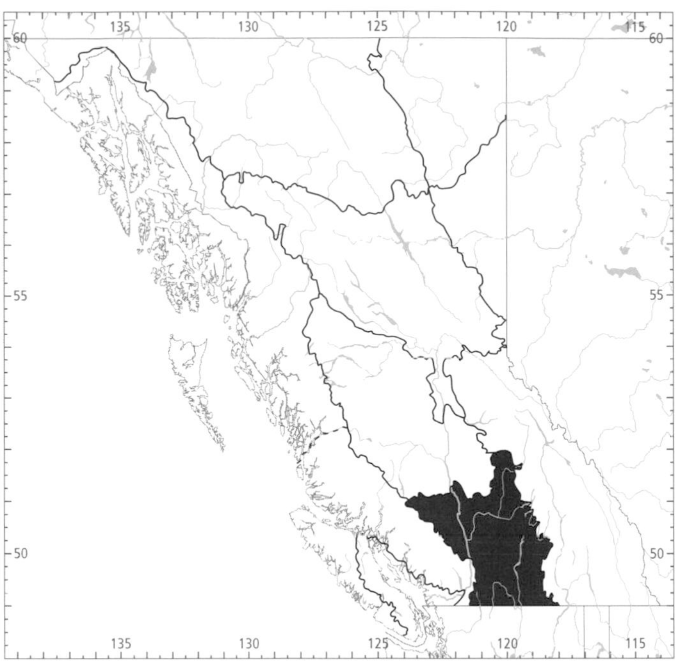

Figure 590. In British Columbia, the highest numbers for the Solitary Vireo in summer occur in the Southern Interior Ecoprovince.

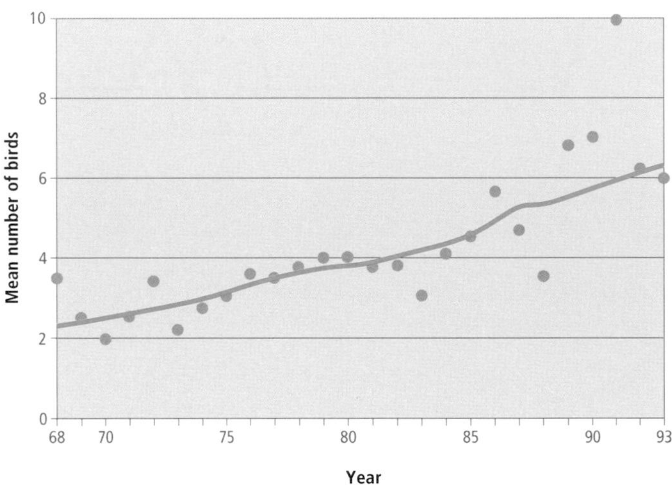

Figure 591. An analysis of Breeding Bird Surveys shows that the number of birds on interior routes increased at an average annual rate of 5% over the period 1968 through 1993 ($P \leq 0.001$).

On the coast, breeding can begin about a month earlier than in the interior. The Solitary Vireo has been recorded breeding on the coast from 20 April (calculated) to 29 July; in the interior, it has been recorded breeding from 23 May (calculated) to 22 July (Fig. 588).

Nests: Most nests were found in trees (94%; n = 47; Figs. 592 and 593); 3 nests were in shrubs. Unlike in the case of the Red-eyed and Warbling vireos, conifers were often used for nest sites (43%), including Douglas-fir (Fig. 592), western redcedar, grand fir, spruce, and ponderosa pine. Deciduous trees used for nest sites (40%) included red alder, maple, willow, arbutus, birch, Garry oak, and horse-chestnut. One nest was in a dead snag. Almost all nests were placed near the ends of branches, usually suspended from diverging twigs. The nest is cup-shaped (Fig. 592) and is constructed of grasses, lichens, spider webs, plant fibres, mosses, bark strips, animal hair, and other soft, pliable materials. The heights of 46 nests ranged from 0.8 to 8.0 m, with 63% between 1.8 and 4.5 m.

Eggs: Dates for 35 clutches ranged from 27 April to 14 July, with 53% recorded between 3 June and 24 June. Calculated dates indicate that eggs can occur as early as 20 April. Sizes of 32 clutches ranged from 1 to 5 eggs (1E-7, 2E-4, 3E-2, 4E-15, 5E-4), with 59% having 4 or 5 eggs (Fig. 593). In Ontario, the incubation period is 13 or 14 days (Peck and James 1987).

Young: Dates for 23 broods ranged from 5 May to 29 July, with 13 dates recorded between 18 June and 17 July. Sizes of 18 broods ranged from 1 to 5 young (1Y-1, 2Y-6, 3Y-5, 4Y-4, 5Y-2), with 11 having 2 or 3 young. The nestling period is poorly known, but is about 14 days (Ehrlich et al. 1988).

Brown-headed Cowbird Parasitism: In British Columbia, 9% of 45 nests recorded with eggs or young were parasitized by the cowbird. Parasitism on the coast was 8% (n = 24); in the interior, it was 10% (n = 21; Fig. 593). Elsewhere in North

Figure 592. In British Columbia, most Solitary Vireo nests were found in coniferous trees and were suspended from diverging twigs near the end of the branch. Note the hanging nest in the middle of the photograph (near Williams Lake, 7 June 1976; Ross D. James).

America, the rate of parasitism varies greatly (Friedmann et al. 1977). In Ontario it was 4.5% (Peck and James 1987), whereas in Colorado it was 49% (Marvil and Cruz 1989).

Nest Success: Of 6 nests found with eggs and followed to a known fate, only 1 produced at least 1 fledgling.

REMARKS: Two subspecies of the Solitary Vireo occur in British Columbia: *V. s. solitarius* occurs northeast of the Rocky Mountains, while *V. s. cassinii* occurs west of the Rocky Mountains (Munro and Cowan 1947). Recent studies indicate that the 2 subspecies behave as separate species (Johnson et al. 1988). If this change in systematic status is recognized by the American Ornithologists' Union, the name *Vireo solitarius* (Wilson) will apply to populations east of the Rocky Mountains, and *Vireo cassinii* (Xantus de Vesey) will apply to populations west of the Rocky Mountains (American Ornithologists' Union 1983; Sibley and Monroe 1990).

See James (1978) for additional information on the nesting behaviour of the Solitary Vireo, and Johnson et al. (1988) for a discussion of relationships within the family Vireonidae.

Figure 593. In British Columbia, 9% of nests, including this one at Burnaby Lake, were parasitized by the Brown-headed Cowbird (11 June 1969; R. Wayne Campbell). Cowbird parasitism of Solitary Vireo nests in North America ranges from 4.5% (Ontario) to 49% (Colorado).

NOTEWORTHY RECORDS

Spring: Coastal – Beacon Hill Park (Victoria) 6 Apr 1959-11, first of year; Mill Hill 23 Mar 1975-1; Shawnigan Lake 5 May 1962-2 nestlings; Duncan 30 Mar 1983-1, 27 Apr 1974-2 eggs; Parksville 30 Mar 1926-1 (MVZ 82681); Tofino 18 May 1974-1; Point Grey (Vancouver) 25 Mar 1983-3; Coquitlam 20 May 1963-4 eggs (RBCM E462); North Vancouver 15 May 1983-10; Thacker Mountain (Hope) 19 Mar 1960-1 (Horvath 1963), 28 May 1963-3 eggs; Cheakamus River 4 May 1969-building nest; Fulmore River 29 Apr 1976-1; Kitlope Lake 3 May 1991-4; Kitimat River estuary 25 Apr 1984-1. **Interior** – Creston 11 May 1980-5; Cawston 17 May 1992-completed nest; Kootenay Bay 8 Apr 1971-1; Naramata 5 Apr 1964-1 (Cannings et al. 1987); Okanagan Landing 28 Apr 1915-1 (Munro and Cowan 1947); Nakusp 19 Apr 1980-1; Hyas Lake road 31 May 1987-4 eggs; Yoho National Park 29 Apr 1977-13; Soda Lake 2 May 1958-1, 31 May 1929-3 eggs; Riske Creek 20 May 1970-4; Meldrum Creek 2 May 1979-1; Williams Lake 3 May 1989-1; Prince George 22 Apr 1980-2; Williston Lake 1 May 1976-2, 31 May 1976-4 eggs (ROM 12288); Taylor 10 May 1983-1, 18 May 1985-5; Fort Nelson 7 May 1974-2; ne Kwokullie Lake 30 May 1982-1; Liard Hot Springs 27 May 1981-2.

Summer: Coastal – Sooke 14 Jul 1956-5 eggs; Jordan Meadows 26 Aug 1947-1; Coquitlam 26 Jul 1967-3 nestlings; North Vancouver 30 Aug 1971-12; Agassiz 8 Jun 1969-4; Sechelt Peninsula 26 Jul 1959-4 nestlings; Mitlenatch Island 24 Aug 1973-1 (BC Photo 327); lower Quinsam Lake 3 Jun 1961-5 nestlings; Loughborough Inlet 29 Aug 1936-1 (Munro and Cowan 1947); Bella Coola 3 Jul 1940-2 (FMNH 175966); Stuie 12 Aug 1982-5; Minette Bay 1 Jul 1975-1; Lakelse Lake 17 Aug 1977-1. **Interior** – Trail 24 Jun 1994-1 egg and 4 nestlings; Kettle River (s Christina Lake) 26 Jun 1993-pair at nest; Goat River 23 Aug 1947-4 (Munro 1957); Okanagan Falls 6 Jun 1980-adult feeding young; Jaffray 22 Jun 1968-15; Summerland 17 Jul 1972-3 nestlings; Slocan River 3 Jul 1983-nestlings; 9 km e Enderby 11 Jun 1962-4 eggs; Tappen to Squilax 3 Jun 1972-8; Sinclair Canyon 22 Jul 1965-2 nestlings (Seel 1965); Shuswap Lake Park 7 Jul 1973-4 eggs, 13 Aug 1973-4; 22 km n Revelstoke 2 Jul 1991-7; Bridge Lake 10 Jun 1959-4 eggs; Horse Lake 21 Jul 1978-3; Canim Lake 1 Aug 1978-2; 20 km e Williams Lake 7 Jun 1976-eggs, 11 Jul 1981-1 young; Murtle Lake road 30 Jun 1968-5; Bouchie Lake 19 Aug 1944-1 (ROM 85599); Topley 18 Jun 1956-7; Willow River 22 Jun 1977-nestlings; 23 km s Ground-birch 17 Jun 1993-1 egg and 3 nestlings; Mackenzie 3 Jun 1976-4 eggs; St. John Creek bridge 6 Jul 1976-3; Beatton Park 14 Aug 1986-3; Cecil Lake 11 Jul 1987-3 nestlings; Kinaskan Lake 20 Jun 1975-1; Haworth Lake 5 Aug 1976-1; Dease Lake 22 Jun 1962-1 (NMC 50080); Fort Nelson 15 Jun 1975-3; Boya Lake 22 Jun 1983-1; Kechika River valley at Coal River 19 Jun 1981-incubating.

Breeding Bird Surveys: Coastal – Recorded from 21 of 27 routes and on 42% of all surveys. Maxima: Squamish 10 Jun 1973-31; Pemberton 14 Jun 1992-24; Gibsons Landing 4 Jul 1993-21. **Interior** – Recorded from 60 of 73 routes and on 71% of all surveys. Maxima: Beaverdell 25 Jun 1977-25; Bridge Lake 21 Jun 1992-20; Chu Chua 10 Jun 1994-19.

Autumn: Interior – Fort Nelson 2 Sep 1985-1, 4 Sep 1986-1; Charlie Lake Park 1 Sep 1982-1; Stoddart Creek 6 Sep 1986-1; St. John Creek 7 Sep 1985-1; Quesnel 3 Sep 1900-1 (MCZ 111149); Chezacut 5 Sep 1933-1; Williams Lake 9 Sep 1983-1; Riske Creek 3 Sep 1978-2; Tiltzarone Lake 4 Sep 1944-1; Horse Lake 18 Sep 1933-1; Mount Revelstoke National Park 19 Sep 1978-1; Seton Lake 11 Oct 1982-1; Mabel Lake 20 Sep 1921-1 (ROM 85600); Okanagan Landing 27 Sep 1938-1 (Cannings et al. 1987); Okanagan Falls 5 Sep 1969-10; Vaseux Lake 26 Sep 1977-1 (Cannings et al. 1987); Trail 22 Sep 1968-1; Creston 17 Sep 1928-1 (CAS 32043). **Coastal** – Port Neville 5 Sep 1975-3; Egmont 5 Sep 1975-1; Comox 5 Oct 1988-1; Qualicum Beach 28 Sep 1983-1; Vancouver 14 Sep 1922-1 (RBCM 7271), 25 Sep 1988-1; Saanich 14 Sep 1983-11; Goldstream Park 27 Sep 1975-1; Oak Bay 24 Sep 1978-1.

Winter: No records.

Hutton's Vireo

Vireo huttoni Cassin

HUVI

RANGE: Resident on Vancouver Island and the southwestern mainland of British Columbia, south along the coastal slope of the United States to northwestern Baja California, Arizona, western New Mexico, and Texas, and along the Mexican highlands to Guatemala.

STATUS: On the coast, an *uncommon* to locally *fairly common* resident on southeastern Vancouver Island and *rare* to *uncommon* in the Fraser Lowland of the Georgia Depression Ecoprovince; *rare* migrant and summer visitant on Western Vancouver Island, becoming *casual* in winter there; *very rare* throughout the year in the southern mainland portions of the Coast and Mountains Ecoprovince. Breeds.

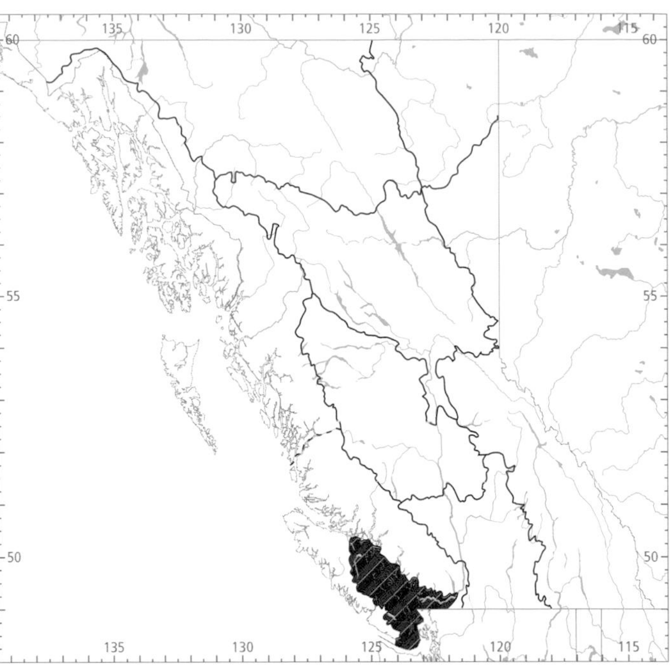

Figure 594. In British Columbia, the highest numbers for the Hutton's Vireo in both winter (black) and summer (red) occur in the Georgia Depression Ecoprovince.

NONBREEDING: In British Columbia, the Hutton's Vireo is restricted to south coastal regions. It occurs from Long Beach on the west coast of Vancouver Island east to Hope, and from Victoria north to Cape Scott on Vancouver Island and the Kingcome River estuary on the mainland. It is most abundant on southeastern Vancouver Island between Comox and Victoria, in forested western portions of the Fraser Lowland, and on the lower mountain slopes above the Fraser River valley. Elsewhere it is more sparsely distributed, especially on northern Vancouver Island. The highest numbers in winter occur in the Georgia Depression (Fig. 594).

The Hutton's Vireo has been recorded at elevations from sea level to 740 m. Most birds occur at lower elevations in coastal Douglas-fir forests with Garry oak, arbutus, and lodgepole pine in drier sites; bigleaf maple and red alder in moister sites (Figs. 595 and 596); or coastal western hemlock with western redcedar and shore pine. For more information on habitat, see BREEDING.

The observability of this vireo changes markedly with the seasons, and these changes mask seasonal migration or altitudinal movements. Although data are inconclusive, part of the population is probably migratory (Figs. 597 and 598). Throughout the range of the Hutton's Vireo in the province, there is a marked increase in numbers recorded when males begin to sing in February and March. With the decline of singing in May, the number of records also declines. On Western Vancouver Island and on the mainland portions of the southern Coast and Mountains Ecoprovince, records show a steep decline in the summer, an increase in September, and a decline in October and November; there are no records in December and January (Fig. 598). The increase in records in September suggests an autumn migration. In addition, some birds seem to wander beyond the normal range after the breeding season (half of the northernmost records are from autumn).

Figure 595. During the nonbreeding seasons, the Hutton's Vireo frequents a variety of habitats on southeastern Vancouver Island, including Douglas-fir and lodgepole pine woodlands with scattered Garry oaks and established shrubby areas (University of Victoria, 30 September 1994; R. Wayne Campbell).

During autumn and winter, the Hutton's Vireo is often found with flocks of Chestnut-backed Chickadees, Ruby-crowned Kinglets, Golden-crowned Kinglets, and Red-breasted Nuthatches. It is the most silent of the assemblage, and often goes unnoticed.

On the coast, the Hutton's Vireo has been recorded throughout the year (Fig. 597).

BREEDING: The Hutton's Vireo probably breeds throughout most of its range in the province, although breeding records are few. About half of our breeding records are from the Langley to Surrey area in the Fraser Lowland, while the other half are from Victoria north to Fanny Bay and Miracle Beach Park on southern Vancouver Island. There are also single breeding records from Vedder Mountain, Hope, and Bamfield.

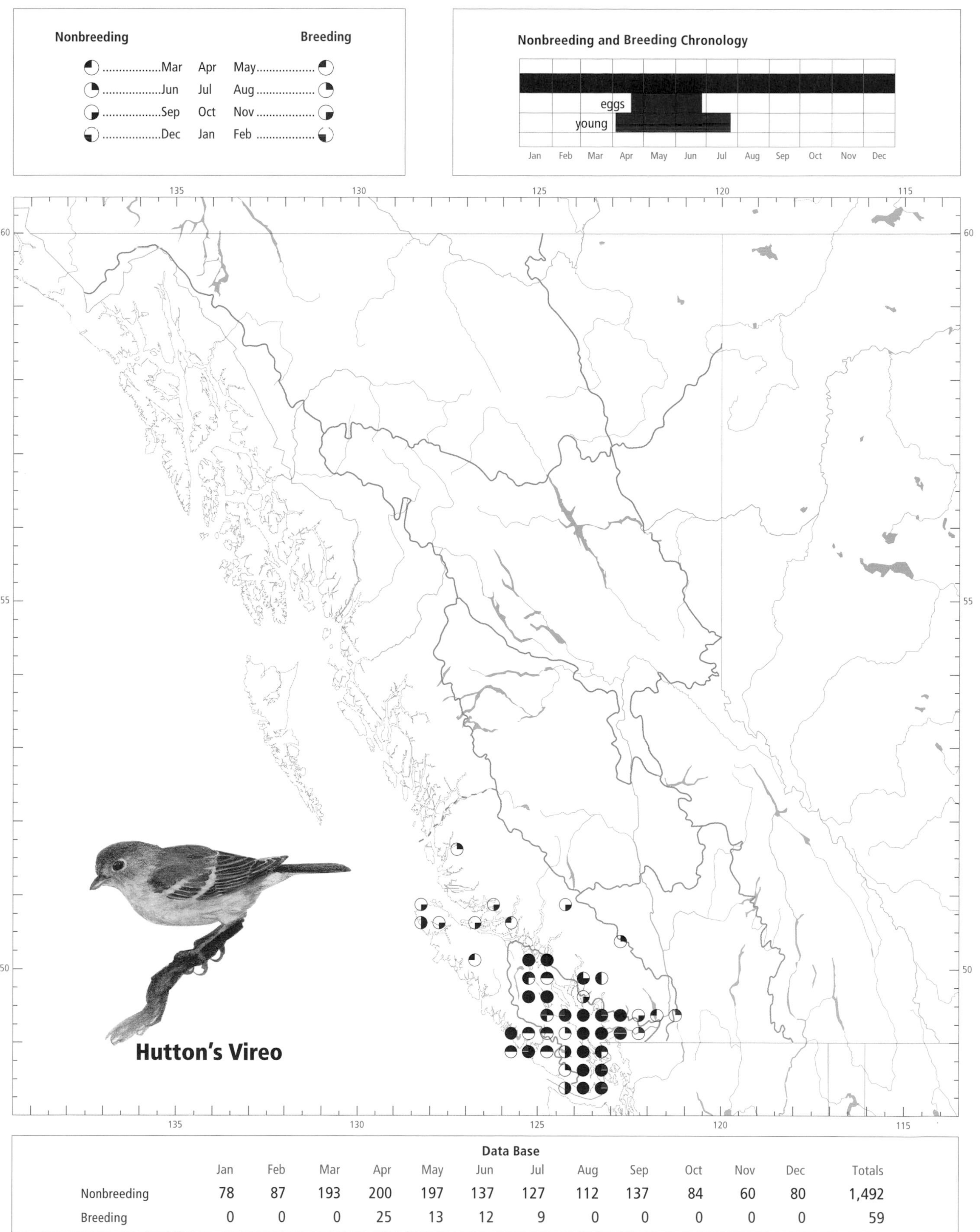

Data Base

	Jan	Feb	Mar	Apr	May	Jun	Jul	Aug	Sep	Oct	Nov	Dec	Totals
Nonbreeding	78	87	193	200	197	137	127	112	137	84	60	80	1,492
Breeding	0	0	0	25	13	12	9	0	0	0	0	0	59

The highest numbers in summer occur in the Georgia Depression (Fig. 594). In the Fraser Lowland, this vireo is fairly common in forested areas such as the Green Timbers Reserve in Surrey, but it is absent as a breeding bird in most urban, suburban, and rural areas. Breeding Bird Surveys for coastal routes for the period 1968 through 1993 contain insufficient data for analysis.

The Hutton's Vireo breeds at elevations from near sea level to 240 m. It seems to prefer mixed conifer-dominated woodland with a moderate to heavy shrub understorey (Fig. 596), and is known to occur and probably breeds in coniferous forests ranging from 30-year-old second growth to old-growth (Bryant et al. 1993). It uses shrub understoreys more than other vireos and tends to occur in open forest or along the forest edge, where the shrub layer is more lush. All nests found were in forested habitats, often along edges associated with roadsides or farm clearings.

On the coast, the Hutton's Vireo has been recorded breeding from 19 March to 24 July (Fig. 597).

Nests: Twelve of 16 nests were in trees, 8 of them in Douglas-firs and 1 each in western redcedar, arbutus, vine maple, and red alder. Four were in shrubs, including hazelnut and ocean-spray. Nests were pendent cups most often suspended from a branch fork well out from the trunk of the tree. Moss was the most frequently used nest material (11 of 13 nests), but grass, feathers, lichens, moth cocoons, and spider webs were also used. The earliest evidence of nest building was a pair carrying nesting material on 2 March. The heights for 17 nests ranged from 1 to 15 m, with 9 nests between 3 and 12 m.

Eggs: Dates for 8 clutches ranged from 25 April to 26 June, with 6 recorded between 25 April and 19 June. Calculated dates indicate that eggs can occur as early as 19 March. Sizes of 7 clutches were 1 or 4 eggs (1E-3, 4E-4). The incubation period for 1 nest in British Columbia was 15 days; elsewhere it is reported as 14 to 16 days (Miller 1953; Harrison 1979).

Young: Dates for 18 broods ranged from 3 April to 24 July, with 66% recorded between 25 April and 20 June. Sizes of 15 broods ranged from 1 to 5 young (1Y-1, 2Y-2, 3Y-6, 4Y-4, 5Y-2), with 10 having 3 or 4 young. Our data suggest that the Hutton's Vireo may be double-brooded. The nestling period for 1 nest in British Columbia was 14 days. In California, Davis (1995) mentions that young leave the nest on the 15th day.

Figure 596. Mixed woodlands of Douglas-fir and scattered red alder with a moderate to heavy shrub understorey of willow and red-osier dogwood are the preferred breeding habitat for the Hutton's Vireo in the Georgia Depression Ecoprovince (Saanich Peninsula, Vancouver Island, 22 May 1985; R. Wayne Campbell).

Brown-headed Cowbird Parasitism: Cowbird parasitism was not found in British Columbia in 18 nests found with eggs or young. There are, however, 2 reports from the province of adult Hutton's Vireos feeding fledged cowbirds. Through 1984, only 13 instances of parasitism were reported for North America (Crowell and Nehls 1973d; Friedmann et al. 1977; Friedmann and Kiff 1985).

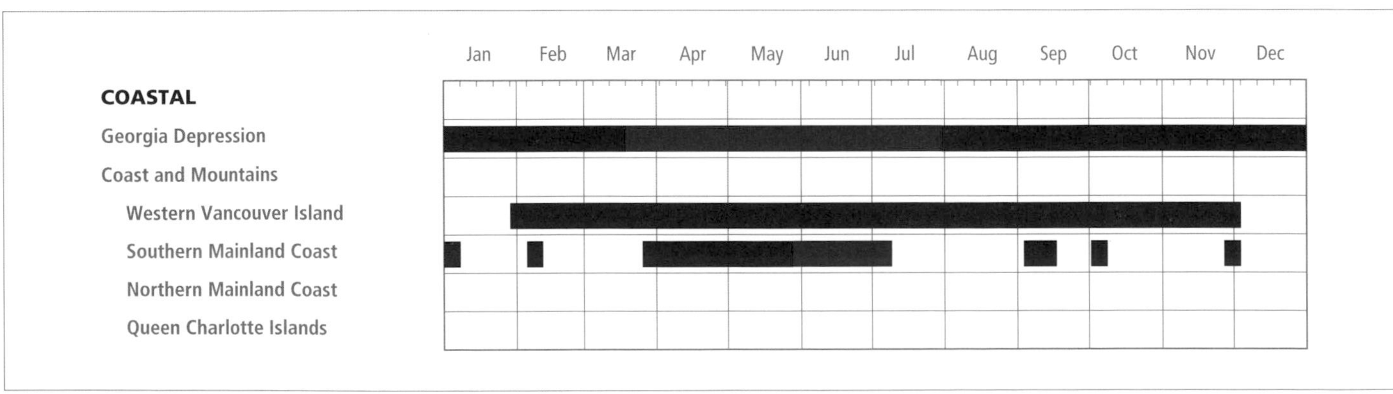

Figure 597. Annual occurrence (black) and breeding chronology (red) for the Hutton's Vireo in coastal ecoprovinces of British Columbia. Records are shown for the week in which they occurred.

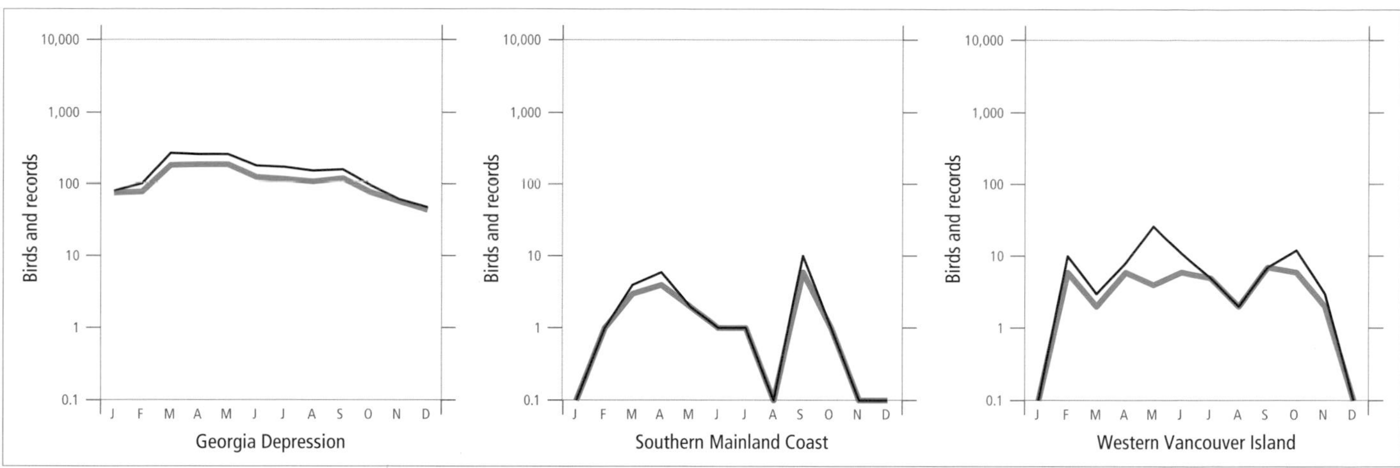

Figure 598. Fluctuations in total number of birds (purple line) and total number of records (green line) for the Hutton's Vireo in coastal ecoprovinces of British Columbia. Christmas Bird Count and nest record data have been excluded.

Nest Success: Insufficient data.

REMARKS: Two subspecies of Hutton's Vireo occur in British Columbia: *V. h. huttoni* is confined to the mainland, whereas *V. h. insularis* occurs on Vancouver Island (Grinnell 1935; American Ornithologists' Union 1957).

There are a number of observations outside the normal range of the Hutton's Vireo, including published records (e.g., near Oliver 23 May 1965-2 [Rogers 1965c]; Osoyoos 28 May 1967-1 [Rogers 1967c]). However, because it is difficult to distinguish between this vireo and the Ruby-crowned Kinglet under some field conditions, we have not accepted observations outside the established range that are not substantiated by photographic or specimen evidence.

The status of the Hutton's Vireo in British Columbia is poorly known outside of the Georgia Depression, especially on western and northern Vancouver Island.

See Davis (1995) for additional information on the life history of the Hutton's Vireo in North America.

NOTEWORTHY RECORDS

Spring: Coastal – Victoria Highlands 31 Mar 1983-11; Somenos 6 May 1984-5; Bamfield 26 May 1991-22 on survey, 1 nest found; Mayne Island 7 Apr 1988-1; Valdez Island 30 Mar 1986-2; Surrey 19 Mar 1968-incubating, 3 Apr 1967 feeding young in nest, 23 Apr 1967-4 nestlings, 25 Apr 1960-5 nestlings, 7 May 1963-4 eggs; Chilliwack May 1905-1 (Brooks 1917); Meares Island 17 Apr 1986-2; Whalley 21 Mar 1970-incubating; Port Alberni 30 Apr 1983-1; Wreck Beach (Pacific Rim National Park) 17 Mar 1981-2; Parksville 28 May 1960-3 nestlings partially feathered; Harrison Hot Springs 13 May 1986-1; Fanny Bay 8 May 1989-4 young had fallen from nest; Alice Lake (Squamish) 19 Apr 1970-3; Miracle Beach Park 11 Apr 1962-2, 17 Mar to 13 Jun 1962-1 pair present and feeding young on 11 Jun (Westerborg 1963). **Interior** – No records.

Summer: Coastal – Sooke 17 Jul 1973-2 (ROM 119570-71); Jordan River 31 Aug 1974-1; Victoria 13 Jul 1973-20, 21 Jul 1984-2 nestlings, 24 Jul 1984-1 nestling; Port Renfrew 11 Jun 1989-5 on Breeding Bird Survey; Willis Island 16 Jul 1971-1 (Hatler et al. 1978); Ucluelet 25 Aug 1987-1 dead on road; Pacific Rim National Park 19 Jul 1972-2 singing (Hatler et al. 1978); Vedder Mountain 26 Jun 1942-eggs (Munro and Cowan 1947); New Westminster 20 Jun 1937-4 eggs (UBC 1719); Port Alberni 10 Jun 1931-2 (UBC 4666-7); Maple Ridge 9 to 11 Jul 1973-7 (ROM 119514, -24, -26, -31, -34-36); North Vancouver 27 Jun 1981-nestlings; Hope 4 Jun 1960-4 eggs, 2 Jul 1960-4 young left nest (Horvath 1963); Roberts Creek 2 Jun 1928-2 (UBC 4659-60); Hornby Island 12 Aug 1977-4; Mitlenatch Island 24 Aug 1965-1; Cortes Island 19 Jul 1979-1; Morte Lake 28 Aug 1977-1; Guildford Island (Shoal Bay) 9 Jul 1993-1; Pemberton 21 Jun 1981-1; San Josef 4 Jul 1933-1; Pine Island 14 Jul 1975-1. **Interior** – No records.

Breeding Bird Surveys: Coastal – Recorded from 13 of 27 routes and on 17% of all surveys. Maxima: Gibsons Landing 7 Jul 1991-7; Port Renfrew 6 Jun 1987-5; Campbell River 4 Jul 1992-4. **Interior** – Not recorded.

Autumn: Interior – No records. **Coastal** – Kingcome River 14 Sep 1936-1 on estuary (NMC 27341); Cape Scott 19 Sep 1935-2 (NMC 26269); Quatsino 6 Oct 1935-5; Gorge Harbour 10 Oct 1977-1; Egmont 7 Oct 1973-1; Mike Lake (Golden Ears Park) 27 Nov 1975-1; Burnaby 7 Oct 1963-2; Marpole 6 Nov 1940-1 (UBC 10427); Bamfield 10 Nov 1979-2; Cowichan Bay 16 Nov 1974-2; Victoria 11 Sep 1977-4; Jordan River 27 Nov 1983-1.

Winter: Interior – No records. **Coastal** – Cortez Island 13 Jan 1979-1; Comox 17 Jan 1933-3 (UMMZ 163382-3; RBCM 13216); Squamish 2 Jan 1988-1; Egmont 8 Feb 1976-1; Redroofs 30 Jan 1987-1 (Campbell 1987b); Sechelt 26 Jan 1986-1 (Campbell 1986b); Lighthouse Park (West Vancouver) 2 Dec 1972-1; Pacific Rim National Park 1 Feb 1983-1; Bamfield 25 Feb 1979-3; North Saanich 26 Feb 1983-8.

Christmas Bird Counts: Interior – Not recorded. **Coastal** – Recorded from 17 of 33 localities and on 29% of all counts. Maxima: Ladner 27 Dec 1988-**16**, all-time Canadian high count (Monroe 1989a); Vancouver 17 Dec 1989-9; White Rock 30 Dec 1989-9; Victoria 20 December 1980-7.

Warbling Vireo

Vireo gilvus (Vieillot)

WAVI

RANGE: Breeds from southeastern Alaska, southeastern Yukon, northern British Columbia, west-central and southwestern Mackenzie, and northern and western Alberta southeast across Saskatchewan, southern Manitoba, western and southern Ontario, southern Quebec, and New Brunswick; south in the west to southern Baja California and in the east to Louisiana and Tennessee. Winters in the mountains from southern Sonora south to Guatemala and El Salvador.

STATUS: On the coast, a *fairly common* migrant and summer visitant in the Georgia Depression Ecoprovince; *uncommon* to locally *fairly common* in the southern and northern mainland of the Coast and Mountains Ecoprovince; *rare* on Western Vancouver Island; absent from the Queen Charlotte Islands.

In the interior, an *uncommon* to *fairly common* (occasionally *common*) migrant and summer visitant in the Southern Interior, Southern Interior Mountains, Central Interior, Sub-Boreal Interior, and Boreal Plains ecoprovinces; *uncommon* in the Northern Boreal Mountains and Taiga Plains ecoprovinces.

Breeds.

Figure 599. The Warbling Vireo occurs throughout most of British Columbia except on the Queen Charlotte Islands (Penticton, 5 July 1981; Stephen R. Cannings).

NONBREEDING: The Warbling Vireo (Fig. 599) has a widespread distribution across most of British Columbia, but is absent from the Queen Charlotte Islands. It is relatively abundant from eastern Vancouver Island east across the southern portions of the province to the Rocky Mountain Trench and north through the interior to the Peace Lowland; further north, it becomes more thinly distributed. It is also sparsely distributed along the outer coast, including Western Vancouver Island, and on islands and the mainland of the central and northern coast. On the mainland north of the Georgia Depression, it has been recorded mainly from the heads of fiords where rivers cut through the mountains from the interior.

The Warbling Vireo has been reported at elevations from sea level to about 300 m on the coast, and from 300 to 1,700 m in the interior. On the coast, it avoids continuous coniferous forests and occurs only in deciduous or mixed forests, which are often restricted to river valleys, wetlands (Fig. 600), or estuaries. In mountainous interior areas, it occurs mainly in valleys, along lower-elevation passes, and on the edges of slides and avalanche chutes where patches of deciduous forest occur. It seldom occurs in conifer-dominated forests found

Figure 600. On the southwest mainland coast of British Columbia, stands of mixed forests, often adjacent to wetlands, attract migrating Warbling Vireos each year (Campbell Valley Park, near Langley, 27 April 1996; R. Wayne Campbell).

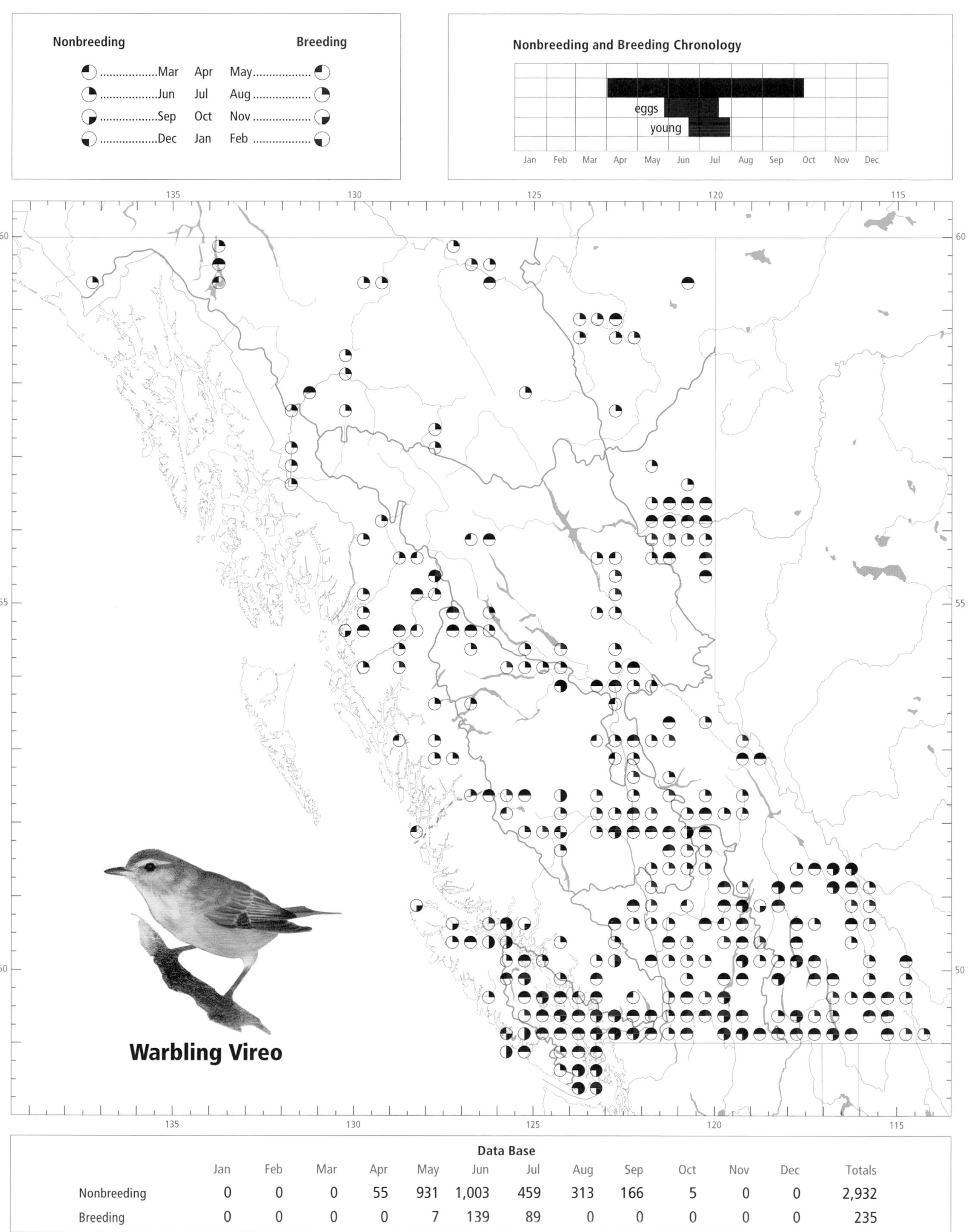

Data Base

	Jan	Feb	Mar	Apr	May	Jun	Jul	Aug	Sep	Oct	Nov	Dec	Totals
Nonbreeding	0	0	0	55	931	1,003	459	313	166	5	0	0	2,932
Breeding	0	0	0	0	7	139	89	0	0	0	0	0	235

on most mountain slopes and plateaus. Spring migration begins in the Georgia Depression in early April (Figs. 601 and 602); numbers increase during late April but the movement does not peak until mid-May. In the southern and central portions of the interior, the first spring migrants arrive in late April and early May, with the movement peaking in mid-May. On the north coast and in the northern interior, spring migration probably begins in early to mid-May and peaks in late May.

In northern British Columbia, the autumn movement begins shortly after the young are independent, probably by late July, and all birds are gone by the end of August. Near Vanderhoof, Munro (1949) noted that the main movement occurred between mid-August and early September. In southern portions of the interior, autumn migration occurs from mid-August to mid-September (Fig. 602). Along the south coast, the autumn migration also begins about mid-August and peaks during the first 2 weeks of September, but most birds have left by the third week of September. There have been only 5 records of the Warbling Vireo in the province in October, all from the southern coast (Vancouver, Saanich, Victoria).

On the coast, the Warbling Vireo has been recorded from 1 April to 10 October; in the interior, it has been recorded from 21 April to 19 September (Fig. 601).

BREEDING: The Warbling Vireo has a widespread breeding distribution throughout most of British Columbia from the Peace Lowland southward. It has not been reported from the Queen Charlotte Islands. Breeding has been confirmed in almost all areas of its summer range except the west coast of Vancouver Island and most of the central and northern mainland coast. In the interior, there are no confirmed breeding records north of Fort St. John. However, an individual was observed carrying nesting material at Liard River, and the Warbling Vireo probably breeds throughout the northern interior.

This vireo reaches its highest numbers in summer in the Sub-Boreal Interior (Fig. 603). Relatively high numbers also occur on southeastern Vancouver Island, the Fraser Lowland, and the valleys of the Kootenay and Columbia rivers. Populations become less abundant north of the Sub-Boreal Interior. An analysis of Breeding Bird Surveys in British Columbia shows that the mean number of birds on interior routes increased at an average annual rate of 3% over the period 1968 through 1993 (Fig. 604); analysis of coastal routes for the same period could not detect a net change in numbers.

On the coast, the Warbling Vireo has been recorded breeding at elevations from near sea level to 300 m; in the interior, from 330 to 1,450 m. It is a conspicuous bird where it occurs during spring and summer, because males are easily detected by their melodious song while they forage in thick foliage in the crowns of trees. It breeds mainly in open deciduous woodlands, but also in mixed woodland with some conifers (Fig. 605). In general, it prefers taller and larger trees in riparian habitats at lower elevations, especially around ponds, sloughs, wet meadows, and lagoons, and at the edges of forest clearings provided by farms, powerlines, and roads. In suburban areas it breeds in parks, along treed boulevards, in well-treed gardens, and in orchards, where a variety of non-native tree species are used.

In the Bulkley valley, Pojar (1995) found the Warbling Vireo in stands of variable age, including sapling, mature, and old-growth trembling aspen and mixed conifer-aspen stands, but mature and old-growth aspen stands had the highest density of singing males (Table 9).

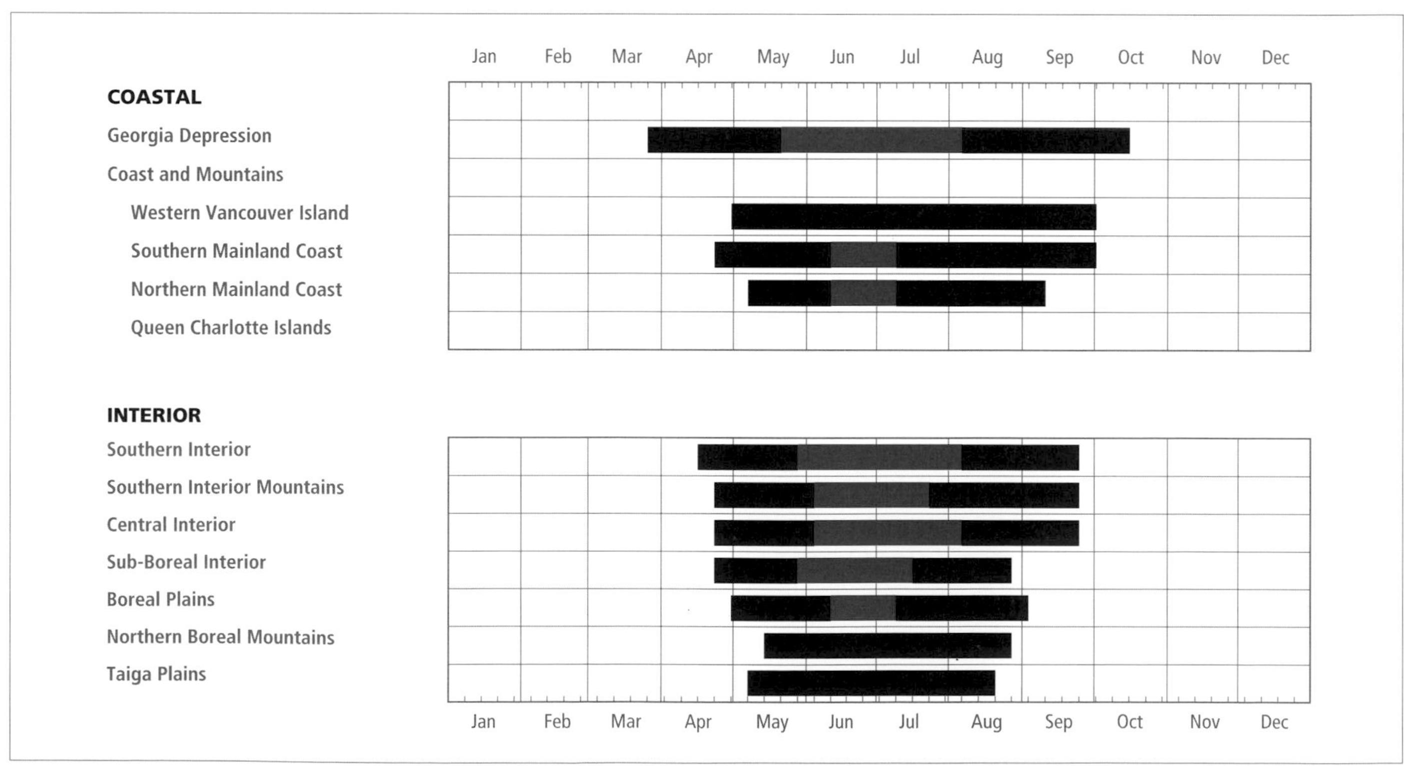

Figure 601. Annual occurrence (black) and breeding chronology (red) for the Warbling Vireo in ecoprovinces of British Columbia. Records are shown for the week in which they occurred.

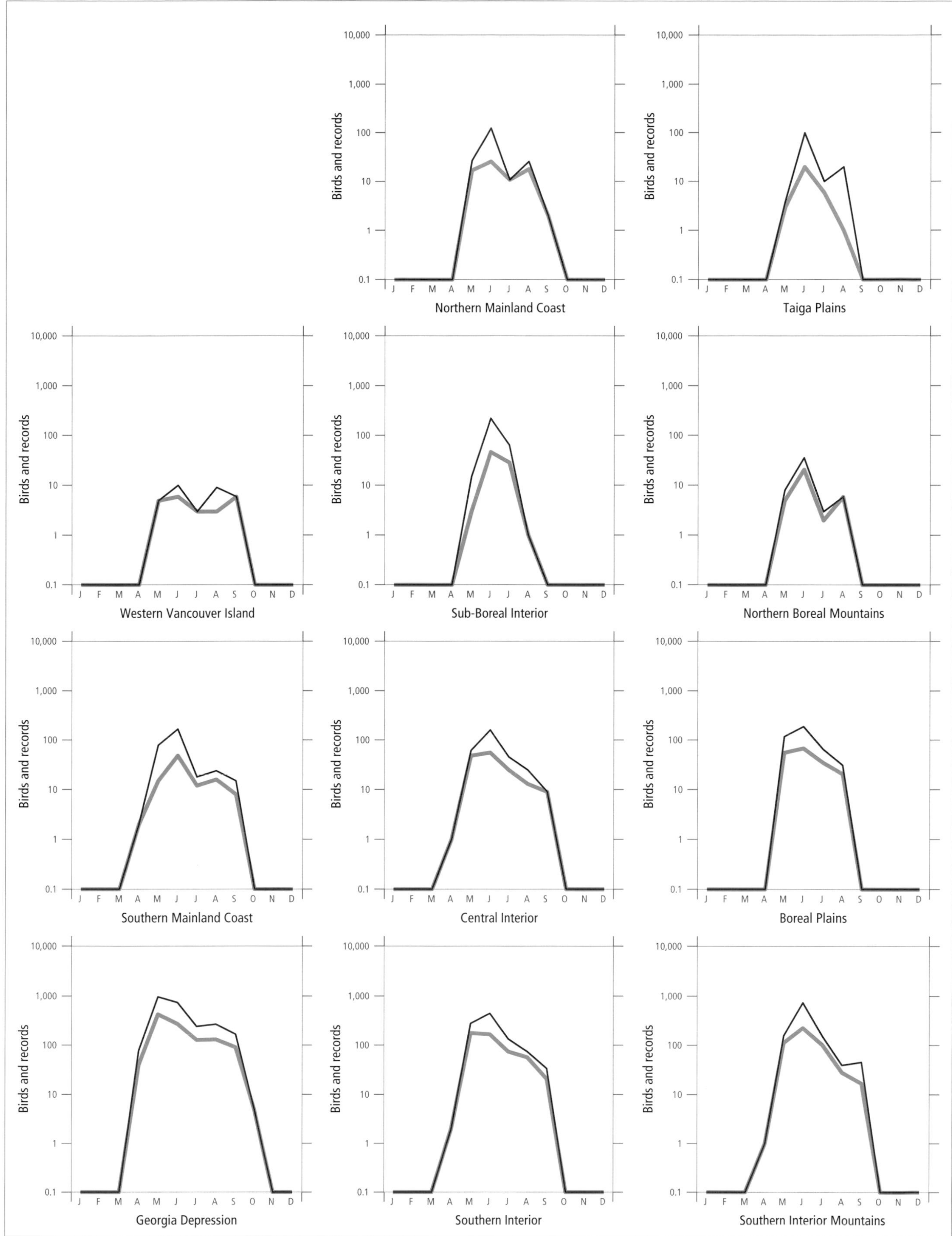

Figure 602. Fluctuations in total number of birds (purple line) and total number of records (green line) for the Warbling Vireo in ecoprovinces of British Columbia. Nest record data have been excluded.

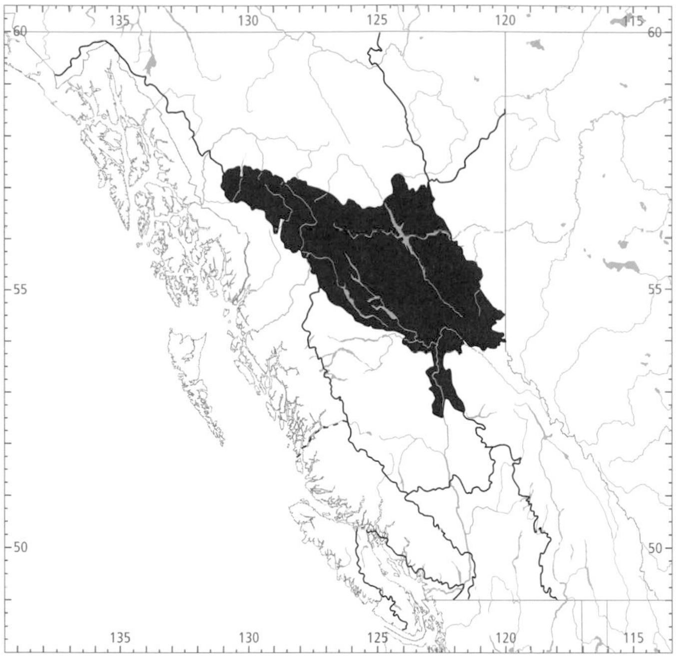

Figure 603. In British Columbia, the highest numbers for the Warbling Vireo in summer occur in the Sub-Boreal Interior Ecoprovince.

Most nesting habitats have been described as forests (60%; n = 75), including mixed (33%; Fig. 605), deciduous (15%), and coniferous (11%) forests. Another 33% of nests were in rural, suburban, or agricultural settings. Most nest sites were near a forest-clearing edge or in relatively open forest parkland.

On the coast, the Warbling Vireo has been recorded breeding from 25 May (calculated) to 28 July; in the interior, it has been recorded from 31 May (calculated) to 3 August (Cannings et al. 1987) (Fig. 601).

Nests: Most nests were found in deciduous trees (76%; n = 118), including trembling aspen (27%), poplar (10%), birch (10%), red alder (8%) black cottonwood (7%), and vine maple; or deciduous shrubs (14%) such as willow, red-osier dogwood, and elderberry. Nests are cup-shaped pendants, generally placed in a fork between 2 twigs or attached to adjacent twigs (Fig. 606). Nest materials included grass (65%; n = 118), plant fibres (28%), plant down (28%), lichens (22%), bark strips (21%), hair (20%), moss (17%), spider webs (13%), and several other materials.

Figure 605. In portions of the east Kootenay, the Warbling Vireo breeds in mixed riparian forests of spruce and alder, with black twinberry as the dominant shrub (Morrissey River, 5 July 1991; R. Wayne Campbell).

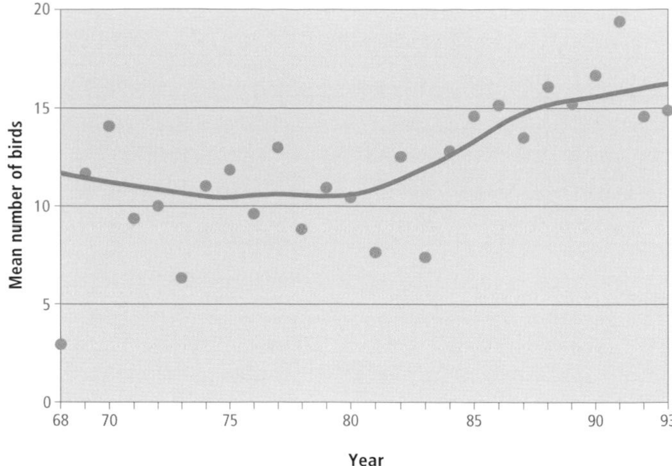

Figure 604. An analysis of Breeding Bird Surveys shows that the number of birds on interior routes increased at an average annual rate of 3% over the period 1968 through 1993 ($P \leq 0.01$).

The heights for 114 nests ranged from 1 to 16 m, with 61% between 2 and 6 m. Higher nests are probably largely undetected by observers; for example, in Manitoba, the Warbling Vireo nested mainly in maples at a mean nest height of 12.6 m (MacKenzie et al. 1982).

Eggs: Dates for 84 clutches ranged from 27 May to 19 July, with 52% recorded between 12 June and 27 June. Calculated dates indicate that eggs can occur as early as 25 May. Sizes of 75 clutches ranged from 1 to 5 eggs (1E-10, 2E-12, 3E-19, 4E-30, 5E-4), with 65% having 3 or 4 eggs. The incubation period has been reported as 12 days (Ehrlich et al. 1988) or 13 to 14 days (Harrison 1979).

Young: Dates for 38 broods ranged from 20 June to 30 July, with 57% recorded between 30 June and 14 July. Calculated dates indicate that young can be found as early as 10 June. Sizes of 31 broods ranged from 1 to 5 young (1Y-6, 2Y-6, 3Y-10, 4Y-8, 5Y-1), with 58% having 3 or 4 young. The nestling period of 2 nests in British Columbia was about 13 days. In southern Ontario, the average nestling period was 12 days, with a maximum of 13 days (Howes-Jones 1985).

Table 9. Density of singing male Warbling Vireos in various forest types and age classes near Smithers, 1991 and 1992 (Pojar 1995). Density is the range of values for all plots in each habitat type for both years; n = number of plots.

Habitat	Density (males/10 ha)	n
Clearcuts	0–12.7	6
Sapling aspen	0–26.6	6
Mature aspen	11.1–27.6	10
Old aspen	14.3–23.9	4
Mixed conifer-aspen	7.3–20.4	6

Brown-headed Cowbird Parasitism: In British Columbia, 49% of 98 nests found with eggs or young were parasitized by the cowbird. Parasitism on the coast was 40% (n = 20); parasitism in the interior was 51% (n = 78). In 37 parasitized nests with eggs, the number of cowbird eggs ranged from 1 to 6 eggs (1E-18, 2E-15, 3E-2, 4E-1, 6E-1). Adult vireos feeding fledged cowbird young were also reported from the interior (n = 10) and coast (n = 1).

In Ontario, the parasitism rate was 11% (Peck and James 1987). The Warbling Vireo is known as a frequent victim of the Brown-headed Cowbird (Friedmann et al. 1977) and appears to be one of the most frequently parasitized species in British Columbia.

Nest Success: Of 23 nests found with eggs and followed to a known fate, 10 produced at least 1 fledgling.

Figure 606. The Warbling Vireo nest is a pendent cup-shaped structure, usually built in the fork of a slender branch (Nine Mile Narrows, Nelson, 27 June 1994; Linda M. Van Damme).

REMARKS: One subspecies, *V. g. swainsoni*, occurs in British Columbia (American Ornithologists' Union 1957); however, see Browning (1974) for a discussion of the possible presence of a second subspecies.

NOTEWORTHY RECORDS

Spring: Coastal – Victoria 27 May 1962-3 eggs; Goldstream River 11 Apr 1962-10; Sidney Island 6 Apr 1988-1; Huntingdon 5 May 1945-7; Grice Bay 1 May 1993-1; Tofino 4 May 1986-1; Surrey 1 Apr 1960-1; Coquitlam 28 May 1939-3 eggs; Stanley Park (Vancouver) 16 May 1975-20; North Vancouver 28 May 1939-3 eggs (RBCM E465); Harrison Hot Springs 26 Apr 1992-1; Miracle Beach Park 4 Apr 1962-1; Fulmore River 29 Apr 1976-1; Swanson Bay 11 May 1936-1 (MCZ 284091); Terrace 30 May 1977-6. **Interior** – Newgate 16 May 1930-1 (Munro and Cowan 1947); Grand Forks 30 Apr 1983-1; Vaseux Lake 26 Apr 1922-1; Jaffray 25 May 1975-6; Okanagan Landing 21 Apr 1940-1 (Cannings et al. 1987), 19 May 1911-9; Nakusp 3 May 1981-1; Leanchoil 29 Apr 1977-1; Riske Creek 29 Apr 1989-1; Tatla Lake 5 May 1990-1; Dragon Lake 31 May 1994-pair building nest; Prince George 28 Apr 1978-1, 29 May 1969-12; Tupper Creek 21 May 1938-1 (RBCM 8083); Le Moray Creek 24 May 1981-2; Taylor 18 May 1985-18; Beatton Park 2 May 1987-1, 6 May 1985-1; Fort Nelson 13 May 1987-1; ne Kwokullie Lake 30 May 1982-2; Liard Hot Springs 23 May 1975-1 (Reid 1975); Atlin 18 May 1934-1 (RBCM 5676).

Summer: Coastal – Lost Lake (Victoria) 28 Jul 1948-3 nestlings; Bamfield 29 Jun 1977-5; Stanley Park (Vancouver) 31 Aug 1980-20; Squamish 8 Jun 1974-1; Brackendale 22 Jun 1916-3 eggs; Alta Lake 23 Aug 1937-3; Port Neville 27 Jun 1985-dead nestling, 30 Aug 1975-6; Kitlope River 6 Jun 1993-several; Kitimat River 27 Jun 1986-4 nestlings; Terrace 30 Jun 1979-3 nestlings; Hazelton 3 Jun 1921-nest building (Swarth 1924); New Hazelton 25 Aug 1917-4 (Spreadborough and Taverner 1917); Flood Glacier 10 Aug 1919-1 (MVZ 40144). **Interior** – Richter Pass 3 Jun 1961-4 eggs plus 2 cowbird eggs (Campbell and Meugens 1971); Creston 19 Aug 1947-10 (Johnstone 1949); Sirdar 10 Jun 1980-4 eggs; Cranbrook 25 Jun 1960-4 eggs; Mount Brent 20 Jul 1977-13; Penticton 30 Jul 1966-5 nestlings; Naramata 25 Jun 1973-6 cowbird eggs, 19 Jul 1966-3 eggs, 30 Jul 1966-5 nestlings, 3 Aug 1966-5 young fledged (Cannings et al. 1987); Nakusp 6 Aug 1988-1 fledgling being fed; Adams Lake 17 Jun 1958-5 eggs; Golden 2 Jul 1976-21; Green Lake (70 Mile House) 11 Jun 1962-4 eggs; 100 Mile House 25 Jul 1976-3 nestlings; Kleena Kleene 11 Jun 1968-incubating eggs; Stillwater Lake (Cariboo) 13 Aug 1982-10; Moose Lake (Mount Robson) 19 Aug 1911-1 (Riley 1912); Sixteen Mile Lake 23 Aug 1944-1 (Munro 1947); Tête Jaune Cache 19 Jul 1987-2 feeding fledgling cowbird; Bentzi Lake 14 Jun 1987-1 singing from nest; Prince George 16 Jun 1969-32 on survey, 10 Jul 1972-3 nestlings fledged; Fort St. James 4 Jun 1889-2 eggs (USNM 238380); 8 km s Dawson Creek 2 Jul 1993-4 nestlings quite well feathered; Dawson Creek 7 Jul 1992-4 nestlings; Moberly Lake 13 Jun 1984-1 egg; e side Pine Pass 16 Jun 1959-incubating; Taylor 24 Jul 1982-2 recent fledglings; Beatton Park 31 Aug 1982-2; Laslui Lake 2 Aug 1976-1; Tetsa River 19 Jun 1992-1; Liard River 22 Aug 1943-1 (NMC 29531; Rand 1944); Lower Laird Crossing 22 Aug 1943-1; Liard Hot Springs 17 Jun 1982-building nest; Atlin 17 Aug 1924-1 (MVZ 104147); Tatshenshini River 7 Jun 1983-1 (RBCM 17823).

Breeding Bird Surveys: Coastal – Recorded from 25 of 27 routes and on 83% of all surveys. Maxima: Kispiox 19 Jun 1994-62; Squamish 8 Jun 1975-45; Kwinitsa 28 Jun 1970-40. **Interior** – Recorded from 66 of 73 routes and on 86% of all surveys. Maxima: Ferndale 30 Jun 1990-81; Telkwa High Road 29 Jun 1991-61; Summit Lake 28 Jun 1991-61; McLeod Lake 16 Jun 1992-52.

Autumn: Interior – Nulki Lake 11 Sep 1953-1 (ROM 83918); Chezacut 17 Sep 1933-1 (MCZ 284079); Riske Creek 12 Sep 1990-1; Tatla Lake 4 Sep 1991-1; Chilcotin River 17 Sep 1991-1; Mount Revelstoke 10 Sep 1982-2; Nicholson 14 Sep 1976-20; Yard Creek Park 16 Sep 1980-1; Summerland 18 Sep 1966-1 (Cannings et al. 1987); Vaseux Lake 19 Sep 1977-2 (Cannings et al. 1987); Oliver 10 Sep 1960-6; Creston 14 Sep 1985-1. **Coastal** – Hazelton 1 Sep 1921-1 (MVZ 42462); Port Simpson 7 Sep 1969-1 (Crowell and Nehls 1970a); Cape Scott 15 Sep 1935-1 (NMC 26059); Hardy Bay 1 Sep 1935-1 (NMC 26244); Loughborough Inlet 4 Sep 1936-1 (NMC 27336; Laing 1942); Port Neville 25 Sep 1977-5; Stanley Park (Vancouver) 5 Oct 1985-1; Vancouver 14 Sep 1931-1 (RBCM 7261); Pacific Rim National Park 28 Sep 1986-1; Saanich 6 Sep 1984-10, 1 Oct 1984-1, 2 Oct 1985-1; Victoria 7 Oct 1981-1, 10 Oct 1976-1.

Winter: No records.

Philadelphia Vireo

PHVI

Vireo philadelphicus (Cassin)

RANGE: Breeds from southeastern Yukon, northeastern British Columbia, northern Alberta, northwestern Saskatchewan, central Manitoba, central Ontario, central Quebec, New Brunswick, and southwestern Newfoundland south to south-central Alberta, central Saskatchewan, North Dakota, extreme southern Manitoba, southern Ontario, southern Quebec, and Minnesota; and east to Maine. Winters from Yucatan south to northern Colombia.

STATUS: In the interior, *uncommon* migrant and summer visitant in the Peace Lowland of the Boreal Plains Ecoprovince; *rare* elsewhere in the Boreal Plains and in the Taiga Plains ecoprovinces. Breeds.

NONBREEDING: The Philadelphia Vireo occurs regularly in British Columbia only in the northeastern corner of the province east of the Rocky Mountains. There it reaches its northernmost distribution in North America. It was first documented in British Columbia at Swan Lake in the Boreal Plains in 1938 (Cowan 1939), and in the Taiga Plains near Fort Nelson in 1974 (Erskine and Davidson 1976). It is widespread only in the Peace Lowland from the Alberta border west to Chetwynd and Hudson's Hope. In the Taiga Plains, there are only a few records from near Fort Nelson, near Kwokullie Lake, at Gutah Creek, and on the Tuchodi River. It is probable that small numbers occur west of its known range along the major river valleys on the east slope of the Rocky Mountains and in scattered pockets of deciduous forest elsewhere in the northeast. It seems to be absent from large areas of mainly coniferous forest between Charlie Lake and Fort Nelson.

The Philadelphia Vireo has been recorded at elevations from 600 to 910 m. In migration and during the breeding period, it frequents deciduous or mixed woodlands (see Figs. 608 and 609), riparian alder and willow thickets, and second-growth deciduous stands on logged or burned areas. In general, it prefers deciduous forest with a rich sapling understorey (Barlow and Rice 1977) and a diversity of foliage and canopy structure provided by natural disturbances or openings (Holmes and Robinson 1981). In British Columbia, foraging habitat includes both the forest canopy and understorey (Enns and Siddle 1996), but the Philadelphia Vireo tends to forage higher in the canopy than other vireos (Holmes and Robinson 1981). Unlike other vireos in British Columbia, the Philadelphia Vireo is seldom seen in suburban or rural habitats. During migration it uses a wider variety of habitats than during the breeding season (Ehrlich et al. 1988).

The Philadelphia Vireo is a relatively late spring migrant. First arrivals reach the Peace Lowland in mid to late May (Fig. 607), with the peak movement probably occurring in late May and early June. Migrants undoubtedly enter the province from Alberta. The autumn movement probably begins in mid July but occurs mainly in August, with all birds gone by the first week of September.

In the northeastern interior, the Philadelphia Vireo has been recorded from 17 May to 6 September (Fig. 607).

BREEDING: The Philadelphia Vireo breeds in the extreme northeastern corner of British Columbia. Breeding has been confirmed only in the Peace Lowland of the Boreal Plains near Tupper, Pouce Coupe, Dawson Creek, and Fort St. John, and in the Fort Nelson Lowland of the Taiga Plains near Fort Nelson. It is probable, however, that it breeds west to Chetwynd and Hudson's Hope, and along the major river valleys on the east side of the Rocky Mountains as far north as the Petitot River. Its breeding distribution is closely associated with the distribution of deciduous forests.

The Philadelphia Vireo reaches its highest numbers in summer in the Peace Lowland of the Boreal Plains (Fig. 610). Breeding Bird Surveys for interior routes for the period 1968 through 1993 contain insufficient data for analysis.

Breeding habitat is primarily deciduous forests that are dominated by trembling aspen (Fig. 608) or balsam poplar. Mixed forests of trembling aspen and white spruce, and

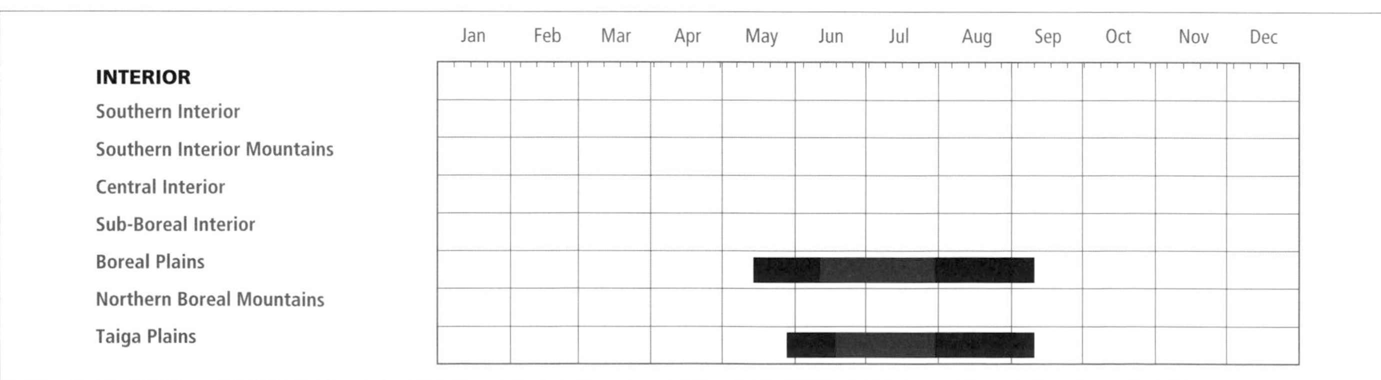

Figure 607. Annual occurrence (black) and breeding chronology (red) for the Philadelphia Vireo in interior ecoprovinces of British Columbia. Records are shown for the week in which they occurred.

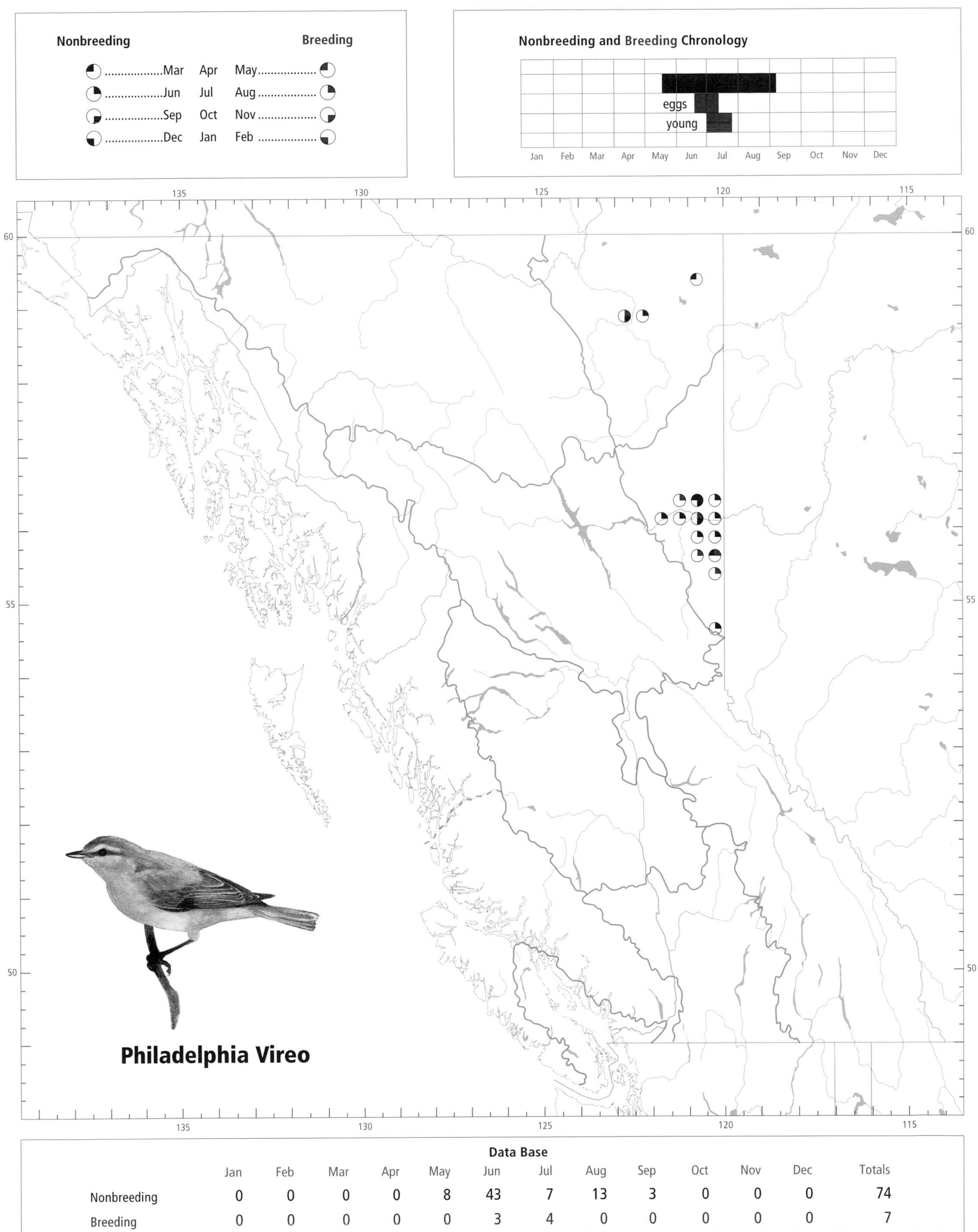

Data Base

	Jan	Feb	Mar	Apr	May	Jun	Jul	Aug	Sep	Oct	Nov	Dec	Totals
Nonbreeding	0	0	0	0	8	43	7	13	3	0	0	0	74
Breeding	0	0	0	0	0	3	4	0	0	0	0	0	7

Figure 608. During migration and in summer, the Philadelphia Vireo frequents trembling aspen woodlands with many branches in the lower canopy (near Pink Mountain, 25 June 1996; R. Wayne Campbell).

Figure 609. Mixed deciduous-coniferous woodlands of mature balsam poplar, paper birch, and mountain alder with an understorey of red-osier dogwood provide breeding habitat for the Philadelphia Vireo in northeastern British Columbia (near Taylor, 12 June 1990; R. Wayne Campbell).

deciduous edges of forest stands, are also used. Both old-growth and second-growth stands are used for breeding, but second-growth stands are probably preferred (Erskine and Davidson 1976; Siddle et al. 1991a; Lance and Phinney 1994). Typical breeding habitat has been described as dense, tall, rapidly growing young aspen stands with an 80% to 100% canopy closure and an understorey of paintbrush, fireweed, highbush-cranberry, Sitka alder, willow, and clover (Enns and Siddle 1996). Siddle (1992d) stressed the importance of riparian forest along the south bank of the Peace River as breeding habitat. In Alberta, breeding habitat has been described as second-growth deciduous riparian thickets (Salt and Wilk 1958); mixed birch, alder, and spruce forest (Salt 1973); and the ecotone between willow uplands and deciduous forest (Francis and Lumbis 1979).

The Philadelphia Vireo has been found breeding in British Columbia from 15 June (calculated) to 27 July (calculated) (Fig. 607).

Nests: Five nests found in the province were in stands of trembling aspen (4) or balsam poplar (1). Both mature and pole-stage trees were used as nest sites. Nests were suspended between diverging twigs that formed a fork at the end of a branch. Nests were composed of fine bark strips, plant down, fine grass, bits of lichen, and leaves. The heights for 4 nests ranged from 1.5 to 18 m.

Eggs: Dates for 3 clutches ranged from 18 June to 12 July. Calculated dates indicate that eggs could be found as early as 15 June. Clutch sizes were 3 and 4 eggs. Elsewhere, the clutch size ranges from 3 to 5 eggs, but is usually 4 eggs (Harrison 1979). The incubation period is about 14 days (Ehrlich et al. 1988).

Young: Dates for 4 broods ranged from 30 June to 25 July. Calculated dates indicate that young can be found between 26 June and 27 July. Sizes of 2 broods ranged from 1 to 3 young (1Y-1, 3Y-1). The nestling period is 12 to 14 days (Ehrlich et al. 1988).

Brown-headed Cowbird Parasitism: In British Columbia, there were no cases of parasitism by the cowbird in 4 nests found with eggs or young. Although cowbird parasitism is seldom reported, the Philadelphia Vireo is probably parasitized more often than records indicate (Friedmann et al. 1977).

Nest Success: Insufficient data.

REMARKS: There are several published reports of the Philadelphia Vireo for which evidence of identification is inconclusive. These reports and all reports after 7 September have been excluded from this account (e.g., Paul 1964; Crowell and Nehls 1974a; Siddle et al. 1991a).

The Philadelphia Vireo is vulnerable in the short term to logging of northeastern hardwood forests (Cooper et al. 1995), as rapid harvesting of deciduous forests is forecast (B.C. Ministry of Forests 1992). However, because it readily uses second-growth deciduous forest, the medium- and long-term impacts of clearcut logging will probably be less serious than for most other songbirds that depend on deciduous forest habitat.

In structurally diverse forests, the territories of the Philadelphia and Red-eyed vireos often overlap because of subtle differences in habitat preference (Rice 1978).

For further information on life history, see Bent (1950), Barlow and Rice (1977), Barlow (1980), Robinson (1981), and Moskoff and Robinson (1996).

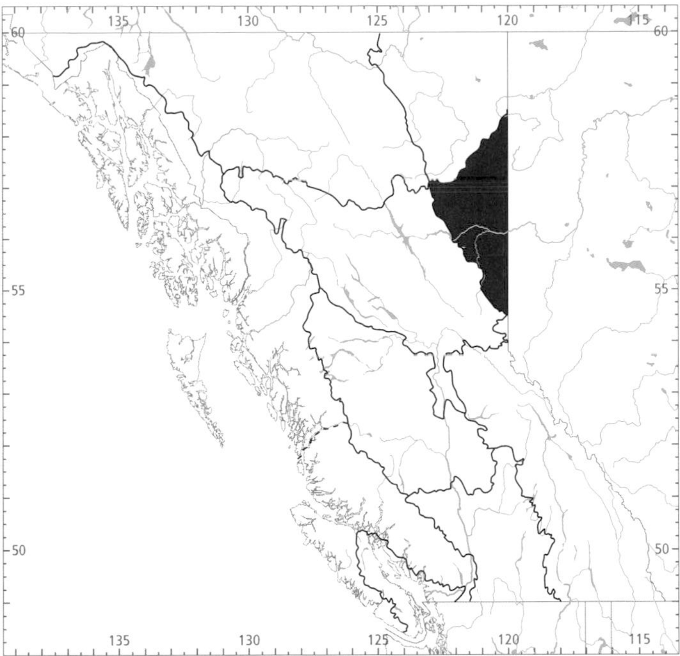

Figure 610. In British Columbia, the highest numbers for the Philadelphia Vireo in summer occur in the Boreal Plains Ecoprovince.

NOTEWORTHY RECORDS

Spring: Coastal – No records. **Interior** – Swan Lake (Tupper) 21 May 1938-1 (RBCM 8076; Cowan 1939), 31 May 1989-2; Stoddart Creek 17 May 1980-1; Beatton Park 25 May 1986-1; ne Kwokullie Lake 30 May 1982-1.

Summer: Coastal – No records. **Interior** – Tupper 28 Jun 1984-pair at nest; Tupper Creek 18 Jun 1941-4 eggs; 10 km wsw of Fellers Heights 12 Jun 1993-male singing from nest, 30 Jun 1993-3 small nestlings; Brassey Creek 18 Jul 1993-nestlings; 3 km w Alcock Lake 18 Jun 1993; 8 km w Chetwynd 5 Jun 1954-1 (NMC 48172; Jobin 1955); Pouce Coupe 7 Jul 1984-2; n Dawson Creek 3 Aug 1975-1; Progress 27 Jun 1993-1; Kiskatinaw Park 19 Jul 1983-1; Hudson's Hope 29 Aug 1979-1; Tupper Creek 18 Jun 1941-4 eggs (WFVZ); Peace Island Park (Taylor) 28 Aug 1984-4; Taylor 1 Jun 1983-1, 23 Jul 1981-fledgling fed by parent, 15 Aug 1987-1; Clayhurst 1 Jul 1978-1 at ferry crossing; Charlie Lake 9 Jun 1938-2 building nest (Cowan 1939), 12 Jul 1962-3 eggs; Muskwa River 30 Jun 1986-1; Fort Nelson 6 Jun 1974-1 (Erskine and Davidson 1976), 14 Jun 1976-1 (RBCM 15362), 3 Jul 1990-1 adult feeding 2 nestlings, 25 Jul 1978-1 adult feeding 2 fledglings, 2 Aug 1978-1 near airport; ne Fort Nelson 2 Jul 1982-1 (Campbell and McNall 1982); Parker Lake 25 Jul 1978-1 adult feeding 2 fledglings, 2 Aug 1978-1.

Breeding Bird Surveys: Coastal – Not recorded. **Interior** – Recorded from 3 of 73 routes and on less than 1% of all surveys. Maxima: Tupper 20 Jun 1994-3; Fort Nelson 19 Jun 1974-1; Hudson's Hope 7 Jun 1976-1.

Autumn: Interior – Fort Nelson 3 Sep 1986-1; Cecil Lake 2 Sep 1984-1; Peace Island Park (Taylor) 6 Sep 1987-1 (Siddle 1988b). **Coastal** – No records.

Winter: No records.

Red-eyed Vireo

Vireo olivaceus (Linnaeus)

REVI

RANGE: Breeds from southeastern Alaska, southeastern Yukon, north-central and northeastern British Columbia, and southern Mackenzie across the Prairie provinces, through southern Ontario and Quebec to Newfoundland and Nova Scotia, and south in the west to northern Oregon, Idaho, and Montana; in the central and eastern United States south to the Gulf coast. Winters in the Amazon Basin from Colombia and Venezuela south to Bolivia and southeastern Peru.

STATUS: On the coast, *uncommon* (occasionally *fairly common*) migrant and summer visitant in the Georgia Depression Ecoprovince and the Southern Mainland Coast and locally in the Northern Mainland Coast of the Coast and Mountains Ecoprovince; *casual* on Western Vancouver Island; absent from the Queen Charlotte Islands.

In the interior, *uncommon* to *fairly common* (occasionally *common*) migrant and summer visitant in the Southern Interior and Southern Interior Mountains ecoprovinces; *uncommon* in the Central Interior Ecoprovince and *rare* to *uncommon* in the Sub-Boreal Interior Ecoprovince from Quesnel to Prince George and Vanderhoof, *very rare* elsewhere in that ecoprovince; *uncommon* to locally *fairly common* migrant and summer visitant in the Boreal Plains and Taiga Plains ecoprovinces; *very rare* in the Liard River area of the Northern Boreal Mountains Ecoprovince.

Breeds.

NONBREEDING: The Red-eyed Vireo is widely distributed across much of the southern third of the province, from southeastern Vancouver Island and the Fraser Lowland to the Rocky Mountains, becoming more scattered through central regions of the province. In the interior, it is a species of valley bottoms and lower mountain slopes, and is widespread in the Kootenay, Columbia, Okanagan, and Thompson river valleys. Its distribution extends northward across the aspen parkland of the Cariboo and Chilcotin areas, but it becomes less abundant north to Williams Lake, Quesnel, and Prince George. A population in the northern Coast and Mountains, mainly along the lower Skeena River valley, appears to be isolated from other populations in the province. In the far north, it has been recorded only east of the Rocky Mountain Trench. It occurs commonly in the Peace Lowland, but becomes more sparsely distributed in the Taiga Plains north to at least Kwokullie Lake.

The Red-eyed Vireo has been recorded at elevations from near sea level to 620 m on the coast; in the interior, it occurs from the valley floor up to 900 m elevation. It avoids areas of continuous coniferous forest and prefers, in both coastal and interior areas, groves of deciduous trees and shrubs in riparian habitats (Fig. 611). The Red-eyed Vireo occurs less frequently in suburban parks and gardens than the Warbling Vireo.

The Red-eyed Vireo is a late spring migrant, arriving after the leaves of deciduous trees have emerged and when the annual horde of leaf-eating insects is thriving. The first migrants reach the Fraser Lowland and the southern portions of the interior in the first week of May (Figs. 613 and 614), and the main movement occurs from late May to early June. First-of-year observations in the Vancouver area range from 1 May (earliest) to 24 May (latest) (Campbell et al. 1972a).

Figure 611. In migration and during the breeding season, the Red-eyed Vireo often frequents riparian habitats with young black cottonwood, willow, and saskatoon (Creston, 2 August 1993; R. Wayne Campbell).

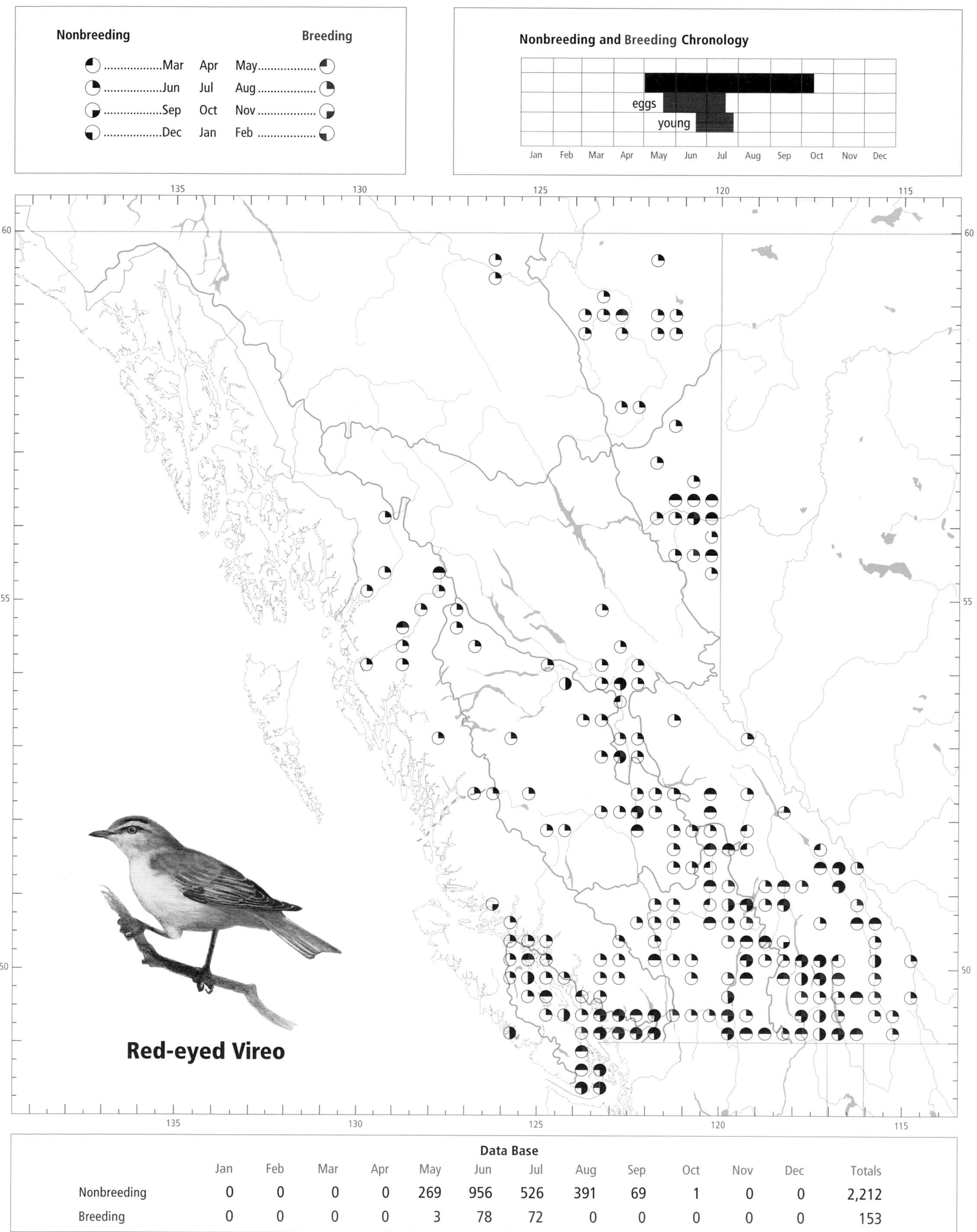

Data Base

	Jan	Feb	Mar	Apr	May	Jun	Jul	Aug	Sep	Oct	Nov	Dec	Totals
Nonbreeding	0	0	0	0	269	956	526	391	69	1	0	0	2,212
Breeding	0	0	0	0	3	78	72	0	0	0	0	0	153

The mean first-arrival date in the Okanagan valley is 25 May, with a range over 23 years of 3 May to 3 June (Cannings et al. 1987). Early migrants seem to arrive in the west Kootenay sooner (early May) than in the east Kootenay (late May). Provincewide, the highest counts are from mid to late June, when most males are on their breeding territories and sing frequently.

Autumn migration in the north occurs in August, and most of the Red-eyed Vireos have left the north by the end of that month. In the southern portions of the province, migration occurs mainly in the second half of August and in early September. Most birds have left the province by mid-September, but a very few remain into early October (Figs. 613 and 614). Migration chronology in British Columbia is similar to that reported for northern Alberta by Salt (1976), and populations in the northeast migrate through Alberta.

On the coast, the Red-eyed Vireo has been recorded from 1 May to 22 September; in the interior, it has been recorded regularly between 2 May and 25 September and irregularly to 12 October (Fig. 613).

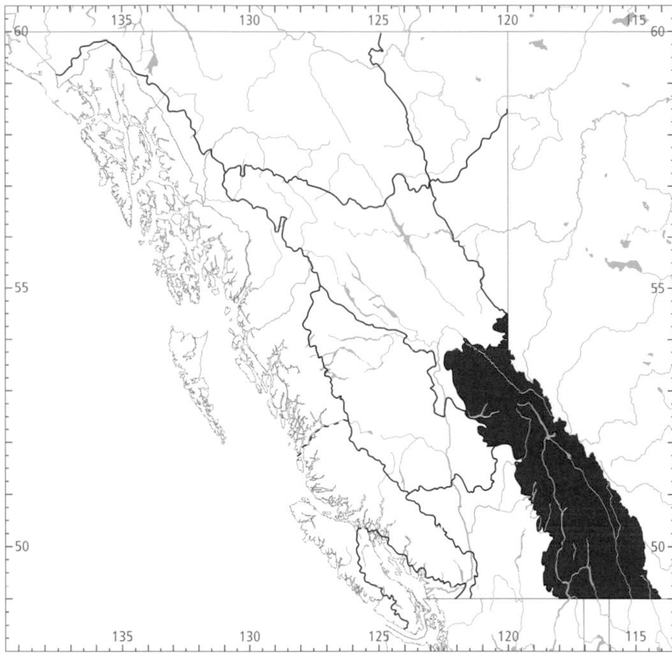

Figure 612. In British Columbia, the highest numbers for the Red-eyed Vireo in summer occur in the valleys of the Kootenay and Columbia rivers in the Southern Interior Mountains Ecoprovince.

BREEDING: In British Columbia, the breeding distribution of the Red-eyed Vireo is concentrated in the southern quarter of the province, from the east coast of Vancouver Island east across the Southern Interior to the Rocky Mountain Trench and north to the North Thompson River valley. On Vancouver Island, the Red-eyed Vireo has been found breeding at Campbell River, Duncan, and the vicinity of Victoria, but not on the west coast. The only confirmed breeding localities in the Central Interior and Sub-Boreal Interior are at Williams Lake and Quesnel, respectively. A small, relatively isolated population is present in summer along the lower Skeena River drainage basin, but Terrace is the only confirmed breeding locality in that region. The Red-eyed Vireo is a relatively common vireo in the Peace Lowland, but there are only a few breeding records. In the Taiga Plains, although it is regularly present from Parker Lake to Smith River, the only breeding

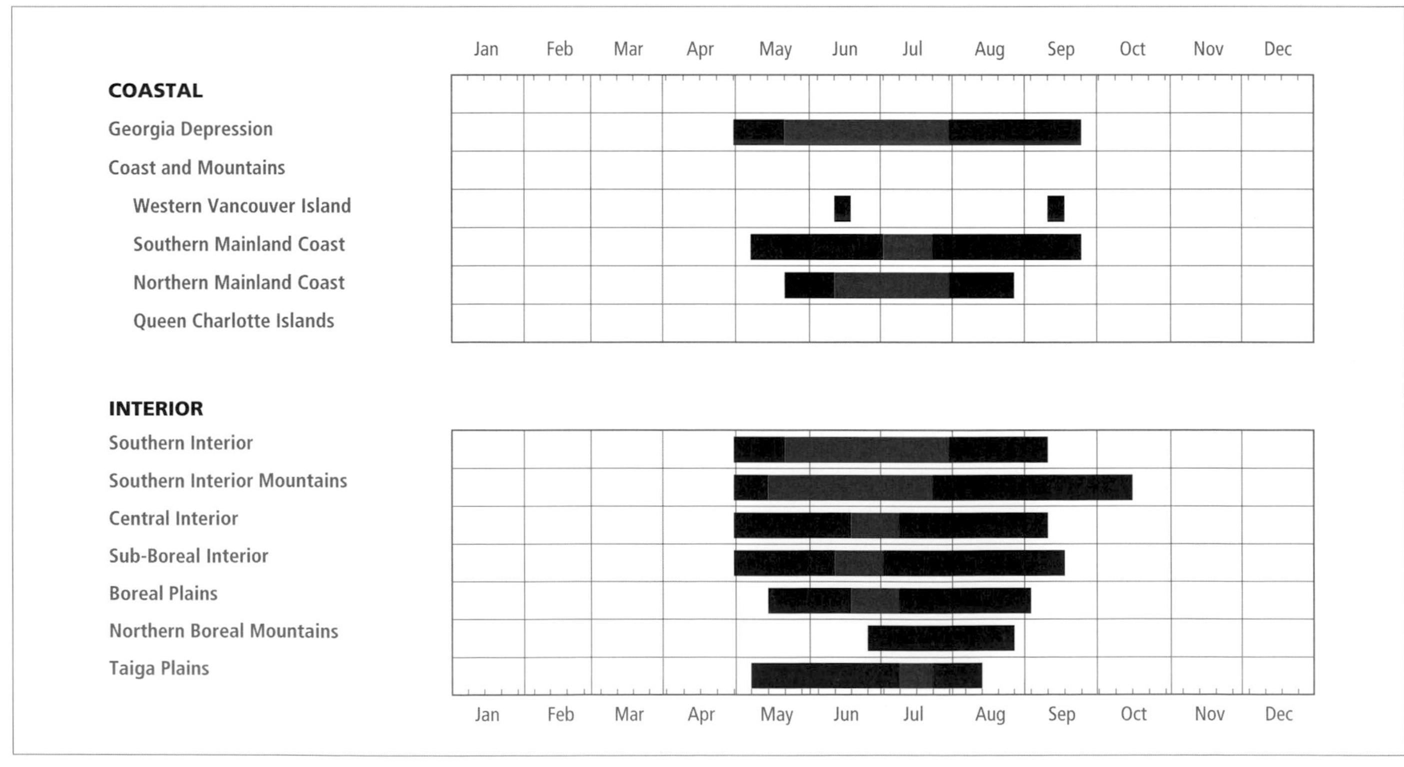

Figure 613. Annual occurrence (black) and breeding chronology (red) for the Red-eyed Vireo in ecoprovinces of British Columbia. Records are shown for the week in which they occurred.

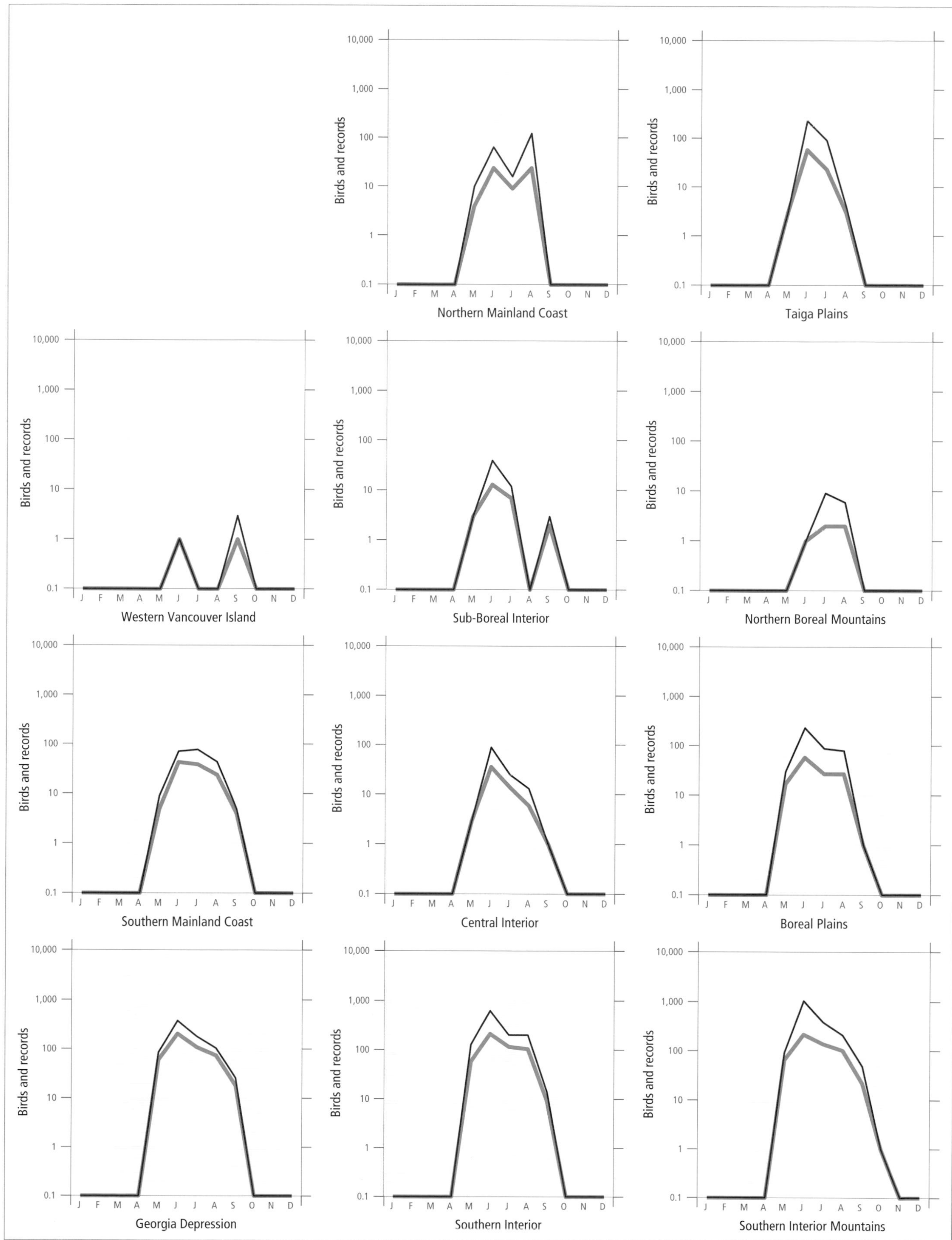

Figure 614. Fluctuations in total number of birds (purple line) and total number of records (green line) for the Red-eyed Vireo in ecoprovinces of British Columbia. Nest record data have been excluded.

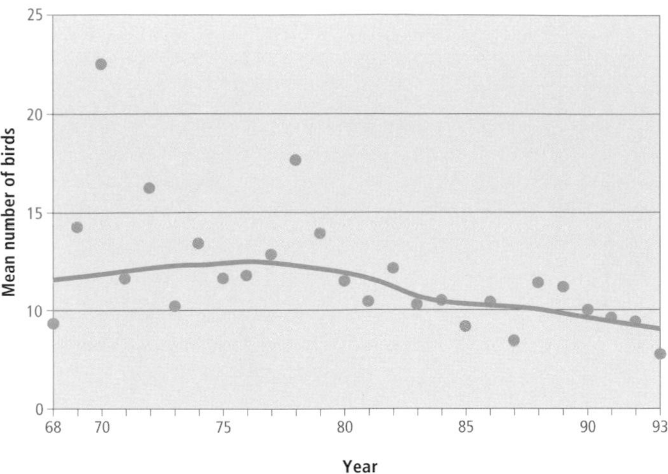

Figure 615. An analysis of Breeding Bird Surveys for the Red-eyed Vireo in British columbia shows that the mean number of birds on interior routes decreased at an average annual rate of 2.0% over the period 1968 through 1993 ($P \leq 0.001$).

records are from near Fort Nelson.

The Red-eyed Vireo occurs in summer, and probably breeds at the heads of some major inlets along the south and central coast of the mainland. Laing (1942) reported it at Bute Inlet, at Loughborough Inlet, at Hagensborg in the Bella Coola valley, and on the estuary of the Kimsquit River. During a recent survey of the Kitlope River, several singing males were found along the lower reaches, but none was recorded in the mid to upper sections. Laing (1942) believed that isolated coastal populations reached the coast from the east by following major river valleys, and that they probably breed there.

Figure 616. In the Taiga Plains Ecoprovince, the Red-eyed Vireo breeds in mixed forests of white spruce, trembling aspen, and paper birch (east of Fort Nelson, 26 June 1996; R. Wayne Campbell).

This vireo reaches its highest numbers in summer in the valleys of the Kootenay and Columbia rivers in the Southern Interior Mountains (see Fig. 612). An analysis of Breeding Bird Surveys shows that the mean number of birds on interior routes decreased at an average annual rate of 2% over the period 1968 to 1993 (Fig. 615); analysis of coastal routes for the same period could not detect a net change in numbers.

The Red-eyed Vireo has been recorded breeding at elevations from near sea level to 200 m on the coast, and from 300 to 800 m in the interior. It breeds in forested habitats, including mixed forests (37%; *n* = 30; Fig. 616), deciduous woodlands (30%), and fragmented woodlands in rural and suburban areas (27%). More specifically, most nesting sites described were in riparian habitat (48%; *n* = 21; Fig. 611); mature, young, or regenerating forest (19%); and gardens (19%). On the coast, the Red-eyed Vireo is most frequently found in mixed deciduous forests as well as plantings of exotic trees such as Lombardy poplar and weeping willow. It is reported infrequently from the Garry oak–arbutus–Douglas-fir forests of southeastern Vancouver Island, but is more common in the drier mixed forests of the northern Gulf Islands. In the interior, this vireo frequents woodlands of deciduous trees and mixed deciduous-coniferous forests, especially those on moist sites.

On the coast, the Red-eyed Vireo has been recorded breeding from 25 May (calculated) to 21 July; in the interior, it has been recorded breeding from 20 May (calculated) to 27 July (Fig. 613).

Nests: Even where the Red-eyed Vireo breeds in mixed forests, all nests reported for British Columbia were built in deciduous vegetation (*n* = 48), either trees (81%) or shrubs (19%). Like other vireos, the Red-eyed Vireo builds a cup-shaped nest suspended between diverging or adjacent twigs. Nest materials included fine grass (74%; *n* = 43), bark strips (40%), plant fibres (35%), moss (16%), lichens (16%), rootlets (14%), plant down (14%), and other fine material, including leaves, paper, twigs, hair, and feathers. The heights for 50 nests ranged from 1 to 15 m, with 64% between 2 and 3 m.

Eggs: Dates for 47 clutches (Fig. 617) ranged from 23 May to 19 July, with 55% recorded between 16 June and 1 July

Figure 617. In British Columbia, nests with eggs have been reported between 23 May and 19 July, with over half of all clutches containing 4 eggs (Victoria, 2 June 1993; R. Wayne Campbell).

(Fig. 618). Calculated dates indicate that eggs can be found as early as 20 May. Sizes of 43 clutches ranged from 1 to 4 eggs (1E-3, 2E-2, 3E-15, 4E-23), with 53% having 4 eggs (Fig. 617). One record in British Columbia suggests that the incubation period lies between 11 and 13 days. Peck and James (1987) give a range of 11 to 15 days for Ontario.

Young: Dates for 21 broods ranged from 20 June to 27 July, with 56% recorded between 2 July and 11 July. Calculated dates indicate that young can occur as early as 3 June. Sizes of 18 broods ranged from 1 to 4 young (1Y-4, 2Y-4, 3Y-5, 4Y-5), with 10 having 3 or 4 young. The nestling period is 10 to 12 days (Ehrlich et al. 1988).

Brown-headed Cowbird Parasitism: In British Columbia, 27% of 56 nests found with eggs or young were parasitized by the cowbird: 6 of 21 nests on the coast, and 9 of 35 nests in the interior. There were also 4 records of adult vireos feeding fledged cowbird young. In Ontario, Peck and James (1987) reported parasitism in 38% of 354 Red-eyed Vireo nests.

This vireo is well known as a victim of the cowbird (Friedmann et al. 1977), and usually abandons its nest if 2 or 3 cowbird eggs are laid in its nest or if all the vireo eggs are removed by cowbirds (Southern 1958).

Nest Success: Of 10 nests found with eggs and followed to a known fate, 4 produced at least 1 fledgling.

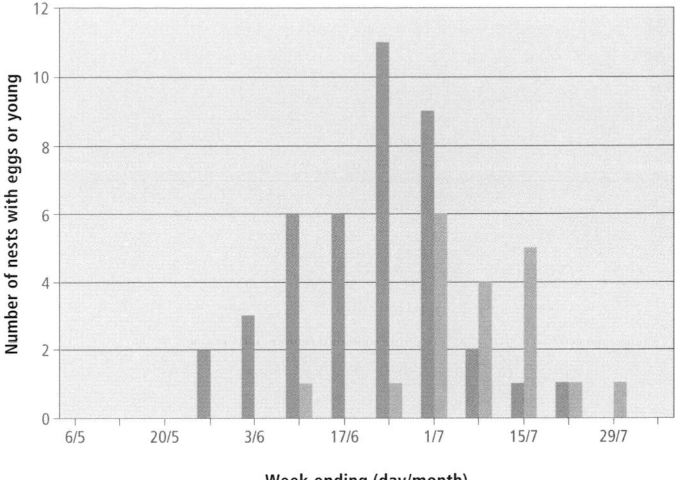

Figure 618. Weekly number of clutches (dark bars) and broods (light bars) for the Red-eyed Vireo in southern British Columbia. Egg laying begins 2 to 3 weeks after the birds arrive. Data are summed from breeding records for the Georgia Depression, Southern Interior, and Southern Interior Mountains ecoprovinces, and are based on the week eggs or young were first found in a nest.

REMARKS: See Lawrence (1953), Barlow and Rice (1977), and Robinson (1981) for additional information on the behaviour and natural history of the Red-eyed Vireo.

NOTEWORTHY RECORDS

Spring: Coastal – Victoria 28 May 1947-4 eggs; Tofino Inlet 10 May 1970-1; Sumas Mountain 1 May 1975-1; Langley 17 May 1972-2; Alouette Lake 8 May 1971-3 (Campbell et al. 1972a); West Vancouver 1 May 1976-1; Black Mountain (West Vancouver) 18 May 1974-10; Hazelton 21 May 1963-1; Terrace 30 May 1976-1. **Interior** – Creston valley 28 May 1980-10; Vaseux Lake 4 May 1922-1 (NMC 17837); White Lake (Okanagan Falls) 30 May 1965-22 on survey; Horn Lake (Olalla) 26 May 1905-4 eggs; Castlegar 2 May 1972-1; St. Mary Lake (Kimberley) 29 May 1941-1 (Johnstone 1949); Kaslo 23 May 1963-4 eggs; Argenta 3 May 1971-1; Coldstream 3 May 1966-1 (Cannings et al. 1987); Botanie Creek 7 May 1966-4; Moberly Marsh 23 May 1994-1; Williams Lake 3 May 1975-1; 15 km n Quesnel 3 May 1992-1 at Cottonwood bridge; Prince George 19 May 1986-1; Tupper Creek 18 May 1938-1 (Cowan 1939), 28 May 1938-1 (RBCM 8081); Taylor 20 May 1988-1; Parker Lake road (Fort Nelson) 11 May 1980-1.

Summer: Coastal – Saanich 1 Jul 1989-1; Cultus Lake 21 Jun 1959-15; Chesterman Beach 12 Jun 1984-1; 6 km e Haney 26 Jun 1961-2 nestlings; Pitt Meadows 4 Aug 1969-4 fledglings; Maple Ridge 21 Jul 1956-3 nestlings in University of British Columbia Research Forest, fledged by 22 Jul; Capilano River 2 Jun 1937-3 eggs; Kawkawa Lake 6 Jul 1942-4 eggs; Qualicum Beach 29 Jun 1963-11; Squamish 15 Aug 1954-8; Stories Beach 23 Aug 1974-15; Campbell River 5 Aug 1978-1 fledgling; Heriot Bay 27 Jul 1973-10; Pemberton Meadows 14 Jul 1985-6; Bute Inlet 17 Aug 1936-1 (NMC 27386); Loughborough Inlet 28 Aug 1936-1 (Laing 1942); Kitlope River 10 Jun 1994-4; Kitimat 10 Jul 1975-4; Lakelse Lake 18 Aug 1977-11, 23 August 1977-2; Kispiox River 26 Jun 1975-15; Terrace 20 Jun 1978-4 eggs, 19 Jul 1979-1 nestling. **Interior** – Richter Pass 2 Jul 1960-3 eggs (Campbell and Meugens 1971); Yahk 9 Jun 1958-4 eggs; Creston 16 Aug 1982-11; Trail 17 Jul 1985-4 fledglings; Gallagher Lake 25 Aug 1964-8; Zincton 17 Jun 1978-54 on Breeding Bird Survey; Creighton Valley 2 Jul 1979-10; Okanagan Landing 29 Aug 1933-1 (USNM 413345); Brisco 19 Jun 1958-3 eggs, 14 Jul 1958-nestlings fledged; Squilax 13 Jun 1973-26 on survey; Shuswap Lake Park 15 Jul 1981-3 nestlings fledged, 27 Jul 1969-nestlings; Revelstoke 1 Jul 1988-1, 4 Aug 1969-3 fledglings; Buffalo Lake 13 Jun 1964-30; Williams Lake 23 Jun 1977-4 eggs, 13 Aug 1971-8; Redstone 30 Jun 1978-1; Horsefly River 1 Jul 1975-6; Prince George 16 Jun 1969-7, 1 Jul 1969-5; Tomslake to Pouce Coupe 15 Jun 1974-14 on survey; Arras 26 Jun 1993-3 eggs; Taylor 5 Jul 1974-10, 29 Jul 1982-1 fledgling, 15 Aug 1986-19; Fort St. John 24 Jul 1982-1 fledgling; Boundary Lake 30 Aug 1982-1; Fort Nelson 21 Jul 1986-2 fledglings recently out of nest; Parker Lake road (Fort Nelson) 7 Jul 1970-10, 7 Aug 1978-1; Kledo Creek 2 Aug 1985-2; Liard River 4 Aug 1943-3 (Rand 1943); Liard Hot Springs 28 Jun 1985-1, 7 to 9 Jul 1992-9.

Breeding Bird Surveys: Coastal – Recorded from 21 of 27 routes and on 49% of all surveys. Maxima: Kispiox 20 Jun 1993-34; Albion 4 Jul 1970-23; Squamish 16 Jun 1985-18. **Interior** – Recorded from 60 of 73 routes and on 77% of all surveys. Maxima: Zincton 26 Jun 1976-68; Salmon Arm 1 Jul 1968-68; Golden 17 Jun 1978-62; Adams Lake 21 Jun 1992-55.

Autumn: Interior – Taylor 2 Sep 1984-1; Nulki Lake 2 Sep 1945-2 (Munro and Cowan 1947); Prince George 8 Sep 1993-2, 14 Sep 1994-1; Williams Lake 9 Sep 1983-1; Golden 25 Sep 1975-last migrant; Yoho National Park 1 Sep 1977-1; Nicholson 12 Sep 1976-3; Shuswap Lake 6 Sep 1959-4; Columbia Lake 2 Sep 1945-2 (Johnstone 1949); Nakusp 13 Sep 1985-1; Brouse 12 Oct 1981-1; Penticton 8 Sep 1978-1 (Cannings et al. 1987); Erie Lake 3 Sep 1971-10. **Coastal** – Pitt Lake 1 Sep 1976-5; Harrison Hot Springs 14 Sep 1984-1, 18 Sep 1986-1; Kent 15 Sep 1977-1; Vancouver 14 Sep 1973-1; Sea Island 22 Sep 1979-1; Tofino 13 Sep 1989-3.

Winter: No records.

Casual, Accidental, Extirpated, and Extinct Species

Acadian Flycatcher ACFL

Empidonax virescens (Vieillot)

RANGE: Breeds from southeastern North Dakota, northern Iowa, southeastern Michigan, extreme southern Ontario, southern New York, and western Massachusetts south to eastern Texas, the Gulf coast, and central Florida. Winters in Central America and northern South America.

STATUS: *Accidental.*

OCCURRENCE: In British Columbia, a male was collected at Leonie Lake (1,920 m elevation), near Barriere, by G.M. Sutton on 9 June 1934. The testes were much enlarged. The specimen remained unrecognized in the bird collections of the Carnegie Museum of Natural History (CMNH 115760) until 1956, when it was correctly identified (Todd 1957).

The Acadian Flycatcher is mostly confined to the eastern United States and breeds no closer to the Pacific coast than the Great Plains (Roberson 1980). The species has been erroneously reported from California (see Unitt 1984) and does not appear on state lists for Oregon (Schmidt 1989) or Washington (Wahl and Paulson 1991).

REMARKS: Phillips et al. (1966), Phillips and Lanyon (1970), and Whitney and Kaufman (1985) provide information on the identification of various species of *Empidonax* flycatchers.

Page and Cadman (1994) recently proposed that the Acadian Flycatcher be added to the endangered species list of Canadian birds.

Black Phoebe BLPH

Sayornis nigricans (Swainson)

RANGE: Resident from extreme southwestern Oregon, southern Nevada, southwestern Utah, central Arizona, south-central Colorado, central New Mexico, and western Texas south to southern Baja California and south to northwestern Argentina.

STATUS: *Casual.*

OCCURRENCE: Two records. On 11 November 1936, a female Black Phoebe was collected at Marpole, south Vancouver (RBCM 6914; Fig. 619; Cowan 1939). On 26 and 27 April 1980, a single bird was seen and photographed at Lost Lagoon in Stanley Park, Vancouver (Weber et al. 1981; BC Photo 658).

REMARKS: There is 1 additional report of a single bird seen on 28 May 1978 in the Botanical Gardens at the University of British Columbia, Vancouver; however, satisfactory documentation is lacking (Anonymous 1978).

Figure 619. Black Phoebe collected at Vancouver on 11 November 1936 (RBCM 6914; Michael C.E. McNall).

Great Crested Flycatcher GCFL

Myiarchus crinitus (Linnaeus)

RANGE: Breeds from east-central Alberta, central and southeastern Saskatchewan, southern Manitoba, western and southern Ontario, southwestern Quebec, central New Brunswick, southern Nova Scotia, and northern Maine south to Texas, the Gulf coast, and southern Florida; and west to the eastern Dakotas, eastern Nebraska, western Kansas, and west-central Oklahoma. Winters in central and southern Florida and Cuba south through Middle America to northern Colombia and Venezuela.

STATUS: *Accidental.*

OCCURRENCE: A single bird was seen on Triangle Island, off the northwestern tip of Vancouver Island, on 29 and 30 September 1995 (Bowling 1995b). The flycatcher arrived during the onset of a storm and departed during a period of sunny weather (Toochin 1995).

The Great Crested Flycatcher has not been recorded in Washington and Oregon (Schmidt 1989; Wahl and Paulson 1991; Gilligan et al. 1994) but is considered an autumn vagrant, between 19 September and 20 October, along the coast of California (Roberson 1980).

REMARKS: This large flycatcher has also been reported from Surrey, Vancouver, 100 Mile House, Radium, and Prince George during spring and autumn migration periods, but without satisfactory documentation.

Thick-billed Kingbird TBKI

Tyrannus crassirostris Swainson

RANGE: Breeds from southeastern Arizona and southwestern New Mexico south to western Guatemala. Winters from Sonora south through the breeding range to Chiapas. Northern populations are migratory.

STATUS: *Accidental.*

OCCURRENCE: A Thick-billed Kingbird appeared at a private residence in Qualicum Beach, on the east coast of Vancouver Island, on 20 October 1974 (Crowell and Nehls 1975a; Dawe 1976). It was subsequently sighted and photographed on 30 October. The bird was seen again on 11 November, and on 12 November it was found dead below a window. It was preserved and identified as an immature male (RBCM 14750; Fig. 620).

The Thick-billed Kingbird occurs mainly as an autumn vagrant along the west coast of North America, a pattern shown to be similar to that of other southern tyrannids (McCaskie 1967; Roberson 1980). In addition, most records are of immature birds.

REMARKS: Dates given in the literature for the British Columbia specimen (see Crowell and Nehls 1975a; Dawe 1976) and on the specimen label (i.e., 11 November) are erroneous.

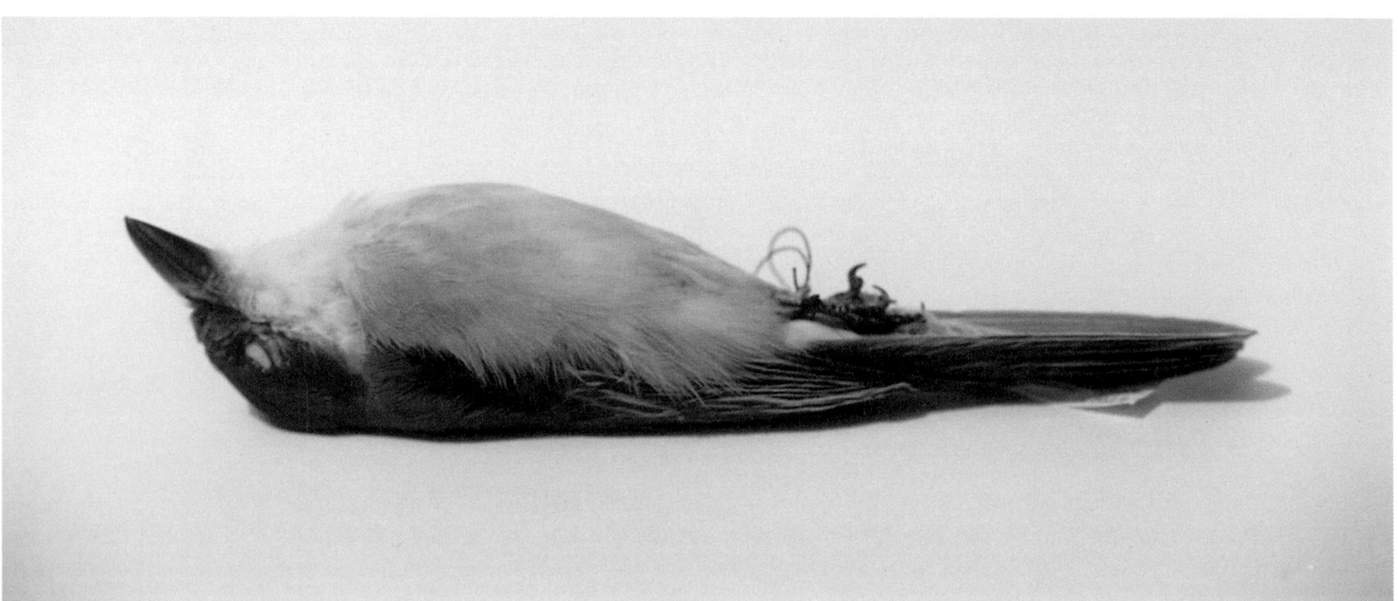

Figure 620. Immature male Thick-billed Kingbird found dead below a window in Qualicum Beach, Vancouver Island, on 12 November 1974 (RBCM 14750; Michael C.E. McNall).

Gray Kingbird

GRKI

Tyrannus dominicensis (Gmelin)

RANGE: Breeds along the Atlantic and Gulf coasts from southeastern South Carolina south to the Florida Keys, and west to Louisiana; south to northern South America. Winters from Hispaniola and Puerto Rico south into Colombia and Venezuela.

STATUS: *Accidental.*

OCCURRENCE: A Gray Kingbird was collected at Cape Beale, near Barkley Sound on southwestern Vancouver Island, on 29 September 1889 (Brooks 1909; Macoun and Macoun 1909).

This species breeds only along a narrow coastal belt of the southern and southeastern United States. The British Columbia specimen is the only record for the west coast of North America (McCaskie 1986; Armstrong 1988; Schmidt 1989; Wahl and Paulson 1991; Gilligan et al. 1994). Roberson (1980) suggests that although the record is extraordinary, "several other species of southern Tyrannids are known to indulge in occasional 'wrong way' flights to the north in the fall."

Western Scrub-Jay

WSJA

Aphelocoma californica (Vigors)

RANGE: Resident from southwestern and south-central Washington south through western and central Oregon to southern Baja California, and from northeastern California, southern Idaho, and southwestern Wyoming south through Nevada, Utah, western Colorado, western New Mexico, and southwestern Texas (locally in central Texas). Also resident in Mexico.

STATUS: *Casual* on the southern mainland coast.

OCCURRENCE: Two records. A single bird was located and photographed in Langley, 40 km southeast of Vancouver, on 8 November 1981 (Campbell 1982a; BC Photo 747). The jay was not seen the following day (9 November) or on 5 subsequent visits from mid-November 1981 through early January 1982 (Campbell 1986e). The date given by Hunn and Mattocks (1982) is erroneous. This represents the first record for Canada.

The second record for the province was of a bird found in the vicinity of Musqueam Nature Park, south Vancouver, on 28 July 1993 (Siddle 1994b). It was present in the general vicinity until 7 April 1994 (Fig. 621).

Coastal records north of its northern breeding limit in central Oregon have become more frequent over the past 2 decades. In Washington the species is regarded as a rare resident in the southwestern part of the state, where it has been recorded since the late 1800s (Jewett et al. 1953). Recent Washington records include a Western Scrub-Jay that spent more than a year in Seattle from late 1977 to early 1979 (Mattocks 1979). There are also several other recent records for the Washington coast.

REMARKS: There are 3 other records for coastal British Columbia between 1980 and 1991, all without supporting documentation. At least 1 of these was a caged bird that was released in Surrey in 1989.

The Western Scrub-Jay was formerly known as Scrub Jay, *Aphelocoma coerulescens* (Bosc.) (American Ornithologists' Union 1995). Pitelka (1951) gives details of the distribution and identification of the former 4 North American races.

Figure 621. Western Scrub-Jay in garden at 6180 Highbury Street, Vancouver, in late November 1993 (Allan Drab).

Blue-gray Gnatcatcher

BGGN

Polioptila caerulea (Linnaeus)

RANGE: Breeds from southern Oregon and Idaho, southwestern Wyoming, southeastern South Dakota, Nebraska, southeastern Wisconsin, southern Michigan, extreme southern Ontario, southwestern Quebec, New York, Vermont, New Hampshire, and Maine south in the west to southern Baja California, through most of Mexico to Chiapas and Belize; also southern Texas and the Gulf coast to Florida. Winters in the southern United States from southern California and Arizona in the west through southern Texas and the Gulf coast, to the Atlantic coast from Virginia to Florida; south through Mexico to Guatemala and Honduras.

STATUS: *Casual* on the south coast.

OCCURRENCE: Three records. A single bird was seen foraging with a flock of Bushtits in a grove of Garry oaks in Uplands Park, Victoria, on 10 and 11 November 1963 (Boggs and Boggs 1964a; Davidson 1966; Roberson 1980) (Fig. 622). The second bird recorded was present from 19 November to 1 December 1981, in the vicinity of Kitsilano Park, Vancouver (Campbell 1982b; Mattocks and Hunn 1982a). It was photographed on 19 November 1981 (BC Photo 775). The third record was of a single bird carefully observed on Bowen Island, 14 km northwest of Vancouver, on 12 June 1994 (Bowling 1995a).

The known range of the Blue-gray Gnatcatcher has been expanding slowly northward. In western North America, there are now scattered breeding records in Oregon, Idaho, and Wyoming (Findholt 1983; White et al. 1983; Gilligan et al. 1994).

Including the 3 British Columbia occurrences, there are at least 6 records of this species on the west coast north of its northernmost breeding range in southern Oregon. All but 1 of them are from the period from late autumn (November) to early spring (March).

REMARKS: See Ellison (1992) for a recent summary of the life history of the Blue-gray Gnatcatcher in North America, and Dunn and Garrett (1987) for details on identifying North American gnatcatchers.

Figure 622. Blue-gray Gnatcatcher in Uplands Park, Victoria, on 10 November 1963 (BC Photo 78; G. Allen Poynter).

Northern Wheatear

NOWH

Oenanthe oenanthe (Linnaeus)

RANGE: Breeds in North America from northern Alaska, northern Yukon, and northwestern Mackenzie south to western and south-coastal Alaska and southern Yukon, and from central Ellesmere Island south to Boothia Peninsula, southeastern Keewatin, eastern and southern Baffin Island, northern Quebec, and Labrador. Winters in the Old World. Most of the distribution of the species is in Eurasia.

STATUS: *Accidental.*

OCCURRENCE: A single bird, in autumn plumage, was present near the airport at Patricia Bay, 32 km north of Victoria on southern Vancouver Island, from 10 to 16 October 1970 (Crowell and Nehls 1971a; Stirling 1971a; Tatum 1971). It was photographed on 11 and 12 October (BC Photos 106 and 125; Fig. 623). The bird spent a lot of its time in the vicinity of a hangar, where it frequently left the roof to capture insects on the bare ground.

There are at least 3 other unpublished sightings for the province, all without convincing details. Two of these are from the Haines Highway area in the Northern Boreal Mountains Ecoprovince. Another published observation, from Gladys Lake in Spatsizi Plateau Wilderness Park in July 1975, is also without adequate documentation (Carswell 1975).

There are very few acceptable records of the Northern Wheatear at lower latitudes in western North America. Through the 1970s, Roberson (1980) lists only 3 other records: 1 for Oregon (22 June 1977) and 2 for California (11 June 1971 and 15 September 1977).

REMARKS: The Northern Wheatear was formerly known as Wheatear (American Ornithologists' Union 1983).

Figure 623. Northern Wheatear, in autumn plumage, near Patricia Bay airport on the Saanich Peninsula, southern Vancouver Island, on 11 October 1970 (Ken R. Summers).

Dusky Thrush

Turdus naumanni Temminck

DUTH

RANGE: Breeds from northern Siberia east to Kamchatka and the Sea of Okhotsk. Winters from Japan and the Ryukyu Islands south to southern China and Taiwan, rarely west to southeast Asia and India. In North America, casual in migration and winter.

STATUS: *Casual.*

OCCURRENCE: Two records. On 2 January 1993 a Dusky Thrush (Fig. 624) was identified as it fed on berries of wild lily-of-the-valley and English holly in southeast Langley, about 40 km east-southeast of Vancouver on the Fraser River delta lowland. The bird remained in the area until 9 April 1993.

During the 93 days the thrush was present, it experienced temperatures as low as –16°C, and, in January, an accumulation of about 50 cm of snow. Through this the bird remained active, and vigorously defended a large holly tree from other berry-eating passerines. The presence of the Dusky Thrush, for the first time on the North American continent south of Alaska, was considered newsworthy by newspapers, radio, and television, and some 1,400 birdwatchers, from as far away as eastern North America and Britain, came to see the bird. Other details of the bird's behaviour, its use of the habitat, and the recreational interest it stimulated have been described by McKay (1993), Siddle (1993b), Siddle and Bowling (1993), and Tyson (1993). Colour and black-and-white photographs were published by Tyson.

The only other record is of a single bird that was carefully studied and documented feeding with American Robins at Alaksen National Wildlife Refuge, Westham Island, near Ladner, on 22 November 1994 (Bowling 1995c).

Previous occurrences of the Dusky Thrush in North America are all from Alaska. These records are from May, June, October, and November (Kessel and Gibson 1978; Tobish 1990; Tobish and Isleib 1990; Gibson 1993).

REMARKS: The 2 subspecies of Dusky Thrush are distinguishable under good field conditions (Dement'ev and Gladkov 1954), and birds occurring in North America appear to be of the northeast Asian race, *T. n. eunomus* (Kessel and Gibson 1978).

There are at least 3 other reports of sightings in British Columbia of what were believed to be the Dusky Thrush. These are from Tofino (Bowling 1995c), Nanaimo, and Surrey. All lack convincing details or do not meet our criteria of specimen, photograph, or at least 2 competent observers.

Figure 624. Dusky Thrush in southeast Langley, about 40 km east-southeast of Vancouver, on 13 January 1993 (Dick McNeely).

Siberian Accentor SIAC

Prunella montanella (Pallas)

RANGE: Breeds in Siberia, mostly north of the Arctic Circle, from the Ural Mountains east to Magadan Province, including Wrangel Island. Winters from southern Manchuria, Korea, and Japan to central China. In migration, occurs on the shores of the Sea of Okhotsk. Casual in North America.

STATUS: *Accidental.*

OCCURRENCE: A single male was located in a private garden at Sunnybrae, Tappen, near Salmon Arm in south-central British Columbia, on 5 March 1994. The following day it was joined by a female. Both birds were seen frequently and remained in the vicinity of the front yard until 10 April (F. Kime pers. comm.; Bowling 1995a) (Fig. 625).

This sighting represents the first occurrence for Canada and the first spring record for North America. Other records, all during autumn along coastal western Alaska, are from Nunivak Island, St. Lawrence Island, and Point Barrow (Roberson 1980).

REMARKS: Two subspecies of Siberian Accentor occupy the range of the species in Siberia. American occurrences are most likely to be the eastern race, *P. m. badia* (Dement'ev and Gladkov 1954).

Another Siberian Accentor was located in Vancouver on 15 December 1993. Unfortunately it was seen by only a single observer (Bowling 1994c).

Figure 625. Siberian Accentor on a Douglas-fir trunk in scrubby forest habitat in Tappen, near Salmon Arm, on 3 April 1994 (Ruth Sullivan).

Yellow Wagtail

YEWG

Motacilla flava Linnaeus

RANGE: In North America, breeds locally in northern and western Alaska, northern Yukon, and northwestern Mackenzie. Also breeds in Eurasia and north Africa. Winters in Africa, southern Eurasia, and occasionally northern Australia.

STATUS: *Casual.*

OCCURRENCE: Three records, all coastal. A single bird was found along Piper Road near Burnaby Lake, east of Vancouver, on 6 October 1985 – the first well-documented record for British Columbia (W.C. Weber pers. comm.). On 1 September 1991, an immature bird was observed feeding among kelp above the high tide line at Sandspit, Queen Charlotte Islands (Siddle 1992a). Another immature was observed feeding in open farm fields on Martindale Flats, Central Saanich, on southern Vancouver Island, from 4 to 7 September 1995.

The Yellow Wagtail is a fairly regular migrant, in small numbers, in the western Aleutian Islands of Alaska (Roberson 1980). There are few records for central and southern Alaska. Its occurrence in British Columbia appears to be that of an aberrant autumn migrant. This coincides with occurrences in California, where the species has been recorded on 10 occasions between 1978 and 1992, all of them between 4 and 21 September (Small 1994).

REMARKS: The taxonomy and relationships of this species and its groups are still uncertain. If the presently recognized *Motacilla flava* proves to include more than 1 species or subspecies, the name *Motacilla tschutschensis*, applied to the eastern Eurasian population, would include those in western North America (American Ornithologists' Union 1983).

There are at least 4 other reports of Yellow Wagtails in British Columbia from the autumn period, including a fairly well documented occurrence of 2 juveniles on Lulu Island, south of Vancouver, on 20 September 1986 (W.C. Weber pers. comm.). Unfortunately these sightings are all from single observers and do not meet the criteria for provincial listing.

Black-backed Wagtail

BBWA

Motacilla lugens Gloger

RANGE: Breeds from southern Ussuriland, Sakhalin, and Kamchatka south to the Kurile Islands and northern Japan. Winters from eastern China, Korea, and Japan south to southeastern China, Taiwan, the Seven Islands of Izu, and Bonin Islands. Casual in North America.

STATUS: *Accidental.*

OCCURRENCE: The only acceptable record is one observed at Ambleside Park in West Vancouver on 18 April 1982 (Mattocks and Hunn 1982b).

REMARKS: The Black-backed Wagtail was formerly classified as a subspecies of the White Wagtail (*Motacilla alba*) (American Ornithologists' Union 1957).

Several other British Columbia wagtail sightings lack sufficient details to determine whether they were Black-backed or White wagtails. These include 1 at the mouth of the Coquitlam River, east of Vancouver, between 2 and 21 March 1973 (Crowell and Nehls 1973b; Jerema 1973; Weber and Shepard 1975; D.R. Paulson pers. comm.); 1 at Pachena Point lighthouse in Pacific Rim National Park, on the west coast of Vancouver Island, on 24 May 1977 (Godfrey 1986); and 1 at Whiffin Spit, Sooke, on southern Vancouver Island on 19 May 1980 (Taylor 1983).

For a discussion of the status and identification of White and Black-backed wagtails, see Morlan (1981) and Howell (1990).

Postscript: A single bird was seen on Triangle Island, 47 km off the northwestern tip of Vancouver Island, from 30 April to 7 May 1996 (A.E. Burger pers. comm.).

Red-throated Pipit

RTPI

Anthus cervinus (Pallas)

RANGE: Breeds in North America in the Bering Strait area of western Alaska, including St. Lawrence Island and the adjacent mainland from Cape Lisburne south to Wales. The main breeding range extends across northern Eurasia from Scandinavia to the Bering Sea coast (Sibley and Monroe 1990). Winters in the Old World from northern Africa east across Asia Minor, Iran, and India to southeastern China; and south to central Africa, southeast Asia, the East Indies, and the Philippines.

STATUS: *Casual.*

OCCURRENCE: Three records, all coastal. A single bird was present near the Boundary Bay airport in Delta, south of Vancouver, from 22 to 28 December 1990 (Siddle 1991b; W.C. Weber pers. comm.). The second record involved 1 to 4 birds, seen several times at Sandspit, Queen Charlotte Islands, between 4 September and 21 November 1991 (Siddle 1992a; M. Bentley pers. comm.). The most recent observation was of a single bird found in Central Saanich, southern Vancouver Island, on 19 September 1992 (Siddle 1993a).

The Red-throated Pipit is a rare spring migrant on the western Aleutian Islands. In autumn, a small number of Red-throated Pipits follow the North American coast southward, and have been recorded as far south as Baja California (American Ornithologists' Union 1983). There are records from Washington (Roberson 1980) and California. Those from California are from 9 September to mid-November, with most occurrences in October (Small 1994).

REMARKS: Two other sightings have been reported. A single bird was found near Burnaby Lake, east of Vancouver, on 5 October 1985, and another was spotted briefly at Tofino, on the west coast of Vancouver Island, on 4 November 1995. Both lack the confirmatory details required for inclusion here (Campbell 1986a).

Dunn (1976), King (1981), and Parkes (1982) provide details useful in the identification of North American pipits.

Appendices

APPENDIX 1

Migration Chronology

In the following table, dates are given for regular spring and/or autumn migrants in 6 coastal and 9 interior locations throughout the province. The data for compiling the migration dates were obtained from the literature as well as from field notebooks of birders. Periods of records vary with area as follows: Victoria (1969-95), Tofino (1968-95), Vancouver (1968-95), Harrison (1974-92), Sechelt (1971-95), Masset (1980-95), Okanagan (1897-95), Nakusp (1975-95), Radium (1979-86), Kamloops (1978-95), Williams Lake (1958-95), Quesnel (1980-92), Prince George (1981-95), Fort St. John (1980-95), and Fort Nelson (1975-86). Dates for peak movements (Fig. 626) are generally given as ranges. A single date indicates a calculated average for all records. Localities and dates were deleted if information was incomplete.

Each species generally falls into 1 of 3 categories: summer visitant, winter visitant, or passage migrant. Earliest arrival, peak movement, and latest departure dates are given for each "season." A completed matrix for each category is as follows:

	SPRING			AUTUMN		
	Early Arrival	Peak Movement	Late Departure	Early Arrival	Peak Movement	Late Departure
Summer Visitant	x	x	–	–	x	x
Winter Visitant	–	x	x	x	x	–
Passage Migrant	x	x	x	x	x	x

Fig. 626. Part of a flock of 230 Violet-green Swallows staging for autumn migration at Emerald Bay, Okanagan Lake, on 9 August 1996 (R. Wayne Campbell).

	SPRING			AUTUMN		
	Early Arrival	Peak Movement	Late Departure	Early Arrival	Peak Movement	Late Departure
Olive-sided Flycatcher						
Victoria	20 Apr	15-30 May	–	–	15-30 Aug	28 Sep
Tofino	2 May	–	–	–	–	6 Sep
Vancouvesr	20 Apr	21 May	–	–	30 Aug	24 Sep
Harrison	9 May	–	–	–	–	5 Sep
Sechelt	30 Apr	20 May	–	–	–	22 Sep
Okanagan	8 May	early Jun	–	–	15-24 Aug	16 Sep
Nakusp	10 May	–	–	–	–	28 Aug
Radium	20 Apr	–	–	–	–	26 Aug
Kamloops	3 May	late May	15 Jun	–	–	7 Sep
Williams Lake	10 May	16-29 May	–	–	–	–
Quesnel	3 May	–	–	–	–	23 Aug
Prince George	4 May	16-24 May	–	–	–	1 Sep
Fort St. John	12 May	18-28 May	–	–	–	31 Aug
Fort Nelson	10 Jun	–	–	–	–	–
Western Wood-Pewee						
Victoria	1 May	20 May	–	–	31 Aug	3 Oct
Tofino	22 Apr	–	–	–	–	11 Sep
Vancouver	24 Apr	23 May	–	–	1 Sep	25 Sep
Harrison	1 May	–	–	–	–	20 Sep
Sechelt	2 May	20 May	–	–	–	19 Sep
Okanagan	25 Apr	15-20 May	–	–	early Sep	26 Sep
Nakusp	7 May	–	–	–	–	7 Sep
Radium	12 May	–	–	–	–	31 Aug
Kamloops	24 Apr	10-15 May	–	–	–	7 Sep
Williams Lake	2 May	12-17 May	–	–	–	30 Sep
Quesnel	7 May	–	–	–	–	1 Sep
Prince George	4 May	20-27 May	–	–	15-27 Aug	31 Aug
Fort St. John	10 May	18-22 May	–	–	–	2 Sep
Fort Nelson	23 May	25-31 May	–	–	–	28 Aug
Yellow-bellied Flycatcher						
Prince George	7 Jun	10-16 Jun	–	–	–	9 Aug
Fort St. John	10 Jun	–	–	–	–	21 Jul
Fort Nelson	15 Jun	–	–	–	–	24 Jul
Alder Flycatcher						
Nakusp	27 May	–	–	–	–	–
Radium	6 Jun	–	–	–	–	–
Williams Lake	31 May	10-12 Jun	–	–	–	–
Prince George	4 Jun	7-10 Jun	–	–	6-13 Aug	13 Aug
Fort St. John	26 May	29 May-3 Jun	–	–	15-20 Aug	30 Aug
Fort Nelson	25 May	6-12 Jun	–	–	–	–
Willow Flycatcher						
Victoria	10 May	23-31 May	–	–	mid-Aug	6 Sep
Vancouver	12 May	27 May	–	–	5 Sep	8 Sep
Harrison	6 May	4-10 Jun	–	–	–	28 Aug
Sechelt	11 May	–	–	–	–	20 Sep
Okanagan	24 May	early Jun	–	–	late Aug	7 Sep
Nakusp	27 May	–	–	–	–	4 Aug
Radium	31 May	–	–	–	–	20 Aug
Kamloops	12 May	–	–	–	20-30 Aug	7 Sep
Williams Lake	22 May	2-4 Jun	–	–	mid-Aug	–
Quesnel	21 May	3 Jun	–	–	–	27 Aug
Least Flycatcher						
Kamloops	17 May	late May	–	–	–	–
Williams Lake	30 Apr	18-22 May	3 Jun	–	–	29 Aug
Quesnel	8 May	–	–	–	–	2 Aug
Prince George	12 May	–	–	–	–	–
Fort St. John	11 May	14-18 May	–	–	15-21 Aug	2 Sep
Fort Nelson	19 May	25-30 May	–	–	–	28 Jul
Hammond's Flycatcher						
Victoria	9 Apr	mid-May	–	–	late Aug	28 Sep
Vancouver	13 Apr	8-12 May	–	–	early Sep	8 Oct
Sechelt	11 Apr	–	–	–	–	21 Sep
Okanagan	15 Apr	early May	–	–	1-7 Sep	28 Sep

►

	SPRING			AUTUMN		
	Early Arrival	Peak Movement	Late Departure	Early Arrival	Peak Movement	Late Departure
◀ *Hammond's Flycatcher*						
Nakusp	1 May	–	–	–	–	20 Sep
Kamloops	26 Apr	–	–	–	–	–
Williams Lake	3 May	6-8 May	12 May	–	early Sep	5 Sep
Quesnel	7 May	–	–	–	–	3 Sep
Prince George	29 Apr	5-15 May	–	–	–	–
Fort Nelson	15 May	–	–	–	–	–
Dusky Flycatcher						
Okanagan	20 Apr	10 May	–	–	late Aug	21 Sep
Kamloops	1 May	3-10 May	–	–	–	18 Sep
Williams Lake	3 May	7-12 May	14 May	–	Aug	18 Aug
Prince George	9 May	13-20 May	–	–	–	–
Fort Nelson	15 May	–	–	–	–	–
Gray Flycatcher						
Okanagan	29 Apr	mid-late May	–	–	–	2 Sep
Western Flycatcher						
Victoria	23 Mar	18-30 May	–	–	1-15 Sep	27 Oct
Tofino	10 May	–	–	–	–	28 Aug
Vancouver	26 Mar	3 May	–	–	19 Sep	30 Sep
Sechelt	19 Apr	–	–	–	–	19 Sep
Masset	25 Apr	–	–	–	–	28 Aug
Okanagan	16 Apr	late May	–	–	–	13 Sep
Kamloops	23 Apr	5-15 May	–	–	–	–
Williams Lake	26 May	15 Jun	22 Jun	–	–	15 Aug
Eastern Phoebe						
Fort St. John	25 Apr	27 Apr-4 May	–	–	15-21 Aug	30 Aug
Fort Nelson	12 May	18-25 May	–	–	–	2 Aug
Say's Phoebe						
Vancouver	9 Mar	14 Apr	17 May	28 Aug	22 Sep	25 Oct
Okanagan	8 Feb	early Mar	–	–	mid-Sep	7 Nov
Kamloops	1 Mar	20 Mar-5 Apr	–	–	1-10 Sep	7 Oct
Williams Lake	29 Feb	20-31 Mar	–	–	19-25 Aug	22 Sep
Prince George	–	–	–	–	17-25 Aug	–
Fort St. John	28 Apr	9-15 May	20 May	15 Aug	15-25 Aug	3 Sep
Fort Nelson	28 Apr	14-19 May	27 May	2 Aug	–	22 Aug
Western Kingbird						
Victoria	15 Apr	–	–	–	–	11 Nov
Vancouver	27 Apr	22 May	–	–	–	3 Oct
Harrison	2 May	–	–	–	–	–
Sechelt	27 Apr	20 May	–	–	–	–
Okanagan	9 Apr	late Apr–early May	–	–	early Aug	9 Sep
Nakusp	4 May	–	–	–	–	10 Sep
Kamloops	9 Apr	5-15 May	–	–	15-25 Aug	4 Sep
Williams Lake	13 Apr	4-10 May	14 May	2 Aug	5-10 Aug	15 Sep
Quesnel	3 May	–	–	–	–	–
Eastern Kingbird						
Vancouver	14 Apr	1 Jun	–	–	4 Sep	10 Sep
Okanagan	2 May	late May	–	–	15-25 Aug	14 Sep
Nakusp	16 May	4-12 Jun	–	–	–	6 Sep
Radium	25 Apr	–	–	–	–	18 Sep
Kamloops	15 May	18-25 May	–	–	20-30 Aug	8 Sep
Williams Lake	16 May	27 May-1 Jun	–	–	15-20 Aug	3 Sep
Prince George	5 May	20-30 May	–	–	–	8 Sep
Fort St. John	9 May	19-26 May	–	–	18-24 Aug	2 Sep
Fort Nelson	30 May	–	–	10 Aug	15-20 Aug	22 Aug
Horned Lark						
Victoria	–	7 Apr-29 May	–	–	2 Sep-30 Oct	–
Vancouver	9 Mar	18 Apr	3 Jun	22 Aug	24 Sep	29 Oct
Harrison	–	–	–	31 Aug	–	5 Oct
Okanagan	6 Feb	6-17 Mar	12 May	1 Sep	Sep	late Oct
Nakusp	21 Mar	–	21 Apr	11 Sep	12-30 Sep	16 Oct

▶

	SPRING			AUTUMN		
	Early Arrival	Peak Movement	Late Departure	Early Arrival	Peak Movement	Late Departure
◀ *Horned Lark*						
Kamloops	20 Feb	25 Feb-15 Mar	7 May	1 Sep	3 Sep-15 Oct	15 Nov
Williams Lake	22 Feb	20 Apr-12 May	–	28 Aug	9-26 Sep	30 Oct
Quesnel	20 Mar	5 Apr	2 May	9 Sep	21 Sep	6 Oct
Prince George	25 Apr	30 Apr	–	9 Sep	9-17 Sep	–
Fort St. John	29 Mar	–	20 Apr	26 Aug	–	18 Sep
Fort Nelson	19 May	–	–	–	–	28 Oct
Purple Martin						
Victoria	4 Apr	–	–	–	–	17 Sep
Vancouver	5 Apr	–	–	–	–	7 Sep
Tree Swallow						
Victoria	11 Feb	late Mar	–	–	late Jul	24 Sep
Vancouver	16 Feb	26 Mar	–	–	23 Oct	12 Nov
Harrison	6 Mar	11-25 Apr	–	–	–	5 Oct
Sechelt	20 Mar	–	–	–	–	11 Aug
Masset	30 Mar	–	–	–	–	15 Aug
Okanagan	27 Feb	late Mar–early Apr	–	–	mid-Aug	27 Sep
Nakusp	27 Mar	–	–	–	–	6 Sep
Kamloops	8 Mar	5-25 Apr	10 May	–	20 Aug-3 Sep	20 Sep
Williams Lake	10 Mar	22 Mar-18 Apr	23 Apr	–	early Aug	25 Aug
Quesnel	22 Mar	6 Apr	30 Apr	–	–	18 Aug
Prince George	23 Mar	–	–	–	25-31 Aug	–
Fort St. John	16 Apr	23-26 Apr	–	–	–	17 Aug
Fort Nelson	27 Apr	12-19 May	–	–	–	–
Violet-green Swallow						
Victoria	5 Feb	late Mar–mid-Apr	–	–	–	3 Nov
Vancouver	21 Feb	late Mar	–	–	early Sep	31 Oct
Harrison	6 Mar	mid-Apr	–	–	20 Aug-6 Sep	16 Sep
Sechelt	6 Mar	12 Apr	–	–	late Jul	23 Aug
Okanagan	19 Feb	mid-Mar–mid-Apr	–	–	mid-Sep	7 Nov
Nakusp	20 Mar	–	–	–	–	7 Sep
Kamloops	11 Mar	Apr	–	–	–	13 Sep
Williams Lake	14 Mar	9-20 Apr	–	–	early Aug	9 Nov
Quesnel	17 Mar	–	–	–	–	21 Sep
Prince George	22 Mar	6-20 Apr	–	–	early Aug	–
Fort St. John	20 Apr	21-27 Apr	–	–	–	–
Northern Rough-winged Swallow						
Victoria	10 Mar	12 Apr	–	–	1 Aug	19 Nov
Tofino	9 Apr	–	–	–	–	–
Vancouver	4 Apr	–	–	–	–	11 Nov
Harrison	25 Mar	15 Apr-10 May	–	–	–	2 Sep
Sechelt	12 Apr	late Apr	–	–	mid-Aug	8 Sep
Okanagan	3 Apr	late Apr	–	–	late Aug	5 Nov
Nakusp	9 Apr	–	–	–	–	7 Nov
Kamloops	11 Apr	15-30 Apr	1 May	–	20 Aug-5 Sep	15 Sep
Williams Lake	2 Apr	16-24 Apr	–	–	late Aug	18 Sep
Quesnel	26 Mar	15 Apr	–	–	–	28 Aug
Prince George	7 Apr	20 Apr-3 May	–	–	1-15 Aug	–
Bank Swallow						
Vancouver	9 May	21 May	6 Jun	–	4 Sep	29 Sep
Harrison	–	–	–	–	–	23 Aug
Okanagan	11 Apr	early May	–	–	mid-Aug	27 Sep
Kamloops	26 Apr	–	–	–	1-10 Sep	13 Sep
Williams Lake	13 Apr	27 Apr-19 May	–	–	15 Aug	2 Sep
Quesnel	15 Apr	–	23 Apr	–	–	5 Sep
Prince George	20 Apr	20-31 May	–	–	13 Aug	19 Aug
Fort St. John	4 May	17-23 May	–	–	–	16 Sep
Fort Nelson	–	2 Jul	–	–	–	–
Cliff Swallow						
Victoria	1 Apr	1 May	–	–	late Jul	9 Nov
Vancouver	22 Mar	1 May	–	–	27 Sep	1 Dec
Harrison	12 Apr	12 Apr-10 May	–	–	24 Jul-6 Aug	7 Sep
Sechelt	26 Apr	–	–	2 Jul	–	30 Aug

▶

	SPRING			AUTUMN		
	Early Arrival	Peak Movement	Late Departure	Early Arrival	Peak Movement	Late Departure
◀ *Cliff Swallow*						
Masset	29 Apr	–	14 May	–	–	–
Okanagan	10 Apr	late Apr–early May	–	–	late Aug	1 Nov
Nakusp	8 Apr	–	–	–	–	8 Nov
Kamloops	8 Apr	25 Apr-15 May	–	–	20 Aug-10 Sep	21 Sep
Williams Lake	14 Apr	20-25 Apr	27 Apr	–	15 Aug	25 Aug
Quesnel	23 Apr	10 May	–	–	–	14 Oct
Prince George	20 Apr	25 Apr-1 May	–	–	23 Aug	1 Sep
Fort St. John	24 Apr	9-15 May	–	–	–	16 Sep
Fort Nelson	5 May	20 May-2 Jun	–	–	–	–
Barn Swallow						
Victoria	18 Mar	15 Apr	–	–	15 Sep	5 Nov
Vancouver	31 Mar	5 May	–	–	15 Sep	16 Dec
Harrison	13 Apr	2-11 May	–	–	25 Aug-16 Sep	20 Oct
Sechelt	15 Apr	early May	–	–	late Aug	3 Oct
Masset	21 Apr	–	–	–	–	16 Oct
Okanagan	11 Apr	late Apr	–	–	early Sep	23 Oct
Nakusp	15 Apr	–	–	–	–	27 Sep
Kamloops	22 Apr	–	–	–	20 Aug-20 Sep	3 Oct
Williams Lake	23 Apr	29 Apr-19 May	28 May	–	early Sep	7 Oct
Quesnel	2 Apr	12 May	19 May	–	5 Sep	29 Sep
Prince George	19 Apr	4-15 May	–	–	25-31 Aug	1 Sep
Fort St. John	4 May	4-8 May	–	–	1-10 Sep	16 Sep
Fort Nelson	16 May	19-25 May	–	–	–	–
American Crow						
Okanagan	15 Feb	early Mar	–	–	early Oct	–
Kamloops	25 Feb	15-30 Mar	–	–	5-25 Sep	1 Nov
Fort St. John	16 Mar	19-24 Mar	–	–	15 Sep-5 Oct	10 Oct
Fort Nelson	19 Apr	–	–	–	–	–
Red-breasted Nuthatch						
Kamloops	–	20 Mar-10 Apr	–	–	–	–
Fort St. John	3 Mar	1 Apr-8 May	–	–	Sep	15 Oct
Brown Creeper						
Kamloops	–	–	–	–	15-30 Oct	–
Williams Lake	–	–	–	1 Sep	Oct-Nov	–
Prince George	–	11 May	–	–	10 Oct	–
Rock Wren						
Okanagan	29 Mar	late Apr	–	–	Sep	–
Kamloops	15 Mar	1-15 May	–	–	15-25 Aug	30 Aug
Williams Lake	–	30 Apr	–	–	–	–
House Wren						
Victoria	7 Apr	1 May	–	–	15 Aug	25 Oct
Okanagan	18 Apr	7-14 May	–	–	early Sep	6 Oct
Kamloops	6 May	–	–	–	–	–
Williams Lake	19 May	–	–	–	–	–
Quesnel	3 May	–	–	–	–	7 Sep
Fort St. John	17 May	22-26 May	–	–	–	12 Sep
Fort Nelson	19 May	1 Jun	–	–	–	1 Sep
Winter Wren						
Kamloops	24 Mar	–	–	–	1-10 Sep	4 Sep
Quesnel	9 Apr	28 Apr	15 May	–	–	14 Oct
Prince George	10 May	15 May	–	–	–	–
Marsh Wren						
Sechelt	22 Mar	–	–	–	–	–
Kamloops	11 Apr	20-30 Apr	–	–	15-30 Sep	28 Nov
Williams Lake	18 Mar	18 Apr-1 May	–	–	early Oct	16 Oct
Quesnel	–	21 May	–	–	–	27 Sep
Fort St. John	2 May	4-16 May	–	–	15 Aug-1 Sep	18 Sep
Fort Nelson	15 Jun	–	–	–	–	–

	SPRING			AUTUMN		
	Early Arrival	Peak Movement	Late Departure	Early Arrival	Peak Movement	Late Departure
American Dipper						
Sechelt	–	–	24 Apr	1 Oct	–	–
Williams Lake	1 Mar	Mar	4 May	2 Nov	8-20 Nov	28 Nov
Quesnel	14 Mar	–	3 May	–	–	4 Nov
Golden-crowned Kinglet						
Kamloops	–	20 Mar-10 Apr	–	–	–	–
Williams Lake	18 Mar	4-20 Apr	–	–	25 Sep-14 Oct	–
Prince George	5 Apr	14 Apr	–	–	1-10 Oct	–
Fort St. John	1 Apr	16-26 Apr	–	–	Sep	18 Oct
Fort Nelson	13 May	–	–	–	–	–
Ruby-crowned Kinglet						
Victoria	–	1 Apr	30 May	24 Aug	Oct	–
Vancouver	–	16 Apr	26 May	7 Sep	22 Sep	–
Harrison	22 Mar	10-29 Apr	11 May	15 Sep	14 Oct	–
Sechelt	–	20 Apr	8 May	3 Sep	1 Oct	–
Okanagan	21 Mar	mid-Apr	–	–	Sep	–
Nakusp	30 Mar	13 Apr-5 May	23 Jun	–	–	18 Oct
Kamloops	27 Mar	15-30 Apr	–	–	8-30 Sep	11 Oct
Williams Lake	1 Apr	9-14 Apr	22 Apr	–	15-27 Sep	5 Nov
Quesnel	28 Mar	16 Apr	–	–	7 Sep	27 Sep
Prince George	2 Apr	13-29 Apr	–	–	20-25 Sep	28 Oct
Fort St. John	11 Apr	16-23 Apr	–	–	5-20 Sep	15 Oct
Fort Nelson	22 Apr	29 Apr-10 May	–	–	–	–
Mountain Bluebird						
Vancouver	6 Mar	13 Apr	1 May	–	–	–
Harrison	7 Mar	27 Mar	18 Apr	–	–	–
Sechelt	20 Mar	–	17 Apr	18 Sep	–	16 Nov
Okanagan	9 Feb	Mar	–	–	Sep	3 Nov
Nakusp	27 Feb	1-20 Apr	–	–	–	–
Kamloops	27 Feb	20 Mar-10 Apr	–	1 Sep	3 Sep-10 Oct	31 Oct
Williams Lake	1 Mar	6-30 Mar	–	5 Aug	15 Aug-15 Sep	5 Nov
Quesnel	31 Mar	21 Apr	9 May	21 Aug	5 Sep	21 Sep
Prince George	18 Mar	30 Mar-10 Apr	–	29 Jul	6 Oct	15 Oct
Fort St. John	11 Apr	–	16 May	–	–	1 Oct
Fort Nelson	12 Apr	22 Apr-12 May	–	–	–	–
Townsend's Solitaire						
Vancouver	10 Apr	21 Apr	2 Jun	18 Aug	13 Sep	–
Harrison	1 Apr	9 Apr-11 May	23 May	–	–	–
Sechelt	17 Mar	18 Apr	11 May	27 Aug	–	18 Sep
Nakusp	17 Mar	20 Apr-12 May	5 Jun	–	–	13 Nov
Kamloops	24 Mar	5-30 Apr	–	–	20-30 Sep	–
Quesnel	2 May	12 May	–	3 Sep	18 Sep	2 Oct
Prince George	1 Apr	12 Apr-10 May	24 May	15 Sep	–	31 Oct
Fort St. John	10 Apr	–	11 May	6 Oct	–	26 Oct
Fort Nelson	28 Apr	–	12 May	8 Sep	–	–
Veery						
Okanagan	8 May	late May	–	–	late Aug	9 Sep
Nakusp	26 May	–	–	–	–	–
Kamloops	24 May	–	–	–	–	20 Aug
Williams Lake	21 May	21 May-4 Jun	–	–	–	13 Aug
Quesnel	14 May	–	–	–	–	30 Aug
Fort St. John	5 Jun	–	–	–	–	–
Swainson's Thrush						
Victoria	23 Apr	15 May	–	–	31 Aug	16 Oct
Vancouver	23 Mar	21 May	–	–	19 Sep	27 Sep
Sechelt	14 May	20 May	–	–	–	27 Sep
Masset	23 May	–	–	–	–	1 Sep
Okanagan	6 May	late May	–	–	late Aug	22 Sep
Nakusp	26 May	–	–	–	–	28 Sep
Kamloops	7 May	–	–	–	25 Aug-1 Sep	6 Sep
Williams Lake	6 May	23-29 May	–	–	27 Aug-9 Sep	21 Sep
Quesnel	–	–	–	–	5 Sep	15 Sep
Prince George	19 May	31 May-4 Jun	–	–	4-10 Sep	17 Sep
Fort St. John	9 May	11-21 May	–	–	24-31 Aug	13 Sep
Fort Nelson	14 May	14-20 May	–	–	16-19 Aug	8 Sep

	SPRING			AUTUMN		
	Early Arrival	Peak Movement	Late Departure	Early Arrival	Peak Movement	Late Departure
Hermit Thrush						
Victoria	–	21 Apr-7 May	17 May	1 Sep	–	15 Oct
Harrison	13 Apr	–	–	6 Sep	–	–
Sechelt	28 Mar	25 Apr	–	–	–	7 Nov
Masset	2 Apr	late May	–	–	–	15 Sep
Okanagan	2 Apr	early May	–	–	late Sep	1 Oct
Nakusp	4 May	–	–	–	–	–
Kamloops	5 May	–	–	–	–	6 Sep
Williams Lake	3 May	5-12 May	–	–	1-10 Sep	–
Quesnel	–	–	–	–	–	6 Sep
Prince George	30 Apr	5-9 May	–	–	–	2 Oct
Fort St. John	30 Apr	3-10 May	–	–	25 Aug-1 Sep	29 Sep
Fort Nelson	5 May	–	–	–	–	–
American Robin						
Victoria	–	Mar	–	–	–	–
Tofino	–	mid-Mar	–	–	mid-Sep	–
Vancouver	–	Mar	–	–	–	–
Harrison	–	late Mar	–	–	–	–
Sechelt	–	10 Mar	–	–	28 Sep	30 Nov
Masset	–	early Apr	–	–	–	–
Okanagan	–	late Mar	–	–	late Sep	–
Nakusp	21 Feb	17 Mar-5 Apr	–	–	9 Sep-9 Oct	21 Nov
Radium	3 Mar	–	–	–	–	19 Nov
Kamloops	16 Feb	10 Mar-10 Apr	–	–	15 Sep-10 Oct	30 Nov
Williams Lake	21 Feb	14-24 Mar	–	–	10 Sep-15 Oct	6 Nov
Quesnel	8 Mar	2 Apr	–	–	19 Sep	4 Nov
Prince George	12 Feb	21 Mar-4 Apr	–	–	14 Aug-25 Sep	–
Fort St. John	15 Feb	5-14 Apr	–	–	5-20 Sep	13 Nov
Fort Nelson	27 Feb	19-23 Apr	–	30 Aug	5-30 Sep	2 Nov
Varied Thrush						
Harrison	–	4-26 Apr	17 May	24 Sep	29 Oct	–
Sechelt	–	–	31 May	15 Sep	–	–
Okanagan	21 Feb	Mar	–	19 Sep	–	–
Nakusp	19 Feb	–	–	–	–	25 Nov
Kamloops	22 Feb	20 Mar-5 Apr	–	–	–	10 Nov
Williams Lake	24 Feb	15 Mar-13 Apr	20 May	–	30 Sep-16 Oct	20 Nov
Quesnel	14 Mar	10 Apr	7 May	–	6 Oct	2 Nov
Prince George	28 Feb	19-28 Mar	–	–	30 Sep	15 Oct
Fort St. John	Apr	21-28 Apr	7 May	6 Sep	–	2 Oct
Fort Nelson	2 Apr	15 Apr-5 May	–	–	–	–
Gray Catbird						
Vancouver	4 May	6 Jun	–	–	31 Aug	10 Oct
Harrison	31 May	–	–	–	–	17 Aug
Okanagan	10 Apr	late May	–	–	14-21 Aug	5 Oct
Nakusp	26 May	–	–	–	–	20 Sep
Kamloops	28 May	10-20 Jun	–	–	–	20 Aug
Williams Lake	18 May	22 May-3 Jun	–	–	early Sep	15 Oct
Quesnel	13 Jun	–	–	–	–	6 Sep
American Pipit						
Victoria	–	30 Apr	30 May	25 Aug	15 Oct	–
Vancouver	–	21 Apr	10 Jun	2 Aug	3 Sep	–
Harrison	17 Apr	1-16 May	24 May	1 Sep	Sep-19 Oct	23 Nov
Sechelt	9 Apr	24 Apr	21 May	7 Sep	16 Sep	8 Nov
Masset	21 Apr	–	5 May	3 Sep	–	30 Oct
Okanagan	10 Apr	15-30 May	5 Jun	26 Aug	late Sep–early Oct	1 Nov
Nakusp	17 Apr	25 Apr-15 May	28 May	3 Sep	6 Sep-6 Oct	31 Oct
Kamloops	8 Apr	20 Apr-10 May	25 May	30 Aug	12 Sep-10 Oct	26 Nov
Williams Lake	14 Apr	28 Apr-16 May	19 May	22 Aug	6-25 Sep	–
Quesnel	26 Apr	14 May	2 Jun	30 Sep	12 Oct	26 Oct
Prince George	20 Apr	28 Apr-16 May	22 May	27 Oct	–	–
Fort St. John	14 Apr	25 Apr-10 May	10 May	–	1-15 Sep	30 Oct
Fort Nelson	25 Apr	30 Apr-19 May	25 May	28 Aug	5 Sep	20 Sep
Bohemian Waxwing						
Victoria	–	–	21 Apr	7 Aug	–	–
Vancouver	–	late Feb	15 Apr	15 Aug	mid-Dec	–

►

	SPRING			AUTUMN		
	Early Arrival	Peak Movement	Late Departure	Early Arrival	Peak Movement	Late Departure
◀ *Bohemian Waxwing*						
Masset	–	–	–	7 Nov	–	13 Jan
Okanagan	–	late Feb–early Mar	19 Apr	7 Oct	Dec	–
Nakusp	–	–	17 Apr	21 Oct	–	–
Kamloops	–	26 Feb-15 Mar	6 Apr	9 Oct	1-10 Nov	–
Williams Lake	–	17-25 Mar	2 Apr	1 Oct	16 Oct-5 Nov	–
Prince George	–	–	14 May	8 Sep	–	–
Fort St. John	–	late Mar–early Apr	20 May	23 Aug	late Oct	–
Fort Nelson	4 Apr	10-25 Apr	–	–	–	–
Cedar Waxwing						
Victoria	–	–	–	–	late Aug	–
Tofino	3 Apr	late May	–	–	early Sep	16 Dec
Vancouver	–	late May	–	–	–	–
Harrison	10 Apr	late May	3 Jun	–	–	7 Nov
Sechelt	20 May	25 May	–	–	–	1 Oct
Okanagan	–	late May	–	–	late Oct	–
Nakusp	27 Apr	early Jun	–	–	late Sep	11 Nov
Kamloops	24 May	1-10 Jun	–	–	5-25 Sep	8 Nov
Williams Lake	18 May	23 May-4 Jun	12 Jun	–	late Sep	11 Nov
Prince George	20 May	mid-Jun	–	–	5-25 Sep	1 Oct
Fort St. John	2 Jun	2-8 Jun	–	–	late Aug	18 Sep
Fort Nelson	–	–	–	–	–	28 Sep
Northern Shrike						
Victoria	–	early Apr	23 May	23 Sep	1-15 Oct	–
Vancouver	–	18 Mar	20 May	23 Sep	3 Oct	–
Sechelt	16 Mar	–	23 Apr	23 Sep	–	–
Okanagan	–	–	4 May	20 Sep	mid-Oct	–
Nakusp	–	22 Mar-16 Apr	16 Apr	27 Oct	–	–
Kamloops	–	1-8 Apr	25 Apr	27 Sep	1-15 Oct	–
Williams Lake	–	6-22 Apr	24 Apr	23 Sep	16 Oct- 4 Nov	–
Prince George	–	–	24 Apr	30 Sep	15 Oct	–
Fort St. John	–	mid-Apr	29 Apr	17 Sep	Oct	–
Fort Nelson	12 Apr	–	24 Apr	29 Oct	–	5 Nov
European Starling						
Kamloops	–	15 Mar-15 Apr	–	–	15 Aug-15 Oct	–
Williams Lake	–	4-9 Mar	–	–	–	–
Prince George	–	27 Mar-7 Apr	–	–	1 Sep	–
Fort St. John	–	9-17 Mar	–	–	Oct	–
Fort Nelson	28 Mar	6 Apr	–	–	–	22 Nov
Solitary Vireo						
Victoria	23 Mar	mid-Apr	–	–	late Aug	27 Sep
Vancouver	25 Mar	5 May	–	–	12 Sep	25 Sep
Harrison	28 Mar	3 May	–	–	–	14 Sep
Sechelt	30 Mar	15 Apr	–	–	3 Sep	20 Sep
Okanagan	5 Apr	1-15 May	–	–	1-10 Sep	27 Sep
Nakusp	19 Apr	1 May-2 Jun	–	–	–	18 Sep
Kamloops	4 May	10-20 May	–	–	–	13 Sep
Williams Lake	2 May	3-13 May	–	–	late Aug–4 Sep	18 Sep
Quesnel	26 Apr	20 May	–	–	–	3 Sep
Prince George	22 Apr	10-15 May	–	–	–	5 Oct
Fort St. John	10 May	12-19 May	–	–	15 Aug-1 Sep	7 Sep
Fort Nelson	7 May	20-24 May	–	–	–	4 Sep
Warbling Vireo						
Victoria	14 Apr	mid-May	–	–	early Sep	10 Oct
Vancouver	21 Apr	12 May	–	–	13 Sep	5 Oct
Harrison	26 Apr	–	–	–	–	–
Sechelt	22 Apr	6 May	–	–	2 Sep	18 Sep
Okanagan	21 Apr	early May	–	–	15 Aug-10 Sep	19 Sep
Nakusp	3 May	–	–	–	–	5 Sep
Kamloops	7 May	10-25 May	–	–	–	13 Sep
Williams Lake	29 Apr	6-14 May	–	–	15-30 Aug	17 Sep
Quesnel	22 Apr	19 May	–	–	24 Aug	22 Sep
Prince George	9 May	12-20 May	–	–	–	25 Aug
Fort St. John	2 May	10-15 May	–	–	20-25 Aug	5 Sep
Fort Nelson	13 May	15-25 May	–	–	–	–

	SPRING			AUTUMN		
	Early Arrival	Peak Movement	Late Departure	Early Arrival	Peak Movement	Late Departure
Philadelphia Vireo						
Fort St. John	17 May	–	–	–	–	6 Sep
Fort Nelson	6 Jun	–	–	–	–	3 Sep
Red-eyed Vireo						
Victoria	17 May	–	–	–	–	14 Sep
Vancouver	1 May	31 May	–	–	31 Aug	22 Sep
Harrison	17 May	–	–	–	–	18 Sep
Sechelt	11 May	24 May	–	–	25 Aug	31 Aug
Okanagan	3 May	25 May-1 Jun	–	–	15 Aug-1 Sep	8 Sep
Nakusp	10 May	–	–	–	–	12 Oct
Kamloops	20 May	–	–	–	–	30 Aug
Williams Lake	5 May	late May	–	–	12-24 Aug	–
Quesnel	–	–	–	–	–	22 Aug
Prince George	3 May	27 May-3 Jun	–	–	–	14 Sep
Fort St. John	19 May	20-25 May	–	–	20-30 Aug	4 Sep
Fort Nelson	24 May	26 May-10 Jun	–	–	–	–

APPENDIX 2

Summary of Christmas Bird Counts in British Columbia: 1957 Through 1993

The following tables summarize, by species, the Christmas Bird Count data for flycatchers through vireos in British Columbia. Only official Christmas Bird Counts (Fig. 627), as published in *Audubon Field Notes, American Birds,* and *Field Notes* from 1957 through 1993, have been used.

Highest provincial, interior, and coastal totals are the highest sums for all official counts in any one year in the province. In the tables, "Count locality" refers to the most recent name under which that count has been published. The time span of count activity for a particular location is shown under "Count years"; discontinuities can be determined by comparing "Count years" with "Total counts." "Frequency of occurrence" indicates the percentage of the counts in a locality on which a species has been recorded.

Figure 627. Map of British Columbia showing locations of official Christmas Bird Counts, 1957-93.

The general abundance of the species over the count periods is shown by the range of values in the last 3 columns. The highest and lowest numbers recorded are given. An asterisk following a "High" number indicates that some caution is required in interpreting the data. Refer to the appropriate species account or to the note at the end of that species' entry in the table.

The median has been chosen as an indicator of the "usual" value when the species does occur. The middle value of the data set, it will lie closer to the low value when the species frequently occurs in small numbers, and closer to the high value when a species rarely occurs in small numbers. In a symmetrically distributed data set, the median is an unbiased and consistent estimate of the mean (Zar 1974). Counts of zero have not been used as the "Low" value, but they can be determined using the values under "Frequency of occurrence" and "Total counts."

For example, there is about a 30% chance of seeing a Horned Lark on the Ladner Christmas Bird Count, and there is a high probability that this is an accurate estimate of the chance of seeing the bird because there is a long history of counts on which to base this estimate. Similarly, if you see the species, you are likely to see about 20 birds, but more than 108 would be exceptional.

See Fig. 627 for the locations of official Christmas Bird Counts, and Table 10 for the number of counts in each locality.

Figure 628. Maurice Ellison counting birds at Trail during the West Kootenay Naturalists Christmas Bird Count in January 1990 (R. Wayne Campbell).

Table 10. Localities holding official Christmas Bird Counts in British Columbia: 1957 through 1993.

Count locality	Number of counts
Coastal	
Bamfield	8
Bella Bella	1
Campbell River	19
Chilliwack	20
Comox	32
Deep Bay	19
Dewdney	1
Duncan	24
Hecate Strait	1
Kitimat	16
Ladner	35
Masset	12
Nanaimo	28
North Saanich	1
Parksville–Qualicum Beach	3
Pender Harbour	3
Pender Islands	28
Pitt Meadows	22
Port Alberni	3
Port Clements	10
Prince Rupert	14
Rose Spit	8
Sayward	12
Skidegate	12
Sooke	11
Squamish	13
Sunshine Coast	14
Surrey	7
Terrace	29
Vancouver	37
Victoria	36
Whistler	3
White Rock	23
TOTAL	505

Count locality	Number of counts
Interior	
Burns Lake–Francois Lake	4
Cranbrook	10
Creston	3
Fauquier	10
Fort St. James	7
Fort St. John	4
Kamloops	10
Kelowna	9
Kimberley	2
Lake Windermere	6
MacKenzie	1
Nakusp	15
Nelson	1
North Pine	11
Oliver-Osoyoos	15
Penticton	18
Prince George	5
Princeton	2
Quesnel	4
Revelstoke	13
Salmon Arm	5
Shuswap Lake Park	22
Smithers	17
Vaseux Lake	20
Vernon	19
Wells Gray Park	6
Yoho National Park	5
TOTAL	244

Say's Phoebe

Highest provincial total: 2 (1976)
Highest interior total: 2 (1976)
Highest coastal total: Not recorded

Count locality	Count years	Total counts	Frequency of occurrence (%)	Range of total counts when species occurred: Low	Median	High
Interior						
Penticton	1974-93	18	6	2	–	2
Vernon	1975-93	19	5	1	–	1

Western Kingbird

Highest provincial total: 1 (1963)
Highest interior total: Not recorded
Highest coastal total: 1 (1963)

Count locality	Count years	Total counts	Frequency of occurrence (%)	Range of total counts when species occurred: Low	Median	High
Coastal						
Deep Bay	1975-93	19	5	1	–	1
Victoria	1958-93	36	3	1	–	1*

Sky Lark

Highest provincial total: 969 (1965)
Highest interior total: Not recorded
Highest coastal total: 969 (1965)

Count locality	Count years	Total counts	Frequency of occurrence (%)	Range of total counts when species occurred: Low	Median	High
Coastal						
Duncan	1970-93	24	21	2	5	11
North Saanich	1960	1	100	65	–	65
Victoria	1958-93	36	86	2	34	969

Horned Lark

Highest provincial total: 408 (1984)
Highest interior total: 408 (1984)
Highest coastal total: 109 (1964)

Count locality	Count years	Total counts	Frequency of occurrence (%)	Range of total counts when species occurred: Low	Median	High
Coastal						
Comox	1961-93	32	3	5	–	5
Kitimat	1974-93	16	6	1	–	1
Ladner	1957-93	35	31	1	22	108
Nanaimo	1963-93	28	4	3	–	3
North Saanich	1960	1	100	2	–	2
Skidegate	1982-93	12	8	2	–	2
Vancouver	1957-93	37	30	1	12	37
Victoria	1958-93	36	14	1	3	32
Interior						
Fauquier	1984-92	10	10	1	–	1
Kamloops	1984-93	10	70	11	40	157
Oliver-Osoyoos	1979-93	15	13	50	175	300
Vernon	1975-93	19	74	5	37	251

Tree Swallow

Highest provincial total: 1 (1970)
Highest interior total: Not recorded
Highest coastal total: 1 (1970)

Count locality	Count years	Total counts	Frequency of occurrence (%)	Low	Range of total counts when species occurred Median	High
Coastal						
Vancouver	1957-93	37	3	1	–	1

Violet-green Swallow

Highest provincial total: 1 (1982)
Highest interior total: Not recorded
Highest coastal total: 1 (1982)

Count locality	Count years	Total counts	Frequency of occurrence (%)	Low	Range of total counts when species occurred Median	High
Coastal						
Victoria	1958-93	36	3	1	–	1

Cliff Swallow

Highest provincial total: 2 (1991)
Highest interior total: Not recorded
Highest coastal total: 2 (1991)

Count locality	Count years	Total counts	Frequency of occurrence (%)	Low	Range of total counts when species occurred Median	High
Coastal						
Vancouver	1957-93	37	3	2	–	2

Barn Swallow

Highest provincial total: 4 (1969)
Highest interior total: Not recorded
Highest coastal total: 4 (1969)

Count locality	Count years	Total counts	Frequency of occurrence (%)	Low	Range of total counts when species occurred Median	High
Coastal						
Vancouver	1957-93	37	8	3	–	3
Ladner	1957-93	35	6	2	3	4

Gray Jay

Highest provincial total: 208 (1991)
Highest interior total: 197 (1991)
Highest coastal total: 37 (1984)

Count locality	Count years	Total counts	Frequency of occurrence (%)	Low	Range of total counts when species occurred Median	High
Coastal						
Campbell River	1972-93	19	5	1	–	1
Chilliwack	1972-93	20	5	1	–	1
Comox	1961-93	32	3	2	–	2
Duncan	1970-93	24	17	1	3	3
Kitimat	1974-93	16	19	1	1	3
Nanaimo	1963-93	28	4	1	–	1
Pender Harbour	1991-93	3	33	8	–	8
Pitt Meadows	1972-93	22	5	5	–	5
Port Alberni	1975-93	3	33	4	–	4

▶

◀ *Gray Jay*

Count locality	Count years	Total counts	Frequency of occurrence (%)	Low	Range of total counts when species occurred Median	High
Sayward	1973-84	12	17	2	2	2
Sooke	1983-93	11	9	2	–	2
Squamish	1980-93	13	100	1	18	35
Sunshine Coast	1979-93	14	14	1	1	1
Terrace	1963-92	29	28	1	2	5
Vancouver	1957-93	37	24	1	3	7
Whistler	1991-93	3	100	6	6	10
Interior						
Burns Lake–Francois Lake	1990-93	4	100	12	13	26
Cranbrook	1984-93	10	70	1	6	11
Creston	1987-89	3	33	8	–	8
Fauquier	1984-92	10	90	1	3	9
Fort St. James	1987-93	7	100	1	10	16
Fort St. John	1975-78	4	50	4	6	8
Kamloops	1984-93	10	20	3	3	3
Kelowna	1981-93	9	78	1	2	6
Kimberley	1992-93	2	100	2	7	11
Lake Windermere	1979-84	6	83	5	10	12
MacKenzie	1982	1	100	6	–	6
Nakusp	1979-93	15	93	2	3	11
Nelson	1985	1	100	7	–	7
North Pine	1983-93	11	100	2	17	34
Oliver-Osoyoos	1979-93	15	67	1	7	17
Penticton	1974-93	18	89	2	9	22
Prince George	1989-93	5	80	3	5	7
Princeton	1992-93	2	100	6	12	18
Revelstoke	1981-93	13	69	1	3	5
Salmon Arm	1988-92	5	80	2	3	5
Shuswap Lake Park	1972-93	22	91	1	5	18
Smithers	1977-93	17	100	8	21	42
Vaseux Lake	1974-93	20	60	2	6	15
Vernon	1975-93	19	74	2	5	11
Wells Gray Park	1984-92	6	100	3	6	7
Yoho National Park	1975-84	5	100	9	18	38

Steller's Jay

Highest provincial total: 3,251 (1992)
Highest interior total: 595 (1992)
Highest coastal total: 2,656 (1992)

Count locality	Count years	Total counts	Frequency of occurrence (%)	Low	Range of total counts when species occurred Median	High
Coastal						
Bamfield	1986-93	8	100	6	17	38
Bella Bella	1976	1	100	2	2	
Campbell River	1972-93	19	100	10	71	198
Chilliwack	1972-93	20	100	8	59	151
Comox	1961-93	32	88	1	47	233
Deep Bay	1975-93	19	89	1	13	46
Dewdney	1967	1	100	4	–	4
Duncan	1970-93	24	100	1	39	293
Kitimat	1974-93	16	100	1	12	60
Ladner	1957-93	35	60	1	5	31
Masset	1982-93	12	83	1	2	13
Nanaimo	1963-93	28	93	2	25	126
North Saanich	1960	1	100	98	–	98
Parksville–Qualicum Beach	1991-93	3	100	9	9	24
Pender Harbour	1991-93	3	100	21	34	38
Pender Islands	1964-93	28	7	1	2	2
Pitt Meadows	1972-93	22	100	43	63	117
Port Alberni	1975-93	3	100	53	70	94
Port Clements	1984-93	10	70	1	2	4
Prince Rupert	1980-93	14	79	3	11	40
Sayward	1973-84	12	100	4	19	63
Skidegate	1982-93	12	100	1	4	15

▶

◀ *Steller's Jay*

Count locality	Count years	Total counts	Frequency of occurrence (%)	Low	Range of total counts when species occurred Median	High
Sooke	1983-93	11	100	29	44	127
Squamish	1980-93	13	100	3	65	124
Sunshine Coast	1979-93	14	100	1	50	78
Surrey	1960-66	7	57	1	2	4
Terrace	1963-92	29	97	1	13	69
Vancouver	1957-93	37	100	7	84	317
Victoria	1958-93	36	97	1	26	659
Whistler	1991-93	3	100	19	46	93
White Rock	1971-93	23	100	14	70	236
Interior						
Burns Lake–Francois Lake	1990-93	4	25	7	–	7
Cranbrook	1984-93	10	100	1	6	15
Creston	1987-89	3	100	2	17	23
Fauquier	1984-92	10	100	8	19	39
Fort St. James	1987-93	7	43	4	8	15
Kamloops	1984-93	10	70	1	2	13
Kelowna	1981-93	9	100	10	33	67
Kimberley	1992-93	2	50	7	–	7
Lake Windermere	1979-84	6	83	4	8	16
MacKenzie	1982	1	100	8	–	8
Nakusp	1979-93	15	100	10	22	79
Nelson	1985	1	100	49	–	49
North Pine	1983-93	11	9	1	–	1
Oliver-Osoyoos	1979-93	15	100	2	21	56
Penticton	1974-93	18	100	5	61	127
Prince George	1989-93	5	60	3	9	33
Princeton	1992-93	2	100	18	19	20
Revelstoke	1981-93	13	100	6	41	119
Salmon Arm	1988-92	5	100	5	11	28
Shuswap Lake Park	1972-93	22	100	2	24	62
Smithers	1977-93	17	82	1	5	19
Vaseux Lake	1974-93	20	100	7	23	79
Vernon	1975-93	19	95	4	20	46
Wells Gray Park	1984-92	6	100	2	11	18
Yoho National Park	1975-84	5	100	4	8	10

Blue Jay

Highest provincial total: 55 (1990)
Highest interior total: 54 (1990)
Highest coastal total: 2 (1986)

Count locality	Count years	Total counts	Frequency of occurrence (%)	Low	Range of total counts when species occurred Median	High
Coastal						
Chilliwack	1972-93	20	5	2	–	2
Ladner	1957-93	35	3	2	–	2
Parksville–Qualicum Beach	1991-93	3	33	1	–	1
Sunshine Coast	1979-93	14	7	1	–	1
Vancouver	1957-93	37	3	2	–	2
Victoria	1958-93	36	3	1	–	1
Interior						
Cranbrook	1984-93	10	20	1	1	1
Creston	1987-89	3	33	9	–	9
Fauquier	1984-92	10	10	1	–	1
Fort St. John	1975-78	4	25	4	–	4
Kamloops	1984-93	10	10	1	–	1
Kimberley	1992-93	2	50	1	–	1
Nakusp	1979-93	15	13	1	2	3
North Pine	1983-93	11	100	4	15	41
Revelstoke	1981-93	13	8	1	–	1
Salmon Arm	1988-92	5	20	1	–	1
Vernon	1975-93	19	11	1	2	2

Clark's Nutcracker

Highest provincial total: 306 (1991)
Highest interior total: 306 (1991)
Highest coastal total: 1 (1972)

Count locality	Count years	Total counts	Frequency of occurrence (%)	Range of total counts when species occurred: Low	Median	High
Coastal						
Chilliwack	1972-93	20	5	1	–	1
Kitimat	1974-93	16	6	1	–	1
Interior						
Cranbrook	1984-93	10	100	4	11	32
Creston	1987-89	3	33	3	–	3
Fort St. James	1987-93	7	14	1	–	1
Kamloops	1984-93	10	90	1	7	19
Kelowna	1981-93	9	100	3	15	54
Kimberley	1992-93	2	100	17	34	51
Lake Windermere	1979-84	6	100	8	24	34
Nakusp	1979-93	15	27	2	4	15
Nelson	1985	1	100	3	–	3
Oliver-Osoyoos	1979-93	15	93	2	11	74
Penticton	1974-93	18	100	10	49	132
Princeton	1992-93	2	100	7	41	74
Revelstoke	1981-93	13	31	1	3	18
Shuswap Lake Park	1972-93	22	18	1	1	12
Smithers	1977-93	17	29	1	1	6
Vaseux Lake	1974-93	20	100	1	21	89
Vernon	1975-93	19	84	2	8	16
Yoho National Park	1975-84	5	100	10	26	30

Black-billed Magpie

Highest provincial total: 1,697 (1993)
Highest interior total: 1,697 (1993)
Highest coastal total: 2 (1960)

Count locality	Count years	Total counts	Frequency of occurrence (%)	Range of total counts when species occurred: Low	Median	High
Coastal						
Campbell River	1972-93	19	5	2	–	2
Comox	1961-93	32	3	1	–	1
Ladner	1957-93	35	3	1	–	1
Pitt Meadows	1972-93	22	5	1	–	1
Vancouver	1957-93	37	3	2	–	2
White Rock	1971-93	23	9	1	1	1
Interior						
Burns Lake–Francois Lake	1990-93	4	25	1	–	1
Cranbrook	1984-93	10	70	1	2	5
Creston	1987-89	3	100	3	3	12
Fort St. James	1987-93	7	14	2	–	2
Fort St. John	1975-78	4	100	23	37	46
Kamloops	1984-93	10	100	68	130	191
Kelowna	1981-93	9	100	113	223	469
Kimberley	1992-93	2	100	8	9	9
Lake Windermere	1979-84	6	100	16	31	43
Nakusp	1979-93	15	7	1	–	1
North Pine	1983-93	11	100	6	38	77
Oliver-Osoyoos	1979-93	15	100	65	169	244
Penticton	1974-93	18	100	171	307	448
Prince George	1989-93	5	100	12	16	33
Princeton	1992-93	2	100	8	11	13
Quesnel	1981-84	4	25	1	–	1
Revelstoke	1981-93	13	8	2	–	2
Salmon Arm	1988-92	5	100	9	35	41
Shuswap Lake Park	1972-93	22	95	3	7	52
Smithers	1977-93	17	12	1	2	2
Vaseux Lake	1974-93	20	100	38	128	205
Vernon	1975-93	19	100	127	292	393
Yoho National Park	1975-84	5	80	2	4	6

American Crow

Highest provincial total: 2,530 (1979)
Highest interior total: 2,036 (1990)
Highest coastal total: 1,877 (1979)

Count locality	Count years	Total counts	Frequency of occurrence (%)	Range of total counts when species occurred Low	Median	High
Coastal						
Chilliwack	1972-93	20	35	263	815	1,446*
Terrace	1963-92	29	100	8	150	470*
Victoria	1958-93	36	3	1	–	1*
Whistler	1991-93	3	33	3	–	3*
Interior						
Burns Lake Francois Lake	1990 93	4	100	61	86	123
Cranbrook	1984-93	10	70	1	8	18
Creston	1987-89	3	67	4	152	300
Fauquier	1984-92	10	60	4	15	50
Fort St. James	1987-93	7	100	34	71	150
Fort St. John	1975-78	4	25	2	–	2
Kamloops	1984-93	10	100	9	39	77
Kelowna	1981-93	9	100	108	175	333
Kimberley	1992-93	2	100	3	4	4
Lake Windermere	1979-84	6	50	1	5	12
MacKenzie	1982	1	100	3	–	3
Nakusp	1979-93	15	100	16	56	142
Nelson	1985	1	100	49	–	49
Oliver-Osoyoos	1979-93	15	87	1	12	65
Penticton	1974-93	18	100	5	37	119
Prince George	1989-93	5	100	83	155	185
Princeton	1992-93	2	100	9	42	74
Quesnel	1981-84	4	75	4	6	32
Revelstoke	1981-93	13	100	26	80	298
Salmon Arm	1988-92	5	100	197	257	523
Shuswap Lake Park	1972-93	22	100	2	53	312
Smithers	1977-93	17	100	27	228	597
Vaseux Lake	1974-93	20	65	1	5	17
Vernon	1975-93	19	100	76	241	723

Northwestern Crow

Highest provincial total: 41,279 (1993)
Highest interior total: Not recorded
Highest coastal total: 41,279 (1993)

Count locality	Count years	Total counts	Frequency of occurrence (%)	Range of total counts when species occurred Low	Median	High
Coastal						
Bamfield	1986-93	8	100	21	212	332
Bella Bella	1976	1	100	159	–	159
Campbell River	1972-93	19	100	234	998	3,374
Chilliwack	1972-93	20	85	25	1,401	3,380
Comox	1961-93	32	100	31	1,428	2,837
Deep Bay	1975-93	19	100	429	542	1,090
Dewdney	1967	1	100	191	–	191
Duncan	1970-93	24	100	146	1,315	3,323
Hecate Strait	1980	1	100	4	–	4
Kitimat	1974-93	16	100	65	130	410
Ladner	1957-93	35	97	82	1,180	3,672
Masset	1982-93	12	100	115	191	422
Nanaimo	1963-93	28	100	13	1,007	1,841
North Saanich	1960	1	100	556	–	556
Parksville–Qualicum Beach	1991-93	3	100	657	963	1,006
Pender Harbour	1991-93	3	100	273	294	355
Pender Islands	1964-93	28	96	40	414	703
Pitt Meadows	1972-93	22	100	517	2,412	7,903
Port Alberni	1975-93	3	100	586	683	1,335
Port Clements	1984-93	10	100	62	88	338
Prince Rupert	1980-93	14	100	74	298	689
Rose Spit	1986-93	8	88	6	12	61
Sayward	1973-84	12	100	68	163	300
Skidegate	1982-93	12	100	178	390	584

▶

◀ Northwestern Crow

Count locality	Count years	Total counts	Frequency of occurrence (%)	Low	Range of total counts when species occurred Median	High
Sooke	1983-93	11	100	400	555	914
Squamish	1980-93	13	100	387	612	830
Sunshine Coast	1979-93	14	100	293	703	1,008
Surrey	1960-66	7	100	3	60	322
Terrace	1963-92	29	69	11	142	664
Vancouver	1957-93	37	100	480	3,477	13,167
Victoria	1958-93	36	100	668	4,460	7,131
White Rock	1971-93	23	100	1,281	2,851	7,380

Common Raven

Highest provincial total: 7,129 (1992)
Highest interior total: 4,856 (1992)
Highest coastal total: 2,447 (1991)

Count locality	Count years	Total counts	Frequency of occurrence (%)	Low	Range of total counts when species occurred Median	High
Coastal						
Bamfield	1986-93	8	100	1	3	5
Bella Bella	1976	1	100	19	–	19
Campbell River	1972-93	19	100	13	77	432
Chilliwack	1972-93	20	90	2	17	226
Comox	1961-93	32	100	7	72	237
Deep Bay	1975-93	19	100	8	14	32
Dewdney	1967	1	100	1	–	1
Duncan	1970-93	24	100	22	38	224
Hecate Strait	1980	1	100	10	–	10
Kitimat	1974-93	16	100	17	92	181
Ladner	1957-93	35	80	1	6	43
Masset	1982-93	12	100	23	116	249
Nanaimo	1963-93	28	100	3	43	108
North Saanich	1960	1	100	11	–	11
Parksville–Qualicum Beach	1991-93	3	100	38	50	55
Pender Harbour	1991-93	3	100	21	44	103
Pender Islands	1964-93	28	100	4	46	105
Pitt Meadows	1972-93	22	100	3	14	59
Port Alberni	1975-93	3	100	29	51	136
Port Clements	1984-93	10	100	6	82	114
Prince Rupert	1980-93	14	100	31	99	267
Rose Spit	1986-93	8	100	3	7	27
Sayward	1973-84	12	100	27	78	135
Skidegate	1982-93	12	100	13	64	221
Sooke	1983-93	11	100	35	56	108
Squamish	1980-93	13	100	23	81	130
Sunshine Coast	1979-93	14	100	4	29	51
Terrace	1963-92	29	100	7	127	323
Vancouver	1957-93	37	100	2	25	119
Victoria	1958-93	36	100	21	110	382
Whistler	1991-93	3	100	74	96	115
White Rock	1971-93	23	100	1	3	14
Interior						
Burns Lake–Francois Lake	1990-93	4	100	19	88	182
Cranbrook	1984-93	10	100	123	172	277
Creston	1987-89	3	100	6	70	114
Fauquier	1984-92	10	100	37	86	130
Fort St. James	1987-93	7	100	62	87	165
Fort St. John	1975-78	4	100	5	32	32
Kamloops	1984-93	10	100	136	214	464
Kelowna	1981-93	9	100	56	594	916
Kimberley	1992-93	2	100	54	194	333
Lake Windermere	1979-84	6	100	65	209	249
MacKenzie	1982	1	100	72	–	72
Nakusp	1979-93	15	100	49	150	308
Nelson	1985	1	100	40	–	40
North Pine	1983-93	11	100	3	31	79

▶

◀ *Common Raven*

Count locality	Count years	Total counts	Frequency of occurrence (%)	Range of total counts when species occurred Low	Median	High
Oliver-Osoyoos	1979-93	15	100	13	85	293
Penticton	1974-93	18	100	38	200	389
Prince George	1989-93	5	100	117	298	334
Princeton	1992-93	2	100	488	611	734
Quesnel	1981-84	4	100	5	43	79
Revelstoke	1981-93	13	100	151	322	573
Salmon Arm	1988-92	5	100	42	83	154
Shuswap Lake Park	1972-93	22	100	10	55	179
Smithers	1977-93	17	100	60	241	481
Vaseux Lake	1974-93	20	100	1	60	196
Vernon	1975-93	19	100	55	239	516
Wells Gray Park	1984-92	6	100	6	23	56
Yoho National Park	1975-84	5	100	8	21	43

Black-capped Chickadee

Highest provincial total: 9,477 (1993)
Highest interior total: 5,361 (1992)
Highest coastal total: 5,704 (1993)

Count locality	Count years	Total counts	Frequency of occurrence (%)	Range of total counts when species occurred Low	Median	High
Coastal						
Chilliwack	1972-93	20	95	54	139	311
Dewdney	1967	1	100	14	–	14
Kitimat	1974-93	16	31	1	2	10
Ladner	1957-93	35	100	9	285	594
Nanaimo	1963-93	28	4	1	–	1*
Pender Islands	1964-93	28	4	3	–	3*
Pitt Meadows	1972-93	22	100	150	359	734
Port Alberni	1975-93	3	33	39	–	39*
Squamish	1980-93	13	100	76	155	238
Sunshine Coast	1979-93	14	21	2	5	91
Surrey	1960-66	7	100	7	41	51
Terrace	1963-92	29	90	3	35	111
Vancouver	1957-93	37	100	164	1,031	2,613
Whistler	1991-93	3	100	3	58	70
White Rock	1971-93	23	100	321	729	1,402
Interior						
Burns Lake–Francois Lake	1990-93	4	100	130	197	1,080
Cranbrook	1984-93	10	100	34	50	108
Creston	1987-89	3	100	28	222	257
Fauquier	1984-92	10	100	9	49	92
Fort St. James	1987-93	7	100	75	152	176
Fort St. John	1975-78	4	100	12	22	42
Kamloops	1984-93	10	100	16	80	114
Kelowna	1981-93	9	100	91	161	256
Kimberley	1992-93	2	100	129	141	153
Lake Windermere	1979-84	6	100	43	163	268
MacKenzie	1982	1	100	31	–	31
Nakusp	1979-93	15	100	18	67	99
Nelson	1985	1	100	126	–	126
North Pine	1983-93	11	100	15	70	224
Oliver-Osoyoos	1979-93	15	100	29	118	397
Penticton	1974-93	18	100	80	124	382
Prince George	1989-93	5	100	453	570	957
Princeton	1992-93	2	100	47	56	65
Quesnel	1981-84	4	100	13	15	53
Revelstoke	1981-93	13	100	16	78	168
Salmon Arm	1988-92	5	100	206	250	315
Shuswap Lake Park	1972-93	22	100	24	210	477
Smithers	1977-93	17	100	201	419	1,165
Vaseux Lake	1974-93	20	100	35	95	308
Vernon	1975-93	19	100	144	351	604
Wells Gray Park	1984-92	6	100	27	66	93
Yoho National Park	1975-84	5	100	14	25	32

Mountain Chickadee

Highest provincial total: 1,499 (1989)
Highest interior total: 1,491 (1989)
Highest coastal total: 14 (1987)

Count locality	Count years	Total counts	Frequency of occurrence (%)	Range of total counts when species occurred: Low	Median	High
Coastal						
Chilliwack	1972-93	20	25	1	2	6
Ladner	1957-93	35	3	2	–	2
Pitt Meadows	1972-93	22	9	3	4	4
Squamish	1980-93	13	23	6	6	7
Sunshine Coast	1979-93	14	14	1	2	2
Terrace	1963-92	29	3	3	–	3
Vancouver	1957-93	37	19	1	1	4
White Rock	1971-93	23	22	1	2	8
Interior						
Burns Lake–Francois Lake	1990-93	4	100	2	11	14
Cranbrook	1984-93	10	100	16	40	57
Creston	1987-89	3	33	132	–	132
Fauquier	1984-92	10	30	1	8	12
Fort St. James	1987-93	7	100	5	7	13
Fort St. John	1975-78	4	25	4	–	4
Kamloops	1984-93	10	100	32	72	95
Kelowna	1981-93	9	100	40	66	186
Kimberley	1992-93	2	100	50	58	65
Lake Windermere	1979-84	6	100	46	122	334
MacKenzie	1982	1	100	9	–	9
Nakusp	1979-93	15	27	1	2	4
Nelson	1985	1	100	2	–	2
Oliver-Osoyoos	1979-93	15	100	42	159	280
Penticton	1974-93	18	100	95	252	519
Prince George	1989-93	5	80	1	3	5
Princeton	1992-93	2	100	62	74	85
Quesnel	1981-84	4	50	2	2	2
Revelstoke	1981-93	13	69	1	3	16
Salmon Arm	1988-92	5	100	25	31	31
Shuswap Lake Park	1972-93	22	100	4	34	83
Smithers	1977-93	17	76	1	10	24
Vaseux Lake	1974-93	20	100	17	150	361
Vernon	1975-93	19	100	20	56	110
Wells Gray Park	1984-92	6	83	1	2	6
Yoho National Park	1975-84	5	100	2	10	13

Boreal Chickadee

Highest provincial total: 54 (1988)
Highest interior total: 41 (1988)
Highest coastal total: 13 (1988)

Count locality	Count years	Total counts	Frequency of occurrence (%)	Range of total counts when species occurred: Low	Median	High
Coastal						
Squamish	1980-93	13	8	2	–	2
Terrace	1963-92	29	17	1	1	13
Vancouver	1957-93	37	3	1	–	1
Interior						
Burns Lake–Francois Lake	1990-93	4	50	2	6	10
Fauquier	1984-92	10	10	20	–	20
Fort St. James	1987-93	7	100	2	5	6
Lake Windermere	1979-84	6	17	2	–	2
North Pine	1983-93	11	55	1	8	14
Oliver-Osoyoos	1979-93	15	27	1	1	3
Penticton	1974-93	18	33	1	3	8
Prince George	1989-93	5	40	1	1	1
Revelstoke	1981-93	13	15	5	7	9
Salmon Arm	1988-92	5	20	1	–	1
Smithers	1977-93	17	65	1	5	13
Vaseux Lake	1974-93	20	25	1	2	2
Vernon	1975-93	19	5	3	–	3
Yoho National Park	1975-84	5	80	6	8	12

Chestnut-backed Chickadee

Highest provincial total: 6,009 (1993)
Highest interior total: 118 (1991)
Highest coastal total: 5,970 (1993)

Count locality	Count years	Total counts	Frequency of occurrence (%)	Range of total counts when species occurred Low	Median	High
Coastal						
Bamfield	1986-93	8	100	12	18	59
Campbell River	1972-93	19	100	75	207	339
Chilliwack	1972-93	20	55	1	8	65
Comox	1961-93	32	100	11	250	545
Deep Bay	1975-93	19	100	31	102	267
Duncan	1970-93	24	100	56	279	832
Kitimat	1974-93	16	81	4	7	38
Ladner	1957-93	35	100	4	19	265
Masset	1982-93	12	100	7	19	108
Nanaimo	1963-93	28	100	14	145	289
North Saanich	1960	1	100	116	–	116
Parksville–Qualicum Beach	1991-93	3	100	155	184	242
Pender Harbour	1991-93	3	100	50	81	102
Pender Islands	1964-93	28	100	24	207	464
Pitt Meadows	1972-93	22	100	2	20	78
Port Alberni	1975-93	3	100	72	85	90
Port Clements	1984-93	10	100	2	7	36
Prince Rupert	1980-93	14	50	2	6	26
Rose Spit	1986-93	8	88	1	4	29
Sayward	1973-84	12	92	3	22	66
Skidegate	1982-93	12	100	7	47	147
Sooke	1983-93	11	100	143	331	492
Squamish	1980-93	13	100	31	112	229
Sunshine Coast	1979-93	14	100	8	96	222
Surrey	1960-66	7	86	2	4	10
Terrace	1963-92	29	90	1	10	64
Vancouver	1957-93	37	100	20	155	412
Victoria	1958-93	36	100	160	550	2,099
Whistler	1991-93	3	100	14	72	103
White Rock	1971-93	23	100	33	109	268
Interior						
Burns Lake–Francois Lake	1990-93	4	50	3	3	3
Cranbrook	1984-93	10	10	2	–	2
Creston	1987-89	3	100	6	7	16
Fauquier	1984-92	10	90	2	16	70
Kelowna	1981-93	9	11	1	1	
Nakusp	1979-93	15	93	4	18	52
Nelson	1985	1	100	11	–	11
Oliver-Osoyoos	1979-93	15	7	1	–	1
Penticton	1974-93	18	6	7	–	7
Revelstoke	1981-93	13	100	1	10	24
Salmon Arm	1988-92	5	20	2	–	2
Shuswap Lake Park	1972-93	22	73	1	2	18
Smithers	1977-93	17	6	2	–	2
Vaseux Lake	1974-93	20	20	1	2	3
Vernon	1975-93	19	5	1	–	1
Wells Gray Park	1984-92	6	17	4	–	4

Bushtit

Highest provincial total: 4,677 (1993)
Highest interior total: Not recorded
Highest coastal total: 4,677 (1993)

Count locality	Count years	Total counts	Frequency of occurrence (%)	Range of total counts when species occurred Low	Median	High
Coastal						
Bamfield	1986-93	8	13	12	–	12
Campbell River	1972-93	19	26	1	20	42
Chilliwack	1972-93	20	85	1	20	330
Comox	1961-93	32	75	3	29	69
Deep Bay	1975-93	19	32	4	15	25
Duncan	1970-93	24	100	2	62	281
Ladner	1957-93	35	97	8	181	560

▶

◀ *Bushtit*

Count locality	Count years	Total counts	Frequency of occurrence (%)	Range of total counts when species occurred Low	Median	High
Nanaimo	1963-93	28	93	1	34	145
North Saanich	1960	1	100	9	–	9
Parksville–Qualicum Beach	1991-93	3	100	17	51	109
Pender Islands	1964-93	28	50	1	11	64
Pitt Meadows	1972-93	22	100	20	101	366
Port Alberni	1975-93	3	100	8	40	47
Sooke	1983-93	11	100	3	32	142
Squamish	1980-93	13	38	2	30	92
Surrey	1960-66	7	43	8	12	29
Vancouver	1957-93	37	86	1	211	1,392
Victoria	1958-93	36	100	29	253	1,578
White Rock	1971-93	23	100	59	299	612

Red-breasted Nuthatch

Highest provincial total: 1,759 (1993)
Highest interior total: 1,277 (1983)
Highest coastal total: 723 (1993)

Count locality	Count years	Total counts	Frequency of occurrence (%)	Range of total counts when species occurred Low	Median	High
Coastal						
Bamfield	1986-93	8	13	1	–	1
Campbell River	1972-93	19	68	1	3	9
Chilliwack	1972-93	20	60	1	2	8
Comox	1961-93	32	69	1	4	31
Deep Bay	1975-93	19	84	1	3	14
Duncan	1970-93	24	100	1	12	45
Kitimat	1974-93	16	6	1	–	1
Ladner	1957-93	35	83	1	7	78
Nanaimo	1963-93	28	79	1	6	24
North Saanich	1960	1	100	9	–	9
Parksville–Qualicum Beach	1991-93	3	67	4	10	16
Pender Harbour	1991-93	3	67	3	3	3
Pender Islands	1964-93	28	96	1	17	169
Pitt Meadows	1972-93	22	50	1	2	10
Port Alberni	1975-93	3	67	1	6	10
Prince Rupert	1980-93	14	7	1	–	1
Rose Spit	1986-93	8	13	1	–	1
Skidegate	1982-93	12	8	1	–	1
Sooke	1983-93	11	100	4	27	198
Squamish	1980-93	13	92	1	6	74
Sunshine Coast	1979-93	14	64	1	7	20
Surrey	1960-66	7	14	2	–	2
Terrace	1963-92	29	21	1	1	2
Vancouver	1957-93	37	89	1	9	75
Victoria	1958-93	36	100	5	39	433
Whistler	1991-93	3	100	2	10	13
White Rock	1971-93	23	83	1	9	54
Interior						
Burns Lake–Francois Lake	1990-93	4	75	1	1	5
Cranbrook	1984-93	10	100	4	13	39
Creston	1987-89	3	100	1	20	26
Fauquier	1984-92	10	90	1	7	15
Fort St. James	1987-93	7	86	1	7	20
Kamloops	1984-93	10	100	2	10	62
Kelowna	1981-93	9	100	18	54	93
Kimberley	1992-93	2	100	26	30	34
Lake Windermere	1979-84	6	100	19	30	129
MacKenzie	1982	1	100	28	–	28
Nakusp	1979-93	15	100	8	14	41
Nelson	1985	1	100	12	–	12
North Pine	1983-93	11	18	4	6	7
Oliver-Osoyoos	1979-93	15	100	4	38	172

▶

◀ *Red-breasted Nuthatch*

Count locality	Count years	Total counts	Frequency of occurrence (%)	Range of total counts when species occurred Low	Median	High
Penticton	1974-93	18	100	4	65	401
Prince George	1989-93	5	100	7	23	85
Princeton	1992-93	2	100	36	60	84
Quesnel	1981-84	4	50	2	2	2
Revelstoke	1981-93	13	100	1	22	39
Salmon Arm	1988-92	5	100	4	18	30
Shuswap Lake Park	1972-93	22	100	2	40	153
Smithers	1977-93	17	71	1	5	34
Vaseux Lake	1974-93	20	95	1	41	359
Vernon	1975-93	19	100	3	26	174
Wells Gray Park	1984-92	6	33	3	5	7
Yoho National Park	1975-84	5	40	1	4	7

White-breasted Nuthatch

Highest provincial total: 144 (1989)
Highest interior total: 144 (1989)
Highest coastal total: 24 (1993)

Count locality	Count years	Total counts	Frequency of occurrence (%)	Range of total counts when species occurred Low	Median	High
Coastal						
Pender Islands	1964-93	28	4	1	–	1
Sooke	1983-93	11	18	1	1	1
White Rock	1971-93	23	4	24	–	24*
Interior						
Burns Lake–Francois Lake	1990-93	4	25	1	–	1
Cranbrook	1984-93	10	70	1	2	6
Fauquier	1984-92	10	40	1	1	2
Fort St. James	1987-93	7	14	2	2	
Kamloops	1984-93	10	100	1	5	14
Kelowna	1981-93	9	89	1	8	12
Kimberley	1992-93	2	100	6	11	16
Lake Windermere	1979-84	6	83	4	7	10
Nakusp	1979-93	15	7	2	–	2
North Pine	1983-93	11	27	1	2	2
Oliver-Osoyoos	1979-93	15	93	1	9	16
Penticton	1974-93	18	100	1	23	64
Prince George	1989-93	5	40	1	3	4
Princeton	1992-93	2	100	2	4	5
Revelstoke	1981-93	13	23	1	2	2
Shuswap Lake Park	1972-93	22	82	1	3	7
Vaseux Lake	1974-93	20	95	3	14	42
Vernon	1975-93	19	100	3	12	45
Yoho National Park	1975-84	5	20	1	–	1

Pygmy Nuthatch

Highest provincial total: 973 (1989)
Highest interior total: 973 (1989)
Highest coastal total: Not recorded

Count locality	Count years	Total counts	Frequency of occurrence (%)	Range of total counts when species occurred Low	Median	High
Interior						
Kamloops	1984-93	10	90	5	15	52
Kelowna	1981-93	9	100	77	133	269
Oliver-Osoyoos	1979-93	15	100	4	41	61
Penticton	1974-93	18	100	90	191	515
Princeton	1992-93	2	100	13	19	25
Revelstoke	1981-93	13	8	1	–	1
Vaseux Lake	1974-93	20	100	44	126	384
Vernon	1975-93	19	100	13	55	169

Brown Creeper

Highest provincial total: 453 (1993)
Highest interior total: 89 (1986)
Highest coastal total: 429 (1993)

Count locality	Count years	Total counts	Frequency of occurrence (%)	Range of total counts when species occurred Low	Median	High
Coastal						
Bamfield	1986-93	8	38	1	2	2
Campbell River	1972-93	19	100	1	4	12
Chilliwack	1972-93	20	70	1	3	6
Comox	1961-93	32	91	1	7	20
Deep Bay	1975-93	19	95	1	3	8
Duncan	1970-93	24	100	5	19	50
Kitimat	1974-93	16	19	1	1	1
Ladner	1957-93	35	94	1	9	39
Masset	1982-93	12	42	1	1	2
Nanaimo	1963-93	28	96	1	6	22
North Saanich	1960	1	100	16	–	16
Parksville–Qualicum Beach	1991-93	3	100	4	7	10
Pender Harbour	1991-93	3	67	2	3	4
Pender Islands	1964-93	28	89	1	7	26
Pitt Meadows	1972-93	22	82	1	3	15
Port Alberni	1975-93	3	100	2	4	6
Port Clements	1984-93	10	50	1	1	2
Prince Rupert	1980-93	14	14	1	1	1
Rose Spit	1986-93	8	13	2	–	2
Sayward	1973-84	12	33	1	1	2
Skidegate	1982-93	12	25	2	2	3
Sooke	1983-93	11	100	4	20	41
Squamish	1980-93	13	100	1	4	17
Sunshine Coast	1979-93	14	86	3	7	10
Surrey	1960-66	7	43	1	1	3
Terrace	1963-92	29	41	1	2	9
Vancouver	1957-93	37	97	2	16	54
Victoria	1958-93	36	97	11	35	171
White Rock	1971-93	23	100	2	16	65
Interior						
Burns Lake–Francois Lake	1990-93	4	50	1	2	2
Cranbrook	1984-93	10	40	1	2	2
Creston	1987-89	3	67	1	3	4
Fauquier	1984-92	10	60	1	2	3
Fort St. James	1987-93	7	100	1	1	3
Kamloops	1984-93	10	40	1	3	3
Kelowna	1981-93	9	100	2	6	16
Lake Windermere	1979-84	6	100	1	2	7
Nakusp	1979-93	15	93	1	2	7
Nelson	1985	1	100	4	–	4
North Pine	1983-93	11	18	1	2	2
Oliver-Osoyoos	1979-93	15	93	1	4	15
Penticton	1974-93	18	89	1	9	22
Prince George	1989-93	5	60	1	2	5
Princeton	1992-93	2	50	1	–	1
Quesnel	1981-84	4	25	1	–	1
Revelstoke	1981-93	13	85	1	4	10
Salmon Arm	1988-92	5	100	2	6	9
Shuswap Lake Park	1972-93	22	95	1	5	13
Smithers	1977-93	17	65	1	3	8
Vaseux Lake	1974-93	20	90	1	5	13
Vernon	1975-93	19	100	1	6	15
Wells Gray Park	1984-92	6	50	1	1	2
Yoho National Park	1975-84	5	40	4	5	6

Rock Wren

Highest provincial total: 2 (1977)
Highest interior total: 2 (1989)
Highest coastal total: 1 (1975)

Count locality	Count years	Total counts	Frequency of occurrence (%)	Low	Range of total counts when species occurred Median	High
Coastal						
Victoria	1958-93	36	6	1	1	1
Interior						
Vaseux Lake	1974-93	20	15	1	1	2
Vernon	1975-93	19	11	1	1	1

Canyon Wren

Highest provincial total: 34 (1983)
Highest interior total: 34 (1983)
Highest coastal total: Not recorded

Count locality	Count years	Total counts	Frequency of occurrence (%)	Low	Range of total counts when species occurred Median	High
Interior						
Oliver-Osoyoos	1979-93	15	87	2	3	11
Penticton	1974-93	18	67	1	2	7
Vaseux Lake	1974-93	20	100	2	8	21

Bewick's Wren

Highest provincial total: 559 (1993)
Highest interior total: Not recorded
Highest coastal total: 559 (1993)

Count locality	Count years	Total counts	Frequency of occurrence (%)	Low	Range of total counts when species occurred Median	High
Interior						
Bamfield	1986-93	8	13	1	–	1
Campbell River	1972-93	19	89	1	5	14
Chilliwack	1972-93	20	75	1	2	10
Comox	1961-93	32	97	1	11	24
Deep Bay	1975-93	19	84	1	3	7
Duncan	1970-93	24	100	12	36	71
Ladner	1957-93	35	100	5	48	155
Nanaimo	1963-93	28	100	1	11	31
North Saanich	1960	1	100	28	–	28
Parksville–Qualicum Beach	1991-93	3	100	2	6	9
Pender Harbour	1991-93	3	33	1	–	1
Pender Islands	1964-93	28	100	1	7	31
Pitt Meadows	1972-93	22	100	1	15	60
Port Alberni	1975-93	3	67	5	7	9
Sooke	1983-93	11	100	3	16	48
Squamish	1980-93	13	8	1	–	1
Sunshine Coast	1979-93	14	79	1	2	5
Surrey	1960-66	7	100	3	5	17
Vancouver	1957-93	37	100	5	36	93
Victoria	1958-93	36	100	21	71	218
White Rock	1971-93	23	100	9	56	89

House Wren

Highest provincial total: 466 (1977)
Highest interior total: 466 (1977)
Highest coastal total: 4 (1979)

Count locality	Count years	Total counts	Frequency of occurrence (%)	Range of total counts when species occurred: Low	Median	High
Coastal						
Campbell River	1972-93	19	5	1	–	1*
Comox	1961-93	32	3	2	–	2*
Ladner	1957-93	35	3	1	–	1*
Nanaimo	1963-93	28	11	1	3	4*
Vancouver	1957-93	37	5	1	1	1*
Victoria	1958-93	36	11	1	1	2*
Interior						
Shuswap Lake Park	1972-93	22	14	1	1	1*

Winter Wren

Highest provincial total: 1,970 (1992)
Highest interior total: 47 (1989)
Highest coastal total: 1,958 (1992)

Count locality	Count years	Total counts	Frequency of occurrence (%)	Range of total counts when species occurred: Low	Median	High
Coastal						
Bamfield	1986-93	8	100	3	16	22
Campbell River	1972-93	19	100	7	28	60
Chilliwack	1972-93	20	100	1	6	33
Comox	1961-93	32	100	3	38	81
Deep Bay	1975-93	19	100	7	26	47
Dewdney	1967	1	100	4	–	4
Duncan	1970-93	24	100	14	46	195
Kitimat	1974-93	16	50	1	2	3
Ladner	1957-93	35	100	8	41	179
Masset	1982-93	12	92	3	7	18
Nanaimo	1963-93	28	100	2	21	58
North Saanich	1960	1	100	29	–	29
Parksville–Qualicum Beach	1991-93	3	100	29	37	38
Pender Harbour	1991-93	3	100	28	35	49
Pender Islands	1964-93	28	100	1	52	124
Pitt Meadows	1972-93	22	100	6	32	71
Port Alberni	1975-93	3	100	36	38	39
Port Clements	1984-93	10	100	2	10	33
Prince Rupert	1980-93	14	86	1	2	12
Rose Spit	1986-93	8	100	2	5	14
Sayward	1973-84	12	100	1	6	12
Skidegate	1982-93	12	100	5	10	31
Sooke	1983-93	11	100	52	102	184
Squamish	1980-93	13	100	11	34	48
Sunshine Coast	1979-93	14	100	5	65	103
Surrey	1960-66	7	100	1	5	10
Terrace	1963-92	29	52	1	3	11
Vancouver	1957-93	37	100	22	96	444
Victoria	1958-93	36	100	29	97	451
Whistler	1991-93	3	33	4	–	4
White Rock	1971-93	23	100	9	58	188
Interior						
Burns Lake–Francois Lake	1990-93	4	25	1	–	1
Creston	1987-89	3	100	1	1	2
Fauquier	1984-92	10	90	1	3	8
Kamloops	1984-93	10	10	1	–	1
Kelowna	1981-93	9	67	1	3	7
Nakusp	1979-93	15	93	1	2	8
Nelson	1985	1	100	2	–	2
Oliver-Osoyoos	1979-93	15	40	1	1	3
Penticton	1974-93	18	83	1	4	9
Princeton	1992-93	2	50	2	–	2
Revelstoke	1981-93	13	69	1	1	5
Salmon Arm	1988-92	5	40	2	5	8

▶

◀ *Winter Wren*

Count locality	Count years	Total counts	Frequency of occurrence (%)	Range of total counts when species occurred: Low	Median	High
Shuswap Lake Park	1972-93	22	82	1	3	8
Smithers	1977-93	17	6	1	–	1
Vaseux Lake	1974-93	20	80	1	2	6
Vernon	1975-93	19	74	1	4	10
Wells Gray Park	1984-92	6	17	1	–	1

Marsh Wren

Highest provincial total: 265 (1991)
Highest interior total: 28 (1991)
Highest coastal total: 237 (1991)

Count locality	Count years	Total counts	Frequency of occurrence (%)	Range of total counts when species occurred: Low	Median	High
Coastal						
Bamfield	1986-93	8	38	1	2	4
Comox	1961-93	32	9	1	1	2
Duncan	1970-93	24	100	1	11	31
Ladner	1957-93	35	91	1	21	102
Nanaimo	1963-93	28	79	1	4	15
Parksville–Qualicum Beach	1991-93	3	33	1	–	1
Pender Harbour	1991-93	3	67	1	3	4
Pender Islands	1964-93	28	7	1	1	1
Pitt Meadows	1972-93	22	82	1	5	27
Port Alberni	1975-93	3	100	6	8	11
Sooke	1983-93	11	73	1	1	2
Squamish	1980-93	13	15	1	1	1
Sunshine Coast	1979-93	14	43	1	1	3
Vancouver	1957-93	37	95	1	11	81
Victoria	1958-93	36	81	1	10	52
White Rock	1971-93	23	78	2	8	56
Interior						
Kamloops	1984-93	10	20	1	2	2
Kelowna	1981-93	9	44	1	2	4
Oliver-Osoyoos	1979-93	15	73	1	2	19
Penticton	1974-93	18	50	1	2	4
Salmon Arm	1988-92	5	20	1	–	1
Vaseux Lake	1974-93	20	65	1	5	10
Vernon	1975-93	19	79	1	2	13

American Dipper

Highest provincial total: 215 (1984)
Highest interior total: 114 (1992)
Highest coastal total: 121 (1984)

Count locality	Count years	Total counts	Frequency of occurrence (%)	Range of total counts when species occurred: Low	Median	High
Coastal						
Bamfield	1986-93	8	88	1	1	3
Campbell River	1972-93	19	95	1	2	6
Chilliwack	1972-93	20	90	1	13	20
Comox	1961-93	32	59	1	3	7
Deep Bay	1975-93	19	79	1	2	3
Duncan	1970-93	24	58	1	2	3
Kitimat	1974-93	16	50	1	1	2
Nanaimo	1963-93	28	39	1	2	5
Parksville–Qualicum Beach	1991-93	3	33	1	–	1
Pender Harbour	1991-93	3	100	2	2	3
Pender Islands	1964-93	28	7	1	2	2
Pitt Meadows	1972-93	22	100	1	5	18
Port Alberni	1975-93	3	100	3	4	9
Port Clements	1984-93	10	30	1	1	1

▶

◀ *American Dipper*

Count locality	Count years	Total counts	Frequency of occurrence (%)	Range of total counts when species occurred Low	Median	High
Prince Rupert	1980-93	14	21	1	1	2
Sayward	1973-84	12	50	1	1	3
Skidegate	1982-93	12	92	1	1	4
Sooke	1983-93	11	73	1	2	4
Squamish	1980-93	13	100	2	24	69
Sunshine Coast	1979-93	14	93	2	3	8
Terrace	1963-92	29	48	1	2	5
Vancouver	1957-93	37	84	1	3	12
Victoria	1958-93	36	89	1	3	11
Whistler	1991-93	3	100	3	4	6
White Rock	1971-93	23	9	1	1	1
Interior						
Burns Lake–Francois Lake	1990-93	4	25	3	–	3
Cranbrook	1984-93	10	100	3	6	14
Creston	1987-89	3	100	1	1	1
Fauquier	1984-92	10	100	4	10	20
Fort St. James	1987-93	7	100	1	2	6
Kamloops	1984-93	10	40	1	2	3
Kelowna	1981-93	9	100	1	9	18
Kimberley	1992-93	2	50	1	–	1
Lake Windermere	1979-84	6	100	5	8	13
MacKenzie	1982	1	100	1	–	1
Nakusp	1979-93	15	100	2	3	8
Nelson	1985	1	100	1	–	1
Oliver-Osoyoos	1979-93	15	100	2	4	9
Penticton	1974-93	18	100	2	3	12
Prince George	1989-93	5	60	1	2	4
Princeton	1992-93	2	100	6	13	20
Quesnel	1981-84	4	25	1	–	1
Revelstoke	1981-93	13	100	5	8	32
Salmon Arm	1988-92	5	60	1	2	2
Shuswap Lake Park	1972-93	22	100	2	6	25
Smithers	1977-93	17	100	1	2	9
Vaseux Lake	1974-93	20	100	1	21	50
Vernon	1975-93	19	95	1	3	10
Wells Gray Park	1984-92	6	67	1	1	2
Yoho National Park	1975-84	5	100	5	7	24

Golden-crowned Kinglet

Highest provincial total: 12,885 (1993)
Highest interior total: 518 (1989)
Highest coastal total: 12,370 (1993)

Count locality	Count years	Total counts	Frequency of occurrence (%)	Range of total counts when species occurred Low	Median	High
Coastal						
Bamfield	1986-93	8	100	12	35	66
Campbell River	1972-93	19	100	16	162	383
Chilliwack	1972-93	20	100	1	68	257
Comox	1961-93	32	100	8	261	742
Deep Bay	1975-93	19	100	84	150	411
Dewdney	1967	1	100	3	–	3
Duncan	1970-93	24	100	19	210	1,103
Kitimat	1974-93	16	69	1	4	33
Ladner	1957-93	35	100	30	196	735
Masset	1982-93	12	100	7	46	104
Nanaimo	1963-93	28	100	6	124	493
North Saanich	1960	1	100	200	–	200
Parksville–Qualicum Beach	1991-93	3	100	265	322	394
Pender Harbour	1991-93	3	100	114	175	184
Pender Islands	1964-93	28	100	16	314	782
Pitt Meadows	1972-93	22	100	57	161	461
Port Alberni	1975-93	3	100	68	240	296
Port Clements	1984-93	10	100	5	30	106
Prince Rupert	1980-93	14	57	3	14	45

▶

◀ *Golden-crowned Kinglet*

Count locality	Count years	Total counts	Frequency of occurrence (%)	Low	Range of total counts when species occurred Median	High
Rose Spit	1986-93	8	88	2	6	16
Sayward	1973-84	12	92	17	26	60
Skidegate	1982-93	12	100	8	36	62
Sooke	1983-93	11	100	194	769	1,721
Squamish	1980-93	13	100	111	241	471
Sunshine Coast	1979-93	14	100	11	271	482
Surrey	1960-66	7	86	14	28	60
Terrace	1963-92	29	62	2	8	78
Vancouver	1957-93	37	100	92	441	1,907
Victoria	1958-93	36	100	168	467	3,337
Whistler	1991-93	3	100	11	11	44
White Rock	1971-93	23	100	152	356	1,029
Interior						
Cranbrook	1984-93	10	40	1	2	2
Creston	1987-89	3	67	3	26	49
Fauquier	1984-92	10	90	7	17	68
Fort St. James	1987-93	7	14	3	–	3
Kamloops	1984-93	10	30	1	3	3
Kelowna	1981-93	9	78	2	9	34
Lake Windermere	1979-84	6	67	1	2	29
Nakusp	1979-93	15	100	4	39	108
Nelson	1985	1	100	15	–	15
North Pine	1983-93	11	9	1	–	1
Oliver-Osoyoos	1979-93	15	100	2	14	66
Penticton	1974-93	18	100	5	30	67
Prince George	1989-93	5	40	1	3	5
Princeton	1992-93	2	50	31	–	31
Revelstoke	1981-93	13	100	2	13	54
Salmon Arm	1988-92	5	100	4	21	93
Shuswap Lake Park	1972-93	22	100	4	43	211
Smithers	1977-93	17	88	1	3	10
Vaseux Lake	1974-93	20	95	2	24	78
Vernon	1975-93	19	95	10	28	72
Wells Gray Park	1984-92	6	67	1	7	11
Yoho National Park	1975-84	5	40	1	3	4

Ruby-crowned Kinglet

Highest provincial total: 988 (1988)
Highest interior total: 7 (1975)
Highest coastal total: 984 (1988)

Count locality	Count years	Total counts	Frequency of occurrence (%)	Low	Range of total counts when species occurred Median	High
Coastal						
Bamfield	1986-93	8	88	1	5	9
Campbell River	1972-93	19	100	1	7	32
Chilliwack	1972-93	20	80	1	6	30
Comox	1961-93	32	97	2	8	57
Deep Bay	1975-93	19	95	1	7	27
Duncan	1970-93	24	100	4	28	113
Kitimat	1974-93	16	13	1	2	2
Ladner	1957-93	35	97	3	32	117
Nanaimo	1963-93	28	93	1	12	30
North Saanich	1960	1	100	31	–	31
Parksville–Qualicum Beach	1991-93	3	100	13	41	45
Pender Harbour	1991-93	3	100	1	1	3
Pender Islands	1964-93	28	93	2	15	44
Pitt Meadows	1972-93	22	95	1	19	99
Port Alberni	1975-93	3	100	2	5	21
Sayward	1973-84	12	25	1	2	5
Sooke	1983-93	11	100	1	51	105
Squamish	1980-93	13	92	2	9	18
Sunshine Coast	1979-93	14	100	1	13	39
Surrey	1960-66	7	86	1	3	5
Terrace	1963-92	29	17	2	5	12

▶

◀ *Ruby-crowned Kinglet*

Count locality	Count years	Total counts	Frequency of occurrence (%)	Range of total counts when species occurred Low	Median	High
Vancouver	1957-93	37	100	20	87	254
Victoria	1958-93	36	100	14	81	353
White Rock	1971-93	23	100	5	41	125
Interior						
Creston	1987-89	3	33	4	–	4
Kelowna	1981-93	9	56	1	2	5
Oliver-Osoyoos	1979-93	15	7	1	–	1
Penticton	1974-93	18	50	1	2	7
Salmon Arm	1988-92	5	20	2	–	2
Shuswap Lake Park	1972-93	22	5	1	–	1
Vaseux Lake	1974-93	20	10	1	1	1

Western Bluebird

Highest provincial total: 73 (1988)
Highest interior total: 73 (1988)
Highest coastal total: 59 (1964)

Count locality	Count years	Total counts	Frequency of occurrence (%)	Range of total counts when species occurred Low	Median	High
Coastal						
Victoria	1958-93	36	11	26	52	59
Interior						
Kelowna	1981-93	9	22	4	9	14
Oliver-Osoyoos	1979-93	15	13	1	2	2
Penticton	1974-93	18	89	2	17	73
Vaseux Lake	1974-93	20	20	4	6	13

Mountain Bluebird

Highest provincial total: 12 (1980)
Highest interior total: 12 (1980)
Highest coastal total: 4 (1990)

Count locality	Count years	Total counts	Frequency of occurrence (%)	Range of total counts when species occurred Low	Median	High
Coastal						
Ladner	1957-93	35	3	2	–	2
Skidegate	1982-93	12	25	1	1	1
Terrace	1963-92	29	3	1	–	1
Victoria	1958-93	36	3	3	–	3
Interior						
Kamloops	1984-93	10	10	1	–	1
Penticton	1974-93	18	6	12	–	12
Vaseux Lake	1974-93	20	10	1	2	3
Vernon	1975-93	19	5	1	–	1

Townsend's Solitaire

Highest provincial total: 124 (1982)
Highest interior total: 123 (1982)
Highest coastal total: 20 (1991)

Count locality	Count years	Total counts	Frequency of occurrence (%)	Range of total counts when species occurred Low	Median	High
Coastal						
Campbell River	1972-93	19	5	1	–	1
Chilliwack	1972-93	20	10	1	1	1
Comox	1961-93	32	9	1	1	1
Deep Bay	1975-93	19	5	1	–	1
Duncan	1970-93	24	8	3	3	3
Ladner	1957-93	35	3	1	–	1
Masset	1982-93	12	8	1	–	1
Nanaimo	1963-93	28	14	1	1	2
Pender Islands	1964-93	28	14	1	1	2
Pitt Meadows	1972-93	22	18	1	1	1
Squamish	1980-93	13	23	1	1	2
Sunshine Coast	1979-93	14	7	1	–	1
Terrace	1963-92	29	3	1	–	1
Vancouver	1957-93	37	51	1	2	8
Victoria	1958-93	36	31	1	1	6
White Rock	1971-93	23	17	1	1	1
Interior						
Cranbrook	1984-93	10	50	1	1	8
Creston	1987-89	3	33	1	–	1
Fauquier	1984-92	10	30	1	4	5
Fort St. James	1987-93	7	29	1	1	1
Kamloops	1984-93	10	100	1	4	14
Kelowna	1981-93	9	100	4	10	17
Kimberley	1992-93	2	50	2	–	2
Lake Windermere	1979-84	6	100	2	27	62
Nakusp	1979-93	15	40	1	2	2
Nelson	1985	1	100	1	–	1
Oliver-Osoyoos	1979-93	15	93	1	4	9
Penticton	1974-93	18	100	2	19	35
Prince George	1989-93	5	40	1	1	1
Princeton	1992-93	2	50	1	–	1
Revelstoke	1981-93	13	15	1	1	1
Salmon Arm	1988-92	5	80	1	1	2
Shuswap Lake Park	1972-93	22	27	1	1	1
Vaseux Lake	1974-93	20	85	3	8	25
Vernon	1975-93	19	100	1	8	28

Swainson's Thrush

Highest provincial total: 2 (1975)
Highest interior total: 1 (1988)
Highest coastal total: 2 (1975)

Count locality	Count years	Total counts	Frequency of occurrence (%)	Range of total counts when species occurred Low	Median	High
Coastal						
Duncan	1970-93	24	4	1	–	1*
Nanaimo	1963-93	28	4	1	–	1*
Pender Islands	1964-93	28	4	1	–	1*
Vancouver	1957-93	37	3	1	–	1*
Victoria	1958-93	36	6	1	1	1*
White Rock	1971-93	23	4	1	–	1*
Interior						
Shuswap Lake Park	1972-93	22	5	1	–	1

Hermit Thrush

Highest provincial total: 46 (1983)
Highest interior total: 1 (1987)
Highest coastal total: 46 (1983)

Count locality	Count years	Total counts	Frequency of occurrence (%)	Range of total counts when species occurred Low	Median	High
Coastal						
Bamfield	1986-93	8	75	1	2	5
Campbell River	1972-93	19	16	1	2	15
Comox	1961-93	32	19	1	2	3
Deep Bay	1975-93	19	11	1	2	2
Duncan	1970-93	24	25	1	2	8
Ladner	1957-93	35	31	1	1	4
Masset	1982-93	12	8	2	–	2
Nanaimo	1963-93	28	4	1	–	1
Pender Islands	1964-93	28	29	1	3	13
Port Alberni	1975-93	3	33	4	–	4
Prince Rupert	1980-93	14	7	1	–	1
Sayward	1973-84	12	8	1	–	1
Skidegate	1982-93	12	42	1	1	1
Sooke	1983-93	11	82	1	4	17
Squamish	1980-93	13	15	1	1	1
Sunshine Coast	1979-93	14	36	1	1	3
Terrace	1963-92	29	7	1	1	1
Vancouver	1957-93	37	62	1	2	6
Victoria	1958-93	36	92	1	5	34
White Rock	1971-93	23	48	1	2	6
Interior						
Oliver-Osoyoos	1979-93	15	7	1	–	1

American Robin

Highest provincial total: 21,834 (1990)
Highest interior total: 2,099 (1987)
Highest coastal total: 21,317 (1990)

Count locality	Count years	Total counts	Frequency of occurrence (%)	Range of total counts when species occurred Low	Median	High
Coastal						
Bamfield	1986-93	8	88	1	4	19
Campbell River	1972-93	19	100	3	18	141
Chilliwack	1972-93	20	100	6	51	354
Comox	1961-93	32	100	1	90	1,258
Deep Bay	1975-93	19	100	5	41	470
Dewdney	1967	1	100	1	–	1
Duncan	1970-93	24	100	29	472	3,471
Kitimat	1974-93	16	88	1	16	92
Ladner	1957-93	35	100	20	974	2,190
Masset	1982-93	12	58	1	3	36
Nanaimo	1963-93	28	100	74	501	3,312
North Saanich	1960	1	100	1,464	–	1,464
Parksville–Qualicum Beach	1991-93	3	100	52	66	320
Pender Harbour	1991-93	3	33	30	–	30
Pender Islands	1964-93	28	100	13	384	2,465
Pitt Meadows	1972-93	22	100	19	119	1,153
Port Alberni	1975-93	3	100	27	79	157
Port Clements	1984-93	10	20	2	3	3
Prince Rupert	1980-93	14	29	1	2	9
Rose Spit	1986-93	8	13	2	–	2
Sayward	1973-84	12	50	2	5	9
Skidegate	1982-93	12	100	1	11	63
Sooke	1983-93	11	100	64	239	826
Squamish	1980-93	13	92	1	8	50
Sunshine Coast	1979-93	14	100	2	14	65
Surrey	1960-66	7	86	4	19	66
Terrace	1963-92	29	66	1	24	208
Vancouver	1957-93	37	100	214	803	3,911
Victoria	1958-93	36	100	1,137	3,433	14,847
White Rock	1971-93	23	100	77	623	1,927

▶

◀ *American Robin*

Count locality	Count years	Total counts	Frequency of occurrence (%)	Range of total counts when species occurred: Low	Median	High
Interior						
Cranbrook	1984-93	10	20	1	2	2
Creston	1987-89	3	100	7	10	17
Fauquier	1984-92	10	10	1	–	1
Kamloops	1984-93	10	80	1	17	22
Kelowna	1981-93	9	100	19	78	330
Lake Windermere	1979-84	6	50	1	1	16
Nakusp	1979-93	15	53	1	1	15
Nelson	1985	1	100	11	–	11
Oliver-Osoyoos	1979-93	15	73	1	3	146
Penticton	1974-93	18	100	6	33	797
Prince George	1989-93	5	60	1	1	2
Princeton	1992-93	2	100	1	9	17
Revelstoke	1981-93	13	69	1	11	61
Salmon Arm	1988-92	5	100	1	1	3
Shuswap Lake Park	1972-93	22	32	1	2	7
Smithers	1977-93	17	29	1	3	18
Vaseux Lake	1974-93	20	70	1	5	56
Vernon	1975-93	19	95	2	34	700

Varied Thrush

Highest provincial total: 4,245 (1983)
Highest interior total: 235 (1990)
Highest coastal total: 4,120 (1983)

Count locality	Count years	Total counts	Frequency of occurrence (%)	Range of total counts when species occurred: Low	Median	High
Coastal						
Bamfield	1986-93	8	100	3	14	166
Campbell River	1972-93	19	89	1	13	418
Chilliwack	1972-93	20	80	1	6	165
Comox	1961-93	32	91	1	20	508
Deep Bay	1975-93	19	100	1	10	91
Dewdney	1967	1	100	14	–	14
Duncan	1970-93	24	100	5	40	581
Kitimat	1974-93	16	50	2	24	82
Ladner	1957-93	35	100	1	29	155
Masset	1982-93	12	100	2	12	69
Nanaimo	1963-93	28	100	1	17	110
North Saanich	1960	1	100	34	–	34
Parksville–Qualicum Beach	1991-93	3	100	3	11	17
Pender Harbour	1991-93	3	100	15	17	19
Pender Islands	1964-93	28	96	1	66	446
Pitt Meadows	1972-93	22	100	3	54	294
Port Alberni	1975-93	3	100	14	83	185
Port Clements	1984-93	10	80	1	6	28
Prince Rupert	1980-93	14	71	1	11	78
Rose Spit	1986-93	8	63	2	7	10
Sayward	1973-84	12	50	1	4	23
Skidegate	1982-93	12	92	3	22	422
Sooke	1983-93	11	100	13	80	352
Squamish	1980-93	13	92	5	21	219
Sunshine Coast	1979-93	14	93	2	46	555
Surrey	1960-66	7	86	7	10	51
Terrace	1963-92	29	45	1	10	156
Vancouver	1957-93	37	100	3	191	1,212
Victoria	1958-93	36	100	12	130	1,243
Whistler	1991-93	3	67	4	5	6
White Rock	1971-93	23	100	2	88	388
Interior						
Creston	1987-89	3	33	2	–	2
Fauquier	1984-92	10	50	1	1	3
Fort St. James	1987-93	7	29	1	1	1
Kamloops	1984-93	10	20	1	2	3
Kelowna	1981-93	9	67	1	5	16

▶

◀ *Varied Thrush*

Count locality	Count years	Total counts	Frequency of occurrence (%)	Low	Range of total counts when species occurred Median	High
Lake Windermere	1979-84	6	17	1	–	1
Nakusp	1979-93	15	67	1	2	58
Oliver-Osoyoos	1979-93	15	47	1	1	3
Penticton	1974-93	18	89	1	4	29
Prince George	1989-93	5	20	1	–	1
Princeton	1992-93	2	50	2	–	2
Revelstoke	1981-93	13	62	1	5	120
Salmon Arm	1988-92	5	40	2	8	14
Shuswap Lake Park	1972-93	22	50	1	5	13
Smithers	1977-93	17	18	1	1	1
Vaseux Lake	1974-93	20	35	1	2	9
Vernon	1975-93	19	42	1	5	20

Gray Catbird

Highest provincial total: 1 (1981)
Highest interior total: 1 (1981)
Highest coastal total: Not recorded

Count locality	Count years	Total counts	Frequency of occurrence (%)	Low	Range of total counts when species occurred Median	High
Interior						
Nakusp	1979-93	15	7	1	–	1
Vaseux Lake	1974-93	20	5	1	–	1

Northern Mockingbird

Highest provincial total: 1 (1972)
Highest interior total: Not recorded
Highest coastal total: 1 (1972)

Count locality	Count years	Total counts	Frequency of occurrence (%)	Low	Range of total counts when species occurred Median	High
Coastal						
Duncan	1970-93	24	4	1	–	1
Ladner	1957-93	35	3	1	–	1
Nanaimo	1963-93	28	4	1	–	1
Vancouver	1957-93	37	3	1	–	1

Brown Thrasher

Highest provincial total: 1 (1976)
Highest interior total: Not recorded
Highest coastal total: 1 (1976)

Count locality	Count years	Total counts	Frequency of occurrence (%)	Low	Range of total counts when species occurred Median	High
Coastal						
Terrace	1963-92	29	3	1	–	1

American Pipit

Highest provincial total: 346 (1962)
Highest interior total: 8 (1993)
Highest coastal total: 346 (1962)

Count locality	Count years	Total counts	Frequency of occurrence (%)	Low	Range of total counts when species occurred Median	High
Coastal						
Campbell River	1972-93	19	5	1	–	1
Chilliwack	1972-93	20	10	9	23	36
Comox	1961-93	32	9	2	3	37
Duncan	1970-93	24	50	1	7	21
Kitimat	1974-93	16	6	1	–	1
Ladner	1957-93	35	74	1	25	317
Masset	1982-93	12	17	16	23	30
Nanaimo	1963-93	28	14	1	2	11
North Saanich	1960	1	100	3	–	3
Parksville–Qualicum Beach	1991-93	3	33	1	–	1
Pender Islands	1964-93	28	4	1	–	1
Pitt Meadows	1972-93	22	23	1	2	11
Port Alberni	1975-93	3	33	5	–	5
Rose Spit	1986-93	8	13	2	–	2
Skidegate	1982-93	12	67	2	20	28
Sooke	1983-93	11	9	2	–	2
Surrey	1960-66	7	14	2	–	2
Vancouver	1957-93	37	24	1	4	40
Victoria	1958-93	36	56	1	28	126
White Rock	1971-93	23	17	1	2	5
Interior						
Kelowna	1981-93	9	11	1	–	1
Oliver-Osoyoos	1979-93	15	13	1	5	8

Bohemian Waxwing

Highest provincial total: 25,829 (1985)
Highest interior total: 24,318 (1990)
Highest coastal total: 3,278 (1985)

Count locality	Count years	Total counts	Frequency of occurrence (%)	Low	Range of total counts when species occurred Median	High
Coastal						
Campbell River	1972-93	19	11	1	2	2
Chilliwack	1972-93	20	25	6	36	101
Deep Bay	1975-93	19	5	1	–	1
Kitimat	1974-93	16	56	1	21	259
Ladner	1957-93	35	11	1	4	15
Masset	1982-93	12	8	1	–	1
Nanaimo	1963-93	28	7	6	44	82
Pitt Meadows	1972-93	22	45	1	6	130
Port Alberni	1975-93	3	33	1	–	1
Sayward	1973-84	12	8	2	–	2
Skidegate	1982-93	12	17	2	5	7
Squamish	1980-93	13	31	7	23	53
Terrace	1963-92	29	62	3	64	3,037
Vancouver	1957-93	37	51	1	32	436
Victoria	1958-93	36	6	1	11	20
White Rock	1971-93	23	13	1	9	74
Interior						
Burns Lake–Francois Lake	1990-93	4	75	20	70	113
Cranbrook	1984-93	10	80	100	382	2,759
Creston	1987-89	3	100	90	117	230
Fauquier	1984-92	10	40	1	30	140
Fort St. James	1987-93	7	57	1	10	33
Kamloops	1984-93	10	100	153	1,115	3,183
Kelowna	1981-93	9	100	1,584	2,815	4,654
Kimberley	1992-93	2	100	463	674	884
Lake Windermere	1979-84	6	100	148	193	635
Nakusp	1979-93	15	73	40	127	380
Nelson	1985	1	100	5,600	–	5,600
North Pine	1983-93	11	45	14	71	434

▶

◀ *Bohemian Waxwing*

Count locality	Count years	Total counts	Frequency of occurrence (%)	Low	Range of total counts when species occurred Median	High
Oliver-Osoyoos	1979-93	15	93	5	696	5,139
Penticton	1974-93	18	100	436	2,320	6,893
Prince George	1989-93	5	100	52	1,193	4,960
Princeton	1992-93	2	50	250	–	250
Quesnel	1981-84	4	25	65	–	65
Revelstoke	1981-93	13	100	7	290	2,634
Salmon Arm	1988-92	5	100	35	241	398
Shuswap Lake Park	1972-93	22	77	21	158	1,207
Smithers	1977-93	17	88	9	40	313
Vaseux Lake	1974-93	20	95	31	322	755
Vernon	1975-93	19	100	250	1,649	11,133
Wells Gray Park	1984-92	6	17	3	–	3
Yoho National Park	1975-84	5	40	3	18	32

Cedar Waxwing

Highest provincial total: 737 (1958)
Highest interior total: 488 (1987)
Highest coastal total: 737 (1958)

Count locality	Count years	Total counts	Frequency of occurrence (%)	Low	Range of total counts when species occurred Median	High
Coastal						
Campbell River	1972-93	19	26	1	1	4
Chilliwack	1972-93	20	25	1	3	53
Comox	1961-93	32	19	1	9	30
Deep Bay	1975-93	19	5	8	–	8
Duncan	1970-93	24	25	1	3	8
Ladner	1957-93	35	49	1	5	29
Nanaimo	1963-93	28	46	2	18	55
Pender Islands	1964-93	28	32	1	8	108
Pitt Meadows	1972-93	22	23	1	2	5
Sayward	1973-84	12	8	5	–	5
Skidegate	1982-93	12	17	3	3	3
Sooke	1983-93	11	27	1	2	6
Surrey	1960-66	7	14	19	–	19
Terrace	1963-92	29	17	6	13	60
Vancouver	1957-93	37	89	2	21	737
Victoria	1958-93	36	92	1	34	602
White Rock	1971-93	23	35	3	16	55
Interior						
Kamloops	1984-93	10	10	16	–	16
Kelowna	1981-93	9	89	12	42	114
Lake Windermere	1979-84	6	17	120	–	120
Nakusp	1979-93	15	7	15	–	15
Nelson	1985	1	100	18	–	18
Oliver-Osoyoos	1979-93	15	40	1	5	79
Penticton	1974-93	18	72	1	13	73
Prince George	1989-93	5	20	28	–	28
Revelstoke	1981-93	13	54	1	34	328
Shuswap Lake Park	1972-93	22	5	3	–	3
Smithers	1977-93	17	35	1	11	18
Vaseux Lake	1974-93	20	25	1	9	49
Vernon	1975-93	19	58	1	11	60

Northern Shrike

Highest provincial total: 152 (1988)
Highest interior total: 107 (1988)
Highest coastal total: 82 (1977)

Count locality	Count years	Total counts	Frequency of occurrence (%)	Range of total counts when species occurred Low	Median	High
Coastal						
Campbell River	1972-93	19	37	1	1	2
Chilliwack	1972-93	20	60	1	2	5
Comox	1961-93	32	69	1	2	5
Deep Bay	1975-93	19	37	1	1	2
Dewdney	1967	1	100	1	–	1
Duncan	1970-93	24	88	1	2	5
Kitimat	1974-93	16	63	1	1	4
Ladner	1957-93	35	97	1	13	30
Masset	1982-93	12	25	1	1	1
Nanaimo	1963-93	28	71	1	2	4
North Saanich	1960	1	100	2	–	2
Parksville–Qualicum Beach	1991-93	3	67	1	1	1
Pender Islands	1964-93	28	18	1	1	3
Pitt Meadows	1972-93	22	100	1	5	12
Port Alberni	1975-93	3	33	1	–	1
Prince Rupert	1980-93	14	14	1	1	1
Sayward	1973-84	12	17	1	1	1
Skidegate	1982-93	12	8	1	–	1
Sooke	1983-93	11	55	1	2	4
Squamish	1980-93	13	54	1	2	2
Surrey	1960-66	7	43	1	2	3
Terrace	1963-92	29	45	1	1	4
Vancouver	1957-93	37	97	1	6	18
Victoria	1958-93	36	86	1	3	11
White Rock	1971-93	23	100	1	6	14
Interior						
Burns Lake–Francois Lake	1990-93	4	50	2	2	2
Cranbrook	1984-93	10	60	1	2	4
Creston	1987-89	3	67	1	2	2
Fauquier	1984-92	10	60	1	2	2
Fort St. James	1987-93	7	14	1	–	1
Fort St. John	1975-78	4	25	1	–	1
Kamloops	1984-93	10	100	3	7	10
Kelowna	1981-93	9	100	3	10	21
Kimberley	1992-93	2	100	2	3	4
Lake Windermere	1979-84	6	67	1	2	3
Nakusp	1979-93	15	60	1	1	2
North Pine	1983-93	11	36	1	1	2
Oliver-Osoyoos	1979-93	15	93	1	11	26
Penticton	1974-93	18	100	2	8	16
Prince George	1989-93	5	100	1	4	6
Princeton	1992-93	2	50	1	–	1
Revelstoke	1981-93	13	46	1	1	3
Salmon Arm	1988-92	5	80	2	4	9
Shuswap Lake Park	1972-93	22	50	1	1	5
Smithers	1977-93	17	94	1	3	5
Vaseux Lake	1974-93	20	95	1	3	8
Vernon	1975-93	19	100	3	12	24
Wells Gray Park	1984-92	6	33	1	1	1

Loggerhead Shrike

Highest provincial total: 2 (1965)
Highest interior total: 1 (1978)
Highest coastal total: 2 (1965)

Count locality	Count years	Total counts	Frequency of occurrence (%)	Range of total counts when species occurred Low	Median	High
Coastal						
Ladner	1957-93	35	3	1	–	1
Terrace	1963-92	29	3	2	–	2*
Vancouver	1957-93	37	5	1	1	1
White Rock	1971-93	23	4	1	–	1*

▶

◀ *Loggerhead Shrike*

Count locality	Count years	Total counts	Frequency of occurrence (%)	Range of total counts when species occurred: Low	Median	High
Interior						
Kamloops	1984-93	10	10	1	–	1
Oliver-Osoyoos	1979-93	15	20	1	1	1
Vernon	1975-93	19	5	1	–	1

European Starling

Highest provincial total: 261,232 (1966)
Highest interior total: 13,189 (1993)
Highest coastal total: 261,232 (1966)

Count locality	Count years	Total counts	Frequency of occurrence (%)	Range of total counts when species occurred: Low	Median	High
Coastal						
Bamfield	1986-93	8	100	5	45	88
Campbell River	1972-93	19	100	118	711	7,497
Chilliwack	1972-93	20	100	874	7221	20,182
Comox	1961-93	32	97	78	1,225	3,893
Deep Bay	1975-93	19	100	94	227	470
Dewdney	1967	1	100	1,177	–	1,177
Duncan	1970-93	24	100	335	1,560	3,979
Hecate Strait	1980	1	100	11	–	11
Kitimat	1974-93	16	94	3	86	230
Ladner	1957-93	35	100	6	10,343	56,019
Masset	1982-93	12	100	33	237	418
Nanaimo	1963-93	28	100	102	1,075	3,577
North Saanich	1960	1	100	3,363	–	3,363
Parksville–Qualicum Beach	1991-93	3	100	154	469	717
Pender Harbour	1991-93	3	100	30	60	78
Pender Islands	1964-93	28	100	5	217	2,486
Pitt Meadows	1972-93	22	100	772	6,607	36,148
Port Alberni	1975-93	3	100	506	585	4,374
Port Clements	1984-93	10	100	9	39	85
Prince Rupert	1980-93	14	86	7	107	536
Rose Spit	1986-93	8	13	3	–	3
Sayward	1973-84	12	100	24	77	140
Skidegate	1982-93	12	100	140	346	955
Sooke	1983-93	11	100	367	661	1,489
Squamish	1980-93	13	100	67	210	601
Sunshine Coast	1979-93	14	100	40	152	370
Surrey	1960-66	7	86	14	26	80
Terrace	1963-92	29	97	3	105	837
Vancouver	1957-93	37	100	10,129	45,450	254,068
Victoria	1958-93	36	97	279	5,642	18,183
Whistler	1991-93	3	100	34	36	39
White Rock	1971-93	23	100	941	7,069	12,381
Interior						
Burns Lake–Francois Lake	1990-93	4	75	34	48	53
Cranbrook	1984-93	10	80	1	40	82
Creston	1987-89	3	100	15	77	251
Fauquier	1984-92	10	80	1	8	56
Fort St. James	1987-93	7	100	19	31	95
Fort St. John	1975-78	4	25	5	–	5
Kamloops	1984-93	10	100	208	380	1,063
Kelowna	1981-93	9	100	1,337	2,493	5,783
Lake Windermere	1979-84	6	33	1	3	4
MacKenzie	1982	1	100	6	–	6
Nakusp	1979-93	15	93	2	40	118
Nelson	1985	1	100	86	–	86
North Pine	1983-93	11	55	3	16	51
Oliver-Osoyoos	1979-93	15	93	102	467	1,887
Penticton	1974-93	18	100	373	1,546	3,558
Prince George	1989-93	5	100	157	276	558
Princeton	1992-93	2	100	6	14	22
Quesnel	1981-84	4	75	6	56	68
Revelstoke	1981-93	13	100	18	90	115

▶

◀ European Starling

Count locality	Count years	Total counts	Frequency of occurrence (%)	Range of total counts when species occurred Low	Median	High
Salmon Arm	1988-92	5	100	117	335	443
Shuswap Lake Park	1972-93	22	86	11	37	328
Smithers	1977-93	17	100	24	84	305
Vaseux Lake	1974-93	20	100	3	166	1,171
Vernon	1975-93	19	100	240	858	2,755
Yoho National Park	1975-84	5	40	2	2	2

Crested Myna

Highest provincial total: 1,774 (1960)
Highest interior total: Not recorded
Highest coastal total: 1,774 (1960)

Count locality	Count years	Total counts	Frequency of occurrence (%)	Range of total counts when species occurred Low	Median	High
Coastal						
Ladner	1957-93	35	80	2	23	300
Nanaimo	1963-93	28	21	4	13	63
Pitt Meadows	1972-93	22	14	1	2	2
Vancouver	1957-93	37	97	8	160	1,632

Hutton's Vireo

Highest provincial total: 43 (1989)
Highest interior total: Not recorded
Highest coastal total: 43 (1989)

Count locality	Count years	Total counts	Frequency of occurrence (%)	Range of total counts when species occurred Low	Median	High
Coastal						
Bamfield	1986-93	8	13	1	–	1
Campbell River	1972-93	19	11	2	3	3
Comox	1961-93	32	9	1	2	4
Deep Bay	1975-93	19	32	1	1	2
Duncan	1970-93	24	46	1	1	4
Ladner	1957-93	35	63	1	4	16
Nanaimo	1963-93	28	25	1	1	5
Parksville–Qualicum Beach	1991-93	3	33	4	–	4
Pender Harbour	1991-93	3	67	1	2	2
Pender Islands	1964-93	28	39	1	2	6
Pitt Meadows	1972-93	22	23	1	1	2
Sooke	1983-93	11	73	1	2	5
Squamish	1980-93	13	15	1	1	1
Sunshine Coast	1979-93	14	50	1	1	4
Vancouver	1957-93	37	68	1	2	9
Victoria	1958-93	36	44	1	3	7
White Rock	1971-93	23	65	1	2	9

APPENDIX 3

Summary of Breeding Bird Surveys in British Columbia: 1969 Through 1994

The following tables summarize, by species, the Breeding Bird Survey data for flycatchers through vireos in British Columbia. These summaries were compiled from data obtained from the Bird Banding Office, Canadian Wildlife Service, Ottawa.

"Survey route" refers to the most recent name under which that survey has been published (Table 11). The time span of survey activity for a particular location is shown under "Survey years"; discontinuities can be determined by comparing "Survey years" with "Total surveys." "Frequency of occurrence" indicates the percentage of the surveys in a locality on which a species has been recorded.

The general abundance of the species over the count periods is shown by the range of values in the last 3 columns. The highest and lowest numbers recorded are given. An

Figure 629. Map of British Columbia showing locations of official Breeding Bird Surveys, 1968-94.

asterisk following a "High" number indicates that some caution is required in interpreting the data. Refer to the appropriate species account or to the note at the end of that species' entry in the table.

The median has been chosen as an indicator of the "usual" value when the species does occur. The middle value of the data set, it will lie closer to the low value when the species frequently occurs in small numbers, and closer to the high value when a species rarely occurs in small numbers. In a symmetrically distributed data set, the median is an unbiased and consistent estimate of the mean (Zar 1974). Surveys of zero have not been used as the "Low" value, but they can be determined using the values under "Frequency of occurrence" and "Total counts."

For example, there is about a 50% chance of seeing an Olive-sided Flycatcher on the Albion Breeding Bird Survey, and there is a high probability that this is an accurate estimate of the chance of seeing the bird because there is a long history of surveys on which to base this estimate. Similarly, if you see the species, you are likely to see about 2 birds, but more than 5 would be exceptional.

See Fig. 629 for the locations of official Breeding Bird Surveys and Table 11 for the number of surveys in each locality.

Figure 630. Since 1977, Richard R. Howie has completed up to 3 Breeding Bird Survey routes each summer in the southern interior of British Columbia (Campbell Range Road, Barnhartvale, 10 August 1996; R. Wayne Campbell).

Table 11. Breeding Bird Survey Routes in British Columbia: 1968 through 1994.

Survey route	Number of surveys
Coastal	
Alberni	6
Albion	21
Campbell River	19
Chemainus	4
Chilliwack	11
Comox Lake	3
Coquitlam	5
Courtenay	2
Elsie Lake	6
Gibsons Landing	10
Kispiox	3
Kitsumkalum	19
Kwinitsa	26
Masset	5
Meziadin	2
Nanaimo River	22
Nass River	1
Pemberton	18
Pitt Meadows	14
Point Grey	11
Port Renfrew	17
Port Hardy	4
Queen Charlotte City	3
Saltery Bay	3
Seabird	16
Squamish	9
Victoria	17
TOTAL	277

Survey route	Number of surveys
Interior	
Adams Lake	6
Alexis Creek	14
Ashnola	2
Atlin Road	1
Beaverdell	24
Bridge Lake	20
Bromley	5
Brookmere	10
Canford	10
Canoe Point	2
Cassiar	2
Chilkat Pass	2
Christian Valley	2
Chu Chua	2
Columbia Lake	7
Creighton Valley	16
Creston	1
Ferndale	15
Fort Nelson	5
Fort St. John	9
Fraser Lake	5
Gerrard	10
Gnat Pass	2
Golden	22
Goldstream Mountain	2
Gosnell	4
Grand Forks	22
Haines Summit	2
Hendrix Lake	1
Horsefly	11
Hudson's Hope	1
Illecilewaet	4
Kamloops	15
Kinaskan Lake	1
Kootenay	10
Kuskonook	21
Lac la Hache	11
Lavington	23
Mabel Lake	19
McLeod Lake	9
Meadowlake	2
Mount Morice	5
Needles	26
Nicola	2
Oliver	18
Osprey Lake	10
Pavilion	6
Pennington	4
Pinkut Creek	1
Pleasant Valley	6
Prince George	20
Princeton	8
Punchaw	9
Riske Creek	11
Salmo	3
Salmon Arm	19
Scotch Creek	6
Shalalth	4
Spillimacheen	17
Steamboat	4
Succour Creek	2
Summerland	21
Summit Lake	8
Syringa Creek	7
Tatlayoko Lake	6
Telkwa High Road	11
Tetsa River	1
Tokay	4
Tupper	6
Wasa	19
Williams Lake	21
Wingdam	3
Zincton	22
TOTAL	663

Olive-sided Flycatcher

Survey route	Survey years	Total surveys	Frequency of occurrence (%)	Range of total surveys when species occurred		
				Low	Median	High
Coastal						
Alberni	1968-79	6	67	1	4	15
Albion	1968-92	21	48	1	2	5
Campbell River	1973-93	19	95	1	2	9
Chemainus	1969-88	4	100	1	3	12
Chilliwack	1973-94	11	36	1	1	3
Comox Lake	1973-75	3	100	3	8	10
Coquitlam	1990-94	5	20	1	–	1
Courtenay	1992-93	2	100	2	4	5
Elsie Lake	1970-94	6	83	2	4	22
Gibsons Landing	1973-94	10	30	1	1	2
Kispiox	1975-94	3	100	2	2	3
Kitsumkalum	1974-94	19	79	1	3	6
Kwinitsa	1968-94	26	19	1	1	2
Meziadin	1975-77	2	50	1	–	1
Nanaimo River	1973-94	22	86	1	2	6
Nass River	1975	1	100	3	–	3
Pemberton	1974-94	18	56	1	2	7
Pitt Meadows	1973-89	14	43	1	1	3
Point Grey	1974-92	11	64	1	1	3
Port Hardy	1982-89	4	100	2	5	17
Port Renfrew	1973-94	17	100	1	4	9
Saltery Bay	1974-76	3	100	1	6	12
Seabird	1969-94	16	75	1	2	10
Squamish	1973-94	9	78	1	2	7
Victoria	1973-94	17	71	1	2	5
Interior						
Adams Lake	1989-94	6	17	1	–	1
Alexis Creek	1973-86	14	93	1	3	8
Ashnola	1973-74	2	100	1	2	2
Beaverdell	1970-94	24	71	1	2	3
Bridge Lake	1974-94	20	90	2	6	11
Bromley	1973-94	5	40	1	2	2
Brookmere	1973-92	10	90	1	4	9
Canford	1974-93	10	90	1	2	5
Cassiar	1972-75	2	50	5	–	5
Christian Valley	1993-94	2	100	1	2	3
Chu Chua	1993-94	2	50	3	–	3
Creighton Valley	1979-94	16	44	1	1	3
Ferndale	1968-94	15	100	1	4	30
Fort Nelson	1974-80	5	40	1	1	1
Fort St. John	1974-90	9	78	1	1	2
Fraser Lake	1968-78	5	60	3	6	6
Gerrard	1976-85	10	100	1	3	5
Gnat Pass	1975-93	2	100	1	1	1
Golden	1973-94	22	5	1	–	1
Gosnell	1969-74	4	75	1	1	3
Grand Forks	1973-94	22	59	1	1	3
Hendrix Lake	1976	1	100	6	–	6
Horsefly	1974-88	11	36	2	3	6
Hudson's Hope	1976	1	100	2	–	2
Kamloops	1974-94	15	67	1	2	3
Kootenay	1981-94	10	80	1	2	9
Kuskonook	1968-91	21	5	1	–	1
Lac la Hache	1973-86	11	91	2	7	17
Lavington	1972-94	24	14	1	1	2
Mabel Lake	1973-94	19	37	1	1	6
McLeod Lake	1968-94	9	56	1	2	8
Meadowlake	1968-70	2	50	5	–	5
Mount Morice	1968-75	5	100	1	6	11
Needles	1968-94	26	50	1	2	4
Oliver	1973-94	18	83	1	3	8
Osprey Lake	1976-94	10	80	1	4	4
Pennington	1987-91	4	50	1	2	2
Pinkut Creek	1975	1	100	8	–	8
Pleasant Valley	1980-93	6	67	2	3	4
Prince George	1973-94	20	75	1	2	9
Princeton	1968-94	8	63	1	2	8
Punchaw	1974-94	9	78	1	3	5
Riske Creek	1973-92	11	91	2	4	11

▶

◀ *Olive-sided Flycatcher*

Survey route	Survey years	Total surveys	Frequency of occurrence (%)	Low	Range of total surveys when species occurred Median	High
Salmo	1969-75	3	67	2	5	7
Salmon Arm	1968-94	19	74	1	1	6
Scotch Creek	1989-94	6	83	1	3	7
Shalalth	1968-94	4	25	1	–	1
Spillimacheen	1976-93	17	76	1	2	5
Succour Creek	1993-94	2	50	3	–	3
Summerland	1973-94	21	29	1	1	4
Summit Lake	1972-94	8	88	2	3	10
Syringa Creek	1986-93	7	43	1	1	3
Tatlayoko Lake	1989-94	6	83	1	1	4
Telkwa High Road	1974-94	11	100	1	8	20
Tokay	1968-75	4	25	2	–	2
Tupper	1974-94	6	100	1	3	4
Wasa	1973-94	19	5	1	–	1
Williams Lake	1973-93	21	67	1	2	4
Wingdam	1968-70	3	100	1	4	6
Zincton	1968-94	22	18	1	1	2

Western Wood-Pewee

Survey route	Survey years	Total surveys	Frequency of occurrence (%)	Low	Range of total surveys when species occurred Median	High
Coastal						
Alberni	1968-79	6	33	2	4	6
Albion	1968-92	21	33	1	1	2
Campbell River	1973-93	19	16	5	5	11
Chemainus	1969-88	4	25	1	–	1
Chilliwack	1973-94	11	73	1	4	6
Coquitlam	1990-94	5	80	1	2	2
Courtenay	1992-93	2	50	3	–	3
Gibsons Landing	1973-94	10	50	1	1	5
Kispiox	1975-94	3	100	5	6	19
Kitsumkalum	1974-94	19	95	1	5	14
Kwinitsa	1968-94	26	15	1	1	2
Nanaimo River	1973-94	22	41	1	1	2
Nass River	1975	1	100	12	–	12
Pemberton	1974-94	18	56	1	2	4
Pitt Meadows	1973-89	14	57	1	1	4
Point Grey	1974-92	11	18	1	1	1
Port Hardy	1982-89	4	25	7	–	7
Port Renfrew	1973-94	17	12	1	1	1
Seabird	1969-94	16	100	1	7	17
Squamish	1973-94	9	67	1	1	7
Interior						
Adams Lake	1989-94	6	100	1	2	4
Alexis Creek	1973-86	14	100	3	10	22
Ashnola	1973-74	2	50	5	–	5
Beaverdell	1970-94	24	100	2	10	21
Bridge Lake	1974-94	20	100	2	13	23
Bromley	1973-94	5	80	3	6	10
Brookmere	1973-92	10	90	3	6	12
Canford	1974-93	10	100	1	6	12
Canoe Point	1993-94	2	100	2	3	4
Cassiar	1972-75	2	50	4	–	4
Christian Valley	1993-94	2	100	2	3	3
Chu Chua	1993-94	2	100	7	9	11
Columbia Lake	1973-83	7	14	2	–	2
Creighton Valley	1979-94	16	88	1	2	5
Creston	1984	1	100	11	11	
Ferndale	1968-94	15	100	2	11	32
Fort Nelson	1974-80	5	100	2	3	5
Fort St. John	1974-90	9	100	2	10	15
Fraser Lake	1968-78	5	100	3	12	29
Gerrard	1976-85	10	100	1	6	9

▶

◀ *Western Wood-Pewee*

Survey route	Survey years	Total surveys	Frequency of occurrence (%)	Range of total surveys when species occurred Low	Median	High
Gnat Pass	1975-93	2	50	4	–	4
Golden	1973-94	22	95	1	7	20
Goldstream Mountain	1993-94	2	50	1	–	1
Gosnell	1969-74	4	100	2	7	9
Grand Forks	1973-94	22	100	3	8	25
Hendrix Lake	1976	1	100	7	–	7
Horsefly	1974-88	11	100	1	4	8
Hudson's Hope	1976	1	100	13	–	13
Illecilewaet	1973-83	4	50	1	2	2
Kamloops	1974-94	15	93	3	8	17
Kinaskan Lake	1975	1	100	1	–	1
Kootenay	1981-94	10	90	1	4	10
Kuskonook	1968-91	21	67	1	4	8
Lac la Hache	1973-86	11	100	6	10	26
Lavington	1972-94	24	100	7	12	21
Mabel Lake	1973-94	19	100	5	9	16
McLeod Lake	1968-94	9	100	1	5	18
Meadowlake	1968-70	2	100	1	2	2
Mount Morice	1968-75	5	100	3	15	19
Needles	1968-94	26	96	1	6	13
Nicola	1973-74	2	100	1	2	3
Oliver	1973-94	18	100	1	6	15
Osprey Lake	1976-94	10	100	2	4	9
Pavilion	1974-94	6	67	2	5	8
Pennington	1987-91	4	50	1	2	3
Pinkut Creek	1975	1	100	6	–	6
Pleasant Valley	1980-93	6	100	2	5	6
Prince George	1973-94	20	100	1	6	16
Princeton	1968-94	8	50	1	2	3
Punchaw	1974-94	9	100	2	11	27
Riske Creek	1973-92	11	100	2	12	21
Salmon Arm	1968-94	19	100	1	4	15
Scotch Creek	1989-94	6	83	1	2	4
Shalalth	1968-94	4	25	1	–	1
Spillimacheen	1976-93	17	100	1	9	12
Steamboat	1974-80	4	75	2	2	5
Succour Creek	1993-94	2	100	1	2	3
Summerland	1973-94	21	100	2	5	11
Summit Lake	1972-94	8	75	5	13	23
Syringa Creek	1986-93	7	100	3	6	8
Tatlayoko Lake	1989-94	6	83	2	3	4
Telkwa High Road	1974-94	11	100	1	21	39
Tokay	1968-75	4	50	5	7	9
Tupper	1974-94	6	100	8	13	18
Wasa	1973-94	19	100	1	3	9
Williams Lake	1973-93	21	100	1	4	9
Wingdam	1968-70	3	67	1	2	3
Zincton	1968-94	22	50	1	1	3

Yellow-bellied Flycatcher

Survey route	Survey years	Total surveys	Frequency of occurrence (%)	Range of total surveys when species occurred Low	Median	High
Interior						
Ferndale	1968-94	15	13	1	2	2

Alder Flycatcher

Survey route	Survey years	Total surveys	Frequency of occurrence (%)	Range of total surveys when species occurred: Low	Median	High
Coastal						
Chilliwack	1973-94	11	9	1	–	1
Kispiox	1975-94	3	100	4	6	19
Kwinitsa	1968-94	26	4	3	–	3
Meziadin	1975-77	2	50	5	–	5
Nass River	1975	1	100	8	–	8
Pemberton	1974-94	18	6	1	–	1
Pitt Meadows	1973-89	14	7	2	–	2
Seabird	1969-94	16	6	3	–	3
Squamish	1973-94	9	11	1	–	1
Interior						
Adams Lake	1989-94	6	17	1	–	1
Alexis Creek	1973-86	14	29	1	1	2
Atlin Road	1989	1	100	3	–	3
Bridge Lake	1974-94	20	20	1	2	2
Canford	1974-93	10	20	1	1	1
Cassiar	1972-75	2	50	4	–	4
Columbia Lake	1973-83	7	14	1	–	1
Ferndale	1968-94	15	60	1	12	30
Fort Nelson	1974-80	5	100	13	50	65
Fort St. John	1974-90	9	100	9	21	36
Gerrard	1976-85	10	40	1	2	2
Gnat Pass	1975-93	2	50	2	–	2
Golden	1973-94	22	64	1	3	7
Goldstream Mountain	1993-94	2	100	11	12	12
Gosnell	1969-74	4	25	9	–	9
Grand Forks	1973-94	22	86	2	7	14
Haines Summit	1993-94	2	50	3	–	3
Horsefly	1974-88	11	55	1	2	4
Hudson's Hope	1976	1	100	30	–	30
Illecilewaet	1973-83	4	25	1	–	1
Lac la Hache	1973-86	11	27	1	2	4
Mabel Lake	1973-94	19	5	4	–	4
McLeod Lake	1968-94	9	56	26	29	38
Mount Morice	1968-75	5	20	7	–	7
Needles	1968-94	26	8	1	1	1
Pennington	1987-91	4	25	2	–	2
Prince George	1973-94	20	65	6	11	23
Punchaw	1974-94	9	100	1	10	21
Riske Creek	1973-92	11	45	1	1	4
Salmon Arm	1968-94	19	11	1	3	4
Scotch Creek	1989-94	6	100	1	1	1
Steamboat	1974-80	4	100	14	29	54
Succour Creek	1993-94	2	100	4	8	11
Summit Lake	1972-94	8	75	6	18	32
Syringa Creek	1986-93	7	14	1	–	1
Tatlayoko Lake	1989-94	6	17	1	–	1
Telkwa High Road	1974-94	11	73	1	4	10
Tupper	1974-94	6	100	15	28	37
Wasa	1973-94	19	32	1	2	2
Williams Lake	1973-93	21	24	1	1	2
Zincton	1968-94	22	27	1	1	2

Willow Flycatcher

Survey route	Survey years	Total surveys	Frequency of occurrence (%)	Range of total surveys when species occurred: Low	Median	High
Coastal						
Alberni	1968-79	6	83	1	4	14
Albion	1968-92	21	95	9	14	24
Campbell River	1973-93	19	95	1	3	12
Chemainus	1969-88	4	75	5	11	22
Chilliwack	1973-94	11	100	5	9	20
Comox Lake	1973-75	3	100	1	2	3

▶

◀ *Willow Flycatcher*

Survey route	Survey years	Total surveys	Frequency of occurrence (%)	Range of total surveys when species occurred Low	Median	High
Coquitlam	1990-94	5	100	4	10	12
Courtenay	1992-93	2	100	10	13	16
Elsie Lake	1970-94	6	67	1	3	7
Gibsons Landing	1973-94	10	100	10	14	45
Kitsumkalum	1974-94	19	100	1	8	20
Kwinitsa	1968-94	26	65	1	2	4
Nanaimo River	1973-94	22	91	1	4	12
Pemberton	1974-94	18	61	1	2	5
Pitt Meadows	1973-89	14	100	2	13	27
Point Grey	1974-92	11	82	1	2	3
Port Renfrew	1973-94	17	65	1	2	31
Saltery Bay	1974-76	3	100	21	22	28
Seabird	1969-94	16	100	4	11	24
Squamish	1973-94	9	100	13	24	31
Victoria	1973-94	17	18	1	2	4
Interior						
Adams Lake	1989-94	6	17	1	–	1
Alexis Creek	1973-86	14	71	1	3	11
Ashnola	1973-74	2	50	1	–	1
Beaverdell	1970-94	24	96	1	4	10
Bridge Lake	1974-94	20	85	1	3	8
Bromley	1973-94	5	100	3	6	7
Brookmere	1973-92	10	100	1	12	20
Canford	1974-93	10	90	5	8	16
Canoe Point	1993-94	2	100	4	6	8
Christian Valley	1993-94	2	100	1	3	4
Chu Chua	1993-94	2	50	1	–	1
Columbia Lake	1973-83	7	29	1	1	1
Creighton Valley	1979-94	16	100	2	3	6
Ferndale	1968-94	15	60	2	7	25
Fraser Lake	1968-78	5	80	2	7	14
Gerrard	1976-85	10	70	1	2	4
Gnat Pass	1975-93	2	50	6	–	6
Golden	1973-94	22	100	2	16	32
Goldstream Mountain	1993-94	2	100	2	3	3
Gosnell	1969-74	4	100	1	4	9
Grand Forks	1973-94	22	14	1	4	9
Hendrix Lake	1976	1	100	1	–	1
Horsefly	1974-88	11	64	1	1	3
Kamloops	1974-94	15	80	1	3	6
Kuskonook	1968-91	21	5	3	–	3
Lac la Hache	1973-86	11	91	3	11	15
Lavington	1972-94	23	87	1	1	5
Mabel Lake	1973-94	19	95	1	4	11
McLeod Lake	1968-94	9	44	3	7	9
Meadowlake	1968-70	2	50	4	–	4
Mount Morice	1968-75	5	60	1	8	12
Needles	1968-94	26	62	1	1	4
Nicola	1973-74	2	100	2	3	3
Oliver	1973-94	18	22	1	1	1
Osprey Lake	1976-94	10	60	1	13	15
Pavilion	1974-94	6	100	1	4	16
Pleasant Valley	1980-93	6	100	12	15	18
Prince George	1973-94	20	35	4	8	16
Princeton	1968-94	8	50	1	1	1
Riske Creek	1973-92	11	82	1	7	14
Salmo	1969-75	3	33	19	–	19
Salmon Arm	1968-94	19	89	1	5	11
Scotch Creek	1989-94	6	100	3	5	6
Spillimacheen	1976-93	17	100	4	7	11
Succour Creek	1993-94	2	100	23	26	29
Summerland	1973-94	21	100	7	12	23
Summit Lake	1972-94	8	25	4	5	6
Syringa Creek	1986-93	7	86	2	3	4
Tatlayoko Lake	1989-94	6	17	3	–	3
Tokay	1968-75	4	25	4	–	4
Wasa	1973-94	19	58	1	2	8
Williams Lake	1973-93	21	95	1	4	8
Wingdam	1968-70	3	100	3	10	10
Zincton	1968-94	22	82	1	4	6

Least Flycatcher

Survey route	Survey years	Total surveys	Frequency of occurrence (%)	Range of total surveys when species occurred: Low	Median	High
Coastal						
Chilliwack	1973-94	11	9	1	–	1
Kispiox	1975-94	3	67	2	5	7
Kitsumkalum	1974-94	19	5	10	–	10
Meziadin	1975-77	2	50	1	–	1
Pemberton	1974-94	18	11	1	2	3
Seabird	1969-94	16	6	1	–	1
Victoria	1973-94	17	6	1	–	1
Interior						
Adams Lake	1989-94	6	17	2	–	2
Alexis Creek	1973-86	14	79	1	2	11
Beaverdell	1970-94	24	13	1	1	1
Bridge Lake	1974-94	20	85	1	1	6
Brookmere	1973-92	10	40	1	1	1
Canoe Point	1993-94	2	50	2	–	2
Cassiar	1972-75	2	50	1	–	1
Chilkat Pass	1975-76	2	50	1	–	1
Christian Valley	1993-94	2	100	3	4	4
Chu Chua	1993-94	2	100	1	2	2
Columbia Lake	1973-83	7	29	1	7	13
Ferndale	1968-94	15	87	1	5	17
Fort Nelson	1974-80	5	100	8	13	16
Fort St. John	1974-90	9	100	7	19	26
Fraser Lake	1968-78	5	20	9	–	9
Gerrard	1976-85	10	10	1	–	1
Gnat Pass	1975-93	2	50	5	–	5
Golden	1973-94	22	95	2	7	29
Gosnell	1969-74	4	25	2	–	2
Grand Forks	1973-94	22	5	3	–	3
Horsefly	1974-88	11	73	1	2	8
Hudson's Hope	1976	1	100	42	–	42
Kamloops	1974-94	15	13	1	1	1
Kuskonook	1968-91	21	38	2	6	14
Lac la Hache	1973-86	11	73	2	3	6
Lavington	1972-94	23	8	1	1	1
Mabel Lake	1973-94	19	21	1	2	2
McLeod Lake	1968-94	9	89	1	3	11
Mount Morice	1968-75	5	40	2	3	4
Needles	1968-94	26	23	1	1	1
Oliver	1973-94	18	6	1	–	1
Pennington	1987-91	4	50	1	1	1
Pinkut Creek	1975	1	100	1	–	1
Prince George	1973-94	20	90	1	4	11
Punchaw	1974-94	9	89	1	3	10
Riske Creek	1973-92	11	73	1	3	6
Salmon Arm	1968-94	19	5	1	–	1
Spillimacheen	1976-93	17	94	2	5	7
Steamboat	1974-80	4	75	1	3	5
Succour Creek	1993-94	2	100	4	4	4
Summerland	1973-94	21	5	5	–	5
Summit Lake	1972-94	8	88	3	6	9
Syringa Creek	1986-93	7	57	1	1	2
Telkwa High Road	1974-94	11	82	1	10	40
Tokay	1968-75	4	25	2	–	2
Tupper	1974-94	6	100	30	43	67
Wasa	1973-94	19	58	1	1	4
Williams Lake	1973-93	21	33	1	1	3
Zincton	1968-94	22	18	1	1	12

Hammond's Flycatcher

Survey route	Survey years	Total surveys	Frequency of occurrence (%)	Range of total surveys when species occurred Low	Median	High
Coastal						
Alberni	1968-79	6	50	3	4	6
Campbell River	1973-93	19	89	1	2	4
Chemainus	1969-88	4	75	7	9	11
Chilliwack	1973-94	11	18	2	2	2
Courtenay	1992-93	2	100	7	10	12
Elsie Lake	1970-94	6	83	2	4	10
Gibsons Landing	1973-94	10	60	1	3	9
Kispiox	1975-94	3	100	1	23	25
Kitsumkalum	1974-94	19	16	1	2	4
Kwinitsa	1968-94	26	8	1	6	11
Nanaimo River	1973-94	22	64	1	2	7
Nass River	1975	1	100	10	–	10
Pemberton	1974-94	18	78	1	3	14
Point Grey	1974-92	11	18	1	1	1
Port Renfrew	1973-94	17	6	1	–	1
Seabird	1969-94	16	38	1	2	7
Squamish	1973-94	9	44	7	9	10
Interior						
Adams Lake	1989-94	6	33	1	3	4
Alexis Creek	1973-86	14	64	1	3	21
Beaverdell	1970-94	24	100	8	33	79
Bridge Lake	1974-94	20	35	2	2	8
Bromley	1973-94	5	40	1	1	1
Brookmere	1973-92	10	90	4	10	20
Canford	1974-93	10	100	3	6	14
Canoe Point	1993-94	2	100	3	7	10
Chilkat Pass	1975-76	2	50	1	–	1
Christian Valley	1993-94	2	100	5	6	7
Chu Chua	1993-94	2	100	16	16	16
Creighton Valley	1979-94	16	81	18	28	41
Creston	1984	1	100	2	–	2
Ferndale	1968-94	15	80	1	15	45
Gerrard	1976-85	10	30	4	18	25
Golden	1973-94	22	45	1	2	19
Goldstream Mountain	1993-94	2	100	8	11	13
Gosnell	1969-74	4	100	5	19	39
Grand Forks	1973-94	22	100	2	6	14
Haines Summit	1993-94	2	100	1	1	1
Horsefly	1974-88	11	73	1	3	4
Illecilewaet	1973-83	4	75	7	14	19
Kamloops	1974-94	15	7	1	–	1
Kinaskan Lake	1975	1	100	1	–	1
Kootenay	1981-94	10	100	4	25	38
Kuskonook	1968-91	21	67	1	4	15
Lac la Hache	1973-86	11	64	1	2	7
Lavington	1972-94	23	100	8	18	29
Mabel Lake	1973-94	19	100	1	16	27
McLeod Lake	1968-94	9	89	22	32	39
Mount Morice	1968-75	5	60	1	6	11
Needles	1968-94	26	92	2	22	31
Oliver	1973-94	18	67	1	2	8
Osprey Lake	1976-94	10	40	1	2	7
Pavilion	1974-94	6	83	2	3	30
Pinkut Creek	1975	1	100	7	–	7
Pleasant Valley	1980-93	6	100	1	2	4
Prince George	1973-94	20	80	1	3	10
Princeton	1968-94	8	38	1	2	3
Punchaw	1974-94	9	44	21	24	40
Riske Creek	1973-92	11	36	1	2	3
Salmon Arm	1968-94	19	68	3	5	14
Scotch Creek	1989-94	6	100	1	3	11
Shalalth	1968-94	4	25	4	–	4
Spillimacheen	1976-93	17	53	1	4	17
Succour Creek	1993-94	2	100	26	32	37
Summerland	1973-94	21	57	1	3	11
Summit Lake	1972-94	8	100	3	14	29
Syringa Creek	1986-93	7	86	3	7	10
Tatlayoko Lake	1989-94	6	67	2	4	4

▶

◀ *Hammond's Flycatcher*

Survey route	Survey years	Total surveys	Frequency of occurrence (%)	Range of total surveys when species occurred Low	Median	High
Telkwa High Road	1974-94	11	45	1	3	15
Tokay	1968-75	4	25	29	–	29
Wasa	1973-94	19	26	1	6	9
Williams Lake	1973-93	21	52	1	1	6
Wingdam	1968-70	3	100	5	36	41
Zincton	1968-94	22	32	11	30	37

Dusky Flycatcher

Survey route	Survey years	Total surveys	Frequency of occurrence (%)	Range of total surveys when species occurred Low	Median	High
Coastal						
Kispiox	1975-94	3	100	2	3	6
Kitsumkalum	1974-94	19	37	1	2	4
Kwinitsa	1968-94	26	62	1	3	13
Pemberton	1974-94	18	17	1	1	14
Seabird	1969-94	16	25	1	1	1
Interior						
Adams Lake	1989-94	6	100	22	30	47
Alexis Creek	1973-86	14	71	5	17	34
Beaverdell	1970-94	24	100	1	5	11
Bridge Lake	1974-94	20	95	4	13	23
Bromley	1973-94	5	80	2	5	5
Brookmere	1973-92	10	90	13	25	44
Canford	1974-93	10	100	11	18	33
Canoe Point	1993-94	2	100	1	8	14
Christian Valley	1993-94	2	100	6	20	34
Chu Chua	1993-94	2	100	21	34	46
Columbia Lake	1973-83	7	86	4	13	13
Creighton Valley	1979-94	16	50	1	1	25
Creston	1984	1	100	1	–	1
Ferndale	1968-94	15	53	1	2	8
Fraser Lake	1968-78	5	20	1	–	1
Gerrard	1976-85	10	90	3	27	34
Gnat Pass	1975-93	2	50	19	–	19
Golden	1973-94	22	77	2	5	12
Goldstream Mountain	1993-94	2	50	2	–	2
Gosnell	1969-74	4	100	1	7	12
Grand Forks	1973-94	22	91	1	4	13
Haines Summit	1993-94	2	100	9	10	11
Hendrix Lake	1976	1	100	1	–	1
Horsefly	1974-88	11	45	1	3	8
Illecilewaet	1973-83	4	25	7	–	7
Kamloops	1974-94	15	93	1	2	6
Kootenay	1981-94	10	30	1	1	2
Kuskonook	1968-91	21	67	1	3	19
Lac la Hache	1973-86	11	64	2	5	18
Lavington	1972-94	23	71	1	2	5
Mabel Lake	1973-94	19	84	1	4	13
McLeod Lake	1968-94	9	44	1	2	3
Mount Morice	1968-75	5	60	1	3	4
Needles	1968-94	26	73	1	2	11
Nicola	1973-74	2	100	4	7	9
Oliver	1973-94	18	89	7	11	21
Osprey Lake	1976-94	10	60	1	7	11
Pavilion	1974-94	6	67	1	5	12
Pleasant Valley	1980-93	6	100	4	5	7
Prince George	1973-94	20	30	1	4	5
Princeton	1968-94	8	63	1	4	8
Punchaw	1974-94	9	44	3	3	6
Riske Creek	1973-92	11	82	1	9	21
Salmon Arm	1968-94	19	37	1	6	12
Scotch Creek	1989-94	6	100	8	17	23

▶

◀ *Dusky Flycatcher*

Survey route	Survey years	Total surveys	Frequency of occurrence (%)	Low	Range of total surveys when species occurred Median	High
Shalalth	1968-94	4	25	2	–	2
Spillimacheen	1976-93	17	100	3	17	32
Succour Creek	1993-94	2	100	9	11	12
Summerland	1973-94	21	100	5	15	25
Summit Lake	1972-94	8	63	3	9	10
Syringa Creek	1986-93	7	100	1	3	6
Tatlayoko Lake	1989-94	6	100	1	3	10
Telkwa High Road	1974-94	11	82	4	9	54
Tokay	1968-75	4	25	25	–	25
Wasa	1973-94	19	53	1	2	6
Williams Lake	1973-93	21	100	10	20	35
Wingdam	1968-70	3	33	1	–	1
Zincton	1968-94	22	82	1	16	35

Gray Flycatcher

Survey route	Survey years	Total surveys	Frequency of occurrence (%)	Low	Range of total surveys when species occurred Median	High
Interior						
Oliver	1973-94	18	28	1	1	2
Summerland	1973-94	21	43	1	3	10

"Western Flycatcher" Complex

Survey route	Survey years	Total surveys	Frequency of occurrence (%)	Low	Range of total surveys when species occurred Median	High
Coastal						
Alberni	1968-79	6	100	3	9	22
Albion	1968-92	21	67	1	3	8
Campbell River	1973-93	19	95	1	15	25
Chemainus	1969-88	4	100	8	18	29
Chilliwack	1973-94	11	45	1	1	2
Comox Lake	1973-75	3	100	8	8	10
Coquitlam	1990-94	5	40	1	1	1
Courtenay	1992-93	2	100	11	14	16
Gibsons Landing	1973-94	10	90	1	7	17
Kispiox	1975-94	3	67	1	2	3
Kwinitsa	1968-94	26	19	1	2	3
Masset	1989-94	5	80	42	61	69
Meziadin	1975-77	2	50	1	–	1
Nanaimo River	1973-94	22	91	1	11	24
Pemberton	1974-94	18	56	1	4	14
Pitt Meadows	1973-89	14	7	1	–	1
Point Grey	1974-92	11	82	1	2	4
Port Hardy	1982-89	4	100	15	46	48
Port Renfrew	1973-94	17	94	7	13	26
Queen Charlotte City	1989-94	3	100	22	37	65
Saltery Bay	1974-76	3	100	1	2	3
Seabird	1969-94	16	75	2	6	13
Squamish	1973-94	9	100	2	5	13
Victoria	1973-94	17	94	1	4	12
Interior						
Adams Lake	1989-94	6	17	1	–	1
Alexis Creek	1973-86	14	21	1	3	4
Beaverdell	1970-94	24	25	1	1	3
Bridge Lake	1974-94	20	10	1	1	1
Bromley	1973-94	5	80	1	3	4
Brookmere	1973-92	10	40	1	1	1

▶

◀ *"Western Flycatcher" Complex*

Survey route	Survey years	Total surveys	Frequency of occurrence (%)	Low	Range of total surveys when species occurred Median	High
Canford	1974-93	10	60	1	1	5
Canoe Point	1993-94	2	100	1	3	4
Christian Valley	1993-94	2	50	1	–	1
Chu Chua	1993-94	2	100	1	2	2
Columbia Lake	1973-83	7	29	1	2	3
Creighton Valley	1979-94	16	6	1	–	1
Gerrard	1976-85	10	10	3	–	3
Golden	1973-94	22	9	1	1	1
Grand Forks	1973-94	22	9	2	6	9
Horsefly	1974-88	11	55	1	2	4
Kamloops	1974-94	15	60	1	1	3
Kootenay	1981-94	10	20	1	1	1
Kuskonook	1968-91	21	67	1	10	18
Lac la Hache	1973-86	11	18	1	2	2
Lavington	1972-94	23	30	1	1	2
Mabel Lake	1973-94	19	95	1	4	12
Meadowlake	1968-70	2	50	7	–	7
Needles	1968-94	26	35	1	1	3
Oliver	1973-94	18	44	1	1	3
Osprey Lake	1976-94	10	40	2	3	3
Pavilion	1974-94	6	33	2	3	4
Pleasant Valley	1980-93	6	100	3	4	5
Princeton	1968-94	8	88	2	5	8
Riske Creek	1973-92	11	64	1	3	6
Salmon Arm	1968-94	19	37	1	1	2
Scotch Creek	1989-94	6	17	3	–	3
Shalalth	1968-94	4	75	2	4	6
Spillimacheen	1976-93	17	47	1	2	6
Summerland	1973-94	21	100	1	5	10
Syringa Creek	1986-93	7	43	1	1	1
Tatlayoko Lake	1989-94	6	17	1	–	1
Telkwa High Road	1974-94	11	9	1	–	1
Wasa	1973-94	19	5	1	–	1
Williams Lake	1973-93	21	14	1	1	2
Wingdam	1968-70	3	67	1	1	1
Zincton	1968-94	22	18	1	1	1

Eastern Phoebe

Survey route	Survey years	Total surveys	Frequency of occurrence (%)	Low	Range of total surveys when species occurred Median	High
Interior						
Fort Nelson	1974-80	5	20	1	–	1
Fort St. John	1974-90	9	44	1	2	3
Fraser Lake	1968-78	5	20	5	–	5*
Mount Morice	1968-75	5	20	10	–	10*
Prince George	1973-94	20	50	1	1	3*
Steamboat	1974-80	4	50	1	2	2
Tupper	1974-94	6	67	1	2	2

Say's Phoebe

Survey route	Survey years	Total surveys	Frequency of occurrence (%)	Low	Range of total surveys when species occurred Median	High
Coastal						
Kwinitsa	1968-94	26	4	1	–	1
Interior						
Beaverdell	1970-94	24	13	1	1	2

▶

◀ *Say's Phoebe*

Survey route	Survey years	Total surveys	Frequency of occurrence (%)	Range of total surveys when species occurred Low	Median	High
Canford	1974-93	10	40	1	1	1
Canoe Point	1993-94	2	50	1	–	1
Chilkat Pass	1975-76	2	50	1	–	1
Columbia Lake	1973-83	7	14	4	–	4
Fraser Lake	1968-78	5	20	3	–	3
Golden	1973-94	22	18	1	1	1
Kamloops	1974-94	15	60	1	1	5
Kuskonook	1968-91	21	5	3	–	3
Lavington	1972-94	23	40	1	1	4
Mabel Lake	1973-94	19	26	1	1	1
Mount Morice	1968-75	5	20	5	–	5
Oliver	1973-94	18	50	1	1	2
Pavilion	1974-94	6	17	1	–	1
Pennington	1987-91	4	50	1	2	3
Salmon Arm	1968-94	19	11	1	1	1
Summerland	1973-94	21	24	1	1	1
Wasa	1973-94	19	5	1	–	1
Williams Lake	1973-93	21	5	1	–	1

Western Kingbird

Survey route	Survey years	Total surveys	Frequency of occurrence (%)	Range of total surveys when species occurred Low	Median	High
Coastal						
Chilliwack	1973-94	11	18	1	1	1
Pemberton	1974-94	18	56	1	3	7
Saltery Bay	1974-76	3	33	1	–	1
Seabird	1969-94	16	38	1	1	2
Interior						
Ashnola	1973-74	2	100	1	3	5
Beaverdell	1970-94	24	29	1	2	6
Bridge Lake	1974-94	20	5	1	–	1
Bromley	1973-94	5	80	1	2	4
Brookmere	1973-92	10	10	1	–	1
Canford	1974-93	10	90	2	5	9
Canoe Point	1993-94	2	100	2	5	8
Chu Chua	1993-94	2	100	3	5	7
Columbia Lake	1973-83	7	86	1	2	3
Creighton Valley	1979-94	16	44	1	2	3
Creston	1984	1	100	4	–	4
Gnat Pass	1975-93	2	50	4	–	4
Golden	1973-94	22	18	1	2	3
Grand Forks	1973-94	22	100	2	5	12
Kamloops	1974-94	15	100	3	7	12
Kuskonook	1968-91	21	5	1	–	1
Lavington	1972-94	23	100	3	10	20
Mabel Lake	1973-94	19	100	1	6	18
Needles	1968-94	26	8	1	1	1
Nicola	1973-74	2	100	10	12	13
Oliver	1973-94	18	100	2	5	14
Pavilion	1974-94	6	100	2	5	12
Pleasant Valley	1980-93	6	67	1	3	4
Princeton	1968-94	8	13	1	–	1
Riske Creek	1973-92	11	9	1	–	1
Salmo	1969-75	3	33	2	–	2
Salmon Arm	1968-94	19	74	1	3	6
Spillimacheen	1976-93	17	82	1	2	5
Summerland	1973-94	21	86	1	2	6
Syringa Creek	1986-93	7	29	2	2	2
Tokay	1968-75	4	25	1	–	1
Wasa	1973-94	19	11	1	2	2
Williams Lake	1973-93	21	33	1	1	2
Zincton	1968-94	22	5	1	–	1

Eastern Kingbird

Survey route	Survey years	Total surveys	Frequency of occurrence (%)	Range of total surveys when species occurred Low	Median	High
Coastal						
Chilliwack	1973-94	11	45	2	2	4
Pemberton	1974-94	18	61	1	2	/
Seabird	1969-94	16	38	1	1	2
Interior						
Alexis Creek	1973-86	14	29	1	3	3
Ashnola	1973-74	2	100	2	3	3
Beaverdell	1970-94	24	42	1	1	3
Bridge Lake	1974-94	20	5	1	–	1
Bromley	1973-94	5	20	2	–	2
Brookmere	1973-92	10	60	1	2	3
Canford	1974-93	10	90	1	2	6
Canoe Point	1993-94	2	50	3	–	3
Chu Chua	1993-94	2	50	1	–	1
Columbia Lake	1973-83	7	71	1	2	3
Creighton Valley	1979-94	16	19	1	1	2
Creston	1984	1	100	5	–	5
Fort St. John	1974-90	9	78	1	3	5
Gnat Pass	1975-93	2	50	5	–	5
Golden	1973-94	22	95	1	4	22
Grand Forks	1973-94	22	68	1	2	7
Horsefly	1974-88	11	36	1	4	5
Hudson's Hope	1976	1	100	1	–	1
Kamloops	1974-94	15	93	1	4	5
Kuskonook	1968-91	21	24	1	2	4
Lac la Hache	1973-86	11	36	1	2	3
Lavington	1972-94	23	100	1	3	15
Mabel Lake	1973-94	19	68	1	2	8
Mount Morice	1968-75	5	20	1	–	1
Needles	1968-94	26	38	1	2	4
Nicola	1973-74	2	50	2	–	2
Oliver	1973-94	18	89	1	5	7
Osprey Lake	1976-94	10	30	1	2	2
Pavilion	1974-94	6	83	2	4	8
Pleasant Valley	1980-93	6	100	2	4	5
Prince George	1973-94	20	10	1	1	1
Princeton	1968-94	8	13	1	–	1
Punchaw	1974-94	9	11	1	–	1
Riske Creek	1973-92	11	27	1	1	4
Salmo	1969-75	3	67	3	4	4
Salmon Arm	1968-94	19	47	1	2	6
Spillimacheen	1976-93	17	94	1	3	9
Succour Creek	1993-94	2	50	1	–	1
Summerland	1973-94	21	95	1	3	5
Syringa Creek	1986-93	7	71	1	3	3
Tokay	1968-75	4	100	2	4	5
Tupper	1974-94	6	17	1	–	1
Wasa	1973-94	19	95	1	4	8
Williams Lake	1973-93	21	19	1	2	2
Zincton	1968-94	22	5	1	–	1

Sky Lark

Survey route	Survey years	Total surveys	Frequency of occurrence (%)	Range of total surveys when species occurred Low	Median	High
Coastal						
Victoria	1973-94	17	59	4	5	17

Horned Lark

Survey route	Survey years	Total surveys	Frequency of occurrence (%)	Range of total surveys when species occurred: Low	Median	High
Interior						
Chilkat Pass	1975-76	2	100	7	9	10
Haines Summit	1993-94	2	100	2	4	6
Kamloops	1974-94	15	13	1	2	3
Meadowlake	1968-70	2	100	10	12	13
Nicola	1973-74	2	50	7	–	7
Pleasant Valley	1980-93	6	100	11	18	24
Riske Creek	1973-92	11	36	1	2	3

Purple Martin

Survey route	Survey years	Total surveys	Frequency of occurrence (%)	Range of total surveys when species occurred: Low	Median	High
Coastal						
Kwinitsa	1968-94	26	4	1	–	1*
Nanaimo River	1973-94	22	5	1	–	1

Tree Swallow

Survey route	Survey years	Total surveys	Frequency of occurrence (%)	Range of total surveys when species occurred: Low	Median	High
Coastal						
Alberni	1968-79	6	17	3	–	3
Albion	1968-92	21	100	2	10	42
Campbell River	1973-93	19	47	2	4	9
Chemainus	1969-88	4	50	4	8	11
Chilliwack	1973-94	11	100	2	10	47
Comox Lake	1973-75	3	100	2	3	8
Coquitlam	1990-94	5	100	6	12	19
Elsie Lake	1970-94	6	67	1	2	3
Gibsons Landing	1973-94	10	10	8	–	8
Kispiox	1975-94	3	67	3	4	4
Kitsumkalum	1974-94	19	47	1	3	10
Kwinitsa	1968-94	26	38	1	6	14
Masset	1989-94	5	100	13	16	44
Meziadin	1975-77	2	50	2	–	2
Nanaimo River	1973-94	22	100	1	7	19
Pemberton	1974-94	18	83	1	5	17
Pitt Meadows	1973-89	14	86	2	7	24
Point Grey	1974-92	11	27	1	2	5
Port Hardy	1982-89	4	50	2	2	2
Port Renfrew	1973-94	17	12	1	3	4
Queen Charlotte City	1989-94	3	100	7	18	20
Saltery Bay	1974-76	3	100	1	5	6
Seabird	1969-94	16	100	2	9	35
Squamish	1973-94	9	67	1	4	9
Victoria	1973-94	17	47	1	2	5
Interior						
Adams Lake	1989-94	6	17	1	–	1
Alexis Creek	1973-86	14	100	1	8	28
Ashnola	1973-74	2	100	1	4	6
Atlin Road	1989	1	100	1	–	1
Beaverdell	1970-94	24	92	1	3	12
Bridge Lake	1974-94	20	100	5	27	58
Bromley	1973-94	5	100	1	27	64
Brookmere	1973-92	10	100	5	14	28
Canford	1974-93	10	90	1	3	4
Canoe Point	1993-94	2	100	36	42	47

▶

◀ *Tree Swallow*

Survey route	Survey years	Total surveys	Frequency of occurrence (%)	Range of total surveys when species occurred Low	Median	High
Cassiar	1972-75	2	100	6	8	9
Christian Valley	1993-94	2	100	6	8	9
Chu Chua	1993-94	2	100	1	3	5
Columbia Lake	1973-83	7	14	3	–	3
Creighton Valley	1979-94	16	88	3	7	20
Creston	1984	1	100	34	–	34
Ferndale	1968-94	15	87	1	6	25
Fort Nelson	1974-80	5	100	1	5	16
Fort St. John	1974-90	9	56	1	1	1
Fraser Lake	1968-78	5	80	6	19	63
Gerrard	1976-85	10	40	1	1	2
Gnat Pass	1975-93	2	100	2	5	8
Golden	1973-94	22	100	2	22	42
Grand Forks	1973-94	22	100	1	8	67
Horsefly	1974-88	11	64	1	2	9
Hudson's Hope	1976	1	100	2	–	2
Kamloops	1974-94	15	100	3	8	25
Kootenay	1981-94	10	30	1	1	1
Kuskonook	1968-91	21	86	1	8	45
Lac la Hache	1973-86	11	91	1	10	53
Lavington	1972-94	23	100	5	9	18
Mabel Lake	1973-94	19	100	2	7	20
McLeod Lake	1968-94	9	78	1	5	18
Meadowlake	1968-70	2	100	2	3	3
Mount Morice	1968-75	5	80	2	11	18
Needles	1968-94	26	77	1	2	6
Nicola	1973-74	2	100	11	22	32
Oliver	1973-94	18	39	1	3	6
Osprey Lake	1976-94	10	80	1	21	47
Pavilion	1974-94	6	83	7	10	11
Pennington	1987-91	4	25	4	–	4
Pinkut Creek	1975	1	100	7	–	7
Pleasant Valley	1980-93	6	100	4	9	24
Prince George	1973-94	20	100	9	48	80
Princeton	1968-94	8	88	1	11	19
Punchaw	1974-94	9	100	1	6	17
Riske Creek	1973-92	11	100	9	16	29
Salmo	1969-75	3	100	2	27	41
Salmon Arm	1968-94	19	95	2	11	57
Scotch Creek	1989-94	6	50	2	2	3
Shalalth	1968-94	4	75	7	8	18
Spillimacheen	1976-93	17	100	1	7	24
Steamboat	1974-80	4	50	1	2	2
Succour Creek	1993-94	2	100	1	1	1
Summerland	1973-94	21	38	1	2	3
Summit Lake	1972-94	8	75	4	6	29
Syringa Creek	1986-93	7	86	1	4	14
Tatlayoko Lake	1989-94	6	100	9	12	16
Telkwa High Road	1974-94	11	100	2	10	39
Tokay	1968-75	4	25	12	–	12
Tupper	1974-94	6	83	2	5	15
Wasa	1973-94	19	95	1	8	24
Williams Lake	1973-93	21	100	5	21	56
Wingdam	1968-70	3	100	9	11	13
Zincton	1968-94	22	82	1	5	26

Violet-green Swallow

Survey route	Survey years	Total surveys	Frequency of occurrence (%)	Range of total surveys when species occurred Low	Median	High
Coastal						
Alberni	1968-79	6	100	2	12	34
Albion	1968-92	21	100	2	12	50
Campbell River	1973-93	19	74	1	5	8
Chemainus	1969-88	4	100	2	25	47

▶

◀ *Violet-green Swallow*

Survey route	Survey years	Total surveys	Frequency of occurrence (%)	Range of total surveys when species occurred Low	Median	High
Chilliwack	1973-94	11	100	5	17	45
Comox Lake	1973-75	3	100	2	7	26
Coquitlam	1990-94	5	100	20	24	36
Courtenay	1992-93	2	100	7	9	10
Elsie Lake	1970-94	6	50	1	1	2
Gibsons Landing	1973-94	10	100	8	12	25
Kispiox	1975-94	3	100	1	9	19
Kwinitsa	1968-94	26	8	2	2	2
Nanaimo River	1973-94	22	86	1	5	13
Pemberton	1974-94	18	83	1	5	13
Pitt Meadows	1973-89	14	100	7	15	68
Point Grey	1974-92	11	91	6	15	48
Port Hardy	1982-89	4	75	2	2	5
Port Renfrew	1973-94	17	76	2	4	9
Saltery Bay	1974-76	3	67	1	4	7
Seabird	1969-94	16	88	1	11	29
Squamish	1973-94	9	100	2	5	15
Victoria	1973-94	17	100	10	36	75
Interior						
Adams Lake	1989-94	6	50	1	1	2
Alexis Creek	1973-86	14	7	1	–	1
Ashnola	1973-74	2	50	1	–	1
Beaverdell	1970-94	24	100	4	9	23
Bridge Lake	1974-94	20	5	8	–	8
Bromley	1973-94	5	100	15	21	50
Brookmere	1973-92	10	80	1	2	7
Canford	1974-93	10	90	2	6	16
Canoe Point	1993-94	2	100	30	34	37
Christian Valley	1993-94	2	50	5	–	5
Chu Chua	1993-94	2	100	3	7	10
Columbia Lake	1973-83	7	100	1	13	55
Creighton Valley	1979-94	16	88	1	3	10
Ferndale	1968-94	15	47	1	2	3
Fort St. John	1974-90	9	11	1	–	1
Fraser Lake	1968-78	5	80	1	5	9
Gerrard	1976-85	10	80	1	2	4
Golden	1973-94	22	100	1	11	65
Grand Forks	1973-94	22	95	5	21	79
Illecilewaet	1973-83	4	100	2	4	6
Kamloops	1974-94	15	100	2	5	17
Kootenay	1981-94	10	60	1	4	8
Kuskonook	1968-91	21	100	3	20	59
Lac la Hache	1973-86	11	64	1	2	9
Lavington	1972-94	23	100	5	16	31
Mabel Lake	1973-94	19	100	2	10	21
McLeod Lake	1968-94	9	11	8	–	8
Mount Morice	1968-75	5	20	2	–	2
Needles	1968-94	26	81	1	2	12
Oliver	1973-94	18	33	1	2	3
Osprey Lake	1976-94	10	60	1	21	48
Pavilion	1974-94	6	100	3	6	20
Pleasant Valley	1980-93	6	100	1	7	26
Prince George	1973-94	20	75	1	6	39
Princeton	1968-94	8	50	1	5	12
Punchaw	1974-94	9	11	1	–	1
Salmo	1969-75	3	100	17	49	68
Salmon Arm	1968-94	19	95	2	7	34
Scotch Creek	1989-94	6	17	1	–	1
Shalalth	1968-94	4	75	2	3	6
Spillimacheen	1976-93	17	76	1	2	10
Succour Creek	1993-94	2	100	2	3	4
Summerland	1973-94	21	29	1	1	2
Summit Lake	1972-94	8	88	3	7	14
Syringa Creek	1986-93	7	100	19	53	66
Telkwa High Road	1974-94	11	73	1	5	26
Tokay	1968-75	4	75	5	11	13
Wasa	1973-94	19	84	1	9	25
Williams Lake	1973-93	21	29	1	1	5
Zincton	1968-94	22	100	2	12	33

Northern Rough-winged Swallow

Survey route	Survey years	Total surveys	Frequency of occurrence (%)	Range of total surveys when species occurred: Low	Median	High
Coastal						
Alberni	1968-79	6	17	4	–	4
Albion	1968-92	21	10	1	1	1
Campbell River	1973-93	19	42	1	1	4
Chemainus	1969-88	4	25	1	–	1
Chilliwack	1973-94	11	45	1	2	8
Courtenay	1992-93	2	50	1	–	1
Elsie Lake	1970-94	6	67	1	2	4
Gibsons Landing	1973-94	10	20	1	2	2
Kispiox	1975-94	3	67	1	1	1
Kitsumkalum	1974-94	19	5	1	–	1
Nanaimo River	1973-94	22	9	1	2	2
Pemberton	1974-94	18	50	1	2	12
Pitt Meadows	1973-89	14	21	2	2	4
Point Grey	1974-92	11	64	1	2	4
Port Renfrew	1973-94	17	24	1	2	2
Saltery Bay	1974-76	3	67	1	2	3
Seabird	1969-94	16	69	1	3	8
Squamish	1973-94	9	67	1	4	7
Victoria	1973-94	17	29	1	2	3
Interior						
Adams Lake	1989-94	6	67	1	3	5
Alexis Creek	1973-86	14	79	1	4	15
Ashnola	1973-74	2	100	4	7	9
Beaverdell	1970-94	24	100	1	5	18
Bridge Lake	1974-94	20	45	1	2	16
Bromley	1973-94	5	60	1	2	2
Brookmere	1973-92	10	60	2	4	11
Canford	1974-93	10	100	2	7	16
Canoe Point	1993-94	2	50	1	–	1
Christian Valley	1993-94	2	100	1	7	13
Chu Chua	1993-94	2	50	9	–	9
Columbia Lake	1973-83	7	43	1	20	36
Creighton Valley	1979-94	16	94	1	2	9
Creston	1984	1	100	10	–	10
Ferndale	1968-94	15	53	2	3	50
Fraser Lake	1968-78	5	20	3	–	3
Gerrard	1976-85	10	100	1	6	12
Golden	1973-94	22	86	1	14	33
Goldstream Mountain	1993-94	2	50	2	–	2
Gosnell	1969-74	4	75	1	6	9
Grand Forks	1973-94	22	91	5	22	61
Horsefly	1974-88	11	100	3	22	62
Illecilewaet	1973-83	4	25	3	–	3
Kamloops	1974-94	15	93	4	14	30
Kootenay	1981-94	10	50	1	4	5
Kuskonook	1968-91	21	71	2	8	16
Lac la Hache	1973-86	11	91	1	12	72
Lavington	1972-94	23	78	1	2	9
Mabel Lake	1973-94	19	95	1	6	16
McLeod Lake	1968-94	9	44	2	3	4
Needles	1968-94	26	69	1	3	8
Oliver	1973-94	18	67	1	2	7
Osprey Lake	1976-94	10	80	1	4	6
Pavilion	1974-94	6	83	2	4	19
Pleasant Valley	1980-93	6	100	2	7	13
Princeton	1968-94	8	88	2	4	7
Punchaw	1974-94	9	22	1	2	2
Riske Creek	1973-92	11	36	1	2	2
Salmo	1969-75	3	67	3	8	12
Salmon Arm	1968-94	19	37	1	1	4
Scotch Creek	1989-94	6	33	1	1	1
Shalalth	1968-94	4	50	1	5	8
Spillimacheen	1976-93	17	88	1	7	32
Succour Creek	1993-94	2	50	3	–	3
Summerland	1973-94	21	100	1	5	16
Summit Lake	1972-94	8	25	1	2	2
Syringa Creek	1986-93	7	100	5	14	17
Tatlayoko Lake	1989-94	6	50	6	6	8

▶

◀ *Northern Rough-winged Swallow*

Survey route	Survey years	Total surveys	Frequency of occurrence (%)	Range of total surveys when species occurred Low	Median	High
Telkwa High Road	1974-94	11	45	1	4	14
Tokay	1968-75	4	50	2	4	5
Wasa	1973-94	19	95	2	13	24
Williams Lake	1973-93	21	57	1	3	7
Zincton	1968-94	22	86	1	6	19

Bank Swallow

Survey route	Survey years	Total surveys	Frequency of occurrence (%)	Range of total surveys when species occurred Low	Median	High
Coastal						
Chilliwack	1973-94	11	9	2	–	2
Kispiox	1975-94	3	33	6	–	6
Nanaimo River	1973-94	22	5	5	–	5*
Pemberton	1974-94	18	11	3	5	6
Pitt Meadows	1973-89	14	7	3	–	3*
Interior						
Alexis Creek	1973-86	14	100	3	18	46
Beaverdell	1970-94	24	13	1	2	9
Bridge Lake	1974-94	20	35	1	3	5
Bromley	1973-94	5	20	15	–	15
Brookmere	1973-92	10	10	1	–	1
Canford	1974-93	10	70	3	12	31
Canoe Point	1993-94	2	100	7	20	32
Chu Chua	1993-94	2	100	5	7	8
Columbia Lake	1973-83	7	29	1	3	4
Creighton Valley	1979-94	16	75	2	4	9
Creston	1984	1	100	18	–	18
Ferndale	1968-94	15	7	30	–	30
Fort St. John	1974-90	9	22	5	7	8
Fraser Lake	1968-78	5	40	28	29	30
Golden	1973-94	22	36	1	36	137
Grand Forks	1973-94	22	100	7	21	65
Horsefly	1974-88	11	82	1	8	20
Hudson's Hope	1976	1	100	3	–	3
Kamloops	1974-94	15	100	1	5	30
Kootenay	1981-94	10	20	2	3	4
Kuskonook	1968-91	21	52	1	3	9
Lac la Hache	1973-86	11	64	2	35	75
Mabel Lake	1973-94	19	32	1	3	20
Nicola	1973-74	2	100	11	27	42
Oliver	1973-94	18	94	3	56	112
Pleasant Valley	1980-93	6	17	2	–	2
Prince George	1973-94	20	5	1	–	1
Punchaw	1974-94	9	11	8	–	8
Riske Creek	1973-92	11	9	1	–	1
Salmo	1969-75	3	67	1	1	1
Salmon Arm	1968-94	19	42	1	3	10
Spillimacheen	1976-93	17	12	2	6	10
Summerland	1973-94	21	100	1	9	34
Summit Lake	1972-94	8	38	2	2	7
Syringa Creek	1986-93	7	29	6	8	10
Tatlayoko Lake	1989-94	6	83	4	8	20
Telkwa High Road	1974-94	11	18	2	14	25
Tokay	1968-75	4	100	33	136	230
Tupper	1974-94	6	17	1	–	1
Wasa	1973-94	19	11	1	3	4
Williams Lake	1973-93	21	57	1	5	24

Cliff Swallow

Survey route	Survey years	Total surveys	Frequency of occurrence (%)	Range of total surveys when species occurred Low	Median	High
Coastal						
Alberni	1968-79	6	17	1	–	1
Albion	1968-92	21	90	1	13	70
Campbell River	1973-93	19	32	1	4	6
Chemainus	1969-88	4	50	2	4	5
Chilliwack	1973-94	11	45	2	10	26
Courtenay	1992-93	2	50	1	–	1
Kispiox	1975-94	3	67	1	21	40
Kwinitsa	1968-94	26	4	2	–	2
Nanaimo River	1973-94	22	68	1	2	8
Pemberton	1974-94	18	22	1	3	7
Pitt Meadows	1973-89	14	7	1	–	1
Point Grey	1974-92	11	45	1	1	2
Seabird	1969-94	16	81	2	7	22
Squamish	1973-94	9	22	1	1	1
Victoria	1973-94	17	71	1	5	19
Interior						
Alexis Creek	1973-86	14	57	2	3	11
Ashnola	1973-74	2	50	2	–	2
Beaverdell	1970-94	24	46	1	4	20
Bridge Lake	1974-94	20	95	6	41	95
Bromley	1973-94	5	60	3	7	7
Brookmere	1973-92	10	90	9	29	74
Canford	1974-93	10	80	1	5	16
Canoe Point	1993-94	2	50	22	–	22
Christian Valley	1993-94	2	100	1	2	2
Columbia Lake	1973-83	7	14	2	–	2
Creighton Valley	1979-94	16	63	1	3	13
Ferndale	1968-94	15	40	2	50	80
Fort Nelson	1974-80	5	20	25	–	25
Fort St. John	1974-90	9	11	1	–	1
Fraser Lake	1968-78	5	80	12	41	64
Gerrard	1976-85	10	50	1	3	7
Gnat Pass	1975-93	2	50	2	–	2
Golden	1973-94	22	50	2	21	84
Grand Forks	1973-94	22	100	2	68	141
Horsefly	1974-88	11	64	1	6	34
Kamloops	1974-94	15	80	1	15	47
Kootenay	1981-94	10	10	2	–	2
Kuskonook	1968-91	21	24	1	7	13
Lac la Hache	1973-86	11	91	5	58	111
Lavington	1972-94	23	82	1	4	26
Mabel Lake	1973-94	19	89	1	8	38
McLeod Lake	1968-94	9	22	4	7	10
Meadowlake	1968-70	2	100	100	101	102
Needles	1968-94	26	15	2	6	67
Nicola	1973-74	2	100	95	165	234
Oliver	1973-94	18	56	1	2	20
Osprey Lake	1976-94	10	50	2	5	18
Pavilion	1974-94	6	17	30	–	30
Pennington	1987-91	4	25	1	–	1
Pleasant Valley	1980-93	6	100	96	143	242
Prince George	1973-94	20	75	1	8	45
Princeton	1968-94	8	13	3	–	3
Punchaw	1974-94	9	44	1	22	31
Riske Creek	1973-92	11	82	2	7	46
Salmo	1969-75	3	100	4	6	14
Salmon Arm	1968-94	19	74	1	4	26
Spillimacheen	1976-93	17	94	1	14	39
Summerland	1973-94	21	52	1	2	15
Summit Lake	1972-94	8	100	1	3	19
Syringa Creek	1986-93	7	100	1	11	38
Tatlayoko Lake	1989-94	6	67	1	5	6
Telkwa High Road	1974-94	11	91	3	25	46
Tokay	1968-75	4	75	1	2	31
Tupper	1974-94	6	67	2	5	11
Williams Lake	1973-93	21	95	2	18	95
Wingdam	1968-70	3	33	1	–	1
Zincton	1968-94	22	59	1	3	25

Barn Swallow

Survey route	Survey years	Total surveys	Frequency of occurrence (%)	Range of total surveys when species occurred: Low	Median	High
Coastal						
Alberni	1968-79	6	83	1	7	10
Albion	1968-92	21	100	42	103	281
Campbell River	1973-93	19	89	1	5	12
Chemainus	1969-88	4	100	4	16	24
Chilliwack	1973-94	11	100	30	46	85
Comox Lake	1973-75	3	100	2	7	9
Coquitlam	1990-94	5	100	49	71	79
Courtenay	1992-93	2	100	1	4	6
Elsie Lake	1970-94	6	83	1	1	3
Gibsons Landing	1973-94	10	100	1	3	10
Kispiox	1975-94	3	100	1	1	1
Kwinitsa	1968-94	26	88	1	6	40
Masset	1989-94	5	100	1	4	8
Nanaimo River	1973-94	22	86	1	9	26
Pemberton	1974-94	18	94	2	10	24
Pitt Meadows	1973-89	14	100	27	67	107
Point Grey	1974-92	11	100	17	42	55
Port Hardy	1982-89	4	75	1	2	2
Port Renfrew	1973-94	17	29	1	2	6
Saltery Bay	1974-76	3	100	2	2	18
Seabird	1969-94	16	100	13	36	67
Squamish	1973-94	9	100	3	10	31
Victoria	1973-94	17	100	4	62	154
Interior						
Adams Lake	1989-94	6	50	1	2	3
Alexis Creek	1973-86	14	86	1	3	14
Ashnola	1973-74	2	50	4	–	4
Beaverdell	1970-94	24	83	1	3	12
Bridge Lake	1974-94	20	100	2	11	38
Bromley	1973-94	5	100	3	6	11
Brookmere	1973-92	10	100	3	8	14
Canford	1974-93	10	70	1	5	16
Canoe Point	1993-94	2	100	35	42	48
Chilkat Pass	1975-76	2	50	3	–	3
Christian Valley	1993-94	2	100	2	3	4
Columbia Lake	1973-83	7	100	4	8	19
Creighton Valley	1979-94	16	94	2	5	13
Creston	1984	1	100	55	–	55
Ferndale	1968-94	15	93	3	8	106
Fort Nelson	1974-80	5	20	1	–	1
Fort St. John	1974-90	9	89	1	3	4
Fraser Lake	1968-78	5	80	6	29	122
Gerrard	1976-85	10	90	1	3	7
Gnat Pass	1975-93	2	50	5	–	5
Golden	1973-94	22	100	2	13	106
Goldstream Mountain	1993-94	2	50	2	–	2
Grand Forks	1973-94	22	100	10	41	114
Hendrix Lake	1976	1	100	2	–	2
Horsefly	1974-88	11	100	1	11	34
Hudson's Hope	1976	1	100	2	–	2
Illecilewaet	1973-83	4	100	4	6	7
Kamloops	1974-94	15	93	1	5	22
Kootenay	1981-94	10	100	2	5	23
Kuskonook	1968-91	21	100	1	11	31
Lac la Hache	1973-86	11	100	9	31	70
Lavington	1972-94	23	100	6	21	38
Mabel Lake	1973-94	19	100	1	16	34
McLeod Lake	1968-94	9	67	1	2	4
Meadowlake	1968-70	2	100	9	19	29
Mount Morice	1968-75	5	60	1	4	11
Needles	1968-94	26	96	1	5	39
Nicola	1973-74	2	100	3	12	20
Oliver	1973-94	18	100	1	5	10
Osprey Lake	1976-94	10	80	4	7	21
Pavilion	1974-94	6	83	1	3	16
Pinkut Creek	1975	1	100	2	–	2
Pleasant Valley	1980-93	6	100	8	11	12
Prince George	1973-94	20	95	5	23	36

▶

◀ Barn Swallow

Survey route	Survey years	Total surveys	Frequency of occurrence (%)	Range of total surveys when species occurred Low	Median	High
Princeton	1968-94	8	38	1	2	2
Punchaw	1974-94	9	78	1	4	15
Riske Creek	1973-92	11	100	1	10	24
Salmo	1969-75	3	100	19	24	28
Salmon Arm	1968-94	19	100	3	6	26
Scotch Creek	1989-94	6	17	1	–	1
Spillimacheen	1976-93	17	100	3	8	32
Steamboat	1974-80	4	50	1	1	1
Succour Creek	1993-94	2	50	1	–	1
Summerland	1973-94	21	52	1	3	8
Summit Lake	1972-94	8	100	3	7	16
Syringa Creek	1986-93	7	100	3	6	12
Tatlayoko Lake	1989-94	6	50	2	2	3
Telkwa High Road	1974-94	11	100	3	11	38
Tokay	1968-75	4	50	1	2	2
Tupper	1974-94	6	67	3	5	9
Wasa	1973-94	19	84	1	2	10
Williams Lake	1973-93	21	90	2	6	16
Wingdam	1968-70	3	67	2	6	9
Zincton	1968-94	22	100	1	9	23

Gray Jay

Survey route	Survey years	Total surveys	Frequency of occurrence (%)	Range of total surveys when species occurred Low	Median	High
Coastal						
Campbell River	1973-93	19	26	1	1	2
Chilliwack	1973-94	11	9	1	–	1
Comox Lake	1973-75	3	33	2	–	2
Kispiox	1975-94	3	33	1	–	1
Kitsumkalum	1974-94	19	26	1	1	2
Meziadin	1975-77	2	50	1	–	1
Pemberton	1974-94	18	17	2	3	6
Port Hardy	1982-89	4	25	1	–	1
Seabird	1969-94	16	6	2	–	2
Interior						
Alexis Creek	1973-86	14	50	1	1	2
Atlin Road	1989	1	100	5	–	5
Beaverdell	1970-94	24	17	1	1	4
Bridge Lake	1974-94	20	80	1	4	9
Bromley	1973-94	5	20	4	–	4
Brookmere	1973-92	10	60	1	3	4
Cassiar	1972-75	2	100	2	4	5
Christian Valley	1993-94	2	100	4	4	4
Chu Chua	1993-94	2	50	2	–	2
Columbia Lake	1973-83	7	14	1	–	1
Creighton Valley	1979-94	16	19	1	2	4
Ferndale	1968-94	15	87	1	3	15
Fort Nelson	1974-80	5	60	1	4	16
Fort St. John	1974-90	9	56	1	2	6
Gerrard	1976-85	10	20	1	1	1
Gnat Pass	1975-93	2	50	3	–	3
Grand Forks	1973-94	22	9	1	1	1
Hendrix Lake	1976	1	100	2	–	2
Horsefly	1974-88	11	36	1	2	3
Hudson's Hope	1976	1	100	1	–	1
Kootenay	1981-94	10	50	1	3	14
Kuskonook	1968-91	21	5	2	–	2
Lac la Hache	1973-86	11	45	2	4	4
Lavington	1972-94	24	8	1	4	6
McLeod Lake	1968-94	9	89	1	4	6
Meadowlake	1968-70	2	50	10	–	10
Mount Morice	1968-75	5	60	1	4	9
Needles	1968-94	26	23	1	2	4

▶

◀ *Gray Jay*

Survey route	Survey years	Total surveys	Frequency of occurrence (%)	Range of total surveys when species occurred Low	Median	High
Nicola	1973-74	2	100	5	6	7
Oliver	1973-94	18	22	1	2	3
Osprey Lake	1976-94	10	50	1	3	4
Pennington	1987-91	4	75	3	5	5
Pinkut Creek	1975	1	100	2	–	2
Prince George	1973-94	20	5	2	–	2
Princeton	1968-94	8	75	1	3	20
Punchaw	1974-94	9	78	3	5	9
Riske Creek	1973-92	11	18	1	2	2
Scotch Creek	1989-94	6	67	1	3	5
Spillimacheen	1976-93	17	35	1	2	3
Steamboat	1974-80	4	100	2	5	13
Summerland	1973-94	21	52	1	1	5
Summit Lake	1972-94	8	13	3	–	3
Tatlayoko Lake	1989-94	6	33	1	1	1
Telkwa High Road	1974-94	11	27	1	1	22
Tetsa River	1974	1	200	4	4	4
Tupper	1974-94	6	50	1	2	9
Wasa	1973-94	19	32	1	1	2
Williams Lake	1973-93	21	5	1	–	1
Wingdam	1968-70	3	67	7	9	11
Zincton	1968-94	22	5	3	–	3

Steller's Jay

Survey route	Survey years	Total surveys	Frequency of occurrence (%)	Range of total surveys when species occurred Low	Median	High
Coastal						
Alberni	1968-79	6	83	1	2	8
Albion	1968-92	21	57	1	1	5
Campbell River	1973-93	19	95	2	5	11
Chemainus	1969-88	4	25	1	–	1
Chilliwack	1973-94	11	64	1	3	3
Comox Lake	1973-75	3	67	1	1	1
Coquitlam	1990-94	5	40	2	3	4
Elsie Lake	1970-94	6	100	2	3	12
Gibsons Landing	1973-94	10	100	1	3	4
Kitsumkalum	1974-94	19	100	1	4	10
Kwinitsa	1968-94	26	100	1	12	23
Meziadin	1975-77	2	50	2	–	2
Nanaimo River	1973-94	22	55	1	2	4
Pemberton	1974-94	18	100	1	5	12
Pitt Meadows	1973-89	14	79	1	2	3
Point Grey	1974-92	11	64	1	2	4
Port Hardy	1982-89	4	100	18	27	31
Port Renfrew	1973-94	17	100	5	15	42
Queen Charlotte City	1989-94	3	33	5	–	5
Saltery Bay	1974-76	3	33	5	–	5
Seabird	1969-94	16	75	1	5	8
Squamish	1973-94	9	56	1	1	5
Interior						
Beaverdell	1970-94	24	4	1	–	1
Bromley	1973-94	5	40	2	3	4
Brookmere	1973-92	10	40	1	1	1
Cassiar	1972-75	2	50	1	–	1
Christian Valley	1993-94	2	50	1	–	1
Chu Chua	1993-94	2	50	1	–	1
Columbia Lake	1973-83	7	14	1	–	1
Creighton Valley	1979-94	16	38	1	1	3
Creston	1984	1	100	3	–	3
Ferndale	1968-94	15	7	2	–	2
Gerrard	1976-85	10	90	1	3	5
Gnat Pass	1975-93	2	50	1	–	1
Goldstream Mountain	1993-94	2	100	2	2	2

▶

◀ *Steller's Jay*

Survey route	Survey years	Total surveys	Frequency of occurrence (%)	Range of total surveys when species occurred Low	Median	High
Gosnell	1969-74	4	75	4	4	10
Grand Forks	1973-94	22	9	1	2	3
Horsefly	1974-88	11	9	1	–	1
Illecilewaet	1973-83	4	100	3	8	10
Kootenay	1981-94	10	10	1	–	1
Kuskonook	1968-91	21	14	1	1	1
Lavington	1972-94	23	26	1	1	8
Mabel Lake	1973-94	19	21	1	1	3
McLeod Lake	1968-94	9	22	1	2	2
Mount Morice	1968-75	5	60	1	1	1
Needles	1968-94	26	58	1	1	2
Nicola	1973-74	2	100	2	2	2
Oliver	1973-94	18	28	1	1	2
Osprey Lake	1976-94	10	20	2	2	2
Pavilion	1974-94	6	33	4	6	8
Salmo	1969-75	3	67	1	3	5
Salmon Arm	1968-94	19	5	1	–	1
Scotch Creek	1989-94	6	83	2	2	3
Spillimacheen	1976-93	17	12	1	1	1
Summerland	1973-94	21	52	1	1	4
Summit Lake	1972-94	8	25	1	1	1
Syringa Creek	1986-93	7	86	1	1	3
Wingdam	1968-70	3	33	1	–	1
Zincton	1968-94	22	23	1	2	2

Blue Jay

Survey route	Survey years	Total surveys	Frequency of occurrence (%)	Range of total surveys when species occurred Low	Median	High
Interior						
Fort St. John	1974-90	9	11	1	–	1
Tupper	1974-94	6	17	3	–	3

Clark's Nutcracker

Survey route	Survey years	Total surveys	Frequency of occurrence (%)	Range of total surveys when species occurred Low	Median	High
Coastal						
Pemberton	1974-94	18	11	1	2	2
Interior						
Ashnola	1973-74	2	50	1	–	1
Beaverdell	1970-94	24	17	1	1	2
Bromley	1973-94	5	60	1	2	2
Brookmere	1973-92	10	30	1	1	3
Canford	1974-93	10	90	2	7	12
Chu Chua	1993-94	2	100	1	5	9
Grand Forks	1973-94	22	14	1	1	2
Kamloops	1974-94	15	60	1	4	8
Kootenay	1981-94	10	20	1	1	1
Mabel Lake	1973-94	19	5	1	–	1
Needles	1968-94	26	4	1	–	1
Nicola	1973-74	2	50	2	–	2
Oliver	1973-94	18	67	1	2	7
Osprey Lake	1976-94	10	30	1	3	4
Pavilion	1974-94	6	50	1	1	2
Pleasant Valley	1980-93	6	83	2	4	10
Princeton	1968-94	8	38	1	2	2
Shalalth	1968-94	4	50	1	2	3

▶

◀ *Clark's Nutcracker*

Survey route	Survey years	Total surveys	Frequency of occurrence (%)	Low	Range of total surveys when species occurred Median	High
Succour Creek	1993-94	2	50	1	–	1
Summerland	1973-94	21	100	1	4	13
Wasa	1973-94	19	37	1	2	4
Williams Lake	1973-93	21	5	1	–	1

Black-billed Magpie

Survey route	Survey years	Total surveys	Frequency of occurrence (%)	Low	Range of total surveys when species occurred Median	High
Interior						
Ashnola	1973-74	2	100	6	6	6
Bromley	1973-94	5	80	1	1	5
Canford	1974-93	10	30	2	3	26
Canoe Point	1993-94	2	50	4	–	4
Chu Chua	1993-94	2	100	3	5	6
Columbia Lake	1973-83	7	100	1	4	20
Fort St. John	1974-90	9	100	7	13	27
Golden	1973-94	22	50	1	2	8
Grand Forks	1973-94	22	86	1	2	24
Haines Summit	1993-94	2	50	3	–	3
Hudson's Hope	1976	1	100	3	–	3
Kamloops	1974-94	15	100	1	17	29
Kuskonook	1968-91	21	5	1	–	1
Lavington	1972-94	23	100	7	18	43
Mabel Lake	1973-94	19	100	1	7	18
Nicola	1973-74	2	100	1	2	2
Oliver	1973-94	18	100	3	11	19
Osprey Lake	1976-94	10	60	1	3	7
Pavilion	1974-94	6	100	2	6	13
Pleasant Valley	1980-93	6	83	3	5	10
Prince George	1973-94	20	100	7	21	30
Riske Creek	1973-92	11	18	1	3	4
Salmon Arm	1968-94	19	58	1	5	12
Spillimacheen	1976-93	17	76	1	2	5
Summerland	1973-94	21	100	1	10	33
Summit Lake	1972-94	8	88	1	2	5
Tokay	1968-75	4	25	1	–	1
Tupper	1974-94	6	100	3	6	17
Wasa	1973-94	19	79	1	3	19
Williams Lake	1973-93	21	62	1	2	4

American Crow

Survey route	Survey years	Total surveys	Frequency of occurrence (%)	Low	Range of total surveys when species occurred Median	High
Coastal						
Chilliwack	1973-94	11	18	55	63	71*
Kispiox	1975-94	3	100	18	20	40*
Kitsumkalum	1974-94	19	79	2	10	28*
Kwinitsa	1968-94	26	92	4	19	54*
Nanaimo River	1973-94	22	5	10	–	10*
Pemberton	1974-94	18	100	11	27	50*
Saltery Bay	1974-76	3	33	7	–	7*
Interior						
Adams Lake	1989-94	6	100	2	4	8
Alexis Creek	1973-86	14	100	8	13	29
Ashnola	1973-74	2	100	2	3	3
Beaverdell	1970-94	24	100	1	11	18

▶

◀ *American Crow*

Survey route	Survey years	Total surveys	Frequency of occurrence (%)	Range of total surveys when species occurred Low	Median	High
Bridge Lake	1974-94	20	100	15	41	91
Bromley	1973-94	5	100	25	31	48
Brookmere	1973-92	10	80	1	6	23
Canford	1974-93	10	100	12	22	36
Canoe Point	1993-94	2	100	32	35	37
Christian Valley	1993-94	2	100	8	8	8
Chu Chua	1993-94	2	100	18	26	34
Columbia Lake	1973-83	7	100	2	19	61
Creighton Valley	1979-94	16	100	3	9	22
Creston	1984	1	100	33	–	33
Ferndale	1968-94	15	80	1	3	9
Fort Nelson	1974-80	5	20	1	–	1
Fort St. John	1974-90	9	100	1	5	8
Fraser Lake	1968-78	5	100	3	36	49
Gnat Pass	1975-93	2	50	12	–	12
Golden	1973-94	22	100	13	61	97
Goldstream Mountain	1993-94	2	50	1	–	1
Gosnell	1969-74	4	100	20	28	53
Grand Forks	1973-94	22	100	41	90	136
Hendrix Lake	1976	1	100	10	–	10
Horsefly	1974-88	11	100	4	20	46
Hudson's Hope	1976	1	100	12	–	12
Illecilewaet	1973-83	4	100	13	16	40
Kamloops	1974-94	15	100	25	40	59
Kootenay	1981-94	10	10	2	–	2
Kuskonook	1968-91	21	33	1	2	15
Lac la Hache	1973-86	11	100	2	19	33
Lavington	1972-94	23	100	5	30	68
Mabel Lake	1973-94	19	100	22	36	89
McLeod Lake	1968-94	9	100	8	24	42
Meadowlake	1968-70	2	100	5	7	9
Mount Morice	1968-75	5	100	4	4	7
Needles	1968-94	26	100	8	19	97
Nicola	1973-74	2	100	26	34	42
Oliver	1973-94	18	83	1	3	9
Osprey Lake	1976-94	10	100	11	18	33
Pavilion	1974-94	6	100	42	48	71
Pinkut Creek	1975	1	100	5	–	5
Pleasant Valley	1980-93	6	100	16	22	34
Prince George	1973-94	20	100	45	77	146
Princeton	1968-94	8	50	4	7	9
Punchaw	1974-94	9	67	1	3	5
Riske Creek	1973-92	11	100	5	15	25
Salmo	1969-75	3	33	2	–	2
Salmon Arm	1968-94	19	100	26	76	142
Shalalth	1968-94	4	75	5	9	13
Spillimacheen	1976-93	17	100	6	15	40
Succour Creek	1993-94	2	100	2	4	5
Summerland	1973-94	21	100	15	22	41
Summit Lake	1972-94	8	100	9	19	34
Syringa Creek	1986-93	7	100	9	11	20
Tatlayoko Lake	1989-94	6	100	16	29	62
Telkwa High Road	1974-94	11	100	7	26	100
Tokay	1968-75	4	100	12	19	41
Tupper	1974-94	6	83	1	4	10
Wasa	1973-94	19	100	2	9	34
Williams Lake	1973-93	21	100	11	29	65
Wingdam	1968-70	3	100	2	3	18
Zincton	1968-94	22	95	3	7	12

Northwestern Crow

Survey route	Survey years	Total surveys	Frequency of occurrence (%)	Low	Range of total surveys when species occurred Median	High
Coastal						
Alberni	1968-79	6	100	12	27	36
Albion	1968-92	21	100	59	117	216
Campbell River	1973-93	19	100	17	25	54
Chemainus	1969-88	4	100	23	30	32
Chilliwack	1973-94	11	91	3	60	120
Comox Lake	1973-75	3	100	2	3	12
Coquitlam	1990-94	5	100	37	90	106
Courtenay	1992-93	2	50	7	–	7
Elsie Lake	1970-94	6	17	1	–	1
Gibsons Landing	1973-94	10	100	29	52	72
Kitsumkalum	1974-94	19	37	1	2	18
Kwinitsa	1968-94	26	62	1	22	78
Masset	1989-94	5	100	4	8	33
Nanaimo River	1973-94	22	95	7	23	37
Pitt Meadows	1973-89	14	100	43	72	134
Point Grey	1974-92	11	100	57	144	182
Port Hardy	1982-89	4	100	2	8	11
Port Renfrew	1973-94	17	100	3	13	28
Saltery Bay	1974-76	3	67	1	8	14
Seabird	1969-94	16	100	21	41	55
Squamish	1973-94	9	100	7	30	46
Victoria	1973-94	17	100	57	122	206

Common Raven

Survey route	Survey years	Total surveys	Frequency of occurrence (%)	Low	Range of total surveys when species occurred Median	High
Coastal						
Alberni	1968-79	6	100	2	5	8
Albion	1968-92	21	52	1	1	5
Campbell River	1973-93	19	95	3	10	50
Chemainus	1969-88	4	75	3	4	6
Chilliwack	1973-94	11	36	1	2	3
Comox Lake	1973-75	3	100	1	2	6
Coquitlam	1990-94	5	20	2	–	2
Courtenay	1992-93	2	50	4	–	4
Elsie Lake	1970-94	6	83	1	3	5
Gibsons Landing	1973-94	10	100	1	3	7
Kispiox	1975-94	3	67	4	7	10
Kitsumkalum	1974-94	19	84	1	6	16
Kwinitsa	1968-94	26	88	1	5	20
Masset	1989-94	5	100	20	24	53
Meziadin	1975-77	2	50	4	–	4
Nanaimo River	1973-94	22	91	1	8	17
Pemberton	1974-94	18	100	2	7	35
Point Grey	1974-92	11	55	1	2	4
Port Hardy	1982-89	4	100	8	11	20
Port Renfrew	1973-94	17	53	1	2	5
Queen Charlotte City	1989-94	3	100	3	9	15
Saltery Bay	1974-76	3	100	3	3	11
Seabird	1969-94	16	69	1	3	12
Squamish	1973-94	9	100	2	3	8
Victoria	1973-94	17	100	1	3	11
Interior						
Adams Lake	1989-94	6	100	4	6	9
Alexis Creek	1973-86	14	64	1	4	5
Atlin Road	1989	1	100	5	–	5
Beaverdell	1970-94	24	100	6	17	38
Bridge Lake	1974-94	20	100	2	17	44
Bromley	1973-94	5	100	4	15	30
Brookmere	1973-92	10	100	1	6	11
Canford	1974-93	10	50	1	2	4
Canoe Point	1993-94	2	100	2	11	19

▶

◀ *Common Raven*

Survey route	Survey years	Total surveys	Frequency of occurrence (%)	Range of total surveys when species occurred Low	Median	High
Cassiar	1972-75	2	100	1	2	2
Chilkat Pass	1975-76	2	50	4	–	4
Christian Valley	1993-94	2	100	18	22	25
Chu Chua	1993-94	2	100	4	7	10
Columbia Lake	1973-83	7	100	1	4	11
Creighton Valley	1979-94	16	100	4	9	15
Creston	1984	1	100	24	–	24
Ferndale	1968-94	15	67	1	9	13
Fort Nelson	1974-80	5	100	6	12	23
Fort St. John	1974-90	9	56	1	2	2
Gerrard	1976-85	10	90	1	3	6
Gnat Pass	1975-93	2	50	11	–	11
Golden	1973-94	22	95	1	7	23
Goldstream Mountain	1993-94	2	50	9	–	9
Gosnell	1969-74	4	100	2	4	5
Grand Forks	1973-94	22	82	1	5	15
Haines Summit	1993-94	2	50	1	–	1
Horsefly	1974-88	11	100	2	9	15
Illecilewaet	1973-83	4	50	5	6	7
Kamloops	1974-94	15	93	1	7	17
Kootenay	1981-94	10	100	2	6	12
Kuskonook	1968-91	21	100	3	15	28
Lac la Hache	1973-86	11	91	6	11	29
Lavington	1972-94	23	92	1	7	47
Mabel Lake	1973-94	19	100	1	13	38
McLeod Lake	1968-94	9	44	1	4	7
Meadowlake	1968-70	2	100	2	4	5
Mount Morice	1968-75	5	20	1	–	1
Needles	1968-94	26	100	7	20	38
Nicola	1973-74	2	100	2	3	3
Oliver	1973-94	18	94	1	4	9
Osprey Lake	1976-94	10	90	1	4	13
Pavilion	1974-94	6	67	1	3	8
Pennington	1987-91	4	100	2	4	11
Pleasant Valley	1980-93	6	83	2	4	4
Prince George	1973-94	20	55	1	4	22
Princeton	1968-94	8	100	5	12	15
Punchaw	1974-94	9	78	3	5	7
Riske Creek	1973-92	11	82	1	3	7
Salmo	1969-75	3	100	7	8	10
Salmon Arm	1968-94	19	100	2	4	19
Scotch Creek	1989-94	6	100	1	4	10
Shalalth	1968-94	4	100	2	5	8
Spillimacheen	1976-93	17	100	8	15	20
Steamboat	1974-80	4	100	1	1	8
Succour Creek	1993-94	2	100	2	4	6
Summerland	1973-94	21	81	1	3	8
Summit Lake	1972-94	8	75	1	4	8
Syringa Creek	1986-93	7	100	9	13	24
Tatlayoko Lake	1989-94	6	67	2	5	9
Telkwa High Road	1974-94	11	73	1	4	19
Tokay	1968-75	4	100	6	8	14
Tupper	1974-94	6	33	1	2	3
Wasa	1973-94	19	100	3	13	39
Williams Lake	1973-93	21	95	1	4	13
Wingdam	1968-70	3	100	4	20	41
Zincton	1968-94	22	100	1	15	40

Black-capped Chickadee

Survey route	Survey years	Total surveys	Frequency of occurrence (%)	Range of total surveys when species occurred Low	Median	High
Coastal						
Albion	1968-92	21	100	5	19	41
Chilliwack	1973-94	11	100	19	28	52

▶

◀ *Black-capped Chickadee*

Survey route	Survey years	Total surveys	Frequency of occurrence (%)	Range of total surveys when species occurred Low	Median	High
Coquitlam	1990-94	5	100	13	40	42
Gibsons Landing	1973-94	10	10	1	–	1
Kispiox	1975-94	3	100	3	14	15
Kitsumkalum	1974-94	19	63	1	6	14
Kwinitsa	1968-94	26	62	3	5	59
Pemberton	1974-94	18	89	1	3	15
Pitt Meadows	1973-89	14	100	28	39	65
Point Grey	1974-92	11	100	12	21	48
Saltery Bay	1974-76	3	33	1	–	1
Seabird	1969-94	16	100	4	18	26
Squamish	1973-94	9	100	1	5	26
Interior						
Adams Lake	1989-94	6	83	2	4	7
Alexis Creek	1973-86	14	100	2	9	14
Ashnola	1973-74	2	50	6	–	6
Atlin Road	1989	1	100	2	–	2
Beaverdell	1970-94	24	100	1	4	11
Bridge Lake	1974-94	20	85	1	8	27
Bromley	1973-94	5	100	5	7	12
Brookmere	1973-92	10	80	1	4	19
Canford	1974-93	10	80	1	3	6
Canoe Point	1993-94	2	100	5	7	8
Cassiar	1972-75	2	50	1	–	1
Christian Valley	1993-94	2	100	4	6	8
Chu Chua	1993-94	2	100	3	6	8
Columbia Lake	1973-83	7	100	1	18	28
Creighton Valley	1979-94	16	94	3	7	13
Creston	1984	1	100	48	–	48
Ferndale	1968-94	15	100	1	5	11
Fort Nelson	1974-80	5	40	6	9	12
Fort St. John	1974-90	9	78	1	7	12
Fraser Lake	1968-78	5	80	2	4	15
Gerrard	1976-85	10	100	2	6	17
Gnat Pass	1975-93	2	50	5	–	5
Golden	1973-94	22	100	1	6	15
Goldstream Mountain	1993-94	2	100	1	4	6
Gosnell	1969-74	4	75	1	1	1
Grand Forks	1973-94	22	100	2	10	34
Haines Summit	1993-94	2	100	1	3	5
Hendrix Lake	1976	1	100	7	–	7
Horsefly	1974-88	11	100	3	6	8
Hudson's Hope	1976	1	100	2	–	2
Illecilewaet	1973-83	4	25	3	–	3
Kamloops	1974-94	15	67	1	2	3
Kootenay	1981-94	10	80	1	3	7
Kuskonook	1968-91	21	100	6	13	37
Lac la Hache	1973-86	11	109	5	13	22
Lavington	1972-94	23	100	1	5	14
Mabel Lake	1973-94	19	95	4	12	17
McLeod Lake	1968-94	9	67	1	4	6
Mount Morice	1968-75	5	80	1	8	40
Needles	1968-94	26	96	3	6	25
Nicola	1973-74	2	100	16	24	31
Oliver	1973-94	18	67	1	2	7
Osprey Lake	1976-94	10	100	3	14	20
Pavilion	1974-94	6	100	1	3	12
Pennington	1987-91	4	100	1	2	8
Pleasant Valley	1980-93	6	33	1	2	2
Prince George	1973-94	20	100	1	7	25
Princeton	1968-94	8	50	2	8	28
Punchaw	1974-94	9	56	1	3	4
Riske Creek	1973-92	11	91	3	6	13
Salmo	1969-75	3	100	5	8	12
Salmon Arm	1968-94	19	100	3	13	62
Scotch Creek	1989-94	6	17	1	–	1
Shalalth	1968-94	4	25	6	–	6
Spillimacheen	1976-93	17	94	2	5	25
Steamboat	1974-80	4	50	1	7	12
Succour Creek	1993-94	2	100	1	3	5

▶

◀ *Black-capped Chickadee*

Survey route	Survey years	Total surveys	Frequency of occurrence (%)	Low	Range of total surveys when species occurred Median	High
Summerland	1973-94	21	67	1	2	3
Summit Lake	1972-94	8	100	3	6	16
Syringa Creek	1986-93	7	100	9	17	21
Tatlayoko Lake	1989-94	6	67	5	7	8
Telkwa High Road	1974-94	11	100	7	19	38
Tokay	1968-75	4	75	9	9	9
Tupper	1974-94	6	83	1	6	11
Wasa	1973-94	19	100	1	4	28
Williams Lake	1973-93	21	100	2	5	14
Wingdam	1968-70	3	67	1	3	5
Zincton	1968-94	22	100	1	6	19

Mountain Chickadee

Survey route	Survey years	Total surveys	Frequency of occurrence (%)	Low	Range of total surveys when species occurred Median	High
Coastal						
Pemberton	1974-94	18	11	1	1	1
Interior						
Adams Lake	1989-94	6	100	1	2	6
Alexis Creek	1973-86	14	100	1	9	20
Beaverdell	1970-94	24	100	1	9	39
Bridge Lake	1974-94	20	100	1	15	43
Bromley	1973-94	5	100	3	5	8
Brookmere	1973-92	10	100	4	13	15
Canford	1974-93	10	100	7	14	17
Canoe Point	1993-94	2	100	12	15	17
Christian Valley	1993-94	2	100	2	2	2
Chu Chua	1993-94	2	100	5	9	13
Columbia Lake	1973-83	7	57	2	2	5
Creighton Valley	1979-94	16	81	1	2	8
Gnat Pass	1975-93	2	50	11	–	11
Golden	1973-94	22	68	1	2	7
Grand Forks	1973-94	22	82	1	2	8
Hendrix Lake	1976	1	100	1	–	1
Horsefly	1974-88	11	36	1	1	5
Illecilewaet	1973-83	4	50	1	5	8
Kamloops	1974-94	15	93	4	10	20
Kootenay	1981-94	10	40	2	3	4
Lac la Hache	1973-86	11	73	2	5	11
Lavington	1972-94	23	96	1	3	8
Mabel Lake	1973-94	19	79	1	3	10
Meadowlake	1968-70	2	100	1	3	5
Mount Morice	1968-75	5	20	4	–	4
Needles	1968-94	26	27	1	3	5
Nicola	1973-74	2	100	17	19	21
Oliver	1973-94	18	94	2	17	31
Osprey Lake	1976-94	10	90	2	7	13
Pavilion	1974-94	6	83	2	3	5
Pleasant Valley	1980-93	6	100	2	8	12
Princeton	1968-94	8	100	1	2	23
Riske Creek	1973-92	11	91	2	9	10
Salmon Arm	1968-94	19	53	1	3	9
Scotch Creek	1989-94	6	67	1	1	5
Shalalth	1968-94	4	100	1	5	8
Spillimacheen	1976-93	17	100	5	11	25
Succour Creek	1993-94	2	100	1	2	2
Summerland	1973-94	21	100	8	15	33
Tatlayoko Lake	1989-94	6	83	1	10	11
Telkwa High Road	1974-94	11	18	1	1	1
Tokay	1968-75	4	100	4	8	13
Wasa	1973-94	19	89	1	3	21
Williams Lake	1973-93	21	76	2	5	10

Boreal Chickadee

Survey route	Survey years	Total surveys	Frequency of occurrence (%)	Range of total surveys when species occurred Low	Median	High
Interior						
Atlin Road	1989	1	100	1	–	1
Brookmere	1973-92	10	10	1	–	1
Cassiar	1972-75	2	50	3	–	3
Creighton Valley	1979-94	16	6	1	–	1
Fort Nelson	1974-80	5	100	1	2	2
Gerrard	1976-85	10	10	2	–	2
Gosnell	1969-74	4	25	2	–	2
Kootenay	1981-94	10	20	2	4	5
Mount Morice	1968-75	5	20	1	–	1
Pennington	1987-91	4	50	2	5	7
Pinkut Creek	1975	1	100	2	–	2
Punchaw	1974-94	9	22	2	3	3
Steamboat	1974-80	4	50	1	3	4
Summerland	1973-94	21	10	1	1	1
Telkwa High Road	1974-94	11	9	3	–	3
Wingdam	1968-70	3	33	1	–	1

Chestnut-backed Chickadee

Survey route	Survey years	Total surveys	Frequency of occurrence (%)	Range of total surveys when species occurred Low	Median	High
Coastal						
Alberni	1968-79	6	100	21	32	58
Albion	1968-92	21	33	1	3	6
Campbell River	1973-93	19	95	19	30	58
Chemainus	1969-88	4	100	34	56	86
Chilliwack	1973-94	11	27	2	2	6
Comox Lake	1973-75	3	100	30	39	44
Coquitlam	1990-94	5	40	3	3	3
Courtenay	1992-93	2	100	10	40	70
Elsie Lake	1970-94	6	83	2	5	11
Gibsons Landing	1973-94	10	100	12	26	34
Kispiox	1975-94	3	67	2	3	3
Kitsumkalum	1974-94	19	58	1	13	51
Kwinitsa	1968-94	26	77	1	8	39
Masset	1989-94	5	100	1	4	31
Meziadin	1975-77	2	50	12	–	12
Nanaimo River	1973-94	22	100	15	31	55
Pemberton	1974-94	18	78	1	3	12
Pitt Meadows	1973-89	14	57	2	5	12
Point Grey	1974-92	11	100	3	8	17
Port Hardy	1982-89	4	100	9	15	22
Port Renfrew	1973-94	17	100	8	16	65
Queen Charlotte City	1989-94	3	100	8	9	17
Saltery Bay	1974-76	3	100	2	3	7
Seabird	1969-94	16	31	2	4	16
Squamish	1973-94	9	100	2	8	21
Victoria	1973-94	17	100	13	50	76
Interior						
Adams Lake	1989-94	6	33	1	2	2
Chilkat Pass	1975-76	2	50	11	–	11
Creighton Valley	1979-94	16	38	1	2	9
Gerrard	1976-85	10	80	1	2	7
Goldstream Mountain	1993-94	2	50	2	–	2
Gosnell	1969-74	4	25	4	–	4
Grand Forks	1973-94	22	5	2	–	2
Illecilewaet	1973-83	4	50	2	4	5
Kuskonook	1968-91	21	14	2	2	6
Needles	1968-94	26	38	1	2	4
Salmo	1969-75	3	33	1	–	1
Scotch Creek	1989-94	6	83	2	3	7
Summerland	1973-94	21	5	1	–	1
Syringa Creek	1986-93	7	29	1	1	1
Zincton	1968-94	22	27	2	2	6

Bushtit

Survey route	Survey years	Total surveys	Frequency of occurrence (%)	Range of total surveys when species occurred		
				Low	Median	High
Coastal						
Albion	1968-92	21	62	2	9	38
Chemainus	1969-88	4	50	2	7	12
Chilliwack	1973-94	11	45	1	4	19
Comox Lake	1973-75	3	33	4	–	4
Coquitlam	1990-94	5	40	4	9	14
Courtenay	1992-93	2	50	5	–	5
Elsie Lake	1970-94	6	17	4	–	4
Nanaimo River	1973-94	22	27	2	5	9
Pemberton	1974-94	18	6	2	–	2
Pitt Meadows	1973-89	14	71	3	10	21
Point Grey	1974-92	11	82	2	5	14
Port Renfrew	1973-94	17	29	1	2	4
Seabird	1969-94	16	6	5	–	5
Victoria	1973-94	17	88	1	10	38

Red-breasted Nuthatch

Survey route	Survey years	Total surveys	Frequency of occurrence (%)	Range of total surveys when species occurred		
				Low	Median	High
Coastal						
Alberni	1968-79	6	67	1	3	3
Albion	1968-92	21	10	1	1	1
Campbell River	1973-93	19	58	1	2	6
Chemainus	1969-88	4	100	1	3	12
Chilliwack	1973-94	11	18	1	1	1
Comox Lake	1973-75	3	33	1	–	1
Courtenay	1992-93	2	50	4	–	4
Elsie Lake	1970-94	6	50	1	1	1
Gibsons Landing	1973-94	10	80	1	3	6
Kispiox	1975-94	3	67	2	5	7
Kitsumkalum	1974-94	19	53	1	2	8
Kwinitsa	1968-94	26	27	1	1	3
Masset	1989-94	5	20	1	–	1
Nanaimo River	1973-94	22	77	1	2	6
Pemberton	1974-94	18	89	1	2	9
Pitt Meadows	1973-89	14	7	2	–	2
Point Grey	1974-92	11	55	1	2	3
Port Hardy	1982-89	4	25	11	–	11
Port Renfrew	1973-94	17	12	1	1	1
Saltery Bay	1974-76	3	33	1	–	1
Seabird	1969-94	16	38	1	2	4
Squamish	1973-94	9	11	5	–	5
Victoria	1973-94	17	88	1	3	10
Interior						
Adams Lake	1989-94	6	100	6	26	32
Alexis Creek	1973-86	14	79	1	2	5
Beaverdell	1970-94	24	100	1	6	21
Bridge Lake	1974-94	20	100	1	10	29
Bromley	1973-94	5	80	2	9	10
Brookmere	1973-92	10	100	2	7	18
Canford	1974-93	10	100	2	6	16
Canoe Point	1993-94	2	100	2	8	14
Christian Valley	1993-94	2	100	11	22	33
Chu Chua	1993-94	2	100	15	18	20
Columbia Lake	1973-83	7	29	1	4	6
Creighton Valley	1979-94	16	88	1	8	17
Creston	1984	1	100	14	–	14
Ferndale	1968-94	15	93	1	9	21
Fort Nelson	1974-80	5	60	1	1	3
Fort St. John	1974-90	9	22	1	1	1
Gerrard	1976-85	10	100	3	6	8
Gnat Pass	1975-93	2	50	9	–	9
Golden	1973-94	22	68	1	3	11
Goldstream Mountain	1993-94	2	100	1	3	4

▶

Red-breasted Nuthatch

Survey route	Survey years	Total surveys	Frequency of occurrence (%)	Low	Range of total surveys when species occurred Median	High
Gosnell	1969-74	4	75	1	1	2
Grand Forks	1973-94	22	91	1	5	12
Haines Summit	1993-94	2	100	1	2	2
Hendrix Lake	1976	1	100	15	–	15
Horsefly	1974-88	11	64	1	3	18
Hudson's Hope	1976	1	100	1	–	1
Illecilewaet	1973-83	4	25	1	–	1
Kamloops	1974-94	15	93	1	4	9
Kootenay	1981-94	10	90	1	3	6
Kuskonook	1968-91	21	90	3	7	17
Lac la Hache	1973-86	11	91	1	9	14
Lavington	1972-94	23	87	1	2	8
Mabel Lake	1973-94	19	95	1	4	12
McLeod Lake	1968-94	9	100	3	15	22
Meadowlake	1968-70	2	100	3	3	3
Mount Morice	1968-75	5	80	1	2	3
Needles	1968-94	26	88	1	5	16
Oliver	1973-94	18	94	1	12	23
Osprey Lake	1976-94	10	40	1	3	8
Pavilion	1974-94	6	83	1	3	9
Pennington	1987-91	4	25	3	–	3
Pleasant Valley	1980-93	6	83	1	3	5
Prince George	1973-94	20	45	1	1	7
Princeton	1968-94	8	100	3	7	9
Punchaw	1974-94	9	100	1	7	22
Riske Creek	1973-92	11	100	1	2	9
Salmo	1969-75	3	100	1	3	5
Salmon Arm	1968-94	19	100	1	6	16
Scotch Creek	1989-94	6	100	3	6	12
Shalalth	1968-94	4	100	3	16	19
Spillimacheen	1976-93	17	100	3	8	19
Steamboat	1974-80	4	50	1	2	3
Succour Creek	1993-94	2	100	7	8	8
Summerland	1973-94	21	76	2	4	17
Summit Lake	1972-94	8	100	1	10	14
Syringa Creek	1986-93	7	100	2	7	12
Tatlayoko Lake	1989-94	6	83	1	7	15
Telkwa High Road	1974-94	11	73	1	5	12
Tokay	1968-75	4	50	4	6	8
Tupper	1974-94	6	17	4	–	4
Wasa	1973-94	19	100	1	5	31
Williams Lake	1973-93	21	100	1	8	17
Wingdam	1968-70	3	67	6	8	10
Zincton	1968-94	22	95	1	6	18

White-breasted Nuthatch

Survey route	Survey years	Total surveys	Frequency of occurrence (%)	Low	Range of total surveys when species occurred Median	High
Interior						
Beaverdell	1970-94	24	4	1	–	1
Bromley	1973-94	5	20	1	–	1
Canford	1974-93	10	60	1	2	3
Chu Chua	1993-94	2	50	2	–	2
Columbia Lake	1973-83	7	14	1	–	1
Grand Forks	1973-94	22	9	1	2	2
Kamloops	1974-94	15	33	1	3	5
Oliver	1973-94	18	78	1	2	3
Osprey Lake	1976-94	10	20	1	3	4
Pavilion	1974-94	6	50	1	1	4
Pleasant Valley	1980-93	6	33	1	2	2
Shalalth	1968-94	4	25	1	–	1
Succour Creek	1993-94	2	50	1	–	1
Summerland	1973-94	21	90	1	2	7
Tokay	1968-75	4	25	2	–	2
Wasa	1973-94	19	37	1	1	6

Pygmy Nuthatch

Survey route	Survey years	Total surveys	Frequency of occurrence (%)	Low	Range of total surveys when species occurred Median	High
Interior						
Bromley	1973-94	5	60	1	5	7
Canford	1974-93	10	60	1	5	12
Kamloops	1974-94	15	33	1	2	13
Lavington	1972-94	23	15	1	1	1
Nicola	1973-74	2	50	1	–	1
Oliver	1973-94	18	56	1	2	5
Osprey Lake	1976-94	10	30	1	3	4
Pleasant Valley	1980-93	6	17	1	–	1
Princeton	1968-94	8	38	1	1	2
Summerland	1973-94	21	81	1	3	7

Brown Creeper

Survey route	Survey years	Total surveys	Frequency of occurrence (%)	Low	Range of total surveys when species occurred Median	High
Coastal						
Alberni	1968-79	6	50	1	1	3
Albion	1968-92	21	5	2	–	2
Campbell River	1973-93	19	37	1	2	4
Chemainus	1969-88	4	100	1	2	4
Chilliwack	1973-94	11	64	1	2	6
Comox Lake	1973-75	3	67	2	3	3
Coquitlam	1990-94	5	20	1	–	1
Courtenay	1992-93	2	100	3	4	5
Elsie Lake	1970-94	6	50	1	1	1
Gibsons Landing	1973-94	10	70	1	2	5
Kitsumkalum	1974-94	19	16	1	1	1
Kwinitsa	1968-94	26	4	1	–	1
Nanaimo River	1973-94	22	73	1	3	14
Pemberton	1974-94	18	56	1	4	9
Pitt Meadows	1973-89	14	43	1	1	2
Point Grey	1974-92	11	36	1	2	3
Port Hardy	1982-89	4	75	1	1	1
Port Renfrew	1973-94	17	6	5	–	5
Seabird	1969-94	16	56	1	2	5
Squamish	1973-94	9	67	1	1	6
Victoria	1973-94	17	76	1	2	8
Interior						
Adams Lake	1989-94	6	83	1	2	4
Alexis Creek	1973-86	14	21	1	1	1
Beaverdell	1970-94	24	4	1	–	1
Bridge Lake	1974-94	20	20	1	2	2
Bromley	1973-94	5	40	1	1	1
Brookmere	1973-92	10	50	1	1	2
Canford	1974-93	10	10	1	–	1
Chu Chua	1993-94	2	50	1	–	1
Creighton Valley	1979-94	16	19	1	2	4
Gerrard	1976-85	10	80	1	1	2
Illecilewaet	1973-83	4	25	3	–	3
Kuskonook	1968-91	21	19	1	1	1
Mabel Lake	1973-94	19	11	1	1	1
Mount Morice	1968-75	5	20	1	–	1
Needles	1968-94	26	8	1	2	2
Nicola	1973-74	2	50	1	–	1
Oliver	1973-94	18	6	1	–	1
Osprey Lake	1976-94	10	50	1	3	4
Princeton	1968-94	8	75	1	6	12
Salmo	1969-75	3	33	1	–	1
Salmon Arm	1968-94	19	5	1	–	1
Scotch Creek	1989-94	6	83	1	2	4
Spillimacheen	1976-93	17	12	1	2	3
Succour Creek	1993-94	2	50	1	–	1
Summerland	1973-94	21	5	1	–	1
Summit Lake	1972-94	8	13	1	–	1
Zincton	1968-94	22	18	1	1	3

Rock Wren

Survey route	Survey years	Total surveys	Frequency of occurrence (%)	Range of total surveys when species occurred Low	Median	High
Coastal						
Seabird	1969-94	16	6	2	–	2
Interior						
Beaverdell	1970-94	24	17	1	1	3
Bromley	1973-94	5	40	2	2	2
Brookmere	1973-92	10	10	1	–	1
Creighton Valley	1979-94	16	6	1	–	1
Grand Forks	1973-94	22	14	1	2	2
Kamloops	1974-94	15	13	1	1	1
Kuskonook	1968-91	21	5	1	–	1
Oliver	1973-94	18	39	1	1	2
Pavilion	1974-94	6	33	1	1	1
Princeton	1968-94	8	13	1	–	1
Summerland	1973-94	21	24	1	2	3
Syringa Creek	1986-93	7	86	1	1	2

Canyon Wren

Survey route	Survey years	Total surveys	Frequency of occurrence (%)	Range of total surveys when species occurred Low	Median	High
Interior						
Needles	1968-94	26	4	2	–	2
Syringa Creek	1986-93	7	43	1	2	3

Bewick's Wren

Survey route	Survey years	Total surveys	Frequency of occurrence (%)	Range of total surveys when species occurred Low	Median	High
Coastal						
Alberni	1968-79	6	50	1	1	1
Albion	1968-92	21	81	1	3	13
Campbell River	1973-93	19	74	1	1	4
Chemainus	1969-88	4	75	9	12	21
Chilliwack	1973-94	11	55	1	2	7
Coquitlam	1990-94	5	60	1	1	1
Courtenay	1992-93	2	100	2	2	2
Gibsons Landing	1973-94	10	30	1	1	2
Nanaimo River	1973-94	22	95	1	8	18
Pemberton	1974-94	18	11	1	2	3
Pitt Meadows	1973-89	14	100	2	6	12
Point Grey	1974-92	11	100	2	5	21
Port Renfrew	1973-94	17	18	1	1	3
Saltery Bay	1974-76	3	67	1	1	1
Seabird	1969-94	16	19	1	1	2
Squamish	1973-94	9	33	1	2	3
Victoria	1973-94	17	100	5	32	57

House Wren

Survey route	Survey years	Total surveys	Frequency of occurrence (%)	Range of total surveys when species occurred Low	Median	High
Coastal						
Alberni	1968-79	6	33	1	2	2
Albion	1968-92	21	5	1	–	1
Chemainus	1969-88	4	25	3	–	3
Chilliwack	1973-94	11	9	1	–	1
Elsie Lake	1970-94	6	17	1	–	1
Nanaimo River	1973-94	22	14	1	1	3
Victoria	1973-94	17	18	1	1	4
Interior						
Alexis Creek	1973-86	14	7	1	–	1
Beaverdell	1970-94	24	46	1	2	5
Brookmere	1973-92	10	90	1	6	10
Canford	1974-93	10	90	1	3	9
Canoe Point	1993-94	2	50	1	–	1
Chu Chua	1993-94	2	100	1	2	2
Creighton Valley	1979-94	16	6	2	–	2
Fort Nelson	1974-80	5	20	3	–	3
Fort St. John	1974-90	9	100	2	4	7
Golden	1973-94	22	18	1	2	2
Grand Forks	1973-94	22	55	1	2	4
Kamloops	1974-94	15	7	3	–	3
Kuskonook	1968-91	21	5	1	–	1
Lac la Hache	1973-86	11	9	2	–	2
Lavington	1972-94	23	100	3	7	17
Mabel Lake	1973-94	19	84	1	2	3
Needles	1968-94	26	8	1	1	1
Oliver	1973-94	18	89	2	4	14
Osprey Lake	1976-94	10	40	1	1	2
Pavilion	1974-94	6	33	1	1	1
Pleasant Valley	1980-93	6	100	1	5	6
Prince George	1973-94	20	5	1	–	1
Princeton	1968-94	8	13	1	–	1
Salmon Arm	1968-94	19	21	1	1	3
Summerland	1973-94	21	90	2	5	9
Telkwa High Road	1974-94	11	9	10	–	10*
Tokay	1968-75	4	75	1	3	7
Tupper	1974-94	6	83	2	8	14

Winter Wren

Survey route	Survey years	Total surveys	Frequency of occurrence (%)	Range of total surveys when species occurred Low	Median	High
Coastal						
Alberni	1968-79	6	100	6	20	28
Albion	1968-92	21	95	3	5	15
Campbell River	1973-93	19	100	9	20	37
Chemainus	1969-88	4	100	8	17	21
Chilliwack	1973-94	11	100	1	4	10
Comox Lake	1973-75	3	100	5	8	15
Coquitlam	1990-94	5	80	2	4	4
Courtenay	1992-93	2	100	24	24	24
Elsie Lake	1970-94	6	100	4	9	26
Gibsons Landing	1973-94	10	100	12	20	32
Kispiox	1975-94	3	67	1	1	1
Kitsumkalum	1974-94	19	100	1	11	34
Kwinitsa	1968-94	26	100	1	18	39
Masset	1989-94	5	100	26	49	88
Meziadin	1975-77	2	100	1	10	19
Nanaimo River	1973-94	22	100	4	8	17
Nass River	1975	1	100	7	–	7
Pemberton	1974-94	18	89	1	3	10
Pitt Meadows	1973-89	14	93	1	2	4
Point Grey	1974-92	11	100	5	11	16
Port Hardy	1982-89	4	100	74	79	81

▶

◀ *Winter Wren*

Survey route	Survey years	Total surveys	Frequency of occurrence (%)	Range of total surveys when species occurred Low	Median	High
Port Renfrew	1973-94	17	100	31	42	60
Queen Charlotte City	1989-94	3	100	25	51	112
Saltery Bay	1974-76	3	67	20	21	21
Seabird	1969-94	16	94	1	9	15
Squamish	1973-94	9	100	6	14	35
Victoria	1973-94	17	71	1	3	6
Interior						
Adams Lake	1989-94	6	100	4	6	19
Beaverdell	1970-94	24	8	1	1	1
Bridge Lake	1974-94	20	5	3	–	3
Brookmere	1973-92	10	90	1	4	6
Canford	1974-93	10	90	1	2	7
Canoe Point	1993-94	2	50	1	–	1
Chilkat Pass	1975-76	2	100	8	8	8
Christian Valley	1993-94	2	50	2	–	2
Chu Chua	1993-94	2	100	1	2	2
Columbia Lake	1973-83	7	14	1	–	1
Creighton Valley	1979-94	16	100	1	5	13
Creston	1984	1	100	1	–	1
Ferndale	1968-94	15	80	1	2	5
Fort Nelson	1974-80	5	20	1	–	1
Gerrard	1976-85	10	100	8	14	33
Goldstream Mountain	1993-94	2	100	7	8	8
Gosnell	1969-74	4	100	7	12	20
Grand Forks	1973-94	22	36	1	1	2
Horsefly	1974-88	11	36	1	1	2
Illecilewaet	1973-83	4	75	1	5	15
Kootenay	1981-94	10	90	1	3	7
Kuskonook	1968-91	21	57	1	3	7
Lavington	1972-94	23	96	1	2	7
Mabel Lake	1973-94	19	84	1	3	10
McLeod Lake	1968-94	9	100	2	3	8
Mount Morice	1968-75	5	60	1	1	2
Needles	1968-94	26	100	1	3	15
Oliver	1973-94	18	56	1	1	3
Osprey Lake	1976-94	10	50	1	2	7
Pavilion	1974-94	6	17	1	–	1
Pennington	1987-91	4	25	3	–	3
Prince George	1973-94	20	5	1	–	1
Princeton	1968-94	8	38	1	1	2
Punchaw	1974-94	9	44	1	2	3
Salmon Arm	1968-94	19	21	1	1	2
Scotch Creek	1989-94	6	100	9	21	33
Spillimacheen	1976-93	17	41	1	1	2
Steamboat	1974-80	4	25	1	–	1
Succour Creek	1993-94	2	100	5	5	5
Summerland	1973-94	21	86	1	3	5
Summit Lake	1972-94	8	38	1	2	3
Syringa Creek	1986-93	7	43	1	1	1
Telkwa High Road	1974-94	11	27	1	2	5
Tokay	1968-75	4	25	4	–	4
Wingdam	1968-70	3	67	1	4	6
Zincton	1968-94	22	91	1	4	12

Marsh Wren

Survey route	Survey years	Total surveys	Frequency of occurrence (%)	Range of total surveys when species occurred Low	Median	High
Coastal						
Albion	1968-92	21	19	1	2	4
Nanaimo River	1973-94	22	32	1	2	4
Pitt Meadows	1973-89	14	7	2	–	2
Seabird	1969-94	16	6	1	–	1
Squamish	1973-94	9	11	1	–	1
Victoria	1973-94	17	6	2	–	2

▶

◀ Marsh Wren

Survey route	Survey years	Total surveys	Frequency of occurrence (%)	Range of total surveys when species occurred Low	Median	High
Interior						
Alexis Creek	1973-86	14	14	1	2	2
Bridge Lake	1974-94	20	35	1	1	2
Gnat Pass	1975-93	2	50	2	–	2
Golden	1973-94	22	59	1	2	5
Grand Forks	1973-94	22	14	1	2	2
Kamloops	1974-94	15	60	1	2	3
Lac la Hache	1973-86	11	45	1	1	2
Mabel Lake	1973-94	19	11	1	1	1
Needles	1968-94	26	4	1	–	1
Pleasant Valley	1980-93	6	67	1	2	3
Riske Creek	1973-92	11	45	1	2	3
Salmon Arm	1968-94	19	21	1	3	4
Spillimacheen	1976-93	17	100	2	3	5
Succour Creek	1993-94	2	100	1	2	3
Wasa	1973-94	19	79	1	5	8
Williams Lake	1973-93	21	62	1	2	4

American Dipper

Survey route	Survey years	Total surveys	Frequency of occurrence (%)	Range of total surveys when species occurred Low	Median	High
Coastal						
Alberni	1968-79	6	17	1	–	1
Campbell River	1973-93	19	16	1	1	1
Chemainus	1969-88	4	25	1	–	1
Meziadin	1975-77	2	50	1	–	1
Pemberton	1974-94	18	28	1	1	1
Seabird	1969-94	16	6	1	–	1
Squamish	1973-94	9	11	1	–	1
Interior						
Christian Valley	1993-94	2	50	1	–	1
Creighton Valley	1979-94	16	19	1	1	2
Grand Forks	1973-94	22	36	1	1	2
Illecilewaet	1973-83	4	25	1	–	1
Kootenay	1981-94	10	40	1	1	2
Kuskonook	1968-91	21	5	2	–	2
Needles	1968-94	26	8	1	1	1
Syringa Creek	1986-93	7	14	1	–	1
Zincton	1968-94	22	14	1	1	1

Golden-crowned Kinglet

Survey route	Survey years	Total surveys	Frequency of occurrence (%)	Range of total surveys when species occurred Low	Median	High
Coastal						
Alberni	1968-79	6	83	11	16	28
Albion	1968-92	21	67	1	1	4
Campbell River	1973-93	19	95	4	13	33
Chemainus	1969-88	4	100	6	18	48
Chilliwack	1973-94	11	55	1	2	14
Comox Lake	1973-75	3	67	4	5	6
Coquitlam	1990-94	5	40	2	2	2
Courtenay	1992-93	2	100	15	16	16
Elsie Lake	1970-94	6	33	3	5	7
Gibsons Landing	1973-94	10	80	3	12	46
Kispiox	1975-94	3	67	5	5	5
Kitsumkalum	1974-94	19	58	1	4	12
Kwinitsa	1968-94	26	54	1	5	12
Masset	1989-94	5	100	9	11	29

▶

◀ *Golden-crowned Kinglet*

Survey route	Survey years	Total surveys	Frequency of occurrence (%)	Range of total surveys when species occurred Low	Median	High
Meziadin	1975-77	2	100	1	4	7
Nanaimo River	1973-94	22	91	1	6	27
Pemberton	1974-94	18	83	1	4	26
Pitt Meadows	1973-89	14	64	2	3	10
Point Grey	1974-92	11	91	1	3	6
Port Hardy	1982-89	4	75	13	14	28
Port Renfrew	1973-94	17	100	2	10	24
Queen Charlotte City	1989-94	3	100	1	13	39
Saltery Bay	1974-76	3	100	11	17	17
Seabird	1969-94	16	69	1	3	7
Squamish	1973-94	9	89	2	6	15
Victoria	1973-94	17	65	1	3	9
Interior						
Adams Lake	1989-94	6	100	3	9	15
Alexis Creek	1973-86	14	21	1	1	1
Ashnola	1973-74	2	50	2	–	2
Beaverdell	1970-94	24	67	1	2	12
Bridge Lake	1974-94	20	80	1	4	8
Bromley	1973-94	5	60	2	4	23
Brookmere	1973-92	10	100	3	4	12
Canford	1974-93	10	80	1	2	4
Canoe Point	1993-94	2	50	10	–	10
Chilkat Pass	1975-76	2	50	3	–	3
Christian Valley	1993-94	2	100	4	8	12
Chu Chua	1993-94	2	50	3	–	3
Creighton Valley	1979-94	16	100	3	11	20
Creston	1984	1	100	8	–	8
Ferndale	1968-94	15	87	2	11	20
Fort Nelson	1974-80	5	20	2	–	2
Gerrard	1976-85	10	100	3	12	34
Golden	1973-94	22	27	1	2	5
Goldstream Mountain	1993-94	2	100	1	4	6
Gosnell	1969-74	4	100	2	8	16
Grand Forks	1973-94	22	50	1	2	8
Haines Summit	1993-94	2	100	1	2	2
Illecilewaet	1973-83	4	75	17	18	26
Kamloops	1974-94	15	13	1	1	1
Kootenay	1981-94	10	100	7	10	25
Kuskonook	1968-91	21	76	1	3	23
Lac la Hache	1973-86	11	55	1	3	4
Lavington	1972-94	23	82	1	2	5
Mabel Lake	1973-94	19	74	1	4	10
McLeod Lake	1968-94	9	89	6	12	17
Mount Morice	1968-75	5	20	3	–	3
Needles	1968-94	26	96	1	8	21
Oliver	1973-94	18	78	1	2	4
Osprey Lake	1976-94	10	90	1	8	15
Pleasant Valley	1980-93	6	17	1	–	1
Prince George	1973-94	20	70	1	1	4
Princeton	1968-94	8	63	2	17	19
Punchaw	1974-94	9	56	2	15	22
Riske Creek	1973-92	11	9	4	–	4
Salmo	1969-75	3	33	3	–	3
Salmon Arm	1968-94	19	79	1	5	9
Scotch Creek	1989-94	6	100	5	12	23
Shalalth	1968-94	4	25	1	–	1
Spillimacheen	1976-93	17	82	1	3	10
Steamboat	1974-80	4	50	1	2	3
Succour Creek	1993-94	2	100	5	6	7
Summerland	1973-94	21	90	1	2	6
Summit Lake	1972-94	8	88	3	9	20
Syringa Creek	1986-93	7	71	2	3	6
Telkwa High Road	1974-94	11	73	1	6	16
Tupper	1974-94	6	17	3	–	3
Wingdam	1968-70	3	100	8	15	17
Zincton	1968-94	22	82	1	8	21

Ruby-crowned Kinglet

Survey route	Survey years	Total surveys	Frequency of occurrence (%)	Range of total surveys when species occurred: Low	Median	High
Coastal						
Albion	1968-92	21	5	1	–	1
Campbell River	1973-93	19	5	2	–	2
Chilliwack	1973-94	11	9	4	–	4
Elsie Lake	1970-94	6	17	1	–	1
Kispiox	1975-94	3	100	4	6	6
Kitsumkalum	1974-94	19	74	1	3	11
Kwinitsa	1968-94	26	77	1	7	26
Nanaimo River	1973-94	22	14	1	1	2
Nass River	1975	1	100	5	–	5
Pemberton	1974-94	18	56	1	2	5
Pitt Meadows	1973-89	14	7	1	–	1
Port Renfrew	1973-94	17	6	1	–	1
Saltery Bay	1974-76	3	33	18	–	18
Seabird	1969-94	16	13	1	1	1
Interior						
Alexis Creek	1973-86	14	100	3	11	20
Ashnola	1973-74	2	100	2	4	5
Beaverdell	1970-94	24	96	1	4	13
Bridge Lake	1974-94	20	100	2	23	42
Bromley	1973-94	5	60	1	1	2
Brookmere	1973-92	10	100	2	11	35
Canford	1974-93	10	80	1	3	6
Cassiar	1972-75	2	50	4	–	4
Chilkat Pass	1975-76	2	100	2	5	7
Christian Valley	1993-94	2	100	8	11	13
Chu Chua	1993-94	2	50	2	–	2
Columbia Lake	1973-83	7	86	1	4	6
Creighton Valley	1979-94	16	88	1	2	11
Creston	1984	1	100	3	–	3
Ferndale	1968-94	15	100	2	14	63
Fort Nelson	1974-80	5	60	1	2	3
Fort St. John	1974-90	9	33	1	1	1
Fraser Lake	1968-78	5	60	4	4	16
Gerrard	1976-85	10	10	1	–	1
Gnat Pass	1975-93	2	100	10	17	23
Golden	1973-94	22	95	1	4	12
Gosnell	1969-74	4	25	1	–	1
Grand Forks	1973-94	22	73	1	1	2
Haines Summit	1993-94	2	100	3	10	16
Hendrix Lake	1976	1	100	1	–	1
Horsefly	1974-88	11	73	1	6	16
Kamloops	1974-94	15	60	1	1	4
Kinaskan Lake	1975	1	100	2	–	2
Kootenay	1981-94	10	90	11	16	31
Kuskonook	1968-91	21	14	2	3	5
Lac la Hache	1973-86	11	100	6	21	40
Lavington	1972-94	23	52	1	2	3
Mabel Lake	1973-94	19	53	1	2	3
McLeod Lake	1968-94	9	100	12	24	51
Mount Morice	1968-75	5	80	8	13	26
Needles	1968-94	26	62	1	2	3
Nicola	1973-74	2	50	2	–	2
Oliver	1973-94	18	100	2	10	15
Osprey Lake	1976-94	10	90	1	11	15
Pavilion	1974-94	6	50	1	2	4
Pennington	1987-91	4	75	4	4	27
Pinkut Creek	1975	1	100	50	–	50
Prince George	1973-94	20	100	7	17	32
Princeton	1968-94	8	100	1	4	6
Punchaw	1974-94	9	100	9	23	55
Riske Creek	1973-92	11	100	2	13	16
Salmo	1969-75	3	33	3	–	3
Salmon Arm	1968-94	19	68	1	2	7
Shalalth	1968-94	4	25	1	–	1
Spillimacheen	1976-93	17	100	2	7	13
Steamboat	1974-80	4	50	5	6	7
Succour Creek	1993-94	2	100	16	22	27
Summerland	1973-94	21	86	1	2	6

►

◀ *Ruby-crowned Kinglet*

Survey route	Survey years	Total surveys	Frequency of occurrence (%)	Low	Range of total surveys when species occurred Median	High
Summit Lake	1972-94	8	100	7	18	36
Syringa Creek	1986-93	7	29	1	1	1
Tatlayoko Lake	1989-94	6	100	1	5	12
Telkwa High Road	1974-94	11	100	3	6	11
Tokay	1968-75	4	75	2	9	9
Tupper	1974-94	6	50	1	1	1
Wasa	1973-94	19	47	1	1	2
Williams Lake	1973-93	21	95	1	6	14
Wingdam	1968-70	3	100	21	28	41
Zincton	1968-94	22	27	1	1	3

Western Bluebird

Survey route	Survey years	Total surveys	Frequency of occurrence (%)	Low	Range of total surveys when species occurred Median	High
Interior						
Beaverdell	1970-94	24	8	1	2	2
Oliver	1973-94	18	44	1	1	3
Osprey Lake	1976-94	10	10	2	–	2
Summerland	1973-94	21	67	1	1	3
Tatlayoko Lake	1989-94	6	17	1	–	1

Mountain Bluebird

Survey route	Survey years	Total surveys	Frequency of occurrence (%)	Low	Range of total surveys when species occurred Median	High
Coastal						
Chilliwack	1973-94	11	9	1	–	1
Seabird	1969-94	16	19	1	1	2
Interior						
Alexis Creek	1973-86	14	79	1	2	7
Beaverdell	1970-94	24	46	1	2	3
Bridge Lake	1974-94	20	20	1	2	2
Bromley	1973-94	5	100	1	4	7
Brookmere	1973-92	10	100	3	5	13
Canford	1974-93	10	90	1	2	4
Canoe Point	1993-94	2	50	1	–	1
Christian Valley	1993-94	2	50	2	–	2
Columbia Lake	1973-83	7	86	2	4	4
Creighton Valley	1979-94	16	31	2	2	4
Ferndale	1968-94	15	7	1	–	1
Gnat Pass	1975-93	2	50	4	–	4
Golden	1973-94	22	23	1	2	5
Grand Forks	1973-94	22	23	1	1	4
Kamloops	1974-94	15	87	1	4	10
Lac la Hache	1973-86	11	18	1	2	3
Lavington	1972-94	23	30	1	2	6
Mabel Lake	1973-94	19	21	1	3	7
Meadowlake	1968-70	2	50	7	–	7
Needles	1968-94	26	19	1	2	2
Nicola	1973-74	2	50	9	–	9
Oliver	1973-94	18	72	1	2	4
Osprey Lake	1976-94	10	70	1	2	5
Pavilion	1974-94	6	67	3	10	13
Pleasant Valley	1980-93	6	67	1	3	5
Prince George	1973-94	20	15	1	1	2
Princeton	1968-94	8	75	1	4	10
Riske Creek	1973-92	11	100	3	11	23

▶

◀ *Mountain Bluebird*

Survey route	Survey years	Total surveys	Frequency of occurrence (%)	Low	Range of total surveys when species occurred Median	High
Salmo	1969-75	3	100	1	3	4
Salmon Arm	1968-94	19	37	1	1	5
Shalalth	1968-94	4	75	7	8	9
Spillimacheen	1976-93	17	71	1	3	8
Succour Creek	1993-94	2	50	1	–	1
Summerland	1973-94	21	95	1	4	8
Summit Lake	1972-94	8	13	1	–	1
Syringa Creek	1986-93	7	57	1	2	2
Tatlayoko Lake	1989-94	6	50	1	1	1
Telkwa High Road	1974-94	11	45	1	3	6
Tokay	1968-75	4	75	5	8	9
Wasa	1973-94	19	79	1	3	7
Williams Lake	1973-93	21	62	1	2	5

Townsend's Solitaire

Survey route	Survey years	Total surveys	Frequency of occurrence (%)	Low	Range of total surveys when species occurred Median	High
Coastal						
Comox Lake	1973-75	3	33	1	–	1
Kitsumkalum	1974-94	19	5	1	–	1
Pemberton	1974-94	18	33	1	2	5
Interior						
Adams Lake	1989-94	6	67	1	1	2
Alexis Creek	1973-86	14	86	1	3	6
Ashnola	1973-74	2	50	2	–	2
Beaverdell	1970-94	24	71	1	3	6
Bridge Lake	1974-94	20	35	1	1	2
Bromley	1973-94	5	80	2	4	8
Brookmere	1973-92	10	90	1	2	7
Canford	1974-93	10	100	1	6	10
Christian Valley	1993-94	2	50	3	–	3
Chu Chua	1993-94	2	100	3	4	4
Columbia Lake	1973-83	7	14	1	–	1
Creighton Valley	1979-94	16	56	1	1	3
Fraser Lake	1968-78	5	20	1	–	1
Golden	1973-94	22	5	2	–	2
Grand Forks	1973-94	22	36	1	2	4
Horsefly	1974-88	11	9	1	–	1
Kamloops	1974-94	15	47	1	2	4
Kootenay	1981-94	10	30	1	1	1
Kuskonook	1968-91	21	10	2	3	3
Lac la Hache	1973-86	11	9	1	–	1
Lavington	1972-94	23	8	2	3	3
Mabel Lake	1973-94	19	16	1	1	2
Meadowlake	1968-70	2	50	1	–	1
Needles	1968-94	26	42	1	1	2
Nicola	1973-74	2	50	2	–	2
Oliver	1973-94	18	100	1	6	11
Osprey Lake	1976-94	10	100	1	2	7
Pavilion	1974-94	6	67	1	2	4
Pleasant Valley	1980-93	6	100	1	3	4
Princeton	1968-94	8	63	2	3	9
Riske Creek	1973-92	11	82	1	3	3
Salmon Arm	1968-94	19	26	1	1	2
Shalalth	1968-94	4	25	1	–	1
Spillimacheen	1976-93	17	35	1	1	2
Summerland	1973-94	21	100	2	11	26
Syringa Creek	1986-93	7	29	1	2	2
Tatlayoko Lake	1989-94	6	83	1	4	5
Wasa	1973-94	19	42	1	2	6
Williams Lake	1973-93	21	62	1	3	5
Zincton	1968-94	22	14	1	1	2

Veery

Survey route	Survey years	Total surveys	Frequency of occurrence (%)	Low	Range of total surveys when species occurred Median	High
Coastal						
Kitsumkalum	1974-94	19	16	1	1	2
Kwinitsa	1968-94	26	23	1	1	2
Pemberton	1974-94	18	100	6	11	26
Interior						
Adams Lake	1989-94	6	50	1	1	3
Alexis Creek	1973-86	14	93	1	3	7
Beaverdell	1970-94	24	100	6	16	31
Bridge Lake	1974-94	20	70	1	1	4
Bromley	1973-94	5	100	8	13	15
Brookmere	1973-92	10	100	1	4	13
Canford	1974-93	10	100	6	19	29
Canoe Point	1993-94	2	50	1	–	1
Chu Chua	1993-94	2	100	8	10	12
Creighton Valley	1979-94	16	19	1	1	2
Ferndale	1968-94	15	7	3	–	3
Fort St. John	1974-90	9	11	1	–	1
Gerrard	1976-85	10	100	4	9	10
Golden	1973-94	22	36	1	1	4
Gosnell	1969-74	4	75	2	3	3
Grand Forks	1973-94	22	100	27	45	64
Horsefly	1974-88	11	64	1	2	16
Illecilewaet	1973-83	4	25	1	–	1
Kamloops	1974-94	15	13	2	2	2
Kuskonook	1968-91	21	10	1	1	1
Lac la Hache	1973-86	11	82	1	6	21
Lavington	1972-94	23	48	1	2	6
Mabel Lake	1973-94	19	95	1	5	18
Needles	1968-94	26	100	1	6	21
Oliver	1973-94	18	94	1	4	8
Osprey Lake	1976-94	10	90	3	5	16
Pavilion	1974-94	6	67	10	25	50
Pleasant Valley	1980-93	6	33	2	3	4
Riske Creek	1973-92	11	73	1	2	7
Salmo	1969-75	3	33	1	–	1
Salmon Arm	1968-94	19	95	1	9	17
Spillimacheen	1976-93	17	12	1	1	1
Succour Creek	1993-94	2	100	3	6	9
Summerland	1973-94	21	100	5	14	26
Syringa Creek	1986-93	7	100	2	7	9
Tatlayoko Lake	1989-94	6	83	1	3	4
Tokay	1968-75	4	75	14	20	21
Wasa	1973-94	19	100	1	4	13
Williams Lake	1973-93	21	52	1	1	4
Zincton	1968-94	22	59	1	2	10

Gray-cheeked Thrush

Survey route	Survey years	Total surveys	Frequency of occurrence (%)	Low	Range of total surveys when species occurred Median	High
Interior						
Cassiar	1972-75	2	50	1	–	1
Chilkat Pass	1975-76	2	50	1	–	1
Gnat Pass	1975-93	2	50	5	–	5
Haines Summit	1993-94	2	100	2	3	3

Swainson's Thrush

Survey route	Survey years	Total surveys	Frequency of occurrence (%)	Range of total surveys when species occurred Low	Median	High
Coastal						
Alberni	1968-79	6	100	20	36	70
Albion	1968-92	21	100	6	31	80
Campbell River	1973-93	19	100	28	45	84
Chemainus	1969-88	4	100	12	40	44
Chilliwack	1973-94	11	100	14	20	34
Comox Lake	1973-75	3	100	25	27	33
Coquitlam	1990-94	5	100	10	14	20
Courtenay	1992-93	2	100	41	45	48
Elsie Lake	1970-94	6	100	38	58	71
Gibsons Landing	1973-94	10	100	40	60	91
Kispiox	1975-94	3	100	57	71	94
Kitsumkalum	1974-94	19	100	17	60	91
Kwinitsa	1968-94	26	100	37	58	90
Masset	1989-94	5	100	18	32	35
Meziadin	1975-77	2	100	34	38	42
Nanaimo River	1973-94	22	100	8	20	39
Nass River	1975	1	100	37	–	37
Pemberton	1974-94	18	100	20	29	53
Pitt Meadows	1973-89	14	100	8	17	29
Point Grey	1974-92	11	100	15	25	31
Port Hardy	1982-89	4	100	23	33	48
Port Renfrew	1973-94	17	100	21	55	195
Queen Charlotte City	1989-94	3	100	57	57	109
Saltery Bay	1974-76	3	100	43	69	97
Seabird	1969-94	16	100	22	41	64
Squamish	1973-94	9	100	38	102	115
Victoria	1973-94	17	82	1	6	18
Interior						
Adams Lake	1989-94	6	100	62	93	118
Alexis Creek	1973-86	14	100	10	25	34
Ashnola	1973-74	2	100	15	20	24
Beaverdell	1970-94	24	100	14	34	75
Bridge Lake	1974-94	20	95	1	15	26
Bromley	1973-94	5	80	1	6	8
Brookmere	1973-92	10	100	5	32	48
Canford	1974-93	10	100	7	13	19
Canoe Point	1993-94	2	100	16	18	19
Cassiar	1972-75	2	100	16	36	56
Chilkat Pass	1975-76	2	100	3	8	12
Christian Valley	1993-94	2	100	68	72	76
Chu Chua	1993-94	2	100	37	40	42
Columbia Lake	1973-83	7	86	1	14	25
Creighton Valley	1979-94	16	100	19	32	47
Creston	1984	1	100	28	–	28
Ferndale	1968-94	15	100	1	62	94
Fort Nelson	1974-80	5	100	19	24	25
Fort St. John	1974-90	9	100	1	6	10
Fraser Lake	1968-78	5	100	19	41	61
Gerrard	1976-85	10	100	23	34	43
Gnat Pass	1975-93	2	100	24	40	55
Golden	1973-94	22	100	4	19	39
Goldstream Mountain	1993-94	2	100	41	53	64
Gosnell	1969-74	4	100	31	44	62
Grand Forks	1973-94	22	100	6	17	38
Haines Summit	1993-94	2	100	6	8	10
Hendrix Lake	1976	1	100	33	–	33
Horsefly	1974-88	11	100	5	16	22
Hudson's Hope	1976	1	100	18	–	18
Illecilewaet	1973-83	4	75	9	15	22
Kamloops	1974-94	15	73	1	2	4
Kinaskan Lake	1975	1	100	55	–	55
Kootenay	1981-94	10	100	12	35	44
Kuskonook	1968-91	21	100	16	32	49
Lac la Hache	1973-86	11	100	22	30	38
Lavington	1972-94	23	100	5	11	30
Mabel Lake	1973-94	19	100	13	22	69
McLeod Lake	1968-94	9	100	62	124	146
Meadowlake	1968-70	2	100	8	14	19

▶

◀ *Swainson's Thrush*

Survey route	Survey years	Total surveys	Frequency of occurrence (%)	Low	Range of total surveys when species occurred Median	High
Mount Morice	1968-75	5	100	52	83	140
Needles	1968-94	26	100	24	42	75
Oliver	1973-94	18	100	9	40	54
Osprey Lake	1976-94	10	100	6	11	20
Pavilion	1974-94	6	67	1	10	10
Pennington	1987-91	4	75	28	52	52
Pinkut Creek	1975	1	100	22	–	22
Pleasant Valley	1980-93	6	100	5	14	18
Prince George	1973-94	20	100	6	18	45
Princeton	1968-94	8	88	18	25	57
Punchaw	1974-94	9	100	29	45	102
Riske Creek	1973-92	11	100	1	3	12
Salmo	1969-75	3	33	26	–	26
Salmon Arm	1968-94	19	100	9	24	33
Scotch Creek	1989-94	6	100	77	94	121
Shalalth	1968-94	4	100	11	14	24
Spillimacheen	1976-93	17	100	20	51	69
Steamboat	1974-80	4	100	31	35	58
Succour Creek	1993-94	2	100	103	106	109
Summerland	1973-94	21	95	1	5	9
Summit Lake	1972-94	8	100	42	66	100
Syringa Creek	1986-93	7	100	19	22	28
Tatlayoko Lake	1989-94	6	100	25	54	58
Telkwa High Road	1974-94	11	100	25	55	89
Tetsa River	1974	1	200	16	16	16
Tokay	1968-75	4	75	14	20	29
Tupper	1974-94	6	100	17	22	30
Wasa	1973-94	19	100	4	11	18
Williams Lake	1973-93	21	100	2	6	15
Wingdam	1968-70	3	100	42	56	67
Zincton	1968-94	22	100	23	33	65

Hermit Thrush

Survey route	Survey years	Total surveys	Frequency of occurrence (%)	Low	Range of total surveys when species occurred Median	High
Coastal						
Alberni	1968-79	6	17	1	–	1
Chilliwack	1973-94	11	9	2	–	2
Kispiox	1975-94	3	33	2	–	2
Kitsumkalum	1974-94	19	89	1	9	36
Kwinitsa	1968-94	26	81	1	7	15
Masset	1989-94	5	100	25	34	38
Meziadin	1975-77	2	100	7	7	7
Pemberton	1974-94	18	22	1	3	16
Port Hardy	1982-89	4	100	4	10	13
Port Renfrew	1973-94	17	65	1	5	9
Queen Charlotte City	1989-94	3	100	15	21	85
Seabird	1969-94	16	6	2	–	2
Squamish	1973-94	9	11	1	–	1
Interior						
Adams Lake	1989-94	6	17	1	–	1
Alexis Creek	1973-86	14	57	1	2	8
Ashnola	1973-74	2	100	1	1	1
Bridge Lake	1974-94	20	100	2	18	33
Bromley	1973-94	5	80	3	3	3
Brookmere	1973-92	10	90	2	6	8
Canford	1974-93	10	20	1	1	1
Cassiar	1972-75	2	100	1	9	16
Chilkat Pass	1975-76	2	100	11	22	32
Columbia Lake	1973-83	7	14	21	–	21
Creighton Valley	1979-94	16	88	1	3	8
Ferndale	1968-94	15	73	1	2	14
Fort St. John	1974-90	9	100	2	4	14

▶

◀ *Hermit Thrush*

Survey route	Survey years	Total surveys	Frequency of occurrence (%)	Low	Range of total surveys when species occurred Median	High
Fraser Lake	1968-78	5	80	2	3	16
Gerrard	1976-85	10	60	1	1	1
Gnat Pass	1975-93	2	50	1	–	1
Golden	1973-94	22	5	2	–	2
Goldstream Mountain	1993-94	2	100	2	4	5
Grand Forks	1973-94	22	14	1	1	1
Haines Summit	1993-94	2	50	2	–	2
Hendrix Lake	1976	1	100	4	–	4
Horsefly	1974-88	11	45	1	3	3
Hudson's Hope	1976	1	100	18	–	18
Illecilewaet	1973-83	4	75	9	12	21
Kinaskan Lake	1975	1	100	2	–	2
Kootenay	1981-94	10	90	1	2	5
Kuskonook	1968-91	21	19	1	4	7
Lac la Hache	1973-86	11	27	2	2	3
McLeod Lake	1968-94	9	44	1	1	3
Meadowlake	1968-70	2	100	3	5	6
Mount Morice	1968-75	5	60	2	3	3
Needles	1968-94	26	58	1	3	9
Oliver	1973-94	18	83	1	6	12
Osprey Lake	1976-94	10	70	1	5	9
Pavilion	1974-94	6	50	1	5	14
Pennington	1987-91	4	75	2	19	41
Pinkut Creek	1975	1	100	2	–	2
Pleasant Valley	1980-93	6	33	1	1	1
Prince George	1973-94	20	90	1	4	14
Princeton	1968-94	8	100	5	7	23
Punchaw	1974-94	9	44	9	11	15
Riske Creek	1973-92	11	55	1	2	3
Salmo	1969-75	3	33	3	–	3
Salmon Arm	1968-94	19	16	1	1	2
Scotch Creek	1989-94	6	100	3	14	18
Shalalth	1968-94	4	75	1	1	3
Steamboat	1974-80	4	25	3	–	3
Succour Creek	1993-94	2	100	5	7	8
Summerland	1973-94	21	90	1	5	10
Summit Lake	1972-94	8	88	1	12	21
Tatlayoko Lake	1989-94	6	100	2	5	11
Telkwa High Road	1974-94	11	82	1	2	13
Tupper	1974-94	6	100	1	10	22
Wasa	1973-94	19	16	1	2	2
Williams Lake	1973-93	21	52	1	1	3
Wingdam	1968-70	3	67	3	5	6
Zincton	1968-94	22	27	1	1	12

American Robin

Survey route	Survey years	Total surveys	Frequency of occurrence (%)	Low	Range of total surveys when species occurred Median	High
Coastal						
Alberni	1968-79	6	100	57	79	126
Albion	1968-92	21	100	89	115	183
Campbell River	1973-93	19	100	52	69	98
Chemainus	1969-88	4	100	103	118	155
Chilliwack	1973-94	11	100	74	108	267
Comox Lake	1973-75	3	100	52	67	90
Coquitlam	1990-94	5	100	53	59	72
Courtenay	1992-93	2	100	73	83	92
Elsie Lake	1970-94	6	100	42	54	63
Gibsons Landing	1973-94	10	100	95	121	163
Kispiox	1975-94	3	100	32	37	60
Kitsumkalum	1974-94	19	100	4	34	79
Kwinitsa	1968-94	26	100	47	76	178
Masset	1989-94	5	100	27	43	66
Meziadin	1975-77	2	100	5	8	11

▶

◀ *American Robin*

Survey route	Survey years	Total surveys	Frequency of occurrence (%)	Range of total surveys when species occurred Low	Median	High
Nanaimo River	1973-94	22	100	66	111	143
Nass River	1975	1	100	1	–	1
Pemberton	1974-94	18	100	58	77	135
Pitt Meadows	1973-89	14	100	36	63	85
Point Grey	1974-92	11	100	53	79	108
Port Hardy	1982-89	4	100	35	40	49
Port Renfrew	1973-94	17	100	48	86	220
Queen Charlotte City	1989-94	3	100	11	31	32
Saltery Bay	1974-76	3	100	37	99	164
Seabird	1969-94	16	100	38	69	100
Squamish	1973-94	9	100	39	75	101
Victoria	1973-94	17	100	52	84	171
Interior						
Adams Lake	1989-94	6	100	20	25	37
Alexis Creek	1973-86	14	100	29	43	53
Ashnola	1973-74	2	100	17	21	24
Atlin Road	1989	1	100	34	–	34
Beaverdell	1970-94	24	100	24	63	91
Bridge Lake	1974-94	20	100	32	49	81
Bromley	1973-94	5	100	42	50	55
Brookmere	1973-92	10	100	22	40	57
Canford	1974-93	10	100	25	41	62
Canoe Point	1993-94	2	100	70	71	72
Cassiar	1972-75	2	100	3	6	8
Chilkat Pass	1975-76	2	100	8	17	26
Christian Valley	1993-94	2	100	65	66	67
Chu Chua	1993-94	2	100	50	54	57
Columbia Lake	1973-83	7	100	56	67	98
Creighton Valley	1979-94	16	100	38	51	61
Creston	1984	1	100	109	–	109
Ferndale	1968-94	15	100	27	60	94
Fort Nelson	1974-80	5	100	14	27	35
Fort St. John	1974-90	9	100	18	28	40
Fraser Lake	1968-78	5	100	7	26	32
Gerrard	1976-85	10	100	9	20	24
Gnat Pass	1975-93	2	50	66	–	66
Golden	1973-94	22	100	31	67	94
Goldstream Mountain	1993-94	2	100	6	8	9
Gosnell	1969-74	4	100	14	29	49
Grand Forks	1973-94	22	100	72	105	148
Haines Summit	1993-94	2	100	11	13	15
Hendrix Lake	1976	1	100	19	–	19
Horsefly	1974-88	11	100	21	40	49
Hudson's Hope	1976	1	100	42	–	42
Illecilewaet	1973-83	4	100	3	10	24
Kamloops	1974-94	15	100	31	47	58
Kinaskan Lake	1975	1	100	4	–	4
Kootenay	1981-94	10	100	25	41	67
Kuskonook	1968-91	21	100	32	45	132
Lac la Hache	1973-86	11	100	45	56	88
Lavington	1972-94	23	100	39	69	105
Mabel Lake	1973-94	19	100	38	66	108
McLeod Lake	1968-94	9	100	28	39	65
Meadowlake	1968-70	2	100	37	40	42
Mount Morice	1968-75	5	100	15	23	26
Needles	1968-94	26	100	25	50	127
Nicola	1973-74	2	100	59	71	83
Oliver	1973-94	18	100	19	40	57
Osprey Lake	1976-94	10	100	25	39	59
Pavilion	1974-94	6	100	26	49	73
Pennington	1987-91	4	100	2	6	12
Pinkut Creek	1975	1	100	5	–	5
Pleasant Valley	1980-93	6	100	18	29	35
Prince George	1973-94	20	100	46	71	96
Princeton	1968-94	8	100	20	25	33
Punchaw	1974-94	9	100	20	33	67
Riske Creek	1973-92	11	100	26	40	48
Salmo	1969-75	3	100	48	71	125
Salmon Arm	1968-94	19	100	24	37	80
Scotch Creek	1989-94	6	100	22	31	37

▶

◀ *American Robin*

Survey route	Survey years	Total surveys	Frequency of occurrence (%)	Low	Range of total surveys when species occurred Median	High
Shalalth	1968-94	4	100	42	44	46
Spillimacheen	1976-93	17	100	46	58	121
Steamboat	1974-80	4	75	1	2	10
Succour Creek	1993-94	2	100	37	46	55
Summerland	1973-94	21	100	22	32	63
Summit Lake	1972-94	8	100	35	67	96
Syringa Creek	1986-93	7	100	33	47	60
Tatlayoko Lake	1989-94	6	100	32	48	58
Telkwa High Road	1974-94	11	100	42	66	156
Tokay	1968-75	4	100	17	49	67
Tupper	1974-94	6	100	21	40	75
Wasa	1973-94	19	100	19	31	46
Williams Lake	1973-93	21	100	27	37	50
Wingdam	1968-70	3	100	36	37	50
Zincton	1968-94	22	100	28	38	132

Varied Thrush

Survey route	Survey years	Total surveys	Frequency of occurrence (%)	Low	Range of total surveys when species occurred Median	High
Coastal						
Alberni	1968-79	6	100	5	7	16
Albion	1968-92	21	38	1	1	2
Campbell River	1973-93	19	95	1	4	12
Chemainus	1969-88	4	75	1	1	9
Chilliwack	1973-94	11	9	1	–	1
Comox Lake	1973-75	3	100	2	3	7
Courtenay	1992-93	2	100	2	2	2
Elsie Lake	1970-94	6	100	7	10	13
Gibsons Landing	1973-94	10	70	1	1	6
Kispiox	1975-94	3	100	6	12	17
Kitsumkalum	1974-94	19	100	12	35	56
Kwinitsa	1968-94	26	100	11	34	59
Masset	1989-94	5	100	47	65	80
Meziadin	1975-77	2	100	3	16	29
Nanaimo River	1973-94	22	45	1	2	12
Nass River	1975	1	100	6	–	6
Pemberton	1974-94	18	100	1	7	46
Point Grey	1974-92	11	27	1	1	1
Port Hardy	1982-89	4	100	27	35	37
Port Renfrew	1973-94	17	100	25	50	66
Queen Charlotte City	1989-94	3	100	50	75	90
Saltery Bay	1974-76	3	100	1	6	10
Seabird	1969-94	16	44	1	5	9
Squamish	1973-94	9	89	1	4	16
Victoria	1973-94	17	6	1	–	1
Interior						
Adams Lake	1989-94	6	83	1	3	6
Alexis Creek	1973-86	14	14	1	2	2
Beaverdell	1970-94	24	29	1	1	2
Bridge Lake	1974-94	20	80	1	2	6
Brookmere	1973-92	10	40	1	5	6
Canford	1974-93	10	10	1	–	1
Canoe Point	1993-94	2	50	1	–	1
Cassiar	1972-75	2	50	1	–	1
Chilkat Pass	1975-76	2	100	12	31	49
Christian Valley	1993-94	2	50	1	–	1
Columbia Lake	1973-83	7	14	2	–	2
Creighton Valley	1979-94	16	100	6	16	29
Creston	1984	1	100	5	–	5
Ferndale	1968-94	15	100	4	12	26
Fort Nelson	1974-80	5	60	1	1	1
Fraser Lake	1968-78	5	20	1	–	1
Gerrard	1976-85	10	100	3	18	37

▶

◀ Varied Thrush

Survey route	Survey years	Total surveys	Frequency of occurrence (%)	Range of total surveys when species occurred Low	Median	High
Gnat Pass	1975-93	2	50	4	–	4
Golden	1973-94	22	18	1	1	2
Goldstream Mountain	1993-94	2	100	9	10	11
Gosnell	1969-74	4	100	14	25	34
Grand Forks	1973-94	22	36	1	1	4
Haines Summit	1993-94	2	100	18	22	26
Horsefly	1974-88	11	9	6	–	6
Illecilewaet	1973-83	4	100	19	40	62
Kinaskan Lake	1975	1	100	3	–	3
Kootenay	1981-94	10	100	6	23	33
Kuskonook	1968-91	21	48	1	3	6
Lac la Hache	1973-86	11	45	1	3	8
Lavington	1972-94	23	82	1	2	7
Mabel Lake	1973-94	19	68	1	2	3
McLeod Lake	1968-94	9	89	11	16	38
Mount Morice	1968-75	5	80	1	3	10
Needles	1968-94	26	81	1	6	12
Oliver	1973-94	18	67	1	3	12
Pennington	1987-91	4	50	2	3	3
Pinkut Creek	1975	1	100	2	–	2
Pleasant Valley	1980-93	6	33	1	1	1
Prince George	1973-94	20	30	1	3	4
Princeton	1968-94	8	25	1	2	3
Punchaw	1974-94	9	100	1	9	17
Scotch Creek	1989-94	6	100	10	22	32
Spillimacheen	1976-93	17	24	1	2	3
Steamboat	1974-80	4	100	1	7	17
Succour Creek	1993-94	2	100	8	16	23
Summerland	1973-94	21	71	1	2	11
Summit Lake	1972-94	8	100	5	14	21
Syringa Creek	1986-93	7	29	1	2	3
Tatlayoko Lake	1989-94	6	17	7	–	7
Telkwa High Road	1974-94	11	100	1	3	8
Tetsa River	1974	1	200	1	1	1
Wingdam	1968-70	3	100	59	84	113
Zincton	1968-94	22	100	1	13	33

Gray Catbird

Survey route	Survey years	Total surveys	Frequency of occurrence (%)	Range of total surveys when species occurred Low	Median	High
Coastal						
Albion	1968-92	21	14	1	1	2
Pemberton	1974-94	18	22	1	3	7
Pitt Meadows	1973-89	14	7	1	–	1
Point Grey	1974-92	11	9	1	–	1
Seabird	1969-94	16	6	1	–	1
Interior						
Alexis Creek	1973-86	14	79	1	2	4
Ashnola	1973-74	2	100	1	2	2
Beaverdell	1970-94	24	50	1	1	2
Bromley	1973-94	5	20	1	–	1
Canford	1974-93	10	30	1	2	2
Chu Chua	1993-94	2	50	1	–	1
Columbia Lake	1973-83	7	14	1	–	1
Creston	1984	1	100	1	–	1
Gerrard	1976-85	10	10	2	–	2
Golden	1973-94	22	86	1	2	12
Grand Forks	1973-94	22	100	1	2	10
Horsefly	1974-88	11	18	1	1	1
Kamloops	1974-94	15	47	1	1	3
Kuskonook	1968-91	21	52	1	1	7
Lac la Hache	1973-86	11	9	1	–	1
Lavington	1972-94	23	52	1	1	4

▶

◀ *Gray Catbird*

Survey route	Survey years	Total surveys	Frequency of occurrence (%)	Low	Range of total surveys when species occurred Median	High
Mabel Lake	1973-94	19	47	1	1	3
Needles	1968-94	26	50	1	1	5
Nicola	1973-74	2	50	6	–	6
Oliver	1973-94	18	33	1	2	2
Pavilion	1974-94	6	50	1	1	2
Pleasant Valley	1980-93	6	33	1	2	3
Riske Creek	1973-92	11	9	1	–	1
Salmo	1969-75	3	67	3	4	5
Salmon Arm	1968-94	19	42	1	2	8
Spillimacheen	1976-93	17	35	1	1	2
Summerland	1973-94	21	48	1	1	4
Syringa Creek	1986-93	7	57	2	4	5
Tokay	1968-75	4	100	2	5	6
Wasa	1973-94	19	79	1	2	5
Williams Lake	1973-93	21	10	1	1	1
Zincton	1968-94	22	32	1	1	2

Brown Thrasher

Survey route	Survey years	Total surveys	Frequency of occurrence (%)	Low	Range of total surveys when species occurred Median	High
Interior						
Tupper	1974-94	6	17	1	–	1

American Pipit

Survey route	Survey years	Total surveys	Frequency of occurrence (%)	Low	Range of total surveys when species occurred Median	High
Interior						
Chilkat Pass	1975-76	2	100	7	8	9

Bohemian Waxwing

Survey route	Survey years	Total surveys	Frequency of occurrence (%)	Low	Range of total surveys when species occurred Median	High
Coastal						
Pemberton	1974-94	18	6	2	–	2
Interior						
Atlin Road	1989	1	100	2	–	2
Cassiar	1972-75	2	50	1	–	1
Chilkat Pass	1975-76	2	50	6	–	6
Columbia Lake	1973-83	7	14	3	–	3
Ferndale	1968-94	15	13	1	2	2
Fort Nelson	1974-80	5	20	1	–	1
Horsefly	1974-88	11	9	2	–	2
Mount Morice	1968-75	5	20	2	–	2
Pennington	1987-91	4	75	1	1	5
Prince George	1973-94	20	35	1	5	8
Punchaw	1974-94	9	11	16	–	16
Steamboat	1974-80	4	75	1	5	8
Wingdam	1968-70	3	33	1	–	1

Cedar Waxwing

Survey route	Survey years	Total surveys	Frequency of occurrence (%)	Range of total surveys when species occurred: Low	Median	High
Coastal						
Alberni	1968-79	6	83	4	6	7
Albion	1968-92	21	100	2	17	51
Campbell River	1973-93	19	63	1	2	4
Chemainus	1969-88	4	100	1	4	14
Chilliwack	1973-94	11	91	13	20	41
Comox Lake	1973-75	3	67	3	3	3
Coquitlam	1990-94	5	100	2	10	20
Courtenay	1992-93	2	50	7	–	7
Elsie Lake	1970-94	6	67	1	3	3
Gibsons Landing	1973-94	10	100	2	16	34
Kispiox	1975-94	3	100	1	3	4
Kitsumkalum	1974-94	19	37	1	3	29
Kwinitsa	1968-94	26	8	1	1	1
Nanaimo River	1973-94	22	82	1	5	14
Pemberton	1974-94	18	100	2	12	45
Pitt Meadows	1973-89	14	100	6	13	23
Point Grey	1974-92	11	82	1	3	8
Port Hardy	1982-89	4	25	5	–	5
Port Renfrew	1973-94	17	71	2	10	18
Saltery Bay	1974-76	3	100	1	3	4
Seabird	1969-94	16	100	3	19	38
Squamish	1973-94	9	100	1	12	27
Victoria	1973-94	17	94	2	5	16
Interior						
Adams Lake	1989-94	6	83	1	8	11
Ashnola	1973-74	2	100	2	5	7
Beaverdell	1970-94	24	83	1	4	21
Bridge Lake	1974-94	20	60	1	4	23
Bromley	1973-94	5	100	2	9	22
Brookmere	1973-92	10	70	1	8	15
Canford	1974-93	10	80	4	7	35
Canoe Point	1993-94	2	100	7	15	22
Christian Valley	1993-94	2	100	10	12	13
Chu Chua	1993-94	2	50	8	–	8
Columbia Lake	1973-83	7	43	1	1	8
Creighton Valley	1979-94	16	75	1	5	10
Creston	1984	1	100	4	–	4
Ferndale	1968-94	15	73	2	5	23
Fort St. John	1974-90	9	44	1	6	6
Fraser Lake	1968-78	5	20	3	–	3
Gerrard	1976-85	10	80	2	4	20
Golden	1973-94	22	95	1	9	37
Goldstream Mountain	1993-94	2	100	4	6	7
Gosnell	1969-74	4	50	1	2	3
Grand Forks	1973-94	22	73	1	9	25
Horsefly	1974-88	11	36	4	5	8
Hudson's Hope	1976	1	100	4	–	4
Kamloops	1974-94	15	93	1	4	35
Kootenay	1981-94	10	40	1	2	8
Kuskonook	1968-91	21	90	1	3	39
Lac la Hache	1973-86	11	64	1	4	19
Lavington	1972-94	23	100	2	9	25
Mabel Lake	1973-94	19	100	1	8	23
McLeod Lake	1968-94	9	67	1	3	7
Needles	1968-94	26	88	1	4	22
Nicola	1973-74	2	100	2	3	4
Oliver	1973-94	18	78	1	3	8
Osprey Lake	1976-94	10	60	2	13	32
Pavilion	1974-94	6	50	1	3	13
Pleasant Valley	1980-93	6	100	4	9	20
Prince George	1973-94	20	45	1	3	11
Princeton	1968-94	8	38	1	2	4
Punchaw	1974-94	9	67	1	5	7
Riske Creek	1973-92	11	36	1	2	7
Salmo	1969-75	3	67	2	6	10
Salmon Arm	1968-94	19	95	1	4	16
Scotch Creek	1989-94	6	100	5	8	10
Shalalth	1968-94	4	25	1	–	1

▶

◀ *Cedar Waxwing*

Survey route	Survey years	Total surveys	Frequency of occurrence (%)	Range of total surveys when species occurred Low	Median	High
Spillimacheen	1976-93	17	65	1	5	12
Steamboat	1974-80	4	25	3	–	3
Succour Creek	1993-94	2	100	16	36	55
Summerland	1973-94	21	62	1	3	10
Summit Lake	1972-94	8	63	2	4	11
Syringa Creek	1986-93	7	86	2	19	24
Tatlayoko Lake	1989-94	6	17	3	–	3
Telkwa High Road	1974-94	11	73	1	4	14
Tokay	1968-75	4	50	6	9	12
Tupper	1974-94	6	33	3	4	5
Wasa	1973-94	19	47	1	3	12
Williams Lake	1973-93	21	29	1	3	10
Wingdam	1968-70	3	67	1	2	2
Zincton	1968-94	22	86	2	11	33

Loggerhead Shrike

Survey route	Survey years	Total surveys	Frequency of occurrence (%)	Range of total surveys when species occurred Low	Median	High
Interior						
Ashnola	1973-74	2	50	1	–	1*

European Starling

Survey route	Survey years	Total surveys	Frequency of occurrence (%)	Range of total surveys when species occurred Low	Median	High
Coastal						
Alberni	1968-79	6	83	5	58	119
Albion	1968-92	21	100	80	229	3,726
Campbell River	1973-93	19	95	1	9	91
Chemainus	1969-88	4	100	13	35	159
Chilliwack	1973-94	11	100	224	376	703
Comox Lake	1973-75	3	100	8	13	33
Coquitlam	1990-94	5	100	60	173	199
Courtenay	1992-93	2	100	17	24	30
Elsie Lake	1970-94	6	50	4	4	25
Gibsons Landing	1973-94	10	100	8	23	113
Kispiox	1975-94	3	100	1	4	4
Kitsumkalum	1974-94	19	5	1	–	1
Kwinitsa	1968-94	26	15	3	4	10
Masset	1989-94	5	60	1	1	1
Nanaimo River	1973-94	22	100	17	38	133
Pemberton	1974-94	18	100	8	29	231
Pitt Meadows	1973-89	14	100	95	176	255
Point Grey	1974-92	11	100	78	142	224
Port Hardy	1982-89	4	100	2	6	30
Port Renfrew	1973-94	17	29	2	3	20
Saltery Bay	1974-76	3	67	3	14	25
Seabird	1969-94	16	100	77	120	230
Squamish	1973-94	9	100	3	18	78
Victoria	1973-94	17	100	72	128	488
Interior						
Alexis Creek	1973-86	14	100	1	10	25
Ashnola	1973-74	2	50	56	–	56
Beaverdell	1970-94	24	92	1	6	27
Bridge Lake	1974-94	20	100	2	20	52
Bromley	1973-94	5	100	26	50	59
Brookmere	1973-92	10	100	2	10	38

▶

◀ *European Starling*

Survey route	Survey years	Total surveys	Frequency of occurrence (%)	Low	Range of total surveys when species occurred Median	High
Canford	1974-93	10	100	7	29	207
Canoe Point	1993-94	2	100	86	105	123
Christian Valley	1993-94	2	100	1	5	8
Chu Chua	1993-94	2	100	6	7	7
Columbia Lake	1973-83	7	100	6	52	104
Creighton Valley	1979-94	16	100	2	12	49
Creston	1984	1	100	59	–	59
Ferndale	1968-94	15	47	1	4	45
Fort Nelson	1974-80	5	100	1	4	29
Fort St. John	1974-90	9	100	4	15	47
Fraser Lake	1968-78	5	80	5	25	97
Gerrard	1976-85	10	60	1	3	6
Gnat Pass	1975-93	2	50	35	–	35
Golden	1973-94	22	100	2	30	148
Grand Forks	1973-94	22	86	1	15	92
Horsefly	1974-88	11	91	3	11	49
Hudson's Hope	1976	1	100	22	–	22
Kamloops	1974-94	15	100	43	120	256
Kuskonook	1968-91	21	67	2	6	14
Lac la Hache	1973-86	11	100	24	86	237
Lavington	1972-94	23	100	37	102	307
Mabel Lake	1973-94	19	100	13	46	241
McLeod Lake	1968-94	9	56	1	3	6
Meadowlake	1968-70	2	100	1	24	46
Mount Morice	1968-75	5	60	1	10	12
Needles	1968-94	26	92	2	11	53
Nicola	1973-74	2	100	9	46	83
Oliver	1973-94	18	100	9	15	210
Osprey Lake	1976-94	10	100	10	39	79
Pavilion	1974-94	6	100	2	7	15
Pleasant Valley	1980-93	6	100	31	88	201
Prince George	1973-94	20	100	42	155	539
Princeton	1968-94	8	75	1	6	13
Punchaw	1974-94	9	11	1	–	1
Riske Creek	1973-92	11	100	10	20	66
Salmo	1969-75	3	100	11	18	28
Salmon Arm	1968-94	19	100	5	49	266
Shalalth	1968-94	4	50	1	2	2
Spillimacheen	1976-93	17	100	3	10	38
Summerland	1973-94	21	95	1	7	56
Summit Lake	1972-94	8	100	3	12	25
Syringa Creek	1986-93	7	100	9	27	37
Tatlayoko Lake	1989-94	6	83	1	1	7
Telkwa High Road	1974-94	11	100	2	24	73
Tokay	1968-75	4	75	3	8	39
Tupper	1974-94	6	83	14	18	28
Wasa	1973-94	19	89	2	15	79
Williams Lake	1973-93	21	100	10	24	79
Wingdam	1968-70	3	33	4	–	4
Zincton	1968-94	22	100	4	17	75

Crested Myna

Survey route	Survey years	Total surveys	Frequency of occurrence (%)	Low	Range of total surveys when species occurred Median	High
Coastal						
Point Grey	1974-92	11	9	1	–	1

Solitary Vireo

Survey route	Survey years	Total surveys	Frequency of occurrence (%)	Range of total surveys when species occurred: Low	Median	High
Coastal						
Alberni	1968-79	6	100	2	3	6
Albion	1968-92	21	5	1	–	1
Campbell River	1973-93	19	95	1	2	19
Chemainus	1969-88	4	100	1	5	8
Chilliwack	1973-94	11	27	2	2	2
Comox Lake	1973-75	3	100	2	2	4
Coquitlam	1990-94	5	20	1	–	1
Courtenay	1992-93	2	100	2	3	3
Elsie Lake	1970-94	6	33	1	1	1
Gibsons Landing	1973-94	10	80	1	7	21
Kispiox	1975-94	3	67	2	3	4
Nanaimo River	1973-94	22	91	1	2	8
Pemberton	1974-94	18	72	1	3	24
Pitt Meadows	1973-89	14	7	1	–	1
Point Grey	1974-92	11	9	1	–	1
Port Hardy	1982-89	4	50	1	2	3
Port Renfrew	1973-94	17	12	1	2	2
Saltery Bay	1974-76	3	100	3	11	13
Seabird	1969-94	16	81	1	2	4
Squamish	1973-94	9	100	1	2	31
Victoria	1973-94	17	12	1	1	1
Interior						
Adams Lake	1989-94	6	100	7	12	15
Alexis Creek	1973-86	14	36	1	1	3
Beaverdell	1970-94	24	100	2	11	25
Bridge Lake	1974-94	20	95	4	7	20
Bromley	1973-94	5	80	2	3	7
Brookmere	1973-92	10	90	3	6	9
Canford	1974-93	10	90	2	5	12
Canoe Point	1993-94	2	100	6	9	12
Christian Valley	1993-94	2	100	6	7	8
Chu Chua	1993-94	2	100	17	18	19
Columbia Lake	1973-83	7	86	1	5	6
Creighton Valley	1979-94	16	94	1	6	9
Creston	1984	1	100	1	–	1
Ferndale	1968-94	15	60	1	3	10
Fort Nelson	1974-80	5	40	1	2	3
Fraser Lake	1968-78	5	20	2	–	2
Gerrard	1976-85	10	50	1	1	1
Gnat Pass	1975-93	2	50	14	–	14
Golden	1973-94	22	82	1	3	10
Goldstream Mountain	1993-94	2	50	1	–	1
Gosnell	1969-74	4	75	1	2	15
Grand Forks	1973-94	22	95	3	6	16
Illecilewaet	1973-83	4	25	1	–	1
Kamloops	1974-94	15	87	1	2	3
Kinaskan Lake	1975	1	100	1	–	1
Kootenay	1981-94	10	90	3	5	13
Kuskonook	1968-91	21	90	1	4	16
Lac la Hache	1973-86	11	27	2	3	10
Lavington	1972-94	23	79	1	2	5
Mabel Lake	1973-94	19	100	1	6	14
McLeod Lake	1968-94	9	89	1	3	6
Meadowlake	1968-70	2	50	2	–	2
Mount Morice	1968-75	5	40	1	1	1
Needles	1968-94	26	88	1	7	10
Oliver	1973-94	18	89	1	6	11
Osprey Lake	1976-94	10	70	1	3	14
Pavilion	1974-94	6	67	2	3	10
Pinkut Creek	1975	1	100	1	–	1
Pleasant Valley	1980-93	6	83	1	3	5
Prince George	1973-94	20	80	1	2	6
Princeton	1968-94	8	63	1	4	8
Punchaw	1974-94	9	44	3	6	8
Riske Creek	1973-92	11	36	2	2	4
Salmo	1969-75	3	33	3	–	3
Salmon Arm	1968-94	19	79	1	3	12
Scotch Creek	1989-94	6	67	2	3	4
Shalalth	1968-94	4	25	1	–	1

▶

◀ *Solitary Vireo*

Survey route	Survey years	Total surveys	Frequency of occurrence (%)	Low	Range of total surveys when species occurred Median	High
Spillimacheen	1976-93	17	100	6	10	18
Steamboat	1974-80	4	25	1	–	1
Succour Creek	1993-94	2	100	7	9	11
Summerland	1973-94	21	81	1	2	6
Summit Lake	1972-94	8	63	3	5	14
Syringa Creek	1986-93	7	86	1	4	7
Tatlayoko Lake	1989-94	6	50	1	2	2
Telkwa High Road	1974-94	11	45	1	3	5
Tokay	1968-75	4	50	9	10	10
Tupper	1974-94	6	83	1	2	3
Wasa	1973-94	19	89	2	6	16
Williams Lake	1973-93	21	52	1	1	3
Zincton	1968-94	22	68	1	3	12

Hutton's Vireo

Survey route	Survey years	Total surveys	Frequency of occurrence (%)	Low	Range of total surveys when species occurred Median	High
Coastal						
Alberni	1968-79	6	17	2	–	2
Albion	1968-92	21	10	1	1	1
Campbell River	1973-93	19	37	1	1	4
Chemainus	1969-88	4	75	1	1	1
Courtenay	1992-93	2	50	1	–	1
Elsie Lake	1970-94	6	17	2	–	2
Gibsons Landing	1973-94	10	100	1	3	7
Nanaimo River	1973-94	22	9	1	2	2
Pemberton	1974-94	18	6	1	–	1
Point Grey	1974-92	11	9	1	–	1
Port Renfrew	1973-94	17	76	1	3	5
Squamish	1973-94	9	11	2	–	2
Victoria	1973-94	17	29	1	1	1

Warbling Vireo

Survey route	Survey years	Total surveys	Frequency of occurrence (%)	Low	Range of total surveys when species occurred Median	High
Coastal						
Alberni	1968-79	6	83	1	8	13
Albion	1968-92	21	76	1	1	2
Campbell River	1973-93	19	100	4	10	23
Chemainus	1969-88	4	75	4	8	21
Chilliwack	1973-94	11	73	1	3	7
Comox Lake	1973-75	3	100	4	7	14
Coquitlam	1990-94	5	20	3	–	3
Courtenay	1992-93	2	100	15	23	30
Elsie Lake	1970-94	6	100	3	8	17
Gibsons Landing	1973-94	10	100	3	11	22
Kispiox	1975-94	3	100	29	60	62
Kitsumkalum	1974-94	19	100	1	14	28
Kwinitsa	1968-94	26	100	1	12	40
Meziadin	1975-77	2	100	2	18	34
Nanaimo River	1973-94	22	95	2	8	16
Nass River	1975	1	100	18	–	18
Pemberton	1974-94	18	78	2	13	31
Pitt Meadows	1973-89	14	50	1	1	2
Point Grey	1974-92	11	82	1	2	4
Port Hardy	1982-89	4	75	1	1	1
Port Renfrew	1973-94	17	100	1	9	15

▶

◀ *Warbling Vireo*

Survey route	Survey years	Total surveys	Frequency of occurrence (%)	Range of total surveys when species occurred Low	Median	High
Saltery Bay	1974-76	3	100	11	11	18
Seabird	1969-94	16	88	1	5	11
Squamish	1973-94	9	89	25	32	45
Victoria	1973-94	17	53	1	2	3
Interior						
Adams Lake	1989-94	6	100	6	14	24
Alexis Creek	1973-86	14	64	1	2	4
Ashnola	1973-74	2	100	1	3	4
Beaverdell	1970-94	24	100	4	18	31
Bridge Lake	1974-94	20	95	5	16	33
Bromley	1973-94	5	80	6	8	11
Brookmere	1973-92	10	90	8	15	23
Canford	1974-93	10	100	4	13	18
Canoe Point	1993-94	2	100	7	13	19
Cassiar	1972-75	2	50	7	–	7
Christian Valley	1993-94	2	100	18	19	19
Chu Chua	1993-94	2	100	13	22	30
Columbia Lake	1973-83	7	43	1	1	4
Creighton Valley	1979-94	16	100	7	21	34
Creston	1984	1	100	5	–	5
Ferndale	1968-94	15	73	1	54	81
Fort Nelson	1974-80	5	80	1	10	26
Fort St. John	1974-90	9	100	3	4	8
Fraser Lake	1968-78	5	80	5	14	27
Gerrard	1976-85	10	100	8	13	21
Gnat Pass	1975-93	2	50	21	–	21
Golden	1973-94	22	100	5	15	44
Goldstream Mountain	1993-94	2	100	40	42	43
Gosnell	1969-74	4	100	11	17	20
Grand Forks	1973-94	22	100	1	5	12
Horsefly	1974-88	11	91	1	3	7
Hudson's Hope	1976	1	100	31	–	31
Illecilewaet	1973-83	4	100	1	19	25
Kamloops	1974-94	15	93	2	5	8
Kinaskan Lake	1975	1	100	7	–	7
Kootenay	1981-94	10	90	2	8	15
Kuskonook	1968-91	21	76	1	8	17
Lac la Hache	1973-86	11	91	1	4	18
Lavington	1972-94	23	100	7	14	29
Mabel Lake	1973-94	19	100	1	5	10
McLeod Lake	1968-94	9	100	6	37	52
Meadowlake	1968-70	2	100	1	2	2
Mount Morice	1968-75	5	80	19	33	44
Needles	1968-94	26	96	2	12	26
Oliver	1973-94	18	89	2	11	19
Osprey Lake	1976-94	10	70	1	13	20
Pavilion	1974-94	6	83	5	11	13
Pinkut Creek	1975	1	100	5	–	5
Pleasant Valley	1980-93	6	100	6	7	10
Prince George	1973-94	20	100	2	8	22
Princeton	1968-94	8	100	3	8	13
Punchaw	1974-94	9	44	18	33	46
Riske Creek	1973-92	11	64	1	3	7
Salmo	1969-75	3	33	5	–	5
Salmon Arm	1968-94	19	53	2	6	9
Scotch Creek	1989-94	6	100	12	26	34
Shalalth	1968-94	4	75	4	5	6
Spillimacheen	1976-93	17	100	10	21	35
Steamboat	1974-80	4	75	1	14	25
Succour Creek	1993-94	2	100	24	30	36
Summerland	1973-94	21	100	1	8	18
Summit Lake	1972-94	8	88	1	23	61
Syringa Creek	1986-93	7	100	2	16	18
Tatlayoko Lake	1989-94	6	100	7	13	15
Telkwa High Road	1974-94	11	91	13	31	61
Tokay	1968-75	4	25	14	–	14
Tupper	1974-94	6	100	6	12	22
Wasa	1973-94	19	63	1	3	9
Williams Lake	1973-93	21	90	1	3	6
Wingdam	1968-70	3	100	14	15	26
Zincton	1968-94	22	82	1	11	27

Philadelphia Vireo

Survey route	Survey years	Total surveys	Frequency of occurrence (%)	Low	Range of total surveys when species occurred Median	High
Interior						
Fort Nelson	1974-80	5	40	1	1	1
Hudson's Hope	1976	1	100	10	–	10
Tupper	1974-94	6	33	1	2	3

Red-eyed Vireo

Survey route	Survey years	Total surveys	Frequency of occurrence (%)	Low	Range of total surveys when species occurred Median	High
Coastal						
Alberni	1968-79	6	67	1	3	3
Albion	1968-92	21	90	1	3	23
Campbell River	1973-93	19	21	1	1	3
Chemainus	1969-88	4	75	1	1	6
Chilliwack	1973-94	11	64	1	1	4
Comox Lake	1973-75	3	67	1	1	1
Coquitlam	1990-94	5	40	1	1	1
Courtenay	1992-93	2	100	1	1	1
Gibsons Landing	1973-94	10	80	1	3	8
Kispiox	1975-94	3	100	15	19	34
Kitsumkalum	1974-94	19	53	1	2	3
Kwinitsa	1968-94	26	12	1	2	3
Nanaimo River	1973-94	22	32	1	1	2
Pemberton	1974-94	18	56	2	4	14
Pitt Meadows	1973-89	14	86	1	2	6
Point Grey	1974-92	11	55	1	1	3
Port Hardy	1982-89	4	25	1	–	1
Saltery Bay	1974-76	3	100	1	1	7
Seabird	1969-94	16	88	1	4	12
Squamish	1973-94	9	78	1	12	18
Victoria	1973-94	17	47	1	1	2
Interior						
Adams Lake	1989-94	6	100	30	46	55
Alexis Creek	1973-86	14	86	1	3	8
Ashnola	1973-74	2	100	1	2	3
Beaverdell	1970-94	24	100	4	11	26
Bridge Lake	1974-94	20	85	2	3	17
Bromley	1973-94	5	40	1	1	1
Brookmere	1973-92	10	80	1	2	5
Canford	1974-93	10	80	1	5	15
Canoe Point	1993-94	2	100	6	12	18
Chu Chua	1993-94	2	100	15	17	19
Columbia Lake	1973-83	7	86	12	27	36
Creighton Valley	1979-94	16	100	1	2	18
Creston	1984	1	100	12	–	12
Ferndale	1968-94	15	53	1	1	8
Fort Nelson	1974-80	5	100	10	24	52
Fort St. John	1974-90	9	100	8	25	30
Fraser Lake	1968-78	5	60	1	6	13
Gerrard	1976-85	10	100	5	16	27
Gnat Pass	1975-93	2	50	7	–	7
Golden	1973-94	22	100	8	31	62
Goldstream Mountain	1993-94	2	100	3	4	5
Gosnell	1969-74	4	75	7	8	10
Grand Forks	1973-94	22	100	5	14	28
Hendrix Lake	1976	1	100	1	–	1
Horsefly	1974-88	11	100	1	7	14
Hudson's Hope	1976	1	100	35	–	35
Kamloops	1974-94	15	13	1	2	2
Kuskonook	1968-91	21	100	3	16	53
Lac la Hache	1973-86	11	55	1	1	5
Lavington	1972-94	23	100	1	5	12
Mabel Lake	1973-94	19	100	10	25	50
McLeod Lake	1968-94	9	78	1	2	2

◀ *Red-eyed Vireo*

Survey route	Survey years	Total surveys	Frequency of occurrence (%)	Range of total surveys when species occurred Low	Median	High
Meadowlake	1968-70	2	50	3	–	3
Mount Morice	1968-75	5	20	1	–	1
Needles	1968-94	26	96	14	31	50
Nicola	1973-74	2	50	2	–	2
Oliver	1973-94	18	56	1	1	5
Osprey Lake	1976-94	10	70	1	3	10
Pavilion	1974-94	6	50	1	5	5
Pleasant Valley	1980-93	6	50	1	1	3
Prince George	1973-94	20	100	2	5	15
Princeton	1968-94	8	75	1	4	7
Punchaw	1974-94	9	22	1	1	1
Riske Creek	1973-92	11	55	2	6	8
Salmo	1969-75	3	33	29	–	29
Salmon Arm	1968-94	19	100	9	20	68
Scotch Creek	1989-94	6	17	1	–	1
Shalalth	1968-94	4	50	1	3	4
Spillimacheen	1976-93	17	100	2	12	19
Steamboat	1974-80	4	100	1	7	14
Succour Creek	1993-94	2	100	1	2	2
Summerland	1973-94	21	100	2	7	16
Summit Lake	1972-94	8	100	1	4	6
Syringa Creek	1986-93	7	100	20	29	34
Telkwa High Road	1974-94	11	27	1	1	2
Tokay	1968-75	4	25	13	–	13
Tupper	1974-94	6	100	3	14	21
Wasa	1973-94	19	79	1	4	9
Williams Lake	1973-93	21	67	1	1	3
Zincton	1968-94	22	95	18	36	68

Fogarty, Ethel
Fogg, J.
Fogg, T.
Fohr, Brian
Folbegg, Joyce
Fomataro, Mark
Fontaine, J.
Fontaine, Lorraine
Fontaine, Marlene
Fooks, A.
Fooks, H.A.
Foote, R.
Foottit, Michael K.
Foottit, Robert G.
Forbes, E.M.
Forbes, Ian
Forbes, Joe
Forbes, L. Scott
Forbes, Robert
Forbes, Susan
Forbes, Ted
Force, Michael P.
Ford, A.H.
Ford, Bruce S.
Ford, John
Ford, Ron
Ford, Victor
Ford, William
Foreman, Barbara
Forer, Barry
Forest, M.W.
Forest, W.H.
Forman, Barry
Forrer, A.
Forrest, Margaret
Forrester, Ed
Forrester, Shelly
Forryan, Doreen
Forsman, Eric D.
Forster, Craig
Forster, Mary
Forster, Nancy
Forsyth, Evelyn
Forsyth, James S.
Forsyth, James B.
Fortin, Shawn
Fortney, Wilf
Forty, Thelma
Foskett, Ann
Foskett, Dudley
Foss, Ray
Foster, A.
Foster, Anthony
Foster, B.
Foster, D.
Foster, Eric
Foster, F.
Foster, G.G.
Foster, Ian
Foster, J. Bristol
Foster, Jack W.
Foster, John
Foster, Mark
Foubister, M.
Foulser, Art
Fowle, D.
Fowle, F.T.
Fowle, J.T.
Fowler, Fran
Fowler, Scott
Fox, G.
Fox, J.
Fox, L.
Fox, Rosemary
Fox, S.D.
Foxall, Roger
Fram, Roland
Francis, Brian
Francis, George
Frank, Floyd
Frank, P.
Franken, John P.
Franklin, D.H.
Franklin, Dan
Franklin, June
Franklin, R.
Franko, G.
Fraser, A.
Fraser, Bill
Fraser, David F.
Fraser, Douglas P.
Fraser, Jack
Fraser, Joan
Fraser, Kitsy
Fraser, M.A.
Fraser, Nancy
Fraser, Tom A.
Fraser, W.H.
Fraser, William
Frazer, Evelyn
Frazer, Frank
Frederick, Bruce G.
Freebairn, Tom
Freeman, Cheryl L.
Freeman, Mrs. John
Freer, Gary
Freer, Jean
Freeze, Ernest
French, Brigitte
Freshwater, N.G.
Frew, Betty
Frew, Gordon
Frewin, M.
Friberg, Sherrie
Fricke, Patricia
Fried, S.
Friedli, E.
Friend, G.B.
Friesen, J.
Friesz, Ron
Friis, Laura K.
Frisby, Alan
Frith, Wendy
Froese, Dave
Froese, Susan
Froimovitch, Mark J.
Frost, Bud
Frost, D. Lorne
Frost, M.L.
Fry, B.
Fry, Kathleen
Fryer, Ralph
Fryer, Ron
Fuhr, Brian I.
Fuhrer, Hans
Fujino, Ken K.
Fulks, Reg
Fuller, Joe
Fulton, Murrey
Fumiss, O.C.
Funk, Maureen
Funk, Phyllis
Fusco, L.
Futur, George
Fyall, Gerrie
Fyfe, Bud
Fyles, Helen
Fyles, James
Fyles, Tim
Fynn, Sonia
Gabreau, Martin
Gadsen, Ron
Gaelick, N.F.
Gage, Kim
Gage, Peter
Gagnon, R.
Gain, Scott
Gain, Mrs. Scott
Gainer, Bob
Gak, Janice
Gak, Joyce
Gak, Marc
Gak, Russ
Galbraith, Florence
Galbraith, J. Douglas
Gale, Alf
Galicz, George
Galliford, J.
Galloway, Phyllis
Galloway, Vera
Galt, Betty
Gamble, Eleanor
Gammer, Anna
Ganges, B.
Ganguin, Reiner
Gant, Jim
Gardiner, Joe
Gardiner, Mark
Gardner, Barbara
Gardner, Gerry
Gardner, Ivan
Gardner, Joe
Gardner, Norah
Gardner, Penny
Gardner, Ted
Gardner, W.E.
Garham, S.W.
Gariett, C.
Garlick, Ella
Garlick, N.F.
Garneau, Larry
Garner, T.
Garnett, C.
Garnett, J.
Garnier, Donald
Garnier, Hattie
Garret, C.B.
Garrett, C.G.
Garrett, O.B.
Garrioch, Hans
Garrioch, Heather M.
Garritt, C.B.
Gaskin, David
Gaskin, Jeff
Gasser, Ellen
Gaston, Anthony J.
Gates, Bryan R.
Gates, Conrad
Gattrell, G.K.
Gaume, S.A.
Gaunt, Sean
Gawn, Mark
Gawn, S.
Gaze, D.
Gebauer, Martin
Gee, Andrea
Gee, Penny
Geernaert, Karen
Geernaert, Tracee O.
Geeroms, Darryn
Gehlert, R.E.
Gehlin, Phil
Geist, V.
George, D.V.
George, H.E.
George, Val
Germyn, D.
Gerow, Dave
Gerow, Helen
Geryn, D.
Gest, Lillian
Gibbard, Fern
Gibbard, H.J.
Gibbard, Les A.
Gibbard, Lorna
Gibbard, P.
Gibbard, Robert T.
Gibbard, S.
Gibbard, Violet
Gibbon, Robert
Gibbons, Bob
Gibbons, Jeanette
Gibbons, Terry
Gibbons, Tim
Gibbs, Andrew E.
Gibbs, Nicholas
Gibbs, Richard E.
Gibson, A.
Gibson, Carlen
Gibson, D.E.
Gibson, Daniel D.
Gibson, George G.
Gibson, Ian
Gibson, Kenneth
Gibson, Kevin
Gibson, Pete
Gibson, W.H.
Gidman, G.
Giesecke, C.
Giesecke, Eve
Gieson, Cyril
Gifford, Bruce
Gifford, Janet
Gifford, Kit
Gilbert, Frank
Gilbert, Laura
Gilbert, Nick
Giles, Lorna
Giliberti, Laura
Gill, Allan M.C.
Gill, Cathy
Gill, Leslie
Gillard, Margaret
Gilles, A.S.
Gilles, Cam
Gilles, Cathy
Gillespie, Grahame E.
Gillespie, Hazel
Gillespie, Jim
Gillies, Barry
Gillingham, Michael
Gillis, M.M.
Gillis, W.M.
Gilmour, Bill
Giovanella, Carlo
Girard, Mark
Gissing, Alwin
Gitt, Margaret
Gladstone, B.
Glanville, Alice
Glasgow, Nancy
Glass, Ted
Glazeer, Bob
Glen, Molacom
Glendenning, R.
Glenny, Jim
Glide, Margaret
Glovanella, Carlo
Glover, Bev
Gobbett, M.
Goble, Edie
Goble, Jim
Godau, Helmut
Goddard, Peter A.
Godfrey, Dudley
Godfrey, Geoff
Godfrey, Judy
Godfrey, Monica
Godfrey, W. Earl
Godin, Tom
Godkin, Sharon
Godlien, Pat
Goff, A.
Goff, D.
Goff, H.
Gold, G.R.
Gold, Nita
Gold, P.
Goldberg, Kim
Golden, Linda
Golden, Sandy
Goldsworthy, M.
Gonzales, B.
Good, Ed
Goodacre, Brian W.
Goodall, Kay
Goodkey, Mrs. C.
Goodman, Elaine
Goodwill, J.E. Victor
Goodwill, Margaret E.
Goodwin, Kent
Goodwin, Lance
Goodwin, Mark
Goodwin, Ruth
Goold, Joan
Goossen, J. Paul
Gorden, John
Gordon, Amelia
Gordon, Bruce
Gordon, Hilary
Gordon, Janette
Gordon, John
Gordon, Keith
Gordon, M.
Gordon, Orville
Gordon, Ruth
Gordon, Sheila
Gorman, Wynne
Gornall, Fred A.
Gorog, K.
Gorsuch, Cecilia V.
Gosling, A.G.
Gosling, Gordon D.
Goss, Kathleen
Goudie, Douglas M.
Gough, C.F.
Gould, Dulce
Gould, Glen
Gould, Lenny
Gould, Lorne
Gould, T.C.
Goulding, Mary
Goulet, Louise
Gow, Bertha
Goward, Trevor
Goysuch, C.
Grabowski, Tony J.
Grady, Glen
Graenager, Earl
Graf, Ronald P.
Graham, David
Graham, Elaine
Graham, J. Douglas
Graham, Jim
Graham, Roy
Graham, Sheila
Graham, Walter
Granger, Ted
Granstrand, Denny
Grant-Duff, Adrian
Grant, James G.
Grant, Peg
Grant, Robert
Grant, S.
Grass, Al
Grass, Jude F.
Grass, R.
Grasser, E.
Grav, David
Grav, Dennis
Gray, Alex
Gray, Chris
Gray, Jim
Gray, N.
Gray, Ronald
Gray, Tom G.
Greaven, Moira
Greber, W.
Green, A.
Green, C. De B.
Green, Daphne
Green, David M.
Green, Herb
Green, Rick
Green, T.R.
Green, William
Greene, R.K.
Greener, Karl
Greenfield, Tony
Greenfield, F.
Greenwood, Eric
Greenwood, Gwen
Gregory, Ann
Gregory, M.S.
Gregory, Patrick T.
Gregson, Jack
Greiner, D.
Greissel, J.
Greissel, M.
Grenager, Earl
Grewer, D.
Grierson, John
Griffee, W.E.
Griffin, Mark
Griffin, R.V.
Griffith, D.
Griffiths, Don
Griffiths, Pele
Grigg, Garry J.
Griggs, Tamar
Grindridge, George
Grinnel, Dick
Grinyer, Chic
Groenveld, Anna
Grogan, Patrick
Gronau, Christian W.
Gronau, Steffi G.
Groseth, Janet
Groseth, Robert
Gross, A.
Grosse, Brigitte
Grosse, Hannes
Grossman, Eric
Grosuch, Cecilia V.
Grotage, Loyd
Groves, Joan
Groves, Sarah
Gruener, Karl
Grunberg, Helmut
Guernsey, Vera
Guest, Catherine
Guest, Harold
Guiguet, Charles J.
Guiguet, Mark
Guiguet, Muriel
Guiguet, Suzanne M.
Guillon, Frank E.
Guinet, Allan
Guinet, Frances
Guinet, Lynn
Guinet, Victor
Gully, P.

Gunther, Jack
Guppy, A.G.
Guppy, G.A.
Guppy, R.
Gurr, Ray
Gustafson, Barbara
Gustafson, Richard
Gutensohn, Joan
Guthrie, David
Guthrie, Don
Guthrie, Doreen
Guthrie, Jim
Guthrie, P.
Gwilliam, John
Gyug, Les W.
Haas, Norma
Haavik, Andre
Haavik, Colleen
Haavik, Miriam
Haavik, Steven
Hack, F.W.
Hack, Mrs. F.W.
Hackett, John
Hackett, Kathy
Hackett, Shannon
Hackman, G.S.
Haddow, Douglas J.
Haddow, W.
Haegart, John
Haering, Penny
Hagen, Barry
Hagen, Betty
Hagen, Catherine
Hagen, L.
Hagen, Patricia
Haggart, Lee
Haggert, Leona
Hagmeier, E.M.
Hahn, Rick J.
Haig-Brown, Roderick
Hainel, P.
Hainel, T.
Halasz, Gabor
Hale, Hilda L.
Hale, T.
Hales, D.
Haley, Gordon
Haley, Kathy
Halfnights, B.
Hall, Audrey
Hall, Bob
Hall, Brian
Hall, D.
Hall, Dorothy
Hall, E.R.S.
Hall, J.
Hall, Ken
Hall, W.A.
Halladay, D. Raymond
Halliday, Erik
Halliday, R.L.
Halliday, Valerie
Halsall, Leah
Haltrecht, Anna
Halverson, Larry R.
Halz, Gabbro
Hambell, O.G.
Hamel, Peter J.
Hames, A.M.
Hames, J.
Hames, Michael
Hamilton, Anthony N.
Hamilton, Daphne
Hamilton, Dulcie
Hamilton, Elsie
Hamilton, I.
Hamilton, John
Hamilton, K.
Hamilton, L.E.
Hamilton, Marla J.
Hamilton, Richard
Hamilton, W.
Hammell, J.
Hammell, Terry
Hammer, H.B.
Hammil, I.
Hammill, Sally
Hammond, D.
Hammond, Elsie
Hammond, Jo
Hammond, S.
Hammond, Vi
Hammonds, Jack
Hanceville, Dorothy
Hancock, David
Handford, Paul
Handley, Catherine
Handley, L.
Hann, Paddy
Hanna, Wilson C.
Hannah, Jack
Hannah, May
Hannon, Susan
Hanrahan, C.
Hanrahan, Tom
Hanry, M.
Hansen, J.
Hansen, Marilyn
Hansen, Ole
Hansen, Ruth
Hansen, Stanley
Hansen, Vicky
Hanson, David
Hanson, Don
Hanson, J.A.
Hanson, L.
Hanson, P.
Hanson, T.A.
Hanson, Wayne G.
Hansvall, Erling
Hansvall, Louise
Hanwell, Barbara
Haraldson, T.
Haras, Moreen
Haras, Willie
Harcombe, Andrew P.
Harcombe, Rick
Hardie, David
Hardie, L.
Hardie, M.
Hardie, W.
Harding, Martha
Harding, Rob
Harding, Tim
Hardley, C.
Hards, Jennifer
Hardstaff, Lynn
Hardwick, S.
Hardy, Bill
Hardy, Chuck
Hardy, David
Hardy, Duncan
Hardy, G.A.
Hardy, Gordon
Hardy, Harry
Hardy, Phyllis
Hardy, W.
Harestad, Alton S.
Hargrave, A. Nairn
Hargrave, Dave
Hargreave, Bob
Hark, F.W.
Harles, R.
Harlock, F.
Harlock, Mrs. F.
Harlow, Susan L.
Harman, Barry C.
Harman, Sam
Harms, W.
Harnell, P.
Harnell, T.
Harper, Charles
Harper, Don
Harper, Fred E.
Harper, John
Harper, Lynn
Harrington, R.F.
Harris, A.C.
Harris, A.E.
Harris, A.M.
Harris, B.J.
Harris, Brian S.
Harris, Christopher G.
Harris, Elizabeth
Harris, Mrs. G.C.
Harris, G.J.
Harris, Margaret
Harris, Nancy
Harris, P.
Harris, R.C.
Harris, R.P.
Harris, Robert D.
Harris, Ron
Harris, Ross E.
Harrison, John
Harrison, Julian D.
Harrison, Linda
Harrison, William
Harrold, C.G.
Hart, A.M.
Hart, Carole
Hart, E.H.
Hart, F. Gordon
Hart, G.A.M.
Hart, Mrs. J.
Hart, J.F.
Hart, J.G.
Hart, J.S.
Hart, John
Hart, Kit
Hart, Lauren
Hart, Mark
Hart, Sue
Hart, Ted
Hartland, D.
Hartland, G.
Hartman, Fay H.
Hartman, Gerald
Hartman, Harold
Hartman, Lisa
Hartman, Mary
Hartt, E.A.
Harttana, G.
Hartwick, E. Brian
Harty, F.G.
Harvard, Peggy
Harvey, Merle
Harvey, Richard
Harvey, Virginnia
Harwell, M.
Harwell, W.
Hasell, D.
Hasell, T.
Haskell, H.
Haskell, K.
Haslam, Cathy
Hassell, Sharon
Hastings, Joanna
Hastings, W.W.
Hatfield, John
Hatler, A.
Hatler, David F.
Hatler, Mareca
Hatler, Mary Eta
Hatter, David
Hatter, Ian
Hatter, James
Haughan, Linda
Haun, Ariel
Hauser, Pearl
Havelin, Louis
Haven, Stoner
Hawes, David M.
Hawes, David B.
Hawes, James
Hawes, Myrnal A.L.
Hawken, J.
Hawken, M.
Hawkes, Bill
Hawkes, Marian
Hawkey, G.
Hawkins, J.
Hawksley, Janet
Hawley, Connie
Haworth, Kent
Haws, C.W.
Hay, Diana
Hay, E.A.
Hay, Heather
Hay, Robert B.
Hayden, M.A.
Hayes, Eric
Hayes, Hugh
Hayes, Lauren
Hayes, Maryann
Hayes, Rachael
Hayes, Richard
Hayhurst, Katie
Haylock, Cliff
Haylock, Linda
Hayman, Gus
Hayman, Tom
Haynard, A.H.
Haynes, Barry
Haynes, J.
Haynes, Muriel
Hayton, B.
Hayton, M.
Hayward, John
Hazeldine, W.R.
Hazelwood, W. Grant
Hazledine, W.R.
Heakes, Todd
Healey, John
Healy, Michael
Hearn, O.
Hearn, David
Hearn, Dorothy
Hearn, Ed B.
Hearn, Martin
Hearne, Georgina
Hearne, Margo E.
Heater, Wendy
Heathman, R.L.
Hebert, Daryl
Hedley, A.F.
Heimstra, H.T.
Heintz, Gretta
Helbig, Marion
Helleiner, Fred
Helleman, O.
Heller, E.
Helm, Charles, W.
Helset, Roy
Henckle, David
Henderson, Bryan A.
Henderson, E.M.
Henderson, Martha
Henderson, Michael
Henderson, Nolan
Henderson, Nonie
Henderson, Otto
Henderson, Phil S.
Henderson, Valerie
Hendra, Isabel
Hendrick, Russell
Hendricks, Allan
Hendricks, Gus
Henkins, Harmon
Henley, O.
Henn, Keith
Hennan, Ed G.
Hennig, Karla
Henning, E.E.
Henning, Mrs. E.E.
Henry, G.
Henry, H.
Henry, John
Henry, Margaret
Henson, Colen
Henson, Gary
Henson, P.
Henson, Simon
Henzig, T.
Hepburn, J.
Heppner, D.
Herbert, B.
Herbert, William S.
Herbison, B.
Herdt, Fred
Heriot, Joan E.
Herne, A.
Herr, G.R.
Herrin, Brian
Hertzberg, D.
Hervieux, Margot
Herzig, Jerry R.
Herzig, Kim
Herzig, R.J.
Herzig, T.
Hesse, Hilde
Hesse, Werner H.
Hetherington, Anne E.
Hett, M.
Hettis, Rob
Hewell, George
Hewson, C.A.
Heybrock, Bill
Heyland, J.B.
Heywood, J.
Hickey, C.
Hickling, Bertha
Hickson, Cathie
Hienstra, H.T.
Higgins, Jno C.
Higginson, T.
Highe, Barbara
Highe, Donald
Hill, Beth
Hill, C.L.
Hill, Cecilia
Hill, Dorothy
Hill, Ivy
Hill, Ken
Hill, Les
Hill, Mark
Hill, Pat
Hill, Robert
Hill, Roy
Hillaby, Bruce
Hillard, R.W.
Hillerman, J.
Hilliar, C.
Hillier, George
Hilligan, D.
Hilligan, S.
Hillington, M.
Hillis, Nancy
Hilton, Jim
Hinckle, David
Hind-Smith, John
Hindson, Mr.
Hindson, Mrs.
Hines, Garfield
Hinton, J.L.
Hippen, H.
Hirrie, A.
Hirschbolz, Heinz
Hirschbolz, Marlene
Hirst, Stanley
Hitchcock, Clare E.
Hitchcock, Gordon
Hitchmough, John
Hlady, Debbie
Hoar, Carol
Hoar, N.J.
Hoar, Rick J.
Hoar, Robin
Hoar, W.
Hobeck, Erika
Hobson, Alan
Hobson, Cam
Hobson, Mrs. J.F.
Hobson, J. Fred
Hobson, James
Hobson, Kerry
Hobson, Marie
Hobson, Mark
Hobson, Shirley
Hochachka, W.
Hocker, Pat
Hocking, H.J.
Hocking, Jack
Hocking, Jennifer H.
Hodgins, Betty
Hodgson, John
Hodgson, K.
Hodson, C.
Hodson, Keith
Hoeg, Steig
Hoek, Jane
Hoffos, Trish
Hogarth, L.
Hogen, Betty
Hogg, Edward H.
Hogg, Lori
Hogg, Ted
Holbigg, J.
Holden, D.
Holdon, Canon M.W.
Holland, A.
Holland, David
Holland, L.
Holland, M.M.
Holland, R.H.
Holland, Stephen
Hollander, J.
Hollands, Grant R.
Hollands, J.
Hollby, M.
Holling, C.
Hollington, Jack
Hollington, Madge J.
Holloway, Lawrence
Holm, Margaret
Holman, John H.
Holman, M.
Holman, Muriel
Holman, P.J.
Holmes, Brian
Holmes, George
Holmes, H.

Wood, Casey A.
Wood, Chauncey
Wood, Christopher S.
Wood, Daryl
Wood, Deirdre
Wood, Douglas
Wood, E.D.
Wood, Eleanor
Wood, Ellie
Wood, F.H.
Wood, Frank W.
Wood, J.
Wood, Lorna
Wood, Murray
Wood, Peter
Wood, Phil
Wood, Rosa
Wood, Sarah
Wood, Trudy
Wood, W.
Woodcock, Don
Woodcock, Joe
Woodcock, Mike
Woods, G.
Woods, John G.
Woods, Marcia E.
Woods, R.
Woods, Susan M.
Woodworth, Freda
Woodworth, John
Wooldridge, Chris E.
Wooldridge, Donald R.
Woolgar, David
Woolgar, Pam
Woolman, Edward
Wootton, A.
Workman, Bob
Worther, C.K.
Worthington, Grant
Wrenshall, Anne
Wright, Dan
Wright, Dorothy
Wright, Eileen
Wright, Gwen W
Wright, James
Wright, Joy
Wright, Kenneth
Wright, Lois
Wright, P.F.
Wright, Mrs. R.T.
Wright, Richard T.
Wright, Rochelle
Wrigley, Bill
Wyatt, H.G.
Wyborn, Margriet
Wyburne, M.
Wydman, Roy
Wye, Doris
Wylie, Bill
Wynja, Mark
Wynn, James
Wynne, E.J.
Wynne, J.
Wynstra, Jack
Wyse, K.
Wysong, Dennis
Yak, John
Yardley, C.
Yarem Ko, Leslie
Yarwood, J.D.
Yates, Roy
Yaunk, Hans
Yellowlees, Jean
Yellowlees, Lou A.
Yellowlees, Mary
Yellowlees, Robin
Yewchan, Karl
Yorke, Paul
Yorkham, S.W.
Youds, J.
Young, C.
Young, Darren
Young, Jim
Young, R.J.
Younger, Dave
Youngs, K.
Youngson, Danny
Youngson, Margie
Younk, Hans
Youwe, A.
Youwe, Phil
Yule, Ian
Zamluk, Joan
Zaniol, Gabriella
Zapf, Rick
Zaremba, Stephan
Zeck, Hal
Zeeman, Richard
Zeeman, Thomas
Zeral, Martin
Zettergreen, Barry
Zielinski, Anne
Zieroth, Dale
Zimmerman, Ellen
Zinkan, Betty
Zinkan, Ted
Zogaris, Stamatis
Zolinski, Ed
Zoyetz, Cynthia
Zroback, Bill
Zroback, Ki
Zurowski, Tim
Zwickel, Fred C.

APPENDIX 5

Structure of the Data Base Files Used for *Birds of British Columbia,* Volume 3

OCCURRENCE DATA BASE FILE

Structure: XXXX.DBF, where "XXXX" refers to a 4-character species code.

Field	Field Name	Type	Width	Dec	Index	Definition
1	SPECIES	Character	4	–	N	Four-letter code identifying the species
2	GRID	Character	6	–	N	A 1:50,000 map grid designation, from the National Topographic Map Series, e.g., 082E05, within which the observation was made
3	MONTH	Numeric	2	–	N	A number between 0 and 12; 0 indicates that the month is unknown
4	DAY	Numeric	2	–	N	A number between 0 and 31; 0 indicates that the day is unknown
5	YEAR	Numeric	4	–	N	Usually a 4-digit number indicating the year; 0 indicates that the year is unknown
6	OBSERVER	Character	30	–	N	The name(s) of the observer(s)
7	TOTAL_BIRD	Numeric	6	–	N	The total number of birds seen
8	TOTAL_AD	Numeric	5	–	N	Number of adult birds
9	TOTAL_IMM	Numeric	5	–	N	Number of immature birds
10	TOTAL_UNAG	Numeric	5	–	N	Number of birds of unknown age
11	ADULT_M	Numeric	5	–	N	Number of adult male birds
12	IMM_M	Numeric	5	–	N	Number of immature male birds
13	UNAGED_M	Numeric	5	–	N	Number of male birds of unknown age
14	ADULT_F	Numeric	5	–	N	Number of adult female birds
15	IMM_F	Numeric	5	–	N	Number of immature female birds
16	UNAGED_F	Numeric	5	–	N	Number of female birds of unknown age
17	LOCATION	Character	20	–	N	Location of the observation, generally the nearest town, lake, mountain, or other gazetted location
18	BIB_SPECMN	Character	5	–	N	The bibliography number or specimen number for observations of museum specimens
19	REF_NUMBER	Character	10	–	N	Reference number for bibliographic or specimen data
20	REMARKS	Character	80	–	N	Relevant information about the observation that does not fit into the above categories
Total			215			

BREEDING DATA BASE FILES

Primary File

Structure: XXXX1B.DBF, where "XXXX" refers to a 4-character species code.

Field	Field Name	Type	Width	Dec	Index	Definition
1	PRIMARY_NO	Numeric	6	–	N	Identification number assigned to the nest or colony
2	SUCESS2	Character	1	–	N	A field to indicate whether or not at least 1 young fledged
3	GRID	Character	6	–	N	A 1:50,000 map grid designation from the National Topographic Map Series, e.g., 082E05, within which the observation was made
4	OBSERVER	Character	30	–	N	The name(s) of the observer(s)
5	NEAR_TOWN	Character	45	–	N	The nearest town or other gazetted location
6	ELEVATION_	Numeric	5	–	N	Elevation in metres
7	HAB_TYPE	Character	6	–	N	General habitat type, e.g., grassland, aquatic, human interference, etc.
8	HAB_CLASS	Character	6	–	N	More specifics about the habitat. For example, under human interference, possibilities would be farmland, urban, etc.
9	HABC_REMRK	Character	40	–	N	Relevant remarks about the specifics concerning the habitat
10	HAB_SPECIF	Character	6	–	N	Specific habitat, e.g., roadside, riparian, etc.
11	HABS_REMRK	Character	40	–	N	Remarks about the specific habitat
12	NEST_HITE	Numeric	4	–	N	Height of the nest above the ground, in metres
13	MATERIAL	Character	5	–	N	The materials used to construct the nest
14	MATERIAL2	Character	5	–	N	
15	MATERIAL3	Character	5	–	N	
16	LINING	Character	5	–	N	The materials used to construct the lining of the nest, or additional nest construction materials
17	LINING2	Character	5	–	N	
18	LINING3	Character	5	–	N	
19	GEN_LOCTN	Character	6	–	N	General location of the nest, e.g., post, tree, or stump
20	SPE_LOCTN	Character	6	–	N	Specific location of the nest, e.g., nestbox, natural cavity, or on a branch
21	NEST_TYPE	Character	5	–	N	Type of nest, e.g., cavity, cup
22	NEST_DIAMI	Numeric	4	–	N	Inside diameter of nest, in centimetres
23	NEST_DIAMO	Numeric	4	–	N	Outside diameter of nest, in centimetres
24	NEST_DPTHI	Numeric	3	–	N	Inside depth of nest, in centimetres
25	NEST_DPTHO	Numeric	3	–	N	Outside depth of nest, in centimetres
26	NEST_SLOPE	Character	2	–	N	Slope of the ground where the nest is located
27	NEST_COMM	Character	80	–	N	Relevant comments regarding the nest that do not fit into the above categories
28	BURO_LNGTH	Numeric	4	–	N	Length of the burrow, in centimetres
29	BURO_ENTR	Numeric	2	–	N	Height of the burrow entrance, in centimetres
30	BURO_ENTRW	Numeric	2	–	N	Width of the burrow entrance, in centimetres
31	NATURAL_	Character	1	–	N	–
32	COWBIRD_	Character	1	–	N	Has this nest been parasitized by the Brown-headed Cowbird?
33	COWBRDHOST	Character	4	–	N	Host species, if parasitized
34	REMARKS	Character	80	–	N	Relevant remarks about the nest
35	JUMP	Character	1	–	N	Field used to correct a data entry anomaly of dBASE
Total			440			

Secondary File

Structure: XXXX2B.DBF, where "XXXX" refers to a 4-character species code.

Field	Field Name	Type	Width	Dec	Index	Definition
1	JUMP	Character	1	–	N	Field used to correct a data entry anomaly of dBASE
2	PRIMARY_NO	Numeric	6	–	N	Identification number assigned to the nest that can be matched with the identification number in the primary breeding file
3	DATE	Date	8	–	N	Date of visit to the nest
4	DATE2	Date	8	–	N	–
5	EGGS_YNG	Character	1	–	N	Indicates one of the following conditions: eggs present (E); young present (Y); nesting activity, but neither eggs nor young present (M); already fledged (F); or unknown (U)
6	NUMBER	Numeric	2	–	N	Number of eggs or young
7	MAX_NO	Logical	1	–	N	Is this the maximum number of eggs or young seen? yes (Y) or no (N)
8	COMMENTS	Character	80	–	N	Relevant information about the observation that does not fit into the above categories
9	COWBIRD	Logical	1	–	N	Evidence of parasitism: yes (Y) or no (N)
Total			109			

Tertiary File

Structure: XXXX3B.DBF, where "XXXX" refers to a 4-character species code.

Field	Field Name	Type	Width	Dec	Index	Definition
1	JUMP	Character	1	–	N	Field used to correct a data entry anomaly of dBASE
2	PRIM_RECNO	Numeric	6	–	N	Identification number assigned to the colony visit record that can be matched with the identification number in the primary breeding file
3	SUCCESS	Numeric	6	–	N	–
4	DATE	Date	8	–	N	Date of visit
5	NO_NESTS	Numeric	6	–	N	Number of nests in the colony on the date of visit
6	EST_PAIRS	Numeric	6	–	N	Estimated number of pairs
7	EMPTY	Numeric	6	–	N	Number of empty nests
8	EGG1	Numeric	6	–	N	Number of nests with no young and 1 egg
9	EGG2	Numeric	6	–	N	Number of nests with no young and 2 eggs
10	EGG3	Numeric	6	–	N	Number of nests with no young and 3 eggs
11	EGG4	Numeric	6	–	N	Number of nests with no young and 4 eggs
12	EGG5	Numeric	6	–	N	Number of nests with no young and 5 eggs
13	EGG6	Numeric	6	–	N	Number of nests with no young and 6 eggs
14	YOUNG1	Numeric	6	–	N	Number of nests with 1 young and no eggs
15	EGG1YNG1	Numeric	6	–	N	Number of nests with 1 young and 1 egg
16	EGG2YNG1	Numeric	6	–	N	Number of nests with 1 young and 2 eggs
17	EGG3YNG1	Numeric	6	–	N	Number of nests with 1 young and 3 eggs
18	EGG4YNG1	Numeric	6	–	N	Number of nests with 1 young and 4 eggs
19	EGG5YNG1	Numeric	6	–	N	Number of nests with 1 young and 5 eggs
20	YNG2	Numeric	6	–	N	Number of nests with 2 young and no eggs
21	EGG1YNG2	Numeric	6	–	N	Number of nests with 2 young and 1 egg
22	EGG2YNG2	Numeric	6	–	N	Number of nests with 2 young and 2 eggs
23	EGG3YNG2	Numeric	6	–	N	Number of nests with 2 young and 3 eggs
24	EGG4YNG2	Numeric	6	–	N	Number of nests with 2 young and 4 eggs
25	YNG3	Numeric	6	–	N	Number of nests with 3 young and no eggs
26	EGG1YNG3	Numeric	6	–	N	Number of nests with 3 young and 1 egg
27	EGG2YNG3	Numeric	6	–	N	Number of nests with 3 young and 2 eggs
28	EGG3YNG3	Numeric	6	–	N	Number of nests with 3 young and 3 eggs
29	YNG4	Numeric	6	–	N	Number of nests with 4 young and no eggs
30	EGG1YNG4	Numeric	6	–	N	Number of nests with 4 young and 1 egg
31	EGG2YNG4	Numeric	6	–	N	Number of nests with 4 young and 2 eggs
32	YNG5	Numeric	6	–	N	Number of nests with 5 young and no eggs
33	EGG1YNG5	Numeric	6	–	N	Number of nests with 5 young and 1 egg
34	YNG6	Numeric	6	–	N	Number of nests with 6 young and no eggs
35	EGG1A	Numeric	3	–	N	For nests whose numbers of eggs and young are not covered by the above, these variables represent the number of eggs and young, and the number of nests containing this combination.
36	YNG1A	Numeric	3	–	N	
37	NO_1A	Numeric	6	–	N	
38	EGG2A	Numeric	3	–	N	Same as EGG1A, YNG1A, NO_1A, but for a second combination of eggs and young
39	YNG2A	Numeric	3	–	N	
40	NO_2A	Numeric	6	–	N	
41	REMARKS	Character	80	–	N	Special remarks about the colony that are not covered in the other information
Total			306			

ECOPROVINCE DATA BASE FILE

Structure: ECO_ONEW.DBF

Field	Field Name	Type	Width	Dec	Index	Definition
1	ECOPROVINC	Character	27	–	N	One of the 13 ecoprovinces in B.C. There are 9 terrestrial ecoprovinces in British Columbia (see Volume 1, p. 61). Because of the wide diversity in the Coast and Mountains Ecoprovince, we divided it into 4 regions: Queen Charlotte Islands, Northern Mainland Coast, Southern Mainland Coast, and Western Vancouver Island.
2	GRID	Character	6	–	Y	1:50,000 National Topographic Map Series map grid designation
3	REC_NO	Numeric	4	–	N	Internal record number to keep track of data
4	COASTINT	Character	8	–	N	Whether GRID is at the coast or in the interior
5	LAT	Character	4	–	N	Latitude of the GRID
6	LONG	Character	5	–	N	Longitude of the GRID
7	LLORDER	Numeric	4	–	N	GRID order from north to south

PASSERINE DATA BASE FILE

Structure: PASSSORT.DBF

Field	Field Name	Type	Width	Dec	Index	Definition
1	SPECIES2	Character	35	–	N	Full species name
2	SPECIES	Character	4	–	N	Species code
3	AOU_NO	Numeric	4	–	N	American Ornithologists' Union species number
4	NUMBER	Character	5	–	N	Taxonomic order of the species
5	INC_PD	Character	2	–	N	Incubation period
6	NEST_PD	Character	2	–	N	Nestling period

WEEK-ENDING DATA BASE FILE

Structure: WEEKEND.DBF

Field	Field Name	Type	Width	Dec	Index	Definition
1	WEEKEND	Character	6		N	Day and month
2	WEEK_NO	Numeric	2		N	Week number
3	DATE	Date	8		N	Weekly dates throughout the year

APPENDIX 6

dBASE Programs Used in Preparing *Birds of British Columbia,* Volume 3

Note that in the descriptions below, "XXXX" refers to a four-character species code.

SYSTEM PROGRAMS

BBSAPPX — Generates the Breeding Bird Survey appendix (Appendix 3) from the Breeding Bird Survey data set.

BBSNOTE — Generates the Breeding Bird Survey summaries for the NOTEWORTHY RECORDS section from the Breeding Bird Survey data set.

CBCAPPX — Generates the Christmas Bird Count appendix (Appendix 2) from the Christmas Bird Count data set.

CBCNOTE — Generates the Christmas Bird Count summaries for the NOTEWORTHY RECORDS section from the Christmas Bird Count data set.

INTERACTIVE PROGRAMS

AVCLTCH — Calculates the average clutch size by ecoprovince, along with statistics that can be tested with AVGCLTST.

AVGCLTST — Calculates a *t* value based on user input generated from the program AVCLTCH.

BIRDYEAR — Calculates the total number of birds and records and mean number of birds by year by selected ecoprovinces. Can help indicate timing of range extensions.

BRCHEK — Allows the verification of the Egg, Young, or Fledged fields in the breeding data base to ensure that no late egg dates (resulting from nestbox cleanings) or fledged young are included in the breeding chronology figures or on the breeding map.

BRCSTINT — Calculates the early and late breeding dates for both the coast and interior. Also back-calculates earlier egg dates from early young records.

BRDGRPH3 — Prepares data file that will allow the seasonal fluctuation figure to be generated with Harvard Graphics 3.0. Creates a new dBASE file: XXXXGRPH.DBF. The program also creates a text file of the results (XXXXSFL.TXT) for use in preparing the final figures.

BRSUCC — Calculates breeding success for the coast, the interior, and British Columbia in general.

CLBRSIZE — Calculates frequency data by week for clutch and/or brood size, and writes them to a new dBASE file: XXXXSE.DBF for eggs and XXXXSY.DBF for young. Data are reported during the week the nest was found.

COLONYTBL — Prepares a dBASE file of colony data and creates a text file of colony data (XXXXCLNY.TXT) for importing into WordPerfect 5.1.

COWBIRD — Appends Brown-headed Cowbird data from host species files to Cowbird file.

COWSUCC2 — Calculates Brown-headed Cowbird parasitism rates for the coast, the interior, and British Columbia in general.

DECADE — Calculates the number of records and mean number of birds by ecoprovince by decade.

ELEV	Displays both occurrence and breeding records that have elevation data.
EYNKD	For nests with eggs and young, calculates early and late dates, the 2nd and 3rd quartile percentages, and the frequency of clutch and brood sizes. Creates a text file (XXXXBR.TXT) for importing into WordPerfect 5.1.
FILE1OCC	Sorts occurrence records by ecoprovince by month by day, and gives the user several choices, including selecting records to be added to the NOTEWORTHY RECORDS in the account. Creates a new dBASE file (XXXXNOTE.DBF) for use with program NOTRECS.
FILE2BRD	Sorts breeding records by ecoprovince by month by day, and gives the user several choices, including selecting records to be added to the NOTEWORTHY RECORDS in the account. Uses the dBASE file XXXXNOTE.DBF that was created with FILE1OCC.
FILE3BRD	Sorts colony breeding records by ecoprovince by month by day, and gives the user several choices, including selecting records to be added to the NOTEWORTHY RECORDS in the account. Uses the dBASE file XXXXNOTE.DBF that was created with FILE1OCC.
GRIDCHEK	Checks both the occurrence and primary breeding data bases for incorrect map grids, and writes the map grids with errors to the file XXXXERR.TXT.
INCFLEDG	Selects records that meet certain criteria that may permit the calculation of incubation and nestling periods by hand.
IRRUPT	Determines possible irruptive activities of birds by ecoprovince.
LAYDATE	Selects records with information known to fall in the laying stage of the nesting cycle, and creates a dBASE file (XXXXELAY.DBF) that can be used to prepare a figure. Needs the dBASE file WEEKEND.DBF.
MAXFLOCK	Determines the number of records by ecoprovince by month with number of birds greater than a value selected by the user. Creates a dBASE file (XXXXMXNO.DBF) that can be used to prepare a figure.
MIGCRON4	Creates a text file for preparing the occurrence and breeding chronology bar graph. The program summarizes data by week and generates the file XXXXMC.TXT.
MIGPEAKS	Allows the user to see the weeks with highest numbers of birds and records, and from this to interpret possible start, peak, and end of movements in an ecoprovince.
NOTRECS	This program is run after the user has finished selecting Noteworthy Records from FILE1OCC, FILE2BRD, and FILE3BRD. Sorts records by season and latitude, and writes them to a text file (XXXXRECS.TXT) for use with WordPerfect 5.1.
OCMXCOMB	Allows the user to append data from a mixed species file to the original data base.
PREDMORT	Selects and displays records with information about predation or mortality.
PROTO	Selects and displays records with information about blowfly infestations.
RECORD	Deletes duplicate records within a species data base file.
SAMECOMB	Allows the user to append data from a file with the same name as the original species data set.
STATUS	Accumulates all the records into status groupings (see Volume 1, p. 148) to help determine the status of a species for each ecoprovince.

REFERENCES CITED

As with the nonpasserine components of *Birds of British Columbia* (Campbell et al. 1990a, 1990b), the passerines are covered in self-contained volumes. Each volume has its own list of references even though this means that repetition occurs.

Citations of most unpublished material contain a reference to the 2-volume *A Bibliography of British Columbia Ornithology* (Campbell et al. 1979b, 1988a). Copies of the papers cited in that work are on file at the Royal British Columbia Museum in Victoria, and are also found in the personal library of R.W. Campbell in Victoria.

Adams, E. and M. Morrison. 1993. Effects of forest stand structure and composition on Red-breasted Nuthatches and Brown Creepers. Journal of Wildlife Management 57:616-629.

Adkisson, C.S. 1966. The nesting and behavior of Mockingbirds in northern lower Michigan. Jack-Pine Warbler 44:102-116.

Ahlquist, J.E., A.H. Bledsoe, F.H. Sheldon, and C.G. Sibley. 1987. DNA hybridization and avian systematics. Auk 104:556-563.

Ainsley, D.T.J. 1991. Vocalizations and nesting behaviour of the Pacific-slope Flycatcher, *Empidonax difficilis*. M.Sc. Thesis, University of Victoria, Victoria, British Columbia. 181 pp.

Alcorn, G.D. 1971. Key to nest and eggs of the birds of the state of Washington. University of Puget Sound Press, Tacoma, Washington.

Aldrich, J.W. 1944. Notes on the races of the White-breasted Nuthatch. Auk 61:592-604.

________. 1968. Population characteristics and nomenclature of the Hermit Thrushes. United States National Museum Bulletin 124:1-33.

Aldrich, J.W. and F.C. James. 1991. Ecogeographic variation in the American Robin (*Turdus migratorius*). Auk 108:230-249.

Alford, C.E. 1928. Field notes on the birds of Vancouver Island. Ibis 4:181-210.

Allen, A.L., S.C. Samis, and E.A. Stanlake. 1977. Urban wildlife in Vancouver and Victoria, British Columbia. British Columbia Fish and Wildlife Branch Unpublished Report, Victoria. 66 pp. (Bibliography 2123).

American Ornithologists' Union. 1957. Check-list of North American birds, 5th edition. Lord Baltimore Press, Maryland. 691 pp.

________. 1973. Corrections and additions to the "thirty-second supplement to the Check-list of North American birds." Auk 90:887.

________. 1982. Thirty-fourth supplement to the American Ornithologists' Union check-list of North American birds. Auk 99:1CC-16CC.

________. 1983. Check-list of North American birds, 6th edition. Allen Press, Lawrence, Kansas. 877 pp.

________. 1985. Thirty-fifth supplement to the American Ornithologists' Union check-list of North American birds. Auk 102:680-686.

________. 1987. Thirty-sixth supplement to the American Ornithologists' Union check-list of North American birds. Auk 104:591-596.

________. 1989. Thirty-seventh supplement to the American Ornithologists' Union check-list of North American birds. Auk 106:532-538.

________. 1993. Thirty-ninth supplement to the American Ornithologists' Union check-list of North American birds. Auk 110:675-682.

________. 1995. Fortieth supplement to the American Ornithologists' Union check-list of North American birds. Auk 112:819-830.

Anderson, E. 1967. The intermediates are busy birders. Vancouver Natural History Society Bulletin 135:9-10.

Anderson, E.A. 1972. Mockingbird visits Salmon Arm. Federation of British Columbia Naturalists Newsletter 10:4.

Anderson, E.M. 1914. Report on birds collected and observed during April, May and June, in the Okanagan Valley, from Okanagan Landing south to Osoyoos. Pages 7-16 *in* Report of the Provincial Museum of Natural History for the year 1913, Victoria, British Columbia.

________. 1915a. Report of E.M. Anderson on Atlin expedition, 1914. Pages F7-F17 *in* Report of the Provincial Museum of Natural History for the year 1914, Victoria, British Columbia.

________. 1915b. Nesting of the Bohemian Waxwing in northern B.C. Condor 17:145-148.

Anderson, J.R. 1920. Bird notes from Victoria, British Columbia. Bird-Lore 22:282-284.

Anderson, R.R. 1976a. Summary of highest counts of individuals for Canada. American Birds 30:643-644.

________. 1976b. Summary of all-time highest counts of individuals for Canada. American Birds 30:645-648.

________. 1977. Summary of highest counts of individuals for Canada. American Birds 31:916-918.

________. 1979. Summary of highest counts of individuals for Canada. American Birds 33:708-709.

________. 1980. Summary of highest counts of individuals for Canada. American Birds 34:708-710.

________. 1981. Summary of highest counts of individuals for Canada. American Birds 35:763-765.

________. 1983. Summary of highest counts of individuals for Canada. American Birds 37:797-799.

Anonymous. 1903. An introduction of English birds to Victoria. Country Life 14(358).

________. 1950. European Starlings. Victoria Naturalist 6:99.

________. 1956. Bird notes. Victoria Naturalist 13:37-38.

________. 1967. Additions to the flora and fauna, Shuswap, 1967. British Columbia Parks Branch Unpublished Report, Victoria. 2 pp. (Bibliography 2154).

________. 1978. Bird sightings. Wandering Tatler 2(1):1-16. (Issued by the Vancouver Natural History Society).

________. 1979. Bird sightings. Wandering Tattler 2(7):1-4. (Issued by the Vancouver Natural History Society).

________. 1991. Notable sightings. Marsh Wrenderings 8(6):1-2. (Issued by the Sechelt Marsh Protective Society).

Antifeau, T. 1977. Wells Gray Park – Moose winter range survey, 1977. British Columbia Wildlife Branch Unpublished Report, Kamloops. 7 pp. (Bibliography 2171).

Arbib, R. 1975. The blue list for 1976. American Birds 29:1067-1072.

________. 1976. The blue list for 1977. American Birds 30:1031-1039.

________. 1979. The blue list for 1980. American Birds 33:830 835.

Armstrong, E.A. and H.L.K. Whitehouse. 1977. Behavioral adaptations of the wren (*Troglodytes troglodytes*). Biological Review 52:235-294.

Armstrong, R.H. 1988. Guide to the birds of Alaska. Alaska Northwest Publishing Company, Anchorage. 332 pp.

Ashby, D. 1940. A record of the Eastern Mockingbird in British Columbia. Condor 42:266.

Auman, G. and J.T. Emlen. 1959. The distribution of Cliff Swallow nesting colonies in Wisconsin. Passenger Pigeon 21:95-100.

Bagg, A.C. 1941. Brown Thrasher in Oregon. Auk 58:99.

Bailey, A.M. 1927. Notes on the birds of southeastern Alaska. Auk 44:351-367.

Bakus, G.J. 1959. Territoriality, movements, and population density of the Dipper in Montana. Condor 61:410-425.

Banks, R.C. 1970. Molt and taxonomy of Red-breasted Nuthatches. Wilson Bulletin 82: 201-205.

Barkley, W.D. 1966. Shuswap Lake nature house – 1966 season report. British Columbia Parks Branch Unpublished Report, Victoria. 15 pp. (Bibliography 2197).

Barlow, J.C. 1980. Patterns of ecological interactions among migrant and resident vireos on the wintering grounds. Pages 79-109 *in* A. Keast and E.S. Morton (editors). Migrant birds in the neotropics: ecology, behavior, distribution and conservation. Smithsonian Institution Press, Washington, D.C.

Barlow, J.C. and W.B. McGillivray. 1983. Foraging and habitat relationships of the sibling species Willow Flycatcher (*Empidonax traillii*) and Alder Flycatcher (*E. alnorum*) in southern Ontario. Canadian Journal of Zoology 61:1510-1516.

Barlow, J.C. and J.C. Rice. 1977. Aspects of the comparative behaviour of Red-eyed and Philadelphia Vireos. Canadian Journal of Zoology 55:528-542.

Beal, F.E.L. 1912. Food of our more important flycatchers. United Sates Department of Agriculture Biology Survey Bulletin No. 44, Washington, D.C. 67 pp.

Beason, R.C. 1995. Horned Lark (*Eremophila alpestris*). *In* A. Poole and F. Gill (editors). The birds of North America, No. 195. The Academy of Natural Sciences, Philadelphia; The American Ornithologists' Union, Washington, D.C.

Beason, R.C. and E.C. Franks. 1974. Breeding behavior of the Horned Lark. Auk 91:65-74.

Beaver, D.L. and P.H. Baldwin. 1975. Ecological overlap and the problem of competition and sympatry in the Western and Hammond's Flycatchers. Condor 77:1-13.

Begg, B. 1989. Skylarks – another little bit of old England. Victoria Naturalist 45(5):15-18.

________. 1990. Skylark, Eurasian – Saanich Peninsula, British Columbia. Winging It 8:7. (Newsletter of the American Birding Association).

________. 1991. Skylark behaviour influenced by snow cover. Victoria Naturalist 47(6):19.

Behle, W.H. 1942. Distribution and variation of the Horned Larks (*Ototcoris alpestris*) of western North America. University of California Publications in Zoology 3:205-316.

________. 1956. A systematic review of the Mountain Chickadee. Condor 58:51-70.

________. 1976. Systematic review, intergradation, and clinal variation in Cliff Swallows. Auk 93:66-77.

Bell, K.M. 1975. Fall flocking and migration. British Columbia Parks Branch Unpublished Report, Victoria. 9 pp. (Bibliography 2211).

Belles-Isles, J.-C. and J. Picman. 1986a. House Wren nest-destroying behavior. Condor 88:190-193.

________. and ________. 1986b. Nesting losses and nest site preference in House Wrens. Condor 88:483-486.

________. and ________. 1986c. Destruction of heterospecific eggs by the Gray Catbird. Wilson Bulletin 98:603-605.

Bendire, C. 1895. Life histories of North American birds, from parrots to the grackles, with special reference to their breeding habits and eggs. United States National Museum Special Bulletin 3:1-518.

Bent, A.C. 1942. Life histories of North American flycatchers, larks, swallows, and their allies. United States National Museum Bulletin No. 179, Washington, D.C. 555 pp.

________. 1946. Life histories of North American jays, crows, and titmice. United States National Museum Bulletin No. 191, Washington, D.C. 495 pp.

________. 1948. Life histories of North American nuthatches, wrens, thrashers, and their allies. United States National Museum Bulletin No. 195, Washington, D.C. 475 pp.

________. 1949. Life histories of North American thrushes, kinglets, and their allies. United States National Museum Bulletin 196, Washington, D.C. 452 pp.

________. 1950. Life histories of North American wagtails, shrikes, vireos, and their allies. United States National Museum Bulletin No. 197, Washington, D.C. 411 pp.

Bertin, R.I. 1977. Breeding habits of the Wood Thrush and Veery. Auk 79:303-311.

Birkhead, T. 1991. The magpies. T & A D Poyser, London, England. 270 pp.

Bishop, L.B. 1900. Birds of the Yukon region with notes on other species. Pages 47-95 *in* Osgood, W.H. Results of a biological reconnaissance of the Yukon River region. North American Fauna No. 19, Washington, D.C.

________. 1930. The Eastern Kingbird breeding on Vancouver Island. Murrelet 11:19.

________. 1931. Descriptions of three new birds from Alaska. Auk 27: 11-12.

Blancher, P.J. and R.J. Robertson. 1982. A double-brooded Eastern Kingbird. Wilson Bulletin 94:212-213.

________. and ________. 1984. Resource use by sympatric kingbirds. Condor 86:305-313.

________. and ________. 1985. A comparison of Eastern Kingbird breeding biology in lakeshore and upland habitats. Canadian Journal of Zoology 63:2305-2312.

Blood, D.A. and M. Chutter. 1978. Raptor nesting survey for Shakwak Highway improvement project. Donald A. Blood and Associates Unpublished Report, Lantzville, British Columbia. 70 pp.

Blood, D.A., M. Chutter, and G. Anweiller. 1981. An annotated list of the birds of the Stikine region, B.C. Donald A. Blood and Associates Unpublished Report, Lantzville, British Columbia. 32 pp. (Bibliography 2278).

Bock, C.E. and L.W. Lepthien. 1972. Winter eruptions of Red-breasted Nuthatches in North America 1950-1970. American Birds 26:558-561.

________. and ________. 1975. Distribution and abundance of the Black-billed Magpie (*Pica pica*) in North America. Great Basin Naturalist 35:269-272.

________. and ________. 1976. Changing winter abundance and distribution of the Blue Jay, 1962-1971. American Midland Naturalist 96:232-236.

________. and ________. 1976a. Synchronous eruptions of boreal seed-eating birds. American Naturalist 110:559-571.

Bock, C.E. and J.F. Lynch. 1970. Breeding bird populations of burned and unburned conifer forest in the Sierra Nevada. Condor 72:182-189.

Boggs, B. and E. Boggs. 1960a. The fall migration – northern Pacific coast region. Audubon Field Notes 14:65-67.

________. and ________. 1960b. The winter season – northern Pacific coast region. Audubon Field Notes 14:334-336.

________. and ________. 1960c. The spring migration – northern Pacific coast region. Audubon Field Notes 14:414-416.

________. and ________. 1960d. The nesting season – northern Pacific coast region. Audubon Field Notes 14:472-474.

________. and ________. 1961a. The fall migration – northern Pacific coast region. Audubon Field Notes 15:68-70.

________. and ________. 1961b. The winter season – northern Pacific coast region. Audubon Field Notes 15:352-353.

________. and ________. 1961c. The spring migration – northern Pacific coast region. Audubon Field Notes 15:433-434.

________. and ________. 1961d. The nesting season – northern Pacific coast region. Audubon Field Notes 15:487-489.

________. and ________. 1962a. The fall migration – northern Pacific coast region. Audubon Field Notes 16:67-69.

________. and ________. 1962b. The winter season – northern Pacific coast region. Audubon Field Notes 16:357-359.

________. and ________. 1962c. The spring migration – northern Pacific coast region. Audubon Field Notes 16: 440-442.

________. and ________. 1962d. The nesting season – northern Pacific coast region. Audubon Field Notes 16:500-502.

________. and ________. 1963a. The fall migration – northern Pacific coast region. Audubon Field Notes 17:58-61.

________. and ________. 1963b. The winter season – northern Pacific coast region. Audubon Field Notes 17:351-353.

________. and ________. 1963c. The spring migration – northern Pacific coast region. Audubon Field Notes 17:427-429.

________. and ________. 1963d. The nesting season – northern Pacific coast region. Audubon Field Notes 17:478-480.

________. and ________. 1964a. The fall migration – northern Pacific coast region. Audubon Field Notes 18:65-68.

________. and ________. 1964b. The winter season – northern Pacific coast region. Audubon Field Notes 18:379-381.

________. and ________. 1964c. The nesting season – northern Pacific coast region. Audubon Field Notes 18:530-532.

Bond, G.M. 1963. Geographic variation in the thrush *Hylocichla ustulata*. Proceedings of the United States National Museum 114:373-387.

Boone, D.D. 1982. Are Tree Swallows colonial? Sialia 4:8-9.

Bowles, J.H. 1898. Notes on the Streaked Horned Lark. Osprey 3:53-54.

________. 1900. Nesting of the Streaked Horned Lark. Condor 2:30-31.

________. 1921. Breeding dates for Washington birds. Murrelet 2:8-12.

Bowles, J. and F.R. Decker. 1930. The ravens of the state of Washington. Condor 32:192-201.

Bowling, J. 1992. Fall, 1992 – British Columbia/Yukon region. American Birds 46:465-469.

________. 1995a. Fall, 1994 – British Columbia/Yukon region. National Audubon Society Field Notes 48:332-334.

________. 1995b. Winter, 1994 – British Columbia/Yukon region. National Audubon Society Field Notes 48:979-981.

________. 1995c. Spring, 1995 – British Columbia/Yukon region. National Audubon Society Field Notes 49:87-92.

________. 1995d. Summer, 1995 – British Columbia/Yukon region. National Audubon Society Field Notes 49:184-189.

Boxall, P.C. 1981. Ruby-crowned Kinglets (*Regulus calendula*) feeding a Brown-headed Cowbird (*Molothrus ater*). Canadian Field-Naturalist 95:99-100.

Boyd, R.L. 1976. Behavioral biology and energy expenditure in a Horned Lark population. Ph.D. Thesis, Colorado State University, Fort Collins.

Brackbill, H. 1977. Eleven-day incubation by an American Robin. Auk 94:607-608.

Braun, C.E., M.F. Baker, R.L. Eng, J.S. Gashwiller, and M.H. Schroeder. 1976. Conservation committee report of effects of alteration of sagebrush communities on the associated avifauna. Wilson Bulletin 88:165-171.

Brawn, J.D. and R.P. Balda. 1988. Population biology of cavity nesters in northern Arizona: do nest sites limit breeding densities? Condor 90:51-57.

Brenchley, A. 1985. Ravens and their effect on sheep production on Saltspring Island, British Columbia. British Columbia Ministry of Agriculture and Food Crop Protection Branch Report, Victoria. 27 pp.

Brennan, L.A. 1989. Comparative use of forest resources by Chestnut-backed and Mountain Chickadees in the western Sierra Nevada. Ph.D. Thesis, University of California, Berkeley. 154 pp.

Brennan, L.A. and M.L. Morrison. 1991. Long-term trends of chickadee populations in western North America. Condor 93:130-137.

Briskie, J.V. 1985. Growth and parental feeding of Least Flycatchers in relation to brood size, hatching order and prey availability. M.Sc. Thesis, University of Manitoba, Winnipeg.

________. 1994. Least Flycatcher (*Empidonax minimus*). *In* A. Poole and F. Gill (editors). The birds of North America, No. 99. The Academy of Natural Sciences, Philadelphia; The American Ornithologists' Union, Washington, D.C.

Briskie, J.V. and S.G. Sealy. 1987a. Polygyny and double-brooding in the Least Flycatcher. Wilson Bulletin 99:492-494.

________. and ________. 1987b. Responses of Least Flycatchers to experimental inter- and intraspecific brood parasitism. Condor 89:899-901.

________. and ________. 1988. Nest re-use and egg burial in the Least Flycatcher, *Empidonax minimus*. Canadian Field-Naturalist 102:729-731.

________. and ________. 1989. Determination of clutch size in the Least Flycatcher. Auk 106:269-278.

British Columbia Ministry of Forests. 1992. Resource management plan for the Fort St. John timber supply area. British Columbia Ministry Forests Report, Victoria. 96 pp.

Brooks, A. 1900. Notes on some of the birds of British Columbia. Auk 17:104-107.

________. 1903. Notes on the birds of the Cariboo district, British Columbia. Auk 20:77-84.

________. 1904. British Columbia notes. Auk 21:289-291.

________. 1905. Notes on the nesting of the Varied Thrush. Auk 34:28-50.

________. 1909. British Columbia supplement. Pages 963-978 *in* William Leon Dawson and John Hooper Bowles. The birds of Washington: a complete, scientific and popular account of the 372 species of birds found in the state. Occidental Publishing Company, Seattle. (Bibliography 2363).

________. 1909a. Three records for British Columbia. Auk 26:313-314.

________. 1909b. Some notes on the birds of the Okanagan, British Columbia. Auk 26:60-63.

________. 1917. Birds of the Chilliwack district. Auk 34:28-50.

________. 1925. Three noteworthy records for British Columbia. Condor 27:211-212.

________. 1927. British Columbia notes. Murrelet 8:43-44.

________. 1942. Status of the Northwestern Crow. Condor 44:166-167.

________. 1942a. Additions to the distributional list of birds of British Columbia. Condor 44:33-34.

Brooks, A. and B. Brooks. 1973. The seventy-third Christmas Bird Count – Pender Island, B.C. American Birds 27:181-182.

Brooks, A. and H.S. Swarth. 1925. A distributional list of the birds of British Columbia. Pacific Coast Avifauna No. 17, Berkeley, California. 158 pp.

Broun, M. 1941. Migration of Blue Jays. Auk 58:262-263.

Brown, C.R. 1986. Cliff Swallow colonies as information centers. Science 234:83-85.

Brown, C.R. and M.B. Brown. 1988. A new form of reproductive parasitism in Cliff Swallows. Nature 331:66-68.

Brown, C.R., A.M. Knott, and E.J. Damrose. 1992. Violet-green Swallow (*Tachycineta thalassina*). *In* A. Poole and F. Gill (editors). The birds of North America, No. 14. The Academy of Natural Sciences, Philadelphia; The American Ornithologists' Union, Washington, D.C.

Browning, M.R. 1974. Taxonomic remarks on recently described subspecies of birds that occur in the northwestern United States. Murrelet 55:32-38.

________. 1976. The status of *Sayornis saya yukonensis* Bishop. Auk 93:843-846.

________. 1979. A review of the geographic variation in continental populations of the Ruby-crowned Kinglet (*Regulus calendula*). Nemouria 21:109.

________. 1992. Geographic variation in *Hirundo pyrrhonota* (Cliff Swallow) from northern North America. Western Birds 23:21-29.

Bruce, J.A. 1961. First record of Eurasian Skylark on San Juan Island, Washington. Condor 63:418.

Bryant, A.A., J.-P.L. Savard, and R.T. McLaughlin. 1993. Avian communities in old-growth and managed forests of western Vancouver Island, British Columbia. Canadian Wildlife Service Technical Report Series No. 167, Delta, British Columbia. 115 pp.

Buckner, C.H., A.J. Erskine, R. Lidstone, B.B. McLeod, and M. Ward. 1975. The breeding bird community of coast forest stands on northern Vancouver Island. Murrelet 56:6-11.

Bull, J. 1974. Birds of New York state. Doubleday/Natural History Press, Garden City, New York. 655 pp.

Bump, S.R. 1986. Yellow-headed Blackbird nest defense: aggressive responses to Marsh Wrens. Condor 86:328-335.

Burcham, J.S. 1904. Notes on the habits of the Water Ousel (*Cinclus mexicanus*). Condor 4:50.

Burleigh, T.D. 1959. Two new subspecies of birds from western North America. Proceedings of the Biological Survey of Washington 72:15-18.

Burns, F.L. 1915. Comparative periods of deposition and incubation of some North American birds. Wilson Bulletin 27:275-286.

Burt, E.H. and R.M. Tuttle. 1983. Effect of timing of banding on reproductive success of Tree Swallows. Journal of Field Ornithology 54:319-323.

Buss, I.O. 1942. A managed Cliff Swallow colony in southern Wisconsin. Wilson Bulletin 54:152-161.

Butler, R.W. 1974. The feeding ecology of the Northwestern Crow on Mitlenatch Island, British Columbia. Canadian Field-Naturalist 88:313-316.

________. 1980. The breeding ecology and social organization of the Northwestern Crow on Mitlenatch Island, British Columbia. M.Sc. Thesis, Simon Fraser University, Burnaby, British Columbia. 100 pp.

________. 1981. The historical and present distribution of the Bushtit in British Columbia. Murrelet 62:87-90.

________. 1988. Population dynamics and migration routes of Tree Swallows, *Tachycineta bicolor,* in North America. Journal of Field Ornithology 59:395-402.

Butler, R.W. and S. Butler. 1976. Naturalist's summer report, Mitlenatch Island (May 19–August 30, 1976). British Columbia Parks Branch Unpublished Report, Victoria. 9 pp.

Butler, R.W. and R.W. Campbell. 1987. The birds of the Fraser River delta: populations, ecology, and international significance. Canadian Wildlife Service Occasional Paper No. 65, Ottawa, Ontario. 73 pp.

Butler, R.W. and J. Toochin. 1984. The Crested Myna in British Columbia. Unpublished Manuscript, Delta, British Columbia. 6 pp.

Butler, R.W., N.A.M. Verbeek, and H. Richardson. 1984. The breeding biology of the Northwestern Crow. Wilson Bulletin 96:408-418.

Butler, R.W., M. Lemon, and M.S. Rodway. 1985. Northwestern Crows in a Rhinoceros Auklet colony: predators and scavengers. Murrelet 66:86-90.

Butler, R.W., B.G. Stushnoff, and E. McMackin. 1986. The birds of the Creston valley and southeastern British Columbia. Canadian Wildlife Service Occasional Paper No. 58, Ottawa, Ontario. 37 pp.

Byrd, G.V., J.L. Trapp, and D.D. Gibson. 1978. New information on Asiatic birds in the Aleutian Islands. Condor 80:309-315.

Cade, T.J. 1967. Ecological and behavioral aspects of predation by a Northern Shrike. Living Bird 6:43-86.

Cade, T.J. and C.M. White. 1973. Breeding of Say's Phoebe in Arctic Alaska. Condor 75:360-361.

Cadman, M.D. 1990. Update status report on the Loggerhead Shrike (*Lanius ludovicianus*) in Ontario. Ontario Ministry of Natural Resources Unpublished Report, Ottawa, Ontario. 30 pp.

________. 1991. Status report on the Loggerhead Shrike (*Lanius ludovicianus*) – eastern population. Committee on the Status of Endangered Wildlife in Canada Report, Ottawa, Ontario. 17 pp.

Campbell, R.W. 1964. An annotated list of birds of Mitlenatch Island (June 5 – August 26, 1964). British Columbia Ministry of Recreation and Conservation (Parks Branch) Unpublished Report, Victoria. 18 pp. (Bibliography 1924).

________. 1964a. Birding on Galiano Island. Vancouver Natural History Society News 125:5-6.

________. 1965. Mitlenatch Island Nature Park, December 19-22, 1965. British Columbia Ministry of Recreation and Conservation (Parks Branch) Unpublished Report, Victoria. 8 pp. (Bibliography 2501).

________. 1965a. An afternoon of birding on Mitlenatch Island. Blue Jay 23:158-160.

________. 1966. Late sightings of Bank Swallows on the West Coast. Victoria Naturalist 23:4-5.

________. 1967. Wickaninnish Provincial Park summer naturalist report, 1967. British Columbia Ministry of Recreation and Conservation (Parks Branch) Unpublished Report, Victoria. 163 pp. (Bibliography 2430).

________. 1967b. Birding in a breeze. Vancouver Natural History Society Bulletin 133:4-5.

________. 1968a. Wickaninnish Provincial Park summer naturalist report, 1968. British Columbia Ministry of Recreation and Conservation (Parks Branch) Unpublished Report, Victoria. 104 pp. (Bibliography 2431).

________. 1968b. Spring migration 1968 – some arrival dates. Vancouver Natural History Society Bulletin 139:10.

________. 1968c. European Starling at Pachena lightstation. Victoria Naturalist 24:55.

________. 1969. Wickaninnish Provincial Park summer naturalist report, 1969. British Columbia Ministry of Recreation and Conservation (Parks Branch) Unpublished Report, Victoria. 85 pp.

________. 1969a. Spring bird observations on Langara Island, British Columbia. Blue Jay 27:155-159.

________. 1969b. Checklist of Vancouver birds, 1969 edition. University of British Columbia, Department of Zoology, Vancouver. Leaflet.

________. 1970a. Bird chatter. Vancouver Natural History Society News 145:7-8.

________. 1970b. Bird chatter. Vancouver Natural History Society News 146:7-9.

________. 1970c. Ornithology section. Vancouver Natural History Society Discovery 148:8-9.

________. 1971. Bird chatter. Vancouver Natural History Society Discovery 152:11-12.

________. 1973. Occurrence and status of the Mockingbird in British Columbia (1931-1971). University of British Columbia Department of Zoology Unpublished Manuscript, Vancouver. 26 pp.

________. 1974. Brown Thrasher on the coast of British Columbia. Canadian Field-Naturalist 88:225.

________. 1978a. Birds observed at the Coal and Liard Rivers and Portage Brule Rapids Ecological Reserve. British Columbia Provincial Museum Unpublished Report, Victoria. 3 pp. (Bibliography 2442).

________. 1978b. Census of waterbirds at Cecil Lake, northeastern British Columbia. British Columbia Provincial Museum Unpublished Report, Victoria. 10 pp. (Bibliography 2441).

________. 1978c. Breeding birds of One Island Lake, northeastern British Columbia. British Columbia Provincial Museum Unpublished Report, Victoria. 4 pp. (Bibliography 2439).

________. 1979a. Proposal for an ecological reserve at "McQueen's Slough," Dawson Creek area. British Columbia Provincial Museum Unpublished Report, Victoria. 6 pp. (Bibliography 2444).

________. 1979b. Birds observed in the vicinity of ecological reserve No. 47 (Parker Lake–Fort Nelson) 1974-1978. British Columbia Provincial Museum Unpublished Report, Victoria. 8 pp. (Bibliography 2445).

________. 1981. Spring migrants observed 16-21 May 1981 at Atlin, British Columbia. British Columbia Provincial Museum Unpublished Report, Victoria. 7 pp. (Bibliography 2447).

________. 1982a. Wildlife atlases progress report. B.C. Naturalist 20(1):8-9.

________. 1982b. Wildlife atlases progress report. B.C. Naturalist 20(2):8-10.

________. 1982c. Wildlife atlases progress report – spring 1982. B.C. Naturalist 20(3):6-8.

________. 1982d. Wildlife atlases progress report – summer 1982. B.C. Naturalist 20(4):5-7.

________. 1983a. Wildlife atlases progress report. B.C. Naturalist 21(1):4-5.

________. 1983b. Wildlife atlases progress report – winter 1982-83. B.C. Naturalist 21(2):4-5.

________. 1983c. Wildlife atlases progress report – spring/summer 1983. B.C. Naturalist 21(3):4-6.

________. 1983d. Wildlife atlases progress report – fall 1983. B.C. Naturalist 21(4):4-6.

________. 1984a. Wildlife atlases progress report – winter 1983. B.C. Naturalist 22(1):4-5.

________. 1984b. Wildlife atlases progress report – winter 1983-84. B.C. Naturalist 22(2):6-7.

________. 1984c. Wildlife atlases progress report – spring/summer 1984. B.C. Naturalist 22(3):6-7.

________. 1984d. Wildlife atlases progress report – fall 1984. B.C. Naturalist 22(4):6-7.

________. 1984e. Wings over the water. Pages 95-117 *in* Islands Protection Society (editors). Islands at the edge: preserving the Queen Charlotte Islands wilderness. Douglas & McIntyre, Toronto, Ontario.

________. 1984f. British Columbia wildlife road mortality census – southern British Columbia, July 8-24, 1984. British Columbia Provincial Museum Unpublished Report, Victoria. 16 pp. (Bibliography 2470).

________. 1985a. Wildlife atlases progress report – winter 1984-85. B.C. Naturalist 23(1):6-7.

________. 1985b. Wildlife atlases progress report – spring 1985. B.C. Naturalist 23(2):6-7.

________. 1985c. Wildlife atlases progress report – summer 1985. B.C. Naturalist 23(3):6-7.

________. 1985d. Wildlife atlases progress report – fall 1985. B.C. Naturalist 23(4):6-7, 9.

________. 1986a. Wildlife atlases progress report – winter 1985-86. B.C. Naturalist 24(1):6-7.

________. 1986b. Wildlife atlases progress report – spring 1986. B.C. Naturalist 24(2):6-7.

________. 1986c. Wildlife atlases progress report – summer 1986. B.C. Naturalist 24(3):6-7.

________. 1986d. Wildlife atlases progress report – autumn 1986. B.C. Naturalist 24(4):6-7.

________. 1986e. First Canadian record of the Scrub Jay, *Aphelocoma coerulescens*. Canadian Field-Naturalist 100:120-121.

________. 1986f. Birds and mammals observed during a cruise of Moresby Island, Queen Charlotte Islands, 11-20 May 1986. Pacific Synergies Limited Unpublished Report, Whistler, British Columbia. 10 pp.

________. 1986g. Birds and mammals observed during a cruise of Moresby Island, Queen Charlotte Islands, 9-19 June 1986. Pacific Synergies Limited Unpublished Report, Whistler, British Columbia. 16 pp.

________. 1987a. British Columbia wildlife – winter report 1986-87. B.C. Naturalist 25(1):6-7.

________. 1987b. British Columbia wildlife – spring report 1987. B.C. Naturalist 25(2):6-7.

________. 1987c. British Columbia wildlife – summer report 1987. B.C. Naturalist 25(3):6-7.

________. 1987d. British Columbia wildlife – autumn report 1987. B.C. Naturalist 25(4):6-7.

________. 1988a. British Columbia wildlife – winter report. B.C. Naturalist 26(1):6-7.

________. 1988b. British Columbia wildlife – spring report. B.C. Naturalist 26(2):6-7.

________. 1988c. British Columbia wildlife – summer report. B.C. Naturalist 26(3):6-7.

________. 1988d. The birds of the Queen Charlotte Islands. Pages 88-94 *in* R.J. Fox (editor). The wildlife of northern British Columbia: past, present, and future. Centennial Wildlife Society of British Columbia Symposium, 27-29 November 1987, Smithers, British Columbia.

________. 1988e. Birds and mammals observed during a cruise of the west coast of Moresby Island, Queen Charlotte Islands, 14-25 May 1988. Pacific Synergies Limited Unpublished Report, Whistler, British Columbia. 11 pp.

________. 1988f. Birds and mammals observed during a cruise of the north and west coasts of Graham Island, Queen Charlotte Islands, 5-25 June 1988. Pacific Synergies Limited Unpublished Report, Whistler, British Columbia. 11 pp.

________. 1989a. British Columbia wildlife – autumn report 1988. B.C. Naturalist 27(1):6-7.

________. 1989b. British Columbia wildlife – winter report 1988-89. B.C. Naturalist 27(2):6-7.

________. 1989c. British Columbia wildlife – spring report 1989. B.C. Naturalist 27(3):6-8.

________. 1989d. British Columbia wildlife – summer report 1989. B.C. Naturalist 27(4):11-12.

________. 1989e. Birds, man, and a Pacific estuary. Bioline 8:1-4.

________. 1990a. British Columbia wildlife – autumn report 1989. B.C. Naturalist 28(1):6-9.

________. 1990b. British Columbia wildlife – winter report 1989-90. B.C. Naturalist 28(2):10-12.

________. 1990c. British Columbia wildlife – spring report 1990. B.C. Naturalist 28(3):12.

________. 1990d. Birds and mammals observed during a cruise of Moresby Island, Queen Charlotte Islands, British Columbia, 22-30 May 1990. Pacific Synergies Limited Unpublished Report, Whistler, British Columbia. 14 pp.

________. 1991a. British Columbia wildlife – spring [autumn] report 1991. B.C. Naturalist 29(3):6-8.

________. 1991b. British Columbia wildlife – spring report 1991. B.C. Naturalist 29(5):8-9.

________. 1992a. British Columbia wildlife – summer report 1991. B.C. Naturalist 30(1):8-10.

Campbell, R.W. and W.J. Anderson. 1968. Mockingbird at Vancouver, British Columbia. Canadian Field-Naturalist 82:227.

Campbell, R.W. and N.K. Dawe. 1991. Field report of the joint Royal British Columbia Museum and Canadian Wildlife Service expedition across southern British Columbia, June 29 to July 13, 1991. Royal British Columbia Museum. Victoria. 287 pp.

________. and ________. 1992. Central British Columbia ornithology field trip (Victoria–Williams Lake–Bella Coola–Prince George–Victoria), 6-19 July 1992. British Columbia Ministry of Environment Unpublished Report, Victoria. 244 pp.

Campbell, R.W. and V. Gibbard. 1981. British Columbia nest records scheme 1979/80. B.C. Naturalist 19:6-10.

________. and ________. 1984. British Columbia nest records scheme twenty-eighth annual report, 1983. B.C. Naturalist 22(1):9-11.

________. and ________. 1986. British Columbia nest records scheme thirtieth annual report – 1985. B.C. Naturalist 24(1):14, 16-17.

Campbell, R.W. and M. Harris. 1990. British Columbia nest records scheme thirty-fourth annual report, 1989. B.C. Naturalist 28(1):18-20.

________. and ________. 1991. British Columbia nest records scheme thirty-fifth annual report – 1990. B.C. Naturalist 29(3):10-11.

Campbell, R.W. and H. Hosford. 1979. Attracting and feeding birds in British Columbia. British Columbia Provincial Museum Methods Manual No. 7, Victoria. 31 pp.

Campbell, R.W. and M.C.E. McNall. 1982. Field report of the Provincial Museum expedition in the vicinity of Kotcho Lake, northeastern British Columbia, June 11 to July 9, 1982. British Columbia Provincial Museum Unpublished Report, Victoria. 307 pp. (Bibliography 2495).

Campbell, R.W. and A.L. Meugens. 1971. The summer birds of Richter Pass, British Columbia. Syesis 4:93-123.

Campbell, R.W. and B.J. Petrar. 1986. Birds observed in the vicinity of Fort St. John, British Columbia, 15-17 July 1986. British Columbia Provincial Museum Unpublished Report, Victoria. 7 pp.

Campbell, R.W. and M.G. Shepard. 1971. Summary of spring and fall pelagic birding trips from Tofino, British Columbia. Vancouver Natural History Society Discovery 150:13-16.

Campbell, R.W. and A.C. Stewart. 1993. British Columbia ornithology field trip (Okanagan-Shuswap-Cariboo-Nechako), May 20-29, 1993. British Columbia Ministry of Environment, Lands and Parks Unpublished Report, Victoria. 256 pp.

Campbell, R.W. and D. Stirling. 1971. A photoduplicate file for British Columbia vertebrate records. Syesis 4:217-222.

Campbell, R.W. and K.R. Summers. In press. Vertebrates of the Brooks Peninsula, British Columbia. Royal British Columbia Museum, Victoria.

Campbell, R.W., M.G. Shepard, and R.H. Drent. 1972a. Status of birds in the Vancouver area in 1970. Syesis 5:180-220.

Campbell, R.W., M.G. Shepard, and W.C. Weber. 1972b. Vancouver birds in 1971. Vancouver Natural History Society Special Publication, Vancouver, British Columbia. 88 pp.

Campbell, R.W., M.G. Shepard, B.A. MacDonald, and W.C. Weber. 1974. Vancouver birds in 1972. Vancouver Natural History Society Special Publication, Vancouver, British Columbia. 96 pp.

Campbell, R.W., R.J. Cannings, S.G. Cannings, and R.A. Cannings. 1979a. A proposal for an ecological reserve at Rock Lake, Becher's Prairie, British Columbia. British Columbia Provincial Museum Unpublished Report, Victoria. 9 pp. (Bibliography 2474).

Campbell, R.W., H.R. Carter, C.D. Shepard, and C.J. Guiguet. 1979b. A bibliography of British Columbia ornithology – Volume 1. British Columbia Provincial Museum Heritage Record No. 7, Victoria. 185 pp.

Campbell, R.W., J.M. Cooper, and M.C.E. McNall. 1983. Field report of the Provincial Museum expedition in the vicinity of Haines Triangle, northwestern British Columbia, May 27 to July 4, 1983. British Columbia Provincial Museum Unpublished Report, Victoria. 351 pp. (Bibliography 2477).

Campbell, R.W., N.K. Dawe, and T.D. Hooper. 1988a. A bibliography of British Columbia ornithology – Volume 2. Royal British Columbia Museum Heritage Record No. 19, Victoria. 591 pp.

Campbell, R.W., N.K. Dawe, I. McT.-Cowan, J.M. Cooper, G.W. Kaiser, and M.C.E. McNall. 1990a. The birds of British Columbia – Volume 1: Nonpasserines (Introduction, loons through waterfowl). Royal British Columbia Museum and Canadian Wildlife Service, Victoria. 535 pp.

________, ________, ________, ________, ________, and ________. 1990b. The birds of British Columbia – Volume 2: Non-passerines (Diurnal birds of prey through woodpeckers). Royal British Columbia Museum and Canadian Wildlife Service, Victoria. 632 pp.

Cannings, R.A., R.J. Cannings, and S.G. Cannings. 1987. Birds of the Okanagan Valley, British Columbia. Royal British Columbia Museum, Victoria. 420 pp.

Cannings, R.J. 1973. Shuswap birds 1973 – annotated list. British Columbia Parks Branch Unpublished Report, Victoria. 15 pp. (Bibliography 2528).

________. 1977. Breeding biology of the Horned Lark (*Eremophila alpestris alpestris* [L.]) at Cape St. Mary's, Newfoundland. M.Sc. Thesis, Memorial University, St. John's, Newfoundland.

________. 1981. Notes on the nesting of Horned Larks on the Chilcotin Plateau of British Columbia. Murrelet 62:21-23.

________. 1987. Gray Flycatcher: a new breeding bird for Canada. American Birds 41:376-378.

________. 1987a. The eighty-seventh Christmas Bird Count – British Columbia (Penticton). American Birds 41:649.

________. 1991. Status report on the Gray Flycatcher (*Empidonax wrightii*) in Canada. Committee on the Status of Endangered Wildlife in Canada Unpublished Report, Ottawa, Ontario. 11 pp.

________. 1992. Status report on the Canyon Wren (*Catherpes mexicanus*) in Canada. Committee on the Status of Endangered Wildlife in Canada Unpublished Report, Ottawa, Ontario. 9 pp.

________. 1992a. Status report on the Sage Thrasher *Oreoscoptes montanus* in Canada. Canadian Wildlife Service Unpublished Report, Ottawa, Ontario. 25 pp.

________. 1994. The ninety-fourth Christmas Bird Count – British Columbia (Victoria). National Audubon Society Field Notes 48:435.

________. 1995. Status of the Gray Flycatcher in British Columbia. British Columbia Ministry of Environment Wildlife Bulletin No. B-76, Victoria. 9 pp.

Cannings, S.G. 1973. Mount Robson vertebrate report, 1973. British Columbia Parks Branch Unpublished Report, Victoria. 18 pp. (Bibliography 2558).

________. 1974. South Okanagan bird report – summer 1974. British Columbia Parks Branch Unpublished Report, Victoria. 27 pp. (Bibliography 2559).

Cannings, S.R. 1972. Brown Thrasher in British Columbia. Canadian Field-Naturalist 86:295.

Carder, J.E.W. 1985. How fares an exotic: skylarks on Vancouver Island. University of Victoria Department of Geography Unpublished Report, Victoria. 37 pp.

Carl, G.C. 1942. British Columbia field notes – Victoria District, Burnaby, Hatzic, Cultus Lake, and Cariboo (Lac la Hache). British Columbia Provincial Museum Unpublished Report, Victoria. 69 pp.

Carl, G.C., C.J. Guiguet, and G.A. Hardy. 1952. A natural history survey of the Manning Park area, British Columbia. British Columbia Provincial Museum Occasional Paper No. 9, Victoria. 130 pp.

Carriger, H.W. 1899. Breeding of the Dusky Horned Lark in eastern Washington. Bulletin of the Cooper Ornithological Club 1:86.

Carswell, R. 1975. Wildlife inventory – Gladys Lake Ecological Reserve. British Columbia Parks Branch Unpublished Report, Victoria. 99 pp.

Castrale, J.S. 1982. Effects of two sagebrush control methods on nongame birds. Journal of Wildlife Management 46:945-952.

Chamberlain-Auger, J.A., P.J. Auger, and E.G. Strauss. 1990. Breeding biology of American Crows. Wilson Bulletin 102:615-622.

Chambers, L.E. 1969. Skylarks near Duncan. Victoria Naturalist 26(4):40.

Chapman, F.M. 1890. Note on *Cyanocitta stelleri* Maynard. Auk 7:91.

Clark, L. 1979. Birding at Cowichan. Victoria Naturalist 36(2):21.

Clay, J.O. 1946. Japanese Starlings. Victoria Naturalist 33:5.

________. 1948. North American waterfowl inventory. Victoria Naturalist 55:4.

________. 1948a. Bush-tits on Vancouver Island. Victoria Naturalist 5:27.

Cleveland, W.S. 1979. Robust locally weighted regression and smoothing scatterplots. Journal of the American Statistics Association 74:829-836.

Collins, B.T. 1990. Using rerandomization tests in route-regression analysis of avian population trends. Pages 63-70 *in* J.R. Sauer and S. Droege (editors). Survey designs and statistical methods for the estimation of avian population trends. United States Department of the Interior Fish and Wildlife Service Biological Report No. 90, Washington, D.C.

Collins, C.T. 1974. Banding worksheet for western birds – Tropical Kingbird (*Tyrannus melancholicus*). Western Bird Banding Association, Cave Creek, Arizona.

Comer, J. 1972. Seventy-second Christmas Bird Count – Duncan, B.C. American Birds 26:174.

________. 1980. The eightieth Audubon Christmas Bird Count – Duncan, B.C. American Birds 34:378.

________. 1981. Birds of the Cowichan Valley, British Columbia. Unpublished manuscript, Duncan, British Columbia. 27 pp.

Conner, R.N. and C.S. Adkisson. 1976. Concentration of foraging Common Ravens along the Trans-Canada Highway. Canadian Field-Naturalist 90:496-497.

Conrad, K.F. and R.J. Robertson. 1993. Patterns of parental provisioning by Eastern Phoebes. Condor 95:57-62.

Cook, F.S. 1947. Notes on some fall and winter birds of the Queen Charlotte Islands, British Columbia. Canadian Field-Naturalist 61:131-133.

Cooke, M.T. 1937. Some returns of banded birds. Bird-Banding 8:144-155.

Cooke, M.T. and P. Knappen. 1941. Some birds naturalized in North America. Transactions of the North American Wildlife Conference 5:176-183.

Cooper, J.M. 1982. A cave-dwelling robin. Discovery 11:3. (Issued by the Vancouver Natural History Society).

________. 1985. Notes on the vertebrates of McClinton Bay and Masset, Queen Charlotte Islands, 15 May – 14 June 1985. Royal British Columbia Museum Unpublished Report, Victoria. 124 pp.

________. 1993. Breeding bird surveys in Airport Reserve on Sea Island, Richmond, British Columbia. LGL Limited Unpublished Report, Sidney, British Columbia. 34 pp.

Cooper, J.M. and M. Adams. 1979. The birds and mammals of Kwadacha Wilderness Park (August 13 – September 8, 1979). Royal British Columbia Museum Unpublished Report, Victoria. 10 pp. (Bibliography 2687).

Cooper, J.M. and D.L.P. Cooper. 1983. A second report on the summer vertebrates of the Fern Lake area of Kwadacha Wilderness Park. British Columbia Provincial Museum Unpublished Report, Victoria. 7 pp. (Bibliography 2688).

Cooper, J.M., K.A. Enns, and M.G. Shepard. 1995. Status of the Philadelphia Vireo (*Vireo philadelphicus*) in British Columbia. British Columbia Ministry of Environment, Lands and Parks Wildlife Branch Unpublished Report, Victoria. 24 pp.

Cowan, I. McT. 1939. The vertebrate fauna of the Peace River district of British Columbia. British Columbia Provincial Museum Occasional Paper No. 1, Victoria. 102 pp.

________. 1939a. Black Phoebe in British Columbia. Condor 41:123.

________. 1940. Bird records from British Columbia. Murrelet 21:69-70.

Cowan, I. McT.- and G. McT.-Cowan. 1961. The Amur Barn Swallow off British Columbia. Condor 63:419.

Cowan, I. McT.- and J.A. Munro. 1946. Birds and mammals of Mount Revelstoke National Park. Canadian Alpine Journal 29:100-119; 29:237-256.

Cramp, S. (editor). 1988. Handbook of the birds of Europe, the Middle East and North Africa: The birds of the Western Palearctic. Volume V – Tyrant flycatchers to thrushes. Oxford University Press, England.

Creaser, C.W. 1925. The egg destroying activity of the House Wren in relation to territorial control. Bird-Lore 27:163-167.

Cresco, C.P. 1960. A sight record of a nesting Canada Jay. Murrelet 41:44-45.

Crowe, R.B. 1963. Recent temperature and precipitation fluctuations along the British Columbia coast. Journal of Applied Meteorology 2:114-118.

Crowell, J.B. and H.B. Nehls. 1966a. The fall migration – northern Pacific coast region. Audubon Field Notes 20:449-453.

________. and ________. 1966b. The spring migration – northern Pacific coast region. Audubon Field Notes 20:539-542.

________. and ________. 1966c. The nesting season – northern Pacific coast region. Audubon Field Notes 20:591-595.

________. and ________. 1967a. The fall migration – northern Pacific coast region. Audubon Field Notes 21:67-72.

________. and ________. 1967b. The winter season – northern Pacific coast region. Audubon Field Notes 21:448-452.

________. and ________. 1967c. The spring migration – northern Pacific coast region. Audubon Field Notes 21:532-535.

________. and ________. 1967d. The nesting season – northern Pacific coast region. Audubon Field Notes 21:596-600.

________. and ________. 1968a. The fall migration – northern Pacific coast region. Audubon Field Notes 22:78-83.

________. and ________. 1968b. The winter season – northern Pacific coast region. Audubon Field Notes 22:468-472.

________. and ________. 1968c. The spring migration – northern Pacific coast region. Audubon Field Notes 22:567-571.

________. and ________. 1968d. The nesting season – northern Pacific coast region. Audubon Field Notes 22:638-642.

________. and ________. 1969a. The fall migration – northern Pacific coast region. Audubon Field Notes 23:94-99.

________. and ________. 1969b. The winter season – northern Pacific coast region. Audubon Field Notes 23:508-513.

________. and ________. 1969c. The spring migration – northern Pacific coast region. Audubon Field Notes 23:615-619.

________. and ________. 1969d. The nesting season – northern Pacific coast region. Audubon Field Notes 23:684-688.

________. and ________. 1970a. The fall migration – northern Pacific coast region. Audubon Field Notes 24:82-88.

________. and ________. 1970b. The winter season – northern Pacific coast region. Audubon Field Notes 24:530-533.

________. and ________. 1970c. The spring migration – northern Pacific coast region. Audubon Field Notes 24:635-638.

________. and ________. 1970d. The nesting season – northern Pacific coast region. Audubon Field Notes 24:708-711.

________. and ________. 1971a. The fall migration – northern Pacific coast region. American Birds 25:94-100.

________. and ________. 1971b. The winter season – northern Pacific coast region. American Birds 25:614-619.

________. and ________. 1971c. The spring migration – northern Pacific coast region. American Birds 25:614-619.

________. and ________. 1971d. The nesting season – northern Pacific coast region. American Birds 25:895-899.

________. and ________. 1972a. The fall migration – northern Pacific coast region. American Birds 26:107-111.

________. and ________. 1972b. The winter season – northern Pacific coast region. American Birds 26:644-648.

________. and ________. 1972c. The spring migration – northern Pacific coast region. American Birds 26:797-801.

________. and ________. 1972d. The nesting season – northern Pacific coast region. American Birds 26:893-897.

________. and ________. 1973a. The fall migration – northern Pacific coast region. American Birds 27:105-110.

________. and ________. 1973b. The winter season – northern Pacific coast region. American Birds 27:652-656.

________. and ________. 1973c. The spring migration – northern Pacific coast region. American Birds 27:809-812.

________. and ________. 1973d. The nesting season – northern Pacific coast region. American Birds 27:908-911.

________. and ________. 1974a. The fall migration – northern Pacific coast region. American Birds 28:93-98.

________. and ________. 1974b. The winter season – northern Pacific coast region. American Birds 28:679-684.

________. and ________. 1974c. The spring migration – northern Pacific coast region. American Birds 28:840-845.

________. and ________. 1974d. The nesting season – northern Pacific coast region. American Birds 28:938-943.

________. and ________. 1975a. The fall migration – northern Pacific coast region. American Birds 29:105-112.

________. and ________. 1975b. The winter season – northern Pacific coast region. American Birds 29:730-735.

________. and ________. 1975c. The spring migration – northern Pacific coast region. American Birds 29:897-902.

________. and ________. 1976a. The fall migration – northern Pacific coast region. American Birds 30:112-117.

________. and ________. 1976b. The winter season – northern Pacific coast region. American Birds 30:755-760.

________. and ________. 1976c. The spring migration – northern Pacific coast region. American Birds 30:878-882.

________. and ________. 1976d. The nesting season – northern Pacific coast region. American Birds 30:992-996.

________. and ________. 1977a. The fall migration – northern Pacific coast region. American Birds 31:212-216.

________. and ________. 1977b. The winter season – northern Pacific coast region. American Birds 31:364-367.

________. and ________. 1977c. The spring migration – northern Pacific coast region. American Birds 31:1037-1041.

Cumming, R.A. 1925. Observations of the Chinese Starling (*Aethiopsar cristatellus*). Canadian Field-Naturalist 39:187-190.

________. 1932. Birds of the Vancouver district, British Columbia. Murrelet 13:1-15.

Daly, M. 1982. First sight record of Alder Flycatcher in Vancouver area. Vancouver Natural History Society Discovery 11(3):107-108.

Darcus, S.J. 1930. Notes on birds of the northern part of the Queen Charlotte Islands. Canadian Field-Naturalist 44:45-49.

________. 1932. The present status of the Sage Thrasher in British Columbia. Murrelet 13:22.

Dare, P.J. 1986. Raven *Corvus corvax* populations in two upland regions of north Wales. Bird Study 33:179-189.

Davidson, A.R. 1953. The birds of Victoria. Victoria Naturalist 10:59-60.

________. 1958. Skylarks. Victoria Naturalist 14(9):113.

________. 1961. The swallow's return. Victoria Naturalist 17(8):117.

________. 1966. Annotated list of birds of southern Vancouver Island. Victoria Natural History Society Mimeographed Report, Victoria. 23 pp.

________. 1967. The skylarks. Victoria Naturalist 23(8):91.

________. 1976. Annotated list of birds of southern Vancouver Island. Victoria Natural History Society Unpublished Report, Victoria. 23 pp. (Bibliography 306).

________. 1978. A rare bird for Victoria. Victoria Naturalist 34:78.

Davidson, G. 1983. The eighty-third Audubon Christmas Bird Count – Nakusp, B.C. American Birds 37:442.

Davidson, G. and C. Siddle. 1991. White-throated Swifts and Canyon Wrens outside the Okanagan valley. B.C. Naturalist 29(1):28.

Davis, C.M. 1980. A nesting study of the Brown Creeper. Living Bird 17:237-264.

Davis, D.E. 1941. The belligerency of the kingbird. Wilson Bulletin 53:157-168.

________. 1954. The breeding biology of Hammond's Flycatcher. Auk 71:164-171.

Davis, J., G.F. Fisler, and B.S. Davis. 1963. The breeding biology of the Western Flycatcher. Condor 65:337-382.

Davis, J.N. 1995. Hutton's Vireo (*Vireo huttoni*). *In* A. Poole and F. Gill (editors). The birds of North America, No. 189. The Academy of Natural Sciences, Philadelphia; The American Ornithologists' Union, Washington, D.C.

Davis, P. 1965. Recoveries of swallows ringed in Britain and Ireland. Bird Study 12:151-169.

Dawe, N.K. 1971. Nature interpretation report on Wasa, Moyie and Jimsmith provincial parks, British Columbia. British Columbia Parks Branch Unpublished Report, Victoria. 97 pp. (Bibliography 1172).

________. 1974. The seventy-fourth Christmas Bird Count – Ladner, B.C. American Birds 28:190.

________. 1975. The seventy-fifth Audubon Christmas Bird Count – Ladner, B.C. American Birds 29:206.

________. 1976. Flora and fauna of the Marshall-Stevenson unit, Qualicum National Wildlife Area, August, 1976. Canadian Wildlife Service Unpublished Report, Delta, British Columbia. 201 pp. (Bibliography 1862).

________. 1980. Flora and fauna of the Marshall-Stevenson unit, Qualicum National Wildlife Area (update to June 1979). Canadian Wildlife Service Unpublished Report, Delta, British Columbia. 149 pp. (Bibliography 2788).

Delius, J.D. 1963. Das Verhalten der Feldlerche. Zeitschrift Fur Tierpsychologie 20:297-348.

________. 1965. A population study of skylarks, *Alauda arvensis*. Ibis 107:466-492.

DellaSalla, D.A. and D.L. Rabe. 1987. Response of Least Fly-catchers *Empidonax minimus* to forest disturbances. Biological Conservation 41:291-299.

DeLong, C., R.M. Annas, and A.C. Stewart. 1991. Pages 237-250 *in* D. Meidinger and J. Pojar (editors). Ecosystems of British Columbia. British Columbia Ministry of Forests Special Report Series 8, Victoria.

Demarchi, D.A. 1993. Ecoregions of British Columbia, 3rd edition. British Columbia Ministry of Environment, Lands and Parks, Victoria. Map (1:2,000,000).

________. 1994. Ecoprovinces of the central North American cordillera and adjacent plains – appendix A. Pages 153-169 *in* L.F. Ruggiero, K.B. Aubry, S.W. Buskirk, L.J. Lyon, and W.J. Zeilinski. The scientific basis for conserving forest carnivores: American marten, fisher, lynx and wolverine, in the western United States. United States Department of Agriculture, Forest Service, Rocky Mountain and Range Experiment Station General Technical Report RM-254, Fort Collins, Colorado.

________. 1995. Ecoregions of British Columbia, 4th edition. British Columbia Ministry of Environment, Lands and Parks, Victoria. Map (1:2,000,000).

Demarchi, D.A., R.D. Marsh, A.P. Harcombe, and E.C. Lea. 1990. The environment. Pages 54-142 *in* R.W. Campbell, N.K. Dawe, I. McT.-Cowan, J.M. Cooper, G.W. Kaiser, and M.C.E. McNall. The birds of British Columbia – nonpasserines (loons through waterfowl). Royal British Columbia Museum, Victoria.

Dement'ev, G.P. and N.A. Gladkov. 1954. The birds of the Soviet Union, Volume 5. Translated 1970 by Israel Program for Scientific Translations, Jerusalem. Clearinghouse for Federal Scientific and Technical Information, Springfield, Virginia.

Derrickson, K.C. and R. Breitwisch. 1992. Northern Mockingbird. *In* A. Poole, P. Stettenheim, and F. Gill (editors). The birds of North America, No. 7. The Academy of Natural Sciences, Philadelphia; The American Ornithologists' Union, Washington, D.C.

DeSante, D.F., N.K. Johnson, R. LeValley, and R.P. Henderson. 1985. Occurrence and identification of the Yellow-bellied Flycatcher on southeast Farallon Island, California. Western Birds 16:153-160.

De Schauensee, R.M. 1984. The birds of China. Smithsonian Institution Press, Washington, D.C.

de Vos, A., A.T. Cringan, J.H. Reynolds, and H.G. Lumsden. 1959. Biological investigations of traplines in northern Ontario, 1951-56. Ontario Department of Lands, Forests and Wildlife Service Publication No. 8, Ottawa, Ontario.

Dickinson, J.C. 1953. Report on the McCabe collection of British Columbia birds. Bulletin of the Museum of Comparative Zoology 109(2):123-211.

Dilger, W.C. 1956. Adaptive modifications and ecological isolating mechanisms in the thrush genera *Catharus* and *Hylocichla*. Wilson Bulletin 68:171-199.

Dixon, K.L. and J.D. Gilbert. 1964. Altitudinal migration in the Mountain Chickadee. Condor 66:61-64.

Dobkin, D.S. 1992. Neotropical migrant landbirds in the northern Rockies and Great Plains. United States Department of Agriculture Forest Service Northern Region Publication No. R1-93-34, Missoula, Montana. 207 pp.

Dodge, H.J. 1965. The association of a bird-roosting site with infection of schoolchildren by *Histoplasma capsulatum*. American Journal of Public Health 55:1203-1211.

Dolbeer, R.A. and R.A. Stehn. 1979. Population trends of blackbirds and starlings in North America. United States Department of the Interior Fish and Wildlife Service Special Scientific Report No. 214, Washington, D.C. 99 pp.

Dow, D.D. 1962. Report on birds observed in Wells Gray Park between May 9 and June 6, 1962. British Columbia Parks Branch Unpublished Report, Victoria. 36 pp. (Bibliography 2826).

________. 1963. A natural history of the southern portion of Garibaldi Provincial Park. British Columbia Parks Branch Unpublished Report, Victoria. 95 pp. (Bibliography 2827).

Drent, R.H. and C.J. Guiguet. 1961. A catalogue of British Columbia seabird colonies. British Columbia Provincial Museum Occasional Papers No. 12, Victoria. 173 pp.

Droege, S. and J. Sauer. 1987. Breeding bird survey – administrative report to cooperators – 1987. Report to United States Department of the Interior Fish and Wildlife Service, Laurel, Maryland. 6 pp.

________ and ________. 1988. North American breeding bird survey annual summary 1988. Report to United States Department of the Interior Fish and Wildlife Service, Laurel Maryland. 12 pp.

Dubois, A.D. 1935. Nests of Horned Larks and longspurs on a Montana prairie. Condor 37:56-72.

Dunn, J. 1976. The identification of pipits. Western Tanager 42:5.

________. 1979. Field Notes – The *Tyrannus* flycatchers. Western Tanager 46(5):5.

________. 1981. The identification of female bluebirds. Birding 13:4-11.

Dunn, J.L. and K.L. Garrett. 1987. The identification of North American gnatcatchers. Birding 19:17-29.

Dunn, P.O. and S.J. Hannon. 1989. Evidence for obligate male parental care in Black-billed Magpies. Auk 106:635-644.

Dwight, J. 1890. The Horned Larks of North America. Auk 7:138-158.

Edson, J.M. 1943. A study of the Violet-green Swallow. Auk 60:396-403.

Edwards, R.Y. 1956. The influence of poisoning for predators on the wildlife of parks. British Columbia Parks Branch Unpublished Report, Victoria. 4 pp.

________. 1963. Twenty-seventh breeding bird census – disturbed Douglas-fir coast forest. Audubon Field Notes 17:499-500.

Edwards, R.Y. and R.W. Ritcey. 1967. The birds of Wells Gray Park, British Columbia. British Columbia Parks Branch, Victoria. 37 pp. (Bibliography 1025).

Egger, M. 1979. Varied Thrushes feeding on Talitrid amphipods. Auk 96:805-806.

Ehrlich, P.R., D.S. Dobkin, and D. Wheye. 1988. The birder's handbook: a field guide to the natural history of North American birds. Simon and Schuster, New York. 785 pp.

________. 1976. The seventy-sixth Christmas Bird Count – White Rock, B.C. American Birds 30:216.

Scriven, D.H. 1989. Bluebird trails in the upper midwest: a guide to successful trail management. Audubon Chapter of Minneapolis, Minnesota. 178 pp.

Schultz, Z.M. 1958a. The fall migration – north Pacific coast region. Audubon Field Notes 12:52-54.

________. 1958b. The winter season – north Pacific coast region. Audubon Field Notes 12:302.

________. 1958c. The spring migration – north Pacific coast region. Audubon Field Notes 12:377-379.

________. 1958d. The nesting season – north Pacific coast region. Audubon Field Notes 12:435.

Scott, D.M. 1977. Cowbird parasitism on the Gray Catbird at London, Ontario. Auk 94:18-27.

________. 1993. On egg-laying times of American Robins. Auk 110:156.

Scott, D.M., J.A. Varley, and A.V. Newsome. 1988. Length of the laying season and clutch size of the Gray Catbird at London, Ontario. Journal of Field Ornithology 59:355-360.

Scott, G. 1973. Avifauna of the Vermilion Pass burn. B.Sc. Thesis, University of Calgary, Alberta. 60 pp.

Scott, V.E. and J.L. Oldemeyer. 1983. Cavity-nesting requirements and response to snag cutting in ponderosa pine. Pages 19-23 *in* Snag management: proceedings of the symposium. United States Department of Agriculture General Technical Report RM-99, Fort Collins, Colorado.

Sealy, S.G. 1974. Ecological segregation of Swainson's and Hermit thrushes on Langara Island, British Columbia. Condor 76:350-351.

Sealy, S.G. and G.C. Biermann. 1983. Timing of breeding and migrations in a population of Least Flycatchers in Manitoba. Journal of Field Ornithology 54:113-124.

Searing, G.F., K.H. Morgan, and C. Engelstoff. 1990a. Starling roost control and dispersal in Vancouver – 1990 implementation. LGL Limited Report, Sidney, British Columbia. 20 pp.

Searing, G.F., D.E. Nowell, S.R. Johnson, and A.D. Sekerak. 1990b. Starling roost control and dispersal in Vancouver, British Columbia. LGL Limited Report, Sidney, British Columbia. 36 pp.

Sedgwick, J.A. 1975. A comparative study of the breeding biology of Hammond's (*Empidonax hammondii*) and Dusky (*Empidonax oberholseri*) Flycatchers. M.S. Thesis, University of Montana, Missoula.

________. 1993a. Reproductive ecology of Dusky Flycatchers in western Montana. Wilson Bulletin 105:84-92.

________. 1993b. Dusky Flycatcher (*Empidonax oberholseri*). *In* A. Poole and F. Gill (editors). The birds of North America, No. 78. The Academy of Natural Sciences, Philadelphia; The American Ornithologists' Union, Washington, D.C.

________. 1994. Hammond's Flycatcher (*Empidonax hammondii*). *In* A. Poole and F. Gill (editors). The birds of North America, No. 109. The Academy of Natural Sciences, Philadelphia; The American Ornithologists' Union, Washington, D.C.

Sedgwick, J.A. and F.L. Knopf. 1988. A high incidence of Brown-headed Cowbird parasitism of Willow Flycatchers. Condor 90:253-256.

________. and ________. 1989. Regionwide polygyny in the Willow Flycatcher. Condor 91:473-475.

Seel, K.E. 1965. The birds of Kootenay National Park, first report, 1965, of field studies. Parks Canada Unpublished Report, Radium Hot Springs, British Columbia. 41 pp. (Bibliography 4123).

Semenchuk, G.P. (editor). 1992. The atlas of breeding birds of Alberta. Federation of Alberta Naturalists, Edmonton. 391 pp.

Sergeant, J. 1974. The seventy-fourth Christmas Bird Count – Chilliwack, B.C. American Birds 29:189.

Shantz, B.R. 1986. A history of the Ellis Bird Farm. Sialia 8:143-146.

Sharp, B. 1985. Guidelines for the management of the Purple Martin Pacific coast population. Sialia 8:9-13.

Shepard, M.G. 1972. The distribution of Cliff Swallow nesting colonies in the Vancouver area. University of British Columbia Department of Zoology Unpublished Report, Vancouver. 11 pp.

________. 1975a. British Columbia birds – spring and summer, 1974. Vancouver Natural History Society Discovery, New Series 3(2):32-38.

________. 1975b. British Columbia birds – spring 1975. Vancouver Natural History Society Discovery, New Series 4(2):41-44.

________. 1976a. British Columbia birds – July to September, 1975. Vancouver Natural History Society Discovery, New Series 4(3):67-69.

________. 1976b. British Columbia birds – October to December, 1975. Vancouver Natural History Society Discovery, New Series 5(1):10-13.

________. 1976c. British Columbia birds – January to March, 1976. Vancouver Natural History Society Discovery, New Series 5(2):30-32.

________. 1976d. British Columbia birds – April to June 1976. Vancouver Natural History Society Discovery, New Series 5(3):48-50.

________. 1976e. British Columbia birds – July to September 1976. Vancouver Natural History Society Discovery, New Series 5(4):65-67.

________. 1977. British Columbia birds – October to December 1976. Vancouver Natural History Society, Discovery New Series 6(1):18-20.

________. 1981. Victoria, British Columbia. Christmas Bird Count. American Birds 35:413.

________. 1995a. Report on a pilot project for migration monitoring at Rocky Point, British Columbia – 1994. Long Point Bird Observatory Report, Port Rowan, Ontario. 35 pp.

________. 1995b. Fall summary report (1995) for the Rocky Point migration monitoring station. Long Point Bird Observatory Report, Port Rowan, Ontario. 36 pp.

Sherman, A.R. 1925. Down with the House Wren boxes. Wilson Bulletin 37:5-13.

Sherry, T.W. 1979. Competitive interactions and adaptive strategies of American Redstarts and Least Flycatchers in a northern hardwood forest. Auk 96:265-283.

Shields, W.M. 1984. Factors affecting nest and site fidelity in Adirondack Barn Swallows (*Hirundo rustica*). Auk 101:102-110.

________. 1990. Information centers and coloniality in Cliff Swallows: statistical design and analysis – a comment. Ecology 71:401-405.

Shields, W.M. and J.R. Crook. 1987. Barn Swallow coloniality: a net cost for group breeding in the Adirondacks. Ecology 68:1373-1386.

Sibley, C.G. and J.E. Ahlquist. 1981. The relationships of the wagtails and pipits (Motacillidae) as indicated by DNA-DNA hybridization. Oiseau 51:189-199.

________. and ________. 1982a. The relationships of the Yellow-breasted Chat (*Icteria virens*), and the alleged "slow-down" in the rate of macromolecular evolution in birds. Postilla 187.

________. and ________. 1982b. The relationships of the swallows (Hirundinidae). Journal of the Yamashina Institute of Ornithology 14:122-130.

________. and ________. 1982c. The relationships of the vireos (Vireoninae) as indicated by DNA-DNA hybridization. Wilson Bulletin 94:114-128.

________. and ________. 1983. The phylogeny and classification of birds based on data of DNA-DNA hybridization. Pages 245-292 *in* R.F. Johnson (editor). Current Ornithology I, Plenum Press, New York.

________. and ________. 1984. The relationships of the starlings (Sturnidae: Sturnini) and the mockingbirds (Sturnidae: Mimini). Auk 101:230-243.

________. and ________. 1985. The phylogeny and classification of the passerine birds based on comparisons of the genetic material, DNA. Pages 83-121 *in* Proceedings of the 18th International Ornithological Congress, Moscow, 1982.

________. and ________. 1986. Reconstructing bird phylogeny by comparing DNA's. Scientific American 254:82-92.

________. and ________. 1987. Avian phylogeny reconstructed from comparisons of genetic material, DNA. Pages 95-121 *in* C. Patterson (editor). Molecules and morphology in evolution: conflict or compromise. Cambridge University Press, London, England.

________. and ________. 1990. Phylogeny and classification of the birds of the world. Yale University Press, New Haven, Connecticut.

Sibley, C.G. and B.L. Monroe. 1990. Distribution and taxonomy of birds of the world. Yale University Press, New Haven, Connecticut. 111 pp.

Siddle, C. 1982. The status of birds in the Peace River area of British Columbia. Unpublished report to the Royal British Columbia Museum, Victoria. 319 pp.

________. 1988a. The winter season – northwestern Canada region. American Birds 42:292-293.

________. 1988b. The autumn migration – northwestern Canada region. American Birds 42:293-295.

________. 1988c. The spring season – northwestern Canada region. American Birds 42:462-463.

________. 1988d. The nesting season – northwestern Canada region. American Birds 42:1316-1317.

________. 1989a. The autumn migration – northwestern Canada region. American Birds 43:134-136.

________. 1989b. The winter season – northwestern Canada region. American Birds 43:340-341.

________. 1990a. The winter season – British Columbia and the Yukon region. American Birds 44:312-317.

________. 1990b. The spring season – British Columbia and the Yukon region. American Birds 44:482-486.
________. 1990c. The nesting season – British Columbia/Yukon region. American Birds 44:1173-1176.
________. 1991a. The autumn migration – British Columbia/Yukon region. American Birds 45:142-145.
________. 1991b. The winter season – British Columbia/Yukon region. American Birds 45:306-308.
________. 1991c. The spring migration – British Columbia/Yukon region. American Birds 45:486-489.
________. 1992a. The autumn migration – British Columbia/Yukon region. American Birds 46:139-142.
________. 1992b. The winter season – British Columbia/Yukon region. American Birds 46:303-306.
________. 1992c. The summer season – British Columbia/Yukon region. American Birds 46:1167-1170.
________. 1993a. The autumn migration – British Columbia region. American Bird 47:136-139.
________. 1993b. The winter season – British Columbia/Yukon. American Birds 47:290-293.
________. 1993c. The spring season – British Columbia/Yukon. American Birds 47:445-447.
________. 1993d. The summer season – British Columbia/Yukon region. American Birds 47:1141-1143.
________. 1994a. The autumn migration – British Columbia/Yukon region. American Birds 48:142-144.
________. 1994b. Spring, 1994 – British Columbia/Yukon region. American Birds 48:142-144.
________. 1994c. Summer, 1994 – British Columbia/Yukon region. National Audubon Society Field Notes 48:240-242.
Siddle, C. and J. Bowling. 1993. Fall, 1993 – British Columbia/Yukon region. American Birds 47:445-447.
Siddle, C., D.F. Fraser, and L.R. Ramsay. 1991. Preliminary species management plan for Philadelphia Vireo in British Columbia. British Columbia Ministry of Environment, Lands and Parks Wildlife Branch, Victoria. 6 pp.
Siddle, C.R., E.L. Walters, and D. Copley. 1991. A status report on the Purple Martin (*Progne subis*) in British Columbia – 1990. British Columbia Ministry of Environment, Lands and Parks Wildlife Branch Unpublished Report, Victoria. 85 pp.
Siderius, J.A. 1994. The relationships between nest defence, nest viability, habitat, and nest success in the Eastern Kingbird (*Tyrannus tyrannus*). Ph.D. Thesis, Simon Fraser University, Burnaby, British Columbia. 150 pp.
Sinclair, P.H. and J.E. Elliott. 1993. A survey of birds and pesticide use in orchards in the south Okanagan/Similkameen region of British Columbia, 1991. Canadian Wildlife Service Technical Report No. 185, Delta, British Columbia.
Sirk, G. 1970. Bird report for the Shuswap Lake region – 1970. British Columbia Parks Branch Unpublished Report, Victoria. 7 pp. (Bibliography 4164).
Sirk, G. and L. Sirk. 1971. Mitlenatch Island Nature Park – annual report, 1971. British Columbia Parks Branch Unpublished Report, Victoria. 12 pp. (Bibliography 4168).
Sirk, G., R.J. Cannings, and M.G. Shepard. 1973. Shuswap Lake Park annual report – 1973. British Columbia Parks Branch Unpublished Report, Victoria. 66 pp. (Bibliography 4169).
Skinner, M.P. 1916. The nutcrackers of Yellowstone Park. Condor 18:62-64.
Slipp, J.W. 1942. Vagrant occurrences of *Tyrannus melancholicus* in North America. Auk 59:310-311.
Small, A. 1974. The birds of California. Winchester Press, New York. 320 pp.
________. 1994. California birds: their status and distribution. Ibis Publishing Company, Vista, California. 342 pp.
Smith, J.P. and R.J. Atkins. 1979. Cowbird parasitism on Common Bushtit nest. Wilson Bulletin 91:122-123.
Smith, K.G. 1978. Range extension of the Blue Jay into western North America. Bird-Banding 49:208-214.
________. 1979. Migrational movements of Blue Jays west of the 100th meridian. North American Bird Bander 4:49-52.
Smith, S.M. 1967. An ecological study of winter flocks of Black-capped and Chestnut-backed Chickadees. Wilson Bulletin 79:130-137.
________. 1991. The Black-capped Chickadee: behavioral ecology and natural history. Cornell University Press, Ithaca, New York. 362 pp.
________. 1993. Black-capped Chickadee. *In* A. Poole and F. Gill (editors). The birds of North America, No. 39. The Philadelphia Academy of Sciences, Philadelphia; The American Ornithologists' Union, Washington, D.C.
Smith, T.S. 1972. Cowbird parasitism of Western Kingbird and Baltimore Oriole nests. Wilson Bulletin 84:497.
Smith, W.J. 1967. Sixty-seventh Christmas Bird Count – Vancouver, B.C. Audubon Field Notes 21:100.
Snapp, B.D. 1976. Colonial breeding in the Barn Swallow (*Hirundo rustica*) and its adaptive significance. Condor 78:471-480.
Snarski, D. 1969. A contribution toward a bibliography on the genus *Parus*. University of Alaska Department of Wildlife Unpublished Report, College. 25 pp. (Bibliography 4220).
Southern, W.E. and L.K. Southern. 1980. A summary of the incidence of cowbird parasitism in northern Michigan from 1911-1948. Jack-Pine Warbler 58:77-84.
Speich, S.M., H.L. Jones, and E.M. Benedict. 1986. Review of the natural nesting of the Barn Swallow in North America. American Midland Naturalist 115:248-254.
Spreadborough, W. 1905. Bird notes – southern British Columbia – Midway, Osoyoos, Similkameen and Hope. National Museum of Canada, Ottawa, Ontario. 142 pp. (Bibliography 4239).
Spreadborough, W. and P.A. Taverner. 1917. Annotated list of the birds of Hazelton district, B.C. National Museum of Canada Unpublished Report, Ottawa, Ontario. 76 pp. (Bibliography 4242).
Sprot, G.D. 1937. Notes on the introduced skylark in the Victoria district of Vancouver Island. Condor 39:24-31.
Stallcup, P.L. 1968. Spatio-temporal relationships of nuthatches and woodpeckers in ponderosa pine forests of Colorado. Ecology 49:831-843.
Stanwell-Fletcher, J.F. and T.C. Stanwell-Fletcher. 1943. Some accounts of the flora and fauna of the Driftwood valley region of north central British Columbia. British Columbia Provincial Museum Occasional Paper No. 4, Victoria. 97 pp.
Steeger, C. and M.M. Machmer. In press. Inventory and description of wildlife trees and primary cavity nesters in selected stands of the Nelson Forest Region. British Columbia Ministry of Forests, Victoria.
Stein, R.C. 1961. Least Flycatchers in northwestern Washington and central British Columbia. Condor 63:181-182.
________. 1963. Isolating mechanisms between populations of Traill's Flycatchers. Proceedings of the American Philosophical Society 107:21-50.
Stevens, G. 1968. Sixty-eighth Christmas Bird Count – Pender Island, B.C. Audubon Field Notes 22:112.
Stevens, G.B. and Mrs. G.B. Stevens. 1966. Sixty-sixth Christmas Bird Count – Pender Islands, B.C. Audubon Field Notes 20:113.
Stevens, T., A. Grass, and G. Sirk. 1970. Shuswap Lake nature house annual report 1970. British Columbia Parks Branch Unpublished Report, Victoria. 23 pp. (Bibliography 4260).
Stevenson, J. 1934. Comments upon systematics of Pacific coast jays of the genus *Cyanocitta*. Condor 36:72-78.
Stewart, A.C. and M.G. Shepard. 1994. Steller's Jay invasion on southern Vancouver Island, British Columbia. North American Bird Bander 19:90-95.
Stewart, R.E. 1975. Breeding birds of North Dakota. Tri-College Center for Environmental Studies, Fargo, North Dakota. 295 pp.
Stiehl, R.B. 1981. Observations of a large roost of Common Raven. Condor 83:8.
Stiles, F.G. 1981. The taxonomy of Rough-winged Swallows (*Stelgidopteryx;* Hirundinidae) in southern Central America. Auk 98:282-293.
Stirling, D. 1960. A sight record of the Say's Phoebe (*Sayornis saya*) on southern Vancouver Island. Victoria Naturalist 17(1):11.
________. 1960a. Sight records of unusual birds in the Victoria area for 1959. Murrelet 41:10-11.
________. 1960b. Two records of the Mockingbird in British Columbia. Canadian Field-Naturalist 74:176.
________. 1961a. Purple Martins. Victoria Naturalist 18:33-34.
________. 1961b. Sixty-first Christmas Bird Count – Victoria, B.C. Audubon Field Notes 15:102.
________. 1962. Birding in the snow. Victoria Naturalist 18(8):102-103.
________. 1962a. Sixty-second Christmas Bird Count – Victoria, B.C. Audubon Field Notes 16:91.
________. 1966. Bird report (Victoria) number four – 1965. Victoria Natural History Society Mimeographed Paper, Victoria, British Columbia. 6 pp. (Bibliography 310).
________. 1968. Sight record of the Scissor-tailed Flycatcher, *Muscivora forficata*, on southern Vancouver Island. Murrelet 49:14.

ABOUT THE AUTHORS

R. Wayne Campbell was born in Edmonton, Alberta. His fields of interest include zoogeography, feeding and breeding ecology of raptors, marine bird populations, and conservation of birds. He graduated from the University of Victoria in 1976 and received his M.S. degree from the University of Washington in 1983.

After high school, he worked as a seasonal naturalist with the British Columbia Parks Branch, and in 1969 joined the staff at the University of British Columbia as Curator of the Cowan Vertebrate Museum in the Department of Zoology. Over the next 4 years, he established the Photo-Records File, a system to document the occurrence of rare vertebrates in the province, and took over administrative and financial responsibilities for the British Columbia Nest Records Scheme, which he continues to oversee today. He also became very active on the executives of several conservation organizations, including the British Columbia Waterfowl Society and the Vancouver Natural History Society.

In 1973 he moved to the then British Columbia Provincial Museum as Curator of Ornithology. Over the next 20 years, he conducted wildlife inventories of remote areas of the province, including the first complete census of breeding seabird colonies. In addition, he amassed an enormous provincial vertebrate data base that includes details for nearly 2 million specimen and sight records, 300,000 breeding records, and hard copies of nearly 10,000 published and unpublished articles on amphibians, reptiles, birds, and mammals.

Wayne has written over 375 scientific papers, popular articles, and government reports, and has co-authored almost 40 books on higher vertebrates. He has served as British Columbia coordinator for the North American Breeding Bird Survey since 1976, and as a select member of the national ornithology group for the Committee on the Status of Endangered Wildlife in Canada since 1981. He belongs to 23 professional and natural history organizations and is a Life Member and Elected Member of the American Ornithologists' Union, Life Member of the Cooper Ornithological Society, and Honorary Life Member of the Vancouver Natural History Society.

He has received numerous honours and awards for lecturing, writing, and conservation activities. In 1989 he received the Award of Excellence in Biology from the Association of Professional Biologists of British Columbia, and two

The authors, from left: John M. Cooper, Michael C.E. McNall, R. Wayne Campbell, Ian McTaggart-Cowan, Neil K. Dawe, and Gary W. Kaiser. (Photo by Mike H. Symons)

years later the Lifetime Service Award from the Federation of British Columbia Naturalists. He was appointed to the Order of British Columbia in 1992, and recently received the Commemorative Medal of Canada in recognition of the 125th anniversary of the Confederation of Canada.

Neil K. Dawe was born in New Westminster, British Columbia. After graduation from high school his interest turned to the world of finance, and banking became his vocation for the next 7 years. In 1970 he returned to the University of British Columbia, where his interest in natural history was inspired by the enthusiasm and encouragement of Wayne Campbell. This interest grew into a career in habitat management and a commitment to maintaining wildlife populations for present and future generations.

In 1971 he worked as a seasonal naturalist for the British Columbia Parks Branch. Later that year, he became Chief Naturalist at the George C. Reifel Bird Sanctuary, where he developed interpretation and education programs for the British Columbia Waterfowl Society until 1975. Since 1975, he has worked for the Canadian Wildlife Service on Vancouver Island, managing their National Wildlife Areas and Migratory Bird Sanctuaries and working to protect migratory birds and their habitat.

Neil is a member of a number of professional, conservation, and scientific organizations, including the Association of Professional Biologists of British Columbia, the Society of Wetland Scientists, and the Federation of British Columbia Naturalists. He served as the first Regional Vice-President of the Federation of British Columbia Naturalists for Vancouver Island, and has sat on the executives of the Vancouver Natural History Society, the Mid Island Wildlife Watch Society, and the Rosewall-Bonnel Land Trust Society. He co-founded the Brant Festival, a major wildlife festival that takes place in the Parksville–Qualicum Beach area each April.

Neil has been the recipient of Environment Canada's regional Citation of Excellence Award for his work in promoting the value of wildlife to Canadians and for the Brant Festival, plus their national Citation of Excellence Award for outstanding achievement in advancing the goals of conservation and protection. He has also received the Ina Mitchell Award of the Tourism Association of Vancouver Island, for bringing wildlife and people together on Vancouver Island.

His primary research interests now focus on the two major factors affecting populations of migratory birds and other wildlife on earth today: too many people and too much consumption.

Neil has written over 50 scientific, technical, and popular papers and articles on birds and ecology. He co-authored the popular children's books *The Bird Book and Bird Feeder* and *The Pond Book and Pond Pail*.

Ian McTaggart-Cowan, born in Edinburgh, Scotland, is a career biologist and educator with special concentration on the systematics, biology, and conservation of birds and mammals. He graduated from the University of British Columbia in 1932 and earned a Ph.D. degree from the University of California (Berkeley) in 1935. He has been awarded D.Sc. degrees by the University of British Columbia and the University of Victoria, LL.D. degrees by the University of Alberta and Simon Fraser University, and a Doctor of Environmental Studies degree by the University of Waterloo.

He was the biologist at the Provincial Museum in Victoria from 1935 to 1940, when he joined the faculty of the University of British Columbia. During 35 years there, he established and taught courses in vertebrate zoology, undertook research in ornithology and mammalogy, and guided the studies and research of some 100 graduate students while serving successively as Professor, Head of the Department of Zoology, and Dean of the Faculty of Graduate Studies. His studies took him to 6 continents and resulted in more than 300 publications, 110 television programs, and 12 teaching films.

His public service related to vertebrate zoology, conservation, and education includes 7 years on the National Research Council of Canada, where he was the first chairman of the Advisory Committee on Wildlife Research, and 14 years on the Fisheries Research Board of Canada. It also includes serving as chairman of such bodies as the Environment Council of Canada, the Advisory Committee on Whales and Whaling, the Habitat Enhancement Fund of the Province of British Columbia, the Board of Governors of the Arctic Institute of North America, The Wildlife Society, and the Pacific Science Association; as vice chairman of the International Union for the Conservation of Nature and Natural Resources; as a member of the Select Committee on National Parks for the United States Secretary of the Interior; and as a director of the Nature Trust of British Columbia and of the National Audubon Society.

Ian McTaggart-Cowan has received numerous honours and awards for his contributions to biology and conservation, including: Officer of the Order of Canada, the Order of British Columbia, Fellow of the Royal Society of Canada, the Leopold Medal of the Wildlife Society, the Fry Medal of the Canadian Society of Zoologists, the Einarsen Award in Conservation from the Northwest Section of the Wildlife Society, and the Dewey Soper Award from the Alberta Society of Professional Biologists.

John M. Cooper, born in New Westminster, British Columbia, is a career wildlife biologist. His early interest in birds and the natural world was stimulated by his parents, Jack and Louise Cooper. Each spring, for 2 decades, the Cooper family travelled throughout British Columbia and Alberta, often with close friends Lorne Frost and Glen Ryder, in search of birds and their nests. His passion for birds, wilderness, and environmental issues was born from those experiences.

John obtained his B.Sc. degree from the University of British Columbia in 1978 and his M.Sc. degree from the University of Victoria in 1993. He worked as a consulting biologist from 1978 to 1981, then joined the Royal British Columbia Museum as the Ornithology Technician. At the museum he realized his dream of working with Wayne Campbell, and travelled to many remote regions of the province to inventory birds and other wildlife. After 10 years at the museum,

John returned to private business with his own consulting company. Since 1991, he has worked on a wide variety of projects for clients in industry, government, and the conservation movement, mainly environmental impact assessments, mitigation of wildlife/development conflicts, wildlife management, and wildlife inventory.

John is a member of many scientific and conservation organizations, but is most active as a fundraiser for Ducks Unlimited Canada, an organization that gave him his first biology-related employment in 1975. For 15 years he has directed or assisted in fundraising activities in Victoria, to aid Ducks Unlimited's wetland enhancement programs. His most poignant career moment came in 1993: he joined a Sierra Club research trip to the Tatshenshini River, his most beloved wilderness area, and while he was on the river, the provincial government announced the creation of the new Tatshenshini park.

John has written over 50 technical reports, popular articles, scientific papers, and books on birds and other wildlife. His most significant research projects include a long-term study of breeding shorebirds on the Queen Charlotte Islands, breeding ecology and the effects of logging on woodpeckers, and an inventory of rare raptors.

Gary W. Kaiser was born in England but grew up in Ottawa. He developed an early interest in natural history through the Macoun Field Club, and went on to earn a B.Sc. degree in biology in 1966 and an M.Sc. degree in 1972 from Carleton University. University field trips to tropical research stations were the first in a long series of teaching and research activities in distant parts of the world. He has conducted field courses and participated in bird-banding projects in the Philippines, Borneo, Australia, Colombia, and Peru.

Gary began his career as a Migratory Bird Biologist with the Canadian Wildlife Service in Ottawa at the beginning of 1968, and moved to a comparable position in British Columbia in 1974. Much of his early work involved waterfowl surveys, wetland studies, and hunting management, but in 1976 he was able to initiate banding studies of migrant sandpipers with the help of local naturalists. In 1980 he began participating in the seabird colony inventory that had been initiated by Wayne Campbell, and he has devoted most of his subsequent career to the conservation of British Columbia's alcids. In 1990 he served on the National Recovery Team for Marbled Murrelets when they were declared a threatened species.

In 1993 he undertook a campaign to exterminate introduced rats on Langara Island, Queen Charlotte Islands, before the last of that island's Ancient Murrelets were destroyed.

His bibliography includes 30 scientific papers and internal government reports.

Michael C.E. McNall, born in Wingham, Ontario, spent much of his early life hunting, fishing, and studying nature. He obtained a diploma in Wildlife Management from Sir Sandford Fleming College, Ontario, in 1971, joined the Ornithology Department of the Royal Ontario Museum, and spent the next 3 years on field expeditions in the West Indies, British Isles, Netherlands, and Iceland, and throughout North America.

While at the Royal Ontario Museum, he was inspired by artist Terry Shortt to carry out his own research. In 1975 and 1976, with guidance and support from Henri Ouellet and Stewart MacDonald of the National Museum of Canada, he carried out a behavioural study of Parasitic and Long-tailed jaegers in the Canadian Arctic.

After his arctic experience, Michael moved to Victoria, and in 1980 joined the staff of the Royal British Columbia Museum. He has travelled throughout the province collecting data for this book, and is currently the Ornithology Collections Manager. In this role, he prepares exhibits and gives talks on bird behaviour and conservation.

Outside of ornithology, Michael is vice president of the Vancouver Island Arms Collectors Association.

Geo. E. John Smith was born in Vancouver, British Columbia. After graduating from high school, he entered the University of British Columbia and received an honours B.Sc. degree in mathematics and physics and an M.A. degree in mathematics. He then pursued studies at the University of Alberta, where he received his Ph.D. degree in mathematics and statistics in 1970.

He then moved to Ottawa, where he joined Statistics Canada and participated in the redesign of the Canadian Labour Survey. In 1972 he joined Environment Canada, where he served as a consulting statistician in the Biometrics Section of the Canadian Wildlife Service. In 1978 he became head of the Biometrics Division. In 1982 John moved back to Vancouver and joined the Pacific Region of the Canadian Wildlife Service, where he was a senior research scientist involved in the statistical design of research projects, including the redesign of the Canadian Waterfowl Harvest and Species Composition Surveys, the Snow Goose population surveys in the eastern Arctic, and the development of a polar bear population model.

After retiring from the Canadian Wildlife Service in 1995, John taught at the University of British Columbia. He is currently an instructor at the British Columbia Institute of Technology, where he teaches technical mathematics and statistics to students in the renewable-resource technologies of forestry, fisheries, and wildlife.

John has authored or co-authored about 30 papers and 20 technical reports and is a member of the Statistical Society of Canada, American Statistical Association, and Biometric Society.